Stereoselective Polymerization
with
Single-Site
Catalysts

Stereoselective Polymerization
with
Single-Site Catalysts

Edited by
Lisa S. Baugh
Jo Ann M. Canich

CRC Press
Taylor & Francis Group
Boca Raton London New York

CRC Press is an imprint of the
Taylor & Francis Group, an **informa** business

On the cover is a space-filling model of an active *rac*-ethylenebis(indenyl)zirconium catalyst with an attached growing polypropylene chain. This represents one of the first discoveries of an *ansa*-metallocene catalyst capable of producing isotactic polypropylene. The transition structure of the modified Green-Rooney propylene insertion mechanism is also depicted. The enantioselectivity of the propylene insertion step requires a catalyst with both appropriate symmetry and stereorigidity.

The cover background photograph shows a reactor in the laboratory of Professor Stephen A. Miller, containing syndiotactic polypropylene prepared by graduate student Levi J. Irwin. The Miller-Bercaw metallocene catalyst employed (catalyst *s*-90 in Chapter 2) is one of the most syndioselective propylene polymerization catalysts currently known. Unlike isotactic polypropylene, which can also be prepared with Ziegler-Natta initiators, syndiotactic polypropylene is only obtained in appreciable quantities using single-site catalysts.

Cover design by Stephen A. Miller

CRC Press
Taylor & Francis Group
6000 Broken Sound Parkway NW, Suite 300
Boca Raton, FL 33487-2742

First issued in paperback 2020

© 2008 by Taylor & Francis Group, LLC
CRC Press is an imprint of Taylor & Francis Group, an Informa business

No claim to original U.S. Government works

ISBN-13: 978-0-367-57760-5 (pbk)
ISBN-13: 978-1-57444-579-4 (hbk)

Library of Congress Cataloging-in-Publication Data

Stereoselective polymerization with single-site catalysts / edited by Lisa S. Baugh and Jo Ann M. Canich.
   p. cm.
   Includes bibliographical references and index.
   ISBN-13: 978-1-57444-579-4 (hardcover : acid-free paper)
   ISBN-10: 1-57444-579-0 (hardcover : acid-free paper)
   1. Polymerization--Research. 2. Polymer engineering--Research. 3. Stereochemistry--Research. 4. Catalysis--Research. I. Baugh, Lisa S. (Lisa Saunders) II. Canich, Jo Ann M.

TP156.P6S74 2008
668.9'2--dc22   2007018734

Visit the Taylor & Francis Web site at
http://www.taylorandfrancis.com

and the CRC Press Web site at
http://www.crcpress.com

Contents

Preface

Polymers are one of the modern world's most intriguing and versatile materials, able to assume a near-infinite variety of shapes and properties that permeate almost every aspect of our lives as consumers. The commutative nature of plastics is not only a function of their melt-processing capabilities, but also a reflection of the wide variety of possible microstructures that each polymer chain (or distribution of chains) can assume as a unique fingerprint of the polymerization mechanism.

Among the many aspects of polymer chain microstructures, tacticity is arguably the most critical for determining polymer properties and uses. For example, atactic polypropylene is a soft, amorphous material, useful only in very limited applications such as adhesives or asphalt modifiers, while isotactic polypropylene is stiff and highly crystalline, possessing good heat resistance and a high melting point. Isotactic polypropylene is one of today's most commercially important polymers, used to make a wide range of products including fibers, food containers, molded auto parts, films, and nonwoven fabrics. These significant property differences arise principally from alignment of the polymer's pendant methyl groups into a regular rather than random arrangement.

Today, the development of a new polymeric material requires a keen understanding of how to manipulate the most intimate features of individual polymer chains—tacticity, branching, comonomer sequence distribution, block length, regioerrors—to obtain desirable physical properties and performance. The modern polymer chemist must possess a good understanding of fundamental microstructural structure–property relationships for any system under study, both from the synthetic perspective (relationships between polymerization catalyst ligand/active site structure, polymerization mechanism, and chain microstructure) and the performance perspective (relationships between chain microstructure, phase behavior, and bulk properties).

With this book, we have sought to create a one-volume reference containing information on the most important classes of tactic polymers and polymerizations carried out using single-site catalysis.[1–5] This book contains 25 individual review chapters on tactic polymers and polymerizations, each contributed by top researchers in the area, grouped into five parts.

The first part describes the basic types of tactic polypropylenes and catalyst systems used to prepare them. These chapters provide the reader with an overview of known catalyst structure–polymer property relationships, and explain the critical ligand modification strategies used to achieve stereocontrol. Newer classes of polypropylene polymerization catalysts, including nonmetallocene and late transition metal precatalysts, are also highlighted in order to exemplify emerging catalyst design techniques. A chapter discussing theoretical site-control studies is included to illustrate how such techniques assist with and complement catalyst design. Introductions to the fundamental concepts of tacticity and polymerization mechanisms are also featured in the earliest chapters. The second part contains chapters highlighting the role of tacticity in the design of polypropylenes with special physical properties (stereoblock, elastomeric, and functional tactic materials). Tacticity considerations in ethylene–propylene copolymers are also examined.

In the third part, tacticity considerations are expanded to other monoolefinic monomers and polymers, with chapters covering tactic α-olefin and styrene polymerizations. Ditactic polymerizations and copolymerizations of cyclic olefinic monomers with two adjacent prochiral centers (cyclopentene and norbornene) are then described. The fourth part addresses the increasingly complex mechanistic and microstructural aspects of diene polymerization, with chapters addressing linear and cyclic conjugated diene monomers, diene cyclo(co)polymerization, tacticity in olefin-containing polymers formed from bicyclic olefin metathesis polymerization (and their hydrogenated derivatives), and the polymerization of acetylenes.

The final part of this book features tactic polymerizations of functional and nonolefinic (ring-opening) monomers—materials for which many aspects of polymer stereochemistry and microstructure control are very different than for simple polyolefins. Acrylate, epoxide, and lactide polymerizations are addressed in this part, along with tactic olefin/carbon monoxide co- and terpolymers. These final chapters provide an expanded view of polymer microstructures and stereocontrol strategies, such as enantiomer-selective polymerization, that may be less familiar to the polyolefin-minded chemist and serve to enhance the reader's overall understanding of stereoregular polymers and polymerization.

Our goal in constructing this book was to create a reference volume that would be valuable both to experts in the area and new researchers or students wishing to learn the fundamentals of an unfamiliar system. To this end, although each review chapter presents exciting recent advances in the field, the chapter content has been carefully written and edited to feature basic and practical information that will continue to be relevant and useful over time. In particular, each chapter includes the following:

- *Introductory material* describing the basic features of the general mechanism, polymer microstructures, and/or catalyst systems concerned, to assist less familiar readers in understanding subsequent material in the chapter
- *Generous references with titles* so that readers can quickly locate additional information for systems of interest
- *Characterization information* explaining how tactic polymers are commonly evaluated, with many spectral examples
- Descriptions of the *physical properties* and *industrial applications* of important tactic polymers presented from a structure–property viewpoint
- Detailed information on important tactic *copolymers* in addition to homopolymers.
- *Cross-references to other chapters* when additional information is available on a topic, allowing readers to find information most quickly and efficiently.

As researchers in the single-site catalyst design/polymer structure–property area, we have found editing this volume to be a tremendous education. We hope that the information in this book will prove equally useful to the readers in their work for many years to come.

Lisa S. Baugh
ExxonMobil Research & Engineering Company
Annandale, New Jersey

Jo Ann M. Canich
ExxonMobil Chemical Company,
a division of Exxon Mobil Corporation
Baytown, Texas

REFERENCES AND NOTES

We would like to point out excellent earlier review articles addressing general aspects of tactic polymerization.

1. Suzuki, N. Stereospecific olefin polymerization catalyzed by metallocene complexes. *Top. Organomet. Chem.* **2004**, *8*, 177–216.
2. Guerra, G.; Cavallo, L.; Corradini, P. Chirality of catalysts for stereospecific polymerizations. In *Materials-Chirality*; Green, M. M., Nolte, R. J. M., Meijer, E. W., Eds; Topics in Stereochemistry 24; Wiley-Interscience: Hoboken, NJ, 2003, pp. 1–69.
3. Coates, G. W. Precise control of polyolefin stereochemistry using single-site metal catalysts. *Chem. Rev.* **2000**, *100*, 1223–1252.
4. Resconi, L.; Cavallo, L.; Fait, A.; Piemontesi, F. Selectivity in propene polymerization with metallocene catalysts. *Chem. Rev.* **2000**, *100*, 1253–1345.

For basic definitions of stereochemical terms relating to tactic polymers and polymerizations, see

5. IUPAC Macromolecular Chemistry Division, UK. Stereochemical definitions and notations relating to polymers. *Pure Appl. Chem.* **1981**, *53*, 733–752.

Editors

Lisa Saunders Baugh is a Research Associate at ExxonMobil's Corporate Strategic Research Laboratory in Annandale, New Jersey, where her research focuses on metal-mediated polymerization, polymer functionalization, and the synthesis of model polyolefins. She received her BS (1991) from the University of Texas, Austin, and her PhD (1996) from the University of California, Berkeley, completing part of her thesis work in the polymer science and engineering department of the University of Massachusetts, Amherst. She has coedited two previous books on metal-mediated polymerization (*Late Transition Metal Polymerization Catalysis* and *Transition Metal Catalysis in Macromolecular Design*) and has authored numerous technical publications and patents (those from 1991 to 2000 under the name of Lisa S. Boffa). Dr. Baugh is a past secretary-general of the American Chemical Society's Catalysis & Surface Science Secretariat, a member-at-large of the ACS Polymeric Materials: Science & Engineering Division, and a member of the editorial advisory board for the ACS magazine *Chemistry*. She is also a member of the National Association of Science Writers, having contributed many feature pieces to college and high school chemistry textbooks.

Jo Ann M. Canich is a Senior Staff Chemist at ExxonMobil's Baytown Technology and Engineering Complex in Baytown, Texas, where she has worked for the last 20 years. Her research interests are based in synthetic organometallic catalysis, with a focus on olefin polymerization and oligomerization catalysts that can produce products ranging from higher molecular weight polymers including linear low-density polyethylene, high-density polyethylene, ethylene-propylene copolymers, and polypropylene to lower molecular weight products such as adhesives, synthetics, additives, and higher α-olefins. She is the author of many technical publications, and the inventor of over 50 issued U.S. patents spanning the areas of organometallic compounds and catalysts, polymerization processes, and related products. In 1998, she received the American Society of Patent Holders Distinguished Corporate Inventors Award. Her educational background includes BS (1981) and MS (1984) degrees from Portland State University, and a PhD (1987) from Texas A&M University.

Contributors

Hiroharu Ajiro
Department of Chemistry and Chemical Biology
Baker Laboratory
Cornell University
Ithaca, New York

Scott D. Allen
Department of Chemistry and Chemical Biology
Baker Laboratory
Cornell University
Ithaca, New York

Michael Arndt-Rosenau
Lanxess Deutschland GmbH
Leverkusen, Germany

David A. Aubry
Department of Chemistry
Texas A&M University
College Station, Texas

Manfred Bochmann
Wolfson Materials and Catalysis Centre
School of Chemical Sciences and Pharmacy
University of East Anglia
Norwich, United Kingdom

Lynn M. Bormann-Rochotte
Motorola, Inc.
Advanced Product Technology Center
Plantation, Florida

Darragh Breen
University College Dublin
School of Chemistry and Chemical Biology
Dublin, Ireland

Vincenzo Busico
Dipartimento di Chimica
Università di Napoli Federico II
Naples, Italy

Carmine Capacchione
Department of Chemistry
University of Salerno
Fisciano (SA), Italy

Malcolm H. Chisholm
Department of Chemistry
The Ohio State University
Columbus, Ohio

T. C. Chung
Department of Materials Science and Engineering
The Pennsylvania State University
University Park, Pennsylvania

Geoffrey W. Coates
Department of Chemistry and Chemical Biology
Baker Laboratory
Cornell University
Ithaca, New York

Cecilia Cobzaru
Department of Inorganic Chemistry II
University of Ulm
Ulm, Germany

Katherine Curran
University College Dublin
School of Chemistry and Chemical Biology
Dublin, Ireland

John J. Esteb
Clowes Department of Chemistry
Butler University
Indianapolis, Indiana

Terunori Fujita
Mitsui Chemicals, Inc.
Research Center
Chiba, Japan

Maurizio Galimberti
Pirelli Pneumatici SpA
Materials Innovation
Milan, Italy

Vernon C. Gibson
Department of Chemistry
Imperial College
London, United Kingdom

Stanislav Groysman
School of Chemistry
Raymond and Beverly Sackler Faculty of Exact
  Sciences
Tel Aviv University
Tel Aviv, Israel

Gaetano Guerra
Dipartimento di Chimica
University of Salerno
Fisciano (SA), Italy

Sabine Hild
Department of Experimental Physics
Central Facility of Electronmicroscopy
University of Ulm
Ulm, Germany

Jiling Huang
Laboratory of Organometallic Chemistry
East China University of Science and Technology
Shanghai, People's Republic of China

Levi J. Irwin
Department of Chemistry
Texas A&M University
College Station, Texas

Walter Kaminsky
Institute of Technical and Macromolecular
  Chemistry
University of Hamburg
Hamburg, Germany

Il Kim
Department of Polymer Science and Engineering
Pusan National University
Busan, Korea

Moshe Kol
School of Chemistry
Raymond and Beverly Sackler Faculty of Exact
  Sciences
Tel Aviv University
Tel Aviv, Israel

Reko Leino
Laboratory of Organic Chemistry
Åbo Akademi University
Åbo, Finland

Shirley Lin
Department of Chemistry
United States Naval Academy
Annapolis, Maryland

Haiyan Ma
Laboratory of Organometallic
  Chemistry
East China University of Science and
  Technology
Shanghai, People's Republic of China

Haruyuki Makio
Mitsui Chemicals, Inc.
Research Center
Chiba, Japan

Edward L. Marshall
Department of Chemistry
Imperial College
London, United Kingdom

Toshio Masuda
Department of Polymer Chemistry
Graduate School of Engineering
Kyoto University
Kyoto, Japan

Stephen A. Miller
The George and Josephine Butler Polymer
  Research Laboratory
Department of Chemistry
University of Florida
Gainesville, Florida

Naofumi Naga
Department of Applied Chemistry
Faculty of Engineering
Shibaura Institute of Technology
Tokyo, Japan

Mitsuru Nakano
Toyota Central Research and Development
  Laboratories, Inc.
Aichi, Japan

Kyoko Nozaki
Department of Chemistry and Biotechnology
Graduate School of Engineering
The University of Tokyo
Tokyo, Japan

Oleg G. Polyakov
Synkera Technologies, Inc.
Longmont, Colorado

Craig J. Price
Department of Chemistry
Texas A&M University
College Station, Texas

Antonio Proto
Department of Chemistry
University of Salerno
Fisciano (SA), Italy

Anthony K. Rappé
Department of Chemistry
Colorado State University
Fort Collins, Colorado

Bernhard Rieger
Department of Chemistry
Technische Universität München
Munich, Germany

Wilhelm Risse
University College Dublin
School of Chemistry and Chemical
    Biology
Dublin, Ireland

Sharon Segal
School of Chemistry
Raymond and Beverly Sackler Faculty of
    Exact Sciences
Tel Aviv University
Tel Aviv, Israel

Pamela J. Shapiro
Department of Chemistry
The University of Idaho
Moscow, Idaho

Masashi Shiotsuki
Department of Polymer Chemistry
Graduate School of Engineering
Kyoto University
Kyoto, Japan

Junichi Tabei
Department of Polymer Chemistry
Graduate School of Engineering
Kyoto University
Kyoto, Japan

Carsten Troll
Department of Chemistry
Technische Universität München
Munich, Germany

Zhiping Zhou
Department of Chemistry
The Ohio State University
Columbus, Ohio

Deanna L. Zubris
Department of Chemistry
Villanova University
Villanova, Pennsylvania

Part I

Polypropylene: Catalysts and Mechanism

1 Isotactic Polypropylene from C_2- and Pseudo-C_2-Symmetric Catalysts

Reko Leino and Shirley Lin

CONTENTS

1.1 INTRODUCTION

Polypropylene (PP) is one of the most highly utilized engineering plastics, with production in the United States and Canada of over 8 million metric tons in 2003.[1] The starting material for PP, propylene, can be transformed into polymers with a range of physical properties by adjusting the relative stereochemistry of the pendant methyl groups on the polymer backbone. For instance, when the methyl groups are placed in a stereorandom configuration, the resulting polymer, known as atactic polypropylene, is amorphous and lacks crystallinity. The vast majority of polypropylene sold is isotactic polypropylene (iPP), a resin valued for its strength, toughness, chemical and heat resistance, and processability. The highly stereoregular structure of iPP, in which the pendant methyl groups reside on the same side of the polymer chain, imparts a crystalline structure to the polymer and allows the material to have a high melting point (160–165 °C).[2,3] A versatile plastic, iPP can be found in a wide range of applications from packaging to fibers to containers. Commercial, highly crystalline iPP is largely synthesized using traditional heterogeneous Ziegler–Natta catalysis, but the advent of homogeneous group 4 metallocenes has enabled polyolefin producers to access a wider range of polymer microstructures and properties. The mechanism of stereoselective polymerization by these catalysts is now well understood; the issues of stereocontrol and influence of the metallocene catalyst structure on the propylene polymerization behavior have been discussed in detail in earlier reviews.[4–6] Excellent review articles are also available on the synthesis of chiral group 4 metallocenes.[7,8] This chapter will introduce some fundamental aspects of metallocene-catalyzed propylene polymerization, discuss the details of how the C_2-symmetry of the catalyst is exploited in the synthesis of iPP, and then survey the general ligand structures that are commonly encountered.

1.1.1 FUNDAMENTALS OF METALLOCENE-CATALYZED PROPYLENE POLYMERIZATION

A brief overview of the necessary steps for polymer formation is discussed here. A more detailed review may be found elsewhere.[6]

1.1.1.1 Activation of Metallocene Precatalysts

The most widely used homogeneous catalysts for the synthesis of iPP are group 4 metallocenes (Scheme 1.1). Metallocene precatalysts typically consist of the metallocene dichlorides. Addition of a Lewis acidic cocatalyst is necessary to generate the catalyst. The cocatalyst replaces the chlorides on the metal with alkyl groups and, in a subsequent step, abstracts one alkyl group to generate the active catalyst. The most common cocatalyst for this purpose is the oligomeric mixed aluminum alkyl/oxide methylaluminoxane (MAO). MAO is obtained from controlled hydrolysis of trimethylaluminum, with the final product being a dynamic mixture of different compounds.[9] Alternatives to MAO include boron-based cocatalysts such as tris(perfluorophenylborane) and tetrakis(perfluorophenylborate)s with various counterions.[9,10]

The metallocene species after activation is a d^0, cationic, 14-electron species with a pseudotetrahedral structure and a highly Lewis acidic metal center. Two of the four coordination sites on the metal are occupied by the ligands, such as η^5-cyclopentadienyl (Cp, C_5H_5) groups or other moieties. The two remaining sites participate directly in the polymerization process.

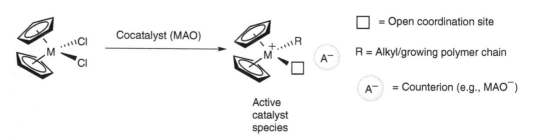

SCHEME 1.1 Formation of an active catalyst species from a metallocene dichloride.

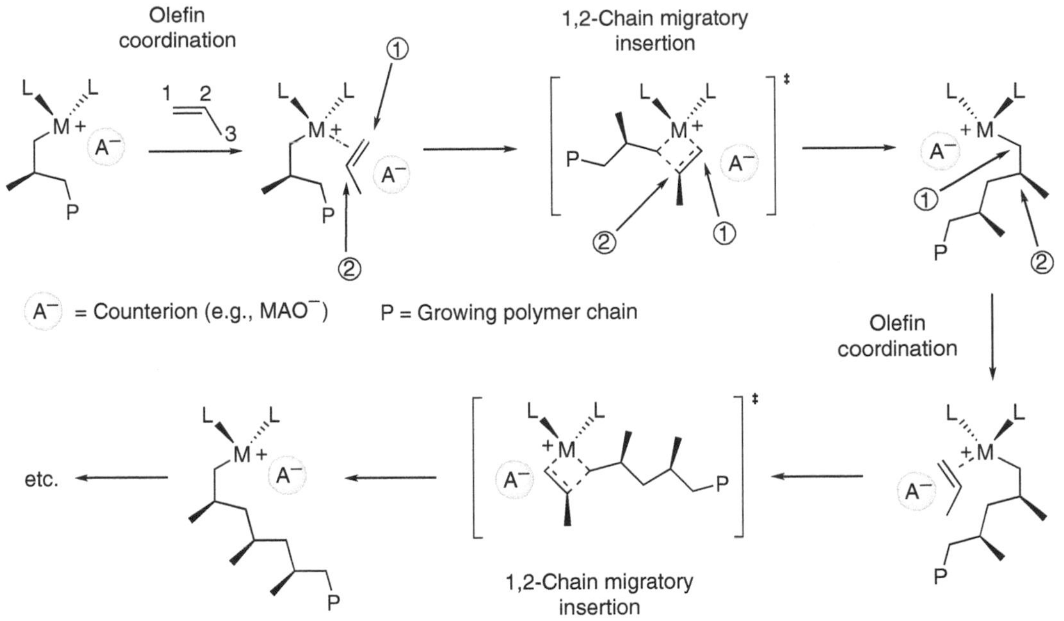

SCHEME 1.2 General mechanism of chain propagation for metallocene-catalyzed propylene polymerization (L = cyclopentadienyl or similar ligand).

1.1.1.2 Mechanism of Chain Propagation

The mechanism of chain propagation in metallocene-catalyzed propylene polymerization has been well studied; the essential steps are shown in Scheme 1.2. Propylene insertion takes place in two steps: (1) olefin coordination with cis opening of the double bond and (2) chain migratory insertion. The monomer inserts primarily with 1,2-regioselectivity with the less substituted carbon bonded to the metal. Following migration of the polymer chain to the site with the coordinated olefin, the monomer coordination site is transformed into the site bearing the growing polymer chain; the site formerly occupied by the polymer chain is now free for monomer coordination. Therefore, in most cases, propylene insertion occurs at each of the two sites in an alternating manner.

The energetics of each step have been deduced using a variety of computational methods. Propylene uptake (coordination) by a coordinatively unsaturated metallocene alkyl species is favorable (~10 kcal/mol) and essentially barrierless unless sterically bulky ligands are involved.[6] The barrier to insertion of propylene for a C_2-symmetric zirconocene catalyst has been calculated at 2–4 kcal/mol.[11]

Four proposed detailed mechanisms for monomer insertion are shown in Scheme 1.3. All mechanisms begin with the coordinatively unsaturated metal center bearing the growing polymer chain. In the Cossee mechanism (Scheme 1.3a),[12–16] olefin coordination to the metal is followed by chain migratory insertion through a four-center metallocycle transition state. The Green–Rooney mechanism (Scheme 1.3b)[17,18] involves an oxidative 1,2-hydrogen shift from the polymer chain onto the metal to form an alkylidene hydride. This alkylidene can react with the monomer to form a metallocyclobutane. Reductive elimination regenerates the three-coordinate metal center. Since the metallocenes discussed here are d^0 species incapable of oxidative addition, this mechanism can be ruled out. The next two mechanisms involve α-agostic interactions: the modified Green–Rooney mechanism (Scheme 1.3c)[19–21] features agostic interactions between a hydrogen atom on the α-carbon of the polymer chain and the metal in the ground and transition states, while the fourth mechanism (Scheme 1.3d) has an agostic interaction of this type only in the transition state.

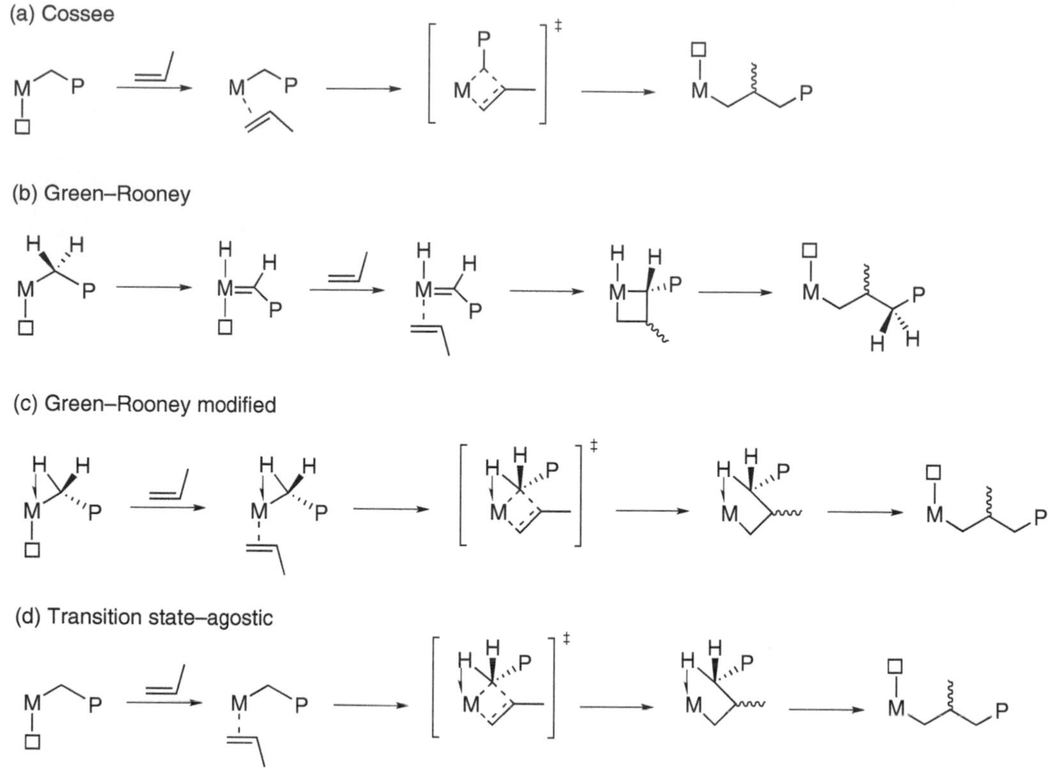

(a) Cossee

(b) Green–Rooney

(c) Green–Rooney modified

(d) Transition state–agostic

SCHEME 1.3 Four mechanisms for propylene insertion (spectator ligands and charge on metal center not shown; P = growing polymer chain; □ = open coordination site; half-arrow represents an agostic interaction). (Adapted with permission from Grubbs, R. H.; Coates, G. W. *Acc. Chem. Res.* **1996**, *29*, 85–93. Copyright 1996 American Chemical Society.)

Ground-state agostic interactions have been observed experimentally in model compounds of active sites;[22,23] growing polymer chain geometries containing β- and γ-agostic interactions are calculated to be more stable than those with α-agostic interactions,[24] and thus are a likely feature of the resting state of the catalyst. The presence of an α-agostic interaction in the transition state has been observed experimentally in mechanistic studies[25–27] and is consistent with the proposed conformation adopted by the growing polymer chain as a means of affecting enantioselective propylene insertion (*vide infra*). These experimental results support the modified Green–Rooney mechanism (Scheme 1.3c) as the most likely mechanism for propylene insertion.

1.1.1.3 Regioirregular Propylene Insertion

Insertion of propylene with 1,2-regioselectivity is known as primary insertion or regioregular insertion because it is the most prevalent type of monomer enchainment for metallocenes. Some metallocenes do insert propylene in other, regioirregular ways[28–30] (Scheme 1.4). The monomer may be inserted with 2,1-regioselectivity (Scheme 1.4a), known as secondary insertion, with the more substituted olefinic carbon bonded to the metal. Following 2,1-insertion, a unimolecular isomerization may take place to give a 3,1-insertion[31–34] (Scheme 1.4b). Alternatively, monomer insertion can occur following a regioirregular insertion. This insertion step is always regioregular (1,2-insertion) and occurs at a slower rate than 1,2-insertion following a regioregular insertion; the reduced rate of chain propagation following 2,1-insertion has led to the suggestion that these sites are effectively dormant.[35] Chain release through β-hydride transfer to monomer (*vide infra*) is another possible

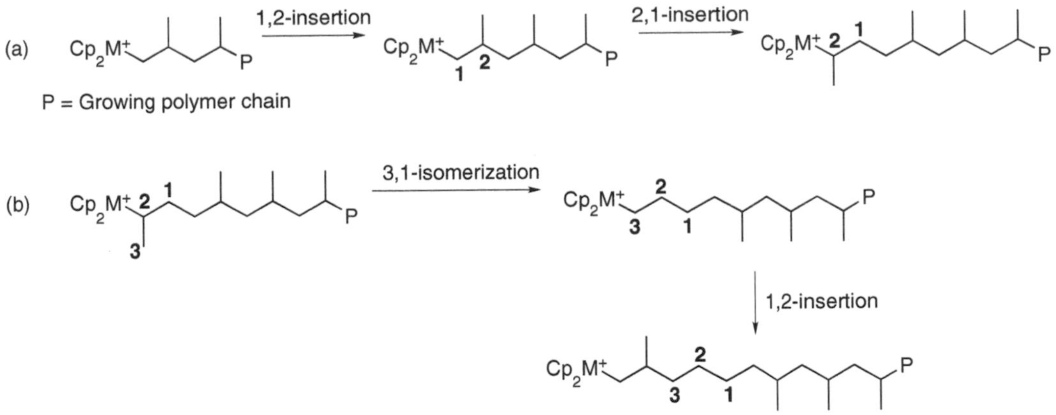

SCHEME 1.4 (a) 2,1-Insertion following 1,2-insertion into a growing polymer chain and (b) isomerization to produce a 3,1-insertion. (Adapted with permission from Resconi, L.; Cavallo, L.; Fait, A.; Piemontesi, F. *Chem. Rev.* **2000**, *100*, 1253–1345. Copyright 2000 American Chemical Society.)

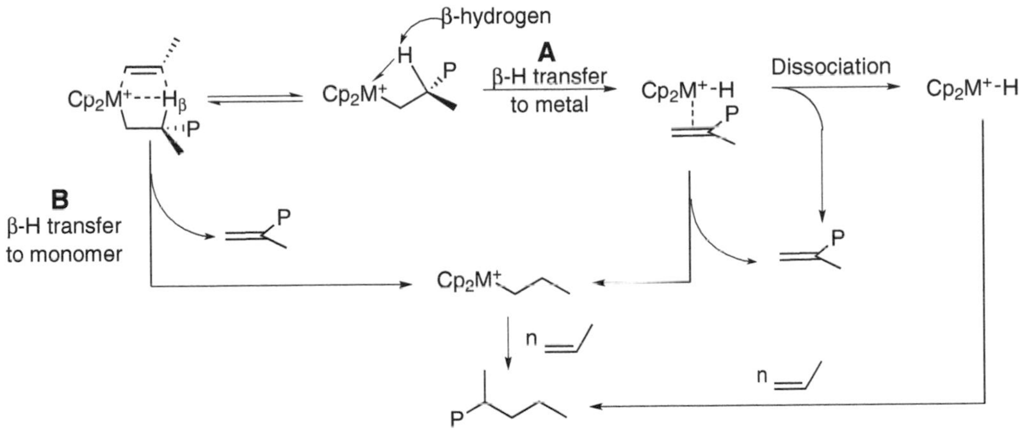

SCHEME 1.5 Hydride transfer chain release mechanisms: (A) β-hydride elimination (transfer to metal); (B) β-hydride transfer to monomer (P = growing polymer chain). (Adapted with permission from Resconi, L.; Cavallo, L.; Fait, A.; Piemontesi, F. *Chem. Rev.* **2000**, *100*, 1253–1345. Copyright 2000 American Chemical Society.)

reaction which is thought to be responsible for the decrease in molecular weights for metallocenes with a high number of 2,1-insertions.[6]

1.1.1.4 Chain Release

One very important feature of metallocene-catalyzed propylene polymerization is that more than one polymer chain is produced by each metal site. This is possible through mechanisms of chain release which follow monomer insertion, both 1,2-insertion and 2,1-insertion. Scheme 1.5 depicts the two most common chain release mechanisms: β-hydride transfer to the metal[36–39] (with or without associative displacement by incoming monomer) and β-hydride transfer to the monomer.[40] Both processes result in identical chain endgroups (vinylidene and *n*-propyl), but can be studied by examining the rate law for chain release.

Another chain release mechanism is β-methyl transfer from the polymer to the metal center (Scheme 1.6a).[37,39] This process occurs when highly substituted Cp ligands are present[41–43] and

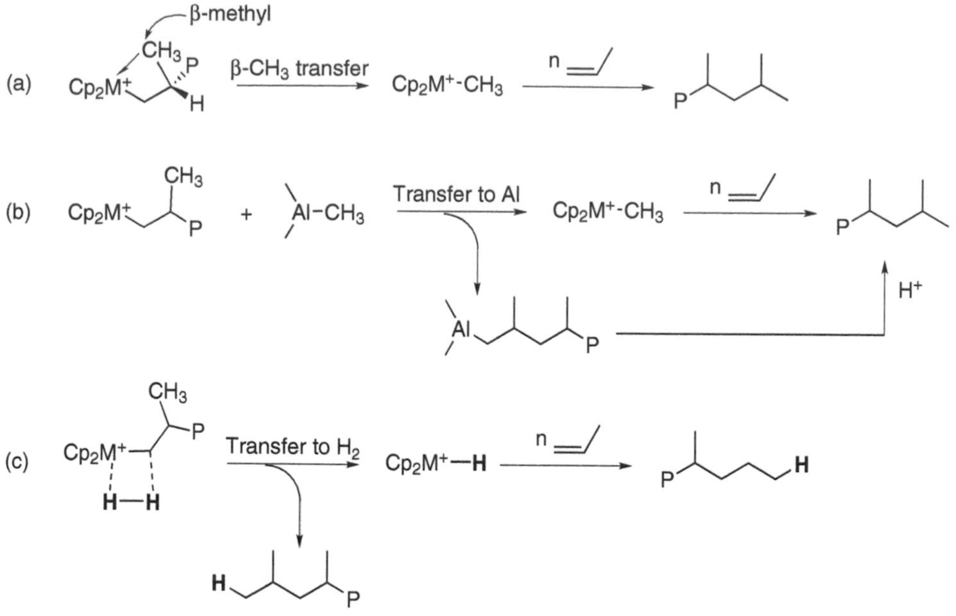

SCHEME 1.6 Other chain release mechanisms: (a) β-methyl elimination (transfer to metal); (b) chain transfer to Al; (c) chain transfer to H_2 (P = growing polymer chain). (Adapted with permission from Resconi, L.; Cavallo, L.; Fait, A.; Piemontesi, F. *Chem. Rev.* **2000**, *100*, 1253–1345. Copyright 2000 American Chemical Society.)

produces allyl and isobutyl chain endgroups. A fourth mechanism is chain transfer to aluminum[44–47] (Scheme 1.6b); trimethylaluminum is a component of the MAO activator required to form the catalyst from the metallocene dichloride. This mechanism is observed at high [Al]/[metal] ratios and results in two saturated isobutyl endgroups on a polymer chain.

An effective way to lower polymer molecular weights is to add hydrogen as a chain transfer agent (Scheme 1.6c).[48,49] This results in saturated *n*-propyl and isobutyl chain ends for the polymer. For catalysts that produce dormant sites through 2,1-insertion (*vide supra*), hydrogen can also serve to increase productivity.[6]

1.2 CONTROL OF POLYPROPYLENE STEREOCHEMISTRY

1.2.1 POLYMER STEREOCHEMISTRY

Figure 1.1 depicts the polymer architectures commonly known to exist for polypropylene. These polymers differ only in the relative stereochemistry of the pendant methyl groups on the polymer backbone. There are two possible configurations for two adjacent methyl groups (dyads): when the methyl groups reside on the same side, they are *meso* to each other, denoted as *m*; two methyl groups on opposite sides of the chain are *racemo*, denoted as *r*. There are three possible sequences of triads (three adjacent methyl groups): *mm*, *rr*, and *mr/rm*. The ^{13}C NMR chemical shift of each methyl group is sensitive to the relative stereochemistry of its neighbors. High-resolution ^{13}C NMR is capable of distinguishing the ten different sets of five stereocenters, known as pentads[50,51] (Figure 1.2). The percentages of the pentads are used as a measure of the degree of stereoregularity of the polymer. For example, perfectly isotactic PP has *mmmm* = 100% whereas statistically atactic PP has *mmmm* = 6.25%.

A spectrum of polymer microstructures exists. There are two crystalline, stereoregular polypropylenes, one in which each methyl group has an *m* configuration with respect to its neighbors

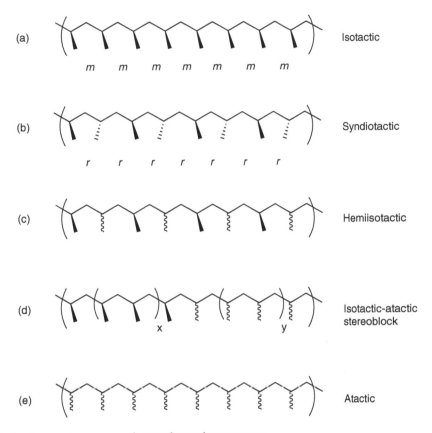

FIGURE 1.1 The most common polypropylene microstructures.

(isotactic, Figure 1.1a) and one in which each methyl group has an *r* configuration with respect to its neighbors (syndiotactic, Figure 1.1b). The completely stereoirregular, amorphous atactic polypropylene (Figure 1.1e) represents the other end of the spectrum. In the middle are hemiisotactic PP (Figure 1.1c), where every other methyl group is on the same side of the polymer chain, and stereoblock PP, where one segment of the polymer is atactic and the other is isotactic (Figure 1.1d).

1.2.2 MECHANISM OF STEREOCONTROL

As discussed in Section 1.1.1.2, metallocene catalysts insert propylene from two sites in an alternating manner (exceptions will be discussed in Section 1.2.2.2). Each insertion of the prochiral propylene monomer into the growing polymer chain creates a new chiral center. The configuration of the new chiral center will be determined by which enantioface of propylene coordinated to the metal prior to insertion. The two enantiofaces, *re* and *si*, are shown coordinated to a metallocene in Figure 1.3.

Since insertion occurs at two different sites on the metal center most often in an alternating fashion, an *m* dyad (two methyl groups on the same side of the polymer chain) results when the same enantioface of propylene is inserted consecutively (*re* after *re* and *si* after *si*); isotactic polymer is produced after many insertions. Similarly, when the two coordination sites of the catalyst prefer opposite enantiofaces (*re* after *si* and *si* after *re*), an *r* dyad (two methyl groups on opposite sides of the polymer chain) is formed and a syndiotactic polypropylene is produced.

It should be noted that a mixture of active catalyst centers that insert either only the *re* or *si* enantioface will still produce iPP. Although the chain start and chain end of each polymer chain are generally not the same group, when iPP molecular weight is high, a pseudo plane of symmetry exists

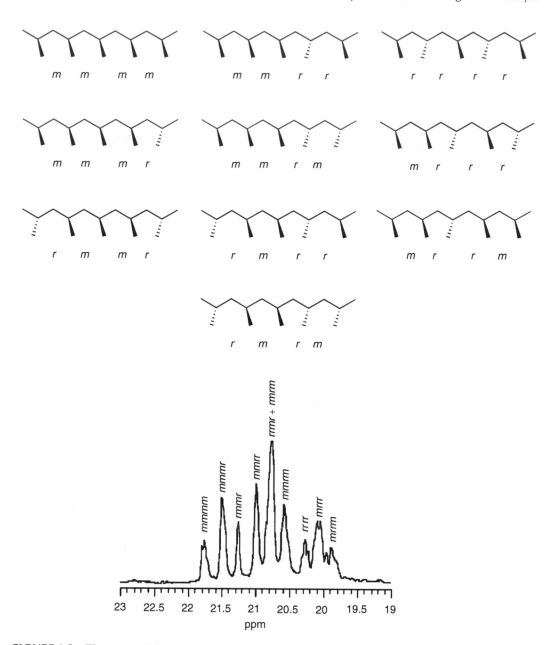

FIGURE 1.2 The ten possible combinations of five adjacent stereocenters (pentads) in polypropylene and the ^{13}C NMR spectrum of the methyl region of atactic polypropylene.

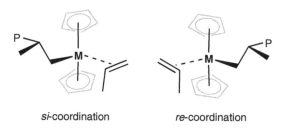

FIGURE 1.3 *SI* and *re* coordination of propylene to a metal center (P = growing polymer chain).

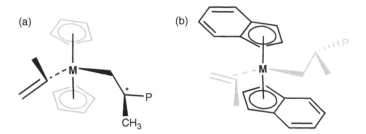

FIGURE 1.4 (a) Chain-end control and (b) Enantiomorphic site control of propylene insertion (P = growing polymer chain).

in each individual polymer chain; a polymer chain created from only *re* insertions is the enantiomer of a polymer chain resulting from only *si* insertions but neither is optically active.

In order for the coordination of one enantioface of propylene to be favored over the other, a chiral environment is required. The two possible sources of this chirality are the stereocenter(s) of the growing polymer chain (chain-end control, Figure 1.4a) and the ligands on the metallocene (enantiomorphic site control, Figure 1.4b). In both mechanisms, enantiofacial selectivity of propylene insertion is achieved through steric interactions between the incoming monomer and the growing polymer chain.

As discussed in the following sections, the two mechanisms can be distinguished by observing the ^{13}C NMR spectra of the polymers produced because the stereochemical misinsertions are distinct. In principle, both mechanisms may be operating for a given catalyst system. But since enantiomorphic site control is far superior to chain-end control in producing highly isotactic polypropylene, the pentads that result from a chain-end control mechanism operating simultaneously with enantiomorphic site control are often too low in intensity to be evaluated.[6]

1.2.2.1 Chain-End Control

In the chain-end control mechanism, the stereochemistry of the last-inserted monomer in the polymer chain influences the enantiofacial selectivity for insertion of the next propylene unit. As suggested by molecular mechanics models for [Cp$_2$Ti(growing polymer chain)]$^+$,[52] the chiral center on the last inserted monomer within the polymer chain causes the chain to favor a particular orientation relative to the ligands of the metallocene (Figure 1.5). The next propylene to be inserted coordinates to the other metal site with its methyl group *anti* to the growing chain. This favors the formation of *m* dyads (i.e., the insertion of an *si*-coordinated monomer favors the *si* enantioface of the subsequent monomer). Therefore, it is possible to generate iPP but not syndiotactic PP using the chain-end control mechanism.

This mechanism will produce perfectly isotactic polypropylene as long as there are no enantiofacial misinsertions. However, the difference in energy between insertion of the *re* and *si* enantiofaces is fairly small (∼2 kcal/mol) for the chain-end control mechanism.[43,53] The stereocenter resulting from each misinsertion will cause the opposite enantioface of the alkene to be preferred for the next insertion; the result is a stereoblock structure with one isolated *r* defect, *mmmmmrmmmm* (Figure 1.6). The signature for a chain-end control mechanism is therefore the presence of the *mmrm* and *mmmr* pentads in a 1:1 ratio in the ^{13}C NMR of the polypropylene synthesized.[54]

Several catalysts have been found to produce iPP through the chain-end control mechanism. These include the Cp$_2'$TiR$_2$/MAO systems (Cp$'$ = C$_5$H$_5$ or substituted Cp; R = Ph or Cl)[53,54] and (pyridyldiimine)FeCl$_2$/modified methylaluminoxane catalysts.[55] Low polymerization temperatures are required to achieve moderate isotacticity with both classes of catalysts. The frequency of misinsertions produces an average stereoblock length of less than 16 units, the minimum required for

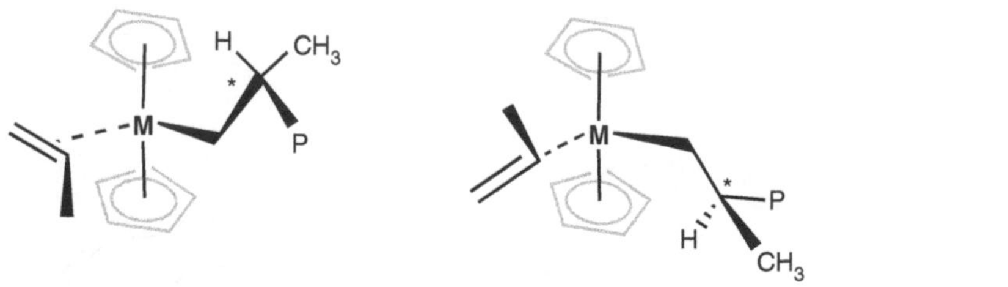

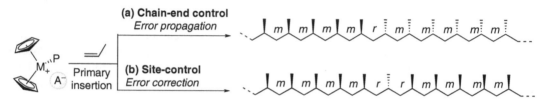

(a) *re*-coordination
to *si*-chain
(disfavored)

(b) *si*-coordination
to *si*-chain
(favored)

FIGURE 1.5   Chain-end control mechanism: (a) *re* coordination following *si* insertion (disfavored); (b) *si* coordination following *si* insertion (favored) (P = growing polymer chain).

(a) Chain-end control
Error propagation

(b) Site-control
Error correction

Primary
insertion

FIGURE 1.6   Microstructures of polypropylene synthesized through (a) chain-end control and (b) enantio-morphic site control (P = growing polymer chain).

crystallite formation.[56,57] Consequently, the iPPs synthesized with $Cp_2'TiR_2/MAO$ have low melting points and crystallinitities.[58,59]

1.2.2.2   Enantiomorphic Site Control

When the structure of the metallocene is responsible for enantiofacial selectivity in propylene inser-tion, the catalyst is operating under enantiomorphic site control. The catalyst environment (the metal and ligands) determines the conformation of the growing polymer chain, and in a mechanism sim-ilar to chain-end control, the incoming propylene monomer coordinates then to the metal with its methyl group *anti* to the growing chain (*vide infra*). In order for a stereoregular polypropylene to be synthesized, the catalyst must consistently place the polymer chain in the same conformation for each of the two metal coordination sites. This type of rigidity is not expected for unbridged cyclopentadienyl-type ligands, which are known to spin freely around the metal-Cp axis, except at low temperatures or in the presence of bulky substituents on the Cp rings (see Section 1.3.2.).

However, metallocene rigidity can be achieved through the introduction of a bridging group between the two rings. For instance, for two identical ligands such as indene, the introduction of a short bridge between the ligands, such as $–CH_2CH_2–$ or $–SiMe_2–$, will result in three diastereomers of the metallocene (Figure 1.7). Two of these catalysts, **A** and **B**, are chiral and exist as a pair of enantiomers with C_2 symmetry; the third, **C**, is achiral with C_s symmetry. All three diastereomeric metallocenes may be formed when the ligands are metallated; the ratio of the diastereomers is determined by their relative stabilities. The bridge prevents the interconversion of the diastereomers under usual polymerization conditions. Typically, the enantiomers **A** and **B** are isolated together as a racemic pair, called *rac*-bis(indenyl)zirconocene dichloride, whereas **C** is obtained as the *meso*-bis(indenyl)zirconocene dichloride. (See Sections 1.3.1.1 and 1.4.1 for further discussion on these types of *ansa*-metallocenes.)

After activation using a Lewis acid cocatalyst such as MAO, catalysts **A**, **B**, and **C** will each have two coordination sites. In **C**, the two coordination sites are diastereotopic; owing to the plane

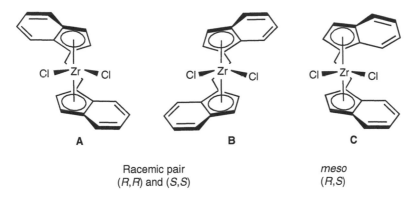

FIGURE 1.7   The three possible diastereomers of ethylene-bridged bis(indenyl)zirconocene dichloride: (a) (R,R)-Et(Ind)$_2$ZrCl$_2$; (b) (S,S)-Et(Ind)$_2$ZrCl$_2$; (c) *meso-(R,S)*-Et(Ind)$_2$ZrCl$_2$.

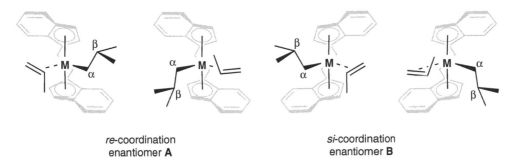

FIGURE 1.8   Lowest energy structures for propylene coordination to a chiral, C_2-symmetric metallocene-isobutyl complex. For both enantiomers, the two sites shown for monomer coordination and chain growth are equivalent.

of symmetry within the metallocene, neither site will preferentially orient the polymer chain in a particular direction so insertion of the *re* or *si* face of the incoming propylene monomer is equally likely at both coordination sites; this results in the formation of atactic polypropylene.

For the chiral racemic pair **A** and **B**, the two sites are equivalent (homotopic) through the C_2 axis of symmetry. Both sites will orient the growing polymer chain away from the benzene portion of the indene ligand. This causes the (R,R)-enantiomer, **A**, to prefer insertion of the *re* enantioface of propylene and the (S,S)-enantiomer, **B**, to prefer the *si* face. The coordination of the two enantiofaces of propylene (*re* and *si*) to metallocenes **A** and **B** is shown in Figure 1.8.

The origin of enantiomorphic site control by C_2-symmetric metallocenes has been well studied. One hypothetical mechanism is the direct influence of the ligand upon the propylene monomer. However, many researchers have demonstrated that the first monomer insertion into the metal–alkyl bond of C_2-symmetric metallocenes proceeds with at the most only slight enantioselectivity when the alkyl group is methyl.[60–65] The energy difference between *re* and *si* coordination of the alkene to this species is < 1 kcal/mol;[63,65] the distance between the methyl group of propylene and the ligand is sufficiently large such that the ligand framework is unable to affect enantioselective coordination.

In contrast, propylene insertion into a metal–alkyl bond is selective when the alkyl group is isobutyl, a model for the growing polymer chain.[61,62,66,67] In the lowest energy structure of the monomer coordinated to the metallocene (Figure 1.8), the polymer chain is directed toward the open sector of the ligand framework (chiral orientation of the growing chain). The propylene then coordinates to the metal with the methyl group *anti* to the β-C atom of the growing chain. The difference in energy between coordination of correct (preferred) versus incorrect (nonpreferred) enantiofaces of propylene by C_2-symmetric metallocenes is typically 3–4 kcal/mol.[68–71]

Since the structure of the catalyst (metal center and ligands) determines the enantiofacial selectivity of propylene insertion, the polymer microstructure of polypropylenes synthesized under enantiomorphic site control is different from the microstructure resulting from chain-end control. An insertion of the opposite enantioface does not change the enantiofacial preference of the following insertion for enantiomorphic site control, and the occurrence of an *r* dyad resulting from misinsertion is not propagated. Rather, the catalyst "corrects itself" by generating a second *r* dyad to generate a *mmmrrmmm* sequence. Therefore, *mmrr* and *mrrm* pentads are observed in the polymer in a 1:1 ratio (Figure 1.6).

In summary, the ability of C_2-symmetric metallocene catalysts to produce iPP can be rationalized given the discussion above. The ligand imposes the same chiral orientation upon the growing polymer chain regardless of which of the two positions (coordination sites) on the metal is occupied by the chain. This results in the insertion of the same enantioface of propylene at both sites. The two coordination sites are therefore homotopic, and each insertion yields a new stereocenter with the same chirality as the previous monomer insertions.

It is important to note that catalysts with C_2-symmetry are not the only metallocenes capable of forming iPP. Metallocenes with C_1-symmetry where one of the coordination sites is very sterically crowded as a result of ligand substitution can also produce polypropylene of low-to-moderate isotacticity. In C_1-symmetric metallocenes, the metal has two different coordination sites, one isoselective and one aselective, that should result in a hemiisotactic PP. However, chain propagation does not occur in a strictly alternating fashion between the two sites on the metal. The extreme steric hindrance at the site closest to the large ligand substituent causes a unimolecular isomerization reaction to occur; the polymer chain "back skips" to the less crowded site without insertion of monomer (called chain back skip or unimolecular site epimerization). Therefore, the majority of monomer insertions take place from the isoselective site. An example of such a catalyst is the isopropyl-bridged (3-*tert*-butyl-Cp)(fluorenyl)zirconocene shown in Scheme 1.7[72] (see Chapter 2 for a further discussion of C_1-symmetric Cp-fluorenyl catalysts and site epimerization).

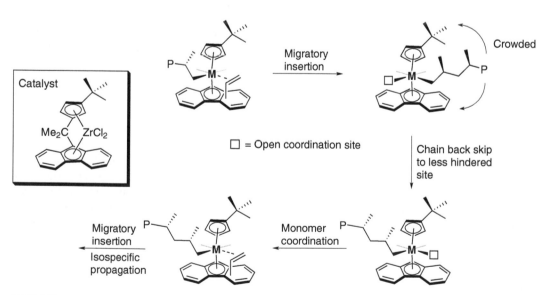

SCHEME 1.7 Synthesis of iPP by a C_1-symmetric catalyst through the chain back skip mechanism (P = growing polymer chain). (Adapted with permission from Resconi, L.; Cavallo, L.; Fait, A.; Piemontesi, F. *Chem. Rev.* **2000**, *100*, 1253–1345. Copyright 2000 American Chemical Society.)

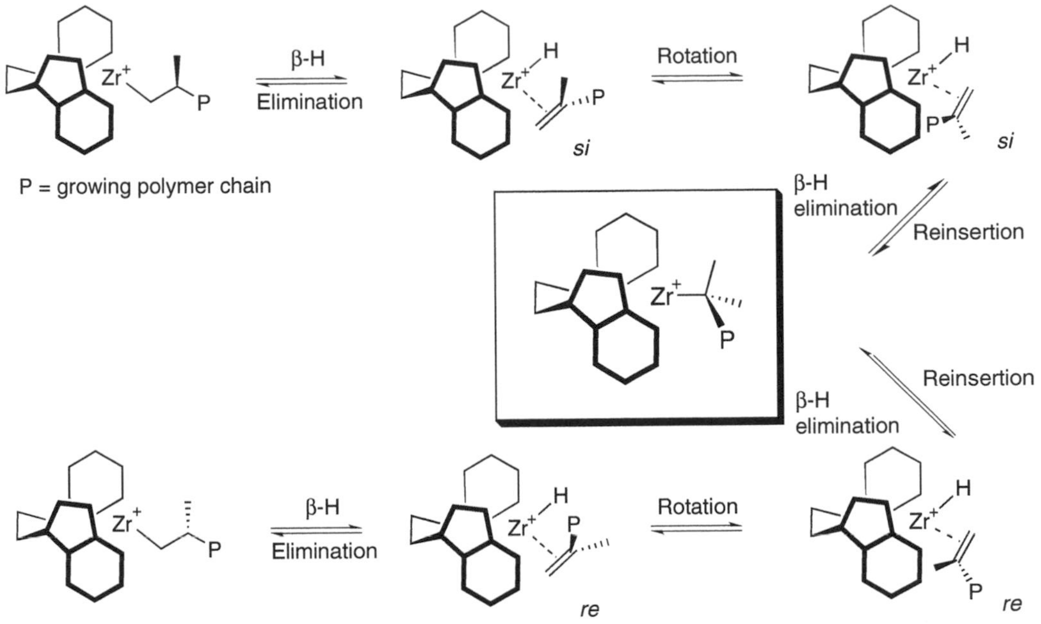

SCHEME 1.8 Mechanism of epimerization of the last stereocenter on the growing polymer chain.

1.2.3 EXPERIMENTAL CONDITIONS THAT AFFECT POLYMER TACTICITY

Polymerization conditions can affect the synthesis of iPP by C_2-symmetric catalysts. A temperature dependence of polymer tacticity has been observed, with higher temperatures resulting in a decrease in the *mmmm* content of the polymer. This observation has been rationalized in terms of the relative energy of enantioface selectivity of propylene insertion.[68–71]

Low monomer concentration can also decrease the isotacticity of polypropylene. The three-coordinate metallocene alkyl species that exists in the absence of a π-coordinated monomer is capable of racemizing the methyl group of the last-inserted monomer unit. Evidence for the occurrence of chain-end epimerization has come from studies using deuterium-labeled propylene;[26,27,73,74] the relevant mechanistic steps are believed to be a series of β-hydride eliminations and subsequent isomerizations with a tertiary alkyl species as an intermediate (Scheme 1.8).

1.3 PROTOTYPICAL C_2- AND PSEUDO-C_2-SYMMETRIC CATALYSTS

In accordance with the mechanistic requirements for enantiomorphic site-controlled formation of iPP from single-site catalysts, the most successful efforts in catalyst development have focused on conformationally rigid systems locked in axial C_2 symmetry. Historically, the milestone development in this area was initiated in the late 1970s by Hans–Herbert Brintzinger's publication of the first chiral C_2-symmetric *ansa*-metallocene, *rac*-trimethylenebis(3-*tert*-butylcyclopentadienyl)titanium dichloride (**1**, Figure 1.9).[75] The complex was initially developed as a stereorigid version of chiral titanocenes for use in asymmetric reactions with prochiral olefins. Remarkably, in this first report, the racemic mixture was already resolved into its single C_2-symmetric enantiomers by selective conversion of one of the enantiomers to the corresponding binaphtholate, a standard technique for resolving chiral racemic *ansa*-metallocenes. The ansa prefix, derived from Latin *bent* or *handle*, had already been used earlier for alkane bridges across arene rings and was now adopted by Hans Brintzinger, "the father of *ansa*-metallocenes," as a short notation for the inter-annular bridge in metallocene complexes. This first report was soon followed by the landmark

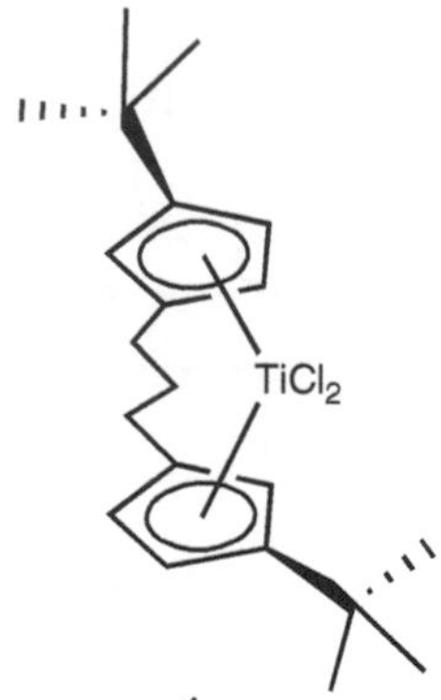

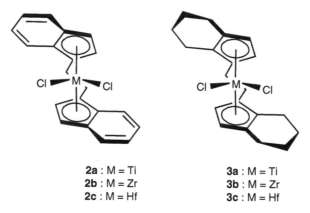

1

FIGURE 1.9   The first chiral *ansa*-metallocene, *rac*-trimethylenebis(3-*tert*-butylcyclopentadienyl)titanium dichloride (**1**). Only one enantiomer of the racemic pair is shown.

2a : M = Ti
2b : M = Zr
2c : M = Hf

3a : M = Ti
3b : M = Zr
3c : M = Hf

FIGURE 1.10   Prototypical group 4 bis(indenyl) and bis(tetrahydroindenyl) *ansa*-metallocenes **2–3**. Only one enantiomer of the racemic pair is shown.

preparations of the first chiral group 4 bis(indenyl) and bis(tetrahydroindenyl) *ansa*-metallocenes, *rac*-ethylenebis(indenyl)titanium dichloride (*rac*-Et(Ind)$_2$TiCl$_2$, **2a**) and its hydrogenated congener *rac*-ethylenebis(4,5,6,7-tetrahydroindenyl)titanium dichloride (*rac*-Et(H$_4$Ind)$_2$TiCl$_2$, **3a**) in 1982 (Figure 1.10).[76] Preparations of the *ansa*-zirconocene analogues **2b** and **3b** were reported in 1985.[77] By 1987, the hafnium analogues **2c** and **3c** had also been described by Ewen.[78]

With these C_2-symmetric homogeneous catalysts at hand, the groups of John Ewen at Exxon and Walter Kaminsky at the University of Hamburg almost simultaneously published revolutionary studies on the formation of isotactic polypropylene upon activation of chiral *ansa*-metallocenes with MAO.[54,79] Ewen used a 56:44 mixture of *rac*- and *meso*-ethylenebis(indenyl)titanium dichlorides to prepare a 63:37 mixture of isotactic and atactic polypropylenes;[54] Kaminsky and Brintzinger obtained exclusively isotactic polypropylene by use of the pure C_2-symmetric *rac* form (as a pair of enantiomers isolated away from the *meso*-diastereomer) of ethylenebis(4,5,6,7-tetrahydroindenyl)zirconium dichloride in combination with MAO.[79] This seminal experiment unambiguously demonstrated the dependence of polypropylene microstructure on the stereochemistry of the metallocene catalyst. The prototypical *ansa*-metallocene indenyl and 4,5,6,7-tetrahydroindenyl catalyst precursors **2a-c** and **3a-c** are shown in Figure 1.10. While huge numbers of structural variants of these parent systems have been described in the literature since the mid-1980s, the racemic ethylene-bridged bis(indenyl)zirconium dichloride compound **2b** remains the landmark prototype against which many of the new catalyst precursors are often referenced. Consequently, considerable efforts have been directed toward synthetic and structural studies on **2b** and its hydrogenated congener **3b**. A detailed investigation on the synthesis, molecular structure, and solution dynamics of these metallocenes has appeared in the literature.[80]

The following discussion highlights the key findings and development of C_2-symmetric catalysts for isoselective propylene polymerization; polymerization results reported here are for MAO-activated catalysts derived from the metallocene dichlorides shown in the figures unless otherwise specified. Most of the catalysts described to date are bridged complexes of the *ansa*-metallocene type. Only a few homogeneous catalyst systems of any significance based on nonmetallocene ligand structures in combination with early or late transition metals have appeared (see Chapter 6).

In addition to the bridged metallocene-based catalysts, a number of unbridged metallocene catalyst systems, exhibiting time-dependent C_2-symmetry due to restricted ligand rotation around the ligand–metal axis, have been reported to produce polypropylenes with a range of tacticities and microstructures. In the context of the present discussion, the unbridged metallocene complexes are considered as *pseudo-C_2-symmetric*.

1.3.1 *ANSA*-METALLOCENES

As reported in Ewen's seminal paper, the racemic ethylene-bridged bis(indenyl)titanocene **2a** yields isotactic polypropylene, albeit with fairly low activity and low stereoselectivity;[54] however, this complex has a limited stability under technical polymerization conditions found in industrial settings. Kaminsky's bis(tetrahydroindenyl)zirconocene analogue **3b**[79] displays a considerably higher polymerization activity, although the properties of the polymers obtained with this catalyst are far inferior to those of commercial-grade isotactic polypropylenes prepared using heterogeneous Ziegler–Natta catalysts. For example, in liquid propylene at 70 °C, **3b** produces iPP with a fourfold decrease in activity as compared to a fourth-generation heterogeneous Ziegler–Natta catalyst of the TiCl$_4$/MgCl$_2$/electron donor type (20 vs. 80 kg PP/mmol metal).[81] The homopolymer weight average molecular weight ($M_w = 15,000$) and melting point ($T_m = 125$ °C) are considerably lower than those of commercial iPP ($M_w = 100,000$–$500,000$, $T_m = 160$–165 °C).[81] Nevertheless, these metallocene catalysts have had a tremendous intellectual impact on the scientific community. With the correlation between polypropylene microstructure and metallocene chirality clearly established, reports of new homogeneous metallocene catalysts for production of syndiotactic,[82] hemiisotactic,[83] and isotactic–atactic stereoblock[84] polypropylene microstructures soon followed. Researchers also pursued the development of metallocene-based catalysts for production of commercial-grade iPPs by structural modifications of the Brintzinger-type catalysts. To date, the most successful attempts at catalyst design for highly isotactic polypropylene have evolved from these C_2-symmetric racemic bis(indenyl) *ansa*-zirconocene-type structures. In a few cases, successful candidates have been obtained by variations of bridged bis(cyclopentadienyl) ligand backbones; attempts using bis(fluorenyl) ligands have been far less fruitful. Selected examples are presented in the discussion below.

1.3.1.1 C_2-Symmetric Bis(indenyl) *ansa*-Metallocenes

Logical starting points for structural variations of bridged bis(indenyl) group 4 metallocenes are as follows: (1) the central metal; (2) the interannular bridge connecting the η^5-indenyl ligands; (3) the substitution pattern of the ligand's five-membered ring; and (4) the substitution pattern of the ligand's six-membered ring. Of these, the last is the most synthetically demanding.

Substantial experimental evidence shows that Zr is clearly the preferred transition metal in terms of cost, activity, stability, and the resulting polymer molecular weight. Hafnium complexes commonly produce polypropylenes with higher molecular weights than the analogous zirconium

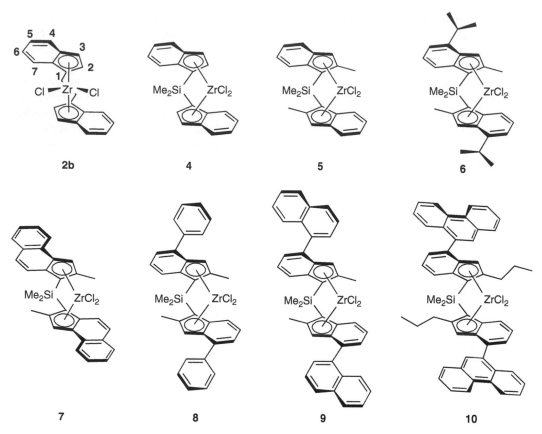

FIGURE 1.11 Representative dimethylsilylene-bridged bis(indenyl) *ansa*-metallocenes **4–10** for isoselective propylene polymerization. Only one enantiomer of the racemic pair is shown.

complexes, owing to the relatively stronger Hf–C bond, while typically suffering from considerably lower polymerization activities. Differences in stereoregulating ability between analogous Zr and Hf catalysts are usually marginal. Most of the titanium analogues reported in the literature are inferior to the corresponding zirconium complexes when catalyst activities, thermal stabilities, and stereoselectivities are compared. Attempts to manipulate catalyst performance by optimizing the ligand structure rather than changing the central metal are thus prevailing in the literature. Key C_2-symmetric bis(indenyl) *ansa*-metallocenes for iPP synthesis are shown in Figure 1.11; a comparison of these catalysts' propylene polymerization characteristics under technical conditions is summarized in Table 1.1.

Soon after the first reports by Ewen[54] and Kaminsky and Brintzinger,[79] it was shown that by changing the two-atom ethylene bridge of **2b** to a one-atom dimethylsilylene bridge (complex **4**), the molecular weight and isotacticity of the iPP produced were enhanced.[85,86] A further increase in polymer molecular weight was obtained by 2-methyl substitution of the indenyl ligands (complexes **5** and **6**).[87]

Truly high-performance metallocene catalysts for the production of isotactic polypropylene have been prepared (**7–9**).[2,3] Key components of these catalysts are the following: (1) a one-atom dimethyl-silylene bridge; (2) alkyl substitution at the 2-position of the indenyl five-membered ring; and (3) a fused aromatic ring on the indenyl six-membered ring (benz[e]indenyl group) or a 4-aryl substituent. For example, in liquid propylene, the 2-methyl-4-naphthyl-substituted catalyst **9** produces highly isotactic polypropylene with an isotactic *mmmm* pentad content of > 99%, a M_w of 920,000, and a T_m

TABLE 1.1

Comparison of Propylene Polymerization Performance for Some Advanced *ansa*-Metallocenes and a Fourth-Generation Ziegler–Natta Catalyst[a]

Catalyst	Year[b]	A[c]	iPP M_w	iPP mmmm (%)	iPP T_m (°C)	References
Ziegler–Natta[d]	1980s	20	900,000	> 99	162	81
rac-Me$_2$Si(2-Me-4-Naph-Ind)$_2$ZrCl$_2$ (**9**)	1994	875	920,000	99.1	161	3
rac-Me$_2$Si(2-Me-4-Ph-Ind)$_2$ZrCl$_2$ (**8**)	1994	755	729,000	95.2	157	3
rac-Me$_2$Si(2-Me-Benz[*e*]Ind)$_2$ZrCl$_2$ (**7**)	1994	403	330,000	88.7	146	3
rac-Me$_2$Si(2-Me-4-*i*Pr-Ind)$_2$ZrCl$_2$ (**6**)	1992	245	213,000	88.6	150	3
rac-Me$_2$Si(2-Me-Ind)$_2$ZrCl$_2$ (**5**)	1992	99	195,000	88.5	145	3
rac-Me$_2$Si(Ind)$_2$ZrCl$_2$ (**4**)	1989	190	36,000	81.7	137	3
rac-Et(H$_4$Ind)$_2$ZrCl$_2$ (**3b**)	1985	80	15,000	na[e]	125	81
rac-Et(Ind)$_2$ZrCl$_2$ (**2b**)	1985	188	24,000	78.5	132	3

[a] Under technical polymerization conditions (70 °C, MAO activator; [Al]:[Zr] = 15,000:1; liquid propylene). Ind = indenyl; H$_4$Ind = 4,5,6,7-tetrahydroindenyl; Naph = naphthyl; Ph = phenyl.

[b] First published.

[c] Activity in kg iPP/mmol metal·h.

[d] Heterogeneous Ziegler–Natta catalyst (Ti/MgCl$_2$/electron donor).

[e] Data not available.

of 161 °C, values essentially identical to those obtained with modern heterogeneous Ziegler–Natta catalysts.[3] The activity of **9**, however, exceeds that of the tabulated Ziegler–Natta catalyst by a factor of 40 in terms of kg PP/mmol metal·h produced (880 versus 20). Further optimization attempts have resulted in the preparation of catalyst precursor **10** having a 2-*n*-propyl-4-phenanthryl substitution pattern, which reportedly provides even higher isotacticities and polymer melting points.[88]

On the basis of these empirical studies on the influence of catalyst structure on polymerization performance, the synergistic influence of 2-alkyl-4-aryl substitution appears to be the most beneficial, albeit through a complicated combination of steric and electronic effects. The 2-alkyl substituent appears to disfavor chain release, even after 2,1-insertion, and consequently increases the polymer molecular weight. The 4-aryl substituent affects the catalyst enantioselectivity of the insertion step, and hence the polymer stereoregularity, apparently through a predominantly steric effect[2,3,6,81,87] (see Chapter 3 for a further discussion of indenyl ligand substitution pattern effects in propylene polymerization).

Notably, the last papers to contain significant conceptual insight into metallocene catalyst development for the synthesis of highly crystalline iPP were published over 10 years ago.[2,3] As envisioned by Resconi, Cavallo, Fait, and Piemontesi in their seminal review,[6] in terms of catalyst efficiency and cost, metallocene catalysts are unlikely to replace the latest generation heterogeneous Ti/MgCl$_2$ Ziegler–Natta catalysts "*in any foreseeable future.*" Instead, the greatest potential of homogeneous catalysts for propylene polymerization appears to have shifted to lower-crystallinity, intermediate-tacticity materials, such as thermoplastic elastomers, not obtainable with the current Ziegler–Natta catalyst technologies.

Of the more recent publications involving catalyst design for highly crystalline iPP, the most notable examples of high-performance C_2-symmetric bis(indenyl) *ansa*-zirconocenes are probably the one-carbon-atom-bridged catalysts with bulky 3-substituents.[68,89] For example, the methylene- and isopropylidene-bridged precursors **11** and **12** (Figure 1.12) produce iPP with superior regioregularity, high isotacticity (mmmm = 95–98%), high melting points (T_m = 154–163 °C), and modest

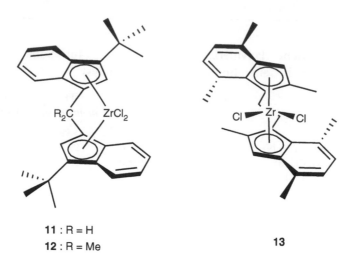

11 : R = H
12 : R = Me

13

FIGURE 1.12 Methylene-, isopropylidene-, and ethylene-bridged bis(indenyl) *ansa*-metallocenes **11–13** for isoselective propylene polymerization. Only one enantiomer of the racemic pair is shown.

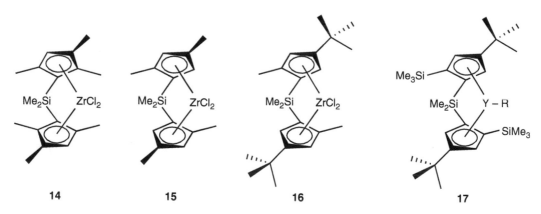

14   **15**   **16**   **17**

FIGURE 1.13 C_2-Symmetric bis(cyclopentadienyl) *ansa*-metallocenes **14–17** for isoselective propylene polymerization. Only one enantiomer of the racemic pair is shown. Note that **17** is a group 3 metallocene that does not require activation by a Lewis acidic species (R = hydride; actual precatalyst structure is a dimer).

to high molecular weights (M_w = 70,000–780,000). The polymer properties and catalyst productivities of **11** and **12** are not quite on the same level as those for the optimized catalysts **8** and **9**, but the very simple ligand design of **11** and **12** may offer some economical advantages in catalyst cost. Isotactic polypropylene with similar properties has also been synthesized using the C_2-symmetric, ethylene-bridged 2,4,7-trimethyl-substituted bis(indenyl) catalyst **13** (Figure 1.12).[90]

1.3.1.2 C_2-Symmetric Bis(cyclopentadienyl) *ansa*-Metallocenes

In comparison to bis(indenyl) *ansa*-metallocenes, there are considerably fewer examples of C_2-symmetric isoselective metallocene catalysts on the basis of bridged bis(cyclopentadienyl) ligand framework. Representative complexes **14–17** are illustrated in Figure 1.13.[3,91–93] Many of the reported chiral bis(cyclopentadienyl) *ansa*-zirconocenes produce highly stereoregular iPP, but with polymerization activies much lower than those of the structurally optimized bis(indenyl)-based catalysts. For example, in liquid propylene at 70 °C, **16** produces iPP with *mmmm* > 94%,

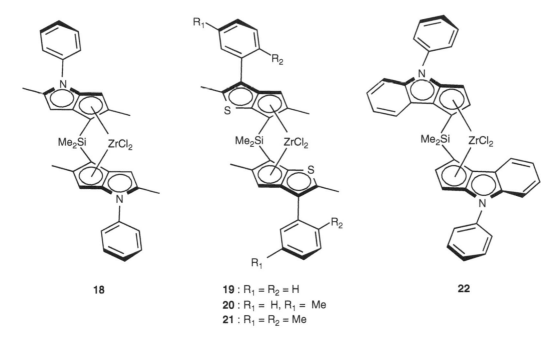

18

19 : R₁ = R₂ = H
20 : R₁ = H, R₁ = Me
21 : R₁ = R₂ = Me

22

FIGURE 1.14 Representative C_2-symmetric *ansa*-"heterocenes" **18–22** for isoselective propylene polymerization. Only one enantiomer of the racemic pair is shown.

$M_w = 19,000$, and $T_m = 155\,°C$ with a polymerization activity of 10 kg PP/mmol Zr·h,[3] as compared to $M_w = 920,000$ and 880 kg PP/mmol Zr·h obtained with the 2-methyl-4-naphthyl-substituted bis(indenyl) catalyst **9** under similar polymerization conditions (see Chapter 3 for a further discussion of cyclopentadienyl ligand substitution pattern effects in propylene polymerization).

The C_2-symmetric neutral yttrocene complex **17** was the first stereoselective single-component catalyst (the catalyst precursor is a dimeric hydride-bridged yttrocene that does not require a Lewis acid activator to insert propylene) reported for propylene polymerization. This catalyst produces highly isotactic polypropylene at ambient temperature ($mmmm = 97\%, T_m = 157\,°C$), albeit with low activity and low polymer molecular weight.[93]

As an entirely new approach, "heterocene"-based C_2-symmetric metallocene catalysts for iPP synthesis have been reported;[94–97] their ligand design incorporates a heterocyclic ring system such as thiophene or pyrrole fused to a cyclopentadienyl ring. Representative complexes **18–22** of this class are shown in Figure 1.14. Essentially, the complexes have been designed to mimic the high-performance bis(indenyl)zirconocenes **8** and **9**.[2,3] Heterocene **19** is reportedly the most active metallocene catalyst for polypropylene reported as of 2001.[95] This catalyst produces highly isotactic PP with $M_w = 122,000$ and $T_m = 157\,°C$ with a polymerization activity of 5000 kg PP/mmol Zr·h (in liquid propylene, 70 °C, in the presence of H₂). The 2-methylcyclopenta[b]indole-derived zirconocene **22** in turn exhibits polymerization characteristics intermediate between those of the high-performance catalysts **7** and **8**.[96]

1.3.1.3 C_2-Symmetric Bis(fluorenyl) *ansa*-Metallocenes

A few C_2-symmetric bis(fluorenyl) *ansa*-metallocenes have been prepared by either desymmetrizing the fluorenyl ligand to generate a geometry analogous to that of bridged bis(indenyl) complexes[98,99] or by desymmetrization of the interannular ethylene bridge with proper substituents (Figure 1.15).[100] These variations result in catalysts with low polymerization activity and stereoselectivity. The methyl- and benzo-fused-substituted complexes **23** and **24** produced atactic

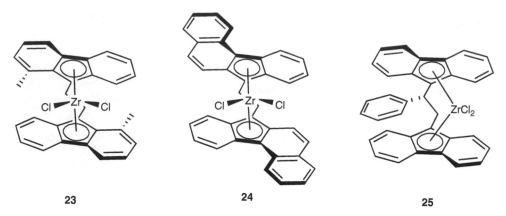

FIGURE 1.15 C_2-Symmetric bis(fluorenyl) *ansa*-metallocenes **23–25** for isoselective propylene polymerization. Only one enantiomer of the racemic pair is shown.

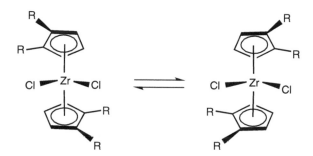

SCHEME 1.9 Temporal C_2 symmetry displayed by unbridged metallocene complexes with identically substituted cyclopentadienyl ligands (a pair of enantiomers is shown).

or slightly isotactic polypropylenes.[98,99] The bridge-substituted catalyst **25** yields polypropylenes with *mmmm* = 32–64% and M_w = 1,800–21,200, depending upon the polymerization temperature. The pentad distributions for **25** indicate stereocontrol by the enantiomorphic site mechanism. The low catalyst activities and the lack of any significant stereocontrol for the bis(fluorenyl) catalysts is attributable to steric crowding on both sides of the active center, resulting in steric hindrance toward monomer coordination/insertion and, even more likely, a failure of the ligands to affect chiral orientation of the growing polymer. Questions concerning the thermal stabilities of group 4 bis(fluorenyl) metallocenes have also been raised.

1.3.2 UNBRIDGED METALLOCENES

Unbridged group 4 metallocene complexes with two identically substituted cyclopentadienyl ligands may display temporal C_2-symmetry (Scheme 1.9), which theoretically can result in stereoselective polymerization by the enantiomorphic site control mechanism. In most cases, however, the barrier against ligand rotation around the cyclopentadienyl-metal axis is low, and time-averaged molecular C_{2v}-symmetry prevails at all temperatures. With carefully selected bulky ligand substituents, these systems may be tuned to produce at least partially isotactic polypropylenes, often at low polymerization temperatures with concomitant decreases in polymerization activities. In such cases, conformational isomers displaying stereoselective monomer coordination and chain propagation may prevail for sufficient time intervals in order to produce isotactic chains.

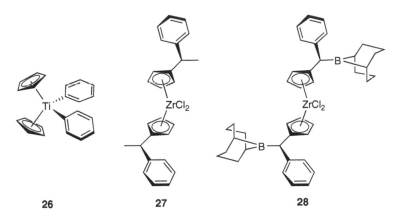

FIGURE 1.16 Unbridged group 4 bis(cyclopentadienyl) complexes **26–28** for isoselective propylene polymerization.

In special cases, this type of catalyst may "oscillate," switching between the two stereoselective C_2-symmetric racemic-like conformations and a nonstereoselective *meso*-like conformation during the propagation of a single polymer chain. The result is a stereoblock polymer microstructure (see Chapter 8). The most-studied catalyst precursor of this type is the unbridged complex bis(2-phenylindenyl)zirconium dichloride (*vide infra*).[84,101] The "oscillating" stereocontrol mechanism was, for the first time, initially proposed for this particular catalyst,[84] although the actual mechanism may be inherently more complex.[102]

1.3.2.1 Pseudo-C_2-Symmetric Bis(cyclopentadienyl) Metallocenes

In 1984, Ewen reported the first chain-end controlled polymerization of propylene to isotactic polypropylene with metallocene catalysts, using the unbridged C_{2v}-symmetric bis(cyclopentadienyl)titanium diphenyl complex **26** at low temperature (Figure 1.16).[54] With this catalyst system, partially isotactic PP with *mmmm* = 52% is obtained at −45 °C.

Enantiomorphic site-controlled formation of iPP with unbridged pseudo-C_2-symmetric bis(cyclopentadienyl) metallocene catalysts has been observed in only a few cases. At −79 °C, racemic 1-phenylethyl-substituted bis(cyclopentadienyl)zirconocene **27** (Figure 1.16) produces low molecular weight iPP with *mmmm* = 60% through a mechanism that is at least partially chain-end controlled.[103] The related racemic 9-boracyclo[3.3.1]nonane-substituted analogue **28** (Figure 1.16) gives slightly higher stereoselectivity, producing iPP having *mmmm* = 75% with predominating enantiomorphic site control at −50 °C.[104]

1.3.2.2 Pseudo-C_2-Symmetric Bis(indenyl) Metallocenes

Certain chirally substituted unbridged bis(indenyl)zirconocenes are active catalysts for isoselective propylene polymerization at low temperatures where ligand rotation is most restricted, and function by enantiomorphic site control (Figure 1.17).[105,106] For example, the catalyst derived from the cholestanyl-substituted bis(indenyl) complex **29** gives, at −30 °C, high molecular weight iPP with *mmmm* = 80%.[105] The neomenthyl- (**30**) and neoisomenthyl-substituted (**31**) zirconocene analogues of this complex (and their tetrahydroindenyl congeners) give iPPs with a range of tacticities from *mmmm* = 16–77% at polymerization temperatures ranging from −30 to −5 °C. At these temperatures, the polymerization activities are fairly low. Higher stereoselectivities are obtained with the neoisomenthyl-substituted complexes as compared to the isomenthyl-substituted catalysts. The observed enantioselectivity of propylene insertion for **29**, **30**, and **31** is likely to result from the predominance of a C_2-symmetric metallocene rotamer.

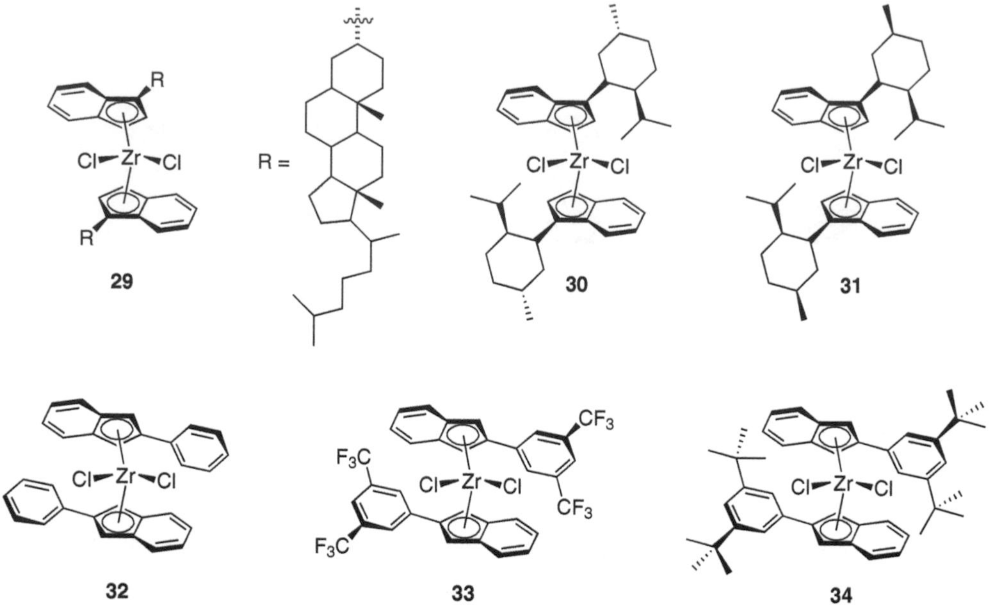

FIGURE 1.17 Unbridged pseudo-C_2-symmetric bis(indenyl) zirconocenes **29–34** for isoselective propylene polymerization.

The 2-arylindenyl zirconocenes developed by Waymouth for the synthesis of stereoblock elastomeric polypropylenes (see Chapter 8) provide de facto a range of tacticities, depending on the polymerization conditions such as temperature and pressure and the ligand substitution pattern.[101] The prototypical catalyst bis(2-phenylindenyl)zirconium dichloride **32** (Figure 1.17) gives stereoblock polypropylene with intermediate tacticity and isotactic pentad content (*mmmm* = 33%) in liquid propylene at 20 °C.[84,101] The sterically more congested 3,5-bis(trifluoromethyl)-[107] and 3,5-(di-*tert*-butyl)-substituted[108] analogues (**33** and **34**, respectively, Figure 1.17) both provide iPPs with *mmmm* = 70% under similar polymerization conditions. The increase in stereoselectivity is believed to arise from the 3,5-substituents' role in restricting ligand rotation, which favors a predominance of the C_2-symmetric chiral rotamers and thus enhances the stereoselectivity of the propagation step.[101,107,108]

1.3.2.3 Pseudo-C_2-Symmetric Bis(fluorenyl) Metallocenes

Among unbridged C_2-symmetric bis(fluorenyl)metallocenes, the racemic 1-methyl-substituted zirconocene **35a** (Figure 1.18) reportedly produces isotactic polypropylene with *mmmm* > 80% upon activation with MAO at unusually high polymerization temperatures (60 °C) for unbridged catalysts.[109] The iPP obtained with **35a** has a T_m of 145 °C, characteristic of moderately high isotacticity, although both the molecular weight of the polymer (M_w = 65,000) and catalyst activity are fairly low. Somewhat lower molecular weights and less stereoregular polymer are obtained with the hafnium analogue **35b** (Figure 1.18). Once again, strongly hindered ligand rotation is responsible for the chiral disposition of the substituted fluorenyl ligands, which in turn affect enantiofacial selectivity in monomer insertion.

1.4 SYNTHETIC CONSIDERATIONS

Of the C_2-symmetric metallocene catalysts used to prepare iPP, the bis(indenyl)- and bis(tetrahydroindenyl)-type complexes form the most important classes. Thus, the synthetic

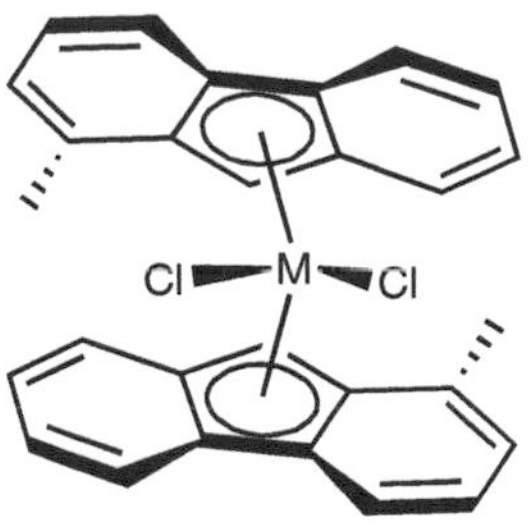

35a : M = Zr
35b : M = Hf

FIGURE 1.18 Bis(1-methylfluorenyl)metallocenes **35a–b** for isoselective propylene polymerization.

considerations illustrated here are focused upon these classes of ligands. In most cases, analogous procedures and strategies apply to the bis(cyclopentadienyl) and bis(fluorenyl) ligand systems as well. Since excellent and detailed review articles are readily available describing the synthesis of various substituted chiral and prochiral cyclopentadienyl, indenyl, and fluorenyl ligands and their corresponding chiral group 4 metallocene complexes,[7,8] only selected synthetic considerations are briefly discussed in the present context, including recently published references. The synthesis of group 4 bis(indenyl) metallocenes containing heteroatom-substituted ligand frameworks (containing N, O, or S) has likewise been reviewed recently.[110]

1.4.1 Separation of *rac* and *meso* Diastereomers

The application of chiral group 4 metallocenes in stereoselective synthesis and polymerization reactions is hampered by their often difficult preparation. Commonly, these complexes are prepared by salt metathesis reactions of the ligand dianions, often lithium or potassium salts, with group 4 metal tetrahalides. Chemical yields may vary to a great extent and issues controlling the chemoselectivity (metallocene versus dinuclear or oligomeric products) and diastereoselectivity (formation of racemic versus *meso* diastereomers) are only poorly understood. In order to optimize yields, which are frequently low, several counterions, transition metal reagents, solvents, reaction temperatures, and work-up procedures may have to be screened in each case. The issue becomes especially critical with multiply-substituted ligand precursors, such as those in complexes **8** and **9**,[2,3,81,87] for which several low-yielding steps are already required for construction of the ligand frameworks.

The most important issue in the synthesis of bridged *ansa*-metallocenes is the separation of the C_s-symmetric *meso* diastereomer from the C_2-symmetric racemic enantiomer pair, as only the latter isomer by definition functions as a stereoselective polymerization catalyst. A simplified schematic presentation of the synthesis of the classic ethylenebis(indenyl)zirconium dichloride **2b** is illustrated in Scheme 1.10; for detailed discussions on the synthesis of this and other chiral metallocenes the reader is referred to the reviews in the literature[7,8] and the original publications.[2,3,68,75–78,80–83,85–89,94–96] In general, a similar situation exists for most other bridged *ansa*-metallocenes as well, and a reaction of the bridged ligand dianion with the appropriate group 4 metal halide yields both racemic and *meso* diastereomers of the corresponding metallocene in variable ratios. The desired *rac* diastereomer is then purified and isolated (as a mixture of enantiomers) by fractional crystallization with variable success. The corresponding tetrahydroindenyl complexes are obtained by catalytic hydrogenation of the parent indenyl complexes, often in nearly quantitative yields.

Various strategies have been devised to maximize the formation of the racemic diastereomers of *ansa*-metallocenes. Most of the hitherto-described methods lack generality and are

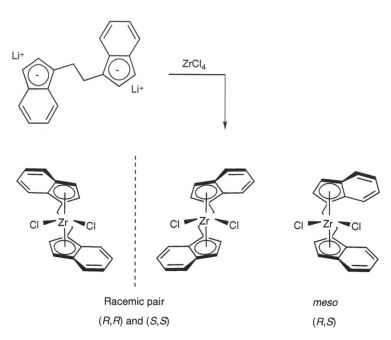

SCHEME 1.10   Generalized synthesis of bis(indenyl) *ansa*-zirconocenes to give a mixture of *rac* and *meso* diastereomers.

applicable in specific cases only.[111] Bulky ligand substituents in the α-position to the interannular bridge, sterically congested or stereogenic bridges, and use of various ligand transfer agents based upon silicon, tin, and aluminum have been employed to increase the *rac* yield with variable results.[7,8] The racemoselective synthesis of several chiral *ansa*-metallocene diamides by the amine elimination reaction of $M(NMe_2)_4$ (M = Zr, Hf) with protic bridged bis(indenes) and bis(cyclopentadienes) has been reported.[112–114] Very recently, the selective synthesis of racemic C_2-symmetric *ansa*-metallocenes by the chelate-controlled reaction of bis(indenyl) ligand dilithium salts with $Zr\{PhN(CH_2)_3NPh\}Cl_2(THF)_2$ (THF = tetrahydrofuran) has been developed and shows considerable promise as a general method for racemate synthesis.[111]

In the case of unbridged bis(indenyl) complexes, a similar selectivity issue exists with unsymmetrically substituted ligand precursors, for which the two π-faces of the ligand are inequivalent (planar chirality; (*pR*) versus (*pS*); Scheme 1.11).[8] If the substituent R is achiral (resulting in enantiotopic ligand π-faces), three stereoisomers can be formed, and a racemic pair consisting of the (*pR*)(*pR*) and (*pS*)(*pS*) combinations must be separated from the (*pR*)(*pS*) *meso*-form by fractional crystallization. If, as for example in the case of the cholestanyl- and menthyl-substituted complexes **29–31**, the substituent R is chiral (resulting in diastereotopic ligand π-faces), three diastereomers are formed; the (*pR*)(*pR*) and (*pS*)(*pS*) combinations no longer share an enantiomeric relationship and are often referred to as "racemic-like" diastereomers apart from the (*pR*)(*pS*) "*meso*-like" diastereomer.[103–106]

For symmetrically substituted ligands (which have equivalent or homotopic ligand π-faces), for example, the 2-arylindenyl metallocenes **32–34**,[84,107,108] the situation becomes simpler because only one stereoisomer complex can be formed regardless of the nature (chiral versus achiral) of the ligand substituent(s).

1.4.2   VARIATIONS ON LIGAND SUBSTITUENTS

Variations in the ligand substitution pattern of the indenyl five-membered ring are easy to perform. Deprotonation of indene with strong bases such as *n*-BuLi generates indenyl anion, which reacts

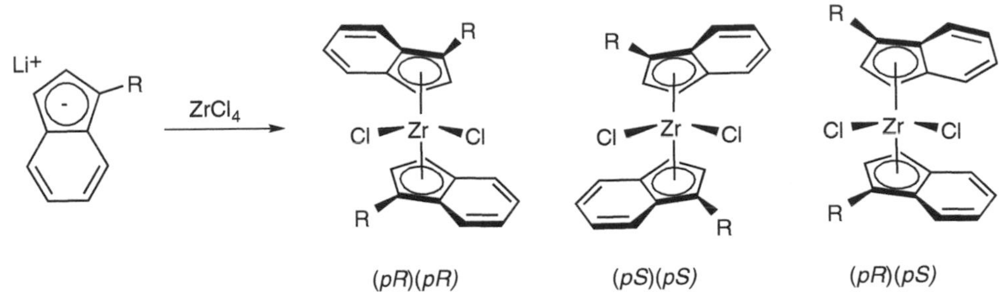

SCHEME 1.11 Synthesis of unbridged bis(indenyl)zirconocenes to give a mixture of *rac* [(*pR*)(*pR*) and (*pS*)(*pS*)] and *meso* [(*pR*)(*pS*)] diastereomers.

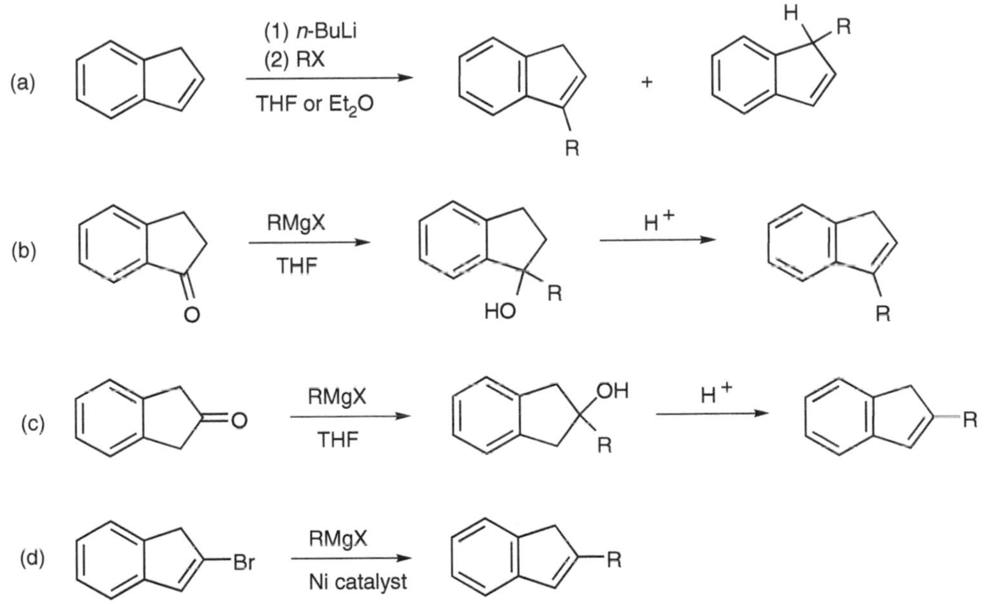

SCHEME 1.12 General methods for the synthesis of 1- and/or 3- and 2-substituted indenes are the following: (a) synthesis of 3-substituted indenes through deprotonation of the indene followed by substitution with an alkyl halide; (b) synthesis of 3-substituted indenes through addition of a Grignard reagent to 1-indanone followed by dehydration; (c) synthesis of 2-substituted indenes through addition of a Grignard reagent to 2-indanone followed by dehydration; (d) synthesis of 2-substituted indenes through metal-catalyzed cross-coupling of a Grignard reagent with 2-bromoindene.

with alkyl and silyl halides to form 1- and/or 3-alkyl/silyl-substituted indenes (Scheme 1.12a). A second, more general method (Scheme 1.12b) is the addition of metal alkyls to 1-indanones; following acid-catalyzed dehydration, 3-alkylindenes are obtained, often in good yields. A similar reaction sequence with 2-indanones (Scheme 1.12c) provides 2-substituted indenes. Another useful synthesis of 2-substituted indenes is based on the metal-catalyzed cross-coupling of alkyl and aryl Grignard reagents with 2-bromoindene (Scheme 1.12d). For a detailed discussion on the generality and applicability of the synthesis methods for these ligand and heteroatom-substituted variants, the reader is referred to recent reviews.[7,8,110]

Substitutions in the six-membered rings of indenes are considerably more difficult to access synthetically. In most cases, construction of the entire indenyl ring system from smaller building blocks is required and the desired ligands are obtained in low overall yields. As an example, the

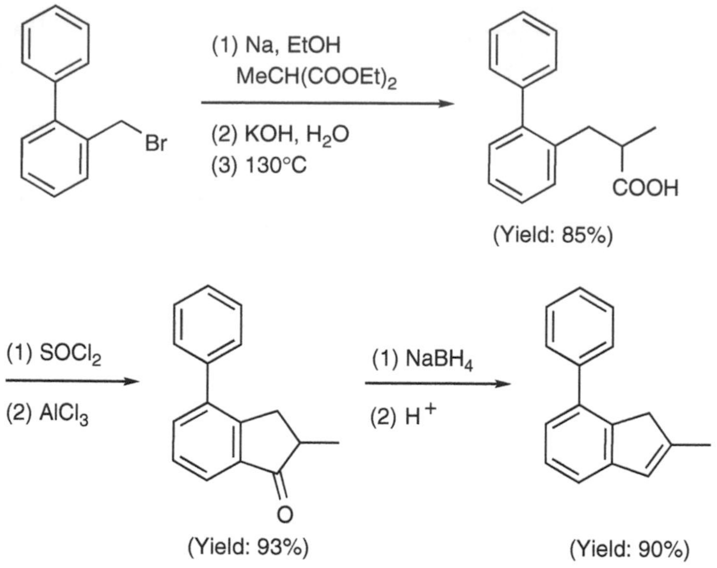

SCHEME 1.13   Synthesis of 2-methyl-7-phenylindene.

synthesis of 2-methyl-7-phenylindene, a building block for the high-performance catalyst **8**, is shown in Scheme 1.13.[3] Related strategies are employed for the synthesis of other multiply substituted indenes.[2,3,8,87]

1.4.3   VARIATIONS IN THE INTERANNULAR BRIDGE

The preparation of ethylene- and dimethylsilylene-bridged indenes generally follows the alkylation routes displayed in Scheme 1.12. Thus, in most cases, the reaction of indenyllithium salts with dibromoethane or dimethyldichlorosilane gives the corresponding bridged bis(indenes) in good yields (Scheme 1.14).[7,8,76–78,80,81,83,86,110] The reaction is equally applicable to substituted indenes.[2,3,87] Considerably less attention has been paid to longer alkyl and silyl bridges, since these variations generally result in a considerable decrease in catalyst efficiency, rendering the complexes either inactive or nonstereoselective toward propylene insertion.[4–6,81,86] These decreases in efficiency have been attributed to increased conformational flexibility of the ligand backbone with the longer bridges, a decrease in the bite angle, and narrowing of the coordination gap.

The one-carbon bridged ligands for the Montell high-performance catalysts **11** and **12** are elegantly constructed in high yield by the base-catalyzed condensation of formaldehyde or acetone with the corresponding 3-substituted indenes.[68,89]

1.5   CONCLUSIONS

The synthesis of iPP from C_2-symmetric metallocenes is a mature field within the discipline of olefin polymerization. The mechanism by which the chirality of the catalyst structure affects enantiofacial insertion of the monomer has been elucidated in great detail. Synthetic efforts have generated many structural variations on the parent bis(cyclopentadienyl), bis(indenyl), and bis(fluorenyl) ligand motifs, both bridged and unbridged. Although C_2-symmetric metallocenes capable of synthesizing highly crystalline iPP have been developed, they are not industrially significant. However, these complexes remain an important milestone in homogeneous catalysis and continue to impact the development of metallocenes for the synthesis of polypropylenes with other microstructures.

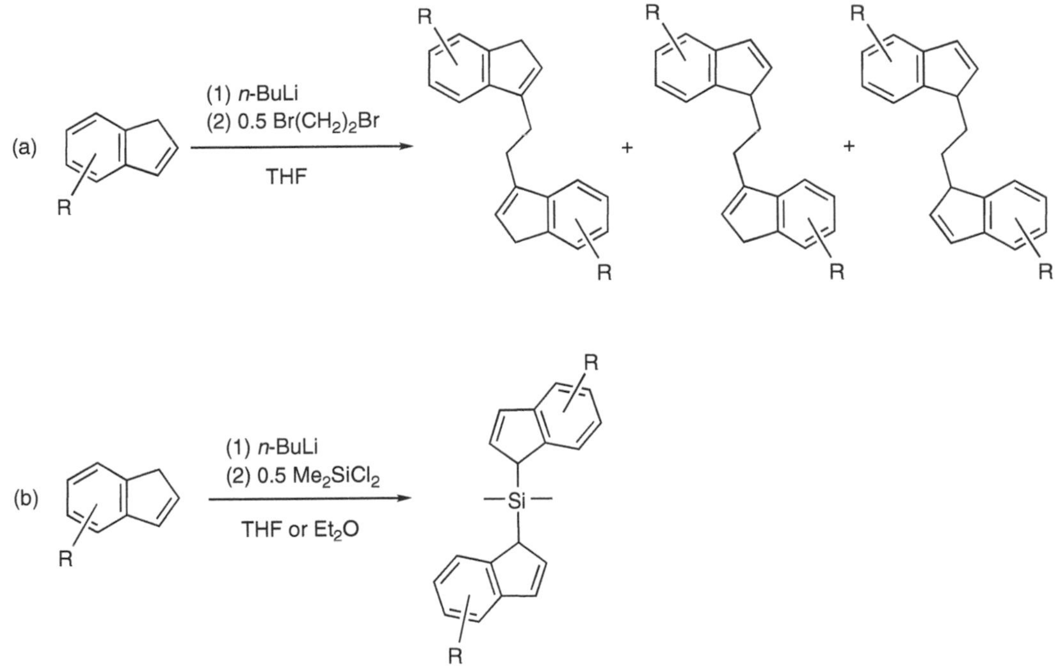

SCHEME 1.14 Synthesis of (a) ethylene-bridged and (b) dimethylsilylene-bridged bis(indenes).

REFERENCES AND NOTES

1. McCoy, M.; Reisch, M. S.; Tullo, A. H.; Short, P. L.; Tremblay, J.-F.; Storck, W. J. Facts and figures for the chemical industry. *Chem. Eng. News* **2004**, *82*, 23–63.

2. Stehling, U.; Diebold, J.; Kirsten, R.; Roell, W.; Brintzinger, H. H.; Juengling, S.; Muelhaupt, R.; Langhauser, F. *ansa*-Zirconocene polymerization catalysts with anelated ring ligands—effects on catalytic activity and polymer chain length. *Organometallics* **1994**, *13*, 964–970.

3. Spaleck, W.; Kueber, F.; Winter, A.; Rohrmann, J.; Bachmann, B.; Antberg, M.; Dolle, V.; Paulus, E. F. The influence of aromatic substituents on the polymerization behavior of bridged zirconocene catalysts. *Organometallics* **1994**, *13*, 954–963.

4. Brintzinger, H. H.; Fischer, D.; Muelhaupt, R.; Rieger, B.; Waymouth, R. M. Stereospecific olefin polymerization with chiral metallocene catalysts. *Angew. Chem., Int. Ed. Engl.* **1995**, *34*, 1143–1170.

5. Coates, G. W. Precise control of polyolefin stereochemistry using single-site metal catalysts. *Chem. Rev.* **2000**, *100*, 1223–1252.

6. Resconi, L.; Cavallo, L.; Fait, A.; Piemontesi, F. Selectivity in propene polymerization with metallocene catalysts. *Chem. Rev.* **2000**, *100*, 1253–1345.

7. Halterman, R. L. Synthesis and applications of chiral cyclopentadienylmetal complexes. *Chem. Rev.* **1992**, *92*, 965–994.

8. Halterman, R. L. Synthesis of chiral titanocene and zirconocene dichlorides. In *Metallocenes—Synthesis, Reactivity and Applications*; Togni, A., Halterman, R. L., Eds.; Wiley-VCH: Weinheim, 1998; pp 455–544.

9. Pedeutour, J.-N.; Radhakrishnan, K.; Cramail, H.; Deffieux, A. Reactivity of metallocene catalysts for olefin polymerization: Influence of activator nature and structure. *Macromol. Rapid Commun.* **2001**, *22*, 1095–1123.

10. Chen, E. Y.-X.; Marks, T. Cocatalysts for metal-catalyzed olefin polymerization: Activators, activation processes and structure-activity relationships. *Chem. Rev.* **2000**, *100*, 1391–1434.

11. Borrelli, M.; Busico, V.; Cipullo, R.; Ronca, S.; Budzelaar, P. H. M. Selectivity of metallocene-catalyzed olefin polymerization: A combined experimental and quantum mechanical study. The *ansa*-Me$_2$Si(Ind)$_2$Zr and *ansa*-Me$_2$C(Cp)(Flu)Zr systems. *Macromolecules* **2003**, *36*, 8171–8177.

12. Arlman, E. J.; Cossee, P. Ziegler–Natta catalysis. III. Stereospecific polymerization of propene with the catalyst system TiCl$_3$-AlEt$_3$. *J. Catal.* **1964**, *3*, 99–104.

13. Cossee, P. Reaction mechanism of ethylene polymerization with heterogeneous Ziegler–Natta catalysts. *Tetrahedron Lett.* **1960**, *17*, 12–16.

14. Cossee, P. Formation of isotactic polypropylene under the influence of Ziegler–Natta catalysts. *Tetrahedron Lett.* **1960**, *17*, 17–21.

15. Cossee, P. Ziegler–Natta catalysis. I. Mechanism of polymerization of α-olefins with Ziegler–Natta catalysts. *J. Catal.* **1964**, *3*, 80–88.

16. Breslow, D. S.; Newburg, N. R. Dicyclopentadienyltitanium dichloride-alkylaluminum complexes as soluble catalysts for the polymerization of ethylene. *J. Am. Chem. Soc.* **1959**, *81*, 81–86.

17. Green, M. L. H. Organic chemistry of molybdenum and related topics. *Pure Appl. Chem.* **1972**, *30*, 373–388.

18. Ivin, K. J.; Rooney, J. J.; Stewart, C. D.; Green, M. L. H.; Mahtab, R. Mechanism for the stereospecific polymerization of olefins by Ziegler–Natta catalysts. *J. Chem. Soc., Chem. Commun.* **1978**, 604–606.

19. Laverty, D. T.; Rooney, J. J. Mechanism of initiation of the ring-opening polymerization and addition oligomerization of norbornene using unicomponent metathesis catalysts. *J. Chem. Soc., Faraday Trans. 1* **1983**, *79*, 869–878.

20. Dawoodi, Z.; Green, M. L. H.; Mtetwa, V. S. B.; Prout, K. A titanium-methyl group containing a covalent bridging hydrogen system: X-ray crystal structure of Ti(Me$_2$PCH$_2$CH$_2$PMe$_2$)MeCl$_3$. *J. Chem. Soc., Chem. Commun.* **1982**, 1410–1411.

21. Brookhart, M.; Green, M. L. H. Carbon-hydrogen-transition metal bonds. *J. Organomet. Chem.* **1983**, *250*, 395–408.

22. Jordan, R. F.; Bradley, P. K.; Baenziger, N. C.; LaPointe, R. E. β-Agostic interactions in (C$_5$H$_4$Me)$_2$Zr(CH$_2$CH$_2$R)(PMe$_3$)$^+$ complexes. *J. Am. Chem. Soc.* **1990**, *112*, 1289–1291.

23. Erker, G.; Froemberg, W.; Angermund, K.; Schlund, R.; Krueger, C. 'Agostic' metal-alkenyl β-CH interaction in (C$_5$H$_4$Me)$_2$ZrCl(CH:CMe)ZrCl(C$_5$H$_5$)$_2$. *J. Chem. Soc., Chem. Commun.* **1986**, 372–374.

24. Lohrenz, J. C. W.; Woo, T. K.; Ziegler, T. A density functional study of the origin of the propagation barrier in the homogeneous ethylene polymerization with Kaminsky-type catalysts. *J. Am. Chem. Soc.* **1995**, *117*, 12793–12780.

25. Krauledat, H.; Brintzinger, H. H. Isotope effect on olefin insertion in catalytic polymerization by zirconocene. Evidence for an α-agostic interaction in the transition state. *Angew. Chem., Int. Ed. Engl.* **1990**, *29*, 1412–1413.

26. Leclerc, M. K.; Brintzinger, H. H. *ansa*-Metallocene derivatives. 31. Origins of stereoselectivity and stereoerror formation in *ansa*-zirconocene-catalyzed isotactic propene polymerization. A deuterium labeling study. *J. Am. Chem. Soc.* **1995**, *117*, 1651–1652.

27. Leclerc, M. K.; Brintzinger, H. H. Zr-alkyl isomerization in *ansa*-zirconocene-catalyzed olefin polymerizations. Contributions to stereoerror formation and chain termination. *J. Am. Chem. Soc.* **1996**, *118*, 9024–9032.

28. Cheng, H. N.; Ewen, J. A. Carbon-13 nuclear magnetic resonance characterization of poly(propylene) prepared with homogeneous catalysts. *Makromol. Chem.* **1989**, *190*, 1931–1943.

29. Grassi, A.; Zambelli, A.; Resconi, L.; Albizzati, E.; Mazzocchi, R. Microstructure of isotactic polypropylene prepared with homogeneous catalysis: Stereoregularity, regioregularity, and 1,3 insertion. *Macromolecules* **1988**, *21*, 617–622.

30. Mizuno, A.; Tsutsui, T.; Kashiwa, N. Stereostructure of regioirregular unit of polypropylene obtained with *rac*-ethylenebis(1-indenyl)zirconium dichloride/methylaluminoxane catalyst system studied by carbon-13-proton shift correlation two-dimensional nuclear magnetic resonance spectroscopy. *Polymer* **1992**, *33*, 254–258.

31. Soga, K.; Shiono, T.; Takemura, S.; Kaminsky, W. Isotactic polymerization of propene with (η-1,1′-ethylenedi-4,5,6,7-tetrahydroindenyl)zirconium dichloride combined with methylaluminoxane. *Makromol. Chem., Rapid Commun.* **1987**, *8*, 305–310.

32. Rieger, B.; Mu, X.; Mallin, D. T.; Chien, J. C. W.; Rausch, M. D. Degree of stereochemical control of racemic ethylenebis(indenyl)zirconium dichloride/methyl aluminoxane catalyst and properties of anisotactic polypropylenes. *Macromolecules* **1990**, *23*, 3559–3368.

33. Schupfner, G.; Kaminsky, W. Microstructure of polypropene samples produced with different homogeneous bridged indenyl zirconium catalysts. Clues on the structure and reactivity relation. *J. Mol. Catal. A: Chem.* **1995**, *102*, 59–65.

34. Prosenc, M.-H.; Brintzinger, H.-H. Zirconium-alkyl isomerizations in zirconocene-catalyzed olefin polymerization: A density functional study. *Organometallics* **1997**, *16*, 3889–3894.

35. Busico, V.; Cipullo, R.; Chadwick, J. C.; Modder, J. F.; Sudmeijer, O. Effects of regiochemical and stereochemical errors on the course of isotactic propene polyinsertion promoted by homogeneous Ziegler–Natta catalysts. *Macromolecules* **1994**, *27*, 7538–7543.

36. Burger, B. J.; Thompson, M. E.; Cotter, W. D.; Bercaw, J. E. Ethylene insertion and β-hydrogen elimination for permethylscandocene alkyl complexes. A study of the chain propagation and termination steps in Ziegler–Natta polymerization of ethylene. *J. Am. Chem. Soc.* **1990**, *112*, 1566–1577.

37. Hajela, S.; Bercaw, J. E. Competitive chain transfer by β-hydrogen and β-methyl elimination for a model Ziegler–Natta olefin polymerization system $[Me_2Si(\eta^5\text{-}C_5Me_4)_2]Sc\{CH_2CH(CH_3)_2\}(PMe_3)$. *Organometallics* **1994**, *13*, 1147–1154.

38. Alelyunas, Y. W.; Guo, Z.; LaPointe, R. E.; Jordan, R. F. Structures and reactivity of $(C_5H_4Me)_2Zr(CH_2CH_2R)(CH_3CN)^{n+}$ complexes. Competition between insertion and β-H elimination. *Organometallics* **1993**, *12*, 544–553.

39. Guo, Z.; Swenson, D. C.; Jordan, R. F. Cationic zirconium and hafnium isobutyl complexes as models for intermediates in metallocene-catalyzed propylene polymerizations. Detection of an α-agostic interaction in $(C_5Me_5)_2Hf(CH_2CHMe_2)(PMe_3)^+$. *Organometallics* **1994**, *13*, 1424–1432.

40. Tsutsui, T.; Mizuno, A.; Kashiwa, N. The microstructure of propylene homo- and copolymers obtained with a bis(cyclopentadienyl)zirconium dichloride and methylaluminoxane catalyst system. *Polymer* **1989**, *30*, 428–431.

41. Eshuis, J. J. W.; Tan, Y. Y.; Meetsma, A.; Teuben, J. H.; Renkema, J.; Evens, G. G. Kinetic and mechanistic aspects of propene oligomerization with ionic organozirconium and -hafnium compounds: Crystal structures of $[Cp*_2MMe(THT)]^+[BPh_4]^-$ (M = zirconium, hafnium). *Organometallics* **1992**, *11*, 362–369.

42. Eshuis, J. J. W.; Tan, Y. Y.; Teuben, J. H.; Renkema, J. Catalytic olefin oligomerization and polymerization with cationic group IV metal complexes $[Cp*_2MMe(THT)]^+[BPh_4]^-$, M = titanium, zirconium and hafnium. *J. Mol. Catal.* **1990**, *62*, 277–287.

43. Resconi, L.; Abis, L.; Franciscono, G. 1-Olefin polymerization at bis(pentamethylcyclopentadienyl) zirconium and -hafnium centers: Enantioface selectivity. *Macromolecules* **1992**, *25*, 6814–6817.

44. Sacchi, M. C.; Barsties, E.; Tritto, I.; Locatelli, P.; Brintzinger, H.-H.; Stehling, U. Stereochemistry of first monomer insertion into metal–methyl bond: A tool for evaluating ligand-monomer interactions in propene polymerization with metallocene catalysts. *Macromolecules* **1997**, *30*, 3955–3957.

45. Sacchi, M. C.; Shan, C.; Locatelli, P.; Tritto, I. Stereochemical investigation of the initiation step in $MgCl_2$-supported Ziegler–Natta catalysts: The Lewis base activation effect. *Macromolecules* **1990**, *23*, 383–386.

46. Zambelli, A.; Sacchi, M. C.; Locatelli, P.; Zannoni, G. Isotactic polymerization of α-olefins: Stereoregulation for different reactive chain ends. *Macromolecules* **1982**, *15*, 211–212.

47. Zambelli, A.; Locatelli, P.; Sacchi, M. C.; Tritto, I. Isotactic polymerization of propene: Stereoregularity of the insertion of the first monomer unit as a fingerprint of the catalytic active site. *Macromolecules* **1982**, *15*, 831–834.

48. Tsutsui, T.; Kashiwa, N.; Mizuno, A. Effect of hydrogen on propene polymerization with ethylenebis (1-indenyl)zirconium dichloride and methylalumoxane catalyst system. *Makromol. Chem., Rapid Commun.* **1990**, *11*, 565–570.

49. Pino, P.; Cioni, P.; Wei, J. Asymmetric hydrooligomerization of propylene. *J. Am. Chem. Soc.* **1987**, *109*, 6189–6191.

50. Bovey, F. A.; Mirau, P. A. *NMR of Polymers*; Academic Press: San Diego, 1996.

51. Cheng, H. N. NMR characterization of polymers. In *Modern Methods of Polymer Characterization*; Barth, H. G., Mayes, J. W., Eds.; Chemical Analysis Series 113; John Wiley and Sons: New York, 1991; pp 409–493.

52. Venditto, V.; Guerra, G.; Corradini, P.; Fusco, R. Possible model for chain end control of stereoregularity in the isospecific homogeneous Ziegler–Natta polymerization. *Polymer* **1990**, *31*, 530–537.

53. Erker, G.; Fritze, C. Selectivity control by temperature variation during formation of isotactic vs. syndiotactic polypropylene on a titanocene/alumoxane catalyst. *Angew. Chem., Int. Ed. Engl.* **1992**, *31*, 199–202.

54. Ewen, J. A. Mechanisms of stereochemical control in propylene polymerizations with soluble group 4B metallocene/methylalumoxane catalysts. *J. Am. Chem. Soc.* **1984**, *106*, 6355–6364.

55. Small, B. L.; Brookhart, M. Polymerization of propylene by a new generation of iron catalysts: Mechanisms of chain initiation, propagation, and termination. *Macromolecules* **1999**, *32*, 2120–2130.

56. Tullock, C. W.; Tebbe, F. N.; Mulhaupt, R.; Ovenall, D. W.; Setterquist, R. A.; Ittel, S. D. Polyethylene and elastomeric polypropylene using alumina-supported bis(arene) titanium, zirconium, and hafnium catalysts. *J. Polym. Sci., Part A: Polym. Chem.* **1989**, *27*, 3063–3081.

57. Collette, J. W.; Tullock, C. W.; MacDonald, R. N.; Buck, W. H.; Su, A. C. L.; Harrell, J. R.; Mulhaupt, R.; Anderson, B. C. Elastomeric polypropylenes from alumina-supported tetraalkyl group IVB catalysts. 1. Synthesis and properties of high molecular weight stereoblock homopolymers. *Macromolecules* **1989**, *22*, 3851–3858.

58. De Candia, F.; Russo, R.; Vittoria, V. Physical behavior of stereoblock-isotactic polypropylene. *Makromol. Chem.* **1988**, *189*, 815–821.

59. De Candia, F.; Russo, R. Crystallization and melting behavior of a stereoblock isotactic polypropylene. *Thermochim. Acta* **1991**, *177*, 221–227.

60. van der Leek, Y.; Angermund, K.; Reffke, M.; Kleinschmidt, R.; Goretzki, R.; Fink, G. On the mechanism of stereospecific polymerization—development of a universal model to demonstrate the relationship between metallocene structure and polymer microstructure. *Chem. Eur. J.* **1997**, *3*, 585–591.

61. Yoshida, T.; Koga, N.; Morokuma, K. A combined *ab initio* MO-MM study on isotacticity control in propylene polymerization with silylene-bridged group 4 metallocenes. C_2-Symmetrical and asymmetrical catalysts. *Organometallics* **1996**, *15*, 766–777.

62. Kawamura-Kuribayashi, H.; Koga, N.; Morokuma, K. An *ab initio* MO and MM study of homogeneous olefin polymerization with silylene-bridged zirconocene catalyst and its regio- and stereoselectivity. *J. Am. Chem. Soc.* **1992**, *114*, 8687–8694.

63. Cavallo, L.; Guerra, G.; Vacatello, M.; Corradini, P. A possible model for the stereospecificity in the syndiospecific polymerization of propene with group 4a metallocenes. *Macromolecules* **1991**, *24*, 1784–1790.

64. Cavallo, L.; Guerra, G.; Oliva, L.; Vacatello, M.; Corradini, P. Steric control in the initiation step of the isospecific homogeneous Ziegler–Natta polymerization of propene and 1-butene. *Polym. Commun.* **1989**, *30*, 16–19.

65. Castonguay, L. A.; Rappe, A. K. Ziegler–Natta catalysis. A theoretical study of the isotactic polymerization of propylene. *J. Am. Chem. Soc.* **1992**, *114*, 5832–5842.

66. Corradini, P.; Guerra, G.; Vacatello, M.; Villani, V. A possible model for site control of stereoregularity in the isotactic specific homogeneous Ziegler–Natta polymerization. *Gazz. Chim. Ital.* **1988**, *118*, 173–177.

67. Guerra, G.; Cavallo, L.; Moscardi, G.; Vacatello, M.; Corradini, P. Enantioselectivity in the regioirregular placements and regiospecificity in the isospecific polymerization of propene with homogeneous Ziegler–Natta catalysts. *J. Am. Chem. Soc.* **1994**, *116*, 2988–2995.

68. Resconi, L.; Balboni, D.; Baruzzi, G.; Fiori, C.; Guidotti, S.; Mercandelli, P.; Sironi, A. *rac*-[Methylene(3-*tert*-butyl-1-indenyl)2]ZrCl2: A simple, high-performance zirconocene catalyst for isotactic polypropene. *Organometallics* **2000**, *19*, 420–429.

69. Resconi, L.; Piemontesi, F.; Camurati, I.; Sudmeijer, O.; Nifant'ev, I. E.; Ivchenko, P. V.; Kuz'mina, L. G. A new class of isospecific, highly regiospecific zirconocene catalysts for the polymerization of propene. *J. Am. Chem. Soc.* **1998**, *120*, 2308–2321.

70. Guerra, G.; Longo, P.; Cavallo, L.; Corradini, P.; Resconi, L. Relationship between regiospecificity and type of stereospecificity in propene polymerization with zirconocene-based catalysts. *J. Am. Chem. Soc.* **1997**, *119*, 4394–4403.

71. Toto, M.; Cavallo, L.; Corradini, P.; Moscardi, G.; Resconi, L.; Guerra, G. Influence of π-ligand substitutions on the regiospecificity and stereospecificity in isospecific zirconocenes for propene polymerization. A molecular mechanics analysis. *Macromolecules* **1998**, *31*, 3431–3438.

72. Kleinschmidt, R.; Reffke, M.; Fink, G. Investigation of the microstructure of poly(propylene) in dependence of the polymerization temperature for the systems iPr[3-RCpFlu]ZrCl₂/MAO, with R = H, Me, Et, iPr, tBu, and iPr[IndFlu]Zr/Cl₂/MAO. *Macromol. Rapid Commun.* **1999**, *20*, 284–288.

73. Sillars, D. R.; Landis, C. R. Catalytic propene polymerization: Determination of propagation, termination, and epimerization kinetics by direct NMR observation of the (EBI)Zr(Me(BC₆F₅)₃)propenyl catalyst species. *J. Am. Chem. Soc.* **2003**, *125*, 9894–9895.

74. Yoder, J. C.; Bercaw, J. E. Chain epimerization during propylene polymerization with metallocene catalysts: Mechanistic studies using a doubly labeled propylene. *J. Am. Chem. Soc.* **2002**, *124*, 2548–2555.

75. Schnutenhaus, H.; Brintzinger, H. H. 1,1′-Trimethylenebis(η⁵-3-*tert*-butylcyclopentadienyl)titanium (IV) dichloride, a chiral *ansa*-titanocene derivative. *Angew. Chem., Int. Ed. Engl.* **1979**, *18*, 777–778.

76. Wild, F. R. W. P.; Zsolnai, L.; Huttner, G.; Brintzinger, H. H. *ansa*-Metallocene derivatives. IV. Synthesis and molecular structures of chiral *ansa*-titanocene derivatives with bridged tetrahydroindenyl ligands. *J. Organomet. Chem.* **1982**, *232*, 233–247.

77. Wild, F. R. W. P.; Wasiucionek, M.; Huttner, G.; Brintzinger, H. H. *ansa*-Metallocene derivatives. VII. Synthesis and crystal structure of a chiral *ansa*-zirconocene derivative with ethylene-bridged tetrahydroindenyl ligands. *J. Organomet. Chem.* **1985**, *288*, 63–67.

78. Ewen, J. A. Crystal structures and stereospecific propylene polymerizations with chiral hafnium metallocene catalysts. *J. Am. Chem. Soc.* **1987**, *100*, 6544–6545.

79. Kaminsky, W.; Külper, K.; Brintzinger, H. H.; Wild, F. R. W. P. Polymerization of propene and butene with a chiral zirconocene and methylalumoxane as cocatalyst. *Angew. Chem., Int. Ed. Engl.* **1985**, *24*, 507–508.

80. Piemontesi, F.; Camurati, I.; Resconi, L.; Balboni, D.; Sironi, A.; Moret, M.; Zeigler, R.; Piccolrovazzi, N. Crystal structures and solution conformations of the meso and racemic isomers of (ethylenebis(1-indenyl))zirconium dichloride. *Organometallics* **1995**, *14*, 1256–1266.

81. Spaleck, W.; Antberg, M.; Aulbach, M.; Bachmann, B.; Dolle, V.; Hafka, S.; Küber, F.; Rohrmann, J.; Winter, A. New isotactic polypropylene via metallocene catalysts. In *Ziegler Catalysts—Recent Scientific Innovations and Technological Improvements*; Fink, G., Mülhaupt, R., Brintzinger, H. H., Eds.; Springer-Verlag: Berlin Heidelberg, 1995; pp 83–97.

82. Ewen, J. A.; Jones, R. L.; Razavi, A.; Ferrara, J. D. Syndiospecific propylene polymerizations with group IVB metallocenes. *J. Am. Chem. Soc.* **1988**, *110*, 6255–6256.

83. Ewen, J. A.; Elder, M. J.; Jones, R. L.; Haspelagh, L.; Atwood, J. L.; Bott, S. G.; Robinson, K. Metallocene/propylene structural relationships: Implications on polymerization and stereochemical control mechanisms. *Makromol. Chem., Macromol. Symp.* **1991**, *48/49*, 253–295.

84. Coates, G. W.; Waymouth, R. M. Oscillating stereocontrol: A strategy for the synthesis of thermoplastic elastomeric polypropylene. *Science* **1995**, *267*, 217–219.

85. Ewen, J. A.; Haspelagh, L.; Elder, M. J.; Atwood, J. L.; Zhang, H.; Cheng, H. N. Propylene polymerization with group 4 metallocene/alumoxane systems. In *Transition Metals and Organometallics as Catalysts for Olefin Polymerization*; Kaminsky, W., Sinn, H., Eds.; Springer-Verlag: Berlin Heidelberg, 1988; pp 281–289.

86. Herrmann, W. A.; Rohrmann, J.; Herdtweck, E.; Spaleck, W.; Winter, A. The first example of a ethylene-selective, soluble Ziegler catalyst of the zirconocene-type. *Angew. Chem., Int. Ed. Engl.* **1989**, *28*, 1511–1512.

87. Spaleck, W.; Antberg, M.; Rohrmann, J.; Winter, A.; Bachmann, B.; Kiprof, P.; Behm, J.; Herrmann, W. A. High molecular weight polypropylene through specifically designed zirconocene catalysts. *Angew. Chem., Int. Ed. Engl.* **1992**, *31*, 1347–1350.

88. Kashiwa, N.; Kojoh, S.-I.; Imuta, J.-I.; Tsutsui, T. Characterization of PP prepared with the latest metallocene and MgCl₂-supported TiCl₄ catalyst systems. In *Metalorganic Catalysts for Synthesis and Polymerization*, Kaminsky, W., Ed.; Springer-Verlag: Berlin Heidelberg, 1999; pp 30–37.

89. Resconi, L.; Pietmontesi, F.; Camurati, I.; Sudmeijer, O.; Nifant'ev, I. E.; Ivchenko, P. V.; Kuz'mina, L. G. Highly regiospecific zirconocene catalysts for the isospecific polymerization of propylene. *J. Am. Chem. Soc.* **1998**, *120*, 2308–2321.

90. Deng, H.; Winkelbach, H.; Taeji, K.; Kaminsky, W.; Soga, K. Synthesis of high-melting, isotactic polypropene with C_2- and C_1-symmetrical zirconocenes. *Macromolecules* **1996**, *29*, 6371–6376.

91. Mise, T.; Miya, S.; Yamazaki, H. Excellent stereoregular isotactic polymerizations of propylene with C_2-symmetric silylene-bridged metallocene catalysts. *Chem. Lett.* **1989**, 1853–1856.

92. Röll, W.; Brintzinger, H. H.; Rieger, B.; Zolk, R. Stereo- and regioselectivity of chiral, alkyl-substituted *ansa*-zirconocene catalysts in methylalumoxane-activated propene polymerization. *Angew. Chem., Int. Ed. Engl.* **1990**, *29*, 279–280.

93. Coughlin, E. B.; Bercaw, J. E. Isospecific Ziegler–Natta polymerization of α-olefins with a single-component organoyttrium catalyst. *J. Am. Chem. Soc.* **1992**, *114*, 7606–7607.

94. Ewen, J. A.; Jones, R. L.; Elder, M. J.; Rheingold, A. L.; Liable-Sands, L. M. Polymerization catalysts with cyclopentadienyl ligands ring-fused to pyrrole and thiophene heterocycles. *J. Am. Chem. Soc.* **1998**, *120*, 10786–10787.

95. Ewen, J. A.; Elder, M. J.; Jones, R. L.; Rheingold, A. L.; Liable-Sands, L. M.; Sommer, R. D. Chiral *ansa* metallocenes with Cp ring-fused to thiophenes and pyrroles: Syntheses, crystal structures, and isotactic polypropylene catalysts. *J. Am. Chem. Soc.* **2001**, *123*, 4763–4773.

96. van Baar, J. F.; Horton, A. D.; De Kloe, K. P.; Kragtwijk, E.; Mkoyan, S. G.; Nifant'ev, I. E.; Schut, P. A.; Taidakov, I. V. *ansa*-Zirconocenes based on *N*-substituted 2-methylcyclopenta[*b*]indoles: Synthesis and catalyst evaluation in liquid propylene polymerization. *Organometallics* **2003**, *22*, 2711–2722.

97. Fisher, R. A.; Temme, R. B. Polymerization catalyst system comprising heterocyclic fused cyclopentadienide ligands. U.S. Patent 6,451,938 B1 (ExxonMobil Chemical Patents Inc.), September 17, 2002.

98. Alt, H. G.; Zenk, R. C_2-symmetric bis(fluorenyl) complexes: Four complex models as potential catalysts for the isospecific polymerization of propylene. *J. Organomet. Chem.* **1996**, *512*, 51–60.

99. Chen, Y.-X.; Rausch, M. D.; Chien, J. C. W. C_{2v}- and C_2-Symmetric *ansa*-bis(fluorenyl)zirconocene catalysts: Synthesis and α-olefin polymerization catalysis. *Macromolecules* **1995**, *28*, 5399–5404.

100. Rieger, B. Stereospecific propene polymerization with *rac*-[1,2-bis(η^5-(9-fluorenyl))-1-phenylethane] zirconium dichloride/methylalumoxane. *Polym. Bull. (Berlin)* **1994**, *32*, 41–46.

101. Lin, S.; Waymouth, R. M. 2-Arylindene metallocenes: Conformationally dynamic catalysts to control the structure and properties of polypropylenes. *Acc. Chem. Res.* **2002**, *35*, 765–773.

102. Busico, V.; Cipullo, R.; Kretchmer, W. P.; Talarico, G.; Vacatello, M.; Van Axel Castelli, V. "Oscillating" metallocene catalysts: How do they oscillate? *Angew. Chem., Int. Ed. Engl.* **2002**, *41*, 505–508.

103. Erker, G.; Nolte, R.; Tsay, Y.-H.; Krüger, C. Double stereodifferentiation in the formation of isotactic polypropylene at chiral $(C_5H_4CHMePh)_2ZrCl_2$/methylalumoxane catalysts. *Angew. Chem., Int. Ed. Engl.* **1989**, *28*, 628–629.

104. Erker, G.; Nolte, R.; Aul, R.; Wilker, S.; Krüger, C.; Noe, R. Cp-substituent additivity effects controlling the stereochemistry of the propene polymerization reaction at conformationally unrestricted $(Cp\text{-}CHR_1R_2)_2ZrCl_2$/methylalumoxane catalysts. *J. Am. Chem. Soc.* **1991**, *113*, 7594–7602.

105. Erker, G.; Temme, B. Use of cholestanylindene-derived nonbridged group 4 bent metallocene/methyl alumoxane catalysts for stereoselective propene polymerization. *J. Am. Chem. Soc.* **1992**, *114*, 4004–4006.

106. Erker, G.; Aulbach, M.; Knickmeier, M.; Wingbermuehle, D.; Krüger, C.; Nolte, M.; Werner, S. The role of torsional isomers of planarly chiral nonbridged bis(indenyl)metal type complexes in stereoselective propene polymerization. *J. Am. Chem. Soc.* **1993**, *115*, 4590–4601.

107. Hauptman, E.; Waymouth, R. M.; Ziller, J. W. Stereoblock polypropylene: Ligand effects on the stereospecificity of 2-arylindene zirconocene catalysts. *J. Am. Chem. Soc.* **1995**, *117*, 11586–11587.

108. Wilmes, G. M.; Lin, S.; Waymouth, R. M. Propylene polymerization with sterically hindered unbridged 2-arylindene metallocenes. *Macromolecules* **2002**, *35*, 5382–5387.

109. Razavi, A.; Atwood, J. L. Isospecific propylene polymerization with unbridged group 4 metallocenes. *J. Am. Chem. Soc.* **1993**, *115*, 7529–7530.

110. Leino, R.; Lehmus, P.; Lehtonen, A. Heteroatom-substituted group 4 bis(indenyl)metallocenes. *Eur. J. Inorg. Chem.* **2004**, 3201–3222.

111. LoCoco, M. D.; Zhang, X.; Jordan, R. F. Chelate-controlled synthesis of *ansa*-zirconocenes. *J. Am. Chem. Soc.* **2004**, *126*, 15231–15244.

112. Christopher, J. N.; Diamond, G. M.; Jordan, R. F.; Petersen, J. F. Synthesis, structure, and reactivity of rac-Me$_2$Si(indenyl)$_2$Zr(NMe$_2$)$_2$. *Organometallics* **1996**, *15*, 4038–4044.
113. Diamond, G. M.; Jordan, R. F.; Petersen, J. L. Efficient synthesis of chiral *ansa*-metallocenes by amine elimination synthesis, structure, and reactivity of rac-(EBI)Zr(NMe$_2$)$_2$. *J. Am. Chem. Soc.* **1996**, *118*, 8024–8033.
114. Diamond, G. M.; Jordan, R. F.; Petersen, J. L. Synthesis of Me$_2$Si-bridged *ansa*-zirconocenes by amine elimination. *Organometallics* **1996**, *15*, 4045–4053.
115. Grubbs, R. H.; Coates, G. W. α-Agostic interactions and olefin insertion in metallocene polymerization catalysts. *Acc. Chem. Res.* **1996**, *29*, 85–93.

2 Fluorenyl-Containing Catalysts for Stereoselective Propylene Polymerization

Craig J. Price, Levi J. Irwin, David A. Aubry, and Stephen A. Miller

CONTENTS

2.1 INTRODUCTION

2.1.1 BACKGROUND

The fortuitous synthesis in 1951[1] and structural elucidation of ferrocene in 1952[2] were monumental steps in what would become the field of organometallic homogeneous polymerization catalysis. Pioneering work in the realm of metallocene synthesis initially focused on the cyclopentadienide ligand, $[C_5H_5]^-$ (Figure 2.1), resulting in a family of structures having the general formula $(C_5H_5)_2MX_n$, where M is a transition, lanthanide, or actinide metal, and X is an anionic ligand (often Cl^- or Br^-) that completes the valence of the overall neutral metallocene.[3] In 1953, the first metallocene

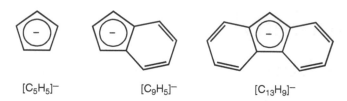

$[C_5H_5]^-$ $[C_9H_5]^-$ $[C_{13}H_9]^-$

FIGURE 2.1 Cyclopentadienide, indenide, and fluorenide are important ligands commonly found in stereoselective olefin polymerization catalysts.

based on indenide (Figure 2.1) was reported and had the formula $(C_9H_7)_2Co$.[4] Progression from "benzocyclopentadienide" to "dibenzocyclopentadienide" in 1958 resulted in the first report of a metallocene based on fluorenide (Figure 2.1), having the formula $(C_{13}H_9)_2Mn$.[5] Since this time the fluorenide ligand—or "fluorenyl" ligand in the neutral formalism—has been incorporated into thousands of organometallic compounds. While the fluorenyl ligand remains the least investigated of these three important cyclopentadiene-based ligands, it is an essential component in a large number of stereoselective olefin polymerization catalysts.[6]

2.1.2 Scope

The focus of this chapter is fluorenyl-containing organometallic species that (upon activation) reportedly catalyze the homogeneous polymerization of propylene with quantifiable stereoregularity in the resultant polymer. Hence, many such species that have been investigated only for ethylene polymerization[7] are excluded. Similarly, certain group 3 species that produce syndiotactic polystyrene[8] will not be discussed (see Chapter 14 for further information on tactic polystyrene). Fluorenyl-containing species that have only been investigated as heterogeneous catalysts following immobilization will not be addressed.[9] Furthermore, a wide variety of fluorenyl-containing species that produce atactic (truly or nearly stereorandom) polymers will not be addressed, but can be categorized into the following groups: nonbridged fluorenyl/cyclopentadienyl species such as $(C_5H_5)(C_{13}H_9)ZrCl_2$;[10] nonbridged fluorenyl/indenyl species such as $(C_9H_7)(C_{13}H_9)ZrCl_2$;[11] nonbridged fluorenyl/fluorenyl C_{2v}-symmetric species such as $(C_{13}H_9)_2ZrCl_2$;[12] bridged fluorenyl/fluorenyl C_{2v}-symmetric species such as $Me_2Si(C_{13}H_8)_2ZrCl_2$, which can yield high molecular weight atactic polypropylene having elastomeric properties;[13] certain fluorenyl-containing species with heteroatom substituents such as $(9-(Me_2N)C_{13}H_8)_2ZrCl_2$;[14] bridged C_1-symmetric species containing one fluorenyl ligand and another bulky ligand such as $(C_{13}H_8)PhCHCH_2(C_{27}H_{16})ZrCl_2$;[15] fluorenyl-containing species that likely do not retain fluorenyl ligation following activation such as $(CH_3)_2C(C_5H_4)(C_{13}H_8)Zr(C_5H_5)Cl$;[16] titanium-based species, such as $Ph_2C(C_5H_4)(C_{13}H_8)TiCl_2$, which are generally difficult to activate[17] or only produce atactic polymers;[18] bridged group 3 C_s-symmetric species such as $Ph_2C(C_5H_4)(C_{13}H_8)ScMe(AlMe_3)$, which produces atactic polypropylene without a cocatalyst;[19] and group 5 compounds, from which there are no clear examples of stereoregular polymers. Finally, it should be noted that monofluorenyl species are generally unstable[20] or otherwise fail to polymerize olefins.[21]

A recent SciFinder® search for fluorenyl-containing zirconium species resulted in 924 Chemical Abstracts Service (CAS) registry numbers in 1137 references. A similar search for fluorenyl-containing hafnium species resulted in 134 CAS registry numbers in 137 references. Still smaller numbers for titanium—103 CAS registry numbers in 95 references—are a result of the scant success found with fluorenyl-containing titanium compounds as olefin polymerization catalysts. Nearly all significantly stereoselective fluorenyl-containing catalysts are based on zirconium and hafnium and thus, such transition metal species that are not excluded for the aforementioned reasons will be the focus of this chapter. While zirconium- and hafnium-based catalysts with identical ligands often

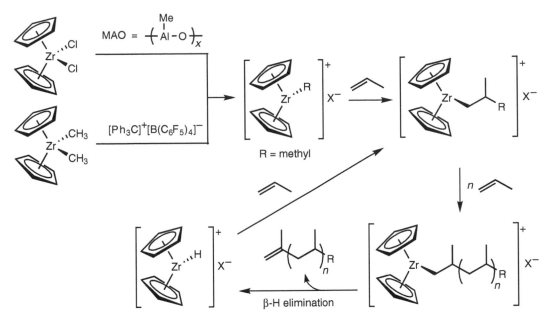

SCHEME 2.1 The activation of a metallocene catalyst precursor initiates a catalytic cycle in which thousands of polymer chains per metal can be synthesized. X^- represents $[B(C_6F_5)_4]^-$ or an aluminoxane-based anion.

differ considerably in terms of catalytic activity, and produce polymers of widely different molecular weights, it is a fair generalization that they display similar stereoselectivities.

2.1.3 MECHANISMS AND NOMENCLATURE

Since reasonably active metallocene-based homogeneous polymerizations were first reported in 1980,[22] a large number of researchers have contributed to the understanding of the mechanism that allows for catalytic stereoselective olefin polymerization. Scheme 2.1 depicts two possible routes to generating the requisite metallocenium cation from a dichloride or dimethyl precatalyst. Methylaluminoxane (MAO, an oligomeric material with a repeat unit of —Al(CH$_3$)O—) serves to alkylate the metallocene and abstract the remaining chloride ligand, generating a metallocene alkyl cation.[23] Boron-based activators are designed to abstract an alkyl anion, R^-, from a dialkyl precatalyst.[24] Iterative migratory insertion of olefinic monomer into the metal–carbon bond constitutes propagation and elongates the pendant polymer until a chain termination event occurs. A number of chain termination modes are possible; a common one, which is depicted, is β-hydride transfer. This results in a new metal–H bond into which an olefin can insert, thereby allowing the catalytic formation of, typically, 500–5000 polymer chains per metal.[25]

The transition state for migratory insertion of an olefin into a metal carbon bond is widely believed to rely on a α-agostic interaction[26] as shown in Figure 2.2a. This interaction (between an α-hydrogen on the polymer chain and the metal center) helps to orient the highly directional sp^3 orbital of the α-carbon toward the π orbital of the incoming olefin, which is known to insert with high 1,2-regioselectivity as depicted. This allows for more bonding in the transition state and lowers the energetic barrier for insertion. The stereochemical implications of the α-agostic interaction are profound and form the basis of stereoregular α-olefin polymerizations. The pendant polymer chain has two α-hydrogens. Selection of one (Figure 2.2b) allows the sterically demanding polymer chain to occupy an open quadrant of the metallocene during the transition state. However, selection of the other α-hydrogen (Figure 2.2c) results in a congested steric interaction wherein the growing polymer chain resides in close proximity to the benzo ring of the fluorenyl ligand during

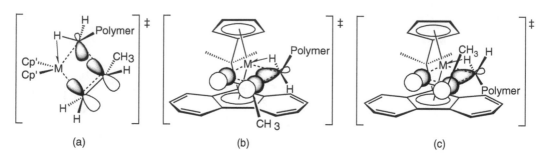

FIGURE 2.2   The α-agostic interaction lowers the energetic barrier for olefin insertion. The stereochemical consequences of this interaction ultimately dictate which enantioface of propylene inserts into the metal–carbon bond; **b** is preferred over **c**.

the insertion. With selection of the "correct" α-hydrogen, the polymer occupies an open quadrant (Figure 2.2b) and the methyl group of the incoming propylene monomer assumes a trans relationship with the growing polymer chain during the transition state for monomer insertion. The absolute stereochemistry of this model has been experimentally established for C_2-symmetric, indenyl-based catalysts.[27] However, for C_s-symmetric, fluorenyl-based catalysts, this stereochemical model is supported only by theoretical calculations (see Chapter 7 for a further discussion on theoretical calculations).[28]

2.1.4   DETERMINATION OF POLYMER STEREOCHEMISTRY

The most important tool for assessing the stereochemistry of a polypropylene chain is ^{13}C NMR spectroscopy.[29] The chemical shift of a methyl carbon is sensitive to the relative stereochemistry of neighboring methine carbons. For example, the methyl carbon of the pentad described as *mmmm* (*m = meso*) resonates at 21.62 ppm while the methyl carbon of the pentad described as *rrrr* (*r = racemo*) resonates at 20.15 ppm, as shown in Figure 2.3. There are a total of ten possible pentads. Since two of these are roughly coincident at 100 MHz, nine peaks are typically observed at this frequency (proton decoupling, {^1H}, is standard). The ^{13}C NMR spectrum for a sample of atactic polypropylene is shown in Figure 2.3. For this stereochemically random polymer, a statistical distribution of pentads is observed with the relative intensities 1:2:1:2:4:2:1:2:1 corresponding to *mmmm:mmmr:rmmr:mmrr:mmrm + rmmr:mrmr:rrrr:rrrm:mrrm*. Generally, a ^{13}C NMR spectrum with one predominant methyl peak at 21.62 ppm (*mmmm*) is considered isotactic and a ^{13}C NMR spectrum with one predominant methyl peak at 20.15 ppm (*rrrr*) is considered syndiotactic. While isotactic and syndiotactic describe limiting structures of the highest stereochemical order for a linear poly(α-olefin), there are many other tacticities that are still considered stereoregular.[30]

2.1.5   OVERVIEW OF POSSIBLE TACTICITIES

A wide variety of stereoregular polymers can be obtained with metallocene-based catalysts.[31] The vast majority of these can be prepared with fluorenyl-containing catalysts, and in this sense, the fluorenyl-based catalysts have displayed a stereoselective versatility not yet demonstrated with catalysts bearing only cyclopentadienyl and/or indenyl ligands.

Figure 2.4 depicts a continuum of polypropylene tacticities that is possible only with fluorenyl-based catalysts. At the extremes of this continuum are isotactic polypropylene and syndiotactic polypropylene. Note the relationship between these two polymers emphasized by the shaded boxes. For both of these polymers, every other stereocenter has the same relative stereochemistry.

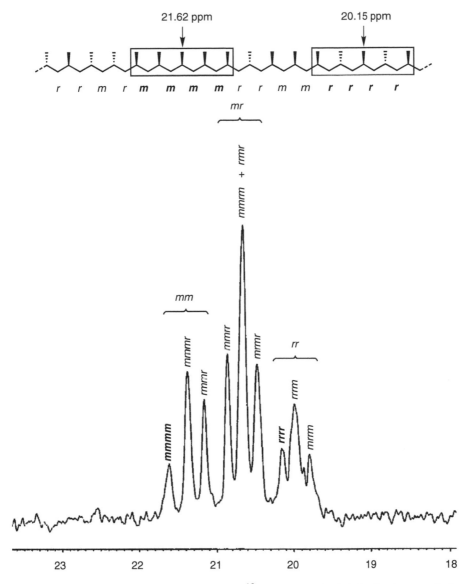

FIGURE 2.3 A pentad consists of five repeat units. The ^{13}C NMR chemical shift of a given methyl carbon is sensitive to the relative stereochemistry of the pentad (m = *meso* and r = *racemo*). A statistical distribution of pentads is observed in the $\{^1\text{H}\}$ ^{13}C NMR spectrum of the methyl region of atactic polypropylene.

The intervening stereocenters align with their neighbors for isotactic polypropylene, whereas the intervening stereocenters do not align with their neighbors for syndiotactic polypropylene. In the middle of this continuum is a tacticity called hemiisotactic, which also has stereoregularity at every other stereocenter; however, the intervening stereocenters are stereorandom. If the intervening stereocenters are partially, but not completely, aligned with their neighbors, then isotactic-hemiisotactic polymer is obtained. Similarly, if the intervening stereocenters are partially, but not completely, oppositely aligned with their neighbors, then syndiotactic-hemiisotactic polymer is obtained. None of these polymers is atactic or stereorandom, but clearly the structure of minimum stereoregularity in this particular continuum lies somewhere between isotactic and syndiotactic.

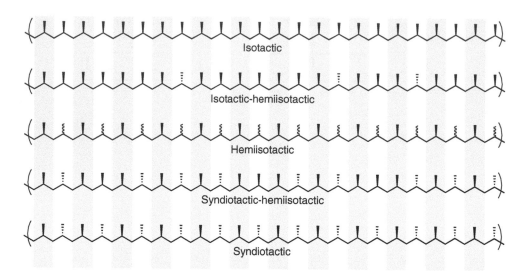

FIGURE 2.4 Every other stereocenter (shaded) is the same in the *hemiisotactic continuum*, which ranges from isotactic to syndiotactic, depending on the stereochemistry of the intervening stereocenters (not shaded).

2.1.6 Quantification of Polymer Stereoregularity

What exactly constitutes stereoregularity in a polymer? It is quite tenable to define isotactic polymers as 100% stereoregular. In addition, most would agree that syndiotactic polymers are also stereoregular—just less so than isotactic ones. But, how much less? Hemiisotactic polymers certainly seem less stereoregular than either isotactic or syndiotactic polymers. However, would it be correct to assert that perfectly hemiisotactic polymers have the minimum stereoregularity among those polymers defined by the continuum of Figure 2.4? Answers to these questions are best formulated by quantifying the stereoregularity of the polymers under discussion. For the following derivation, it is useful to note that any tacticity described in Figure 2.4 can be defined by a single stereochemical parameter.[32] The m dyad composition $[m]$, which ranges from 0% to 100%, is equal to the probability that an intervening stereocenter is aligned with its two neighbors. This is true because such alignment results in an mm triad, but the rr triad is formed when the intervening stereocenter is not aligned with its neighbors.

The *absolute stereoregularity* (Ψ_{abs}) of a polymer can be defined by quantifying the likelihood that identical stereocenters occur in succession along a polymer chain. Mathematically, there are several possible ways to proceed, but Ψ_{abs} in Table 2.1 averages the probability that adjacent (α,β and β,γ) and nearly adjacent (α,γ and β,δ) stereocenters are identical. For the hemiisotactic continuum, the probability that two adjacent stereocenters are the same is simply $[m]$. The probability that nearly adjacent stereocenters are the same is either 1.00 (for α,γ) or $[m]^2 + (1-[m])^2$ (for β,δ), depending on which pair of stereocenters is considered. Thus $\Psi_{abs} = \{1 + [m]^2\}/2$ and will vary between 0.50 and 1.00.

Similarly, the *relative stereoregularity* (Ψ_{rel}) of a polymer can be defined by quantifying the likelihood that identical dyads occur in succession along a polymer chain. Ψ_{rel} in Table 2.1 averages the probability that adjacent (α,β and β,γ) and nearly adjacent (α,γ and β,δ) dyads are identical. For the hemiisotactic continuum, the probability that two adjacent dyads are identical is either 1.00 (for α,β) or $[m]^2 + (1-[m])^2$ (for β,γ), depending on which pair of dyads is considered. The probability that two nearly adjacent dyads are the same is simply $[m]^2+(1-[m])^2$. Thus, $\Psi_{rel} = \{2-3[m]+3[m]^2\}/2$ and will vary parabolically between 0.625 and 1.00, with the minimum at 0.625 when $[m] = 0.50$.

The average of Ψ_{abs} and Ψ_{rel} is Ψ (Table 2.1) and provides a quantitative measure of a polymer's stereoregularity. If only Ψ_{abs} were considered, then a syndiotactic polymer ($\Psi_{abs} = 0.50$) would be

TABLE 2.1
Mathematical Quantification of the Absolute (Ψ_{abs}), Relative (Ψ_{rel}), and Total (Ψ) Stereoregularity of a Polymer in the Hemiisotactic Continuum

Relationship			
Probability of identical stereocenters	$[m]$	$[m]$	$[m]^2 + (1-[m])^2$
Average = Ψ_{abs}	$\{2[m] + 1 + [m]^2 + (1-[m])^2\}/4 = \{1+[m]^2\}/2$	1.00	$[m]^2 + (1-[m])^2$
Relationship			
Probability of identical dyads	1.00	$[m]^2 + (1-[m])^2$	$[m]^2 + (1-[m])^2$
Average = Ψ_{rel}	$\{1 + 3([m]^2 + (1-[m])^2)\}/4 = \{2 - 3[m] + 3[m]^2\}/2$		
Stereoregularity, $\Psi = (\Psi_{rel} + \Psi_{abs})/2$	$\{2 + 2[m] + 4([m]^2 + (1-[m])^2)\}/8 = \{3 - 3[m] + 4[m]^2\}/4$		

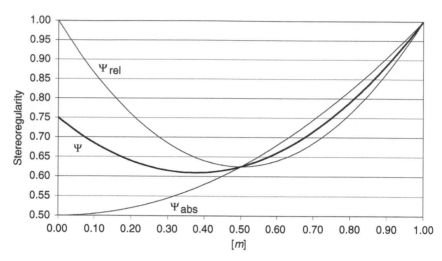

FIGURE 2.5 The stereoregularity (Ψ) of a polymer in the hemiisotactic continuum can be defined as the average of the absolute stereoregularity (Ψ_{abs}) and the relative stereoregularity (Ψ_{rel}).

less stereoregular than a purely hemiisotactic one ($\Psi_{abs} = 0.625$). If only Ψ_{rel} were considered, then a syndiotactic polymer ($\Psi_{rel} = 1.00$) would have the same perfect stereoregularity as an isotactic one ($\Psi_{rel} = 1.00$). Therefore, both the absolute and relative components are needed to provide a realistic quantification of polymer stereoregularity. Figure 2.5 plots Ψ_{abs}, Ψ_{rel}, and Ψ as a function of $[m]$. It is satisfying to note that the syndiotactic ($\Psi = 0.75$) and isotactic ($\Psi = 1.00$) polymers represent local maxima on the stereoregularity curve. In the hemiisotactic continuum, the polymer with the least stereoregularity occurs when $[m] = 0.375$ ($\Psi = 0.609375$). This mathematical analysis of stereoregularity can be applied to polymers outside the scope of this chapter. For example, a purely atactic Bernoullian polymer,[30] which should be the least stereoregular polymer possible, will have $\Psi_{abs} = \Psi_{rel} = \Psi = 0.50$.

2.2 SYNDIOTACTIC POLYPROPYLENE

2.2.1 BRIDGED FLUORENYL/CYCLOPENTADIENYL METALLOCENES

In 1988, Ewen et al. reported the first organometallic olefin polymerization catalysts that contained the fluorenyl ligand.[33] The zirconium and hafnium precatalysts $(CH_3)_2C(C_5H_4)(C_{13}H_8)MCl_2$ were activated with MAO and yielded syndiotactic polypropylene with $[rrrr] = 86\%$ for M = Zr and $[rrrr] = 74\%$ for M = Hf. This initial report opened the door to a detailed mechanistic understanding of the syndioselective olefin polymerization mechanism that is common to many fluorenyl-containing metallocene catalysts,[34] and even a class of doubly bridged metallocene catalysts that do not rely on the fluorenyl ligand (see Chapter 4 for further information on doubly bridged metallocene catalysts).[35]

Scheme 2.2 illustrates the commonly accepted stereochemical mechanism for syndioselective propylene polymerization.[36] Two principal types of stereoerrors can be detected (Figure 2.6). A unimolecular site epimerization event,[37] which disrupts the alternating employment of enantiotopic coordination sites, can lead to the creation of an isolated m dyad in an otherwise perfect sequence of r dyads (…$rrmrrr$…). An enantiofacial misinsertion, in which the wrong enantiotopic face of the incoming monomer is used for coordination and migratory insertion, can lead to the creation of an isolated mm triad in an otherwise perfect sequence of r dyads (…$rrmmrrr$…). As shown in Figure 2.6, these two kinds of stereoerrors generate different pentad signatures and can thus be differentiated and quantified.

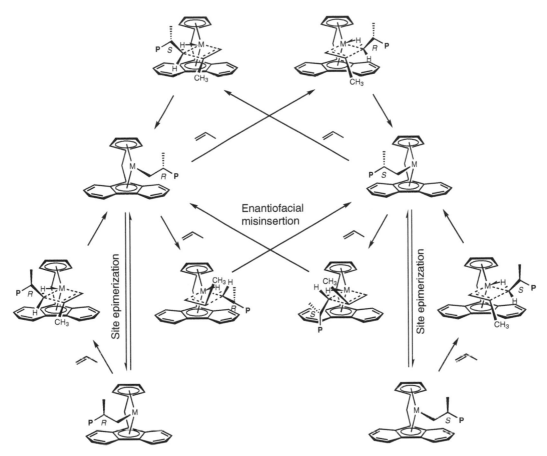

SCHEME 2.2 Site epimerization and enantiofacial misinsertion are two principal sources of stereoerrors commonly observed in syndioselective polymerizations. The bridge substituents have been omitted for clarity; **P** represents the growing polymer chain; and M = cationic Zr^+ or Hf^+, generally.

Since 1988, many fluorenyl/cyclopentadienyl variants related to the original Ewen/Razavi catalyst (s-**1**) have been prepared. Figure 2.7 depicts essentially all of these in chronological order of their appearance in the literature.[38–40] For simplicity, only the reported zirconocene dichlorides are shown; but in many cases, derivatives—such as the zirconocene dimethyls—are known and have also been investigated. Where the corresponding hafnium dichloride species have been reported, a parenthetical note (Hf) in Figure 2.7 is provided. None of the related titanium-based metallocenes is known to produce significantly stereoregular polypropylene.[18]

Several recent reviews chronicle the polymerization behavior displayed by the most prominent of these catalysts.[36,41] The following observations and general trends are noteworthy.

Not all syndioselective precatalysts are C_s-symmetric; indeed, many of the metallocenes in Figure 2.7 are C_1-symmetric. Usually symmetry is broken by distal substituents on the fluorenyl ring or on the bridge. But, even proximally substituted C_1-symmetric species s-**70** and s-**83**, with cyclopentadienyl substitution, give syndiotactic polypropylene with melting points (T_ms) of 129 °C ([r] = 86.4%)[42] and 134 °C,[43] respectively. Even double cyclopentadienyl substitution, as in C_s-symmetric s-**93** ([rrrr] = 90.5%)[44] or C_1-symmetric s-**95** ([r] = 66%),[45] does not preclude syndioselectivity. The cyclopentaphenanthrene derivative s-**6** exhibits diminished stereoselectivity ([rrrr] = 72%),[46] in line with its relative inability to accommodate the methyl group of the inserting propylene monomer (see Figure 2.2b). The 4-methyl-substituted catalyst s-**18** shows a similarly decreased syndioselectivity ([rrrr] = 67%).[40]

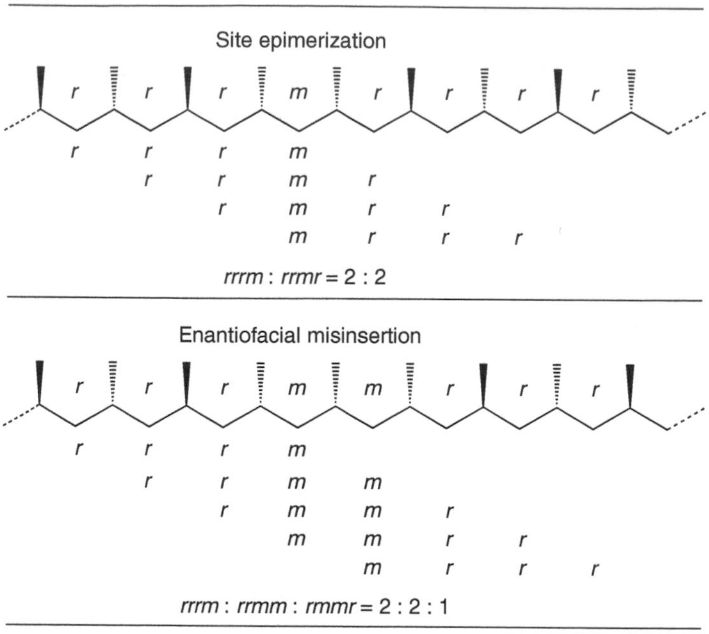

FIGURE 2.6 Site epimerization and enantiofacial misinsertion produce defective pentads in characteristic ratios. (Reprinted with permission from Miller, S. A.; Bercaw, J. E. *Organometallics* **2004**, *23*, 1777–1789. Copyright 2004 American Chemical Society.)

Two-atom bridges, such as those found in s-**12**, s-**27**, s-**50**, and s-**79**, generally result in a less rigid, less stereoselective catalyst; for example, s-**12** has a [*rrrr*] pentad fraction (83.0%) that is 3% less than that of the parent catalyst s-**1**.[47] Catalysts bearing dialkyl-silicon or diaryl-silicon bridges usually afford syndiotactic polymer plagued with site epimerization stereoerrors. For example, the silicon bridge of s-**7** decreases the [*rrrr*] fraction to 24% from the 72% afforded by its carbon-bridged analogue s-**6**.[46] Depending on the polymerization conditions, silicon-bridged s-**29** can produce atactic polymer with [*rr*] = 12% [48] or syndiotactic polymer with [*rr*] = 76.2%.[49] The catalyst bearing the diphenyl germanium bridge (s-**52**) suffers similarly ([*rr*] = 65.5%),[49] but the catalyst bearing a phenyl phosphorus bridge (s-**72**) ([*rrrr*] = 81%, T_m = 128 °C, T_p (polymerization temperature) = 50 °C) fares comparably to the parent catalyst, s-**1**.[50] Diarylmethylidene bridges typically allow for a higher molecular weight than isopropylidene bridges.[51,52] The increase is usually between 200% and 500%.

The inclusion of heteroatom substituents on the metallocene ligand has varied consequences. A generalization found for species s-**19** through s-**26** is that electron-donating substituents in the 1-, 2-, or 3-position of the fluorenyl ring (as in s-**19**, s-**20**, and s-**21**, respectively) decrease catalytic activity and polymer molecular weight. In contrast, such substituents in the 4-position of the fluorenyl ring (as in s-**22**) increase catalytic activity and polymer molecular weight. For example, s-**21** provides a catalytic activity of 3,000 g PP/(g cat.·h) with a viscosity average molecular weight (M_v) of 114,000, whereas s-**22** provides 38,000 g PP/(g cat.·h) with M_v = 295,000 under comparable conditions. Stereoselectivity operates on a steric basis with a 4-substitutent being more disruptive: s-**21**, [*rrrr*] = 87%, s-**22**, [*rrrr*] = 80%.[40]

Inspection of the ligand structure of more recently developed catalysts reveals considerable effort directed toward the incorporation of steric bulk into the fluorenyl plane of the ligand framework. Indeed, this has proven to be the most successful strategy toward increasing catalyst syndioselectivity. When compared to other catalysts under identical reaction conditions, one of the most stereoselective catalysts in Figure 2.7 is s-**90**, which produces high molecular weight syndiotactic polypropylene with

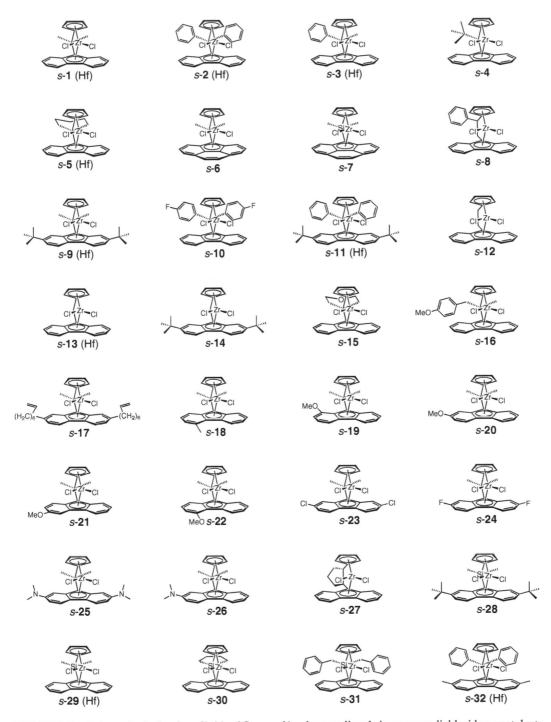

FIGURE 2.7 A chronological series of bridged fluorenyl/cyclopentadienyl zirconocene dichloride precatalysts that have been employed for syndioselective propylene polymerization. Known hafnium dichloride analogues are indicated by (Hf).

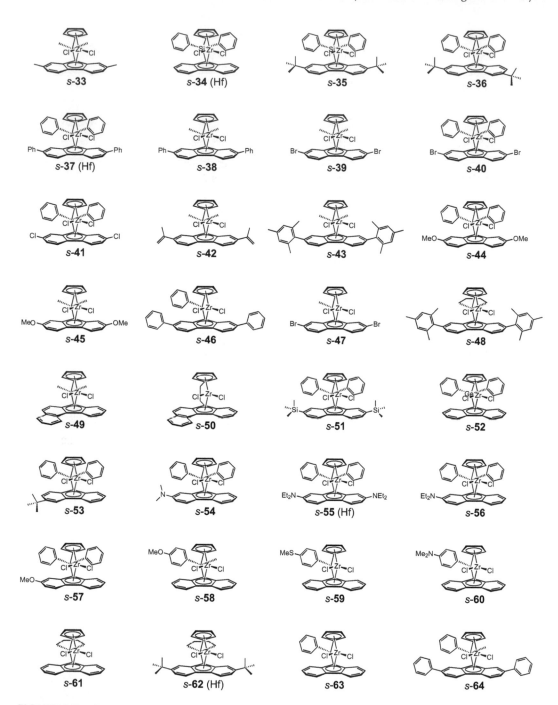

FIGURE 2.7 Continued.

M_w (weight average molecular weight) = 535,000, $[r]$ = 97.5%, and T_m = 154 °C (T_p = 0 °C).[53]
Interestingly, catalytic activity generally increases with the added steric bulk. A tentative explanation
has been posited that an extended ligand framework enforces a greater ion pair separation, decreasing
the energetic barrier to counteranion dissociation, which can be the rate-determining step[54] before
monomer coordination and insertion.[53]

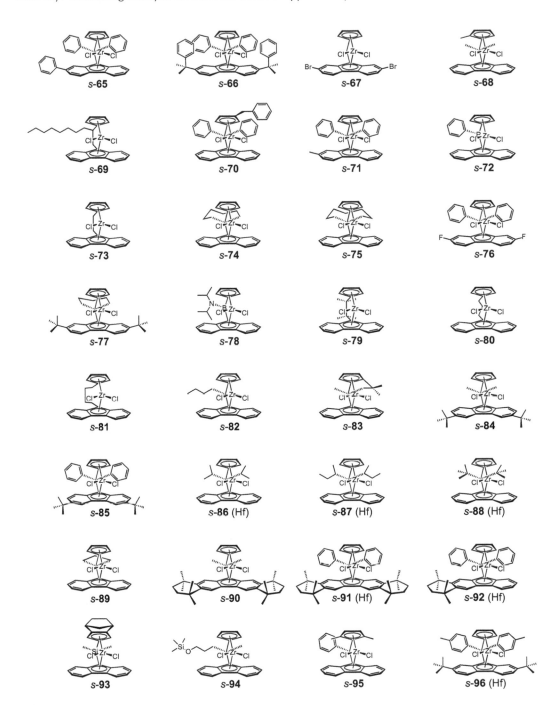

FIGURE 2.7 Continued.

Of all the hafnium compounds listed in Figure 2.7, none excels their zirconium counterparts in syndioselectivity. This observation is sometimes attributed to hafnium's decreased rate of bimolecular propagation relative to unimolecular site epimerization. Lower overall activities with hafnium catalysts corroborate this argument. The most syndioselective hafnium catalyst appears to be C_1-symmetric s-**92**-Hf, which produces syndiotactic polypropylene with $[r] = 92.2\%$ and $T_m = 141\,°C$ ($T_p = 0\,°C$).[53]

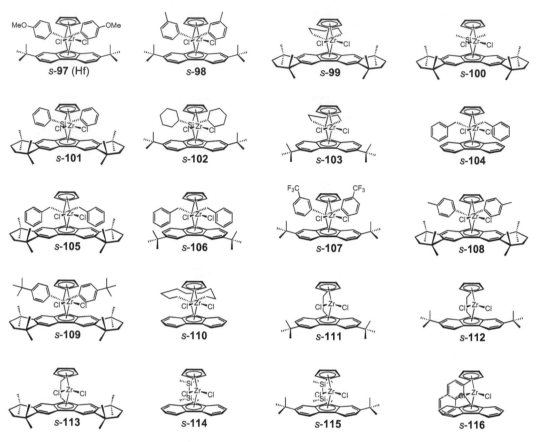

FIGURE 2.7   Continued.

2.2.2 BRIDGED FLUORENYL/INDENYL METALLOCENES

A few bridged fluorenyl/indenyl metallocenes show a tendency toward modest syndioselectivity (Figure 2.8).[55] These are rare, but a particularly syndioselective example is s-**120**, which produces syndiotactic polypropylene having $[r] = 93.8\%$, $[rr] = 89.4\%$, and $[rrrr] = 74.4\%$ $(T_p = 0\,^\circ C)$.[56] It has been proposed that this catalyst operates with alternating stereodifferentiation mechanisms at the two coordination sites. In addition to the normal stereochemical mechanism wherein the ligand directs the growing polymer chain (see Figure 2.2), monomer insertion at the other coordination site is controlled directly by a ligand/monomer interaction that favors placement of propylene's methyl group away from the 4-methyl substituent of the indenyl ring. Lower monomer concentrations or higher polymerization temperatures result in less stereoregular polymer, as might be expected with increasing site epimerization.

2.2.3 BRIDGED FLUORENYL/FLUORENYL METALLOCENES

Members of a small class of fluorenyl/fluorenyl catalysts, shown in Figure 2.9, exhibit weak syndioselectivity. Several of these produce nearly atactic polypropylene with $[rr]$ values near 25%, but are included here for comparison. Steric perturbation at the 4- and 5-positions of one fluorenyl ring serves to increase the syndioselectivity through weak enantiomorphic site control that is in competition with site epimerization.[55] Note that the direct comparison of s-**124** with s-**124**-Hf provides an unusual example for which the hafnium analogue is more stereoselective than the zirconium version. The most syndioselective of these is s-**130**, which arguably has the greatest size discrepancy between

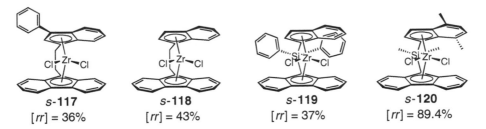

FIGURE 2.8 Rare examples of syndioselective bridged fluorenyl/indenyl precatalysts.

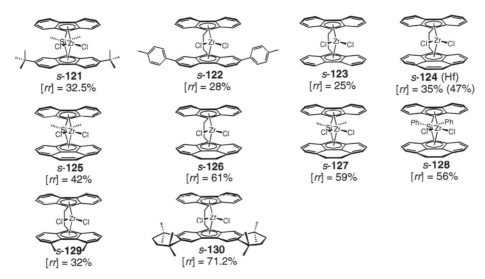

FIGURE 2.9 Weakly syndioselective bridged fluorenyl/fluorenyl precatalysts.

the fluorenyl moieties and can generate syndiotactic polypropylene with $[r] = 81.3\%$, $[rr] = 71.2\%$, and $[rrrr] = 49.1\%$ ($T_p = 0\,°C$, no T_m, amorphous).[53] The enantiomorphic site control of this catalyst operates in contrast to the completely aselective nature of the unsubstituted, C_{2v}-symmetric parent compound $(CH_2CH_2)(C_{13}H_8)_2ZrCl_2$ s-**123** ($[r] = 50.2\%$, $[rr] = 25.1\%$, $[rrrr] = 6.3\%$).[53,57]

2.2.4 Bridged Fluorenyl/Amido Half-Metallocenes

A number of fluorenyl/amido half-metallocenes or "constrained geometry catalysts" (CGCs) have been investigated for ethylene/α-olefin copolymerizations.[58–60] However, such CGCs typically exert little or no stereocontrol in α-olefin homopolymerizations. Figure 2.10 depicts fluorenyl-containing exceptions to this rule. The zirconium-based species s-**131**[61] is reportedly syndioselective ($[r] = 91.8\%$, $[rrrr] = 77.4\%$, $T_m = 148.9\,°C$, $T_p = 20\,°C$) when activated by MAO/Al(i-butyl)₃, but isoselective when activated by $[HMe_2N(C_6H_5)]^+[B(C_6F_5)_4]^-/Al(i\text{-butyl})_3$ ($[m] = 96.3\%$, $[mmmm] = 92.0\%$, $T_m = 160.5\,°C$, $T_p = 20\,°C$).[62] The syndiotactic polymer is generated with enantiomorphic site control, but the corresponding titanium catalyst s-**131**-Ti reportedly operates with moderately syndioselective chain-end control ($[rr] = 55$–64%, $[rrrr] = 30$–38%, $T_p = 40\,°C$), regardless of the activator.[63] In nonpolar solvents, living syndioselective polymerization is possible with the dimethyl variant of s-**131**-Ti; a syndioselectivity of $[rrrr] = 42\%$ is typical.[64] The 3,6-di-$tert$-butylated s-**133** is considerably syndioselective ($[rr] = 94.5\%$, $[rrrr] = 86.7\%$, $T_m = 151.2\,°C$, $T_p = 40\,°C$)[36,65] and the titanium analogue s-**133**-Ti has been shown to employ a weak enantiomorphic site control mechanism,[66] suggesting that this is the case for s-**131**-Ti as well. Repeated

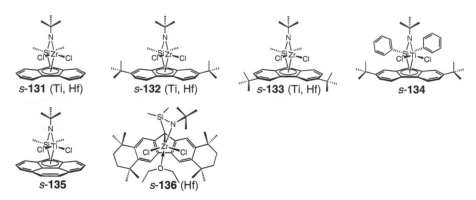

FIGURE 2.10 Syndioselective bridged fluorenyl/amido half-metallocene precatalysts. Titanium or hafnium dichloride analogues are indicated by (Ti) or (Hf). Unusual η^1 binding to zirconium is observed for s-**136**.

polymerizations with s-**133**-Ti/MAO at 30 °C revealed the catalyst's propensity to undergo site epimerization, making it less syndioselective ([*rrrr*] = 62.9–78.9%) than the parent metallocene catalyst, s-**1**. Activation of the dimethyl version of s-**133**-Ti with borate activators resulted in a dramatic decline in stereoselectivity ([*rrrr*] = 20.7–35.6%).[66]

The zirconium-based catalyst s-**136**/MAO[67] is highly syndioselective ([*r*] > 99%, $T_p = 0$ °C) and apparently provides the highest-melting syndiotactic polypropylene known.[68] The unannealed T_m for the polypropylene is 164 °C, and the annealed T_m is 174 °C. Stereoerrors have not been detected in this polymer by ^{13}C NMR, but pentads presumably attributable to site epimerization (single *m* mistakes) can be found if polymerizations are conducted in dilute propylene at elevated temperatures.

2.3 ISOTACTIC POLYPROPYLENE

2.3.1 BRIDGED FLUORENYL/CYCLOPENTADIENYL METALLOCENES

Shortly following the report of Ewen, Razavi, et al.[33] demonstrating the MAO-cocatalyzed formation of syndiotactic polypropylene with fluorenyl-containing metallocenes of the type $(CH_3)_2C(C_5H_4)(C_{13}H_8)MCl_2$ (e.g., M = Zr, s-**1**), several authors—principally Ewen,[46,69] Spaleck,[70] Razavi,[71] and Fink[72]—prepared cyclopentadienyl-substituted variants of the parent C_s-symmetric metallocene. Incorporation of a substituent at the 3-position of the cyclopentadienyl ring lowers the symmetry of the metallocene to C_1. As a consequence, the obtained polymers are no longer syndiotactic, but display alternative tacticities depending on the nature of the substituent. When the substituent is *tert*-butyl (*i*-**1**), isotactic polypropylene is obtained having [*m*] = 91.9%, [*mmmm*] = 78.0%, and $T_m = 129$ °C ($T_p = 40$ °C).[71] Several isoselective fluorenyl/cyclopentadienyl precatalysts are shown in Figure 2.11.

The stereochemical mechanism responsible for the isoselectivity of C_1-symmetric metallocene catalysts has been a topic of considerable debate. There are two limiting mechanisms possible for the formation of isotactic polypropylene with such C_1-symmetric catalysts having diastereotopic coordination sites. These are the *site epimerization* mechanism and the *alternating* mechanism, as shown in Scheme 2.3.

The vast majority of published reports for isotactic polypropylene formation with metallocenes based on *i*-**1** invoke the site epimerization mechanism[37] to account for the observed isoselectivity.[69–72] When the growing polymer chain occupies the coordination site distal to the *tert*-butyl group (Scheme 2.3, site α), it is directed away from the benzo substituent of the fluorenyl ligand in the transition state for monomer insertion. The methyl group of the incoming monomer is

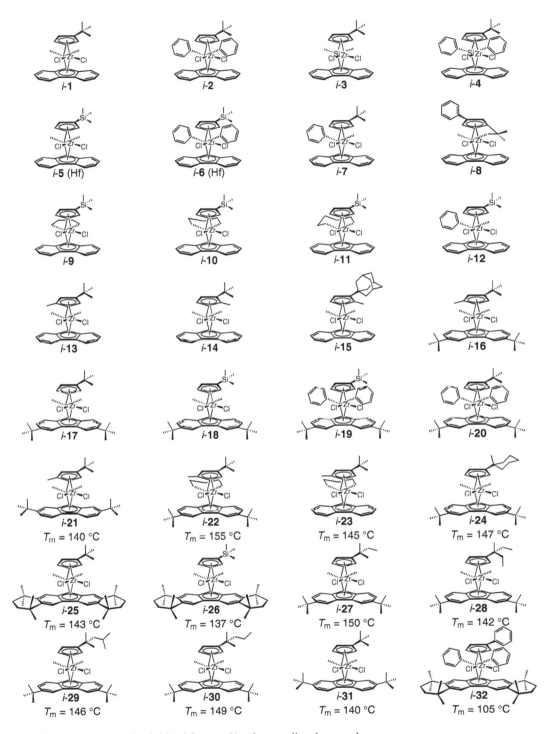

FIGURE 2.11 Isoselective bridged fluorenyl/cyclopentadienyl precatalysts.

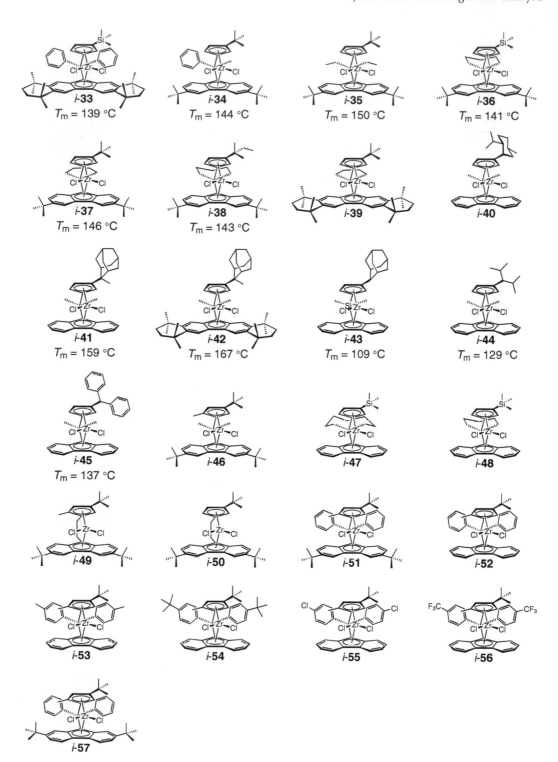

FIGURE 2.11 Continued.

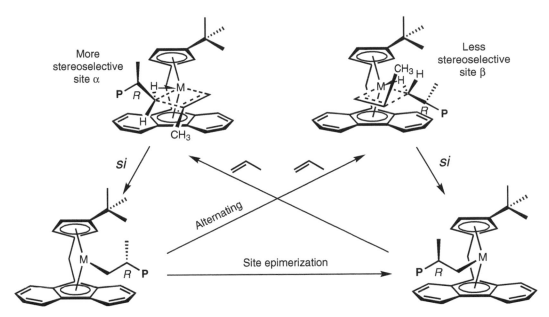

SCHEME 2.3 There are two limiting mechanisms for isoselective polymerization with C_1-symmetric catalysts: the site epimerization mechanism and the alternating mechanism. The bridge substituents have been omitted for clarity; **P** represents the growing polymer chain; and M = cationic Zr^+ or Hf^+, generally. (Reprinted with permission from Miller, S. A.; Bercaw, J. E. *Organometallics* **2006**, *25*, 3576–3592. Copyright 2006 American Chemical Society.)

directed in a trans fashion away from the growing polymer chain. Following migratory insertion, the growing polymer chain briefly occupies the coordination site proximal to the *tert*-butyl substituent (Scheme 2.3, bottom left). Epimerization at the metal moves the polymer chain away from the bulky *tert*-butyl group in a unimolecular process that regenerates the original coordination site for monomer insertion (Scheme 2.3, bottom right). The site epimerization process is usually presented as occurring only in the direction shown. Hence, only the α coordination site of the metallocene is employed for monomer insertion according to a strict site epimerization model. This results in repeated employment of the same monomer enantioface for insertion, and isotactic polypropylene is thereby formed.

Also depicted in Scheme 2.3 is a second limiting mechanism, the alternating mechanism. Following monomer insertion at the more stereoselective α site, the β site becomes available for monomer coordination. In the transition state for insertion at the less stereoselective site (β), the growing polymer chain is directed competitively by both the *tert*-butyl group and the benzo substituent on that half of the metallocene. In order for the resulting polymer to be isotactic, the *tert*-butyl group must prevail and the growing polymer chain is preferentially directed toward the benzo substituent. Insertion ensues with a trans arrangement between the polymer chain and the methyl group of the inserting monomer; this regenerates the original coordination site (Scheme 2.3, bottom right). In contrast to the site epimerization mechanism, the alternating mechanism employs two coordination sites (α and β sites) for monomer insertion. Note that both sites must exhibit the same general monomer enantioface selectivity for isotactic polypropylene to result.

Efforts to differentiate between these two models through statistical analysis of the pentad distributions have been inconclusive. Figure 2.12 summarizes one such investigation with *i*-**1**,[73] which concluded that the site epimerization model and the alternating model converged to the same enantiofacial selectivity parameter upon minimization of the RMS (root mean square) error in comparison with the observed pentad populations. If the site epimerization model applies, then

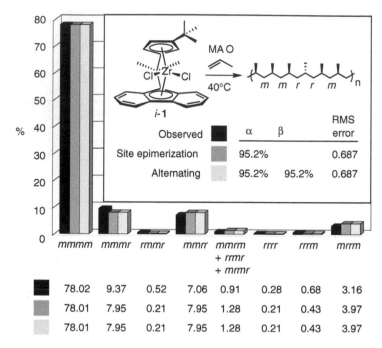

The figure's data table (below the chart):

	mmmm	mmmr	rmmr	mmrr	mmrm + rrmr + mrmr	rrrr	rrrm	mrrm
■	78.02	9.37	0.52	7.06	0.91	0.28	0.68	3.16
▨	78.01	7.95	0.21	7.95	1.28	0.21	0.43	3.97
▢	78.01	7.95	0.21	7.95	1.28	0.21	0.43	3.97

Legend within figure:

	α	β	RMS error
Observed			
Site epimerization	95.2%		0.687
Alternating	95.2%	95.2%	0.687

FIGURE 2.12 A statistical analysis of isotactic polypropylene from i-**1**/MAO cannot conclusively differentiate between the site epimerization model and the alternating model. (Reprinted with permission from Miller, S. A.; Bercaw, J. E. *Organometallics* **2006**, *25*, 3576–3592. Copyright 2006 American Chemical Society.)

the enantioselectivity of the one-site catalyst is $\alpha = 95.2\%$. If the alternating model applies, then the enantioselectivities of the two-site catalyst are the same at both sites, $\alpha = 95.2\%$ and $\beta = 95.2\%$.

An early investigation[71] found that the stereoregularity and polymer melting temperature of polypropylene derived from i-**1** both decreased slightly as the polymerization temperature increased from 20 to 80 °C (in liquid monomer). The [mmmm] fraction decreased from 79.2% to 76.8%, as did the melting temperature, from 133 to 127 °C. These results are not wholly consistent with either model. If the alternating model were operative at low temperatures, one would predict that an increase in polymerization temperature would increase the likelihood of the unimolecular site epimerization process relative to bimolecular propagation, thereby accessing the more stereoselective insertion to a greater degree—resulting in an increase in [mmmm] and T_m. Moreover, if the site epimerization model is operative at low temperatures, then only one coordination site for monomer insertion is employed and therefore, the stereoselectivity is predicted to be essentially invariant to an increase in polymerization temperature.

A more recent set of experiments favors the alternating mechanism.[73] If polymerizations are conducted in dilute monomer (10% by volume in toluene), the melting temperatures and [mmmm] fractions exceed those obtained from bulk polymerizations. For example, $T_m = 135$ °C and [mmmm] = 88.8% are obtained in 10% monomer solution as compared to $T_m = 129$ °C and [mmmm] = 78.0% in neat monomer at the same $T_p (40 °C)$. In addition, under these dilute conditions, [mmmm] is found to increase from 82.2% to 89.4% as the polymerization temperature increases over the range of 0–60 °C. A similar effect was reported with 2 bar of propylene monomer in toluene; the [mmmm] fraction increased from 83.5% to 87.7% as T_p increased from 10 to 50 °C.[74] These results are consistent with an alternating mechanism that increasingly yields to a site epimerization mechanism as the monomer concentration decreases or the polymerization temperature increases. This conclusion is contrary to the popular view in the literature that invokes the site epimerization mechanism under all conditions.

One of the primary explanations offered in support of an exclusive site epimerization mechanism is that the *tert*-butyl side of the metallocene is too sterically crowded to accommodate the growing polymer chain. Calculations by Morokuma[75] suggest that the bulky substituent forbids the growing chain to reside nearby, necessitating a site epimerization following every insertion. Calculations by Fink[28d,76] and Corradini,[77] however, are more forgiving and allow the insertion with the polymer chain proximal to the *tert*-butyl group.

Convincing experimental evidence that such an insertion occurs has been found with a 4-methyl substituted variant of this catalyst, $(CH_3)_2C(3\text{-}tert\text{-butyl-4-methyl-}C_5H_2)(C_{13}H_8)ZrCl_2$ (*ih*-**5**).[73] Attempts to fit the pentad distribution ($T_p = 20\,°C$, Figure 2.13) to a one-site (site epimerization) model generally fail, whereas a two-site (alternating) model fits the data quite satisfactorily (RMS

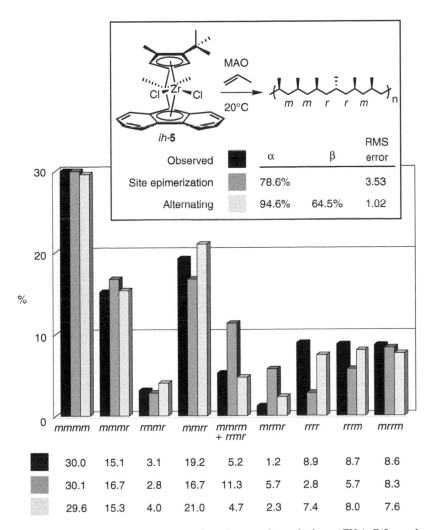

FIGURE 2.13 The pentad distribution of polymer formed by $(CH_3)_2C(3\text{-}tert\text{-butyl-4-methyl-}C_5H_2)(C_{13}H_8)ZrCl_2/MAO$ (*ih*-**5**) at $T_p = 20\,°C$ is poorly described by a one-site enantiomorphic site control model. A superior fit is obtained with a two-site model that approximates hemiisotactic polypropylene. (Reprinted with permission from Miller, S. A.; Bercaw, J. E. *Organometallics* **2006**, *25*, 3576–3592. Copyright 2006 American Chemical Society.)

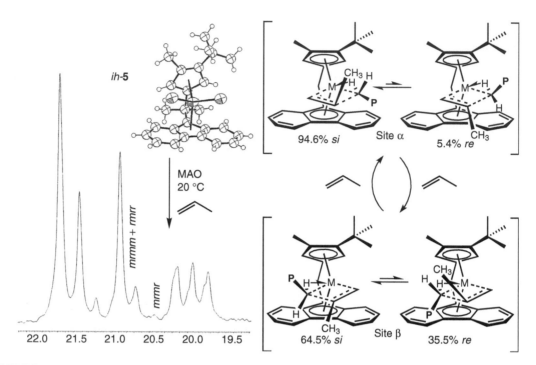

FIGURE 2.14   A statistical analysis of the pentad distribution predicts that $(CH_3)_2C(3\text{-}tert\text{-butyl-4-methyl-}$ $C_5H_2)(C_{13}H_8)ZrCl_2/MAO$ (*ih*-**5**) operates through a two-site mechanism having enantiofacial selectivities of 94.6% (site α) and 64.5% (site β). The bridge substituents have been omitted for clarity; **P** represents the growing polymer chain; and M = cationic Zr^+. (Reprinted with permission from Miller, S. A.; Bercaw, J. E. *Organometallics* **2006**, *25*, 3576–3592. Copyright 2006 American Chemical Society.)

error = 1.02). The formed polymer is essentially hemiisotactic polypropylene having the characteristically low intensities of pentads with isolated *m* and *r* dyads (*mrmm*, *rmrr*, and *mrmr*),[30] and therefore must have arisen through a two-site mechanism wherein the growing polymer chain resides proximal to either the methyl group or the *tert*-butyl group in an alternating fashion. The statistical analysis indicates that the more stereoselective site (α) is 94.6% enantioselective whereas the less stereoselective site (β) is 64.5% stereoselective toward the same enantioface of the coordinating propylene monomer (Figure 2.14).

With the understanding that C_1-symmetric metallocene catalysts can operate through a two-site, alternating mechanism, it has been possible to engineer more highly isoselective monosubstituted (pre)catalysts of this symmetry. The identification and modification of three important catalyst features have allowed for a greater isoselectivity.[73]

First, the 3-cyclopentadienyl substituent R should be larger than *tert*-butyl. If the stereocontrol mechanism proposed in Figure 2.14 is correct, then a group larger than *tert*-butyl should enforce greater enantiofacial selectivity at the more stereoselective site (α). This substituent has a greater purpose than simply effecting site epimerization; it competes with the opposing benzo substituent in repelling the growing polymer chain during the transition state for monomer insertion at one of the two operative sites. The diphenyl methyl substituent (*i*-**45**, [*m*] = 86.1%, T_m = 137 °C) and the 2-methyl-2-adamantyl substituent (*i*-**41**, [*m*] > 99%, T_m = 159 °C) are phenomenologically larger than *tert*-butyl (*i*-**1**, [*m*] = 79.5%, T_m = 126 °C) and allow for greater isoselectivity (comparative data for T_p = 0 °C).[73] However, the group should not be so large that it retards polymerization activity altogether. Apparently, R = 2-phenyl-2-adamantyl or R = 2-$(CH_2Si(CH_3)_3)$-2-adamantyl are extremely large substituents that do not allow the catalyst to accommodate both the polymer chain and monomer in the coordination sphere. The corresponding zirconocene dichloride/MAO species are essentially inactive for both propylene and ethylene homopolymerizations.[73]

Second, factors that encourage site epimerization should lead to polymers of higher isotacticity since the more stereoselective site can be used preferentially. "Turning on" the site epimerization mechanism can be accomplished, in principle, by altering polymerization conditions (e.g., increasing temperature, decreasing monomer concentration)[78] or by the inclusion of a silicon-based bridge. It is well documented for C_s-symmetric complexes that a disubstituted silylene bridge increases the site epimerization/propagation quotient compared to an isopropylidene bridge.[48,49] Silicon-bridged i-3 ($T_m = 161\,°C$ for $T_p = 30\,°C$; $T_m = 148\,°C$ for $T_p = 60\,°C$) exhibits greater isoselectivity than carbon-bridged i-1 ($T_m = 130\,°C$ for $T_p = 30\,°C$; $T_m = 125\,°C$ for $T_p = 60\,°C$) as gauged by the melting temperatures of the obtained isotactic polymers.[79] Even a catalyst with a poorly directing 3-substituent can provide moderately isotactic polymer if the site epimerization mechanism has been turned on.[70c] Such is the case with the silicon-bridged 2-adamantyl catalyst i-43 ($[m] = 75.3\%$, $T_m = 109\,°C$); its isopropylidene-bridged counterpart is less isoselective ($[m] = 62.3\%$, ~amorphous) because it largely avoids site epimerization.[80]

Third, to the extent that the catalyst system preferentially utilizes the more stereoselective site for monomer insertion, enhancement of the stereoselectivity at that site will lead to higher isotacticity. Thus, efforts to increase the size of the fluorenyl ring have led to enhanced isoselectivity, presumably because a substituted benzo moiety is more repulsive in the transition state than the benzo group itself. Catalysts derived from *tert*-butylated fluorenes such as i-16 ($[mmmm] = 95.7\%$, $T_m = 153\,°C$) successfully demonstrate this approach.[81]

One of the most isoselective C_1-symmetric catalysts known is i-42, which encompasses all three strategies for improving isoselectivity. It contains the bulky 2-methyl-2-adamantyl substituent. While it does not bear a silicon-based bridge, it is nonetheless plausible that site epimerization is facile because of extreme steric crowding contributed by both the 2-methyl-2-adamantyl substituent and the opposing octamethyloctahydrodibenzofluorenyl ligand. The directing ability of the sterically enhanced fluorenyl ligand results in greater stereoselectivity at the more stereoselective site. With i-42/MAO, highly isotactic polypropylene is obtained, as stereoerrors are virtually absent by ^{13}C NMR analysis ($[mmmm] > 99\%$). The polymers prepared at 0 and 20 °C have high melting temperatures (167.0 and 162.7 °C, respectively) and large enthalpies of melt (92.0 J/g and 87.5 J/g, respectively). Apparently, the highest previously reported melting temperature for an isotactic polypropylene made through homogeneous catalysis is 166 °C.[82] The high isoselectivity of i-42/MAO suggests that it may employ a single propagative transition state for monomer insertion.[73]

2.3.2 Bridged Fluorenyl/Indenyl Metallocenes

The vast majority of isoselective metallocene catalysts are based on bridged C_2-symmetric indenyl/indenyl species.[34] However, it has been demonstrated that certain bridged C_1-symmetric fluorenyl/indenyl metallocenes are isoselective as well. Examples are depicted in Figure 2.15, along with the highest $mmmm$ pentad fractions reported. Generally, the isoselectivities are poorer than the related indenyl/indenyl C_2-symmetric catalysts and are quite dependent on the reaction conditions (e.g., MAO activation versus borate-based activation of the dimethyl species). However, specific substitution of the indenyl ring and optimization of the reaction conditions—neither follow an obvious trend—can lead to a very high isoselectivity, as in i-63/MAO ($[mmmm] = 98.0\%$).[83] Note also that the nature of the bridge is important; the dimethyl silylene variant of i-63, i-65, is markedly less isoselective ($[mmmm] = 71.7\%$).[83] The mechanism is generally thought to be a combination of the alternating and site epimerization pathways;[56,84,85] hence isoselectivity is strongly dependent on the polymerization temperature and monomer concentration.[85–87] With some exceptions, isotacticity increases with decreasing monomer concentration and increasing polymerization temperature because of increased utilization of the site

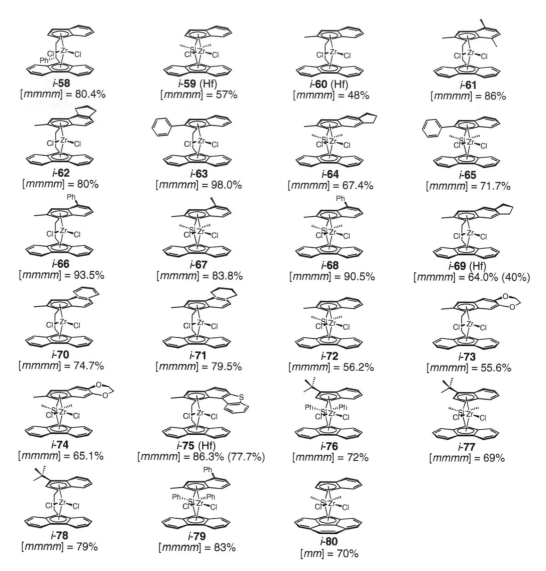

FIGURE 2.15   Isoselective bridged fluorenyl/indenyl precatalysts. Pentad fractions [*mmmm*] (or triad fraction [*mm*]) correspond to the highest reported value.

epimerization mechanism; these trends are opposite those found with C_2-symmetric indenyl/indenyl catalysts.[88]

2.3.3 FLUORENYL/FLUORENYL METALLOCENES

Several fluorenyl/fluorenyl metallocenes are reportedly useful for the preparation of moderately isotactic polypropylene (Figure 2.16). One strategy has been to prepare bridged C_2-symmetric fluorenyl/fluorenyl metallocenes, in analogy to *ansa*-C_2-symmetric indenyl/indenyl metallocenes. This has led to catalysts that operate with modest enantiomorphic site control, with *i*-**83** and *i*-**84** exhibiting [*m*] = 88.9 % and [*m*] = 89.5%, respectively.[89,90] Bridged C_1-symmetric species have also been prepared (*i*-**82** and *i*-**87**), but these are less isoselective.[55,91] Complex *i*-**81** still stands as perhaps the only nonbridged fluorenyl-containing metallocene that is sufficiently stereoselective ([*mmmm*] = 82.9%) to make high-melting polypropylene (145 °C).[57,92]

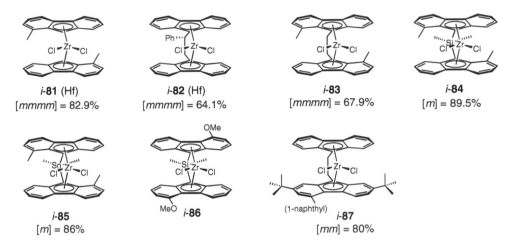

FIGURE 2.16 Isoselective nonbridged and bridged fluorenyl/fluorenyl precatalysts. Dyad, triad, and pentad fractions correspond to the highest reported value.

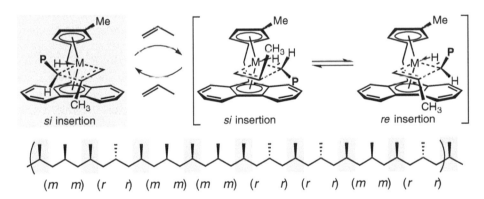

SCHEME 2.4 A hemiisoselective mechanism (shown here for catalyst *h*-1) relies on alternating employment of diastereotopic coordination sites—one that is highly enantioselective (in gray) and one that is aselective (in brackets). The bridge substituents have been omitted for clarity; **P** represents the growing polymer chain; and M = cationic Zr^+ or Hf^+, generally.

2.4 HEMIISOTACTIC POLYPROPYLENE

Farina et al. first prepared hemiisotactic polypropylene in 1982 through the hydrogenation of isotactic poly(2-methylpenta-1,3-diene).[93] The first direct, stereoselective synthesis of hemiisotactic polypropylene was reported by Ewen, et al. in 1991[46] with the methyl-substituted zirconocene $(CH_3)_2C(3-CH_3-C_5H_3)(C_{13}H_8)ZrCl_2$ (*h*-1) activated with MAO.[57,70c,71b,72] Scheme 2.4 depicts the proposed stereochemical mechanism. Modeling and calculations suggest that one of the diastereotopic coordination sites is highly enantioselective whereas the other site is aselective.[28] To the degree that site epimerization is minimal, ideal hemiisotactic polypropylene should result, providing only sequential *mm* and *rr* triads, as depicted in Scheme 2.4.

Naturally, this catalyst is not perfect and stereoerrors are invariably present in the formed polymer. The principal stereoerrors arise from enantiofacial misinsertion at the more stereoselective site and site epimerization. Such mistakes give rise to *mr* and *rm* triads and *m* and *r* dyads in an otherwise consistent procession of *mm* and *rr* triads (Figure 2.17). Thus, perfect hemiisotactic polypropylene

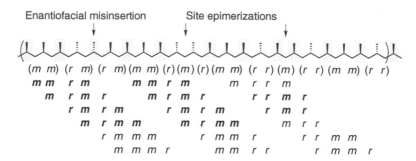

FIGURE 2.17   Enantiofacial misinsertion and site epimerization stereoerrors give rise to pentads containing isolated *m* and *r* dyads (in bold); such pentads are forbidden in perfectly hemiisotactic polypropylene.

is contaminated with the three forbidden pentads that contain isolated *m* and *r* dyads (*mmrm*, *rrmr*, and *mrmr*).

Figure 2.18 shows typical deviations from the ideal 3:2:1:4:0:0:3:2:1 pentad distribution that is expected from perfectly hemiisotactic polypropylene.[30,46,71,80] An increase in the polymerization temperature effects an increase in the prevalence of stereoerror pentads, *mmrm*, *rrmr*, and *mrmr*. Note that a site epimerization can skip an insertion at either one of the two coordination sites. If the site epimerization effects polymer chain migration *away* from the methyl group of the cyclopentadienyl substituent, then only (*mm*)(*m*)(*mm*), (*mm*)(*m*)(*rr*), (*rr*)(*m*)(*mm*), and (*rr*)(*m*)(*rr*) stereoerrors can result. None of these sequences gives rise to the *mrmr* pentad, which can form through a site epimerization involving chain migration toward the methyl group or through enantiofacial misinsertion. Hence, the failure of the [*mrmr*] pentad fraction to grow as rapidly as the [*mmrm*] + [*rrmr*] pentad fractions with increasing polymerization temperature (Figure 2.18) suggests that site epimerization of the chain migrating away from the methyl group is the dominant stereochemical mistake. For *h*-1/MAO, the best statistical match with ideal hemiisotactic polypropylene is obtained at 20 °C (RMS error = 2.36, 1.46, and 2.19 for T_p = 0, 20, and 60 °C, respectively).[70c,80]

Figure 2.19 depicts several reported hemiisoselective fluorenyl-containing metallocenes. The parent catalyst, *h*-**1**, appears to be the most hemiisoselective with [*m*] = 49.6% at T_p = 20 °C.[80] Any deviation in catalyst structure tends to effect less hemiisotactic polymers. For example, ethyl substitution (*h*-**5**) instead of methyl substitution (*h*-**1**) results in a measurable increase in syndioselectivity. The indenyl variant *h*-**4** commits a considerable fraction of stereoerrors ([*mmrm*] + [*rrmr*] + [*mrmr*] = 7.5% at T_p = 10 °C).[74] In addition, the hafnium complex *h*-**1**-Hf is more isoselective than the parent zirconium complex *h*-**1** under identical conditions ([*mmmm*] = 24.2% versus [*mmmm*] = 15.1% at T_p = 60 °C).[71b]

2.5   ISOTACTIC-HEMIISOTACTIC POLYPROPYLENE

Isotactic-hemiisotactic polymers are variants of hemiisotactic polymers wherein the intervening stereocenters exhibit partial alignment with their stereoregular neighbors. The general approach to this kind of polymer is to employ a steric variant of the methyl-substituted hemiisoselective metallocene $(CH_3)_2C(3\text{-}CH_3\text{-}C_5H_3)(C_{13}H_8)ZrCl_2$ (*h*-**1**). As predicted by the stereochemical mechanism in Scheme 2.5, 3-cyclopentadienyl substituents that are considerably larger than the methyl group should more effectively repel the growing polymer chain than the opposing benzo group. Hence, the less stereoselective coordination site will preferentially employ the same monomer enantioface as the more stereoselective site. If site epimerization is minimized, an isotactic-hemiisotactic polymer will result.

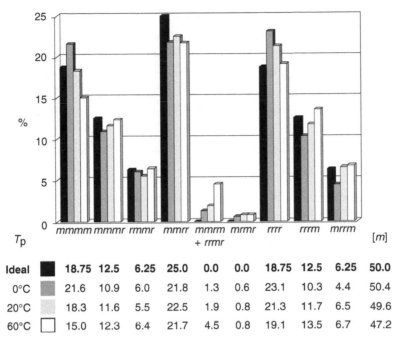

T_p		$mmmm$	$mmmr$	$rmmr$	$mmrr$	$mmrm + rrmr$	$mrmr$	$rrrr$	$rrrm$	$mrrm$	$[m]$
Ideal	■	18.75	12.5	6.25	25.0	0.0	0.0	18.75	12.5	6.25	50.0
0°C		21.6	10.9	6.0	21.8	1.3	0.6	23.1	10.3	4.4	50.4
20°C		18.3	11.6	5.5	22.5	1.9	0.8	21.3	11.7	6.5	49.6
60°C	□	15.0	12.3	6.4	21.7	4.5	0.8	19.1	13.5	6.7	47.2

FIGURE 2.18 Ideal (bold) and actual pentad distributions for hemiisotactic polypropylene obtained with $(CH_3)_2C(3\text{-}CH_3\text{-}C_5H_3)(C_{13}H_8)ZrCl_2/MAO$ (*h*-1/MAO) at varying polymerization temperatures.

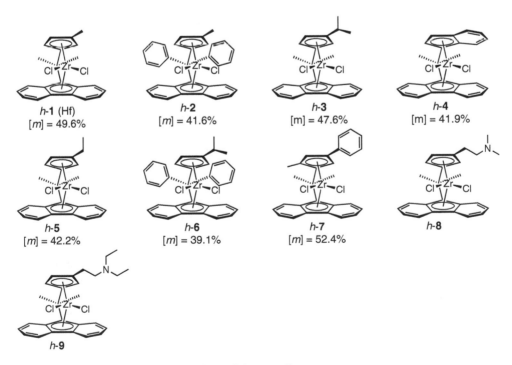

FIGURE 2.19 Hemiisoselective fluorenyl-containing metallocenes.

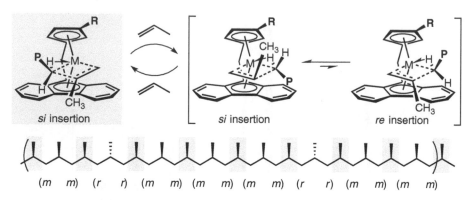

SCHEME 2.5 Isotactic-hemiisotactic polymers should arise from a two-site catalyst that has one highly enantioselective coordination site (in gray) and one coordination site that is moderately enantioselective for the same monomer enantioface (in brackets). **R** is substantially larger than methyl; the bridge substituents have been omitted for clarity; **P** represents the growing polymer chain; and M = cationic Zr^+ or Hf^+, generally.

Isotactic-hemiisotactic polypropylene has been obtained from the metallocenes shown in Figure 2.20, which have the general structure $(R')_2C(3\text{-R-}C_5H_3)(C_{13}H_8)MCl_2$.[80,94] Figure 2.21 depicts the pentad distributions obtained for each of the corresponding zirconocene/MAO polymerizations performed at 0 °C. As mechanistically prescribed, the *mmrm*, *rrmr*, and *mrmr* pentads are minimally abundant. Not surprisingly, the 3,4-disubstituted metallocene *ih*-**5** commits the most stereoerrors. The more stereoselective insertion with this catalyst pits a *tert*-butyl group against a benzo substituent, resulting in increased enantiofacial misinsertion as compared to the stereoselectivity when benzo competes against a simple hydrogen substituent.

As the size of the cyclopentadienyl substituent increases, the less stereoselective site is more likely to choose the same monomer enantioface as the more stereoselective site. The [*m*] dyad fraction increases accordingly. Figure 2.22 plots the observed polymer melting temperature as a function of the [*m*] dyad composition.[94] Several polymers exhibit no strong endotherm via differential scanning calorimetry (DSC) and are essentially amorphous. Above [*m*] ≈ 70%, normal DSC traces are obtained and the melting temperature is proportional to the isotacticity as quantified by [*m*].

An interesting case arises for *ih*-**9**, which contains the sterically expanded octamethyloctahydrodibenzofluorenyl ligand.[94] Its diminished isoselectivity ([*m*] = 89.2%), compared to that of similarly substituted *ih*-**6** ([*m*] = 94.7%), suggests that this expanded fluorenyl ligand competes effectively against the opposing diphenylmethyl group during monomer insertion at the less stereoselective site. In short, the diphenylmethyl substituent of *ih*-**9** and *ih*-**6** is a potent polymer-directing group, but its ability to repel the growing polymer chain is attenuated when it competes against the octamethyloctahydrodibenzofluorenyl component. Note that site epimerization is likely not prevalent with *ih*-**9**. If it were, *ih*-**9** should equal or excel the isoselectivity of *ih*-**6** because of an equal or enhanced isoselectivity at the more stereoselective site.

Ewen, et al. have also reported the synthesis of isotactic-hemiisotactic polypropylene from a catalyst that employs a sulfur-containing cyclopentadithiophene moiety instead of the fluorenyl component.[95] This "heterocene" catalyst, $Me_2C(3\text{-}(2\text{-adamantyl})\text{-}C_5H_3)(C_9H_4S_2)ZrCl_2$, is comparable to *ih*-**1** in its stereoselectivity at 0 °C ([*m*] = 57.9%), but engages in significant site epimerization at higher polymerization temperatures, and is slightly more prone to enantiofacial misinsertion at the more stereoselective site.

A stereochemical analysis has predicted the distribution of isotactic stereoblocks present in isotactic-hemiisotactic polypropylene.[32] For an ideal case (without site epimerization and with perfect enantiofacial selectivity at the more stereoselective site), the derived equations depend only on

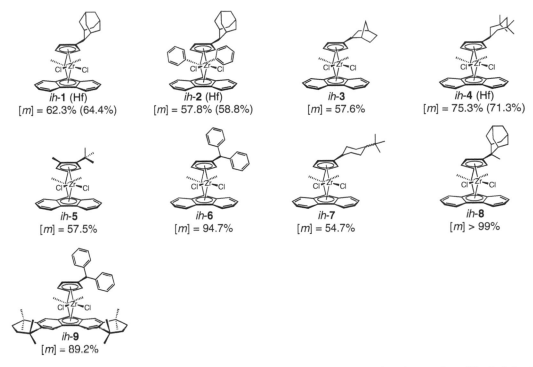

FIGURE 2.20 Metallocenes used for the preparation of isotactic-hemiisotactic polypropylene. The [*m*] dyad fractions are for MAO-cocatalyzed polymerizations performed in liquid propylene at 0 °C. Hafnocene [*m*] values are given in parentheses.

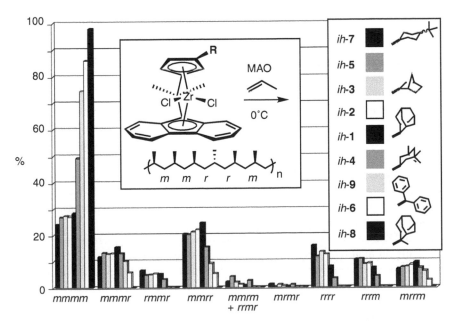

FIGURE 2.21 A ^{13}C NMR pentad analysis of nine different isotactic-hemiisotactic polypropylenes reveals an increase in catalyst isoselectivity with increasing 3-substituent (R) size, but a consistently low level of stereoerror pentads, *mmrm*, *rrmr*, and *mrmr*.

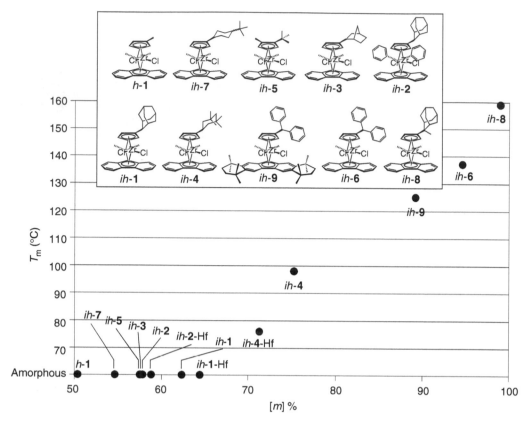

FIGURE 2.22 The [m] dyad fraction and the polymer melting temperature of isotactic-hemiisotactic polypropylenes can be controlled by manipulation of the catalyst steric framework.

the [m] dyad content and the number average molecular weight (M_n). Thus, the elastomeric properties observed for certain isotactic-hemiisotactic polypropylenes can be rationalized.[80] As illustrated by the Monte Carlo simulation in Figure 2.23, chains with [m] $\approx$ 60% and $M_n \approx$ 100,000 are predicted to have an average of 2.9 isotactic blocks of length 21 monomer units or greater, a length presumed sufficient for cocrystallization with other isotactic segments. Polymers with [m] near 50% are poor elastomers because there is an average of less than one isotactic block per chain—an insufficient number to create a substantially cross-linked network. Finally, polymers with [m] $\approx$ 75% are predicted to have properties that are dominated by the crystalline phase because fully 19% of the monomers reside in isotactic blocks of length 21 or greater. Indeed, ih-**4** and ih-**4**-Hf produce rigid, yet flexible polymers that are the first in Figure 2.22 along the continuum of increasing [m] to exhibit normal, well-defined melting endotherms by DSC.

A more careful DSC analysis shows that isotactic-hemiisotactic polymers generally have a melting temperature higher than that predicted by simple consideration of the [mmmm] pentad content. Figure 2.24 shows several exemplary data points (black squares) and their relationship to data collected from typical isoselective catalysts (gray circles) that range from moderately isoselective[96] to highly isoselective.[34] For isotactic-hemiisotactic polymers with [mmmm] below 40%, the melting temperatures are obtained from DSC endotherms with small melting enthalpies. These peaks, the location of which depends on the thermal history of the sample, are attributed to the crystalline fraction of the polymer sample made up of cocrystallized isotactic segments from multiple chains present in an otherwise amorphous medium.[80]

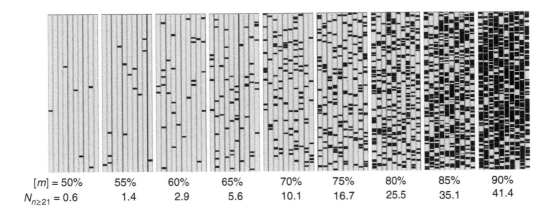

$[m] = 50\%$	55%	60%	65%	70%	75%	80%	85%	90%
$N_{n\geq21} = 0.6$	1.4	2.9	5.6	10.1	16.7	25.5	35.1	41.4

FIGURE 2.23 Ninety simulated isotactic-hemiisotactic polypropylene chains ($M_n = 100,000$; degree of polymerization $= 2376$) with ten chains per value of $[m]$. The average number of isotactic stereoblocks—denoted by the black regions—of length 21 or greater per chain is given by $N_{n\geq21}$. The gray regions denote amorphous, hemiisotactic stereoblocks (calculated according to equations in Reference 32). (Reprinted with permission from Miller, S. A. *Macromolecules* **2004**, *37*, 3983–3995. Copyright 2004 American Chemical Society.)

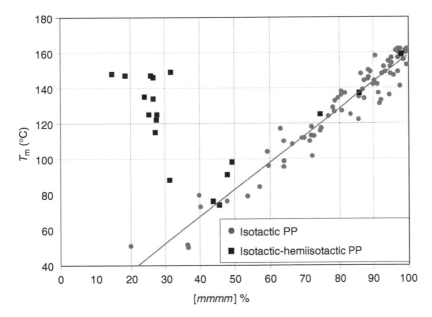

FIGURE 2.24 Typical isotactic polymers generally adhere to a linear relationship between the $[mmmm]$ pentad fraction and melting temperature (gray circles), whereas isotactic-hemiisotactic polymers can exhibit faint melting endotherms at considerably higher temperatures (black squares) than predicted by this relationship.

2.6 SYNDIOTACTIC-HEMIISOTACTIC POLYPROPYLENE

Syndiotactic-hemiisotactic polymers are variants of hemiisotactic polymers wherein the intervening stereocenters, on average, have stereochemistry opposite that of their stereoregular neighbors. An initial approach to this kind of polymer has been reported[94] through the use of steric variants of the methyl-substituted hemiisoselective metallocene $(CH_3)_2C(3\text{-}CH_3\text{-}C_5H_3)(C_{13}H_8)ZrCl_2$ (*h*-1). Whereas the strategy for making isotactic-hemiisotactic polymers involves increasing the

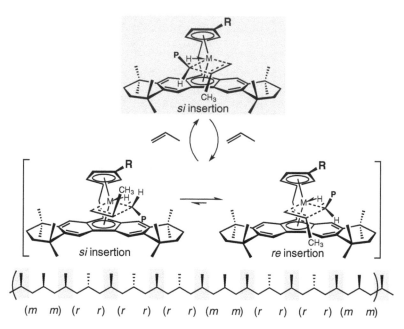

SCHEME 2.6 Syndiotactic-hemiisotactic polymers result from a two-site catalyst that has one highly enantioselective coordination site (in gray) and one coordination site that is moderately enantioselective for the opposite monomer enantioface (in brackets). The bridge substituents have been omitted for clarity; **P** represents the growing polymer chain; and M = cationic Zr^+ or Hf^+, generally.

steric size of the cyclopentadienyl substituent, the strategy for making syndiotactic-hemiisotactic polymers involves increasing the steric size of the fluorenyl substituent. Scheme 2.6 shows the general stereochemical mechanism when the sterically expanded octamethyloctahydrodibenzo-fluorenyl component is employed. As long as the cyclopentadienyl substituent is comparatively small, then the less stereoselective site prefers the monomer enantioface opposite that which is chosen at the more stereoselective site. The alternating use of these two coordination sites results in syndiotactic-hemiisotactic polypropylene.

Syndiotactic-hemiisotactic polypropylene with $[m] < 50\%$ has been obtained from the metallocenes shown in Figure 2.25.[94] According to the stereochemical model, the growing polymer chain prefers to reside near the cyclopentadienyl substituent during monomer insertion at the less stereoselective site. This is quite plausible for *sh*-**1** and *sh*-**2** because of the presumed steric repulsion of the octamethyloctahydrodibenzofluorenyl ligand. However, it is somewhat surprising that this is true for *sh*-**3**, *sh*-**4**, and *sh*-**5** because this suggests that the 2-methylcyclohexyl, cyclohexyl, and 2-norbornyl substituents are smaller than the methyl group of the parent hemiisoselective catalyst (*h*-**1**), which affords polypropylene with $[m] \approx 50\%$. A possible explanation for this behavior invokes an attractive interaction between the chain and the alkyl substituent of the cyclopentadienyl ring during the transition state for monomer insertion.[94] The attractive interaction may derive from cumulative van der Waals dispersion forces.

Figure 2.26 illustrates the pentad composition of the corresponding polypropylenes. Enantiofacial misinsertion and site epimerization are largely suppressed in liquid propylene at 0 °C, as evidenced by a consistently low level of stereoerror pentads, *mmrm*, *rrmr*, and *mrmr*. However, further analysis suggests that site epimerization, with chain migration away from the cyclopentadienyl substituent, is considerably more frequent than enantiofacial misinsertion. The *mrmr* pentad, which is essentially absent, only results from enantiofacial misinsertion or the unlikely event of site epimerization with chain migration toward the cyclopentadienyl substituent.

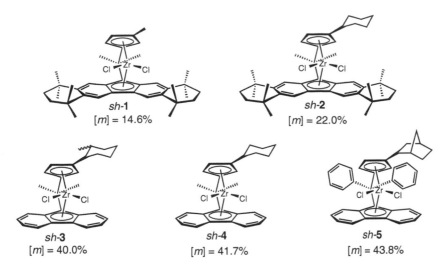

FIGURE 2.25 Metallocenes used for the preparation of syndiotactic-hemiisotactic polypropylene. The [*m*] dyad fractions are for MAO-cocatalyzed polymerizations performed in liquid propylene at 0 °C.

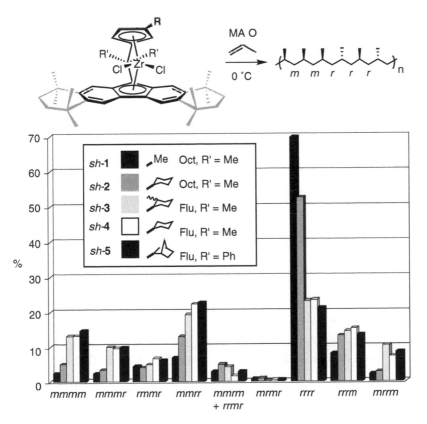

FIGURE 2.26 A ^{13}C NMR pentad analysis of five different syndiotactic-hemiisotactic polypropylenes reveals a strong dependence on the steric substitution of the metallocene precatalyst. A consistently low level of stereoerror pentads, *mmrm*, *rrmr*, and *mrmr*, is observed. Flu = fluorenyl; Oct = octamethyloctahydrodibenzofluorenyl.

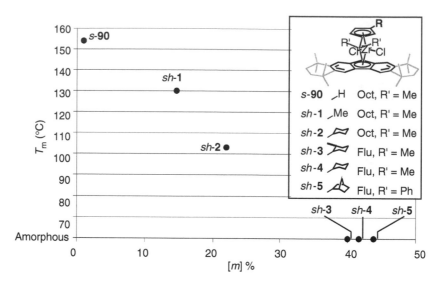

FIGURE 2.27 The [m] dyad fraction and the polymer melting temperature of syndiotactic-hemiisotactic polypropylenes can be controlled by manipulation of the catalyst steric framework. Flu = fluorenyl; Oct = octamethyloctahydrodibenzofluorenyl.

Figure 2.27 plots the observed polymer melting temperature as a function of the [m] dyad composition. The syndiotactic-hemiisotactic polypropylenes with [m] > 30% exhibit no strong endotherm through DSC and are essentially amorphous. Below [m] ≈ 25%, normal DSC traces are obtained and melting temperature is proportional to the syndiotacticity as quantified by 1 − [m]. For comparison, a data point for highly syndiotactic polypropylene from s-**90** (R = H)[53] has been included.

2.7 CONCLUSIONS

The fluorenyl ligand has proven to be an essential component in many homogeneous, stereoselective olefin polymerization catalysts. Since the first fluorenyl-containing metallocene was employed in 1988 for syndioselective propylene polymerization, other tacticities have been accessed: isotactic, hemiisotactic, isotactic-hemiisotactic, and syndiotactic-hemiisotactic. Indeed, fluorenyl-containing metallocenes are remarkably versatile, valuable for mechanistic inquiry, and functional in the synthesis of polymers with novel structures and properties.

2.8 ACKNOWLEDGMENTS

Our research in olefin polymerization is supported by grants from The Robert A. Welch Foundation (No. A-1537), the Texas Advanced Technology Program (No. 010366-0196-2003), and The Dow Chemical Company.

REFERENCES AND NOTES

Table 2.2, Table 2.3 and Table 2.4 tabulate all of the fluorenyl-containing compounds in this chapter along with their respective reference citations.

TABLE 2.2
Syndioselective Compounds and References

#	References	#	References	#	References	#	References	#	References	#	References
s-1	33	s-24	40	s-47	133	s-70	43	s-93	44	s-116	97
s-2	51	s-25	40	s-48	52	s-71	98	s-94	39	s-117	55
s-3	99, 100	s-26	40	s-49	38	s-72	50	s-95	45	s-118	55
s-4	101	s-27	102	s-50	38	s-73	103	s-96	104, 105	s-119	55
s-5	52, 130	s-28	49	s-51	106	s-74	107	s-97	104, 105	s-120	56
s-6	46	s-29	48, 108	s-52	49	s-75	107	s-98	104, 105	s-121	55
s-7	46	s-30	109	s-53	110	s-76	111	s-99	112	s-122	55
s-8	113	s-31	109	s-54	114	s-77	115	s-100	112	s-123	57
s-9	116, 117	s-32	40, 118	s-55	114	s-78	119	s-101	112	s-124	55
s-10	120	s-33	40	s-56	114	s-79	121	s-102	112	s-125	55
s-11	116	s-34	49, 122	s-57	114	s-80	123	s-103	112	s-126	55
s-12	47	s-35	49	s-58	124	s-81	123	s-104	112	s-127	55
s-13	125, 126	s-36	127	s-59	124	s-82	123	s-105	112	s-128	55
s-14	126	s-37	40, 116	s-60	124	s-83	43	s-106	112	s-129	55
s-15	128	s-38	40	s-61	38	s-84	81	s-107	112	s-130	53
s-16	128	s-39	40	s-62	38, 52	s-85	41b	s-108	112	s-131	61
s-17	129	s-40	40	s-63	38	s-86	130	s-109	112	s-132	41b, 131
s-18	40	s-41	40	s-64	38	s-87	130	s-110	132	s-133	36, 65
s-19	40	s-42	133	s-65	40	s-88	130	s-111	134	s-134	41b, 131
s-20	40	s-43	40	s-66	40	s-89	135	s-112	134	s-135	136
s-21	40	s-44	40	s-67	125	s-90	53	s-113	134	s-136	67, 68
s-22	40	s-45	40	s-68	137	s-91	53	s-114	138		
s-23	40	s-46	52	s-69	139	s-92	53	s-115	138		

TABLE 2.3
Isoselective Compounds and References

#	References	#	References	#	References	#	References	#	References	#	References
i-1	46, 140	i-16	81	i-31	145	i-46	141	i-61	84	i-76	142
i-2	143	i-17	81	i-32	145	i-47	132	i-62	96	i-77	142
i-3	144	i-18	81	i-33	145	i-48	132	i-63	83	i-78	142
i-4	70d	i-19	81	i-34	145	i-49	138	i-64	83	i-79	142
i-5	65	i-20	81	i-35	145	i-50	138	i-65	83	i-80	55
i-6	42	i-21	145	i-36	145	i-51	146	i-66	87	i-81	92, 47
i-7	147	i-22	145	i-37	145	i-52	146	i-67	87	i-82	91, 148
i-8	43	i-23	145	i-38	145	i-53	146	i-68	87	i-83	89
i-9	107	i-24	145	i-39	145	i-54	146	i-69	86, 149	i-84	90
i-10	107	i-25	145	i-40	150	i-55	146	i-70	86	i-85	90
i-11	107	i-26	145	i-41	73, 151	i-56	146	i-71	86	i-86	152
i-12	107	i-27	145	i-42	73, 151	i-57	146	i-72	86	i-87	55
i-13	153	i-28	145	i-43	73, 151	i-58	85	i-73	154		
i-14	153b	i-29	145	i-44	73, 151	i-59	48, 155	i-74	154		
i-15	156	i-30	145	i-45	73, 151	i-60	84	i-75	157		

TABLE 2.4

Hemiisoselective Compounds and References

#	References	#	References	#	References	#	References	#	References	#	References
h-1	71, 80	*h*-5	74	*h*-9	159	*ih*-4	80	*ih*-8	73, 151	*sh*-3	94, 151
h-2	42	*h*-6	42	*ih*-1	80	*ih*-5	80	*ih*-9	73, 151	*sh*-4	73, 151
h-3	74	*h*-7	158	*ih*-2	80	*ih*-6	73, 151	*sh*-1	73, 151	*sh*-5	94, 151
h-4	74	*h*-8	159	*ih*-3	80	*ih*-7	94, 151	*sh*-2	73, 151		

1. (a) Kealy, T. J.; Paulson, P. L. A new type of organo-iron compound. *Nature* **1951**, *168*, 1039–1040. (b) Miller, S. A.; Tebboth, J. A.; Tremaine, J. F. Dicyclopentadienyliron. *J. Chem. Soc.* **1952**, 632–635.
2. (a) Wilkinson, G.; Rosenblum, M.; Whiting, M. C.; Woodward, R. B. The structure of iron biscyclopentadienyl. *J. Am. Chem. Soc.* **1952**, *74*, 2125–2126. (b) Fischer, E. O.; Pfab, W. Cyclopentadiene-metallic complex, a new type of organo-metallic compound. *Z. Naturforsch.* **1952**, *7b*, 377–379. (c) Wilkinson, G. Iron sandwich. Recollection of the first four months. *J. Organomet. Chem.* **1975**, *100*, 273–278.
3. Wilkinson, G.; Pauson, P. L.; Birmingham, J. M.; Cotton, F. A. Bis-cyclopentadienyl derivatives of some transition elements. *J. Am. Chem. Soc.* **1953**, *75*, 1011–1012.
4. (a) Fischer, E. O.; Seus, D.; Jira, R. Metal complexes of indene with cobalt. *Z. Naturforsch.* **1953**, *8b*, 692–693. (b) Fischer, E. O.; Seus, D. Diindenyl iron. *Z. Naturforsch.* **1953**, *8b*, 694.
5. Shapiro, H.; De Witt, E. G.; Brown, J. E. Cyclomatic manganese compounds. U.S. Patent 2,839,552 (Ethyl Corporation), June 17, 1958.
6. Alt, H. G.; Samuel, E. Fluorenyl complexes of zirconium and hafnium as catalysts for olefin polymerization. *Chem. Soc. Rev.* **1998**, *27*, 323–329.
7. (a) Alt, H. G.; Köppl, A. Effect of the nature of metallocene complexes of group IV metals on their performance in catalytic ethylene and propylene polymerization. *Chem. Rev.* **2000**, *100*, 1205–1221. (b) Licht, E. H.; Alt, H. G.; Karim, M. M. Mixed substituted zirconocene dichloride complexes as catalyst precursors for homogeneous ethylene polymerization. *J. Mol. Catal. A: Chem.* **2000**, *59*, 273–283. (c) Rodriguez, G.; Crowther, D. J. Olefin copolymerization process with bridged hafnocenes. PCT Int. Pat. Appl. WO 2000/024792 A1 (Exxon Chemical Patents Inc.), May 4, 2000. (d) Schertl, P.; Alt, H. G. Synthesis and polymerization properties of substituted *ansa*-bis(fluorenylidene) complexes of zirconium. *J. Organomet. Chem.* **1999**, *582*, 328–337.
8. Kirillov, E.; Lehmann, C. W.; Razavi, A.; Carpentier, J.-F. Highly syndiospecific polymerization of styrene catalyzed by allyl lanthanide complexes. *J. Am. Chem. Soc.* **2004**, *126*, 12240–12241.
9. (a) Peifer, B.; Milius, W.; Alt, H. G. Self-immobilized metallocene catalysts. *J. Organomet. Chem.* **1998**, *553*, 205–220. (b) Alt, H. G.; Jung, M. C_2-bridged metallocene dichloride complexes of the types $(C_{13}H_8\text{-}CH_2CHR\text{-}C_9H_{6-n}R'_n)ZrCl_2$ and $(C_{13}H_8\text{-}CH_2CHR\text{-}C_{13}H_8)MCl_2$ (n = 0, 1; R = H, alkenyl; R' = alkenyl, benzyl; M = Zr, Hf) as self-immobilizing catalyst precursors for ethylene polymerization. *J. Organomet. Chem.* **1999**, *580*, 1–16. (c) Hlatky, G. G. Heterogeneous single-site catalysts for olefin polymerization. *Chem. Rev.* **2000**, *100*, 1347–1376.
10. Schmid, M. A.; Alt, H. G.; Milius, W. Unbridged cyclopentadienyl-fluorenyl complexes of zirconium as catalysts for homogeneous olefin polymerization. *J. Organomet. Chem.* **1995**, *501*, 101–106.
11. Surprisingly, $(C_9H_7)(C_{13}H_9)ZrCl_2$ does not have a CAS number. There is also no indication from the literature that substituted non-bridged indenyl/fluorenyl metallocenes can make stereoregular polymers.
12. Despite reports and confirmation (S. A. Miller, unpublished results) that $(C_{13}H_9)_2ZrCl_2$ is unstable in the solid state, even at –30 °C, the X-ray crystal structure has been solved: (a) Kowala, C.; Wunderlich, J. A. The crystal and molecular structure of (pentahaptofluorenyl) (trihaptofluorenyl) dichlorozirconium(IV). *Acta Crystallogr.* **1976**, *B32*, 820–823. MAO-cocatalyzed polymerizations are generally not reproducible, but can yield small amounts of atactic polypropylene. (b) Alt, H. G.; Zenk, R. C_2-symmetric bis(fluorenyl) complexes: Four complex models as potential catalysts for the isospecific polymerization of propylene. *J. Organomet. Chem.* **1996**, *512*, 51–60.

13. Resconi, L.; Jones, R. L.; Rheingold, A. L.; Yap, G. P. A. High-molecular-weight atactic polypropylene from metallocene catalysts. 1. $Me_2Si(9\text{-}Flu)_2ZrX_2$ (X = Cl, Me). *Organometallics* **1996**, *15*, 998–1005.

14. Miller, S. A.; Bercaw, J. E. Aminofluorenyl-pentamethylcyclopentadienyl and bis(aminofluorenyl) derivatives of group 4 metals. *Organometallics* **2000**, *19*, 5608–5613.

15. Repo, T.; Jany, G.; Hakala, K.; Klinga, M.; Polamo, M.; Leskelä, M.; Rieger, B. Metallocene dichlorides bearing acenaphthyl substituted cyclopentadienyl rings: Preparation and polymerization behavior. *J. Organomet. Chem.* **1997**, *549*, 177–186.

16. Diamond, G. M.; Chernega, A. N.; Mountford, P.; Green, M. L. H. New mono- and bi-nuclear *ansa*-metallocenes of zirconium and hafnium as catalysts for the polymerization of ethene and propene. *J. Chem. Soc., Dalton Trans.* **1996**, *6*, 921–938.

17. (a) Okumura, Y.; Ishikawa, T. Polypropylene having low isotacticity and relatively high melting point. PCT Int. Pat. Appl. WO 2002/038634A2 (Basell Technology Company B.V.), May 16, 2002. (b) Surprisingly, such titanium-based species are nearly absent from the literature. The first structural characterization via X-ray crystallography for any fluorenyl/cyclopentadienyl titanocene was performed on $Ph_2C(3\text{-}(CHPh_2)\text{-}C_5H_3)(C_{13}H_8)TiCl_2$. Under a variety of conditions, it is essentially inactive for propylene polymerization. Al-Bahily, K. Design and synthesis of new C_1 and C_2-symmetric *ansa*-metallocene catalysts for isotactic polypropylene formation. Masters Thesis, Texas A&M University, College Station, TX, December 2004.

18. Grisi, F.; Longo, P.; Zambelli, A.; Ewen, J. A. Group 4 C_s-symmetric catalysts and 1-olefin polymerization. *J. Mol. Catal. A: Chem.* **1999**, *140*, 225–233.

19. Zubris, D. L. Investigations of the origin of stereocontrol in syndiospecific Ziegler–Natta polymerizations. Ph.D. Thesis, California Institute of Technology, Pasadena, CA, July 2000.

20. Virtually all monofluorenyl species are thermally unstable. However, their decomposition products can sometimes make atactic polypropylene. Van der Zeijden, A. A. H.; Mattheis, C.; Frohlich, R. The first monofluorenylzirconium trichloride. *Chem. Ber. Recl.* **1997**, *130*, 1231–1234.

21. Knjazhanski, S. Y.; Cadenas, G.; Garcia, M.; Perez, C. M.; Nifant'ev, I. E.; Kashulin, I. A.; Ivchenko, P. V.; Lyssenko, K. A. (Fluorenyl)titanium triisopropoxide and bis(fluorenyl)titanium diisopropoxide: A facile synthesis, molecular structure, and catalytic activity in styrene polymerization. *Organometallics* **2002**, *21*, 3094–3099.

22. (a) Sinn, H.; Kaminsky, W. Ziegler–Natta catalysis. *Adv. Organomet. Chem.* **1980**, *18*, 99–149. (b) Sinn, H.; Kaminsky, W.; Vollmer, H. J.; Woldt, R. "Living polymers" with Ziegler catalysts of high productivity. *Angew. Chem.* **1980**, *92*, 396–402.

23. Jordan, R. F. Chemistry of cationic dicyclopentadienyl metal-alkyl complexes. *Adv. Organomet. Chem.* **1991**, *32*, 325–387.

24. Chen, E. Y.-X.; Marks, T. J. Cocatalysts for metal-catalyzed olefin polymerization: Activators, activation processes, and structure-activity relationships. *Chem. Rev.* **2000**, *100*, 1391–1434.

25. (a) Brintzinger, H.-H.; Fischer, D.; Mülhaupt, R.; Rieger, B.; Waymouth, R. M. Stereospecific olefin polymerization with chiral metallocene catalysts. *Angew. Chem., Int. Ed. Engl.* **1995**, *34*, 1143–1170. (b) Chum, P. S.; Kruper, W. J.; Guest, M. J. Materials properties derived from INSITE metallocene catalysts. *Adv. Mater.* **2000**, *12*, 1759–1767.

26. Grubbs, R. H.; Coates, G. W. α-Agostic interactions and olefin insertion in metallocene polymerization catalysts. *Acc. Chem. Res.* **1996**, *29*, 85–93.

27. Dahlmann, M.; Erker, G.; Nissinen, M.; Fröhlich, R. Direct experimental observation of the stereochemistry of the first propene insertion step at an active homogeneous single-component metallocene Ziegler catalyst. *J. Am. Chem. Soc.* **1999**, *121*, 2820–2828.

28. (a) Cavallo, L.; Guerra, G.; Vacatello, M.; Corradini, P. A possible model for the stereospecificity in the syndiospecific polymerization of propene with group 4a metallocenes. *Macromolecules* **1991**, *24*, 1784–1790. (b) Castonguay, L. A.; Rappe, A. K. Ziegler–Natta catalysis. A theoretical study of the isotactic polymerization of propylene. *J. Am. Chem. Soc.* **1992**, *114*, 5832–5842. (c) Kawamura-Kuribayashi, H.; Koga, N.; Morokuma, K. An *ab initio* MO and MM study of homogeneous olefin polymerization with silylene-bridged zirconocene catalyst and its regio- and stereoselectivity. *J. Am. Chem. Soc.* **1992**, *114*, 8687–8694. (d) van der Leek, Y.; Angermund, K.; Reffke, M.; Kleinschmidt, R.; Goretzki, R.; Fink, G. On the mechanism of stereospecific polymerization—development of a universal model to demonstrate the relationship between metallocene structure and polymer microstructure.

Chem. Eur. J. **1997**, *3*, 585–591. (e) Borrelli, M.; Busico, V.; Cipullo, R.; Ronca, S.; Budzelaar, P. H. M. Selectivity of metallocene-catalyzed olefin polymerization: A combined experimental and quantum mechanical study. The *ansa*-Me$_2$Si(Ind)$_2$Zr and *ansa*-Me$_2$C(Cp)(Flu)Zr Systems. *Macromolecules* **2003**, *36*, 8171–8177.

29. (a) Busico, V.; Cipullo, R.; Corradini, P.; Landriani, L.; Vacatello, M.; Segre, A. L. Advances in the ^{13}C NMR microstructural characterization of propene polymers. *Macromolecules* **1995**, *28*, 1887–1892. (b) Busico, V.; Cipullo, R.; Monaco, G.; Vacatello, M. Full assignment of the ^{13}C NMR spectra of regioregular polypropylenes: Methyl and methylene region. *Macromolecules* **1997**, *30*, 6251–6263. (c) Busico, V.; Cipullo, R. Microstructure of polypropylene. *Prog. Polym. Sci.* **2001**, *26*, 443–533. (d) Busico, V.; Mannina, L.; Segre, A. L.; Van Axel Castelli, V. Recent advances in the NMR description of polypropylene. In *NMR Spectroscopy of Polymers in Solution and in the Solid State*; Cheng, H. N., English, A. D., Eds.; ACS Symposium Series 834; American Chemical Society: Washington, DC, 2003; pp 192–207.

30. Farina, M. The stereochemistry of linear macromolecules. *Top. Stereochem.* **1987**, 17,1–111.

31. (a) Coates, G. W. Precise control of polyolefin stereochemistry using single-site metal catalysts. *Chem. Rev.* **2000**, *100*, 1223–1252. (b) Ewen, J. A. Symmetry rules and reaction mechanisms of Ziegler–Natta catalysts. *J. Mol. Catal. A: Chem.* **1998**, *128*, 103–109. (c) Ewen, J. A. New chemical tools to create plastics. *Scientific Am.* **1997**, *276*, 86–91.

32. Miller, S. A. Isotactic block length distribution in polypropylene: Bernoullian vs hemiisotactic. *Macromolecules* **2004**, *37*, 3983–3995.

33. Ewen, J. A.; Jones, R. L.; Razavi, A.; Ferrara, J. D. Syndiospecific propylene polymerizations with group 4 metallocenes. *J. Am. Chem. Soc.* **1988**, *110*, 6255–6256.

34. Resconi, L.; Cavallo, L.; Fait, A.; Piemontesi, F. Selectivity in propene polymerization with metallocene catalysts. *Chem. Rev.* **2000**, *100*, 1253–1345.

35. (a) Herzog, T. A.; Zubris, D. L.; Bercaw, J. E. A new class of zirconocene catalysts for the syndio-specific polymerization of propylene and its modification for varying polypropylene from isotactic to syndiotactic. *J. Am. Chem. Soc.* **1996**, *118*, 11988–11989. (b) Miyake, S.; Bercaw, J. E. Doubly [SiMe$_2$]-bridged C_s- and C_{2v}-symmetric zirconocene catalysts for propylene polymerization. Synthesis and polymerization characteristics. *J. Mol. Catal. A: Chem.* **1998**, *128*, 29–39. (c) Veghini, D.; Henling, L. M.; Burkhardt, T. J.; Bercaw, J. E. Mechanisms of stereocontrol for doubly silylene-bridged C_s- and C_1-symmetric zirconocene catalysts for propylene polymerization. Synthesis and molecular structure of Li$_2$[(1,2-Me$_2$Si)$_2${C$_5$H$_2$-4-(1R,2S,5R-menthyl)}{C$_5$H-3,5-(CHMe$_2$)$_2$)}]·3THF and [(1,2-Me$_2$Si)$_2${η^5-C$_5$H$_2$-4-(1R,2S,5R-menthyl)}{η^5-C$_5$H-3,5-(CHMe$_2$)$_2$}]ZrCl$_2$. *J. Am. Chem. Soc.* **1999**, *121*, 564–573. (d) Zubris, D. L.; Veghini, D.; Herzog, T. A.; Bercaw, J. E. Reactivity and mechanistic studies of stereocontrol for Ziegler–Natta polymerization utilizing doubly-silylene bridged group 3 and group 4 metallocenes. In *Olefin Polymerization*; Arjunan, P., McGrath, J. E., Hanlon, T. L., Eds.; ACS Symposium Series 749; American Chemical Society: Washington, DC, 2000; pp 2–14.

36. Razavi, A.; Baekelmans, D.; Bellia, V.; De Brauwer, Y.; Hortmann, K.; Lambrecht, M.; Miserque, O.; Peters, L.; Slawinski, M.; Van Belle, S. Syndiotactic specific structures, symmetry considerations, mechanistic aspects. In *Organometallic Catalysts and Olefin Polymerization: Catalysts for a New Millennium*; Blom, R., Follestad, A., Rytter, E., Tilset, M., Ystenes, M., Eds.; Springer: Berlin, 2001; pp 267–279.

37. The term "site epimerization" is used in preference to other terms found in the literature. The term "epimerization" (Eliel, E. L.; Wilen, S. H. *Stereochemistry of Organic Compounds*; Wiley-Interscience: New York, 1994; p 1198) refers to the inversion of a single stereocenter when two or more are present, and can be correctly applied to metallocene polymerizations when the stereochemistry of the metal is inverted while the stereocenters present on the polymer chain remain unchanged. Reference 73 cites several colloquial terms found in the literature, including: Consecutive addition; skipped insertion; skipped-out insertions; chain migratory catalyst isomerization; back-skip mechanism; retention mechanism; isomerization without monomer insertion; site-to-site chain migration; side-to-side swing; back swing; and chain stationary insertion mechanism.

38. Technically, the compounds are listed in order of increasing CAS number. For the benzo substituted metallocenes *s*-**49** and *s*-**50**, large numbers of related catalysts have been reported in the patent literature and thus, only representative examples are depicted: (a) Zenk, R.; Alt, H. G.; Welch, M. B. Organometallic fluorenyl compounds, preparation, and use. U.S. Patent 5,451,649 (Phillips Petroleum

Company), September 19, 1995. (b) Zenk, R.; Alt, H. G.; Welch, M. B. Organometallic fluorenyl compounds, preparation, and use. European Patent 0672675 B1 (Phillips Petroleum Company), July 24, 2002.

39. For the silyl ether metallocene s-**94**, large numbers of related catalysts have been reported in the patent literature and thus, only a representative example is depicted. Nunez, M. F. M.; Lafuente, A. M.-E.; Garcia, B. P.; Canas, P. L. Single-carbon bridged biscyclopentadienyl compounds and metallocene complexes thereof. U.S. Patent 6,534,665 B1 (Repsol Quimica S.A.), March 18, 2003.

40. For the heteroatom-substituted metallocenes (s-**19** through s-**26**), large numbers of related catalysts have been reported in the patent literature and thus, only representative examples are depicted. Several of these have not been assigned a CAS number. Ewen, J. A.; Reddy, B. R.; Elder, M. J. Method for controlling the melting points and molecular weights of syndiotactic polyolefins using metallocene catalyst systems. U.S. Patent 5,710,222 (Fina Technology, Inc.), January 20, 1998.

41. (a) Razavi, A.; Bellia, V.; De Brauwer, Y.; Hortmann, K.; Peters, L.; Sirole, S.; Van Belle, S.; Thewalt, U. Syndiotactic- and isotactic specific bridged cyclopentadienyl-fluorenyl based metallocenes; structural features, catalytic behavior. *Macromol. Chem. Phys.* **2004**, *205*, 347–356. (b) Razavi, A.; Bellia, V.; De Brauwer, Y.; Hortmann, K.; Peters, L.; Sirole, S.; Van Belle, S.; Thewalt, U. Fluorenyl based syndiotactic specific metallocene catalysts structural features, origin of syndiospecificity. *Macromol. Symp.* **2004**, *213*, 157–171. (c) Zambelli, A.; Sessa, I.; Grisi, F.; Fusco, R.; Accomazzi, P. Syndiotactic polymerization of propylene: Single-site vanadium catalysts in comparison with zirconium and nickel. *Macromol. Rapid Commun.* **2001**, *22*, 297–310. (d) Shiomura, T.; Uchikawa, N.; Asanuma, T.; Sugimoto, R.; Fujio, I.; Kimura, S.; Harima, S.; Akiyama, M.; Kohno, M.; Inoue, N. Metallocene-catalyzed syndiotactic polypropylene: Preparation and properties. In *Metallocene-Based Polyolefins: Preparation, Properties and Technology*; Scheirs, J., Kaminsky, W., Eds.; John Wiley & Sons Ltd.: Chichester, UK, 2000; pp 437–465. (e) Makio, H.; Kashiwa, N.; Fujita, T. FI catalysts: A new family of high performance catalysts for olefin polymerization. *Adv. Synth. Catal.* **2002**, *344*, 477–493. (f) Razavi, A.; Bellia, V.; De Branwer, Y.; Hortnam, K.; Lambrecht, M.; Miseque, O.; Peters, L.; Van Belle, S. Syndiotactic and isotactic specific metallocene catalysts with hapto-flexible cyclopentadienyl-fluorenyl ligand. In *Metalorganic Catalysts for Synthesis and Polymerization*; Kaminsky, W., Ed.; Springer: Berlin, 1999; pp 236–248.

42. Yano, A.; Kaneko, T.; Sato, M.; Akimoto, A. Propylene polymerization with $Ph_2C(3-RCp)(Flu)ZrCl_2$ [R = Me, *i*-Pr, $PhCH_2$, Me_3Si] catalysts activated with MAO and $Me_2PhNH \cdot B(C_6F_5)_4/i$-$Bu_3Al$. *Macromol. Chem. Phys.* **1999**, *200*, 2127–2135.

43. Kaminsky, W.; Scholz, V.; Werner, R. Progress of olefin polymerization by metallocene catalysts. *Macromol. Symp.* **2000**, *159*, 9–17.

44. Gentil, S.; Dietz, M.; Pirio, N.; Meunier, P.; Gallucci, J.C.; Gallou, F.; Paquette, L. A. A silylene-bridged (isodicyclopentadienyl)(fluorenyl) complex of zirconium for homogeneous olefin polymerization. *Organometallics* **2002**, *21*, 5162–5166.

45. Won, Y. C.; Kwon, H. Y.; Lee, B. Y.; Park, Y.-W. Fulvene having substituents only on 1-, 4-, and 6-positions: A key intermediate for novel *ansa*-metallocene complexes. *J. Organomet. Chem.* **2003**, *677*, 133–139.

46. Ewen, J. A.; Elder, M. J.; Jones, R. L.; Haspeslagh, L.; Atwood, J. L.; Bott, S. G.; Robinson, K. Metallocene/polypropylene structural relationships: Implications on polymerization and stereochemical control mechanisms. *Makromol. Chem., Macromol. Symp.* **1991**, *48/49*, 253–295.

47. Razavi, A.; Peters, L.; Nafpliotis, L. Geometric flexibility, ligand and transition metal electronic effects on stereoselective polymerization of propylene in homogeneous catalysis. *J. Mol. Catal. A: Chem.* **1997**, *115*, 129–154.

48. Chen, Y.-X.; Rausch, M. D.; Chien, J. C. W. Silylene-bridged fluorenyl-containing ligands and zirconium complexes with C_1 and C_s symmetry: General synthesis and olefin polymerization catalysis. *J. Organomet. Chem.* **1995**, *497*, 1–9.

49. Patsidis, K.; Alt, H. G.; Milius, W.; Palackal, S. J. The synthesis, characterization and polymerization behavior of *ansa* cyclopentadienyl fluorenyl complexes; the X-ray structures of the complexes $[(C_{13}H_8)SiR_2(C_5H_4)]ZrCl_2$ (R = Me or Ph). *J. Organomet. Chem.* **1996**, *509*, 63–71.

50. Schaverien, C. J.; Ernst, R.; Terlouw, W.; Schut, P.; Sudmeijer, O.; Budzelaar, P. H. M. Phosphorus-bridged metallocenes: New homogeneous catalysts for the polymerization of propene. *J. Mol. Catal. A: Chem.* **1998**, *128*, 245–256.

51. Razavi, A.; Atwood, J. L. Preparation and crystal structures of the cyclopentadienylfluorenyldiphenyl-methane zirconium and hafnium complexes (η^5-$C_5H_4CPh_2$-η^5-$C_{13}H_8$)MCl_2 (M = Zr, Hf) and the catalytic formation of high molecular weight high tacticity syndiotactic polypropylene. *J. Organomet. Chem.* **1993**, *459*, 117–123.

52. Alt, H. G.; Zenk, R. Syndiospecific polymerization of propylene: New metallocene complexes of type ($C_{13}H_{8-n}R_nCR'R''C_5H_4$)$MCl_2$ (n = 0,2; R = Alkyl, Aryl, Hal; R', R'' = H, Alkyl, Aryl; M = Zr, Hf) with special regard for different bridge substituents. *J. Organomet. Chem.* **1996**, *518*, 7–15.

53. Miller, S. A.; Bercaw, J. E. Highly stereoregular syndiotactic polypropylene formation with metallocene catalysts via influence of distal ligand substituents. *Organometallics* **2004**, *23*, 1777–1789.

54. (a) Deck, P. A.; Beswick, C. L.; Marks, T. J. Highly electrophilic olefin polymerization catalysts. Quantitative reaction coordinates for fluoroarylborane/alumoxane methide abstraction and ion-pair reorganization in Group 4 metallocene and "constrained geometry" catalysts. *J. Am. Chem. Soc.* **1998**, *120*, 1772–1784. (b) Lanza, G.; Fragalà I. L.; Marks, T. J. Energetic, structural, and dynamic aspects of ethylene polymerization mediated by homogeneous single-site "constrained geometry catalysts" in the presence of cocatalyst and solvation: An investigation at the *ab initio* quantum chemical level. *Organometallics* **2002**, *21*, 5594–5612.

55. Siedle, A. R.; Theissen, K. M.; Stevens, J. A measure of metallocene catalyst shape asymmetry. *J. Mol. Catal. A: Chem.* **2003**, *191*, 167–175.

56. Fan, W.; Waymouth, R. M. Sequence and stereoselectivity of the C_1-symmetric metallocene $Me_2Si(1$-$(4,7$-$Me_2Ind))(9$-$Flu)ZrCl_2$. *Macromolecules* **2003**, *36*, 3010–3014.

57. Razavi, A.; Vereecke, D.; Peters, L. Den Dauw, K.; Nafpliotis, L.; Atwood, J. L. Manipulation of the ligand structure as an effective and versatile tool for modification of active site properties in homogeneous Ziegler–Natta catalyst systems. In *Ziegler Catalysts, Recent Scientific Innovations and Technological Improvements*; Fink, G., Mülhaupt, R., Brintzinger, H.-H., Eds.; Springer: Berlin, 1995; pp 111–147.

58. McKnight, A. L.; Waymouth, R. M. Group 4 *ansa*-cyclopentadienyl-amido catalysts for olefin polymerization. *Chem. Rev.* **1998**, *98*, 2587–2598.

59. Okuda, J.; Schattenmann, F. J.; Wocadlo, S.; Massa, W. Synthesis and characterization of zirconium complexes containing a linked amido-fluorenyl ligand. *Organometallics* **1995**, *14*, 789–795.

60. Dias, H. V. R.; Wang, Z.; Bott, S. G. Preparation of group 4 metal complexes of a bulky amido-fluorenyl ligand. *J. Organomet. Chem.* **1996**, *508*, 91–99.

61. (a) Canich, J. A. M. Process for producing crystalline poly-α-olefins with a monocyclopentadienyl transition metal catalyst system. U.S. Patent 5,026,798 (Exxon Chemical Patents Inc.), June 25, 1991. (b) Dias, H. V. R.; Wang, Z. Syntheses and characterization of zirconium and hafnium complexes of amido-fluorenyl ligands $[(NBu^t)SiMe_2CH_2(C_{13}H_8)]^{2-}$ and $[(NPr^i)SiMe_2CH_2(C_{13}H_8)]^{2-}$. *J. Organomet. Chem.* **1997**, *539*, 77–85. (c) Alt, H. G.; Fottinger, K.; Milius, W. Synthesis, characterization and polymerization properties of bridged half-sandwich complexes of titanium, zirconium and hafnium; molecular structure of $[C_{13}H_8$-$SiMe_2$-$N^tBu]ZrCl_2$. *J. Organomet. Chem.* **1999**, *572*, 21–30. (d) McKnight, A. L.; Masood, Md. A.; Waymouth, R. M.; Straus, D. A. Selectivity in propylene polymerization with group 4 Cp-amido catalysts. *Organometallics* **1997**, *16*, 2879–2885.

62. (a) Shiomura, T.; Asanuma, T.; Inoue, N. Inversion of stereoselectivity in a metallocene catalyst. *Macromol. Rapid Commun.* **1996**, *17*, 9–14. (b) Shiomura, T.; Asanuma, T.; Sunaga, T. Effect of cocatalysts on the character of a constrained geometry catalyst. *Macromol. Rapid Commun.* **1997**, *18*, 169–173.

63. Hagihara, H.; Shiono, T.; Ikeda, T. Syndiospecific polymerization of propene with [t-BuNSiMe$_2$Flu]TiMe$_2$-based catalysts by chain-end controlled mechanism. *Macromolecules* **1997**, *30*, 4783–4785.

64. (a) Hasan, T.; Ioku, A.; Nishii, K.; Shiono, T.; Ikeda, T. Syndiospecific living polymerization of propene with [t-BuNSiMe$_2$Flu]TiMe$_2$ using MAO as cocatalyst. *Macromolecules* **2001**, *34*, 3142–3145. (b) Nishii, K.; Shiono, T.; Ikeda, T. A novel synthetic procedure for stereoblock poly(propylene) with a living polymerization system. *Macromol. Rapid Commun.* **2004**, *25*, 1029–1032.

65. Razavi, A.; Thewalt, U. Preparation and crystal structures of the complexes (η^5-C_5H_3TMS-CMe_2-η^5-$C_{13}H_8$)MCl_2 and [3,6-di^{t}Bu$C_{13}H_6$-$SiMe_2$-N^tBu]MCl_2 (M=Hf, Zr or Ti): Mechanistic aspects of the catalytic formation of a isotactic-syndiotactic stereoblock-type polypropylene. *J. Organomet. Chem.* **2001**, *621*, 267–276.

66. Busico, V.; Cipullo, R.; Cutillo, F.; Talarico, G.; Razavi, A. Syndiotactic poly(propylene) from [Me$_2$Si(3,6-di-*tert*-butyl-9-fluorenyl)(N-*tert*-butyl)]TiCl$_2$-based catalysts: Chain-end or enantiotopic-sites stereocontrol? *Macromol. Chem. Phys.* **2003**, *204*, 1269–1274.

67. (a) Irwin, L. J.; Reibenspies, J. H.; Miller, S. A. A sterically expanded "constrained geometry catalyst" for highly active olefin polymerization and copolymerization: An unyielding comonomer effect. *J. Am. Chem. Soc.* **2004**, *126*, 16716–16717. (b) Irwin, L. J.; Reibenspies, J. H.; Miller, S. A. Synthesis and characterization of sterically expanded *ansa*-η^1-fluorenyl-amido complexes. *Polyhedron* **2005**, *24*, 1314–1324.

68. Irwin, L. J.; Miller, S. A. Unprecedented syndioselectivity and syndiotactic polyolefin melting temperature: Polypropylene and poly(4-methyl-1-pentene) from a highly active, sterically expanded η^1-fluorenyl-η^1-amido zirconium complex. *J. Am. Chem. Soc.* **2005**, *127*, 9972–9973.

69. (a) Ewen, J. A. Catalyst for producing hemiisotactic polypropylene. European Patent 0423101B1 (Fina Technology, Inc.), January 26, 2000. (b) Ewen, J. A. Symmetrical and lopsided zirconocene pro-catalysts. *Macromol. Symp.* **1995**, *89*, 181–196. (c) Ewen, J. A.; Elder, M. J. Isospecific pseudo-helical zirconocenium catalysts. In *Ziegler Catalysts, Recent Scientific Innovations and Technological Improvements*; Fink, G., Mülhaupt, R., Brintzinger, H.-H., Eds.; Springer: Berlin, 1995; pp 99–109.

70. (a) Dolle, V.; Rohrmann, J.; Winter, A.; Antberg, M.; Klein, R. Process for preparing a syndio-isoblockpolymer. European Patent 0399347B1 (Hoechst Aktiengesellschaft), August 14, 1996. (b) Antberg, M.; Dolle, V.; Klein, R.; Rohrmann, J.; Spaleck, W.; Winter, A. Propylene polymerization by stereorigid metallocene catalysts: Some new aspects of the metallocene structure/polypropylene microstructure correlation. In *Catalytic Olefin Polymerization*; Keii, T., Soga, K., Eds.; Kokansha: Tokyo, 1990; pp 501–515. (c) Spaleck, W.; Antberg, M.; Aulbach, M.; Bachmann, B.; Dolle, V.; Haftka, S.; Kuber, F.; Rohrmann, J.; Winter, A. New isotactic polypropylenes via metallocene catalysts. In *Ziegler Catalysts, Recent Scientific Innovations and Technological Improvements*; Fink, G., Mülhaupt, R., Brintzinger, H.-H., Eds.; Springer: Berlin, 1995; pp 83–97. (d) Spaleck, W.; Aulbach, M.; Bachmann, B.; Küber, F.; Winter, A. Stereospecific metallocene catalysts: Scope and limits of rational catalyst design. *Macromol. Symp.* **1995**, *89*, 237–247.

71. (a) Razavi, A.; Atwood, J. L. Synthesis and characterization of the catalytic isotactic-specific metallocene complex (η^5-C$_5$H$_3$C$_4$H$_9$-CMe$_2$-η^5-C$_{13}$H$_8$)ZrCl$_2$. Mechanistic aspects of the formation of isotactic polypropylene, the stereoregulative effect of the distal substituent and the relevance of C_2 symmetry. *J. Organomet. Chem.* **1996**, *520*, 115–120. (b) Razavi, A.; Atwood, J. L. Preparation and crystal structure of the complexes (η^5-C$_5$H$_3$Me-CMe$_2$-η^5-C$_{13}$H$_8$)MCl$_2$ (M = Zr or Hf): Mechanistic aspects of the catalytic formation of a syndiotactic-isotactic stereoblock-type polypropylene. *J. Organomet. Chem.* **1995**, *497*, 105–111. (c) Razavi. A.; Peters, L.; Nafpliotis, L.; Vereecke, D.; Dendauw, K.; Atwood, J. L.; Thewald, U. The geometry of the site and its relevance for chain migration and stereospecificity. *Macromol. Symp.* **1995**, *89*, 345–367.

72. Herfert, N.; Fink, G. Hemiisotactic polypropylene through propene polymerization with the ^{i}Pr[3-MeCpFlu]ZrCl$_2$/MAO catalyst system: A kinetic and microstructural analysis. *Makromol. Chem., Macromol. Symp.* **1993**, *66*, 157–178.

73. Miller, S. A.; Bercaw, J. E. Mechanism of isotactic polypropylene formation with C_1-symmetric metallocene catalysts. *Organometallics* **2006**, *25*, 3576–3592. Portions of this text have been reprinted with permission. Copyright 2006 American Chemical Society.

74. Kleinschmidt, R.; Reffke, M.; Fink, G. Investigation of the microstructure of poly(propylene) in dependence of the polymerization temperature for the systems ^{i}Pr[3-RCpFlu]ZrCl$_2$/MAO, with R = H, Me, Et, ^{i}Pr, ^{t}Bu, and ^{i}Pr[IndFlu]ZrCl$_2$/MAO. *Macromol. Rapid Commun.* **1999**, *20*, 284–288.

75. Yoshida, T.; Koga, N.; Morokuma, K. A combined *ab initio* MO-MM study on isotacticity control in propylene polymerization with silylene-bridged group 4 metallocenes. C_2 symmetrical and asymmetrical catalysts. *Organometallics* **1996**, *15*, 766–777.

76. Angermund, K.; Fink, G.; Jensen, V. R.; Kleinschmidt, R. The role of intermediate chain migration in propene polymerization using substituted {^{i}Pr(CpFlu)}ZrCl$_2$/MAO catalysts. *Macromol. Rapid Commun.* **2000**, *21*, 91–97.

77. Corradini, P.; Cavallo, L.; Guerra, G. Molecular modeling studies on stereospecificity and regiospecificity of propene polymerization by metallocenes. In *Metallocene-Based Polyolefins*; Scheirs, J., Kaminsky, W., Eds.; John Wiley & Sons Ltd.: Chichester, UK, 2000; pp 3–36.

78. In addition to low monomer concentration and high polymerization temperatures, the addition of methylene chloride has been observed to increase the relative rate of the site epimerization process for *s*-**1**/MAO: Fink, G.; Herfert, N.; Montag, P. The relationship between kinetics and mechanisms. In *Catalysts, Recent Scientific Innovations and Technological Improvements*; Fink, G., Mülhaupt, R., Brintzinger, H.-H., Eds.; Springer: Berlin, 1995; pp 159–179.

79. Ewen, J. A.; Elder, M. J. Process and catalyst for producing isotactic polyolefins. European Patent 0537130B1 (Fina Technology, Inc.), September 18, 1996.

80. Miller, S. A.; Bercaw, J. E. Isotactic-hemiisotactic polypropylene from C_1-symmetric *ansa*-metallocene catalysts: A new strategy for the synthesis of elastomeric polypropylene. *Organometallics* **2002**, *21*, 934–945.

81. Razavi, A. Polyolefin production. PCT Int. Pat. Appl.WO 2000/049029A1 (Fina Research S.A.), August 24, 2000.

82. Ewen, J. A.; Elder, M. J.; Jones, R. L.; Rheingold, A. L.; Liable-Sands, L. M.; Sommer, R. D. Chiral *ansa* metallocenes with Cp ring-fused to thiophenes and pyrroles: Syntheses, crystal structures, and isotactic polypropylene catalysts *J. Am. Chem. Soc.* **2001**, *123*, 4763–4773.

83. Kukral, J.; Lehmus, P.; Feifel, T.; Troll, C.; Rieger, B. Dual-side *ansa*-zirconocene dichlorides for high molecular weight isotactic polypropene elastomers. *Organometallics* **2000**, *19*, 3767–3775.

84. Thomas, E. J.; Chien, J. C. W.; Rausch, M. D. Influence of alkyl substituents on the polymerization behavior of asymmetric ethylene-bridged zirconocene catalysts. *Organometallics* **1999**, *18*, 1439–1443.

85. Rieger, B.; Jany, G.; Fawzi, R.; Steimann, M. Unsymmetric *ansa*-zirconocene complexes with chiral ethylene bridges: Influence of bridge conformation and monomer concentration on the stereoselectivity of the propene polymerization reaction. *Organometallics* **1994**, *13*, 647–653.

86. Kukral, J.; Rieger, B. High molecular weight polypropene elastomers via "dual-side" zirconocene dichlorides. *Macromol. Symp.* **2002**, *177*, 71–86.

87. Thomas, E. J.; Rausch, M. D.; Chien, J. C. W. New C_1 symmetric Ziegler–Natta type zirconocenes for the production of isotactic polypropylene. *Organometallics* **2000**, *19*, 4077–4083.

88. Thomas, E. J.; Rausch, M. D.; Chien, J. C. W. Substituent effects on the stereospecificity of propylene polymerization by novel asymmetric bridged zirconocenes. A mechanistic discussion. *Macromolecules* **2000**, *28*, 1546–1552.

89. Chen, Y.-X.; Rausch, M. D.; Chien, J. C. W. C_{2v}- and C_2-symmetric *ansa*-bis(fluorenyl)zirconocene catalysts: Synthesis and α-olefin polymerization catalysis. *Macromolecules* **1995**, *28*, 5399–5404.

90. (a) Patsidis, K.; Alt, H. G.; Welch, B. M.; Chu, P. P. Bis fluorenyl metallocenes and use thereof. European Patent 0628565B1 (Phillips Petroleum Company), April 26, 2000. (b) Alt, H. G.; Patsidis, K.; Welch, M. B.; Chu, P. P. Process of polymerizing olefins using diphenylsilyl or dimethyl tin bridged 1-methylfluorenyl metallocenes. U.S. Patent 5,631,335 (Phillips Petroleum Company), May 20, 1997.

91. Rieger, B. Stereospecific propene polymerization with *rac*-[1,2-bis(η^5-(9-fluorenyl))-1-phenylethane] zirconium dichloride/methylalumoxane. *Polym. Bull.* **1994**, *32*, 41–46.

92. (a) Razavi, A.; Atwood, J. L. Isospecific propylene polymerization with unbridged group 4 metallocenes. *J. Am. Chem. Soc.* **1993**, *115*, 7529–7530. (b) Razavi, A. Process and catalyst for producing crystalline polyolefins. U.S. Patent 5,304,523 (Fina Technology, Inc.), April 19, 1994.

93. (a) Farina, M.; Di Silvestro, G.; Sozzani, P. Hemitactic polypropylene: An example of a novel kind of polymer tacticity. *Macromolecules* **1982**, *15*, 1451–1452. (b) Farina, M.; Di Silvestro, G.; Sozzani, P.; Savaré, B. Hemitactic polymers. *Macromolecules* **1985**, *18*, 923–928. (c) Farina, M.; Di Silvestro, G.; Sozzani, P. Hemiisotactic polypropylene: A key point in the elucidation of the polymerization mechanism with metallocene catalysts. *Macromolecules* **1993**, *26*, 946–950. (d) Farina, M.; Di Silvestro, G.; Sozzani, P. Hemitactic polymers. *Prog. Polym. Sci.* **1991**, *16*, 219–238.

94. Miller, S. A. Metallocene-mediated olefin polymerization: The effects of distal ligand perturbations on polymer stereochemistry. Ph.D. Thesis, California Institute of Technology, Pasadena, CA, November 1999.

95. Ewen, J. A.; Jones, R. L.; Elder, M. J.; Camurati, I.; Pritzkow, H. Stereoblock isotactic-hemiisotactic poly(propylene)s and ethylene/propylene copolymers obtained with *ansa*-cyclopenta[1,2-b;4, 3-b']dithiophene catalysts. *Macromol. Chem. Phys.* **2004**, *205*, 302–307.

96. Dietrich, U.; Hackmann, M.; Rieger, B.; Klinga, M.; Leskelä, M. Control of stereoerror formation with high-activity "dual-side" zirconocene catalysts: A novel strategy to design the properties of thermoplastic elastic polypropenes. *J. Am. Chem. Soc.* **1999**, *121*, 4348–4355.

97. Alt, H. G.; Baker, R. W.; Dakkak, M.; Foulkes, M. A.; Schilling, M. O.; Turner, P. Zirconocene dichloride complexes with a 1,2-naphthylidene bridge as catalysts for the polymerization of ethylene and propylene. *J. Organomet. Chem.* **2004**, *689*, 1965–1977.

98. Miyata, H.; Ikeda, R. Ethylene-α-olefin copolymer, its production method and its composition. *Jpn. Kokai Tokkyo Koho.* JP 09183817A2 (Tosoh Corp.), July 15, 1997.

99. Winter, A.; Rohrmann, J.; Antberg, M.; Dolle, V.; Spaleck, W. Process and catalysts for the manufacture of syndiotactic polyolefins. *Ger. Offen.* DE 3907965A1 (Hoechst A.-G.), September 13, 1990.

100. Winter, A.; Rohrmann, J.; Antberg, M.; Dolle, V.; Spaleck, W. Catalysts for the syndiotactic polymerization of olefins. *Ger. Offen.* DE 3907964A1 (Hoechst A.-G.), September 13, 1990.

101. Fierro, R.; Yu, Z.; Rausch, M. D.; Dong, S.; Alvares, D.; Chien, J. C. W. Syndioselective propylene polymerization catalyzed by *rac*-2,2-dimethylpropylidene(1-η^5-cyclopentadienyl)(1-η^5-fluorenyl)dichlorozirconium. *J. Polym. Sci., Part A: Polym. Chem.* **1994**, *32*, 661–673.

102. Rieger, B.; Jany, G.; Steimann, M.; Fawzi, R. Synthesis of ethylene bridged biscyclopentadiene ligand precursor compounds and some of their *ansa*-zirconocene derivatives via chiral epoxides: A synthetic strategy of high variability. *Z. Naturforsch., B: Chem. Sci.* **1994**, *49*, 451–458.

103. Kim, I.; Kim, K.-T.; Lee, M. H.; Do, Y.; Won, M.-S. Syndioselective propylene polymerization: Comparison of Me$_2$C(Cp)(Flu)ZrMe$_2$ with Et(Cp)(Flu)ZrMe$_2$. *J. Appl. Polym. Sci.* **1998**, *70*, 973–983.

104. Kaminsky, W.; Hopf, A.; Arndt-Rosenau, M. Efficient and tailored polymerization of olefins and styrene by metallocene catalysts. *Macromol. Symp.* **2003**, *201*, 301–307.

105. Kaminsky, W.; Hopf, A.; Piel, C. C_s-symmetric hafnocene complexes for synthesis of syndiotactic polypropene. *J. Organomet. Chem.* **2003**, *684*, 200–205.

106. Inoue, N.; Jinno, M.; Shiomura, T. Novel transition metal compounds for polymerization of olefins with improved efficiency and polymerization of olefins using them. *Jpn. Kokai Tokkyo Koho.* JP 07247309A2 (Mitsui Toatsu Chemicals), September 26, 1995.

107. Alt, H. G.; Jung, M. C_1-Bridged fluorenylidene cyclopentadienylidene complexes of the type (C$_{13}$H$_8$-CR$_1$R$_2$-C$_5$H$_3$R)ZrCl$_2$ (R$_1$, R$_2$=alkyl, phenyl, alkenyl; R=H, alkyl, alkenyl, substituted silyl) as catalyst precursors for the polymerization of ethylene and propylene. *J. Organomet. Chem.* **1998**, *568*, 87–112.

108. Brinen, J. L.; Cozewith, C. Polymerization process with mixed catalyst compositions. PCT Int. Pat. Appl. WO 2002/060957A2 (ExxonMobil Chemical Patents Inc.), August 8, 2002.

109. Rohrmann, J.; Brekner, M.-J.; Kueber, F.; Osan, F.; Weller, T. Process for preparing cyclo-olefin polymers. European Patent 0610843B1 (Ticona GmbH), May 19, 1999.

110. Alt, H. G.; Zenk, R.; Milius, W. Syndiospecific polymerization of propylene: 3-, 4-, 3,4- and 4,5-substituted zirconocene complexes of the type (C$_{13}$H$_{8-n}$R$_n$CR$_2'$C$_5$H$_4$)ZrCl$_2$ (n = 1, 2; R = alkyl, aryl; R' = Me, Ph). *J. Organomet. Chem.* **1996**, *514*, 257–270.

111. Yano, A.; Hasegawa, S.; Kaneko, T.; Sone, M.; Sato, M.; Akimoto, A. Ethylene/1-hexene copolymerization with Ph$_2$C(Cp)(Flu)ZrCl$_2$ derivatives. Correlation between ligand structure and copolymerization behavior at high temperature. *Macromol. Chem. Phys.* **1999**, *200*, 1542–1553.

112. Tohi, Y.; Endo, K.; Kaneyoshi, H.; Urakawa, N.; Yamamura, Y.; Kawai, K. Crosslinked metallocene compound for olefin polymerization and method of polymerizing olefin with the same. PCT Int. Pat. Appl. WO 2004/029062A1 (Mitsui Chemicals, Inc.), April 8, 2004.

113. Rieger, B.; Repo, T.; Jany, G. Ethylene-bridged *ansa*-zirconocene dichlorides for syndiospecific propene polymerization. *Polym. Bull.* **1995**, *35*, 87–94.

114. Sone, M.; Hasegawa, S.; Yamada, S.; Yano, A. Olefin polymerization catalyst and process for producing olefin polymer. European Patent Application 0709393A2 (Tosoh Corporation), May 1, 1996.

115. Sunaga, T.; Michiue, K.; Yamashita, M.; Ishii, Y. Metallocene compound, and process for preparing polyolefin by using it. European Patent 0955305B1 (Mitsui Chemicals, Inc.), October 26, 2005.

116. Alt, H. G.; Zenk, R. Syndiospecific polymerization of propylene: 2- and 2,7-substituted metallocene complexes of type (C$_{13}$H$_{8-n}$R$_n$CR$_2'$C$_5$H$_4$)MCl$_2$ (n = 1, 2; R = alkoxy, alkyl, aryl, hal; R' = Me, Ph; M = Zr, Hf). *J. Organomet. Chem.* **1996**, *522*, 39–54.

117. Matsukawa, T.; Kiba, R.; Ikeda, T. Manufacture of stereospecific vinyl chloride-based polymers. *Jpn. Kokai Tokkyo Koho.* JP 08208736A2 (Chisso Corp.), August 13, 1996.

118. Crowther, D. J.; Folie, B. J.; Walzer, J. F., Jr.; Schiffino, R. S. High temperature olefin polymerization process. PCT Int. Pat. Appl. WO 99/45041A1 (Exxon Chemical Patents Inc.), September 10, 1999.

119. Ashe, A. J., III; Fang, X.; Hokky, A.; Kampf, J. W. The C_s-symmetric aminoboranediyl-bridged zirconocene dichloride [(η-9-$C_{13}H_8$)-BN(^{i}Pr)$_2$(η-C_5H_4)]$ZrCl_2$: Its synthesis, structure, and behavior as an olefin polymerization catalyst. *Organometallics* **2004**, *23*, 2197–2200.

120. Inoue, N.; Jinno, M.; Sonobe, Y.; Mizutani, K.; Shiomura, T. Catalysts for manufacture of syndiotactic propene polymers. *Jpn. Kokai Tokkyo Koho.* JP 04366107A2 (Mitsui Toatsu Chemicals, Inc.), December 18, 1992.

121. Mülhaupt, R.; Heinemann, J.; Reichert, P.; Geprägs, M.; Queisser, J. Polyolefin nanocomposites. *Ger. Offen.* DE 19846314A1 (BASF A.-G.), April 13, 2000.

122. Izmer, V. V.; Agarkov, A. Y.; Nosova, V. M.; Kuz'mina, L. G.; Howard, J. A. K.; Beletskaya, I. P.; Voskoboynikov, A. Z. *ansa*-Metallocenes with a Ph$_2$Si bridge: Molecular structures of HfCl$_2$[Ph$_2$Si(η^5-$C_{13}H_8$)(η^5-C_5H_4)] and HfCl$_2$[Ph$_2$Si($C_{13}H_9$)(η^5-C_5H_4)]$_2$. *J. Chem. Soc., Dalton Trans.* **2001**, *7*, 1131–1136.

123. Köppl, A.; Babel, A. I.; Alt, H. G. Homopolymerization of ethylene and copolymerization of ethylene and 1-hexene with bridged metallocene/methylaluminoxane catalysts: The influence of the bridging moiety. *J. Mol. Catal. A: Chem.* **2000**, *153*, 109–119.

124. Razavi, A.; Vereecke, D. Bridged metallocenes for use in catalyst systems for the polymerization of olefins. PCT Int. Pat. Appl. WO 96/16069A1 (Fina Research S.A.), May 30, 1996.

125. Alt, H. G.; Zenk, R. Syndiospecific polymerization of propylene: Synthesis of CH$_2$- and CHR-bridged fluorenyl-containing ligand precursors for metallocene complexes of type ($C_{13}H_{8-n}$R$'_n$CHR-C_5H_4)$ZrCl_2$ (n = 0, 2; R = H, alkyl; R$'$ = H, Hal). *J. Organomet. Chem.* **1996**, *526*, 295–302.

126. Alt, H. G.; Palackal, S. J.; Patsidis, K.; Welch, M. B.; Geerts, R. L.; Hsieh, E. T.; McDaniel, M. P.; Hawley, G. R.; Smith, P. D. Fluorenyl-containing metallocenes, their preparation and use as polymerization catalysts, and polymers prepared thereby. *Can. Pat. Appl.* CA 2067525C (Phillips Petroleum Co.), September 15, 1998.

127. Sugimoto, R.; Ooe, T. Preparation of polyolefins with controlled molecular weight. *Jpn. Kokai Tokkyo Koho.* JP 07062013A2 (Mitsui Toatsu Chemicals), March 7, 1995.

128. Inoue, N.; Jinno, M.; Shiomura, T. Preparation of transition metal compounds as catalysts for polymerization of α-olefins. *Jpn. Kokai Tokkyo Koho.* JP 04275293A2 (Mitsui Toatsu Chemicals, Inc.), September 30, 1992.

129. Shamshoum, E.; Reddy, B. Metallocene catalyst component with good catalyst efficiency after aging. European Patent 0574370B1 (Fina Technology, Inc.), March 11, 1998.

130. Szul, J. F.; Kwalk, T. H.; Schreck, D. J.; Mawson, S.; McKee, M. G.; Terry, K. A.; Goode, M. G.; Whiteker, G. T.; Lucas, E. Catalyst composition, method of polymerization and polymer therefrom. PCT Int. Pat. Appl. WO 2002/046250A2 (Univation Technologies, LLC), June 13, 2002.

131. Razavi, A. Syndiotactic/atactic block polyolefins, catalysts and processes for producing the same. PCT Int. Pat. Appl. WO 98/02469A1 (Fina Research S.A.), January 22, 1998.

132. Dai, Z.; Jing, Z.; Zhou, H.; Liu, W. QSAR studies of C_1-bridged Flu-Cp complexes of zirconium metallocene. *Comput. Chem. Eng.* **2003**, *15B*, 1170–1174.

133. Alt, H. G.; Palackal, S. J.; Zenk, R.; Welch, M. B.; Schmid, M. Organometallic fluorenyl compounds, preparation, and use. European Patent 0666267B1 (Phillips Petroleum Company), April 18, 2001.

134. Tohi, Y.; Urakawa, N.; Endo, K.; Kawai, K.; Okawa, K.; Tsutsui, T. Process for preparing low molecular weight olefin (co)polymer and polymerizing catalyst used therefore. European Patent Application 1416000A1 (Mitsui Chemicals, Inc.), May 6, 2004.

135. Tsai, J.-C.; Wu, M.-Y.; Hsieh, T.-Y.; Wei, Y.-Y. Catalyst compostion for preparing olefin polymers. U.S. Patent Application 2002/0147104A1 (Industrial Technology Research Institute), October, 10, 2002.

136. Hahn, S. F.; Redwine, O. D.; Shankar, R. B.; Timmers, F. J. Ethylene and/or alpha-olefin/vinyl or vinylidene aromatic interpolymer compositions. PCT Int. Pat. Appl. WO 2000/078831A1 (The Dow Chemical Company), December 28, 2000.

137. Ewen, J. A. Stereospecific metallocene catalysts with stereolocking α-*CP* substituents. U.S. Patent 5,631,202 (Montell Technology Company B.V.), May 20, 1997.

138. Urakawa, N.; Dohi, Y.; Endo, H.; Kawai, K. Bridged metallocene compounds, polymerization catalysts containing them, and olefin polymerization using them. *Jpn. Kokai Tokkyo Koho*. JP 2004175707A2 (Mitsui Chemicals Inc.) June 24, 2004.

139. Oh, J. S.; Park, T. H.; Lee, B. Y.; Jeong, S. W. Process for the preparation of olefinic polymers using metallocene catalyst. PCT Int. Pat. Appl. WO 97/19960A1 (LG Chemical Ltd.), June 5, 1997.

140. Kaminsky, W.; Rabe, O.; Schauwienold, A.-M.; Schupfner, G. U.; Hanss, J.; Kopf, J. Crystal structure and propene polymerization characteristics of bridged zirconocene catalysts. *J. Organomet. Chem.* **1995**, *497*, 181–193.

141. Kawahara, N.; Kawai, K.; Kashiwa, N. Olefin polymerization catalysts and polymerization of olefins therewith. *Jpn. Kokai Tokkyo Koho*. JP 2004027163A2 (Mitsui Chemicals Inc.) January 29, 2004.

142. Esteb, J. J.; Chien, J. C. W.; Rausch, M. D. Novel C_1 symmetric zirconocenes containing substituted indenyl moieties for the stereoregular polymerization of propylene. *J. Organomet. Chem.* **2003**, *688*, 153–160.

143. Brekner, M. J.; Osan, F.; Rohrmann, J.; Antberg, M. Process for the preparation of chemically homogenous cycloolefin copolymers. European Patent 0503422B1 (Targor GmbH), June 3, 1998.

144. Ewen, J. A.; Elder, M. J. Process and catalyst for producing isotactic polyolefins. European Patent 0537130B1 (Fina Technology, Inc.), September 18, 1996.

145. Kawai, K.; Yamashita, M.; Tohi, Y.; Kawahara, N.; Michiue, K.; Kaneyoshi, H.; Mori, R. Metallocene compound, process for producing metallocene compound, olefin polymerization catalyst, process for producing polyolefin, and polyolefin. PCT Int. Pat. Appl. WO 2001/027124A1 (Mitsui Chemicals, Inc.), April 19, 2001.

146. Ikenaga, S.; Okada, K.; Takayasu, H.; Inoue, N.; Hirota, N.; Kaneyoshi, H.; Funaya, M. et al. Propylene copolymer, polypropylene composition, use thereof, transition metal compounds, and catalysts for olefin polymerization. PCT Int. Pat. Appl. WO 2004/087775A1 (Mitsui Chemicals, Inc.), October 14, 2004.

147. Reddy, B. R.; Shamshoum, E. S.; Lopez, M. Production of E-B copolymers with a single metallocene catalyst and a single monomer. U.S. Patent 5,753,785 (Fina Technology, Inc.), May 19, 1998.

148. Jany, G.; Fawzi, R.; Steimann, M.; Rieger, B. Synthesis of enantiomerically pure ethylene-bridged *ansa*-zirconocene and -hafnocene complexes bearing fluorenyl, indenyl, octahydrofluorenyl, and tetrahydroindenyl ligands. *Organometallics* **1997**, *16*, 544–550.

149. Rieger, B.; Troll, C.; Preuschen, J. Ultrahigh molecular weight polypropene elastomers by high activity "dual-side" hafnocene catalysts. *Macromolecules* **2002**, *35*, 5742–5743.

150. Halterman, R. L.; Fahey, D. R.; Marin, V. P.; Dockter, D. W.; Khan, M. A. Synthesis, characterization and polymerization properties of isopropylidene(η^5-3-neomenthylcyclopentadienyl)(η^5-fluorenyl) zirconium dichloride. *J. Organomet. Chem.* **2001**, *625*, 154–159.

151. (a) Miller, S. A.; Bercaw, J. E. Catalyst system for the polymerization of alkenes to polyolefins. U.S. Patent 6,469,188B1 (California Institute of Technology), October 22, 2002. (b) Miller, S. A.; Bercaw, J. E. Catalyst system for the polymerization of alkenes to polyolefins. U.S. Patent 6,693,153B2 (California Institute of Technology), February 17, 2004.

152. (a) Reddy, B. R. Stereorigid bis-fluorenyl metallocenes. U.S. Patent 6,313,242B1 (Fina Technology, Inc.), November 6, 2001. (b) Reddy, B. R. Stereorigid bis-fluorenyl metallocenes. U.S. Patent 5,945,365 (Fina Technology, Inc.), August 31, 1999.

153. (a) Debras, G.; Dupire, M.; Michel, J. Production of polypropylene. European Patent Application 1195391A1 (Atofina Research), April 10, 2002. (b) Razavi, A.; Bellia, V. Metallocene catalyst component for use in producing isotactic polyolefins. European Patent Application 0881236A1 (Fina Research S.A.), December 2, 1998.

154. Kukral, J.; Lehmus, P.; Klinga, M.; Leskelä, M.; Rieger, B. Oxygen-containing, asymmetric "dual-side" zirconocenes: Investigations on a reversible chain transfer to aluminum. *Eur. J. Inorg. Chem.* **2002**, *2002*, 1349–1356.

155. Rohrmann, J. Bridged chiral metallocenes, process for their preparation and their use as catalysts. European Patent 0528287B1 (Targor GmbH), November 11, 1998.

156. Ivchenko, N. B.; Ivchenko, P. V.; Nifant'ev, I. E. Synthesis of 1-(cyclopentadienyl)adamantane and the corresponding zirconium complexes. *Russ. Chem. Bull.* **2000**, *49*, 508–513.

157. Deisenhofer, S.; Feifel, T.; Kukral, J.; Klinga, M.; Leskelä, M.; Rieger, B. Asymmetric metallocene catalysts based on dibenzothiophene: A new approach to high molecular weight polypropylene plastomers. *Organometallics* **2003**, *22*, 3495–3501.
158. Razavi, A.; Hortmann, K. Polyolefin production. European Patent Application 0965603A1 (Fina Research S.A.), December 22, 1999.
159. Jutzi, P.; Muller, C.; Neumann, B.; Stammler, H.-G. Dialkylaminoethyl-functionalized *ansa*-zirconocene dichlorides: Synthesis, structure, and polymerization properties. *J. Organomet. Chem.* **2001**, *625*, 180–185.

3 Substituted Indenyl and Cyclopentadienyl Catalysts for Stereoselective Propylene Polymerization

John J. Esteb

CONTENTS

3.1 INTRODUCTION

Metallocenes based on group 4 metals were first synthesized by Wilkinson et al. in 1953;[1] however, the great utility of these compounds as catalyst precursors for the stereoselective polymerization of α-olefins was not realized until the mid-1980s when two key discoveries were made. The first of these was the discovery that alkyl substituents placed on the cyclopentadienyl framework of the metallocene can significantly influence the performance and behavior of the catalyst.[2] The second was the discovery that enantioselectivity in the insertion step of an α-olefin polymerization could result if one made metallocene catalysts with the appropriate chirality and stereorigidity.[3]

Since these early discoveries, an enormous amount of effort has been put forth to understand and improve upon stereoselective catalyst systems. Because of the excellent research that has been accomplished over the past 20+ years, the ability to influence the mechanistic details of each step of the polymerization process from insertion to chain release now exists. This chapter will focus on how the "tuning" of metallocene polymerization catalyst structures can be carried out to influence stereoregularity in the resulting polymer.

For the purpose of the present chapter, the definition of metallocene catalysts is limited to the bis(cyclopentadienyl)-based complexes of the group 4 transition metals (titanium, zirconium, or hafnium). The focus is exclusively on group 4 metallocenes since they uniquely have demonstrated the ability to influence stereocontrol over the entire range of polymer microstructures and molecular weights. Furthermore, the discussion is limited to bridged frameworks, that is, the so called "*ansa*-metallocenes" containing only substituted indenyl or cyclopentadienyl ligands.

3.2 CATALYST STRUCTURE AND SYMMETRY

Ewen first demonstrated a correlation between a catalyst's molecular symmetry and the stereoregularity of the polymer produced by studying the two isomers (*rac* and *meso*) of the *ansa*-metallocene $Et(Ind)_2TiCl_2$ (Ind = indenyl).[3a] Ewen found that the achiral *meso* form (**1**) produced atactic polypropylene, whereas the C_2-symmetric *rac* form (**2**) produced polypropylene with moderate isotacticity (Figure 3.1). Both (**1**) and (**2**) suffered from poor catalytic activity owing to the instability of the titanium species.

Shortly after the initial discovery by Ewen had been disclosed, Brintzinger and Kamisky demonstrated that a similar metallocene, C_2-symmetric *rac*-$Et(H_4Ind)_2ZrCl_2$ (H_4Ind = tetrahydroindenyl) (**3**), was able to produce a much greater yield of isotactic polypropylene (iPP) relative to the similar titanocene species (**2**).[3b] These two discoveries led the way for the development of several different families of symmetry-based metallocene catalysts. In almost every metallocene example, zirconium is selected as the metal of choice because it usually produces the most active catalyst (relative to the hafnium and titanium analogues). However, hafnium metallocenes do tend to produce higher molecular weights in the resulting polymers than do the corresponding zirconium or titanium analogues.

This chapter will focus on four classes of *ansa*-metallocene catalysts: C_2-symmetric bis-Cp metallocenes, C_2-symmetric bis(indenyl) metallocenes, C_1-symmetric Cp-indenyl mixed metallocenes, and lastly, C_1-symmetric bis(indenyl) metallocenes containing two different indenyl moieties (Figure 3.2). There are several other classes of catalysts that have been shown to produce stereoregular polypropylene. These systems will not be reviewed herein, but are discussed in other chapters of this book.

3.2.1 C_2-SYMMETRIC *ANSA*-METALLOCENES

C_2-symmetric *ansa*-metallocenes greatly outnumber all of the other catalyst structures studied to date. This is partly due to the flexibility offered by the cyclopentadienyl ligands in these systems, which can be readily substituted to change the structural framework of the resulting metallocene catalyst. In addition, this structural variety can and has lead to a whole spectrum of molecular weights and tacticities available in the polymers produced by these catalysts. In other words, metallocene catalysts can be "tuned" to produce polypropylene with tacticity ranging from completely atactic to perfectly isotactic by changing the ligand framework around the group 4 metals.

The symmetry of an *ansa*-metallocene arises from the ligand framework of the compound and is preserved by the bridge, which is the group that links the two Cp moieties together and prevents their rotation. In terms of bis-Cp *ansa*-metallocene nomenclature, the position where connection to the bridge (Z) takes place is labeled as position 1. The bridging group (Z) is usually a CH_2CH_2-, R_2C-,

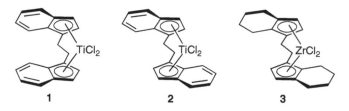

FIGURE 3.1   Examples of *ansa*-bis(indenyl) metallocenes.

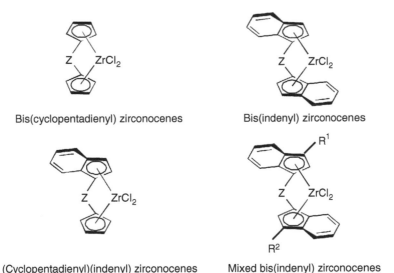

Bis(cyclopentadienyl) zirconocenes   Bis(indenyl) zirconocenes

(Cyclopentadienyl)(indenyl) zirconocenes   Mixed bis(indenyl) zirconocenes

FIGURE 3.2   The four classes of cyclopentadienyl/indenyl metallocenes reviewed in this chapter. Z represents a bridge, for example, CH_2CH_2 or Me_2Si.

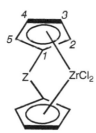

FIGURE 3.3   Numbering system for biscyclopentadienyl *ansa*-metallocenes.

or R_2Si- fragment (R = alkyl, aryl, or H), although several other bridge types have been investigated. The "front" positions of the Cp are labeled as positions 3 and 4, which bear the α-substituents. The "back" positions, which bear the β-substituents, are labeled as positions 2 and 5. The terms α and β refer to the position of each substituent relative to the bridging position (Figure 3.3).

Bis(indenyl) metallocenes employ a slightly different numbering system owing to the presence of the fused aromatic ring within the Cp framework. In these instances, a traditional classification for fused ring compounds is adopted where one begins numbering along the periphery of the fused ring system beginning at the first nonfused atom contained within the 5-membered ring. Fused carbon atoms are not included in this numbering system. The bridge in most cases is located at position one, but as will be discussed later, the bridge can theoretically exist at any location (Figure 3.4).

3.2.1.1 Bis(cyclopentadienyl) Systems

There are only a handful of examples of C_2-symmetric bis-Cp *ansa*-metallocenes, in contrast to the numerous existing examples of analogous bis(indenyl) systems. The known bis-Cp complexes demonstrate a definite correlation between the Cp substituent (both position and type) and the physical properties of the polypropylene produced using these systems.[4] Table 3.1 lists some representative examples of bis-Cp catalysts and their polymerization behaviors.[4–6] In the series of compounds characterized by the general formula *rac*-Z-$(R_nCp)_2ZrCl_2$ (where *n* indicates the number

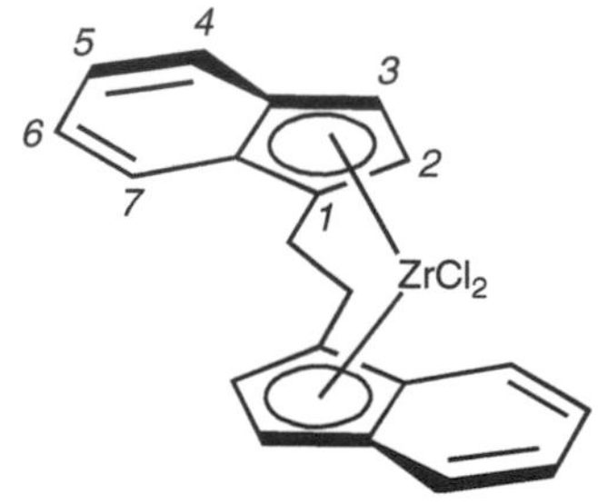

FIGURE 3.4   Numbering system for bis(indenyl) *ansa*-metallocenes.

TABLE 3.1

Propylene Polymerization Data from Representative Substituted Bis(cyclopentadienyl) Zirconocene Precatalysts Activated with MAO

Precatalyst	Al/Zr Ratio	T_p^a (°C)	Activity[b]	M_w	T_m (°C)	*mmmm* (%)	References
(**4**) *rac*-CH$_2$CH$_2$(3-MeCp)$_2$ZrCl$_2$	2,000	40	5.8	19,600	133	92.2	4
(**5**) *rac*-CH$_2$CH$_2$(3-*i*-PrCp)$_2$ZrCl$_2$	2,000	40	4.5	19,400	136	94.6	4
(**6**) *rac*-CH$_2$CH$_2$(3-*t*-BuCp)$_2$ZrCl$_2$	2,000	40	1.0	17,400	141	97.6	4
(**7**) *rac*-Me$_2$Si(3-MeCp)$_2$ZrCl$_2$	10,000	30	16.3	13,700	148	92.5	5
(**8**) *rac*-Me$_2$Si(3-*t*-BuCp)$_2$ZrCl$_2$	10,000	30	0.3	9,500	149	93.4	5
(**9**) *rac*-Me$_2$C(3-*t*-BuCp)$_2$ZrCl$_2$	3,000	50	15	17,000[c]	153	99.5[d]	6
(**10**) *rac*-Me$_2$Si(2-Me-4-*t*-BuCp)$_2$ZrCl$_2$	15,000	70	10	19,000	155	94.3	7
(**11**) *rac*-Me$_2$C(2-Me-4-*t*-BuCp)$_2$ZrCl$_2$	8,000	50	73	103,000	162	99.5[d]	6

[a] Polymerization temperature.
[b] kg PP/mmol Zr·h.
[c] Recorded as M_v.
[d] Value on primary (1,2) insertions only.

---

of R substituents), changing the substituent has a distinct effect on the tacticity and the molecular weight of the polypropylene produced.

As a general trend, the alkyl substituent at C-3 appears to have a direct influence on the isotacticity of the polymer whereas the substituent at C-2 appears to have more influence on the polymer molecular weight. The effect of the substituent at C-3 can be seen most clearly by comparing data for the ethylene-bridged series *rac*-CH$_2$CH$_2$(3-RCp)$_2$ZrCl$_2$ as R is varied from a methyl to an isopropyl to a *tert*-butyl substituent (**4–6**). The methyl analogue (**4**) gives rise to a polypropylene with an *mmmm* value of 92.2%. However, as the size of the substituent is increased, the value increases to 94.6% and 97.6% for the isopropyl (**5**) and *tert*-butyl (**6**) analogues, respectively.

Similar influence of alkyl substitution around the catalyst framework has also been demonstrated with both Me$_2$C- and Me$_2$Si-bridged analogues of the bis(cyclopentadienyl) zirconcocenes.[5,6] For example, the introduction of an alkyl substituent at C-2 causes a net increase in the molecular weight of the resulting polymer relative to derivatives without substitution at that position. This observation can be explained by the fact that alkyl substitution at C-2 hinders the bimolecular chain release by β-hydride transfer to the monomer. This hypothesis is also supported by the observation that the molecular weight effects become more pronounced in liquid propylene, which would not be the case if the substituent were simply hindering the unimolecular β-hydride transfer to the metal. In addition, replacement of a Me group (**7**) with a bulkier *t*-Bu group (**8**) at the C-3 position of the Me_2Si- metallocene worsens catalyst activity, and produces polypropylene of lower molecular weight.

To date, optimal catalyst performance has been achieved by placing alkyl substituents at both C-2 and C-4, as was demonstrated with precatalysts (10) and (11). Each of these precatalysts contains a Me substituent at C-2 and a bulky t-Bu substituent at C-4, producing a synergistic effect and leading to improved catalyst activity and molecular weights and isotacticities of the resulting polymers far superior to the unsubstituted analogues.

Attempts to further increase catalyst performance by making alternative modifications to the bridge have also been made. Most notably, Brintzinger and coworkers[8,9] created a series of differently bridged bis-Cp zirconocene analogues, including complexes having spirosilane (($(CH_2)_3Si$-) or 2,2'-biphenyl bridges.[8,9a,b] Unfortunately, these catalysts performed poorly relative to the analogues listed in Table 3.1.

3.2.1.2 Bis(indenyl) Systems

The C_2-symmetric *ansa*-metallocenes containing a bis(indenyl) ligand are the most common and best studied of all of the metallocene catalysts.[10] The chemistry involved in synthesizing the metallocene ligands has been very creative, oftentimes elegant, and inventive, which has lead to a plethora of ligand structures and metallocene frameworks for study. The effect of ligand substitution has been studied at every position of the indenyl ring to gain a better understanding of the substituents' ability to control polymerization behavior. In addition, multitudes of bridges have been investigated to probe more completely the role that the bridge (both position and type) has in influencing the polymerization behavior of the catalysts.

When placing alkyl substituents on the bis(indenyl) framework, the most significant effects are obtained when the substituents are placed on the C-2, C-3, and C-4 positions. There have been numerous modeling and experimental studies performed that reveal the significance of each of these positions.[10a] Table 3.2 lists representative examples of alkyl-substituted bis(indenyl) catalysts and their propylene polymerization behavior.

TABLE 3.2

Propylene Polymerization Data from Representative Substituted Bis(indenyl) Zirconocene Precatalysts Activated with MAO

Precatalyst	Al/Zr Ratio	T_p[a] (°C)	Activity[b]	M_w	T_m (°C)	*mmmm* (%)	References
(12) *rac*-$Me_2Si(Ind)_2ZrCl_2$	n/a[c]	70	190	36,000	137	81.7	13
(13) *rac*-$Me_2Si(H_4Ind)_2ZrCl_2$	8,000	50	54	30,300	148	94.9	14
(14) *rac*-$Me_2Si(benz[e]Ind)_2ZrCl_2$	15,000	70	274	270,000	138	80.5	7
(15) *rac*-$CH_2CH_2(Ind)_2ZrCl_2$	8,000	50	140	33,600[d]	134	87.4	17a
(16) *rac*-$Me_2Si(2-MeInd)_2ZrCl_2$	15,000	70	99	195,000	145	88.5	7
(17) *rac*-$Me_2Si(2-MeInd-H_4)_2ZrCl_2$	15,000	70	40	55,000	144	87.4	7
(18) *rac*-$Me_2Si(2-PhInd)_2ZrCl_2$	1,000	20	2.4	444,000	139	87.0	11
(19) *rac*-$Me_2Si(2-Me-benz[e]Ind)_2ZrCl_2$	15,000	70	403	330,000	146	88.7	7
(20) *rac*-$CH_2CH_2(4,7-Me_2Ind)_2ZrCl_2$	2,000	40	7.8	5,800	130	92.2	15
(21) *rac*-$CH_2CH_2(2,4,7-Me_3Ind)_2ZrCl_2$[e]		30	21	n/a	158	90.6	12

[a] Polymerization temperature.
[b] kg PP/mmol Zr·h.
[c] Not available.
[d] M_v.
[e] $[Ph_3C][B(C_6F_5)_4]/Al(i-Bu)_3$; Zr/B = 1:1, Al/Zr = 100:1.

As can be seen by comparing the 2-alkyl substituted analogues (**16–19**) with their unsubstituted parent catalysts at the same polymerization temperature (**12–14**), the introduction of an alkyl substituent at C-2 increases both the molecular weight and isotacticity of the resulting polypropylene.[7,11–15] Furthermore, substitution with methyl groups at the C-4 and C-7 positions of the indene (as demonstrated in **20** and **21**) increases the stereoselectivity slightly, but is detrimental to both the molecular weight and regioregularity (2,1- versus 1,2-insertions) of the resulting polymer.[15] The low molecular weight of these polymers is due to the large number of regioerrors associated with these polymerization catalysts since 2,1-insertions frequently lead to a chain termination event.

The most interesting and promising substitution pattern has been obtained by placing substituents at both C-2 and C-4. This 2,4-disubstituted arrangement has produced some of the most active metallocene catalysts known to date, and the polypropylenes produced rank amongst the highest in regioregularity, stereoregularity, and molecular weight.[7,16] The observed catalyst performance can be attributed to the substituent at C-2 blocking a coordination sight and hence preventing/reducing the possibility of a misinsertion and thus producing a highly regioregular polymer. In addition, the bulky substituent at C-4 directs the face that the monomer will use to coordinate to the metal center leading to a highly stereoregular arrangement of the resulting polymer. Some representative examples of these 2,4-disubstituted metallocenes (and some of their counterparts without a substituent at C-2) are shown in Table 3.3.

For example, when comparing rac-Me$_2$Si(2-MeBenz[e]Ind)$_2$ZrCl$_2$ (**19**) to its unsubstituted parent, rac-Me$_2$Si(Benz[e]Ind)$_2$ZrCl$_2$ (**14**), one observes a doubling of the catalyst activity and a significant increase in both molecular weight and isotacticity of the resulting iPP (Table 3.2). Similarly, when comparing the 2,4-disubstituted rac-Me$_2$Si(2-Me-4-PhInd)$_2$ZrCl$_2$ (**24**) with rac-Me$_2$Si(4-PhInd)$_2$ZrCl$_2$ (**22**), which lacks the corresponding substituent at C-2, the activity of the 2,4-disubstitued complex is over 15 times greater than that for the less-substituted analogue. There is a correspondingly impressive jump in molecular weight and isotacticity for this system as well. Table 3.3 lists several other examples of 2,4-disubstituted indenyl systems that exhibit even higher activities, molecular weights, and stereoregularity than similar complexes lacking the 2,4-disubstitution pattern.

Substitution at C-3 can also have a dramatic effect on catalyst activity, but only under very specific conditions (Table 3.4). For example, the metallocene series rac-Z-(3-MeInd)$_2$ZrCl$_2$ where Z is a CH$_2$CH$_2$- (**28**), Me$_2$Si- (**29**), or Me$_2$C- (**30**) bridge produces polypropylene of low molecular weights and tacticity.[13,17–19] However, by changing the substituent at C-3 from a methyl group (**30**) to a bulkier $tert$-butyl (**31**) or trimethylsilyl group (**32**) in the isopropylidene-bridged analogues, the catalysts produce highly isotactic polypropylene of moderate molecular weights exhibiting $mmmm$ values of close to 95% from the former catalyst (**31**) and 86% from the latter catalyst (**32**).[17] Interestingly, the same increase in polymer properties is not observed in either rac-Me$_2$Si(3-t-BuInd)$_2$ZrCl$_2$ (**33**) or rac-CH$_2$CH$_2$(3-(Me$_3$Si)Ind)$_2$ZrCl$_2$ (**34**).[13,19] The improved performance therefore arises synergistically from both the presence of the bulky alkyl group at C-3, and the higher rigidity and larger bite angle that results from the isopropylidene bridge relative to the Me$_2$Si- bridge or CH$_2$CH$_2$- bridge. This change in angle allows the substituent group at C-3 to influence the environment typically influenced by substituents at C-2 and C-4 when the smaller bite angle is present, which thus produces a marked increase in polymer regioregularity, stereoregularity, and molecular weight.

3.2.1.3 Effect of Changing the Bridging System

In addition to the extensive studies performed on the substitution patterns in bis(indenyl) $ansa$-metallocenes, the nature of the bridging moiety has also been exhaustively probed. The wide array of bridging groups investigated ranges from simple one-atom bridges to complex organic molecules. Although the effect of varying the bridge is reviewed elsewhere, some generalizations regarding the effects of the most common bridging fragments on polymerization behavior of the catalysts are noted herein.

TABLE 3.3

Propylene Polymerization Data from Representative 2,4-Disubstituted Bis(indenyl) Zirconocene and Their Related Parent Precatalysts Activated with MAO

Precatalyst	Al/Zr Ratio	T_p^a (°C)	Activity[b]	M_w	T_m (°C)	$mmmm$ (%)	References
(12) rac-Me$_2$Si(Ind)$_2$ZrCl$_2$	n/a^c	70	190	36,000	137	81.7	13
(16) rac-Me$_2$Si(2-MeInd)$_2$ZrCl$_2$	15,000	70	99	195,000	145	88.5	7
(22) rac-Me$_2$Si(4-PhInd)$_2$ZrCl$_2$	15,000	70	48	42,000	148	86.5	7
(23) rac-Me$_2$Si(4-α-naphthyl-Ind)$_2$ZrCl$_2$	15,000	70	875	920,000	161	99.1	7
(24) rac-Me$_2$Si(2-Me-4-PhInd)$_2$ZrCl$_2$	15,000	70	755	729,000	157	95.2	7
(25) rac-Me$_2$Si(2-Me-4-i-PrInd)$_2$ZrCl$_2$	15,000	70	245	213,000	150	88.6	7
(26) rac-Me$_2$Si(2-Me-4-α-naphthyl-Ind)$_2$ZrCl$_2$	350	50	22.5	380,000	156	98.6	16
(27) rac-Me$_2$Si(2-n-Pr-4-phenanthryl-Ind)$_2$ZrCl$_2$	350	50	46.8	400,000	160	99.2	16

[a]Polymerization temperature.
[b]kg PP/mmol Zr·h.
[c]Not available.

TABLE 3.4

Influence of Alkyl Substitution at C-3 on Bis(indenyl) Zirconocene Precatalysts Activated with MAO in Propylene Polymerization

Precatalyst	Al/Zr Ratio	$T_p{}^a$ (°C)	Activity[b]	M_w	$T_m{}^{}$ (°C)	mmmm (%)	References
(28) rac-CH$_2$CH$_2$(3-MeInd)$_2$ZrCl$_2$	8,000	50	28	15,800	n/a[c]	19.9	17a
(29) rac-Me$_2$Si(3-MeInd)$_2$ZrCl$_2$	15,000	70	33	28,000	n/a	n/a	17a
(31) rac-Me$_2$C(3-t-BuInd)$_2$ZrCl$_2$	8,000	50	125	89,400	152	94.8	17a
(32) rac-Me$_2$C(3-(Me$_3$Si)Ind)$_2$ZrCl$_2$	3,000	50	74	70,900	135	85.9	17a
(33) rac-Me$_2$Si(3-t-BuInd)$_2$ZrCl$_2$	n/a	70	0.5	700	n/a	10.5	19
(34) rac-CH$_2$CH$_2$(3-(Me$_3$Si)Ind)$_2$ZrCl$_2$	2,000	1	2.6	5,000	n/a	75.5 (mm)	13

[a]Polymerization temperature.
[b]kg PP/mmol Zr·h.
[c]Not available.

TABLE 3.5

Influence of the Bridge on Bis(indenyl) Zirconocene Precatalysts Activated with MAO in Propylene Polymerization

Precatalyst	Al/Zr Ratio	$T_p{}^a$ (°C)	Activity[b]	M_v	T_m (°C)	mmmm (%)	References
(35) rac-CH$_2$(Ind)$_2$ZrCl$_2$	4,000	50	62	5,300	110	71.4	17c
(36) rac-Me$_2$C(Ind)$_2$ZrCl$_2$	3,000	50	66	11,000	127	80.6	17a
(12) rac-Me$_2$Si(Ind)$_2$ZrCl$_2$	3,000	50	17	56,000	144	90.3	13,17a
(15) rac-CH$_2$CH$_2$(Ind)$_2$ZrCl$_2$	8,000	50	140	33,600	134	87.4	17a
(37) rac-Ph$_2$Si(Ind)$_2$ZrCl$_2$	n/a[c]	70	40	42,000	136	80.5	13
(38) rac-Me$_2$SiMe$_2$Si(Ind)$_2$ZrCl2	n/a	70	3	9,000[d]	73	40.2	13

[a]Polymerization temperature.
[b]kg PP/mmol Zr·h.
[c]Not available.
[d]Recorded as M_w.

The most popular and best studied bridging moieties are the H$_2$C-, Me$_2$C-, C$_2$H$_4$-, and Me$_2$Si- groups. In general, the molecular weight and isotacticity of the polymer produced increases as the catalyst bridge is changed from H$_2$C- (**35**) to Me$_2$C- (**36**) to C$_2$H$_4$- (**15**) to Me$_2$Si- (**12**) (Table 3.5), although there are exceptions (e.g., the rac-Z-(3-MeInd)$_2$ZrCl$_2$ series (**28–30**) discussed above). The effect of changing the methyl groups to other alkyl groups in the R$_2$C- and R$_2$Si-bridged bis(indenyl) framework has also been thoroughly investigated and the effects vary depending on the system. Last, several longer bridges[20] and heteroatom bridges[21] have been investigated, but generally these analogues suffer from poorer catalyst performance as compared to the previously mentioned carbon- and silicon-based bridges.

3.2.1.4 Influence of Changing the Position of the Bridge

Almost all of the known examples of ansa-bis(indenyl) metallocenes consist of two indenyl ligands bridged at the C-1 position of each indenyl ring. There are three major variations to this structural

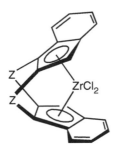

FIGURE 3.5 Generalized structure of a doubly bridged zirconocene. Each Z represents a bridge, for example, CH_2CH_2 or Me_2Si.

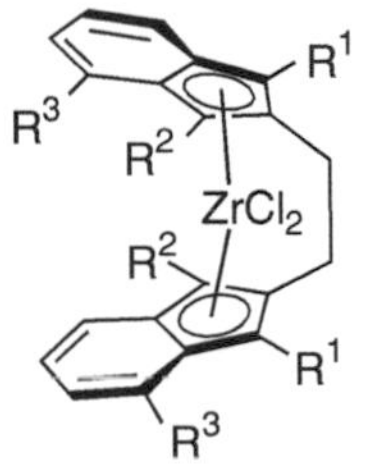

39: R^1, R^2, R^3 = H
40: R^1 = Me, R^2, R^3 = H
41: R^1 = Et, R^2, R^3 = H
42: R^1, R^2 = H, R^3 = Ph
43: R^1 = Me, R^2 = H, R^3 = Ph
44: R^1, R^2 = Me, R^3 = H

FIGURE 3.6 Generalized structure of a 2,2′-bridged *ansa*-metallocene.

framework. The first variation involves the formation of doubly bridged systems that involve tethering the indenyl moieties together at both the C-1 and C-2 positions (Figure 3.5).[22,23] To date, mainly CH_2CH_2- and Me_2Si-bridged analogues have been synthesized; however very little polymerization work has been performed. Not only do these species suffer from a challenging synthetic route, but they also have thus far proven to be very poor catalysts that decompose quite readily to singly bridged systems.[23] (For additional information on doubly bridged bis(indenyl) metallocenes, see Chapter 4.)

A second approach has involved moving the position of the bridge from C-1 to C-2. Several of these substituted 2,2′-bridged bis(indenyl) complexes of Ti and Zr have been prepared by both Nantz[24] and Schaverien[25] and their polymerization behaviors have been studied. The bridge in each case was an ethylene moiety and substitution, when present, was located at a combination of C-1, C-3, or C-4 (**39–44**) (Figure 3.6). Although these systems were active for ethylene polymerization, the attempted polymerization of propylene resulted only in oligomers (M_n = 400–900 Da) with very little stereoregularity along the backbone (Table 3.6).[25]

Halterman,[26] and Bosnich[27] have also synthesized some specialized biaryl strapped 2,2′-bis(indenyl) metallocenes (**45–48**) (Figure 3.7), but the routes are quite demanding synthetically and very little has been done in the way of polymerization studies. The polymerization studies that have been done have generally shown these derivatives to display poor activities and to produce oligomers of low stereoregularity similar to the other C-2 bridged systems shown in Table 3.6.

Lastly, the most successful modification to the position of the bridge has involved moving the bridge from the C-1 to the C-4 or C-7 positions of the indenyl ligands (**49–55**) (Figure 3.8).[28,29] The best reported catalyst of this type, *rac*-$Me_2Si\{4,4′$-$(3$-Me-1-PhInd)$\}_2ZrCl_2$ (**50**), has been shown to produce iPP with an *mmmm* value of 98% and a molecular weight of 60,000 Da. This catalyst also exhibits activities comparable to those for several well-known C-1 bridged indenyl analogues.[28]

TABLE 3.6

C-2-Bridged Bis(indenyl) Zirconocene Precatalysts Activated with MAO in Propylene Polymerization

Precatalyst	Al/Zr Ratio	$T_p{}^a$ (°C)	Activity[b]	M_n	mm (%)	References
(39) $CH_2CH_2(2\text{-Ind})_2ZrCl_2$	10,000	50	530	400	n/a[c]	25
(40r) rac-$CH_2CH_2(1\text{-Me-2-Ind})_2ZrCl_2$	40,200	50	6,240	850	n/a	25
(40m) $meso$-$CH_2CH_2(1\text{-Me-2-Ind})_2ZrCl_2$	40,200	50	4,160	1,800	22.0	25
(41r) rac-$CH_2CH_2(1\text{-Et-2-Ind})_2ZrCl_2$	40,000	50	1,320	1,600	36	25
(41m) $meso$-$CH_2CH_2(1\text{-Et-2-Ind})_2ZrCl_2$	40,000	50	5,080	1,700	23	25
(42) rac-$CH_2CH_2(4\text{-Ph-2-Ind})_2ZrCl_2^d$	40,000	50	330	1,100	62.2	25
(43) rac-$CH_2CH_2(1\text{-Me-4-Ph-2-Ind})_2ZrCl_2^d$	40,000	50	1,080	1,000	27.8	25
(43m) $meso$-$CH_2CH_2(1\text{-Me-4-Ph-2-Ind})_2ZrCl_2$	40,000	50	900	1,000	26.3	25
(44) $CH_2CH_2(1,3\text{-Me}_2\text{-2-Ind})_2ZrCl_2$	4,000	50	5,040	1,900	40.3	25
(45) $1,1'$-biphenyl$(2\text{-Ind})_2ZrCl_2$	40,195	50	114	1,700	44.2	25

[a] Polymerization temperature.
[b] kg PP/g Zr·h.
[c] Not available.
[d] Polymerizations performed with a 4:1 rac:$meso$ ratio.

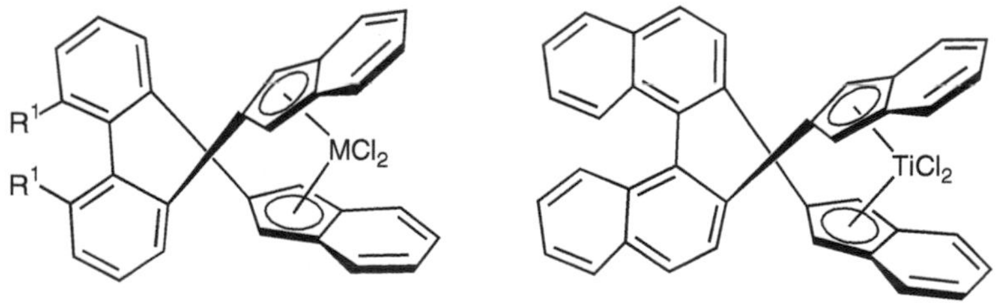

45: R^1 = H, M = Zr
46: R^1 = H, M = Ti
47: R^1 = Me, M = Ti

48

FIGURE 3.7 Representative structures of biaryl strapped 2,2′-bridged *ansa*-metallocenes.

3.2.2 C_1-SYMMETRIC *ANSA*-METALLOCENES

By definition, a C_1-symmetric compound will lack any symmetry elements, which allows for the possibility of a wide range of structures. C_1-symmetric metallocenes like their C_2-symmetric analogues can produce all degrees of molecular weight and tacticities. However, C_1-metallocenes offer an even greater degree of ligand structure variation, which should in turn allow for greater "tunability" of the resulting polymer, including the production of materials with hemiisotactic microstructures and elastomeric properties.

For the purpose of this chapter, two types of C_1-symmetric metallocenes are considered. The first is a "mixed ligand metallocene" in which the metal is bound to a Cp ligand and an indenyl ligand connected to one another through a bridging unit. The second is a metallocene in which the metal is bound to two indenyl ligands (linked by a bridging unit), but the substituents on each indenyl ligand differ from one another, thus eliminating the symmetry in the molecule. A third type of C_1 *ansa*-metallocene containing a fluorene ligand has been extensively studied and is reviewed elsewhere. (For additional information on fluorene containing metallocenes, see Chapter 2.)

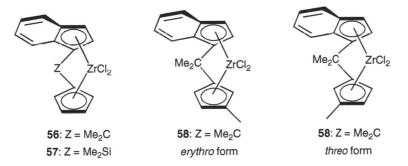

4,4'-bridged metallocenes

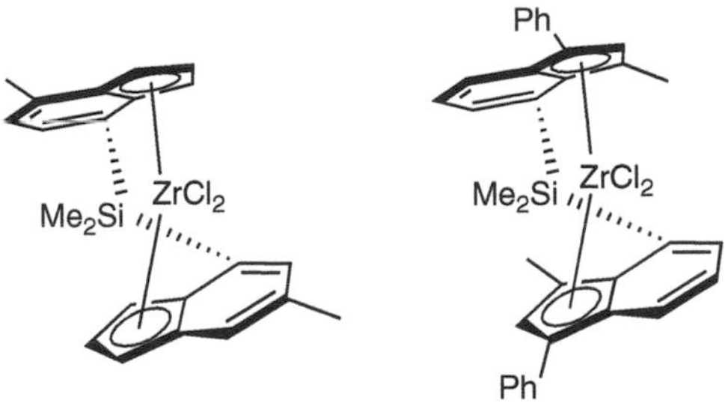

49   50

7,7'-bridged metallocenes

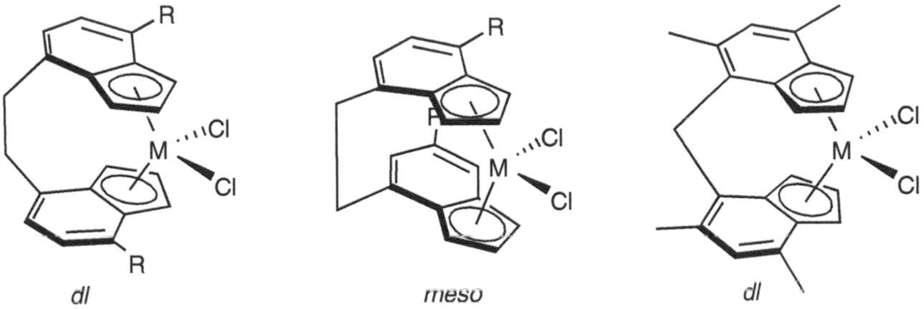

dl   *meso*   *dl*

51: R = Me   **53**: R = Me   **55**
52: R = *i*-Pr   **54**: R = *i*-Pr

FIGURE 3.8   Some representative examples of 4,4' and 7,7'-bridged *ansa*-metallocenes.

56: Z = Me$_2$C   **58**: Z = Me$_2$C   **58**: Z = Me$_2$C
57: Z = Me$_2$Si   *erythro* form   *threo* form

FIGURE 3.9 C_1-Symmetric Cp-indenyl mixed metallocene catalysts showing poor propylene polymerization behavior.

3.2.2.1   Cp-Indenyl Metallocenes

The simplest known examples of the class of mixed indenyl-Cp metallocenes are the isopropylidene-(Z = Me$_2$C-, **56**) and dimethylsilylene-bridged (Z = Me$_2$Si-, **57**) cyclopentadienyl-indenyl zirconocene dihalides (Figure 3.9). Many polymerization studies have been performed on these unsubstituted systems, but unfortunately, the published studies demonstrate poor catalytic behavior for propylene polymerizations.[30] Even the substituted analogue of the isopropylidene-bridged complex, *erythro/threo*-Me$_2$C(3-MeCp)(Ind)ZrCl$_2$ (**58**),[31] exhibits the same low catalytic activity and low polypropylene molecular weight observed for the unsubstituted parent system.

Miyake et al.[19] have had some success developing mixed metallocenes that produce iPP ($mm =$ 51–99.6%), albeit with low molecular weights (ca. 10–100 kDa). These catalysts, with the general formula Z-(3-t-BuCp)(3-t-BuInd)MCl$_2$ (where Z = Me$_2$C, or Me$_2$Si- and M = Ti, Zr, Hf (**59–62**) (Figure 3.10) possess bulky *tert*-butyl groups on both the indenyl and cyclopentadienyl moieties of the metallocene framework which leads to increased isoselectivity of the resulting iPP. An unusual feature of these catalysts is that the molecular weight of the resulting iPP exhibits a strong dependence on the polymerization temperature decreasing from 105,000 Da at 1°C to 9,000 Da at 60 °C while the isoselectivity remains relatively unchanged (Table 3.7).[19]

Interestingly, the Ti analogue **60** is more active (1950 g PP/mmol Zr·h) than the analogous zirconocene **59t** (620 g PP/mmol Zr·h), yet retains very high isoselectivity (mm = 99.6% for each species). This observation is unusual, since most zirconocene catalysts are more active than their titanocene analogues. Furthermore, the analogous hafnocene **61** follows the typical trend seen for most metallocene catalysts, exhibiting a large decrease in polymerization activity (30 g PP/mmol Zr·h) relative to the zirconocene analogue **59t**.[19] Last, when the bridging moiety is changed from an

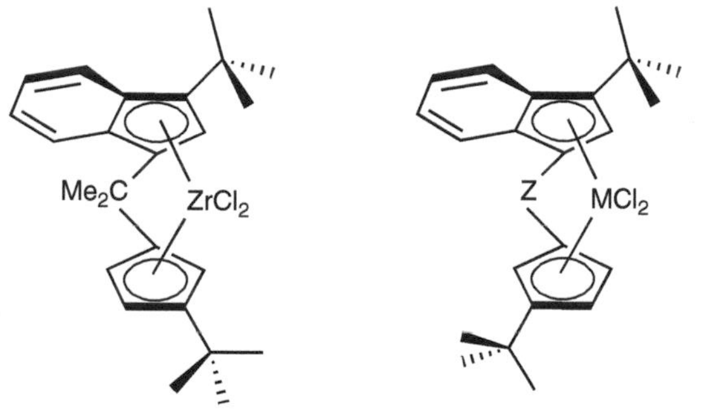

erythro-form

59e

threo-form

59t: Z = Me$_2$C, M = Zr
60: Z = Me$_2$C, M = Ti
61: Z = Me$_2$C, M = Hf
62: Z = Me$_2$Si, M = Zr

FIGURE 3.10 C_1-Symmetric isoselective (3-*tert*-butyl)Cp-(3-*tert*-butyl)indenyl mixed metallocene catalysts.

TABLE 3.7
Unsymmetrical Metallocene Precatalysts Activated with MAO in Propylene Polymerization

Precatalyst	Al/Zr Ratio	T_p[a] (°C)	Activity[b]	M_n	mm (%)	References
(**59t**) *threo*-Me$_2$C(3-t-BuCp)(3-t-BuInd)ZrCl$_2$	2,000	1	620	105,000	99.6	19
(**59t**) *threo*-Me$_2$C(3-t-BuCp)(3-t-BuInd)ZrCl$_2$	2,000	60	42,000	9,000	99.2	19
(**59e**) *erythro*-Me$_2$C(3-t-BuCp)(3-t-BuInd)ZrCl$_2$	2,000	1	60	9,000	51.8	19
(**60t**) *threo*-Me$_2$C(3-t-BuCp)(3-t-BuInd)TiCl$_2$	2,000	1	1,950	34,100	99.6	19
(**61t**) *threo*-Me$_2$C(3-t-BuCp)(3-t-BuInd)HfCl$_2$	2,000	1	30	39,000	99.5	19
(**62t**) *threo*-Me$_2$Si(3-t-BuCp)(3-t-BuInd)ZrCl$_2$	2,000	1	110	28,000	98.6	19
(**56**) Me$_2$C(Cp)(Ind)ZrCl$_2$	2,000	1	730	1,100	35.2	19

[a] Polymerization temperature.
[b] g PP/mmol Zr·h.

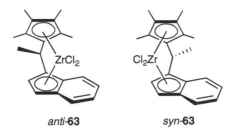

anti-**63**   syn-**63**

FIGURE 3.11 Example syn/anti C_1-symmetric mixed Cp/indenyl metallocene precatalysts for the polymerization of thermoplastic-elastomeric polypropylene.

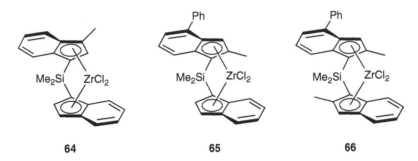

64   **65**   **66**

FIGURE 3.12 C_1-Symmetric bis(indenyl) mixed metallocenes.

isopropylidene **59t** to dimethylsilylene group **62**, a corresponding decrease in tacticity, molecular weight, and activity is observed (Table 3.7).[19]

As was mentioned earlier, C_1-symmetric catalysts allow for the unique opportunity to produce novel polypropylenes of varying microstructures and properties. This control of polymer properties has been nicely demonstrated by Rausch et al. in the synthesis of thermoplastic-elastomeric polypropylene (TPE-PP).[32] Novel TPE-PP consisting of isotactic and atactic blocks of polypropylene was produced by using the *syn* and *anti* isomers of the mixed metallocene catalyst, $(CH_3)CH(Me_4Cp)(Ind)ZrCl_2$ (**63**) (Figure 3.11). The polypropylene that is produced has a very low crystallinity (low isotacticity) and melting temperature (*mmmm* = 40%; T_m = 50–65 °C).[32a] These initial studies by Rausch et al. have led to the development of several related C_1-symmetric, mixed metallocene catalysts capable of producing elastomeric polypropylene. (For additional information on elastomeric polypropylene, see Chapter 8.)

3.2.2.2 Mixed Bis(indenyl) Metallocenes

There are only a few examples of metallocenes having two differently substituted indenyls as part of their ligand framework. Remarkably, these complexes have demonstrated decent catalytic behavior for the polymerization of propylene. For example, anti-*rac*-$Me_2Si(Ind)(3-MeInd)ZrCl_2$ (**64**) (Figure 3.12) has been shown to produce elastomeric PP with *mmmm* contents ranging from 30–50% depending on polymerization temperature and monomer concentration.[33]

Spaleck et al. have produced the best catalysts of this type to date. The metallocenes, *rac*-$Me_2Si(Ind)(2-Me-4-PhInd)ZrCl_2$ (**65**) and *rac*-$Me_2Si(2-MeInd)(2-Me-4-PhInd)ZrCl_2$ (**66**),[34] combine individual indenyl moieties that have previously demonstrated excellent catalytic properties in C_2-symmetric metallocenes. These systems give rise to polypropylene having both high molecular weight and high isotacticity. For example, (**66**) produces iPP with molecular weights greater than 500,000 Da and *mm* values exceeding 96%. All of these findings can be fully explained by molecular modeling analyses.[35]

3.3 CONCLUSIONS

This chapter has provided a brief overview of factors that one should consider when designing substituted indenyl and cyclopentadienyl catalysts for stereoselective propylene polymerization. Symmetry of the catalyst precursor, substitution patterns around the cyclopentadienyl framework, position and type of bridging moiety, and type of ligand employed all appear to be critical factors in tuning the properties of the resulting polypropylene. The most studied and best performance catalysts to date belong to the *ansa*-bis(indenyl) class of metallocenes. Both C_1- and C_2-symmetric analogues of this class have been prepared which have been shown to be highly active for the polymerization of polypropylene. Polymer tacticities ranging from completely atactic to isotactic are possible and the specific tuning of the polymer properties can be accomplished through modification of the groups surrounding the metal framework. Bridging at the 1 position of the indenyl rings has proved to be the most successful strategy using either a one- or two-atom bridge. Lastly, the most significant effects on the properties of the resulting polymer can be obtained through modification of the substituent groups, particularly in the 2, 3, 4, and 7 positions of the indenyl ring.

REFERENCES AND NOTES

1. Wilkinson, G.; Pauson, P. L.; Birmingham, J. M.; Cotton, F. A. Biscyclopentadienyl derivatives of some transition elements. *J. Am. Chem. Soc.* **1953**, *75*, 1011–1012.

2. (a) Ewen, J. A. Ligand effects on metallocene catalyzed Ziegler-Natta polymerizations. In *Catalytic Polymerization of Olefins, Studies in Surface Science and Catalysis*; Keii, T., Soga, K., Eds.; Elsevier: New York, 1986; pp 271–292. (b) Giannetti, E.; Nicoletti, G.; Mazzocchi, R. Homogeneous Ziegler-Natta catalysis. II. Ethylene polymerization by IVB transition metal complexes/methyl aluminoxane catalyst systems. *J. Polym. Sci. Part A: Polym. Chem.* **1985**, *23*, 2117–2134.

3. (a) Ewen, J. A. Mechanisms of stereochemical control in propylene polymerizations with soluble Group 4B metallocene/methylalumoxane catalysts. *J. Am. Chem. Soc.* **1984**, *106*, 6355–6364. (b) Kaminsky, W.; Külper, K.; Brintzinger, H.; Wild, F. Polymerization of propene and butene with a chiral zirconocene and methylaluminoxane as cocatalyst. *Angew. Chem., Int. Ed. Engl.* **1985**, *24*, 507–508.

4. Collins, S.; Gauthier, W. J.; Holden, D. A.; Kuntz, B. A.; Taylor, N. J.; Ward, D. G. Variation of polypropylene microtacticity by catalyst selection. *Organometallics* **1991**, *10*, 2061–2068.

5. Mise, T.; Miya, S.; Yamazaki, H. Excellent stereoregular isotactic polymerizations of propylene with C_2-symmetric silylene-bridged metallocene catalysts. *Chem. Lett.* **1989**, 1853–1856.

6. Resconi, L.; Piemontesi, F.; Nifant'ev, I.; Ivchenko, P. Metallocene compounds, process for their preparation, and their use in catalysts for the polymerization of olefins. PCT Int. Pat. Appl. WO 96/22995 (Montell), August 1, 1996.

7. Spaleck, W.; Kuber, F.; Winter, A.; Rohrmann, J.; Bachmann, B.; Antberg, M.; Dolle, V.; Paulus, E. The influence of aromatic substituents on the polymerization behavior of bridged zirconocene catalysts. *Organometallics* **1994**, *13*, 954–963.

8. Mansel, S.; Rief, U.; Prosenc, M.-H.; Kirsten, R.; Brintzinger, H.-H. *ansa*-Metallocene derivatives. XXXII. Zirconocene complexes with a spirosilane bridge: Syntheses, crystal structures and properties as olefin polymerization catalysts. *J. Organomet. Chem.* **1996**, *512*, 225–236.

9. (a) Huttenloch, M. E.; Dorer, B.; Rief, U.; Prosenc, M.-H.; Schmidt, K.; Brintzinger, H.-H. *ansa*-Metallocene derivatives. XXXIX. Biphenyl-bridged metallocene complexes of titanium, zirconium, and vanadium: Syntheses, crystal structures and enantioseparation. *J. Organomet. Chem.* **1997**, *541*, 219–232. (b) Huttenloch, M. E.; Diebold, J.; Rief, U.; Brintzinger, H.-H.; Gilbert, A. M.; Katz, T. J. *ansa*-Metallocene derivatives. 26. Biphenyl-bridged metallocenes that are chiral, configurationally stable, and free of diastereomers. *Organometallics* **1992**, *11*, 3600–3607.

10. For some leading reviews see (a) Resconi, L.; Cavallo, L.; Fait, A.; Piemontesi, F. Selectivity in propene polymerization with metallocene catalysts. *Chem. Rev.* **2000**, *100*, 1253–1345. (b) Alt, H. G.; Köppl, A. Effect of the nature of metallocene complexes of Group IV metals on their performance in catalytic ethylene and propylene polymerization. *Chem. Rev.* **2000**, *100*, 1205–1221. (c) Halterman, R. L. Synthesis and applications of chiral cyclopentadienylmetal complexes. *Chem. Rev.* **1992**, *92*, 965–994. (d) Halterman, R. L. Synthesis of chiral titanocene and zirconocene dichlorides. In *Metallocenes*;

Halterman, R. L., Togni, A., Eds.; Wiley-VCH: New York, 1998; pp 455–544. (e) Chirik, P. J.; Bercaw, J. E. Group 3 metallocenes. In *Metallocenes*; Togni, A., Halterman, R. L., Eds.; Wiley-VCH: New York, 1998; pp 111–152. (f) Cardin, D. J.; Lappert, M. F.; Raston, C. L. *Chemistry of Organo-Zirconium and -Hafnium Compounds*; Wiley: New York, 1986.

11. Maciejewski Petoff, J. L.; Agoston, T.; Lal, T. K.; Waymouth, R.M. Elastomeric polypropylene from unbridged 2-arylindenyl zirconocenes: Modeling polymerization behavior using *ansa*-metallocene analogs. *J. Am. Chem. Soc.* **1998**, *120*, 11316–11322.

12. Deng, H.; Winkelbach, H.; Taeji, K.; Kaminsky, W.; Soga, K. Synthesis of high-melting, isotactic polypropene with C_2- and C_1-symmetrical zirconocenes. *Macromolecules* **1996**, *29*, 6371–6376.

13. Spaleck, W.; Antberg, M.; Aulbach, M.; Bachmann, B.; Dolle, V.; Haftka, S.; Küber, F.; Rohrmann, J.; Winter, A. New isotactic polypropylenes via metallocene catalysts. In *Ziegler Catalysts*; Fink, G., Mülhaupt, R., Brintzinger, H.-H., Eds.; Springer-Verlag: Berlin, 1995; pp 83–97.

14. Resconi, L.; Camurati, I.; Sudmeijer, O. Chain transfer reactions in propylene polymerization with zirconocene catalysts. *Top. Catal.* **1999**, *7*, 145–163.

15. Lee, I.- M.; Gauthier, W. J.; Ball, J. M.; Iyengar, B.; Collins, S. Electronic effects in Ziegler–Natta polymerization of propylene and ethylene using soluble metallocene catalysts. *Organometallics* **1992**, *11*, 2115–2122.

16. Kashiwa, N.; Kojoh, S.; Imuta, J.; Tsutsui, T. Characterization of PP prepared with the latest metallocene and $MgCl_2$-supported $TiCl_4$ catalyst systems. In *Metalorganic Catalysts for Synthesis and Polymerization*; Kaminsky, W., Ed.; Springer-Verlag: Berlin, 1999; pp 30–37.

17. (a) Resconi, L.; Piemontesi, F.; Camurati, I.; Sudmeijer, O.; Nifant'ev, I. E.; Ivchenko, P. V.; Kuz'mina, L. G. A new class of isospecific, highly regiospecific zirconocene catalysts for the polymerization of propene. *J. Am. Chem. Soc.* **1998**, *120*, 2308–2321. (b) Spaleck, W.; Antberg, M.; Dolle, V.; Klein, R.; Rohrmann, J.; Winter, A. Stereorigid metallocenes: Correlations between structure and behavior in homopolymerizations of propylene. *New J. Chem.* **1990**, *14*, 499–503. (c) Balboni, D.; Moscardi, G.; Nifant'ev, I.; Baruzzi, G.; Angeli, D.; Resconi, L. C_2-Symmetric zirconocenes for high molecular weight amorphous polypropylene. *Polym. Prepr. (Am. Chem. Soc., Div. Polym. Chem.)* **2000**, *41*, 1920–1921.

18. Resconi, L.; Balboni, D.; Baruzzi, G.; Fiori, C.; Guidotti, S. *rac*-[Methylene(3-*tert*-butyl-1-indenyl)$_2$]ZrCl$_2$: A simple, high-performance zirconocene catalyst for isotactic polypropene. *Organometallics* **2000**, *19*, 420–429.

19. Miyake, S.; Okumura, Y.; Inazawa, S. Highly isospecific polymerization of propylene with unsymmetrical metallocene catalysts. *Macromolecules* **1995**, *28*, 3074–3079.

20. For some representative examples see: (a) Han, T. K.; Woo, B. W.; Park, J. T.; Do, Y.; Ko, Y. S.; Woo, S. I. Ethylene and propylene polymerization over chiral *ansa*-dichloro[*o*-phenylenedimethylenebis(η^5-1-indenyl)]zirconium [Zr{C$_6$H$_4$(CH$_2$-1-C$_9$H$_6$)$_2$-1,2}Cl$_2$]. *Macromolecules* **1995**, *28*, 4801–4805. (b) Antberg, M.; Spaleck, W.; Rohrmann, J.; Lueker, H.; Winter, A. Catalysts for polymerization of ethylene with propylene. Ger. Offen. DE 3836059 (Hoechst), May 3, 1990.

21. For some representative examples see: (a) Schaverien, C. J.; Ernst, R.; Terlouw, W.; Schut, P.; Sudmeijer, O.; Budzelaar, P. H. M. Phosphorus-bridged metallocenes: New homogeneous catalysts for the polymerization of propene. *J. Mol. Catal. A: Chem.* **1998**, *128*, 245–256. (b) Reetz, M. T.; Willuhn, M.; Psiorz, C.; Goddard, R. Donor complexes of bis(1-indenyl)phenylborane dichlorozirconium as isospecific catalysts in propene polymerization. *Chem. Commun.* **1999**, 1106–1107. (c) Ashe, A. J., III.; Fang, X.; Kampf, J. W. Aminoboranediyl-bridged zirconocenes: Highly active olefin polymerization catalysts. *Organometallics* **1999**, *18*, 2288–2290. (d) Alt, H. G.; Jung, M. PPh-bridged metallocene complexes of the type (C$_{13}$H$_8$-PPh-C$_{13}$H$_8$)MCl$_2$ (M=Zr, Hf). *J. Organomet. Chem.* **1998**, *568*, 127–131.

22. Halterman, R. L.; Tretyakov, A.; Combs, D.; Chang, J.; Khan, M. Synthesis and structure of C_2-symmetric, doubly bridged bis(indenyl)titanium and -zirconium dichlorides. *Organometallics* **1997**, *16*, 3333–3339.

23. Mengele, W.; Diebold, J.; Troll, C.; Röll, W.; Brintzinger, H.-H. *ansa*-Metallocene derivatives. 27. Chiral zirconocene complexes with two dimethylsilylene bridges. *Organometallics* **1993**, *12*, 1931–1935.

24. Hitchcock, S. R.; Situ, J. J.; Covel, J. A.; Olmstead, M. M.; Nantz, M. H. Synthesis of *ansa*-titanocenes from 1,2-bis(2-indenyl)ethane and structural comparisons in the catalytic epoxidation of unfunctionalized alkenes. *Organometallics* **1995**, *14*, 3732–3740.

25. (a) Schaverien, C. J.; Ernst, R.; van Loon, J.-D.; Dall'Occo, T. Bridged zirconocene compounds, process for their preparation, and their use as catalyst components in the polymerization of olefins. European Patent 0941997 B1 (Montell), December 18, 2002. (b) Schaverien, C. J.; Ernst, R.; Schut, P.; Dall'Occo, T. Ethylene bis(2-indenyl)zirconocenes: A new class of diastereomeric metallocenes for the (co)polymerization of α-olefins. *Organometallics* **2001**, *20*, 3436–3452.

26. Halterman, R. L.; Ramsey, T. M. Asymmetric synthesis of a sterically rigid binaphthyl-bridged chiral metallocene: Asymmetric catalytic epoxidation of unfunctionalized alkenes. *Organometallics* **1993**, *12*, 2879–2880.

27. Ellis, W. W.; Hollis, T. K.; Odenkirk, W.; Whelan, J.; Ostrander, R.; Rheingold, A. L.; Bosnich, B. Synthesis, structure, and properties of chiral titanium and zirconium complexes bearing biaryl strapped substituted cyclopentadienyl ligands. *Organometallics* **1993**, *12*, 4391–4401.

28. (a) Kato, T.; Uchino, H.; Iwama, N.; Imaeda, K.; Kashimoto, M.; Osano, Y.; Sugano, T. Synthesis of novel complex with bridged bis(indenyl)ligand and its polymerization behavior of propylene. In *Metalorganic Catalysts for Synthesis and Polymerization*; Kaminsky, W., Ed.; Springer-Verlag: Berlin, 1999; pp 192–199. (b) Kato, T.; Uchino, H.; Iwama, N.; Osano, Y.; Sugano, T. Synthesis of novel *ansa*-metallocene complexes with bridged bis(indenyl) ligand and its application for olefin polymerization. *Studies Sur. Sci. Catalysis* **1999**, *121* (*Sci. Technol. Catalysis* 1998), 473–476. (c) Schulze, U.; Arndt, M.; Freidanck, F.; Beulich, I.; Pompe, G.; Meyer, E.; Jehnichen, D.; Pionteck, J.; Kaminsky, W. Structure and properties of ethene copolymers synthesized by metallocene catalysts. *J. Macromol. Sci. ,Pure and Applied Chemistry* **1998**, *A35*(7 and 8), 1037–1044. (d) Halterman, R. L.; Combs, D.; Khan, M. A. Synthesis of *C*7,*C*7′-ethylene- and *C*7,*C*7′-methylene-bridged *C*2-symmetric bis(indenyl)zirconium and -titanium dichlorides. *Organometallics* **1998**, *17*, 3900–3907.

29. Schaverien, C.; Ernst, R.; Schut, P.; Skiff, W.; Resconi, L.; Barbassa, E.; Balboni, D. et al. A new class of chiral bridged metallocene: Synthesis, structure, and olefin (co)polymerization behavior of *rac*- and *meso*-1,2-CH₂CH₂{4-(7-Me-indenyl)}₂ZrCl₂. *J. Am. Chem. Soc.* **1998**, *120*, 9945–9946.

30. (a) Fierro, R.; Chien, J. C. W.; Rausch, M. D. Asymmetric zirconocene precursors for catalysis of propylene polymerization. *J. Polym. Sci., Part A: Polym. Chem.* **1994**, *32*, 2817–2824. (b) Gauthier, W. J.; Collins, S. Elastomeric Poly(propylene): Propagation models and relationship to catalyst structure. *Macromolecules* **1995**, *28*, 3779–3786. (c) Gauthier, W. J.; Corrigan, J. F.; Taylor, N. J.; Collins, S. Elastomeric poly(propylene): Influence of catalyst structure and polymerization conditions on polymer structure and properties. *Macromolecules* **1995**, *28*, 3771–3778. (d) Gauthier, W. J.; Collins, S. Preparation of elastomeric polypropylene using metallocene catalysts: Catalyst design criteria and mechanisms for propagation. *Macromol. Symp.* **1995**, *98*, 223–231. (e) Montag, P.; van der Leek, Y.; Angermund, K.; Fink, G. Mechanistic study of propene polymerization with the catalytic system [2,4-Cyclopentadien-1-ylidene(methylethylidene)-1H-inden-1-ylidene]zirconium dichloride-methylaluminoxane. *J. Organomet. Chem.* **1995**, *497*, 201–209. (f) Antberg, M.; Dolle, V.; Klein, R.; Rohrmann, J.; Spaleck, W.; Winter, A. Propylene polymerization by stereorigid metallocene catalysts: Some new aspects of the metallocene structure/polypropylene microstructure correlation. In *Catalytic Olefin Polymerization, Studies in Surface Science and Catalysis*; Keii, T., Soga, K., Eds.; Kodansha-Elsevier: Tokyo, 1990; pp 501–515. (g) Green, M. L. H.; Ishihara, N. New *ansa*-metallocenes of the group 4 transition metals as homogeneous catalysts for the polymerization of propene and styrene. *J. Chem. Soc., Dalton Trans.* **1994**, 657–665.

31. Kaminsky, W.; Engehausen, R.; Zoumis, K.; Spaleck, W.; Rohrmann, J. Standardized polymerizations of ethylene and propene with bridged and unbridged metallocene derivatives: A comparison. *Makromol. Chem.* **1992**, *193*, 1643–1651.

32. (a) Mallin, D. T.; Rausch, M. D.; Lin, Y.-G.; Dong, S.; Chien, J. C. W. *rac*-[Ethylidene(1-η⁵-tetramethylcyclopentadienyl)(1-η⁵-indenyl)]dichlorotitanium and its homopolymerization of propylene to crystalline-amorphous block thermoplastic elastomers. *J. Am. Chem. Soc.* **1990**, *112*, 2030–2031. (b) Chien, J. C. W.; Llinas, G. H.; Rausch, M. D.; Lin, Y.-G.; Winter, H. H.; Atwood, J. L.; Bott, S. G. Metallocene catalysts for olefin polymerization. XXIV. Stereoblock propylene polymerization catalyzed by *rac*-[anti-ethylidene(1-η⁵-tetramethylcyclopentadienyl)(1-η⁵-indenyl)dimethyltitanium: A two-state propagation. *J. Polym. Sci., Part A: Polym. Chem.* **1992**, *30*, 2601–2617. (c) Llinas, G. H.; Chien, J. C. W. Comparison of polymerization catalyzed by the *syn* and *anti* diastereomers of [ethylidene(1-η⁵-tetramethylcyclopentadienyl)(1-η⁵-indenyl)]titanium dichloride and methylaluminoxane. *Polym. Bull.* **1992**, *28*, 41–45. (d) Chien, J. C. W.; Llinas, G. H.;

Rausch, M. D.; Lin, Y.-G.; Winter, H. H. Two-state propagation mechanism for propylene polymerization catalyzed by *rac*-[anti-ethylidene(1-η^5-tetramethylcyclopentadienyl)(1-η^5-indenyl)] dimethyltitanium. *J. Am. Chem. Soc.* **1991**, *113*, 8569–8570. (e) Llinas, G. H.; Day, R. O.; Rausch, M. D.; Chien, J. C. W. Ethylidene(1-η^5-tetramethylcyclopentadienyl)(1-η^5-indenyl)dichlorozirconium: Synthesis, molecular structure, and polymerization catalysis. *Organometallics* **1993**, *12*, 1283–1288. (f) Babu, G. N.; Newmark, R. A.; Cheng, H. N.; Llinas, G. H.; Chien, J. C. W. Microstructure of elastomeric polypropylenes obtained with nonsymmetric *ansa*-titanocene catalysts. *Macromolecules* **1992**, *25*, 7400–7402.

33. Bravakis, A. M.; Bailey, L. E.; Pigeon, M.; Collins, S. Synthesis of elastomeric poly(propylene) using unsymmetrical zirconocene catalysts: Marked reactivity differences of "*rac*"- and "*meso*"-like diastereomers. *Macromolecules* **1998**, *31*, 1000–1009.

34. Spaleck, W.; Kuber, F.; Bachmann, B.; Fritze, C.; Winter, A. New bridged zirconocenes for olefin polymerization: Binuclear and hybrid structures. *J. Mol. Catal. A: Chem.* **1998**, *128*, 279–287.

35. (a) Corradini, P.; Cavallo, L.; Guerra, G. Molecular modeling studies of stereospecificity and regiospecificity of propene polymerizations by metallocences. In *Metallocene-Based Polyolefins: Preparation, Properties, and Technology*; Kaminsky, W., Scheirs, J., Eds.; Wiley: New York, 2000; Vol. 2, pp 3–36. (b) Yoshida, T.; Koga, N.; Morokuma, K. A combined *ab initio* MO-MM study on isotacticity control in propylene polymerization with silylene-bridged Group 4 metallocenes. C_2 symmetrical and asymmetrical catalysts. *Organometallics* **1996**, *15*, 766–777. (c) van der Leek, Y.; Angermund, K.; Reffke, M.; Kleinschmidt, R.; Goretzki, R.; Fink, G. On the mechanism of stereospecific polymerization—development of a universal model to demonstrate the relationship between metallocene structure and polymer microstructure. *Chem. Eur. J.* **1997**, *3*, 585–591.

4 Doubly Bridged Metallocenes for Stereoselective Propylene Polymerization

Deanna L. Zubris

CONTENTS

4.1 DOUBLY BRIDGED METALLOCENES

Many academic and industrial researchers have probed the utility of group 3 and group 4 metallocenes as catalyst precursors for the polymerization of ethylene and α-olefins such as propylene.[1–4] In particular, much emphasis has been placed on the use of an interannular bridge, or linking group, between the two cyclopentadienyl rings in these metallocenes. These complexes are known as *ansa*-metallocenes, a term that Brintzinger introduced in his landmark work in this field.[5] The term *ansa* (Latin, for "bent handle") may be used to describe metallocenes with either one or two interannular bridges (Figure 4.1).

The presence of an interannular bridge causes a geometric, and hence, steric distortion in these catalysts as compared to unlinked metallocenes. *ansa*-Metallocenes are also electronically different from their unlinked counterparts; these differences were recently treated in the literature.[6] Variations in geometry and electronics can affect reactivity, and in many cases, also affect activity and stereo-control in the polymerization of α-olefins. A recent review article summarizes the broad array of interannular bridges that have been incorporated in group 3 and 4 metallocenes; consequences for reactivity are also treated.[7] While many group 3 and 4 singly bridged *ansa*-metallocenes are known,

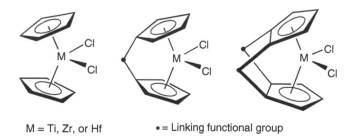

M = Ti, Zr, or Hf   • = Linking functional group

FIGURE 4.1 Group 4 metallocenes can be unlinked, singly linked, or doubly linked. These metallocene dichloride complexes are common precatalysts for olefin polymerization.

examples of doubly bridged *ansa*-metallocenes are relatively rare. A review of doubly bridged bis(cyclopentadienyl) metal complexes, with a structural focus, was published in 2003.[8]

Few doubly bridged group 3 and 4 *ansa*-metallocenes have been tested as precatalysts for the polymerization of ethylene or α-olefins. These complexes fall into the following three categories: (1) the precatalyst can produce polyethylene, but tests for α-olefin polymerization are not reported; (2) the precatalyst can produce polyethylene, but does not polymerize α-olefins; and (3) the precatalyst polymerizes both ethylene and α-olefins. Some complexes in this third class serve as outstanding precatalysts for making polypropylene with regard to catalyst activity, polymer molecular weight (M_w or M_n), polydispersity index (PDI, M_w/M_n) and tacticity.

This chapter will discuss all known group 3 and 4 doubly bridged *ansa*-metallocenes made to date. When polymerization data is unavailable, comments will be made on the perceived viability of the compounds as precatalysts for α-olefin polymerization based on their structure and symmetry. For the α-olefin polymerization precatalysts described herein, the correlation between catalyst structure and polymer tacticity will be discussed. Further, the correlation between catalyst structure and regiocontrol, polymerization activity, and polymer molecular weight will be addressed when there is pertinent data present for a given precatalyst.

4.1.1 GENERAL DESCRIPTION OF BONDING IN DOUBLY BRIDGED *ANSA*-METALLOCENES

Structural information is known for many doubly bridged *ansa*-metallocenes. The presence of one or two interannular bridges causes distortion from unlinked metallocene geometry (*vide infra*). In part, these distortions may be quantified by the angles depicted in Figure 4.2. Angles α, β, τ, and γ are often reported.[6] The interplanar cyclopentadienyl ring angle (dihedral angle between the ring planes) is represented by α. The angle between the normals to the ring centroids (Cp_{norm}–Cp_{norm}) is designated β, where $α + β = 180°$. The Cp_{cent}–metal–Cp_{cent} angle (where Cp_{cent} is the cyclopentadienyl centroid) is referred to as γ.

Ring slippage, defined as a cyclopentadienyl coordination mode of less than $η^5$, can be ascertained by comparison of β and γ. Specifically, the tilt angle can be used to verify whether or not a linked metallocene exhibits ring slippage when compared to its unlinked counterpart. The tilt angle, τ, is defined as $(γ − β)/2$. Angle α is also related to τ by the following expression: $α = (2τ − γ + 180°)$. A linked metallocene will tend to have a larger value of τ than its unlinked counterpart. In addition, a doubly linked metallocene will tend to have a larger value of τ than its singly linked counterpart does. There are two caveats when using τ as a measure of ring slippage: (1) two-atom bridges (or greater) tend to provide τ values that are comparable to unlinked metallocenes, and (2) substituents on the cyclopentadienyl ligands can lead to unexpected variations in τ values. When assessing ring slippage for a given *ansa*-metallocene, it is best to make a comparison to a similarly substituted, unlinked analog.

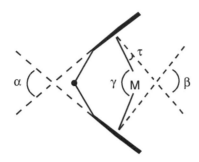

FIGURE 4.2 Angles α, β, τ, and γ are often used to describe the bonding in *ansa*-metallocenes with two interannular linkers.

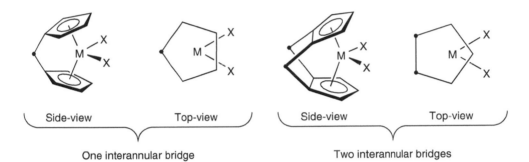

FIGURE 4.3 The presence of one versus two interannular bridges affects the steric environment of the metallocene wedge.

Some generalities can be made when comparing metallocenes with one interannular bridge to those with two. Doubly bridged metallocenes typically have increased α values, decreased γ values, and increased τ values as compared to singly bridged metallocenes. In this chapter, α, β, τ, and γ values are provided in Table 4.1 for doubly bridged metallocenes with reported X-ray crystallographic data. When available, R–M–R' angles are also reported in Table 4.1, where R and R' are noncyclopentadienyl ligands located in the metallocene wedge.

Another important consequence of interannular linkers is the resulting orientation of cyclopentadienyl rings themselves (Figure 4.3). When one interannular bridge is present, it forces two cyclopentadienyl carbon atoms to be eclipsed in the narrow rear of the metallocene. The central coordination site of the metallocene wedge "sees" a carbon–carbon bond of a cyclopentadienyl ring directly above and below. In contrast, the presence of two interannular bridges forces two sets of two carbon atoms to be eclipsed in the narrow rear of the metallocene. The central coordination site of the metallocene wedge "sees" a carbon atom (not a carbon–carbon bond) of a cyclopentadienyl ring directly above and below. This has important consequences for coordination of the cyclopentadienyl rings to the metal, that is, for M–C bond distances. In doubly bridged metallocenes, the *ipso*-carbon atoms are closer to the metal than the three remaining distal carbon atoms. In the extreme, this can be considered an η^2-allyl coordination mode.

The presence of two interannular linkers changes the substitution patterns required to achieve various metallocene symmetries. Though the data is somewhat limited, doubly bridged metallocenes tend to follow metallocene symmetry/polymer tacticity relationships that have been established for singly bridged metallocenes.[3] There are examples (*vide infra*) where the steric congestion of the metallocene wedge can hinder activity so dramatically that metallocene symmetry/polymer tacticity relationships cannot be addressed. Regardless, metallocene symmetry is used as an organizing principle for this chapter.

TABLE 4.1

Structural Data for Selected Doubly Bridged Compounds with Representative Unlinked and Singly Linked Metallocenes Included for Comparison Purposes

	α (°)	β (°)	γ (°)	τ (°)	R-M-R' (°)	Reference
Representative unlinked metallocenes						
Cp₂TiCl₂ᵃ	51.7 [51.3]	128.3 [128.7]	130.7 [130.7]	1.2 [1.0]	94.43(6) [94.62(6)]	57
Cp₂ZrCl₂	53.5	126.5	129.2	1.4	97.0(1)	58
Representative singly linked metallocenes						
[(Me₂Si)(C₅H₄)₂]TiCl₂	56.2	123.8	128.7	2.1	95.75(4)	59
[(Me₂Si)(C₅H₄)₂]ZrCl₂	60.1	119.9	125.4	2.8	97.98(4)	59
[(CH₂CH₂)(C₅H₄)₂]ZrCl₂	56.4	123.6	125.8	1.1	97.43(8)	6
[(Me₂SiOSiMe₂)(C₅H₄)₂]ZrCl₂	51.1	128.9	130.8	0.95	98.71(3)	23
Two -CH₂CH₂- interannular linkers						
[(CH₂CH₂)₂(C₅H₃)₂]TiCl₂ (**1a**)	57.8	122.2	124.5	1.2	94.8(1)	9
[(CH₂CH₂)₂(C₅H₃)₂]ZrCl₂ (**1b**)ᵇ	62.5	117.5	120.0	1.3	97.8(1)	9
rac-[(CH₂CH₂)₂(indenyl)₂]ZrCl₂ (**20b**)ᵃ	{62.9} —ᶜ	—	120.2(1) [120.4(1)]	—	{97.95(5)} 93.88(4) [96.43(4)]	10, 11 38
Spiro- interannular linkers						
[*rac*-(R)(3-MeCp)₂]TiBr₂ (**3c**), where R=(2,3,6,7)-(bicyclo[3.3.1]nonane)	—	—	124.2	<5	95.4(1)	12
bis(spirosilacyclohexane-[*b*]-*i*-PrCp₂)ZrCl₂ (**16a**)	61.1	118.9	126.6	3.9	99.1(1)	36
bis(spirosilacyclohexane-[*b*]-*t*-BuCp₂)ZrCl₂ (**16b**)	63.7	116.3	126.2	5.0	98.0(2)	36
Two -Me₂SiOSiMe₂- interannular linkers						
[(Me₂SiOSiMe₂)₂(4-*t*-BuCp₂)]Sm(THF)₂ (**12**)	55.4	124.6	133.4	4.4	99.6(3)	25
[(Me₂SiOSiMe₂)₂(C₅H₃)₂]ZrCl₂ (**10b**)	44.1	135.9	137.33	0.7	96.20	22
One -Me₂SiOSiMe₂- and one Me₂Si- interannular linker						
[(Me₂Si)(Me₂SiOSiMe₂)(4-*t*-BuCp₂)]Sm(THF)₂ (**11a**)	76.0	104	116.5	6.3	94.4(8)	23
[(Me₂Si)(Me₂SiOSiMe₂)(4-*t*-BuCp₂)]YCl(THF) (**11c**)	—	—	120.8	—	89.7(5)	26

Two Me₂Si- interannular linkers

Compound						
[(Me₂Si)₂{3,5-i-Pr₂Cp)(4-t-BuCp)}Sc(μ-CH₃)₂Al(CH₃)₂ (**38b**)	75.68(7)	104.32	123.53	9.6	88.48(8)	49
[(Me₂Si)₂{3,5-i-Pr₂Cp)(4-t-BuCp)}YCH(SiMe₃)₂ (**39**)	80.7(1)	99.3	116.1	8.4	na^d	49
rac-[(Me₂Si)₂(3-i-Pr-5-MeCp)₂]TiCl₂ (**7a**)	69.7	110.3	128.1	8.9	98.28(1)	20
[(Me₂Si)₂(C₅H₃)₂]TiMe₂ (**4c**)^a	63.5(2)	116.5	126.8	5.2	93.0(2)	14
	[64.1(2)]	[115.9]	[126.7]	[5.4]		
[(Me₂Si)₂(C₅H₃)₂]ZrCl₂ (**4b**)	69.6(1)	110.4	120.6	5.1	99.64(4)	13
[(Me₂Si)₂(4-t-BuCp)₂]ZrCl₂ (**5a**)	73(4)	107	—	—	101.9(1)	16
[(Me₂Si)₂(3,4-Me₂Cp)₂]ZrCl₂ (**18**)	72.1	107.9	121.8	7.0	98.7(1)	37
[(Me₂Si)₂(4,5,6,7-tetrahydroindenyl)₂]ZrCl₂ (**19**)	72.9	107.1	122.5	7.7	99.7(1)	37
[(Me₂Si)₂(3,5-i-Pr₂Cp)(4-i-PrCp)]ZrCl₂ (**35b**)	73.9(2)	106.1	120.3	7.1	100.38(5)	43
[(Me₂Si)₂(3,5-i-Pr₂Cp)(4-i-PrCp)]Zr(CH₂Ph)₂ (**36a**)	75.08	104.92(21)	121.67(3)	8.4	96.5(2)	44
[(Me₂Si)₂{3,5-i-Pr₂Cp}{4-(1R,2S,5R-menthyl)-Cp}]ZrCl₂ (**40c**)^e	—	—	122	—	105	47
[(Me₂Si)₂{3,5-i-Pr₂Cp}{4-((S)-CHMeCMe₃)-Cp}]ZrCl₂ ((S)-**41a**)^a	72.05	107.95(25)	121.7	6.9	102.15(5)	53
	[72.19]	[107.81(23)]	[122.2]	[7.2]	[101.80(5)]	
[(Me₂Si)₂{3,5-i-Pr₂Cp}{4-((S)-CHMeCMe₃)-Cp}]Zr(SPh)₂ ((S)-**42b**)	75.76(9)	104.24	121.9	8.8	104.70(3)	53
[(Me₂Si)₂{3,5-(CHEt₂)₂Cp}{4-((S)-CHMeCMe₃)-Cp}]ZrCl₂ ((S)-**41b**)	—	—	—	—	101.32(2)	53
[(Me₂Si)₂{3,5-Cy₂-Cp}{4-((S)-CHMeCMe₃)-Cp}]ZrCl₂ ((S)-**41c**)^a	—	—	—	—	103.01(4)	53
					[102.09(4)]	

(Cy = cyclohexyl)

a Two molecules are present in the unit cell. The bracketed data represents the second molecule in the unit cell.

b There are two reported sets of data for this compound. The data in curly brackets represents the second set of data.

c —: not available.

d na: not applicable.

e Six molecules are present in the unit cell. Average bond angles are given.

4.1.2 Metallocenes without Reported Polymerization Data

Some metallocenes with two interannular bridges have appeared in the literature, but with no polymerization data included in the reports of their synthesis and structure. Most of these examples possess C_{2v}-symmetry. These types of precatalysts are expected to produce atactic polypropylene under enantiomorphic site control conditions (the most common scenario). Tacticity may vary if chain-end control is obeyed (rarer). Perhaps the authors did not publish polymerization data for propylene because of the expectation of obtaining atactic polymer, considered to be less industrially useful than its isotactic counterpart. (For more information on enantiomorphic site control and chain-end control, see Chapter 1.)

Two ethanediyl interannular linkers have been used to produce C_{2v}-symmetric titanocene and zirconocene dichloride complexes (**1a** and **1b**, respectively; Figure 4.4). In 1994, Brintzinger described the synthesis and X-ray crystallographic analysis of **1a** and **1b**[9] and Hafner described a distinct synthesis for **1b**[10,11] along with crystallographic data (see Table 4.1). For both complexes, α (the dihedral angle between the ring planes) was found to be only slightly larger than that of the corresponding complexes with a single ethanediyl bridge (a similar phenomenon has been described for metallocenes with a single ethanediyl bridge versus the corresponding unlinked metallocenes[6]). The two cyclopentadienyl rings in both **1a** and **1b** are nearly eclipsed, yielding a slight distortion from ideal C_{2v}-symmetry. Thus, these complexes display approximate C_2-symmetry.

Also in 1994, Buchwald[12] described the synthesis of a ligand with two fused interannular linkers comprised entirely of carbon and hydrogen (**2**, Figure 4.4). This ligand enforces a C_2-symmetry in *ansa*-titanocene complexes (**3a–c**, Figure 4.4). The bicyclo[3.3.1]nonane skeleton present in these complexes is structurally analogous to norbornadiene. Combination of **2** with either $ZrCl_4(THF)_2$ or $HfCl_4(THF)_2$ did not provide the desired zirconocene or hafnocene complexes. Compound **3c** was analyzed by X-ray crystallography (see Table 4.1). The authors propose that the failed metallation of **2** with both Zr and Hf is due to the compact nature of ligand **2**. Namely, the Ti-C(cyclopentadienyl) and Ti-cyclopentadienyl centroid (Cp_{cent}) distances are reported to be relatively short in comparison to a titanocene with a single ethanediyl bridge. Angles of α and β were not reported for **3c**; it should be noted that the cyclopentadienyl rings are canted relative to one another (at the back of the metallocene wedge, the C—C bonds of each cyclopentadienyl ring are not parallel).

Examples of metallocenes with two dimethylsilyl interannular linkers are more numerous than those composed strictly of C—C and C—H bonds. Royo first reported the synthesis of *ansa*-metallocenes with a $[\{(CH_3)_2Si\}_2(C_5H_3)_2]$ ligand array.[13–15] Relevant metallocene dichloride and dimethyl complexes **4a–d** are depicted in Figure 4.5. X-ray crystal structures have been obtained for dichloride complexes **4a** and **4b**[13] and dimethyl complex **4c**[14] (see Table 4.1). In all three structures, the cyclopentadienyl rings are eclipsed. Despite a wealth of synthetic and crystallographic data,

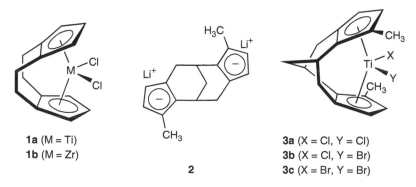

1a (M = Ti)
1b (M = Zr)

2

3a (X = Cl, Y = Cl)
3b (X = Cl, Y = Br)
3c (X = Br, Y = Br)

FIGURE 4.4 Metallocenes and ligands with two carbon-based interannular bridges.

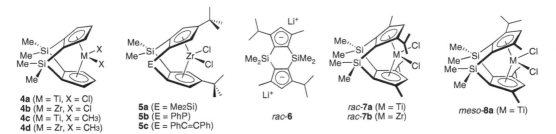

FIGURE 4.5 Metallocenes and ligands with one or two Me$_2$Si- interannular bridges.

Royo has not put forward α-olefin polymerization results for precatalysts **4a–d**. However, Lang has tested **4a** for polymerization of ethylene and propylene (*vide infra*).[34]

Bulls and Bercaw employed two dimethylsilyl interannular linkers in the preparation of [(Me$_2$Si)$_2$\{η^5-C$_5$H$_2$-4-C(CH$_3$)$_3$\}$_2$]ZrCl$_2$ (**5a**, Figure 4.5).[16] The most notable aspect of the X-ray crystal structure for **5a** is the dihedral angle (α) of 73(4)° formed by the cyclopentadienyl ring planes (see Table 4.1). This value is high in comparison to both singly linked and unlinked zirconocenes.

Owing to the C_{2v}-symmetry of complexes **1–5a**, it is anticipated that these precatalysts would produce atactic polypropylene through an enantiomorphic site control mechanism. Furthermore, the distortion of **1a** and **1b** from C_{2v}-symmetry is so slight that atactic polypropylene would be anticipated under a range of polymerization conditions.

In the Zubris laboratories, the synthesis of two C_s-symmetric, *meso* complexes of the form [(Me$_2$Si)(E)\{η^5-C$_5$H$_2$-4-C(CH$_3$)$_3$\}$_2$]ZrCl$_2$ (E = PhP, **5b**; E = PhC=CPh, **5c**; Figure 4.5) is currently being pursued.[17] Syntheses of the singly linked ligands, [(PhP)(*t*-Bu-C$_5$H$_4$)$_2$] and a cis/trans mixture of [(PhC=CPh)(*t*-Bu-C$_5$H$_4$)$_2$], were recently reported.[18,19] Zirconocenes **5b** and **5c** with two distinct interannular linkers have been targeted to examine the steric requirements for regioselectivity in α-olefin polymerization. It is anticipated that these precatalysts may favor 2,1-olefin insertion, a mode that is not common for most zirconocenes. Synthetic methods to prepare **5b** and **5c** are currently under development.

In 1998, Miyake and Bercaw reported the synthesis of a zirconocene and two titanocenes, each with two dimethylsilyl interannular linkers.[20] Notably, they described a novel intermediate and an isomerization reaction for doubly bridged metallocenes. C_2-symmetric *rac*-**7b** was prepared by reaction of the ligand, *rac*-**6**, with ZrCl$_4$ in dichloromethane (Figure 4.5). The ^{1}H NMR spectrum of isolated *rac*-**7b** was consistent with C_2-symmetry; no signals were evident for the *meso* isomer. In contrast, reaction of *rac*-**6** with TiCl$_3$(THF)$_3$ in toluene followed by treatment with 1 equivalent of PbCl$_2$ provided a roughly 1:1 mixture upon workup of *rac*-**7a** and *meso*-**8a** (Figure 4.5). Since pure *rac*-**6** was used as a starting material, the authors postulate that a *rac*-to-*meso* interconversion was occurring under the reaction conditions. Furthermore, pure *rac*-**7a** was found to equilibrate with *meso*-**8a** rapidly in hydrocarbon solvent in the dark at ambient temperature. The authors propose an isomerization mechanism involving an intermediate with both Si atoms bound to the same sp^3 hybridized C of Cp and with Ti coordinated to Cp in an η^1-fashion (via an adjacent sp^2 hybridized C) (Scheme 4.1). This is the first example of an intermediate and isomerization reaction requiring a change in cyclopentadienyl face for a doubly bridged metallocene.

Single crystals of *rac*-**7a** were obtained and analyzed by X-ray diffraction (see Table 4.1). While the angle γ (128.1°) is moderately less than unlinked or singly linked metallocenes (see Table 4.1 for representative data), α (69.7°) is considerably greater than for related unlinked or singly linked metallocenes. A relatively large value of τ (8.9°) also indicates that the titanium-ring carbon distances in the back of the wedge are much shorter than those to the front ring carbons.

The rapid *rac-meso* interconversion for [(Me$_2$Si)$_2$\{3-(CHMe$_2$)-5-Me-C$_5$H\}$_2$]TiCl$_2$ leads to complications in predicting tacticity preferences for this precatalyst. In the absence of *rac-meso*

rac-7a   Intermediate   meso-8a

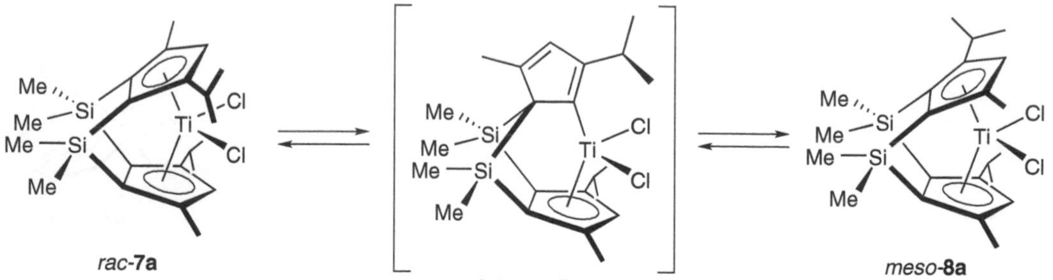

SCHEME 4.1   Proposed mechanism for conversion of *rac*-**7a** to *meso*-**8a**.

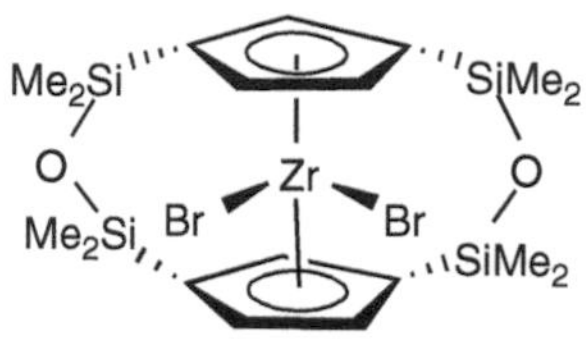

9

FIGURE 4.6   A zirconocene with two -Me$_2$SiOSiMe$_2$- interannular bridges in a 1,3-relationship relative to one another.

interconversion, C_2-symmetric *rac*-**7a** would be expected to yield isotactic polypropylene, and C_s-symmetric *meso*-**8a** would be expected to yield atactic polypropylene through an enantiomorphic site control mechanism. Owing to the absence of rapid *rac-meso* interconversion, C_2-symmetric zirconocene *rac*-**7b** would be expected to produce isotactic polypropylene through an enantiomorphic site control mechanism.

When two transannular linkers are present in an *ansa*-metallocene, they do not necessarily adopt positions that are adjacent to one another (in a 1,2-relationship). One such example of a nonadjacent complex was reported by Deck (**9**, Figure 4.6). This zirconocene dibromide has two tetramethyldisiloxane bridges in a 1,3-relationship relative to one another.[21] This complex was characterized by ^{1}H NMR, ^{13}C NMR, and elemental analysis but not by X-ray crystallography. Owing to the C_{2v}-symmetry of **9**, it is anticipated that this precatalyst would yield atactic polypropylene. However, the steric bulk of the tetramethyldisiloxane bridges may hinder the coordination of propylene to zirconium through a 1,2-coordination mode.

4.1.3   POLYETHYLENE CATALYSTS WITHOUT PROPYLENE POLYMERIZATION DATA

Some metallocenes with two interannular linkers have been used as precatalysts for polyethylene formation although their reactivity towards α-olefins has not been reported. Noh described the synthesis of group 4 metallocenes with two disiloxanediyl interannular linkers, [(Me$_2$SiOSiMe$_2$)$_2$(C$_5$H$_3$)$_2$]MCl$_2$, (M = Hf, **10a**; M = Zr, **10b**; Figure 4.7).[22] Complex **10b** was methylated to provide the corresponding dimethyl complex (**10c**), and benzylated to provide the corresponding bis(benzyl) complex (**10d**). Complex **10b** was analyzed by single-crystal X-ray diffraction. This compound displays C_2-symmetry in the solid state and the two Cp rings are nearly eclipsed, as expected. The interannular linkers give this complex its C_2-symmetry. The angle α (44.1°) was found to be smaller than in the related compound containing a single disiloxane bridge (51.1°)[23] (see Table 4.1). Notably, this is the opposite of the trend for other doubly bridged versus singly bridged metallocenes, perhaps owing to the bridge length.

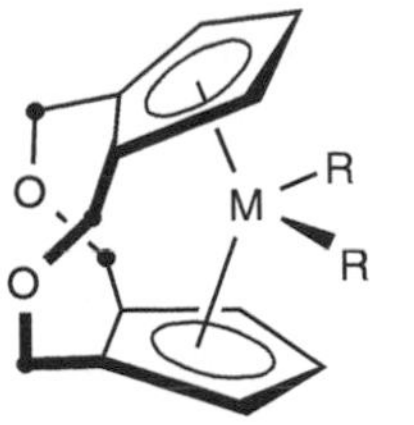

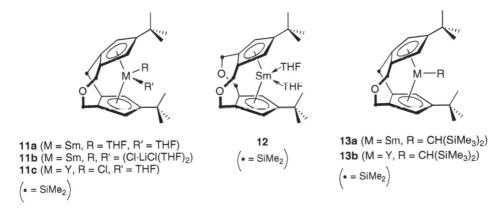

10a (M = Hf, R = Cl)
10b (M = Zr, R = Cl)
10c (M = Zr, R = CH$_3$)
10d (M = Zr, R = CH$_2$C$_6$H$_5$)

$\left(\bullet = \text{SiMe}_2 \right)$

FIGURE 4.7   Group 4 metallocenes with two -Me$_2$SiOSiMe$_2$- interannular bridges in a 1,2-relationship can serve as effective precatalysts for ethylene polymerization.

11a (M = Sm, R = THF, R′ = THF)
11b (M = Sm, R, R′ = (Cl·LiCl(THF)$_2$)
11c (M = Y, R = Cl, R′ = THF)

$\left(\bullet = \text{SiMe}_2 \right)$

12

$\left(\bullet = \text{SiMe}_2 \right)$

13a (M = Sm, R = CH(SiMe$_3$)$_2$)
13b (M = Y, R = CH(SiMe$_3$)$_2$)

$\left(\bullet = \text{SiMe}_2 \right)$

FIGURE 4.8   Samarocenes and yttrocenes with one or two -Me$_2$SiOSiMe$_2$- interannular bridges and one or zero Me$_2$Si- interannular bridges do not polymerize α-olefins (THF = tetrahydrofuran; complexes **11a** and **12** are divalent while complexes **11b**, **11c**, **13a**, and **13b** are trivalent).

Complex **10b** serves as an efficient ethylene polymerization catalyst when activated with modified methylaluminoxane (MMAO) at 70 °C in an ethylene-saturated toluene solution. The analogous Hf compound (**10a**) was found to be less active. Complex **10b** displayed a higher activity than Cp$_2$ZrCl$_2$, Ind$_2$ZrCl$_2$, and Et(Ind)$_2$ZrCl$_2$ (Ind = indenyl) under the polymerization conditions studied. Polymerization attempts using α-olefins were not reported. Owing to the C_2-symmetry of **10a–d**, it is anticipated that the precatalysts would yield isotactic polypropylene through an enantiomorphic site control mechanism.

4.1.4 Polyethylene Catalysts that Do not Polymerize Propylene

Yasuda and Kai have prepared samarocenes with one disiloxanediyl bridge and one dimethylsilyl bridge (**11a**, **11b**, **13a**), yttrocenes with one disiloxanediyl bridge and one dimethylsilyl bridge (**11c**, **13b**), and a samarocene with two disiloxanediyl bridges (**12**) (Figure 4.8).[24–26] Compounds **11a**, **11c**, and **12** have been characterized by X-ray crystallography (see Table 4.1).[24,25]

The solid-state structure of samarocene **11a** confirms that it is a *meso* compound with C_s-symmetry. In comparison to two related *ansa*-metallocenes with only a single dimethylsilyl bridge[25] the (Cp$_{cent}$)–Sm–(Cp$_{cent}$) bite angle (γ) changed only slightly (<1 °) as a result of adding the second disiloxanediyl linker. Samarocene **12** displays C_{2v}-symmetry both in solution (as evidenced by ^{1}H NMR spectroscopy) and the solid state.[25] The value of α for **12** is fairly small compared to other doubly bridged metallocenes (Table 4.1); in fact, the long bridging groups make this metallocene geometrically similar to an unbridged complex, Cp*$_2$Sm(THF)$_2$ (Cp* = C$_5$Me$_5$; γ = 136°), according to the authors. In the solid-state structure of yttrocene **11c**, it appears that the cyclopentadienyl rings are not "planar," making the dihedral angle between the cyclopentadienyl ring planes, α (112.4(6)°), a number of little utility. And while γ (120.8°) was reported to be less than for unlinked yttrocenes (γ = 134.4(4)° for Cp*$_2$YCH(SiMe$_3$)$_2$ and γ = 132.2(2)° [132.4(2)°] for Cp*$_2$YN(SiMe$_3$)$_2$),[27] no comparison was made to singly bridged yttrocenes.

Compound **11a** acts as an initiator for ethylene polymerization at 23 °C with 1 atm of ethylene in toluene solution, although the polydispersity index of the resulting polymers is somewhat broad (3.29–3.49) depending on reaction time. No cocatalyst (e.g., methylaluminoxane, MAO) was added to **11a** to initiate polymerization: this divalent complex undergoes a different initiation mechanism than group 4 or group 3 d^0 metal alkyls or hydrides. The authors propose an initiation and propagation mechanism for **11a** analogous to that reported by Evans for Cp*$_2$Sm(THF)$_2$.[28] For Cp*$_2$Sm(THF)$_2$, the polymerization mechanism involves coordination of ethylene to two Cp*$_2$Sm units accompanied by one electron transfer from each Sm to ethylene. This first step generates Cp*$_2$Sm-(CH$_2$CH$_2$)-SmCp*$_2$, which subsequently initiates and propagates from both sides. While the initiation mechanism is distinct, the propagation mechanism is proposed to be analogous to group 4 and group 3 d^0 metal alkyls or hydrides. Similarly, **12** also catalyzes ethylene polymerization, although it is much less active than **11a** and produces polymer having a narrower PDI (3.04). The same initiation and propagation mechanism is suggested for polymerizations involving **12**. Both **11a** and **12** are inactive for the polymerization of α-olefins such as 1-pentene and 1-hexene.

Compound **11c** was not tested as a polymerization catalyst; this compound would not be expected to serve as a Ziegler–Natta catalyst owing to its ligands (R and R′). Trivalent alkyl complexes **13a** and **13b** were tested for ethylene polymerization, but no evidence of polymerization was observed.[26] The steric bulk of the CH(SiMe$_3$)$_2$ alkyl ligand was provided as an explanation for the inactivity of these compounds. This phenomenon is typical for group 3 metallocene alkyls bearing this bulky alkyl substituent; for example, complexes of the type, Cp*$_2$LnCH(SiMe$_3$)$_2$ (Ln = La, Nd) are inactive toward ethylene polymerization.[29] In contrast, trivalent lanthanocene methyl complexes (R = Me), which are isoelectronic with group 4 metallocene alkyl cations, are known to function as single-component polymerization catalysts for ethylene; these methyl complexes are also known to oligomerize α-olefins, such as propylene.[30–32] Trivalent lanthanocene methyl complexes are capable of acting as polymerization initiators without any cocatalyst present, such as MAO. However, these authors do not describe the synthesis of trivalent lanthanocene methyl complexes with their ligand arrays.

4.2 PROPYLENE POLYMERIZATION CATALYSTS

Polypropylene tacticity data for selected precatalysts from this section is summarized in Table 4.2.

4.2.1 C_{2v}-SYMMETRIC (ASPECIFIC)

Lang has described the synthesis and polymerization testing of C_{2v}-symmetric precatalysts with two dimethylsilyl interannular linkers (**4a**, Figure 4.5; **14b–d**, Figure 4.9).[33,34] While compound **4a** was first synthesized by Royo (*vide supra*), Lang was the first to describe its polymerization activity. Ethylene and propylene polymerization data have been reported for precatalysts **4a** and **14b–c**; only ethylene polymerization data has been reported for **14d**.[33] Titanocene **14d** was also characterized by X-ray crystallography.[34] The crystallographic data for **14d** is typical for a titanocene with two dimethylsilyl interannular linkers.

TABLE 4.2

Representative Propylene Tacticity Data for Selected Doubly Bridged Precatalysts[a]

Compound	mmmm	mm (%)	m (%)	r (%)[b]	rrrr (%)[c]	Reference (%)
C2v symmetric						
[(Me2Si)2(C5H3)2]TiCl2 (**4a**)	—[d]	42.0	—	—	—	33
[(Me2Si)2(4-SiMe3-Cp)2]TiCl2 (**14b**)	—	25.7	—	—	—	33
[(Me2Si)2(4-SiMe3-Cp)2]ZrCl2 (**14c**)	—	8.7	—	—	—	33
[(Me2Si)2(3,5-i-Pr2Cp)2]ZrCl2 (**15**)	—	—	—	61.6	12.9	35
C2 symmetric						
[bis(spirosilacyclohexane-[b]-i-PrCp)2]ZrCl2 (**16a**)[e]	90	—	—	—	—	36
[bis(spirosilacyclohexane-[b]-t-BuCp)2]ZrCl2 (**16b**)[e]	97	—	—	—	—	36
[bis(spirosilacyclohexane-[b]-Cp)2]ZrCl2 (**16c**)[e]	6	—	—	—	—	36
[(Me2Si)2(3,4-Me2Cp)2]ZrCl2 (**18**)[e]	38	—	—	—	—	37
[(Me2Si)2(4,5,6,7-tetrahydroindenyl)2]ZrCl2 (**19**)[e]	80	—	—	—	—	37
[(CH2CH2)2(indenyl)2]ZrCl2 (**20b**)[f]	—	—	isotactic	—	—	38
Cs symmetric						
[(Me2Si)2(3,5-i-Pr2Cp)(Cp)]ZrCl2 (**35a**)	0.0	1.7	6.0	94.0	83.7	40
[(Me2Si)2(3,5-i-Pr2Cp)(4-i-PrCp)]ZrCl2 (**35b**)	0.0	0.1	0.4	99.6	98.9	40
[(Me2Si)2(3,5-i-Pr2Cp)(4-SiMe3-Cp)]ZrCl2 (**35c**)	0.0	0.0	1.0	99.0	95.9	40
[(Me2Si)2(3,5-i-Pr2Cp)(4-t-BuCp)]ZrCl2 (**35d**)	0.0	—	3.1	96.9	90.5	40
[(Me2Si)2(3,5-(SiMe3)2-Cp)(4-i-PrCp)]ZrCl2 (**37**)	0.0	2.2	6.3	93.7	75.4	35
[(Me2Si)2(3,5-i-Pr2Cp)(4-t-BuCp)]Sc(μ-CH3)2Al(CH3)2 (**38b**)	1.44	15.35	41.35	58.66	11.34	48,49
[(Me2Si)2(3,5-i-Pr2Cp)(4-i-PrCp)]Sc(μ-CH3)2Al(CH3)2 (**38c**)	4.97	24.73	51.14	48.84	5.18	48,49
C1 symmetric						
[(Me2Si)2{3,5-i-Pr2Cp}{4-(CHMeCMe3)-Cp}]ZrCl2 (**40b**)	5.6	16.1	26.5	73.5	41.8	40
[(Me2Si)2{3,5-i-Pr2Cp}{4-(1R,2S,5R-menthyl)-Cp}]ZrCl2 (**40c**)	3.3	13.8	26.0	74.0	45.4	47
[(Me2Si)2{3-(S)-CHMeCy-5-i-PrCp}{4-(CHMeCMe3)-Cp}]ZrCl2 (**43**)	—	—	—	91.5	—	52

[a] Unless otherwise noted, polymerizations were carried out using neat propylene.

[b] $[m] = [mm] + 1/2[mr]$.

[c] $[r] = [rr] + 1/2[mr]$.

[d] —; not available.

[e] Polymerization carried out using 2 bar propylene in toluene solution.

[f] No polymerization details provided.

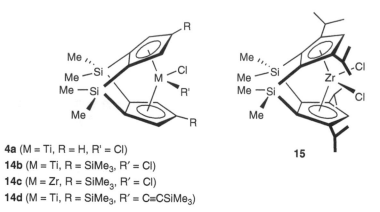

4a (M = Ti, R = H, R' = Cl)
14b (M = Ti, R = SiMe$_3$, R' = Cl)
14c (M = Zr, R = SiMe$_3$, R' = Cl)
14d (M = Ti, R = SiMe$_3$, R' = C≡CSiMe$_3$)

15

FIGURE 4.9 C_{2v}-symmetric titanocenes and zirconocenes with two Me$_2$Si- interannular bridges.

Upon activation with MAO, precatalysts **4a** and **14b–d** showed activity for ethylene polymerization at 10 bar ethylene in toluene solution with reaction temperatures of either 30 or 40 °C. Polymerization activities increased as the amount of MAO or the temperature was increased (from 30 to 40 °C). The activities of precatalysts **4a** and **14b–d** were fairly comparable to one another.

Precatalysts **4a** and **14b–c** do yield polypropylene when activated with MAO, although the activities are fairly low. Polymerizations were carried out in neat propylene at either 40 or 50 °C. All three catalysts yield essentially atactic polypropylene; low *mm* triad values were reported (42.0%, 25.7%, and 8.7%, respectively). The polypropylene sample formed from precatalyst **4a** had an *mmmm* pentad value of 15.3%. Tacticity data for polypropylenes formed by precatalysts **4a** and **14b–c** is summarized in Table 4.2. The authors postulate that **4a** may degrade under polymerization conditions, leading to an *mm* triad value that is only marginally higher than the ideal atactic polypropylene's *mm* triad value of 25%. The polypropylene sample formed from precatalyst **14b** has a large PDI value (14.4), suggesting the presence of more than one active catalyst under reaction conditions. Precatalyst **14c** yields polypropylene with the highest activity of the three precatalysts; no PDI is given for this polymer sample. In summary, precatalysts **4a** and **14b–c** form primarily atactic polypropylene, consistent with what is expected for C_{2v}-symmetric precatalysts operating under an enantiomorphic site control mechanism.

Miyake and Bercaw have synthesized another C_{2v}-symmetric precatalyst with two dimethylsilyl interannular linkers, [(Me$_2$Si)$_2$(3,5-(CHMe$_2$)$_2$-C$_5$H}$_2$]ZrCl$_2$ (**15**, Figure 4.9).[35] No X-ray crystallographic data was reported for **15**. In combination with MAO in liquid propylene at 0 °C, precatalyst **15** forms atactic polypropylene and displays a low activity. The resulting polymer has an M_w value of 430,000 g/mol and a PDI value of 2.11. This polypropylene sample is predominantly atactic, with an [*rrrr*] value of 12.9% and an [*r*] value of 61.6%. Tacticity data is summarized in Table 4.2. The authors postulate that the 3,5-substitution pattern of both cyclopentadienyl rings may hinder polymer chain movement during the migratory insertion step of chain propagation (see Chapter 1 for more information on migratory insertion).

4.2.2 C_2-SYMMETRIC (ISOSELECTIVE)

When enantiomorphic site control is operating, C_2-symmetric precatalysts with two interannular linkers are predicted to yield isotactic polypropylene. In most cases, this has been borne out by experiment, but the steric environment of the metallocene wedge (as illustrated in Figure 4.3) has important implications for both activity and regiocontrol. In particular, substituents in the 4-position of the cyclopentadienyl ring can be influential.

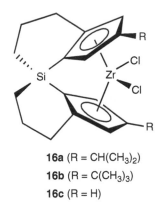

16a (R = CH(CH$_3$)$_2$)
16b (R = C(CH$_3$)$_3$)
16c (R = H)

FIGURE 4.10 Zirconocenes with a spirosilane bridge may be considered a hybrid of metallocenes with one and two interannular bridges.

Brintzinger prepared a set of zirconocenes with a spirosilane bridge that could be considered hybrids—metallocenes with properties somewhere in between those of complexes with one and two interannular bridges (**16a–c**, Figure 4.10).[36] Three chiral *ansa*-zirconocenes were synthesized with a trimethylene linker between the silicon bridge atom and the α-position of each cyclopentadienyl ring. Complexes **16a** and **16b** were analyzed crystallographically (see Table 4.1). Indeed, these complexes do not adopt the idealized geometries of either singly or doubly bridged metallocenes; this is the most notable aspect of both crystal structures. In **16a** and **16b**, the cyclopentadienyl rings are not eclipsed; this is in contrast to metallocenes with either one or two interannular bridges (see Figure 4.3 for the idealized eclipsed orientations).

Compounds **16a–c** have been activated with MAO and used for propylene polymerization. Polymerizations were carried out with a constant propylene pressure of 2 bars in toluene solution at 50 °C. Surprisingly, precatalyst **16c** produces essentially atactic polypropylene ([*mmmm*] = 6%) of low molecular weight (M_w = 3700). The authors suggest that the absence of substituents on the Cp ligand β to the silicon linker (i.e., since R = H) caused this low level of isotacticity. **16a** and **16b** provide highly isotactic polypropylene, with [*mmmm*] values of 90% and 97%, respectively. The molecular weights for these polymers were moderate (M_w = 8400 and 9800, respectively). Tacticity data for polypropylenes formed by precatalysts **16a–c** is summarized in Table 4.2.

A notable aspect of the polymer formed by precatalyst **16b** is the high number of 1,3-insertions in the polymer microstructure (approximately 2.5%, compared to <0.5% for **16a**). When a 2,1-insertion (a regioerror) occurs, the resulting secondary metal-alkyl species can rearrange to the terminally bound isomer before the next olefin inserts; this is called a 1,3-insertion. Regioerrors can cause melting point depression in polymers of high tacticity. This is consistent with the different observed melting points for the polypropylene samples generated by **16a** and **16b** (138 versus 128 °C, respectively). The authors carried out molecular mechanics analysis of the transition states for 1,2- versus 2,1-insertions for **16b**; this analysis indicates that the energetic difference between the two modes is approximately 4 kJ/mol, significantly smaller than the estimated 13 kJ/mol energetic difference for an analogue with only one dimethylsilyl interannular linker. Overall, these calculations are consistent with the increased frequency of 1,3-misinsertions for precatalyst **16b**.

Brintzinger has described the synthesis of two chiral, C_2-symmetric zirconocenes having two dimethylsilyl interannular linkers.[37] Metallation of dilithio salt *rac*-**17a** with ZrCl$_4$ provides a 3:1 ratio of *rac* and *meso* zirconocene; the desired *rac*-zirconocene (**18**) was isolated by fractional crystallizations (Figure 4.11). Similarly, metallation of *rac*-**17b** provides a 3:2 ratio of *rac* and *meso* zirconocene; the desired *rac*-zirconocene (**19**) was also isolated by fractional crystallizations (Figure 4.11). Crystal structures were reported for complexes **18** and **19** (see Table 4.1). The dihedral angle α is fairly large in both cases (72.1 ° and 72.9 °, respectively). The authors state that α is

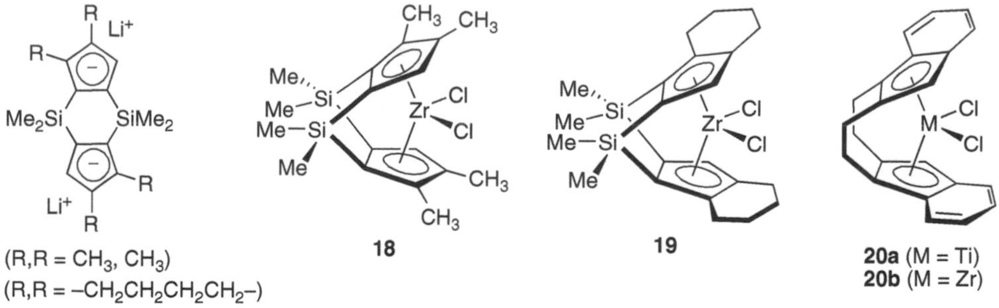

rac-**17a** (R,R = CH₃, CH₃)
rac-**17b** (R,R = –CH₂CH₂CH₂CH₂–)

18

19

20a (M = Ti)
20b (M = Zr)

FIGURE 4.11 When suitably activated, C_2-symmetric group 4 metallocenes with two dimethylsilyl bridges provide isotactic polypropylene.

10 °–12 ° larger for **18** and **19** than for related zirconocenes with a single dimethylsilyl interannular linker. The crystallographic data verifies that **18** and **19** exhibit C_2-symmetry in the solid state.

Both **18** and **19** were used as precatalysts for polypropylene formation, in combination with MAO as a cocatalyst. Polymerizations were carried out with 2 bar of propylene in toluene solution. The tacticities of the resulting polymers are [*mmmm*] = 38% and [*mmmm*] = 80%, respectively, with the former polymer an oily wax and the latter polymer a solid. Pertinent tacticity data is provided in Table 4.2. The time profiles for these polymerizations are noteworthy; catalysts **18**/MAO and **19**/MAO display minimal activity for the first 10–30 min, and then the activity increases substantially over the next several hours. At extended reaction times (> 16 h), activities decline. The authors propose that **18** and **19** are prone to MAO-promoted decomposition and supported this assertion with a control experiment. In this experiment, 5–10 mg samples of **18** and **19** wcre dissolvcd in 5 mL of a 10% MAO solution (in toluene) and incubated at 50 °C for 8 h. The singly bridged zirconocene, Me₂Si(2-Me-4-*t*-Bu-C₅H₂)₂ZrCl₂, was incubated under the same conditions for comparison purposes. After incubation, the solutions were hydrolyzed with CH₃OH/HCl, evaporated to dryness, and the residues extracted with CDCl₃. ¹H NMR analysis of the residues revealed that the sample of the singly bridged zirconocene remained largely unchanged. In contrast, the residues from doubly bridged zirconocenes showed that **18** and **19** accounted for only one-half of the recovered material. The authors suggest that the doubly bridged ligand framework degrades in the presence of MAO, and furthermore, these decomposition products are proposed to serve as the active catalytic species. The isotacticity of the polypropylenes obtained suggests that the decomposition products maintain their C_2-symmetry.

Halterman has reported the synthesis of C_2-symmetric, doubly bridged bis(indenyl)titanium and zirconium dichloride compounds with two ethane-1,2-diyl interannular linkers (**20a–b**, Figure 4.11).[38] Only the *rac*- isomer was isolated in each case, owing to the geometric constraints imposed by the two -CH₂CH₂- interannular linkers. A single crystal of **20b** was analyzed by X-ray diffraction studies; two independent molecules of slightly different conformations were found in the unit cell (see Table 4.1). While **20b** displays C_2-symmetry in the solid state, the cyclopentadienyl rings are not quite eclipsed. The Cp$_{cent}$-Zr-Cp$_{cent}$ angle (γ) of 120.4 ° [120.8 °] is comparable to **1a** and **1b** (Figure 4.4). While no details are provided, Halterman states that **20b** promotes the isotactic polymerization of propylene.

4.2.3 CALCULATIONS FOR C_2-SYMMETRIC (ISOSELECTIVE) AND C_s-SYMMETRIC (SYNDIOSELECTIVE) ZIRCONOCENES

Cavallo has applied *ab initio* methods and density functional calculations to examine the geometry of coordination of zirconium to a doubly bridged ligand based on a norbornadiene skeleton.[39] C_{2v}-symmetric **21** (Figure 4.12) was evaluated, revealing that a ligand array with a norbornadiene skeleton is computationally viable for generating an *ansa*-zirconocene. The two Cp rings are pentahapto

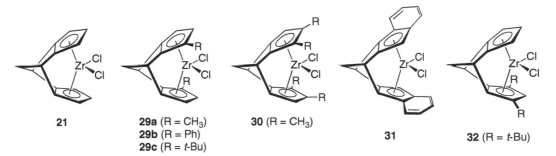

21   **29a** (R = CH₃)   **30** (R = CH₃)   **31**   **32** (R = t-Bu)
         29b (R = Ph)
         29c (R = t-Bu)

FIGURE 4.12 *Ab initio* methods and density functional calculations were used to analyze these doubly bridged zirconocene dichloride complexes with norbornadiene-skeleton linkers.

coordinated to zirconium in each case, and the bis(cyclopentadienyl) ligand is not greatly strained. *Ab initio* methods and density functional calculations predicted similar geometries for **21**. The synthesis of complex **21** has not been reported to date.

Cavallo also targeted a range of substituted doubly bridged zirconocenes based on a norbornadiene skeleton for enantioselectivity studies using molecular mechanics calculations.[39] Specifically, ligands (**22a–c**, **23**, **24**, and **25**, Figure 4.13) were chosen which theoretically yield the corresponding zirconocene dichloride complexes **29a–c**, **30**, **31**, and **32** (Figure 4.12). These substitution patterns were employed to generate both C_2- and C_s-symmetric complexes (C_2: **29a–c**, **30**, **31**; C_s: **32**). The synthesis of complexes **29a–c**, **30**, **31**, and **32** has not been reported to date. Ligands with a single interannular linker (**26–28**) were also used for comparison purposes in these molecular mechanics studies (Figure 4.13). The authors used molecular mechanics methods to study preinsertion intermediates along with pseudotransition states for monomer insertion; these intermediates and pseudotransition states were chosen to glean information about potential enantioselectivity of these zirconocenes in propylene polymerization. A priori, these model systems were predicted to be enantioselective, with C_2-symmetric complexes favoring formation of isotactic polyolefins and C_s-symmetric complexes favoring formation of syndiotactic polyolefins.

For the C_2-symmetric zirconocenes, only **29c** was calculated to be highly enantioselective, and thus, have a strong iso preference. Complex **31** was calculated to exhibit some iso preference, while **29a**, **29b**, and **30** were calculated to show poor enantioselectivity, and thus a poor iso preference. C_s-symmetric **32** was calculated to be highly enantioselective (and thus have a high syndio preference), although the predicted enantioselectivity was lower than that for its isoselective analogue, **29c**. In comparison with zirconium analogues having a single interannular linker (derived from **26–28**), the metallocenes with two interannular linkers were always calculated to be less enantioselective.

4.2.4 C_s-SYMMETRIC (SYNDIOSELECTIVE)

In 1996, Bercaw[40] first described the use of C_s-symmetric zirconocenes with two dimethylsilyl interannular linkers for the syndioselective polymerization of propylene. Importantly, this communication clarified important metallocene design features for high syndioselectivity. Prior to this work, singly bridged C_s-symmetric [(Me₂C){(fluorenyl)(Cp)}]ZrCl₂ (**33**) was synthesized by Ewen and Razavi in 1988 (Figure 4.14) and found to be a highly syndioselective precatalyst for propylene polymerization.[41] These authors attributed syndioselectivity to regularly alternating propylene insertions occurring at each enantiotopic side of the metallocene wedge. Furthermore, the authors postulated that both the growing polymer chain and the methyl group of an incoming propylene monomer would be directed away from the larger fluorenyl ligand in their catalyst system.

To further probe these two hypotheses of Ewen and Razavi, Bercaw prepared singly bridged C_s-symmetric **34** (Figure 4.14) and tested it for propylene polymerization. Combination of **34** with MAO in liquid propylene at 0 °C provided polypropylene that was essentially atactic (55% *r* dyads).[40]

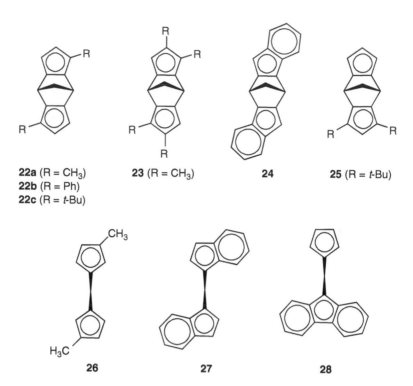

22a (R = CH₃)   **23** (R = CH₃)   **24**   **25** (R = *t*-Bu)
22b (R = Ph)
22c (R = *t*-Bu)

26   **27**   **28**

FIGURE 4.13 C_2- (**22–24**) and C_s- (**25**) symmetric doubly bridged ligands based on a norbornadiene skeleton, and comparative singly bridged ligands (**26–28**).

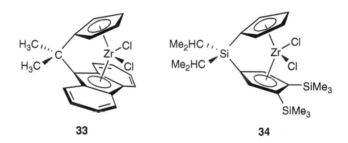

33   **34**

FIGURE 4.14   Singly bridged, C_s-symmetric *ansa*-metallocenes.

This led the authors to believe that an open region, or "pocket," was necessary for the bulky cyclopentadienyl ring to accommodate the methyl group of an incoming propylene monomer and control tacticity. In this model, it follows that the methyl group of propylene would adopt a trans relationship with respect to a growing polymer chain during polypropylene formation. To test this model, a novel C_s-symmetric zirconocene framework was targeted, as illustrated in Figure 4.15.[42]

C_s-symmetric precatalysts **35a–d** (Figure 4.16) and C_1-symmetric precatalysts **40a–b** (*vide infra*, Figure 4.19) were synthesized and tested for propylene polymerization. Precatalysts **35a–d** were activated with MAO and found to be highly syndioselective for the polymerization of propylene. Polymerizations were run in either neat propylene or dilute propylene (40 psig in toluene solution), with reaction temperatures of either 0 or 25 °C, respectively. One trial was carried out in neat propylene at 60 °C. Zirconocene **35a** displays the highest activity of the series; at 60 °C in neat propylene, an activity of 74,200 g polymer/(g cat·h) was found, along with high syndioselectivity ([$rrrr$] = 76.0% and [r] = 92.6%). Zirconocene **35b** is the most syndioselective at 0 °C in neat propylene, tacticities

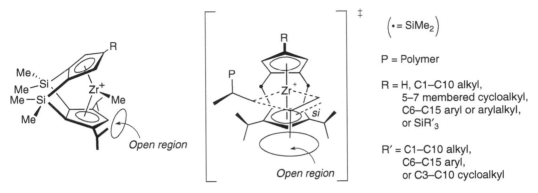

FIGURE 4.15 The open region of this zirconocene with two Me₂Si- interannular linkers can accommodate the alkyl group of the incoming α-olefin, such as the methyl group of propylene. It is thought to help give these catalysts a high syndio preference for polymerization of α-olefins.

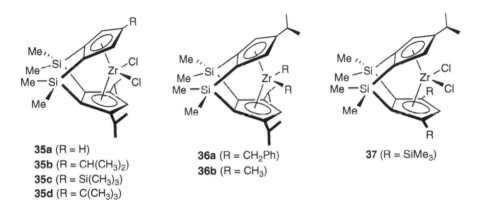

35a (R = H)
35b (R = CH(CH₃)₂)
35c (R = Si(CH₃)₃)
35d (R = C(CH₃)₃)

36a (R = CH₂Ph)
36b (R = CH₃)

37 (R = SiMe₃)

FIGURE 4.16 When suitably activated, C_s-symmetric group 4 metallocenes with two dimethylsilyl bridges provide syndiotactic polypropylene.

of [*rrrr*] = 98.9% and [*r*] = 99.6% are obtained for the resulting polypropylene. Representative polypropylene tacticity data for precatalysts **35a–d** appears in Table 4.2.

The X-ray crystal structure of **35b** was reported elsewhere (see Table 4.1).[43] Similar to other substituted metallocenes with two dimethylsilyl interannular linkers, the tilt angle τ is fairly large (7.1 °). This is reflected by the longer Zr–C distances to C-4 of each cyclopentadienyl ring, versus C-1 and C-2 (in the rear of the metallocene wedge). Representative bond lengths appear in Table 4.3.

In another study by Bercaw, compound **35b** was tested as a precatalyst for polymerization of 1-pentene, along with a zirconocene dibenzyl complex (**36a**) and a zirconocene dimethyl complex (**36b**) with the same ligand array (Figure 4.16).[44] Zirconocene dibenzyl complex **36a** was analyzed by X-ray crystallography. What is most notable is the value of τ for this complex (8.4 °) (see Table 4.1). Longer Zr–C distances to the C-4 position (labeled C3 and C8) of the cyclopentadienyl rings, versus to C-1 and C-5, C-6 and C-10 (in the rear of the metallocene wedge) are observed. Representative bond lengths appear in Table 4.3.

In polymerization tests, **35b** was activated with MAO, **36b** was activated with [CPh₃]⁺[B(C₆F₅)₄]⁻, and **36a** was activated with [CPh₃]⁺[B(C₆F₅)₄]⁻ or B(C₆F₅)₃. The authors carried out these experiments to investigate whether fluorinated borane or borate activators would provide comparable stereoselectivity and productivity as compared to MAO as a cocatalyst. All three compounds were found to be active precatalysts for 1-pentene polymerization. Polymerizations were

TABLE 4.3

Representative Bond Lengths for [(Me$_2$Si)$_2$(3,5-i-Pr$_2$Cp)(4-i-PrCp)ZrCl$_2$ (35b) and [(Me$_2$Si)$_2$(3,5-i-Pr$_2$Cp)(4-i-PrCp)]Zr(CH$_2$Ph)$_2$ (36a)^a

35b		36a	
Chemical bond	**Bond length (Å)**	**Chemical Bond**	**Bond length (Å)**
Zr-C1 (CpA)	2.441(5)	**Zr-C1** (CpA)	2.441(6)
Zr-C2 (CpA)	2.428(5)	**Zr-C5** (CpA)	2.441(6)
Zr-C4 (CpA)	2.652(5)	**Zr-C3** (CpA)	2.686(6)
Zr-C6 (CpB)	2.415(5)	**Zr-C6** (CpB)	2.419(6)
Zr-C7 (CpB)	2.418(5)	**Zr-C10** (CpB)	2.426(6)
Zr-C9 (CpB)	2.673(5)	**Zr-C8** (CpB)	2.713(6)

^{a}CpA is the "top" Cp ring and CpB is the "bottom" Cp ring in Figure 4.16.

carried out in neat 1-pentene at $-10\,^\circ$C, 6.6 M 1-pentene in toluene solution at $-10\,^\circ$C, or 3 M 1-pentene in toluene solution at $-10\,^\circ$C. Catalyst systems **35b**/MAO and **36a**/[CPh$_3$]$^+$[B(C$_6$F$_5$)$_4$]$^-$ were found to be highly syndioselective, with only the *rrrr* pentad apparent by ^{13}C NMR spectroscopy; **36b**/[CPh$_3$]$^+$[B(C$_6$F$_5$)$_4$]$^-$ was found to be slightly less syndioselective, with peaks for other pentads present in addition to the *rrrr* pentad. Owing to overlapping signals for most pentads in the ^{13}C NMR spectra for **36b**/[CPh$_3$]$^+$[B(C$_6$F$_5$)$_4$]$^-$, quantitative poly(1-pentene) tacticity data (e.g., [*rrrr*] percentages) was not reported. The authors attribute the decreased syndioselectivity of **36b**/[CPh$_3$]$^+$[B(C$_6$F$_5$)$_4$]$^-$ to insufficient reaction temperature control for this exothermic polymerization reaction.

The effect of monomer concentration on activity and molecular weight distributions varied for these catalysts. For precatalysts **36a** and **36b**, dilute polymerization conditions (6.6 M 1-pentene or 3 M 1-pentene) provided either negligible amounts of polymer or a polymerization appearing to have more than one active species (large PDI values and multimodal molecular weight distributions as evidenced by gel permeation chromatography). The authors postulate that decomposition of the starting material under reaction conditions serves as an explanation for these results.

C_s-symmetric [(SiMe$_2$)$_2${3,5-(SiMe$_3$)$_2$-C$_5$H}{4-CHMe$_2$-C$_5$H$_2$}]ZrCl$_2$ (37) was also synthesized by Bercaw and tested as a precatalyst for propylene polymerization (Figure 4.16).[35] In the presence of neat propylene at 0 $^\circ$C, **37**/MAO is syndioselective for propylene polymerization. While the M_w of the resulting polymer is comparable to that formed from **35b**/MAO under similar reaction conditions (1,000,000 versus 1,100,000, respectively), the syndioselectivity is slightly lower. For **37**/MAO, values of [*rrrr*] = 75.4% and [*r*] = 93.7% were reported (representative data is provided in Table 4.2); under similar reaction conditions for **35b**/MAO, values of [*rrrr*] = 98.9% and [*r*] = 99.6% were reported.

4.2.4.1 Mechanisms for Formation of Stereoerrors

For the C_s-symmetric zirconocenes that appear in Figure 4.16 (**35a–d**, **36a–b**, and **37**), there are three potential mechanisms for error formation in the polymerization of propylene; these three mechanisms are illustrated in Scheme 4.2. The first, enantiofacial misinsertion, generates an *mm* triad and is independent of propylene concentration. The second, site epimerization (chain movement from one side of the wedge to the other without monomer insertion), generates an *m* dyad, and is dependent on propylene concentration. The third, chain epimerization (inversion of the chirality of the β-C of the polymer chain) can generate either an *m* dyad or *mm* triad, and is also dependent on propylene concentration. (If there is retention of configuration at the metal center during chain

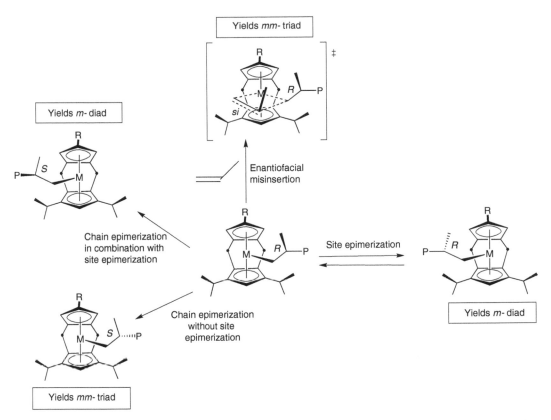

SCHEME 4.2 Mechanisms for stereoerror formation in propylene polymerization: enantiofacial misinsertion, site epimerization, and chain epimerization (with or without site epimerization). (P represents the polymer chain.)

epimerization, an *mm* triad is produced; if there is inversion of configuration at the metal center during chain epimerization, an *m* dyad is produced.) Busico[45,46] first put forth a mechanism for chain epimerization consisting of the following steps: (1) β-hydrogen elimination, (2) 180° rotation of the resulting geminally disubstituted olefin, (3) secondary insertion to give a tertiary alkyl complex, (4) 120° rotation about the M—C bond for this tertiary alkyl, (5) β-hydrogen elimination, (6) 180° rotation of the resulting geminally disubstituted olefin, and (7) primary insertion to regenerate a primary alkyl complex. These steps alone will generate an *mm* triad; if these steps are accompanied by a site epimerization, an *m* dyad is produced.

For precatalyst **37**, Bercaw postulates that the bulky trimethylsilyl groups at the 3 and 5 positions of the lower cyclopentadienyl ring (Figure 4.16) may encourage the polymer chain to rest near the center of the metallocene wedge, making site epimerization more prevalent. Site epimerization interrupts the regular alternation of monomer insertions at the two sides of the metallocene wedge, thus lowering syndioselectivity.

The pentad distribution for polypropylene formed by **37**/MAO was examined to support this hypothesis for lowered syndioselectivity. Both single *m* defects (...*rrrrmrrr*...) and double *m* defects (...*rrrrmmrr*.....) were detected. However, the *mrmr* pentad was not detected. The authors used two pentads to estimate the probability of these two errors types: $mm_{defect} = [rmmr]$ and $m_{defect} = 1/2[rmrr]$. With these estimates, the amount of site epimerization for **37**/MAO was reported as 2.1% (versus 1.8% for **35b**/MAO) and the amount of enantiofacial misinsertion for **37**/MAO was reported as 0.3% (versus 0.1% for **35b**/MAO). Detailed mechanisms for site epimerization and enantiofacial misinsertion appear in Schemes 4.3 and 4.4, respectively.

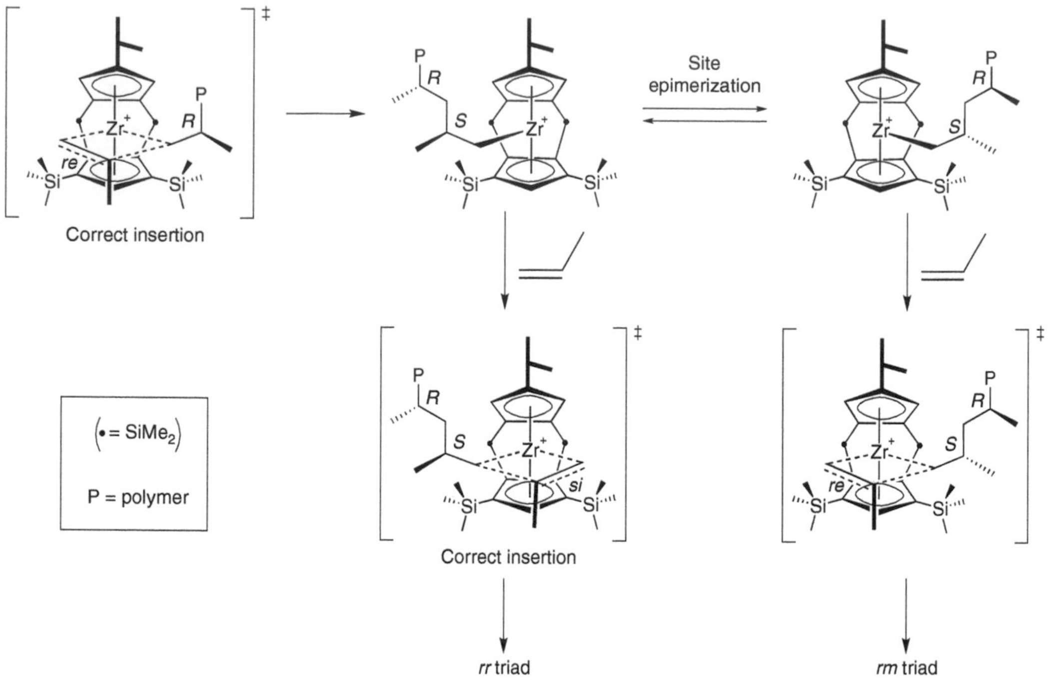

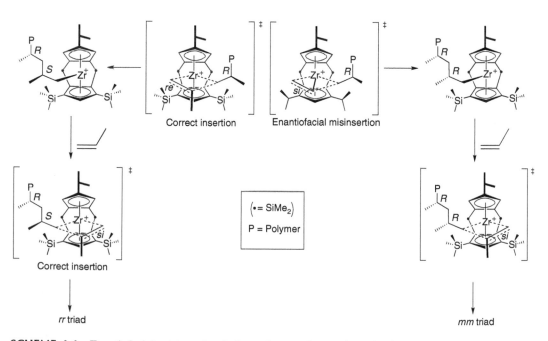

SCHEME 4.3 Site epimerization in isotactic propylene polymerization generates single *m* errors (i.e., *rm* triads).

SCHEME 4.4 Enantiofacial misinsertion in isotactic propylene polymerization generates double *m* errors (i.e., *mm* triads).

Bercaw examined **35a–c** under a range of polymerization conditions (varying propylene concentration and temperature) to determine the error-forming mechanisms for these precatalysts.[47] Polymer molecular weight and polydispersity index were also evaluated as a function of propylene concentration and temperature. For catalysts **35a–c**/MAO, as the concentration of propylene decreases (from neat propylene down to 0.8 M propylene in toluene solution), polypropylene molecular weight decreases. As temperature increases (from 20 up to 70 °C), polypropylene molecular weight decreases and PDI increases. Overall, **35a** was found to be more active for propylene polymerization than **35b** or **35c**. For catalysts **35a–c**/MAO, no regioerrors (2,1-insertions or 1,3-insertions) were observed through ^{13}C NMR spectroscopy.

With regard to tacticity, as the concentration of propylene decreases for catalysts **35a–c**/MAO, *r* dyad content decreases and *m* dyad content increases. This propylene dependence suggests that enantiofacial misinsertion is not an important error-forming mechanism for **35a–c**/MAO. At low propylene concentrations, 2–3% *mm* triads were detected for all three precatalysts. These *mm* triads were attributed to chain epimerization. At higher propylene concentrations, <1% *mm* triads were detected for **35a**/MAO and negligible *mm* triads were observed for **35b**/MAO and **35c**/MAO. This provides further support for the assertion that enantiofacial misinsertion is not significant for these catalysts, accounting for less than 1% of stereoerrors.

At low propylene concentrations, isolated *m* stereoerrors were also observed for **35a–c**/MAO and found to be more numerous than *mm* stereoerrors. The isolated *m* stereoerrors were found to have a strong dependence on propylene concentration. This gave further support for site epimerization as the major stereoerror-forming mechanism for **35a–c**/MAO. However, since chain epimerization occurring with simultaneous site epimerization can also lead to isolated *m* stereoerrors, the authors undertook a deuterium labeling study to probe the importance of chain epimerization as a potential mechanism for forming both single *m* and double *mm* stereoerrors.

For catalyst **35b**/MAO, a kinetic isotope effect experiment was devised using the monomers d_0-propylene and 2-d_1-propylene for polymerization. 2-d_1-Propylene was chosen for this experiment since β-hydride elimination is a component of the mechanism for chain epimerization. Polymerizations were carried out at 25 °C in toluene solution with 1 atm of propylene (0.8 M) (Figure 4.17). The deuterated polymer was examined by ^2D NMR spectroscopy; this technique showed that most of the deuterium was incorporated as –CD(CH$_3$), with 2–3% incorporation as –CH(CH$_2$D). The latter can be attributed to chain epimerization, which appears to be consistent with the 2–3% *mm* triads observed under similar conditions for propylene polymerization with **35a–c**/MAO.

Both polypropylene and poly(2-d_1-propylene) displayed similar tacticities ($r = 83.3\%$ and 84.2%, respectively), but the deuterated monomer produced higher molecular weight polymer (M_n 32,840 versus 56,830, respectively). Based on these molecular weights, a k(β-H)/k(β-D) value of 1.6 can be calculated, suggesting that β-H elimination is the predominant chain termination pathway. This result is consistent with the expectation that β-H elimination would be slower for the polymerization of 2-d_1-propylene than for polypropylene.

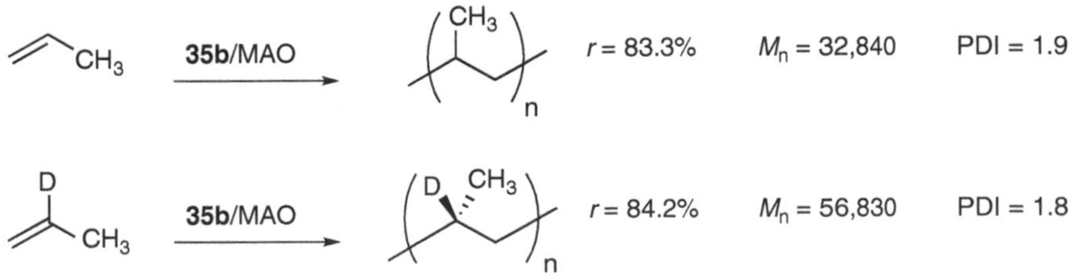

FIGURE 4.17 Use of **35b**/MAO as a catalyst for polymerization of d_0-propylene and 2-d_1-propylene.

The effect of temperature on stereoerror formation for catalysts **35a–c**/MAO was also studied. Neat propylene polymerizations were carried out at 20, 50, and 70 °C. In general, as temperature increased, the r dyad content decreased. Besides, at 70 °C, isolated m stereoerrors were more common than mm stereoerrors. These observations are consistent with site epimerization (a first order process) being in competition with propagation through migratory insertion (a second order process). Migratory insertion has a more negative $\Delta S^{\ddagger}$ and a smaller $\Delta H^{\ddagger}$ than site epimerization; thus, at higher temperature, site epimerization becomes more significant.

In summary, experiments at reduced propylene concentration and at elevated temperatures support the assertion that site epimerization is the major stereoerror-forming mechanism for precatalysts **35a–c**. Chain epimerization also appears to play a role as an error-forming mechanism, though to a lesser extent.

In addition to zirconocenes, C_s-symmetric scandocenes and yttrocenes with two dimethylsilyl interannular linkers (**38a–c**, **39**, Figure 4.18) have been synthesized and tested as polymerization catalysts for propylene and 1-pentene.[48,49] Owing to the electronics of these group 3 catalysts (14-electron, d^0 electronic configuration) no cocatalysts are required. However, metallocene chlorides cannot be used for polymerization; metallocene alkyl or hydride compounds are necessary.

Three tetramethylaluminate complexes were prepared (**38a–c**) along with an yttrocene complex with a bis(trimethylsilyl)methyl group (**39**). Scandocene **38b** and yttrocene **39** were analyzed through X-ray crystallography (see Table 4.1). As observed for related zirconocenes such as **35b**, the value of γ is large for **38b** and **39** (9.6 ° and 8.4 °, respectively). In the crystal structure of **39**, the bis(trimethylsilyl)methyl group sits to one side of the metallocene wedge, with an apparent interaction between the yttrium atom and one of the silicon carbon bonds of the alkyl ligand. This type of interaction has been observed for other yttrocenes (e.g., for $[Me_2Si(\eta^5\text{-}C_5Me_4)_2]YCH(SiMe_3)_2)$.[50]

Reaction of **39** with hydrogen gas $(H_{2(g)})$ to form the corresponding yttrocene hydride complex was also achieved. Hydrogenation of a benzene-d_6 solution of **39** provides a mixture of two dimeric yttrocene hydride complexes, as evidenced by ^{1}H NMR spectroscopy. One of these complexes selectively crystallized from benzene-d_6 solution and was found to be a "fly-over" dimer, in which each doubly bridged bis(cyclopentadienyl) ligand spans two yttrium atoms. While the other dimethylsilyl yttrocene hydride complex could not be characterized by X-ray crystallography, ^{1}H NMR data suggested that it was also a "fly-over" dimer. A mixture of these two dimeric yttrocene hydride complexes did not react with ethylene or α-olefins (such as 1-butene) over the course of several days at 22 °C.

Yttrocene **39** did not polymerize ethylene or α-olefins such as 1-pentene over the course of several days at 22 °C. For this reason, in situ activation of **39** with $H_{2(g)}$ was required to generate a viable

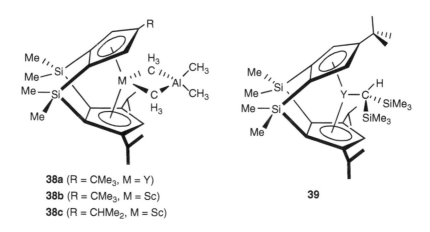

38a (R = CMe₃, M = Y)
38b (R = CMe₃, M = Sc)
38c (R = CHMe₂, M = Sc)

39

FIGURE 4.18 C_s-symmetric group 3 metallocenes do not provide syndiotactic polypropylene.

polymerization catalyst. Presumably, in situ activation of **39** with $H_{2(g)}$ generates a small amount of a monomeric yttrocene hydride complex that can react readily with an olefin. Notably, the reaction time for polymerization of 1-pentene using **39**/$H_{2(g)}$ ($\approx$1 week) is much longer than the reaction time for polymerization of 1-pentene using either **38a** or **38b**.

Yttrocene tetramethylaluminate complex **38a** was found to be less reactive towards ethylene and α-olefins than the corresponding scandocene, **38b**. This reactivity trend is consistent with the observed reactivity for **39**/$H_{2(g)}$. Compounds **38a–c** and **39**/$H_{2(g)}$ were tested for the polymerization of propylene and 1-pentene. 1-pentene polymerizations were carried out in neat monomer, either at $0\,°C$ or at room temperature ($\approx$22 °C). Propylene polymerizations were carried out either in neat monomer or with a 50/50 (v/v) mixture with toluene, all at $-5\,°C$. In general, scandocenes **38b** and **38c** provided higher polymer M_n values and were more active for polymerization than yttrocenes **38a** and **39**/$H_{2(g)}$.

Representative poly(1-pentene) tacticity data is provided in Table 4.4; representative polypropylene tacticity data is provided in Table 4.2. Poly(pentene) was produced using neat 1-pentene; poly(propylene) was produced using neat propylene or a 50/50 (v/v) mixture with toluene. These scandocenes and yttrocenes provide atactic polymers. The tacticity values range from $[r] = 28.62\%$ to $[r] = 52.12\%$ for poly(1-pentene) samples, and from $[r] = 48.77\%$ to $[r] = 61.13\%$ for poly(propylene) samples. Within experimental error, polymerization temperature and monomer dilution appear to exert a minimal effect on polymer tacticity.

The tacticity data was found to be consistent with a chain-end control mechanism, where the relative energies of diastereomeric transition states for olefin insertion dictate tacticity. Theoretical work by Bierwagen[51] suggested that these catalysts might be nonsyndioselective, owing to an energetic preference for an alkyl substituent (a model of a polymer chain) to rest in the center of the metallocene wedge for neutral, group 3 metallocenes. This is in contrast to the energetic preference for cationic group 4 metallocenes, in which an alkyl substituent has a preference to rest on the side of the metallocene wedge. For catalysts **38a–c** and **39**/$H_{2(g)}$, if the growing polymer chain prefers to rest in the center of the metallocene wedge after an olefin insertion, both sides of the metallocene wedge are comparable in accessibility for the following olefin insertion. Thus, regularly alternating migratory insertion is lost. Alternatively, if site epimerization effectively competes with regular migratory insertions, the same outcome would be realized. Scheme 4.5 illustrates the chain-end control mechanism for polymerization of α-olefins using a group 3 metallocene with two dimethylsilyl interannular linkers. In summary, these C_s-symmetric yttrocenes and scandocenes with two dimethylsilyl interannular linkers do not provide syndiotactic poly(α-olefin) and do not follow an enantiomorphic site control mechanism.

TABLE 4.4

Representative 1-pentene Polymerization Data for 38a–c and 39/$H_{2(g)}$[a]

Compound	Symmetry	Reaction Conditions	m (%)	r (%)	M_n (g/mol)	PDI
38b	C_S	0 °C, 10.75 h	48.31	51.73	9300	2.9
38b	C_S	20 °C, 2 h	47.92	52.12	4500	2.6
38c	C_S	0 °C, 9 h	62.76	37.26	—[b]	—
38c	C_S	21 °C, 2 h	61.69	28.62	3900	1.9
38a	C_S	22 °C, 7.5 d	64.39	35.51	2900	1.9
39/$H_{2(g)}$	C_S	22 °C, 7.5 d	59.23	40.81	—	—

[a] All polymerizations were carried out with neat 1-pentene.
[b] —: not available.

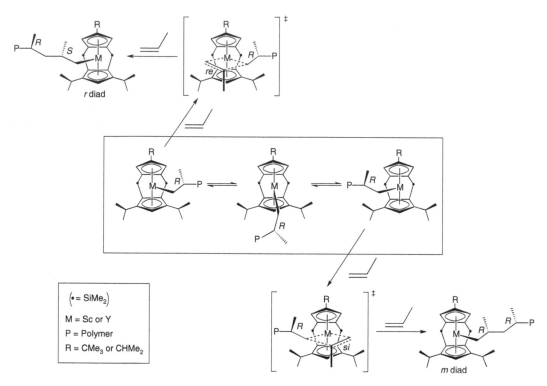

SCHEME 4.5 Chain-end control mechanism for stereocontrol of doubly bridged group 3 metallocenes (M is Sc or Y).

4.2.5 C_1-Symmetric (Varying Stereoselectivity)

When the synthesis of C_s-symmetric zirconocenes **35a–c** was first described, two C_1-symmetric zirconocenes (**40a–b**, Figure 4.19) were also noted.[40] In combination with MAO, both **40a** and **40b** were used as catalysts for propylene polymerization. Pertinent tacticity data for **40a–b**/MAO appears in Table 4.2. At 0 °C in liquid propylene, **40a** behaves similar to C_s-symmetric **35b** in terms of activity, though it is slightly less syndioselective ($[rrrr] = 83.1\%$ and $[r] = 94.4\%$ for **40a** versus $[rrrr] = 98.9\%$ and $[r] = 99.6\%$ for **35b**). However, when polymerization is carried out at 25 °C in dilute propylene (40 psig propylene in toluene solution), the syndioselectivity of precatalyst **40a** decreases dramatically, to $[rrrr] = 20.0\%$ and $[r] = 62.0\%$. Precatalyst **40b** yields even more dramatic results, at 0 °C in liquid propylene $[rrrr] = 41.8\%$ and $[r] = 73.5\%$ and at 25 °C in dilute propylene (40 psig propylene in toluene solution) $[rrrr] = 0.0\%$ and $[r] = 14.6\%$. When polymerization is carried out with precatalyst **40b** at 25 °C in dilute propylene (either 40 psig or 10 psig propylene in toluene solution), the resulting polymers are moderately isotactic with $mmmm$ values of 61.2% and 58.5%, respectively (and m values of 85.4% and 82.4%, respectively).

Site epimerization is used to account for the isoselectivity of **40b** at 25 °C at dilute propylene concentrations. When site epimerization is competitive with olefin insertion for a C_1-symmetric catalyst, it follows that olefin coordination and insertion from one side of the metallocene wedge will be energetically favored in comparison to the other side. Repetitive olefin insertions from one side of the metallocene wedge with the same enantioface will lead to a sequence of m dyads. The mechanism for site epimerization is illustrated in Scheme 4.3 (*vide supra*); this concept illustrated for C_s-symmetric **37** can be extrapolated to this case.

In 1999, Bercaw synthesized C_1-symmetric zirconocene **40c**[47] that has a (+)-menthyl substituent ($1R,2S,5R$-menthyl) on the "top" cyclopentadienyl ring (Figure 4.19). Complex **40c** was

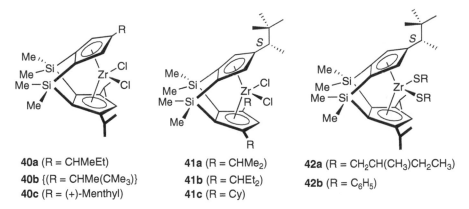

40a (R = CHMeEt)
40b {(R = CHMe(CMe₃)}
40c (R = (+)-Menthyl)

41a (R = CHMe₂)
41b (R = CHEt₂)
41c (R = Cy)

42a (R = CH₂CH(CH₃)CH₂CH₃)
42b (R = C₆H₅)

FIGURE 4.19 C_1-symmetric zirconocene precatalysts where -SCH₂CH(CH₃)CH₂CH₃ is (S)-2-methyl-1-butanethiolate.

characterized by X-ray crystallography, and six independent molecules were found in the unit cell (see Table 4.1). Average angles of 122 ° for γ and 105 ° for the Cl–Zr–Cl bond angle were reported.

Both **40b** and **40c** were tested as precatalysts for propylene polymerization in combination with MAO cocatalyst. The concentration of propylene was varied to examine the effect on polymer molecular weight, PDI, and tacticity. When the propylene concentration was decreased from neat propylene down to dilute conditions (0.5 M propylene in toluene solution), the resulting polypropylene was found to change from moderately syndiotactic, to imperfect hemiisotactic, to moderately isotactic. In addition, a nearly linear decrease of [r] was observed with decreasing propylene concentration. These tacticity results can be attributed to a site epimerization process competing with regularly alternating migratory insertion. Since site epimerization is a unimolecular process and migratory insertion is a bimolecular process, it follows that the relative rates of these two processes will change as a function of propylene concentration. For these catalysts, migratory insertion appears to be favored at high propylene concentration, thus providing moderately syndiotactic polymer. At low propylene concentration, site epimerization dominates. For catalyst **40b**/MAO at low propylene concentration (0.5 M propylene in toluene solution), isoselective enantiomorphic site control is followed. These results are consistent with the mechanistic description provided for **40a–b**/MAO (*vide supra*).

At intermediate propylene concentrations, where the resultant polypropylene displays $r \approx m \approx 50\%$, the polymers are not atactic. In fact, the pentad data suggests that these polymers are hemiisotactic. The following is the characteristic pentad distribution for hemiisotactic polymers: *mmmm*: *mmmr*: *rmmr*: *mmrr*: (*rmrr* + *mrmm*): *rmrm*: *rrrr*: *rrrm*: *mrrm* = 3:2:1:4:0:0:3:2:1. In these polymers, the *rmrm* pentad is absent, and the amount of (*rmrr* + *mrmm*) is quite low (for **40b–c**/MAO, it is present in 9.1% and 11.4%, respectively, at 3.4 M propylene in toluene solution). These results can also be attributed to the site epimerization mechanism. If chain swinging to the less hindered side of the metallocene wedge (for energetically preferred olefin insertion) occurs in competition with regularly alternating migratory insertion, hemiisotactic polymer can form.

Interestingly, this same propylene dependence for tacticity is not observed for the Ewen catalyst, [Me₂C{(3-Me-C₅H₄)(fluorenyl)}]ZrCl₂/MAO.[52] This precatalyst is a C_1-symmetric, methyl-substituted version of **33** (Figure 4.14) and was chosen by the authors for comparison purposes since it was reported to produce polypropylene with a microstructure that approximates hemiisotactic. For [Me₂C{(3-Me-C₅H₄)(fluorenyl)}]ZrCl₂/MAO, hemiisotactic polypropylene is formed over a range of propylene concentrations; it appears that site epimerization does not compete with propagation. This is a noteworthy example of singly bridged and doubly bridged metallocenes displaying different mechanisms for stereoerror formation in propylene polymerization.

Reaction temperature was also varied for propylene polymerization using **40b–c**/MAO to examine the effect on polymer tacticity. In general, as reaction temperature decreased, [r] content increased. For example, polypropylene formed by **40c**/MAO in neat propylene at 25 °C had an [r] content of 78.7%; polypropylene formed at 0 °C had an [r] content of 90.4%. This is consistent with unimolecular site epimerization occurring in competition with bimolecular chain propagation where the latter is favored at lower temperature.

The effect of propylene concentration on polymer molecular weight was inspected for C_s-symmetric precatalysts **35a–c** and C_1-symmetric precatalysts **40b–c**, all activated with MAO. For **35a–c**, as propylene concentration increases, there is a roughly linear increase in polymer molecular weight. For **40b–c** (in particular, **40b**), as propylene concentration increases, polymer molecular weight increases; however, there is slightly less than first order dependence of polymer molecular weight on propylene concentration.

Based on ^{1}H NMR analysis of low molecular weight polypropylene formed by **40b**/MAO at 25 °C with a propylene concentration of 0.5 M (in toluene solution), the predominant chain termination pathway appears to be β-H elimination (i.e., vinylidene end groups are present). The primary kinetic isotope effect observed for **35b**/MAO (*vide supra*) is also indicative of β-H elimination operating as the predominant chain termination pathway. No β-CH₃ elimination was observed for these catalysts; specifically, no vinylic end groups are present in the ^{1}H NMR spectra. For **35a–c** and **40c**, chain transfer was found to be primarily unimolecular, derived from β-H elimination. For **40b**, a small bimolecular contribution (chain transfer to monomer) was also suggested, owing to the less than first order dependence of molecular weight on propylene concentration.

4.2.5.1 Kinetic Resolution of Chiral Olefins

Recently, Bercaw described the synthesis of optically active, C_1-symmetric *ansa*-zirconocene precatalysts for the kinetic resolution of chiral α-olefins.[53] Compounds of the form [(SiMe₂)₂{3,5-R₂-C₅H}{4-((S)-CHMeCMe₃)-C₅H₂}]ZrCl₂, where R is CHMe₂ (**41a**), CHEt₂ (**41b**), or cyclohexyl (**41c**) were prepared (Figure 4.19). The stereochemistry for these enantiopure zirconocenes is derived from an enantiopure methylneopentyl substituent on the "upper" cyclopentadienyl ring. The parent compound, **41a**, was derivatized by reaction with two equivalents of lithium (S)-2-methyl-1-butanethiolate, thus forming (S,S,S)-[(SiMe₂)₂{3,5-(CHMe₂)₂-C₅H}{4-((S)-CHMeCMe₃)-C₅H₂}]Zr(SCH₂CH(CH₃)CH₂CH₃)₂ (**42a**) in greater than 98% optical purity (Figure 4.19). Besides, the parent compound, **41a**, was derivatized by reaction with two equivalents of Li(SC₆H₅), thus forming [(SiMe₂)₂{3,5-(CHMe₂)₂-C₅H}{4-((S)-CHMeCMe₃)-C₅H₂}]Zr(SC₆H₅)₂ (**42b**, Figure 4.19); this compound was prepared solely for characterization through X-ray crystallography.

Compounds **41a–c** and **42b** were characterized through X-ray crystallography (see Table 4.1). In the solid state, the enantiopure methylneopentyl substituent adopts a confirmation with the *tert*-butyl group directed nearly perpendicular to the cyclopentadienyl ring plane, and thus the methine hydrogen and methyl group encroach upon the metallocene wedge. This orientation of substituents is depicted in Figure 4.19. The authors propose that the steric environment presented by complexes **41a–c** is fairly similar, since both the R = CHEt₂ and R = Cy (where Cy is cyclohexyl) substituents adopt similar orientations on the "lower" cyclopentadienyl ring as compared to the parent compound (**35b**, where R = CHMe₂). The authors discuss the solid-state orientation of the thiophenoxy ligands of **42b** in some detail.

In addition, diastereomerically pure precatalysts were prepared with the stereochemistry derived from one enantiopure 1-cyclohexylethyl substituent on the "lower" cyclopentadienyl ring (**43** and **44**, Figure 4.20). Crystallographic data was not provided for **43** or **44**.

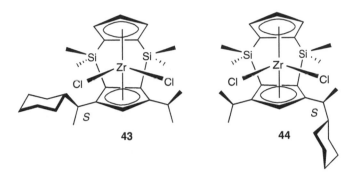

FIGURE 4.20 Diastereomerically pure C_1-symmetric precatalysts.

3-MP1   3-MH1   3,5,5-TMH1   3,4-DMP1   4-MH1

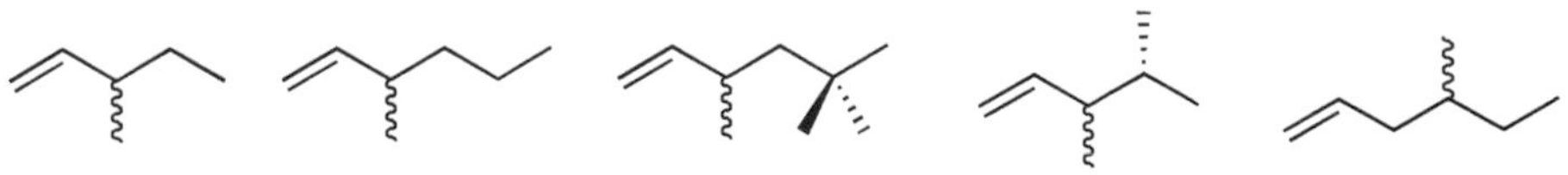

FIGURE 4.21 Chiral α-olefins employed for kinetic resolution experiments.

The authors carried out polymerizations with bulky, racemic α-olefins with substituents in either the 3- or 4-positions, or both. The monomers tested in this report appear in Figure 4.21: 3-methyl-1-pentene (3-MP1); 3-methyl-1-hexene (3-MH1); 3,5,5-trimethyl-1-hexene (3,5,5-TMH1); 3,4-dimethyl-1-pentene (3,4-DMP1); and 4-methyl-1-hexene (4-MH1). It is notable that these doubly bridged precatalysts are capable of polymerizing α-olefins with steric bulk in these positions; most singly bridged metallocenes are incapable of doing so. Only singly bridged C_s-symmetric [(Me$_2$C){(fluorenyl)(Cp)}]ZrCl$_2$ (Figure 4.14, **33**) and racemic C_2-symmetric metallocenes *rac*-[Me$_2$Si(Ind)$_2$]ZrCl$_2$, *rac*-[Me$_2$Si(BenzInd)$_2$]ZrCl$_2$ (BenzInd = benz[*e*]indenyl), *rac*-[Me$_2$Si(2-Me-BenzInd)$_2$]ZrCl$_2$, and *rac*-[C$_2$H$_4$(Ind)$_2$]ZrCl$_2$ have been said to polymerize 3-methyl-1-pentene prior to this report;[54,55] in all cases, MAO/Al(^{13}CH$_3$)$_3$ was used as the cocatalyst. These C_s- and C_2-symmetric precatalysts are not capable of kinetic resolution owing to their symmetry.[54,55]

Precatalysts **41a–c** and **44** were activated with MAO and tested for kinetic resolution. Tetradecane was used as a solvent for these polymerizations at 25 °C. Kinetic resolution was reported by using stereoselectivity factors, or *s* values, where *s* = (rate of fast reacting enantiomer)/(rate of slow reacting enantiomer). Experimentally, *s* may be calculated by using the following equation: $s = \ln[(1-c)(1-ee)]/\ln[(1-c)(1+ee)]$, where ee is the enantiomeric excess of the recovered olefin and c is the fraction conversion.[56] If no kinetic resolution is achieved, *s* = 1. The authors assayed the fraction conversion, c, by gas chromatography (GC) analysis of two aliquots for each polymerization run: (1) an aliquot removed immediately before the start of polymerization (i.e., immediately before the addition of zirconocene catalyst) and (2) an aliquot removed after the desired conversion was reached; in all cases, tetradecane was used as the internal standard.

For the parent compound, **41a**, an *s* value of 2.6(2) was obtained for 3-MP1 polymerization. The best kinetic resolution using **41a** was obtained for 3,4-DMP1 with an *s* value of 16.8(8). Addition of another methyl group to the monomer at the 5-position (i.e., 3,5,5-TMH1) decreased the level of kinetic resolution to *s* = 2.1(1). The presence of a single substituent in the 5-position (i.e., 4-MH1) lead to poor kinetic resolution with *s* = 1.1. In all cases, the *S* enantiomer of the racemic olefin is preferentially incorporated in the polymer. Experiments using **41b** and **41c** as precatalysts revealed the same trends. For these two complexes, 3,4-DMP1 yielded the best kinetic resolution, followed by 3-MP1 with slightly poorer kinetic resolution. As observed when using **41a**, 4-MH1 gave poor kinetic resolution when **41b** or **41c** were used.

For precatalyst **44**, the best kinetic resolution was found for 3,4-DMP1 ($s = 2.6$). The kinetic resolution for other α-olefins was negligible (s values from 1.0 to 1.2). In all cases, the R enantiomer was preferentially polymerized using precatalyst **44**. This is consistent with the authors' comments regarding ^{1}H NMR data and molecular mechanics calculations for **44**; it appears that the axial orientation of the cyclohexyl group means that it has a minimal role in steric interactions with coordinated monomer or the growing polymer. The authors go on to predict that both **43** and **44** would be expected to behave like an achiral, C_s-symmetric system.

Polymer tacticity was assayed by using parent compound **41a** as a precatalyst for polymerization. Poly(3-methyl-1-pentene) made by using **41a**/MAO is predominantly isotactic. Quantitative measures of isotacticity (e.g., [*mmmm*]) were not provided. The authors suggest that the isotactic microstructure results from a rapid site epimerization process. It appears that the rate of site epimerization greatly exceeds the rate of propagation; thus, the olefin is repetitively enchained on one side of the metallocene wedge, with the same olefin enantioface presented each time. The authors suggest an indirect mechanism for selectivity, where the methylneopentyl substituent on the "top" cyclopentadienyl ring dictates the preferred side of the metallocene wedge for the polymer chain. The polymer chain then directs the incoming monomer to the opposite side of the metallocene wedge. The proposed favored transition state for olefin insertion is illustrated in Figure 4.22. As shown, the orientation of the methylneopentyl substituent favors the polymer chain resting on the left side of the metallocene wedge. This directs the S enantiomer of the chiral α-olefin to the right side of the metallocene wedge. Coordination and subsequent insertion occurs by presentation of the *si* olefin face to the metal.

To provide further support for this proposed transition state, molecular mechanics calculations (using CAChe: Computer-Aided Chemistry Modeling Software on Windows, version 4.9, Fujitsu America, Inc.) were carried out to study the olefin binding intermediate using ([(SiMe$_2$)$_2$\{3,5-(CHMe$_2$)$_2$-C$_5$H\}\{4-((S)-CHMeCMe$_3$)-C$_5$H$_2$\}]ZrCH$_3$)$^+$. The authors did not specify the level of theory used for these calculations. For 3-MP1, coordination of the S enantiomer on the right side of the metallocene wedge is slightly favored ($\Delta E = 0.8$ kcal/mol versus the R enantiomer). However, for 3,4-DMP1, the R enantiomer is favored by more than 1.5 kcal/mol for the right side of the metallocene wedge; a result that is apparently inconsistent with the experimental data. To account for the experimental data, the authors conclude that the ΔE of the diastereomeric transition states for alkyl migration (not evaluated by CAChe) must determine olefin enantiomer preferences. In addition, the importance of the asymmetric β and γ centers of the migrating polymer chain in the diastereomeric transition states for alkyl migration was noted.

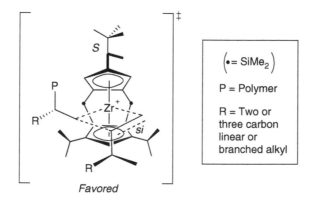

FIGURE 4.22   Proposed favored olefin insertion transition state for activated **41a**.

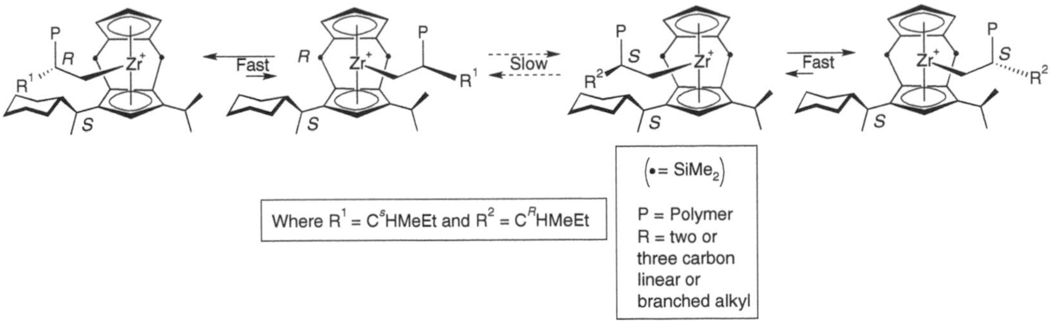

SCHEME 4.6   Proposed polymerization mechanism for activated **43**.

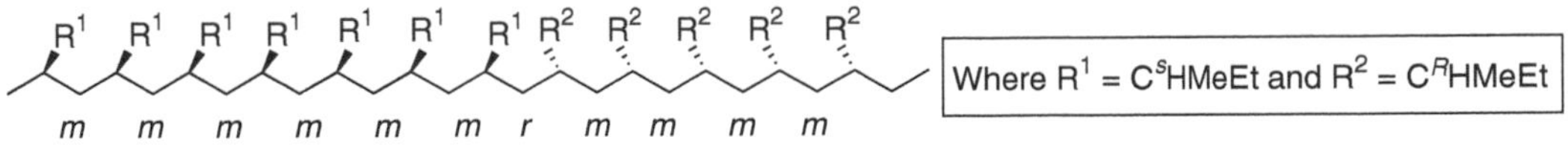

FIGURE 4.23   The resulting polymer microstructure formed by suitably activated **43**.

Polymer tacticity was also assayed by using compound **43** as a precatalyst for polymerization. The prediction for **43** was that it would behave like an achiral, C_s-symmetric system (*vide supra*). This prediction was borne out by using **43** as a precatalyst for polymerization of neat propylene; a predominately syndiotactic polymer was formed with $[r] = 91.5\%$. Pertinent tacticity data appears in Table 4.2. However, when **43** is used to polymerize 3-methyl-1-pentene, the resulting polymer is predominantly isotactic (similar to the results from precatalyst **41a**). Quantitative tacticity data was not given.

The authors justify these results, along with the corresponding low level of kinetic resolution of 3- and 4-substituted α-olefins, by a chain-end mediated site epimerization mechanism (see Scheme 4.6). In this mechanism, site epimerization controls which diastereomeric active site is favored for a β-R or β-S chain end. For example, a polymer chain containing a β-R chain end will favor the less hindered side of the metallocene wedge (the right side), and thus coordination of the S olefin enantiomer will be favored on the left side of the metallocene wedge. It follows that a polymer chain containing a β-S chain end will favor the more hindered side of the metallocene wedge (the left side), and thus coordination of the R olefin enantiomer will be favored on the right side of the metallocene wedge.

Site epimerization is estimated to be fast relative to olefin uptake and insertion. Thus, repetitive enchainment of one olefin enantiomer will occur, producing an isotactic block of poly(α-olefin) until a syndiotactic enchainment error occurs. When this happens, olefin insertion will be switched to the opposite side, and another block of isotactic poly(α-olefin) will be produced. Thus, the resulting polymer is predominantly isotactic with isolated r dyads; stereoblocks of the two olefin enantiomers surround each isolated r dyad (Figure 4.23).

The authors conclude that the importance of polymer chain-end chirality (especially at the β and γ positions) is largely unknown. Further, the relative importance of enantiomorphic site control versus polymer chain-end chirality is not known. In the Bercaw laboratories, copolymerization experiments with chiral and achiral olefins are underway to address these issues.

4.3   CONCLUSIONS

Doubly bridged *ansa*-metallocenes are a relatively small yet significant category of group 3 and 4 metallocenes. More experimentation is needed to supplement what has already been learned about how precatalyst structure and symmetry affects α-olefin polymerization activity, catalyst decomposition, regioselectivity, and stereoselectivity. As described in this chapter, stereoselectivity and stereoerrors formation can deviate from what is observed for analogous singly bridged *ansa*-metallocenes. Recent successes in producing highly syndio- and regio-selective catalysts along with advances in polymerizing sterically hindered α-olefins (with steric bulk in the C_3 and C_4 positions) suggest that doubly bridged group 3 and 4 *ansa*-metallocenes will receive continuing attention.

REFERENCES AND NOTES

1. Brintzinger, H. H.; Fischer, D.; Mülhaupt, R.; Rieger, B.; Waymouth, R. M. Stereospecific olefin polymerization with chiral metallocene catalysts. *Angew. Chem., Int. Ed. Engl.* **1995**, *34*, 1143–1170.

2. Britovsek, G. J. P.; Gibson, V. C.; Wass, D. F. The search for new-generation olefin polymerization catalysts: Life beyond metallocenes. *Angew. Chem., Int. Ed.* **1999**, *38*, 428–447.

3. Coates, G. W. Precise control of polyolefin stereochemistry using single-site metal catalysts. *Chem. Rev.* **2000**, *100*, 1223–1252.

4. Resconi, L.; Cavallo, L.; Fait, A.; Piemontesi, F. Selectivity in propene polymerization with metallocene catalysts. *Chem. Rev.* **2000**, *100*, 1253–1346.

5. Smith, J. A.; Seyerl, J. V.; Huttner, G.; Brintzinger, H. H. *ansa*-Metallocene derivatives: Molecular structure and proton magnetic resonance spectra of methylene- and ethylene-bridged dicyclopentadienyltitanium compounds. *J. Organomet. Chem.* **1979**, *173*, 175–185.

6. Zachmanoglou, C. E.; Docrat, A.; Bridgewater, B. M.; Parkin, G.; Brandow, C. G.; Bercaw, J. E.; Jardine, C. N.; Lyall, M.; Green, J. C.; Keister, J. B. The electronic influence of ring substituents and *ansa*-bridges in zirconocene complexes as probed by infrared, spectroscopic, electrochemical, and computational studies. *J. Am. Chem. Soc.* **2002**, *124*, 9525–9546.

7. Shapiro, P. J. The evolution of the *ansa*-bridge and its effect on the scope of metallocene chemistry. *Coord. Chem. Rev.* **2002**, *231*, 67–81.

8. Zhu, B. L.; Wang, B. Q.; Xu, S. S.; Zhou, X. Z. Progress on the doubly bridged bis(cyclopentadienyl) metal complexes. *Youji Huaxue* **2003**, *23*, 1049–1057; *Chem. Abstr.* **2003**, 830385.

9. Dorer, B.; Prosenc, M. -H.; Rief, U.; Brintzinger, H. H. Synthesis and structure of titanocene, zirconocene, and vanadocene dichloride complexes with two ethanediyl bridges. *Organometallics* **1994**, *13*, 3868–3878.

10. Hafner, K.; Mink, C.; Lindner, H. J. Synthesis and structure of the first [2$_2$]metallocenophanes. *Angew. Chem., Int. Ed. Engl.* **1994**, *33*, 1479–1480.

11. Mink, C.; Hafner, K. Synthesis and reactions of tricyclo[9.3.0.0$^{4.8}$]tetradeca-4,7,11,14-tetraene. *Tetrahedron Lett.* **1994**, *35*, 4087–4090.

12. Grossman, R. B.; Tsai, J. -C.; Davis, W. M.; Gutiérrez, A.; Buchwald, S. L. Synthesis and structure of a C_2-symmetric, doubly bridged *ansa*-titanocene complex. *Organometallics* **1994**, *13*, 3892–2896.

13. Cano, A.; Cuenca, T.; Gómez-Sal, P.; Royo, B.; Royo, P. Double-dimethylsilyl-bridged dicyclopentadienyl group 4 metal complexes. X-ray molecular structure of $M[(Me_2Si)_2(\eta^5-C_5H_3)_2]Cl_2$ (M = Ti, Zr) and $(TiCl_3)_2\{\mu-[(Me_2Si)_2(\eta^5-C_5H_3)_2]\}$. *Organometallics* **1994**, *13*, 1688–1694.

14. Cano, A.; Cuenca, T.; Gómez-Sal, P.; Manzanero, A.; Royo, P. Dicyclopentadienyl titanium and zirconium complexes with the double bridged bis(dimethylsilanodiyl)dicyclopentadienyl $[(Me_2Si)_2(\eta^5-C_5H_3)_2]^{2-}$ ligand: X-ray molecular structure of $[Ti\{(SiMe_2)_2(\eta^5-C_5H_3)_2\}Me_2]$. *J. Organomet. Chem.* **1996**, *526*, 227–235.

15. Royo, P. Early transition-metal compounds with doubly silyl-bridged dicyclopentadienyl ligands. *New J. Chem.* **1997**, *21*, 791–796.

16. Bulls, A. R. Carbon-hydrogen bond activation by peralkylhafnocene and peralkylscandocene derivatives. Ph.D. Thesis, California Institute of Technology, Pasadena, CA, 1988.

17. Daundikhed, N.; Chong, A. A.; Zubris, D. L. Villanova University, Villanova, PA. Unpublished work, 2005.

18. Chong, A. A.; Zubris, D. L. Progress toward the synthesis of silicon/phosphorous doubly-bridged group 4 metallocenes. Poster presentation at the 32nd Northeast Regional Meeting of the American Chemical Society, Rochester, NY, October 31–November 3, 2004; Paper GEN-081.

19. Daunikhed, N. Progress towards the synthesis of a doubly chelating metallocene analog and a doubly bridged *ansa*-metallocene for the polymerization of α-olefins. M.S. Thesis, Villanova University, Villanova, PA, August 2005.

20. Miyake, S.; Henling, L. M.; Bercaw, J. E. Synthesis, molecular structure, and racemate-meso interconversion for *rac*-(Me$_2$Si)$_2$\{η^5-C$_5$H-3-(CHMe$_2$)-5-Me\}$_2$MCl$_2$ (M = Ti and Zr). *Organometallics* **1998**, *17*, 5528–5533.

21. Deck, P. A.; Fisher, T. S.; Downey, J. S. Boron-silicon exchange reactions of boron trihalides with trimethylsilyl-substituted metallocenes. *Organometallics* **1997**, *16*, 1193–1196.

22. Jung, J.; Noh, S. K.; Lee, D.; Park, S. K.; Kim, H. Synthesis and characterization of group 4 metallocene complexes with two disiloxanediyl bridges. *J. Organomet. Chem.* **2000**, *595*, 147–152.

23. Ciruelos, S.; Cuenca, T.; Gomez-Sal, P.; Manzanero, A.; Royo, P. New silyl-substituted cyclopentadienyl titanium and zirconium complexes. X-ray molecular structures of [TiCl$_2$\{μ-(OSiMe$_2$-η^5-C$_5$H$_4$)\}]$_2$ and [ZrCl$_2$\{μ-[(η^5-C$_5$H$_4$)SiMe$_2$OSiMe$_2$(η^5-C$_5$H$_4$)]\}]. *Organometallics* **1995**, *14*, 177–185.

24. Ihara, E.; Nodono, M.; Yasuda, H.; Kanehisa, N.; Kai, Y. Single site polymerization of ethylene and 1-olefins initiated by rare earth metal complexes. *Macromol. Chem. Phys.* **1996**, *197*, 1909–1917.

25. Ihara, E.; Nodono, M.; Katsura, K.; Adachi, Y.; Yasuda, H.; Yamagashira, M.; Hashimoto, H.; Kanehisa, N.; Kai, Y. Synthesis and olefin polymerization catalysts of new divalent samarium complexes with bridging bis(cyclopentadienyl) ligands. *Organometallics* **1998**, *17*, 3945–3956.

26. Ihara, E.; Yoshioko, S.; Furo, M.; Katsura, K.; Yasuda, H.; Mohri, S.; Kanehisa, N.; Kai, Y. Synthesis and olefin polymerization catalysts of new trivalent samarium and yttrium complexes with bridging bis(cyclopentadienyl) ligands. *Organometallics* **2001**, *20*, 1752–1761.

27. Haan, K. H.; Boer, J. L.; Teuben, J. H.; Speck, A. L.; Kojic-Prodic, B.; Hays, G. R.; Huis, R. Synthesis of monomeric permethyl yttrocene derivatives. The crystal structures of Cp$_2$*YN(SiMe$_3$)$_2$ and Cp$_2$*YCH(SiMe$_3$)$_2$. *Organometallics* **1986**, *5*, 1726–1733.

28. Evans, W. J.; DeCoster, D. M.; Greaves, J. Field desorption mass spectrometry studies of the samarium-catalyzed polymerization of ethylene under hydrogen. *Macromolecules* **1995**, *28*, 7929–7936.

29. Jeske, G.; Lauke, H.; Mauermann, H.; Swepston, P. N.; Schumann, H.; Marks, T. J. Highly reactive organolanthanides. Synthesis, chemistry, and structures of 4f hydrocarbyls and hydrides with chelating bis(polymethylcyclopentadienyl) ligands. *J. Am. Chem. Soc.* **1985**, *107*, 8103–8110.

30. Watson, P. L. Ziegler–Natta polymerization: The lanthanide model. *J. Am. Chem. Soc.* **1982**, *104*, 337–339.

31. Watson, P. L.; Parshall, G. W. Organolanthanides in catalysis. *Acc. Chem. Res.* **1985**, *18*, 51–56.

32. Edelmann, F. T. Lanthanide metallocenes in homogeneous catalysis. *Top. Curr. Chem.* **1996**, *179*, 247–267.

33. Weiss, K.; Neugebauer, U.; Blau, S.; Lang, H. Untersuchungen von polymerisations- und metathesereaktionen, XXIII. Einfach und zeifach dimethylsilylen-verbrückte metallocendichloride des Ti, Zr, und Hf in der ethen- und propen- polymerization. *J. Organomet. Chem.* **1996**, *520*, 171–179.

34. Lang, H.; Blau, S.; Pritzkow, H.; Zsolnai, L. Synthesis and reaction behavior of the novel mono(σ-alkynyl)titanocene chloride [(η^5-C$_5$H$_2$SiMe$_3$)SiMe$_2$]$_2$Ti(Cl)(CCSiMe$_3$). *Organometallics* **1995**, *14*, 1850–1854.

35. Miyake, S.; Bercaw, J. E. Doubly [SiMe$_2$]-bridged C_s- and C_{2v}-symmetric zirconocene catalysts for propylene polymerization. Synthesis and polymerization characteristics. *J. Mol. Catal. A: Chem.* **1998**, *128*, 29–39.

36. Mansel, S.; Rief, U.; Prosenc, M. H.; Kirsten, R.; Brintzinger, H. H. *ansa*-Metallocene derivatives XXXII. Zirconocene complexes with a spirosilane bridge: Syntheses, crystal structures and properties as olefin polymerization catalysts. *J. Organomet. Chem.* **1996**, *512*, 225–236.

37. Menegele, W.; Diebold, J.; Troll, C.; Röll, W.; Brintzinger, H. H. *ansa*-Metallocene derivatives. 27. Chiral zirconocene complexes with two dimethylsilylene bridges. *Organometallics* **1993**, *12*, 1931–1935.

38. Halterman, R. L.; Tretyakov, A.; Combs, D.; Chang, J.; Khan, M. A. Synthesis and structure of C_2-symmetric, doubly bridged bis(indenyl)titanium and –zirconium dichlorides. *Organometallics* **1997**, *16*, 3333–3339.

39. Cavallo, L.; Corradini, P.; Guerra, G.; Resconi, L. Doubly bridged *ansa*-zirconocenes based on the norbornadiene skeleton: A quantum mechanical and molecular mechanics study. *Organometallics* **1996**, *15*, 2554–2263.

40. Herzog, T. A.; Zubris, D. L.; Bercaw, J. E. A new class of zirconocene catalysts for the syndiospecific polymerization of propylene and its modification for varying polypropylene from isotactic to syndiotactic. *J. Am. Chem. Soc.* **1996**, *188*, 11988–11989.

41. Ewen, J. A.; Jones, R. L.; Razavi, A.; Ferrara, J. D. Syndiospecific propylene polymerizations with group IVB metallocenes. *J. Am. Chem. Soc.* **1988**, *110*, 6255–6256.

42. Bercaw, J. E.; Herzog, T. A. Stereospecific metallocene polymerization catalysts for olefins. U.S. Patent 5,708,101 A (California Institute of Technology), August 12, 1999.

43. Herzog, T. A. Deuterium isotope effects as evidence for alpha-agostic assistance in Ziegler–Natta catalysts, design, synthesis, and reactivity of a new class af highly syndiospecific Zeigler–Natta polymerization catalysts. Ph.D. Thesis, California Institute of Technology, Pasadena, CA, 2007.

44. Veghini, D.; Day, M. W.; Bercaw, J. E. Preparation of doubly-silylene-bridged zirconocene alkyl complexes, $(Me_2Si)_2\{\eta^5\text{-}C_5H_2\text{-}4\text{-}CHMe_2\}(\eta^5\text{-}C_5H\text{-}3,5\text{-}(CHMe_2)_2\}ZrR_2$ (R = CH_3, CH_2Ph) and investigations of their activity in 1-pentene polymerization. Molecular structure of $(Me_2Si)_2\{\eta^5\text{-}C_5H_2\text{-}4\text{-}CHMe_2\}(\eta^5\text{-}C_5H\text{-}3,5\text{-}(CHMe_2)_2\}Zr(CH_2Ph)_2$. *Inorg. Chim. Acta* **1998**, *280*, 226–232.

45. Busico, V.; Cipullo, R. Influence of monomer concentration on the stereospecificity of 1-alkene polymerization promoted by C_2−symmetric *ansa*-metallocene catalysts. *J. Am. Chem. Soc.* **1994**, *116*, 9329–9330.

46. Busico, V.; Cipullo, R. Growing chain isomerizations in metallocene-catalyzed Ziegler–Natta 1-alkene polymerization. *J. Organomet. Chem.* **1995**, *497*, 113–118.

47. Veghini, D.; Henling, L. M.; Burkhardt, T. J.; Bercaw, J. E. Mechanisms of stereocontrol for doubly silylene-bridged C_s- and C_1-symmetric zirconocene catalysts for propylene polymerization. Synthesis and molecular structure of $Li_2[(1,2\text{-}Me_2Si)_2\{C_5H_2\text{-}4\text{-}(1R,2S,5R\text{-}menthyl)\}\{C_5H\text{-}3,5\text{-}(CHMe_2)_2)] \cdot$ 3THF and $[(1,2\text{-}Me_2Si)_2\{\eta^5\text{-}C_5H_2\text{-}4\text{-}(1R,2S,5R\text{-}menthyl)\}\{\eta^5\text{-}C_5H\text{-}3,5\text{-}(CHMe_2)_2)\}]ZrCl_2$. *J. Am. Chem. Soc.* **1999**, *121*, 564–573.

48. Zubris, D. L.; Veghini, D.; Herzog, T. A.; Bercaw, J. E. Reactivity and mechanistic studies of stereocontrol for Ziegler–Natta polymerization utilizing doubly-silylene bridged group 3 and group 4 metallocenes. In *Olefin Polymerization: Emerging Frontiers*; Arjunan, P., McGrath, J. E., Hanlon, T. L., Eds.; ACS Symposium Series 749; American Chemical Society: Washington, DC, 2000; pp 2–14.

49. Zubris, D. L. Investigations of the origin of stereocontrol in syndiospecific Ziegler–Natta polymerizations. Ph.D. Thesis, California Institute of Technology, Pasadena, CA, July 2001.

50. Coughlin, E. B.; Henling, L. M.; Bercaw, J. E. Synthesis and structural characterization of $\{(\eta^5\text{-}C_5Me_4)_2SiMe_2\}YCH(SiMe_3)_2$. Hydrogenation to $[\{(\eta^5\text{-}C_5Me_4)_2SiMe_2\}Y]_2(\mu_2\text{-}H_2)$ and its facile ligand redistribution to $Y_2[\mu_2\text{-}\{(\eta^5\text{-}C_5Me_4)SiMe_2(\eta^5\text{-}C_5Me_4)\}]_2(\mu_2\text{-}H_2)$. *Inorg. Chim. Acta* **1996**, *242*, 205–210.

51. Bierwagen, E. P.; Bercaw, J. E.; Goddard, W. A. III. Theoretical studies of Ziegler–Natta catalysis: Structural variations and tacticity control. *J. Am. Chem. Soc.* **1994**, *116*, 1481–1489.

52. Ewen, J. A.; Elder, M. J.; Jones, R. L.; Haspeslagh, L.; Atwood, J. L.; Bott, S. G.; Robinson, K. Metallocene/polypropylene structural relationships: implications on polymerization and stereochemical control mechanisms. *Makromol. Chem., Macromol. Symp.* **1991**, *48–49*, 253–295.

53. Baar, C. R.; Levy, C. J.; Min, E. Y. -J.; Henling, L. M.; Day, M. W.; Bercaw, J. E. Kinetic resolution of chiral α-olefins using optically active *ansa*-zirconocene polymerization catalysts. *J. Am. Chem. Soc.* **2004**, *126*, 8216–8231.

54. Oliva, L.; Longo, P.; Zambelli, A. ^{13}C-Enriched end groups of poly(3-methyl-1-pentene) prepared in the presence of metallocene catalysts. *Macromolecules* **1996**, *29*, 6383–6385.

55. Sacchi, M. C.; Barsties, E.; Tritto, I.; Locatelli, P.; Brintzinger, H. H.; Stehling, U. Stereochemistry of first monomer insertion into a metal-methyl bond: Enantioselectivity and diastereoselectivity in 3-methyl-1-pentene polymerization with metallocene catalysts. *Macromolecules* **1997**, *30*, 1267–1271.

56. Eliel, A. L.; Wilen, S. H.; Mander, L. N. *Stereochemistry of Organic Compounds*; John Wiley & Sons, Inc.: New York, 1994; pp 266–268.

57. Clearfield, A.; Warner, D. K.; Saldarriaga-Molina, C. H.; Ropal, R.; Bernal, I. Structural studies of $(\pi\text{-}C_5H_5)_2MX_2$ complexes and their derivatives. The structure of bis(π-cyclopentadienyl)titanium dichloride. *Can. J. Chem.* **1975**, *53*, 1622–1629.

58. Corey, J. Y.; Zhu, X. H.; Brammer, L.; Rath, N. P. Zirconocene Dichloride. *Acta Crystallogr.* **1995**, *C*51, 565–567.

59. Bajgur, C. S.; Tikkanen, W. R.; Petersen, J. L. Synthesis, structural characterization, and electrochemistry of [1]metallocenophane complexes, [Si(alkyl)$_2$(C$_5$H$_4$)$_2$]MCl$_2$, M = Ti, Zr. *Inorg. Chem.* **1985**, *24*, 2539–2546.

5 Metallocenes with Donor/Acceptor and Other Heteroatom Bridges for Stereoselective Propylene Polymerization

Pamela J. Shapiro

CONTENTS

5.1 INTRODUCTION

The demonstration of single-site control over polymer stereochemistry by C_2-symmetric *ansa*-titanocene and zirconocene catalysts in 1984 and 1985 ushered in a new era in alkene polymerization catalysis.[1,2] Since then, a plethora of *ansa*-metallocene ligand designs have been developed for the purpose of controlling the stereochemistry and molecular weight of the polymer product and for increasing the activity of the catalyst. Besides numerous ring substitution patterns as well as the replacement of one ring with another moiety (such as an amide ligand), different types of bridges have been used to alter both the geometry and electronic properties of the *ansa*-metallocene. For the most part, these bridges have consisted of relatively inert carbon- or silicon-based moieties. Apart from a few phosphorus-[3,4] and arsenic[5]-bridged *ansa*-zirconocene complexes, examples of heteroatom-bridged group 4 *ansa*-metallocenes were scarce until about a decade ago, when the potential of the heteroatom to expand the functionality of the *ansa*-metallocene ligand and increase its influence on the polymerization properties of the catalyst was recognized. (Note: The term "heteroatom" is generally associated with any p-block element other than carbon; however, in this chapter it refers to any p-block element outside of group 14.)

So far, only boron, phosphorus, and arsenic have served as single heteroatom bridges in group 4 *ansa*-metallocene complexes, probably owing to the greater synthetic accessibility of these systems. *ansa*-Metallocenes with a single oxygen or nitrogen atom joining the π ligands have not been developed, although these atoms have been combined with one or two other heteroatoms[6–9] or have been flanked by carbon and silicon to form the backbone of the bridge.[10–15] This chapter will focus only on *ansa*-metallocenes containing heteroatoms within the backbone of the bridge. Heteroatom-bridged half sandwich complexes of the "constrained-geometry" type, in which the π-ligand is linked to an amide ligand,[16–18] are also outside the scope of this chapter.

The following abbreviations are used for the π-ligands of the *ansa*-metallocene complexes: cyclopentadienyl (Cp), substituted cyclopentadienyl (Cp^R), indenyl (Ind), 2-methyl-4-phenyl-indenyl (2-Me-4-PhInd), and fluorenyl (Flu).

5.2 GROUP 4 BORON-BRIDGED *ANSA*-METALLOCENE COMPLEXES

A boron bridge offers more functionality than silicon or carbon bridges to the chemistry of group 4 *ansa*-metallocene complexes because it can reversibly bind a variety of Lewis basic moieties that are designed to influence the electronic and stereochemical properties of the complex. By coordinating an anionic Lewis base, the boron can potentially serve as an internal, counteranion to the catalytically active, cationic group 4 metal alkyl, as in the hypothetical zwitterionic complex shown in Figure 5.1. The polymerization activity of the zwitterionic catalyst should benefit from the remote location of the counteranion from the active site, where it will not compete with the alkene monomer for a coordination site on the metal.

5.2.1 Y(Cp)₂MX₂ (M = Ti, Zr)

Very little has been reported about the propylene polymerization properties of boron-bridged dicyclopentadienyl group 4 complexes. In most cases, only the ethylene polymerization behavior of the complex has been examined. This is owing to the fact that *ansa*-metallocenes containing two unsubstituted cyclopentadienyl rings produce atactic polypropylene (PP) and are typically less active propylene polymerization catalysts than their bis(indenyl) counterparts. Nevertheless, the investigation of a new type of *ansa*-bridge is usually initiated with the cyclopentadienyl ligand, which is less costly and often easier to handle than indenyl and fluorenyl ligands. Furthermore, dicyclopentadienyl complexes are useful prototypes for the more synthetically intensive stereoselective *ansa*-metallocenes. Their polymerization activities (even if only for polyethylene) indicate whether or not the pursuit of more elaborate *ansa*-metallocene designs is worthwhile. For this reason, an overview of the chemistry of boron-bridged Y(Cp)₂MX₂ (M = Ti, Zr, Hf) complexes is presented here as prelude to the discussion of their stereoselective

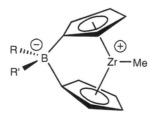

FIGURE 5.1 Hypothetical boron-bridged zwitterionic zirconocene complex.

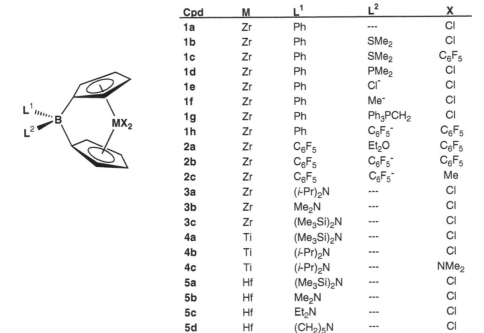

Cpd	M	L^1	L^2	X
1a	Zr	Ph	---	Cl
1b	Zr	Ph	SMe$_2$	Cl
1c	Zr	Ph	SMe$_2$	C$_6$F$_5$
1d	Zr	Ph	PMe$_2$	Cl
1e	Zr	Ph	Cl$^-$	Cl
1f	Zr	Ph	Me$^-$	Cl
1g	Zr	Ph	Ph$_3$PCH$_2$	Cl
1h	Zr	Ph	C$_6$F$_5$$^-$	C$_6$F$_5$
2a	Zr	C$_6$F$_5$	Et$_2$O	C$_6$F$_5$
2b	Zr	C$_6$F$_5$	C$_6$F$_5$$^-$	C$_6$F$_5$
2c	Zr	C$_6$F$_5$	C$_6$F$_5$$^-$	Me
3a	Zr	(i-Pr)$_2$N	---	Cl
3b	Zr	Me$_2$N	---	Cl
3c	Zr	(Me$_3$Si)$_2$N	---	Cl
4a	Ti	(Me$_3$Si)$_2$N	---	Cl
4b	Ti	(i-Pr)$_2$N	---	Cl
4c	Ti	(i-Pr)$_2$N	---	NMe$_2$
5a	Hf	(Me$_3$Si)$_2$N	---	Cl
5b	Hf	Me$_2$N	---	Cl
5c	Hf	Et$_2$N	---	Cl
5d	Hf	(CH$_2$)$_5$N	---	Cl

FIGURE 5.2   Boron-bridged (Cp)$_2$MX$_2$ complexes (M = Zr, Ti, Hf). Compounds **1e, 1f, 1h, 2b,** and **2c** are anionic species (counter ions not shown), while compound **1g** is zwitterionic.

analogues: Y(Ind)$_2$ZrX$_2$, Y(2-Me-4-PhInd)$_2$ZrCl$_2$, and Y(Flu)(Cp)ZrCl$_2$. Figure 5.2 presents a complete list of these complexes. The first example, PhB(Cp)$_2$ZrCl$_2$ (**1a**), was prepared independently by Rufanov[19] and by Reetz.[20]

The Rufanov synthesis involved the reaction between ZrCl$_4$ and PhB{(SnMe$_3$)C$_5$H$_4$}$_2$, which was formed in situ from the 2:1 reaction between (SnMe$_3$)$_2$C$_5$H$_4$ and PhBCl$_2$. A 40% yield of **1a** was reported; however, the synthesis has not been reproducible.[21] The Reetz synthesis was similar but instead used PhB{(SiMe$_3$)C$_5$H$_4$} as a ligand source and afforded **1a** in only 7% yield. A more reliable approach to the phenylboron-bridged *ansa*-zirconocene system was developed by Shapiro and coworkers,[22,23] who prepared **1b** by reacting ZrCl$_4$(SMe$_2$)$_2$ with PhB{(SiMe$_3$)(SnMe$_3$)C$_5$H$_3$}$_2$. Complex **1b** forms selectively even if two equivalents of ZrCl$_4$(SMe$_2$)$_2$ are reacted with the ligand precursor. The Me$_2$S donor in **1b** is weakly coordinated and dissociates readily from the complex in solution, leaving the boron bridge vulnerable to attack by strong nucleophiles. Consequently, efforts to alkylate the zirconium center of **1b** using various alkyllithium, Grignard, dialkylzinc, and trialkylaluminum reagents result in decomposition of the complex.[21] The involvement of boron–cyclopentadienyl bond cleavage in the decomposition was indicated by signals characteristic of nonbridged CpZr moieties in the ^{1}H NMR spectra of the decomposition products. When the SMe$_2$ adduct in **1b** was replaced with more tightly coordinating PMe$_3$ (**1d**), alkylation of the zirconium center by alkyllithium and Grignard reagents proceeded smoothly.[21]

An exception to the decomposition of **1b** by alkylating reagents was discovered by Lancaster and Bochmann,[24] who selectively alkylated the zirconium center of **1b** with two equivalents of (C$_6$F$_5$)Li while leaving the Me$_2$S adduct on the boron bridge undisturbed to form complex **1c**. Displacement of the Me$_2$S with additional Li(C$_6$F$_5$) produced the borate-bridged species **1h**. The opposite selectivity was observed in the reaction between **1b** and (C$_5$Me$_5$)$_2$AlMe. Methyl anion transfer to boron instead of zirconium occurred to produce **1f**.[25] Thus, the outcome of these alkylation reactions is very

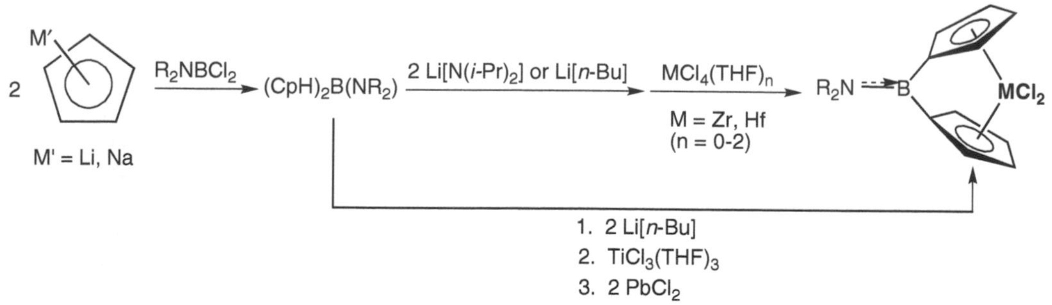

SCHEME 5.1   Synthetic routes to $R_2NB(Cp)_2MCl_2$ complexes (M = Ti, Zr, Hf).

sensitive to the choice of alkylating reagent, and the stability of the alkylborate-bridge appears to depend on the identities of both the anionic donor and its countercation.

R_2NB-bridged group 4 *ansa*-metallocene complexes **3a–c**, **4a–c** and **5a–d**, were prepared in the research groups of Ashe[26–28] and Braunschweig.[29–33] These complexes are considerably more robust than the aforementioned $Ph(L^2)B$-bridged systems, presumably owing to π-bonding between the amide nitrogen and boron. Moreover, the ligand syntheses and metallations are accomplished using standard nucleophilic displacement and acid/base reactions without compromising the boron bridge (Scheme 5.1).

Selected olefin polymerization data for these boron-bridged zirconocene complexes are presented in Table 5.1. Since precise comparisons cannot be made among data acquired under different experimental conditions by different researchers, an effort is made whenever possible to compare data arising from a single source. Otherwise, conservative comparisons are made among data from different sources when important trends are indicated.

A comparison of the ethylene polymerization activities of **1d–1f** (Table 5.1, entries 2–4) and a separate comparison of the ethylene polymerization activities of **1h** and **2b–c** (Table 5.1, entries 6–10) indicate that the identity of the donor ligand (L^2) does not have a substantial influence on the reactivities of the $Ph(L^2)B$-bridged complexes. Complexes **1b** and **1g** are exceptions to this observation. Complex **1b** exhibited no polymerization activity whatsoever (Table 5.1, entry 1), presumably owing to its decomposition by methylaluminoxane (MAO). It is not possible to compare **1g** directly with the other $Ph(L^2)B$-bridged complexes using the available data; however, in a parallel screening of the ethylene/octene copolymerization properties of **1g**, **3a**, Cp_2ZrCl_2 (**C1**), and $Me_2Si(Cp)_2ZrCl_2$ (**C2**) at 140 °C, **1g** exhibited remarkably high ethylene polymerization activity compared to the other complexes (Table 5.1, entries 5, 11, 18, 19).[34] Significantly, **1g** was over 30 times more active than **3a**. The 140 °C polymerization temperature indicates that it is also surprisingly thermally robust. A bis(indenyl)zirconium analogue of **1g** was pursued (see Section 5.2.3) with the hope that it would be as thermally robust as **1g** and exhibit similarly high activity for polypropylene polymerization. Unfortunately, the indenyl analogue proved to be much less thermally stable than the cyclopentadienyl complex.

Braunschweig reported that his R_2NB-bridged *ansa*-zirconocene complexes are approximately seven times more active than their hafnocene analogues toward ethylene polymerization, while the hafnocene complexes produce higher molecular weight polymer (Table 5.1, entries 12 and 15).[33] This is consistent with the alkene polymerization behavior of other zirconocene and hafnocene systems.

More recently, Braunschweig and coworkers reported the synthesis and characterization of the $B(NR_2)$–$B(NR_2)$-bridged group 4 metallocenes shown in Figure 5.3.[9] The molecular structures of these complexes indicate that a bridge containing two boron atoms exerts significantly less strain on metallocene geometry than a bridge containing one boron atom. In fact, the geometric parameters for **6b** are remarkably similar to those of nonbridged Cp_2HfCl_2. Both **6a** and **6b** are reasonably active

TABLE 5.1

Polymerization Data^a for Boron-Bridged Cp$_2$MCl$_2$ Complexes (M = Zr, Hf)

Entry	Complex	Activator (equivalent)	$T_p{}^b$ (°C)	P (bar)	Activity^c	References
1	**1b**	MAO (50)	25	~1	none	21
2	**1d**	MAO (1000)	25	2.2	3.6	25
3	**1e** [(Ph$_2$P)$_2$N]$^+$ counter ion	MAO (1000)	25	2.2	1.2	25
4	**1f**[Cp*Al]$^+$ counter ion^d	MAO (1000)	25	2.2	2.0	25
5	**1g**	MAO (1000)	140	34	2490 (C$_2^=$/C$_8^=$)	34
			70	nr	7.4 (C$_3^=$)	
6	**1h** ([NEt$_4$]$^+$ counter ion)	MAO (500)	60	1	0.80	24
7	**2b** ([Li(Et$_2$O)$_4$]$^+$ counter ion)	MAO (600)	60	1	1.10	24
8	**2b** ([Li(Et$_2$O)$_4$]$^+$ counter ion)	[Ph$_3$C]$^+$[B(C$_6$F$_5$)$_4$]$^-$ (2) + Al(i-Bu)$_3$ (10)	60	1	2.13	24
9	**2c** ([Li(Et$_2$O)$_4$]$^+$ counter ion)	MAO (1000)	60	1	0.69	24
10	**2c** ([Li(Et$_2$O)$_4$]$^+$ counter ion)	[Ph$_3$C]$^+$[B(C$_6$F$_5$)$_4$]$^-$ (1) + Al(i-Bu)$_3$ (10)	60	1	0.37	24
11	**3a**	MAO (1000)	140	34	12.6 (C$_2^=$/C$_8^=$)	34
			70	150 g^e	3.8 (C$_3^=$)	
12	**3b**	MAO (4500)	nr^f	150 g	1.8	30
13	**3c**	MAO (1000)	20	5	1.7	32
		MAO (650)	60	5	4.2	
14	**4b**	MAO (100)	20	5	1.9	32
15	**5b**	MAO (4500)	nd^g	nr	0.26	33
16	**6a**	MAO (2000)	70–85	20	28	35
17	**6b**	MAO (2000)	70–85	20	27	35
18	**C1** (Cp$_2$ZrCl$_2$)	MAO (1000)	140	34	72 (C$_2^=$/C$_8^=$)	34
				nr	2.0 (C$_3^=$)	
19	**C2** (Me$_2$Si(Cp)$_2$ZrCl$_2$)	MAO (1000)	140	34	3.4 (C$_2^=$/C$_8^=$)	34
				nr	6.2 (C$_3^=$)	

^{a} Data is for ethylene polymerization, unless otherwise indicated.

^{b} T_p polymerization temperature.

^{c} Activity in kg polymer/mmol Zr·h. Notation in parentheses indicates alternative monomers used in the polymerization where C$_2^=$/C$_8^=$ is an ethylene/1-octene copolymerization and C$_3^=$ is propylene homopolymerization.

^{d} Cp* is pentamethylcyclopentadienyl.

^{e} Mass of polypropylene used in the polymerization.

^{f} nr = not reported.

^{g} The authors report that a "slightly elevated" temperature was used.

ethylene polymerization catalysts (Table 5.1, entries 16 and 17) and surprisingly similar in their activity. Both systems also produce very high-molecular-weight polymers ($M_v = 896,000$ g/mol and $1,600,000$ g/mol for **6a** and **6b**, respectively) compared to complexes containing a single boron atom in the bridge ($M_w < 20,000$). The authors attributed the lower chain-termination rates of complexes **6a** and **6b** to the smaller tilt angles of the cyclopentadienyl rings. Analogous complexes containing modified cyclopentadienyl rings (i.e., indenyl and fluorenyl) for stereoselective catalysis are undoubtedly under development.

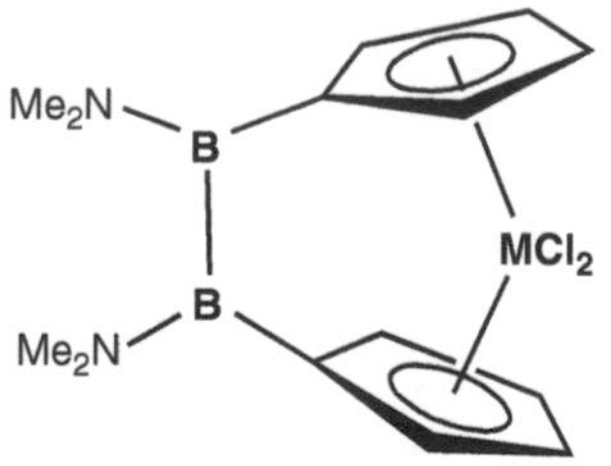

FIGURE 5.3   Braunschweig's diboron-bridged group 4 *ansa*-metallocene complexes (**6a** and **6b**).

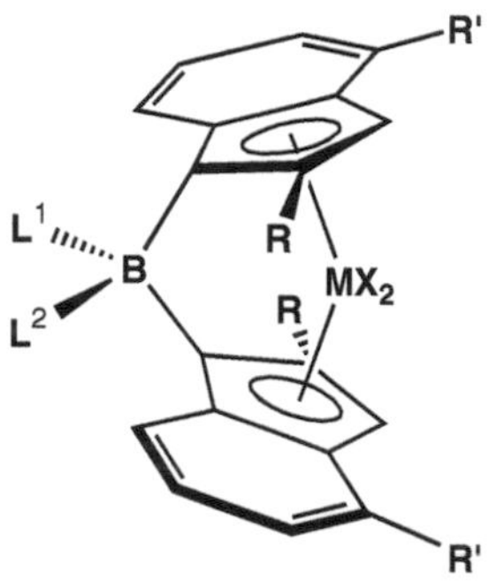

Cpd	M	L^1	L^2	R	R'	X
7a	Zr	Ph	Et$_2$O	H	H	Cl
7b	Zr	Ph	THF	H	H	Cl
7c	Zr	Ph	PMe$_3$	H	H	Cl
8	Zr	Ph	Me$_2$S	Me	Ph	Cl
8a	Zr	Ph	Ph$_3$PCH$_2$	Me	Ph	Cl
9a	Zr	(*i*-Pr$_2$)N	—	H	H	Cl
9b	Zr	(*i*-Pr$_2$)N	—	H	H	NMe$_2$
10a	Ti	(Me$_3$Si)$_2$N	—	H	H	Cl
10b	Zr	(Me$_3$Si)$_2$N	—	H	H	Cl
10c	Hf	(Me$_3$Si)$_2$N	—	H	H	Cl

FIGURE 5.4   Boron-bridged (Ind)$_2$MX$_2$ complexes (M = Zr, Ti, Hf). Complex **8a** is zwitterionic.

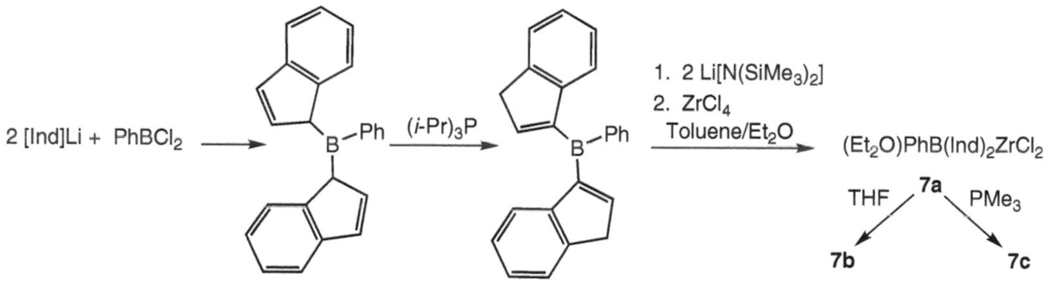

SCHEME 5.2   Synthetic route to (L^2)PhB(Ind)$_2$ZrCl$_2$ complexes (**7a**, **7b** and **7c**).

5.2.2   Y(Ind)$_2$MX$_2$ (M = Ti, Zr, Hf)

ansa-Bis(indenyl)zirconium complexes are generally more efficient propylene polymerization cata-lysts than their cyclopentadienyl counterparts, and the C_2-symmetric *rac* isomers are preferred over the *meso* isomers for their selectivity. Figure 5.4 shows the *rac* boron-bridged (Ind)$_2$Zr complexes that have been reported to date.[36,37,26]

Complexes **7a–c** was reported by Reetz et al. in 1999.[35] Complex **9a** was reported by Ashe et al. in the same year.[26] The synthetic route to **7a–c** is shown in Scheme 5.2. Base-catalyzed isomerization of PhB(HInd)$_2$ to a more stable isomer with vinylic instead of allylic B–C bonds before metallation of the ligand improved the yield of **7a**. Replacement of the Et$_2$O ligand on boron with PMe$_3$ afforded **7c** as a mixture of *rac* and *meso* isomers, from which the *rac* isomer was selectively crystallized in 12% yield.

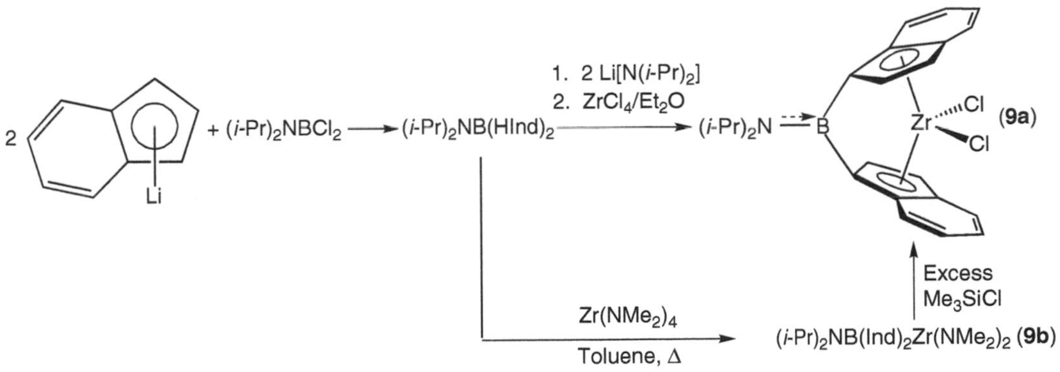

SCHEME 5.3 Synthetic routes to $(i\text{-Pr})_2\text{NB(Ind)}_2\text{ZrCl}_2$ (**9a**).

TABLE 5.2
Propylene Polymerization Data for Racemic Y(Ind)$_2$ZrCl$_2$ Complexes[a]

Entry	Y	MAO (equivalent)	T_p^{b}(°C)	Activity[c]	mmmm (%)	T_m^{d}(°C)	PPMw ($\times 10^{-5}$g/mol)	PDI[e]	References
1	(PMe$_3$)PhB (**7c**)	1,000	40	1.05	90	nd[f]	1.29	nr[g]	35
2	(PMe$_3$)PhB (**7c**)	1,000	60	0.174	85	nd	0.62	nr	35
3	i-Pr$_2$NB (**9a**)[h]	1,000	70	2.4	72[i]	113	nr	nr	26
4	Me$_2$C (**C3**)	8,000	70	27	69.7	124	0.15	nr	39
5	Me$_2$Si (**C4**)[j]	15,800	50	14	89	142	0.61	1.7	41
6	Me$_2$Si (**C4**)	15,000	70	190	81.7	137	0.36	2.5	38, 40
7	CH$_2$CH$_2$ (**C5**)	15,000	70	188	78.5	132	0.24	nr	38, 40

[a] Unless otherwise indicated, polymerizations were performed in liquid propylene.
[b] T_p = polymerization temperature.
[c] Activity in kg PP/mmol Zr·h.
[d] T_m = polymer melting point.
[e] PDI = molecular weight polydispersity (M_w/M_n).
[f] nd = not determined.
[g] nr = not reported.
[h] Polymerization conditions: 23.3 atm. propylene, Isopar-E^{TM} solvent.
[i] Polymer tacticity also reported as 82% mm:12% mr:6% rr.
[j] Polymerization conditions: 2 bar propylene, toluene solvent.

Two different synthetic approaches were used to prepare complex **9a** (Scheme 5.3). The aminolysis approach, modeled after the method developed by Jordan and coworkers,[37] afforded **9b** with a higher selectivity for the *rac* isomer (*rac:meso* ratio of 7:1) than the salt metathesis route to form **9a** (*rac:meso* ratio of 3:2).

The propylene polymerization activities of **7c**, **9a**, and their nonheteroatom-bridged racemic analogues[38–41] Me$_2$C(Ind)$_2$ZrCl$_2$ (**C3**), Me$_2$Si(Ind)$_2$ZrCl$_2$ (**C4**), and CH$_2$CH$_2$(Ind)$_2$ZrCl$_2$ (**C5**) are compared in Table 5.2. The polymerization activities of the boron-bridged complexes (Table 5.2, entries 1–3) were 1–2 orders of magnitude lower than the activities of **C3**, **C4**, and **C5** (Table 5.2, entries 4–7). The polymerization activity of **7a** (L^2 = Et$_2$O), as a mixture of *meso* and *rac* isomers, was also examined. This species exhibited much poorer activity than **7c** and produced low molecular

TABLE 5.3

Selected Structural Parameters and Stereoselectivities of Y(Ind)$_2$ZrX$_2$ Complexes

Entry	Y	$\theta(°)$	$\gamma(°)$	$\alpha(°)$	$\tau(°)$	d_{cc}(Å)	% $mmmm^a$ (°C)	References
1	i-Pr$_2$NB (**9a**)	104.9	121.6	66.8	4.2	2.517	72 (70)	26
2	Me$_2$C (**C3**)	100.3	118.3	70.8	4.55	2.354	69.7 (70)	43
3	CH$_2$CH$_2$ (**C5**)	—	125.31	63.47	4.39	2.763	78.5 (70)	46
4	Me$_2$Si (**C4**)	94.57	127.81	61.94	4.875	2.749	81.7 (70)	40
5	(PMe$_3$)PhB (**7c**)	99.4	121.6	66.7	4.15	2.483	85 (60)	35

^{a}Polymer tacticity measured as % $mmmm$ from polymerizations conducted at 70 or 60 °C as indicated by the number in parentheses.

weight, atactic polymer. The higher activity of **7c** relative to **7a** can probably be attributed to the greater protection afforded to the boron bridge by the more tightly coordinated PMe$_3$.

The tacticities of the PP produced by the racemic Y(Ind)$_2$ZrCl$_2$ complexes are listed in Table 5.3 along with selected metric parameters (defined in Figure 5.5) from the X-ray crystal structures of these complexes. The parameters α, τ, γ, and d_{cc} are commonly used to evaluate the tilt of the π-ligands on the metal.[42] These parameters reveal a direct correlation between the degree of ring tilt in Y(Ind)$_2$ZrCl$_2$(Me$_2$C > (i-Pr)$_2$NB $\approx$ (PMe$_3$)PhB > Me$_2$Si $\approx$ CH$_2$CH$_2$) and the size of Y (Me$_2$C < (i-Pr)$_2$NB $\approx$ (PMe$_3$)PhB < Me$_2$Si < CH$_2$CH$_2$). In general, the smaller the bridging atom, the greater the ring tilt angle.

Theoretical and experimental models indicate that the selectivity of the *rac ansa*-metallocenes for isotactic PP arises primarily from nonbonding interactions between the indenyl rings and the growing polymer.[44,45] Logically, the steric influence of the indenyl rings at the active site, and hence the stereoselectivity of the metallocene catalyst, should decrease with increasing ring tilt angle. The relative stereoselectivities of the racemic Y(Ind)$_2$ZrCl$_2$ complexes (Me$_2$C < (i-Pr)$_2$NB < CH$_2$CH$_2$ < (PMe$_3$)PhB $\approx$ Me$_2$Si) follow roughly the expected trend; however, a less than perfect correspondence between stereoselectivity and ring tilt indicates the influence of other factors as well. For example, **7c** and Me$_2$Si(Ind)$_2$ZrCl$_2$ (**C4**) exhibit comparable stereoselectivities (Table 5.3, entries 4 and 5), taking into account differences in polymerization temperatures (T_p), despite the greater ring tilt angle of the former species. Furthermore, **7c** exhibits a higher stereoselectivity than **9a** (Table 5.3, entry 1), which has nearly identical ring tilt parameters. This last comparison is somewhat mitigated, however, by the different polymerization conditions that were employed for **7c** and **9a**.

The rigidity of the bridge must also be considered and is responsible for the higher stereose-lectivity of the Me$_2$Si-bridged **C4** relative to the CH$_2$CH$_2$-bridged **C5** (Table 5.3, entries 3 and 4).[39] The more flexible CH$_2$CH$_2$ bridge allows **C5** to rearrange dynamically in solution between indenyl-forward and indenyl-backward conformations.[46] Evidently, the greater conformational mobility of the ligand framework in **C5** reduces its control over the stereochemistry of the polymerization.

5.2.3 Y(2-Me-4-PhInd)$_2$ZrCl$_2$

Zwitterionic (Ph$_3$PCH$_2$)PhB(2-Me-4-PhInd)$_2$ZrCl$_2$ complexes **8a–c** (Figure 5.6) were recently reported by Shapiro et al.[34] The precursor, (Me$_2$S)PhB(2-Me-4-PhInd)$_2$ZrCl$_2$(**8**, Figure 5.4), was formed as a 2:1 mixture of *meso:rac* isomers by using ZrCl$_4$(SMe$_2$)$_2$ in a synthetic procedure similar to the one developed by Reetz (Scheme 5.3).[35] Displacement of the Me$_2$S adduct by the ylide base produced a mixture of three diastereomers, the *rac* isomer and two *meso*-like isomers with

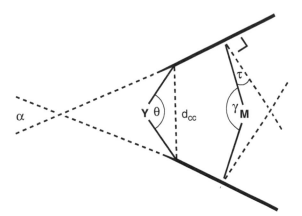

FIGURE 5.5 Metric parameters for describing *ansa*-metallocene geometry.

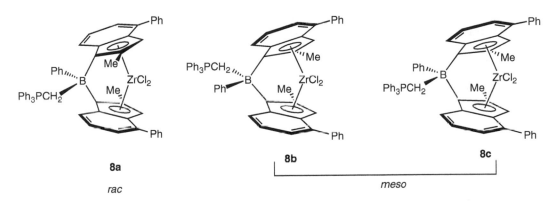

FIGURE 5.6 *rac* and *meso* isomers of $(Ph_2CH_2)PhB(2-Me-4-PhInd)_2ZrCl_2$.

different boron configurations (Figure 5.6). Efforts to separate the *rac* isomer from the mixture were unsuccessful; however, the authors managed to isolate single crystals of the *rac* isomer and one of the *meso* isomers for X-ray crystal structure determinations. Unlike the zwitterionic dicyclopentadienyl analogue, **1g** (Section 5.2.1), **8a–c** are thermally unstable. Consequently, the propylene polymerization activity of **8a–c** decayed from a fairly high level at 70 °C to a low level at 115 °C compared to the complex *rac*-Me$_2$Si(2-Me-4-PhInd)$_2$Zr(-(Ph)C=CHCH=C(Ph)-)[47] (**C6**, Table 5.4, entries 1–8), which was used as a benchmark owing to its high thermal stability and high selectivity for isotactic PP. A similar decay in propylene polymerization activity with increasing polymerization temperature (T_p) was observed for Reetz's phosphine adducts, **7c** (Table 5.2, entries 1 and 2). The PP produced by **8a–c** (as a 2:1 *meso:rac* mixture) was moderately isotactic (% *mmmm* = 52.4), indicating that the *rac* isomer is slightly more active than the *meso* isomers. It is not uncommon for *rac*-zirconocene complexes to be as much as 10–20 times more active than their *meso* isomers.[48,49]

5.2.4 Y(Flu)(Cp)ZrCl$_2$

Ashe and coworkers also developed the C_s-symmetric boron-bridged complex, $(i\text{-Pr})_2NB(Flu)(Cp)$ ZrCl$_2$ (**11**, Figure 5.7)[50] which, like the analogous Me$_2$Si(Flu)(Cp)ZrCl$_2$ (**C8**),[51] Me$_2$C(Flu)(Cp)ZrCl$_2$ (**C9**),[52] CH$_2$CH$_2$(Flu)(Cp)ZrCl$_2$ (**C10**)[52] and PhP(Flu)(Cp)ZrCl$_2$ (**12**)[48] complexes, is selective for syndiotactic PP. The data in Table 5.5 indicate the following order of stereoselectivity for the series

TABLE 5.4
Propylene Polymerization Data for Racemic $Y(2\text{-Me-4-PhInd})_2ZrX_2$[a] Complexes

Entry	Y	MAO (equivalent)	T_p[b](°C)	Activity[c]	*mmmm* (%)	T_m[d](°C)	M_w[e] ($\times 10^{-5}$ g/mol)	PDI[f]	References
1	$(Ph_3PCH_2)PhB$ (8a–c)[g]	1,000	70	197	52.4	156	3.73	2.0	33
2	$(Ph_3PCH_2)PhB$ (8a–c)[g]	1,000	85	186	nd[h]	155	1.96	2.2	33
3	$(Ph_3PCH_2)PhB$ (8a–c)[g]	1,000	100	130	nd	154	1.32	2.1	33
4	$(Ph_3PCH_2)PhB$ (8a–c)[g]	1,000	115	61	nd	153	0.32	2.1	33
5	Me_2Si (C6)[i]	1,000	70	82	nd	159	5.13	1.8	33
6	Me_2Si (C6)[i]	1,000	85	92	nd	157	3.09	2.0	33
7	Me_2Si (C6)[i]	1,000	100	152	nd	154	1.84	3.3	33
8	Me_2Si (C6)[i]	1,000	115	164	nd	154	0.80	2.8	33
9	PhP (15a)	37,000	67	57.6	98	156	0.64[j]	nr	48
10	*i*-PrP (16a)	37,600	50	126	99.6	160	nr[k]	nr	48
11	Me_2Si (C7)	15,000	70	755	95.2	157	7.3	nr	38

[a] X = Cl, and complexes are racemic unless otherwise indicated.
[b] T_p = polymerization temperature.
[c] Activity in kg PP/mmol Zr·h.
[d] T_m = polymer melting point.
[e] M_w = PP weight average molecular weight unless otherwise indicated.
[f] PDI = molecular weight polydispersity (M_w/M_n).
[g] Used as a 2:1 mixture of *meso:rac* isomers.
[h] nd = not determined.
[i] $X_2 = -(Ph)C{=}CHCH{=}C(Ph)-$.
[j] Polymer molecular weight as the polymer viscosity average molecular weight (M_v).
[k] nr = not reported.

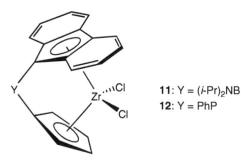

11: Y = $(i\text{-Pr})_2$NB
12: Y = PhP

FIGURE 5.7 $Y(Flu)(Cp)ZrCl_2$ complexes.

of $Y(Flu)(Cp)ZrCl_2$ complexes: Me_2C > PhP > R_2NB ≈ CH_2CH_2 > Me_2Si. A different relative order for polymerization activity is indicated: R_2NB > Me_2C > CH_2CH_2 > PhP > Me_2Si. The high ethylene polymerization activity exhibited by **11** at 140 °C indicates that the complex is thermally robust, again illustrating the stabilizing influence of π-electron donation from the amide nitrogen to boron.

TABLE 5.5
Propylene Polymerization Data for Y(Flu)(Cp)ZrCl$_2$ Complexes[a]

Entry	Y	MAO (equivalent)	T_p[b](°C)	Activity[c]	rrrr (%)	T_m[d](°C)	PPM_w ($\times 10^{-5}$g/mol)	PDI[e]	References
1	i-Pr$_2$NB (**11**)[f]	1000	70	77	71[g]; 81 (rr)	112	1.1	1.9	50
2	PhP (**12**)	1000	50	9	81	128	3.0	nd[h]	48
3	PhP (**12**)	1000	67	14	nd	119	1.3	nd	48
4	Me$_2$C (**C9**)	1000	67	37	85	134	1.2	2.0	49
5	Me$_2$C (**C9**)	nr[i]	60	17	84	136	0.09	nr	52
6	Me$_2$Si (**C8**)	1000	67	1.6	51	nr	nr	nr	48
7	Me$_2$Si (**C8**)	1000	70	25	76 (rr)	nr	nr	2.3	51
8	Me$_2$Ge (**C11**)	1000	70	8	65 (rr)	nr	2.9	2.3	51
9	CH$_2$CH$_2$ (**C10**)	nr	60	4.5	74.3	111	1.7	nr	52

[a]Unless otherwise stated, all polymerizations were performed in liquid propylene.
[b]T_p = polymerization temperature.
[c] Activity in kg PP/mmol Zr·h.
[d]T_m = polymer melting point.
[e]PDI = molecular weight polydispersity (M_w/M_n).
[f] Polymerization conditions: Isopar-E^{TM} solution, 34 atm. propylene.
[g]Polymer tacticity also reported as 4% *mm*:15% *mr*:81% *rr*.
[h]nd = not determined.
[i]nr = not reported.

5.3 GROUP 4 PHOSPHORUS-BRIDGED *ANSA*-METALLOCENE COMPLEXES

5.3.1 Y(Cp^R)$_2$MCl$_2$(R = H, *t*-Bu; M = Ti, Zr)

Similar to the boron bridge, the phosphorus bridge possesses more functionality than a carbon or silicon bridge because of its Lewis basicity and its ability to be either three or four coordinate. Although the first phosphorus-bridged titanocene complexes were reported as early as 1983[4] and 1988,[3] it was not until 1996 that the first application of phosphorus-bridged *ansa*-zirconocene complexes to olefin polymerization was described in a patent application by Brintzinger et al.[53] Of the two complexes shown in Figure 5.8, only the phosphonium-bridged species (**14**) was active toward propylene polymerization in the presence of MAO, producing polypropylene with $M_w = 42,500$ and $M_n = 19,800$ (polydispersity index [PDI] $= 2.15$). The polymer microstructure was not described; however, its melting point of 151.2 °C indicates that it was isotactic-rich.

5.3.2 Y(Flu)(Cp)ZrCl$_2$

Schaverien[48] prepared a series of phosphorus-bridged (Flu)(Cp)ZrCl$_2$, (Ind)$_2$ZrCl$_2$, and (Flu)$_2$ZrCl$_2$ complexes and performed a comprehensive study of their performance as propylene polymerization catalysts. The performance of the C_s-symmetric complexes Me$_2$C(Flu)(Cp)ZrCl$_2$ (**C9**), PhP(Flu)(Cp)ZrCl$_2$ (**12**), and Me$_2$Si(Flu)(Cp)ZrCl$_2$ (**C8**)) were compared under identical reaction conditions. The data is presented in Table 5.5 along with polymerization data collected by other groups on complexes **C9** and **C8**. Polymerization data for complexes **11** and CH$_2$CH$_2$(Flu)(Cp)ZrCl$_2$ (**C10**)[52] are also included for comparison.

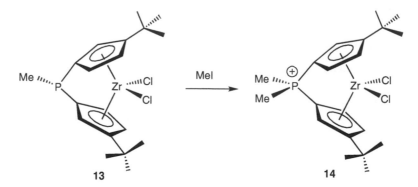

FIGURE 5.8 Tertiary and quaternary phosphorus-bridged bis(*tert*-butylcyclopentadienyl)zirconocene complexes.

TABLE 5.6

Selected Structural Parameters and Stereoselectivities of Y(Flu)(Cp)ZrCl$_2$ Complexes

Entry	Y	d_{cc}(Å)	θ(°)	γ(°)	α(°)	% *rrrr*[a](°C)	References
Calculated Structures							
1	H$_2$C	2.34	99.3	121.5	nr[b]	—	48
2	H$_2$Si	2.73	94.3	129.4	nr	—	48
3	HP	12.56	87.7	127.2	nr	—	48
4	PhP (**12**)	2.57	87.8	127.4	nr	81 (50)	48
Crystallographically Determined Structures							
5	Me$_2$C (**C9**)	2.336	99.4	118.6	72.0	85 (67)	55
6	*i*-Pr$_2$NB (**11**)	2.504	104.7	122.1	68.1	71 (70)	50
7	Me$_2$Si (**C8**)	2.756	93.4	127.9	62.5	51 (67)	51
9	CH$_2$CH$_2$ (**C10**)	nr	—	127.2	nr	74.3 (60)	52

[a] Polymer tacticity measured as % *rrrr* from polymerizations conducted at 70 or 60 °C as indicated by the number in parentheses.

[b] nr = not reported.

In order to explain the variations in the stereoselectivity of complexes **C9**, **11**, and **C8**, Schaverien[48] sought structural clues from calculated (*ab initio*) structures of model complexes (Table 5.6). As can be seen, there is excellent agreement between the calculated structures of the H$_2$C- and H$_2$Si-bridged complexes and the molecular structures of **C9** and **C8**. The authors attributed the relative syndioselectivities of the complexes to the effect of ligand "bite angle" (which is the inverse of ring tilt angle) on the barrier to inversion of stereochemistry (epimerization) at the metal center. (See Chapter 2 for further information on epimerization of bridged (Flu)(Cp) complexes.) The smaller the "bite angle" (greater ring tilt), the higher the barrier to epimerization of the metal center through a shifting of the growing polymer chain in the equatorial plane from one coordination site to the other.[54] Since the syndioselectivity of the polymerization is believed to depend on the regular alternation of olefin insertion from side to side in the equatorial plane,[55,56] a higher inversion barrier should translate into fewer stereoerrors. The authors found that the relative syndioselectivities of complexes **C9**, **C8**, and **12** (from highest to lowest: **C9** > **12** > **C8**) correspond to their ligand "bite angles" (from smallest to largest: **C9** < **12** < **C8**), which they based on the structural parameter d_{cc}. Notably, the stereoselectivity and "bite angle" of boron-bridged species **11** falls between that of **C9**

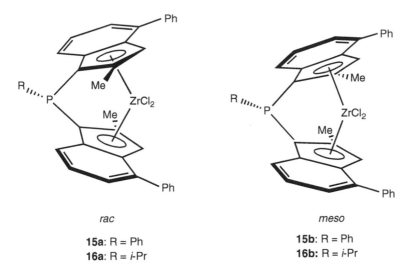

rac

15a: R = Ph
16a: R = i-Pr

meso

15b: R = Ph
16b: R = i-Pr

FIGURE 5.9 *rac* and *meso* isomers of RP(2-Me-4-PhInd)$_2$ZrCl$_2$ (R = Ph, *i*-Pr).

and **C8**, as predicted by the model. The lower stereoselectivity exhibited by Me$_2$Ge(Flu)(Cp)ZrCl$_2$ (**C11**) versus **C8** is consistent with expectation since the larger Me$_2$Ge bridge should produce a larger "bite angle" (smaller tilt angle) between the metallocene rings (Table 5.5, entries 7–8). The stereoselectivity of **C10** is greater than expected judging from its ring centroid-Zr-ring centroid angle (γ), which is narrower than that of **C8**. This may be somehow associated with the greater flexibility of its CH$_2$CH$_2$ bridge.

5.3.3 Y(2-Me-4-PhInd)$_2$ZrCl$_2$

The same authors examined the propylene polymerization properties of PhP(2-Me-4-PhInd)$_2$ZrCl$_2$ (**15a–b**, Figure 5.9).[48] Since they were not able to separate the *rac* from the *meso* isomers, they used the 1:2 *rac:meso* mixture directly and obtained highly isotactic PP (*mmmm* = 97.5–98%) along with 4–10% atactic PP, which originated from the *meso* isomers and could be largely removed from the bulk polymer by extraction with hot xylene. The PP was also highly regioregular, with 99.2–99.8 mole% 1,2 insertions and 0.2–0.5 mole% and 0.03–0.2 mole% of 2,1-and 1,3-regiodefects, respectively. The presence of primarily isobutyl end groups in the PP implicated chain transfer to aluminum as the primary chain transfer pathway. The activity of the system increased more than twofold (from 200 to 576 kg PP/g Zr·h) when the polymerization temperature was increased from 50 to 67 °C. This activity is comparable to that of Me$_2$Si(2-Me-4-PhInd)$_2$ZrCl$_2$ (**C7**, Table 5.4), which was measured by Spaleck et al. under similar conditions (liquid propylene, 70 °C).[38,40] Interestingly, changing the phenyl substituent on the phosphorus bridge to an isopropyl substituent (**16a–b**) increased the activity of the system three to fivefold.

5.3.4 Y(Flu)$_2$ZrCl$_2$

Schaverien[48] and Alt[57] independently reported the synthesis of PhP(Flu)$_2$ZrCl$_2$ (**17**, Figure 5.10), which has average C_{2v} symmetry. This system exhibited much lower ethylene and propylene polymerization activity than the Me$_2$Si(Flu)$_2$ZrCl$_2$ (**C12**) and CH$_2$CH$_2$(Flu)$_2$ZrCl$_2$ (**C13**) analogues, although the molecular weight and tacticity of the PP produced by **17** were comparable to that of the **C12** (Table 5.7).[48,58] As expected, C_{2v}-symmetric Y(Flu)$_2$ZrCl$_2$ complexes are not especially

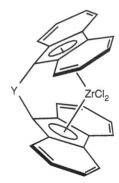

17: Y = PhP
C12: Y = Me$_2$Si
C13: Y = CH$_2$CH$_2$

FIGURE 5.10 Y(Flu)$_2$ZrCl$_2$ complexes.

TABLE 5.7
Polymerization Data for Racemic Y(Flu)$_2$ ZrCl$_2$ Complexes[a]

Entry	Y	MAO (equivalent)	T_p[b] (°C)	Activity[c]	Tacticity (mm/mr/rr)	Polymer M_w ($\times 10^{-5}$ g/mol)	References
1	PhP (**17**)	1,000	50	0.09 (C$_3^=$)	15.7/49.5/34.8	1.9 (M_v)	48
2	PhP (**17**)	17,000	60	28 (C$_2^=$)[d]	na[e]	4.3 (M_n)	57
3	Me$_2$Si (**C12**)	1,000	50	2.0 (C$_3^=$)	19/49/32	2.3 (M_v)[f]	48
4	Me$_2$Si (**C12**)	1,000	50	1.6 (C$_3^=$)	3.6/1 rrrr/mmmm	2.0 (M_n)[g]	58
5	CH$_2$CH$_2$ (**C13**)	200	70	nr (C$_3^=$)[h]	32/24/44	0.35 (M_w)	52
6	CH$_2$CH$_2$ (**C13**)	1,000	50	2.0 (C$_3^=$)	nr	0.71 (M_n)[g]	58

[a] All propylene polymerizations were performed in liquid propylene (C$_3^=$).
[b] T_p = polymerization temperature.
[c] Activity in kg polymer/mmol Zr·h; C$_3^=$ indicates a propylene polymerization while C$_2^=$ indicates an ethylene polymerization.
[d] Polymerization conditions: 10 bar ethylene (C$_2^=$), pentane solvent.
[e] na = not applicable.
[f] Calculated from a viscosity (η) = 2.2 dl/g using the same formula as the authors in reference 48 did for entry 1.
[g] Calculated from P$_n$ value (number average polymerization degree) reported by authors.
[h] nr = not reported.

stereoselective; however, the oily PP products are not completely atactic. Rather, they are slightly enriched in syndiotactic sequences.

Alt attributed the low activity of **17** versus **C13** to the greater electron donating capability of the phosphorus bridge, which should increase electron density at the metal, strengthen the carbon–metal bond, and thereby reduce the rate of alkene insertion. This argument, although logical, is contradicted by the relative polymerization activities of **15a–b** versus **16a–b** mentioned in Section 5.3.3.

5.4 GROUP 4 DONOR-ACCEPTOR-BRIDGED *ANSA*-METALLOCENE COMPLEXES

Starzewski and coworkers[6,7,59–61] developed a highly versatile *ansa*-metallocene design in which a donor–acceptor (D/A) bridge is formed by a Lewis acidic boryl substituent on one π-ligand and either a phosphoryl or amine substituent on the other. A wide range of D/A-bridged Cp_2MCl_2 (M = Ti, Zr) (**18a–d, 19a–g**), (Ind)(Cp)ZrCl$_2$(**20a–f**), Ind$_2$ZrCl$_2$(**21a–f**), (Flu)(Cp)ZrCl$_2$ (**22a–c**), and (Flu)(Ind)ZrCl$_2$ (**23a–b**) complexes, as well as Cp(phosphole)TiCl$_2$ (**24**) and Cp(pyrrole)TiCl$_2$ (**25**) complexes with in-ring donors have been prepared (Figure 5.11).

The reversibility of the D/A interaction of the bridge (Figure 5.12) affords additional control over the polymer microstructure for stereoselective metallocenes. By controlling the equilibrium between bridged and nonbridged catalysts, using the temperature of the polymerization and the strength of the D/A interaction as variables, one can vary the properties of the PP from elastomeric material, with alternating atactic and isotactic blocks, to high melting, isotactic material.

Table 5.8 lists propylene polymerization data for some of the complexes shown in Figure 5.12. Varieties of other polymer products have been prepared using the D/A-bridged complexes. For example the (Flu)(Cp)ZrCl$_2$ complexes (**22a–b**) produce ultra-high molecular weight (UHMW) polyethylene with intrinsic viscosities as high as 13.5 d/g. The D/A metallocenes are also effective catalysts for the copolymerization of α-olefins with ethylene to produce medium to ultra-low density polyethylene products.[7]

5.5 CONCLUSIONS

It is striking how pronounced an effect varying the bridging atom and its substituents can have on the polymerization activities and stereoselectivities of group 4 *ansa*-metallocene complexes. Even subtle changes to the bridge have a significant influence on catalyst activity. The electronic reasons for these effects are currently not well understood. The electronic effect of the bridging moiety on the strength of the metal–carbon bond has been suggested as a possible explanation;[62] however, contradictory results indicate that the situation is more complex. Some rationales have been put forth concerning the effects of the heteroatom bridges on catalyst stereoselectivity. In general, bridges that enforce large ring tilt angles are detrimental to the isoselectivity of C_2-symmetric bis(indenyl) complexes because they reduce the steric influence of the indenyl rings over the polymerization. The opposite effect of ring tilt applies to C_s symmetric (Flu)(Cp) complexes. Increasing ring tilt (decreasing "bite angle") increases the barrier to epimerization at the metal center.

Whereas Ph(L^2)B-bridged metallocenes are too chemically and thermally sensitive to be practical for industrial use, (R$_2$N)B-bridged metallocenes are relatively robust and exhibit high stereoselectivities and activities that sometimes exceed that of their single carbon- and silicon-bridged analogues. It will be interesting to see how the stereoselectivity of *rac*-bis(indenyl)zirconium and hafnium complexes containing a -B(NR$_2$)B(NR$_2$)- bridge compares to that of other Y(Ind)$_2$MX$_2$ analogues. The trends identified in Section 5.2.2 indicate that the larger size of this bridge relative to single-atom bridges and its greater rigidity relative to the CH$_2$CH$_2$ bridge should favor a comparatively high selectivity for isotactic PP.

The phosphorus-bridged metallocene complexes exhibit high stereoselectivities and, in some cases, relatively competitive polymerization activities. It would be worthwhile to examine the effect of quaternizing the phosphorus bridge on the activities of these complexes in light of the significant effect that this has on the relative polymerization activity of complex **13** versus **14** (see Section 5.3.1).

Starzewski's D/A-bridged metallocene complexes illustrate the enormous versatility that heteroatom bridges can bring to the chemistry of *ansa*-metallocenes. The modular nature of the syntheses of these complexes and the medley of donors and acceptors available for varying the strength of the bridge make it possible to tailor these systems toward a remarkable variety of polymer products.

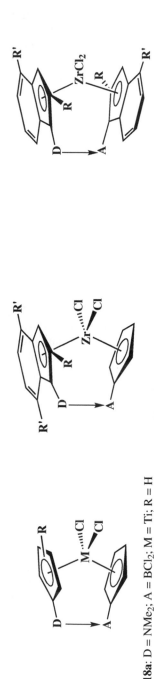

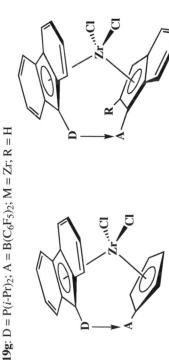

21a: D = PPh$_2$; A = BCl$_2$; R = R' = H
21b: D = Et$_2$P; A = BCl$_2$; R = R' = H
21c: D = Me$_2$P; A = BCl$_2$; R = R' = H
21d: D = Ph$_2$P; A = BMe$_2$; R = R' = H
21e: D = Et$_2$P; A = B(C$_6$F$_5$)$_2$; R = Me; R' = Ph
21f: D = Me$_2$P; A = B(C$_6$F$_5$)$_2$; R = Me; R' = H

20a: D = PMe$_2$; A = B(C$_6$F$_5$)$_2$; R = Me; R' = H
20b: D = PMe$_2$; A = BCl$_2$; R = Me; R' = H
20c: D = PEt$_2$; A = B(C$_6$F$_5$)$_2$; R = Me; R' = H
20d: D = PPh$_2$; A = BCl$_2$; R = R' = H
20e: D = PEt$_2$; A = BCl$_2$; R = R' = H
20f: D = P(i-Pr)$_2$; A = BCl$_2$; R = H; R' = Me

18a: D = NMe$_2$; A = BCl$_2$; M = Ti; R = H
18b: D = NMe$_2$; A = BMe$_2$; M = Ti; R = H
18c: D = PPh$_2$; A = BCl$_2$; M = Ti; R = H
18d: D = P(i-Pr)$_2$; A = BMe$_2$; M = Ti; R = H
19a: D = PPh$_2$; A = BCl$_2$; M = Zr; R = H
19b: D = PPh$_2$; A = B(C$_6$H$_{11}$)$_2$; M = Zr; R = SiMe$_3$
19c: D = PMe$_2$; A = BCl$_2$; M = Zr; R = H
19d: D = PMe$_2$; A = BCl$_2$; M = Zr; R = H
19e: D = PMe$_2$; A = B(C$_6$F$_5$)$_2$; M = Zr; R = H
19f: D = PEt$_2$; A = B(C$_6$F$_5$)$_2$; M = Zr; R = H
19g: D = P(i-Pr)$_2$; A = B(C$_6$F$_5$)$_2$; M = Zr; R = H

23a: D = PEt$_2$; A = BPh$_2$; R = H
23b: D = PEt$_2$; A = B(C$_6$F$_5$)$_2$; R = Me

22a: D = PEt$_2$; A = BPh$_2$
22b: D = PEt$_2$; A = B(C$_6$F$_5$)$_2$
22c: D = PEt$_2$; A = BEt$_2$

FIGURE 5.11 D/A-bridged group 4 *ansa*-metallocene complexes.

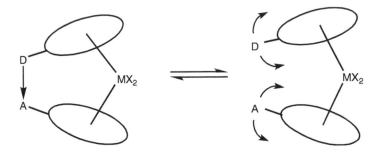

FIGURE 5.12 Reversible cleavage of the D/A bridge.

TABLE 5.8
Propylene Polymerization Data for D/A Complexes

Entry	Complex	$T_p{}^a$ (°C)	Activity[b]	Polymer Tacticity (%)	$T_m{}^c$ (°C)	PP M_w (×10^{-5} g/mol)	PDI[d]	References
1	21b[e]	rt[f]	7	*mmmm* (92)	161	4.2 (M_v)	nr[g]	6
2	21b[e]	50	nr	*mmmm* (81)	158	4.3 (M_v)	nr	6
3	21b[h]	rt	nr	*mmmm* (98)	165	20 (M_v)	nr	6
4	22a[i]	20–22	0.014	*mm/mr/rr* (5/21/74)	87	0.059 (M_v)	nr	60
	22b[j]	20–24	0.017	*mm/mr/rr* (15/41/45)	nr	5.3 (M_w)	2.3	61
5	22c[k]	0	0.0004	*rrrr* (63) *mm/mr/rr* (3/16/82)	100	μ = 10.8 dl/g[l]	nr	59
6	20a[k]	20	0.009	*mm/mr/rr* (37/41/22)	−4 (T_g)[m]	1.6 (M_w)	1.8	61
7	21e[k]	55–60	40.3	*mm/mr/rr* (99/1/0)	161	μ = 2.01 dl/g	nr	61
8	21f[e]	20–25	0.009	*mm/mr/rr* (93/5/2)	156	12.9 (M_w)	4.1	61
9	23b[k]	20–23	0.014	*mm/mr/rr* (68/21/11)	148	μ = 1.04 dl/g	nr	61

[a] T_p = polymerization temperature.
[b] Activity in kg PP/mmol Zr·h.
[c] T_m = polymer melting point.
[d] PDI = molecular weight polydispersity (M_w/M_n).
[e] 2 bar propylene, toluene solvent, $(i\text{-Bu})_3\text{Al}$ (TIBA), $[\text{Me}_2\text{HNC}_6\text{H}_5]^+[\text{B}(\text{C}_6\text{F}_5)_4]^-$ activator.
[f] rt = room temperature.
[g] nr = not reported.
[h] Liquid propylene used.
[i] 1 mL propylene, toluene solvent, 10,000 equivalent MAO activator.
[j] 1 mole propylene, toluene solvent, 10,000 equivalent MAO activator.
[k] 200 g propylene, toluene solvent, 10,000 equivalent MAO activator.
[l] μ = polymer viscosity.
[m] T_g = polymer glass transition temperature.

REFERENCES AND NOTES

1. Ewen, J. Mechanisms of stereochemical control in propylene polymerizations with soluble group 4B metallocene/methylalumoxane catalysts. *J. Am. Chem. Soc.* **1984**, *106*, 6355–6364.

2. Kaminsky, W.; Külper, K.; Brintzinger, H. H.; Wild, F. R. W. P. Polymerization of propene and butene with a chiral zirconocene and methylalumoxane as cocatalyst. *Angew. Chem. Int. Ed. Engl.* **1985**, *204*, 507–508.

3. Anderson, G. K.; Lin, M. Preparation of phenylplatinum-zirconium complexes with $Ph_2PC_5H_4$ or $PhP(C_5H_4)_2$ bridging ligands and their reactions with carbon monoxide. *Organometallics* **1988**, *7*, 2285–2288.

4. Köpf, H.; Klouras, N. Das erste [1]titanocenophan mit phosphor-brücke. *Monatsh. Chem.* **1983**, *114*, 243–247; *Chem. Abstr.* **1983**, *98*, 179546.

5. Klouras, N. Z. $PhAs(C_5H_4)_2TiCl_2$: The first arsenic-bridged titanocenophane. *Z. Naturforsch, B: Anorg. Chem.* **1990**, *46*, 647–649; *Chem. Abstr.* **1991**, *115*, 49871.

6. Starzewski, K. A. O.; Kelly, W. M.; Stumpf, A.; Freitag, D. Donor/acceptor metallocenes: A new structure principle in catalyst design. *Angew. Chem. Int. Ed. Engl.* **1999**, *38*, 2439–2443.

7. Starzewski, A. O. D/A-Metallocenes: The new dimension in catalyst design. *Macromol. Symp.* **2004**, *213*, 47–55.

8. Reetz, M. T.; Brümmer, H.; Kessler, M.; Kuhnigk, J. Preparation and catalytic activity of boron-substituted zirconocenes. *Chimia* **1995**, *49*, 501–503.

9. Braunschweig, H.; Gross, M.; Kraft, M.; Kristen, M. O.; Leusser, D. [2]Borametallocenophanes of Zr and Hf: Synthesis, structure, and polymerization activity. *J. Am. Chem. Soc.* **2005**, *127*, 3282–3283.

10. Alt, H. G.; Foettinger, K.; Milius, W. Synthesis, characterization and polymerization potential of *ansa*-metallocene dichloride complexes of titanium, zirconium and hafnium containing a Si-N-Si bridging unit. *J. Organomet. Chem.* **1998**, *564*, 109–114.

11. Alt, H. G.; Koeppl, A. Effect of the nature of metallocene complexes of group IV metals on their performance in catalytic ethylene and propylene polymerization. *Chem. Rev.* **2000**, *100*, 1205–1221.

12. Qian, C.; Zhu, D. Studies on organolanthanide complexes. Part 55. Synthesis of furan-bridged bis(cyclopentadienyl)lanthanide and yttrium chlorides, and ligand and metal tuning of reactivity of organolanthanide hydrides (*in situ*). *J. Chem. Soc., Dalton Trans.* **1994**, 1599–1603.

13. Damrau, W. E.; Paolucci, G.; Zanon, J.; Siebel, E.; Fischer, D. R. $[Y\{2,6-(C_5H_4CH_2)_2C_5H_3N\}(\mu-OH)]_2$: A dinuclear 5-*ansa*-yttrocene hydroxide containing the 2,6-dimethylenepyridyl bridge. *Inorg. Chem. Commun.* **1998**, *1*, 424–426.

14. Lofthus, O. W.; Slebodnick, C.; Deck, P. A. Electrophile-functionalized metallocene intermediates. Application in the diastereoselective synthesis of a tetramethyldisiloxane-bridged C_2-symmetric *ansa*-zirconocene dibromide. *Organometallics* **1999**, *18*, 3702–3708.

15. Deck, P. A.; Fischer, T. S.; Downey, J. S. Boron-silicon exchange reactions of boron trihalides with trimethylsilyl-substituted metallocenes. *Organometallics* **1997**, *16*, 1193–1196.

16. Braunschweig, H.; von Koblinski, C.; Englert, U. Synthesis and structure of the first boron-bridged constrained geometry complexes. *Chem. Commun.* **2000**, 1049–1050.

17. Braunschweig, H.; von Koblinski, C.; Breitling, F. M.; Racacki, K.; Hu, C.; Wesemann, L.; Marx, T.; Patenburg, I. Synthesis and structure of $[Zr\{\eta^5 : \eta^1-C_9H_6B(NiPr_2)[NPh\}_2]$: A new complex with a boron-bridged amido-indenyl ligand. *Inorg. Chim. Acta* **2003**, *350*, 467–474.

18. Braunschweig, H.; Breitlung, F. M.; von Koblinski, C.; White, A. J. P.; Williams, D. J. Synthesis and structure of boron-bridged constrained geometry complexes of titanium. *J. Chem. Soc., Dalton Trans.* **2004**, 938–943.

19. Rufanov, K. A.; Kotov, V. V.; Kazennova, N. B.; Lemenovskii, D. A.; Avtomonov, E. V.; Lorberth, J. Hetero-*ansa*-metallocenes: I. Synthesis of the novel [1]-borylidene-bridged-zirconocene dichloride. *J. Organomet. Chem.* **1996**, *525*, 287–289.

20. Reetz, M. T.; Bruemmer, H.; Psiorz, C.; Willuhn, M. Preparation and use of borylated cyclopentadienyl ligand containing zirconocene and hafnocene compounds. PCT Int. Pat. Appl. WO 1997/15581 (Studiengesellschaft Kohle Mbh, Germany), May 1, 1997.

21. Stelck, D. S. Advancements in boryl-bridged *ansa*-zirconocene chemistry. Ph.D. Thesis, University of Idaho, 2001.

22. Stelck, D. S.; Shapiro, P. J.; Basickes, N. Novel *ansa*-Metallocenes with a single boron atom in the bridge: Syntheses, reactivities, and X-ray structures of {Ph(L)B(η^5-C$_5$H$_5$)$_2$}ZrCl$_2$(L = SMe$_2$, PMe$_3$). *Organometallics* **1997**, *16*, 4546–4550.

23. Shapiro, P. J. Boron-bridged group 4 *ansa*-metallocene complexes. *Eur. J. Inorg. Chem.* **2001**, 321–326.

24. Lancaster, S. J.; Bochmann, M. Anionic *ansa*-zirconocenes with pentafluorophenyl-substituted borato bridges. *Organometallics* **2001**, *20*, 2093–2101.

25. Burns, C. T.; Stelck, D. S.; Shapiro, P. J.; Vij, A.; Kunz, K.; Kehr, G.; Concolino, T.; Rheingold, A. L. Stable borate-bridged *ansa*-zirconocene complexes. Preparation and X-ray crystallographic characterization of [Cp$_2^*$Al]$^+$[Me(Ph)B(η^5-C$_5$H$_4$)$_2$ZrCl$_2$]$^-$ and [PPN]$^+$[Cl(Ph)B(η^5-C$_5$H$_4$)$_2$ZrCl$_2$]$^-$. *Organometallics* **1999**, *18*, 5432–5434.

26. Ashe, A. J.; Fang, X.; Kampf, J. W. Aminoboranediyl-bridged zirconcenes: Highly active olefin polymerization catalysts. *Organometallics* **1999**, *18*, 2288–2290.

27. Ashe, A. J.; Devore, D. D.; Fang, X.; Green, D. P.; Frazier, K. A.; Patton, J. T.; Timmers, F. J. Bridged metal complex catalysts for olefin polymerization. PCT Int. Pat. Appl. WO 2000/020426A1(The Dow Chemical Company, USA; The Regents of the University of Michigan), April 13, 2000.

28. Ashe, A. J.; Yang, H.; Timmers, F. J., Azaborolyl group 4 metal complexes, catalysts and olefin polymerization process. U.S. Patent Application 2005/0119114A1 (The Dow Chemical Company, USA; The Regents of the University of Michigan), June 2, 2005.

29. Braunschweig, H.; von Koblinski, C.; Wang, R. Synthesis and structure of the first [1]boratitanocenophanes. *Eur. J. Inorg. Chem.* **1999**, 69–73.

30. Braunschweig, H.; von Koblinski, C.; Mamuti, M.; Englert, U.; Wang, R. Synthesis and structure of [1]borametallocenophanes of titanium, zirconium, and hafnium. *Eur. J. Inorg. Chem.* **1999**, 1899–1904.

31. Braunschweig, H.; Breitling, F. M.; Gullo, E.; Kraft, M. The chemistry of [1]borametallocenophanes and related compounds. *J. Organomet. Chem.* **2003**, *680*, 31–42.

32. Kristen, M. O.; Braunschweig, H.; von Koblinski, C. Metallocene complexes suitable as olefin polymerization catalysts. PCT Int. Pat. Appl. WO 2000/035928A1 (BASF AG), June 22, 2000.

33. Braunschweig, H.; Kraft, M.; Radacki, K.; Stellwag, S. [1]Borahafnocenophanes: Synthesis, structure and catalytic activity. *Eu. J. Inorg. Chem.* **2005**, 2754–2759.

34. Shapiro, P. J.; Jiang, F.; Jin, X.; Twamley, B.; Patton, J. T.; Rheingold, A. L. Zwitterionic phosphorus ylide adducts of boron-bridged *ansa*-zirconocene complexes as precatalysts for olefin polymerization. *Eur. J. Inorg. Chem.* **2004**, 3370–3378.

35. Reetz, M. T.; Willuhn, M.; Psiorz, C.; Goddard, R. Donor complexes of bis(1-indenyl)phenylborane dichlorozirconium as isospecific catalysts in propene polymerization. *Chem. Commun.* **1999**, 1105–1106.

36. Rufanov, K.; Avtomonov, E.; Kazennova, N.; Kotov, V.; Khvorost, A.; Lemenovskii, D.; Lorberth, J. Polyelement substituted cyclopentadienes and indenes—novel ligand precursors for organotransition metal chemistry. *J. Organomet. Chem.* **1997**, *536–537*, 361–373.

37. Diamond, G. M.; Rodewald, S.; Jordan, R. F. Efficient Synthesis of *rac*-(ethylenebis(indenyl)ZrX$_2$ complexes via amine elimination. *Organometallics* **1995**, *14*, 5–7.

38. Spaleck, W.; Antberg, M.; Aulbach, M.; Bachmann, B.; Dolle, V.; Haftka, S.; Kueber, F.; Rohrmann, J.; Winter, A. New isotactic polypropylenes via metallocene catalysts. In *Ziegler Catalysts*; Fink, G., Muelhaupt, R., Brintzinger, H. H., Eds.; Springer: Berlin, 1995; pp 83–97 and references therein.

39. Spaleck, W.; Antberg, M.; Rohrmann, J.; Winter, A.; Bachmann, B.; Kiprof, P.; Behm, J.; Herrmann, W. A. High-molecular-weight polypropylene via mass-tailored zirconocene-type catalysts. *Angew. Chem., Int. Ed. Engl.* **1992**, *31*, 1347–1350.

40. Spaleck, W.; Kueber, F.; Winter, A.; Rohrmann, J.; Bachmann, B.; Antberg, M.; Dolle, V.; Paulus, E. F. The influence of aromatic substituents on the polymerization behavior of bridged zirconocene catalysts. *Organometallics* **1994**, *13*, 954–963.

41. Stehling, U.; Diebold, J.; Kirsten, R.; Roll, W.; Brintzinger, H.-H.; Jiingling, S.; Muelhaupt, R.; Langhauser, F. *ansa*-Zirconocene polymerization catalysts with annelated ring ligands-effects on catalytic activity and polymer chain length. *Organometallics* **1994**, *13*, 964–970.

42. Zachmanoglou, C. E.; Docrat, A.; Bridgewater, B. M.; Parkin, G.; Brandow, C. G.; Bercaw, J. E.; Jardine, C. N.; Lyall, M.; Green, J. C.; Keister, J. B. The electronic influence of ring substitutents and *ansa* bridges in zirconocene complexes as probed by infrared spectroscopic, electrochemical, and computational studies. *J. Am. Chem. Soc.* **2002**, *124*, 9525–9546.

43. Voskoboynikov, A. Z.; Agarkov, A. Y.; Chernyshev, E. A.; Beletskaya, I. P.; Churakov, A. V.; Kuz'mina, L. G. Synthesis and X-ray crystal structures of *rac*- and *meso*-2,2'-propylidenebis(1-indenyl)zirconium dichlorides. *J. Organomet. Chem.* **1997**, *530*, 75–82.

44. Corradini, P.; Guerra, G.; Cavallo, L.; Moscardi, G.; Vacatello, M. Models for the explanation of the stereospecific behavior of Ziegler–Natta catalysts. In *Ziegler Catalysts*; Fink, G., Muelhaupt, R., Brintzinger, H. H., Eds.; Springer: Berlin, 1995; pp 237–249.

45. Gilchrist, J. H.; Bercaw, J. E. NMR spectroscopic probe of the absolute stereoselectivity for metal-hydride and metal-alkyl additions to the carbon-carbon double bond. Demonstration with a single-component, isospecific Ziegler–Natta olefin polymerization catalyst. *J. Am. Chem. Soc.* **1996**, *118*, 12021–12028.

46. Piemontesi, F.; Camurati, I.; Resconi, L.; Balboni, D.; Sironi, A.; Moret, M.; Ziegler, R.; Piccolrovazzi, N. Crystal structures and solution conformations of the meso and racemic isomers of (ethylenebis(1-indenyl))zirconium dichloride. *Organometallics* **1995**, *14*, 1256–1266.

47. Chen, E. Y.-X.; Campbell, R. E.; Devore, D. D.; Green, D. P.; Link, B.; Soto, J.; Wilson, D. R.; Abboud, K. A. Divalent *ansa*-zirconocenes: Stereoselective synthesis and high activity for propylene polymerization. *J. Am. Chem. Soc.* **2004**, *126*, 42–43.

48. Schaverien, C. J. Phosphorus-bridged metallocenes: New homogeneous catalysts for the polymerization of propene. *J. Mol. Cat. A: Chem.* **1998**, *128*, 245–256.

49. Collins, S.; Gauthier, W. J.; Holden, D. A.; Kuntz, B. A.; Taylor, N. J.; Ward, D. G. Variation of polypropylene microtacticity by catalyst selection. *Organometallics* **1991**, *10*, 2061–2068.

50. Ashe, A. J.; Fang, X.; Hokky, A.; Kampf, J. W. The C_s-symmetric aminoboranediyl-bridged zirconocene dichloride [(η-9-$C_{13}H_8$)-BN(i-Pr)$_2$(η-C_5H_4)]ZrCl$_2$: Its synthesis, structure, and behavior as an olefin polymerization catalyst. *Organometallics* **2004**, *23*, 2197–2200.

51. Patsidis K.; Alt, H. G.; Milius, W.; Palackal, S. J. The synthesis, characterization and polymerization behavior of cyclopentadienyl fluorenyl complexes; the X-ray structures of the complexes [($C_{13}H_8$)SiR$_2$(C_5H_4)]ZrCl$_2$ (R – Me or Ph). *J. Organomet. Chem.* **1996**, *509*, 63–71.

52. Razavi, A.; Vereecke, D.; Peters, L.; Den Auw, K.; Nafpliotis, L.; Atwood, J. L. Manipulation of the ligand structure as an effective and versatile tool for modification of active site properties in homogeneous Ziegler–Natta catalyst systems. In *Ziegler Catalysts*; Fink, G., Muelhaupt, R., Brintzinger, H. H., Eds.; Springer: Berlin, 1995; pp 111–147.

53. Brintzinger, H. H.; Langhauser, F.; Leyser, N.; Fischer, D. R.; Schweier, G. Metallocene complexes with cationic bridges and their preparation and use as polymerization catalysts. PCT Int. Pat. Appl. WO 1996/26211A1 (BASF A.-G., Germany), August 29, 1996.

54. Bierwagen, E. P.; Bercaw, J. E.; Goddard, W. A. Theoretical studies of Ziegler–Natta catalysis: Structural variations and tacticity control. *J. Am. Chem. Soc.* **1994**, *116*, 1481–1489.

55. Ewen, J. A.; Jones, R. L.; Razavi, A. Syndiospecific propylene polymerizations with group 4 metallocenes. *J. Am. Chem. Soc.* **1988**, *110*, 6255–6256.

56. Cavallo, L.; Guerra, G.; Vacatello, M.; Corradini, P. A possible model for the stereospecificity in the syndiospecific polymerization of propene with group 4A metallocenes. *Macromolecules* **1991**, *24*, 1784–1790.

57. Alt, H. G.; Jung, M. PPh-bridged metallocene complexes of the type ($C_{13}H_8$-PPh-$C_{13}H_8$)MCl$_2$ (M = Zr, Hf). *J. Organomet. Chem.* **1998**, *568*, 127–131.

58. Resconi, L.; Jones, R. L.; Rheingold, A. L.; Yap, G. P. A. High-molecular-weight atactic polypropylene from metallocene catalysts. 1. Me$_2$Si(9-Flu)$_2$ZrX$_2$ (X = Cl, Me) *Organometallics* **1996**, *15*, 998–1005.

59. Starzewski, K.-H. A.; Steinhauser, N. Method for Producing Homopolymers, Copolymers and/or Block Copolymers Comprising Metallocenes with a Donor–Acceptor Interaction, According to a Living Polymerisation Process. German Patent DE 10244213 A1 (Bayer AG), April 1, 2004.

60. Starzewski, K.-H. A. O.; Steinhauser, N. *Transition metal compounds having a donor–acceptor interaction and a specific substitution pattern for polymerization catalysts.* U.S. Patent Application 2004/059073 A1 (Bayer AG), March 25, 2004.

61. Starzewski, K.-H. A. O.; Xin, B. *Catalysts with a donor–acceptor interaction.* U.S. Patent US2003/0036474 A1 (Bayer AG), March 20, 2003.

62. Spaleck, W.; Antberg, M.; Dolle, R. K.; Rohrmann, J.; Winter, A. Sterorigid Metallocenes: Correlations between structure and behavior in homopolymerizations of propylene. *New J. Chem.* **1990**, *14*, 499–503.

6 Stereoselective Propylene Polymerization with Early and Late Transition Metal Catalysts

Haruyuki Makio and Terunori Fujita

CONTENTS

6.1 INTRODUCTION

The advent of homogeneous metallocene catalysis has allowed chemists to make significant progress in understanding the steric, electronic, and dynamic structures of the active organometallic species in olefin polymerizations, as described in previous chapters. The polymer structures obtained using these well-defined homogeneous systems can be qualitatively related to the structures of the metallocene catalysts, and vice versa. This significant achievement has stimulated many researchers to develop nonmetallocene olefin polymerization catalysts, that is, catalysts that do not have any η^5-cyclopentadienyl (Cp) type ligands. Drawing on the knowledge obtained from metallocene catalysts, many electrophilic (cationic) nonmetallocene complexes based on both early and late transition metals have been investigated as potential olefin polymerization catalysts.[1,2] Propylene polymerization has been examined in those cases where the steric properties of the catalyst were expected to provide stereoselective polymerization. These nonmetallocene catalysts generally have ligands containing oxygen and/or nitrogen (sometimes sulfur and phosphorus), and these heteroatoms are usually incorporated into a chelate ligand structure. Even though many well-defined nonmetallocene catalysts have been developed since the late 1980s, their use to achieve propylene polymerization with comparable stereoselectivity to heterogeneous Ziegler–Natta and chiral *ansa*-metallocene catalysts has remained a challenge, until quite recently.

In this chapter, recent advances in stereoregular propylene polymerizations with nonmetallocene complexes are briefly reviewed. The greatest emphasis is placed on phenoxy-imine ligated group 4 complexes, which have been used to achieve both highly isotactic and highly syndiotactic (living) propylene polymerization.

6.2 RECENT ADVANCES IN NONMETALLOCENE CATALYSTS FOR STEREOSELECTIVE PROPYLENE POLYMERIZATION

6.2.1 NONMETALLOCENE CATALYSTS BASED ON EARLY TRANSITION METALS

In 1962, Natta discovered that syndiotactic polypropylene, which was first separated from the largely isotactic crude polypropylene products prepared by heterogeneous titanium catalysts, could be produced in a selective manner using vanadium compounds such as VCl_4 and $V(acac)_3$ (acac = acetylacetonate) combined with R_2AlX (R = alkyl; X = halogen) at subambient polymerization temperatures (e.g., $-78\,^{\circ}C$).[3] These systems may be qualified as the first single-site catalysts to mediate stereoregular propylene polymerization, because these catalysts are soluble in the polymerization medium and the product polymers have narrow molecular weight distributions (represented by M_w/M_n, also referred to as polydispersity indexes, PDIs) of around 2, or sometimes even close to 1. These narrow PDIs are indicative of single-site catalysis, and those close to 1 are suggestive of a living polymerization.[4] The propylene monomer is inserted in a 2,1-manner and the syndioselectivity (which can reach up to 76% rr) arises from the asymmetry of the last-inserted monomer unit (chain-end control).[5–8] The actual catalytic active species, although not well characterized, are assumed to be in the trivalent state[9] and have octahedral configurations around vanadium which resemble the geometries of many nonmetallocene group 4 complexes recently developed as stereoselective propylene polymerization catalysts.

Eisen and coworkers reported one such example of these recently developed octahedral complexes. Benzamidinate zirconium complexes (**1** and **2**, Figure 6.1) activated by methylaluminoxane

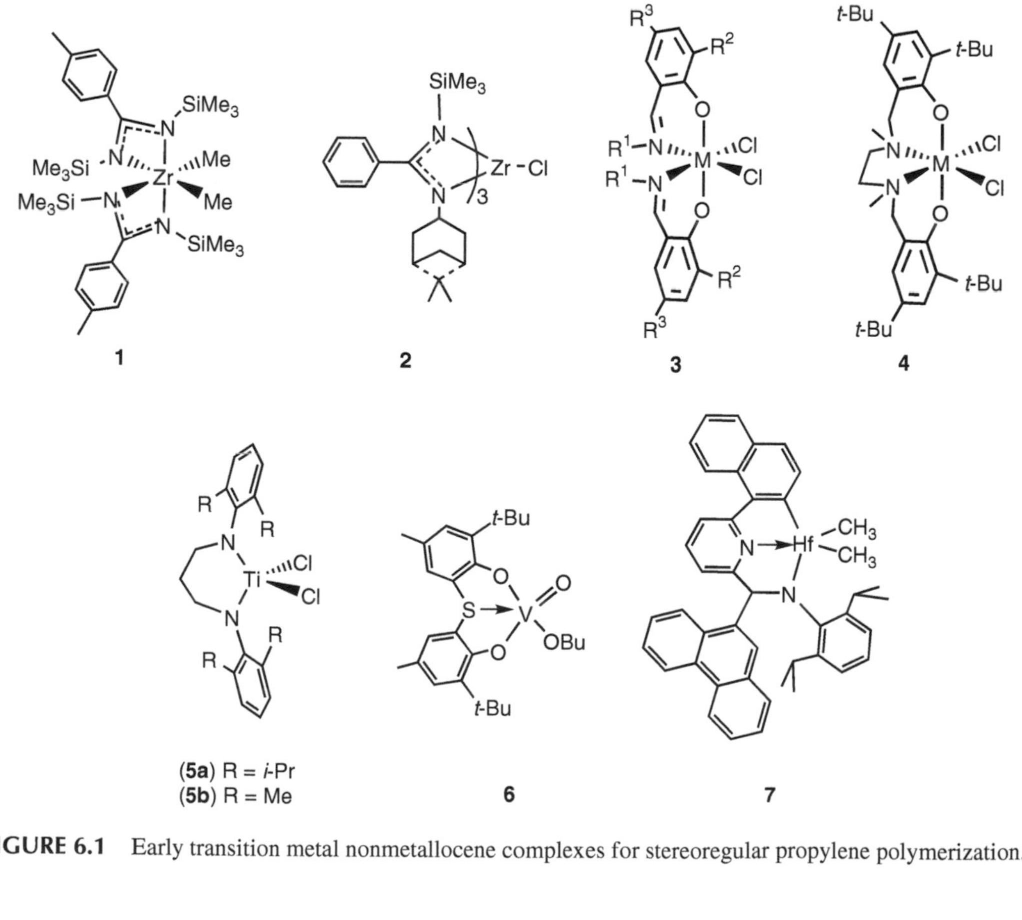

FIGURE 6.1   Early transition metal nonmetallocene complexes for stereoregular propylene polymerization.

(MAO) can polymerize propylene in a highly isotactic fashion at 25 °C in CH_2Cl_2 under high pressure (9.2 atm propylene). As expected from their C_2-symmetric structure, isotactic polypropylene was produced with up to 98% *mmmm* and a polymer melting point (T_m) of 149 °C.[10–12] The polymerization reaction propagates through 1,2-insertions, with occasional 2,1- and 1,3-insertions, and is terminated exclusively by β-methyl elimination. The isotacticity observed is sensitive to monomer concentration (pressure), such that the isotacticity is rapidly decreased at lower propylene monomer pressure. This is because the unimolecular chain-end epimerization reaction (which racemizes the last inserted propylene unit) is more pronounced than the bimolecular propagation reaction (which produces isotactic linkages) at low monomer concentration. (For further information on the mechanisms of propylene insertion, see Chapter 1.)

Other examples of octahedral complexes that can be used for stereoselective propylene polymerization are group 4 complexes incorporating two bidentate phenoxy-imine ligands (**3**, Figure 6.1) developed by scientists at Mitsui as a family of highly versatile olefin polymerization catalysts.[13–16] When a metal, a ligand, and a cocatalyst are chosen appropriately, these complexes can afford either highly isotactic[17] or highly syndiotactic[18] polypropylene with high melting temperatures. The chemistry and proposed mechanisms for stereoregulation of these catalysts are discussed in subsequent sections of this chapter.

Complexes featuring a tetradentate bisphenoxy-bisamine [ONNO] ligand (**4**, Figure 6.1) were found by Kol and coworkers to promote the living polymerization of 1-hexene with high isoselectivity. These complexes are C_2-symmetric in an octahedral framework with a *cis*-N, *trans*-O disposition.[19] The complexes are also capable of polymerizing propylene with high isoselectivity. For example, upon activation with MAO at 25 °C, complex **4** polymerizes propylene into isotactic polypropylene with 80% *mmmm* through an enantiomorphic site-control mechanism involving 1,2-insertion.[20]

The tetrahedral titanium diamide complex, [Ar-N(CH$_2$)$_3$N-Ar]TiCl$_2$ (Ar = 2,6-(*i*-Pr)$_2$-C$_6$H$_3$) (**5a**, Figure 6.1), yields a mixture of isotactic and atactic polypropylene upon activation with R$_3$Al (R = *i*-Bu, C$_6$H$_{13}$, C$_8$H$_{17}$)/[Ph$_3$C]$^+$[B(C$_6$F$_5$)$_4$]$^-$.[21,22] The weight fraction of the isotactic polymer increases as the monomer concentration increases or with the addition of cyclohexene to the system at low concentration of propylene monomer, although the *mmmm* values of the isotactic fraction are almost constant at around 80%. From the fact that the isoselective propagation is roughly second order in the propylene monomer, and that cyclohexene is scarcely incorporated into the resulting polymer, a species involving two coordinating monomer molecules was proposed (i.e., a coordinating propylene or cyclohexene might assist the isoselective insertion of propylene monomer). For reasons that are yet to be elucidated, the isotactic fraction vanishes when the alkyl groups on the Al cocatalyst are methyl- or ethyl-groups, or if the substituents on the aromatic ring of the ligand are methyl groups (**5b**) instead of isopropyl groups. Judging from the ^{13}C NMR spectrum of the isotactic fraction of the polymer, having stereoerrors in [*mmmr*]:[*mmrr*]:[*mrrm*] = 2:2:1 an enantiomorphic site-control mechanism seems to be operating, although the authors did not specifically mention it in their work. The regiochemistry of monomer insertion was not identified in this study.

A vanadium complex ligated by a tridentate thiobisphenoxy group (**6**, Figure 6.1) was reported by scientists at Sumitomo Chemical to produce isotactic polypropylene with 68% *mm* and a T_m of 138 °C (MAO, 20 °C, 1 h, propylene bulk polymerization).[23]

More recently, zirconium and hafnium complexes incorporating a C_1-symmetric pyridyl-amine-C(aryl) ligand (**7**, Figure 6.1) were discovered that produced highly isotactic polypropylene at polymerization temperatures (T_p) high enough for the catalyst to be used in a solution polymerization process (T_m in the range 150–155 °C; T_p was not specified although a high T_p of about 100 °C or above is usually required to dissolve isotactic polypropylene in organic solvents).[24–26] The combination of ligand substituents required in order to realize high isotacticity is a subtle balance, which, according to the authors, would have been difficult to achieve without the high-throughput screening used in the study. The Hf complexes give higher activity, higher isotacticity, and higher molecular weight polypropylene than the analogous Zr complexes under identical polymerization conditions.

8a: $R^1 = R^2 = R^3 = R^4 = i\text{-Pr}$, $R^5 = H$ (C_{2v})
8b: $R^1 = R^2 = R^3 = R^4 = Me$, $R^5 = H$ (C_{2v})
8c: $R^1 = R^3 = Me$, $R^2 = R^4 = R^5 = t\text{-Bu}$ (C_s)
8d: $R^1 = R^4 = i\text{-Pr}$, $R^2 = R^3 = Me$, $R^5 = H$ (C_2)
8e: $R^1 = R^4 = t\text{-Bu}$, $R^2 = R^3 = Me$, $R^5 = H$ (C_2)
8f: $R^1 = R^4 = t\text{-Bu}$, $R^2 = R^3 = Me$, $R^5 = t\text{-Bu}$ (C_2)

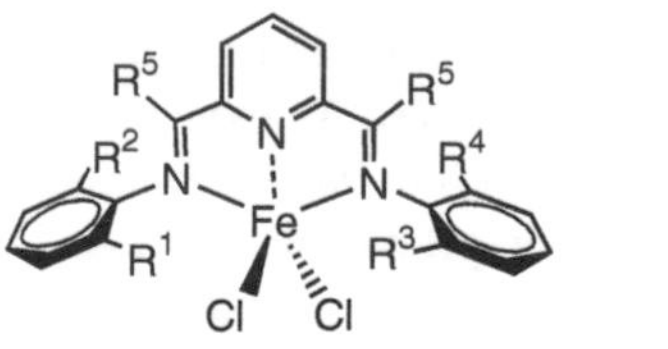

9a: $R^1 = R^2 = R^3 = R^4 = i\text{-Pr}$, $R^5 = Me$ (C_{2v})
9b: $R^1 = R^2 = R^3 = R^4 = R^5 = Me$ (C_{2v})
9c: $R^1 = R^3 = H$, $R^2 = R^4 = t\text{-Bu}$, $R^5 = Me$ (C_s/C_2)
9d: $R^1 = t\text{-Bu}$, $R^2 = H$, $R^3 = R^4 = R^5 = Me$ (C_1)
9e: $R^1 = t\text{-Bu}$, $R^2 = H$, $R^3 = R^4 = i\text{-Pr}$, $R^5 = Me$ (C_1)
9f: $R^1 = R^3 = i\text{-Pr}$, $R^2 = R^4 = R^5 = Me$ (C_s)
9g: $R^1 = R^2 = Me$, $R^3 = R^4 = i\text{-Pr}$, $R^5 = Me$ (C_s)
9h: $R^1 = R^3 = Me$, $R^2 = R^4 = i\text{-Pr}$, $R^5 = H$ (C_s/C_2)
9i: $R^1 = R^2 = R^3 = i\text{-Pr}$, $R^4 = R^5 = Me$ (C_1)

FIGURE 6.2   Late transition metal nonmetallocene complexes for stereoregular propylene polymerization. For complex **3**, R^1 to R^3 can be a variety of substituents. Examples can be found in Table 6.1.

The isotactic polymerization proceeds through an enantiomorphic site-control mechanism with successive 1,2-insertions. Theoretical studies demonstrate that each diastereotopic polymerization site exhibits opposite stereoselectivity, suggesting that only one of the two sites is used for the isotactic polymerization.

6.2.2   NONMETALLOCENE CATALYSTS BASED ON LATE TRANSITION METALS

In late transition metal catalyzed olefin polymerization, generally fast β-hydrogen elimination relative to propagation results in low molecular weight products with terminal double bonds. This disadvantage was successfully overcome by the ingenious molecular design of Ni(II) and Pd(II) diimine complexes (e.g., **8**, Figure 6.2), where the axial positions of the square planar complexes are effectively blocked by bulky 2,6-disubstituted aryl groups to prevent associative displacement of a coordinating oligomeric olefin from the olefin-hydride species by an ethylene molecule, or to avoid chain-transfer of the metal-alkyl species to ethylene monomer.[27] Because the spatial environment of the catalytic center can be modulated amongst C_{2v}, C_2, and C_s symmetry by changing these aryl substituents, propylene polymerization was examined to see if the relationship between the symmetry of the Ni complex and the stereochemistry of the polypropylene as observed for metallocene catalysts still holds true.[28,29] Upon activation with MAO, polymerization proceeds by 1,2-insertion of propylene with a considerable amount of regioerrors, including 1,3-insertions. The polypropylene obtained at a polymerization temperature of −45 °C is prevailingly syndiotactic (61–75% rr; 1–8% mm) through chain-end control for C_{2v} and C_s complexes (**8a–8c**, Figure 6.2), whereas the polymer formed by complexes with C_2 symmetry (**8d–8f**, Figure 6.2) exhibits clearly intensified mm triad signals (25–34% rr; 23–41% mm), suggesting that the chain-end control and enantiomorphic site-control mechanisms operate simultaneously.

The steric environment for iron complexes with tridentate pyridine-bis(imine) ligands is similar to that for the Ni diimine complexes, except that the two metal-bound halogen atoms, which are presumably transformed into a polymerization site upon activation, are located in the plane perpendicular to the Fe-N_3 plane (**9**, Figure 6.2), while they are in the Ni-N_2 plane of the square planar Ni complexes.[30,31] In the iron complex–catalyzed polymerizations, the propylene monomer is inserted in a highly regioregular 2,1-fashion and exclusively yields 1-propenyl chain ends. The polypropylene that is produced is prevailingly isotactic (up to 67% $mmmm$ at −20 °C; 69% mm at 0 °C) irrespective of the catalyst symmetry (**9a–9i**, Figure 6.2), indicating that the stereochemistry is dictated by chain-end control.[32,33]

6.3 BIS(PHENOXY-IMINE) GROUP 4 COMPLEXES (FI CATALYSTS)

A series of group 4 transition metal complexes discovered by scientists at Mitsui incorporating two phenoxy-imine chelate ligands[15] (FI catalysts, named after the Japanese pronunciation of the ligand[13]) can be readily varied in their structures and catalytic properties by systematically changing the substituents (R^1 to R^3 of **3**, Figure 6.1) of their phenoxy-imine ligands, which can be synthesized in a straightforward way from readily available chemicals.[13–16] An enormous ligand library has been built and studied around the polymerization properties of the various combinations with the central metals and the cocatalysts. The existence of this diversified library makes the FI catalyst family extremely versatile, such that polyolefins with specific desired properties (molecular weight, molecular weight distribution, comonomer content, chain-end structure, tacticity, etc.) can sometimes be obtained in a predictable way by selecting appropriate combinations of ligands, metal, and cocatalyst.

Peculiar propylene polymerization behaviors of FI catalysts including syndioselective living polymerization were first disclosed in the patent application by Mitsui Chemicals, Inc. filed in January 2000[16] and have been published in the subsequent scientific papers by the same group. Coates and coworkers at Cornell University independently conducted intensive work on Ti-FI catalysts, and reported their results on syndioselective propylene polymerization in June 2000[34] and on syndioselective living propylene polymerization in February 2001.[35] Since then, a great deal of work aiming at the design and synthesis of new and superior FI catalysts and at understanding polymerization mechanisms for FI catalysts, has been carried out, and is discussed below.

6.3.1 PROPYLENE POLYMERIZATION WITH FI CATALYSTS ACTIVATED WITH MAO

Theoretically, the FI catalysts can possess five isomers, among which a C_2-symmetric *cis*-N, *trans*-O, *cis*-Cl configuration shown in Figure 6.1 (complex **3**) is the most stable, although a C_1-symmetric *cis*-N, *cis*-O, *cis*-Cl configuration is sometimes observable depending on the ligand structure and the group 4 metal centers.[13–16,37,51] Despite the fact that most of the FI catalysts have a C_2-symmetric *cis*-N, *trans*-O, *cis*-Cl configuration, the FI catalysts behave in quite a different way from the isoselective C_2-symmetric metallocene complexes for propylene polymerization. Ti-FI catalysts with MAO used as the cocatalyst polymerize propylene in a moderately to highly syndiotactic manner,[18,34–38] whereas the corresponding Zr-FI and Hf-FI catalysts used in conjunction with MAO afford atactic to slightly syndiotactic polypropylene.[39,40] The stereoerrors in the syndiotactic polypropylenes obtained with Ti-FI catalysts are of the isolated *m*-type (...*rrrrmrrrr*...), indicating that the syndioselective enchainment is mediated through chain-end control.[34–38] The molecular weights of the polypropylenes that are obtained are also strongly affected by the identity of the group 4 metal center. Low molecular weight polymers or oligomers with double bonds at one chain end (formed through β-H transfer) are principally obtained when Zr-FI and Hf-FI catalysts are used,[39,40] while much higher molecular weight polypropylenes are produced with Ti-FI catalysts.[34–38] When R^1 is perfluorophenyl, the nature of Ti-FI catalysts to form high molecular weight polymers is further pronounced, and the catalysts exhibit a robust living nature for both propylene polymerization and ethylene polymerization,[41] with much enhanced propylene syndioselectivity relative to the corresponding nonfluorinated catalysts ($R^1 = C_6H_5$).[42] The difference in polymerization characteristics between Ti-FI and Zr-FI/Hf-FI catalysts when activated with MAO may be attributed to differences in the regiochemistry of propylene insertion (Scheme 6.1).[43–46,29] Regarding Ti-FI catalysts, a propylene molecule is inserted into the Ti-R bond exclusively in a 1,2-fashion (primary insertion) when R is -H or -CH$_3$. For ethylene/propylene copolymerization, primary insertion of propylene is also

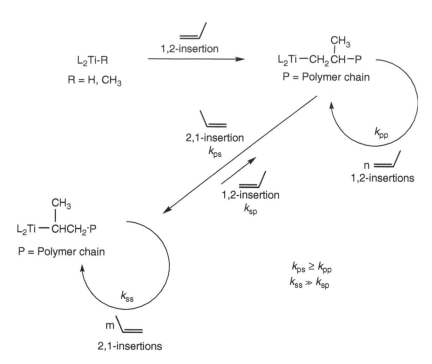

SCHEME 6.1 Regiochemistry of propylene polymerization mediated by Ti-FI Catalysts activated with MAO (L = phenoxy-imine ligand). Two subscript characters attached to k stand for regiochemistry of the last-inserted propylene unit (left) and that of a coordinating propylene monomer (right). For example, k_{ps} is the rate constant of a secondary (2,1-) insertion to a primary chain end.

favored after ethylene insertion, that is, R = -CH$_2$CH$_2$-. However, when R becomes -CH$_2$CH(CH$_3$)- (as after the primary insertion of a propylene monomer), primary propylene insertion to this primary chain end (k_{pp}) seems to be energetically comparable with or slightly less favorable than 2,1-insertion (secondary insertion; $k_{ps} \geq k_{pp}$), depending on the catalyst structures and polymerization conditions. Once a secondary insertion occurs, which sets a -CH(CH$_3$)CH$_2$- structure at the chain end of the active species, the subsequent insertion is preferentially secondary ($k_{ss} \gg k_{sp}$). On the whole, the regioselectivity described above ends up comprising prevailingly secondary insertion with an exclusive 1,2-insertion at the initiating chain-end. Propylene polymerization through secondary insertion is highly unusual for group 4 transition metal-mediated olefin polymerization. In contrast, Zr-FI and Hf-FI catalysts polymerize propylene through ordinary primary insertion with a considerable amount of regioerrors.[39,40]

The behavior of apparently C_2-symmetric Ti-FI catalysts is obviously different from that of isoselective C_2-symmetric metallocenes that polymerize propylene by 1,2-insertion through an enantiomorphic site-control mechanism. Instead, it is similar to that of soluble vanadium catalysts for which the propylene insertion is secondary, and where polypropylene with low to moderate syndiotacticity is formed through a chain-end control mechanism (see Section 6.2.1).[5–8] The well-defined nature of FI catalysts relative to classical vanadium catalysts allows systematical change of their molecular structures, which leads to intriguing results on a stereoregulating mechanism. The steric modulation of R^2 substituents gives a sharp increase in syndioregularity that is proportional to the volume of R^2, producing up to 94% of *rr* triads when R^2 = Me$_3$Si (Table 6.1).[18,34–38,47] In a comparison between R^1 = C$_6$H$_5$ and R^1 = C$_6$F$_5$, the fluorinated complex tends to give higher syndioregularity when R^2 and R^3 are the same for both catalysts (Table 6.1, runs 4, 5).[34–38,42] All of these results are rather unexpected in view of traditional chain-end control, in which stereoselectivity is governed by the asymmetry of the last inserted monomer unit, whereas the chain-end

TABLE 6.1

Effects of the R^2 Substituent of Ti-FI Catalysts (3, Figure 6.1) on Propylene Polymerization[36,38,42]

Run	R^1	R^2	R^3	Activity[a]	$M_n/10^{4}$[b]	M_w/M_n[b]	T_m[c]	rr[d]
1[e]	C_6F_5	H	H	30.7	18.9	1.51	n.d.[f]	43
2[e]	C_6F_5	Me	H	68.8	26.0	1.22	n.d.[f]	50
3[e]	C_6F_5	i-Pr	H	31.1	15.4	1.16	n.d.[f]	75
4[e]	C_6F_5	t-Bu	H	3.70	2.85	1.11	137	87
5[g]	C_6H_5	t-Bu	H	0.95	0.43	1.38	97	62.9[h]
6[e]	C_6F_5	SiMe$_3$	H	5.86	4.70	1.08	152	93

[a] kg polymer/mol Ti·h.

[b] Number average molecular weight (M_n) and weight average molecular weight (M_w) measured by gel permeation chromatography (GPC, polypropylene calibration).

[c] Polymer melting point measured by differential scanning calorimetry (DSC).

[d] % rr dyads unless otherwise noted as measured by ^{13}C NMR.

[e] Toluene (250 mL), 0.1 MPa of propylene, 25 °C, 5 h, Ti-FI Catalyst = 10 μmol, MAO = 2.5 mmol.

[f] Not detected.

[g] Toluene (350 mL), 0.37 MPa of propylene, 1 °C, 6 h, Ti-FI Catalyst = 0.1 mmol, MAO = 15 mmol.

[h] % $rrrr$.

controlled syndioselectivity observed with the Ti-FI catalysts is obviously affected by the catalyst ligand structure (ligand-directed chain-end control). The chain-end controlled syndioselectivity that was observed was explained theoretically using the hypothesis that the Ti-FI catalysts are fluxional between the Δ and Λ enantiomers of their *cis*-N, *trans*-O, *cis*-Cl configuration (Scheme 6.2).[48,49] According to the proposed mechanism, the octahedral active species are subject to fluxional isomerization between the Λ and Δ forms at a rate comparable to or faster than chain propagation, driven by the stereochemistry of the α-carbon of the last 2,1-inserted propylene unit to ease steric interference between the specific polymer configuration and an R^1 substituent. The energetically favored diastereomers (*re*-chain/Λ-site, **E** or **F** and *si*-chain/Δ-site, **A** or **B**) exert a strong stereoselectivity on propylene coordination and insertion, which essentially functions as a result of steric repulsion between the ligand R^2 substituents, and the methyl group of the coordinating propylene (Λ-site/*si*-face, **E** and Δ-site/*re*-face, **B**). Thus, the stereochemistry of the α-carbon dictates the face-selectivity of the prochiral propylene monomer (*re*-chain preferred *si*-face, and *si*-chain preferred *re*-face) using the ligand of the catalyst as an intermediary. This scenario, which was originally proposed for the soluble vanadium catalysts,[50] is in good accord with the observation that the steric bulk of the Ti-FI catalyst R^2 substituent has a pronounced effect upon syndioselectivity. Although the postulated inversion between the Δ and Λ forms is not experimentally supported, FI catalysts are likely to be fluxional in solution, as has been reported for some Zr-FI catalysts[51] and some other related complexes,[52] probably through dissociation of an M–N bond.

6.3.2 PROPYLENE POLYMERIZATION WITH FI CATALYSTS ACTIVATED WITH i-Bu$_3$Al/[Ph$_3$C]$^+$[B(C$_6$F$_5$)$_4$]$^-$

When FI catalysts are activated by i-Bu$_3$Al/[Ph$_3$C]$^+$[B(C$_6$F$_5$)$_4$]$^-$, the polymerization characteristics are strikingly different from those observed for MAO activation. This is because the imine groups of the FI catalysts, -CH=NR^1, are susceptible to attack by electrophiles and are easily

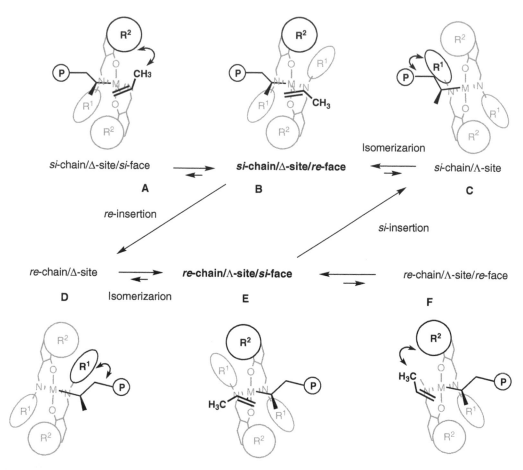

SCHEME 6.2 A proposed mechanism for the chain-end controlled syndioselective propylene polymerization by Ti-FI Catalysts activated with MAO (P = polymer chain). The species A/F, B/E, and C/D are pairs of enantiomers.

reduced, in this case by i-Bu$_3$Al to -CH$_2$-NR^1-Al-i-Bu$_2$ groups with the concurrent formation of isobutylene (Scheme 6.3, lower path).[53] The catalyst species with reduced ligands are, in general, lower in polymerization activity than the corresponding MAO-activated species, but tend to polymerize olefins into extremely high molecular weight polymers. Propylene polymerization characteristics of these species with reduced ligands can be varied by changing the metal center and the ligand substituents in an apparently similar manner to that for the MAO-activated FI catalysts, but with opposite selectivity. A Ti-FI catalyst, bis[N-3-*tert*-butylsalicylidene)anilinato]titanium (IV) dichloride, yields ultra-high molecular weight atactic polypropylene with a significant amount of regioirregular units upon activation with i-Bu$_3$Al/[Ph$_3$C]$^+$[B(C$_6$F$_5$)$_4$]$^-$.[54] On the other hand, the corresponding Zr-FI and Hf-FI catalysts afford rather isotactic polypropylenes, $mmmm$ = 36% for Zr and $mmmm$ = 56% for Hf.[54] In these polymerizations, the polymer product obtained with the Ti catalyst exhibits a relatively broad molecular weight distribution (M_w/M_n = 4.15), indicating the presence of multiple active species, whereas the Zr and Hf catalysts demonstrate single-site polymerization characteristics (M_w/M_n = 2.42, and 2.15, respectively). The isoselectivity of the Zr- and Hf-FI catalysts activated with i-Bu$_3$Al/[Ph$_3$C]$^+$[B(C$_6$F$_5$)$_4$]$^-$ can be improved by changing the ligand substituents. The bulkiness of the R^2 substituents has a tremendous impact on the isotacticity of the polypropylene that is produced. The best result thus far have been obtained when M = Hf, R^1 = cyclohexyl, R^2 = adamantyl, and R^3 = CH$_3$. Although this catalyst exhibits

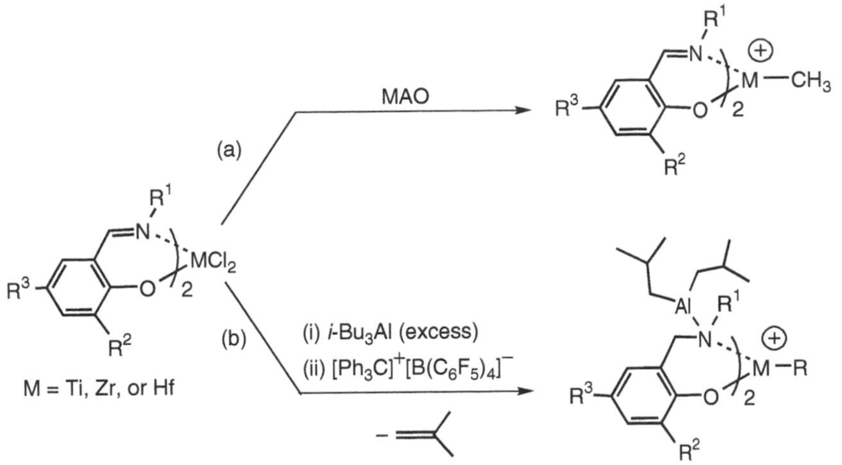

SCHEME 6.3 Activation of FI Catalysts (a) with MAO and (b) with i-Bu$_3$Al/[Ph$_3$C]$^+$[B(C$_6$F$_5$)$_4$]$^-$ and formation of putative active species for olefin polymerizations. R^1 to R^3 are, for example, as defined in Table 6.1. R is presumably the isobutyl group.

multiple-site character ($M_w/M_n = 14.6$), isotacticity reaches 97% *mmmm* and the melting temperature of the polypropylene is 165 °C after removal of low stereoselective polymers by fractionation with boiling hexane.[17] Microstructural analyses of these isotactic polypropylenes suggest that the polymerization proceeds through an enantiomorphic site-control mechanism with the primary (1,2-) insertion of propylene.[17] Although the structure of the active species of FI catalysts activated with i-Bu$_3$Al/[Ph$_3$C]$^+$[B(C$_6$F$_5$)$_4$]$^-$ remains elusive, there are three possibilities that might rationalize the observed isoselectivity. The first possible explanation is that the presumed phenoxy-amine catalyst species possesses a C_2-symmetric, octahedral *cis*-N, *trans*-O, *cis*-Cl configuration similar to that of the nonreduced phenoxy-imine. However, the fluxional inversion of chirality assumed to occur with the phenoxy-imine catalysts (Section 6.3.1) may be inhibited by the large substituents on the amine nitrogen. The second possibility is incomplete reduction of the imines by i-Bu$_3$Al, which may result in a C_1-symmetric (phenoxy-imine) (phenoxy-amine)M-P species (P = polymer chain). The third possibility is that both imines are reduced into amines but one of the amine-metal bonds is elongated or dissociated owing to the steric bulk and the reduced basicity of Al-coordinated amine moiety, which leads to a species with C_1-symmetric structure. In any case, an isoselective polymerization through the primary insertion of propylene monomers with these Zr- or Hf-FI catalysts can occur through a similar mechanism to that operative for metallocene catalysts having C_2 and C_1 symmetry.

6.4 CONCLUSIONS

Rapid progress in the design and synthesis of well-defined organometallic complexes and the rational interpretation of the polymerization reaction mechanisms has made highly stereoregular propylene polymerizations using nonmetallocene catalysts a reality. Two important outcomes may arise from the recent development of stereoselective nonmetallocene polymerization catalysts. First, the dynamic character of these complexes (such as fluxional isomerization) may present another useful strategy for realizing highly stereoselective reactions, which is in a sharp contrast with the sterically rigid ligand frameworks that have been used to improve stereoregularity in metallocene-based chemistry. Secondly, new stereoselective catalysts with novel molecular structures (which may, unfortunately, be beyond rational design or imagination for most of us) may be identified with combinatorial chemistry.

REFERENCES AND NOTES

1. Gibson, V. C.; Spitzmesser, S. K. Advances in non-metallocene olefin polymerization catalysis. *Chem. Rev.* **2003**, *103*, 283–315.

2. Britovsek, G. J. P.; Gibson, V. C.; Wass, D. F. The search for new-generation olefin polymerization catalysts: Life beyond metallocenes. *Angew. Chem. Int. Ed.* **1999**, *38*, 428–447.

3. Natta, G.; Pasquon, I.; Zambelli, A. Stereospecific catalysts for the head-to-tail polymerization of propylene to a crystalline syndiotactic polymer. *J. Am. Chem. Soc.* **1962**, *84*, 1488–1490.

4. Doi, Y.; Ueki, S.; Keii, T. "Living" coordination polymerization of propene initiated by the soluble V(acac)$_3$-Al(C$_2$H$_5$)$_2$Cl system. *Macromolecules* **1979**, *14*, 814–819.

5. Zambelli, A.; Gatti, G.; Sacchi, C.; Crain, Jr. W. O.; Roberts, J. D. Nuclear magnetic resonance spectroscopy. Steric control in α-olefin polymerization as determined by ^{13}C spectra. *Macromolecules* **1971**, *4*, 475–477.

6. Zambelli, A.; Tosi, C.; Sacchi, C. Polymerization of propylene to syndiotactic polymer. VI. Monomer insertion. *Macromolecules* **1972**, *5*, 649–654.

7. Zambelli, A.; Wolfsgruber, C.; Zannoni, G.; Bovey, F. A. Polymerization of propylene to syndiotactic polymer. VIII. Steric control forces. *Macromolecules* **1974**, *7*, 750–752.

8. Bovey, F. A.; Sacchi, M. C.; Zambelli, A. Polymerization of propylene to syndiotactic polymer. IX. Ethylene perturbation of syndiotactic propylene polymerization. *Macromolecules* **1974**, *7*, 752–754.

9. Lehr, M. H. The active oxidation state of vanadium in soluble monoolefin polymerization catalysts. *Macromolecules* **1967**, *1*, 178–184.

10. Averbuj, C.; Tish, E.; Eisen, M. S. Stereoregular polymerization of α-olefins catalyzed by chiral group 4 benzamidinate complexes of C_1 and C_3 Symmetry. *J. Am. Chem. Soc.* **1998**, *120*, 8640–8646.

11. Volkis, V.; Shmulinson, M.; Averbuj, C.; Lisovskii, A.; Edelmann, F. T.; Eisen, M. S. Pressure modulates stereoregularities in the polymerization of propylene promoted by *rac*-octahedral heteroallylic complexes. *Organometallics* **1998**, *17*, 3155–3157.

12. Volkis, V.; Nelkenbaum, E.; Lisovskii, A.; Hasson, G.; Semiat, R.; Kapon, M.; Botoshansky, M.; Eishen, Y.; Eisen, M. S. Group 4 octahedral benzamidinate complexes: Syntheses, structures, and catalytic activities in the polymerization of propylene modulated by pressure. *J. Am. Chem. Soc.* **2003**, *125*, 2179–2194.

13. Makio, H.; Kashiwa, N.; Fujita, T. FI Catalysts: A new family of high performance catalysts for olefin polymerization. *Adv. Synth. Catal.* **2002**, *344*, 477–493.

14. Mitani, M.; Saito, J.; Ishii, S.; Nakayama, Y.; Makio, H.; Matsukawa, N.; Matsui, S. et al. FI Catalysts: New olefin polymerization catalysts for the creation of value-added polymers. *Chem. Rec.* **2004**, *4*, 137–158.

15. Fujita, T.; Tohi, Y.; Mitani, M.; Matsui, S.; Saito, J.; Nitabaru, M.; Sugi, K.; Makio, H.; Tsutsui, T. Olefin polymerization catalysts, transition metal compounds, processes for olefin polymerization, and alpha-olefin/conjugate diene copolymers. European Patent 0874005 B1 (Mitsui Chemicals, Inc.), July 23, 2003.

16. Mitani, M.; Yoshida, Y.; Mohri, J.; Tsuru, K.; Ishii, S.; Kojoh, S.; Matsugi, T. et al. Olefin polymers and production processes thereof. PCT Int. Pat. Appl. WO 2001/55231 A1 (Mitsui Chemicals, Inc.), August 2, 2001.

17. Prasad, A. V.; Makio, H.; Saito, J.; Onda, M.; Fujita, T. Highly isospecific polymerization of propylene with bis(phenoxy-imine) Zr and Hf complexes using ^{i}Bu$_3$Al/Ph$_3$CB(C$_6$F$_5$)$_4$ as a cocatalyst. *Chem. Lett.* **2004**, 250–251.

18. Mitani, M.; Furuyama, R.; Mohri, J.; Saito, J.; Ishii, S.; Terao, H.; Kashiwa, N.; Fujita, T. Fluorine- and trimethylsilyl-containing phenoxy-imine Ti complex for highly syndiotactic living polypropylenes with extremely high melting temperatures. *J. Am. Chem. Soc.* **2002**, *124*, 7888–7889.

19. Tshuva, E. Y.; Goldberg, I.; Kol, M. Isospecific living polymerization of 1-hexene by a readily available nonmetallocene C_2-symmetrical zirconium catalyst. *J. Am. Chem. Soc.* **2000**, *122*, 10706–10707.

20. Busico, V.; Cipullo, R.; Ronca, S.; Budzelaar, P. H. M. Mimicking Ziegler–Natta catalysts in homogeneous phase. Part 1. C_2-symmetric octahedral Zr(IV) complexes with tetradentate [ONNO]-type ligands. *Macromol. Rapid Commun.* **2001**, *22*, 1405–1410.

21. Tsubaki, S.; Jin, J.; Ahn, C.-H.; Sano, T.; Uozumi, T.; Soga, K. Synthesis of isotactic poly(propylene) by titanium based catalysts containing diamide ligands. *Macromol. Chem. Phys.* **2001**, *202*, 482–487.

22. Uozumi, T.; Tsubaki, S.; Jin, J.; Sano, T.; Soga, K. Isospecific propylene polymerization using the [ArN(CH$_2$)$_3$NAr]TiCl$_2$/Al(iBu)$_3$/Ph$_3$CB(C$_6$F$_5$)$_4$ catalyst system in the presence of cyclohexene. *Macromol. Chem. Phys.* **2001**, *202*, 3279–3283.

23. Takaoki, K.; Miyatake, T. Titanium and vanadium based non-metallocene catalysts for olefin polymerization. *Macromol. Symp.* **2000**, *157*, 251–257.

24. Stevens, J. C.; Boone, H.; VanderLende, D.; Boussie, T.; Diamond, G. M.; Goh, C.; Hall, K. et al. New high activity group 4 C_1-symmetric catalysts for isotactic-selective high temperature solution polymerization of propene, and copolymerization of propene with ethene. 1-Synthesis, process and products. *Abstract of Papers*, European Polymer Conference on Stereospecific Polymerization and Stereoregular polymers. Milano, Italy, June 8–12, 2003; pp 79–80.

25. Busico, V.; Cipullo, R.; Talarico, G.; Stevens, J. C. New high activity group 4 C_1-symmetric catalysts for isotactic-selective high temperature solution polymerization of propene, and copolymerization of propene with ethene. 2-Polymer microstructure and polymerization mechanism. *Abstract of Papers*, European Polymer Conference on Stereospecific Polymerization and Stereoregular polymers. Milano, Italy, June 8–12, 2003; pp 81–82.

26. Boussie, T. R.; Diamond, G. M.; Goh, C.; Hall, K. A.; Lapointe, A. M.; Leclerc, M. K.; Lund, C. Substituted pyridyl amine ligands, complexes, catalysts and processes for polymerizing and polymers. PCT Int. Pat. Appl. WO 2002/38628 A2 (Symyx Technologies, Inc.), May 16, 2002.

27. Johnson, L. K.; Killian, C. M.; Brookhart, M. New Pd(II)- and Ni(II)-based catalysts for polymerization of ethylene and α-olefins. *J. Am. Chem. Soc.* **1995**, *117*, 6414–6415.

28. Pappalardo, D.; Mazzeo, M.; Antinucci, S.; Pellecchia, C. Some evidence of a dual stereodifferentiation mechanism in the polymcrization of propene by α-diimine nickel catalysts. *Macromolecules* **2000**, *33*, 9483–9487.

29. Lamberti, M.; Mazzeo, M.; Pappalardo, D.; Zambelli, A.; Pellecchia, C. Polymerization of propene by post-metallocene catalysts. *Macromol. Symp.* **2004**, *213*, 251–257.

30. Small, B. L.; Brookhart, M.; Bennett, A. M. A. Highly active iron and cobalt catalysts for the polymerization of ethylene. *J. Am. Chem. Soc.* **1998**, *120*, 4049–4050.

31. Britovsek, G. J. P.; Gibson, V. C.; Kimberley, B. S.; Maddox, P. J.; McTavish, S. J.; Solan, G. A.; White, A. J. P.; Williams, D. J. Novel olefin polymerization catalysts based on iron and cobalt. *Chem. Commun.* **1998**, 849–850.

32. Pellecchia, C.; Mazzeo, M.; Pappalardo, D. Isotactic-specific polymerization of propene with an iron-based catalyst: Polymer end groups and regiochemistry of propagation. *Macromol. Rapid Commun.* **1998**, *19*, 651–655.

33. Small B. L.; Brookhart, M. Polymerization of propylene by a new generation of iron catalysts: Mechanisms of chain initiation, propagation, and termination. *Macromolecules* **1999**, *32*, 2120–2130.

34. Tian, J.; Coates, G. W. Development of a diversity-based approach for the discovery of stereoselective polymerization catalysts: Identification of a catalyst for the synthesis of syndiotactic polypropylene. *Angew. Chem. Int. Ed.* **2000**, *39*, 3626–3629.

35. Tian, J.; Hustad, P. D.; Coates, G. W. A new catalyst for highly syndiospecific living olefin polymerization: Homopolymers and block copolymers from ethylene and propylene. *J. Am. Chem. Soc.* **2001**, *123*, 5134–5135.

36. Saito, J.; Mitani, M.; Mohri, J.; Ishii, S.; Yoshida, Y.; Matsugi, T.; Matsui, S. et al. Highly syndiospecific living polymerization of propylene using a titanium complex having two phenoxy-imine chelate ligands. *Chem. Lett.* **2001**, 576–577.

37. Mason, A. F.; Tian, J.; Hustad, P. D.; Lobkovsky, E. B.; Coates, G. W. Syndiospecific living catalysts for propylene polymerization: Effect of fluorination on activity, stereoselectivity, and termination. *Israel J. Chem.* **2002**, *42*, 301–306.

38. Mitani, M.; Furuyama, R.; Mohri, J.; Saito, J.; Ishii, S.; Terao, H.; Nakano, T.; Tanaka, H.; Fujita, T. Syndiospecific living propylene polymerization catalyzed by titanium complexes having fluorine-containing phenoxy-imine chelate ligands. *J. Am. Chem. Soc.* **2003**, *125*, 4293–4305.

39. Makio, H.; Tohi, Y.; Saito, J.; Onda, M.; Fujita, T. Regio- and stereochemistry in propylene polymerization catalyzed with bis(phenoxy-imine) Zr and Hf complexes/MAO systems. *Macromol. Rapid Commun.* **2003**, *24*, 894–899.

40. Lamberti, M.; Gliubizzi, R.; Mazzeo, M.; Tedesco, C.; Pellecchia, C. Bis(phenoxyimine)zirconium and titanium catalysts affording prevailingly syndiotactic polypropylenes via opposite modes of monomer insertion. *Macromolecules* **2004**, *37*, 276–282.
41. Mitani, M.; Mohri, J.; Yoshida, Y.; Saito, J.; Ishii, S.; Tsuru, K.; Matsui, S. et al. Living polymerization of ethylene catalyzed by titanium complexes having fluorine-containing phenoxy-imine chelate ligands. *J. Am. Chem. Soc.* **2002**, *124*, 3327–3336.
42. Furuyama, R.; Saito, J.; Ishii, S.; Mitani, M.; Matsui, S.; Tohi, Y.; Makio, H.; Matsukawa, N.; Tanaka, H.; Fujita, T. Ethylene and propylene polymerization behavior of a series of bis(phenoxy-imine)titanium complexes. *J. Mol. Catal. A: Chem.* **2003**, *200*, 31–42.
43. Saito, J.; Mitani, M.; Onda, M.; Mohri, J.; Ishii, S.; Yoshida, Y.; Nakano, T. et al. Microstructure of highly syndiotactic "living" poly(propylene)s produced from a titanium complex with chelating fluorine-containing phenoxyimine ligands (an FI Catalyst). *Macromol. Rapid Commun.* **2001**, *22*, 1072–1075.
44. Lamberti, M.; Pappalardo, D.; Zambelli, A.; Pellecchia, C. Syndiospecific polymerization of propene promoted by bis(salicylaldiminato)titanium catalysts: Regiochemistry of monomer insertion and polymerization mechanism. *Macromolecules* **2002**, *35*, 658–663.
45. Hustad, P. D.; Tian, J.; Coates, G. W. Mechanism of propylene insertion using bis(phenoxyimine)-based titanium catalysts: An unusual secondary insertion of propylene in a group IV catalyst system. *J. Am. Chem. Soc.* **2002**, *124*, 3614–3621.
46. Talarico, G.; Busico, V.; Cavallo, L. Origin of the regiochemistry of propene insertion at octahedral column 4 polymerization catalysts: Design or serendipity? *J. Am. Chem. Soc.* **2003**, *125*, 7172–7173.
47. Mitani, M.; Nakano, T.; Fujita, T. Unprecedented living olefin polymerization derived from an attractive interaction between a ligand and a growing polymer chain. *Chem. Eur. J.* **2003**, *9*, 2396–2403.
48. Milano, G.; Cavallo, L.; Guerra, G. Site chirality as a messenger in chain-end stereocontrolled propene polymerization. *J. Am. Chem. Soc.* **2002**, *124*, 13368–13369.
49. Corradini, P.; Guerra, G.; Cavallo, L. Do new century catalysts unravel the mechanism of stereocontrol of old Ziegler-Natta catalysts? *Acc. Chem. Res.* **2004**, *37*, 231–241.
50. Corradini, P.; Guerra, G.; Pucciariello, R. New model of the origin of the stereospecificity in the synthesis of syndiotactic polypropylene. *Macromolecules* **1985**, *18*, 2030–2034.
51. Tohi, Y.; Matsui, S.; Makio, H.; Onda, M.; Fujita, T. Polyethylenes with uni-, bi-, and trimodal molecular weight distributions produced with a single bis(phenoxy-imine)zirconium complex. *Macromolecules* **2003**, *36*, 523–525.
52. Bei, X.; Swenson, D. C.; Jordan, R. F. Synthesis, structures, bonding, and ethylene reactivity of group 4 metal alkyl complexes incorporating 8-quinolinolato ligands. *Organometallics* **1997**, *16*, 3282–3302.
53. Matsui, S.; Mitani, M.; Saito, J.; Tohi, Y.; Makio, H.; Matsukawa, N.; Takagi, Y. et al. A family of zirconium complexes having two phenoxy-imine chelate ligands for olefin polymerization. *J. Am. Chem. Soc.* **2001**, *123*, 6847–6856.
54. Saito, J.; Onda, M.; Matsui, S.; Mitani, M.; Furuyama, R.; Tanaka, H.; Fujita, T. Propylene polymerization with bis(phenoxy-imine) group-4 catalysts using $^{i}Bu_3Al/Ph_3CB(C_6F_5)_4$ as a cocatalyst. *Macromol. Rapid Commun.* **2002**, *23*, 1118–1123.

7 Catalyst Structure and Polymer Tacticity: The Theory Behind Site Control in Propylene Polymerization

Anthony K. Rappé, Oleg G. Polyakov, and Lynn M. Bormann-Rochotte

CONTENTS

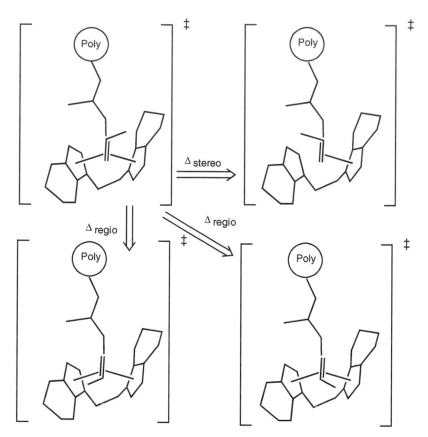

SCHEME 7.1 Stereochemical and regiochemical conformational alternatives for a representative *ansa*-metallocene catalyst shown with growing polypropylene chain.

7.1 INTRODUCTION

This chapter presents a review of theoretical studies of stereo- and regiocontrol in propylene polymerization. The working hypothesis of researchers in this field is that an enhanced understanding of the molecular sources of stereo- and regiocontrol will lead to the development of improved polymerization catalysts. The intellectual challenge of this and other theoretical catalytic studies is that understanding selectivity is an exercise in transition state energetic differentiation. For site control in propylene polymerization, this means computing the energetic difference between transition states that are merely conformeric alternatives for the methyl group of the approaching propylene (Scheme 7.1). In this scheme, the upper left conformation illustrates stereonormal 1,2-insertion, whereas the upper right illustrates stereoerror 1,2-insertion. The two lower conformations illustrate 2,1-regioerror conformations. Given that heroic efforts are needed to compute bond energies or excitation energies to "chemical accuracy" for even the smallest of organic molecules,[1] one must ask the question of how meaningful stereo- and regiodifferentiations can be computed for the large complexes of olefin polymerization. In this chapter, the limitations of alternative theoretical methodologies are described, and a roadmap for computing the necessary small energy differences are provided along with validating examples.

7.2 METHODOLOGIES

As discussed earlier,[2] a range of theoretical approaches (including simple sketches, through rigid docking studies, and complete conformational searches including all electronic and nuclear degrees

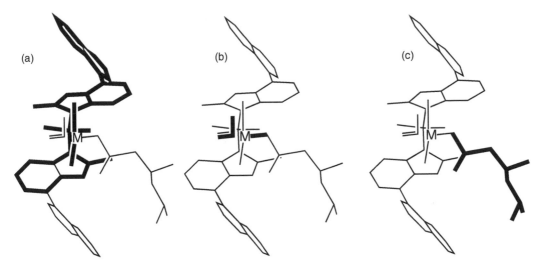

FIGURE 7.1 Fragment analysis of a metallocene-based olefin insertion transition state (a = transition metal + associated ligand; b = olefin + α carbon of polymer chain; c = remainder of polymer chain).

of freedom) have been used to study regio- and stereodifferentiation in olefin polymerization. Each theoretical approach plays a useful role in this process, but has its own limitations. In this section, a number of areas of concern are summarized, and a number of theoretical methodologies are reviewed.

As sketched in Figure 7.1, the analysis of the challenges associated with modeling stereo- and regiodifferentiation breaks the problem into three structural fragments. For the first fragment, the transition metal and associated ligand (Figure 7.1a) is considered to be rather rigid and is common to each regio- and stereoconfiguration. For the second fragment, the olefin and α carbon (C_α) of the polymer chain (Figure 7.1b) along with the metal center are the focus of the electronic reorganization that occurs during the reaction. For the final fragment, the remaining polymer chain (Figure 7.1c) is the source of conformational flexibility and nonbonded/steric interactions.

7.2.1 Electronic Reorganization Errors

A number of years ago, it was recognized that the geometry of the olefin and α carbon of the polymer chain in an olefin polymerization transition state (Figure 7.1b) is remarkably similar to the transition state geometry for addition of a methyl radical to an olefin (compare Figure 7.2a and b).[3] This structural similarity led Polyakov to carry out a large-scale quantum mechanics study of both the addition of a methyl radical to ethylene, and the 1,2- and 2,1-additions of a methyl radical to propylene (Figure 7.2c). The geometrically similar model system was studied because more computationally intense and accurate methodologies could be applied to this much smaller system. Both hybrid density functional (B3LYP) and wave function Projected Møller Plessset perturbation theory to second order (PMP2) quantum mechanical (QM) techniques were systematically used as a function of basis set size.[4] Figure 7.3 summarizes key results. In this study, the Gaussian basis set was systematically improved through the addition of shells of Gaussian functions[5,6] as displayed from left to right in the figure for both the PMP2[7] and B3LYP[8] calculated activation energy barriers. The PMP2 barriers systematically drop as the basis set is improved, whereas the B3LYP barriers systematically increase. At the large basis limit, the density functional (B3LYP) barriers are roughly 1.2 kcal/mol higher than the wave function (PMP2) barriers. This error would have no impact on stereo- or regiodifferentiation were it not for the additional 1.4 kcal/mol differential between 1,2- and 2,1-addition. This 1.4 kcal/mol additional error will have a profound impact on regiodifferentiation for propylene homopolymerization as well as for ethylene–propylene copolymerization. A probable explanation for this 1.4 kcal/mol error between the two methods can be extracted from an

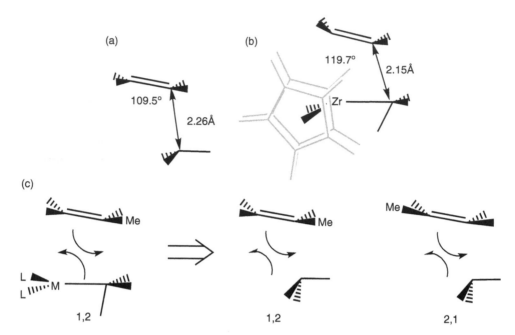

FIGURE 7.2 (a) MP2 geometry for the methyl radical plus ethylene transition state; (b) MP2 geometry for the ethylene plus bis(cyclopentadienyl)zirconium methyl cation transition state; (c) schematic of the correlation between the 1,2 propylene insertion transition state and corresponding 1,2 and 2,1 methyl radical transition states.

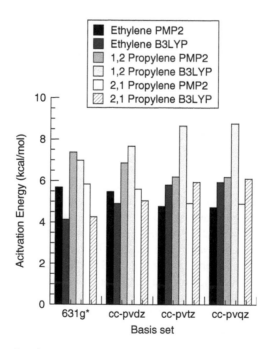

FIGURE 7.3 PMP2 (wave function) and B3LYP (hybrid density functional) activation energies for methyl radical reaction with ethylene, and for 1,2 and 2,1 approach to propylene, as a function of Gaussian basis set. Energies are in kcal/mol.

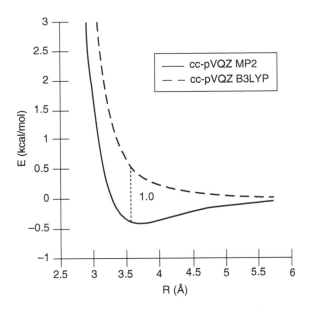

FIGURE 7.4 Methane dimer potential curve showing the van der Waals binding energy (E) as a function of the C–C distance (R) for MP2 wave function (solid line) and B3LYP density functional (dashed line) approaches.

analysis of methane dimer. Dispersion, the attractive part of a van der Waals potential, is absent in modern density functional methods such as B3LYP; they simply do not account for the dispersive attraction between nonbonded electrons.[9–11] Hence, the interaction between two methane molecules is computed by B3LYP to be purely repulsive. In contrast, MP2 does provide the type of electron correlation associated with dispersion,[11] and a binding energy of roughly 0.5 kcal/mol is computed; potential curves are provided in Figure 7.4. At the methane–methane equilibrium distance, $R_e \sim 3.6$ Å (the methyl–methyl distance at the transition state for the 2,1 approach of methyl to propylene modeled in Figure 7.3), the error between the density functional and wave function computations is roughly 1 kcal/mol. For $R_{Me-Me} = 2.7$ Å (the methyl–methyl distance for the 1,2 methyl approach to propylene transition state modeled in Figure 7.3), the error is 2.7 kcal/mol. Thus, a differential dispersion error of 1.7 kcal/mol is anticipated for density functional-computed methyl-ethylene regiodifferentiation.

7.2.2 POLYMER CHAIN ISSUES

The growing polymer chain (Figure 7.1c) is generally acknowledged to provide the bulk of the stereodifferentiation observed in *ansa*-metallocene propylene polymerization. Early computational studies of stereodifferentiation[12] arose from chemists' pencil and paper models that assumed, *a la* Ockham's razor, that the polymer chain could only adopt a single conformational orientation.[13] If conformational flexibility of the polymer chain is acknowledged, then several computational issues arise regarding the growing polymer chain. First, what is the orientation of the α carbon directly attached to the metal (C_α)? Secondly, what is the minimally appropriate length of the chain? Thirdly, what is the conformational flexibility of the chain, that is, how many low-lying conformations are there? And lastly, what, if any, is the interaction of the growing polymer chain with the associated counteranion?

7.2.2.1 C_α Orientation

Starting with the first electronic structure calculations on olefin polymerization,[14–16] it has been observed that there are two classes of transition state orientation for C_α that are nearly isoenergetic

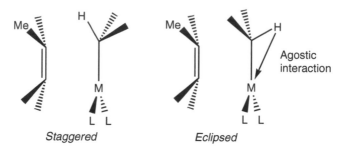

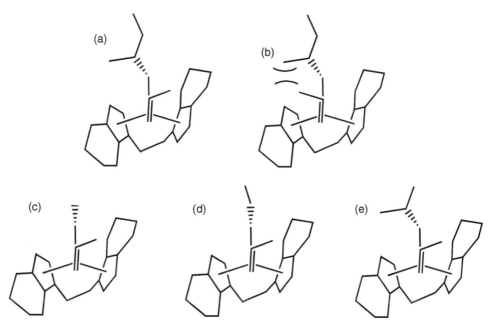

FIGURE 7.5 C_α transition state conformational orientations for 1,2 approach of propylene to a L_2MR^+ catalyst active site.

FIGURE 7.6 Pino–Corradini Stereocontrol model: (a) stereonormal insertion, (b) stereoerror insertion with associated steric repulsions, (c) ethyl, (d) propyl, and (e) isobutyl chain models.

(within 1 kcal/mol) (Figure 7.5). In the first class, the methyl adopts a staggered orientation with respect to the olefin substituents consistent with the orientation of the product polymer chain. In the second class, the methyl adopts an eclipsed orientation relative to the olefin substituents. This orientation provides a metal to C–H bond agostic interaction, but also introduces repulsive interactions with the olefin. For both orientations, the C_α hydrogen shown in Figure 7.5 is in the plane defined by the olefin C–C bond and the M–C_α bond. As discussed in Section 7.3.1, the orientation that is ultimately favored is strongly dependent on ligand structure, stereoconfiguration of the propylene molecule, stereoconfiguration of the polymer chain, as well as the presence or absence of a counteranion. For any given polymerization system, both classes of C_α orientation must be considered.

7.2.2.2 Minimal Chain Length

Dating from the 1980s, Corradini's extension[13] of Pino's active site model[12] defined the minimum polypropylene chain length needed to account for stereocontrol. As shown in Figure 7.6, stereodifferentiation arises from an interaction between the propylene methyl and a methyl group attached to the β carbon of the polymer chain (C_β); compare Figure 7.6a and b. Despite its use in the literature as a polymer chain model, an ethyl chain is simply too short[17] to provide predictive stereodifferentiation. The preferred conformation of an ethyl substituent places it in the *ansa*-metallocene equatorial plane away from the propylene (Figure 7.6c). This symmetric placement cannot provide significant stereodifferentiation. Constraining the ethyl chain so as to provide stereodifferentiation is certainly artificial—the generally accepted Corradini–Pino model does this by ascribing stereodifferentiation to a pendant methyl of the polymer chain. Extension of the chain by one carbon to a propyl chain again cannot provide stereodifferentiation—equatorial placement of the chain is again preferred (Figure 7.6d). Constraining the propyl chain into a conformation that provides steric interaction does not necessarily mean that it is the proper conformation; it merely presupposes an answer and eliminates the possibility of contributing to the understanding of why a given catalyst yields a particular level of stereodifferentiation. Addition of another carbon, generating an isobutyl chain (Figure 7.6e), again cannot provide stereodifferentiation because the methyl groups of an isobutyl chain are equivalent. The ligand backbone can, in principle, provide some orientational bias; however, proper conformational sampling should minimize this bias. The smallest chain model that can, in principle, provide chain-induced stereodifferentiation is the pentyl chain (Figure 7.6a). Even this chain should provide only minimal differentiation owing to the minor steric difference between its methyl and ethyl substituents.

7.2.2.3 Chain Flexibility

Considering the pentyl chain of Figure 7.6 with a pendant polymer group, there are four conformationally active torsions, each with three torsional minima giving rise to 81 conformational minima (Figure 7.7). When the two classes of C_α orientation are considered, the number of conformational minima doubles to 162. Based on the trans-gauche energy difference in butane of 0.7 kcal/mol and a carbon sp^3–sp^3 rotation barrier of roughly 3 kcal/mol, a large fraction of these 162 conformations are kinetically and thermally accessible. This conformational space needs to be coupled with the four propylene orientations of Figure 7.3 since chain conformational preferences are likely coupled with propylene orientation. A few conformations are eliminated by the "pentane effect" that describes the relative inaccessibility of the *gauche+*-*gauche−* and equivalent *gauche−*-*gauche+* conformations owing to methyl–methyl repulsions; in the idealized structure, the nonbonded C–C distance is 2.5 Å. An additional number of conformations are excluded by the ligand framework—the precise number being dictated by the ligand fragment. The calculational challenge lies in knowing which conformations are accessible/preferred for a given ligand. This can be accomplished through a systematic sampling of the entire conformational space, through a Monte Carlo search, or through an annealed molecular dynamics procedure.[18]

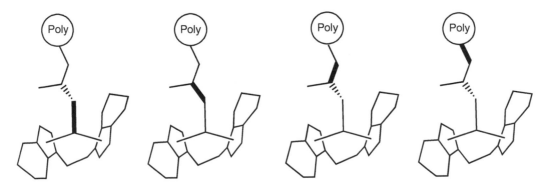

FIGURE 7.7 The four conformationally active torsions of a pentyl chain model.

7.2.2.4 Chain–Anion Interaction

As described in other chapters, single-site olefin polymerization is a rather complex task. Typically, a well-defined organometallic complex is mixed with an activator complex, usually a Lewis acidic main group organometallic complex, to generate an active catalytic system composed of a cationic metal complex and an anionic main group complex. In the primarily nonpolar polymerization medium, the cation and anion form an ion pair wherein the counteranion is not an innocent spectator. The strongly interacting anion certainly must sterically restrict or distort the available conformational space of the polymer chain.

The most common activator is known as MAO (methylaluminoxane). MAO is a complex mixture of chemical species, but has the rough C:Al:O stoichiometry of 1:1:1. MAO is prepared from the careful reaction of trimethylaluminum with water. As described earlier, the Rappé group has developed a MAO model based on structural studies and analogy to $AlR(NR')$ clusters. The model consists of a $(AlMeO)_9$ cluster that has abstracted a Me^- and has had the resulting two-coordinate oxygen coordinated by the Lewis acidic $AlMe_3$. In the following discussion, we refer to this counteranion model as **MAO9**.

7.2.3 Diastereomeric Energy Differences

Practical stereodifferentiation corresponds to activation energy differentiation of 1–2 kcal/mol. In studies on stereodifferentiation,[19,20] computational chemists have relied on the empirical observation that stereodifferentiating transition state energy differences are more accurate than the computation of the absolute barrier heights—the energy difference between the reactants and transition state. If the monomer's olefinic atoms are roughly in the right place, errors in intramolecular energetic terms will likely cancel for the four possible conformeric transition states of Scheme 7.1, and energetic differentiation will arise predominantly from intermolecular effects. Hence, errors in energetic differentiation will arise from errors in the computation of intermolecular interactions. The limitations highlighted in Section 7.2.1 are of particular concern for density functional studies.

Olefin polymerization is not a simple single-step olefin insertion process; it involves complex kinetics and multiple equilibria. Insertion is preceded by olefin complexation[21,22] (Scheme 7.2a) or olefin complexation coupled with counterion displacement (Scheme 7.2b). In view of this complexity, how can the use of barrier differences be justified? The use of barrier height differences arises from the Curtin–Hammett principle.[23] The Curtin–Hammett principle is a kinetic analysis that describes product distributions for reactions involving a pair of equilibrating reactants or intermediates, each capable of forming a product. The stereodifferentiation of general isotactic polymerization can be

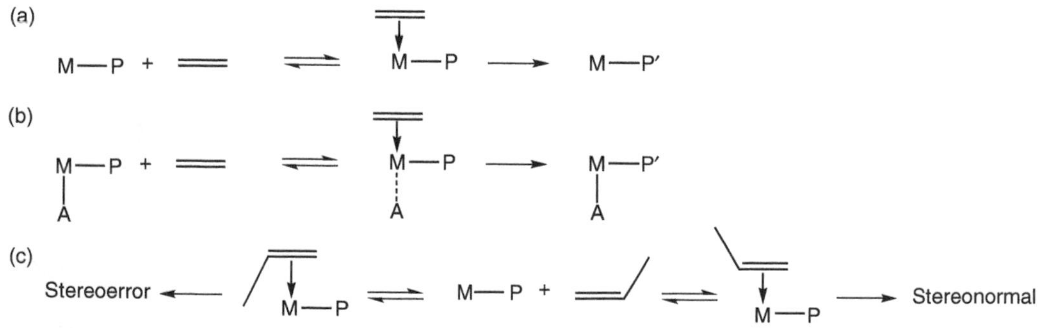

SCHEME 7.2   Olefin polymerization reaction schemes; (a) conventional Cossee–Arlman reaction sequence; (b) counteranion displaced reaction sequence; (c) stereodifferentiation reaction sequence. P and P′ represent polymer chains of length n and n + 1, respectively.

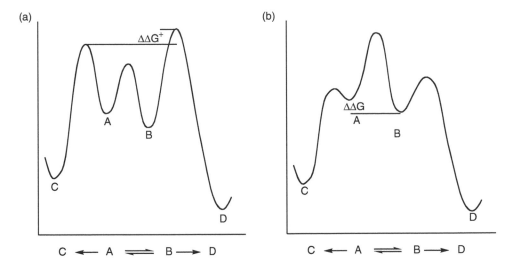

FIGURE 7.8   Potential curves for Curtin–Hammett/Winstein–Holness kinetics limiting cases (a) **L1** limit where the free energy difference between barriers dictates product distribution and (b) **L2** limit where the free energy difference between equilibrating intermediates dictates the product distribution.

described by a set of reactions in Scheme 7.2c. Starting in the middle with free olefin and a metal active site, reversible coordination by each enantioface occurs. Insertion through the right pathway leads to stereonormal chain growth. Insertion through the left pathway leads to a stereoerror. If the olefin complexation equilibria are rapidly maintained relative to product formation, then the **L1** potential curve of Figure 7.8a applies, and the energy differences between transition states dictate the polymer stereochemistry. If instead olefin complexation equilibria is rate limiting, then the **L2** potential curve of Figure 7.8b applies, and the energy difference between intermediates **A** and **B** dictates polymer stereochemistry.

7.2.4   Models

The primary goals of site-control theoretical modeling efforts are to understand the molecular basis for stereo- and regiocontrol and to suggest ways to enhance differentiation. As reviewed elsewhere,[2] these modeling efforts range from simple pencil and paper sketches to large-scale electronic structure studies. Each approach has both utility limitations.

7.2.4.1   Conceptual Models

Over the years, advances in the stereochemical control of polymerization catalysis have arisen from conceptual models of the 3-dimensional shape of the polymerization active site. The original stereocontrol model of Pino et al.[12] was based on crystal structure of the dichloride catalyst precursor. The chlorides of the crystal structure were removed, and a propylene monomer and ethyl chain visually added. The utility of the model was established by its success in explaining the stereochemistry of propylene hydrooligomerization. As discussed in Section 7.2.4.2, Pino's model was quickly replaced by a model that more properly incorporated the shape of the polymer chain. Lacking an energy representation, it is not possible for conceptual models to estimate the magnitude of stereo- or regiodifferentiation of a given model. The usefulness of a conceptual model rests in its utility for explaining experimental observations and for formulating additional experiments.

7.2.4.2 Molecular Mechanics

Molecular mechanics (MM)[18] is a simple computational molecular model primarily used for understanding conformational energy differences and structural deviations due to steric interactions in isolated molecules. It is also used to understand nonbonded interactions between molecules. Because bonded atoms are typically held together by nonphysical harmonic potentials in MM, the bond breaking/bond making events of chemical reactivity are not within its natural arena. Since the late 1980s, the stereo- and regiochemical differentiation questions of single-site propylene polymerization have been cast as conformational questions, albeit involving somewhat distorted structures. Early and useful efforts relied on rigid transition state models that varied only a few torsional degrees of freedom to relieve steric repulsions. Corradini et al.[13] used such a model to confirm Pino's active site model, and to enhance it to include the impact of the stereochemistry of the growing polymer chain. Using MM, Corradini and coworkers found that steric interactions between the propylene methyl groups and the methyl substituent of the last inserted propylene were important for stereocontrol. They suggested that the dominant interaction was that between the polymer chain and the propylene methyl group, rather than direct interactions between the propylene methyl group and the active site. The active site was proposed to cause the polymer chain to adopt a conformation leading to the chain-monomer repulsion. In 1991, this hypothesis received experimental support from Erker et al's observation[24] of double stereodifferentiation; both enantiomorphic-site control and chain-end control were found to contribute to stereodifferentiation. (For additional information on enantiomorphic-site control and chain-end control see Chapter 1.)

More recent MM studies[25–28] have permitted more complete geometric relaxation to release steric strain and moderate the magnitudes of steric effects. Quite recently, full transition state geometric relaxation including polymer chain conformational sampling has been achieved, and stereo- and regiodifferentiation events comparable in magnitude to experiment have been computed.[29]

7.2.4.3 Electronic Structure Models

Modern electronic structure or QM technologies do not rely on any knowledge of molecular structure or preconceived ideas about bonding.[30] The geometry of unknown as well as known complexes can be determined. Electronic structure methods can be straightforwardly used to study new compositions of matter, and hence, novel catalysts. They can also be employed to characterize the transition states of chemical transformations. This freedom comes at the expense of an increased computational effort. As discussed in Section 7.2.1, density-functional theory (DFT), the most popular of the electronic structure methodologies, lacks dispersion and hence cannot properly describe the important intermolecular effects associated with stereodifferentiation.

7.2.4.4 Combined Methods

The time demands of modern electronic structure techniques have limited researchers to electronic structure studies on small model complexes. Unfortunately, the interesting and important questions in single-site catalyzed polymerization concern the large steric encumbrances that chemists have added to achieve molecular control. To begin to address these important questions, theorists have begun to develop and apply hybrid methods.[31–35] These methods are a combination of electronic structure (QM) methodologies and MM techniques. In the most widely used QM/MM technique, ONIUM[31,32] modern electronic structure theory is used for the atoms involved in electronic reorganization, and force field methods are used for the steric periphery. In another hybrid technique termed the reaction force field (RFF),[33–35] electronic structure methods and bonding concepts are used to develop the shapes of potential surfaces,[33] and QM resonance is used to couple potential surfaces together to describe reactions.[34,35] Recent theoretical reports[36–41] and the results in Section 7.3 highlight the dramatic role that the "real" ligands play in olefin binding as well as in stereocontrol.

7.3 EXAMPLES

The issues raised in Section 7.2 are best illustrated by considering several real-world examples. Isotactic polymerization is considered first, followed by syndiotactic polymerization. These classic examples are followed by two studies on novel polymerization systems. The third case study concerns a single-site metallocene system that produces a stereoblocky material with novel elastomeric properties. The last example involves an ethylene–propylene copolymerization catalyst that exhibits nonrandom monomer incorporation. Common themes of these four examples include monomer–polymer chain control, van der Waals attraction, and the importance of the counteranion.

7.3.1 ISOTACTIC POLYPROPYLENE

A number of groups have contributed to the development of computational models of *ansa*-metallocene-based isotactic polypropylene catalysts. Their work will be briefly reviewed here.

In the early 1990s, Castonguay and Rappé,[42] Hart and Rappé,[43] Guerra et al.,[28] and Yu and Chien[25] reported MM investigations of stereoselectivity for *ansa*-metallocene catalysts, predominantly zirconium indenyl and tetrahydroindenyl *ansa*-metallocene bare cation models. By adopting somewhat rigid transition state models, stereodifferentiation magnitudes in line with experiment were computed. Despite agreement with experiment, it is not possible to assess whether or not the agreement was due to the author's precise choice of transition state model.

In addition in the early 1990s, Kawamura-Kuribayashi et al. published[44] a combined *ab initio*-MM (QM/MM) study of a range of substituted Ti, Zr, and Hf indenyl bare cation *ansa*-metallocenes that radically overestimated the energetic differentiation for stereo- and regiodifferentiation in isotactic propylene polymerization. Here, regiodifferentiation greater than 20 kcal/mol was computed for catalysts with experimentally observed regiodifferentiations of at most a few kcal/mol. More refined work from the Morokuma and coworkers in 1996[45] dropped the regiodifferentiation energies to ~ 10 kcal/mol, though the computed stereodifferentiation rose to 16 kcal/mol for a bare cation model of an aryl substituted bis(indenyl) catalyst with an experimentally observed stereodifferentiation of roughly 4 kcal/mol.[46,47] In this study, the positions of the transition state core (the two olefinic carbons, C_α, and zirconium) were held fixed based on the electronic structure optimization of the ethylene-zirconium methyl system.

In 1994, Spaleck et al. reported one of the major achievements in single-site catalyst evolution.[46,47] They found that by making appropriate bis(indenyl) *ansa*-metallocene ligand modifications, they were able to achieve enhanced stereocontrol, increased molecular weight, and improved catalyst productivity in isotactic propylene polymerization. A portion of this experimental data is collected in Figure 7.9. Normally, attempts to improve catalyst performance result in a trade off between increasing activity and increasing selectivity. Adding large ligand substituent groups to control the reaction, as for example in going from **1** to **2**, also usually slows the polymerization. The aryl-substituted system **2** has been studied twice; first as discussed above by the Morokuma and coworkers,[45] and more recently by Bormann–Rochotte.[29]

Findings by Bormann–Rochotte relevant to the discussion in Section 7.2.2 include (1) the importance of van der Waals interactions, (2) the importance of chain conformational sampling, and (3) the importance of the counteranion. This work is summarized below.

7.3.1.1 van der Waals Interactions

One of the most intriguing computational observations of the Bormann–Rochotte work was the computed decrease in the stereonormal propylene insertion barrier caused by the addition of the 4-naphthyl substituent to each indenyl ring. In agreement with the experiment, addition of a steric encumbrance decreased the barrier. Visual examination of the insertion saddle point provides an

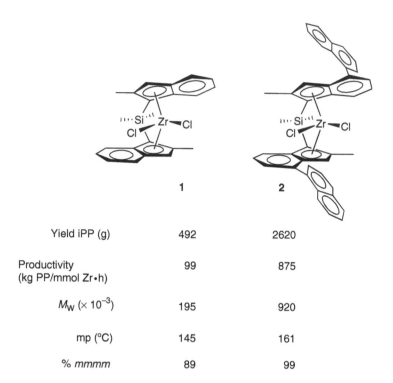

	1	2
Yield iPP (g)	492	2620
Productivity (kg PP/mmol Zr·h)	99	875
M_W ($\times 10^{-3}$)	195	920
mp (°C)	145	161
% *mmmm*	89	99

FIGURE 7.9 As studied by Spaleck et al., 4-aryl substitution impacts catalyst performance in producing isotactic polypropylene. Shown are precatalysts dimethylsilylene bis(2-methylindenyl)zirconium dichloride (**1**) and dimethylsilylene bis(2-methyl-4-naphthylindenyl)zirconium dichloride (**2**). Liquid propylene polymerization at 70 °C, Zr:Al 1:15,000, methylaluminoxane cocatalyst.

explanation for this observation (Figure 7.10). At the insertion saddle point, the methyl group of the propylene is found to be within van der Waals contact of the aromatic ring, and is placed in the attractive well of the interaction rather than in the inner repulsive wall (3.5 Å). In the reactants, free propylene and the alkylated cation of **2**, this stabilizing van der Waals interaction is absent; thus, the saddle point is differentially stabilized.

7.3.1.2 Chain Orientation

As discussed in Section 7.2.2.1, there are two classes of transition state chain orientation—staggered and eclipsed with respect to the monomer olefin substituents. For both classes, at the insertion transition state, there is a C_α–H bond nearly in the plane defined by the metal–C_α and olefinic C–C bonds. Bormann–Rochotte computed the C_α–H bond to be 31 ° out of plane for the staggered orientation and 23 ° out of plane for the eclipsed orientation. The remaining two C_α substituents including the polymer chain occupy out-of-plane positions. These staggered and eclipsed chain orientations are illustrated in Figure 7.11 for **2**. For the staggered orientation, the C_1–Zr–C_α–C_β dihedral angle of complex **2** is 119 °, very near the idealized value of 120 °. In the eclipsed orientation, the C_1–Zr–C_α–C_β dihedral angle is 104 °, opened up significantly from the idealized angle of 60 °. For **2**, the staggered orientation has a 140 ° Zr–C_α–C_β–C_γ dihedral angle. The eclipsed orientation places the Zr–C_α–C_β–C_γ dihedral angle at 92 °. Earlier computational studies have assumed the eclipsed, "least motion" polymer chain orientation for the transition state.[25–28,48] In the past, this "least motion" pathway was assumed because it led to maximal stereodifferentiation.

The eclipsed and staggered chain orientations were found by Bormann–Rochotte to be nearly isoenergetic for the transition state for stereoregular isotactic propylene insertion into a stereoregular

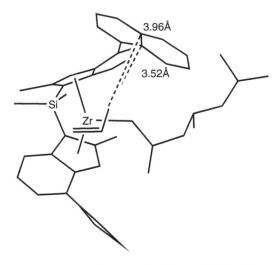

FIGURE 7.10 Stereonormal insertion transition state for **2** as studied by Bormann-Rochotte; attractive van der Waals interactions between the propylene methyl group and the naphthyl substituent aryl carbons are highlighted. The C–C distances provided demonstrate that the carbons are in the attractive part of the curve as illustrated in Figure 7.4.

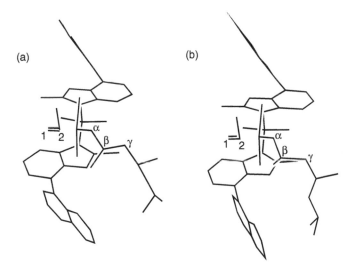

FIGURE 7.11 The (a) staggered and (b) eclipsed chain orientations of the stereonormal isotactic propylene insertion transition state for **2**. The olefinic carbons are labeled 1 and 2. The α, β, and γ polymer chain carbons are labeled as well.

polymer chain (see Figure 7.12). For the stereonormal insertion, the staggered orientation was favored by 1.8 kcal/mol for a bare cation model of **1**, and by 0.6 kcal/mol for a methylaluminoxane ion pair (**1·MAO9**) (gray bars in Figure 7.12). For **2**, the eclipsed orientation was favored by 0.2 kcal/mol for a bare cation model of the stereonormal insertion, and the staggered orientation was favored by 1 kcal/mol for the ion pair model (**2·MAO9**). Thus for bare cations, the staggered chain orientation was favored for **1**, but for **2**, the eclipsed orientation was favored. For ion pair models, the staggered chain orientation was favored for both.

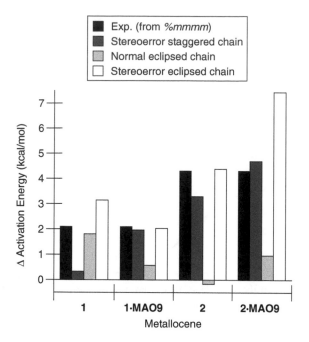

FIGURE 7.12 Olefin insertion activation energy differences relative to the isotactic propylene insertion transition state for a stereonormal staggered chain orientation as compared to a stereonormal eclipsed chain orientation, a stereoerror staggered chain orientation, and a stereoerror eclipsed chain orientation. Results are reported for **1** and **2** bare cations and **1·MAO9** and **2·MAO9** ion pair models. The experimental estimate was derived from the %*mmmm* distribution on an experimental polymer.

For the stereoerror insertion transition state, a syndiotactic 1,2-addition to an isotactic 1,2-ended chain, there were more significant energetic consequences due to the chain orientation. For the bare cation model **1**, the eclipsed conformation stereoerror insertion transition state is 2.8 kcal/mol higher than the staggered conformation stereoerror insertion transition state. This conformational differentiation has mechanistic consequences. For the bare cation model **1** for the energetically favored staggered chain orientation, a stereodifferentiation of 0.3 kcal/mol was predicted, whereas for the higher barrier eclipsed chain orientation, a stereodifferentiation of 1.3 kcal/mol was predicted. If only the eclipsed conformation were studied, it would be tempting to say that the observed stereodifferentiation was reproduced and the counterion plays no role. However, for the ion pair model **1·MAO9**, the energetically favored staggered chain orientation model predicted a 2 kcal/mol stereodifferentiation; for the higher barrier eclipsed chain orientation a 1.5 kcal/mol stereodifferentiation was computed. For the eclipsed chain orientation, the counterion only had a 0.2 kcal/mol impact, whereas for the staggered chain orientation, the counterion raised the stereodifferentiation 1.7 kcal/mol. This chain orientation dependence carries over to **2**. For the bare cation model **2**, the eclipsed stereoerror insertion transition state is 1.1 kcal/mol higher than the staggered stereoerror insertion transition state. For the bare cation model **2** for the slightly higher barrier staggered chain orientation, a stereodifferentiation of 3.3 kcal/mol was predicted, whereas for the energetically favored eclipsed chain orientation, a stereodifferentiation of 4.6 kcal/mol was predicted which is in good agreement with the experimental estimate of 4.3 kcal/mol. Again, if only the eclipsed conformation were studied, it would be tempting to say that the observed stereodifferentiation was reproduced and the counterion plays no role. If both chain orientations are considered (the lowest energy stereonormal eclipsed chain compared to the lowest energy stereoerror chain, i.e., the staggered chain), as is proper, the bare cation stereodifferentiation is 3.5 kcal/mol. For the ion pair model **2·MAO9**, the energetically favored staggered chain orientation model predicted a 4.7 kcal/mol stereodifferentiation, in good

(a)　　　　　　　　　　　(b)

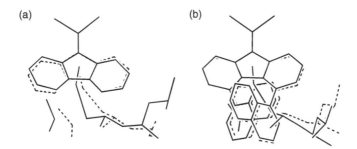

FIGURE 7.13 Graphical impact of counteranion on chain orientation for (a) **1** and (b) **2** for a stereonormal isotactic propylene insertion. The dashed line structure is for the bare cation; the solid line structure is for the ion pair (**1·MAO9** or **2·MAO9**).

agreement with the experimental estimate of 4.3 kcal/mol; for the higher energy eclipsed chain orientation, a 6.5 kcal/mol stereodifferentiation was computed. Hence, the counterion raises the barrier by 1.2 kcal/mol.

7.3.1.3 Counteranion

As discussed above, the counteranion has a significant impact on the stereoselectivity of **1** during isotactic polymerization, raising the relative stereodefect barrier to nearly 2 kcal/mol. As shown in Figure 7.13a, this increase is due to a counteranion-induced polymer chain and propylene movement. Repositioning the polymer chain causes greater interaction with the methyl substituent of the incoming propylene monomer, raising the energy of the stereodefect transition state. The polymer chain reorganization computed for **2** is significantly smaller (see Figure 7.13b), though the energetic impact is comparable.

7.3.2 Syndiotactic Polypropylene

The model required for syndiotactic polymerization is a bit more complex than models needed for isotactic polymerization. Rather than requiring repeated reaction by the same olefin enantioface, olefin enantioface alternation is needed to produce a syndiotactic polymer. In 1988, Ewen et al.[49] reported the discovery of a metallocene catalyst, isopropyl(cyclopentadienyl)(9-fluorenyl)zirconium dichloride (**3**) that produced syndiotactic polypropylene when activated with MAO. In contrast to catalysts such as **1** and **2** that are C_2-symmetric catalyst precursors, **3** is C_s-symmetric. Ewen proposed that in order for this catalyst to produce syndiotactic polymer, the active site must epimerize, flipping the polymer from one active site side to the other during/after each insertion step (Scheme 7.3). During normal-stereoselective polymerization, polymerization was suggested to proceed through **a** and **b** (Scheme 7.3), with the polymer flipping from the left to the right side upon monomer insertion. Stereoerrors were suggested to be due to either reaction with the incorrect olefin enantioface, as in conventional isotactic polymerization **c**, or by the metal site undergoing site epimerization (polymeryl chain inversion at the stereogenic metal center, which "swings" the chain from one site to the other without inserting a monomer unit),[50] also referred to as chain "back-skipping" **d**. Epimerization is usually observed in situations where one site is more energetically favorable than the other, therefore a chain that has "flipped" into the less favored site utilizes epimerization to return to the more favored site. The need to think about site epimerization is unique to C_s- and C_1-symmetric active site catalysts. For C_2-symmetric catalyst precursors, site epimerization is mechanistically invisible because the two active sites are stereochemically equivalent. (For additional information on epimerization of C_s-symmetric catalysts, see Chapter 2.)

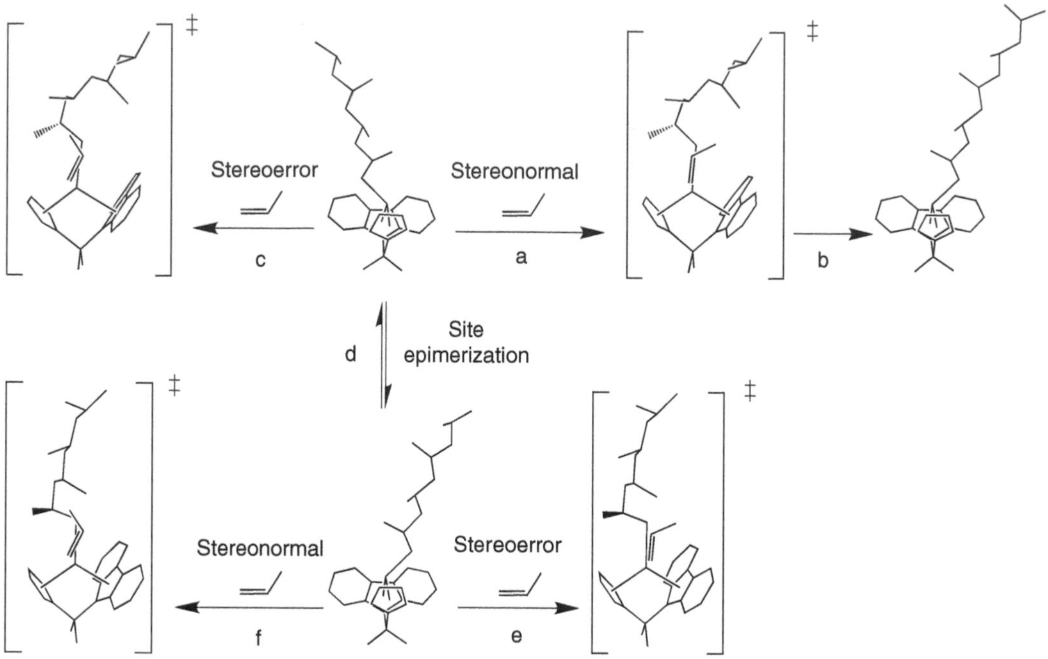

SCHEME 7.3   Stereochemical alternatives for metallocene-based syndiotactic propylene polymerization.

This model of syndiotactic polymerization involving C_s-symmetric active sites implicitly assumes that site epimerization follows Curtin–Hammett **L2** kinetics (Figure 7.8b) wherein the intermediate interconversion barrier is significantly larger than the propagation barriers. Furthermore, if epimeric defects are infrequent, then by the **L2** model, the epimerized chain must be significantly higher in energy than the normal chain.

In 1991, Guerra and coworkers[51] reported a few-degrees-of-freedom force field study on metallocene syndiotactic polymerization with **3**. Their modeling work supported Ewen's mechanistic proposal, and suggested again that the polymer chain was the primary source of stereocontrol. The catalyst active site shape was responsible for providing the preferred orientation for the polymer chain. In contrast to the isoselective C_2-symmetric active sites in which the polymer chain and propylene monomer can each be placed in a steric "hole," for syndioselective C_s-symmetric active sites, the steric demands of the polymer chain induce the propylene methyl group to be pointed towards the more substituted fluorenyl ring. This initial study was followed by more complete QM/MM studies on **3**[17,28,48] wherein stereodifferentiation consistent with experiment was reported without considering epimerization.

The site epimerization feature (Scheme 7.3d) has prompted several computational efforts. Starting with the work of Jolly and Marynick,[52] theoretical studies[53] have found that group 14 metallocenium ions are pyramidal at the metal center. This gives rise to the stereogenic character of the metal center. However, as discussed earlier,[2] computed barriers to inversion at the metal are quite small, and therefore not consistent with retention of active site stereochemistry. Recently, Busico, Budzelaar, and coworkers,[17] based on a DFT study of propylene insertion with an ethyl polymer chain model, suggested that polymer chains longer than methyl should yield larger inversion barriers. However, Chen and Marks reported[54] an anion-dependent experimental barrier of ~ 20 kcal/mol for the equilibration (epimerization) step **d** for a methyl chain. If the methyl bare cation has a small barrier and the methyl ion pair has a large barrier to inversion, then the counteranion should be the source of the large barrier. This raises the question, if the counteranion does prevent site epimerization during polymerization, how can the normal site isomerization that occurs upon monomer insertion

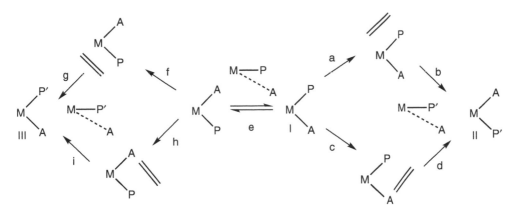

SCHEME 7.4 Stereochemical alternatives for syndiotactic propylene polymerization, incorporating the impact of the counteranion (A). P and P′ represent polymer chains of length n and n + 1, respectively.

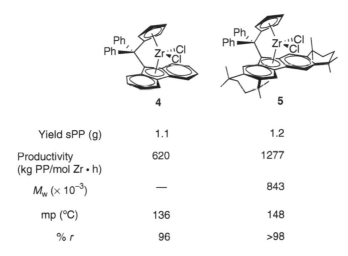

	4	5
Yield sPP (g)	1.1	1.2
Productivity (kg PP/mol Zr · h)	620	1277
M_w ($\times 10^{-3}$)	—	843
mp (°C)	136	148
% r	96	>98

FIGURE 7.14 Distal ligand elaboration impacts catalyst performance for the production of syndiotactic polypropylene. Shown are precatalysts diphenylmethylene(9-fluorenyl)(cyclopentadienyl)zirconium dichloride (4) and diphenylmethylene(octamethyloctahydrodibenzofluorenyl)(cyclopentadienyl)zirconium dichloride (5). (Liquid propylene polymerization at 20 °C, Zr:Al 1:1000 and 1:2000, methylaluminoxane cocatalyst.

(Scheme 7.3a and b) occur with a low barrier? What is the nature of the choreography of the anion dance? This dilemma is summarized in Scheme 7.4. Starting in the middle of Scheme 7.4 with **I**, olefin approach can either occur opposite the anion (through **a**), or from the same face as the anion (through **c**). For "normal" site isomerization to occur somehow during the insertion process, the anion must concertedly switch places with the polymer chain (**b** or **d**) to form **II**. If site epimerization occurs through **e**, the frontside and backside olefin approaches are depicted as **f** and **h**, respectively. Again, during the insertion process, the anion must somehow concertedly switch places with the polymer chain to form **III**. It is difficult to envision an anion-switching insertion process occurring with a barrier lower than simple site epimerization. Unfortunately, the literature provides little guidance; the impact of the counteranion on site epimerization or isomerization has not been computationally investigated since the early work of Castonguay and Rappé.[42]

This anion dilemma in conjunction with the recent report of an exceptionally productive highly ligand-substituted syndiotactic catalyst by Miller and Bercaw,[50] (5, Figure 7.14), prompted Rappé to initiate an RFF study of syndiotactic polymerization. As with Spaleck et al's observation[46,47] for **2**, Miller and Bercaw reported that an increase in distal steric bulk, going from **4** (a diphenylmethane-bridged analogue of **3**) to **5**, led to an increase in productivity as well as an increase in stereocontrol (as evidenced by an enhanced melting point).

Consistent with Bormann–Rochotte observations described in Section 7.3.1, the counteranion was found by Rappé to play a critical role in stereocontrol, and van der Waals interactions are again one source of the observed increase in productivity. This work is summarized below.

7.3.2.1 Counteranion

As for isotactic propylene polymerization, bare cation models are found to provide very little stereodifferentiation in syndiotactic polymerization; both eclipsed and staggered C_α chain models for **3** yield insertion stereodifferentiation of less than 0.3 kcal/mol. In addition, a barrier of 7 kcal/mol is computed for active site epimerization. As reported by Bormann–Rochotte,[29] addition of a MAO counteranion results in an increase in stereodifferentiation. For the ion pair **3·MAO9**, the barrier for propylene approach from the same side as the counteranion **c** of Scheme 7.4 is 2.5 kcal/mol higher than propylene approach from the opposite side, **a** of Scheme 7.4. Both the stereoerror and stereonormal transition states adopt eclipsed C_α conformations. Addition of a counteranion only raises the active site epimerization barrier by 2 kcal/mol. Thus, the ion pair site epimerization barrier is only 9 kcal/mol owing to a significant stabilizing interaction between the polymer chain and the counteranion that is not lost in the transition state. Active site epimerization equilibration should therefore occur prior to chain propagation. But if active site epimerization occurs rapidly, how does stereoregular polymerization proceed? The implicitly assumed Curtin–Hammett **L2** behavior cannot apply. If epimerization occurs rapidly, then epimerized and normal pathways must be considered, and Curtin–Hammett **L1** behavior must be considered. In Scheme 7.5, these barriers for **3·MAO9** are presented relative to stereonormal insertion (**N**), wherein the propylene methyl is directed toward the fluorenyl ring (see Figure 7.16). Stereoerror insertion, where the propylene methyl is pointed toward the cyclopentadienyl ring, is denoted **S**. Insertion with the propylene methyl directed toward the fluorenyl ring, but following metal site epimerization, is labeled **E**. Insertion with the propylene methyl pointed toward the cyclopentadienyl ring following metal site epimerization is denoted **ES**. In isolation, stereo (**S**) and epimerization (**E**) insertion errors both lead to the formation of an *m* defect; subsequent insertion dictates whether an *m* or an *mm* defect has occurred. Stereonormal insertion (**N**) is favored over defect insertions by roughly 2 kcal/mol (Scheme 7.5a). Both stereoerror (**S**) and epimerization (**E**) barriers are higher by roughly 2.5 kcal/mol for an epimerized chain. Subsequent insertion into the defect chain follows a different energetic course (Scheme 7.5b). In this situation, insertion following site epimerization (**E**) is favored, leading to the favorable production of an overall isolated *m* defect. Barriers for the two pathways that lead to an *mm* triad (**N** and **ES**) are each roughly 1 kcal/mol above the **E** barrier. The preference for **E** is not surprising, because the chain end for this transition state is the same as the chain end for the defect-free chain—configurational differences occur at the penultimate monomer unit. In order to test the feasibility of this new model, a pentad distribution has been computed. Comparison of this new model to experimental data for **3·MAO9** and a previous computational study[17] is provided in Figure 7.15. The magnitudes of the classical epimerization errors (*rrmr*) and classical stereoerrors (*mmrr*) are both reproduced well by the current model. The earlier literature model did not include an epimerization pathway, so underestimation of *rrmr* pentads is expected.

The normal stereoregular propylene insertion for **3·MAO9** adopts a unique C_α conformation that places the polymer chain in the equatorial plane. The counteranion is positioned above the olefin, corresponding to a frontside displacement. In the corresponding stereoerror transition state, the polymer chain adopts a staggered C_α conformation with respect to the monomer olefin substituents. The counteranion is positioned above the polymer chain, corresponding to a backside displacement. The full set of lowest energy C_α conformations and anion positions is summarized in Figure 7.16. Active site epimerization effectively interchanges positions of the polymer chain β carbon substituents.

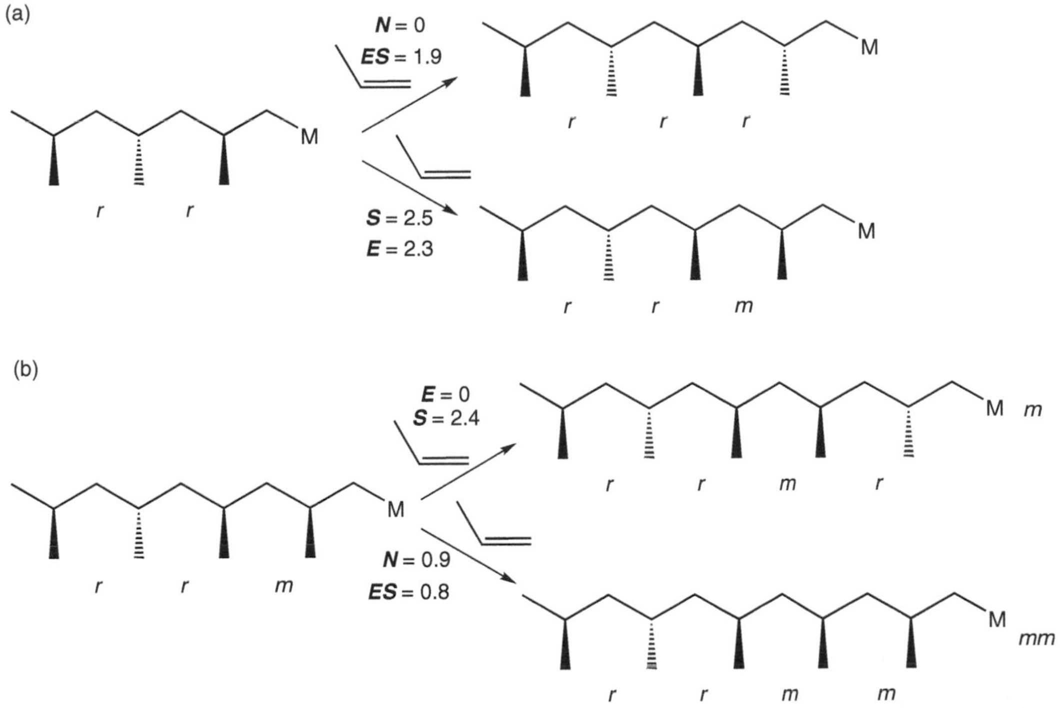

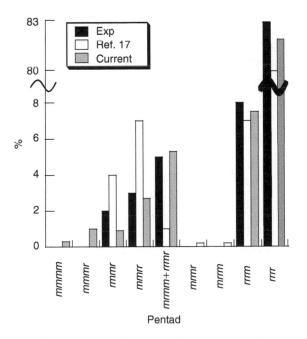

SCHEME 7.5 New model for stereocontrol in metallocene-based syndiotactic propylene polymerization with **3·MAO9** where (a) = normal insertion pathway; (b) defect chain insertion pathway; N = stereonormal insertion; S = stereoerror insertion; E = insertion following epimerization; ES = insertion following epimerization with the propylene methyl pointed in the opposite orientation as E. Numeric values represent transition state energetic differences relative to the lowest insertion transition state, in kcal/mol.

FIGURE 7.15 Experimental (Exp), literature (Reference 17), and new model (current) pentad distributions for propylene polymerization with **3·MAO9**.

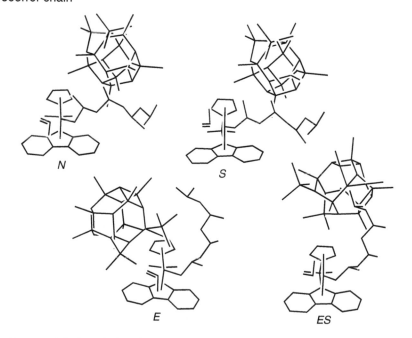

FIGURE 7.16   Structural characteristics of syndiotactic propylene insertion transition states for **3**.

Following substituent interchange, for example, of the methyl substituent with the polymer chain, the resulting polymer chain rotates to reduce active site steric interactions. This exchange-rotation leads to the methyl attached to the β carbon being pointed toward the counteranion rather than the approaching propylene as it is in the nonepimerized chain. This structural reorganization increases repulsions between the approaching propylene monomer and the polymer chain.

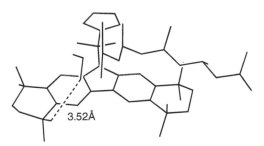

FIGURE 7.17 Stereonormal insertion transition state for **5·MAO9**, showing attractive van der Waals interaction between the propylene monomer and the substituted fluorenyl ring (dashed line). Counterion is not shown for clarity.

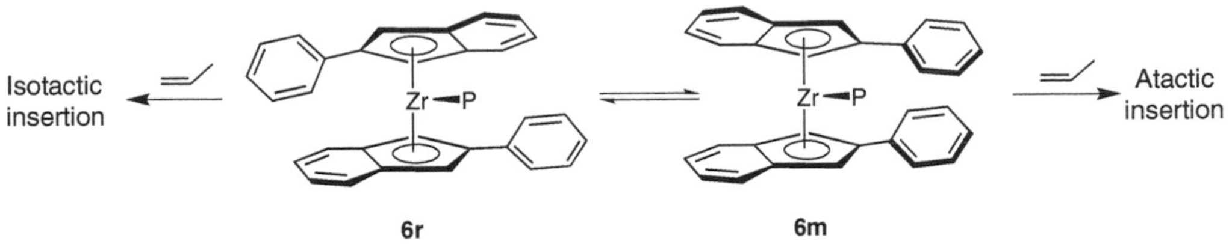

Isotactic insertion ⟵ 6r ⇌ 6m ⟶ Atactic insertion

SCHEME 7.6 Site alternation control for EHPP catalysis (cationic active species shown).

7.3.2.2 van der Waals Stabilization

The stereonormal insertion transition state for **5·MAO9** provides a possible explanation for its increased productivity as compared to **4·MAO9**. The methyl group of the propylene monomer is found to be within van der Waals contact of the distal ring of the substituted fluorenyl ligand at the insertion saddle point of **5·MAO9** (Figure 7.17). This places it in the attractive well of the interaction rather than the inner repulsive wall (3.5 Å). In the reactants, free propylene and **5·MAO9**, this stabilizing van der Waals interaction is absent, thus the saddle point (relative to reactants) for **5·MAO9** is differentially stabilized relative to **4·MAO9**, in this case by 1.4 kcal/mol. The van der Waals attraction between the propylene methyl and the substituted cyclohexyl substituent in **5·MAO9** present only in the insertion transition state, lowers the barrier and hence increases the propagation rate relative to **4·MAO9**.

7.3.3 OSCILLATING-SITE POLYPROPYLENE

In 1995, Coates and Waymouth[55,56] reported a catalyst that produced polypropylene containing blocks of atactic polypropylene and isotactic polypropylene. This novel elastomeric material was referred to as elastomeric homopolypropylene (EHPP). The unit cell of the crystal structure of the EHPP catalyst precursor, bis(2-phenylindenyl)zirconium dichloride (**6**), was observed to contain two distinct diastereomeric conformers for the structures and reaction model (Scheme 7.6). In one conformer, **6m**, the phenyl rings of the substituted indenyl ligands were postulated to be on the same side of the active site yielding a *meso* diastereomer, whereas in the other conformer, **6r**, the indenyl ligands were postulated to be on opposite sides of the active site producing a *rac* enantiomeric pair. The authors suggested that the isotactic blocks of the polypropylene were produced from the *rac* active site, and atactic blocks from the *meso* active site. This model implicitly assumes **L2** Curtin–Hammett kinetics wherein the **6r–6m** equilibration barrier is higher than the propagation barriers.

In a 1996 paper, Cavallo et al.[57] found the experimental conformations **6m** and **6r** (dichloride precursor) to be nearly isoenergetic and suggested that site alternation was controlled by steric interactions present in the *rac-meso* isomerization transition state.

In 1996 Pietsch and Rappé, using MM on the dichloride precursor, suggested[58] that switching between the *rac* and *meso* active site shapes of **6m** and **6r** was controlled by ground state-stabilizing π-stacking interactions between the phenyl substituents in the *rac* conformer and between phenyl rings and the benzo groups of the indenyls in the *meso* conformer. This π-stacking model and a **L2** Curtin–Hammett kinetics model, wherein the *rac-meso* energy differentiation dominates the block structure, have been used at BP–Amoco as an important part of the EHPP catalyst development effort.[59] In this model, isotactic block growth was assumed to occur from *rac* active sites, atactic block growth from *meso* active sites, and block termination through *rac-meso* active site equilibration.

Despite the practical utility of the model outlined in Scheme 7.6, it cannot explain the atomic force microscopy (AFM) images of EHPPs prepared using **6** as reported by Kravchenko et al.,[60] nor the magnitude of their crystallite-induced elasticity. The AFM images show pronounced hard wormlike crystalline features distinct from the soft amorphous background. The thickness of the crystalline regions (12±4 nm) corresponds to roughly 54 monomer units. The AFM images and elasticity require a significant concentration of isotactic blocks of approximately 50 monomers in length. Digestion of a sample of EHPP polymeric material in hot HNO_3 yielded an oligomeric material with a degree of polymerization of 120 and a polydispersity (M_w/M_n) of 1.06. Polydispersities nearing 1.0 are consistent with living block growth and a Poisson distribution of block lengths. To aid the discussion, generic Shulz–Flory and Poisson M_n and M_w distributions are given in Figure 7.18. For the active site model of Scheme 7.6, the required site switching equilibrium provides a Shulz–Flory block termination step. This site switching step terminates the synthesis of an isotactic or atactic block along a chain, and initiates the synthesis of a subsequent block having the other tacticity. Isotactic block growth consistent with the model presented in Scheme 7.6 must yield the exponentially decaying block length distribution of Shulz–Flory statistics. In a Shulz–Flory polymer with an average isotactic block M_w of 100, only 7.7% of the blocks actually fall in the M_w range between 90 and 110. The observed bulk material is 25% *mmmm*, so only 2% of the isotactic blocks would have the correct length of 100 or at least 100—not enough to produce the wormlike crystalline features in the AFM images observed.

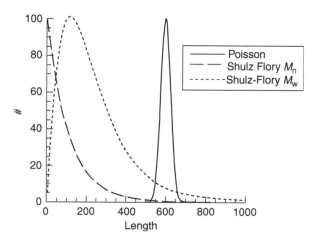

FIGURE 7.18   Generic Schulz–Flory and Poisson model number-average (M_n) and weight-average (M_w) molecular weight distributions; # = number of chains of a given length; length = number of monomer units in a given chain. For the Poisson distribution, the M_n and M_w distributions are identical.

In order to rationalize the disagreement between the model of Scheme 7.6 and observation, Polyakov developed[61] a more computable chain/block length distribution model to treat the continuum from a living polymerization (Poisson distribution) to a terminating polymerization (Shulz–Flory distribution).[62,63]

The population distribution of the living chains in this generalized Flory model is given by Equation 7.1. In this equation, x is the polymerization degree or chain length, α is the first-order propagation rate constant, β is the chain termination rate constant, and t is time. The analogous distribution of terminated chains in provided in Equation 7.2. As β approaches 0, P_x the population distribution of terminated chains, approaches zero, and C_x the population distribution of living chains, is given by Equation 7.3.

$$C_x = e^{-(\alpha+\beta)\cdot t} \cdot \frac{(\alpha \cdot t)^{x-1}}{(x-1)!} \tag{7.1}$$

$$P_x = \frac{\beta}{\alpha} \cdot \left(\frac{\alpha}{\alpha+\beta}\right)^x \cdot \left[1 - e^{-(\alpha+\beta)\cdot t} \cdot \sum_{i=0}^{x-1} \frac{[(\alpha+\beta)\cdot t]^i}{i!}\right], \quad x = 1, 2, 3, \ldots \tag{7.2}$$

$$C_x = e^{-\alpha\cdot t} \cdot \frac{(\alpha \cdot t)^{x-1}}{(x-1)!} \tag{7.3}$$

This model can be directly applied to the structure of individual blocks in EHPP oscillating-site polymerization. The time evolution of the molecular weight distribution provided by this model is displayed in Figure 7.19. Independent of model parameters α and β, Polyakov's Generalized Flory distribution model yields a Poisson molecular weight distribution early in the polymerization process, but inevitably yields a Shulz–Flory distribution at completion. Since there are multiple isotactic blocks within each polymer chain and each block terminates by a block terminating site switching equilibrium, an assumption of low conversion is untenable. While not providing an explanation of the observed isotactic block-length distribution in the EHPP prepared with **6**, this generalized Flory distribution model does provide a kinetic framework for assessing

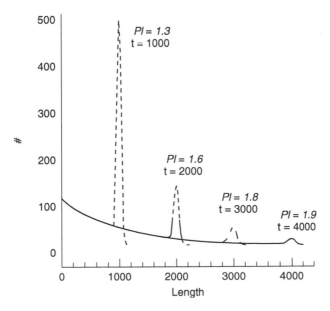

FIGURE 7.19 Number-based generalized Flory molecular weight distribution as a function of time (t); # = number of chains of a given length; length = number of monomer units in a given chain; PI = polydispersity index.

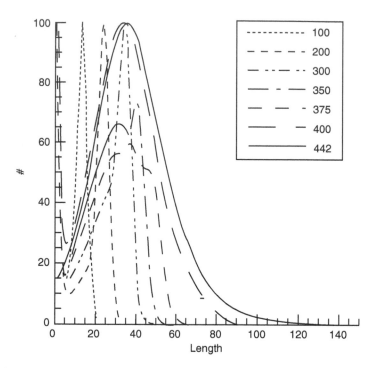

FIGURE 7.20 Number-based molecular weight distribution as a function of scaled time (t) for a generalized Flory model that includes a chain/block length dependent propagation–termination rate ratio; # = number of blocks/chains of a given length; length = number of monomer units in a given block/chain; legend is scaled time.

alternative unconventional mechanistic scenarios. To date, a single scenario has been found that is consistent with a peak in the number-based distribution for block-terminating kinetics (Figure 7.20). The curves in Figure 7.20 trace the evolution in the block length, or M_n distribution. The initial $t = 100$ dotted line curve depicts a distribution that is dominantly Poisson with a small Shulz–Flory feature at small length. At $t = 200$, the Poisson peak has shifted toward longer length. As time progresses, the Poisson feature moves to longer length, but it and the Shulz–Flory feature at short length get swamped by the peak at 50. In this proposed scenario, the difference in activation energy between block propagation and block termination varies, and is dependent upon the growing block length. In the curves shown in Figure 7.20, the energy difference between propagation barrier and the higher termination barrier linearly drops 0.02 kcal/mol per monomer unit. Rather than having a propagation rate *constant*, the model assumes that as the block grows, the propagation rate slows, and termination thus becomes relatively more favorable. The precise position of the peak in the final distribution is dependent upon the propagation α and termination β parameters as well as how fast α changes with block length. Given the decreased conformational flexibility of an isotactic block relative to an atactic block, it is plausible that as an isotactic block increases in length, the associated insertion barrier will increase in height.

To provide a molecular basis for a block length-dependent activation energy for isotactic block propagation, RFF computations of the insertion transition state for propylene polymerization with **6r** and **6r·MAO9** have been carried out (Figure 7.21). The polarizable isotactic polypropylene chain is found to be attracted to the counteranion, leading to a chain distortion. The bare cation model adopts a C_α chain eclipsed conformation with a C_1–Zr–C_α–H dihedral angle of 164°, whereas the ion pair model adopts a more intermediate conformation with a C_1–Zr–C_α–H dihedral angle of 140°. This distortion is at odds with the preferred helical growth of an isotactic chain. As the isotactic block increases in length, the conformational strain induced by the counteranion increases, leading to an increase in insertion barrier while not impacting the site isomerization barrier.

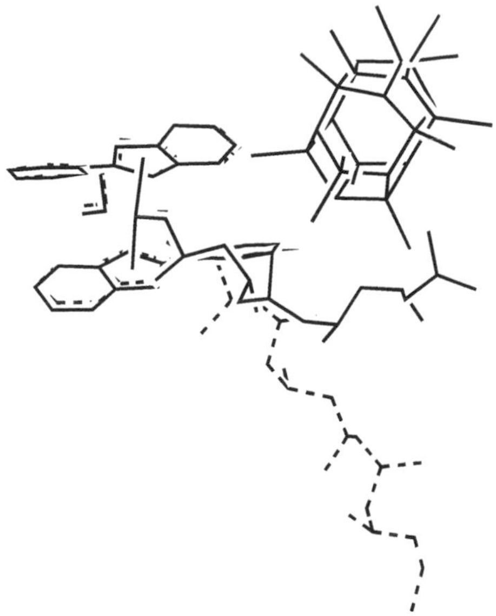

FIGURE 7.21   Superposition of ion pair (solid line) and bare cation (dashed line) insertion transition state models for isotactic polypropylene block formation with **6r/6r·MAO9**.

This theory has been supported by experimental results. In 2002, Wilmes et al. reported[64] that for EHPP polymerization with **6**, catalyst productivity and stereocontrol depended upon the nature of the cocatalyst, with MAO yielding the highest stereocontrol and *tris*(pentafluorophenyl)borane the lowest.

7.3.4 ALTERNATING ETHYLENE/PROPYLENE COPOLYMERIZATION

Ethylene/propylene copolymerization is of significant commercial importance. Elastomers formed from random ethylene/propylene copolymers (EP) possess a number of valuable properties[65–67] including a high plateau modulus (~1.6 MPa), which permits a higher filler loading and more cost-effective compounding.[65] Furthermore, of the major hydrocarbon-based rubbers, EP is by far the least reactive with oxygen and ozone.

A major distinction between heterogeneous Ziegler–Natta ethylene polymerization and single-site polymerization is the greater facility with which comonomers are incorporated into the chain with single-site catalysts.[68] Adding steric encumbrances to a single-site catalyst has been shown to enhance this effect. With the parent bis(cyclopentadienyl)zirconium dichloride activated with MAO, ethylene polymerizes 25 times faster than propylene; with the more bulky catalyst precursor *rac*-ethylene bis(tetrahydroindenyl)zirconium dichloride activated with MAO, ethylene is polymerized only 10 times faster than propylene;[69] and with bridged bis(2-methyl-4-aryl-indenyl)zirconium dichloride complexes such as **2** activated with MAO, ethylene and propylene are polymerized at comparable rates.[46,47] It is reasonable to hypothesize that this differential enhancement of polymerization rate for propylene versus ethylene with increasing steric bulk is due to favorable van der Waals interactions between the ligand and the propylene methyl substituent.

In 1997, Schneider et al. reported[70] that benzannelation of the indenyl ring in **1** forming *rac*-dimethylsilylene bis(2-methylbenz[*e*]indenyl)zirconium dichloride (**7**) enhances the incorporation of octene in ethylene–octene copolymerization. In order to explain this remarkable observation, they carried out a MM study and indeed found that **7** has a smaller transition state energy difference between ethylene insertion and octene insertion than the parent complex **1**.

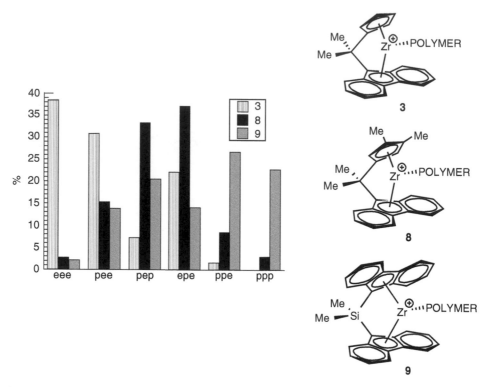

FIGURE 7.22 Experimental ethylene–propylene triad distributions for ethylene/propylene copolymers made using MAO-activated *ansa*-metallocene catalysts, **3**, **8**, and **9**.

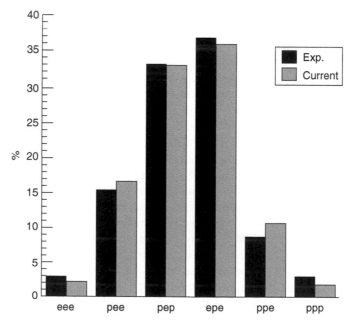

FIGURE 7.23 Experimental (Exp.) and computed (Current) triad distributions for ethylene/propylene copolymerization with **8** show good agreement.

Leclerc and Waymouth have exploited this apparently general differential rate enhancement arising from steric bulk to produce a highly alternating copolymer of ethylene and propylene.[71] Copolymers with ethylene–propylene–ethylene (EPE) and propylene–ethylene–propylene (PEP) alternating triad contents of up to 70% have been achieved.[71,72] Triad distributions are collected in Figure 7.22 for an EP made from MAO-activated catalysts **3**, **8**, and **9**. The triad distribution produced by **3** is that of random propylene incorporation into polyethylene, produced in an ethylene-rich feed. The distribution produced by **9** is that of random ethylene incorporation into polypropylene, produced in a propylene-rich feed. For **8**, there is a pronounced peak for the alternating EPE and PEP sequences. Remarkably, this high degree of alternation was achieved with a C_s-symmetric catalyst, **8**. Leclerc and Waymouth proposed that, in analogy to other metallocene control mechanisms, monomer alternation was chain-end controlled. As with other metallocenes, the orientation of the chain end was induced by the ligand framework. For **8**, site and hence monomer differentiation is not obvious. Which of the two symmetrically equivalent sites prefers ethylene and which favors propylene? The structural nature and relative contributions of the various structural factors leading to alternating polymerization remains an open question.

In order to understand the structural source of alternation for ethylene–propylene copolymerization, Polyakov carried out an RFF study.[61] Transition states for ethylene and propylene insertion into ethylene- and propylene-terminated chains were found, and triad distributions were computed. Experimental and computed triad distributions are compared in Figure 7.23 for a bare cation model of **8**. The agreement is remarkable. Relatively speaking, an ethylene-terminated chain prefers to react with propylene while a propylene-terminated chain has a preference for ethylene. The source of this olefin-based differentiation is shown in Figure 7.24. For the ethylene-terminated chain, the chain orientations for ethylene and propylene insertions are remarkably similar (Figure 7.24a and b). Independent of whether ethylene or propylene is approaching, the ethylene-terminated chain adopts a C_α staggered conformation owing to a β-agostic interaction between the chain and the zirconium center. For the propylene-terminated chains, the two orientations for ethylene and propylene approach are significantly different (Figure 7.24c and d). In this instance, the propylene methyl substituent prevents the propylene-terminated chain from adopting its preferred conformation. For ethylene approach, the propylene-terminated chain adopts an eclipsed conformation for C_α. For the propylene approach transition state, the propylene-terminated chain adopts a staggered C_α conformation. Conversion of the ethylene monomer fragment in the conformation shown in Figure 7.24c into propylene without allowing geometric relaxation leads to the situation depicted in Figure 7.24e, where there is a short C–C contact of 2.64 Å between the propylene methyl and the first methyl of the polymer chain.

7.4 CONCLUSIONS

The energetics associated with stereo-, regio-, and monomer selectivity can be reproduced by modern theoretical methodologies. The counteranion, the polymer conformation, and the metallocene active site shape and van der Waals attraction are each computed to play a role in dictating selectivity.

Reactive force field methodologies such as RFF, QM/MM approaches, and DFT methodologies continue to be used to study polymerization, though the level of activity has dropped since the 1990s. Perhaps this decline is due to a lack of agreement with experiment in efforts initiated, but not published. The lack of dispersion in DFT summarized in Section 7.2.1, the polymer chain conformational issues discussed in Section 7.2.2, and the difficulty in accounting for the counteranion in Section 7.2.2.4 are the most probable sources of disagreement with experiment.

The simplest energetic differentiation question is the stereodifferentiation discussed in this chapter where bonding reorganization is nearly identical for each of the transition states of interest. More challenging questions involve significant electronic differentiation. The methyl radical plus ethylene regiochemical differentiation discussed in Section 7.2.1 demonstrates the level of uncertainty to be expected. Particularly important processes, where detailed model calibration

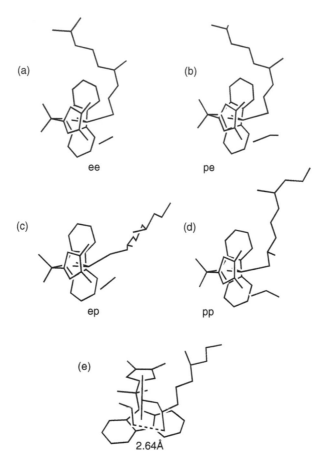

FIGURE 7.24 Ethylene/propylene copolymerization insertion transition states of **8** for: (a) ethylene insertion into an ethylene-terminated chain (ee); (b) propylene insertion into an ethylene-terminated chain (pe); (c) ethylene inserting into a propylene-terminated chain (ep); and (d) propylene insertion into a propylene-terminated chain (pp). The steric impact of a propylene methyl substituent on the geometry of the ep ethylene insertion transition state in (c) is shown in (e). The C–C distance of 2.64 Å is indicative of substantial steric repulsion.

studies have not been carried out, include a comparison of β-hydride and β-methyl termination pathways with the propagation insertion pathway and the related composite chain isomerization process. Because of the change in metal-polymer chain bonding, chain transfer to counteranion as well as chain transfer to alternative metal sites, are of even greater uncertainty.

REFERENCES AND NOTES

1. *Quantum Mechanical Electronic Structure Calculations With Chemical Accuracy* (Understanding Chemical Reactivity, Vol. 13); Langhoff, S. R., Ed.; Kluwer Academic Publishers: Dordrecht, Netherlands, 1995.

2. Rappé, A. K.; Skiff, W. M.; Casewit, C. J. Modeling metal catalyzed olefin polymerization. *Chem. Rev.* **2000**, *200*, 1435–1456.

3. Rappé, A. K.; Upton, T. H. Sigma metathesis reactions involving group 3 and 13 metals. Cl_2MH plus H_2 and Cl_2MCH_3 plus CH_4, M = Al and Sc. *J. Am. Chem. Soc.* **1992**, *114*, 7507–7517.

4. Jensen, F. *Introduction to Computational* Chemistry; John Wiley & Sons: New York, 1998.

5. Dunning, T. H. Gaussian basis sets for use in correlated molecular calculations. I. The atoms boron through neon and hydrogen. *J. Chem. Phys.* **1989**, *90*, 1007–1023.

6. Kendall, R. A.; Dunning, T. H. Electron affinities of the first-row atoms revisited. Systematic basis sets and wave functions. *J. Chem. Phys.* **1992**, *96*, 6796–6806.

7. Schlegel, H. B.; Sosa, C. *Ab initio* molecular orbital calculations on atomic fluorine + molecular hydrogen → hydrogen fluoride + atomic hydrogen and hydroxyl + molecular hydrogen → water + atomic hydrogen using unrestricted Moeller-Plesset perturbation theory with spin projection. *Chem. Phys. Lett.* **1988**, *145*, 329–333.

8. Becke, A. D. Density-functional exchange-energy approximation with correct asymptotic behavior. *Phys. Rev. A.* **1988**, *38*, 3098–3100.

9. Meijer, E. J.; Sprik, M. A density-functional study of the intermolecular interactions of benzene. *J. Chem. Phys.* **1996**, *105*, 8684–8689.

10. Tsuzuki, S.; Uchimaru, T.; Tanabe, K. Intermolecular interaction potentials of methane and ethylene dimers calculated with the Moller-Plesset, coupled cluster and density functional methods. *Chem. Phys. Lett.* **1998**, *287*, 202–208.

11. Rappé, A. K.; Bernstein, E. R. *Ab initio* calculation of nonbonded interactions: Are we there yet? *J. Phys. Chem.* **2002**, *104*, 6117–6128.

12. Pino, P.; Cioni, P.; Wei, J. Asymmetric hydrooligomerization of propylene. *J. Am. Chem. Soc.* **1987**, *109*, 6189–6191.

13. Corradini, F.; Guerra, G.; Vacatello, M.; Villani, V. A possible model for site control of stereoregularity in the isotactic specific homogeneous Ziegler–Natta polymerization. *Gazz. Chim. Ital.* **1988**, *118*, 173–177.

14. Fujimoto, H.; Yamasaki, T.; Mizutani, H.; Koga, N. A theoretical study of olefin insertions into Ti-C and Ti-H Bonds. An analysis by paired interacting orbitals. *J. Am. Chem. Soc.* **1985**, *107*, 6157–6161.

15. Jolly, C. A.; Marynick D. S. The direct insertion mechanism in Ziegler–Natta polymerization: A theoretical study of $Cp_2TiCH_3^+ + C_2H_4 - > Cp_2TiC_3H_7^+$. *J. Am. Chem. Soc.* **1989**, *111*, 7968–7974.

16. Kawamura-Kuribayashi, H.; Koga, N.; Morokuma, K. An *ab initio* MO study on ethylene and propylene insertion into the Ti-C Bond in $CH_3TiCl_2^+$ as a model of homogeneous olefin polymerization. *J. Am. Chem. Soc.* **1992**, *114*, 2359–2366.

17. Borrelli, M.; Busico, V.; Cipullo, R.; Ronca, S.; Budzelaar, P. H. M. Selectivity of metallocene-catalyzed olefin polymerization: A combined experimental and quantum mechanical study. The *ansa*-$Me_2Si(Ind)_2$Zr and *ansa*-$Me_2C(Cp)(Flu)$Zr systems. *Macromolecules* **2003**, *36*, 8171–8177.

18. Rappé, A. K.; Casewit, C. J. *Molecular Mechanics Across Chemistry*; University Science Books: Sausalito, CA, 1997.

19. Lipkowitz, K. B.; Demeter, D. A.; Zegarra, R.; Larter, R.; Darden, T. A protocol for determining enantioselective binding of chiral analytes on chiral chromatographic surfaces. *J. Am. Chem. Soc.* **1988**, *110*, 3446–3452.

20. Lipkowitz, K. B.; Baker, B.; Zegarra, R. Theoretical studies in molecular recognition: Enantioselectivity in chiral chromatography. *J. Comp. Chem.* **1989**, *10*, 718–732.

21. Cossee, P. Ziegler–Natta Catalysis I. Mechanism of polymerization of α-olefins with Ziegler–Natta catalysts. *J. Catal.* **1964**, *3*, 80–88.

22. Arlman, E. J.; Cossee, P. Ziegler–Natta catalysis III. Stereospecific polymerization of propene with the catalyst system $TiCl_3$-$AlEt_3$. *J. Catal.* **1964**, *3*, 99–104.

23. Seeman, J. I. Effect of conformational change on reactivity in organic chemistry. Evaluations, applications, and extensions of Curtin–Hammett/Winstein–Holness kinetics. *Chem. Rev.* **1983**, *83*, 83–134.

24. Erker, G.; Nolte, R.; Aul, R.; Wilker, S.; Krüger, C.; Noe, R. Cp-substituent additivity effects controlling the stereochemistry of the propene polymerization reaction at conformationally unrestricted $(Cp-CHR^1R^2)_2ZrCl_2$/methylalumoxane catalysts. *J. Am. Chem. Soc.* **1991**, *113*, 7594–7602.

25. Yu, Z.; Chien, J. C. W. Molecular mechanics study of zirconocenium catalyzed isospecific polymerization of propylene. *J. Polym. Sci., Part A: Polym. Chem.* **1995**, *33*, 125–135.

26. Cavallo, L.; Guerra, G. A density functional and molecular mechanics study of β-hydrogen transfer in homogeneous Ziegler–Natta catalysis. *Macromolecules* **1996**, *29*, 2729–2737.

27. Fait, A.; Resconi, L.; Guerra, G.; Corradini, P. A possible interpretation of the nonlinear propagation rate laws for insertion polymerizations: A kinetic model based on a single-center, two-state catalyst. *Macromolecules* **1999**, *32*, 2104–2109.

28. Guerra, G.; Longo, P.; Cavallo, L.; Corradini, P.; Resconi, L. Relationship between regiospecificity and type of stereospecificity in propene polymerization with zirconocene-based catalysts. *J. Am. Chem. Soc.* **1997**, *119*, 4394–4403.

29. Bormann-Rochotte, L. M. Molecular Modeling Applied to an Isotactic Polypropylene Catalyst. Ph. D. Thesis, Colorado State University, Fort Collins Colorado, 1998. *Diss. Abstr. Int. B.* **1998**, *59*, 2228; *Chem. Abstr.* **1998**, *129*, 260917.

30. Hehre, W. J.; Radom, L.; v. R. Schleyer, P.; Pople, J. A. *Ab Initio Molecular Orbital Theory*; Wiley-Interscience: New York, 1986.

31. Svensson, M.; Humbel, S.; Froese, R. D. J.; Matsubara, T.; Sieber, S.; Morokuma, K. ONIOM: A multi-layered integrated MO + MM method for geometry optimizations and single point energy predictions. A test for Diels-Alder reactions and Pt(P(t-Bu)$_3$)$_2$ + H$_2$ oxidative addition. *J. Phys. Chem.* **1996**, *100*, 19357–19363.

32. Woo, T. K.; Cavallo, L.; Ziegler, T. Implementation of the IMOMM methodology for performing combined QM/MM molecular dynamics simulations and frequency calculations. *Theor. Chem. Acc.* **1998**, *100*, 307–313.

33. Rappé, A. K.; Pietsch, M. A.; Wiser, D. C.; Hart, J. R.; Bormann-Rochotte, L. M.; Skiff, W. M. RFF, conceptual development of a full periodic table force field for studying reaction potential surfaces. *Mol. Eng.* **1997**, *7*, 385–400.

34. Pietsch, M. A.; Rappé, A. K. Transition state energetics for catalytic processes, incorporating resonance through valence bond theory. Book of Abstracts, 212th National Meeting of the American Chemical Society, Orlando, FL, August 25–29, 1996; American Chemical Society: Washington, DC, 1996; COMP-023. *Chem. Abstr.* **1996**, *1996*, 413403.

35. Pietsch, M. A.; Rappé, A. K. Molecular mechanics studies of catalytic polymerization transition states. *Polym. Mater. Sci. Eng.* **1996**, *74*, 423–424.

36. Griffiths, E. A. H.; Britovsek, G. J. P.; Gibson, V. C.; Gould I. R. Highly active ethylene polymerization catalysts based on iron: An *ab initio* study. *Chem. Commun.* **1999**, 1333–1334.

37. Deng, L.; Margl, P.; Ziegler, T. Mechanistic aspects of ethylene polymerization by iron(II)-bisimine pyridine catalysts: A combined density functional theory and molecular mechanics study. *J. Am. Chem. Soc.* **1999**, *121*, 6479–6487.

38. Deng, L.; Woo, T. K.; Cavallo, L.; Margl, P. M.; Ziegler, T. The role of bulky substituents in Brookhart-type Ni(II) diimine catalyzed olefin polymerization: A combined density functional theory and molecular mechanics study. *J. Am. Chem. Soc.* **1997**, *119*, 6177–6186.

39. Froese, R. J.; Musaev, D. G.; Morokuma, K. Theoretical study of substituent effects in the diimine-M(II) catalyzed ethylene polymerization reaction using the IMOMM method. *J. Am. Chem. Soc.* **1998**, *120*, 1581–1587.

40. Musaev, D. G.; Froese, R. D. J.; Morokuma, K. Molecular orbital and IMOMM studies of the chain transfer mechanisms of the diimine-M(II)-catalyzed (M = Ni, Pd) ethylene polymerization reaction. *Organometallics* **1998**, *17*, 1850–1860.

41. Froese, R. D. J.; Musaev, D. G.; Morokuma, K. Theoretical studies of the factors controlling insertion barriers for olefin polymerization by the titanium-chelating bridged catalysts. A search for more active new catalysts. *Organometallics* **1999**, *18*, 373–379.

42. Castonguay, L. A.; Rappé, A. K. Ziegler–Natta Catalysis. A theoretical study of the isotactic polymerization of propylene. *J. Am. Chem. Soc.* **1992**, *114*, 5832–5842.

43. Hart, J. R.; Rappé, A. K. Predicted structure selectivity trends: Propylene polymerization with sub-stituted *rac*-(1,2-ethylenebis(η^5-indenyl))zirconium(IV) catalysts. *J. Am. Chem. Soc.* **1993**, *115*, 6159–6164.

44. Kawamura-Kuribayashi, H.; Koga, N.; Morokuma, K. An ab initio MO and MM study of homogeneous olefin polymerization with silylene-bridged zirconocene catalyst and its regio- and stereoselectivity. *J. Am. Chem. Soc.* **1992**, *114*, 8687–8694.

45. Yoshida, T.; Koga, N.; Morokuma, K. A combined ab initio MO-MM study on isotacticity control in propylene polymerization with silylene-bridged group 4 metallocenes. C_2 symmetrical and asymmetrical catalysts. *Organometallics* **1996**, *15*, 766–777.

46. Spaleck, W.; Küber, F.; Winter, A.; Rohrmann, J.; Bachmann, B.; Antberg, M.; Dolle, V.; Paulus, E. F. The influence of aromatic substituents on the polymerization behavior of bridged zirconocene catalysts. *Organometallics* **1994**, *13*, 954–963.

47. Spaleck, W.; Küber, F.; Bachmann, B.; Fritze, C.; Winter, A. new bridged zirconocenes for olefin polymerization: Binuclear and hybrid structures. *J. Mol. Catal. A: Chem.* **1998**, *128*, 279–287.

48. Angermund, K.; Fink, G.; Jensen, V. R.; Kleinschmidt, R. Toward quantitative prediction of stereospecificity of metallocene-based catalysts for alpha-olefin polymerization. *Chem. Rev.* **2000**, *100*, 1457–1470.

49. Ewen, J. A.; Jones, R. L.; Razavi, A. Syndiospecific propylene polymerizations with group IVB metallocenes. *J. Am. Chem. Soc.* **1988**, *110*, 6255–6256.

50. Miller, S. A.; Bercaw, J. E. Highly stereoregular syndiotactic polypropylene formation with metallocene catalysts via influence of distal ligand substituents. *Organometallics* **2004**, *23*, 1777–1789.

51. Cavallo, L.; Guerra, G.; Vacatello, M.; Corradini, P. A possible model for the stereospecificity in the syndiospecific polymerization of propene with group 4a metallocenes. *Macromolecules* **1991**, *24*, 1784–1790.

52. Jolly, C. A.; Marynick, D. S. Ground-state geometries and inversion barriers for simple complexes of early transition metals. *Inorg. Chem.* **1989**, *28*, 2893–2895.

53. Bierwagen, E. P.; Bercaw, J. E.; Goddard III, W. A. Theoretical studies of Ziegler–Natta catalysis: Structural variations and tacticity control. *J. Am. Chem. Soc.* **1994**, *116*, 1481–1489.

54. Chen, M.-C.; Marks, T. J. Strong ion pairing effects on single-site olefin polymerization: Mechanistic insights in syndiospecific propylene enchainment. *J. Am. Chem. Soc.* **2001**, *123*, 11803–11804.

55. Coates, G. W.; Waymouth, R. M. Oscillating stereocontrol: A strategy for the synthesis of thermoplastic elastomeric polypropylene. *Science* **1995**, *267*, 217–219.

56. Hauptman, E.; Waymouth, R. M. Stereoblock polypropylene: Ligand effects on the stereospecificity of 2-arylindene zirconocene catalysts. *J. Am. Chem. Soc.* **1995**, *117*, 11586–11587.

57. Cavallo, L.; Guerra, G.; Corradini, P. Molecular mechanics analysis and oscillating stereocontrol for the propene polymerization with metallocene-based catalysts. *Gazz. Chim. Ital.* **1996**, *126*, 463–467.

58. Pietsch, M. A.; Rappé, A. K. π-Stacking as a control element in the (2-PhInd)$_2$Zr elastomeric polypropylene catalyst. *J. Am. Chem. Soc.* **1996**, *118*, 10908–10909.

59. Golab, J. T. Making industrial decisions with computational chemistry. *Chemtech.* **1998**, *28*, 17–23.

60. Kravchenko, R. L.; Sauer, B. B.; McLean, R. S.; Keating, M. Y.; Cotts, P. M.; Kim, Y. H. Morphology investigation of stereoblock polypropylene elastomer. *Macromolecules* **2000**, *33*, 11–13.

61. Polyakov, O. G.; Rappé, A. K. Alternating ethylene/propylene copolymerization: RFF molecular mechanics force field prognosticates effective zirconocene catalysts. Book of Abstracts, 218th National Meeting of the American Chemical Society, New Orleans, LA, Aug 22–26, 1999; American Chemical Society: Washington, DC, 1999; INOR-153. *Chem. Abstr.* **1999**, *1999*, 541786.

62. McLaughlin, K. W. Ziegler–Natta catalyzed polymerization kinetics: Origin of the molecular weight distribution. Ph. D. Thesis, Texas A&M University, College Station, TX, May 1986. *Chem. Abstr.* **1986**, *1986*, 609462.

63. McLaughlin, K. W.; Bertolucci, C. M.; Pierson, S.; Latham, D. D.; Kang, D. Fundamental chemical kinetics study of catalytic polymerizations: Influence of reaction mechanism on product distribution for Ziegler–Natta catalysis. *Polym. Prepr. (Am. Chem. Soc., Div. Polym. Chem.)* **1990**, *31*(2), 669–670.

64. Wilmes, G. M.; Polse, J. L.; Waymouth, R. M. Influence of cocatalyst on the stereoselectivty of unbridged 2-phenylindenyl metallocene catalysts. *Macromolecules* **2002**, *35*, 6766–6772.

65. *Polymer Data Handbook*; Mark, J. E., Ed.; Oxford University Press: New York, 1999.

66. *Encyclopedia of Material Science and Engineering*; Bever, M. B., Ed.; Pergamon Press: Cambridge, Massachusetts, 1986.

67. *Handbook of Industrial Materials*, 2nd ed.; Elsevier Advanced Technology: Oxford, UK, 1992.

68. Brintzinger, H. H.; Fischer, D.; Mülhaupt, R.; Rieger, B.; Waymouth, R. Stereospecific olefin polymerization with chiral metallocene catalysts. *Angew. Chem., Int. Ed. Engl.* **1995**, *34*, 1143–1170.

69. Kaminsky, W.; Külper, K.; Brintzinger, H. H. Wild, F. R. W. P. Polymerization of propene and butene with a chiral zirconocene and methylaluminoxane as cocatalyst. *Angew. Chem., Int. Ed. Engl.* **1985**, *24*, 507–508.
70. Schneider, M. J.; Suhm, J.; Muelhaupt, R.; Prosenc, M.-H.; Brintzinger, H.-H. Influence of indenyl ligand substitution pattern on metallocene-catalyzed ethene copolymerization with 1-octene. *Macromolecules* **1997**, *30*, 3164–3168.
71. Leclerc, M. K.; Waymouth, R. M. Alternating ethene/propene copolymerization with a metallocene catalyst. *Angew. Chem., Int. Ed. Engl.* **1998**, *37*, 922–925.
72. Fan, W.; Leclerc, M. K.; Waymouth, R. M. Alternating stereospecific copolymerization of ethylene and propylene with metallocene catalysts. *J. Am. Chem. Soc.* **2001**, *123*, 9555–9563.

Part II

Polypropylene: Applications of Tacticity

8 Stereoblock Polypropylene

Vincenzo Busico

CONTENTS

8.1 INTRODUCTION

Stereoblock polypropylenes are complicated, intriguing, and challenging polymers. It may be worthy to open this chapter by recalling that, according to the IUPAC *Glossary of Basic Terms in Polymer Science*,[1] a *block* is "a portion of a macromolecule comprising *many* constitutional units, that has at least one feature which is not present in the adjacent portions"; a *block macromolecule* is "a macromolecule which is composed of blocks in *linear* sequence"; a *stereoblock macromolecule* is "a block macromolecule composed of stereoregular, and possibly non-stereoregular, blocks." On paper, applying these definitions to propylene polymers is straightforward with the notable exception of the "*stereoblock-isotactic*" case, which will be discussed specifically in the last part of this section. On the other hand, demonstrating that these structures exist for a real polypropylene sample is a formidable microstructural problem; in fact, of the many cases claimed in the scientific and patent literature, those rigorously proven to be stereoblock polypropylenes can be counted on the fingers of one hand.

The stereoblocks present in these polymers can be all crystallizable (e.g., isotactic and syndiotactic), or crystallizable and noncrystallizable (e.g., isotactic or syndiotactic and atactic). In the latter case, predominantly amorphous materials can behave as thermoplastic elastomers (TPEs), because the stereoregular chain segments trapped in crystalline domains act as physical crosslinks, and induce an elastic recovery after deformation.[2,3] Producing a polyolefin-based TPE from a single monomer greatly simplifies the manufacturing process, and correspondingly reduces cost. This is the reason why elastomeric polypropylene has always been a target of industrial research.[4] Of course, the relatively high glass transition temperature (T_g) of polypropylene chains (ca. $-10\,°C$, irrespective of the tacticity) is a limitation, but for certain applications (e.g., as a replacement for the less environmentally friendly plasticized poly(vinyl chloride)) it is not necessarily a severe drawback.

At the molecular level, stereoblock propylene polymerization is truly fascinating. For a single-center catalyst, it requires a controlled fluxionality of the active species, reversibly changing the local environment of the transition metal like a molecular switch. Alternatively, one can imagine using mixtures of catalysts with diverse stereoselectivities in the presence of a "chain shuttling" agent, so that each polymer chain grows at intervals in environments with different stereocontrol. Unfortunately, both strategies are difficult to achieve in a clean and usefully manageable way, as we shall see in the following sections.

This chapter aims to introduce the reader to the various aspects of the stereoblock polypropylene field. Rather than presenting a detailed list of catalysts and polymers, the approach is to provide the tools for a critical reading of the literature, which already includes a number of recent reviews.[4–7] After a short section on the methods of polymer characterization, the various possible catalytic routes into stereoblock polypropylene are presented and discussed. This is followed by some general conclusions and a perspective outlook.

To begin with, to avoid ambiguity, we call the reader's attention to two important definitions. As is well known,[5,6] catalytic olefin polymerization occurs at transition metal centers with two available coordination sites: one for the growing polymer chain, another for the inserting monomer. With only a few (albeit important) exceptions (e.g., C_2-symmetric catalysts), the said two sites are *not* equivalent, and owing to the chain migratory insertion mechanism both can host the M–C σ bond (active site). Therefore, rather than "*single-site*," we believe that (homogeneous) catalysts with one active species (as opposed to typical heterogeneous catalysts with multiple active species) should be referred to as "*single-center*"; this more rigorous definition is adopted throughout the chapter.

As far as polymer stereochemistry is concerned, a controversial issue is what should be defined as "*stereoblock-isotactic*." Isotactic polypropylene is usually obtained as a result of site control (i.e., the preference of an intrinsically chiral transition metal active species to react with one of the two enantiofaces of the prochiral monomer).[5,6] In the case of a simple C_2-symmetric single-center catalyst with homotopic active sites, if we denote as σ the probability that the monomer inserts with a given enantioface at an active site of given chirotopicity, the fractions $[m]$ and $[r]$ of *meso* and *racemo* diads in the polymer are given by Equations 8.1 and 8.2

$$[m] = \sigma^2 + (1 - \sigma)^2 \tag{8.1}$$

$$[r] = 2\sigma(1 - \sigma) \tag{8.2}$$

Extension of these relationships to longer sequences is trivial. Inevitable failures of enantioselection result in …*mmmrrmmm*… stereodefects (Figure 8.1a), which can be viewed as the signature of site control in the chains.[5,6] When the catalyst is used in racemic form (which is the norm), the two enantiomers select opposite monomer enantiofaces ("enantiomorphic-sites control"[8]); however, as long as their configuration is invariant throughout the polymerization process, polymer chains generated at transition metal centers of opposite chirality are indistinguishable and equally described by the simple Equations 8.1 and 8.2.

Less common, but well-documented, is isotactic propagation under chain-end control.[5,6] In this case, it is the configuration of the last-inserted monomer unit that drives the selection of monomer enantioface in such a way that an *m* diad is formed (1,3-*like* asymmetric induction). Intuitively, the predominant stereodefects in chain-end-controlled isotactic polypropylene are isolated *r* diads (…*mmmrmmm*…, Figure 8.1b), because occasional faults of enantioselection invert the chirality of the stereocontrolling agent, which then tends to be perpetuated. This can be expressed in terms of Bovey's well-known chain-end statistical model of propagation,[9] assuming a first-order Markov distribution of configurations (Equations 8.3 and 8.4)

$$[m] = P_m = P_{(RR)} = P_{(SS)} \tag{8.3}$$

$$[r] = P_r = 1 - P_m = P_{(RS)} = P_{(SR)} \tag{8.4}$$

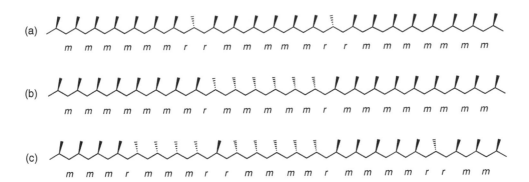

(a)

m m m m m m r r m m m m m r r m m m m m m m m

(b)

m m m m m m m r m m m m m r m m m m m m m m

(c)

m m m r m m m r r m m m m r m m m m r r m m

FIGURE 8.1 Sawhorse representations of site-controlled isotactic (a), chain-end-controlled isotactic (b), and stereoblock-isotactic (c) polypropylene sequences.

where $P_{(XY)}$ is the probability that a monomeric unit of configuration X is followed by another of configuration Y (X and $Y = R$ or S). Isotactic polypropylenes with this microstructure have been obtained, for example, with Cp_2TiX_2/MAO (Cp = cyclopentadienyl; X = Cl, alkyl (R) or aryl (Ar); MAO = methylalumoxane), and more recently with some late transition metal catalysts, but always at very low polymerization temperatures and with modest degree of stereoregularity.[5,6]

The microstructure shown in Figure 8.1b is often referred to as stereoblock-isotactic, instead of chain-end controlled isotactic, to emphasize the alternation of isotactic strands with opposite configurations along the individual chains.[10] We are of the opinion that this should be avoided, mainly for the following two reasons:

The definition is redundant.

In all known cases, as already noted before, the extent of chain-end control is modest. As a result, the average length of the $(m)_n$ strands spanned by two r diads is very low (typically, n = 3–6), which is inconsistent with the IUPAC definition of block.

It was only recently that true stereoblock-isotactic polypropylenes were reported, as a result of enantiomorphic site control at a fluxional catalytic species, which reversibly inverts its configuration during individual chain growth.[11–13] It is important to note that, in addition to isolated r diads at the stereoblock junctures, rr triads *inside the blocks* are also present owing to occasional faults of enantioselection in monomer insertion (Figure 8.1c). A simple Coleman–Fox-type model[14] reproducing the ^{13}C NMR stereosequence distribution and compact matrix multiplication algorithms for its application have been described.[12,13] The model has two adjustable parameters: site enantioselectivity, σ (where $\sigma(1) = 1 - \sigma(2)$ as defined in Equations 8.1 and 8.2 for the two enantiomorphous catalyst states 1 and 2), and the conditional probability of their interconversion, $P_{12}(= P_{21})$. The number-average stereoblock length, $\langle L_{iso} \rangle$, is given by the relationship in Equation 8.5

$$\langle L_{iso} \rangle = \frac{(1 - P_{12})}{P_{12}} \tag{8.5}$$

8.2 CHARACTERIZATION OF STEREOBLOCK POLYPROPYLENE

A direct proof that a certain sample of polypropylene contains stereoblock chains is ^{13}C NMR identification of the (stereo)chemical junctures between the blocks. In most cases, though, this is very difficult to achieve, if not impossible.

Taking the apparently straightforward example of sequential isotactic (...*mmmmmm*...) and syndiotactic (...*rrrrrr*...) blocks, in the ideal case (Figure 8.2a), any "mixed" stereosequences (e.g., *mmmr*, *mmrr*, *mrrr*) would be the sign of a ...*mmmmrrrr*... juncture. Real polypropylenes,

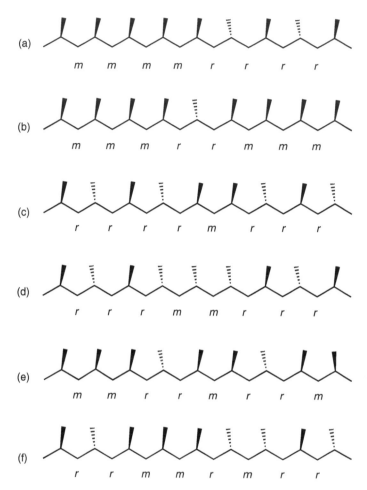

(a)   *m*   *m*   *m*   *m*   *r*   *r*   *r*   *r*

(b)   *m*   *m*   *m*   *r*   *r*   *m*   *m*   *m*

(c)   *r*   *r*   *r*   *r*   *m*   *r*   *r*   *r*

(d)   *r*   *r*   *r*   *m*   *m*   *r*   *r*   *r*

(e)   *m*   *m*   *r*   *r*   *m*   *r*   *r*   *m*

(f)   *r*   *r*   *m*   *m*   *r*   *m*   *r*   *r*

FIGURE 8.2   Sawhorse representations of an ideal isotactic/syndiotactic stereoblock polypropylene juncture (a), and of stereodefective isotactic, and syndiotactic polypropylene sequences (b–f).

however, invariably contain stereodefects, which are usually randomly distributed *rr* triads (...*mmmrrmmm*..., Figure 8.2b) in isotactic chains, and *m* diads (...*rrrmrrr*..., Figure 8.2c) or *mm* triads (...*rrrmmrrr*..., Figure 8.2d) in syndiotactic ones.[5,6] Unless the stereoregularity is very high, adjacent stereodefects (e.g., ...*mmrrmrrm*... or ...*rrmmrmrr*..., Figure 8.2e and f) are also present in nonnegligible amounts. Therefore, the ^{13}C NMR detection of "mixed" sequences is not immediately diagnostic for a stereoblock nature, and only a quantitative statistical analysis of the stereosequence distribution can allow one to recognize the extra presence of block junctures.

For this, good quality ^{13}C NMR data (in terms of sensitivity and resolution), the availability of adequate codes for the simulation and analysis of the spectrum, and previous information (or at least an educated guess) regarding the origin of the stereocontrol during isotactic and syndiotactic chain propagation are equally important prerequisites. An ambiguous verdict, though, is to be expected whenever the fraction of hypothetical junctures is (much) lower than that of the stereodefects.

In the case of inherently more complicated stereoblock polypropylenes containing (close-to-)atactic blocks, with all possible stereosequences present in comparable amounts, the requirements for a meaningful microstructural analysis are even more stringent. Routine ^{13}C NMR characterization of propylene polymers gives access to the stereosequence distribution at the pentad level, which is greater than what is achievable for any other vinyl polymer,[5] but not enough for this problem.[15] Unfortunately, statistical analyses at higher resolution are still a rarity. This is despite the fact that they have been shown to be technically feasible, and that the full assignment of the 150 MHz ^{13}C NMR spectrum of regioregular polypropylene at the heptad to undecad level has been reported in the literature[16] along with mathematical codes based on matrix multiplication techniques suited to handle the high amount of experimental data inherent in this approach.[5,16]

As will be seen in this chapter, in the vast majority of cases the diagnosis of a stereoblock microstructure is indirect and, as such, presumptive. Solvent fractionation is a widely used tool for this purpose. Nonstereoregular (atactic or poorly tactic) and stereoregular sequences found in insoluble polypropylene fractions are commonly assumed to be chemically bound, because only the latter can develop crystallinity. This conclusion is sound, if the fractionation is accurate (exhaustive in case of solvent extraction; slow enough not to be under kinetic control for temperature rising elution fractionation (TREF) or fractional crystallization). For polymers of very high average molecular mass, flawed results are not rare. An additional, more serious drawback of this approach is that it cannot be used (or is inconclusive) when the stereoblocks are all crystallizable and have similar melting temperatures, or—at the other extreme—when they are all amorphous.

Physicomechanical properties have also been proposed as indirect indicators of a stereoblock structure. As we have already noted in the introduction, stereoblock polypropylenes containing crystallizable and noncrystallizable blocks are expected to behave as TPEs. The problem here is that intimate physical blends of semicrystalline and amorphous polypropylenes can also have elastomeric properties, particularly when small amounts of truly stereoblock chains are also present and act as phase compatibilizers.[4]

In our opinion, in the absence of definitive ^{13}C NMR data, any diagnosis of a stereoblock nature for polypropylene-based materials should be regarded as potentially flawed and subjected to a critical analysis.

8.3 CATALYTIC ROUTES TO STEREOBLOCK POLYPROPYLENE

8.3.1 HETEROGENEOUS MULTICENTER ZIEGLER–NATTA CATALYSTS

Although this book is focused on single-center olefin polymerization catalysts, this chapter would be incomplete if it did not mention that the roots of stereoblock polypropylene are deep within heterogeneous Ti-based systems, and date back to the very early days of Ziegler–Natta catalysis.[4,5] In fact, samples of polypropylene obtained with the original Ziegler catalyst ($TiCl_4$ + $AlEt_3$), as well as with combinations of "violet" $TiCl_3$ and aluminum trialkyls, contained more or less significant amounts of a weakly crystalline fraction, insoluble in boiling diethyl ether and soluble in boiling heptane, which was referred to by Natta et al. as [isotactic/atactic] stereoblock fraction.[17] Rather than being based on direct microstructural evidence (still with years to go for the development of NMR spectroscopy), this definition was based on the intermediate solubility of the fraction ("isotactic" and "atactic" polypropylene being defined at that time as the polymer fractions insoluble in boiling heptane and soluble in diethyl ether, respectively), and on its observed elastomeric properties.

Today, we know that Natta's intuition was correct, and also that stereoblock chains are present in commercial polypropylenes prepared with industrial Ziegler–Natta catalysts (including latest-generation $MgCl_2$-supported catalysts).[4,5] The active species of these multicenter heterogeneous systems have been grouped into three basic classes according to the type of stereocontrol that they offer: highly isotactic, poorly isotactic ("isotactoid"), and syndiotactic.[5] Notably, the corresponding polymerization products (Figure 8.3a through c, respectively) can be found within individual polypropylene chains as linear sequences of stereoblocks.[5,18]

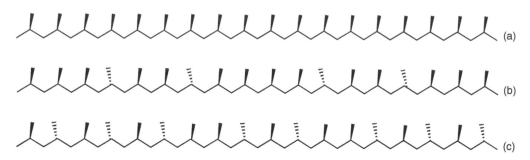

FIGURE 8.3 Sawhorse representations of highly isotactic (a), poorly isotactic ("isotactoid") (b), and syndiotactic (c) sequences/blocks found in polypropylenes produced with heterogeneous Ziegler–Natta catalysts.

Recent high-field ^{13}C NMR studies[18] have shown that the individual fractions obtained from a given polypropylene sample, for example, by solvent extraction or TREF, differ mainly in their relative amounts of the three types of building stereosequences; with increasing melting temperature, the fractions become richer in highly isotactic blocks, but even highly crystalline ones can contain minor amounts of isotactoid and syndiotactic blocks.[18,19] It follows that most stereodefects in the polymer are *not* distributed at random, but instead confined in the isotactoid and syndiotactic blocks. This explains why isotactic polypropylene samples produced with heterogeneous Ziegler–Natta catalysts invariably melt at a higher temperature than samples with the same average degree of isotacticity prepared with homogeneous single-center catalysts[5,20] (Figure 8.4). In the latter case, the stereodefects are present in a random distribution, which corresponds to the highest possible disturbance to the crystallization (i.e., to the minimum average length of the perfectly isotactic strands).[5,20,21]

To account for these observations, a model of catalytic species has been proposed (Figure 8.5).[18] It assumes that the coordination geometry of all active Ti centers is the octahedral, locally C_2-symmetric arrangement originally postulated by Cossee and Arlman.[22] The model traces the three possible types of stereocontrol (isotactic, isotactoid, and syndiotactic) to the presence or absence of "ligands" L1, L2 (*vide infra*) bound to adjacent coordinatively unsaturated metal atoms (Ti or Mg), and at close nonbonded contact with the first C–C bond of the growing polypropylene chain. It is only when at least one such ligand is present that the chiral information of the active Ti center can be carried over to an inserting propylene molecule, with the intermediacy of the growing chain (first chain C–C bond pointing away from L1 or L2, and propylene reacting faster with the enantioface which directs the methyl substituent *anti* to the said C–C bond; Figure 8.6).[5,23]

Experimental and computer modeling results[18,24] indicate that the extent of the asymmetric induction can be modulated by the nature and bulkiness of L1 and L2 (e.g., a surface Cl atom, or a Lewis base molecule purposely added to enhance stereoselectivity[4,5]). In particular, when L1 and L2 are both present (Figure 8.5a) and bulky (e.g., Lewis bases) chain propagation is predicted to be highly isotactic; losing one (Figure 8.5b) or both (Figure 8.5c) ligands would result in isotactoid or (chain-end-controlled) syndiotactic chain propagation, respectively. Well-defined single-center catalysts mimicking the different environments of Figure 8.5 have been prepared, and their stereoselectivity in propylene polymerization found to be consistent with the above assumptions.[25]

What is most relevant for the purposes of this chapter is the concept of active center fluxionality inherent in this model. The three basic surface structures in Figure 8.5 are proposed to be in equilibrium through Lewis base chemisorption/desorption.[18] In the presence of strongly coordinating Lewis base molecules (e.g., chelating alkoxysilanes), the equilibrium is shifted largely toward structure 8.5a, which is consistent with the very high isotactic selectivities of catalyst systems modified with such donors.[4,5] On the other hand, the use of certain labile donors results in the formation of elastomeric polypropylene samples or fractions rich in isotactoid and syndiotactic blocks along with

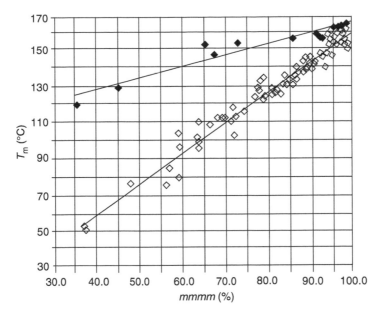

FIGURE 8.4 Correlation between degree of isotacticity (expressed as % of *mmmm* pentad) and DSC melting temperature (T_m; 2nd heating scan) for (predominantly) isotactic polypropylenes ($\diamondsuit$) = samples made with metallocene catalysts; ($\blacklozenge$) = fractions obtained by Temperature Rising Elution Fractionation of samples made with heterogeneous Ziegler–Natta catalysts. The straight lines through the data points are only for orientation. (Reprinted with permission from Busico, V.; Cipullo, R. *Prog. Polym. Sci.* **2001**, *26*, 443–533. Copyright 2001 Elsevier Science Ltd.)

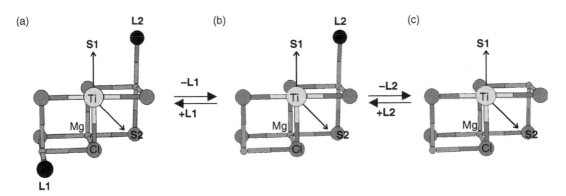

FIGURE 8.5 General three-site dynamic model of active species for heterogeneous Ti-based Ziegler–Natta catalysts: (a) highly isotactic stereocontrol; (b) isotactoid stereocontrol; (c) syndiotactic stereocontrol. L1 and L2 = Cl, AlR_xCl_{3-x}, or Lewis base. (Reprinted with permission from Busico, V.; Cipullo, R. *Prog. Polym. Sci.* **2001**, *26*, 443–533. Copyright 2001 Elsevier Science Ltd.; Busico, V.; Cipullo, R.; Monaco, G.; Talarico, G.; Vacatello, M.; Chadwick, J. C.; Segre, A.; Sudmeijer, O. *Macromolecules* **1999**, *32*, 4173–4182. Copyright 1999 American Chemical Society.)

minor amounts of highly isotactic ones; this points to reversible interconversions between structures 8.5a, 8.5b, and 8.5c.[18]

Several patents claim the production of stereoblock polypropylene-based TPEs with suitably modified heterogeneous Ziegler–Natta systems.[26,27] Unfortunately, control of the distribution of catalytic species in these systems is limited, and the polymerization products are complicated

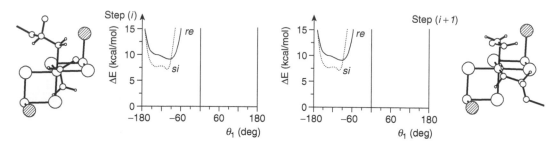

FIGURE 8.6 Model of C_2-symmetric active species (**a** in Figure 8.5) for heterogeneous Ti-based Ziegler–Natta catalysts. According to Molecular Mechanics calculations (see internal energy maps shown for L1 = L2 = Cl), at the two homotopic active sites the growing polymer chain is constrained into a chiral orientation by the repulsive interaction with the nearest-in-space of ligands L1, L2 (generically represented with shaded spheres). This favors, in turn, propylene coordination and 1,2-insertion with the enantioface orienting the methyl group *anti* to the first chain C–C bond (as shown); therefore, chain propagation is isotactic. (Reprinted with permission form Corradini, P.; Busico, V.; Cavallo, L.; Guerra, G.; Vacatello, M.; Venditto, V. *J. Mol. Catal.* **1992**, *74*, 433–442. Copyright 1992 Elsevier Science Ltd.)

blends containing not only a truly elastomeric fraction, but a completely amorphous and/or a highly crystalline fraction as well.

8.3.2 UNBRIDGED METALLOCENE CATALYSTS

Other chapters of this book discuss the manner in which well-defined stereorigid metal complexes with chirotopic sites can be converted into highly efficient catalysts for the stereoselective polymerization of propylene. For *ansa*-metallocenes, in particular, the relationship between precursor symmetry and catalyst stereoselectivity has been deeply studied and is now well-understood.[5,6]

Unbridged metallocene catalysts are more complicated. Catalysts derived from unsubstituted bis(cyclopentadienyl) group 4 metal complexes, having C_{2v} symmetry and nonchirotopic sites, afford purely atactic polypropylene unless polymerization is carried out at a temperature low enough to effect a weak isotactic chain-end stereocontrol (Figure 8.1b).[5,6] An upper limit of $P_m = 0.85$ (Equation 8.3) was reported by Ewen[28] for the Cp_2TiPh_2/MAO catalyst system at $T \leq -45\,°C$, corresponding to an average isotactic strand length of between 5 and 6 monomeric units.

The picture changes for catalyst systems with substituted cyclopentadienyl ligands. Bulky substituents, in particular, hinder the rotation of the aromatic cyclopentadienyl rings, and conformational isomerism becomes an issue. Torsional isomers with different symmetries (C_2, C_s, C_1) have been found in the solid state, and detected (or proposed to exist) in solution, for complexes of the type $(R-Cp)_2MX_2$,[29,30] $(1-R-indenyl)_2MX_2$,[31] $(1-R-4,5,6,7-tetrahydroindenyl)_2MX_2$,[31] and $(2-Ar-indenyl)_2MX_2$[6,32,33] (R = sterically bulky alkyl, such as menthyl; Ar = aryl including phenyl and substituted phenyl; M = Zr, Hf; X = Cl, Me). The steric interference between substituents, minimized in *rac*-like (C_2-symmetric) conformations, can be significant in *meso*-like (C_s-symmetric) conformations, which in fact have never been identified by means of NMR in solution.[31,34–36] The local environment of the transition metal in *rac*-like conformers (Figures 8.7a and b) is very similar to that of stereorigid *ansa*-metallocene homologues in a *rac*-configuration. Therefore, it is conceivable that unbridged *rac*-like active cations can produce site-controlled isotactic polypropylene blocks, or even whole isotactic polypropylene chains, when their conformational dynamics are (much) slower than monomer insertion.

Kaminsky and Buschermoehle[29] and Erker et al.[30,31] were the first to report that isotactically enriched propylene polymers are obtained with MAO-activated $(R-Cp)_2ZrCl_2$, $(1-R-indenyl)_2ZrCl_2$, or $(1-R-4,5,6,7-tetrahydroindenyl)_2ZrCl_2$ catalysts bearing sterically demanding substituents (such as R = (neo/iso)menthyl). Thorough studies aimed at correlating precursor solid-state structures,

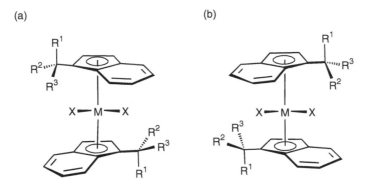

FIGURE 8.7   Possible C_2-symmetric (*rac*-like) conformations (a,b) of unbridged (1-R-indenyl)$_2$MX$_2$ complexes. (Adapted from Erker, G.; Aulbach, M.; Knickmeier, M.; Wingbermuehle, D.; Krueger, C.; Nolte, M.; Werner, S. *J. Am. Chem. Soc.* **1993**, *115*, 4590–4601. Copyright 1993 American Chemical Society.)

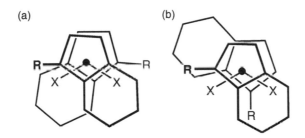

FIGURE 8.8   Possible C_2-symmetric (a) and C_1-symmetric (b) conformations of unbridged (1-R-indenyl)$_2$MX$_2$ complexes. Adapted from reference. (Reprinted with permission from Erker, G.; Aulbach, M.; Knickmeier, M.; Wingbermuehle, D.; Krueger, C.; Nolte, M.; Werner, S. *J. Am. Chem. Soc.* **1993**, *115*, 4590–4601. Copyright 1993 American Chemical Society.)

conformational statistics in solution, and catalyst stereoselectivities were carried out for complexes with 1-menthyl-substituted indenyl and tetrahydroindenyl moieties.[31,34] Single crystal X-ray structure analyses of such complexes revealed the possible occurrence of two conformational types, differing in molecular symmetry: (1) C_2-symmetric, with the bulky terpenyl substituents oriented antiperiplanarly in the lateral sectors of the bent metallocene wedge (Figure 8.8a), (2) C_1-symmetric, with only one substituent oriented laterally, and the second directed toward the metallocene front (Figure 8.8b). Variable-temperature NMR studies in solution gave evidence for the C_2-symmetric species, either alone or in rapid interconversion with minor amounts of C_1-symmetric species. Notably, in the former case of only C_2-symmetric species being present, precursor activation by MAO led to catalysts producing polypropylenes with microstructures close to the purely site-controlled isotactic, and [*mmmm*] values of up to 77% (Figure 8.9a). In the other case, instead, polymers of much lower stereoregularity and complex microstructure were produced. The methyl region "fingerprint" (Figure 8.9b) is seemingly a mixture of site-controlled isotactic sequences (*rr* stereodefects, Figure 8.1a and Equations 8.1 and 8.2; *mmmr*:*mmrr*:*mrrm* = 2:2:1) and of chain-end-controlled isotactic sequences (*r* stereodefects, Figure 8.1b and Equations 8.3 and 8.4; *mmmr*:*mmrm* = 1:1). As a matter of fact, attempts to reproduce the normalized pentad distributions in terms of a linear combination of the corresponding statistical models of chain propagation ended up with reasonably good fits.[30,31]

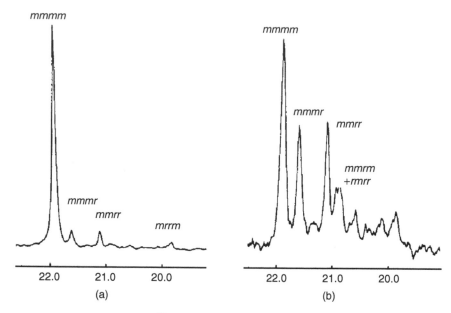

FIGURE 8.9 Methyl region of the ^{13}C NMR spectra of polypropylene samples prepared at $-50\,°C$ with catalyst systems (1-neoisomenthylindenyl)$_2$ZrCl$_2$/MAO (a) and (1-neomenthylindenyl)$_2$ZrCl$_2$/MAO (b). (Reprinted with permission from Erker, G.; Aulbach, M.; Knickmeier, M.; Wingbermuehle, D.; Krueger, C.; Nolte, M.; Werner, S. *J. Am. Chem. Soc.* **1993**, *115*, 4590–4601. Copyright 1993 American Chemical Society.)

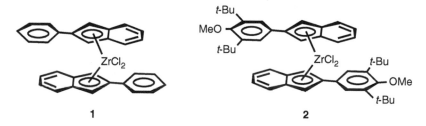

FIGURE 8.10 The structure of precursor complexes **1** and **2**.

Therefore, assuming that conformational equilibria similar to those observed for the precursor complexes occur for the active cations as well, Erker et al. proposed[31] that (1) C_2-symmetric conformers can produce enantiomorphic-sites-controlled isotactic polypropylene with *rr* stereodefects, (2) C_1-symmetric conformers are (much) less stereoselective, and (3) reversible catalyst switches between the different accessible conformations can lead to stereoblock polypropylene microstructures, including the stereoblock-isotactic one (Figure 8.1c).

Another important finding of these studies is that there is no necessary relationship between the "frozen" conformation(s) found in the solid state and those dynamically generated in solution. As an example, (1-neoisomenthylindenyl)$_2$ZrCl$_2$, which has a C_1-symmetric conformation in the crystal, appears as a C_2-symmetric species in solution.[31]

Similar results were reported a few years later by Waymouth and Coates for (2-Ar-indenyl)$_2$MX$_2$ complexes.[7,32] The first and most simple member of this class, (2-phenylindenyl)$_2$ZrCl$_2$ (Figure 8.10, **1**) was found to crystallize as a mixture of C_2-symmetric (*rac*-like) and C_s-symmetric (*meso*-like) configurational isomers. The catalyst obtained by activation of this complex with MAO is reasonably active for propylene polymerization, and produces polymers that appear to be blends of isotactic and atactic polypropylene when looked at by ^{13}C NMR. This is in line with the observation

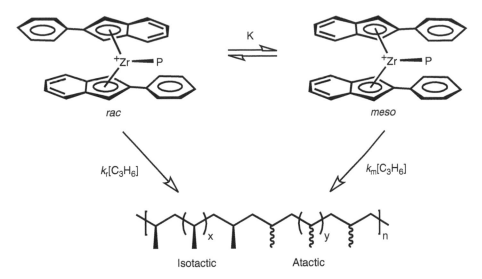

SCHEME 8.1 Proposed mechanism of isotactic/atactic stereoblock propylene polymerization with **1**/MAO, entailing an "oscillation" of the active cation between a *rac*-like (isotactic-selective) and a *meso*-like (nonstereoselective) conformation. (Reprinted with permission from Coates, G. W.; Waymouth, R. M. *Science* **1995**, *267*, 217–219. Copyright 1995 American Association for the Advancement of Science.)

of comparatively low X-ray crystallinities concomitantly with high differential scanning calorimetry (DSC) melting points (of up to 150 °C). However, the physical properties of these polymers, which are those of TPEs, led the authors to postulate[32] an isotactic/atactic stereoblock structure, and to propose that this resulted from an "oscillation" of the active cation between the two conformations found in the solid state (Scheme 8.1) at an average frequency intermediate between those of monomer insertion and chain transfer. Based on well-known structure/property relationships for bridged metallocene homologues,[37] the *rac*-like species was proposed to be isotactic-selective, and the *meso*-like one nonstereoselective.

An intensive research effort led to the synthesis of a large number of (2-Ar-indenyl)$_2$MX$_2$ complexes (M = Zr or Hf) with different aryl substituents, including very bulky ones (see, for example, Figure 8.10, **2**).[7] After activation with MAO, most of these compounds were reported to produce elastomeric polypropylenes; hence the mechanism of Scheme 8.1 was extended to the whole class of "oscillating" metallocenes.

Although simple and captivating, this interpretation is not in line with the previously discussed results obtained by Kaminsky and Erker for unbridged metallocenes having 1-substituted-bis(indenyl) ligands, particularly with respect to the unprecedented (and undemonstrated) hypothesis of a significant presence of *meso*-like species in solution. Common sense suggests that the *meso*-like conformations of the (2-Ar-indenyl)$_2$MX$_2$ complexes are higher in energy than the *rac*-like conformations, and that the bulkier the aryl group, the greater the energy difference. This hypothesis was confirmed by computer modeling.[38–40] In addition, both experimental[34–36,41] and theoretical[39,40] estimates of the rotational barriers for a large variety of (2-Ar-indenyl)$_2$MX$_2$ led to the disturbing conclusion that, in general, these energy barriers are too low to justify the observed ability of the catalysts to produce long (high-melting) isotactic blocks.

High-field ^{13}C NMR studies by Busico et al. ultimately led to a better understanding of the polymer microstructure and to a substantial revision of the mechanism of stereocontrol, which was found to be critically dependent on the steric hindrance of the aryl substituent on the indenyl rings.[11,12]

The methyl region of the 150 MHz ^{13}C NMR spectrum of a polypropylene sample produced with a catalyst bearing a very bulky substituent (Ar = 3,5-di-*tert*-butyl-4-methoxyphenyl) is shown in Figure 8.11. The active catalyst was obtained by reacting precatalyst **2** with MAO, and the

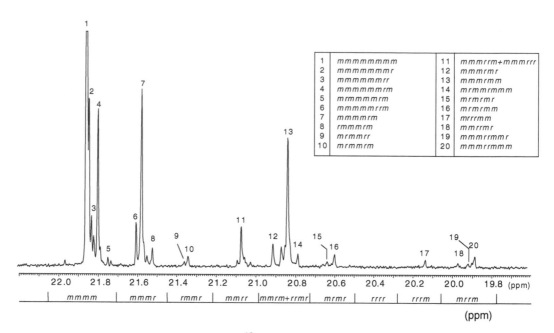

1	mmmmmmmm	11	mmmrrm+mmmrrr
2	mmmmmmmr	12	mmmrmr
3	mmmmmmrr	13	mmmrmm
4	mmmmmmrm	14	mrmrmrmm
5	mmmmmrm	15	mrmrmr
6	mmmmmrrm	16	mrmrmm
7	mmmmrm	17	mrrrmm
8	rmmmrm	18	mmmrmr
9	mrmmrr	19	mmmrrmmr
10	mrmmrm	20	mmmrrmmm

FIGURE 8.11 Methyl region of the 150 MHz ^{13}C NMR spectrum (in tetrachloroethane-1,2-d_2 at 90 °C) of a stereoblock-isotactic polypropylene sample prepared with **2**/MAO at 25 °C and [C_3H_6] = 6.7 M in toluene. The chemical shift scale is in ppm downfield of TMS. (Reprinted with permission from Busico, V.; Van Axel Castelli, V.; Aprea, P.; Cipullo, R.; Segre, A.; Talarico, G.; Vacatello, M. *J. Am. Chem. Soc.* **2003**, *125*, 5451–5460. Copyright 2003 American Chemical Society.)

polymerization was conducted in toluene at 25 °C at high propylene concentration ([C_3H_6] = 6.7 M). The spectrum reveals that the polymer is moderately isotactic, with a large predominance of single-r stereodefects (...*mmmrmmm*...) over rr (...*mmmrrmmm*...) stereodefects (see peak attributions in the same Figure 8.11); quite surprisingly, this rather obvious feature was overlooked in previous ^{13}C NMR studies at lower field. The unprecedented heptad/nonad-level resolution at 150 MHz, along with the use of adequate tools for spectral simulation and statistical analysis, made it possible to demonstrate that the observed microstructure, while deceptively similar to chain-end-controlled isotactic (Figure 8.1b), is instead site-controlled stereoblock-isotactic (Figure 8.1c). This is consistent with a mechanism of stereocontrol entailing an oscillation of the active cation between the two minimum-energy enantiomorphous *rac*-like conformations (Scheme 8.2).

In Table 8.1, the experimental stereosequence distribution, as obtained from the spectrum of Figure 8.11, is compared with best-fit calculated stereosequence distributions according to the chain-end model (Equations 8.3 and 8.4) and to the Coleman–Fox version of the enantiomorphic-sites model (see Section 8.1). The fit is poor ($\chi^2_{red} = 17$) in the former case, owing to the systematic tendency of the model to underestimate the low but nonnegligible amounts of rr stereodefects internal to the isotactic blocks, and idiosyncratic to site control. On the other hand, very good agreement ($\chi^2_{red} = 1.3$) was obtained with the Coleman–Fox model using best-fit values of $\sigma = 0.985$ (Equations 8.1 and 8.2) and $P_{12} = P_{21} = P_{osc} = 0.086$. The calculated number-average stereoblock length is $\langle L_{iso} \rangle = (1 - P_{osc})/P_{osc} \approx 12$ monomeric units (Equation 8.5). The mechanism of Scheme 8.2 predicts that P_{osc} increases with decreasing [C_3H_6], and tends asymptotically to 0.5 for [C_3H_6] → 0 (in other words, monomer insertion and catalyst oscillation tend to be equiprobable). This limit corresponds to purely atactic propagation, irrespective of the value of σ. Such a trend was confirmed experimentally.[11,12]

With decreasing steric hindrance of the aryl substituent, the fluxionality of the catalyst complex increases. For example, at 20 °C ligand rotation for (2-phenylindenyl)$_2$ZrBn$_2$ (Bn = benzyl) was

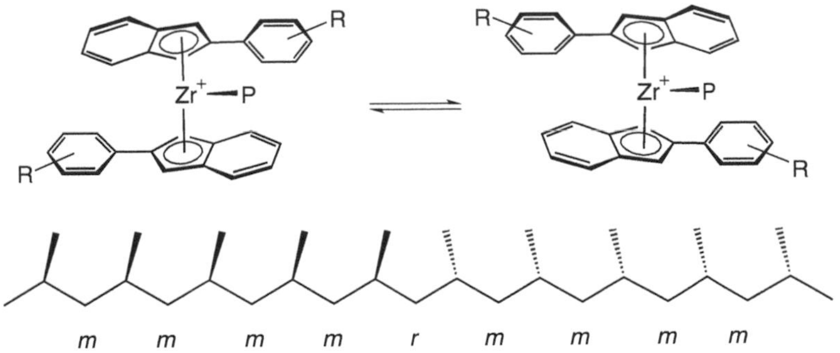

SCHEME 8.2 Mechanism of stereoblock-isotactic propylene polymerization with **2**/MAO, through reversible interconversions of the active cation between the two enantiomorphous *rac*-like conformations. (Reprinted with permission from Busico, V.; Van Axel Castelli, V.; Aprea, P.; Cipullo, R.; Segre, A.; Talarico, G.; Vacatello, M. *J. Am. Chem. Soc.* **2003**, *125*, 5451–5460. Copyright 2003 American Chemical Society.)

TABLE 8.1
150 MHz ^{13} C NMR Stereosequence Distribution of the Polypropylene Sample Shown in Figure 8.11 and Best-Fit Calculated Stereosequence Distributions in Terms of the Chain-End (CE) or Enantiomorphic-Sites (ES) Stochastic Model of Chain Propagation[12]

Stereosequence	Fractional Abundance		
	Experimental	Calculated CE Model	Calculated ES Model
mmmm	0.6446 (125)	0.6735	0.6495
mmmmmm	0.5178 (127)	0.5527	0.5267
mmmmmr	0.1082 (30)	0.1148	0.1165
mmmr	0.1419 (22)	0.1399	0.1436
mmmmrr	0.0255 (20)	0.0119	0.0267
rmmr	0.0105 (20)	0.0073	0.0079
mmrr	0.0342 (20)	0.0145	0.0329
mmrm+rmrr	0.1336 (20)	0.1414	0.1303
mmmrmm	0.0991 (20)	0.1148	0.1025
mmmrmr	0.0147 (22)	0.0119	0.0116
rmrm	0.0162 (20)	0.0145	0.0143
mrmrmm	0.0098 (20)	0.0119	0.0102
rrrr	0.0009 (20)	0.0001	0.0005
rrrm	0.0045 (20)	0.0015	0.0053
mrrm	0.0145 (20)	0.0073	0.0156
mmrrmr	0.0029 (20)	0.0012	0.0028
mmrrmm	0.0116 (20)	0.0060	0.0127
		$\chi^2_{red} = 17$	$\chi^2_{red} = 1.3$
		$P_m = 0.906(5)$	$\sigma = 0.985(1)$
			$P_{osc} = 0.086(1)$

P_m = probability to generate a *meso* diad.
σ = probability to select a given monomer enantioface at an active species of given chirality.
P_{osc} = probability of enantiomorphic sites interconversion ("oscillation").

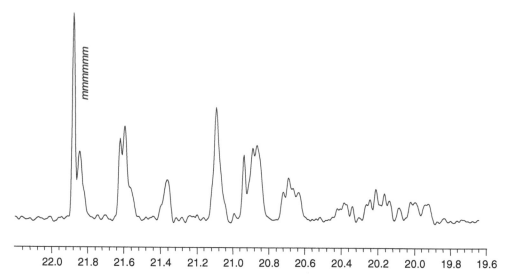

FIGURE 8.12 Methyl region of the 50 MHz ^{13}C NMR spectrum (in tetrachloroethane-1,2-d_2 at 120 °C) of the amorphous (diethylether-soluble) fraction of a typical polypropylene sample prepared with **2**/[HMe$_2$NC$_6$H$_5$]$^+$[B(C$_6$F$_5$)$_4$]$^-$/Al(i-butyl)$_3$ at 25 °C and [C$_3$H$_6$] = 6.7 M in toluene. The chemical shift scale is in ppm downfield of TMS. (Reprinted with permission from Busico, V.; Van Axel Castelli, V.; Aprea, P.; Cipullo, R.; Segre, A.; Talarico, G.; Vacatello, M. *J. Am. Chem. Soc.* **2003**, *125*, 5451–5460. Copyright 2003 American Chemical Society.)

reported to be about 100 times faster than that for (2-(3,5-di-*tert*-butylphenyl)indenyl)$_2$ZrBn$_2$.[41] Therefore, it is not surprising that polypropylenes produced with **1**/MAO are largely atactic ($P_{osc} \approx 0.5$) even for polymerizations carried out at high monomer concentration.[5–7,11,12,32] However, as first noted by Coates and Waymouth,[32] these polymers can also contain low amounts of high-melting isotactic sequences. Quite unexpectedly, in such cases a highly crystalline (albeit not completely isotactic) fraction can be separated from the amorphous part of the polymer sample by solvent extraction.[6,7,42,43] This strongly suggests the coexistence of different catalytic species, and that in the case they are in equilibrium, their average lifetimes are (much) longer than the average polymer chain growth time. Another sign of a nonsingle-center catalyst nature is the frequent observation of comparatively broad polymer molecular weight distributions for these polymers ($M_w/M_n > 2.0$).[6,7]

Even more puzzling is the fact that the stereoselectivity of this catalyst is affected by the nature of the cocatalyst/activator and the solvent.[11,12] For polymerizations in toluene, the semicrystalline fraction in the polymer products was found to be significantly more abundant when using **1**/[HMe$_2$NC$_6$H$_5$]$^+$[B(C$_6$F$_5$)$_4$]$^-$/Al(i-butyl)$_3$ rather than **1**/MAO. On the other hand, the semicrystalline fraction was completely absent, regardless of activator choice, for polymers prepared in moderately polar solvents such as bromobenzene or 1,2-dichlorobenzene.

A thorough high-field ^{13}C NMR microstructural analysis of polypropylene samples obtained with this complicated catalyst system under various conditions and subjected to solvent fractionation was recently reported.[12] The amorphous polymer fraction was confirmed to be atactic-like, albeit with an excess of *m* diads *in sequence* for samples polymerized in a nonpolar medium such as toluene (Figure 8.12).

In contrast, the semicrystalline fraction was found to consist mainly of isotactic blocks, exhibiting the *rr* stereodefects typical of site control and separated by *very short* stereoirregular sequences. Owing to the unprecedented resolution of the 150 MHz ^{13}C NMR spectra, it was possible to demonstrate that the relative configuration of adjacent isotactic blocks is the same (Figure 8.13a). This rules out the hypothesis that the microstructure arises from an oscillation of the active species between

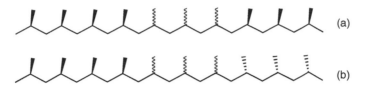

FIGURE 8.13 Sawhorse representation of two isotactic polypropylene blocks with same (a) or opposite (b) relative configurations separated by a short stereoirregular sequence.

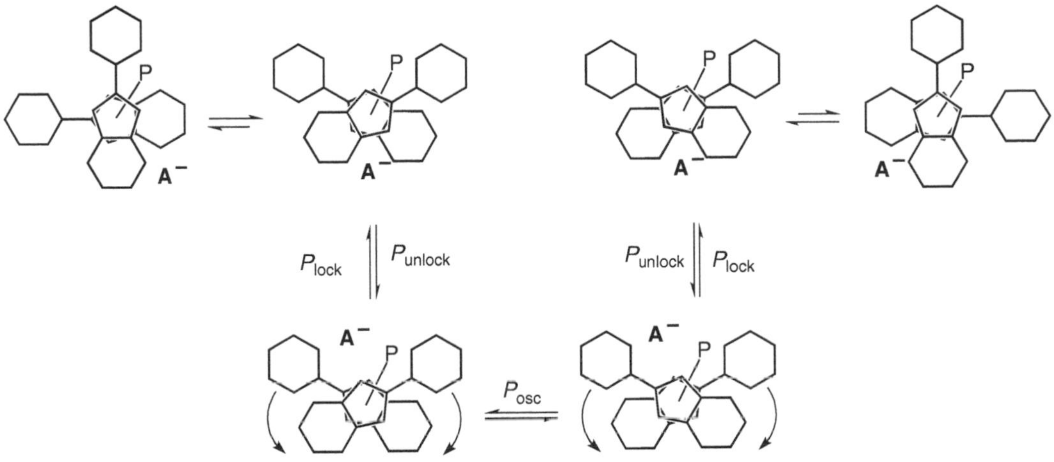

SCHEME 8.3 Proposed equilibria in solution for an active cation of **1** (A^- = counteranion). (Reprinted with permission from Busico, V.; Van Axel Castelli, V.; Aprea, P.; Cipullo, R.; Segre, A.; Talarico, G.; Vacatello, M. *J. Am. Chem. Soc.* **2003**, *125*, 5451–5460. Copyright 2003 American Chemical Society.)

a *rac*-like (isotactic-selective) and a *meso*-like (nonstereoselective) conformation, as originally proposed[32] (Scheme 8.1), because in such a case, equal amounts of adjacent isotactic blocks having the same configuration (Figure 8.13a) and the opposite configuration (Figure 8.13b) are expected.

An alternative mechanism was therefore suggested (Scheme 8.3),[11,12] which disregards the less stable *meso*-like conformation and assumes that the active cation is mainly *rac*-like. Two dynamic regimes, however, are still possible in this scenario: "oscillating" and "conformationally locked." In the former, the two enantiomorphous *rac*-like conformations are in fast interconversion ($P_{osc} = 0.5$), and chain propagation is nonstereoselective. In the latter, the oscillation is frozen, and the active species retains its configuration for a sufficient time to produce crystallizable isotactic blocks. Based on the observed cocatalyst and solvent effects, the transition from the oscillating to the locked regime was proposed to be induced by the association of the active cation with the counterion in the form of a sterically hindered contact ion pair. In this state, the anion would play a role analogous to that of the covalent bridge in an *ansa*-metallocene, albeit presumably with a lower ability to freeze ligand conformation. In particular, there could be room for transient and limited distortions from a *rac*-like to a T-shaped C_1-symmetric conformation, without crossing the barrier to the enantiomorphous *rac*-like conformation; results of computer modeling support this view (Figure 8.14).[12] The microstructure of Figure 8.13a would then be explained by tracing the short stereoirregular sequences to the transient C_1-symmetric catalyst conformer.

The discussed reinterpretation of (2-Ar-indenyl)$_2$MX$_2$-based catalysts[11,12] unifies the mechanistic picture for all substituted unbridged metallocenes. However, the scientific debate is still open.[44,45]

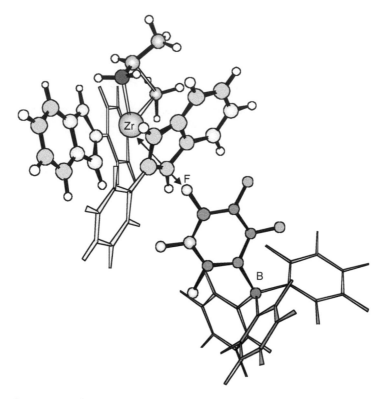

FIGURE 8.14 Quantum Mechanics/Molecular Mechanics (QM/MM) optimized structure of the transition state for propylene insertion into the Zr-Me bond of a putative $[rac\text{-}(2\text{-phenyl-indenyl})_2 Zr(Me)]^+ [B(C_6F_5)_4]^-$ ion couple with the borate anion sterically interlocked at the catalyst rear (Zr $\cdots$ F distance, 0.32 nm). The fragments treated at MM level are shown in wire frame. (Reprinted with permission from Busico, V.; Van Axel Castelli, V.; Aprea, P.; Cipullo, R.; Segre, A.; Talarico, G.; Vacatello, M. *J. Am. Chem. Soc.* **2003**, *125*, 5451–5460. Copyright 2003 American Chemical Society.)

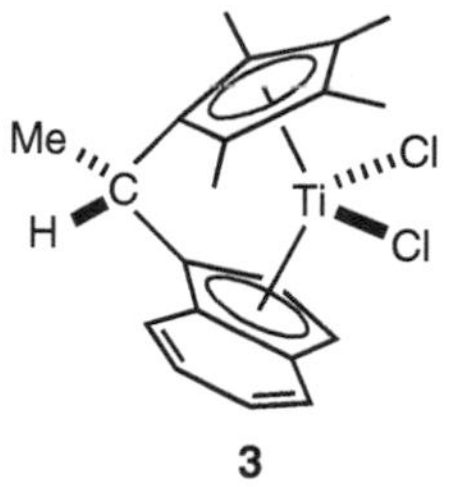

3

FIGURE 8.15 The structure of precursor complex **3**.

8.3.3 STEREORIGID BRIDGED METALLOCENE CATALYSTS

The synthesis of stereoblock polypropylene has been claimed not only from fluxional unbridged metallocene catalysts, but also—more intriguingly—from some stereorigid *ansa*-metallocene catalysts. In particular, a series of papers by Chien and coworkers[46] reported that, upon activation with MAO, *rac*-[ethylidene(tetramethylcyclopentadienyl)(1-indenyl)]TiCl$_2$ (Figure 8.15, **3**) and its dimethyl homologue produced polypropylenes with an isotactic/atactic multiblock stereostructure. This claim was based on (1) the physical properties of the material, which were typical of a TPE; (2) a ^{13}C NMR microstructural characterization at pentad level, revealing the copresence of predominantly

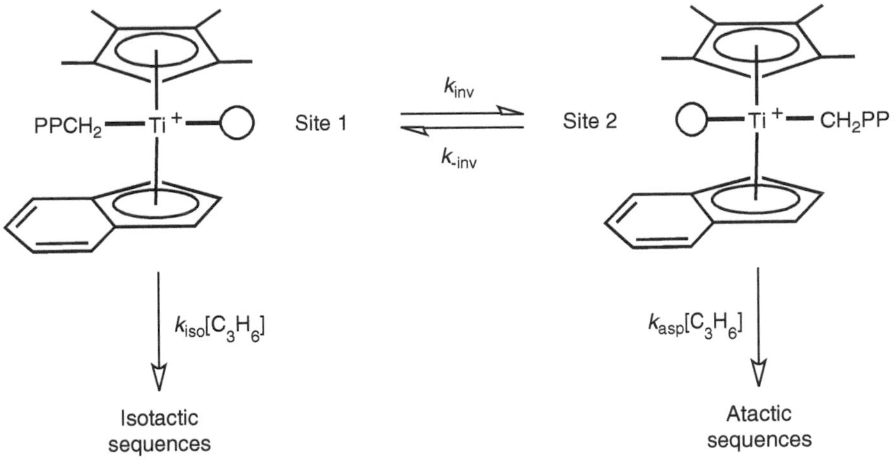

SCHEME 8.4 Proposed mechanism of isotactic/atactic stereoblock propylene polymerization at the active cation of **3**.[46] (Reprinted with permission from Gauthier, W. J.; Corrigan, J. F.; Taylor, N. J.; Collins, S. *Macromolecules* **1995**, *28*, 3771–3778. Copyright 1995 American Chemical Society.)

isotactic and close-to-atactic sequences; (3) solvent fractionation studies documenting the inability to completely separate the two types of stereosequences, which should be possible for physical mixtures of isotactic and atactic polypropylene.

The C_1-symmetric active cation derived from **3** has two diastereotopic sites. To explain their observations, the authors postulated that propylene insertion is enantioselective at only one of the sites, and isotactic/atactic stereoblock chain propagation results from long sequences of monomer insertions at the two sites ($k_{iso}[C_3H_6] \gg k_{inv}$ and $k_{asp}[C_3H_6] \gg k_{-inv}$ in Scheme 8.4).

A variety of Ti, Zr, and Hf analogues of **3** with different bridges were subsequently prepared and investigated by Collins and coworkers, and found to behave very similarly.[15,47] However, these authors underlined that the mechanism proposed in Scheme 8.4 is in obvious conflict with the known tendency of propylene insertion to occur in a chain migratory fashion.[5,6] They also demonstrated, by means of thorough mathematical modeling of stereosequence distributions, that it is not possible to discriminate unambiguously between isotactic/atactic stereoblock polypropylene and a physical blend of isotactic and atactic polypropylene based on ^{13}C NMR data at pentad level.[15] To date, this case has not been reconsidered at higher resolution.

It should be noted here that elastomeric polypropylenes with predominantly isotactic or syndiotactic structures have been obtained using a variety of C_2-, C_s-, and C_1-symmetric *ansa*-metallocene catalysts of comparatively poor stereoselectivity.[48] In such polymers, the stereodefects are distributed nearly at random, and the average length of the stereoregular sequences is very close to the minimum length needed for crystallization (11–15 monomeric units[15]). This results in the formation of a three-dimensional network in which large amorphous domains are tied up by low amounts of (very) small crystalline domains acting as physical crosslinks. It is possible, at least in some cases, that polymers with this structure have been claimed to be stereoblock polypropylenes.

8.3.4 FLUXIONAL BRIDGED METALLOCENE CATALYSTS

A strategy for stereoblock propylene polymerization that can be viewed as a hybrid between those described in Sections 8.3.2 and 8.3.3 consists of using metallocene catalysts with "hapto-flexible" cyclopentadienyl-aryl ligands.[49,50] The catalyst precursors in these systems are Ti(IV) or Zr(IV) complexes bearing a (substituted) cyclopentadienyl or indenyl ligand bridged to a neutral aromatic moiety such as a (substituted) phenyl or naphthyl. A typical example is **4** (Scheme 8.5), which

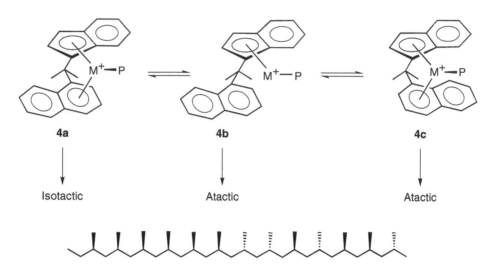

SCHEME 8.5 Proposed mechanism of isotactic/atactic stereoblock propylene polymerization with **4/MAO**, entailing reversible interconversions between three different active cations. (Reprinted with permission from De Rosa, C.; Auriemma, F.; Circelli, T.; Longo, P.; Boccia, A. C. *Macromolecules* **2003**, *36*, 3465–3474. Copyright 2003 American Chemical Society.)

has been claimed to yield isotactic/atactic stereoblock polypropylene owing to the fluxionality of the active cation.[50] This supposedly entails an equilibrium between two metallocene-like species, with the two aromatic moieties coordinated in a *rac*-like or a *meso*-like configuration (Scheme 8.5, **4a** and **4c**, respectively), and a half-metallocene in which the neutral ligand is not coordinated (Scheme 8.5, **4b**). The *rac*-like metallocene cation would be responsible for the formation of the isotactic polypropylene sequences, whereas both the *meso*-like metallocene and the half-metallocene cation would be nonstereoselective. Once again, it should be noted that the microstructural analysis of the polymer is inconclusive, and that the mechanism of stereocontrol is only presumptive; therefore, caution should be taken in the evaluation of the aforementioned results.

8.3.5 NONMETALLOCENE CATALYSTS

The recent literature shows a rapidly growing interest in the development of new nonmetallocene ("postmetallocene") single-center olefin polymerization catalysts.[51] Among these, MAO-activated octahedral bis(benzamidinate) complexes of group 4 metals (Scheme 8.6 for M = Zr) have been claimed by Eisen and coworkers to afford, in some cases, elastomeric isotactic/atactic stereoblock polypropylenes.[52–54] For a number of these complexes, the stereoselectivity in propylene polymerization at a given temperature was found to decrease from isotactic to atactic with decreasing propylene concentration. This was attributed[52] to a competition between site-controlled isotactic chain propagation and an intramolecular racemization of the last-inserted monomeric unit (growing chain epimerization), similar to what occurs with C_2-symmetric *ansa*-zirconocenes.[55] By making use of a special reactor in which high and low propylene partial pressures could be attained with very fast cycling (up to 100 Hz), isotactic/atactic multiblock chain propagation was claimed.[53] Although impractical for large scale operation and not well-proven with respect to the real microstructure of the polymerization products, this path to stereoblock polypropylene deserves citation as an ingenious example of unconventional thinking.

More recently, the same authors have proposed that stereoblock polymerization with these catalysts might be due to a fluxional behavior of the active species, oscillating between an isotactic-selective C_2-symmetric form, with both benzamidinate ligands in η^2 coordination, and a

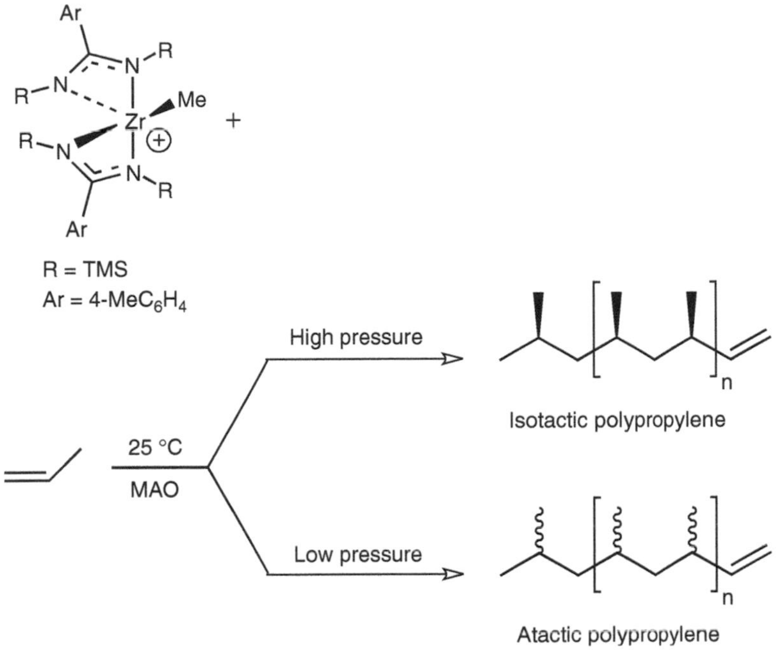

SCHEME 8.6 Proposed mechanism of isotactic/atactic stereoblock propylene polymerization with bis(benzamidinate) Zr(IV) complexes.[52] Chain propagation at high monomer concentration would be site-controlled isotactic-selective, whereas at low monomer concentration it would become nonstereoselective owing to intramolecular growing chain epimerization. (Reprinted with permission from Volkis, V.; Nelken-baum, E.; Lisovskii, A.; Hasson, G.; Semiat, R.; Kapon, M.; Botonshansky, M.; Eishen, Y.; Eisen, M. S. *J. Am. Chem. Soc.* **2003**, *125*, 2179–2194. Copyright 2003 American Chemical Society.)

nonstereoselective C_1-symmetric form in which one benzamidinate ligand is bound to the transition metal in an η^1 fashion.[54]

8.3.6 STEREOBLOCK POLYPROPYLENE VIA EXCHANGE OF GROWING POLYMER CHAINS

A conceptually different route to the synthesis of stereoblock polypropylenes entails the repeated exchange of growing polymer chains between active species of different stereoselectivity. Two pathways are conceivable:

1. Reversible trans-alkylation by a main group metal-alkyl cocatalyst (such as AlR_3 or ZnR_2)

$$L_y M_{(1)} - P + R_x M - R \rightarrow L_y M_{(1)} - R + R_x M - P \qquad (8.6)$$

$$L_y M_{(2)} - R + R_x M - P \rightarrow L_y M_{(2)} - P + R_x M - R \qquad (8.7)$$

($M_{(1)}$ and $M_{(2)}$ = transition metal; M = main group metal; R = alkyl; P = Polymeryl)

2. Direct polymeryl exchange through dinuclear transition metal species.

Pathway (1) has recently found an interesting demonstration in the related area of ethylene/1-alkene multiblock copolymerization, with ZnR_2 acting as the "chain shuttling" agent.[56] In propylene homopolymerization, presumptive evidence of isotactic/atactic stereoblock chain formation with a similar mechanism was reported by Chien et al.[57] for polypropylenes obtained with mixtures of a C_2-symmetric *ansa*-zirconocene, such as *rac*-ethylene-bis(1-indenyl)ZrCl$_2$ or its dimethylsilyl-bridged homologue, and a C_{2v}-symmetric metallocene, such as ethylene-bis(9-fluorenyl)ZrCl$_2$, in combination with Al(*i*-butyl)$_3$ and the triphenylcarbenium salt of tetrakis-perfluorophenylborate. In spite of exhaustive extraction of the polymers with boiling hexane, appreciable amounts of atactic sequences were found in the insoluble residue; in contrast, complete separation of isotactic and atactic chains by the same boiling hexane extraction procedure was achieved for physical blends of isotactic and atactic polypropylene prepared independently with the same catalysts. Isotactic/atactic stereoblock chains were estimated to represent up to 10 wt% of the raw polymer, and were suggested to act as phase compatibilizers between the isotactic and atactic polypropylene parts; this would explain the remarkable elastomeric properties of the material.

More recently, the same strategy was attempted to prepare isotactic/syndiotactic stereoblock polypropylene by using mixtures of *rac*-dimethylsilyl-bis(2-methyl-benz[*e*]-1-indenyl)ZrCl$_2$ and diphenylcarbenyl(Cp)(9-fluorenyl)ZrCl$_2$ in combination with MAO. Morphological studies by means of atomic force microscopy on such materials documented exclusive microphase separation between the isotactic and syndiotactic domains. This behavior is at odds with physical blends of isotactic and syndiotactic polypropylene, for which macrophase separation was observed.[58]

A diligent study on the possibility of achieving Al-trialkyl-mediated stereoblock polypropylenes using mixed *ansa*-metallocene catalysts of various symmetries was carried out by Lieber and Brintzinger.[59] These authors studied propylene polymerization using the zirconocene complexes **5–8** (Figure 8.16), activated either with MAO (with or without added AlMe$_3$) or with Al(*i*-butyl)$_3$ and triphenylcarbenium tetrakis-perfluorophenylborate. These catalysts were first studied individually regarding their tendency toward alkyl-polymeryl exchange with the respective alkylaluminum activators, and then pairwise regarding their capability to generate polymers with a stereoblock structure. A direct ^{13}C NMR identification of stereoblock junctures was not achieved. However, the combined results of the trans-alkylation study and of the solvent fractionation and pentad-level microstructural analysis of the polymers led the authors to conclude with reasonable confidence that growing chain exchange between different active cations can occur, provided that an efficient trans-alkylating agent is present. In particular, stereoblock chain formation was not observed for any metallocene pair in

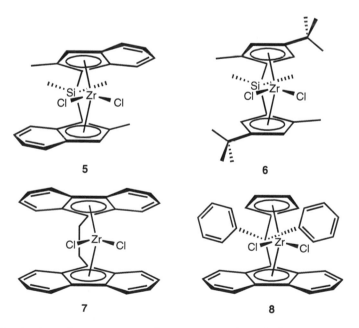

FIGURE 8.16 The structure of precursor *ansa*-zirconocene complexes **5–8**.

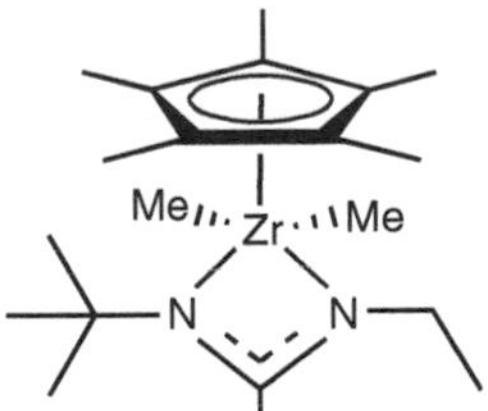

9

FIGURE 8.17   The structure of precursor complex **9**.

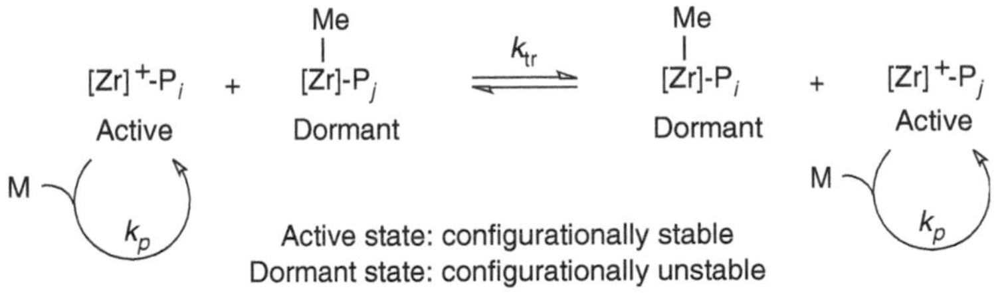

Active state: configurationally stable
Dormant state: configurationally unstable

SCHEME 8.7   Mechanism of isotactic/atactic stereoblock propylene polymerization with **9**, through degenerative chain transfer.[61,62] (Reprinted with permission from Harney, M. B.; Zhang, Y.; Sita, L. R. *Angew. Chem. Int. Ed.* **2006**, *45*, 6140–6144. Copyright 2006 John Wiley & Sons, Inc.)

combination with Al(*i*-butyl)$_3$ (a negative result which seems to be in conflict with the aforementioned previous results by Chien[57,58]), whereas in the presence of AlMe$_3$ (in equilibrium with MAO), chain transfer from the active species of **6** to that of **7** (and possibly also to that of **8**) occurred. This clearly points to an AlMe$_3$-mediated process (Equations 8.6 and 8.7; $R_xM–R = AlMe_3$), which seems to be favored when steric hindrance is relieved, that is, when a growing polymeryl is transferred from a crowded zirconocene cation (like that derived from **6**) to a less congested one (like that of **7**). A recent thorough investigation of reversible trans-alkylation by ZnEt$_2$ in catalytic ethylene polymerizations[60] is substantially in line with Brintzinger's conclusions.[59]

Concerning pathway (2), in turn, nice examples of stereoblock and "stereogradient" propylene polymerizations were reported by Sita and coworkers.[61,62] In previous papers,[63] they had demonstrated that the (η^5-pentamethylcyclopentadienyl)Zr-amidinate, (Cp*){N(*t*-Bu)C(Me)N(Et)}ZrMe$_2$ (Figure 8.17, **9**), can be converted by methide abstraction (e.g., with [HMe$_2$N(C$_6$H$_5$)]$^+$[B(C$_6$F$_5$)$_4$]$^-$) into a catalyst (initiator) for the isotactic Ziegler–Natta polymerization of 1-alkenes, and that at low temperature ($\leq -10\,°C$) chain propagation is living. The stereoselectivity is controlled by the intrinsic chirality of the metal center in the configurationally stable [(Cp*){N(*t*-Bu)C(Me)N(Et)}Zr(P)]$^+$ active cation (P = Polymeryl). Neutral dialkyls, on the other hand, undergo racemization owing to fast rotation of the amidinate ligand. Importantly, [(Cp*){N(*t*-Bu)C(Me)N(Et)}Zr(P)]$^+$ and (Cp*){N(*t*-Bu)C(Me)N(Et)}Zr(Me)(P) species interconvert through facile methyl exchange;[63] this results into degenerative chain transfer,[64] that is, a condition of direct exchange between active (cationic) and "dormant" (neutral) Zr-Polymeryls (Scheme 8.7).

With elegant and ingenious manipulations of the system, and in particular of the precatalyst activation chemistry, it was possible to adjust the relative values of k_p and k_{tr} of Scheme 8.7, and by this means to fine-tune the stereoselectivity of chain growth between the two extremes of isotactic (for $k_p \gg k_{tr}$) and atactic (for $k_p \ll k_{tr}$).[61,62] The living nature of the catalyst made it possible to change such values stepwise or—even—continuously during individual chain growth, with the resulting formation, respectively, of isotactic/atactic stereoblock or of "stereogradient" polypropylene (a novel microstructural type, in which the tacticity of the chain changes progressively from one end to the other). Unfortunately, important drawbacks for practical application are a rather poor catalyst productivity and stereoselectivity; in particular, in spite of the low polymerization temperature ($-10\,°C$) the upper limit of polypropylene stereoregularity was modest ($[mmmm] = 71\%$).

8.4 CONCLUDING REMARKS AND PERSPECTIVE OUTLOOK

The literature results reviewed in this chapter prove the existence of several catalytic routes (either direct or cocatalyst-mediated) to stereoblock polypropylenes with practically all conceivable microstructures, but also demonstrate that a complete control over the very many process variables is—euphemistically—difficult to achieve.

The catalytic route closest to being completely managed and understood is stereoblock-isotactic polymerization in the presence of bis(2-Ar-indenyl) group 4 metallocenes with very bulky aryl substituents. The simple and clean mechanism of stereocontrol entails an oscillation of the active cation between the two enantiomorphous *rac*-like conformations. Unfortunately, the resulting polymers have fairly high crystallinity and thermoplastic properties which are of rather limited practical interest, compared with those in which the alternation of crystallizable and amorphous blocks results in materials behaving as TPEs.

Amongst the latter TPE-like materials, elastomeric polypropylenes containing isotactic/atactic stereoblock chains have been obtained with a number of single- and multicenter catalyst systems, but unfortunately always as a mixture with isotactic and/or atactic polypropylene and with limited process management at both the molecular and the macroscopic scale. One decade after the first optimistic literature announcements,[32] the idea of well-behaved and versatile *rac*-like/*meso*-like oscillating metallocenes has largely evaporated, leaving behind a new and more critical awareness of the persisting obstacles to rational catalyst design.[65]

The demonstration of the viability of "chain shuttling" between different active species in mixed catalyst systems[56] may offer alternative and possibly easier routes into well-controlled stereoblock polypropylenes and block (co)polymers in general. The fundamentals of this process still need to be better understood, but the fact that existing catalysts with a known behavior can be employed makes the search for effective "chain shuttling" agents and conditions potentially of very broad scope and perspective.

Whatever the synthetic strategy, it is highly advisable that the modern tools of high-field ^{13}C NMR microstructural analysis are used for stereoblock polypropylene evaluation.[5] Until recently, with few exceptions, polymer microstructures have been determined at an inadequate level of detail; as a result, many structure/property studies are flawed at the origin.[66]

There is a clear trend in current polyolefin research toward the synthesis of novel and well-controlled polymer architectures, leading to materials with special properties and higher added value.[67] Stereoblock polypropylenes can still represent a promising area for development; it is up to those who invest in it to take full advantage of the modern experimental and theoretical toolkits, and to transform the traditional, substantially "blind" approach into a more conscious (and by no means less exciting) scientific adventure.

ACKNOWLEDGMENTS

Part of the experimental and theoretical results reviewed in this chapter has been obtained in the author's laboratory. For these, the author thanks, for their competent and enthusiastic contribution, the students and colleagues involved, namely (in alphabetical order): Paola Aprea, Prof. Roberta Cipullo, Dr. Winfried P. Kretschmer, Prof. Annalaura Segre, Prof. Giovanni Talarico, Prof. Michele Vacatello, and Dr. Valeria Van Axel Castelli. The Dutch Polymer Institute (DPI) is gratefully acknowledged for nurturing the project titled "*Catalysts that produce 'blocky' homopolymers and copolymers*"

(1999–2003), carried out in collaboration by the author's group and that of Profs. Jan Teuben and Bart Hessen at the University of Groningen (The Netherlands).

REFERENCES AND NOTES

1. Jenkins, A. D.; Kratochvil, P.; Stepto, R. F. T.; Suter, U. W. Glossary of basic terms in polymer science. *Pure Appl. Chem.* **1996**, *68*, 2287–2311.
2. *Thermoplastic Elastomers*, 2nd ed.; Holden, G., Legge, N. R., Quirk, R. P., Schroeder, H. E., Eds.; Hanser Publishers: Munich, Vienna, New York, 1996.
3. *Handbook of Thermoplastic Elastomers*, 2nd ed.; Walker, B. M., Rader, C. P., Eds.; Van Nostrand Reinhold: New York, 1998.
4. *Polypropylene Handbook*; Moore, E. P., Jr., Ed.; Hanser Publishers: Munich, Vienna, New York, 1996.
5. Busico, V.; Cipullo, R. Microstructure of polypropylene. *Prog. Polym. Sci.* **2001**, *26*, 443–533.
6. Resconi, L.; Cavallo, L.; Fait, A.; Piemontesi, F. Selectivity in propene polymerization with metallocene catalysts. *Chem. Rev.* **2000**, *100*, 1253–1345.
7. Lin, S.; Waymouth, R. M. 2-Arylindene metallocenes: Conformationally dynamic catalysts to control the structure and properties of polypropylenes. *Acc. Chem. Res.* **2002**, *35*, 765–773.
8. Shelden, R. A.; Fueno, T.; Tsunetsugu, T.; Furukawa, J. *J. Polym. Sci., Polym. Lett. Ed.* **1965**, *3*, 23–26.
9. Bovey, F. A.; Tiers, G. V. D. Polymer nuclear spin resonance spectroscopy. II. The high-resolution spectra of methyl methacrylate polymers prepared with free radical and anionic initiators. *J. Polym. Sci.* **1960**, *44*, 173–182.
10. For example, see: Kaminsky, W. New polymers by metallocene catalysis. *Macromol. Chem. Phys.* **1996**, *197*, 3907–3945.
11. Busico, V.; Cipullo, R.; Kretschmer, W. P.; Talarico, G.; Vacatello, M.; Van Axel Castelli, V. "Oscillating" metallocene catalysts: How do they oscillate? *Angew. Chem. Int. Ed.* **2002**, *41*, 505–508.
12. Busico, V.; Van Axel Castelli, V.; Aprea, P.; Cipullo, R.; Segre, A.; Talarico, G.; Vacatello, M. "Oscillating" metallocene catalysts: What stops the oscillation? *J. Am. Chem. Soc.* **2003**, *125*, 5451–5460.
13. Busico, V.; Cipullo, R.; Talarico, G.; Van Axel Castelli, V. "Chain-end-controlled isotactic" and "stereoblock-isotactic" polypropylene: Where is the difference? *Israel J. Chem.* **2002**, *42*, 295–299.
14. Coleman, B. D.; Fox, T. G. Multistate mechanism for homogeneous ionic polymerization. I. The diastereosequence distribution. *J. Chem. Phys.* **1963**, *38*, 1065–1075.
15. Gauthier, W. J.; Collins, S. Propagation models and relationship to catalyst structure. *Macromolecules* **1995**, *28*, 3779–3786.
16. (a) Busico, V.; Cipullo, R.; Monaco, G.; Vacatello, M.; Segre, A. Full assignment of the ^{13}C NMR spectra of regioregular polypropylenes: Methyl and methylene region. *Macromolecules* **1997**, *30*, 6251–6263. (b) Busico, V.; Cipullo, R.; Monaco, G.; Vacatello, M.; Bella, J.; Segre, A. Full assignment of the ^{13}C NMR spectra of regioregular polypropylenes: Methine region. *Macromolecules* **1998**, *31*, 8713–8719.
17. (a) Natta, G.; Mazzanti, G.; Crespi, G.; Moraglio, G. Stereoblock and isotactic propylene polymers. *Chim. Ind. (Milan)* **1957**, *39*, 275–283. (b) Natta, G. Properties of isotactic, atactic, and stereoblock homopolymers, random and block copolymers of α-olefins. *J. Polym. Sci.* **1959**, *34*, 531–549.
18. Busico, V.; Cipullo, R.; Monaco, G.; Talarico, G.; Vacatello, M.; Chadwick, J. C.; Segre, A.; Sudmeijer, O. High-resolution ^{13}C NMR configurational analysis of polypropylene made with MgCl$_2$-supported Ziegler–Natta catalysts. 1. The "model" system MgCl$_2$/TiCl$_4$-2,6-dimethylpyridine/Al(C$_2$H$_5$)$_3$. *Macromolecules* **1999**, *32*, 4173–4182.
19. Randall, J. C.; Alamo, R. G.; Agarwal, P. K.; Ruff, C. J. Crystallization rates of matched fractions of MgCl$_2$-supported Ziegler–Natta and metallocene isotactic poly(propylene)s. 2. Chain microstructures from a supercritical fluid fractionation of a MgCl$_2$-supported Ziegler–Natta isotactic poly(propylene). *Macromolecules* **2003**, *36*, 1572–1584.
20. Alamo, R. G.; Blanco, J. A.; Agarwal, P. K.; Randall, J. C., Crystallization rates of matched fractions of MgCl$_2$-supported Ziegler Natta and metallocene isotactic poly(propylene)s. 1. The role of chain microstructure. *Macromolecules* **2003**, *36*, 1559–1571.

21. Alamo, R. G.; Kim, M.-H.; Galante, M. J.; Isasi, J. R.; Mandelkern, L. Structural and kinetic factors governing the formation of the γ polymorph of isotactic polypropylene. *Macromolecules* **1999**, *32*, 4050–4064.

22. (a) Cossee, P. Mechanism of polymerization of α-olefins with Ziegler–Natta catalysts. *J. Catal.* **1964**, *3*, 80–88. (b) Arlman, E. J.; Cossee, P. Ziegler–Natta catalysis. III. Stereospecific polymerization of propene with the catalyst system TiCl$_3$-AlEt$_3$. *J. Catal.* **1964**, *3*, 99–104.

23. Corradini, P.; Busico, V.; Cavallo, L.; Guerra, G.; Vacatello, M.; Venditto, V. Structural analogies between homogeneous and heterogeneous catalysts for the stereospecific polymerization of 1-alkenes. *J. Mol. Catal.* **1992**, *74*, 433–442 and references therein.

24. (a) Busico, V.; Cipullo, R.; Polzone, C.; Talarico, G.; Chadwick, J. C. Propene/ethene-[1-^{13}C] copolymerization as a tool for investigating catalyst regioselectivity. 2. The MgCl$_2$/TiCl$_4$-AlR$_3$ system. *Macromolecules* **2003**, *36*, 2616–2622. (b) Busico, V.; Chadwick, J. C.; Cipullo, R.; Ronca, S.; Talarico, G. Propene/ethene-[1-^{13}C] copolymerization as a tool for investigating catalyst regioselectivity. MgCl$_2$/internal donor/TiCl$_4$-external donor/AlR$_3$ systems. *Macromolecules* **2004**, *37*, 7437–7443.

25. (a) Busico, V.; Cipullo, R.; Ronca, S.; Budzelaar, P. H. M. Mimicking Ziegler–Natta catalysts in homogeneous phase. Part 1. C$_2$-Symmetric octahedral Zr(IV) complexes with tetradentate [ONNO]-type ligands. *Macromol. Rapid Commun.* **2001**, *22*, 1405–1410. (b) Busico, V.; Cipullo, R.; Friederichs, N.; Ronca, S.; Talarico, G.; Togrou, M.; Wang, B. Block copolymers of highly isotactic polypropylene via controlled Ziegler–Natta polymerization. *Macromolecules* **2004**, *37*, 8201–8203. (c) Busico, V.; Cipullo, R.; Pellecchia, R.; Ronca, S.; Roviello, G.; Talarico, G. Design of stereoselective Ziegler–Natta propene polymerization catalysts. *Proc. Natl. Acad. Sci.* **2006**, *103*, 15321–15326.

26. Job, R.C. Process for the production of elastomeric primarily isotactic polyolefins and catalysts for use in said process. U.S. Patent 5,118,649 (Shell Oil Company), June 2, 1992.

27. (a) Collette, J. W.; Tullock, C. W. Elastomeric polypropylene. U.S. Patent 4,335,225 (E. I. du Pont de Nemours and Co.), June 15, 1982. (b) Collette, J. W.; Tullock, C. W.; MacDonald, R. N.; Buck, W. H.; Su, A. C. L.; Harrel, J. R.; Muelhaupt, R.; Anderson, B. C. Elastomeric polypropylenes from alumina-supported tetraalkyl Group IVB catalysts. 1. Synthesis and properties of high molecular weight stereoblock homopolymers. *Macromolecules* **1989**, *22*, 3851–3858. (c) Collette, J. W.; Ovenall, D. W.; Buck, W. H.; Ferguson, R. C. Elastomeric polypropylenes from alumina-supported tetraalkyl group IVB catalysts. 2. Chain microstructure, crystallinity, and morphology. *Macromolecules* **1989**, *22*, 3858–3866.

28. Ewen, J. A. Mechanisms of stereochemical control in propylene polymerizations with soluble Group 4B metallocene/methylalumoxane catalysts. *J. Am. Chem. Soc.* **1984**, *106*, 6355–6364.

29. Kaminsky, W.; Buschermoehle, M. Stereospecific polymerization of olefins with homogeneous catalysts. In *Recent Advances in Mechanistic and Synthetic Aspects of Polymerization*; Fontanille, M., Guyot, A., Eds.; NATO ASI Series C, Math. Phys. Sci. 215. Reidel: New York, 1987; pp 503–514.

30. Erker, G.; Nolte, R.; Aul, R.; Wilker, S.; Krueger, C.; Noe, R. Cp-substituent additivity effects controlling the stereochemistry of the propene polymerization reaction at conformationally unrestricted (Cp-CHR^{1}R^2)$_2$ZrCl$_2$/methylalumoxane catalysts. *J. Am. Chem. Soc.* **1991**, *113*, 7594–7602.

31. Erker, G.; Aulbach, M.; Knickmeier, M.; Wingbermuehle, D.; Krueger, C.; Nolte, M.; Werner, S. The role of torsional isomers of planarly chiral nonbridged bis(indenyl)metal type complexes in stereoselective propene polymerization. *J. Am. Chem. Soc.* **1993**, *115*, 4590–4601.

32. Coates, G. W.; Waymouth, R. M. Oscillating stereocontrol: A strategy for the synthesis of thermoplastic elastomeric polypropylene. *Science* **1995**, *267*, 217–219.

33. Dreier, T.; Erker, G.; Froehlich, R.; Wibbeling, B. 2-Hetaryl-substituted bis(indenyl)zirconium complexes as catalyst precursors for elastomeric polypropylene formation. *Organometallics* **2000**, *19*, 4095–4103.

34. Knickmeier, M.; Erker, G.; Fox, T. Conformational analysis of nonbridged bent metallocene Ziegler-catalyst precursors-detection of the third torsional isomer. *J. Am. Chem. Soc.* **1996**, *118*, 9623–9630.

35. Knueppel, S.; Fauré, J.; Erker, G.; Kehr, G.; Nissinen, M.; Froehlich, R. Probing the dynamic features of bis(aminocyclopentadienyl) and bis(aminoindenyl) Zirconium complexes. *Organometallics* **2000**, *19*, 1262–1268.

36. Schneider, N.; Schaper, F.; Schmidt, K.; Kirsten, R.; Geyer, A.; Brintzinger, H. H. Zirconocene complexes with cyclopenta[l]phenanthrene ligands: Syntheses, structural dynamics, and properties as olefin polymerization catalysts. *Organometallics* **2000**, *19*, 3597–3604.

37. Petoff, J. L. M.; Agoston, T.; Lal, T. K.; Waymouth, R. M. Elastomeric polypropylene from unbridged 2-arylindenyl zirconocenes: Modeling polymerization behavior using *ansa*-metallocene analogs. *J. Am. Chem. Soc.* **1998**, *120*, 11316–11322.

38. Cavallo, L.; Guerra, G.; Corradini, P. Molecular mechanics analysis and oscillating stereocontrol for the propene polymerization with metallocene-based catalysts. *Gazz. Chim. Ital.* **1996**, *126*, 463–467.

39. Pietsch, M. A.; Rappé, A. K. p-Stacking as a control element in the $(2\text{-PhInd})_2\text{Zr}$ elastomeric polypropylene catalyst. *J. Am. Chem. Soc.* **1996**, *118*, 10908–10909.

40. Maiti, A.; Sierka, M.; Andzelm, J.; Golab, J.; Sauer, J. Combined Quantum Mechanics: Interatomic potential function investigation of *rac-meso* configurational stability and rotational transition in zirconocene-based Ziegler–Natta catalysts. *J. Phys. Chem. A* **2000**, *104*, 10932–10938.

41. Wilmes, G. M.; France, M. B.; Lynch, S. R.; Waymouth, R. M. Rotation rates of bis(2-arylindenyl)zirconocenes: Effects of ligands and implications for formation of stereoblock polypropylene. *Abstract of Papers*, 224th National Meeting of the American Chemical Society, Boston, MA, August 18–22, 2002; American Chemical Society: Washington, DC, 2002 paper INOR-278.

42. Busico, V.; Cipullo, R.; Segre, A.; Talarico, G.; Vacatello, M.; Van Axel Castelli, V. "Seeing" the stereoblock junctions in polypropylene made with oscillating metallocene catalysts. *Macromolecules* **2001**, *34*, 8412–8415.

43. Hu, Y.; Krejchi, M. T.; Shah, C. D.; Myers, C. L.; Waymouth, R. M. Elastomeric polypropylenes from unbridged (2-phenylindene)zirconocene catalysts: Thermal characterization and mechanical properties. *Macromolecules* **1998**, *31*, 6908–6916.

44. For example, see: Lincoln, A. L.; Wilmes, G. M.; Waymouth, R. M. Dynamic NMR studies of cationic bis(2-phenylindenyl)zirconium pyridyl complexes: Evidence for syn conformers in solution. *Organometallics* **2005**, *24*, 5828–5835 and references therein.

45. For example, see: Lyakin, O. Y.; Bryliakov, K. P.; Semikolenova, N. V.; Lebedev, A. Y.; Voskoboynikov, A. Z.; Zakharov, V. A.; Talsi, E. P. ^1H and ^{13}C NMR studies of cationic intermediates formed upon activation of "oscillating" catalyst $(2\text{-PhInd})_2\text{ZrCl}_2$ with MAO, MMAO, and $\text{AlMe}_3/[\text{CPh}_3]^+[\text{B}(\text{C}_6\text{F}_5)_4]^-$. *Organometallics* **2007**, *26*, 1536–1540 and references therein.

46. (a) Mallin, D. T.; Rausch, M. D.; Lin, Y. G.; Dong, S.; Chien, J. C. W. *rac*-[Ethylidene(1-η^5-tetramethylcyclopentadienyl)(1-η^5-indenyl)]dichlorotitanium and its homopolymerization of propylene to crystalline-amorphous block thermoplastic elastomers. *J. Am. Chem. Soc.* **1990**, *112*, 2030–2031. (b) Chien, J. C. W.; Llinas, G. H.; Rausch, M. D.; Lin, G. Y.; Winter, H. H.; Atwood, J. L.; Bott, S. G. Two-state propagation mechanism for propylene polymerization catalyzed by *rac*-[anti-ethylidene (1-η^5-tetramethylcyclopentadienyl)(1-η^5-indenyl)] dimethyltitanium. *J. Am. Chem. Soc.* **1991**, *113*, 8569–8570. (c) Llinas, G. H.; Dong, S. H.; Mallin, D. T.; Rausch, M. D.; Lin, Y. G.; Winter, H. H.; Chien, J. C. W. Homogeneous Ziegler–Natta catalysts. 17. Crystalline-amorphous block polypropylene and nonsymmetric *ansa*-metallocene catalyzed polymerization. *Macromolecules* **1992**, *25*, 1242–1253. (d) Babu, G. N.; Newmark, R. A.; Cheng, H. N.; Llinas, G. H.; Chien, J. C. W. Microstructure of elastomeric polypropylenes obtained with nonsymmetric *ansa*-titanocene catalysts. *Macromolecules* **1992**, *25*, 7400–7402.

47. Gauthier, W. J.; Corrigan, J. F.; Taylor, N. J.; Collins, S. Elastomeric poly(propylene): Influence of catalyst structure and polymerization conditions on polymer structure and properties. *Macromolecules* **1995**, *28*, 3771–3778.

48. For example, see: (a) Bravakis, A. M.; Bailey, L. E.; Pigeon, M.; Collins, S. Synthesis of elastomeric poly(propylene) using unsymmetrical zirconocene catalysts: Marked reactivity differences of "*rac*"- and "*meso*"-like diastereomers. *Macromolecules* **1998**, *31*, 1000–1009. (b) Dietrich, U.; Hackmann, M.; Rieger, B.; Klinga, M.; Leskelae, M. Control of stereoerror formation with high-activity "dual-side" zirconocene catalysts: A novel strategy to design the properties of thermoplastic elastic polypropenes. *J. Am. Chem. Soc.* **1999**, *121*, 4348–4355. (c) Siedle, A. R.; Misemer, D. K.; Kolpe, V. V.; Duerr, B. F. Elastomeric polypropylenes and catalysts for their manufacture. PCT Int. Pat. Appl. WO 99/20664 (Minnesota Mining and Manufacturing Co.), April 29, 1999. (d) De Rosa, C.; Auriemma, F.; Ruiz de Ballesteros, O.; Resconi, L.; Fait, A.; Ciaccia, E.; Camurati, I. Synthesis and

characterization of high-molecular-weight syndiotactic amorphous polypropylene, *J. Am. Chem. Soc.* **2003**, *125*, 10913–10920. (e) De Rosa, C.; Auriemma, F.; Di Capua, A.; Resconi, L.; Guidotti, S.; Camurati, I.; Nifant'ev, I. E.; Laishevtsev, I. P. Structure-property correlations in polypropylene from metallocene catalysts: Stereodefective, regioregular isotactic polypropylene. *J. Am. Chem. Soc.* **2004**, *126*, 17040–17049.

49. Longo, P.; Amendola, A. G.; Fortunato, E.; Boccia, A. C.; Zambelli, A. Group 4 metallocene catalysts with hapto-flexible cyclopentadienyl-aryl ligand. *Macromol. Rapid Commun.* **2001**, *22*, 339–344.

50. De Rosa, C.; Auriemma, F.; Circelli, T.; Longo, P.; Boccia, A. C. Stereoblock polypropylene from a metallocene catalyst with a hapto-flexible naphthyl-indenyl ligand. *Macromolecules* **2003**, *36*, 3465–3474.

51. Gibson, V. C.; Spitzmesser, S. Advances in non-metallocene olefin polymerization catalysis. *Chem. Rev.* **2003**, *103*, 283–315.

52. Volkis, V.; Shmulinson, M.; Averbuj, C.; Lisovskii, A.; Edelmann, F. T.; Eisen, M. S. Pressure modulates stereoregularities in the polymerization of propylene promoted by *rac*-octahedral heteroallylic complexes. *Organometallics* **1998**, *17*, 3155–3157.

53. Eisen, M.; Volkis, V.; Shmulinson, M.; Averbuj, C.; Tish, E. Process for the production of stereoregular polymers and elastomers of alpha-olefins and certain novel catalysts therefor. PCT Int. Pat. Appl. WO 2001/55227 A2 (Technion Research and Development Foundation Ltd.), August 2, 2001.

54. Volkis, V.; Nelkenbaum, E.; Lisovskii, A.; Hasson, G.; Semiat, R.; Kapon, M.; Botonshansky, M.; Eishen, Y.; Eisen, M. S. Group 4 octahedral benzamidinate complexes: Syntheses, structures, and catalytic activities in the polymerization of propylene modulated by pressure. *J. Am. Chem. Soc.* **2003**, *125*, 2179–2194.

55. Busico, V.; Cipullo, R.; Caporaso, L.; Angelini, G.; Segre, A.L. C$_2$-symmetric *ansa*-metallocene catalysts for propene polymerization: Stereoselectivity and enantioselectivity. *J. Mol. Catal., Part A.* **1998**, *128*, 53–64 and references therein.

56. Arriola, D. J.; Carnahan, E. M.; Hustad, P. D.; Kuhlman, R. L.; Wenzel, T. T. Catalytic production of olefin block copolymers via chain shuttling polymerization. *Science* **2006**, *312*, 714–719.

57. Chien, J. C. W.; Iwamoto, Y.; Rausch, M. D.; Wedler, W.; Winter, H. H. Homogeneous binary zirconocenium catalyst systems for propylene polymerization. 1. Isotactic/atactic interfacial compatibilized polymers having thermoplastic elastomeric properties. *Macromolecules* **1997**, *30*, 3447–3458.

58. Thomann, R.; Thomann, Y.; Muelhaupt, R.; Kressler, R.; Busse, K.; Lilge, D.; Chien, J. C. W. Morphology of stereoblock polypropylene. *J. Macromol. Sci., Phys.* **2002**, *B41*, 1079–1090.

59. Lieber, S.; Brintzinger, H. H. Propene polymerization with catalyst mixtures containing different ansa-zirconocenes: Chain transfer to alkylaluminum cocatalysts and formation of stereoblock polymers. *Macromolecules* **2000**, *33*, 9192–9199.

60. van Meurs, M.; Britovsek, G. J. P.; Gibson, V. C.; Cohen, S. A. Polyethylene chain growth on zinc catalyzed by olefin polymerization catalysts: A comparative investigation of highly active catalyst systems across the transition series. *J. Am. Chem. Soc.* **2005**, *127*, 9913–9923.

61. Harney, M. B.; Zhang, Y.; Sita, L. R. Discrete, multiblock isotactic-atactic stereoblock polypropene microstructures of differing block architectures through programmable stereomodulated living Ziegler–Natta polymerization. *Angew. Chem. Int. Ed.* **2006**, *45*, 2400–2404.

62. Harney, M. B.; Zhang, Y.; Sita, L. R. Bimolecular control over polypropene stereochemical microstructure in a well-defined two-state system and a new fundamental form: Stereogradient polypropene. *Angew. Chem. Int. Ed.* **2006**, *45*, 6140–6144.

63. For example, see: Zhang, Y.; Keaton, R. J.; Sita, L. R. Degenerative transfer living Ziegler–Natta polymerization: Application to the synthesis of monomodal stereoblock polyolefins of narrow polydispersity and tunable block length. *J. Am. Chem. Soc.* **2003**, *125*, 9062–9069 and references therein.

64. Mueller, A. H. E.; Zhuang, R.; Yan, D.; Litvinenko, G. Kinetic analysis of "living" polymerization processes exhibiting slow equilibria. 1. Degenerative transfer (direct activity exchange between active and "dormant" species). Application to Group Transfer Polymerization. *Macromolecules* **1995**, *28*, 4326–4333.

65. For example, see: Talarico, G.; Busico, V.; Cavallo, L. Origin of the regiochemistry of propene insertion at octahedral column 4 polymerization catalysts: Design or serendipity? *J. Am. Chem. Soc.* **2003**, *125*, 7172–7173.

66. For a recent example of stereoblock-isotactic polypropylenes assumed to be isotactic/atactic stereoblock polypropylenes, see: De Rosa, C.; Auriemma, F.; Circelli, T.; Waymouth, R. M. Crystallization of the α and γ forms of isotactic polypropylene as a tool to test the degree of segregation of defects in the polymer chains. *Macromolecules* **2002**, *35*, 3622–3629.

67. For example, see: (a) Coates, G. W.; Hustad, P. D.; Reinartz, S. Catalysts for the living insertion polymerization of alkenes: Access to new polyolefin architectures using Ziegler–Natta chemistry. *Angew. Chem. Int. Ed. Engl.* **2002**, *41*, 2236–2257 and references therein. (b) Kaneko, H.; Matsugi, T.; Kojoh, S.-I.; Kawahara, N.; Matsuo, S.; Kashiwa, N. Creation of new polyolefin-based polymers with unique topologies and compositions: Application of polyolefin macromonomer. *Polym. Mat. Sci. Eng.* **2003**, *89*, 664–665 and references therein. (c) Busico. V. Catalytic olefin polymerization is a mature field. Isn't it? *Macromol. Chem. Phys.* **2007**, *208*, 26–29 and references therein.

9 Elastomeric Homo-Polypropylene: Solid State Properties and Synthesis via Control of Reaction Parameters

Sabine Hild, Cecilia Cobzaru, Carsten Troll, and Bernhard Rieger

CONTENTS

9.1 INTRODUCTION

Tailoring polymer properties such as stiffness, strength, or processing requires control of the molecular architecture of the polymers. For the polymerization of olefins like propylene, the development of metallocene catalysts gives access to new polymer microstructures. During the last decade, major advances have been made in metallocene catalysis for polypropylenes, providing higher levels of control over composition, molecular mass distributions, and stereoregularity. This leads to a large variety of polypropylenes with varying types and degrees of stereoregularity,[1–4] from amorphous or low-crystallinity types to highly stereoregular ones such as isotactic and syndiotactic polypropylene.[4] Owing to the regular arrangement of methyl groups, these stereoregular polypropylenes are semicrystalline materials with melting points above 180 °C exhibiting typical thermoplastic deformation behavior (Thermoplast, Figure 9.1), but isotactic and syndiotactic polypropylene differ in elastic modulus and impact strength.[5] In atactic polypropylenes, the statistically random arrangement of the methyl groups prevents crystallization. Therefore, atactic polypropylenes are soft, amorphous polymers ranging from oils to soft, waxy materials (Wax, Figure 9.1).[6]

Isotactic polypropylenes with lower levels of stereoregularity show physical and mechanical properties different from those of stiff, plastic, highly isotactic polypropylene (iPP). When the *mmmm* content decreases to a level below 50%, the materials become more flexible than iPP. In the stress–strain curves no yield points appear (Plastomer, Figure 9.1), and the specimens do not show distinct necking zone. Instead, the samples can be homogeneously deformed revealing a continuous increase in stress until the samples fail. Higher elastic recovery of the samples, as compared to pure thermoplastic polypropylene, can be found after stretching the samples to a given elongation and releasing the stress. Such behavior is typical of so-called "plastomer" behavior.[7,8]

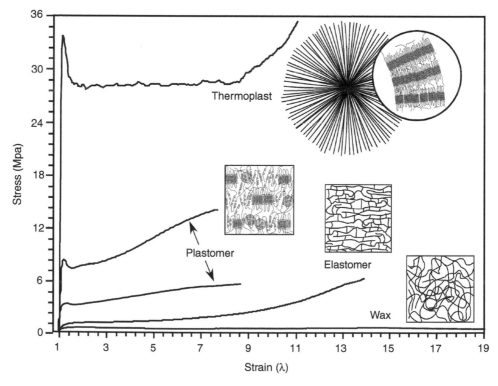

FIGURE 9.1 Stress–strain curves of semicrystalline polypropylenes allow a classification based on their mechanical properties.

Decreasing the *mmmm* content further leads to transparent materials with improved toughness that may present elastomeric properties such as high elongation ratios and improved elastic recovery (Elastomer, Figure 9.1). Owing to this interesting variation in mechanical properties, low-crystalline polypropylenes have become an increasing point of interest.

Natta discovered the first elastomeric polypropylene in 1950.[9,10] He explained the elastic properties in terms of a heterogeneous chain microstructure. Such chains consist of alternating domains of more regular isotactic sequences, which are able to crystallize, and stereoirregular, noncrystallizable sequences. This leads to a phase-separated morphology where crystalline domains provide physical crosslinks of the amorphous chain segments.[11] During the last 30 years a broad variety of microstructures with different sizes and distributions of isotactic sequences have been synthesized, such as polypropylenes with an intermediate isotactic-atactic microstructure,[12–17] stereoblock isotactic-atactic polypropylenes,[18–22] and stereoblock isotactic-hemiisotactic polypropylenes[23] showing elastomeric properties. All of these low-isotactic polypropylenes have semicrystallinity in common.

In the solid state, the crystalline morphology and the crystallinity of such polymers are affected by three major factors: (1) the amount and size of isotactic sequences, (2) the molecular weight of the polymer, and (3) the crystallization conditions. In the case of low-isotactic polypropylenes, little is known about the correlation between the chain microstructures, the morphologies they will generate, and the resulting mechanical properties. It is therefore of particular interest to control the chain microstructures of low-isotactic polypropylenes and to analyze how and to what extent they crystallize.

9.2 THERMOPLASTIC ELASTOMERS

Polymers can be broadly classified on the basis of their macroscopic properties.[24] *Thermoplastic* polymers are linear or branched polymers that can be reversibly melted or solidified, because the polymer chains of these polymers are not crosslinked. The polymer chains can slide by each other. Thus, they are irreversibly deformed under mechanical stress. *Elastomers* are polymers showing a large deformability with essentially high recovery. They consist of long main chains, which form wide-meshed, physically or chemically crosslinked polymer networks. Owing to the relatively low crosslinking density of the polymers and the presence of flexible segments that can alter their arrangement and extension response to external stress, elastomers display rubber elastic behavior.[11] Conventional rubbers are chemically crosslinked polymers. These materials can be reversibly deformed, but reprocessing through softening or melting is impossible because chemical crosslinking is an irreversible process, making it impossible for the chains to flow along each other at higher temperatures. About 60 years ago, a new class of elastomeric materials was described that could be molded and remolded again and again.[25–28] These polymers feature elastomeric behavior due to the formation of reversible crosslinks. Reversible crosslinks use noncovalent (secondary) interactions between the polymer chains to bind them together in a physically crosslinked network. Such crosslinks can easily be opened at increased temperatures. Therefore, transitions between the elastomeric and melt states can be initiated by temperature variations. Since these polymers possess physical properties similar to vulcanized rubber and processing characteristics similar to thermoplastics, they are called *thermoplastic elastomers* (TPEs).[11]

9.2.1 MORPHOLOGY OF THERMOPLASTIC-ELASTIC POLYPROPYLENE

Conventional TPEs consist of block copolymers with defined ABA building units. This block structure is responsible for their phase-separated morphology, where one segment of the polymer chain forms the soft phase responsible for the elastomeric behavior. The other (stiffer) segments form the

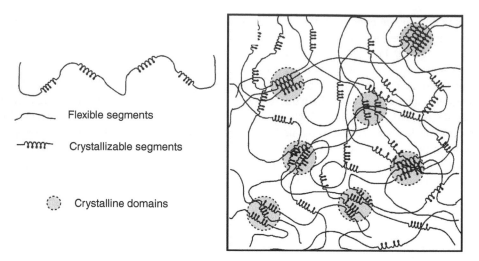

FIGURE 9.2   In thermoplastic-elastic polypropylenes, the network is formed by crystallization of stereoregular chain segments.

hard phase, which consists either of chain segments having a glass transition temperature (T_g) far above room temperature or of crystallizable chain segments.

This hard phase exists as a physical crosslinked network at room temperature, but becomes fluid when the temperature exceeds either the T_g or the melting temperature (T_m) of this phase. Low-isotactic polypropylenes contain only of one kind of monomer, but owing to variations in their stereoregular arrangement, a heterogeneous chain microstructure can be obtained. The chains consist of alternating domains of stereoirregular, noncrystallizable sequences and more regular isotactic sequences that are able to crystallize (Figure 9.2, left). The stereoregular, isotactic sequences of different polymer chains can cocrystallize. These crystalline domains dispersed in the amorphous matrix are thought to provide the physical crosslinks for the amorphous segments of the chain (Figure 9.2, right). Thus, low-isotactic polypropylenes reveal a phase-separated morphology and exhibit elastomeric behavior. These polypropylenes are called thermoplastic-elastic polypropylenes (TPE-PPs).[11]

9.2.2 CRYSTALLINE STRUCTURES

The crystalline aggregates formed by isotactic chain sequences are proposed to be necessary for the elastic behavior of TPE-PP.[29] It is therefore of particular interest to understand how polypropylenes crystallize. Stereoregular, highly isotactic polypropylene chains form a 3_1 helix (Figure 9.3). Owing to stereoerrors present in the chains, these helices can crystallize through chain folding in a lamellar morphology. Bensason et al.[30] studied the morphology of polyethylene-based TPEs. They revealed that polyethylenes with densities below 0.89 g/cm^3 form a granular morphology where the individual granules have diameter of about 5–10 nm.[31] The granules are made of bundled crystals or fringed micelles.[32] Thus, most probably, low-isotactic polypropylenes will crystallize through the parallel alignment of helical chain segments in a granular structure, forming fringed micelles (Figure 9.3). When the crystallinity—and therefore the isotacticity—increases, stacked lamellae may be formed in a manner similar to the process observed for polyethylene.

Zhu et al.[33] showed that polypropylene crystallization is not induced until the isotactic block length exceeds a critical number of sequential isotactic propylene units, $n_{iso\,crit}$. Based on the Doi–Edwards theory for isotropic-to-nematic transitions of liquid crystals[34–37] and the proposal of Imai et al.[38–40] that a parallel order of polymer segments induces a spinodal composition-type microphase separation prior to crystallization, they calculated the persistence length L of the helical

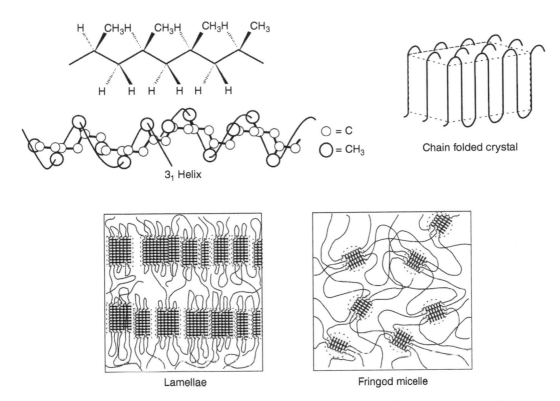

FIGURE 9.3 Crystallization of isotactic polypropylene chains leads to lamellae and fringed micelles.

sequences necessary for crystallization. If the persistence length L is smaller than the critical value, no crystallization occurs because the melted state is stable. Alternatively, the melt becomes instable and the parallel alignment of helices starts to increase when the helical sequences are larger in length than the critical value. The critical value L for increasing order can be calculated by Equation 9.1[33,34,40]

$$L = \frac{4.19M_0}{bl_0\rho N_A} \tag{9.1}$$

where b is the diameter of the polymer segment, ρ is the density, N_A is Avogadro's number, and M_0 and l_0 are the molecular weight and the length of the enchained monomer repeat unit, respectively. For an isotactic polypropylene melt, M_0, b, l_0, and ρ can be taken to be 42 g/mol, 0.665 nm, 0.217 nm, and 0.85 g/cm^3, respectively. With these parameters, the critical length L was found to be 2.38 nm.[33] This suggests that when the helical sequence length exceeds 2.38 nm, the level of parallel ordering of helix structures starts to grow, and crystallization occurs. Since the crystalline isotactic polypropylene has a 3_1 helix conformation and the c-axis dimension of the repeating unit is 0.665 nm, this value corresponds to 11 monomer units in an isotactic sequence ($n_{iso\ crit} = 11$). This calculation implies that low-isotactic polypropylenes do not crystallize until their isotactic segment length n_{iso} is above 11 isotactic repeating units.

Isotactic polypropylene can crystallize in different forms (modifications), such as α, β, γ, and smectic, which differ by their unit cell type and thus by their packing density.[41–47] The most common are the α-form and the γ-form. The α-modification is the preferred crystalline form of polypropylenes synthesized by conventional Ziegler–Natta catalysts.[48–52] High molecular weight isotactic polypropylenes prepared by metallocene catalysts preferentially crystallize in the γ-form.[44,49,50,53–59] The different polymorphic behaviors of metallocene and Ziegler–Natta samples can be related to the different distributions of defects in the polymeric chains generated by these different kinds of catalytic systems.[49–53] Whereas in metallocene-made isotactic polypropylenes the distribution of defects along the chains is random, in Ziegler–Natta iPP samples the majority of defects are segregated in small fractions of poorly crystallizable macromolecules or in the more irregular portions of the chain. Therefore, much longer fully isotactic sequences can be produced in Ziegler–Natta iPP samples, leading to the crystallization of the α-form even for a relatively high overall concentration of defects. In contrast, the presence of interruptions in the regular sequences of isotactic polypropylene favors the development of the γ-form.[51,53,60] Alamo et al.[53] discovered a linear correlation between the content of the γ-modification and the average isotactic segment length n_{iso}. Fischer and Mülhaupt[60] showed for higher-isotactic polypropylenes that polymers having an isotactic block length n_{iso} of below 40 monomer units crystallize exclusively in the γ-modification. De Rosa et al.[51] confirmed these results by showing that isotactic polypropylenes will crystallize preferentially in the γ-modification when increasing amount of rr stereoerrors are present, leading to shorter isotactic sequences.[51] Nevertheless, both of the two modifications (the α-form[54] as well as the γ-form[55,62]) have been found, in addition to mixed α/γ-modifications.[55,56,63] This indicates that, depending on the catalytic system used to prepare the polypropylene, the isotactic block length can vary over a broad range.

The different crystalline forms have different unit cells with different crystalline densities. The monoclinic α-phase (Figure 9.4a) has a higher crystalline density than the γ-phase.[41] The latter crystalline form exhibits a less densely packed orthorhombic unit cell, which may also contain some stereodefects.[64–67] The γ-phase also represents a unique packing arrangement,[58,59,68–70] wherein the orthorhombic unit cell is composed of bilayers of two parallel helices (Figure 9.5a). The directions of the chain axes in adjacent bilayers are tilted at an angle of 81° to each other.[68] In contrast, only parallel-arranged lamellae are observed in α-phase crystals.[68] The different unit cells providing the basis for the different crystalline forms can be discriminated using wide-angle X-ray scattering (WAXS) experiments. The α-modification reveals a WAXS peak at $\Theta = 9.3°$ (Figure 9.4b). For the γ-modification, this peak is missing and a WAXS peak at $\Theta = 10°$ can be found (Figure 9.5b).[58] For the case of a mixed crystal phase, both peaks are present. Here, the fractions of the corresponding α- and γ-modifications can be estimated by fitting the individual peaks using a Lorentzian function.

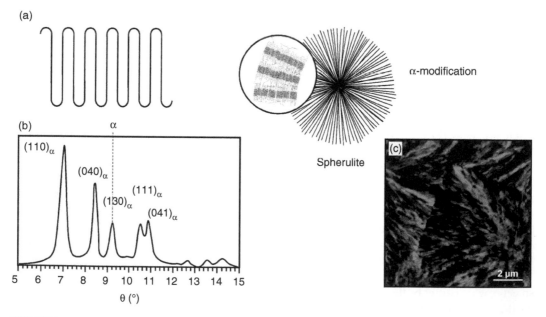

FIGURE 9.4 Polypropylene crystallized in the monoclinic α-form consisting of parallel-aligned lamellae (a) reveals a characteristic peak at $\Theta = 9.3°$ in its WAXS pattern (b); the lamellae will aggregate and form spherulitic superstructures as shown in a Scanning Force Microscopy phase image (c).

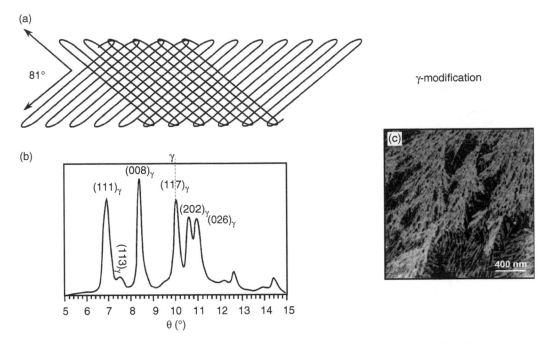

FIGURE 9.5 Polypropylene crystallized in the γ-form consisting of bilayers of two parallel helices (a) reveals a characteristic peak at $\Theta = 10°$ in its WAXS pattern (b); lamellae will aggregate and form arborescent lamellar superstructures as shown in a Scanning Force Microscopy phase image (c).

On a microscopic scale, discrimination owing to morphological differences arising from different lamellar arrangements is also possible. For α-form lamellae, various morphologies have been observed, such as single ribbon-like lamellae, lamellar stacks, and spherulites (Figure 9.4c).[43,68,69] When isotactic polypropylene crystallizes, α-phase lamellae are initially generated. These lamellae serve as nuclei for the edge-on crystallization of short, secondary lamellae appearing as thin branches. When secondary lamellae crystallize in the α-form, the angle between the primary and the secondary lamellae is about 81°. This results in extended crosshatch structure.[61,70,71] In contrast, for the γ-modification only an arborescent arrangement of lamellae can be observed (Figure 9.5c).[44,46,53,72,73] This morphology can be explained by the fact that γ-phase lamellae are only generated by epitaxial growth on the surface of primarily formed α-phase lamellae, leading to a branching angle of about 40°.[46]

9.3 SYNTHESIS OF THERMOPLASTIC-ELASTIC POLYPROPYLENES

9.3.1 POLYPROPYLENES THROUGH METALLOCENE CATALYSIS

Transition metal-catalyzed polymerization reactions allow the formation of polymeric materials with a unique relationship between catalyst structure and material properties. In contrast to conventional heterogeneous Ziegler–Natta systems, which comprise multiple active sites with different stereoselectivities, metallocenes are uniform catalysts with a defined molecular structure at a single active site. In 1984, Ewen[74] first proved the correlation between metallocene chirality and polymer tacticity with the *ansa*-titanocene $C_2H_4(1\text{-Ind})_2TiCl_2$[75] (Ind = indenyl; *ansa*-metallocenes are complexes containing a bridging group that links their two cyclopentadienyl (or similar) ligands, blocking their rotation[4]). The C_2-symmetric racemic form of this metallocene yields isotactic polypropylene (up to 71% *mm*), but the *meso* form produces atactic polymer with low molecular weight. One year

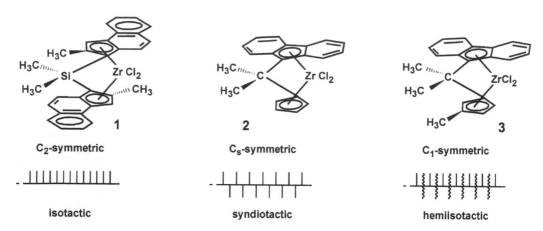

FIGURE 9.6 Correlation of polypropylene microstructure with metallocene precatalyst symmetry.

later, Kaminsky and Brintzinger showed that much higher yields of isotactic polypropylene could be synthesized with a similar C_2-symmetric zirconocene, rac-$C_2H_4(1$-$H_4Ind)_2ZrCl_2$.[76]

From then on, metallocene catalysts were no longer considered purely as model systems for heterogeneous Ziegler–Natta systems; rather, it became obvious that they could be the key to preparing tailor-made polymers with new, fascinating properties. This was a consequence of their ability to provide efficient control of polymer regio- and stereoregularity, monomer incorporation, molecular weight, and molecular weight distribution. Thus, metallocene catalyst structures can be correlated with a product polymer's properties, such as chain microstructure, and based on this correlation, also with a polymer's crystallization behavior and mechanical properties.[4,6,77] Some representative examples of different types of the metallocene precatalysts illustrating the facts outlined above are shown in Figure 9.6.

A prochiral monomer such as propylene offers two faces for coordination to a metal center. The steric environment at the active site, formed by the coordinated ligands and the growing polymer chain after activation with a cocatalyst, determines the orientation of the incoming monomer. In this case, the mechanism of stereoselection is referred to as enantiomorphic site control.[78] The stereochemistry of the polymer is thus determined by the chirality relationship of the two coordination sites of the catalyst. However, every monomer insertion generates a new stereogenic center. As a consequence, chiral induction (enantioface preference) arises from the last-inserted monomer unit in the growing polymer chain. This mechanism is referred to as chain-end control[78] (see Chapter 1 for an introduction to chain-end and enantiomorphic site control mechanisms in iPP synthesis).

In C_2-symmetric catalysts (e.g., **1**, Figure 9.6) having bifold symmetry of rotation around a horizontal axis, both polymerization sites are identical and therefore possess equal selectivity for the coordination of the prochiral monomer. All coordinations on either monomer enantioface or at either coordination site of the metal lead to identical stereoselective insertions, and thus an isotactic polypropylene chain is produced by a chain migratory insertion mechanism (enantiomorphic site control). Nevertheless, the resulting polypropylene's tacticity can be decreased by means of catalyst framework and polymerization conditions. For less stereoselective C_2-symmetric catalysts, the magnitude of chain-end control operating during polymerization can be comparable to that of enantiomorphic site control, thus leading to a decline in polymer tacticity.[78] Instead, at conditions of lower propylene concentration, unimolecular primary-growing-chain-end epimerization (Scheme 9.1) accounts for the decreased stereoregularity of the polymers.

Complexes with C_s-symmetry (e.g., **2**, Figure 9.6) have an internal vertical mirror plane bisecting the ligand from back to front.[78] The two coordination sites formed after activation are mirror images, and therefore show opposite selectivity for the coordination and insertion of the prochiral monomer. This means that the preferred propylene face for coordination changes after every insertion step, which affords a syndiotactic polypropylene microstructure.

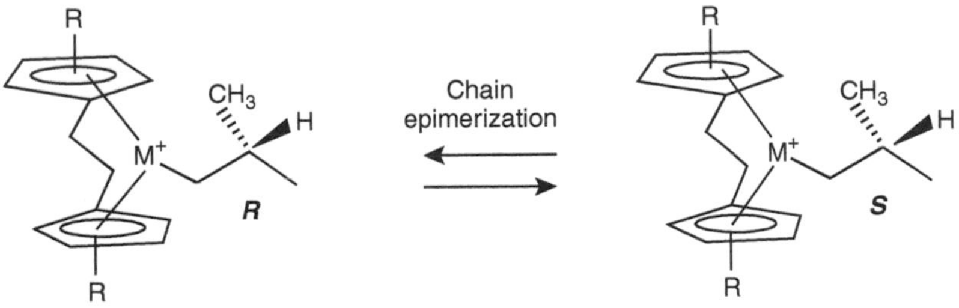

SCHEME 9.1 Epimerization (R = ligand substituent).

Reducing the symmetry of the C_s-symmetric complex, for example by introducing a methyl substituent in the 3-position on the cyclopentadienyl ring of **2**, leads to an asymmetric catalyst with no mirror plane or rotation axis, having C_1-symmetry (e.g., **3**, Figure 9.6). This particular species possesses a selective (isospecific, more sterically hindered) site and a nonselective (atactic, less sterically hindered) site. Every second insertion (assuming polymer chain migration to the site previously occupied by the coordinated propylene monomer) is thus random, and a hemiisotactic polymer is obtained.[6] However, asymmetric C_1 catalysts can not only produce hemiisotactic polymers, but depending on their ligand framework, can also produce isotactic microstructures. Replacing the symmetry-breaking substitutent residing on one of the cyclopentadienyl distal positions (i.e., the 3-methyl substituent at the β-position to the bridgehead C atom in **3**) with a bulkier one leads to a strong nonbonded, repulsive interaction between this bulky substituent and the growing polymer chain. The bulkier substituent hinders chain migration to the coordination site underneath the β-substitutent, which is therefore only available for propylene coordination.[79–81] The formation of an isotactic polymer can thus be explained only by assuming back-skip of the growing chain to its initial, less hindered position after each propylene insertion[82] (*vide infra*). The exclusive availability of only one site for monomer coordination at each active center, and its preference for only one kind of propylene enantioface, are the reasons for the formation of isotactic polypropylene with such systems.[80] This is a consequence of enantiomorphic site control over the polymerization stereochemistry (see Chapter 2 for a further discussion of C_1- and C_s-symmetric propylene polymerization catalysts).

9.3.2 DESIGNED CATALYSTS FOR CONTROLLING THE POLYMER TACTICITY

In recent years, polypropylenes with lower levels of stereoregularity have become an increasing point of interest owing to their elastic properties. In 1990, more than 30 years after Natta isolated the first polypropylene elastomer,[9] Chien reported a homogenous asymmetric C_1 metallocene catalyst, [1-(η^5-indenyl)-1-(η^5-tetramethylcyclopentadienyl)ethane]TiCl$_2$ (**4**, Figure 9.7), capable of producing elastomeric polypropylene.[83,84] Chien attributed the elastomeric properties of the isolated polypropylenes to atactic and isotactic blocks statistically distributed along the polymer chain.[15,85–87] The relative abundance of the isotactic and atactic blocks is based on a competitive two-site model made by mixing a chain-end-controlled site (to model the atactic blocks) and an enatiomorphic site (for the isotactic blocks).

A modification by Collins and coworkers[88,89] of the system introduced by Chien, to a dimethylsilylene-bridged zirconocene (**5**, Figure 9.7), afforded an improved activity (10,400 kg PP/mol Zr) and higher molecular weight polypropylenes (weight average molecular weight (M_w) $\approx$ 100,000 g/mol). This catalyst has only one sterically demanding β-carbon ligand substituent (CH$_3$ on top indenyl ring) allowing facile monomer coordination at each catalyst side. Consequently, only the polymerization temperature influences the polymer stereoregularity. In these cases, low-isotactic polypropylenes with low T_ms or amorphous polypropylenes having a more homogeneous distribution of stereoerrors are obtained.

More successful for producing elastomeric polypropylenes were Resconi et al.,[90] who have reported a chiral C_2-symmetric *ansa*-zirconocene catalyst (**6**, Figure 9.7). This catalyst produces

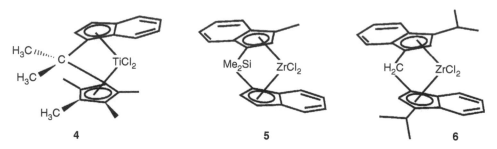

FIGURE 9.7 Asymmetric catalysts investigated by different research groups for the production of elastomeric polypropylene.

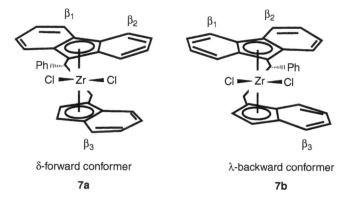

FIGURE 9.8 δ-Forward (**7a**) and λ-backward (**7b**) conformers of $[1\text{-}\eta^5\text{-}9\text{-fluorenyl})\text{-}1\text{-}(R,S)\text{-phenyl-2-}(\eta^5\text{-}1\text{-}(R)\text{-indenyl})$ethane]$ZrCl_2$.

highly flexible, transparent, nonsticky, amorphous polypropylenes with medium molecular weights and preferably isotactic microstructures. In these cases, the resultant polymers have a random distribution of stereoerrors. The presence of short isotactic sequences allows the development of a small level of crystallinity in some of these materials, which show some elastic properties.[90]

9.3.3 C_1-SYMMETRIC CATALYSTS

Rieger[91] introduced a new generation of C_1-symmetric catalysts based on ethylene-bridged (Ind-CH$_2$CH(Ph)-Flu) (Flu = 9-fluorenyl) ligands having two opposed β-substituents, which exhibited, for the first time, a strong dependence of stereoselectivity on monomer concentration. Two of the four possible forms of seven are sketched in Figure 9.8: $[1\text{-}(\eta^5\text{-}9\text{-fluorenyl})\text{-}1\text{-}(R)\text{-phenyl-}2\text{-}(\eta^5\text{-}1\text{-}(R)\text{-indenyl})$ethane]$ZrCl_2$ (**7a**) and $[1\text{-}(\eta^5\text{-}9\text{-fluorenyl})\text{-}1\text{-}(S)\text{-phenyl-2-}(\eta^5\text{-}1\text{-}(R)\text{-indenyl})$ ethane]$ZrCl_2$ (**7b**). The bulky phenyl groups occupy the energetically favored equatorial positions of the metallacycles[92] in both complexes. This leads to preferred conformations of the chelate rings, depending on the nature of the stereogenic backbone center. A (R)-configuration at the bridge carbon bearing the phenyl substituent causes a δ-conformation (**7a**), whereas the (S)-conformation gives rise to the λ-conformer (**7b**), while the stereochemistry of the indenyl fragments remains unchanged (R).[93] The different bridge twists result in staggered arrangements of the fluorenyl and indenyl units within **7a** and **7b**. This allows control of the relative positions of the ligand β-carbon C<u>H</u> substituents (fluorenyl-5H (β_2) and indenyl-4H (β_3), Figure 9.8) to each other, which are assumed to play a major role in the enantiofacial discrimination of the inserting propylene monomer.[94] According to a notation introduced by Brintzinger and coworkers,[95] the δ-conformation places the

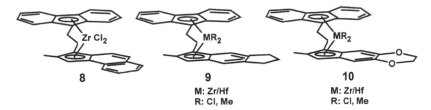

FIGURE 9.9 Ethylene-bridged (Ind-H-Flu) asymmetric catalysts (**9a**: M = Zr, R = Cl; **9b**: M = Hf, R = Me; **10a**: M = Zr, R = Cl; **10b**: M = Zr, R = Me).

opposite-standing substituents β^1 and β^3 in a forward position, minimizing the distance between the groups. Consequently, a backward arrangement characterizes the λ-conformation, with a maximum distance between β^1 and β^3.[96]

The δ-forward conformer **7a** is by far the most selective catalyst. The polymers produced with **7a**/MAO (MAO = methylaluminoxane) are crystalline materials with defined T_ms at all applied polymerization temperatures. Inversion of the backbone twist to the λ-backward conformer results in a nearly complete loss of stereoselectivity; **7b** produces atactic oils or waxes with *mmmm* values ranking between 26.5% and 36.0%.

For **7a** and **7b**, both diastereomers (forward and backward conformers) are asymmetric with comparatively small differences between their ligand arrangements. The main difference between them is that **7a** provides a tight chiral coordination cage, owing to the forward position of the β-substituents. To account for the stereoregularity of polypropylenes prepared with **7a**, one could assume that the stereorigidity of this template favors one particular geometry of the chain and the propylene monomer. It would then be easy to visualize that the opened cage in **7b** allows a less rigid arrangement with more than one possible olefin insertion transition state. The importance of obtaining optimal coordination gap geometry for the design of highly selective catalysts was recently pointed out by Brintzinger and Hortmann.[97]

Further research by Rieger and coworkers[98–100] using complexes **8–10** (Figure 9.9) showed that this concept can be used for control of the regular/nonregular segments within an isotactic polymer chain, and hence to induce phase separation phenomena leading to stereoregular polyolefins with elastomeric properties.

The highly active precatalyst *rac*-[1-(9-η^5-fluorenyl)-2-(5,6-cyclopenta-2-methyl-1-η^5-indenyl) ethane]zirconium dichloride (**9a**) was the first asymmetric complex used for the production of high molecular weight isotactic polypropylene with controllable amounts of isolated stereoerrors for achieving and adjusting elastic properties.[98] It was found that the 2-methyl group and the 5,6-substitution on the indenyl fragment are necessary requirements to obtain a high enough molecular weight and a sufficient amount of stereoerrors for the formation of elastic, isotactic polypropenes. Beside high molecular weight and stereocontrol of the lengths of the isotactic blocks, increased activities were achieved (32,020 kg PP/mol Zr·[C₃]·h). By using hafnium rather than zirconium as the active metal center (**9b**), even ultrahigh molecular weight polypropylenes (4.9×10^6 g/mol) can be obtained, which are X-ray-amorphous materials.[101]

9.3.3.1 Polymerization Mechanism

To get a closer insight into the polymerization mechanism responsible for the strong dependence of stereoselectivity on monomer concentration observed for this type of asymmetric catalyst, the pentad distributions of the polypropylenes prepared with **9a** were investigated using ^{13}C nuclear magnetic resonance (NMR) spectroscopy.[98] The *mmmm* pentad content was observed to decline with increasing monomer concentration; this relationship was attributed to the existence of two coordination sites in these "dual-side" complexes, which show different stereoselectivities for monomer coordination and insertion. Guerra et al. have supported this hypothesis in a theoretical study.[102]

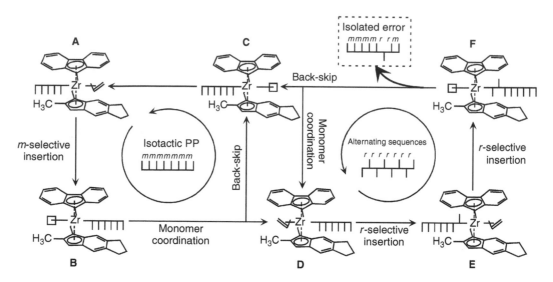

SCHEME 9.2 Proposed back-skip mechanism for the formation of isotactic polypropylenes with isolated stereoerrors obtained with C_1-symmetric catalyst **9a** ($\square$ = empty coordination site).

Beside the chain migratory insertion mechanism operative for metallocene catalysts, a chain stationary mechanism (known as chain "back-skip")[98,101,103] was proposed to be responsible for the statistical distribution of stereoerrors in isotactic polypropylene chains prepared using a C_1-symmetric polymerization catalyst of type **9a**. The two sites available for monomer coordination in the active center form of **9a** are depicted in Scheme 9.2, along with the proposed reaction pathway for explaining the formation of isotactic polypropylenes with variable degrees of stereoerrors. The incoming monomer can be coordinated between the sterically demanding Flu-Ind moieties (site A) or at the less hindered site of the catalyst (site D). Isotactic *mmmm* sequences are produced when the polymerization reaction is performed in the order A → B → C, that is, repeated migratory insertion of the monomer coordinated at site A (A → B) and consecutive back-skip of the growing polymer chain to the less crowded site (B → C). The difference between the activation energies for the back-skip and for the formation of the high-energy alkene-coordinated intermediate D is a decisive factor in determining the probability of the back-skipping process of the polymer chain.

At low propylene monomer concentrations, the back-skip of the polymer chain (B → C) is faster than monomer coordination. This leads to high isotacticities at low propylene concentrations and elevated temperatures. At higher propylene monomer concentrations, however, coordination of monomer at the less hindered site D is favored over back-skip of the polymer chain from site B to site C. Subsequent coordination of the monomer at site D followed by migratory insertion leads to the formation of a stereoerror (D → E). Insertion from E → F proceeds in a stereoselective way, similar to the process A → B, but instead leads to the formation of an *rr* triad owing to the previous nonselective insertion (D → E). At low propylene monomer concentrations, the back-skip of the polymer chain to the less encumbered site (F → C) is favored over monomer coordination (F → D). At site C, the catalyst follows the isotactic cycle A → B → C.

The formation of single isolated stereoerrors is supported by the fact that the *mrmr* pentad, characteristic for atactic sequences,[104–106] is absent in the polymer samples or only detectable in minor concentrations (below 1%) (Figure 9.10). Owing to the occurrence of consecutive selective and nonselective insertions, the same pentad is also absent in hemiisotactic polypropylene. However, polymers products prepared with catalyst **9a** do not fit the pentad distribution required for hemiisotactic polypropylene.[107,108] They can best be characterized as isotactic with variable amounts of stereoerrors that fuse into longer stereoerror sequences as isotacticity decreases.

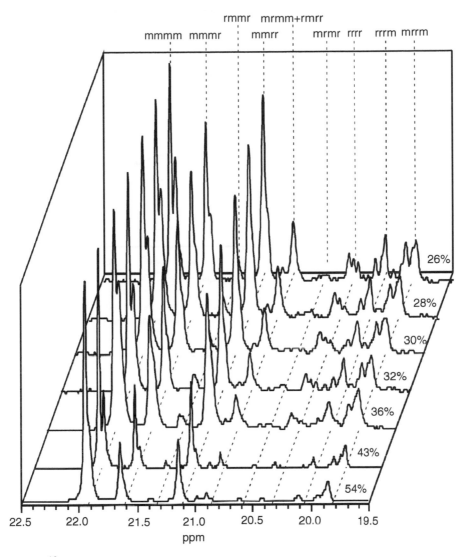

FIGURE 9.10 ^{13}C NMR spectra for polypropylenes with 26–52% *mmmm* prepared with **9a**; the *mrmr* pentad characteristic of atactic sequences is detectable only in minor concentrations.

Furthermore, the formation of single isolated stereoerrors can be proved by the distribution of pentads characteristic of isolated stereoerrors (*mmmr*, *mmrr* and *mrrm*)[109,110] in relation to *mmmm* content (Figure 9.11). When *mmmm* is above 40%, the *mmmr*, *mmrr*, and *mrrm* pentads all decrease continuously with increasing *mmmm* content. A pentad distribution of *mmmr/mmrr/mrrm* = 2:2:1 also indicates the formation of isolated stereoerrors.[109,111] Below 40% *mmmm*, the *mrrr* (not shown in Figure 9.11), *rrrr*, and *mmrr* signals increase overproportionally, owing to the fusion of isolated stereoerrors into longer stereoerror sequences. However, the content of the *rrrr* pentads is not high enough to characterize the polymers as block isotactic–syndiotactic, since it is present only in minor quantities relative to the *mmmm* sequences.

These results show that using catalyst **9a** for polypropylene synthesis results in isotactic polypropylenes not in the manner of stereoblock polymers, but as iPPs with statistically distributed stereoerrors.

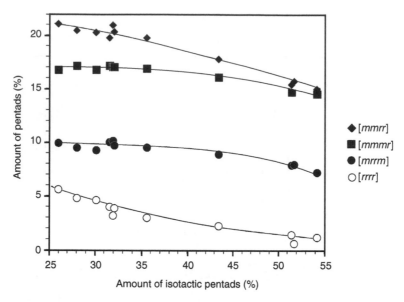

FIGURE 9.11 Variation of stereoerrors (*mmmr, mmrr, mrrm, rrrr*) in polypropylenes with reduced isotacticity (as isotactic pentads, % *mmmm*) as a fingerprint for the polymerization mechanism.

TABLE 9.1

Selected Propylene Polymerization Results Obtained with the Catalysts 10a/MAO and 10b/[Ph₃C]⁺[B(C₆F₅)₄]⁻

Entry	Precatalyst	Cocatalyst	Al/Zr	T_p[a] (°C)	$[C_3]$[b]	Catalyst Activity[c]	M_w[d] (g/mol)	*mmmm*[e]
1	**10a**	MAO	300	30	3.0	1.2	60,000	33.5
2	**10a**	MAO	1000	30	3.0	0.7	45,000	34.7
3	**10a**	MAO	300	30	5.1	1.5	115,000	35.4
4	**10b**	Borate[f]	—	30	1.2	5.1	135,000	55.6
5	**10b**	Borate[f]	—	30	3.0	2.9	150,000	42.5
6	**10b**	Borate[f]	—	30	5.1	3.1	160,000	36.8

[a] Polymerization temperature.

[b] Propylene monomer concentration in mol/L.

[c] In units of $(10^3$ kg PP/mol Zr·$[C_3]$·h).

[d] By gel permeation chromatography (GPC) at 145 °C in 1,2,4-trichlorobenzene, versus polypropylene standards.

[e] In %.

[f] Borate = $[Ph_3C]^+[B(C_6F_5)_4]^-$.[78,98–101]

9.3.3.2 Reversible Chain Transfer to Aluminum

In addition to differences in polymer microstructures caused by cocatalyst (activation with MAO versus borate activation), some specific particularities are noticed with regards to the influence of additional parameters on polymerization results using different cocatalysts. For the 5,6-ethoxy-substituted indenyl zirconium complexes **10a** and **10b** (Figure 9.9), where **10a** is activated with MAO and **10b** with trityl tetrakis(pentafluorophenyl)borate, $[Ph_3C]^+[B(C_6F_5)_4]^-$, it has been found that the *mmmm* pentad concentration of the polymer products varies over a broad range (33–56%) depending on temperature and monomer concentration but is independent of the Al/Zr ratio (Table 9.1).[99,100,103]

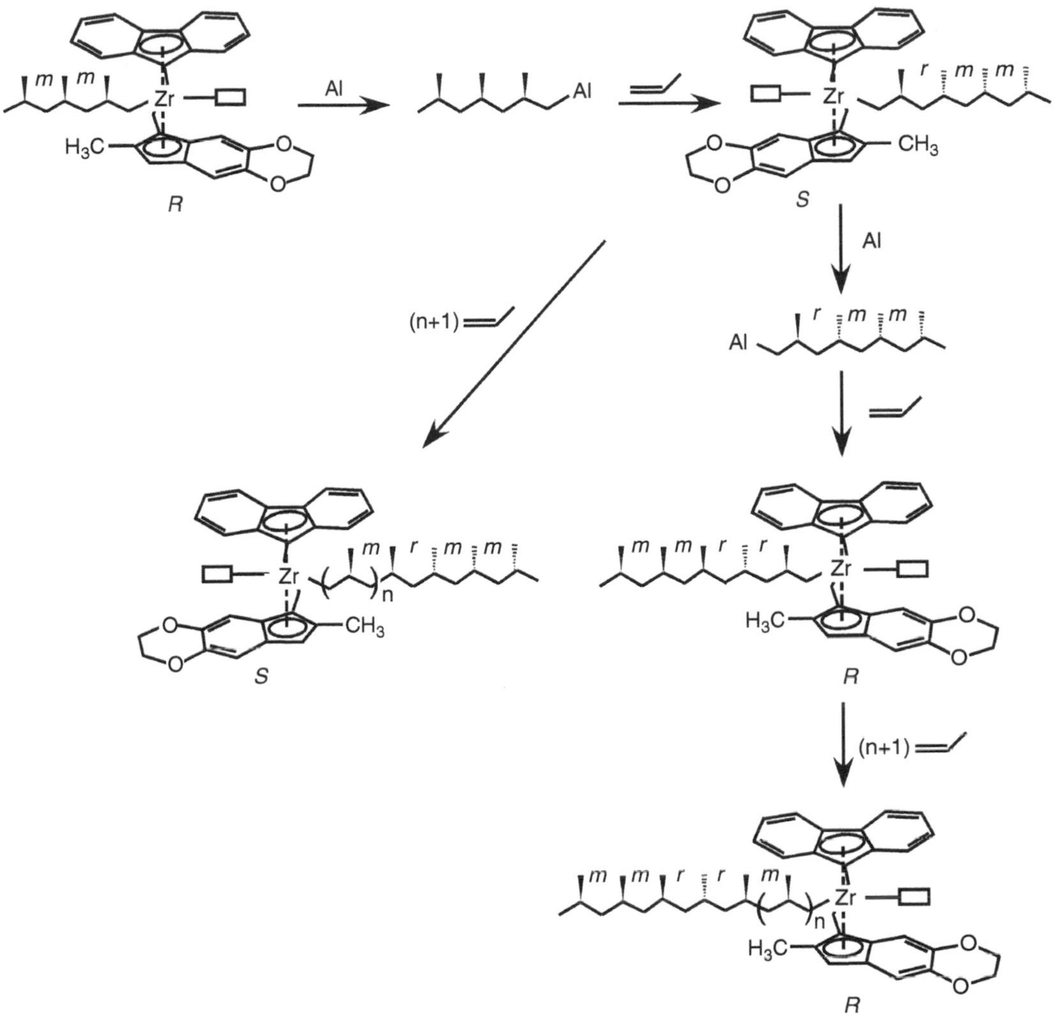

SCHEME 9.3 Proposed mechanism of reversible chain transfer to aluminum for propylene polymerization with catalyst **10a** (□ = empty coordination site).

For typical MAO-activated polymerizations, increased activities are observed at higher aluminum contents. In contrast, reduced activities at higher aluminum contents were found with **10a**. This indicates a direct interaction between MAO and the Zr^{IV} center. In situ activation with [Ph$_3$C]$^+$[B(C$_6$F$_5$)$_4$]$^-$ led to increased molecular weights and also to increased isotacticities. This proves that, when using MAO as the cocatalyst, the aluminum content influences the overall stereoselectivity of the polymerization reaction.

As a result of the particularities of the MAO activation method noticed so far, it has been concluded that, besides chain "back-skip," a different mechanism occurs when using this cocatalyst. One possible explanation might be a reversible chain transfer reaction between the cocatalyst and the active species.[7] As a result of the intrinsic chirality at the metal center, the catalytic system consists of two enantiomers[112] (**S** and **R**, Scheme 9.3). Under different polymerization conditions (i.e., different Al/Zr ratios), the coordination and insertion of the monomer can take place at the metal center of either of the two enantiomers. At higher Al/Zr ratios, a unidirectional transfer of polymer chains from Zr^{IV} (enantiomer **R**, for example) to aluminum can be suggested, because reduced molecular weights of the polymer products have been found. Relocation of the chain from aluminum to the other enantiomer of the C_1-symmetric catalyst species (enantiomer **S**, Scheme 9.3) and then back to **R** through a similar transfer process would lead to the formation of a single stereoerror *mrrm* and could—at enhanced frequencies—account for the observed reduction of isotacticity for the MAO-activated complexes.[113–116] If the insertion of the new monomer unit takes place at the Zr center of enantiomer **S**, stereoerrors of type *mrmm* are found along the polymer chain. The proposed reversible chain transfer to aluminum does not exclude the chain back-skip mechanism valid for the asymmetric catalysts, but could take place in addition to it at high Al/Zr ratios.

9.3.3.3   Influence of the Cocatalyst Nature on Polymer Microstructure

Since it has been shown that the cocatalyst nature has an influence on the distribution of stereoerrors along the polymer chain, this parameter (cocatalyst identity) was probed in propylene polymerization experiments using either MAO (Table 9.2, **A** samples) or $[Ph_3C]^+[B(C_6F_5)_4]^-$/TIBAL ("borate activation," TIBAL = triisobutylaluminum; Table 9.2, **B** samples) to activate catalyst **9a**.[117] Polymer samples prepared in the presence of MAO showed a strong decline in molecular weight when the *mmmm* content is increased, for example, $M_w \leq 7.5 \times 10^4$ g/mol at *mmmm* contents greater than 40%. When borate activation is used, the molecular weights of the polymers also decrease at higher *mmmm* contents, but remains above 1.2×10^5 g/mol. A comparison of the samples obtained with these two cocatalysts under similar polymerization conditions reveals lower values of the molecular weights for samples prepared in the presence of MAO (see also the results in Table 9.1 for **10a/10b**). Similar increases in isotacticity are noticed for samples obtained with MAO and borate cocatalysts under similar conditional variations, but samples prepared using borate activation reveal about 3–4% higher *mmmm* values than the analogous MAO-activated samples (e.g., **B29** versus **A26** and **B34**

TABLE 9.2

Selected Propylene Polymerization Results Obtained with Precatalyst 9a after MAO or $[Ph_3C]^+[B(C_6F_5)_4]^-$/TIBAL Activation

Sample Number	Precatalyst	Cocatalyst[a]	T_p[b] (°C)	C_3 pressure (bar)[c]	M_w[d] (g/mol)	M_w/M_n[d]	*mmmm*[e]
A26	9a	MAO	30	7	153,000	2.2	26
A28	9a	MAO	30	6	110,000	1.5	28
A30	9a	MAO	30	5	110,000	2.2	30
A32	9a	MAO	35	7	121,000	2.1	32
A36	9a	MAO	35	6	160,000	1.8	36
A43	9a	MAO	30	3	71,000	1.9	43
A51	9a	MAO	40	3	48,300	2.0	51
A54	9a	MAO	50	5	75,000	2.3	54
B29	9a	Borate	30	7	200,000	2.2	29
B34	9a	Borate	30	5	205,000	2.1	34
B39	9a	Borate	35	5	133,500	2.2	39
B46	9a	Borate	40	5	160,000	2.0	46
B51	9a	Borate	45	5	120,000	1.9	51

[a] Borate = $[Ph_3C]^+[B(C_6F_5)_4]^-$/TIBAL; Al/Zr ratio for MAO activation = 2000:1; Al/Zr ratio for borate activation (from TIBAL) = 200:1.
[b] Polymerization temperature.
[c] Propylene monomer pressure in bar.
[d] By gel permeation chromatography (GPC) at 145 °C in 1,2,4-trichlorobenzene, versus polypropylene standards.
[e] In %.

---

versus **A30**, Table 9.2; the numeric portion of each polymer sample name corresponds to its % *mmmm*).

In order to identify differences in the polymer microstructures, polypropylenes containing similar amounts of isotactic *mmmm* pentads but prepared with either MAO (**A26–A54**, Table 9.2) or borate (**B29–B51**, Table 9.2) cocatalyst have been subjected to further analysis. No comparable differences were found in the samples' ^{13}C NMR spectra or in their isotactic block lengths; however, WAXS and differential scanning calorimetry (DSC) experiments revealed specific particularities corresponding to each activation method.

9.3.3.3.1 Isotactic Block Lengths

The distribution of pentads allows an estimation of the percentage of blocks containing four or more monomers units in an isotactic sequence. Based on the findings of Collette et al.,[118] the average isotactic block length n_{iso} between two isolated stereoirregular insertions can be estimated by Equation 9.2

$$n_{iso} = 4 + \frac{2[mmmm]}{[mmmr]} \qquad (9.2)$$

where *mmmm* = the percentage amount of isotactic pentads and *mmmr* = the percentage of pentads containing one syndiotactic stereoerror. The amounts of *mmmm* and *mmmr* fractions estimated by ^{13}C NMR (Table 9.3)[114] are used to determine the isotactic block length of the polypropylenes synthesized with **9a**.

Within the accuracy of the method, the determined isotactic block lengths are equal for samples made using different cocatalysts but having comparable amounts of isotactic pentads (Figure 9.12). Thus, the average isotactic block length seems to be independent of the cocatalyst nature.

TABLE 9.3
^{13}C NMR Pentad Distribution of Polypropylenes Synthesized with 9a

Sample Number	*mmmm*[a]	*mmmr*[b]	n_{iso}[c]
A26	26.0	16.8	7.1
A28	28.0	17.2	7.3
A30	30.1	16.8	7.6
A32	32.1	17.0	7.8
A36	35.6	16.9	8.2
A43	43.4	16.1	9.4
A51	51.4	14.7	11.0
A54	54.2	14.5	11.5
B29	29.1	16.4	7.5
B34	34.0	17.1	8.0
B39	39.0	17.2	8.5
B46	45.7	16.2	9.6
B51	50.9	15.3	10.7

[a,b] In %.
[c] Number of monomer units in isotactic sequence.

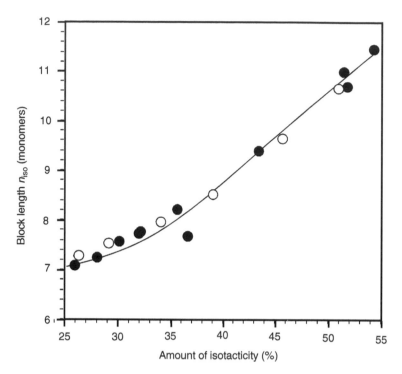

FIGURE 9.12 Isotactic block lengths (in monomer units) for polymers prepared with precatalyst **9a**, determined from the pentad distribution and plotted versus % *mmmm*, show no difference between MAO-activated samples (filled circles) and borate-activated samples (open circles).

9.3.3.3.2 Crystallization Properties

For semicrystalline polymers, melting temperature (T_M) is correlated to the thickness of crystalline lamellae. If all other variables are held constant, thinner lamellae will melt at lower temperatures than thicker ones.[119] Thus, the melting temperature T_M, defined as the maximum temperature within a melting regime M, can be used to approximate the lamellar thickness. Within the first DSC run curves[120] of polypropylenes prepared by **9a**, typically several melting transitions, M_1, M_2, and M_3, with maxima T_{M_1}, T_{M_2}, and T_{M_3}, can be seen (Figure 9.13).

It is likely that the various melting transitions indicate the formation of more than one distinct crystallite size, owing to a nonequal distribution of isotactic block lengths.[121] This distribution may limit the number of lamellae of a given thickness L that can be formed. If isotactic sequences are assumed to generate the lamellae, the melting temperature T_M is controlled by the average segment length n_{iso}.[119,122] Thus, the detected melting temperature correlated to an isotactic block length n_{iso} is given by Equation 9.3[118,119]

$$l = \frac{2\sigma_e T_M^0 M}{\rho_c \Delta H_f^0 (T_M^0 - T_M)} + \frac{k T_M}{b_0 \sigma_s} \tag{9.3}$$

where l = average lamellar thickness in nm; σ_e = specific fold surface free energy = 100 mJ/m²; T_M^0 = equilibrium melting temperature = 460.7 K; M = molecular weight of repeating unit = 42 g/mol; ρ_c = isotactic crystal density = 0.94 g/cm³; ΔH_f^0 = molar heat of fusion = 8.79 kJ/mol; T_M = observed melting temperature (maximum temperature within the melting regime); k = Boltzmann's constant = 1.38×10^{-23} J/K; b_0 = single layer thickness = helix diameter = 0.65 nm; and σ_s = specific side surface free energy. The melting temperature, molar heat of fusion, crystal

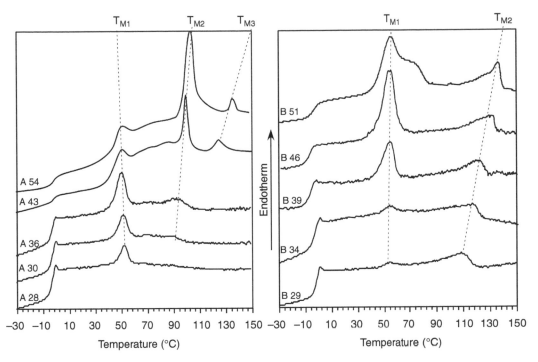

FIGURE 9.13 DSC curves of elastomeric polypropylenes prepared with **9a** using different cocatalysts (A = MAO, left; B = borate, right) obtained during the first DSC run reveal several melting transitions indicating multiple block lengths (n_{iso}s).

density, and σ_e were suggested as given in the literature.[118] The value σ_s was arbitrarily taken to be the same as for polyethylene.[122] The resulting block lengths for the different low-isotactic polypropylenes are plotted in Figure 9.14.

All samples show their first melting transition T_{M_1} at about 50 °C (Figure 9.13). This temperature can be correlated to the melting of the thinnest lamellae L_1 consisting of chain sequences with a block length n_{iso} of about 22 consecutive isotactic monomers (Figure 9.14a, squares).[117,118] This indicates that within the temperature range and monomer concentration used for polymerization (Table 9.2), the block length of the thinnest lamellae seems to be independent of the polymerization conditions. In addition, the influence of the two cocatalysts used for activation (MAO and borate) can be neglected.[117]

The integral area under first melting peak, $\Delta H_{T_{M_1}}$, can be attributed to amount of energy necessary to melt the crystalline fraction of thin lamellae L_1 (n_{iso} = 22).[123] Thus, the ratio $\Delta H_{T_{M_1}}/\Delta H_{tot}$, where ΔH_{tot} is equal to the total heat of fusion obtained after adding all the melt transitions together, characterizes the percentage of short blocks present in samples (as a fraction of all blocks) produced under different polymerization conditions. Plotting this ratio versus the total amount of isotactic pentads (% *mmmm*, Figure 9.14b) reveals differences due to the cocatalyst used. When samples are synthesized in the presence of MAO, the relative amounts of short blocks decreases when % *mmmm* increases. Taking the polymerization conditions (Table 9.2, series A) into account, this gives evidence that at higher polymerization temperatures, the formation of thicker lamellae is favored compared to the formation of thinner ones. In contrast, increased percentages of short blocks can be observed at increasing % *mmmm* when borate activation is used. This refers to an influence of the cocatalyst on the obtained block length distribution: in the absence of MAO, the formation of shorter isotactic blocks seem to be favored.

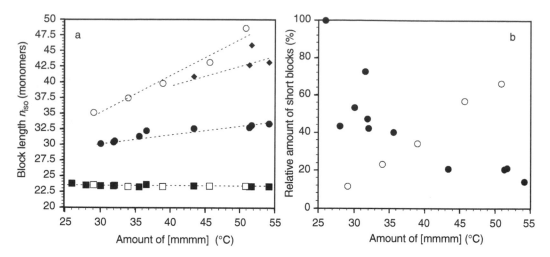

FIGURE 9.14   (a) Isotactic blocks lengths determined from the melting temperatures of polypropylenes prepared with catalyst **9a** (T_{M_1} = squares, T_{M_2} = circles, T_{M_3} = diamonds) reveal differences when different cocatalysts are used (filled = MAO, unfilled = borate). (b) The relative amount of short isotactic blocks as a function of isotacticity (% *mmmm*) in polypropylenes synthesized with **9a** also varies by cocatalyst (filled = MAO, unfilled = borate).

In addition, melting peaks appearing at higher temperatures (Figure 9.13, T_{M_2} and T_{M_3}) indicated that longer isotactic blocks were also formed, whose length n_{iso} increases linearly with increasing *mmmm* content (Figure 9.14a, circles and diamonds). The chain "back-skip" mechanism is proposed to be responsible for this increase in isotactic block length, owing to the fusion of shorter blocks to form longer ones.[98] The eye-catching fact in the DSC curves shown in Figure 9.13 is that for samples prepared using borate activation, only one melting transition M₂ appears at higher temperatures (T_{M_2} > 85 °C) referring to the formation of isotactic blocks with a minimum block length n_{iso} = 36 monomers. In contrast, two melting regimes M₂ and M₃ with melting transitions T_{M_2} (ranging from 90 to 95 °C) and T_{M_3} (>103 °C) can be observed for MAO-activated samples containing more than 40% *mmmm* pentads. This indicates the formation of two crystalline fractions with different block lengths. The lamellae melting at T_{M_3} consist of isotactic sequences of least 36 monomers and fit to the lamellar fraction obtained for the samples prepared using borate activation. The lamellar fraction of M₂ contains blocks where n_{iso} ranges between 28 and 32 monomers. Comparing samples containing similar amounts of *mmmm* pentads, but prepared by different cocatalysts (Figure 9.14a), longer block lengths for the borate-activated samples have been determined. The differences in isotactic block length for samples with similar isotacticity, as well as the formation of several populations of block length, cannot be explained by the chain "back-skip" mechanism, but provides strong evidence that a different mechanism occurs when MAO is used as a cocatalyst as compared to when borate activation is used.

As was shown in Section 9.2.2, the crystalline form of iPP can be correlated to isotactic block length. Based on this, the samples prepared using both MAO and borate activation (Table 9.2) should mainly crystallize in the γ-form. To study the presence of the different crystalline forms in these MAO- and borate-activated low-crystalline polypropylene samples, WAXS analysis was performed as shown in Figure 9.15. Both series **A** and **B** display WAXS peaks at Θ = 9.3°, characteristic for the presence of the α-modification, and at Θ = 10°, characteristic for the presence of the γ-modification, pointing out that lamellae will crystallize in a mixed α/γ-form.

By fitting individual peaks using a Lorentzian function, as shown in the small inset diagram in Figure 9.15a, the fractions of the corresponding α- and γ-modifications can be estimated. Since it

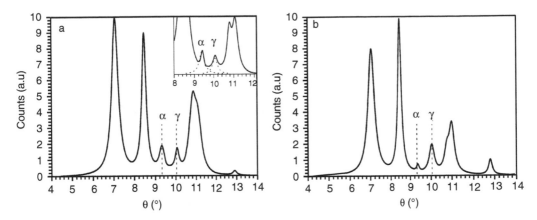

FIGURE 9.15 WAXS diffraction patterns of (a) **9a**/MAO-derived polypropylenes and (b) **9a**/borate-derived polypropylenes having similar *mmmm* contents show different amounts of the α-and γ-forms.

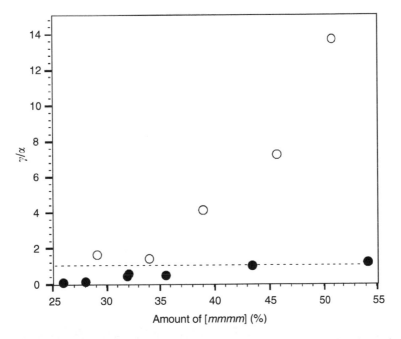

FIGURE 9.16 Ratios of the α- and γ-forms (γ/α) in polymers prepared with **9a**, determined by WAXS analysis, show that the γ-modification is preferred in borate-activated samples (open circles); MAO-activated samples (filled circles) show both modifications in comparable amounts (dotted line indicates γ/α = 1).

has been shown that the formation of γ-phase lamellae requires the previous formation of α-form crystals,[46] the γ/α-modification ratio was estimated and plotted versus the *mmmm* content of the samples (Figure 9.16). For both series **A** and **B**, an increasing amount of the γ-phase is seen as a constant increase in the curves at increasing *mmmm* content. When MAO is used as the cocatalyst, the α-modification is the preferred crystalline unit cell for samples with *mmmm* contents below 40%. For these materials, nearly equal portions of both modifications were found; thus, the ratio α/γ ≈ 1 (Figure 9.16, dotted line). In polymers synthesized using borate activation, lamellae crystallize preferentially in the γ-modification, resulting in the ratio γ/α > 1. According to the DSC experiments, where higher amounts of thinner lamellae L_1 were detected in samples prepared using borate activation, these results corroborate the findings for higher-isotactic polypropylenes that the prevailing crystalline form can be correlated to the isotactic block length. Furthermore, these results confirm the assumption that different types of catalyst activation influence the chain microstructure.

9.4 SOLID STATE PROPERTIES OF ELASTIC POLYPROPYLENES WITH VARIABLE TACTICITY

Low-isotactic polypropylenes belong to the group of semicrystalline polymers. Thus, their crystalline morphologies depend on the amount and size of isotactic sequences and the molecular weight of the polymer. Although a broad variety of elastomeric polypropylenes have been synthesized, little is known about the interplay of chain microstructure, morphology, and mechanical properties. It is therefore of particular interest to analyze and understand how, and to what extent, low-isotactic polypropylenes crystallize. Typically, the morphology of semicrystalline polymers is studied using polarized optical microscopy (POM), but the size of the crystalline aggregates formed in low-isotactic polypropylenes is below 500 nm, which is below the resolution limit of an optical microscope. Electron microscopy (EM) studies revealed that these polypropylenes crystallize as short individual lamellae embedded in an amorphous matrix.[63,118] The disadvantage of this technique is that only conducting samples can be imaged. Thus, the polymer samples have to be metal coated. In 1986, Binnig et al. developed a complimentary technique, called scanning force microscopy (SFM),[124] which requires no special sample treatment and enables imaging of the surface topography at scales ranging from micrometers to nanometers. Here, a presumably atomically sharp silicon or silicon nitride probe attached to the bottom side of a cantilever with an approximate length of 100 μm is moved across a sample surface at a constant distance to the surface. The probe is moved by piezo crystals in the x-, y-, and z-directions, and its bending is detected by a laser beam focused on the top side of the cantilever. Surface topographic information is obtained from the feedback signal, which monitors the deflection of the cantilever as a result of the changing force due to height variations.[125] The resulting height plot is depicted in a color-coded map (topography image), where elevated areas appear bright (e.g., Figure 9.17a, *vide infra*). The development of intermittent contact techniques, such as tapping mode,[126,127] where the tip is not continuously in contact with the samples, enables the imaging of polymeric surfaces with the nanometer-level resolution needed for the investigation of small crystalline structures. In the tapping mode, the height data are complemented with simultaneously measured phase shift data, and these phase data are very useful to map domains of varying material properties at or near the surface.[128,129] Typically, imaging conditions are adjusted

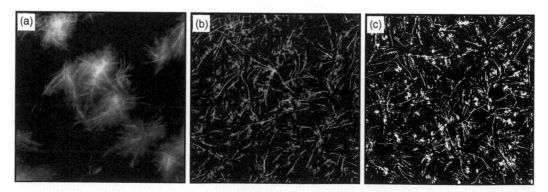

FIGURE 9.17   (a) Topography and (b) phase shift TM-SFM images reveal the phase-separated morphology of low-isotactic polypropylene dip coated thin films prepared with MAO cocatalyst (sample **A28**). Converting the phase shift image into a binary picture (c) enables the determination of the amount of hard phase.

in such a way[129] that harder domains appear bright in the phase shift image (e.g., Figure 9.17b, *vide infra*).

Tapping Mode SFM (TM-SFM) has been successfully used to image the crystalline morphology of highly isotactic iPP crystallized in the α- or γ-modifications.[130–135] Also, the phase-separated morphology of low-isotactic polypropylenes can be evidenced. Schönherr et al.[61] found that elastic stereoblock isotactic-atactic polypropylenes with *mmmm* contents of about 30%, obtained by fractionation, generate morphologies reminiscent of classical semicrystalline polymers such as lamellae, crosshatching, hedrites, and spherulites. Kravchenko et al.[136] showed that elastomeric polypropylenes with *mmmm* contents between 25% and 30%, derived from unbriged bis(2-arylindenyl) metallocenes, crystallize as short individual lamellae embedded in an amorphous matrix. This morphology was attributed to the extremely low crystallinity of the polymer. These results support earlier findings obtained from EM studies.[63,118]

Nevertheless, the morphology of low-isotactic polypropylenes with statistically distributed stereoerrors is sparsely explored. To get a better understanding of the morphology of low-isotactic polypropylenes, samples prepared with C_1-symmetric catalyst **9a** were investigated using TM-SFM. Thin films dip coated on silicon from a hot polypropylene solution in toluene have been used to study the polymer morphology. Depending on the amount of isotactic sequences and on the cocatalyst used (MAO or borate), different morphologies can be observed. Since dip coated samples reveal an increased crystallinity due to the surface–polymer interaction, bulk morphology has also been studied using microtome cuts.

9.4.1 Polypropylenes Prepared by a C_1-symmetric Catalyst Using Methylaluminoxane as Cocatalyst

9.4.1.1 Basic Structure

In dip coated thin films of polypropylenes prepared with catalyst **9a** in the presence of MAO (**A26**–**A54**, Table 9.2), rodlike features were found to be the basic structure (Figure 9.17). The correlation of topography (Figure 9.17a) and phase shift images (Figure 9.17b) points out that the rods turning up as elevated (brighter) features in the topography image correspond to brighter, and thus harder, domains in the phase image. Considering the imaging conditions, the contrast in the tapping mode phase image is assumed to be based on differences in stiffness of the crystals and the amorphous matrix. Comparing the tensile modulus of the amorphous fraction, $E_{am} = 1.73$ MPa,[41] to the one of the crystalline fraction, $E_{lam} = 1.0–1.7$ GPa,[137] a difference of three orders of magnitude can be determined. This difference in elastic modulus is responsible for the pronounced contrast in the TM-SFM phase image. Thus, most likely the rodlike structures can be attributed to ordered chain segments or crystalline lamellae, which are likely caused by the finite isotactic blocks and the low-crystalline segment fractions. This refers to sufficiently long and perfectly stereoregular sequences that are able to crystallize.

To determine the amount of hard phase, the phase shift image was converted into a binary picture (Figure 9.17c), where the brighter domains appear white and the soft matrix black. For the shown sample (**A28**, 28% *mmmm*) a relative amount of $14 \pm 2\%$ of white phase was found.[117] Thus, the surface is covered by a similar amount of rodlike lamellae. This value exceeds the detected DSC crystallinity,[138] but the deviation can be explained by experimental considerations: (1) in dip coated films used for SFM experiments, crystallization is more facile as compared to bulk samples; (2) it is not a priori clear whether, or to what extent, the fibrils protrude from the surface; and (3) the tip radius of the cantilever used limits the resolution in a manner such that hard domains appear larger than they really are. Bearing these discrepancies in mind, the amount of hard domains (Figure 9.18, circles) determined for a series of polypropylenes with increasing *mmmm* content (**A** and **B** series, Table 9.2) shows a good correlation with the crystallinity determined from the heat of fusion ΔH_{tot} (Figure 9.18,

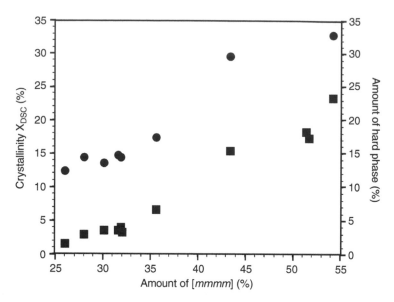

FIGURE 9.18 Per cent crystallinity (squares) and the amount of hard phase determined from dip coated film TM-SFM phase shift images (circles) of polypropylenes prepared with **9a**/MAO show the same trend but different absolute values.

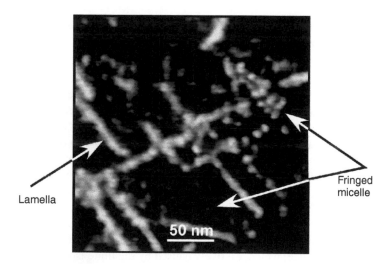

FIGURE 9.19 A TM-SFM phase shift image of a dip coated film of low-isotactic polypropylene having 28% *mmmm* prepared with **9a**/MAO (sample **A28**) shows that either homogeneous lamellae or small crystalline blocks (fringed micelles) are present.

squares). Thus, the bright domains in phase shift images can be attributed to the crystalline phase of the polymer.

An image taken at higher resolution (Figure 9.19) reveals lamellae with an average thickness l of 7 ± 3 nm. This thickness is in good agreement with values reported for crystalline lamellae observed in low-isotactic polypropylenes.[61,63,118] Assuming that parallel-aligned helices generate the crystalline structures, their thickness can be used to estimate the length of isotactic sequences present in the harder (crystalline) domains. Since crystalline isotactic polypropylene has a 3_1 helix conformation and the c-axis of the repeating unit is 0.665 nm, the lamellar l can be attributed to

32 ± 3 monomer units in isotactic sequence. This block length is in good agreement with the isotactic block length $n_{iso} > 28$ monomers obtained from the melting temperatures detected in DSC curves for the thicker lamellar fraction of $M_{2,3}$ (it should be noted that whenever such small structures are imaged, the tip radius of the cantilever used must be taken into account. On the basis of imaging a calibration grid, the radii of the tips used in these studies were 5–15 nm. A quantitative deconvolution of the SFM images was not attempted, however, owing to the uncertainty of the indentation depth and the experiment geometry).

A closer look at the rodlike structures (Figure 9.19) enables discrimination between two different types of rods: (1) longer, homogeneous rods with a lateral extension of up to 500 nm, which might be attributed to crystalline lamellae; and (2) shorter rods with a length of about 150 nm, revealing a bead-like structure consisting of individual blocks with a lateral extension of 5–10 nm separated by amorphous domains. Owing to the size of the individual blocks and their higher stiffness, these blocks are most probably fringed micelles.[139] The appearance of such beads of crystalline blocks can be explained using a crystallization model proposed by Strobl and coworkers.[140,141] Crystallization from the melt produces, in a first step, imperfect "native" crystals where small lamellar blocks are aligned in a parallel fashion. A structural relaxation process stabilizes these imperfect lamellae, subsequently leading to more perfect lamellae. Owing to the low amount of regular sequences in low-isotactic polypropylenes, only some of the lamellae can undergo this second step. More often, the native lamellar state seems to remain stable.

9.4.1.2 Crystallization Properties

Hot stage TM-SFM can be used to study the crystallization behavior of polymers in situ.[142,143] Crystallization studies on a hot stage TM-SFM (Figures 9.20 and 9.21) confirm the mixed α/γ-modification of polypropylenes prepared by **9a** using MAO activation proposed from the WAXS studies.[144] When a polypropylene with an *mmmm* content of 36% (**A36**, Table 9.2) crystallizes from the melt at 80 °C, stable lamellar structures with a lateral extension of up to 500 nm are formed (Figure 9.20a). Based on the DSC results, these fibrils can be assigned to lamellae composed of longer isotactic sequences ($n_{iso} > 28$). Upon cooling further to room temperature (Figure 9.20b), the contrast in the phase shift image is improved, but the size of the lamellar crystals is still unchanged. Thus, no further crystallization occurs.

This low-isotactic polypropylene sample shows a low crystallization rate at room temperature, which allows one to follow the crystallization in situ (Figure 9.21). The initially formed lamellar

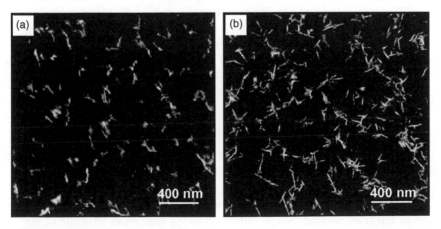

FIGURE 9.20 Temperature-dependent TM-SFM phase shift images of a dip coated film of polypropylene having 36% *mmmm* prepared with **9a**/MAO (sample **A36**) cooled down from the melt to (a) = 80 °C and (b) = 25 °C.

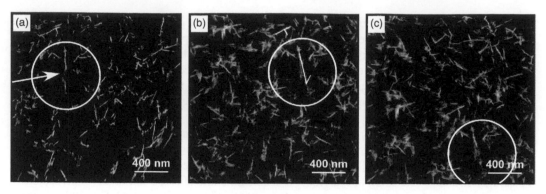

FIGURE 9.21   The time-dependent development of crystalline domains in a dip coated film of polypropylene having 36% *mmmm* prepared with **9a**/MAO (sample **A36**) can be tracked by obtaining TM-SFM phase shift images at room temperature after (a) 0 min, (b) 120 min, and (c) 280 min.

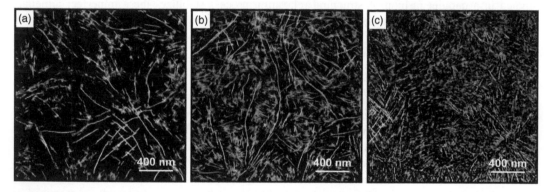

FIGURE 9.22   TM-SFM phase shift images of dip coated polypropylene films derived from **9a**/MAO having different amounts of isotacticity reveal the underlying lamellar structure: (a) sample **A28**, 28% *mmmm*; (b) sample **A36**, 36% *mmmm*; (c) sample **A54**, 54% *mmmm*.

crystals (Figure 9.21a, circled rod) develop and act as stable nuclei for lamellae growing edge-on with two preferential angles (Figure 9.21b). An angle of about $80 \pm 5°$ indicates crystallization in the α-form,[46] whereas the additionally found angle of $40 \pm 3°$ refers to secondary lamellae crystallization in the γ-form.[46] These secondary lamellae are thinner and shorter that the primary lamellae. Thus, they are most probably formed by the shorter isotactic sequences with $n_{iso} = 22$ monomers determined from DSC experiments. At increasing times of crystallization (Figure 9.21c), the amount of slow-growing lamellae increases until nearly 18% of the surface is covered with crystalline lamellae.

9.4.1.3   Morphology at Various *mmmm* Contents

The lamellar features described above represent the basic structure of polypropylenes produced with catalyst **9a** having *mmmm* contents of between 25% and 54%, as shown in Figure 9.22. However, depending on the amount of isotactic sequences, the arrangement of the lamellar phase changes. When the isotacticity is below 30% (**A28**, 28% *mmmm*, Figure 9.22a), individual fibrils with a lateral extension up to 1 μm are formed. Shorter fibrils growing edge-on at an angle of 80° to the main fibrils give strong evidence that these are lamellae crystallized in the monoclinic α-modification.[46] In addition, diffuse gray phase domains appearing close to the lamellae can most probably be attributed to a crystalline phase consisting of less ordered γ-modification crystals. With increasing amount

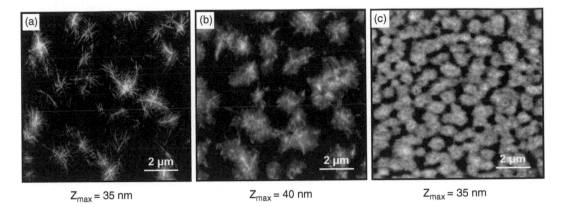

FIGURE 9.23 TM-SFM topography images of dip coated polypropylene films derived from **9a**/MAO having different amounts of isotacticity show the transition from a structure consisting of individual lamellae to circular aggregated lamellae: (a) sample **A28**, 28% *mmmm*; (b) sample **A36**, 36% *mmmm*; (c) sample **A54**, 54% *mmmm*.

of *mmmm* (**A36**, 36% *mmmm*, Figure 9.22b), rodlike lamellae growing from different aggregates meet and join at various angles. With increasing amount of isotactic sequences, the crystalline structures (fibrils as well as the diffuse phase) accumulate until a densely packed structure has been formed. When the isotacticity exceeds 50% (**A54**, 54% *mmmm*, Figure 9.22c), in addition to the edge-on lamellae, a crosshatching pattern[46] can be found, confirming the presence of the monoclinic α-modification.

The increase in lamellar density visible in the phase shift images is accompanied by a change in topography (Figure 9.23). The samples with *mmmm* < 30% reveal individual lamellae joining at various angles. Their intersections are the center of star-like aggregates appearing elevated (brighter) in topography images (**A28**, 28% *mmmm*, Figure 9.23a). When the isotactic content exceeds 30%, disk-like structures with a diameter of about 1 μm, which seem to consist of more densely packed radially arranged lamellae, can be found in the topography image (**A36**, 36% *mmmm*, Figure 9.23b). When the *mmmm* content exceeds 40%, the circular features appear more compact and disks with an average diameter of about 1 μm are formed (not shown in Figure 9.23). At *mmmm* contents above 50%, circles with densely packed radially arranged lamellae are generated, exhibiting diameters of about 2 μm (**A54**, 54% *mmmm*, Figure 9.23c). Similar but larger circular structures were observed on highly isotactic dip coated polypropylene films. The structures are assigned to two-dimensional spherulites.[145] Thus, most likely the circular features observed in the topography images shown in Figure 9.23 can be suggested to be a spherulitic pre-state.

Although dip coated films reflect the morphology of semicrystalline polymers, the bulk structure of melt-pressed samples can differ from those of dip coated films owing to a reduced degree of crystallization in thicker specimens. This necessitates the additional investigation of bulk samples. To elucidate the bulk morphology of the polypropylenes, phase shift images have been performed on the surface of melt-pressed samples after microtoming. Such samples are known to more realistically mirror the true phase-separated morphology.

Bulk samples (Figure 9.24) reveal the different crystalline morphologies at increasing *mmmm* content more clearly. To begin with, when the *mmmm* content is about 25%, small crystalline blocks are formed (not shown in Figure 9.24), which are arranged parallel to each other forming bead-like aggregates. With an increasing amount of isotactic sequences in the polymer chain, the structure partially passes into a lamellar morphology (**A28**, 28% *mmmm*, Figure 9.24a). When the isotacticity exceeds 30%, the lamellae aggregate and join at various angles forming a star-like morphology (**A36**, 36% *mmmm*, Figure 9.24b). The branching angle between two lamellae (line marks, Figure 9.24b) is about 40 ± 3°. Therefore, these structures can be assigned to cocrystallized α/γ-form lamellae.

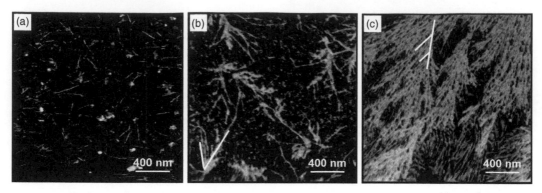

FIGURE 9.24   TM-SFM phase shift images of microtomed, melt-pressed polypropylene films derived from **9a**/MAO having different amounts of isotacticity show the change in morphology from individual lamellae (a = **A28**, 28% *mmmm*) to branched lamellar aggregates (b = **A36**, 36% *mmmm*) forming arborescent features at higher *mmmm* content (c = **A54**, 54% *mmmm*).

Here, initially, longer isotactic chains crystallize in α-form lamellae. These lamellae act as nuclei for the epitaxial growth of γ-form crystals.[46] In addition, in between the lamellae, a fine, bright (thus hard), granular phase appears. Most probably, this corresponds to fringed micelles, which, according to WAXS and DSC measurements, consist of short chain segments crystallized in the γ-form. Joint lamellae in combination with the fringed micelles provide a crystalline network. Such a network is required for low-crystalline polymers to behave like elastomers.[31] When the *mmmm* amount exceeds 40%, the samples form more extended lamellar structures appearing in an arborescent shape (**A54**, 54% *mmmm*, Figure 9.24c). Marking the direction of the primary lamellae and some of the secondary lamellae (line marks, Figure 9.24c) clearly shows an angle of 40 ± 3° between these features. This confirms the formation of branched α/γ-form lamellae.[46,58,63]

9.4.2 Polypropylenes Prepared Using a Perfluorophenyl Borate as Cocatalyst

The differences in chain microstructure observed for the samples prepared using different catalyst activation (**A** and **B** series, Table 9.2) may lead to a change in the polymer morphology. The differences are obvious when comparing TM-SFM images of dip coated polypropylene films of MAO-activated samples (Figure 9.22) to film samples produced using borate activation (**B29–B51**, Table 9.2). The phase shift images of these samples show a rodlike basic structure similar to the MAO-activated samples, but specimens with comparable *mmmm* contents show characteristic differences (Figure 9.25). The eye-catching fact is that when the *mmmm* content is below 50%, lamellae with an average length of 500 nm are formed, independent of the isotaciticity (Figure 9.25a and b). None of the lamellae show secondary crystallization. Even when the amount of fibrils increases (Figure 9.25c), they do not join.

The topography images of samples with *mmmm* contents below 35% (**B29**, 29% *mmmm*, Figure 9.26a) clearly show the formation of lamellae, which do not join to form circular aggregates. When the *mmmm* content exceeds 39% (**B39**, 39% *mmmm*, Figure 9.26b), disk-like structures appear in the topography images, although no lamellar aggregation had been observed in the phase shift images (**B39**, 39% *mmmm*, Figure 9.25b). The diameter of the circular features is always in the range of 500 nm. Unlike for the MAO-activated samples (Figure 9.23), the diameter of the disk-like structures is independent of the isotacticity, but the number of these aggregates increases with increasing isotacticity (**B51**, 51% *mmmm*, Figure 9.26c). Even though the phase shift images show no preferential ordering of the fibrils, the topography images show an in-line alignment of circular aggregates.

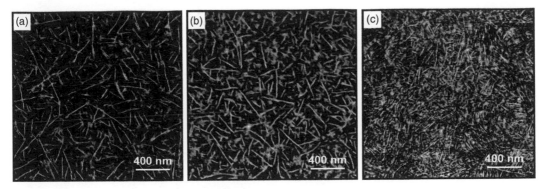

FIGURE 9.25 TM-SFM phase shift images of dip coated polypropylene films derived from **9a**/borate having different amounts of isotacticity do not show secondary crystallization: (a) sample **B29**, 29% *mmmm*; (b) sample **B39**, 39% *mmmm*; (c) sample **B51**, 51% *mmmm*.

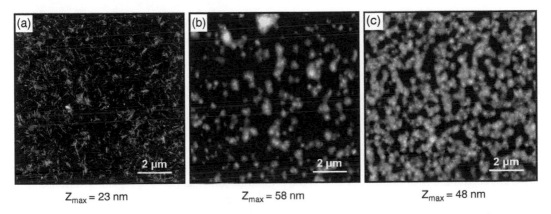

$Z_{max} = 23$ nm $Z_{max} = 58$ nm $Z_{max} = 48$ nm

FIGURE 9.26 TM-SFM topography images of dip coated polypropylene films derived from **9a**/borate having different amounts of isotacticity show the transition from a structure consisting of individual lamellae to small circular aggregates: (a) sample **B29**, 29% *mmmm*; (b) sample **B39**, 39% *mmmm*; (c) sample **B51**, 51% *mmmm*.

The differences in morphology between polypropylenes derived from catalyst **9a** with different activators become more palpable in the volume structures displayed in phase shift images taken from microtomed bulk samples of borate-activated polymer series **B** (Figure 9.27), as compared to samples from MAO-activated series **A** (Figure 9.24). Most striking are the isolated linear lamellae with short edge-on lamellae, which turn up in similar amounts in all samples independent of isotacticity. Since the angle between longer primary and shorter secondary lamellae is about $80 \pm 5°$, these structures are most probably lamellae crystallized in the α-modification. The thickness of the lamellae is about 15 nm. Thus, these are lamellae formed owing to the crystallization of isotactic sequences with lengths of about 60–70 monomers. Within the limits of resolution, this value corresponds to isotactic block length n_{iso} estimated by DSC experiments for the crystalline fraction of M_2, which melts at higher temperatures T_{M_2}.

In between the lamellae a second structure can be found, which varies with the amount of *mmmm* content. Samples with *mmmm* contents below 35% show bright granular features with dimensions of 5–10 nm^2 in between the lamellae (**B29**, 29% *mmmm*, Figure 9.27a) giving evidence for the presence of larger amounts of fringed micelles. The granular structure extends to form thin fibrils when the *mmmm* content increases (**B39**, 39% *mmmm*, Figure 9.27b). The amount of fibrils increases until a crystalline network has been formed (**B51**, 51% *mmmm*, Figure 9.27c). Since the diameter of these lamellae is a third of the width of the α-modification lamellae, it is most likely that short isotactic

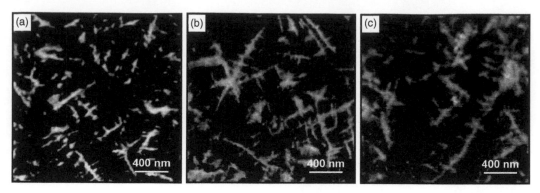

FIGURE 9.27   TM-SFM phase shift images of microtomed, melt pressed polypropylene films derived from **10a**/borate having different amounts of isotacticity show only individual lamellae crystallized in the α-form: (a) sample **B29**, 29% *mmmm*; (b) sample **B39**, 39% *mmmm*; (c) sample **B51**, 51% *mmmm*.

sequences (n_{iso} = 22) generate this substructure. Based on the WAXS and DSC experiments,[117] it can be suggested that this intermediate structure comprises lamellae crystallized in the γ-modification.

9.4.3 COMPARING THE PROPERTIES OF POLYPROPYLENES PREPARED USING DIFFERENT COCATALYSTS

9.4.3.1 Morphology

The SFM investigations[117] clearly show that the different chain microstructures not only form different crystalline modifications, but that the crystalline lamellae also aggregate in various super-structures. When samples are prepared in the presence of MAO, crystalline lamellae and an intermediate granular substructure are generated. The crystalline lamellae are not homogeneously distributed within the bulk samples. When the *mmmm* content increases to above 50%, the discontinuous lamellar structure seems to be converted into a continuous lamellar structure. In contrast, the morphology of in situ borate-activated samples reflects the bimodal distribution of crystalline lamellac proposed from DSC experiments. Thicker, isolated α-phase lamellae and fringed micelles or thin lamellae, most probably formed by γ-phase crystals, coexist. This suggests that the more regular chain microstructures obtained with borate activation may result in homogeneously distributed crystalline crosslinks. Chains with a less regular distribution of isotactic chain segments obtained using MAO activation are preferably arranged in larger, not homogeneously distributed crystalline domains. These differences in morphology should cause variance in the elastic properties of the samples.

9.4.3.2 Mechanical Properties

The different morphologies generated in the bulk polymer samples prepared using catalyst **10a** in combination with either MAO or borate activation (**A** and **B** series, Table 9.2) can be implicated in variations in mechanical behavior. To study the influence of the morphology on macroscopic mechanical behavior, melt-pressed samples were subjected to uniaxial stretching until they failed (**A** series, Figure 9.28; **B** series, Figure 9.29). For *mmmm* contents below 40%, stress–strain curves are observed that are characteristic of elastomeric materials.

As long as the elongation ratio is below $\lambda = 1.5$ ($\lambda = l/l_0$, where l = the actual sample length and l_0 = the initial sample length) a linear increase with stress can be seen, characteristic

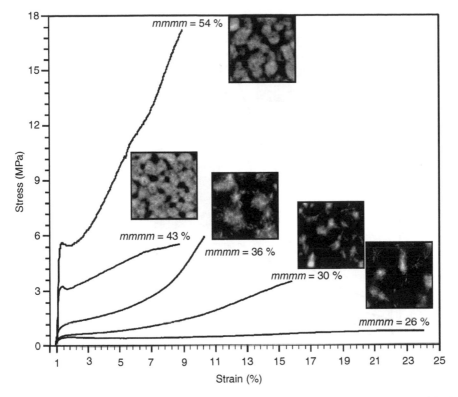

FIGURE 9.28 Stress–strain curves of melt-pressed polypropylene samples prepared with **9a**/MAO show a transition from elastomeric to plastomeric deformation behavior when the *mmmm* content rises above 40%.

of elastic deformation.[146] The slope of this line section is correlated to Young's modulus, which ranges between 2 and 4 MPa for the samples (Figure 9.30a). When the elongation ratio increases, no further increase in stress or the formation of a yield point can be seen; the stress stays constant and a so-called elastic plateau is developed ranging from $\lambda = 1.5$ to $\lambda = 9$. Plateau stresses in the range between 0.5 and 1.5 MPa are low. At elongation ratios above $\lambda = 9$, the stress increases again continuously with increasing deformation until the samples fail at maximum stresses σ_{max} in the range of 4–8 MPa. This indicates that at higher elongation ratios a strain-hardening region occurs, most probably owing to strain-induced crystallization. Nevertheless, high maximum deformation ratios of between 10 and 20 times the original length have been detected for these samples. When the *mmmm* content rises above 40%, the stress–strain curves change from elastic to plastomeric. This change in mechanical behavior is supported[7] by the following: (1) maximum elongation drops below $\lambda = 9$, (2) the maximum stress σ_{fail} increases by a factor of three, (3) the elastic plateau region becomes smaller ($\lambda = 1.5$–3) and a distinct region of strain hardening can be observed. This assumption is supported by a strong increase in the Young's modulus (Figure 9.30a), which increases about 8 times when the *mmmm* contents exceed 40%. The improvement of mechanical parameters, such as the increase in Young's modulus, can be explained by the presence of a larger quantity of crystalline lamellae, leading to a higher crosslinking density of the materials.

Comparing the stress–strain curves obtained for the two series of polypropylenes **A** and **B**, similar Young's moduli were found for samples with similar *mmmm* contents. This can be explained by similar amounts of crystalline aggregates formed within the samples. Nevertheless, a closer look at the stress–strain curves reveals a clear difference in mechanical behavior for samples with similar *mmmm* contents, but showing different chain microstructures (e.g., **A43**, 43% *mmmm*, Figure 9.28

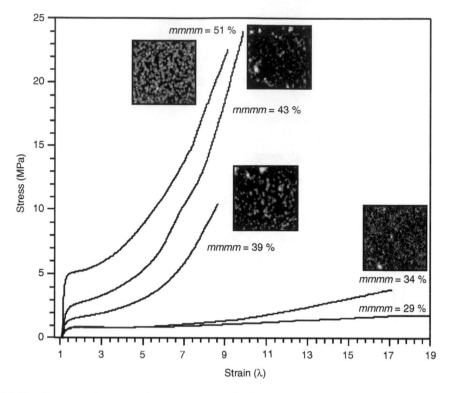

FIGURE 9.29 Stress–strain curves of **9a**/borate-derived polypropylene samples show a transition from elastic to plastomeric deformation behavior when the *mmmm* content rises above 40%.

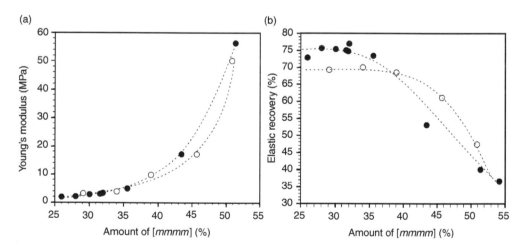

FIGURE 9.30 Polypropylenes prepared with **9a**/MAO (filled circles) or **9a**/borate (open circles) show an increase in mechanical strength, detected as an increasing Young's modulus (a), when their *mmmm* content exceeds 40%; at the same time their elastic recovery (b) decreases.

versus **B39**, 39% *mmmm*, Figure 9.29; **A54**, 54% *mmmm*, Figure 9.28 versus **B51**, 51% *mmmm*, Figure 9.29). Small yield points were developed in the MAO-activated samples with *mmmm* > 40%, but none of the samples show a yielding zone characteristic of thermoplastic deformation. This refers to irreversible changes in the lamellar structure occurring at small elongation ratios. In contrast, stress–strain curves of the borate-activated samples exhibit no yield point, indicating elastic deformation.

At higher crosslinking densities (as evidenced by higher Young's moduli), the materials are reinforced and their viscous properties are reduced, enhancing elastic properties. To demonstrate[147] variations in elastic properties, the reversible deformation (ε_{rev}) after elongation to a given strain rate λ_{cyc} can be determined using Equation 9.4

$$\varepsilon_{rev} = \left(\frac{\lambda_{cyc} - \lambda_{rel}}{\lambda_{cyc} - 1} \right) \cdot 100 \tag{9.4}$$

where $\lambda_{rel} = l_{rel}/l_0$; $l_0 =$ the initial length of the sample; and $l_{rel} =$ the length of the relaxed sample after stretching to λ_{cyc}.[147] Thus, improved elasticity can be correlated to an increase in the reversible deformation ε_{rev}.

Now, differences between samples having similar *mmmm* contents but different microstructures are more obvious. Both polymer series **A** and **B** display the lowest set at *mmmm* contents below 40%, but for samples prepared using MAO cocatalyst, a higher elastic recovery was observed as compared to samples prepared by borate activation (Figure 9.30b). Within this range of isotacticity, the samples preferably form a discontinuous crystalline network consisting of mixed crystalline blocks and lamellar aggregates. This suggests that the overall amount of crystalline crosslinks determines the macroscopic elastic properties of the samples. Thus, most probably the differences in elastic recovery are due to changes in the total amount of crystalline domains generated.

If *mmmm* content exceeds 40%, the elastic recovery strongly decreases to values below 50%, indicating reduced elastic recovery. This decline in elastic recovery is in accordance with the observed change from elastomeric to plastomeric deformation behavior and can be explained as a result of a change in morphology. With increasing *mmmm* content, the lamellar networks become less flexible because the amount of the granular domains is increased and additional lamellae are formed. This leads to an increase in mechanical strength, but, on the other hand, reduces the elastic recovery. When lamellae are extended, they are irreversibly transformed at the yield point into a fibrillar structure.[148] Since the samples still show elastic recovery, it might be concluded that flexible domains that can be reversibly deformed are still present. Nevertheless, the results show that, when a continuous lamellar network is generated, the distribution of the crystalline lamellae determines the elastic recovery. Smaller, more regular distributed lamellar structures, like those formed in samples of series **B** (cf. Figure 9.27), provide improved mechanical properties such as higher mechanical strength and increased flexibility as compared to materials where extended branched lamellae are formed (cf. **A** series, Figure 9.24).

9.5 CONCLUSIONS

The combination of polymerization and structural studies of a series of low-isotactic polypropylenes derived from a C_1-symmetric metallocene provided insights not only into the polymerization mechanism and the resulting chain structures, but also into polymer microstructure and the resulting macroscopic properties. Polymerization experiments with an *ansa*-asymmetric metallocene catalyst were used to prepare low-isotactic polypropylenes with different chain microstructures. The existence of two active sites having different selectivities in the catalyst framework proved to be a useful tool for varying the stereoregularity of the resultant polymers. Control over the stereoregularity is gained by manipulating two parameters, polymerization temperature and propylene pressure. A chain "back-skip" mechanism is suitable to explain the abovementioned facts, but this mechanism is not sufficient to explain an increase in molecular weight or higher isotacticity when $[Ph_3C]^+[B(C_6F_5)_4]^-$/TIBAL is used to activate the catalyst **9a** instead of MAO. Experiments at different MAO/catalyst ratios show that the catalytic activities and molecular weights of the polymers are sensitive to the aluminum content provided by the activators. This dependence suggests an additional reversible chain transfer to aluminum as mechanistic event when activation is performed with MAO. In the case of activation with $[Ph_3C]^+[B(C_6F_5)_4]^-$/TIBAL, lower contents of aluminum are provided in the polymerization system, and the only mechanism occurring is the chain "back skip."

The influence of cocatalyst nature on polymer microstructure was investigated by preparing two series of polypropylenes having increasing amounts of *mmmm* pentads, which showed different polymer microstructures for samples prepared with MAO and borate cocatalysts. The differences for polymers prepared with different activators can best be explained by a change in stereoerror distribution, leading to different block length distributions of the isotactic chain segments. These differences in chain microstructure are reflected in the polymers' thermal and mechanical behaviors, and sustain the proposed reversible chain transfer process proposed for MAO activation.

Different chain microstructures lead to different crystalline modifications, facilitating the formation of various polymer morphologies. Elastic behavior can be found for morphologies providing a flexible crystalline network based on a granular (fringed micelle) structure. In the case that extended lamellar superstructures are formed, such as a lamellar network, branched lamellar aggregates, or spherultites, the elastic properties of the polymer are deteriorated. Taking into consideration the correlation between polymer microstructure and material properties, it was demonstrated that *ansa*-asymmetric catalysts possess a high potential for tailoring the polymer microstructures of low-isotactic polypropylenes, leading to new materials with tunable elastic properties.

ACKNOWLEDGMENTS

Thanks are due to the Margarethe-von-Wrangell foundation for supporting one of the authors (S. H.). Thanks are also due to Prof. O. Marti and the members of the Department of Experimental Physics, especially to M. Kienzle for performing hundreds of microtome cuts.

REFERENCES AND NOTES

1. Ewen, J. A.; Elder, M. J.; Jones, R. L.; Haspeslagh, L.; Atwood, J. L.; Bott, S. J.; Robinson, K. Metallocene/polypropylene structural relationships: Implications on polymerization and stereochemical control mechanisms. *Makromol. Chem., Macromol. Symp.* **1991**, *48/49*, 253–295.

2. Brintzinger, H. H.; Fischer, D.; Mülhaupt, R.; Rieger, B.; Waymouth, R. M. Stereospecific olefin polymerization with chiral metallocene catalysts. *Angew. Chem., Int. Ed. Engl.* **1995**, *34*, 1143–1170.

3. Kaminsky, W. New polymers by metallocene catalysis. *Macromol. Chem. Phys.* **1996**, *197*, 3907–3945.

4. Resconi, L.; Cavallo, L.; Fait, A.; Piemontesi, F. Selectivity in propene polymerization with metallocene catalysts. *Chem. Rev.* **2000**, *100*, 1253–1345.

5. Rätzsch, M. PP-Entwicklung ungebremst. *Kunststoffe* **1998**, *88*, 1108–1116.

6. Hackmann M.; Rieger, B. Metallocene catalysts in stereoselective synthesis. *CaTTech* **1997**, *2*, 79–97.

7. Cobzaru, C.; Hild, S.; Boger, A.; Troll, C.; Rieger, B. Dual-side catalysts for high and ultrahigh molecular weight homopolypropylene elastomers and plastomers. *Coord. Chem. Rev.* **2005**, *250*, 189–211.

8. Cobzaru, C.; Deisenhofer, S.; Harley, A.; Troll, C.; Hild, S.; Rieger, B. Novel high and ultrahigh molecular weight polypropylene plastomers by asymmetric hafnocene catalysts. *Macromol. Chem. Phys.* **2005**, *206*, 1231–1240.

9. Natta, G.; Mazzanti, G.; Crespi, G.; Moraglio, G. Stereoblock and isotactic propylene polymers. *Chim. Ind. (Milan)* **1957**, *39*, 275–283; *Chem. Abstr.* **1957**, *51*, 69238.

10. Natta, G.; Crespi, G. Block polymers of alpha-olefines, processes for producing the same, and mixtures thereof with isotactic polyolefines. U.S. Patent 3,175,999 (Montecatini), March 30, 1965.

11. Holden, G. Thermoplastic elastomers. In *Polymeric Materials Encyclopedia*; Salamone, J. C., Ed.; CRC Press: Boca Raton, 1996; Vol. 11, pp 8343–8358.

12. Bravakis, A. M.; Bailey, L. E.; Pigeon, M.; Collins, S. Synthesis of elastomeric poly(propylene) using unsymmetrical zirconocene catalysts: Marked reactivity differences of "*rac*"- and "*meso*"-like diastereomers. *Macromolecules* **1998**, *31*, 1000–1009.

13. Gauthier, W. J.; Collins, S. Preparation of elastomeric polypropylene using metallocene catalysts: Catalyst design criteria and mechanism for propagation. *Macromol. Symp.* **1995**, *98*, 223–231.

14. Gauthier, W. J.; Collins, S. Elastomeric poly(propylene): Propagation models and relationship to catalyst structure. *Macromolecules* **1995**, *28*, 3779–3786.

15. Chien, J. C. W.; Llinas, G. H.; Rausch, M. D.; Lin, Y. G.; Winter, H. H.; Atwood, J. L.; Bott, S. G. J. Metallocene catalysts for olefin polymerization catalyzed by *rac*-[*anti*-ethylidene(1-η^5-tetramethylcyclopentadienyl)(1-η^5-indenyl)]dimethyltitanium: A two-state propagation. XXIV. Stereoblock propylene polymerization. *J. Polym. Sci., Part A: Polym. Chem.* **1992**, *30*, 2601–2617.

16. Llinas, G. H.; Dong, S. H.; Mallin, D. T.; Rausch, M. D.; Lin, Y. G.; Winter, H. H.; Chien, J. C. W. Crystalline-amorphous block polypropylene and nonsymmetric *ansa*-metallocene catalyzed polymerization. *Macromolecules* **1992**, *25*, 1242–1253.

17. Chien, J. C. W.; Rausch, M. D. Polypropylene and other olefin polymer thermoplastic elastomers, novel catalysts for preparing the same, and method of preparation. U.S. Patent 5,756,614 (Academy of Applied Science), May 26, 1998.

18. Coates, G. W.; Waymouth, R. M. Oscilllating stereocontrol: A strategy for the synthesis of thermoplastic elastomeric polypropylene. *Science* **1995**, *267*, 217–219.

19. Waymouth, R. M.; Coates, G. W.; Hauptman, E. M. Elastomeric olefin polymers, method of production and catalysts therefore. U.S. Patent 5,594,080 (Stanford University), January 14, 1997.

20. Bruce, M. D.; Coates, G. W.; Hauptman, E.; Waymouth, R. M.; Ziller, J. W. Effect of metal on the stereospecificity of 2-arylindene catalysts for elastomeric polypropylene. *J. Am. Chem. Soc.* **1997**, *119*, 11174–11182.

21. Carlson, E. D.; Krejchi, M. T.; Shah, C. D.; Terakawa, T.; Waymouth, R. M.; Fuller, G. G. Rheological and thermal properties of elastomeric polypropylene. *Macromolecules* **1998**, *31*, 5343–5351.

22. Kravchenko, R.; Masood, A.; Waymouth, R. M.; Myers, C. L. J. Strategies for the synthesis of elastomeric polypropylene: Fluxional metallocenes with C_1-symmetry. *J. Am. Chem. Soc.* **1998**, *120*, 2039–2046.

23. Miller, S. A.; Bercaw, J. E. Isotactic-hemiisotactic polypropylene from C_1-asymmetric *ansa*-metallocene catalysts: A new strategy for the synthesis of elastomeric polypropylene. *Organometallics* **2002**, *21*, 934–945.

24. *Physical Properties of Polymers*; Mark, J. E.; Eisenberg, A.; Graessley, W. W.; Mandelkern, L., Eds; ACS Professional Reference Book; American Chemical Society: Washington, DC, 1993.

25. *Thermoplastic Elastomers*; Holden, G.; Legge, N. R.; Quirk, R.; Schroeder, H. E., Eds; Carl Hanser Verlag: München, 1996.

26. Kresge, E. N. Polyolefin based thermoplastic elastomers. In *Thermoplastic Elastomers*; Holden, G., Legge, N. R., Quirk, R., Schroeder, H. E., Eds.; Carl Hanser Verlag: München, 1996; pp 101–104.

27. School, R. J. *Markets for thermoplastic elastomers into the new millennium*. Presented at the 154th Fall Technical Meeting of the Rubber Division of the American Chemical Society, 1998; American Chemical Society: Nashville, 1998; Paper 65.

28. Falbe, J.; Regitz, M. *Römpp Lexikon Chemie*; Georg Thieme Verlag: Stuttgart, 1992.

29. Natta, G.; Pino, P.; Corradini, P.; Danusso, F.; Mantica, E.; Mazzanti, G.; Moragli, G. Crystalline high polymers of α-olefins. *J. Am. Chem. Soc.* **1955**, *77*, 1708–1710.

30. Bensason, S.; Minick, J.; Moet, A.; Chum, S.; Hiltner, A.; Baer, E. Classification of homogenous ethylene-octene copolymers based on comonomer content. *J. Polym. Sci., Part B: Polym. Phys.* **1996**, *34*, 1301–1315.

31. Minick, J.; Moet, A.; Hiltner, A.; Baer, E.; Chum, S. P. Crystallization of very low density copolymers of ethylene with α-olefins. *J. Appl. Polym. Sci.* **1995**, *58*, 1371–1384.

32. Suhm, J.; Heinemann, J.; Thomann, Y.; Thomann, R.; Maier, R.-D.; Schleis, T.; Okuda, J.; Kressler, J.; Mühlhaupt, R. New molecular and supermolecular polymer architectures via transition metal catalyzed alkene polymerization. *J. Mater. Chem.* **1998**, *84*, 553–563.

33. Zhu, X.; Yan, D.; Fang, Y. In situ FTIR spectroscopic study of the conformational change of isotactic polypropylene during the crystallization process. *J. Phys. Chem. B* **2001**, *105*, 12461–12463.

34. Doi, M.; Edwards, S. F. *The Theory of Polymer Dynamics*; Oxford University Press: New York, 1986; Chapter 10.

35. Shimada, T.; Doi, M.; Okano, K. Concentration fluctuation of stiff polymers. I. Static structure factor. *J. Chem. Phys.* **1988**, *88*, 2815–2821.

36. Doi, M.; Shimada, T.; Okano, K. Concentration fluctuation of stiff polymers. II. Dynamical structure factor of rod-like polymers in the isotropic phase. *J. Chem. Phys.* **1988**, *88*, 4070–4075.

37. Shimada, T.; Doi, M.; Okano, K. Concentration fluctuation of stiff polymers. III. Spinodal decomposition. *J. Chem. Phys.* **1988**, *88*, 7181–7186.

38. Imai, M.; Mori, K.; Mizukami, T.; Kaji, K.; Kanaya, T. Structural formation of poly(ethylene terephthalate) during the induction period of crystallization. 1. Ordered structure appearing before crystal nucleation. *Polymer* **1992**, *33*, 4451–4456.

39. Imai, M.; Mori, K.; Mizukami, T.; Kaji, K.; Kanaya, T. Structural formation of poly(ethylene terephthalate) during the induction period of crystallization. 2. Kinetic analysis based on the theories of phase separation. *Polymer* **1992**, *33*, 4457–4462.

40. Matsuba, G.; Kaji, K.; Nishida, K.; Kanaya, T.; Imai, M. T. Conformational change and orientation fluctuations of isotactic polystyrene prior to crystallization. *Polym. J.* **1991**, *31*, 722–727.

41. *Polymer Handbook*, 3rd ed.; Brandrup, J., Immergut, E. H., Eds.; John Wiley and Sons: New York, 1989; p V29.

42. Natta, G.; Corradini, P.; Cesari, M. Crystalline structure of isotactic polypropylene. *Atti Accad. Nazl. Lincei, Rend., Classe Sci. Fis., Mat. e Nat.* **1956**, *21*, 365–372; *Chem. Abstr.* **1957**, *51*, 65132.

43. Keith, H. D.; Padden, F. J. Jr. Spherulitic crystallization in polypropylene. *J. Appl. Phys.* **1959**, *30*, 1479–1484.

44. Keith, H. D.; Padden, F. J. Jr.; Walter, N. M.; Wyckoff, H. W. Evidence for a second crystal form of polypropylene. *J. Appl. Phys.* **1959**, *30*, 1479–1484.

45. Turner-Jones, A.; Aizlewood, J. M.; Beckett, D. R. Crystalline forms of isotactic polypropylene. *Makromol. Chem.* **1964**, *75*, 134–158.

46. Brückner, S.; Meille, S. V.; Petracone, V.; Pirozzi, B. Polymorphism in isotactic polypropylene. *Prog. Polym. Sci.* **1991**, *16*, 361–404.

47. Auriemma, F.; Ruiz de Ballesteros, O.; De Rosa, C.; Corradini, P. Structural disorder in the α form of isotactic polypropylene. *Macromolecules* **2000**, *33*, 8764–8774.

48. Turner-Jones, A. Development of the γ-crystal form in random copolymers of propylene and their analysis by differential scanning calorimetry and X-ray methods. *Polymer* **1971**, *12*, 487–508.

49. Alamo, R. G.; Blanco, J. A.; Agarwal, P. K.; Randall, J. C. Crystallization rates of matched fractions of MgCl$_2$-supported Ziegler–Natta and metallocene isotactic poly(propylene)s. 1. The role of chain microstructure. *Macromolecules* **2003**, *36*, 1559–1571.

50. Randall, J. C.; Alamo, R. G.; Agarwal, P. K.; Russ, C. J. Crystallization rates of matched fractions of MgCl$_2$-supported Ziegler–Natta and metallocene isotactic poly(propylene)s. 2. Chain microstructures from a supercritical fluid fractionation of a MgCl$_2$-supported Ziegler–Natta isotactic poly(propylene). *Macromolecules* **2003**, *36*, 1572–1584.

51. De Rosa, C.; Auriemma, F.; Di Capua, L.; Resconi, S.; Guidotti, L.; Camurati, I. E.; Nifant'ev, I. P.; Laishevtev, I. P. Structure-property correlations in polypropylene from metallocene catalysts: Stereodefective, regioregular isotactic polypropylene. *J. Am. Chem. Soc.* **2004**, *126*, 17040–17049.

52. De Rosa, C.; Auriemma, F.; Perretta, C. Structure and properties of elastomeric polypropylene from C_2 and C_{2v}-symmetric zirconocenes. The origin of crystallinity and elastic properties in poorly isotactic polypropylene. *Macromolecules* **2004**, *37*, 6843–6855.

53. Alamo, R. G.; Kim, M.-H.; Galante, M. J.; Isasi, J. R.; Mandelkern, L. Structural and kinetic factors governing the formation of the γ polymorph of isotactic polypropylene. *Macromolecules* **1999**, *32*, 4050–4064.

54. Auriemma, F.; De Rosa, C. Crystallization of metallocene-made isotactic polypropylene: Disordered modifications intermediate between the α and γ forms. *Macromolecules* **2002**, *35*, 9057–9068.

55. Auriemma, F.; De Rosa, C.; Boscato, T.; Corradini, P. T. The oriented γ-form of isotactic polypropylene. *Macromolecules* **2001**, *34*, 4815–4821.

56. De Rosa, C.; Auriemma, F.; Spera, C.; Talarico, G.; Gahleitner, M. Crystallization properties of elastomeric polypropylene from alumina-supported tetraalkyl zirconium catalysts. *Polymer* **2004**, *45*, 5875–5888.

57. De Rosa, C.; Auriemma, F. Stereoblock polypropylene from a metallocene catalyst with a hapto-flexible naphthyl-indenyl ligand. *Macromolecules* **2003**, *36*, 3465–3474.

58. Meille, S. V.; Brückner, S.; Porzio, W. γ-Isotactic polypropylene. A structure with nonparallel chain axes. *Macromolecules* **1990**, *23*, 4114–4121.

59. Lotz, B.; Wittmann, J. C.; Lovinger, A. J. Structure and morphology of poly(propylenes): A molecular analysis. *Polymer* **1996**, *37*, 4979–4992.

60. Fischer, D.; Mülhaupt, R. The influence of regio- and stereoirregularities on the crystallization behavior of isotactic poly(propylene)s prepared with homogeneous group IVa metallocene/methylaluminoxane Ziegler–Natta catalysts. *Macromol. Chem. Phys.* **1994**, *195*, 1433–1441.

61. Schönherr, H.; Wiyatno, W.; Pople, J.; Frank, C. W.; Fuller, G. G.; Gast, A. P.; Waymouth, R. M. Morphology of thermoplastic elastomers: Elastomeric polypropylene. *Macromolecules* **2002**, *35*, 2654–2666.

62. Balboni, D.; Moscardi, G.; Barruzzi, G.; Braga, V.; Camurati, I.; Piemontesi, F.; Resconi, L.; Nifantév, I. E.; Vanditto, V.; Antinucci, S. C_2-Symmetric zirconocenes for high molecular weight amorphous polypropylene. *Macromol. Chem. Phys.* **2001**, *202*, 2010–2028.

63. Morrow, D. R.; Newman, B. A. Crystallization of low molecular weight polypropylene fractions. *J. Appl. Phys.* **1968**, *39*, 4944–4950.

64. Natta, G.; Corradini, P. Structure and properties of isotactic polypropylene. *Nuovo Cimento, Suppl.* **1960**, *15*, 40–47.

65. Bassett, D. C.; Olley, R. H. On the lamellar morphology of isotactic polypropylene spherulites. *Polymer* **1984**, *25*, 935–943.

66. Addink, E. J.; Beintema, J. Polymorphism of crystalline polypropylene. *Polymer* **1961**, *2*, 185–193.

67. Chatani, Y.; Ueda, Y.; Tadokoro, H. (Title unavailable.) Abstracts of the Annual Meeting of The Society of Polymer Science, Japan; The Society of Polymer Science, Japan: Tokyo, 1977, pp 1326–1328.

68. Bassett, D. C. *Principles of Polymer Morphology*; Cambridge University Press: Cambridge, U.K., 1981.

69. Olley, R. H.; Bassett, D. C. On the development of polypropylene spherulites. *Polymer* **1989**, *30*, 399–409.

70. Padden, F. J.; Keith, H. D. Crystallization in thin films of isotactic polypropylene. *J. Appl. Phys.* **1966**, *37*, 4013–4020.

71. Lotz, B.; Wittmann, J. C. The molecular origin of lamellar branching in the alpha (monoclinic) form of isotactic polypropylene. *J. Polym. Sci., Part B: Polym. Phys.* **1986**, *24*, 1541–1548.

72. Thomann, R.; Kressler, J.; Setz, S.; Wang, C.; Mülhaupt, R. Morphology and phase behavior of blends of syndiotactic and isotactic polypropylene: 1. X-ray scattering, light microscopy, atomic force microscopy, and scanning electron microscopy. *Polymer* **1996**, *37*, 2627–2634.

73. Thomann, R.; Wang, C.; Kressler, J.; Mülhaupt, R. On the γ-phase of isotactic polypropylene. *Macromolecules* **1996**, *29*, 8425–8434.

74. Ewen, J. A. Mechanisms of stereochemical control in propylene polymerizations with soluble group 4B metallocene/methylalumoxane catalysts. *J. Am. Chem. Soc.* **1984**, *106*, 6355–6364.

75. Wild, F.; Zsolnai, L.; Huttner, G.; Brintzinger, H.-H. *ansa*-Metallocene derivatives. IV. Synthesis and molecular structures of chiral *ansa*-titanocene derivatives with bridged tetrahydroindenyl ligands. *J. Organomet. Chem.* **1982**, *232*, 233–247.

76. Kaminsky, W.; Külper, K.; Brintzinger, H. H.; Wild, F. R. W. P. Polymerization of propene and butene with a chiral zirconocene and methylalumoxane as cocatalyst. *Angew. Chem., Int. Ed. Engl.* **1985**, *24*, 507–508.

77. Mülhaupt, R.; Rieger, B. Transition metal catalysts for the polymerization of olefins. Part 2. Preparation of stereoregular poly(1-olefins). *Chimia* **1996**, *50*, 10–19.

78. Resconi, L.; Cavallo, L.; Fait, A.; Piemontesi, F. Selectivity in propene polymerization with metallocene catalysts. *Chem. Rev.* **2000**, *100*, 1253–1345.

79. Ewen, J. A. Symmetrical and lopsided zirconocene pro-catalysts. *Macromol. Symp.* **1995**, *89*, 181–196.

80. Razavi, A.; Vereecke, D.; Peters, L.; Dauw, K. D.; Nafpliotis, L.; Atwood, J. L. Manipulation of the ligand structures as an effective and versatile tool for modification of active site properties in homogeneous Ziegler–Natta catalyst systems. In *Ziegler Catalysts: Recent Scientific Innovations and Technological Improvements*; Fink, G., Mülhaupt, R., Brintzinger, H. H., Eds.; Springer-Verlag: Berlin, 1995; pp 111–147.

81. Ewen, J. A.; Elder, M. J. Isospecific pseudo-helical zirconocenium catalysts. In *Ziegler Catalysts: Recent Scientific Innovations and Technological Improvements*; Fink, G., Mülhaupt, R., Brintzinger, H. H., Eds.; Springer-Verlag: Berlin, 1995; pp 99–109.

82. Albizzati, E.; Giannini, U.; Collina, G.; Noristi, L.; Resconi, L. Catalysts and Polymerizations. In *Polypropylene Handbook: Polymerization, Characterization, Properties, Processing, Applications*; Moore, E. P., Ed.; Hanser: Munich, 1996; pp 11–111.

83. Mallin, D. T.; Rausch, M. D.; Lin, Y.-G.; Dong, S.; Chien, J. C. W. *rac*-[Ethylidene(1-η^5-tetramethylcyclopentadienyl)(1-η^5-indenyl)]dichlorotitanium and its homopolymerization of propylene to crystalline-amorphous block thermoplastic elastomers. *J. Am. Chem. Soc.* **1990**, *112*, 2030–2031.

84. Chien, J. C. W.; Rieger, B.; Sugimoto, R.; Mallin, D. T.; Rausch, M. D. Homogeneous Ziegler–Natta catalysts and synthesis of anisotactic and thermoplastic elastomeric poly(propenes). *Stud. Surf. Sci. Catal.* **1990**, *56*, 535–574.

85. Chien, J. C. W.; Llinas, G. H.; Rausch, M. D.; Lin, G.-Y.; Winter, H. H. J. Two-state propagation mechanism for propylene polymerization catalyzed by *rac*-[*anti*-ethylidene(1-η^5-tetramethylcyclopentadienyl)(1-η^5-indenyl)]dimethyltitanium. *J. Am. Chem. Soc.* **1991**, *113*, 8569–8570.

86. Babu, G. N.; Newmark, R. A.; Cheng, H. N.; Llinas, G. H.; Chien, J. C. W. Microstructure of elastomeric polypropylenes obtained with nonsymmetric *ansa*-titanocene catalysts. *Macromolecules* **1992**, *25*, 7400–7402.

87. Cheng, H. N.; Babu, G. N.; Newmark, R. A.; Chien, J. C. W. Consecutive two-state statistical polymerization models. *Macromolecules* **1992**, *25*, 6980–6987.

88. (a) Gauthier, W. J.; Corrigan, F. J.; Nicholas, N. J.; Collins, S. Elastomeric poly(propylene): Influence of catalyst structure and polymerization conditions on polymer structure and properties. *Macromolecules* **1995**, *28*, 3771–3778. (b) Gauthier, W. J.; Collins, S. Elastomeric poly(propylene): Propagation models and relationship to catalyst structure. *Macromolecules* **1995**, *28*, 3779–3785.

89. Bravakis, A. M.; Bailey, L. E.; Pigeon, M.; Collins, S. Synthesis of elastomeric poly(propylene) using unsymmetrical zirconocene catalysts: Marked reactivity differences of "*rac*"- and "*meso*"-like diastereomers. *Macromolecules* **1998**, *31*, 1000–1009.

90. Resconi, L.; Moscardi, G.; Silvestri, R.; Balboni, D. Process for the preparation of substantially amorphous alpha-olefin polymers and compositions containing them and process for the preparation of bridged ligand. PCT Int. Pat. Appl. WO 00/01738 (Montell), January 13, 2000.

91. Rieger, B.; Jany, G.; Fawzi, R.; Steimann, M. Unsymmetric *ansa*-zirconocene complexes with chiral ethylene bridges: Influence of bridge conformation and monomer concentration on the stereoselectivity of the propene polymerization reaction. *Organometallics* **1994**, *13*, 647–653.

92. Buckingham, D. A.; Sargeson, A. M. Conformational analysis and steric effects in metal chelates. *Top. Stereochem.* **1971**, *6*, 219–277.

93. For the notation of chiral metallacycles, see: Corey, E. J.; Ballar, J. C. Jr. The stereochemistry of complex inorganic compounds. XXII stereospecific effects in complex ions. *J. Am. Chem. Soc.* **1959**, *81*, 2620–2629.

94. Rieger, B.; Reinmuth, A.; Röll, W.; Brintzinger, H. H. Highly isotactic polypropene prepared with *rac*-dimethylsilyl-bis(2-methyl-4-*tert*-butyl-cyclopentadienyl) zirconiumdichloride: An NMR investigation of the polymer microstructure. *J. Mol. Catal.* **1993**, *82*, 67–73.

95. (a) Schäfer, A.; Karl, E.; Zsolnai, L.; Huttner, G.; Brintzinger, H. H. *ansa*-Metallocene derivatives. XII. Diastereomeric derivatisation and enantiomer separation of ethylenebis(tetrahydroindenyl)-titanium and -zirconium dichlorides. *J. Organomet. Chem.* **1987**, *328*, 87–99. (b) Brintzinger, H. H. Synthesis and characterization of chiral metallocene cocatalysts for stereospecific α-olefin polymerizations. In *Transition Metals and Organometallics as Catalysts for Olefin Polymerization*; Kaminsky, W., Sinn, H.-J., Eds.; Springer-Verlag: Berlin, 1988; pp 249–256.

96. Rieger, B.; Jany, G.; Fawzi, R.; Steimann, M. Unsymmetric *ansa*-zirconocene complexes with chiral ethylene bridges: Influence of bridge conformation and monomer concentration on the stereoselectivity of the propene polymerization reaction. *Organometallics* **1994**, *13*, 647–653.

97. Hortmann, K.; Brintzinger, H. H. Steric effects in *ansa*-metallocene-based Ziegler–Natta catalysts: Coordination gap aperture and obliquity angles as parameters for structure-reactivity correlations. *New J. Chem.* **1992**, *16*, 51–55.

98. Dietrich, U.; Hackmann, M.; Rieger, B.; Klinga, M.; Leskelä, M. Control of stereoerror formation with high activity "dual-side" zirconocene catalysts: A novel strategy to design the properties of thermoplastic elastic polypropenes. *J. Am. Chem. Soc.* **1999**, *121*, 4348–4355.

99. Kukral, J.; Lehmus, P.; Feifel, T.; Troll, C.; Rieger, B. Dual-side *ansa*-zirconocene dichlorides for high molecular weight isotactic polypropene elastomers. *Organometallics* **2000**, *19*, 3767–3775.

100. Kukral, J.; Lehmus, P.; Klinga, M.; Leskelä, M.; Rieger, B. Oxygen-containing asymmetric "dual-side" zirconocenes: Investigations on a reversible chain transfer to aluminium. *Eur. J. Inorg. Chem.* **2002**, 1349–1356.

101. Rieger, B.; Troll, C.; Preuschen, J. Ultrahigh molecular weight polypropene elastomers by high activity dual-side hafnocene catalysts. *Macromolecules* **2002**, *35*, 5742–5743.

102. Guerra, G.; Cavallo, L.; Moscardi, G.; Vacatello, M.; Corradini, P. Back-skip of the growing chain at model complexes for the metallocene polymerization catalysis. *Macromolecules* **1996**, *29*, 4834–4845.

103. Müller, G.; Rieger, B. Propene based thermoplastic elastomers by early and late transition metal catalysts. *Prog. Polym. Sci.* **2002**, *27*, 815–853.

104. Schilling, F. C.; Tonelli, A. E. Carbon-13 nuclear magnetic resonance of atactic polypropylene. *Macromolecules* **1980**, *13*, 270–275.

105. Tonelli, A. E. *NMR Spectroscopy and Polymer Microstructure: The Conformational Connection*; VCH: New York, 1989; Chapter 6.

106. Zakharov, V. A.; Bukatov, G. P.; Yermakov, Y. I. On the mechanism of olefin polymerization by Ziegler–Natta catalysts. *Adv. Polym. Sci.* **1983**, *51*, 61–100.

107. Farina, M.; Di Silvestro, G.; Sozzani, P. Hemitatctic polypropylene: An example of a novel kind of polymer tacticity. *Macromolecules* **1982**, *15*, 1451–1452.

108. Di Silvestro, G.; Sozzani, P.; Savare, B.; Farina, M. Resolution of the carbon-13 nuclear magnetic resonance spectrum of hemiisotactic polypropylene at the decade and undecade level. *Macromolecules* **1985**, *18*, 928–932.

109. Bovey, F. A. *Chain Structure and Conformation of Macromolecules*; Academic Press: New York, 1982; pp 75–82.

110. Ewen, J. A. Metallocene polymerization catalysts: Past, present and future. In *Metallocene-Based Polyolefins—Preparation, Properties and Technology*; Scheirs, J., Kaminsky, W., Eds.; John Wiley and Sons: Chichester, 2000; Vol. 1, pp 3–31.

111. Ewen, J. A.; Elder, M. J.; Jones, R. L.; Haspeslagh, L.; Atwood, J. L.; Bott, S. J.; Robinson, K. Metallocene/polypropylene structural relationships: Implications on polymerization and stereochemical control mechanisms. *Makromol. Chem., Macromol. Symp.* **1993**, *66*, 179–190.

112. Stanley, K.; Baird, M. C. Demonstration of controlled asymmetric induction in organoiron chemistry. Suggestions concerning the specification of chirality in pseudotetrahedral metal complexes containing polyhapto ligands. *J. Am. Chem. Soc.* **1975**, *97*, 6598–6599.

113. Lieber, S.; Brintzinger, H.-H. Propene polymerization with catalyst mixtures containing different *ansa*-zirconocenes: Chain transfer to alkylaluminum cocatalysts and formation of stereoblock polymers. *Macromolecules* **2000**, *33*, 9192–9199.

114. Chien, J. C. W.; Iwamoto, Y.; Rausch, M. D.; Wedler, W.; Winter, H. H. Homogeneous binary zirconocenium catalyst systems for propylene polymerization. 1. Isotactic/atactic interfacial compatibilized polymers having thermoplastic elastomeric properties. *Macromolecules* **1997**, *30*, 3447–2445.

115. Chien, J. C. W.; Iwamoto, Y.; Rausch, M. D. Homogeneous binary zirconocenium catalysts for propylene polymerization. II. Mixtures of isospecific and syndiospecific zirconocene systems. *J. Polym. Sci., Part A: Polym. Chem.* **1999**, *37*, 2439–2445.

116. Przybyla, C.; Fink, G. Two different, on the same silica supported metallocene catalysts, activated by various trialkylaluminums. A kinetic and morphological study as well as an experimental investigation for building stereoblock polymers. *Acta Polym.* **1999**, *50*, 77–83.

117. Hild, S.; Cobzaru, C.; Troll, C.; Rieger, B. Elastomeric polypropylene from "dual-side" metallocenes: Reversible chain transfer and its influence in polymer microstructure. *Macromol. Chem. Phys.* **2006**, *207*, 665–683.

118. (a) Collette, J. W.; Tullock, C. W.; MacDonald, R. N.; Buck, W. A.; Su, A. C. L.; Harrell, R.; Mülhaupt, R.; Anderson, B. C. Elastomeric polypropylenes from aluminium-supported tetraalkyl group IVB catalysts. 1. Synthesis and properties of high molecular weight stereoblock homopolymers.

Macromolecules **1989**, *22*, 3858–3866. (b) Collette, J. W.; Tullock, C. W.; MacDonald, R. N.; Harrell, R.; Aaron, C. L.; Buck, W. A.; Mülhaupt, R.; Burton, C. A. Elastomeric polypropylenes from aluminum-supported tetraalkyl group IVB catalysts. 2. Chain microstructures, crystallinity, and morphology. *Macromolecules* **1989**, *22*, 3858–3866.

119. Wunderlich, B. *Macromolecular Physics*; Academic Press: New York, 1976; Vol. 1 and 2.

120. Low-isotactic polypropylenes show slow crystallization; thus, melting peaks appear in DSC curves when the specimens are crystallized for at least two weeks.

121. Alizadeh, A.; Richardson, L.; Xu, J.; McCartney, S.; Marand, H.; Cheung, Y. W.; Chum, S. Influence of structural and topological constraints on the crystallization and melting behavior of polymers. 1. Ethylene/1-octene copolymers. *Macromolecules* **1999**, *32*, 6221–6235.

122. Wunderlich, B. *Macromolecular Physics*; Academic Press: New York, 1976; Vol. 1, Chapters 5 and 6.

123. Elias, H.-G. *Macromolecules: Volume 1: Chemical Structure and Synthesis*; Wiley-VCH: Weinheim, 2005.

124. Binnig, G.; Quate, C. F.; Gerber, C. Atomic Force Microscope. *Phys. Rev. Lett.* **1986**, *56*, 930–933.

125. Magonov, S. N.; Reneker, D. H. Characterization of polymer surfaces with atomic force microscopy. *Ann. Rev. Mater. Sci.* **1997**, *27*, 175–222.

126. Zhong, Q.; Inniss, D.; Kjoller; K.; Ellings, V. B. Fractured polymer/silica fiber surface studied by tapping mode atomic force microscopy. *Surf. Sci.* **1993**, *290*, L688–L692.

127. Quist, A. P.; Ahlborn, J.; Reimann, C. T.; Sundqvist, B. U. R. Scanning force microscopy studies of surface defects induced by incident energetic macromolecular ions. *Nucl. Instrum. Methods Phys. Res., Sect. B* **1994**, *88*, 164–169.

128. Magonov, S. N.; Elings, V. B.; Whangbo, M.-H. Phase imaging and stiffness in tapping-mode atomic force microscopy. *Surf. Sci.* **1997**, *375*, L385–L391.

129. Bar, G.; Thomann, Y.; Brandsch, R.; Cantow, H.-J.;Whangbo, M.-H. Factors affecting the height and phase images in tapping mode atomic force microscopy. Study of phase-separated polymer blends of poly(ethene-*co*-styrene) and poly(2,6-dimethyl-1,4-phenylene oxide). *Langmuir* **1997**, *13*, 3807–3812.

130. Vancso, J. Morphology and nanostructure of polypropylenes by atomic force microscopy. In *Polypropylene: An A-Z Reference*; Karger-Kocsis, J., Ed.; Kluwer Academic Press: Dordrecht, 1999; pp 510–533.

131. Schönherr, H.; Snetivy, D.; Vancso, G. J. A nanoscopic view at the spherulitic morphology of isotactic polypropylene by atomic force microscopy. *Polym. Bull. (Berlin)* **1993**, *30*, 567–574.

132. Haeringen, D. T.-V.; Varga, J.; Ehrenstein, G. W.; Vancso, G. J. Features of the hedritic morphology of β-isotactic polypropylene studied by atomic force microscopy. *J. Polym. Sci., Part B: Polym. Phys.* **2000**, *38*, 672–681.

133. Stocker, W.; Magonov, S. N.; Cantow, H. J.; Wittmann, J. C.; Lotz, B. Contact faces of epitaxially crystallized α- and γ-phase isotactic polypropylene observed by atomic force microscopy. *Macromolecules* **1993**, *26*, 5915–5923.

134. Snetivy, D.; Guillet, J. E.; Vancso, G. J. Atomic force microscopy of polymer crystals. 4. Imaging of oriented isotactic polypropylene with molecular resolution. *Polymer* **1993**, *34*, 429–431.

135. Stocker, W.; Schumacher, M.; Graff, S.; Thierry, A.; Wittmann, J.-C.; Lotz, B. Epitaxial crystallization and AFM investigation of a frustrated polymer structure: Isotactic poly(propylene), γ phase. *Macromolecules* **1998**, *31*, 807–814.

136. Kravchenko, R. L.; Sauer, B. B.; McLean, R. S.; Keating, M. Y.; Cotts, P. M.; Kim, Y. H. Morphology investigation of stereoblock polypropylene elastomer. *Macromolecules* **2000**, *33*, 11–13.

137. Wiyatno, W.; Pople, J.; Gast, A. P.; Waymouth, R. M. Dynamic response of stereoblock elastomeric polypropylene studied by rheooptics and X-ray scattering. 1. Influence of isotacticity. *Macromolecules* **2002**, *35*, 8488–8498.

138. Assuming that the heat of fusion ΔH_{perf} given for a perfect infinite crystal of the crystalline polymer is equal to that of the crystalline fraction ΔH_{tot} of a low crystalline polymer, the relation between the overall heat of fusion ΔH_{tot} and the enthalpy of fusion ΔH_{perf} gives the crystallinity x_{DSC} for an individual sample.

139. Bensason, S.; Minick, J.; Moet, A.; Chum, S. Classification of homogenous ethylene-octene copolymers based on comonomer content. *J. Polym. Sci., Part B: Polym. Phys.* **1996**, *34*, 1301–1315.

140. Schmidtke, J.; Strobl, G.; Thurn-Albrecht, T. A four-state scheme for treating polymer crystallization and melting suggested by calorimetric and small angle X-ray scattering experiments on syndiotactic polypropylene. *Macromolecules* **1997**, *30*, 5804–5821.

141. Hugel, T.; Strobl, G.; Thomann, R. Building lamellae from blocks: The pathway followed in the formation of crystallites of syndiotactic polypropylene. *Acta Polym.* **1999**, *50*, 214–217.

142. Schönherr, H.; Frank, C. W. Ultrathin films of poly(ethylene oxides) on oxidized silicon. 2. In situ study of crystallization and melting by hot stage AFM. *Macromolecules* **2003**, *36*, 1199–1208.

143. Schönherr, H.; Waymouth, C. J.; Frank, C. W. Nucleation and crystallization of low-crystallinity polypropylene followed in situ by hot stage atomic force microscopy. *Macromolecules* **2003**, *36*, 2412–2418.

144. Hild, S.; Boger, A.; Troll, C.; Rieger, B. Nucleation and crystallization of low crystalline polypropylenes with statistically distributed stereoerrors. Submitted to *Polymer*.

145. Schönherr, H.; Snétivy, D.; Vancso, G. J. A nanoscopic view at the spherulitic morphology of isotactic polypropylene by atomic force microscopy. *Polym. Bull.* **1993**, *30*, 567–576.

146. Strobl. G. *The Physics of Polymers*; Springer: Berlin, 1996.

147. The elastic recovery was determined from cyclic deformation experiments, where the samples were stretched three times to $\lambda_{cyc} = 8$ (where $\lambda_{cyc} = l_{cyc}/l_0$ and $l_{cyc} = $ the length of the sample stretched to $\lambda = 8$), and the stress was released to be zero. Subsequently, the samples were removed from the stretching device and stored without strain for 24 hours to allow them to recover before measuring the final sample length used in Equation 9.4.

148. Chum, S. P.; Kao, C. I.; Knight, G. W. Structure, properties and preparation of polyolefines produced by single-site catalyst technology. In *Metallocene-Based Polyolefins—Preparation, Properties and Technology*; Scheirs, J., Kaminsky, W., Eds.; John Wiley and Sons: Chichester, 2000; Vol. 1, pp 261–287.

10 Control of Molecular Weight and Chain-End Group in Tactic Polypropylenes Using Chain Transfer Agents

T. C. Chung

CONTENTS

10.1 INTRODUCTION

The intense research interest in tactic control of metallocene-mediated propylene polymerization has provided a tremendous amount of detail regarding the mechanism of polymerization,[1-4] including active site structure, the regioselectivity and stereoselectivity of propylene coordination and insertion, and the relationship between catalyst symmetry and polymer microstructure and properties. Nevertheless, chain release mechanisms, usually involving various chain transfer reactions, have received less attention. Some reports[5-7] focus on the control of catalyst activity, polymer molecular weight, and saturated polymer chain-end structure by introducing hydrogen transfer agents using different metallocene catalysts.

Considering a propylene polymerization having a typical catalyst activity ($>5 \times 10^4$ kg polymer/mol Zr·h), each catalytic site on average produces hundreds and thousands of polymer chains before deactivation. In other words, each active site experiences chain transfer, chain release, and reinitiation of a new polymer chain hundreds and thousands of times during the polymerization process. Chain transfer reactions clearly play a key role in governing the active site concentration, and affect catalyst activity, polymer molecular weight, chain-end structure, and even chain microstructure. On the other hand, chain transfer reactions also provide an excellent opportunity to introduce a terminal functional group at the polymer chain end. With a frequent chain transfer reaction, a suitable chain transfer agent (CTA), and a metallocene catalyst, it is possible to prepare chain-end functionalized polymers in situ with high polymer yield (no change in catalyst activity) during metallocene-mediated propylene polymerization.

Since the discovery of Ziegler–Natta catalysts in the early 1950s, the functionalization of polyolefins has been a scientifically interesting and technologically important subject.[8–11] The lack of functionality (i.e., heteroatom-containing groups) in polypropylene (PP), resulting in its poor compatibility with other materials, has been a major drawback for many applications,[12–14] particularly those in which adhesion, dyeability, printability, or compatibility with other polymers is paramount. Unfortunately, polypropylene is a particularly difficult substrate for functionalization reactions. Common free-radical-mediated processes usually involve many undesirable side reactions such as degradation and crosslinking.[15]

Chain-end functionalized polypropylene has a very attractive polymer structure that possesses an unperturbed polymer chain with desirable physical properties (such as high melting temperature, crystallinity, good mechanical properties, etc.) almost the same as those of pure polypropylene. Nevertheless, the terminal reactive group at the polymer chain end has good mobility to diffuse to the surface and can provide a useful reactive site for many applications, such as adhesion to substrates, reactive blending, and the formation of block copolymers.

The preparation of chain-end functionalized polymers has been largely limited to living polymerization techniques, used in combination with a controlled initiation or termination (functionalization) reaction.[16,17] Unfortunately, there are only a few transition metal coordination catalysts that exhibit living polymerization behavior, and the utility of most of these is limited to the preparation of polyethylene.[18–20] Furthermore, living polymerization only produces one polymer chain per initiator, which presents a very low rate of catalyst activity for a typical polyolefin preparation.

This chapter will review most of the well-studied chain transfer reactions in metallocene-mediated propylene polymerization, starting with regular chain release mechanisms that occur without the addition of any external CTAs. Subsequently, polymerization reactions using hydrogen to control polymer molecular weight and chain-end structure will be discussed, as well as the effects of hydrogen on catalyst activity and polymer microstructure. Special attention will then be focused on the chain-end functionalization of polypropylene involving various specially designed CTAs.

10.2 CHAIN RELEASE MECHANISMS

10.2.1 β-Hydride and β-Methyl Transfer Reactions

When propylene is polymerized using homogeneous metallocene/methylaluminoxane (MAO) catalysts, several chain transfer mechanisms occur to release free polymer chains. Their relative frequencies are dependent on the polymerization conditions and the catalyst structure. Three chain transfer mechanisms are identified that form chain-end unsaturated polypropylene. These are: (1) β-hydride transfer to metal after a primary (1,2-) propylene insertion, (2) β-hydride transfer to monomer after a primary propylene insertion, and (3) β-hydride transfer after a secondary (2,1-) propylene insertion. The formations of these chain ends and associated polymer head groups (chain starts) are shown in Scheme 10.1. Mechanism (2) is commonly referred to as chain transfer to monomer.

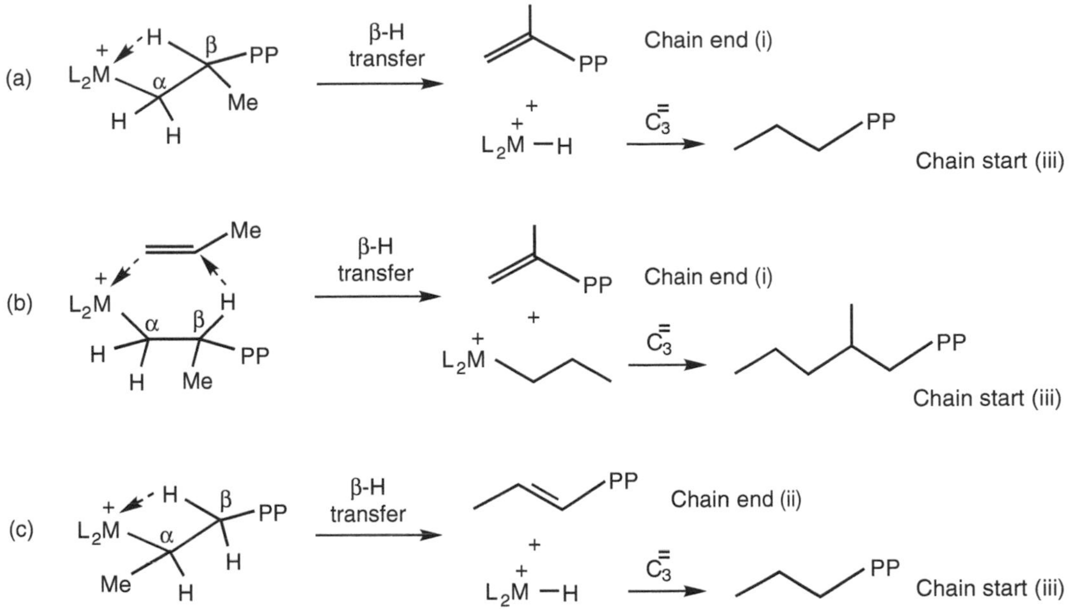

SCHEME 10.1 β-Hydride chain transfer reactions. M = metal, for example, Ti, Zr, Hf (a generic cationic group 4 complex is shown); L = ligand, for example, cyclopentadienyl, indenyl, fluorenyl; PP = polypropylene chain.

The predominant chain transfer mechanism in propylene polymerization is β-hydrogen transfer to metal after a primary propylene insertion (Scheme 10.1a). This occurs when the active site metal (M) abstracts a hydrogen atom from the β-carbon of the growing polymer chain, forming a M–H bond and a polymer chain with a vinylidene unsaturated end group (i). The newly formed metal-hydride active site reinitiates polymerization of propylene to incorporate an *n*-propyl head group (chain start) at the beginning of polymer chain (iii). Bercaw[21,22] and Jordan[23,24] have shown that both neutral group 3 and cationic group 4 metallocene alkyl complexes readily undergo spontaneous β-hydride transfer, forming the corresponding metal hydrides and releasing vinylidene-terminated polypropylene. This β-hydrogen elimination process can also occur simultaneously during monomer insertion (β-hydrogen transfer to monomer, Scheme 10.1b),[25–27] which may reduce the energy barrier for the reaction. This bimolecular mechanism also results in the same PP chain-end structures (a vinylidene group at the chain end (i) and an *n*-propyl group at the chain start of the newly formed chain (iii)) without generating a metal hydride active site. A metal-alkyl (M-CH$_2$CH$_2$CH$_3$) species is generated instead. This reaction is first order in propylene, and is a common chain transfer mechanism during the polymerization of propylene with heterogeneous catalysts.

On the other hand, β-hydrogen transfer after a regioirregular secondary propylene insertion (Scheme 10.1c) produces a polypropylene chain with a 2-butenyl end group (ii),[6,28,29] along with a M–H active site that initiates growth of a new polymer chain with an *n*-propyl chain start (iii). In several catalyst systems, including *rac*-Et(4,7-Me$_2$-Ind)$_2$ZrCl$_2$/MAO (Ind = 1-indenyl) and *rac*-Me$_2$Si(Benz[*e*]-Ind)$_2$ZrCl$_2$/MAO, propylene polymerization only yields a low molecular weight polymer with reduced catalyst activity, owing to a great majority (up to 90%) of the active sites being trapped in a "dormant" state after secondary propylene insertion occurs.[30,31] The methyl group in the 4-position of the ligand in the *rac*-Et(4,7-Me$_2$-Ind)$_2$ZrCl$_2$ system increases the tendency for secondary propylene insertion,[6] owing to the steric interaction between the 4-methyl substituent and the methyl group of the incoming primary propylene monomer; the methyl group in the 2-position of the ligand in the *rac*-Me$_2$Si(2-Me-Benz[*e*]-Ind)$_2$ZrCl$_2$ system[6] suppresses this insertion mechanism.

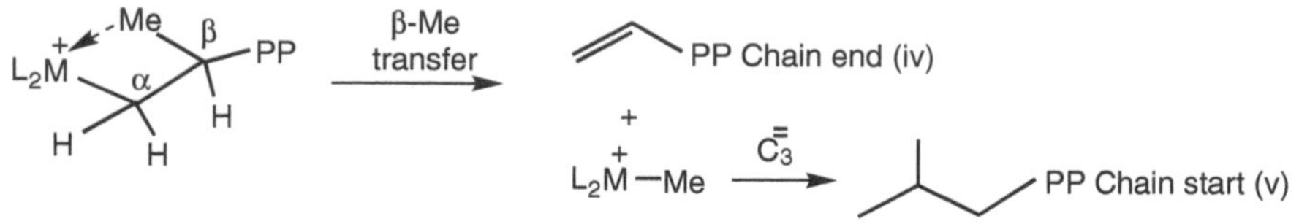

SCHEME 10.2 β-Methyl group chain transfer reaction.

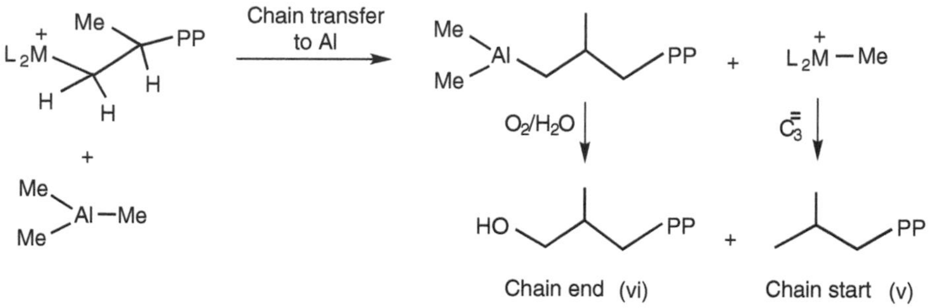

SCHEME 10.3 Chain transfer to aluminum cocatalyst.

The steric congestion at the active site after 2,1-propylene insertion significantly diminishes the rate of subsequent propylene insertions. To release steric hindrance in this dormant state, the active site either engages in isomerization to form a linear 1,3-propylene unit[28] in the polymer chain or undergoes β-hydride transfer reactions, which reduce the average polymer molecular weight.

There is another chain transfer mechanism, first observed by Watson[32] in the polymerization of propylene using a Cp*₂LuMe (Cp* = C₅Me₅) catalyst, which involves β-methyl group migration to metal center. As illustrated in Scheme 10.2, the metal center abstracts a β-CH₃ group, instead of a β-hydrogen atom, at the β-carbon of the growing polymer chain, thus forming a M–CH₃ bond at the active center and leaving the polymer with a vinyl end group (iv). This remarkable C–C bond activation is a unimolecular process.[22,24] Reinitiation of polymerization at the M-CH₃ active site results in an isobutyl group at the beginning of the PP chain (v). This β-CH₃ transfer mechanism was later found to be quite common in group 4 catalysts when the metal center bears highly substituted cyclopentadienyl (Cp; C₅H₅) ligands, such as indenyl and fluorenyl (Flu) ligands.[33–35]

Chain transfer to an alkylaluminum cocatalyst is common with heterogeneous Ziegler–Natta catalyst systems. It also occurs with some metallocene systems that utilize a high concentration of MAO cocatalyst. The growing polymer chain exchanges with the methyl group of AlMe₃ present in the MAO molecule to form a M-CH₃ active species and a Me₂Al-terminated polymer chain,[36–39] as illustrated in Scheme 10.3. Subsequent propylene insertion into the M-CH₃ active site produces an isobutyl head group at the beginning of the new PP chain (v); oxidation/hydrolysis of the Me₂Al terminus of the released chain results in the formation of polypropylene with an OH end group (vi).

10.2.2 INFLUENCE OF CATALYST STRUCTURE AND REACTION CONDITIONS ON CHAIN TRANSFER REACTIONS

All chain transfer reactions are influenced by the nature of the metallocene catalyst (steric/electronic factors) and polymerization conditions (temperature/pressure). Table 10.1 summarizes the effect of the ligand structure in the metallocene active site on chain transfer mechanisms and the structure of the growing polymer chain in propylene polymerization.

TABLE 10.1

Effect of Metallocene Ligand Structure on Chain Transfer Reactions in Propylene Polymerization at 40–60 °C with MAO Cocatalyst

Metallocene	Symmetry	Primary Chain Transfer Reaction	Type of Polymer
Cp_2ZrCl_2	C_{2v}	$(\beta\text{-H})_p$	Oligomers[40]
Cp_2HfCl_2	C_{2v}	$(\beta\text{-H})_p$	Oligomers[40]
$Cp*_2ZrCl_2$	C_{2v}	$\beta\text{-Me}$	Oligomers[41]
$Cp*_2HfCl_2$	C_{2v}	$\beta\text{-Me}$	Oligomers[41]
Ind_2ZrCl_2	C_{2v}	$\beta\text{-Me}$	Oligomers[42]
meso-$Et(Ind)_2ZrCl_2$	C_s	$\beta\text{-H}$	aPP
rac-$Et(Ind)_2ZrCl_2$	C_2	$\beta\text{-H}$	iPP[43]
rac-$Me_2Si(Ind)_2ZrCl_2$	C_2	$\beta\text{-H}$	iPP[44]
rac-$Me_2Si(2\text{-Me-Ind})_2ZrCl_2$	C_2	$\beta\text{-H}$	iPP[44]
rac-$Me_2Si(Benz[e]\text{-Ind})_2ZrCl_2$	C_2	$\beta\text{-H}$	iPP[6]
rac-$Me_2Si(2\text{-Me-Benz}[e]\text{-Ind})_2ZrCl_2$	C_2	$(\beta\text{-H})_p$	iPP[6]
rac-$Et(4,7\text{-Me}_2\text{-Ind})_2ZrCl_2$	C_2	$(\beta\text{-H})_s$	iPP[43]
$Me_2C(Cp)(Flu)ZrCl_2$	C_s	$(\beta\text{-H})_p$	sPP[40]
$Et(Flu)_2ZrCl_2$	C_{2v}	$\beta\text{-Me}$	aPP[42]
$Me_2Si(Flu)_2ZrCl_2$	C_{2v}	$\beta\text{-Mc}$	aPP[42]

$(\beta\text{-H})_p$ = β-hydride transfer from a primary growing chain.

$(\beta\text{-H})_s$ = β-hydride transfer from a secondary growing chain.

β-H = β-hydride transfer from both primary and secondary growing chains.

β-Me = β-methyl transfer from a primary growing chain.

Cp = cyclopentadienyl (C_5H_5); Cp* = C_5Me_5; Ind = 1-indenyl; 2-Me-Ind = 2-methyl-1-indenyl; 4,7-Me_2-Ind = 4,7-dimethyl-1-indenyl; Flu = 9-fluorenyl; iPP = isotactic polypropylene; aPP = atactic polypropylene; sPP = syndiotactic polypropylene.

Steric effects around the metal atom in group 4 metallocene catalysts can cause a significant increase in the energy of the transition state associated with β-H elimination, owing to the nonbonded repulsion between the polymer chain and the periphery of the bulky ligand. In order to attain the transition state for β-H elimination, it is necessary to rotate the polymer chain about the C_α–C_β bond, so that the filled C_β-H orbital overlaps with one of the vacant orbitals on the transition metal atom. This effect increases with the steric bulk of the ligand. Furthermore, as the ligands around the electron-deficient metal center become more electron releasing, the thermodynamic driving force for β-H elimination diminishes. The electron releasing ability of commonly used metallocene ligands increases in the order Cp < Ind < Flu. Table 10.1 shows the influence of metallocene ligand substitution pattern on chain transfer reactions and polypropylene molecular weight. It is clearly shown that β-H transfer after primary insertion is the most commonly recurring chain transfer pathway for zirconocenes with unsubstituted Cp ligands; for chiral bis(indenyl) zirconocenes, β-H transfer occurs from both primary and secondary chains. On the other hand, β-Me transfer becomes predominant in the case of highly substituted metallocenes, such as $Cp*_2ZrCl_2$ and zirconocenes with fluorenyl ligands. Although the methyl substitution in the 4-position of the indenyl ligand of rac-$Et(4,7\text{-Me}_2\text{-Ind})_2ZrCl_2$ increases the tendency of secondary insertion,[43] the methyl group in the 2-position of the indenyl ligand in the rac-$Me_2Si(2\text{-Me-Benz}[e]\text{-Ind})_2ZrCl_2$ system[6] suppresses the same insertion mechanism.

Chain transfer reactions have higher activation energies[4] than monomer insertions; consequently, a change in temperature strongly affects the relative rate of chain transfer as compared to propagation, which, in turn, affects polymer molecular weight. Soga[45] has reported a living propylene polymerization utilizing a $Cp_2ZrMe_2/B(C_6F_5)_3$ catalyst at $-78\ °C$ or a Cp_2HfMe_2/trioctylaluminum catalyst at $-50\ °C$ for which chain transfer is so slow that polymer molecular weight is only a function of polymerization time.

10.3 CONTROL OF POLYPROPYLENE MOLECULAR WEIGHT BY HYDROGEN

Hydrogen is the most commonly added CTA in both heterogeneous Ziegler–Natta[46–48] and metallocene-mediated[49–52] propylene polymerization. It is used to control polymer molecular weight, increase catalyst activity, and to produce polymer with saturated chain ends. The effect of hydrogen is strongly dependent upon the specific catalyst structure used and the polymerization conditions (i.e., monomer concentration, temperature, and H_2 pressure). In some cases, a substantial increase (>10 times) of catalyst activity is observed upon hydrogen addition. The most likely mechanism of the H_2 chain transfer reaction is the direct insertion of H_2 into the metal–carbon (M–C) bond of the growing polymer chain after either primary (Scheme 10.4a) or secondary (Scheme 10.4b) propylene insertion.

Kashiwa[49] described the influence of hydrogen on the catalytic efficiency of the isoselective catalyst rac-Et(Ind)$_2$ZrCl$_2$/MAO in toluene at $30\ °C$ and low propylene concentration (~ 1 atm propylene). On the basis of the strong activating effect of hydrogen on this catalyst, along with the presence of n-butyl end groups (viii) and the disappearance of 1,3-regioerrors, they concluded that the activating effect arises from the hydrogenolysis of secondary (2,1-inserted) "dormant" chains (Scheme 10.4b). The M–H active site regenerated after chain transfer reinitiates propylene polymerization, incorporating an n-propyl group at the beginning of the new polymer chain.

Mülhaupt[6] examined two highly isoselective metallocenes, rac-Me$_2$Si(Benz[e]-Ind)$_2$ZrCl$_2$ and rac-Me$_2$Si(2-Me-Benz[e]-Ind)$_2$ZrCl$_2$, which showed a limited catalyst activation by hydrogen (38% and 17%, respectively, for polymerization at $40\ °C$ in toluene, 0.35 bar H_2, and 1.9 bar pressure of propylene monomer). An appreciable amount of internal regioerrors (head-to-head enchainments) were observed in the isotactic polypropylene (iPP) samples prepared with both catalysts, showing that at the low concentrations of hydrogen used, propylene insertion into a secondary growing chain is still competitive with hydrogenolysis. All of the effects of hydrogen on the polymerization, including a reduction of 2,1-inserted head-to-head monomer sequences to one-third of their original content, formation of n-butyl end groups (viii, Scheme 10.4b) (with a 1:1 n-butyl:n-propyl end group molar ratio, both increasing in parallel with the reduction of incorporated 2,1-units), and a strong hydrogen response of polymer molecular weight reduction to hydrogen concentration, indicate hydrogen-mediated chain transfer at secondary growing chains. Since no isobutyl end groups (vii,

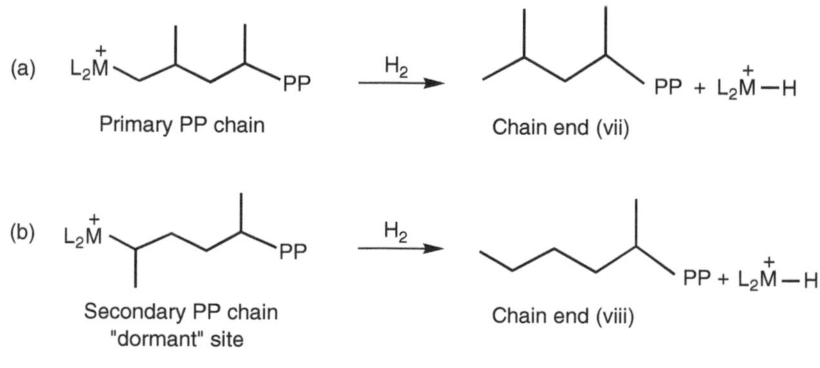

SCHEME 10.4 Hydrogen chain transfer reactions.

Scheme 10.4a) were detected, the hydrogen-mediated chain transfer at secondary growing chains is much faster than transfer at primary chains for these two zirconocenes.

10.4 CHAIN-END FUNCTIONALIZED POLYPROPYLENE

A polymer containing a terminal functional group presents a unique opportunity to serve as a building block for constructing multisegmented polymers and for forming composites with fillers. For polypropylene, the opportunity is even more interesting owing to its commercial importance and the scientific challenge in finding suitable routes to prepare functional PPs[8,53] with desirable functional groups. In general, there are three methods used for the preparation of polymers with terminal functional groups. These are: (1) living polymerization[54–56] with a functional initiator or controlled termination with a specific functional reagent, (2) in situ chain transfer reactions carried out during polymerization (*vide infra*), and (3) chemical modification[57–59] of a polymer having an unsaturated chain end. The use of in situ chain transfer reactions is by far the most convenient and effective method for introducing a functional group to the polymer chain end. With the combination of a well-defined metallocene catalyst and a suitable CTA, it is possible to have a chain transfer reaction take place with little change in the base polymerization rate, and with each polymer chain produced bearing a terminal functional group (the residue of the CTA). Usually, polymer molecular weight is inversely proportional to the [CTA]/[monomer] ratio.

Several interesting CTAs have been studied over the past two decades, with the major objective of this work being the introduction of functional (polar) groups at the polypropylene chain end (see Chapter 11 for a discussion of the synthesis of tactic polypropylenes having functionalities at locations other than the chain end). As discussed previously, an aluminum alkyl (MAO/Me$_3$Al) was reported to act as both a coinitiator and a CTA in propylene polymerization. Zinc alkyls were also reported to act with dual function in the heterogeneous Ziegler–Natta polymerization of propylene. Recently, two other hydride CTAs, silanes and boranes containing Si-H or B-H moieties, respectively, have been successfully used in this manner in metallocene-mediated polymerization (*vide infra*). For the latter, this chemistry is very general for most metallocene systems and can be applied to the polymerization of most α-olefin monomers and their mixtures. The resulting silane- and borane-terminated polyolefins, having relatively well-defined molecular structures, can be used as intermediates for the preparation of more elaborate polymer structures. It is very intriguing that the versatility of metallocene catalysis also allows this chain transfer reaction principle to be extended to styrenic and allylic molecules, which typically serve instead as monomers. These new routes allow for the preparation of chain-end functional polypropylenes containing terminal functional groups, such as OH and NH$_2$, through a one-pot polymerization process.

10.4.1 ZINC ALKYL-TERMINATED POLYPROPYLENE

For a long time, zinc alkyls have been known to have a dual function[60] as both coinitiators (alkylating agents) and CTAs in Ziegler–Natta polymerizations. The addition of a large amount of ZnEt$_2$ to the TiCl$_3$/AlEt$_3$ propylene polymerization system causes a marked decrease in polymer molecular weight. The resulting EtZn-terminated iPP[61] can be converted into various functional groups, such as hydroxy, carboxyl, halogen, or vinyl, in approximately 50% yield. The remaining 50% of the polymers contain no functional group.

In the early 1990s, Soga[62,63] studied several bis(ω-alkenyl)zinc compounds, such as bis(3-butenyl)zinc and bis(4-methyl-3-pentenyl)zinc, as CTAs in propylene polymerization in an attempt to prepare polypropylenes having a terminal alkylzinc group on one chain end and an alkenyl group on the other. As expected for a heterogeneous TiCl$_3$ catalyst system with multiple active sites, the reaction forms a mixture of polymer structures with zinc groups both in the side chains and on the chain ends. In fact, bis(ω-alkenyl)zinc serves three roles in the reaction: as coinitiator, comonomer

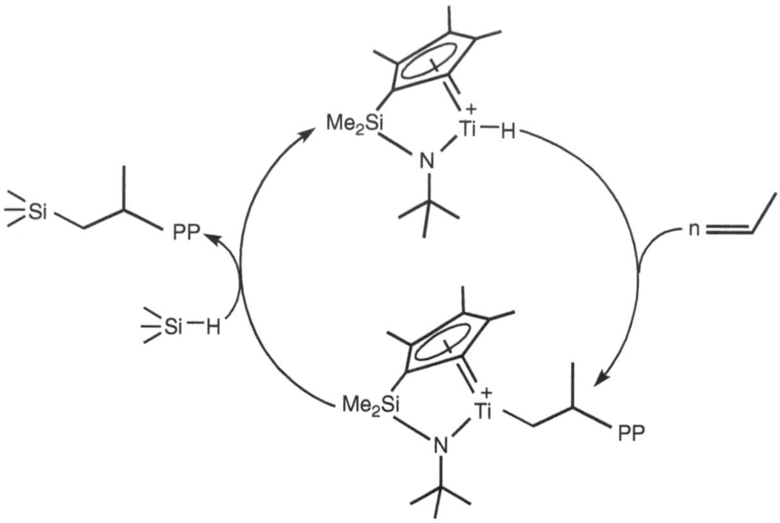

SCHEME 10.5   Catalytic cycle for $[Me_2Si(Me_4C_5)(t\text{-}BuN)TiMe]^+[B(C_6F_5)_4]^-$-mediated propylene polymerization using $PhSiH_3$ chain transfer agent to form a silyl-terminated atactic polypropylene product $(aPP\text{-}SiPhH_2)$.

(through polymerization of the ω-alkenyl group), and CTA. Overall, the polymer product has both broad molecular weight and composition distributions.

10.4.2   SILANE-TERMINATED POLYPROPYLENE

Marks[64] reported that organosilanes containing Si-H groups serve as CTAs in lanthanocene-mediated olefin polymerizations to afford silyl-terminated ethylene polymers and copolymers, including poly(ethylene-co-1-hexene) and poly(ethylene-co-styrene). This chemistry was expanded to group 4 metallocene catalysts[65,66] and other monomers, including propylene. The primary silane $PhSiH_3$ was shown to be an efficient CTA in propylene polymerization systems using a catalyst with a relatively sterically open active site. An example is the polymerization of propylene in the presence of $PhSiH_3$ with the constrained geometry catalyst $[Me_2Si(Me_4C_5)(t\text{-}BuN)TiMe]^+[B(C_6F_5)_4]^-$, illustrated in Scheme 10.5.

The molecular weight of the silyl-terminated atactic PP product $(aPP\text{-}SiPhH_2)$ is inversely proportional to $PhSiH_3$ concentration. It is interesting to note that only one of the three Si-H moieties in $PhSiH_3$ participates in the chain transfer reaction with the $[Me_2Si(Me_4C_5)(t\text{-}BuN)TiMe]^+[B(C_6F_5)_4]^-$ system. However, the remaining two Si-H moieties in $aPP\text{-}SiPhH_2$ are reactive toward other catalyst systems having more open active sites, such as $[(Me_5C_5)TiMe_2]^+[B(C_6F_5)_4]^-$. In one study,[66] $aPP\text{-}SiPhH_2$ was used as a polymeric CTA for the polymerization of styrene with a single-Cp Ti catalyst, $[(Me_5C_5)TiMe_2]^+[B(C_6F_5)_4]^-$. Diblock copolymers of atactic polypropylene-syndiotactic polystyrene were produced.

10.4.3   BORANE-TERMINATED POLYPROPYLENE

Chung[67] has investigated organoboranes (containing the B-H moiety) as CTAs for olefin polymerizations. Given the facile ligand exchange between B–H bonds and most metal-alkyl groups, it is logical to expect a fast chain transfer reaction to take place in most metallocene catalyst systems.[68–70] Indeed, the B-H chain transfer reaction also occurs with $[Me_2Si(2\text{-}Me\text{-}4\text{-}Ph\text{-}Ind)_2ZrMe]^+[B(C_6F_5)_4]^-$, for which hydrogen and Si-H chain transfer reactions are ineffective. As shown in Scheme 10.6, the resulting borane-terminated polyolefin is a very versatile intermediate, and can be converted into a broad

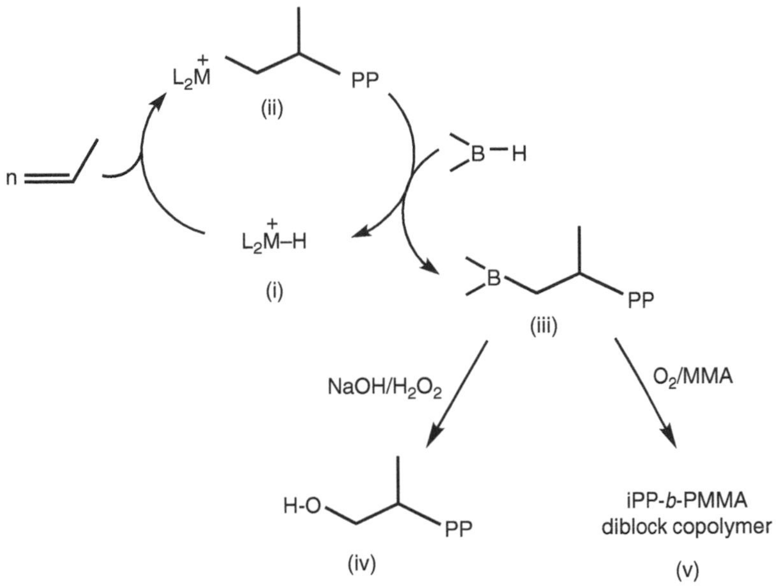

SCHEME 10.6 Reaction mechanism for B-H chain transfer reactions and subsequent transformations to prepare chain-end functionalized iPP and iPP-*b*-PMMA diblock copolymer (MMA = methyl methacrylate; PMMA = poly(methyl methacrylate)).

range of polar group-terminated polyolefins and diblock copolymers containing both polyolefin and functional polymer segments.

In the presence of a dialkylborane containing a B-H group, the metallocene-mediated propagating isotactic PP chain (ii) engages in a facile ligand exchange reaction between the B–H and C–M bonds, owing to the favorable acid–base interaction between the cationic metal center and anionic hydride. This ligand exchange reaction results in a borane-terminated polyolefin (iii) and a new active site (i) that can reinitiate polymerization. Ideally, this chain transfer reaction should not change the overall catalyst activity, and each polymer will contain a terminal borane group with the average polymer molecular weight inversely proportional to the molar ratio of [CTA]/[olefin].[67] The resulting borane-terminated polyolefin (iii) can be converted into a hydroxy-terminated polyolefin (iv) by treatment with $NaOH/H_2O_2$, or can be selectively transformed to a polymeric peroxide containing a peroxylborane (C-O-O-BR_2) moiety by a simple oxygen auto-oxidation reaction. This peroxylborane is a reactive radical initiator even at ambient temperature. Spontaneous homolytic cleavage of the peroxide group generates a reactive alkoxy radical (C-O*) and a stable, dormant (owing to the back-donation of electron density to the empty π-orbital of boron) boronate radical (*O-BR_2). The alkoxy radical then initiates the radical polymerization of functional (polar) monomers such as methyl methacrylate (MMA) at ambient temperature (chain extension). The dormant boronate radical serves as an endcapping agent during radical polymerization, forming a weak, reversible bond with the growing chain end. This process minimizes undesirable radical chain transfer and termination reactions between the growing chain ends. On the whole, this process produces functional polyolefin diblock copolymers (v) with the nature and concentration of the functional groups in the polymer determined by the functional monomers introduced during the chain extension reaction.[67]

There are two potential side reactions that can derail the catalytic cycle shown in Scheme 10.6, namely, hydroboration of the olefin monomer by borane and ligand exchange reactions between the borane and cocatalysts containing aluminum alkyl groups. Fortunately, borane compounds containing B-H groups usually form stable dimers (Figure 10.1) that are unreactive toward olefins in typical olefin polymerization solvents (e.g., hexane, toluene). To eliminate the second process, however,

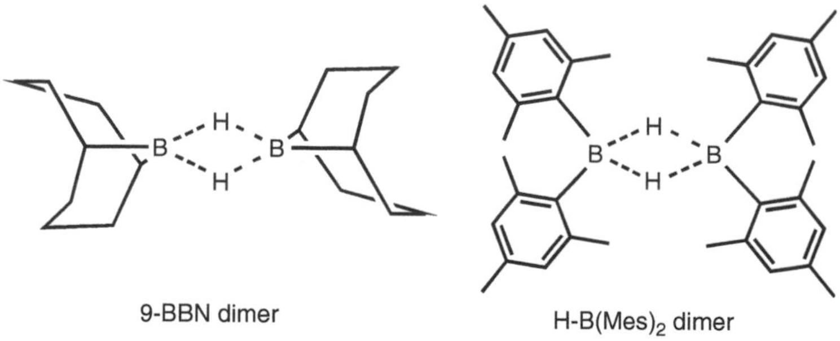

FIGURE 10.1 Two borane dimers used in chain transfer reactions (9-BBN = 9-borabicyclo[3.3.1]nonane; H-B(Mes)$_2$ = dimesitylborane).

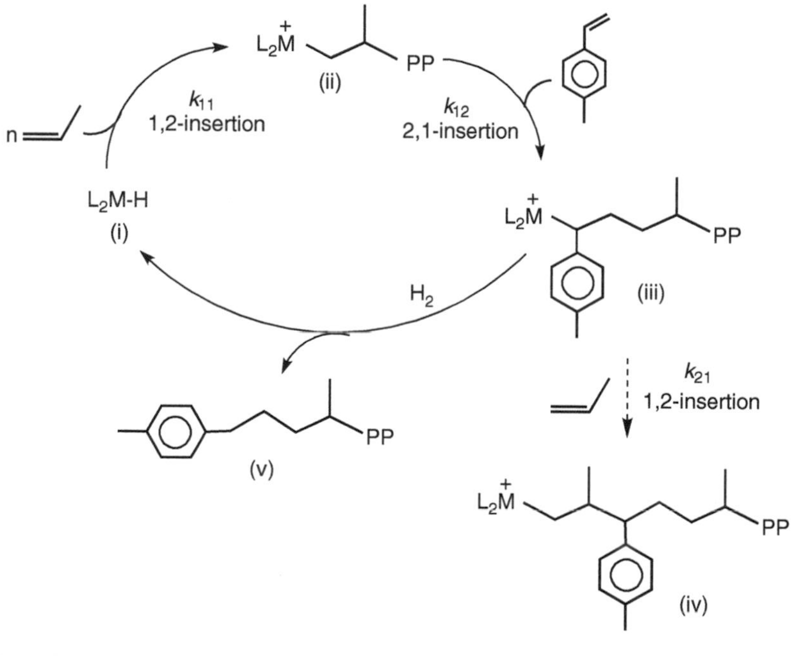

SCHEME 10.7 Catalytic cycle during metallocene-mediated propylene/p-MS copolymerization, with p-MS/H$_2$ chain transfer reaction (monomer 1 = propylene; monomer 2 = p-MS).

a perfluoroborate cocatalyst (stable toward most alkylborane compounds) must be used instead of MAO or aluminum alkyl activators.

10.4.4　Styrene-Terminated Polypropylene

It is most intriguing to expand chain transfer methodology to styrenic molecules that usually serve as monomers during polymerization reactions. Chung[71,72] has studied a new two-component chain transfer system based on the combination of styrene derivatives and hydrogen. This system performs most effectively with certain stereoselective metallocene catalysts, including isoselective ones used for propylene polymerization. Ironically, the metallocene catalysts best suited for propylene polymerization usually show very poor styrene incorporation. In fact, this research stemmed from several observations[73] made during propylene/p-methylstyrene (p-MS) copolymerization using an isoselective rac-Me$_2$Si(2-Me-4-Ph-Ind)$_2$ZrCl$_2$/MAO catalyst system, as illustrated in Scheme 10.7.

The copolymerization reaction became completely stalled at the very beginning of the copolymerization process. Catalyst deactivation was speculated to result from steric interference during the process of 1,2-propylene insertion (k_{21}; monomer 1 = propylene; monomer 2 = p-MS) following a previous 2,1-insertion of p-MS (k_{12}). The combination of this unfavorable 1,2-insertion of propylene and the lack of p-MS homopolymerization (k_{22} reaction) at the propagating site (iii, Scheme 10.7) drastically reduces catalyst activity. This hypothesis was supported by fact that addition of a small amount of ethylene dramatically improves catalyst activity. The sluggish propagating chain end (iii) more readily allows the insertion of ethylene, which reenergizes the propagation process.

Assuming the above mechanism to be operative, the dormant propagating site (iii) may be advantageously reacted with hydrogen, a process that not only regenerates the M–H catalytic site (i) but also produces iPP with a terminal p-MS group (v) (iPP-p-MS). Ideally, the chain transfer reaction does not significantly affect the rate of polymerization, but reduces the average molecular weight of the resulting polymer. The molecular weight of the iPP-p-MS should be linearly proportional to the concentration ratio of propylene to p-MS, and independent of the propylene to hydrogen ratio.

Table 10.2 summarizes a systematic study of propylene polymerization using rac-Me$_2$Si(2-Me-4-Ph-Ind)$_2$ZrCl$_2$/MAO in the presence of p-MS/hydrogen.[72] The in situ chain transfer reaction is evidenced by its comparison with control reactions carried out under similar reaction conditions without hydrogen and/or p-MS. A small amount of p-MS used alone (control 2) effectively stops the polymerization of propylene. The introduction of hydrogen (runs 1–4) restores the catalyst activity; run 4 exhibits about 85% of the catalyst activity of control 1 (a polymerization without any CTAs).

TABLE 10.2

A Summary of PP-p-MS Polymers Prepared by Batch Reaction Using rac-Me$_2$Si(2-Me-4-Ph-Ind)$_2$ZrCl$_2$/MAO Catalyst and p-MS/Hydrogen Chain Transfer Agent[a]

Run	p-MS (mol/L)	H$_2$ (psi)	Yield (g)	Catalyst Activity[b]	p-MS in PP (mol%)[c]	p-MS Conversion (%)	M_n[d]	M_w/M_n[d]
Control 1	0	0	26.94	86,208	0	–	77,600	2.9
Control 2	0.0305	0	0.051	163	0.16	0.05	59,700	3.4
1	0.0305	2	3.80	12,160	0.14	8.30	55,500	2.4
2	0.0305	6	8.04	25,728	0.15	18.83	54,800	2.5
3	0.0305	12	12.04	38,528	0.15	28.19	55,400	2.3
4	0.0305	35	24.67	78,944	0.13	50.05	34,600	2.8
Control 3	0.076	0	0	–	–	–	–	–
5	0.076	6	0.91	2,912	0.40	2.33	27,600	2.1
6	0.076	12	1.69	5,408	0.41	4.33	25,900	2.3
7	0.076	20	8.81	28,192	0.43	23.65	20,500	2.3
8	0.076	35	10.52	33,664	0.41	26.86	25,800	2.3
Control 4	0.153	0	0	–	–	–	–	–
9	0.153	12	0.35	1,120	0.66	0.72	10,000	2.0
10	0.153	20	3.81	12,192	0.61	7.26	11,700	2.0
11	0.153	35	4.41	14,112	0.63	8.67	9,700	1.9

[a] Reaction conditions: 50 mL toluene, 100 psi propylene, [Zr] = 1.25×10^{-6} mol/L, [MAO]:[Zr] = 3000:1, temperature = 30 °C, time = 15 min.

[b] Catalyst activity in kg PP/mol Zr·h.

[c] Determined by ^{1}H NMR.

[d] Measured by gel permeation chromatography; M_n = number average molecular weight; M_w = weight average molecular weight.

Source: Reproduced with permission from Chung, T. C.; Dong, J. Y. *J. Am. Chem. Soc.* **2001**, *123*, 4871–4876. Copyright 2001 American Chemical Society.

Hydrogen is clearly needed to complete the chain transfer cycle during polymerization. The higher the concentration of p-MS, the lower the molecular weight of the resulting polymer (e.g., comparison of runs 4, 8, and 11). Polymer with very low molecular weight ($M_n \sim 2000$) has been obtained by further increasing p-MS concentration to 0.45 mol/L, and the molecular weight distributions of the polymers are generally narrow, consistent with a well-defined single-site polymerization process.[1–3] The catalyst activity is also proportionally depressed with increasing p-MS concentration, which reflects the slow chain transfer and reinitiation process.

Three comparative reaction sets (runs 1–4, runs 5–8, and runs 9–11) were conducted under the same reaction conditions, varying hydrogen pressure between the runs in each set from 0 to 35 psi. Changing the hydrogen concentration does not significantly affect the polymer molecular weight and molecular weight distribution, but has a profound effect on catalyst activity. Therefore, it can be concluded that hydrogen does not engage in the initial chain transfer reaction, but rather assists in the completion of the reaction cycle (as shown in Scheme 10.7). A sufficient concentration of hydrogen, as compared to the p-MS concentration, is needed to maintain high catalyst activity and p-MS conversion.

The effects of the chain transfer reaction are further revealed by the existence of terminal p-MS groups. Figure 10.2 shows the ^{13}C NMR spectrum of an iPP-p-MS sample (M_n 4600; M_w/M_n

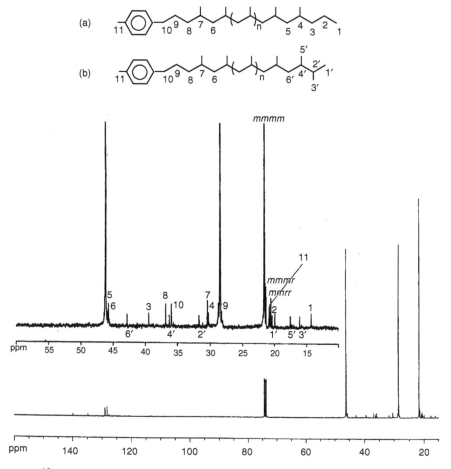

FIGURE 10.2 ^{13}C NMR spectra of iPP-p-MS (M_n 4600; M_w/M_n 1.7) with insets for the expanded aliphatic region and for two types of chain-end groups (CDCl$_2$CDCl$_2$, 110 °C). (Reproduced with permission from Chung, T. C.; Dong, J. Y. *J. Am. Chem. Soc.* **2001**, *123*, 4871–4876. Copyright 2001 American Chemical Society.)

1.7) prepared with *rac*-Me$_2$Si(2-Me-4-Ph-Ind)$_2$ZrCl$_2$/MAO, with an inset of the expanded aliphatic region. In addition to three major peaks (δ 21.6, 28.5, and 46.2 ppm) corresponding to the **CH$_3$** (*mmmm*), **CH$_2$**, and **CH** groups in the PP backbone, the spectrum exhibits all of the carbon chemical shifts associated with both chain ends. Two types of chain start structures (a and b in the inset of Figure 10.2) are seen, owing to the initiation reaction of Zr-H species (i, Scheme 10.7) with 1,2- and 2,1-insertions of propylene, respectively. Although both insertion modes are allowed in the initiation step, it is generally accepted that the propagation step of isotactic propylene polymerization with zirconocene catalysts takes place by a regioselective 1,2-insertion of the propylene monomer at the Zr-C active center (ii, Scheme 10.7). The peak intensity ratio indicates that the *p*-MS and alkyl chain ends are present in about a 1:1 molar ratio, and the iPP-*p*-MS molecular weights derived from NMR and gel permeation chromatography (GPC) are in good agreement. It is important to note that there are no detectable vinyl groups (associated with conventional β-H elimination chain transfer processes), nor any chemical shifts for enchained pendant *p*-MS units (associated with propylene/*p*-MS copolymerization).

10.4.5 FUNCTIONALIZED STYRENE-TERMINATED POLYPROPYLENE

It is very desirable to extend this styrene chain transfer reaction to the direct preparation of polypropylenes with useful terminal functional groups[74] such as Cl, OH, and NH$_2$. In other words, the ideal polypropylene functionalization reaction is a one-pot in situ polymerization process where no chain-end functionalization would be needed after the polymerization reaction. The basic idea in the direct preparation of the chain-end functionalized iPP is to use a functionalized styrenic CTA (St-f) that can engage in metallocene-mediated propylene polymerization/chain transfer reactions, similar to *p*-MS. Three St-f molecules have been investigated: *p*-chlorostyrene (St-Cl), trialkylsilane-protected *p*-vinylphenol (St-OSi(Me)$_2$(*t*-Bu)), and trialkylsilane-protected *p*-ethylaminostyrene (St-N(SiMe$_3$)$_2$) (Figure 10.3). No external protecting agent is needed for St-Cl in the *rac*-Me$_2$Si(2-Me-4-Ph-Ind)$_2$ZrCl$_2$/MAO system. However, the cationic metallocene active site is very sensitive to both the OH and NH$_2$ functionalities. The trialkylsilane groups provide effective protection for the OH and NH$_2$ groups during metallocene-mediated polymerization, and can be completely removed by aqueous HCl solution during the polymer workup procedure. The overall reaction especially benefits from the very small quantity of the CTA needed in the preparation of high-molecular-weight polymers. Therefore, the additional protection–deprotection step causes almost no change in the polymerization conditions and procedures.

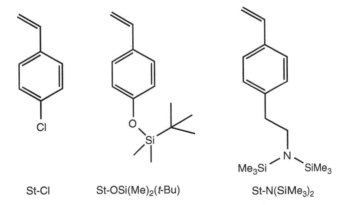

FIGURE 10.3 Three functionalized styrenic chain transfer agents (St-f) used in *rac*-Me$_2$Si(2-Me-4-Ph-Ind)$_2$ZrCl$_2$/MAO-mediated propylene polymerization to prepare chain-end functionalized iPP polymers.

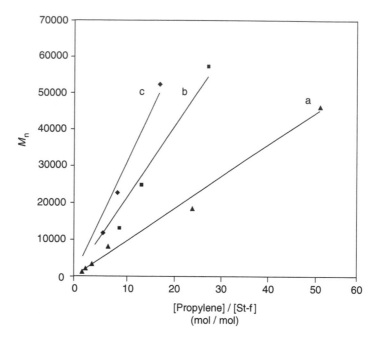

FIGURE 10.4 Plots of number average molecular weight (M_n) of iPP-St-f polymers versus the concentration ratio of [propylene]/[St-f] using (a) St-Cl/H$_2$, (b) St-OSi(Me)$_2$(t-Bu)/H$_2$, and (c) St-N(SiMe$_3$)$_2$/H$_2$ chain transfer agents with the rac-Me$_2$Si[2-Me-4-Ph(Ind)]$_2$ZrCl$_2$ catalyst system. (Reproduced with permission from Dong, J. Y.; Wang, Z. M.; Han, H.; Chung, T. C. $Macromolecules$ **2002**, 35, 9352–9359. Copyright 2002 American Chemical Society.)

In a typical reaction, a 450 mL autoclave equipped with a mechanical stirrer was charged with 50 mL of toluene and 4.5 mL of MAO solution (10 wt% in toluene) before purging with hydrogen (20 psi). Then, the desired amount of St-f was injected into the reactor and 100 psi (3.24 M concentration) of propylene was applied, bringing the total pressure to 120 psi at ambient temperature. About 1.25×10^{-6} mol of rac-Me$_2$Si[2-Mc-4-Ph(Ind)]$_2$ZrCl$_2$ catalyst in toluene solution was then syringed into the rapidly stirred solution under propylene pressure to initiate the polymerization reaction. Additional propylene was fed continuously into the reactor to maintain a constant pressure (120 psi) during the course of the polymerization. After 15 min of reaction at 30 °C, the reaction solution was quenched with methanol and filtered.

Figure 10.4 shows plots of iPP polymer molecular weight (M_n) versus the concentration ratio of propylene to St-f for the St-Cl/H$_2$, St-OSi(Me)$_2$(t-Bu)/H$_2$, and St-N(SiMe$_3$)$_2$/H$_2$ chain transfer systems.[74] In general, the average polymer molecular weight and the concentration ratio of propylene to St-f are linearly proportional. It is clear that the chain transfer reaction to St-f (rate constant k_{tr}) is the dominant chain-termination process, and that it competes with propagation (rate constant k_p). The degree of polymerization (X_n) follows a simple comparative equation, $X_n = k_p$[propylene]/k_{tr}[p-MS], with chain transfer constants (k_{tr}/k_p) of 1/21 for St-Cl/H$_2$, 1/48 for St-OSi(Me)$_2$(t-Bu)/H$_2$, and 1/34 for St-N(SiMe$_3$)$_2$/H$_2$. It is intriguing that the k_{tr}/k_p values are significantly lower than those seen for styrene/H$_2$ and p-MS/H$_2$ CTAs[72] under similar reaction conditions. The relatively large steric bulk of the protected functional groups may reduce the frequency of the chain transfer reaction. As shown in Figure 10.4, chain-end functionalized iPP polymers with very low molecular weights ($M_n \sim 3000$) have been obtained, and the molecular weight distributions of the polymers are quite narrow ($M_w/M_n \sim 2$), which is generally consistent with a single-site polymerization process.[1–3]

The terminal functional group at the chain end provides direct evidence for the chain transfer reaction. Figure 10.5 (left) shows the ^{1}H NMR spectra (with inset of magnified region and chemical shift

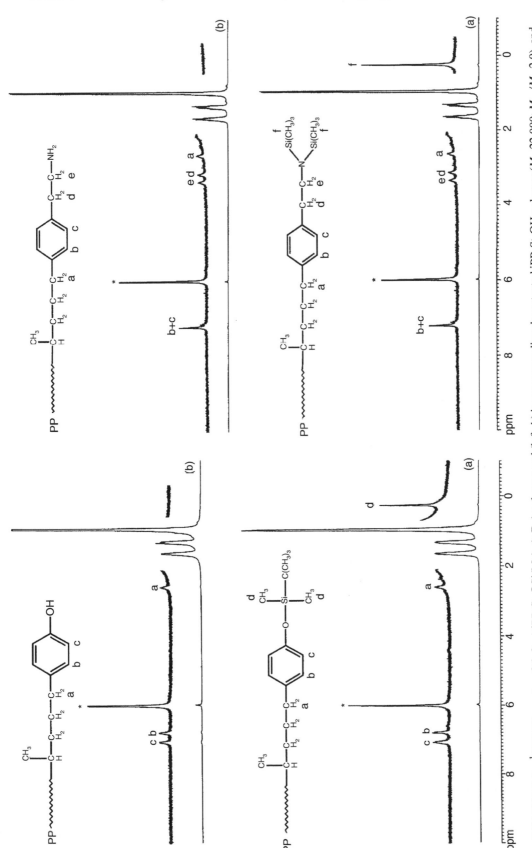

FIGURE 10.5 ¹H NMR spectra of (left, a) iPP-St-OSi(Me)₂(t-Bu) polymer and (left, b) its corresponding deprotected iPP-St-OH polymer (M_n 22,000; M_w/M_n 2.0), and (right, a) iPP-St-N(SiMe₃)₂ polymer and (right, b) its corresponding deprotected iPP-St-NH₂ polymer (M_n 24,200; M_w/M_n 2.3) (CDCl₂CDCl₂, 110 °C). (Reproduced with permission from Dong, J. Y.; Wang, Z. M.; Han, H.; Chung, T. C. *Macromolecules* **2002**, 35, 9352–9359. Copyright 2002 American Chemical Society.)

assignments) of iPP-St-OSi(Me)$_2$(t-Bu) (a) prepared with rac-Me$_2$Si[2-Me-4-Ph(Ind)]$_2$ZrCl$_2$/MAO catalyst and the corresponding deprotected iPP-St-OH polymer (b) (M_n 22,000; M_w/M_n 2.0).[74] For iPP-St-OSi(Me)$_2$(t-Bu), in addition to three major peaks (δ 0.95, 1.35, and 1.65 ppm) for the **CH$_3$**, **CH$_2$**, and **CH** groups in the iPP backbone, there are three minor chemical shifts at 0.25, 2.61, and 6.75–7.18 ppm (with an intensity ratio near 6:2:4) shown in Figure 10.5 (left, a), corresponding to the -OSi(**CH$_3$**)$_2$(t-Bu), -**CH$_2$**-St-OSi, and -**CH$_2$**-**C$_6$H$_4$**-OSi groups, respectively. The chemical shift for the silane protecting group completely disappears in Figure 10.5 (left, b), indicating complete deprotection. The equally split chemical shifts for the phenyl protons (labeled b and c) in both spectra, combined with no detectable side products, further indicate clean formation of the terminal p-alkylphenol moiety. Figure 10.5 (right) compares the ^{1}H NMR spectra[74] of similarly prepared iPP-St-N(SiMe$_3$)$_2$ (a) and the corresponding deprotected iPP-St-NH$_2$ polymer (b) (M_n 24,200; M_w/M_n 2.3). In addition to the chemical shifts for iPP polymer, Figure 10.5 (right, a) shows all of the chemical shifts associated with the bis(trimethylsilyl)amino terminal group connected to the p-dialkylbenzene moiety. All four phenyl protons merge into a single chemical shift at 7.22 ppm. Figure 10.5 (right, b) shows an almost identical spectrum, except for the disappearance of the trimethylsilane protecting groups at 0.24 ppm.

10.4.6 ALLYL MONOMER-TERMINATED POLYPROPYLENE

Shiono[75] recently reported that olefin monomers bearing allylic OH or NH$_2$ groups serve as CTAs during propylene polymerization using the rac-Me$_2$Si(Ind)$_2$ZrCl$_2$/MAO catalyst system. To prevent catalyst poisoning, the allyl monomers were pretreated with alkylaluminums, including trimethyl- and triisobutylaluminum, before polymerization. In general, the catalyst activity was very low, less than 10% of that for the corresponding propylene homopolymerization under similar reaction conditions. The incorporation of allyl alcohol (protected with trimethyl- and triisobutylaluminum) was extraordinarily low (0.04–0.08 mol%); the Al(CH$_3$)$_3$-treated allylamine was found to be somewhat better incorporated (0.65 mol%). The polymer ^{13}C NMR spectra indicated that the position of the allylamine unit in the polymer was influenced by the nature of the alkylaluminum protecting group. When Al(CH$_3$)$_3$ was used, 83% of the iPP polymer obtained contain a terminal NH$_2$ group. However, when Al(i-Bu)$_3$ was used, the incorporated allylamine units are located both at the chain end and within the main chain (see Chapter 11 for a further discussion of the use of this system to prepare side-chain-functionalized iPP).

As illustrated in Scheme 10.8, the amino group-terminated polypropylene is probably obtained through a special chain transfer mechanism that may involve a "dormant" species (iii) formed after insertion of the alkylaluminum-treated allylamine. The nitrogen atom of the inserted masked allylamine is located near the Zr cation and results in a stable complex (iii), to which nucleophilic attack from Al(CH$_3$)$_3$ can occur at the chain end to produce a polymeric intermediate (iv). With HCl treatment during workup, this intermediate (iv) is converted into iPP-NH$_2$ (v).[75]

10.5 APPLICATIONS OF CHAIN-END FUNCTIONALIZED POLYPROPYLENE

10.5.1 DIBLOCK COPOLYMERS

Diblock copolymers are known to be the most effective compatibilizers[76] for improving the interfacial interactions between two polymers that are immiscible. This is particularly interesting for iPP, since its lack of functionality and the poor compatibility between iPP and other materials have imposed limitations for iPP applications in many areas, including polymer blends and composites. The synthesis of iPP with terminal functional groups (OH, NH$_2$, etc.) offers a good opportunity to carry out chain extensions through simple coupling reactions with suitable polymers. These may be carried out in solution or in the polymer melt. Reactive extrusion of two chain-end reactive polymers

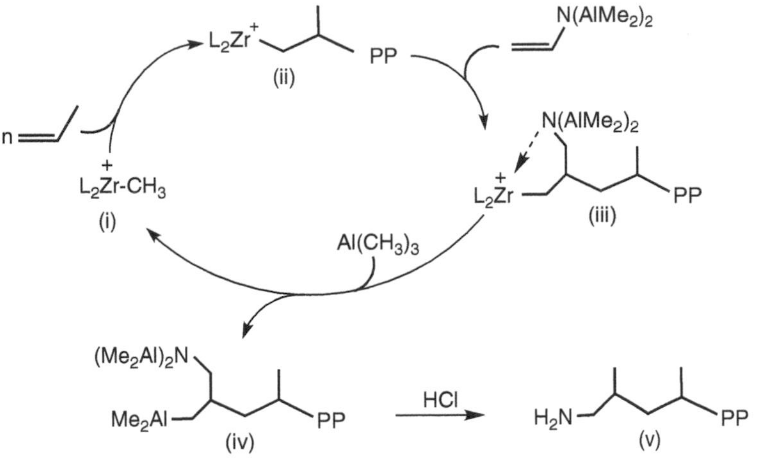

SCHEME 10.8 The postulated mechanism for *rac*-Me$_2$Si(Ind)$_2$ZrCl$_2$/MAO-mediated propylene polymerization with Al(CH$_3$)$_3$-treated allylamine chain transfer agent, forming iPP-NH$_2$ polymer.

can be used to form small amounts of diblock copolymers in situ at the interfaces of a polymer blend; these in situ-formed diblocks can then serve as compatibilizers for the polymer blend.

One example of diblock formation through the coupling of chain-end functional polymers is the coupling reaction between iPP-St-NH$_2$ (M_n 110,000; M_w/M_n 2.0) and poly(ε-caprolactone) (PCL; M_n 50,000; M_w/M_n 2.0) containing a terminal COOH group (1:1 molar ratio, reacted in refluxing toluene).[77] The resulting iPP-*b*-PCL diblock copolymer (with an ester linkage) was subjected to vigorous Soxhlet extraction with boiling acetone to remove any unreacted PCL homopolymer. The diblock nature of the resultant acetone-insoluble polymer sample is evidenced by several characterization techniques. The intrinsic viscosity after the coupling reaction changes from 1.03 dL/g for the iPP-St-NH$_2$ precursor to 1.42 dL/g for the iPP-*b*-PCL diblock. The GPC trace of the copolymer shows a single peak with an increased polymer molecular weight and a relatively narrow molecular weight distribution. The ^{1}H NMR spectrum of the material (in 1,1,2,2-dichloroethane-d_2 at elevated temperature) indicates a composition of about 30 mol% PCL, which is consistent with the expected iPP-*b*-PCL composition.

This iPP-*b*-PCL diblock copolymer has served as a compatibilizer in iPP/PCL polymer blends.[77] Two polymer blends—an iPP/PCL homopolymer blend (70:30 weight ratio) and a blend comprised of a 70:30:10 weight ratio of homo-iPP, homo-PCL, and iPP-*b*-PCL, respectively—were prepared by homogeneous mixing in dichlorobenzene solution at 180 °C and precipitation into hexane at ambient temperature. Films were then press-molded at 180 °C. Figure 10.6 compares scanning electron microscopy (SEM) images of the cross sections of cryofractured films (prepared at liquid N$_2$ temperature) of the homopolymer blend (a) and the blend containing the diblock compatibilizer (b), which are indicative of the bulk morphologies.

In the homopolymer blend (Figure 10.6a), the polymers are grossly phase separated, as can be seen by a PCL minor component that exhibits nonuniform, poorly dispersed domains, and voids at the fracture surface. This "ball and socket" topography is indicative of poor interfacial adhesion between the iPP and PCL domains and represents PCL domains that are pulled out of the iPP matrix. Such pullout indicates that no stress transfer takes place between the phases during fracture. Upon blending iPP and PCL with the in situ-formed iPP-*b*-PCL compatibilizer, a drastic change in morphology occurs. The compatibilized blend (Figure 10.6b) no longer displays distinct PCL globules, and has a rather flat, featureless surface, indicating very small domain sizes. The addition of the diblock copolymer to the system stabilizes the interfaces and increases the interfacial adhesion between the iPP and PCL microdomains. Similar improvements for iPP/Nylon 11 and iPP/PEO (PEO = polyethylene oxide) blends using iPP-*b*-polyamide (Nylon 11)[59] and iPP-*b*-PEO[78] compatibilizers

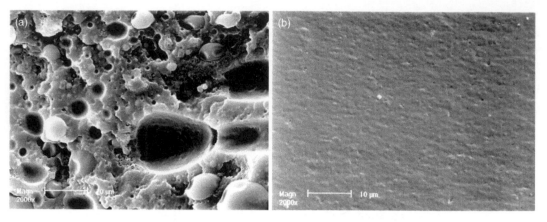

FIGURE 10.6 SEM micrographs of (a) 70:30 iPP/PCL homopolymer blend and (b) 70:30:10 iPP/PCL/iPP-*b*-PCL blend (2000x). (Reproduced with permission from Dong, J. Y.; Wang, Z. M.; Han, H.; Chung, T. C. *Macromolecules* **2002**, *35*, 9352–9359. Copyright 2002 American Chemical Society.)

(prepared, respectively, through the coupling reaction between iPP-St-NH$_2$ and chain-end functional Nylon 11 and the chain extension of iPP-OH with ethylene oxide) were also observed. In fact, iPP-St-NH$_2$ itself has been used directly as a component in the reactive blending between iPP and Nylon 11 in a melt mixer (Brabender). The in situ-formed iPP-*b*-Nylon 11 diblock copolymer, located right at the iPP/Nylon 11 interfaces in the blend, dramatically reduces domain size and improves morphology in PP/Nylon 11 blends.

10.5.2 POLYPROPYLENE/CLAY NANOCOMPOSITES

Recently, polymer/clay nanocomposites (having a nanodispersion of inorganic silicate layers in a polymer matrix)[79] have attracted great attention. This nanocomposite structure simultaneously increases the tensile strength, flex modulus, and impact toughness of a polymer, contributes a general flame retardant character, and affords a dramatic improvement in barrier properties that cannot be realized by conventional fillers.[80] The availability of a broad range of well-defined, functionalized iPP polymers provides a great advantage for evaluating the applications of functionalized iPPs in iPP/clay nanocomposites.[81] Ammonium group-terminated iPP (iPP-NH$_3^+$Cl$^-$) polymers have been prepared by simple mixing between iPP-NH$_2$ and HCl. Both pristine Na$^+$-montmorillonite clay (Na$^+$-mmt), with an ion-exchange capacity of ca. 0.95 meq/g, and a dioctadecylammonium-modified montmorillonite organophilic clay (2C18-mmt) (Southern Clay Products) have been used for the preparation of nanocomposites through static melt intercalation.

Figure 10.7a compares the X-ray diffraction (XRD) patterns[81] before and after static annealing of a physical mixture (90:10 weight ratio) of iPP-NH$_3^+$Cl$^-$ (M_n 58,900; M_w 135,500; melting point (T_m) 158.2 °C) and pristine Na$^+$-mmt clay. Simple mixing of dried iPP-NH$_3^+$Cl$^-$ powder and Na$^+$-mmt, ground together by mortar and pestle at ambient temperature, creates an XRD pattern (Figure 10.7a, top) with an (001) peak at $2\theta \cong 7°$, corresponding to the characteristic Na$^+$-mmt d-spacing of ca. 1.2 nm. The mixed powder was then annealed under static conditions at 190 °C for 2 h under vacuum. The resulting iPP-NH$_3^+$/mmt hybrid shows a featureless XRD pattern (Figure 10.7a, bottom), indicating the formation of an exfoliated clay structure. This corresponds to the thermodynamically stable state of the nanocomposite, as the ammonium terminus of the iPP-NH$_3^+$Cl$^-$ exchanges with the alkali (Na$^+$) cations at the mmt surfaces. It is clear that an organic surfactant is not needed to promote compatibility between iPP-NH$_3^+$Cl$^-$ and pristine Na$^+$-mmt clay. Besides any economic benefits, elimination of the mmt's organic surfactant also offers some significant materials advantages. For example, it eliminates two major concerns relating to thermal stability of

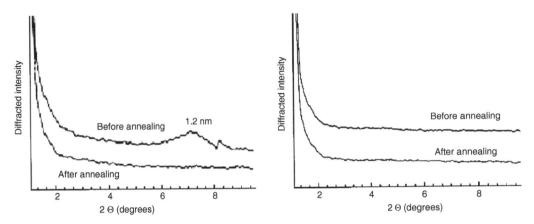

FIGURE 10.7 (a) X-ray diffraction patterns of iPP-$NH_3^+Cl^-$/Na^+-mmt (90:10 weight ratio); top = physical mixture made by simple powder mixing at ambient temperature; bottom = same mixture after static melt-intercalation (iPP-NH_3^+/mmt hybrid). (b) X-ray diffraction patterns of a 50:50 mixture by weight of neat iPP and exfoliated iPP-NH_3^+/mmt (90:10 weight ratio); top = before melt-annealing; bottom = after melt-annealing. (Reproduced with permission from Wang, Z. M.; Nakajima, H.; Manias, E.; Chung, T. C. *Macromolecules* **2003**, *36*, 8919–8922. Copyright 2002 American Chemical Society.)

organic surfactants during high temperature melt processing and long-term stability (migration and degradation) of organic surfactants in the polymer/clay nanocomposite under various application conditions.

The binary iPP-NH_3^+/mmt hybrid was subsequently blended further (at a 50:50 weight ratio) with unfunctionalized iPP (M_n 110,000; M_w 250,000). Figure 10.7b (top) shows the XRD pattern of a physical mixture of exfoliated iPP-NH_3^+/mmt and neat iPP (50:50 weight ratio). The featureless XRD pattern is maintained after further annealing (Figure 10.7b, bottom). This exfoliated structure was also directly observed by transmission electron microscopy (TEM).[82] Apparently, the added iPP is compatible (cocrystallizable) with the backbone of the iPP-NH_3^+ polymer, and the iPP polymer chains serve largely as diluents in the ternary iPP-NH_3^+/mmt/iPP system, with the thermodynamically stable iPP-NH_3^+/mmt exfoliated structure dispersed in an iPP matrix.

For comparison with these results obtained using terminally functionalized iPPs, several function-alized iPP polymers containing functional groups (including OH and succinic anhydride) randomly distributed as side chains or lumped together in a block copolymer microstructure were also evaluated for the synthesis of iPP/mmt nanocomposites.[81] Similar static melt-intercalation procedures were followed with these comparative functional iPPs, except that alkylammonium-modified montmorillonites were used for improved miscibility with iPP: 2C18-mmt (2C18 = dioctadecylammonium) for random copolymers and C18-mmt (C18 = octadecylammonium) for block copolymers. All XRD patterns of the nanocomposites prepared using these comparative functional iPPs show a (001) peak near $2\theta \sim 3°$, indicating intercalated structures with an interlayer d-spacing of about 3 nm (d-spacing of 2C18-mmt = 2 nm).

These comparative results between chain end and side-chain-functionalized iPP clearly demonstrate the advantages of chain-end functionalized iPPs (iPP-NH_3^+s) for nanocomposite synthesis. These polymers seem to adopt a unique molecular structure atop the clay surfaces, which results in an exfoliated montmorillonite structure (Figure 10.8a). The terminal hydrophilic NH_3^+ functional group anchors the iPP chains (through ion exchange) to the inorganic surfaces, and the hydrophobic high molecular weight, semicrystalline iPP "tail" effectively exfoliates the clay platelets to give a structure that is maintained even after further mixing with neat iPP. In contrast, side-chain-functionalized and block copolymer functional iPPs form multiple contacts with each of the clay surfaces along each chain (Figure 10.8b). This results not only in the alignment of the polymer chains

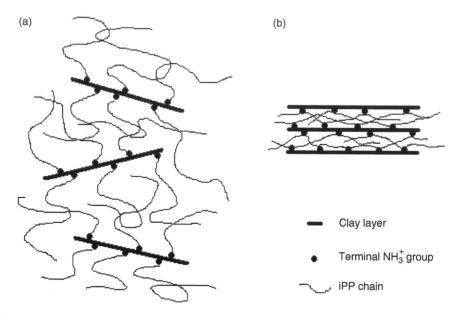

FIGURE 10.8 Illustration of the iPP/clay nanocomposite structures using (a) chain-end functionalized iPPs and (b) side-chain-functionalized iPPs. (Reproduced with permission from Wang, Z. M.; Nakajima, H.; Manias, E.; Chung, T. C. *Macromolecules* **2003**, *36*, 8919–8922. Copyright 2002 American Chemical Society.)

parallel to the clay surfaces but also in bridging of neighboring clay platelets. This bridging promotes intercalated structures, especially for higher lateral size montmorillonites.

10.6 CONCLUSIONS

This chapter summarizes several known chain transfer mechanisms operative in metallocene-mediated propylene polymerization. In the absence of added CTAs, normal propylene chain transfer processes involve β-hydride and β-methyl groups and form polypropylenes having various unsaturated chain ends, the exact identities of which are determined by catalyst structure and reaction conditions. In the presence of added hydrogen, catalyst activity usually increases owing to hydrogen-mediated release of the polymer chain from "dormant" active sites generated by the 2,1-insertion of propylene. This is very significant for catalysts prone to 2,1-regioerror formation. Hydrogen-mediated chain transfer also controls polymer molecular weight and produces polypropylenes having fully saturated chain ends.

It is scientifically interesting and industrially important to use the chain transfer process to form chain-end functionalized polypropylene polymers. Several specially designed CTAs have been discussed. The in situ simultaneous chain transfer/functionalization process shows many advantages, including simplicity, retention of high catalyst activity, good yields of chain-end functional groups in the polymer products, and utility for the subsequent preparation of diblock copolymers. Polypropylenes with reactive terminal functional groups (such as -OH and -NH₂) are also ideal interfacial agents for reactive polymer blending and for the synthesis of exfoliated polypropylene/clay nanocomposites.

ACKNOWLEDGMENTS

The author would like to thank Mitsubishi Chemical Company and the office of Naval Research for financial support.

REFERENCES AND NOTES

1. Kaminsky, W.; Külper, K.; Brintzinger, H. H.; Wild, F. R. W. P. Polymerization of propene and butene with a chiral zirconocene and MAO as cocatalyst. *Angew. Chem., Int. Ed. Engl.* **1985**, *24*, 507–508.

2. Kaminsky, W. Olefin polymerization catalyzed by metallocenes. *Adv. Catal.* **2001**, *46*, 89–159.

3. Ewen, J. A. Mechanisms of stereochemical control in propylene polymerizations with soluble group 4B metallocene/methylalumoxane catalysts. *J. Am. Chem. Soc.* **1984**, *106*, 6355–6364.

4. Resconi, L.; Cavallo, L.; Fait, A.; Piemontesi, F. Selectivity in propene polymerization with metallocene catalysts. *Chem. Rev.* **2000**, *100*, 1253–1346.

5. Carvill, A.; Tritto, I.; Locatelli, P.; Sacchi, M. C. Polymer microstructure as a probe into hydrogen activation effect in *ansa*-zirconocene/methylaluminoxane catalyzed propene polymerizations. *Macromolecules* **1997**, *30*, 7056–7062.

6. Jüngling, S.; Mülhaupt, R.; Stehling, U.; Brintzinger, H. H.; Fischer, D.; Langauser, F. Propene polymerization using homogeneous MAO-activated metallocene catalysts: $Me_2Si(Benz[e]indenyl)_2ZrCl_2$/MAO vs. $Me_2Si(2-Me-Benz[e]indenyl)_2ZrCl_2$/MAO. *J. Polym. Sci., Part A: Polym. Chem.* **1995**, *33*, 1305–1317.

7. Tsutsui, T.; Kashiwa, N.; Mizuno, A. Effect of hydrogen on propene polymerization with ethylenebis (1-indenyl)zirconium dichloride and methylalumoxane catalyst system. *Makromol. Chem., Rapid Commun.* **1990**, *11*, 565–570.

8. Chung, T. C. *Functionalization of Polyolefins*; Academic Press: London, 2002.

9. Chung, T. C. Synthesis of polyalcohols via Ziegler–Natta polymerization. *Macromolecules* **1988**, *21*, 865–869.

10. Purgett, M. D.; Vogl, O. Functional polymers. XLVIII. Polymerization of ω-alkenoate derivatives. *J. Polym. Sci., Part A: Polym. Chem.* **1988**, *26*, 677–700.

11. Stehling, U. M.; Stein, K. M.; Kesti, M. R.; Waymouth, R. M. Metallocene/borate-catalyzed polymerization of amino-functionalized α-olefins. *Macromolecules* **1998**, *31*, 2019–2027.

12. Clark, K. J.; Powell, T. Polymers of halogen-substituted 1-olefins. *Polymer* **1965**, *6*, 531–534.

13. Lee, L. H. *Adhesion and Adsorption of Polymers*; Plenum Press: New York, 1980.

14. Felix, J. M.; Gatenholm, P. The nature of adhesion in composites of modified cellulose fibers and polypropylene. *J. Appl. Polym. Sci.* **1991**, *42*, 609–620.

15. Ruggeri, G.; Aglietto, M.; Petragnani, A.; Ciardelli, F. Some aspects of polypropylene functionalization by free radical reactions. *Eur. Polym. J.* **1983**, *19*, 863–866.

16. Bywater, S. Anionic polymerization. *Prog. Polym. Sci.* **1975**, *4*, 27–69.

17. Yamagishi, A.; Szwarc, M. Kinetics of styrene addition in benzene solution to living lithium polymers terminated by 1,1-diphenylethylene units. The effect of mixed dimerization of monomeric polymers. *Macromolecules* **1978**, *11*, 504–506.

18. Yasuda, H.; Furo, M.; Yamamoto, H.; Nakamura, A.; Miyake, S.; Kibino, N. New approach to block copolymerizations of ethylene with alkyl methacrylates and lactones by unique catalysis with organolanthanide complexes. *Macromolecules* **1992**, *25*, 5115–5116.

19. Brookhart, M.; DeSimone, J. M.; Grant, B. E.; Tanner, M. J. Cobalt(III)-catalyzed living polymerization of ethylene: Routes to end-capped polyethylene with a narrow molar mass distribution. *Macromolecules* **1995**, *28*, 5378–5380.

20. Shea, K. J.; Walker, J. W.; Zhu, H.; Paz, M.; Greaves, J. Polyhomologation. A living polymethylene synthesis. *J. Am. Chem. Soc.* **1997**, *119*, 9049–9050.

21. Burger, B. J.; Thompson, M. E.; Cotter, W. D.; Bercaw, J. E. Ethylene insertion and β-hydrogen elimination for permethylscandocene alkyl complexes. A study of the chain propagation and termination steps in Ziegler–Natta polymerization of ethylene. *J. Am. Chem. Soc.* **1990**, *112*, 1566–1577.

22. Hajela, S.; Bercaw, J. E. Competitive chain transfer by β-hydrogen and β-methyl elimination for a model Ziegler–Natta olefin polymerization system $[Me_2Si(\eta^5-C_5Me_4)_2]Sc\{CH_2CH(CH_3)_2\}(PMe_3)$. *Organometallics* **1994**, *13*, 1147–1154.

23. Alelyunas, Y. W.; Guo, Z.; LaPointe, R. E.; Jordan, R. F. Structures and reactivity of $(C_5H_4Me)_2Zr(CH_2CH_2R)(CH_3CN)_n^+$ complexes. Competition between insertion and β-H elimination. *Organometallics* **1993**, *12*, 544–553.

24. Guo, Z.; Swenson, D.; Jordan, R. Cationic zirconium and hafnium isobutyl complexes as models for intermediates in metallocene-catalyzed propylene polymerizations. Detection of an α-agostic interaction in $(C_5Me_5)_2Hf(CH_2CHMe_2)(PMe_3)^+$. *Organometallics* **1994**, *13*, 1424–1432.

25. Tsutsui, T.; Mizuno, A.; Kashiwa, N. The microstructure of propylene homo- and copolymers obtained with a Cp_2ZrCl_2 and methylaluminoxane catalyst system. *Polymer* **1989**, *30*, 428–431.

26. Kashiwa, N.; Yoshitake, J. Kinetic study on propylene polymerization by a high activity catalyst system: $MgCl_2/TiCl_4/PhCO_2Et$-$AlEt_3/PhCO_2Et$. *Polym. Bull.* **1984**, *11*, 479–484.

27. Cavallo, L.; Guerra, G.; Corradini, P. Mechanisms of propagation and termination reactions in classical heterogeneous Ziegler–Natta catalytic systems: A nonlocal density functional study. *J. Am. Chem. Soc.* **1998**, *120*, 2428–2436.

28. Resconi, L.; Fait, A.; Piemontesi, F.; Colonnesi, M.; Rychlicki, H.; Zeigler, R. Effect of monomer concentration on propene polymerization with the *rac*-[ethylenebis(1-indenyl)]zirconium dichloride/methylaluminoxane catalyst. *Macromolecules* **1995**, *28*, 6667–6676.

29. Schneider, M. J.; Mulhaupt, R. Influence of indenyl ligand substitution pattern on metallocene-catalyzed propene copolymerization with 1-octene. *Macromol. Chem. Phys.* **1997**, *198*, 1121–1129.

30. Kashiwa, N.; Kioka, M. Study on the nature of active sites activity enhancement by the addition of hydrogen in olefin polymerization. *Polym. Mater. Sci. Eng.* **1991**, *64*, 43–44.

31. Busico, V.; Cipullo, R.; Corradini, P. Ziegler–Natta oligomerization of 1-alkenes: A catalyst's "fingerprint." 1. Hydrooligomerization of propene in the presence of a highly isospecific $MgCl_2$-supported catalyst. *Makromol. Chem.* **1993**, *194*, 1079–1093.

32. Watson, P. L.; Roe, D. C. β-Alkyl transfer in a lanthanide model for chain termination. *J. Am. Chem. Soc.* **1982**, *104*, 6471–6473.

33. Eshuis, J. J. W.; Tan, Y. Y.; Teuben, J. H.; Renkema, J. Catalytic olefin oligomerization and polymerization with cationic group IV metal complexes $[Cp*_2MMe(THT)]^+[BPh_4]^-$, M = Ti, Zr and Hf. *J. Mol. Catal.* **1990**, *62*, 277–287.

34. Yang, X.; Stern, C. L.; Marks, T. J. Cationic zirconocene olefin polymerization catalysts based on the organo-Lewis acid tris(pentafluorophenyl)borane. A synthetic, structural, solution dynamic, and polymerization catalytic study. *J. Am. Chem. Soc.* **1994**, *116*, 10015–10031.

35. Resconi, L.; Piemontesi, F.; Camurati, I.; Sudmeijer, O.; Nifant'ev, I. E.; Ivchenko, P. V.; Kuz'mina, L. G. Highly regiospecific zirconocene catalysts for the isospecific polymerization of propene. *J. Am. Chem. Soc.* **1998**, *120*, 2308–2321.

36. Zambelli, A.; Sacchi, M. C.; Locatelli, P.; Zannoni, G. Isotactic polymerization of alpha-olefins: Stereoregulation for different reactive chain ends. *Macromolecules* **1982**, *15*, 211–212.

37. Sacchi, M. C.; Shan, C.; Locatelli, P.; Tritto, I. Stereochemical investigation of the initiation step in $MgCl_2$-supported Ziegler–Natta catalysts: The Lewis base activation effect. *Macromolecules* **1990**, *23*, 383–386.

38. Chien, J. C. W.; Kuo, C. I. Magnesium chloride supported high-mileage catalyst for olefin polymerization. IX. Molecular weight and distribution and chain transfer processes. *J. Polym. Sci., Part A: Polym. Chem.* **1986**, *24*, 1779–1818.

39. Chien, J. C. W.; Wang, B. P. Metallocene-methylaluminoxane catalysts for olefin polymerization. I. Trimethylaluminum as coactivator. *J. Polym. Sci., Part A: Polym. Chem.* **1988**, *26*, 3089–3102.

40. Kaminsky, W.; Ahlers, A.; Moller-Lindenhof, N. Asymmetric oligomerization of propene and 1-butene with a zirconium/alumoxane catalyst. *Angew. Chem., Int. Ed. Engl.* **1989**, *28*, 1216–1218.

41. Resconi, L.; Piemontesi, F.; Franciscono, G.; Abis, L.; Fiorani, T. J. Olefin polymerization at bis(pentamethylcyclopentadienyl)zirconium and -hafnium centers: Chain-transfer mechanisms. *J. Am. Chem. Soc.* **1992**, *114*, 1025–1032.

42. Resconi, L.; Jones, R. L.; Rheingold, A. L.; Yap, G. High-molecular-weight atactic polypropylene from metallocene catalysts. 1. $Me_2Si(9$-$Flu)_2ZrX_2$ (X = Cl, Me). *Organometallics* **1996**, *15*, 998–1005.

43. Resconi, L.; Piemontesi, F.; Camurati, I.; Balboni, D.; Sironi, A.; Moret, M.; Rychlicki, H.; Zeigler, R. Diastereoselective synthesis, molecular structure, and solution dynamics of *meso*- and *rac*-[ethylenebis(4,7-dimethyl-η^5-1-indenyl)]zirconium dichloride isomers and chain transfer reactions in propene polymerization with the *rac* isomer. *Organometallics* **1996**, *15*, 5046–5059.

44. Spaleck, W.; Kuber, F.; Winter, A.; Rohrmann, J.; Bachmann, B.; Antberg, M.; Dolle, V.; Paulus, E. The influence of aromatic substituents on the polymerization behavior of bridged zirconocene catalysts. *Organometallics* **1994**, *13*, 954–963.

45. Fukui, Y.; Murata, M.; Soga, K. Living polymerization of propylene and 1-hexene using bis-Cp type metallocene catalysts. *Macromol. Rapid. Commun.* **1999**, *20*, 637–640.
46. Chien, J. C. W.; Kuo, C. I. Magnesium chloride supported high-mileage catalysts for olefin polymerization. X. Effect of hydrogen and catalytic site deactivation. *J. Polym. Sci., Part A: Polym. Chem.* **1986**, *24*, 2707–2727.
47. Sun, L.; Soga, K. Effect of hydrogen on propene polymerization with Solvay-type TiCl$_3$/Cp$_2$TiMe$_2$ catalyst. *Makromol. Chem.* **1989**, *190*, 3137–3142.
48. Spitz, R.; Bobichon, C.; Guyot, A. Synthesis of polypropylene with improved MgCl$_2$-supported Ziegler–Natta catalysts, including silane compounds as external bases. *Makromol. Chem.* **1989**, *190*, 707–716.
49. Tsutsui, T.; Kashiwa, N.; Mizuno, A. Effect of hydrogen on propene polymerization with ethylenebis(1-indenyl)zirconium dichloride and methylalumoxane catalyst system. *Makromol. Chem. Rapid. Commn.* **1990**, *11*, 565–570.
50. Carvill, A.; Tritto, I.; Locatelli, P.; Sacchi, M. C. Polymer microstructure as a probe into hydrogen activation effect in *ansa*-zirconocene/methylaluminoxane catalyzed propene polymerizations. *Macromolecules* **1997**, *30*, 7056–7062.
51. Busico, V.; Cipullo, R.; Talarico, G. Highly regioselective transition-metal-catalyzed 1-alkene polymerizations: A simple method for the detection and precise determination of regioirregular monomer enchainments. *Macromolecules* **1998**, *31*, 2387–2390.
52. Ramos, J.; Cruz, V.; Munoz-Escalona, A.; Martinez-Salazar, J. *Ab initio* study of hydrogenolysis as a chain transfer mechanism in olefin polymerization catalyzed by metallocenes. *Polymer* **2000**, *41*, 6161–6169.
53. Chung, T. C. Synthesis of functional polyolefin copolymers with graft and block structures. *Prog. Polym. Sci.* **2002**, *27*, 39–85.
54. Yasuda, H.; Furo, M.; Yamamoto, H. New approach to block copolymerizations of ethylene with alkyl methacrylates and lactones by unique catalysis with organolanthanide complexes. *Macromolecules* **1992**, *25*, 5115–5116.
55. Brookhart, M.; DeSimone, J. M.; Grant, B. E.; Tanner, M. J. Cobalt(III)-catalyzed living polymerization of ethylene: Routes to end-capped polyethylene with a narrow molar mass distribution. *Macromolecules* **1995**, *28*, 5378–5380.
56. Hagihara, H.; Shiono, T.; Ikeda, T. Living polymerization of propene and 1-hexene with the [*t*-BuNSiMe$_2$Flu]TiMe$_2$/B(C$_6$F$_5$)$_3$ catalyst. *Macromolecules* **1998**, *31*, 3184–3188.
57. Mülhaupt, R.; Duschek, T.; Rieger, B. Functional polypropylene blend compatibilizers. *Makromol. Chem., Macromol. Symp.* **1991**, *48/49*, 317–332.
58. Shiono, T.; Kurosawa, H.; Ishida, O.; Soga, K. Synthesis of polypropylenes functionalized with secondary amino groups at the chain ends. *Macromolecules* **1993**, *26*, 2085–2089.
59. Lu, B.; Chung, T. C. New maleic anhydride modified PP copolymers with block structure: Synthesis and application in PP/polyamide reactive blends. *Macromolecules* **1999**, *32*, 2525–2533.
60. Burfield, D. R. The synthesis of low molecular weight hydroxy-tipped polyethylene and polypropylene by the intermediacy of Ziegler–Natta catalysts. *Polymer* **1984**, *25*, 1817–1822.
61. Shiono, T.; Soga, K. Synthesis of terminally aluminum-functionalized polypropylene. *Macromolecules* **1992**, *25*, 3356–3361.
62. Shiono, T.; Kurosawa, H.; Soga, K. Synthesis of isotactic polypropylene functionalized with a primary amino group at the initiation chain end. *Macromolecules* **1994**, *27*, 2635–2637.
63. Shiono, T.; Kurosawa, H.; Soga, K. Isospecific polymerization of propene over TiCl$_3$ combined with bis(ω-alkenyl)zinc compounds. *Macromolecules* **1995**, *28*, 437–443.
64. Fu, P. F.; Marks, T. J. Silanes as chain transfer agents in metallocene-mediated olefin polymerization. Facile in situ catalytic synthesis of silyl-terminated polyolefins. *J Am. Chem. Soc.* **1995**, *117*, 10747–10748.
65. Koo, K.; Marks, T. J. Silanolytic chain transfer in Ziegler–Natta catalysis. Organotitanium-mediated formation of new silapolyolefins and polyolefin architectures. *J. Am. Chem. Soc.* **1998**, *120*, 4019–4020.
66. Koo, K.; Marks, T. J. Silicon-modified Ziegler–Natta polymerization. Catalytic approaches to silyl-capped and silyl-linked polyolefins using "single-site" cationic Ziegler–Natta catalysts. *J. Am. Chem. Soc.* **1999**, *121*, 8791–8802.

67. Chung, T. C.; Xu, G. Process for preparing polyolefin diblock copolymers involving borane chain transfer reaction in transition metal-mediated olefin polymerization. U.S. Patent 6,248,837 B1 (Penn State Research Foundation), June 19, 2001.

68. Xu, G.; Chung, T. C. Borane chain transfer agent in metallocene-mediated olefin polymerization. Synthesis of borane-terminated polyethylene and diblock copolymers containing polyethylene and polar polymer. *J. Am. Chem. Soc.* **1999**, *121*, 6763–6764.

69. Xu, G.; Chung, T. C. Synthesis of syndiotactic polystyrene (*s*-PS) containing a terminal polar group and diblock copolymers containing *s*-PS and polar polymers. *Macromolecules* **1999**, *32*, 8689–8692.

70. Chung, T. C.; Xu, G.; Lu, Y.; Hu, Y. Metallocene-mediated olefin polymerization with B-H chain transfer agents; synthesis of chain-end functionalized polyolefins and diblock copolymers. *Macromolecules* **2001**, *34*, 8040–8050.

71. Chung, T. C.; Dong, J. Y. Polyolefin containing a terminal phenyl or substituted phenyl group and process for preparing same. U.S. Patent 6,479,600 B2 (Penn State Research Foundation), November 12, 2002.

72. Chung, T. C.; Dong, J. Y. A novel consecutive chain transfer reaction to *p*-methylstyrene and hydrogen during metallocene-mediated olefin polymerization. *J. Am. Chem. Soc.* **2001**, *123*, 4871–4876.

73. Lu, H. L.; Hong, S.; Chung, T. C. Synthesis of poly(propylene-*co-p*-methylstyrene) copolymers and functionalization. *J. Polym. Sci., Part A: Polym. Chem.* **1999**, *37*, 2795–2802.

74. Dong, J. Y.; Wang, Z. M.; Han, H.; Chung, T. C. Synthesis of isotactic polypropylene containing a terminal Cl, OH, or NH$_2$ group via metallocene-mediated polymerization/chain transfer reaction. *Macromolecules* **2002**, *35*, 9352–9359.

75. Hagihara, H.; Tsuchihara, K.; Sugiyama, J.; Takeuchi, K.; Shiono, T. Copolymerization of propylene and polar allyl monomer with zirconocene/methylaluminoxane catalyst: Catalytic synthesis of amino-terminated isotactic polypropylene. *Macromolecules* **2004**, *37*, 5145–5148.

76. Riess, G.; Periard, J.; Bonderet, A. *Colloidal and Morphological Behavior of Block and Graft Copolymers*; Plenum Press: New York, 1971.

77. Lu, Y. Y.; Hu, Y. L.; Chung, T. C. Synthesis of diblock copolymers polyolefin-*b*-poly(ε-caprolactone) and their applications as the polymeric compatilizer. *Polymer* **2005**, *46*, 10585–10591.

78. Lu, Y. Y.; Wang, Z. M.; Manias, E.; Chung, T. C. Synthesis of new amphiphilic diblock copolymers containing poly(ethylene oxide) and poly(α-olefin). *J. Polym. Sci., Part A: Polym. Chem.* **2002**, *40*, 3416–3425.

79. Usuki, A.; Kojima, Y.; Kawasumi, M.; Okada, A.; Fukushima, Y.; Kurauchi, T.; Kamigaito. O. Synthesis of nylon-6-clay hybrid. *J. Mater. Res.* **1993**, *8*, 1179–1184.

80. Vaia, R. A.; Ishii, H.; Giannelis, E. P. Synthesis and properties of two-dimensional nanostructures by direct intercalation of polymer melts in layered silicates. *Chem. Mater.* **1993**, *5*, 1694–1696.

81. Wang, Z. M.; Nakajima, H.; Manias, E.; Chung, T. C. Exfoliated PP/clay nanocomposites using ammonium-terminated PP as the organic modification montmorillonite. *Macromolecules* **2003**, *36*, 8919–8922.

82. Chung, T. C.; Manias, E.; Wang, Z. M. Exfoliated polyolefin/clay nanocomposites using chain end functionalized polyolefin as the polymeric surfactant. U.S. Patent 7,241,829 B2 (Penn State Research Foundation), July 10, 2007.

11 Functionalization of Tactic Polypropylenes

Manfred Bochmann

CONTENTS

11.1 INTRODUCTION

Polyolefin materials occur in almost every walk of life. They are inexpensive and nontoxic, and their properties can now be tailored to suit a wide range of applications. However, there are some properties that these materials lack owing to their hydrophobic surfaces. These include compatibility with polar compounds (including other polymers) and adhesion to materials such as fillers, printing inks, and pigments, as well as compatibility with microorganisms and cells that would be important in biological and medical applications. For this reason their chemical structure, or at least the structure of the surfaces of polymer films, needs to be modified by functional groups.[1] There are various ways of achieving functionalization. Surface treatment of premolded polypropylene (PP) components with oxygen or CO_2 plasma is industrially important but beyond the scope of this chapter. Other approaches are the modification of preformed PP by radical grafting, reactions of endgroups, and copolymerization of propylene with functionalized or functionalizable comonomers.

The functionalization of PP has a significant influence on its surface energy, and is therefore important in the compatibilization of PP polymer blends. For example, polypropylene functionalized with primary or secondary amino groups gives compatibilized blends with thermoplastic polyurethanes, based on the reaction of the amine functionalities with urethane linkages or traces of isocyanates ("reactive compatibilization"). This process results in much finer domain size and more stable blend morphology, with improvements in tear and abrasion resistance of the composite material.[2] Maleic anhydride grafting is particularly widely applied to improve PP blend compatibility.[3] The presence of functionalized PP in blends with isotactic PP (iPP) reduces the crystallinity and the crystallite size of PP, but increases crystallization rate and increases the crystallization temperature compared to PP homopolymer.[4] Dye-labeled samples of chlorinated maleated iPP ("CPO", 21.8 wt% chlorine) were used as tracers to study the miscibility and the morphology of polyolefin blends, such as the thickness of the interface between functionalized PP and other blend components.[5]

Other applications of functionalized PP include PP composites with inorganic fillers and tougheners, such as glass fibers. Thus, maleic anhydride was grafted onto PP and the resulting material blended with styrene/ethylene-*co*-butene/styrene block copolymers and injection molded with short glass fibers.[6] Phenol-modified iPP was tested for the same purpose.[7] The functionalized PP was found to improve the adhesion between the glass fibers and the polymer blend, and increased the mechanical strength of the composite. Chlorinated iPP was shown to bind covalently to multi-walled carbon nanotubes and to lead to a large stress transfer, giving composite materials with an increased Young's modulus, tensile strength, and toughness.[8] Similarly, addition of maleated iPP as a compatibilizer to iPP blends allows the incorporation of organically modified layer silicates.[9]

Polypropylene-based ion exchange membranes can be prepared from iPP nonwoven fabrics functionalized with sulfonate groups on pendant sidechains.[10] Polypropylene modified by radiation grafting with 2,3-epoxypropylmethacrylate and subsequent ring-opening functionalization of the epoxide (hydroxylation, iminodiacetation, sulfonation, phosphonation, and amination) gave films that showed improved blood compatibility; phosphoric-acid-group- and heparin-introduced PP films were especially good in this respect.[11] The chemistry leading to such functionalized PP materials will be discussed in the sections below, with emphasis on PP produced by metallocene-catalyzed reactions.

11.2 RADICAL GRAFTING OF TACTIC POLYPROPYLENES

There are various ways for functionalizing PP by radical methods, involving either ultraviolet (UV) or gamma radiation or radical initiators. The advantage of the latter technique is that it can be conducted in the polymer melt and during the extrusion process. The functional molecule is added to the chain statistically by radical-induced C-H activation, which may also cause polymer degradation. Sulfonated, chlorinated, carboxylated, and hydroperoxidated PP can be prepared, while modifying PP with polar monomers such as styrene, acrylates, methacrylates, and maleic anhydride allows the attachment of polar monomer sidechains. Aspects of this chemistry have recently been surveyed.[12] In early work, Datta and Lohse showed that iPP chain-end functionalized with succinic anhydride reacts with ethylene–propylene–norbornene (EPNB) copolymers bearing pendant amino groups either in solution or in the melt state, giving graft copolymers with iPP pendant arms. These materials act as compatibilizers for blends of iPP and ethylene–propylene (EP) copolymers (Scheme 11.1).[13]

Radical functionalization of iPP with acrylic acid (AA) gives iPP-AA, which acts as a compatibilizer for PP/liquid crystal polyblend fibers and increases fiber crystallinity and interfacial adhesion.[14] The reaction of PP with maleic anhydride in xylene at elevated temperatures (120 °C) in the presence of benzoyl peroxide gives maleated iPP. Co-melting of this material with poly(ε-caprolactone) (PCL) for 6 h at 120 °C gave a PP-*g*-PCL graft copolymer.[15] Similar methods have been employed for the grafting of *N*-phenylmaleimide and maleimido benzoic acid onto iPP oligomers in order to

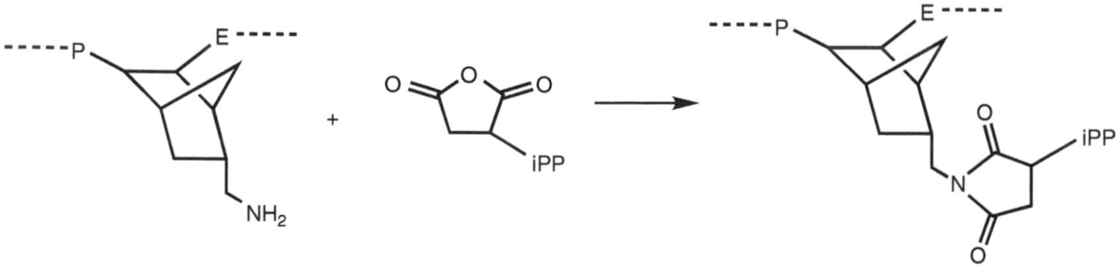

SCHEME 11.1 Reaction of amino-functionalized ethylene-propylene copolymers with maleated iPP.

probe the efficiency of the grafting reactions. Grafting was conducted in the presence of a peroxide radical initiator, with progress monitored by various spectroscopic methods. Grafting yields were essentially quantitative with up to 10 mol% of maleic imide but dropped at higher maleic imide loadings. Grafting of polymers was done by reactive extrusion and products were analyzed by infrared spectroscopy. Optimal grafting yields could be achieved with 1 wt% monomer and 0.5 wt% peroxide concentration.[16] Other applications of radical grafting with peroxide initiators include PP functionalization with 4-allylphenols,[17] itaconic acid esters,[18] maleic anhydride/vinyl acetate mixtures,[19] and isopropenyl-substituted benzylic isocyanates.[20] Recent reports on radiation grafting include the functionalization of PP films with acryloyl chloride under gamma radiation[21] and the attachment of glycidyl methacrylate in the vapor phase under UV irradiation with benzophenone as photoinitiator.[22] UV irradiation of PP in the presence of 1-fluoro-2-nitro-4-azidobenzene allowed the immobilization of enzymes such as horseradish peroxidase and glucose oxidase on PP surfaces.[23]

11.3 COPOLYMERIZATIONS OF PROPYLENE WITH FUNCTIONAL MONOMERS

The copolymerization of simple 1-alkenes such as propylene with comonomers bearing functional groups is, in principle, the simplest route to functionalized PP or poly(α-olefin) polymers. The disadvantage of this approach is the sensitivity of high-activity polymerization catalysts to heteroatoms. This can be minimized by three strategies: (1) the protection of the functional group with bulky substituents, (2) by a long spacer between the C=C double bond of the comonomer and the functionality, or (3) by incorporating a non-heteroatom-containing comonomer that can be functionalized easily in a second, separate step. Both methods 1 and 3 require post-polymerization reactions.

Most olefin copolymerizations are conducted using group 4 metallocene catalysts. However, as Marques has pointed out,[25] poisoning by heteroatoms is less prevalent with late transition metals, and nickel catalysts proved to be several times more active than group 4 metallocenes for the co- and terpolymerization of ethylene and propylene with 1-alkenes bearing terminal –OH and –COOH functional groups. The use of methylaluminoxane (MAO) as activator proved to be an advantage, since the polar functionalities were protected by the stoichiometric reaction with trimethylaluminum and easily deprotected during subsequent aqueous workup of the copolymer product to regenerate the –OH and –COOH moieties. Diimine-ligated nickel complexes (**1**, Figure 11.1) were found to be highly active for such copolymerizations, although they do not give stereoselectivity in propylene copolymerizations. The presence of 5-hexen-1-ol had little effect on catalyst activity. Copolymers with 5-hexen-1-ol, 10-undecen-1-ol, and 10-undecenoic acid were produced. Ethylene/propylene/10-undecen-1-ol terpolymerizations catalyzed by either **1**/MAO (R$_1$ = R$_2$ = Me) or by the group 4 catalyst system *rac*-Et(Ind)$_2$ZrCl$_2$/MAO (Ind = 1-indenyl) (**2**/MAO, Figure 11.1) gave typically 3–8 mol% 10-undecen-1-ol enchainment.[24,25]

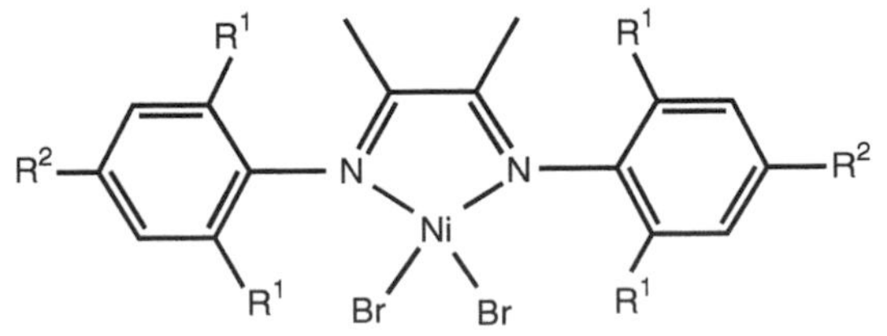

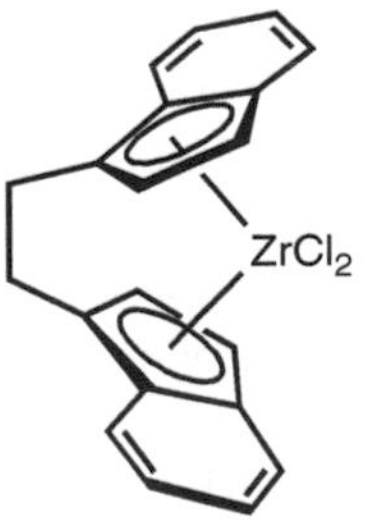

FIGURE 11.1   Structures of catalyst precursor complexes **1** (R$_1$, R$_2$ = H, alkyl) and **2**.

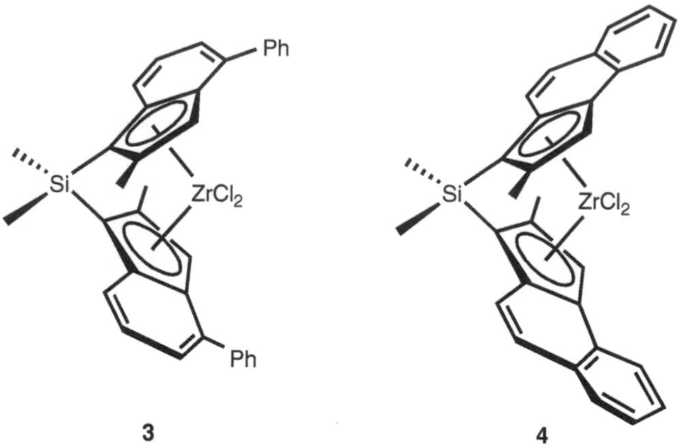

FIGURE 11.2 Structures of catalyst precursor complexes **3** and **4**.

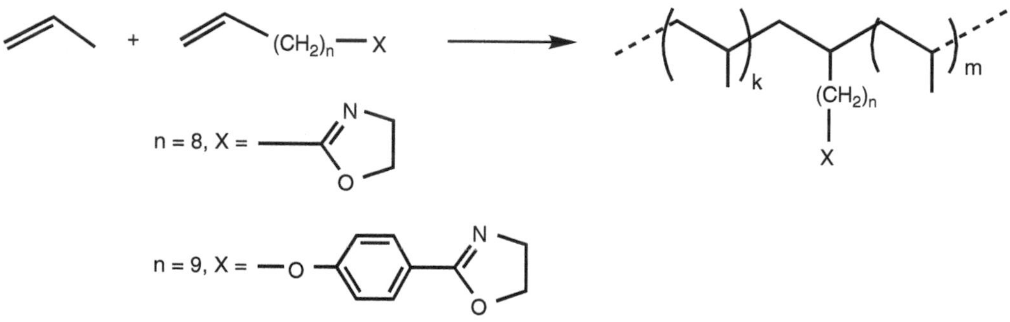

SCHEME 11.2 Copolymerization of propylene with oxazoline-functionalized monomers.

A similar strategy, that is, using long spacers between the comonomer C=C bond and the functionality, was employed by Deffieux et al., who used 11-chloro-1-undecene in terpolymerizations with ethylene and propylene catalyzed by **2**/MAO. The termonomer incorporation levels were low (1–2 mol% 11-chloro-1-undecene). The copolymer product was converted into sidechain-functionalized azides and benzoic esters.[26] The copolymerization of propylene with 10-undecen-1-ol using the sterically hindered zirconocene system **2**/MAO has also been reported.[27] Sterically congested catalysts such as **3** and **4** (Figure 11.2), which are optimized for the production of highly isotactic PP, are poorly suited to higher α-olefin comonomer incorporation. In agreement with this, the incorporation rates of 10-undecen-1-ol into iPP with these catalysts were found to be low (0.1–0.9 mol%). The copolymers were used as blend compatibilizers.[27] Long chain 1-alkenes bearing terminal oxazoline functional groups copolymerize with propylene in the presence of catalysts **2**/MAO and **4**/MAO at high Al/Zr ratios (10,000:1) and mild conditions (Scheme 11.2). As was the case for **3** with 10-undecen-1-ol, the incorporation levels of the functional comonomer are low, typically <0.5 mol%, and the productivities for copolymerization are about two orders of magnitude lower than for propylene homopolymerizations. The copolymers are, however, highly isotactic, with *mmmm* pentad intensities of 80%–90% for copolymers derived from **2** and 95.5% for copolymers derived from **4**.[28]

In this context, it may be noted that the copolymerization of ethylene (but not propylene) with allylic alcohol, that is, of a comonomer with only *one* CH₂ spacer between the functional group and the double bond, is possible with **5**/MAO, whereas **6**/MAO and the constrained-geometry catalyst **7**/MAO (Figure 11.3) were inactive for the copolymerization. The allylic alcohol was predominantly

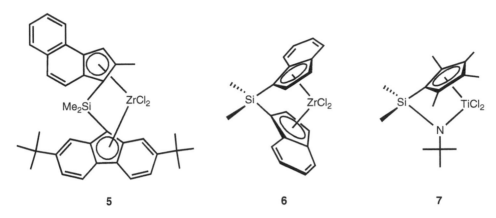

FIGURE 11.3 Structures of catalyst precursor complexes **5–7**.

incorporated as an end group in the polyethylene.[29,30] The copolymerization of propylene with allylic alcohol or allyl amine is, however, possible using catalyst **6**/MAO. Not unexpectedly, the comonomer incorporation rate is low (up to 0.65 mol%). The polymer is isotactic (*mmmm* 81%–84%). Protection of the monomer functional group with trimethylaluminum led mainly to PP with terminal amino end groups.[31] The analogous copolymerization of propylene with $CH_2=CH(CH_2)_2OAlR_2$ (from a 1:1 mixture of 3-buten-1-ol and AlR_3; R = Me or *i*-Bu) catalyzed by **6**/MAO gives copolymer with 3-buten-1-ol incorporated into the main chain at an incorporation level of 0.6–3.8 mol%. Only about 12%–34% of the OH groups were terminal. The addition of excess trimethylaluminum increased the proportion of hydroxyl-terminated copolymer versus copolymer with random hydroxyl enchainment. Catalyst **3**/MAO also gave PP with 3-buten-1-ol incorporated almost exclusively into the main chain (as opposed to the chain ends) and *mm* = 87%–95%. The activities and molecular weights were modest and depended on the amount of trialkylaluminum added. Whereas the triisobutylaluminum-protected system produced a copolymer containing 3-buten-1-ol in the main chain, protection with excess $AlMe_3$ gave end-hydroxylated polypropylenes. The end-functionalized polymer was probably formed by chain transfer reactions with $AlMe_3$[32] (see Chapter 10 for a further discussion of the synthesis of polypropylenes with terminal functionalities).

Conceptually similar is the copolymerization of propylene or other 1-alkenes with unsaturated amines of medium chain length. For example, Waymouth reported the copolymerization of an amine with a three-carbon spacer, 5-*N,N*-diisopropylamino-1-pentene, with either 1-hexene or 4-methyl-1-pentene using *rac*-Et(H$_4$Ind)$_2$ZrMe$_2$ (H$_4$Ind = 4,5,6,7-tetrahydroindenyl) activated with dimethylanilinium tetrakis(pentafluorophenyl)borate ([HNMe$_2$Ph]$^+$[B(C$_6$F$_5$)$_4$]$^-$). The resulting copolymers were isotactic. In 1-hexene/aminopentene copolymerizations the incorporation rate was a linear function of the comonomer feed concentration, and the product of the reactivity ratios $r_1r_2 \approx 0.99$ indicated almost ideal random copolymerization behavior ($r_1r_2 = 1.0$).[33]

For copolymerization with propylene, good incorporation rates were also achieved with the alkoxyamine-functionalized alkene **A**, a precursor for radical reactions (Scheme 11.3). Copolymerization of **A** with propylene catalyzed by *rac*-Et(H$_4$Ind)$_2$ZrMe$_2$/[HNMe$_2$Ph]$^+$[B(C$_6$F$_5$)$_4$]$^-$ afforded a range of copolymers with a copolymer content that corresponded well to the feed ratio. The alkoxyamine side chains were used for subsequent grafting, for example, to attach polystyrene chains by "living" free radical techniques.[34]

The catalyst system **3**/MAO catalyzes the polymerization of propylene in the presence of small amounts of *para*-functionalized styrenes, $H_2C=CH-C_6H_4X$ (X = Cl, OSiMe$_3$, or CH$_2$CH$_2$N(SiMe$_3$)$_2$) and hydrogen.[35] Silyl protection of –OH and –NH$_2$ as –OSiMe$_3$ and –N(SiMe$_3$)$_2$ functionalities is required during polymerization, although the trimethylsilyl groups are readily removed during aqueous acidic workup of the polymer. The styrenes act as chain transfer

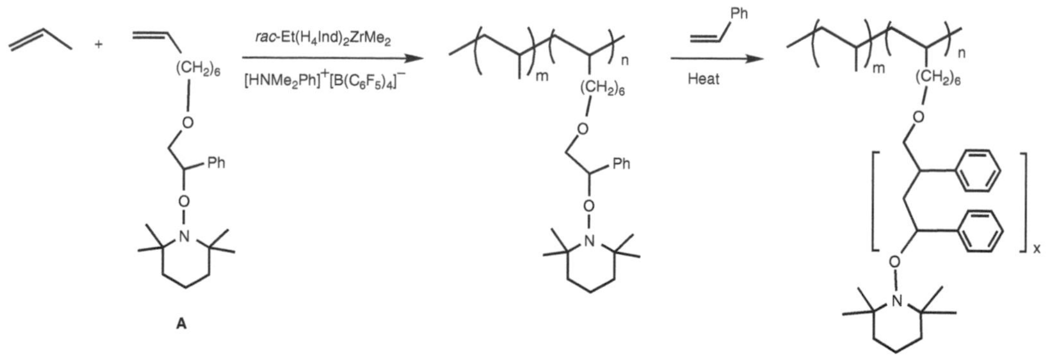

SCHEME 11.3 Random copolymerization of propylene with TEMPO-functionalized alkenes (TEMPO = tetramethylpiperidine N-oxide). The product is suitable for heat-induced radical grafting of styrene.

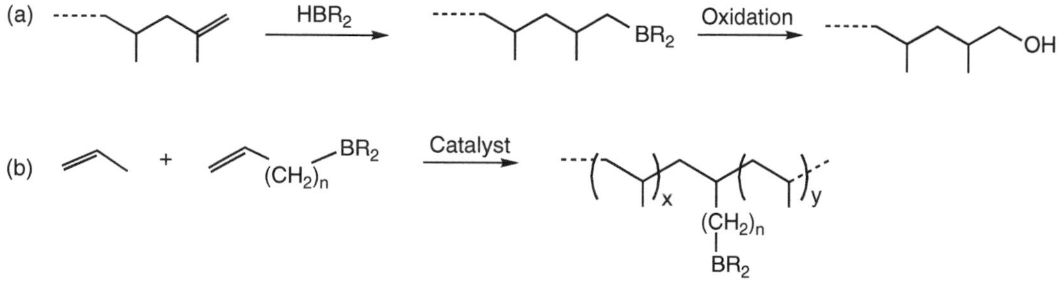

SCHEME 11.4 Preparation of polypropylenes with (a) terminal and (b) pendant borane functional groups (R = alkyl).

agents; since they add to the growing polymer chain in a 2,1- rather than a 1,2-fashion, they block further monomer insertion until the chain is released by chain transfer to hydrogen (hydrogenolysis). The product is iPP carrying a functionalized styrene chain end (iPP-Cl, iPP-OH, or iPP-NH$_2$). iPP-NH$_2$ was coupled with the -COOH functionality of poly(ε-caprolactone) to give iPP-b-PCL diblock copolymers that act as effective compatibilizers for iPP/PCL blends.[35]

11.4 COPOLYMERIZATIONS OF PROPYLENE WITH BORANE AND SILANE REAGENTS

Unlike molecules containing electron-rich heteroatoms, boron compounds do not poison Ziegler–Natta or metallocene polymerization catalysts. Borane-containing olefin comonomers are therefore well suited to produce olefin copolymers while retaining good catalyst activity. The resulting polymers are suitable for subsequent conversion into a variety of functional groups. In principle, two approaches are possible: (1) hydroboration of the terminal double bond (formed by typical chain transfer processes) of a preformed polyolefin, and (2) direct copolymerization of propylene or a 1-alkene with an alkenyl borane (Scheme 11.4).

Early work by Mülhaupt et al.[36] and Shiono et al.[37] demonstrated the first of these strategies. The hydroboration of the vinylidene endgroups of atactic or isotactic polypropylene was achieved using borane reagents in benzene, followed by the addition of 1-hexene to convert remaining B-H functionalities into boron alkyls. Treatment with BCl$_3$ resulted in the conversion of the boron alkyls into -BCl$_2$ groups, which upon reaction with 1-butylazide and subsequent hydrolysis, gave amino-terminated

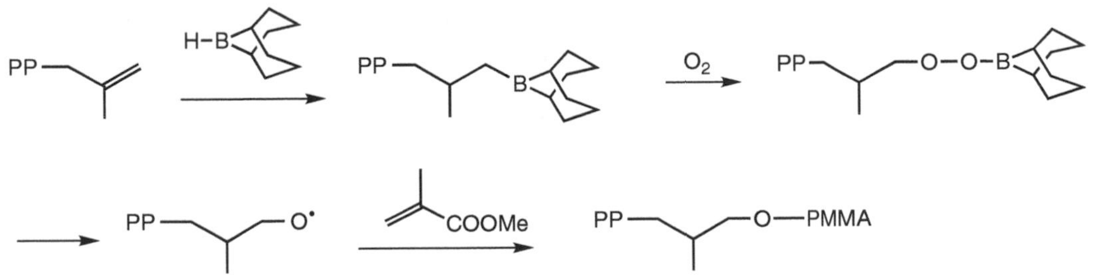

SCHEME 11.5 Hydroboration—oxidation reaction sequence leading to PP-PMMA block copolymers by radical grafting.

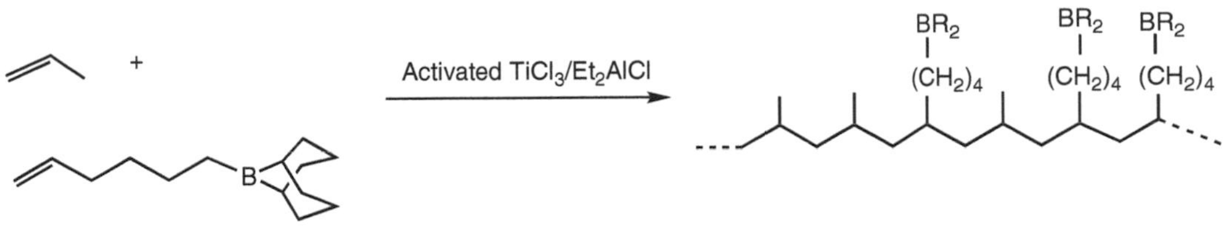

SCHEME 11.6 Copolymerization of propylene with alkenylboranes catalyzed by Ziegler–Natta catalysts.

polypropylenes (PP-NHBu).[37] In a series of papers, Chung exploited this reaction to prepare chain-end functionalized polymers suitable as precursors for the synthesis of graft copolymers. For example, hydroboration of the terminal unsaturations of metallocene-generated iPP with 9-borabicyclononane (9-BBN) gives boryl-terminated iPP (iPP-BR$_2$; R$_2$ = bicyclo[3.3.1]nonane) (Scheme 11.5). Reaction of this material with O$_2$ under controlled conditions affords iPP-O-O-BR$_2$. Since O–O bond cleavage is facile, iPP-O radicals are formed, which upon subsequent treatment with methyl methacrylate (MMA) generate iPP-O-PMMA diblock polymers (PMMA = poly(methyl methacrylate)).[38–40] The first reaction principle can be applied to prepare multifunctionalized polymers if the polypropylene contains unsaturated side chains. For example, the copolymerization of ethylene with 1,4-hexadiene gives products with 2-butenyl side chains, which can be subsequently hydroborated with 9-BBN and oxidized with H$_2$O$_2$/NaOH to afford side-chain-OH-containing polymers.[41–43]

5-Hexenyl-9-BBN can be polymerized with a variety of olefins, including propylene, to give copolymers with pendant –(CH$_2$)$_4$-9-BBN side chains. While ethylene copolymerizations with this comonomer proceeded well using homogeneous zirconocene catalysts, heterogeneous isospecific Ziegler catalysts such as TiCl$_3$/Et$_2$AlCl proved more effective for its copolymerization with higher 1-alkenes. High molecular weight propylene copolymers containing up to 6 mol% of 5-hexenyl-9-BBN were obtained by copolymerization at 25 °C (Scheme 11.6). The 9-BBN-containing side chains appeared to be concentrated at the chain ends, giving "brush-like" structures surrounding crystalline iPP domains, so that the physical characteristics of pure iPP are essentially preserved.[44]

Suspending this borane-functionalized iPP powder in aqueous NaOH/H$_2$O$_2$ proved an effective method of converting the material into a polyhydroxylated iPP. This material cocrystallizes with iPP and is suitable for the preparation of iPP/polyhydroxylated iPP membranes having controllable pore structures, with the OH functional groups apparently concentrated on the surface. These membranes offer good performance in ultrafiltration.[45] With a MgCl$_2$-supported TiCl$_4$/AlEt$_3$ catalyst, the copolymerization of propylene with 5-hexenyl-9-BBN results in spherical PP particles, which upon treatment with NaOH/H$_2$O$_2$ give PP granules with a high surface concentration of OH groups. These granules have been used for the immobilization of Cp$_2$ZrCl$_2$/MAO (Cp = cyclopentadienyl) catalysts for the slurry polymerization of ethylene.[46]

Apart from acting as a post-polymerization hydroboration agent, boranes added during the polymerization process can act as chain transfer agents, giving boron-terminated polyolefins (Scheme 11.7). The amount of borane added is an effective way of controlling the polymer molecular weight, while at the same time introducing a terminal functionality. Different boranes show different reactivities, that is, 9-BBN is a more reactive chain transfer agent than dimesitylborane. This principle has been applied to ethylene polymerization, styrene polymerization, and ethylene/1-alkene copolymerizations catalyzed by a variety of metallocene complexes.[47]

Apart from boranes, other main group hydrides can be used as chain transfer agents in metallocene-catalyzed alkene polymerizations. Most notably, silanes can be used to give silyl-capped linear polymers (Scheme 11.8a and b) and silyl-linked star polymers (Scheme 11.8c). In this case, the precise structure of the catalyst is important. For the polymerization of propylene in the presence of $PhSiH_3$ to give phenylsilane-capped atactic PP (aPP) (Scheme 11.8a), the constrained-geometry catalyst $[Me_2Si(C_5Me_4)N\text{-}t\text{-}BuTiMe]^+[B(C_6F_5)_4]^-$ was 2800 times more active than $[Me_2Si(C_5Me_4)N\text{-}t\text{-}BuTiMe]^+[MeB(C_6F_5)_3]^-$ under the same conditions, owing to differences in the strength of anion coordination to the active site. On the other hand, zirconocene and hafnocene complexes, including the analogous Zr catalysts $[Me_2Si(C_5Me_4)N\text{-}t\text{-}BuZrMe]^+[X]^-$ ($X = B(C_6F_5)_4$ or $MeB(C_6F_5)_3$), gave predominantly or exclusively polyolefins devoid of silyl caps, together with dehydrogenative silane coupling products $(PhSiH_2)_n$.

Disilanes such as Me_2SiH_2 and Et_2SiH_2 can also be used as endcapping reagents. The latter tends to give mixtures of silyl-capped polymers and uncapped terminally unsaturated polymers when used with the Ti constrained-geometry catalysts, whereas the more open half-sandwich catalyst $[(C_5Me_5)TiMe_2]^+[B(C_6F_5)_4]^-$ gives silyl-capped polymers derived from propagating species with primary and secondary alkyl chains (Scheme 11.8b). iPP with $-SiH_2Ph$ endgroups is obtained with $[rac\text{-}Et(Ind)_2TiMe]^+[B(C_6F_5)_4]^-$ at low temperature ($-45\ °C$), whereas analogous *ansa*-zirconocenes give uncapped iPP. With di- and trisilane endcapping reagents and

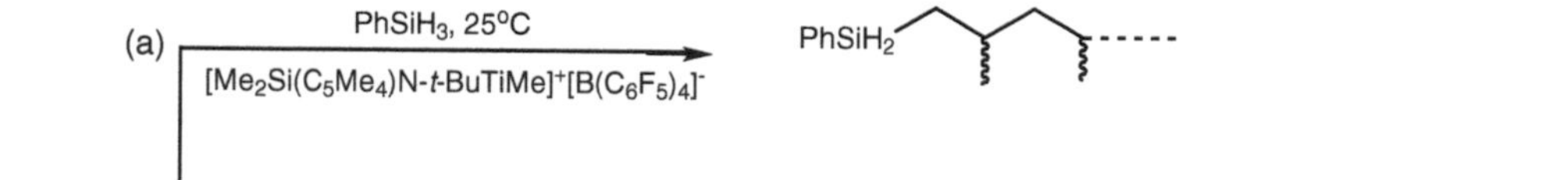

SCHEME 11.7 Boranes as chain transfer agents (R = alkyl; $M(L)_n$ = metallocene).

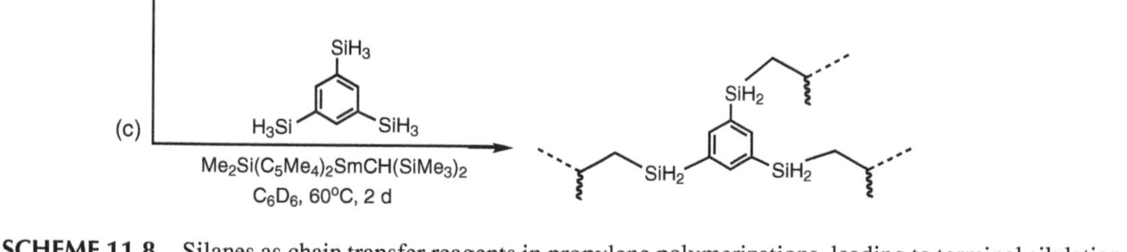

SCHEME 11.8 Silanes as chain transfer reagents in propylene polymerizations, leading to terminal silylation (a,b) or star polymers (c).

bis(cyclopentadienyl) samarium catalysts, silane-coupled and star-like polymer architectures are accessible (Scheme 11.8c).[48]

11.5 POST-POLYMERIZATION FUNCTIONALIZATION OF TACTIC POLYPROPYLENES BY NON-RADICAL-GRAFTING METHODS

Various non-radical-grafting methods have been employed to functionalize premade PP, including oxidation of PP films, solution oxidation of PP (including catalytic oxidations), functionalization of the terminal unsaturated endgroups of premade PP, and copolymerization of propylene with diolefinic monomers that readily undergo post-polymerization chemistry.

The treatment of PP films with ozone under UV irradiation leads to the formation of ether linkages, ketones, and carboxylate groups on the polymer surface. Oxidation of the PP surface is proposed to occur through two alternate mechanisms: (1) insertion of a singlet oxygen atom to form ether linkages, or (2) hydrogen abstraction by triplet oxygen followed either by crosslinking or by reaction with oxygen species to form carbonyl and carboxyl functional groups. The extent of functionalization can, to some degree, be controlled by reaction parameters such as the UV intensity. The product functionalized PP material shows increased wettability.[49] The manganese(III) porphyrin complex 8 (Figure 11.4) catalyzes the potassium peroxymonosulfate oxidation of alternating ethylene-*alt*-propylene copolymers in dichloromethane or chloroform in the presence of imidazole and a phase transfer agent under ambient conditions, without polymer degradation. The primary oxidation products are ketones and tertiary alcohols.[50] An even simpler method used to functionalize iPP is the oxidation of the polymer powder at 80–130 °C with concentrated nitric acid. Nitration, preferably at tertiary positions, is observed together with the formation of –COOH functionalities. The latter appear to be located predominantly at the chain ends, possibly arising from polymer degradation by C–C bond scission. The –COOH groups can be transformed into acyl chlorides, esters, amides, and hydroxyls.[51]

The chain-end functionalization of PP to give anhydrides, esters, amines, carboxylic acids, silanes, boranes, alcohols, and thiols has been briefly summarized by Mülhaupt et al. These materials were obtained by converting the terminal C=C bonds of the propylenes (formed during chain transfer) into functional groups using conventional organic synthetic methods.[36] The functionalization of metallocene-derived ethylene-propylene copolymers with terminal vinylidene groups is also possible, by treatment of the vinylidenes with maleic anhydride in the absence of free radical initiators. The reaction is accompanied by partial double bond isomerization of the vinylidene to an internal olefin; both types of double bond can react to give maleated products.[52] Diblock PP-*b*-PMMA

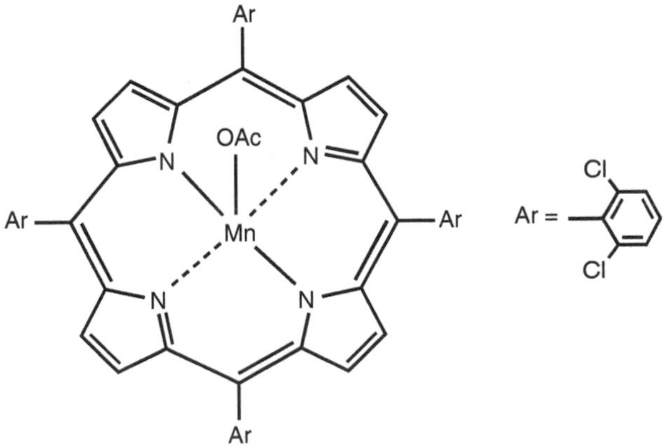

FIGURE 11.4 Structure of manganese(III) oxidation catalyst **8** (Ac = acetyl).

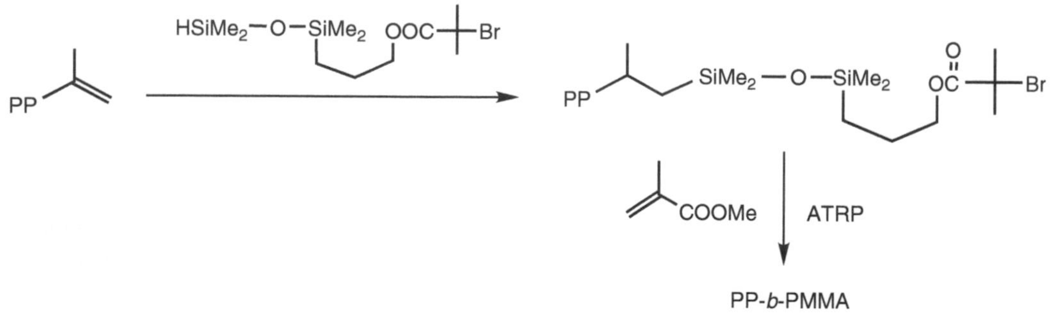

SCHEME 11.9 Hydrosilylation of polypropylene endgroups suitable for subsequent grafting of poly(methyl methacrylate) by atom transfer radical polymerization (ATRP) methods.

copolymers (Scheme 11.9) were prepared by first carrying out hydrosilylation of the terminal C=C bonds of low-molecular weight PP ($M_n \approx 3100$) with 1-(2-bromoisobutyryloxy)propyltetramethyldisiloxane. These functionalized PPs were then used as macroinitiators for the atom transfer radical polymerization (ATRP) of MMA, giving diblock copolymers with a narrow molecular weight distribution ($M_w/M_n = 1.14$). Unreacted PP homopolymer could be removed from the diblock by ether extraction.[53]

para-Methylstyrene (*p*-MS) is a versatile comonomer for copolymerization with 1-alkenes, including propylene, since the C–H bonds of its benzylic methyl group can readily be converted into a variety of functional groups. Thus, the copolymerization of propylene with *p*-MS was conducted using two metallocene catalysts, Et(H$_4$Ind)$_2$ZrCl$_2$/MAO and **2**/MAO, as well as two heterogeneous Ziegler catalysts, TiCl$_4$/MgCl$_2$/electron donor/AlEt$_3$ and "TiCl$_3$ AA"/Et$_2$AlCl. The metallocene catalysts showed low activities and low *p*-MS incorporations in the product PPs (0.9–2.3 mol%), most probably owing to steric jamming by 2,1-inserted *p*-MS units (formed by *p*-MS insertion after 1,2-insertion of propylene). The polymers showed melt transitions (T_ms) from 75 to 143 °C. The heterogeneous catalysts, on the other hand, polymerize both monomers exclusively through 1,2-insertion, and therefore copolymerization is not retarded by *p*-MS incorporation. The TiCl$_4$/MgCl$_2$ catalyst is by far the most active and does not form *p*-MS homopolymer. However, *p*-MS incorporation rates were still modest with this catalyst, up to 0.8 mol%, while the polymer showed improved tacticity (T_m 155–158 °C).[54]

Nonconjugated diene comonomers containing both a terminal (vinylic) double bond and a more highly substituted (vinylene, vinylidene, or trisubstituted) double bond are suitable for the synthesis of copolymers with unsaturated sidechains that can subsequently be functionalized. Only the less substituted double bond is incorporated into the polymer chain (typically a vinyl group is incorporated, although in some instances the incorporated olefin can be a 1,2-disubstitued vinylene). For example, Rieger et al. have used **2**/MAO for the isotactic copolymerization of propylene with dienes **B–D** (Figure 11.5) with good catalyst activities. Comonomer incorporation increased linearly with diene feed concentration and could be used to provide control over copolymer crystallinity and tacticity. Up to 15.6 mol% isocitronellene (**D**) was incorporated at 30 °C. Raising the polymerization temperature to 60 °C halved the diene content in the copolymer; under such conditions a fluffy product was obtained that showed a glass transition (T_g) of −18 °C and a T_m of 71.9 °C. The resulting copolymers were quantitatively epoxidized or brominated at the pendant olefin sites, using chloroform solutions of *m*-chloroperoxybenzoic acid (mCPBA) or bromine, respectively. Irradiation of hexene solutions of the copolymer in the presence of perfluorohexyl iodide led to radical grafting of the C$_6$F$_{13}$ moiety onto the pendant C=C bond.[55]

The copolymerization of propylene with 7-methyl-1,6-octadiene (**C**, "MOD") catalyzed by **2**/MAO was also investigated in detail, with and without controlled addition of the comonomer, to determine the influence of the comonomer feed concentration on the polymer composition

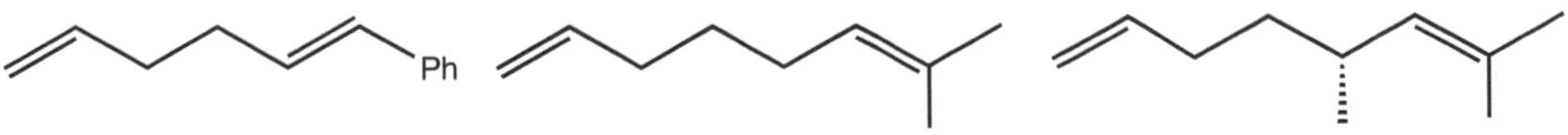

FIGURE 11.5 Diene comonomers for vinyl-selective copolymerizations with propylene.

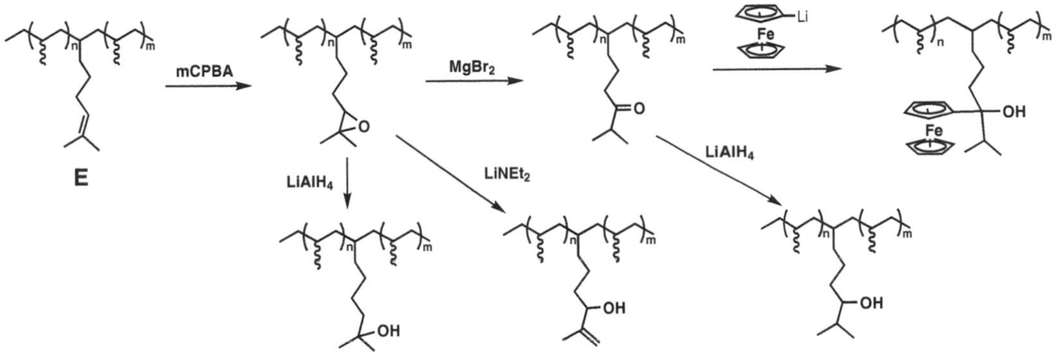

SCHEME 11.10 Functionalization of aPP/7-methyl-1,6-octadiene copolymers by epoxidation and subsequent transformations to ketones and alcohols.

as a function of time. Working at constant comonomer concentration by controlling the rate of comonomer addition spectroscopically resulted in a uniform distribution of **MOD** throughout the polymer chain.[56]

The copolymerization of **MOD** with propylene, catalyzed by $(C_5Me_5)TiMe_3/B(C_6F_5)_3$ in toluene under 1 bar of propylene, gives elastomeric atactic PP (aPP) copolymers with up to ~20 mol% incorporation of the comonomer and high molecular weights (copolymer **E**, Scheme 11.14; M_w 70,000–380,000). With *ansa*-zirconocenes such as **2**/MAO, copolymers with up to 3.8 mol% **MOD** content were prepared. The copolymerizations were conducted by adding the total amount of comonomer at the start of the polymerization. Even though these conditions did not provide for controlled addition to keep the diene concentration constant, under these conditions the comonomer incorporation was a linear function of the diene feed ratio. The catalyst $(C_5Me_5)TiMe_3/B(C_6F_5)_3$ showed a significantly higher affinity for **MOD** than the more sterically hindered zirconocene **2**; for copolymerization using the latter catalyst, comonomer incorporation ranged from 0.4 to 3.8 mol% and the copolymer molecular weights were lower ($M_w \approx 35,000$). Subsequent mCPBA epoxidation of the unsaturated sidechain followed by LiAlH$_4$ reduction led to OH-functionalized copolymers, whereas addition of MgBr$_2$ caused isomerization of the epoxide to a ketone. Other functional groups available through the epoxidized copolymer are shown in Scheme 11.10.

A more versatile functionalization method (as compared to epoxidation) was the treatment of dissolved or chloroform-swollen iPP/**MOD** copolymers with ozone under controlled conditions. Ozonolysis at −78 °C followed by treatment with PPh$_3$ gave the aldehyde-functionalized copolymer, while at room temperature the carboxylic acid derivative was obtained (Scheme 11.11, copolymers **F** and **G**, respectively). These products were converted into a number of derivatives using standard organic synthesis methods. Monitoring the progress of the reactions by ^{1}H nuclear magnetic resonance (NMR) spectroscopy showed that most of these transformations proceeded with essentially quantitative conversion. This methodology also allowed for the covalent binding of dyes. Functionalization, in particular the impregnation of the copolymer with hydrophilic dye molecules, drastically increased the solubility of even iPPs in polar solvents such as acetone or methanol.[57]

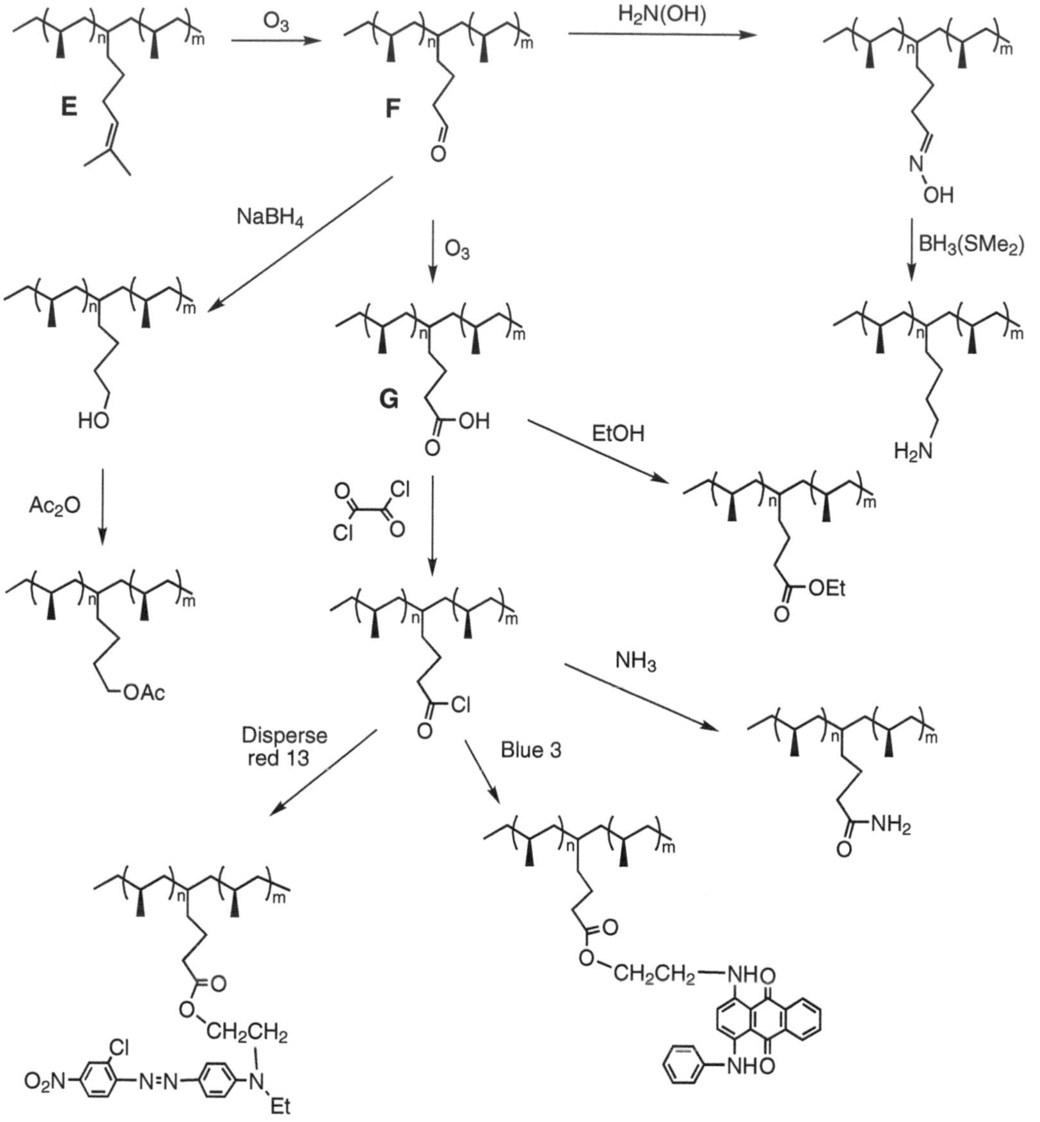

SCHEME 11.11 Functionalization of iPP/7-methyl-1,6-octadiene copolymers through ozonolysis of pendant C=C bonds for the synthesis of highly functionalized polymer derivatives. (Adapted from Song, F.; Pappalardo, D.; Johnson, A. F.; Rieger, B.; Bochmann, M. *J. Polym. Sci., Part A: Polym. Chem.* **2002**, *40*, 1484–1497.)

The copolymerization of propylene and 5-vinyl-2-norbornene (VNB) is possible using precatalysts **2, 4,** or **6** activated with [Ph₃C]⁺[B(C₆F₅)₄]⁻/Al(*i*-Bu)₃ or MAO. VNB was selectively incorporated through its endocyclic double bond (Scheme 11.12). Catalysts **2**/[Ph₃C]⁺[B(C₆F₅)₄]⁻/Al(*i*Bu)₃ and **6**/[Ph₃C]⁺[B(C₆F₅)₄]⁻/Al(*i*-Bu)₃ yielded iPP copolymers with M_w 4,000–52,000 and VNB contents as high as 41.0 mol%, with good catalyst activities. The VNB incorporation levels increased linearly with the feed concentration as long as the VNB fraction in the feed was below 30 mol%; above this value, increasing the initial feed concentration of VNB did not result in substantially increased incorporation levels. In contrast to catalysts **2** and **6**, the more sterically hindered catalyst **4**/MAO displayed low productivity at room temperature and only moderate incorporation of VNB at 60 °C. Surprisingly, in spite of their closely related structures, catalysts **2** and **6** showed very different reactivities toward VNB, with **2** having a greater affinity

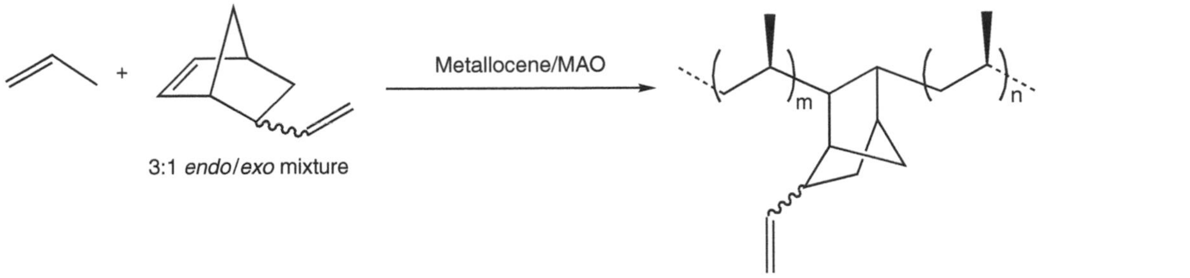

SCHEME 11.12 Copolymerization of propylene with vinylnorbornene catalyzed by zirconocenes.

for VNB than for propylene ($r_{VNB} > r_P$). The resulting PP-*co*-VNB polymers were quantitatively converted into ester- and epoxy-functionalized copolymers by means of post-polymerization functionalization reactions of the VNB vinyl groups; thus, ozonolysis followed by addition of PPh$_3$ and methanol/HCl afforded methyl esters, while addition of mCPBA gave peroxides.[58]

11.6 CONCLUSIONS

The functionalization of tactic polypropylenes follows several strategies. Surface functionalization of premolded PPs does not affect the physical bulk properties of these materials, but increases their adhesive properties and allows for the attachment of dyes and pigments to the polymer surface. Bulk functionalization of PP is best carried out by free radical methods that allow for modifications to be conducted in the melt, for example, during the resin extrusion process, even though polymer chain degradation may be a side reaction. There have been some recent successes involving oxidative methods for the functionalization of PP that utilize two-phase reactions capable of introducing functional groups without the need to dissolve or melt the polymer. However, these reactions are unlikely to give uniform material.

More selective chemical methods for PP functionalization, such as epoxidation or hydroboration, allow for a much higher degree of control but generally need to be conducted in solution. Some of these functionalization reagents, such as boranes and silanes, have little catalyst poisoning effect and can act as chain transfer agents to give end-functionalized PPs. Alternatively, the unsaturated endgroups of PPs can be used to advantage to produce a large number of chain-end functionalized PP derivatives, including block copolymers. These materials are primarily of interest as blend compatibilizers. Polyfunctionalized PPs bearing side chain functionalities are accessible through copolymerizations of propylene, either with functionalized comonomers or, alternatively, with dienes. The latter lead to copolymers with sidechains bearing C=C bonds, which may be regarded as protected functional groups. The disadvantage of the first (functionalized comonomer) strategy is the sensitivity of early transition metal catalysts to heteroatoms, requiring the use of protecting group chemistry. The second (diene) strategy often gives good catalytic activities and maximum control of chemical structure in the resulting copolymers, but may present additional challenges on a larger scale. However, such functionalized materials may be of interest for specialty uses, for example, in medical applications where improved cell adhesion and good biocompatibility, coupled with the low toxicity and good durability of PP, are important.

REFERENCES AND NOTES

1. Desai, S. M.; Singh, P. R. Surface modification of polyethylene. *Adv. Polym. Sci.* **2004**, *169*, 231–293.
2. Lu, Q. W.; Macosko, C. W.; Horrion, J. Compatibilized blends of thermoplastic polyurethane and polypropylene. *Macromol. Symp.* **2003**, *198*, 221–232.

3. Colbeaux, A.; Fenouillot, F.; Gerard, J. F.; Taha, M.; Wauthier, H. Compatibilization of a polyolefin blend through covalent and ionic coupling of grafted polypropylene and polyethylene. *J. Appl. Polym. Sci.* **2004**, *93*, 2237–2244 and references therein.

4. Guan, Y.; Wang, S. H.; Zheng, A.; Xiao, H. N. Crystallization behaviors of polypropylene and functional polypropylene. *J. Appl. Polym. Sci.* **2003**, *88*, 872–877.

5. Ma, Y.; Farinha, J. P. S.; Winnik, M. A.; Yaneff, P. V.; Ryntz, R. A. Compatibility of chlorinated polyolefin with the components of thermoplastic polyolefin: A study by laser scanning confocal fluorescence microscopy. *Macromolecules* **2004**, *37*, 6544–6552.

6. Tjong, S. C.; Xu, S. A.; Li, R. K. Y.; Mai, Y. W. Mechanical behavior and fracture toughness evaluation of maleic anhydride compatibilized short glass fiber/SEBS/polypropylene hybrid composites. *Compos. Sci. Technol.* **2002**, *62*, 831–840.

7. Etelaaho, P.; Jarvela, P.; Stenlund, B.; Wilen, C. E.; Nicolas, R. Phenol-modified polypropylene as adhesion promoters in glass-fiber-reinforced polypropylene composites. *J. Appl. Polym. Sci.* **2002**, *84*, 1203–1213.

8. Coleman, J. N.; Cadek, M.; Blake, R.; Nicolosi, V.; Ryan, K. P.; Belton, C.; Fonseca, A.; Nagy, J. B.; Gun'ko, Y. K.; Blau, W. J. High-performance nanotube-reinforced plastics: Understanding the mechanism of strength increase. *Adv. Funct. Mat.* **2004**, *14*, 791–798.

9. Marchant, D.; Jayaraman, K. Strategies for optimizing polypropylene-clay nanocomposite structure. *Ind. Eng. Chem. Res.* **2002**, *41*, 6402–6408.

10. Bondar, Y.; Kim, H. J.; Yoon, S. H.; Lim, Y. J. Synthesis of cation-exchange adsorbent for anchoring metal ions by modification of poly(glycidyl methacrylate) chains grafted onto polypropylene fabric. *React. Funct. Polym.* **2004**, *58*, 43–51.

11. Kwon, O. H.; Nho, Y. C.; Chen, J. Surface modification of polypropylene film by radiation-induced grafting and its blood compatibility. *J. Appl. Polym. Sci.* **2003**, *88*, 1726–1736.

12. Borsig, E. Polypropylene derivatives. *J. Macromol. Sci., Pure Appl. Chem.* **1999**, *A36*, 1699–1715.

13. Datta, S.; Lohse, D. J. Graft copolymer compatibilizers for blends of isotactic polypropylenes and ethene-propene copolymers. *Macromolecules* **1993**, *26*, 2064–2076.

14. Miller, M. M.; Cowie, J. M. G.; Tait, J. G.; Brydon, D. L.; Mather, R. R. Fibres from polypropylene and liquid-crystal polymer blends using compatibilizing agents. 1. Assesment of functional and non-functional polypropylene-acrylic acid compatibilizers. *Polymer* **1995**, *36*, 3107–3112.

15. Zhang, Q. J.; Wen, J.; Luo, X. L. Synthesis of polypropylene-*graft*-poly(ε-caprolactone). *Macromol. Chem. Phys.* **1995**, *196*, 1221–1228.

16. Sulek, P.; Knaus, S.; Liska, R. Grafting of functional maleimides onto oligo- and polyolefins. *Macromol. Symp.* **2001**, *176*, 155–165.

17. Oktem, Z.; Cetin, S.; Akin-Oktem, G. Functionalization of low-molecular weight atactic polypropylene, Part I. Spectroscopic studies. *Polym. Bull.* **1999**, *43*, 239–246.

18. Yazdani-Pedram, M.; Vega, H.; Quijada, R. Melt functionalization of polypropylene with methyl esters of itaconic acid. *Polymer* **2001**, *42*, 4751–4758.

19. Zhang, L. F.; Guo, B. H.; Zhang, Z. M. Synthesis of multifunctional polypropylene via solid-phase cografting and its grafting mechanism. *J. Appl. Polym. Sci.* **2002**, *84*, 929–935.

20. Karmarkar, A.; Aggarwal, P.; Modak, J.; Chanda, M. Grafting of *m*-isopropenyl-α,α-dimethylbenzyl isocyanate onto isotactic polypropylene: Synthesis and characterization. *J. Polym. Mat.* **2003**, *20*, 101–107.

21. Bucio, E.; Cedillo, G.; Burillo, G.; Ogawa, T. Radiation-induced grafting of functional acrylic monomers onto polyethylene and polypropylene films using acryloyl chloride. *Polym. Bull.* **2001**, *46*, 115–121.

22. Hwang, T. S.; Park, J. W. UV-induced graft polymerization of polypropylene-*g*-glycidyl methacrylate membrane in the vapor phase. *Macromol. Res.* **2003**, *11*, 495–500.

23. Naqvi, A.; Nahar, P.; Gandhi, R. P. Introduction of functional groups onto polypropylene and polyethylene surfaces for immobilization of enzymes. *Anal. Biochem.* **2002**, *306*, 74–78.

24. Marques, M. M.; Correia, S. G.; Ascenso, J. R.; Ribeiro, A. F. G.; Gomes, P. T.; Dias, A. R.; Foster, P.; Rausch, M. D.; Chien, J. C. W. Polymerization with TMA-protected polar vinyl comonomers. I. Catalyzed by group 4 metal complexes with η^5-type ligands. *J. Polym. Sci., Part A: Polym. Chem.* **1999**, *37*, 2457–2469.

25. Correia, S. G.; Marques, M. M.; Ascenso, J. R.; Ribeiro, A. F. G.; Gomes, P. T.; Dias, A. R.; Blais, M.; Rausch, M. D.; Chien, J. C. W. Polymerization with TMA protected polar vinyl copolymers. II. Catalyzed by nickel complexes containing alpha-diimine type ligands. *J. Polym. Sci., Part A: Polym. Chem.* **1999**, *37*, 2471–2480.

26. Bruzaud, S.; Cramail, H.; Duvignac, L.; Deffieux, A. ω-Chloro-α-olefins as co-and termonomers for the synthesis of functional polyolefins. *Macromol. Chem. Phys.* **1997**, *198*, 291–303.

27. Hippi, U.; Korhonen, M.; Paavola, S.; Seppälä, J. Compatibilization of poly(propylene)/polyamide 6 blends with functionalized poly(propylene)s prepared with metallocene catalyst. *Macromol. Mat. Engin.* **2004**, *289*, 714–721.

28. Kaya, A.; Jakisch, L.; Komber, H.; Pompe, G.; Piontek, J.; Voit, B.; Schulze, U. Synthesis of oxazoline functionalized polypropene using metallocene catalysts. *Macromol. Rapid Commun.* **2000**, *21*, 1267–1271.

29. Imuta, J.; Kashiwa, N.; Toda, Y. Catalytic regioselective introduction of allyl alcohol into nonpolar polyolefins: Development of one-pot synthesis of hydroxyl-capped polyolefins mediated by a new metallocene IF catalyst. *J. Am. Chem. Soc.* **2002**, *124*, 1176–1177.

30. Imuta, J.; Toda, A.; Tsutsui, T.; Hachimori, T.; Kashiwa, N. Development of polyethylene copolymers manufacturing technologies and synthesis of new functionalized polyolefins with designed catalysts. *Bull. Chem. Soc. Jpn.* **2004**, *77*, 607–615.

31. Hagihara, H.; Tsuchihara, K.; Sugiyama, J.; Takeuchi, K.; Shiono, T. Copolymerization of propylene and polar allyl monomer with zirconocene/methylaluminoxane catalyst: Catalytic synthesis of amino-terminated isotactic polypropene. *Macromolecules* **2004**, *37*, 5145–5148.

32. Hagihara, H.; Tsuchihara, K.; Sugiyama, J.; Takeuchi, K.; Shiono, T. Copolymerization of 3-buten-1-ol and propylene with an isospecific zirconocene/methylaluminoxane catalyst. *J. Polym. Sci., Part A: Polym. Chem.* **2004**, *42*, 5600–5607.

33. Stehling, U. M.; Stein, K. M.; Fischer, D.; Waymouth, R. M. Metallocene/borate catalyzed copolymerization of 5-*N*,*N*-diisopropylamino-1-pentene with 1-hexene or 4-methylpentene. *Macromolecules* **1999**, *32*, 14–20.

34. Stehling, U. M.; Malström, E. E.; Waymouth, R. M.; Hawker, C. J. Synthesis of poly(olefin) graft copolymers by a combination of metallocene and "living" free radical polymerization techniques. *Macromolecules* **1998**, *31*, 4396–4398.

35. Dong, J. Y.; Wang, Z. M.; Hong, H.; Chung, T. C. Synthesis of isotactic polypropylene containing a terminal Cl, OH, or NH$_2$ group via metallocene-mediated polymerization/chain transfer reaction. *Macromolecules* **2002**, *35*, 9352–9359.

36. Mülhaupt, R.; Duschek, T.; Rieger, B. Functional polypropylene blend compatibilizers. *Makromol. Chem., Macromol. Symp.* **1991**, *48/49*, 317–332.

37. Shiono, T.; Kurosawa, H.; Ishida, O.; Soga, K. Synthesis of polypropylene functionalized with secondary amino groups at the chain ends. *Macromolecules* **1993**, *26*, 2085–2089.

38. Chung, T. C.; Lu, H. L. Functionalization and block reactions of polyolefins using metallocene catalysts and borane reagents. *J. Mol. Catal. A: Chem.* **1997**, *115*, 115–127.

39. Chung, T. C. Functionalized polyolefins prepared by the combination of borane monomers and transition metal catalysts. *Macromol. Symp.* **1995**, *89*, 151–162.

40. Chung, T. C.; Lu, H. L.; Janvikul, W. A novel synthesis of PP-*b*-PMMA copolymers via metallocene catalysis and borane chemistry. *Polymer* **1997**, *38*, 1495–1502.

41. Chung, T. C.; Lu, H. L.; Li, C. L. Synthesis and functionalization of unsaturated polyethylene: Poly(ethylene-*co*-1,4-hexadiene). *Macromolecules* **1994**, *27*, 7533–7537.

42. Chung, T. C.; Lu, H. L.; Li, C. L. Functionalization of polyethylene using borane reagents and metallocene catalysts. *Polym. Int.* **1995**, *37*, 197–205.

43. Chung, T. C. Recent developments in the functionalization of polyolefins utilizing borane-containing copolymers. *Trends Polym. Sci.* **1995**, *3*, 191–198.

44. Chung, T. C.; Janvikul, W. Borane-containing polyolefins: Synthesis and applications. *J. Organomet. Chem.* **1999**, *581*, 176–187.

45. Chung, T. C.; Lee, S. H. New hydrophilic polypropylene membranes: Fabrication and evaluation. *J. Appl. Polym. Sci.* **1997**, *64*, 567–575.

46. Liu, J.; Dong, J. Y.; Cui, N.; Hu, Y. Supporting a metallocene on functional polypropylene granules for slurry ethylene polymerization. *Macromolecules* **2004**, *37*, 6275–6282.

47. Chung, T. C.; Xu, G.; Lu, Y.; Hu, Y. Metallocene-mediated olefin polymerization with B-H chain transfer agents: Synthesis of chain-end functionalized polyolefins and diblock copolymers. *Macromolecules* **2001**, *34*, 8040–8050.

48. Koo, K.; Marks, T. J. Silicon-modified Ziegler–Natta polymerization. Catalytic approaches to silyl-capped and silyl-linked polyolefins using "single-site" cationic Ziegler–Natta catalysts. *J. Am. Chem. Soc.* **1999**, *121*, 8791–8802.

49. Macmanus, L. F.; Walzak, M. J.; McIntyre, N. S. Study of ultraviolet light and ozone surface modification of polypropylene. *J. Polym. Sci., Part A: Polym. Chem.* **1999**, *37*, 2489–2501.

50. Boaen, N. K.; Hillmyer, M. A. Selective and mild oxyfunctionalization of model polyolefins. *Macromolecules* **2003**, *36*, 7027–7034.

51. Fu, P. F.; Tomalia, M. K. Carboxyl-terminated isotactic polypropylene. Preparation, characterization, kinetics and reactivities. *Macromolecules* **2004**, *37*, 267–275.

52. Toyota, A.; Mizuno, A.; Tsutsui, T.; Kaneko, H.; Kashiwa, N. Synthesis and characterization of metallocene-catalyzed propylene-ethylene copolymer with end-capped functionality. *Polymer* **2002**, *43*, 6351–6355.

53. Matyjaszewski, K.; Saget, J.; Pyun, J.; Schlogl, M.; Rieger, B. Synthesis of polypropylene-poly(meth)acrylate block copolymers using metallocene catalyzed processes and subsequent atom transfer radical polymerization. *J. Macromol. Sci., Pure Appl. Chem.* **2002**, *A39*, 901–913.

54. Lu, H. L.; Hong, S.; Chung, T. C. Synthesis of polypropylene-*co*-*p*-methylstyrene copolymers by metallocene and Ziegler–Natta catalysts. *J. Polym. Sci., Part A: Polym. Chem.* **1999**, *37*, 2795–2802.

55. Hackmann, H.; Repo, T.; Jany, G.; Rieger, B. Zirconocene-MAO catalyzed homo- and copolymerizations of linear asymmetrically substituted dienes with propene: A novel strategy to functional (co)poly(alpha-olefin)s. *Macromol. Chem. Phys.* **1998**, *199*, 1511–1517.

56. Hackmann, M.; Rieger, B. Functional olefin copolymers: Uniform architectures of propene/7-methyl-1,6-octadiene copolymers by ATR-FTIR spectroscopy control of monomer composition. *Macromolecules* **2000**, *33*, 1524–1529.

57. Song, F.; Pappalardo, D.; Johnson, A. F.; Rieger, B.; Bochmann, M. Derivatization of propene/methyloctadiene copolymers: A flexible approach to side-chain-functionalized polypropenes. *J. Polym. Sci., Part A: Polym. Chem.* **2002**, *40*, 1484–1497.

58. Sarazin, Y.; Fink, G.; Hauschild, K.; Bochmann, M. Copolymerization of propene and 5-vinyl-2-norbornene: A simple route to functionalized polypropenes. *Macromol. Rapid Commun.* **2005**, *26*, 1208–1213.

12 Influence of Tacticity of Propylene Placement on Structure and Properties of Ethylene/Propylene Copolymers

Maurizio Galimberti and Gaetano Guerra

CONTENTS

12.1 INTRODUCTION

Following the discovery of the stereoselective polymerization of vinyl monomers,[1-4] the relevance of tacticity in the placement of monomeric units along the macromolecular chain was immediately recognized.[2,3] In fact, the average length of tactic sequences of monomeric units in the polymer chain is the main factor determining the occurrence and the degree of crystallinity of polymeric materials, and hence, their physical properties. It is also well-recognized that tacticity can also be relevant for copolymers. This is not only true for the obvious case of block copolymers, but also for random copolymers when the formation of crystallizable sequences requires a tactic placement of monomeric units, for instance for propylene-rich[4] or 1-butene-rich[5] random copolymers with a minor amount of ethylene.

Much less recognized is the possible influence of tacticity on copolymer properties when α-olefin monomer units are a minor component, and crystallinity is not based on a tactic α-olefin sequence but on a different comonomer such as ethylene. In this chapter, this tacticity effect is shown for ethylene-rich ethylene/propylene (EP) copolymers, where the crystallizable sequences are based on ethylene, that is, a comonomer that does not have tacticity requirements. In particular, this chapter describes in detail the microstructure of EP copolymers having industrially relevant compositions (ethylene content 80–55 mol%), with particular focus on the placement of propylene units along the ethylene-based macromolecular chains and their influence on copolymer properties. This subject is, of course, related to the industrial relevance of EP copolymers and ethylene/propylene/diene monomer terpolymers (EPDMs) (collectively referred to as EP(D)Ms), which presently represent the most widely produced saturated rubbers.[6]

In this chapter, copolymers of propylene with ethylene are discussed, giving consideration to the relative distribution of ethylene and propylene along the chain and the regio- and stereochemistry of propylene placement.[2] The products of monomer reactivity ratios are used to assess the intramolecular distribution of comonomers, and the stereo- and regiochemistry of propylene insertion are quantitatively examined. Catalytic systems suitable for preparation of the abovementioned copolymers are presented, and in particular, the implications of the structure of the catalytic sites on the important molecular parameters of the copolymers are discussed. Copolymer property discussions are focused on the crystallinity and elasticity of samples having compositions suitable for elastomeric behavior. It is thus worth reminding the reader here that an elastomer, according to the American Society for Testing and Materials (ASTM) standard definition, is "a natural or synthetic polymer, which at room temperature, can be stretched repeatedly to at least twice its original length and which after removal of the tensile load will immediately and forcibly return to its original length."[7] In this chapter, the effect of propylene placement tacticity on the elasticity of EP elastomers is rationalized.

12.2 SYNTHESIS OF ETHYLENE/PROPYLENE COPOLYMERS BY ZIEGLER–NATTA-TYPE CATALYSTS

EP copolymers are obtained through polymerization promoted by Ziegler–Natta-type catalysts.[8] A Ziegler–Natta catalyst polymerizes olefins by inserting them into a metal–carbon bond. As a consequence, the exact structure of the catalytic site, determined by the interaction between the organometallic complex and the growing polymer chain, has a dramatic influence on the insertion reaction and hence on copolymer characteristics.

Three families of catalytic systems have been used to date for the preparation of copolymers containing ethylene and propylene over a wide compositional range. They are based on vanadium,[9] titanium (heterogeneous),[9e,10] and so-called "single-site" catalysts (SSCs), which are typically metallocenes[11] or cyclopentadienyl (Cp)-amido group 4 dihalides often referred to as "constrained geometry" catalysts (CGCs).[12] All three families of catalysts have industrial relevance. Vanadium-based catalysts are largely used for the industrial production of elastomeric ethylene copolymers, mainly EP(D)Ms. Titanium-based catalysts are used for the production of crystalline propylene copolymers with a minor amount of ethylene. EPM (ethylene/propylene) grades made in this manner were once commercially available from DUTRAL (now Polimeri Europa)[10h,13] but are no longer industrially produced. In the family of SSCs, those based on Cp-amido and metallocene catalysts have found applications on a commercial scale for the production of polyolefin elastomers. The concept of single-site catalysis (and hence the name) is correlated with the homogeneous nature of the organometallic transition metal active species.[14] These kinds of catalysts produce polymers with narrow molecular weight distributions and copolymers with narrow composition distributions. It is also worth noting that the class of vanadium-based catalysts discussed in this chapter also behaves as SSCs. It would therefore be more correct to consider single-site metallocene and CGCs as a new generation of SSCs.

12.2.1 VANADIUM-BASED CATALYSTS

Vanadium-based catalytic systems for EP(D)M synthesis are comprised of a vanadium compound ("catalyst precursor"), a chlorinated aluminum alkyl ("cocatalyst"), and a chlorinated ester ("promoter"). Typical components of a vanadium-based catalytic system are the following:

Vanadium compound	$VOCl_3$, $V(acac)_3$ (acac = acetylacetonate)
Chlorinated aluminum alkyl	$(C_2H_5)_3Al_2Cl_3$, aluminum sesquichloride
Chlorinated promoter	Ethyl trichloroacetate, n-butyl perchlorocrotonate, α,α,α-trichlorotoluene

The reaction of the vanadium salt with the aluminum alkyl leads to displacement of the vanadium ligands, with the formation of V–C bonds.[15] As a consequence, the catalytic site is not sterically demanding. As mentioned above, vanadium-based catalysts behave as SSCs for olefin polymerization. Although the structure of their active catalytic sites is still unknown, the structures of polymers produced with vanadium systems are typical of those from SSCs. The majority of commercially available EP(D)M grades are from vanadium-based catalysts.[6]

12.2.2 TITANIUM-BASED CATALYSTS

The components of a typical heterogeneous titanium-based catalytic system are a titanium compound ("catalyst precursor"), an aluminum alkyl ("cocatalyst"), and an organic substance acting as electron donor substantially to titanium (as widely accepted in the literature).[16] The titanium compound is typically a titanium halide and is usually supported on $MgCl_2$. Electron donors are present as internal donors, that is, ones presupported on $MgCl_2$, and as external donors, that is, those added along with the aluminum alkyls.[16] The cooperation between the support and the electron donor provides a steric environment at the titanium center that is responsible for the regio- and stereoselective control of propylene insertion. A characteristic of titanium-based catalysts is their multiplicity of catalytic sites, which is typical for heterogeneous catalysts.

12.2.3 "SINGLE-SITE" CATALYSTS (SSCs)

Among the new generations of Ziegler–Natta catalysts, two families of SSCs have been well-studied from a scientific point of view and have been used for commercial applications: metallocenes[11] and CGCs.[12]

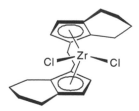

FIGURE 12.1 A prototypical metallocene suitable for the preparation of EP copolymers, *rac*-[ethylenebis(4,5,6,7-tetrahydroindenyl)]zirconium dichloride (*rac*-[C$_2$H$_4$(H$_4$Ind)$_2$]ZrCl$_2$).

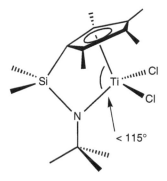

FIGURE 12.2 A prototypical constrained geometry catalyst, [dimethylsilylene(tetramethylcyclopentadienyl)(*tert*-butylamido)]titanium dichloride (Me$_2$Si(Me$_4$Cp)(N-*t*-Bu)TiCl$_2$).

12.2.3.1 Metallocenes

A metallocene is a coordination compound in which a transition metal is π-bonded to two cyclopentadienyl-type ring ligands. The structure of a prototypical metallocene suitable for the preparation of EP copolymers, *rac*-ethylenebis(4,5,6,7-tetrahydroindenyl)zirconium dichloride (*rac*-[C$_2$H$_4$(H$_4$Ind)$_2$]ZrCl$_2$), is shown in Figure 12.1. This isoselective metallocene is a suitable catalyst for the preparation of the whole range of polyolefins and is particularly interesting for the synthesis of EP(D)Ms for two basic reasons: it gives high molecular mass elastomers, and it promotes good conversions of 5-ethylidene-2-norbornene (ENB), the diene termonomer that is almost exclusively used on the industrial scale. On the market, examples of EP(D)M and plastomers from metallocenes are respectively the Vistalon^{TM} and the Exact^{TM} grades from ExxonMobil. Mitsui recently announced the production for mid-2006 of metallocene propylene-based elastomers with the trade name of Tafmer XM^{TM}.

12.2.3.2 Constrained Geometry Catalysts (CGCs)

"Constrained geometry" catalysts of the group 4 metals were first reported in 1990 by both Dow Chemical Company and Exxon Chemical Company.[12] A typical CGC, [dimethylsilylene(tetramethylcyclopentadienyl)(*tert*-butylamido)]titanium dichloride (Me$_2$Si(Me$_4$-Cp)(N-*t*-Bu)TiCl$_2$), is shown in Figure 12.2. The "constrained geometry" terminology refers to the ligand bite angle (N–Ti–Cp centroid angle) as being smaller than the analogous angle in metallocenes (Cp–Ti–Cp centroid angle). For metallocenes, this angle is generally found to be much larger than 115°. From the definition point of view, a CGC is not a metallocene but rather a half sandwich Cp or substituted Cp complex. However, CGCs are often referred to as "metallocenes."

A peculiar feature of CGCs is that they are quite aselective catalysts; out of the many structures known, only a few are able to produce stereoregular polypropylenes (PPs). However, CGCs based on fluorenyl-amido structures have been reported as having stereoselectivity, giving rise to either syndiotactic or isotactic PP, as a function of the activator. In particular, the syndiotactic PP with the highest known melting point (T_m) was obtained with this family of CGCs[17] (see Chapter 2). CGCs can also be used to prepare atactic PPs (with slightly prevalent syndiotactic placement) with relatively high molecular masses. CGCs used for ethylene/propylene copolymerization produce copolymers characterized by low reactivity ratio products ($r_1 r_2$ values, *vide infra*), and hence by very short propylene sequences, that have been reported as atactic.[11p,12c]

The use of CGCs to make ethylene-based polymers gives rise to long-chain branching. In this process, CGCs used at very high polymerization temperatures ($>100\ °C$) are able to produce macromolecular chains terminated with double bonds and then to insert the double bond termini of these chains into a new growing chain. CGCs produce EP copolymers with a random comonomer distribution.[11p,12c] So far, they have found applications at the industrial level for the production of both polyolefin plastomers and polyolefin elastomers. The plastomer materials are ethylene/1-octene copolymers, commercialized by Dow under the trade name of Engage™. Ethylene/styrene copolymer samples (produced industrially by Dow) were available in the late 1990s but are no longer present on the market. The elastomer materials are EPDMs based on ENB as the diene termonomer, commercialized first by Dow Dupont Elastomers and now by Dow under the trade name of Nordel™ IP.

12.3 COMONOMERS IN ETHYLENE/PROPYLENE COPOLYMERS

Ethylene and propylene are molecules that possess different symmetry (Figure 12.3). Ethylene has two C_2 rotation axes and a σ mirror plane. Propylene, owing to the presence of the methyl group, possesses only a σ mirror plane (the double bond plane). The first consequence of this difference is that propylene has two chiral faces with respect to the plane defined by the double bond. In this respect, each face should be regarded as an *enantioface*. In Figure 12.3, the enantiofaces are indicated using the *si* and *re* nomenclature.[18]

The reaction of the propylene double bond with the metal–carbon bond, that is, the insertion reaction, produces an enchained monomeric unit with a tertiary carbon atom whose configuration is determined by the identify of the reacting enantioface (Figure 12.4). The *stereoselectivity* of a catalyst is its ability to discriminate between the two different monomer enantiofaces.

In more precise terms, propylene is a *prochiral* molecule, and a reaction involving addition to its double bond generates a chiral center, the configuration of which depends upon the face involved when insertion occurs. Furthermore, the two unsaturated carbon atoms of propylene are substituted

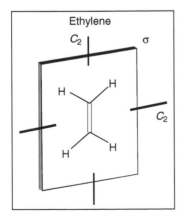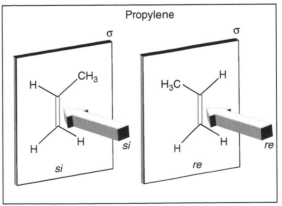

FIGURE 12.3 Ethylene and propylene monomer symmetry.

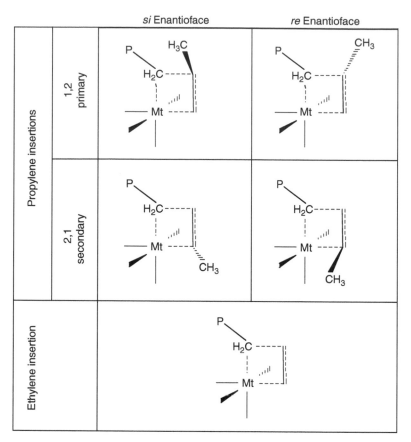

FIGURE 12.4   Insertion of ethylene and propylene into a metal–growing chain bond (P = growing polymer chain).

differently and this gives rise to two possible insertion modes having different regiochemistry. When propylene addition occurs with formation of a metal–CH_2 bond, the insertion is called primary (or 1,2) (Figure 12.4, top). When the propylene addition occurs with formation of a metal–CH bond, the insertion is called secondary (or 2,1) (Figure 12.4, middle). The *regioselectivity* of a catalyst is its ability to discriminate between primary and secondary monomer insertions. The symmetry of ethylene makes all possible approaches of the monomer to the catalyst identical, and only one type of insertion occurs (Figure 12.4, bottom).

12.3.1   ANALYTICAL CHARACTERIZATION OF ETHYLENE/PROPYLENE COPOLYMERS

The regiochemistry of propylene insertion is characterized by counting the number of methylene groups ($-CH_2-$) in sequence along the produced macromolecular chains (Figure 12.5). In a regioregular propylene homopolymer, only isolated methylenes are detected, whereas the occurrence of a misinsertion leads to a sequence of two methylene groups. In EP copolymers with a regioregular placement of propylene, sequences of an odd number of ($-CH_2-$) units between two $-CH(CH_3)-$ units are observed, for example, [metal-(CH_2)-$CH(CH_3)$-(CH_2)-(CH_2)-(CH_2)-$CH(CH_3)$-]. On the other hand, sequences having an even number of ($-CH_2-$) units between two $-CH(CH_3)-$ units, for example, metal-$CH(CH_3)$-(CH_2)-(CH_2)-(CH_2)-(CH_2)-$CH(CH_3)$-, are diagnostic of regioirregular propylene placements. Even or odd ($-CH_2-$) sequences can be analytically discriminated by ^{13}C nuclear magnetic resonance (NMR) spectroscopy and counted, although only to a degree of approximation. It is

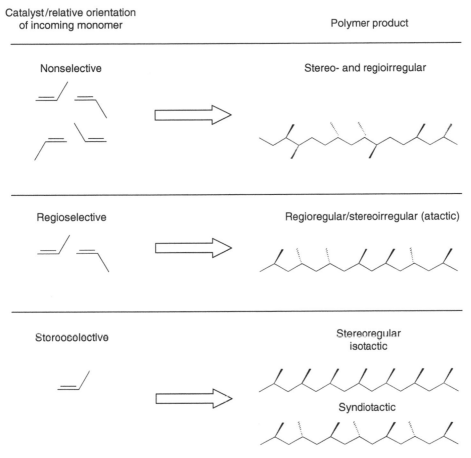

FIGURE 12.5 Catalyst selectivity, relative orientation of incoming monomer, and resultant polymer microstructure for propylene homopolymerization.

thus possible to use the methylene sequences to quantify the regioregularity of propylene placement in EP copolymers.[19]

12.3.2 FIVE DIFFERENT REPEAT UNITS IN ETHYLENE/PROPYLENE COPOLYMERS

The five different monomer insertions shown in Figure 12.4 (ethylene and the four different propylene combinations of 1,2- or 2,1-insertion with the *re* or *si* enantioface) can be considered to be five different repeat units (five different comonomer units) present in the macromolecular chain. Scheme 12.1 summarizes these five comonomer insertions for an EP copolymer.

12.4 INTRAMOLECULAR DISTRIBUTION OF THE COMONOMERS: REACTIVITY RATIOS r_1 AND r_2

12.4.1 GENERAL REMARKS

The relative reactivity of comonomers during copolymerization is usually expressed through the reactivity ratios r_1 and r_2.[20] The reactions involved in a copolymerization are shown in Figure 12.6. Generally speaking, the value k_{ij} is the rate constant for addition of monomer j to a growing chain bearing comonomer i as the last inserted monomer, where i and j equal either 1 or 2. Here, 1 and 2 stand

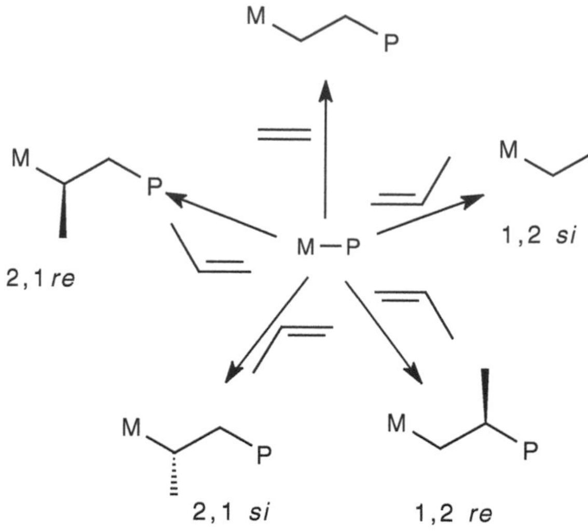

SCHEME 12.1 The five comonomer insertion events in ethylene/propylene copolymerization (M = metal center; P = growing polymer chain).

$$M\text{-}E + E \longrightarrow M\text{-}E\text{-}E \qquad R_{11} = k_{11}[M\text{-}E][E]$$

$$M\text{-}E + P \longrightarrow M\text{-}P\text{-}E \qquad R_{12} = k_{12}[M\text{-}E][P]$$

$$M\text{-}P + E \longrightarrow M\text{-}E\text{-}P \qquad R_{21} = k_{21}[M\text{-}P][E]$$

$$M\text{-}P + P \longrightarrow M\text{-}P\text{-}P \qquad R_{22} = k_{22}[M\text{-}P][P]$$

$$r_1 = \frac{k_{11}}{k_{12}} \qquad r_2 = \frac{k_{22}}{k_{21}}$$

FIGURE 12.6 Definition of the reactivity ratios r_1 and r_2 for ethylene/propylene copolymerization (E = ethylene = monomer 1; P = propylene = monomer 2; R = rate of indicated insertion process).

for ethylene (E) and propylene (P), respectively. In simple words, r_1 represents the relative probability of a homo-addition of ethylene (an ethylene insertion followed by another ethylene insertion, k_{11}) versus the hetero-addition of propylene (an ethylene insertion followed by a propylene insertion, k_{12}). The value r_2 represents the relative probability of a homo-addition of propylene (a propylene insertion followed by another propylene insertion, k_{22}) versus the hetero-addition of propylene (a propylene insertion followed by a ethylene insertion, k_{21}). When evaluating r_1 and r_2 ratios for EP copolymers, the insertion mode of propylene (1,2 versus 2,1 and *re* versus *si*) is not taken into account and propylene is considered as one comonomer.

12.4.2 THE $r_1 r_2$ PRODUCT

The way in which ethylene and propylene distribute themselves along the macromolecular chain is expressed by the product of the reactivity ratios r_1 and r_2 (Figure 12.7). In particular, if the comonomer insertion:

1. Is not influenced by the last inserted comonomer unit, then $k_{11}/k_{12} = k_{21}/k_{22}$, and the relative probabilities of insertion of comonomers 1 and 2 are the same, regardless

$r_1 r_2 = 1$	Random copolymerization	PPEPEPEPPPEPPPEEPEEPE
$r_1 r_2 > 1$	Blocky copolymerization	PPPPEEEEEPPPEEEEEPP
$r_1 r_2 = 0$	Alternating copolymerization	EPEPEPEPEPEPEPEPEP

FIGURE 12.7 Product of reactivity ratios $r_1 r_2$ and comonomer distribution in EP copolymers.

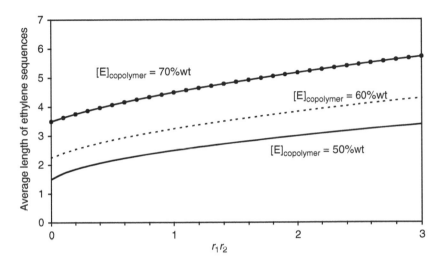

FIGURE 12.8 Average length of ethylene sequences in EP copolymers versus $r_1 r_2$. [E]$_{copolymer}$ is the concentration of ethylene in the copolymer, expressed as wt%.

of whether the last inserted comonomer was 1 or 2. This means that $r_1 r_2 = 1$. The copolymerization is called *random*.

2. Is influenced by the last inserted comonomer unit(s), then $k_{11}/k_{12} \neq k_{21}/k_{22}$, and the relative probabilities of insertion of comonomers 1 and 2 depend on the identity of the last inserted comonomer unit. This means that $r_1 r_2 \neq 1$. In particular, values of $r_1 r_2 > 1$ indicate that the catalytic system favors homosequences rather than an alternating distribution of comonomers. Conversely, alternating sequences are more abundant in copolymers obtained with values of $r_1 r_2 < 1$. It is important to emphasize that the way ethylene and propylene distribute themselves along the macromolecular chain is expressed only by the product of reactivity ratios $r_1 r_2$, no matter what the individual values of r_1 and r_2 values are.

It is thus clear that (1) in a random copolymer (from a random copolymerization), homosequences of both comonomers can be present, with a length depending on chemical composition; and (2) in a copolymer with $r_1 r_2 \gg 1$, the probability of having long homosequences of both comonomers increases.

Figure 12.8 shows the correlation between values of $r_1 r_2$ and the average length of ethylene sequences in an EP copolymer for three levels of ethylene content ([E]$_{copolymer}$). Figure 12.9 shows the effect of $r_1 r_2$ on the probability (fraction) of ethylene sequences with 10 or more ethylene units (E$_{10}$), a microstructural feature that influences the copolymer crystallinity. For copolymers with a given comonomer composition, crystallization ability is generally maximized by blocky constitutions ($r_1 r_2 > 1$) and minimized by alternating constitutions ($r_1 r_2 \rightarrow 0$). As a consequence, minimum

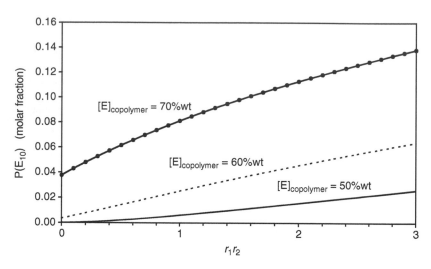

FIGURE 12.9 Probability (fraction) of ethylene sequences with ten or more ethylene units (E_{10}) in EP copolymers versus $r_1 r_2$.

crystallinity, and hence structural order, generally corresponds to copolymers having alternating tendencies rather than to random copolymers ($r_1 r_2 \approx 1$) with a minimum of microstructural order.

12.5 ANALYTICAL EVALUATION OF PROPYLENE PLACEMENT IN THE COPOLYMER CHAINS

The stereoregularity and regioregularity of propylene sequences in EP copolymers can be analytically evaluated by ^{13}C NMR.[19] Tacticity is obtained as the percentage of *mm* triads from the signals of the central tertiary CH carbon in the propylene triad (PPP) sequences ($T_{\beta\beta}$), using Equation 12.1

$$mm(\%) = \frac{100 \times T_{\beta\beta}(mm)}{[T_{\beta\beta}(mm) + T_{\beta\beta}(mr + rr)]} \qquad (12.1)$$

According to Randall[19b,c] the lower limit of regioirregular unit content can be estimated from the relative amount of sequences containing 2 and 4 methylene groups, as quantified by Equation 12.2 (where $S_{\alpha\alpha}$, $S_{\alpha\beta}$, $S_{\beta\beta}$, $S_{\beta\gamma}$, $S_{\gamma\delta}$, and $S_{\gamma\gamma}$ represent secondary CH_2 chain carbons α, β, or γ to the nearest tertiary propylene CH carbon on either side)

$$\frac{\sum [-(CH_2)_2-] + [-(CH_2)_4-]}{\sum [-(CH_2)-]} = \frac{(100)\left(\frac{1}{2} S_{\alpha\beta} + \frac{1}{2} S_{\beta\gamma}\right)}{\left(S_{\alpha\alpha} + \frac{1}{2} S_{\alpha\beta} + S_{\beta\beta} + \frac{1}{2} S_{\beta\gamma} + S_{\gamma\gamma} + \frac{1}{2} S_{\gamma\delta}\right)} \qquad (12.2)$$

This represents a lower limit of regioirregular units, because sequences containing six or more methylenes are indistinguishable from those formed by sequences of at least three ethylene units. Since the reactivity ratios r_1 and r_2 are obtained only from triads containing regioregular propylene units, the presence of a relatively large amount of regioirregular propylene placements introduces uncertainties in their evaluation.

TABLE 12.1

Nature of Catalytic Sites and Distribution of Copolymer Molecular Properties for Common Ethylene/Propylene Copolymerization Systems

Catalytic System	Catalytic Site[a]	Catalyst Precursor		Distribution of Molecular Properties
		Structure	Ligand Stability[b]	
Titanium (heterogeneous)[10o,11j,p]	Multi	Unknown	(No)	Broad
Vanadium[9f,11j,p]	Single	Unknown	No	Narrow
	Multi	Unknown	No	Broad
Metallocene[11j,l,o,p]	Single	Known	Yes	Narrow
Constrained geometry[11p,12c]	Single	Known	Yes	Narrow

[a] Number of active sites.
[b] During the polymerization.

12.6 CATALYTIC SYSTEMS AND COPOLYMER CONSTITUTION AND CONFIGURATION

The behavior of different families of catalysts in the polymerization of ethylene and propylene is described as follows.

12.6.1 NATURE OF CATALYST ACTIVE SITE AND INTERMOLECULAR DISTRIBUTION OF MOLECULAR PROPERTIES

Table 12.1 summarizes some features of common catalyst systems used for ethylene/propylene copolymerizations. SSCs, either based on vanadium salts[9f,11j,p] or on organometallic compounds such as metallocenes[11j,o,p] and constrained geometry complexes,[11p,12c] are able to produce a narrow distribution of copolymer properties such as chemical composition and molecular weight distribution, distinguishing these catalysts from heterogeneous titanium-based catalysts. This ability of SSCs is of fundamental importance not only from an applications point of view, but also because it allows for accurate determination of the microstructures of EP copolymers and prevents misleading evaluations of mixtures of macromolecules having different molecular properties, in particular different chemical compositions. With a multisite catalyst, such as a heterogeneous titanium-based species, macromolecules with a wide range of chemical compositions are produced, from almost pure polyethylene (PE) to almost pure PP.

Vanadium-based catalysts used on a commercial scale are definitely single site, as demonstrated by the resultant intermolecular distribution of copolymer molecular properties. However, it is possible to have multisite vanadium catalysts, for example, by supporting a vanadium salt on an inorganic carrier such as $MgCl_2$,[9f] changing the nature of the aluminum-based cocatalyst (i.e., substituting the chlorinated aluminum alkyl with an aluminum trialkyl),[9f] or modulating the Al:V ratio (with higher ratios producing narrower molecular property distributions).[11j] A broad intermolecular distribution of copolymer properties can also be obtained for metallocene-based copolymers, for example, by using a mixture of metallocenes.

What is really key for the "SSC family" based on metallocenes and CGCs is the opportunity to have a catalyst precursor of defined chemical structure. For example, in the case of metallocenes, the π-bound cyclopentadienyl-type ligands remain coordinated to the transition metal atom during the course of polymerization. This allows for the following: (1) the control of polymerization behavior and polymer microstructure by modifying the structure of the ligands, and, as a consequence, (2) establishment of a correlation between ligand structure and polymer microstructure.

TABLE 12.2
Intermolecular Distribution of Molecular Mass and Chemical Composition for EP Copolymers Prepared with Various Catalyst Systems

Catalytic System	Molecular Weight Distribution[a]	Chemical Composition Distribution[b]
Titanium (heterogeneous)[10o,11p]	5–15	10–25
Vanadium (single site)[9f,11j,p]	2–2.5	0.5–2
Metallocene[11j,l,o,p]	2-2.5	0.5–2
Constrained geometry[11p,12c]	2–2.5	0.5–2

[a] M_w/M_n.
[b] Variation coefficient, V (see text).

Table 12.2 shows a numerical evaluation of the intermolecular molecular weight distribution and chemical composition distribution in EP copolymers made with different catalyst systems. The ratio between the polymer weight average molecular weight and number average molecular weight (M_w/M_n, generally measured by gel permeation chromatography) is used to describe the distribution of molecular masses, and a statistical parameter called the variation coefficient (V) is used to characterize the chemical composition distribution. The variation coefficient is a relative dispersion index, as given by Equation 12.3, where the arithmetic mean and standard deviation refer to numerical data indicating the chemical composition (comonomer content) of copolymer fractions obtained through fractionation:[9f]

$$V = \left(\frac{\text{standard deviation}}{\text{arithmetic mean}} \right) \times 100 \qquad (12.3)$$

Copolymers analyzed were fractionated using solvents (or mixtures of solvents) having increasing solubility power, at their boiling points. Dissolved polymer fractions were subsequently collected, their chemical composition determined through NMR, and arithmetic mean and standard deviation of these data were then calculated. Thus, the range of ethylene content in an EP copolymer is given by Equation 12.4

Ethylene content = (average ethylene content) $\pm$ [(V / 100)(average ethylene content)]   (12.4)

where the average ethylene content expressed as either mol% or wt% is the arithmetic mean obtained from the fractions (and is coincident with the value obtained for the copolymer before fractionation). Copolymers prepared with single-site vanadium, metallocene, and CGCs thus have a high homogeneity of molecular properties. This allows an easier investigation of the correlation between molecular and physical properties.

12.6.2 CONTROL OF PROPYLENE INSERTION

Table 12.3 summarizes stereochemical control in propylene homopolymerization for different catalytic systems. Metallocenes are the only SSCs currently able to produce the entire variety of possible propylene sequence microstructures (isotactic, syndiotactic, and hemiisotactic). As mentioned above,

TABLE 12.3
Polypropylene Tacticities Achievable with Different Catalyst Systems

	Stereochemical Control of Propylene Insertion			
	iso-	syndio-	hemi-	a-
Titanium (heterogeneous)	Yes	No	No	Yes
Vanadium (single site)	No	Yes	No	Yes
Metallocene	Yes	Yes	Yes	Yes
Constrained geometry	No	Yes[a]	No	Yes

[a] See text.

essentially atactic (actually slightly syndiotactic) PPs are obtained with most CGCs. Vanadium-based catalysts can be used to prepare either atactic or syndiotactic PPs. Heterogeneous titanium-based catalysts are the source of isotactic PPs available on a commercial scale.

It is worth commenting on the regioregularity of propylene insertion for various systems. Heterogeneous titanium-based catalysts are well-known for being extremely regioregular; propylene insertion occurs only through 1,2-enchainments, and constitutional mistakes cannot be detected in most commercial products. In contrast, syndiotactic PP from vanadium-based catalysts is formed at a low polymerization temperature (-78 °C) through the 2,1-insertion of propylene.[15,21a] In EP copolymers from vanadium-based catalysts, the 1,2-insertion of propylene is largely prevailing. As discussed below, a large content of regioirregularities, formed through 2,1-insertion, is the traditional fingerprint of an EP copolymer obtained from a vanadium-based catalyst. Up to 20 mol% of 2,1-inserted propylene units can be detected (in a commercial product such as DUTRAL® CO034 from Polimeri Europa, regioirregularities were found to be about 14 mol%).[11p] Metallocenes present a variety of situations regarding regioregularity, the isoselective metallocenes being the most regioirregular. However, the low regioselectivity of vanadium-based catalysts has not been equaled by any new SSC.

There is a general agreement in the field of Ziegler–Natta catalysis that both the stereo- and regioselectivity of all catalyst families can be rationalized mainly on the basis of steric interactions. Refined models have been proposed for the heterogeneous titanium and SSCs to correlate the steric structure of the catalyst active site to the microstructure of the resultant PP.[21] Conversely, the displacement of the ligands in vanadium-based catalysts appears to be responsible for the low stereoregularity and very low level of regioregularity in the PPs formed.

12.6.3 Ethylene/Propylene Copolymers: Propylene Placement

Table 12.4 presents a summary of the propylene placement in EP(D)M copolymers, in terms of tacticity and regioregularity, for the catalyst families discussed. The values reported are a collection of those available in the scientific literature and determined by the authors on commercial samples.

Heterogeneous titanium-based catalysts give rise to isotactic propylene sequences with negligible regioirregularities. One can thus say that these catalysts produce EP copolymers with the highest possible degree of order in the propylene sequences. As previously mentioned, the crucial aspect determining the behavior of titanium-based catalysts is the multiplicity of catalytic sites, which leads to a multiplicity of macromolecular chains. Atactic sequences and a remarkable amount of regioirregularities are brought about by vanadium catalysts, introducing the highest degree of disorder in propylene sequences.

TABLE 12.4

Propylene Placement in EP(D)M as a Function of the Catalyst System Used for Polymerization

Catalytic System	Example[a]	Tacticity (% mm)	R.I.[b] (mol %)	Comonomer Repeat Units, n[c]	References
Titanium (heterogeneous)	MgCl$_2$/donor (isoselective system)	~100	n.d.[d]	2	10o,11j,p
Vanadium (single site)	VOCl$_3$, V(acac)$_3$	30	up to 20	5	11j,p
Metallocene	rac-Et(H$_4$Ind)$_2$ZrCl$_2$	~100	0	2	11j,l,p
	Ind$_2$ZrCl$_2$	30	n.d.[d]	3	11j,l,p
Constrained geometry	Me$_2$Si(Me$_4$Cp)(N-t-Bu)TiCl$_2$	30	6	5	11p,12c

[a] Ind = indenyl; H$_4$Ind = tetrahydroindenyl; Cp = cyclopentadienyl.
[b] Propylene regioirregularity values corresponding to the lower limits calculated as (CH$_2$)$_2$ + (CH$_2$)$_4$ using ^{13}C NMR.
[c] Number of different types of repeat units present out of the five possible structures (see Section 12.3.2 and Figure 12.4).
[d] Not detectable.

Metallocenes offer a unique opportunity, unmatched by traditional catalysts as well as by other classes of SSCs, to control the placement of propylene. Sequences achievable range from atactic to isotactic. Syndiotactic sequences are not observed in metallocene-based EP copolymers, because the syndioselective metallocenes are characterized by low values of $r_1 r_2$ and thus do not allow the formation of propylene sequences long enough to discriminate between atactic and syndiotactic placement. Only a minor amount of regioirregular propylene placements are detectable in EP copolymers prepared with isoselective metallocenes. It should be noted that presently available aselective metallocenes are not able to prepare EP copolymers with sequences that are simultaneously stereo- and regioirregular.

CGCs give rise to atactic propylene sequences and clearly detectable amounts of regioirregularities. CGCs are thus closest to vanadium-based catalysts in their behavior, although by comparison only a minor amounts of regioerrors are present in CGC-derived copolymers.

As previously discussed, these different placements of propylene can be considered to be different comonomer units in the macromolecular chain. This is a crucial point, since the stereo- and regioselectivity of the catalytic system determines propylene placement and hence the microstructure of the resultant copolymers.

12.6.4 Ethylene/Propylene Copolymers: Intramolecular Distribution of the Comonomers

Table 12.5 collects literature values for r_1 and $r_1 r_2$ for different catalyst systems. With vanadium-based catalysts, a correct evaluation of $r_1 r_2$ is hindered by the remarkable presence of regioirregularities.[9] One could say, however, that short comonomer sequences are obtained. With heterogeneous titanium-based catalysts, blocky copolymers are prepared. A correct determination of $r_1 r_2$ is, however, hindered by the multisite nature of the catalyst, that is, by the presence of (many) different polymer fractions resulting from many catalyst sites with different $r_1 r_2$ values that could in principle be responsible for an apparent higher blockiness.

Metallocene-based catalysts are the only catalysts that can be used to prepare EP copolymers with a broad range of $r_1 r_2$ values, from almost perfectly alternating structures to blocky EP(D)M materials. In fact, metallocenes are the only class of catalysts that can produce EP copolymers with microstructures ranging from long sequences of both comonomers ($r_1 r_2 \gg 1$) to an alternating

TABLE 12.5

Type of Catalyst Systems and Reactivity Ratios in Ethylene/ Propylene Copolymerization (Monomer 1 = Ethylene; Monomer 2 = Propylene)

Catalytic System	$r_1 r_2$	r_1	$r_1 r_2$ Control	References
Titanium (heterogeneous)	>1	2–4	No	10o, 11j,p
Vanadium (single site)	0.1–1	3–26	No[a]	11j,p
Metallocene	0.001–6.3	1.3–250	Yes	11j,l,p
Constrained geometry	1–1.5	1.4	No	11p, 12c

[a]Different values of $r_1 r_2$ can be obtained by, for example, changing the Al:V ratio.

TABLE 12.6

Properties of Commercial EP Copolymers Prepared with Constrained Geometry Catalysts

Commercial Grade Based on CGC[a]	Propylene Content (mol%)	$r_1 r_2$	Tacticity (mol %)	R.I.[b](mol %)	References
Nordel[TM] IP 4770	18.8	1.84	20.1	3.3	11p
Nordel[TM] IP 3430	43.9	1.52	11.9	6	11p

[a]From DuPont Dow Elastomers.
[b]Propylene regioirregularity values corresponding to the lower limits calculated as $(CH_2)_2 + (CH_2)_4$ using ^{13}C NMR.

distribution of comonomers ($r_1 r_2 \to 0$). The only example of a truly random EP copolymer ($r_1 r_2 = 1$) is from a CGC.[12c] Moreover, the values of r_1 and r_2 for this system are very similar to each other, that is, ethylene and propylene show similar reactivities.

12.7 EXAMPLES OF ETHYLENE/PROPYLENE COPOLYMERS FROM SINGLE-SITE CATALYSTS

In this section, examples of EP copolymers prepared with the three families of SSC systems discussed above (vanadium, metallocene, and CGC) are described. Commercial EP copolymer samples based on vanadium typically have an ethylene content of about 20%. By applying the abovementioned methods for the determination of $r_1 r_2$ and regioirregularities, averages of about $r_1 r_2 = 0.6$ and 12 mol% regioirregularities can be obtained. However, these numbers have to be considered only as qualitative indications, because a precise interpretation of the copolymer NMR spectra is inhibited by the large amount of regioirregularities. Table 12.6 gives examples of commercial samples of EP copolymers made using CGCs. These copolymers are characterized by a limited amount of regioirregularities, and thus can be analyzed to give a reasonably complete characterization of comonomer distribution and propylene placement.

Table 12.7 provides a comparison of EP copolymers having different $r_1 r_2$ values (<1, ≈1, and >1) and different propylene sequences (isotactic without regioirregularities, atactic without regioirregularities, and atactic with regioirregularities). These materials were prepared using the SSC types

TABLE 12.7

Composition, Microstructure, Reactivity Ratio Product, and Melting Enthalpy for EP Copolymers Prepared Using Various Single-Site Catalysts

Number	Catalyst[a]	E[b] (mol %)	R.I.[c]	m (%)	PPP (%)	PPE (%)	EPE (%)	PEP (%)	EEP (%)	EEE (%)	$r_1 r_2$	P (E_{10})[d]	ΔH_m[e] (J/g)
1	rac-(C$_2$H$_4$)(H$_4$Ind)$_2$ZrCl$_2$	69.7	n.d.[f]	>99	2.5	8.5	19.2	8.8	30.4	30.5	0.48	0.017	20.3
2	(4,7-Me$_2$Ind)$_2$ZrCl$_2$[g]	71.1	<1	≈20	6.8	6.3	15.8	7.0	28.9	35.2	0.82	0.031	18.8
3	Me$_2$Si(Me$_4$Cp)(N-t-Bu)TiCl$_2$	71.4	2.7	11	4.4	13.1	11.1	4.2	26.1	41.1	1.81	0.059	14.3
4	(4,7-Me$_2$Ind)$_2$ZrCl$_2$[g]	71.8	<1	18	7.8	9.6	10.9	2.5	22.8	46.5	1.90	0.096	27.2
5	V(acac)$_3$[h]	72.7	14	23	2.4	7.4	17.5	5.8	30.0	36.8	0.57	0.033	13.2
6	$meso$-(C$_2$H$_4$)(4,7-Me$_2$Ind)$_2$ZrCl$_2$	73.7	<1	n.d.[i]	0	4.4	21.9	7.2	32.5	34.0	0.20	0.022	17.0
7	rac-Me$_2$C(3-t-BuInd)$_2$ZrCl$_2$	74.0	<1	>99	4.2	9.5	12.3	3.7	25.8	44.6	1.42	0.073	26.2
8	rac-(C$_2$H$_4$)(4,7-Me$_2$Ind)$_2$ZrCl$_2$	76.0	n.d.[j]	>99	4.3	10.6	9.1	3.0	24.5	48.5	2.60	0.103	33.2
9	rac-Me$_2$Si(Ind)$_2$ZrCl$_2$	76.4	<1	n.d.[k]	0	4.0	19.7	5.4	30.4	40.5	0.24	0.043	23.4
10	V(acac)$_3$[h]	80.0	8	36	1.3	4.4	14.3	3.9	24.9	51.2	0.68	0.106	10.0

[a] Experimental conditions and analysis details given in Reference 22. Metallocenes and CGCs were activated with alumoxanes; V(acac)$_3$ was activated with AlEt$_2$Cl/n-butyl perchlorocrotonate. Ind = indenyl; H$_4$Ind = tetrahydroindenyl; Cp = cyclopentadienyl; acac = acetylacetonate.

[b] E = ethylene; P = propylene.

[c] Propylene regioirregularity values corresponding to the lower limits calculated as (CH$_2$)$_2$ + (CH$_2$)$_4$ using ^{13}C NMR.

[d] Probability (fraction) of sequences of at least ten ethylene units.

[e] Enthalpy of melting.

[f] n.d. = not determined; however, the catalyst system is highly regioselective for propylene homopolymerization.

[g] Different experimental conditions allowed preparation of different microstructures with the same catalyst.

[h] Polymerizations performed under very similar polymerization conditions to obtain copolymers with different chemical compositions.

[i] n.d. = not determined; however, the catalytic system is regioselective for propylene homopolymerization.

[j] n.d. = not determined; however, the catalyst system is aselective for propylene homopolymerization.

[k] n.d. = not determined; however, the catalyst system is isoselective for propylene homopolymerization.

previously described. In particular, metallocenes capable of different stereo- and regiocontrol are compared. Experimental conditions for the preparation of these polymers are reported elsewhere.[22] The probability (fraction) of ethylene homosequences with 10 or more ethylene units and the enthalpy of fusion (ΔH_m) of the copolymers' T_ms are reported as well.

12.8 CRYSTALLINITY OF ETHYLENE/PROPYLENE COPOLYMERS

This section deals with general aspects of crystalline organization and crystallization behavior of EP copolymers, although all of the EP data shown in the Figures discussed below refer only to samples obtained from a rac-$(C_2H_4)(H_4Ind)_2ZrCl_2$-based system. The influence of the regio- and stereoregularity of propylene insertion (i.e., of the microstructural features discussed in the previous sections) on EP crystallization behavior will be discussed in Section 12.9.

The crystallinity present in EP copolymers has been extensively studied by X-ray diffraction.[23] The bulk of these crystallinity studies have been devoted to copolymers useful for the production of industrially relevant crosslinked elastomers having ethylene contents generally lower than 85 mol%. It is well established that propylene units enter into the lattice of orthorhombic PE,[24] gradually increasing the disorder in the crystalline phase but leaving the *trans*-planar conformation of the chains substantially unaltered. In fact, the dimension of the a axis of the unit cell of PE increases almost proportionally to the propylene content of the copolymer, whereas the b and the c axes practically retain the same dimensions found in PE (Figure 12.10).[23c,h]

At high propylene content, a becomes nearly equal to $b\sqrt{3}$ and, hence, the unit cell becomes pseudohexagonal[23c] (Figure 12.11). X-ray diffraction studies on oriented EP copolymer samples have shown that this pseudohexagonal form presents a long-range order only in the hexagonal arrangement

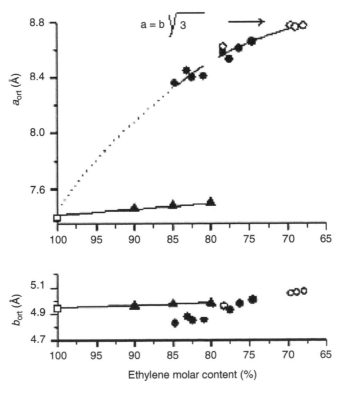

FIGURE 12.10 Dependence of the a (top) and b (bottom) unit cell parameters on ethylene molar content for EP (circles) and ethylene/styrene (triangles) copolymer samples.

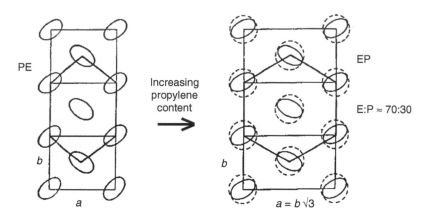

FIGURE 12.11 Schematic presentation of the transition from the orthorhombic crystal structure typical of PE to the pseudohexagonal crystal structure typical of EP copolymers with less than 70 mol% of ethylene.

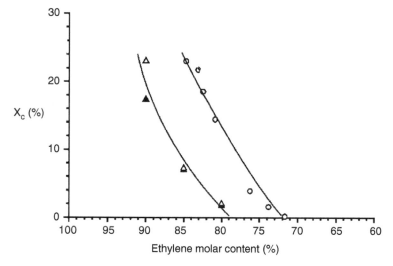

FIGURE 12.12 Degree of crystallinity for EP (circles) and ethylene/styrene (triangles) copolymer samples as a function of ethylene molar content as evaluated from X-ray diffraction patterns at room temperature.

of the axes of nearly *trans*-planar chains; a large packing disorder (rotational and translational as well as conformational) is associated with the inclusion of methyl groups of the propylene comonomer units in the crystalline phase.[23g]

It is worth noting that behavior of ethylene copolymers containing bulkier α-olefins is completely different from that of EP copolymers. In fact, bulkier comonomer units are excluded from the crystalline phase, which remains orthorhombic for high comonomer contents, giving rise only to small increases of the *a* and *b* unit cell parameters (as shown for ethylene/styrene random copolymers in Figure 12.10).[23h] Moreover, as shown in Figure 12.12, as the ethylene molar content in EP copolymers decreases, the concomitant decrease in crystallinity is less rapid for EP copolymers than for ethylene/styrene and similar ethylene copolymers containing bulkier α-olefins. This is due to the inclusion of propylene units and the exclusion of bulkier units from the crystalline phase.[23h]

The data in Figure 12.12 also show that the room temperature crystallinity of EP copolymers generally becomes null for ethylene contents lower than 70–75 mol%. It is worth noting, however, that EP copolymers, although amorphous at room temperature, can crystallize at low temperatures.

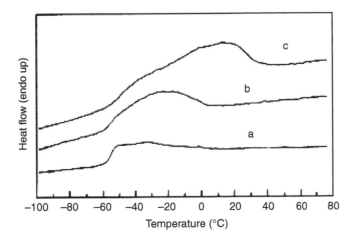

FIGURE 12.13 DSC scans of unannealed compression molded EP copolymer samples with ethylene molar contents of (a) 63.6%, (b) 65%, and (c) 74%.

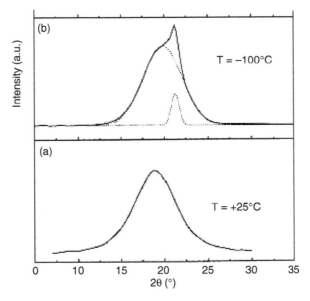

FIGURE 12.14 X-ray diffraction patterns of a compression molded EP copolymer sample (74 mol% ethylene, DSC curve c, Figure 12.13) at (a) room temperature; (b) −100 °C.

This can be shown by differential scanning calorimetry (DSC), which can detect broad endothermic peaks below room temperature.[25] This is exemplified in Figure 12.13 for EP samples with ethylene molar contents in the range 63–74 mol%.[25c]

The occurrence of crystallization phenomena at low temperatures can be also shown by comparing X-ray diffraction patterns taken at different temperatures. For example, for an EP copolymer with 74 mol% ethylene, the X-ray diffraction pattern taken at room temperature presents only a broad diffraction halo (Figure 12.14a), as typical of fully amorphous samples, whereas the X-ray diffraction pattern taken at −100 °C (Figure 12.14b) additionally presents a narrower diffraction peak (at $2\theta° = 21.3°$) associated with formation of the crystalline phase.[25c]

It is worth adding that EP samples that are amorphous at room temperature (e.g., the material in Figure 12.14a) can also crystallize as a consequence of the molecular orientation associated with stretching.[23c] In addition, for copolymer samples with ethylene contents greater than 70 mol%, long-term room temperature annealing generally produces crystallite reorganization phenomena. These improve the quality (perfection and size) of the imperfect crystallites present, as deduced from the appearance of a narrower DSC melting peak in the temperature range 40–50 °C and the appearance of a well-defined crystalline peak in the room temperature X-ray diffraction pattern, analogous to that observed in Figure 12.14b (although shifted to $2\theta = 20.7$ ° due to the temperature difference).[25c]

12.9   INFLUENCE OF REGIO- AND STEREOREGULARITY OF PROPYLENE INSERTION ON CRYSTALLIZATION BEHAVIOR AND ELASTICITY OF ETHYLENE/PROPYLENE COPOLYMERS

It is well-known that the crystallinity of EP copolymers mainly depends on comonomer content (composition) and comonomer distribution (constitution). It is the objective of this section to demonstrate that the tacticity of propylene, that is, the regio- and stereoregularity of propylene insertion, also plays a fundamental role in controlling crystallinity. As discussed earlier in detail, all of these microstructural features in turn essentially depend on the catalytic system and on polymerization conditions. ^{13}C NMR-derived microstructural information for several EP copolymer samples, obtained using different SSC systems and having constitutions and compositions in the range of industrial interest, has been collected in Table 12.7.[22]

To achieve a significant comparison between EPs having different placements of propylene units, the different copolymer constitutions (varying from nearly alternating to almost blocklike) must be taken into account. A standardized parameter for comparison can be obtained by reporting the copolymer melting enthalpy versus the fraction (probability) of PE sequences longer than a fixed value (as shown in Figure 12.15 for sequences of 10 or more ethylene units). In Figure 12.15, the experimental data collected in Table 12.7 are approximately fitted to three lines corresponding to the three different general placement types of propylene comonomer units: a lower line for regio-/stereoirregular copolymers, an intermediate line for regioregular/stereoirregular copolymers, and an upper line for regio-/stereoregular copolymers.[22]

These results are in qualitative agreement with simple energy calculations made using packing models of EP copolymers.[22] Figure 12.16 shows side- and along-the-chain views of *trans*-planar EP copolymer chains (75 mol% ethylene), having regio-/stereoirregular, regioregular/stereoirregular, or regio-/stereoregular placements of the propylene units. The placement of the comparatively less bulky regio-/stereoregular chains into pseudohexagonal crystals involves the least steric interactions. Hence, the molecular modeling clearly indicates that EP chains with regio-/stereoregular enchainments of the propylene units are more easily accommodated into the crystallites.[22]

In summary, for copolymers having equivalent amounts and distributions of propylene comonomer units, a higher disturbance of PE crystallinity is achieved in the case of regio- and stereoirregular placements of propylene units. Moreover, both experimental data and calculations indicate that regioirregularity of propylene insertion is more effective in disturbing crystallinity than stereoirregularity.

12.10   MOLECULAR PICTURE OF ELASTICITY: DIENE- AND ETHYLENE-BASED ELASTOMERS

To obtain a good synthetic elastomer, macromolecular chains must be characterized by: (1) linearity and very high molecular weight to prevent viscous creep; (2) the absence of strong intermolecular interactions that could reduce macromolecular mobility, such as interactions owing to the presence of polar groups or hydrogen bonding; and (3) sequences of monomeric units

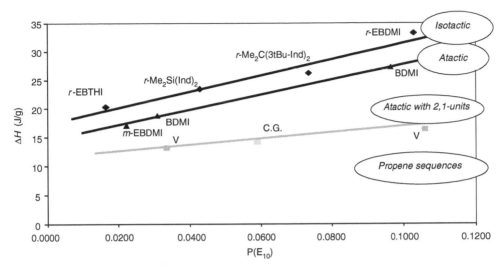

FIGURE 12.15 Melting enthalpy of the EP copolymers shown in Table 12.7 versus the probability (fraction) of copolymer ethylene sequences with ten or more ethylene units (keyed to entries in Table 12.7: r-EBTHI $= 1$; BDMI $= 2,4$; C.G. $= 3$; V $= 5,10$; m-EBDMI $= 6$; r-Me$_2$C(3tBu-Ind)$_2 = 7$; r-EBDMI $= 8$; r-Me$_2$Si(Ind)$_2 = 9$.) Considered copolymers present regio-/stereoregular (diamonds), regioregular/stereoirregular (triangles), or regio-/stereoirregular (squares) placements of propylene co-units.

(a)

(b)

Regioregular propylene insertion

(c)

Regio-and stereoregular propylene insertion

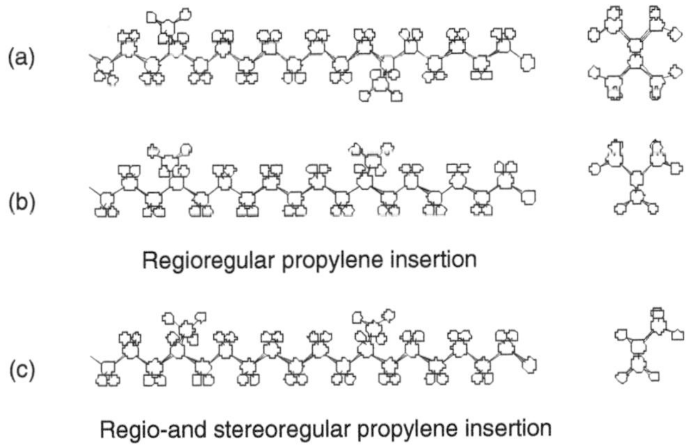

FIGURE 12.16 Side- (left) and along-the-chain (right) views of *trans*-planar EP copolymer chains (75:25 E:P) presenting regio-/stereoirregular (a), regioregular/stereoirregular (b), and regio-/stereoregular (c) placements of propylene co-units.

with constitutional and configurational irregularities to render crystallization at rest difficult. For elastomeric properties, however, it is advantageous to have partial and reversible crystallization of the macromolecular chains with stretching. It is well-known that unsaturated macromolecular chains such as *cis*-1,4-polyisoprene and *cis*-1,4-polybutadiene are ideal elastomers because they possess the abovementioned molecular features. In particular, *cis*-1,4-polyisoprene presents a negligible crystallization rate at rest, but can easily crystallize under stretching, thus leading to the physical crosslinking responsible for the elasticity in the uncured state of natural rubber.[26] Among saturated polymer chains, those based on ethylene meet most of the basic requirements for being elastic. In fact, since PE can be completely linear, it potentially possesses a very high macromolecular flexibility, as demonstrated by its very low glass transition (T_g) values (always reported in the literature as $< -60\,°C$). However, rubber materials are not based on PE because of its high crystallinity at rest.

Elastomeric copolymers based on ethylene were first obtained by Giulio Natta through the copolymerization of ethylene and propylene. The reasoning behind the decision to insert propylene comonomer units into ethylene-based polymers was explained by Natta in a paper prepared for the "XVI Congres Internationale de Chimie Pure at Appliqué," held in Paris in 1957.[27] Natta stated, "If the regularity of the polyethylene chain could be reduced so as to prevent crystallization, without introducing any polar group and without reducing the flexibility of the chain and the rotation capacity of the C–C linkage of the chain, a chain should be obtained that would be suitable as the basic material for obtaining a good rubber. The problem has been solved with the ethylene alpha-olefin copolymers (for example, ethylene–propylene). Those having a content of alpha-olefin, such as to make them not crystalline in the state of rest, have proved to be interesting rubbers. The possibility of maintaining not too short portions of ordered ethylenic units, capable of crystallizing when the molecules are oriented by stretching, is a favorable aspect of the problem and can make it possible to realize better mechanical behaviors." The role of propylene in preventing PE crystallization and the importance of a suitable length of ethylene sequences was already clear to Natta in 1957. Regarding the placement of propylene in EP copolymers, Ver Strate[6b] reported that "In the case of an EP(D)M, stereoregular placement of ethylene and propylene contribute to crystallinity, where crystallinity is not desired." This was a clear indication of the preferred use of copolymerization catalysts that produce regio- and stereoirregular propylene sequences.

12.11 ETHYLENE/PROPYLENE COPOLYMERS: ELASTICITY IN THE UNCURED STATE

Microcrystalline domains in EP copolymers have a strong influence on the physical properties of uncrosslinked materials. In particular, as shown in Figure 12.17, the elongation at break at room temperature of EP copolymer samples derived from rac-$C_2H_4(H_4Ind)_2ZrCl_2$, which have crystalline domains already present in the unstretched state (generally, samples with ethylene contents of >72 mol%), is relatively small, and is smaller still for fully amorphous samples that are noncrystallizing under stretching (generally, those with ethylene contents of <65 mol%). However, elongation at break is much larger for samples that are essentially amorphous in the unstretched state and partially crystallizing under stretching (generally, those with ethylene contents between 65 and 72 mol%). However, the most relevant effect of crystallinity on the physical properties of EP copolymers concerns the elastomeric behavior of uncured samples, as shown in Figure 12.18.

Figure 12.18 demonstrates that ethylene-rich EP copolymers presenting stereo- and regioregular placements of propylene (such as those from rac-$Et(H_4Ind)_2ZrCl_2$) show elasticity in the uncured state.[22,25c,d] In fact, they can be repeatedly stretched, even at high elongations ($>200\%$), and repeatedly recover their initial size and shape, thus showing very low tension set values.[7,25c] It is clear from the discussion above that the highest degree of order in propylene placement allows for (1) the formation of disordered microcrystalline domains at rest, in a chemical composition range that depends on the r_1r_2 product (from about 68–77 mol% ethylene for rac-$C_2H_4(H_4Ind)_2ZrCl_2$[25c,d]); and (2) the development of further crystallinity under stretching. Microcrystalline domains act as physical crosslinks for the prevailingly amorphous samples and account for the elastomeric behavior observed in the uncured state. Elasticity in the uncured state is unusual for EP copolymers, and copolymers prepared using rac-$C_2H_4(H_4Ind)_2ZrCl_2$ were the first reported case.[25d] Commercially available vanadium-based EPM and EPDM do not show this behavior.[10n] In fact, when the ethylene content in EP copolymers is low enough to avoid any crystallinity at rest, the macromolecules undergo viscous creep as the strain increases. On the contrary, an ethylene content high enough to avoid viscous creep under stress renders the copolymers crystalline at rest, and therefore high

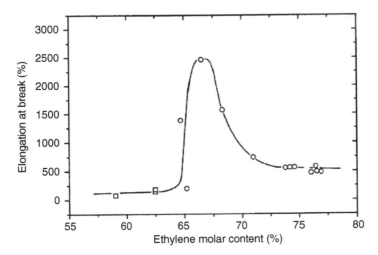

FIGURE 12.17   Elongation at break of EP copolymer samples (inherent viscosity $= 3.6 < \eta < 6.0$) as a function of copolymer composition.

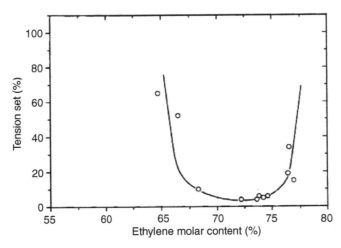

FIGURE 12.18   Tension set of EP copolymer samples (inherent viscosity $= 3.6 < \eta < 6.0$) as a function of copolymer composition.

values of tension set are obtained. It thus appears that it is insufficient simply to have (in Natta's words) "not too short portions of ordered ethylenic units"[27] to have reversible crystallization under stretching in EP copolymers and hence elastic behavior. The tacticity of the propylene sequences plays a fundamental role and the combination of highly stereo- and regioregular sequences with "not too long portions of ordered ethylenic units" seems to be a favorable molecular feature.

Thus, while the "stereoregular placement of propylene"[6b] should be prevented in order to produce a fully amorphous EP copolymer and a very soft elastomer, the opposite direction (greater regularity) is desired in order to achieve elastomeric behavior in the uncured state (i.e., Natta's "better mechanical behaviors"[27]).

The stress-strain behavior of uncured EP copolymers with highly stereo- and regioregular propylene placements resembles that of natural rubber. Both elastomers possess all the requisites discussed in Section 12.10. In particular, they have negligible crystallinity at rest and are able to develop crystallinity under stretching.

12.12 ELASTOMERS BASED ON ETHYLENE: SELECTION OF CATALYTIC SYSTEMS FOR THEIR PRODUCTION

From the above discussion, it can be concluded that in an EP copolymer of industrial relevance, a residual, almost negligible, crystallinity at rest in the unstretched state may be desired to effect thermoplastic elastomeric behavior. This could be achieved, for a given ethylene content, by enhancing the tacticity of propylene sequences. However, when a highly stereo- and regioselective catalyst is used, the only other strategy for controlling crystallinity is changing the relative contents of ethylene and propylene. A high propylene content must be avoided, mainly to obtain low T_gs and to produce a minimum content of tertiary carbon atoms (potential degradation sites) along the macromolecular chain.

EP elastomers were first obtained by Natta and coworkers using vanadium-based catalysts in the early 1960s. Since that time, the key structural features of vanadium-based EP(D)M materials have allowed for the commercial production of millions of tons of elastomeric products. These features are high copolymer molecular weight, a narrow distribution of copolymer molecular properties (mainly chemical composition distribution due to the single-site nature of the catalysts), and high conversion of the diene termonomer. Regarding control of EP copolymer crystallinity and elasticity, it is now clear that the most critical feature brought about by vanadium-based catalysts—more important than the so-called random distribution of the comonomers—is actually the high degree of disorder of propylene placement, particularly the high level of propylene regioirregularities. This in turn can be reasonably attributed to vanadium ligand displacement after reaction with the aluminum cocatalyst, which provides sufficient room for propylene misinsertions. Moreover, there is also enough space at the vanadium center for insertion of a bulky diene such as ENB, whose presence is a *condicio sine qua non* for sulfur-based vulcanization processes.

A new generation of ethylene-based elastomers is now industrially produced using borane-activated CGCs. These catalysts also give rise to copolymers with high molecular weights and narrow molecular weight and composition distributions, and affect reasonable conversions of diene termonomers. However, the longer average length of ethylene sequences and the lower amount of regioirregularities produced by these catalysts as compared to vanadium catalysts requires a higher amount of propylene to avoid crystallinity.

12.13 CONCLUSIONS

The crystallinity and physical properties of EP copolymers mainly depend on comonomer content (composition) and comonomer distribution (constitution). However, for industrially relevant ethylene-rich copolymers, the regio- and stereoregularity of propylene insertion also plays a fundamental role in controlling both crystallinity and elastic properties. Highly stereo- and regioregular propylene units are at the origin of elasticity in the uncured state. These microstructural features are further dependent upon both the catalytic system employed and the polymerization conditions.

REFERENCES AND NOTES

1. (a) Ziegler, K. Consequences and development of an invention. Nobel Prize Lecture, December 12, 1963. Nobel Foundation Web Site. http://nobelprize.org/nobel_prizes/chemistry/laureates/1963/ziegler-lecture.html (accessed Sept 2006). (b) Natta, G. From the stereospecific polymerization to the asymmetric autocatalytic synthesis of macromolecules. Nobel Prize Lecture, December 12, 1963. Nobel Foundation Web Site. http://nobelprize.org/nobel_prizes/chemistry/laureates/1963/natta-lecture.html (accessed Sept 2006). (c) Natta, G.; Corradini, P. Crystal structure of a new type of polypropylene. *Atti Accad. Nazl. Lincei, Mem. Classe sci. fis., mat. e nat.* **1955**, *4 (Sez. II)*, 73–80. (d) Natta, G.; Corradini, P. Crystal structure of isotactic polystyrenes. *Makromol. Chem.* **1955**, *16*, 77–80.

2. (a) Corradini, P.; Guerra, G.; Pirozzi, B. Present status of the configurational and conformational analysis of stereoregular polymers. In *Preparation and Properties of Stereoregular Polymers*; Lenz, R.W., Ciardelli, F., Eds.; D. Reidel: Dordrecht, 1979; pp 317–352. (b) Pino, P.; Mülhaupt, R. The stereospecific polymerization of propylene: A survey 25 years after its discovery. *Angew. Chem.* **1980**, *92*, 869–887. (c) Farina, M. The stereochemistry of linear macromolecules. *Top. Stereochem.* **1987**, *17*, 1–111.

3. (a) De Rosa, C.; Auriemma, F.; Ruiz de Ballesteros, O. Mechanical properties and elastic behavior of high-molecular-weight poorly syndiotactic polypropylene. *Macromolecules* **2003**, *36*, 7607–7617. (b) De Rosa, C.; Auriemma, F.; Perretta, C. Structure and properties of elastomeric polypropylene from C_2 and C_{2v}-symmetric zirconocenes. The origin of crystallinity and elastic properties in poorly isotactic polypropylene. *Macromolecules* **2004**, *37*, 6843–6855.

4. (a) Juengling, S.; Muelhaupt, R.; Fischer, D.; Langhauser, F. Modified syndiotactic polypropylene. Propylene-octene copolymers and blends with atactic oligopropylene. *Angew. Makromol. Chem.* **1995**, *229*, 93–112. (b) Shin, Y.-W.; Uozumi, T.; Terano, M.; Nitta, K.-H. Synthesis and characterization of ethylene-propylene random copolymers with isotactic propylene sequence. *Polymer* **2001**, *42*, 9611–9615. (c) De Rosa, C.; Buono, A.; Caporaso, L.; Oliva, L. Structural characterization of syndiotactic propylene-styrene-ethylene terpolymers. *Macromolecules* **2003**, *36*, 7119–7125.

5. Al-Hussein, M., Strobl, G. Strain-controlled tensile deformation behavior of isotactic poly(1-butene) and its ethylene copolymers. *Macromolecules* **2002**, *35*, 8515–8520.

6. For reviews on EP(D)M see: (a) Karpeles, R.; Grossi, A. V. EPDM rubber technology. In *Handbook of Elastomers*; Bhowmick, A. K., Stephens, H. L., Eds.; Marcel Dekker: New York, **2001**; pp 845–876. (b) Baldwin, F. P.; Ver Strate, G. Polyolefin elastomers based on ethylene and propylene. *Rubber Chem. Technol.* **1972**, *45*, 709–881. (c) Ver Strate, G. Ethylene–propylene elastomers. In *Encyclopedia of Polymer Science and Engineering*, 6th ed.; Kroschwitz, J. I., Ed.; John Wiley and Sons: New York, 1986; pp 552–564. (d) Easterbrook, E. K.; Allen, R. Ethylene-propylene rubber. In *Rubber Technology*, 3rd ed.; Morton, M., Ed.; Van Nostrand Reinhold: New York, 1987; pp 260–283.

7. Mark, J. E.; Erman, B.; Eirich F. R. *The Science and Technology of Rubber*, 3rd ed., Elsevier Academic Press: San Diego, 2005.

8. (a) Boor, J. Jr. *Ziegler–Natta Catalysts and Polymerizations*. Academic Press: New York, 1979. (b) Tait, P. J. T.; Berry, I. G. Monoalkene polymerization: Copolymerization. In *Comprehensive Polymer Science*; Eastmond, G. C., Ledwith, A., Russo, S., Sigwalt, P., Eds.; Pergamon Press: Oxford, 1989; Vol. 4, pp 575–584. (c) Pino, P.; Giannini, U.; Porri, L. Insertion polymerization. In *Encyclopedia of Polymer Science and Engineering*, 2nd ed.; Wiley: New York, 1987; Vol. 8, p 147. (d) Guerra, G.; Cavallo, L.; Corradini, P. Chirality of catalysts for stereospecific polymerizations. *Top. Stereochem.* **2003**, *24*, 1–69.

9. For EP(D)M from vanadium-based catalysts see: (a) Natta, G.; Dall'Asta, G.; Mazzanti, G.; Pasquon, I.; Valvassori, A.; Zambelli, A. Copolymers of ethylene, higher α-olefins, and monocyclomonoolefins or alkyl derivatives thereof. U.S. Patent 3,505,301 (Montecatini Edison S.p.A.), April 7, 1970. (b) Zambelli, A.; Tosi, C.; Sacchi, C. Polymerization of propylene to syndiotactic polymer. VI. Monomer insertion. *Macromolecules* **1972**, *5*, 649–654. (c) Locatelli, P.; Immirzi, A.; Zambelli, A.; Palumbo, R.; Maglio, G. Orientation of propylene units in polypropylene and ethylene/propylene copolymers. *Makromol. Chem.* **1975**, *176*, 1121–1128. (d) Carman, C. J.; Harrington, R. A.; Wilkes, C. E. Monomer sequence distribution in ethylene-propylene rubber measured by ^{13}C NMR. 3. Use of reaction probability mode. *Macromolecules* **1977**, *10*, 536–544. (e) Zucchini, U.; Dall'Occo, T.; Resconi, L. Ziegler–Natta catalysis for the polyolefin industry: Present status and perspectives. *Indian J. Tech.* **1993**, *31*, 247–262. (f) Dall'Occo, T.; Galimberti, M.; Balbontin, G. Ethylene/propylene copolymers with vanadium-based catalysts: Cocatalyst effect. *Polym. Adv. Technol.* **1993**, *4*, 429–434 and references therein.

10. For EP(D)M from heterogeneous titanium-based catalysts see: (a) Corbelli, L.; Milani, F.; Fabbri, R. New ethylene-propylene elastomers produced with highly active catalysts. *Kautsch. Gummi Kunstst.* **1981**, *34*, 11–15. (b) Doi, Y.; Ohnishi R.; Soga, K. Monomer sequence distribution in ethylene-propylene copolymers prepared with a silica-supported magnesium chloride/titanium tetrachloride catalyst. *Makromol. Chem., Rapid. Commun.* **1983**, *4*, 169–174. (c) Ammendola, P.; Oliva, L.; Gianotti, G.; Zambelli, A. Ethylene-propylene copolymerization. Monomer reactivity and reaction mechanism. *Macromolecules* **1985**, *18*, 1407–1409. (d) Ammendola, P.; Zambelli, A.; Oliva, L.;

Tancredi, T. Polymerization of α-olefins in the presence of δ-titanium trichloride/trimethylaluminum: Chain propagation rate constants for ethylene and propylene. *Makromol. Chem.* **1986**, *187*, 1175–1188. (e) Abis, L.; Bacchilega, G.; Milani, F. Carbon-13 NMR characterization of a new ethylene-propene copolymer obtained with a high-yield titanium catalyst. *Makromol. Chem.* **1986**, *187*, 1877–1886. (f) Soga, K.; Sano, T.; Ohnishi, R.; Kawata, T.; Ishii, K.; Shiono, T.; Doi, Y. Synthesis of EP rubber using titanium catalysts. *Stud. Surf. Sci. Catal.* **1986**, *25*, 109–122. (g) Cozewith, C. Interpretation of carbon-13 NMR sequence distribution for ethylene-propylene copolymers made with heterogeneous catalysts. *Macromolecules* **1987**, *20*, 1237–1244. (h) Maglio, G.; Milani, F.; Musto, P.; Riva, F. "Residual" crystallinity and superreticular order in ethylene-propylene rubbers prepared with different catalytic systems. *Makromol. Chem., Rapid. Commun.* **1987**, *8*, 589–593. (i) Yechevskaya, L. G.; Bukatov, G. D.; Zakharov, V.; Nosov, A. Study of the molecular structure of ethylene-propylene copolymers obtained with catalysts of different composition. *Makromol. Chem.* **1987**, *188*, 2573–2583. (j) Bukatov, G. D.; Yechevskaya, L. G.; Zakharov, V. A. Copolymerization of ethylene with α-olefins by highly active supported catalysts of various composition. In *Transition Metals and Organometallics as Catalysts for Olefin Polymerization*; Kaminsky, W., Sinn, H., Eds.; Springer: Berlin, **1988**; pp 101–108. (k) Kakugo, K.; Noito, Y.; Mizunuma, K.; Miyatake, T. Characterization of ethylene-propene copolymers prepared with heterogeneous titanium-based catalysts. *Makromol. Chem.* **1989**, *190*, 849–864. (l) Soga, K.; Uozomi, T.; Park, J. R. Effect of catalyst isospecificity on olefin copolymerization. *Makromol. Chem.* **1990**, *191*, 2853–2864. (m) Zakharov, V. A.; Yechevskaya, L. G.; Bukatov, G. D. Copolymerization of ethylene with propene in the presence of titanium-magnesium catalysts of different compositions. *Makromol. Chem.* **1991**, *192*, 2865–2874. (n) Galimberti, M.; Martini, E.; Piemontesi, F.; Sartori, F.; Camurati, I.; Resconi, L.; Albizzati, E. New polyolefins from metallocenes. *Macromol. Symp.* **1995**, *89*, 259–275.

11. For EP(D)M from metallocenes see: (a) Kaminsky, W.; Miri, M. Ethylene propylene diene terpolymers produced with a homogeneous and highly active zirconium catalyst. *J. Polym. Sci., Polym. Chem. Ed.* **1985**, *23*, 2151–2164. (b) Ewen, J. A. Ligand effects on metallocene catalyzed Ziegler–Natta polymerizations. *Stud. Surf. Sci. Catal.* **1986**, *25*, 271–292. (c) Kaminsky, W.; Schlobohm, M. Elastomers by atactic linkage of α-olefins using soluble Ziegler catalysts. *Makromol. Chem., Macromol. Symp.* **1986**, *4*, 103–118. (d) Chien, J. C. W.; He, D. Olefin copolymerization with metallocene catalysts. IV. Metallocene/methyl aluminoxane catalyzed olefin terpolymerization. *J. Polym. Sci., Part A: Polym. Chem.* **1991**, *29*, 1609–1613. (e) Zambelli, A.; Grassi, A.; Galimberti, M.; Mazzocchi, R.; Piemontesi, F. Copolymerization of ethylene with propylene in the presence of homogeneous catalytic systems based on group 4 metallocenes and methylalumoxane: Implications of the reactivity ratios on the reaction mechanism. *Makromol. Chem., Rapid Commun.* **1991**, *12*, 523–528. (f) Uozumi, T.; Soga, K. Copolymerization of olefins with Kaminsky-Sinn-type catalysts. *Makromol. Chem.* **1992**, *193*, 823–831. (g) Herfert, N.; Montag, P.; Fink, G. Elementary processes of the Ziegler catalysis. 7. Ethylene, α-olefin and norbornene copolymerization with the stereorigid catalyst systems iPr[FluCp]ZrCl₂/MAO and Me₂Si[Ind]₂ZrCl₂/MAO. *Makromol. Chem.* **1993**, *194*, 3167–3182. (h) Randall, J. C.; Rucker, S. P. Markovian statistics for finite chains: Characterization of end group structures and initiation, chain propagation, and chain-transfer probabilities in poly(ethylene-*co*-propylene). *Macromolecules* **1994**, *27*, 2120–2129. (i) Galimberti, M.; Piemontesi, F.; Fusco, O.; Camurati, I.; Destro, M. Ethylene/propylene copolymerization with high product of reactivity ratios from a single center, metallocene-based catalytic system. *Macromolecules* **1998**, *31*, 3049–3416. (j) Galimberti, M.; Baruzzi, G.; Camurati, I.; Fusco, O.; Piemontesi, F.; Vianello, M. Die bedeutung der metallorganichen Katalystoren für die kommerzielle Entwicklung von EP(D)M. *Gummi, Fasern, Kunstst.* **1998**, *51*, 570–578. (k) Kravchenko, R.; Waymouth, R. M. Ethylene-propylene copolymerization with 2-arylindene zirconocenes. *Macromolecules* **1998**, *31*, 1–6. (l) Galimberti, M.; Piemontesi, F.; Fusco, O. Metallocenes as catalysts for the copolymerization of ethylene with propylene and dienes. In *Metallocene-Based Polyolefins: Preparation, Properties, and Technology*; Scheirs, J., Kaminsky, W., Eds.; John Wiley and Sons: Chichester, 1999; Vol. 1, pp 309–344. (m) Polo, E.; Galimberti, M.; Mascellani, N.; Fusco, O.; Mueller G.; Sostero, S. Ethene/propene copolymerizations with *rac*-EBTHIZrR₂/alumoxane: σ-ligands effect. *J. Mol. Catal. A: Chem.* **2000**, *160*, 229–236. (n) Kim, J. D.; Soares, J. B. P. Copolymerization of ethylene and alpha-olefins with combined metallocene catalysts. III. Production of polyolefins with controlled microstructures. *J. Polym. Sci., Part A: Polym.*

Chem. **2000**, *38*, 1427–1432. (o) Galimberti, M.; Piemontesi, F.; Baruzzi, G.; Mascellani, N.; Camurati, I.; Fusco, O. Ethene/propene copolymerizations from metallocene-based catalytic systems: The role of comonomer concentration. *Macromol. Chem. Phys.* **2001**, *202*, 2029–2037. (p) Galimberti, M. EP(D)M Elastomers. Presented at Elastomers 2001, First European Polymer Federation School, XXIII M. Farina—AIM School, Gargnano, Italy, June 4–8, 2001. (q) Kaminsky, W. New elastomers by metallocene catalysis. *Macromol. Symp.* **2001**, *174*, 269–276. (r) Fan, W.; Leclerc, M. K.; Waymouth, R. M. Alternating stereospecific copolymerization of ethylene and propylene with metallocene catalysts. *J. Am. Chem. Soc.* **2001**, *123*, 9555–9563. (s) Fan, W.; Waymouth, R. M. Alternating copolymerization of ethylene and propylene: Evidence for selective chain transfer to ethylene. *Macromolecules* **2001**, *34*, 8619–8625. (t) Longo, P; Siani, E; Pragliola, S; Monaco, G. Copolymerization of ethene and propene in the presence of C_S symmetric group 4 metallocenes and methylaluminoxane. *J. Polym. Sci., Part A: Polym. Chem.* **2002**, *40*, 3249–3255. (u) Reybuck, S. E.; Meyer, A., Waymouth, R. M. Copolymerization behavior of unbridged indenyl metallocenes: Substituent effects on the degree of comonomer incorporation. *Macromolecules* **2002**, *35*, 637–643. (v) Busico, V.; Cipullo, R.; Segre, A. L. Advances in the ^{13}C NMR characterization of ethene/propene copolymers, 1—C_2-symmetric *ansa*-metallocene catalysts. *Macromol. Chem. Phys.* **2002**, *203*, 1403–1412. (w) Hung, J.; Cole, A. P.; Waymouth, R. M. Control of sequence distribution of ethylene copolymers: Influence of comonomer sequence on the melting behavior of ethylene copolymers. *Macromolecules* **2003**, *36*, 2454–2463. (x) Reybuck, S. E.; Waymouth, R. M. Investigation of bridge and 2-phenyl substituent effects on ethylene/α-olefin copolymerization behavior with 1,2'-bridged bis(indenyl)zirconium dichlorides. *Macromolecules* **2004**, *37*, 2342–2347. (y) Heuer, B.; Kaminsky, W. Alternating ethene/propene copolymers by C_1-symmetric metallocene/MAO catalysts. *Macromolecules* **2004**, *38*, 3054–3059. (z) Baughman, T. W.; Sworen, J. C.; Wagener, K. B. Sequenced ethylene-propylene copolymers: Effects of short ethylene run lengths. *Macromolecules* **2006**, *39*, 5028–5036.

12. (a) Stevens, J. C.; Timmers, F. J.; Wilson, D. R.; Schmidt, G. F.; Nickias, P. N.; Rosen, R. K.; Knight, G. W.; Lai, S. Y. Constrained geometry addition polymerization catalysts, processes for their preparation, precursors therefore, methods of use, and novel polymers formed therewith. European Patent Application EP 416,815 A2 (Dow Chemical Company), August 30, 1990. (b) Canich, J. M. Olefin polymerization catalysts. European Patent Application EP 420,436 A1 (Exxon Chemical Patents, Inc., USA), September 10, 1990. (c) Galimberti, M.; Mascellani, N.; Piemontesi, F.; Camurati, I. Random ethylene/propylene copolymerisation from a catalyst system based on a "constrained geometry" half sandwich complex. *Macromol. Rapid Commun.* **1999**, *20*, 214–218.

13. The first industrial production of EPM from heterogeneous titanium catalysts was in 1983; four commercial grades were initially available. From 1986 on, CO053TX was the only grade on the market. Production was then stopped in the late 1980s.

14. For a detailed discussion on the usage of "single center" rather than "single site" nomenclature see: Resconi, L.; Fait, A.; Piemontesi, F.; Colonnesi, M.; Rychlicki, H.; Ziegler, R. Effect of monomer concentration on propylene polymerization with the *rac*-[ethylenebis(1-indenyl)]zirconium dichloride/methylaluminoxane catalyst. *Macromolecules* **1995**, *28*, 6667–6676. Although we consider "single center" as the correct nomenclature we use the "single site" nomenclature herein, basically because of more common usage.

15. (a) Bovey, F. A.; Sacchi, M. C.; Zambelli, A. Polymerization of propylene to syndiotactic polymer. IX. Ethylene perturbation of syndiotactic propylene polymerization. *Macromolecules* **1974**, *7*, 752–753. (b) Zambelli, A.; Sessa, I.; Grisi, F.; Fusco, R.; Accomazzi, P. Syndiotactic polymerization of propylene: Single-site vanadium catalysts in comparison with zirconium and nickel. *Macromol. Rapid Commun.* **2001**, *22*, 297–310.

16. (a) Albizzati, E.; Galimberti, M.; Giannini, U.; Morini, G. The chemistry of magnesium chloride-supported catalysts for polypropylene. *Makromol. Chem., Macromol. Symp.* **1991**, *48/49*, 223–238. (b) Albizzati, E.; Giannini, U.; Collina, G.; Noristi, L.; Resconi, L. Catalysts and polymerizations. In *Polypropylene Handbook*; Moore, E. P., Jr., Ed.; Hanser: Munich, 1996; pp 11–98. (c) Albizzati, E.; Galimberti, M. Catalysts for olefins polymerization. *Catal. Today* **1998**, *41*, 159–168. (d) Cecchin, G.; Morini, G.; Piemontesi, F. Ziegler–Natta catalysts. In *Kirk-Othmer Encyclopedia of Chemical Technology Online*; John Wiley and Sons: New York, 2001 (accessed April 2006).

17. (a) Irwin, L. J.; Miller, S. A. Unprecedented syndioselectivity and syndiotactic polyolefin melting temperature: Polypropylene and poly(4-methyl-1-pentene) from a highly active, sterically expanded η^1-fluorenyl-η^1-amido zirconium complex. *J. Am. Chem. Soc.* **2005**, *127*, 9972–9973. (b) Miller, S. A.; Irwin, L. J. Novel catalyst system for high activity and stereoselectivity in the homopolymerization and copolymerization of olefins. U.S. Patent Application Publ. 2006/0025299 A1 (Texas A & M University), February 2, 2006.

18. Hanson, K. R. Applications of the sequence rule. I. Naming the paired ligands g,g at a tetrahedral atom Xggij. II. Naming the two faces of a trigonal atom Yghi. *J. Am. Chem. Soc.* **1966**, *88*, 2731–2742.

19. For reviews on ^{13}C NMR analysis of EP copolymers see, for example: (a) Bovey, F. A.; Mirau, P. A. *NMR of Polymers*. Academic Press: San Diego, 1996. (b) Randall, J. C. A review of high-resolution liquid carbon-13 nuclear magnetic resonance characterizations of ethylene-based polymers. *J. Macromol. Sci., Rev. Macromol. Chem. Phys.* **1989**, *C29*, 201–317. (c) Randall, J. C. *Polymer Sequence Determination*. Academic Press: New York, 1977. (d) For ^{13}C NMR analysis on EP copolymers from metallocenes see, for example: Tritto, I.; Fan, Z. Q.; Locatelli, P.; Sacchi, M. C.; Camurati, I.; Galimberti, M. ^{13}C NMR studies of ethylene-propylene copolymers prepared with homogeneous metallocene-based Ziegler–Natta catalysts. *Macromolecules* **1995**, *28*, 3342–3350.

20. For general information on copolymerization see, for example: (a) Odian, G. *Principles of Polymerization*, 3rd ed.; John Wiley and Sons: New York, 1991. (b) Ham, G. E. *Copolymerization*. Interscience: New York, 1964.

21. (a) Corradini, P.; Guerra, G.; Pucciariello, R. New model of the origin of the stereospecificity in the synthesis of syndiotactic polypropylene. *Macromolecules* **1985**, *18*, 2030–2034. (b) Corradini, P.; Cavallo, L.; Guerra, G. Molecular modeling studies on stereospecificity and regiospecificity of propylene polymerisation by metallocenes. In *Metallocene-Based Polyolefins*; Scheirs, J., Kaminsky, W., Eds.; John Wiley and Sons: Chichester, 2000; Vol. 2, pp 1–36. (c) Corradini, P.; Guerra, G.; Cavallo, L. Do new century catalysts unravel the mechanism of stereocontrol of old Ziegler–Natta catalysts? *Acc. Chem. Res.* **2004**, *37*, 231–241.

22. Guerra, G.; Galimberti, M.; Piemontesi, F.; Ruiz de Ballesteros, O. Influence of regio- and stereoregularity of propylene insertion on crystallization behavior and elasticity of ethylene-propylene copolymers. *J. Am. Chem. Soc.* **2002**, *124*, 1566–1567.

23. (a) Wunderlich, B.; Poland, D. Thermodynamics of crystalline linear high polymers. II. The influence of copolymer units on the thermodynamic properties of polyethylene. *J. Polym. Sci., Part A* **1963**, *1*, 357–372. (b) Baker, C. H.; Mandelkern, L. Crystallization and melting of copolymers. II. Variations in unit-cell dimensions in polymethylene copolymers. *Polymer* **1966**, *7*, 71–83. (c) Bassi, I. W.; Corradini, P.; Fagherazzi, G.; Valvassori, A. Crystallization of high ethylene EPDM [ethylene-propylene-diene] terpolymers in the stretched state. *Eur. Polym. J.* **1970**, *6*, 709–718. (d) Ver Strate, G., Wilchinsky, Z. W. Ethylene-propylene copolymers: Degree of crystallinity and composition. *J. Polym. Sci., Part A-2: Polym. Phys.* **1971**, *9*, 127–142. (e) Crespi, G.; Valvassori, A.; Zamboni, V.; Flisi, U. Relation between structure and properties of ethylene-propylene elastomers. *Chim. Ind. (Milan)* **1973**, *55*, 130–141. (f) Scholtens, B. J. R.; Riande, E.; Mark, J. E. Crystallization in stretched and unstretched EPDM elastomers. *J. Polym. Sci., Polym. Phys. Ed.* **1984**, *22*, 1223–1238. (g) Ruiz de Ballesteros, O.; Auriemma, F.; Guerra, G.; Corradini, P. Molecular organization in the pseudo-hexagonal crystalline phase of ethylene-propylene copolymers. *Macromolecules* **1996**, *29*, 7141–7148. (h) Antinucci, S.; Guerra, G.; Oliva, L.; Ruiz de Ballesteros, O.; Venditto, V. Pseudo-hexagonal crystallinity in ethylene-styrene random copolymers. *Macromol. Chem. Phys.* **2001**, *202*, 382–387.

24. Bunn, C. W. Crystal structure of long-chain normal paraffin hydrocarbons. "Shape" of the methylene group. *Trans. Faraday Soc.* **1939**, *35*, 482–491.

25. (a) Guerra, G.; Ilavský, M.; Biroš, J.; Dušek, K. The viscoelastic and equilibrium rheooptical behavior of crosslinked ethylene-propylene copolymers. *Coll. Polym. Sci.* **1981**, *259*, 1190–1197. (b) Koivumäki, J.; Seppälä, J. V. Copolymerization of ethylene and propylene in liquid propylene with $MgCl_2/TiCl_4$, $VOCl_3$ and Cp_2ZrCl_2 catalyst systems—a comparison of products. *Eur. Polym. J.* **1994**, *30*, 1111–1115. (c) Guerra, G.; Ruiz de Ballesteros, O.; Venditto, V.; Galimberti, M.; Sartori, F.; Pucciariello, R. Pseudohexagonal crystallinity and thermal and tensile properties of ethene-propene copolymers. *J. Polym. Sci., Part B: Polym. Phys.* **1999**, *37*, 1095–1103. (d) Galimberti, M.; Resconi, L.; Martini, E.; Guglielmi, F.; Albizzati, E. Elastic copolymer of ethylene with alpha-olefins. European Patent EP 586,658 B1 (Montell Technology Company), January 28, 1998.

26. Eng, A. H.; Ong, E. L. Hevea natural rubber. In *Handbook of Elastomers*; Bhowmick, A. K., Stephens, H. L., Eds.; Marcel Dekker: New York, 2001; pp 29–59.

27. (a) Natta, G. (Title unavailable). Presented at the XVI Congres International de Chimie Pure et Appliqué, Paris, France, 1957. (b) Natta, G.; Crespi, G.; Valvassori, A.; Sartori, G. Polyolefin elastomers. *Rubber Chem. Technol.* **1963**, *36*, 1583–1668.

Part III

Other Monoolefin Monomers

13 Stereocontrol in the Polymerization of Higher α-Olefin Monomers

Moshe Kol, Sharon Segal, and Stanislav Groysman

CONTENTS

13.1 SCOPE OF CHAPTER

This chapter gives a perspective on the evolution of catalyst systems for the stereoregular polymerization of higher α-olefins (CH_2=CHR, where R ≥ CH_2CH_3). Heterogeneous systems, metallocene systems, and nonmetallocene systems (including amine-phenolate catalyst systems introduced by the Kol research group) are reviewed. In general, the mechanisms controlling chain growth, regioregularity, and stereoregularity that apply to propylene polymerization also apply for higher α-olefin polymerization, so they will not be discussed in detail in this chapter.

13.2 SIGNIFICANCE OF STEREOREGULAR POLYMERIZATION OF HIGHER α-OLEFINS

Stereoregular α-olefin polymerization by early transition metal Ziegler-type compounds was first reported in 1955 by Natta et al.[1] Since then, major industrial and academic research efforts have focused mainly on the development of highly efficient, stereoselective catalyst systems for propylene polymerization. Concurrently, the polymerization of higher α-olefins was also investigated, with the hope of revealing new opportunities in the field of polymeric materials.

Natta first reported the stereoselective (isotactic and syndiotactic) polymerization of α-olefins using heterogeneous catalysts.[1-5] In addition, he was the first to describe the solid state structures

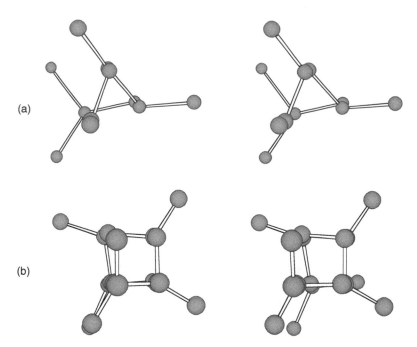

FIGURE 13.1   Stereoviews along the main chain of helical structures exhibited by isotactic poly(α-olefins), with a single atom representing the side chains (a) a threefold helix (3_1-polypropylene) and (b) a fourfold helix (4_1-polyvinyl cyclohexane).

of isotactic poly(α-olefins).[6] Among others, the threefold (3_1-helix) structures of poly(1-butene) and poly(5-methylhexene), the fourfold (4_1-helix) structures of poly(vinylcyclohexane) and poly(3-methylbutene), and the sevenfold (7_2-helix) structure of poly(4-methylpentene) were described (Figure 13.1). On the basis of crystal structure data, Natta concluded that the shape of the R substituent of the α-olefin (CH$_2$-CHR) repeat unit was responsible for the different conformations of these isotactic helices. These findings seem to indicate that, bulkier olefins, that is, olefins in which the branch is closer to the double bond, give rise to a longer repeat unit of the helix. More significantly, the shape of particular R groups may cause lack of crystallinity in the isotactic polymer, which in turn can have substantial influence on polymer properties. For instance, isotactic poly(*meta*-methylstyrene) and poly(*ortho*-methylstyrene) are crystalline, whereas isotactic poly(*para*-methylstyrene) is amorphous.

13.3   STEREOCONTROLLED α-OLEFIN POLYMERIZATION BY HETEROGENEOUS CATALYSTS

Very early on in poly(α-olefin) research, it became clear that for a given monomer, the nature of the polymeric material obtained was determined largely by the nature of catalyst used. Several heterogeneous Ziegler-type catalytic systems were examined by Natta for α-olefin polymerization and were shown to produce polymers with different stereochemical properties. In general, the catalysts studied were comprised of an early transition metal complex (mostly in a form of a metal halide) combined with a main-group compound (cocatalyst, generally in a form of alkyl or alkyl-halide compound). Different crystalline forms of TiCl$_3$ (α, β, and γ) were studied as the early transition metal components in heterogeneous α-olefin polymerization systems. Among these, the α (violet) form of TiCl$_3$ produced the most isotactic poly(α-olefin)s.[7] The β (brown) form of TiCl$_3$ and similarly studied Ti alkoxides produced polymers having much lower tacticities. High valency vanadium compounds, when used as the early transition metal component, produced stereoblock polymers.[8] In terms of the main group organometallic compounds (cocatalysts) used, their effects on stereoregularity were found to depend mainly upon the ionic radii of the main-group metal. Employing the α (violet) modification of $TiCl_3$, cocatalysts that were organometallic derivatives of highly electropositive metals (Al, Be) with small radii resulted in highly isotactic polymers,[8] whereas cocatalysts featuring higher atomic radii (Mg) or less electropositive metals (Zn) led to polymers with a lower portion of an isotactic fraction.[7]

The chirality of the active sites in the heterogeneous Ziegler–Natta system $TiCl_3/AlR_3$ was postulated to be the origin of the stereoselectivity shown in the polymerization of α-olefins ("enantiomorphic site control").[2] Although heterogeneous systems are not easily accessible for mechanistic study, Pino and others conducted a series of elegant experiments using $TiCl_4/Al(i\text{-}C_4H_9)_3$ that involved polymerization of chiral α-olefins possessing an asymmetric carbon atom at the 3- (α to the double bond) or 4- (β to the double bond) positions.[9] Polymerization of racemic monomer mixtures, led to polymer chains that were enriched in specific enantiomers (a stereoselective process) even when the stereoregularity of the chains was very low (as revealed from measurement of optical rotation of fractions of the polymer separated on a chiral stationary phase). These experiments confirmed the intrinsic chirality (dissymmetry) of the heterogeneous active centers in the Ziegler–Natta catalyst, and implied that stereocontrol in α-olefin polymerization is governed by an enantiomorphic site control mechanism. The stereodiscrimination ability of Ziegler–Natta catalysts in the polymerization of chiral α-olefins (having substituents in the 3- and 4-positions, such as racemic 3-methylpentene and 4-methylhexene) was further demonstrated by Giardelli and coworkers. Thus, they induced the elective polymerization of one enantiomer of a racemic olefin by reacting the initiators ($TiCl_4$ or $TiCl_3$) with an optically active metal alkyl, or alternatively, by employing an optically active third component such as $Zn((S)\text{-}2\text{-methylbutyl})_2$.[10] In contrast, a study of 1-butene polymerization in the presence of $MgCl_2$-supported Ziegler–Natta catalysts, conducted by Busico et al., gave evidence for the formation of stereoblock polymers, demonstrating that the syndiopreference in this system results from a chain-end control mechanism.[11] (See Chapter 1 for further information on enantiomorphic site control and chain-end control mechanisms.)

13.4 STEREOCONTROLLED α-OLEFIN POLYMERIZATION BY METALLOCENE CATALYSTS

The development of single-site metallocene catalysts provided investigators with useful tools for controlling stereoregularity and for studying the mechanistic aspects of α-olefin polymerization. The use of metallocene polymerization systems has shown, for the first time, a clear relationship between the structure and the symmetry of a well-defined catalyst and the microstructure of the resulting polymer. In 1985, Kaminsky and Brintzinger reported the highly isotactic polymerization of both propylene and 1-butene using the chiral C_2-symmetric *ansa*-zirconocene precatalyst (**1**, see Figure 13.2) activated by methylaluminoxane (MAO).[12] The chirality of the catalytic site induced by the ligand wrapping around the metal induces an enantiomorphic site control mechanism able to differentiate between the two enantiotopic faces of the incoming olefin. The C_2-symmetry ensures that a possible flip of the growing polymeryl chain will not affect the stereocontrol. Similarly, Ewen's C_s-symmetric catalyst (**2**) produced highly syndiotactic poly(1-butene) via a site control mechanism.[13] In this case, the olefin may approach the metal from two mirror-reflected directions being enantiotopic to one another. An alternating olefin approach through its opposite enantiotopic faces will lead to the syndiotactic polymer. Significantly, even a complex of higher symmetry, that is, the C_{2v}-symmetric $(Cp^*)_2ZrCl_2$/MAO (Cp^* = pentamethylcyclopentadienyl) catalyst (**3**) was able to produce predominantly syndiotactic poly(1-butene), operating, however, through a chain-end control mechanism, according to the Bernoullian value ($B = 4[mm][rr]/[mr]^2$) of ≈ 1. Stereoselectivity was lost when a bulkier α-olefin (4-methylpentene) was employed.[14] The chain-end

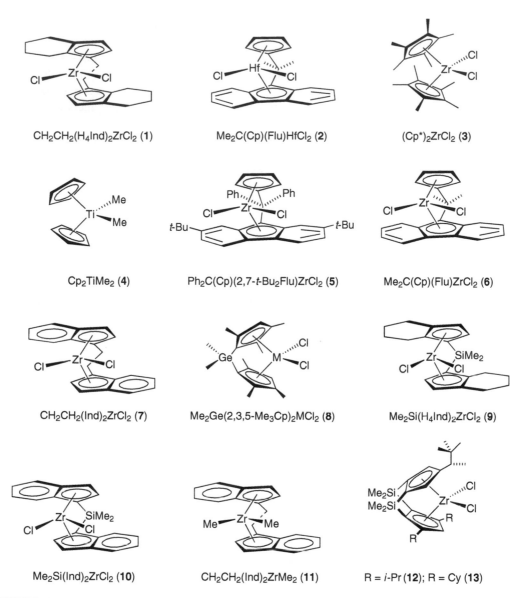

FIGURE 13.2 Various metallocene group 4 complexes employed as precatalysts or cocatalysts in stereoregular polymerizations.

control requires discrimination between the steric bulk of the C-2 substituent of the last inserted unit and the rest of the growing polymer chain. The loss of stereoselectivity (for 4-methylpentene polymerization) was attributed to a similar steric bulk between the former (*i*-Bu group) and the latter. Notably, the combination of a Solvay-type $TiCl_3$ with a nonstereoselective C_{2v}-symmetric Cp_2TiMe_2 (**4**) led to the highly isoselective polymerization of various α-olefins.[15] However, in this case, the metallocene served as a cocatalyst for the actual catalyst being the heterogeneous $TiCl_3$.

The homopolymerization of several important α-olefins by C_2- and C_s-symmetric *ansa*-metallocenes was thoroughly investigated in the past decade. Kaminsky and Grumel developed highly active syndioselective *ansa*-zirconocene catalysts (**5**, **6** respectively) for the polymerization of 1-pentene, 1-hexene, and 1-octene, and studied the effects of temperature on polymer molecular weight and microstructure.[16,17] It was found that polymer syndiotacticity depended strongly upon polymerization temperature, rapidly decreasing at elevated temperatures. Deffieux and coworkers observed that the syndiotacticity of poly(1-hexene) produced by the i-Pr(Cp)(Flu)ZrCl$_2$/MAO catalyst system, **6**/MAO, (Cp = cyclopentadienyl; Flu = 9-fluorenyl) depends strongly on the polymerization media, being high when aromatic solvents are used and much lower when polar chlorinated media are employed.[18] In contrast, the use of chlorinated solvents for polymerization did not decrease the isotacticity of poly(1-hexene) produced by the C_2-symmetric catalyst system, rac-CH$_2$CH$_2$(Ind)$_2$ZrCl$_2$/MAO, **7**/MAO (Ind = indenyl). However, chlorinated solvents did increase the activity of the catalyst, and enabled the use of a lower MAO/precatalyst molar ratio.[19] Yamaguchi et al. studied the influence of high pressure on the stereoselectivity of 1-hexene polymerization by various C_2-symmetric $ansa$-metallocene catalysts, such as **8**. They found that, employing high pressures of the olefin (100–500 MPa) increased catalyst activity and poly(1-hexene) molecular weights, while maintaining the high isoselectivity of the catalysts.[20] Finally, studying the polymerization of high α-olefins (1-pentene, 1-hexene, 1-octene, 1-decene) with rac-Me$_2$Si(2-Me-4-t-BuCp)$_2$Zr(NMe$_2$)$_2$/Al(i-Bu)$_3$/[Ph$_3$C]$^+$[B(C$_6$F$_5$)$_4$]$^-$, Kim et al. observed that the isotacticity of the resultant poly(α-olefins) (as measured by the percentage of mm triads in the polymer ^{13}C NMR spectrum) decreased with increasing bulk of the α-olefin side chain. This relationship was explained by an increased hindering of the rearrangement of the polymeryl unit (to allow the next monomer insertion) in enantiomorphic site control mechanism.[21]

Several research groups have investigated the mechanism of α-olefin polymerization through metallocene-based catalysts, using ^{13}C NMR spectroscopy as a major diagnostic tool. As reported by Rossi and others, end group analysis of poly(1-butene) produced with an $ansa$-zirconocene/MAO catalyst, **9**/MAO, revealed that chain propagation proceeds through 1,2-enchainment in this system, and that termination takes place through β-hydride transfer, chain transfer to aluminum, or β-alkyl transfer.[22] (See Chapter 1 for further information of chain release.) Busico and coworkers, using rac-CH$_2$CH$_2$(Ind)$_2$ZrCl$_2$ (**7**) and rac-Me$_2$Si(Ind)$_2$ZrCl$_2$ (**10**), have observed that 1-butene as compared to propylene is substantially less reactive after 2,1-insertions (regioerrors), and that the resulting "dormant" sites isomerize to give 4,1-regioregular units.[23,24] Chien and coworkers studied the end groups of poly(1-hexene) produced by $ansa$-zirconocene-based precatalysts rac-CH$_2$CH$_2$(Ind)$_2$ZrCl$_2$ (**7**) and AlEt$_3$/[Ph$_3$C]$^+$[B(C$_6$F$_5$)$_4$]$^-$ as cocatalyst using ^{13}C NMR spectroscopy. They found that β-hydride elimination (transfer) was the prevailing chain termination pathway, resulting in the formation of a disubstituted vinylidene endgroup. In addition, trisubstituted vinylidene end groups were observed; these groups were postulated to arise from zirconium migration along the carbon backbone.[25] A detailed kinetic study of the initiation, propagation, and termination steps of isoselective 1-hexene polymerization by the system rac-CH$_2$CH$_2$(Ind)$_2$ZrMe$_2$ (**11**) activated with B(C$_6$F$_5$)$_3$ was reported by Landis and coworkers.[26] The initiation and propagation processes were found to be of the first order in olefin and in catalytic species, and the initiation step was slower than the propagation step by a factor of 70. β-Hydride elimination, following both 1,2- and 2,1-insertions, was found to be the main chain termination process.

Stereoselective metallocene-based systems have enabled the efficient preparation of a large number of stereoregular poly(α-olefins), and the properties of these polymers have been thoroughly investigated. In agreement with the early observations of Natta, the properties of stereoregular poly(α-olefins) were found by later researchers to depend on the nature of the side chain. Wahner and coworkers carried out a comparative study of the thermal properties of poly(α-olefins) possessing linear side chains of varying length.[27,28] The melting point (T_m) of the stereoregular (isotactic) poly(α-olefins) was found to drop from polypropylene to poly(1-pentene), and then to increase from poly(1-decene).[27-29] Whereas poly(1-pentene) is a relatively high-melting material, both poly(1-hexene) and poly(1-octene) are amorphous polymers without a distinct melting point. Further, two melting points are generally detected for the poly(α-olefins) with long side chains (starting from poly(1-pentene)). For example, poly(1-pentene) displays $T_{m_1} = 53\ °C$ and $T_{m_2} = 95\ °C$, and poly(1-decene) displays $T_{m_1} = 10\ °C$ and $T_{m_2} = 26\ °C$. Several explanations have been proposed for this phenomenon. Among those are the melting of the crystals formed by the side chains and the main chains (for poly(1-decene) and higher α-olefins), or the different crystal modifications (for poly(1-pentene)).[28-30]

In 1999, the polymerization of 1-hexene by a system showing both isoselective and living character was reported for the first time by Fukui, Soga, and coworkers, who utilized a *rac*-CH$_2$CH$_2$(Ind)$_2$ZrMe$_2$ (**11**) precatalyst, activated by B(C$_6$F$_5$)$_3$ at -78 °C. The living character of this polymerization was demonstrated by the narrow molecular weight distributions of the poly(1-hexene) ($M_w/M_n < 1.3$). However, only a small fraction of the zirconium complex was actually activated for polymerization (ca. 10%, assuming a living polymerization), and the activity of the system at -78 °C was very low.[31]

Very recently, Bercaw and coworkers described the kinetic resolution of chiral α-olefins using enantiomerically pure *ansa*-zirconocenes.[32] Several C_1-symmetric, doubly bridged zirconocenes (e.g., **12** and **13**) were prepared in enantiomerically pure form. Subsequently, they were activated by MAO for the polymerization of bulky racemic α-olefin monomers substituted in the 3- and 4-positions (such as 3- and 4-Me-1-pentene). Preferential uptake of one enantiomer over the other during polymerization occurred to an extent dependent upon the exact monomer and catalyst structures; subsequently, the product polymer was easily separated from the unreacted monomer, now enriched in the slower reacting enantiomer (enantiomeric excesses of up to 58.6% were obtained). Although only a low differential between polymerization rates of the individual enantiomer pairs was found for most of the monomers, a rate ratio of $>15:1$ was found for the two enantiomers of 3,4-dimethyl-1-pentene. The ease of polymer separation from the remaining enantiomerically enriched monomer, and the lack of useful chiral transformations for the relatively simple α-olefins, reveal the potential importance of this efficient route toward enantiomerically pure α-olefins through kinetic resolution, as a source of chiral starting materials in organic synthesis.

13.5 STEREOCONTROLLED POLYMERIZATION OF α-OLEFINS BY NONMETALLOCENE SYSTEMS

13.5.1 BIPHENOLATE GROUP 4 TYPE CATALYSTS

The first significant diversion from the metallocene catalyst strategy for tactic higher α-olefin polymerization was reported in 1995.[33] Schaverien and coworkers described a series of chelating biphenolate and binaphtholate complexes of titanium and zirconium and their catalytic use including 1-hexene polymerization/oligomerization in the presence of MAO. It was found that the steric bulk of the ligands had substantial effects on both the molecular weight and the tacticity of the resulting oligo/poly(1-hexene). Thus, zirconium dichloride complexes ligated with binaphtholates having bulky substituents, for example, *ortho*-trialkylsilyl groups (**14**, see Figure 13.3) gave high-molecular weight poly(1-hexene)s having relatively narrow molecular weight distributions and high isotacticities ($> 90\%$) through a regioregular 1,2-insertion process. In contrast, titanium complexes of sulfur-bridged biphenolate ligands with bulky *ortho-tert*-butyl substituents (**15**) gave atactic polymers of low molecular weight, obtained by regioregular 1,2-insertion.

13.5.2 CHELATING DIAMIDO GROUP 4 TYPE CATALYSTS

A noteworthy development in the area of higher α-olefin polymerization catalysis was McConville's 1996 report of a new catalyst system (**16**, Figure 13.4): a dialkyltitanium complex featuring a chelating, bulky diamide ligand that upon activation with B(C$_6$F$_5$)$_3$ afforded the living polymerization of neat 1-hexene and other higher α-olefins at room temperature.[34,35] Although these poly(α-olefins) were atactic, a dichlorotitanium complex incorporating this ligand and related ligands led to isotactic propylene polymerization when activated with Al(i-Bu)$_3$/[Ph$_3$C]$^+$[B(C$_6$F$_5$)$_4$]$^-$ at 40 °C as reported by Uozumi and coworkers.[36,37]

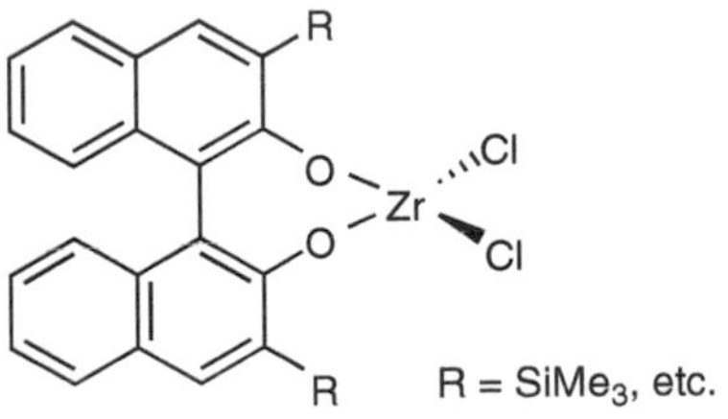

FIGURE 13.3   Chelating bis(naphtholate) (**14**) and bis(phenolate) (**15**) group 4 polymerization precatalysts exhibit different behavior for 1-hexene polymerization when activated with MAO.

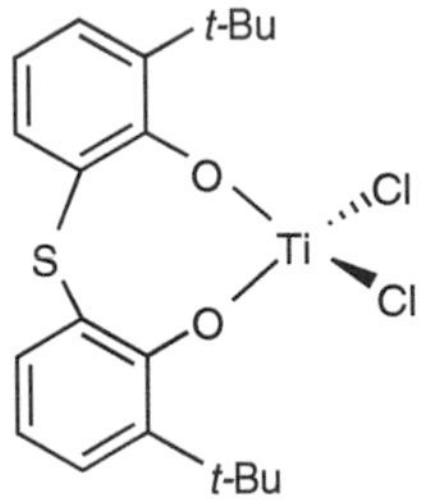

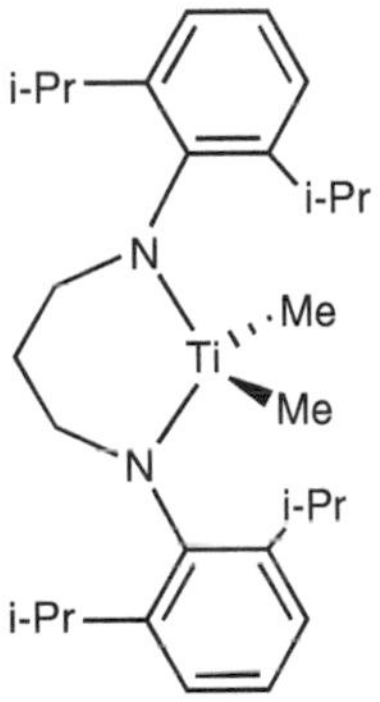

FIGURE 13.4   McConville's precatalyst (**16**), when activated with B(C$_6$F$_5$)$_3$ performs living polymerizations of higher α-olefins at room temperature; bis(imine-phenolate) complexes (**17**) lead to stereoirregular and regioirregular polymerization of high olefins.

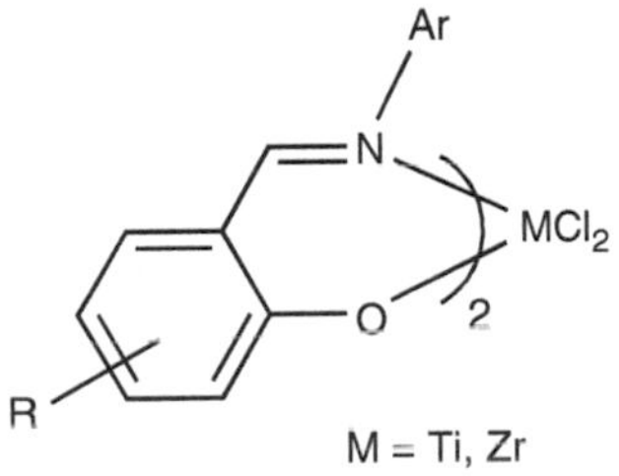

13.5.3   Bis(phenoxy-imine) Group 4 Type Catalysts

Another noteworthy α-olefin polymerization development was the use of MAO activated Ti and Zr catalysts bearing bis(phenoxy-imine)s ligand (**17**, Figure 13.4) or related bidentate monoanionic ligands. These systems were first introduced by Fujita and coworkers from the Mitsui company, and subsequently also studied by Coates.[38,39] These systems have been observed to carry out ultrahigh activity ethylene polymerization and the living syndiotactic polymerization of propylene. Although these systems were also found to be highly active for the polymerization of 1-hexene and other high α-olefins, the high-molecular weight poly(α-olefins) obtained were both stereoirregular and regioirregular.[40]

13.5.4   Cp-Acetamidinate Group 4 Type Catalysts

A versatile, efficient family of higher α-olefin polymerization catalysts was introduced by Sita and Jayaratne in 2000 (**18a-b**, Figure 13.5).[41] These C_1-symmetric zirconium dimethyl precatalysts feature a cyclopentadienyl or pentamethylcyclopentadienyl ligand and a chelating acetamidinate ligand having two different N-alkyl substituents. Upon activation with [PhNMe$_2$H]$^+$[B(C$_6$F$_5$)$_4$]$^-$ at -10 °C, a catalyst of this family (**18a**: Cp*; R^1 = Et; R^2 = t-Bu) effected the highly isotactic polymerization of 1-hexene (>95% *mmmm*). This process was living as evidenced by very narrow M_w/M_ns (as narrow as 1.03), a linear relationship between polymer molecular weight and polymerization time, and the successful consumption of a second portion of monomer after complete

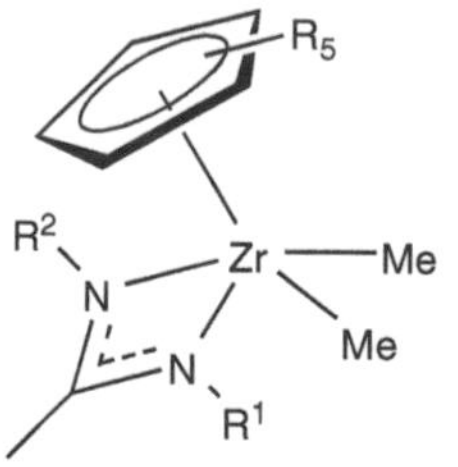

18a: R^1 = Et, R^2 = t-Bu, R = Me
18b: R^1 = Et, R^2 = t-Bu, R = H

FIGURE 13.5 Sita's cyclopentadienyl-acetamidinate precatalysts (**18a and 18b**), when activated with [PhNMe$_2$H]$^+$[B(C$_6$F$_5$)$_4$]$^-$, leads to living polymerization of higher α-olefins and isotactic polymer.

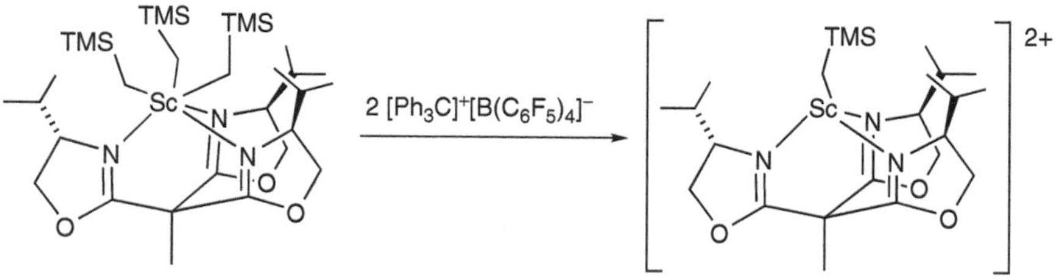

SCHEME 13.1 Activation of a scandium complex forming a dicationic, C$_3$-symmetric scandium catalyst that is capable of polymerizing 1-hexene to isotactic poly(1-hexene) (TMS = trimethylsilyl).

consumption of an initial portion. The high isotacticity found with the system was attributed to a relatively high barrier to racemization of the chiral metal center of the active cationic species relative to the low barrier found for the neutral dialkyl precursor. Subsequent studies have shown that these systems may be employed in block-copolymerization of 1,5-hexadiene and 1-hexene;[42] that systems based on Cp ligand are more active than systems based on Cp* ligand and may be employed for isoselective polymerization of vinylcyclohexane by a chain-end control mechanism (the C$_s$-symmetric catalyst yielded isotactic poly(vinylcyclohexane));[43] and revealed mechanistic insight regarding the structure of the active species.[44]

13.5.5 TRIS(OXAZOLINE) SCANDIUM TYPE CATALYSTS

Very recently, a scandium complex based on a C$_3$-symmetric, neutral tris(oxazoline) ligand introduced by Gade and coworkers was reported to lead to active 1-hexene polymerization catalysis (**19**, Scheme 13.1).[45] This complex produced either monocationic or (proposed) dicationic species (**20**) upon activation with [Ph$_3$C]$^+$[B(C$_6$F$_5$)$_4$]$^-$, depending on the number of equivalents of cocatalyst added. Whereas the monocationic catalyst exhibited low activity polymerization and variable tacticity control, the proposed dicationic complex was highly active, giving high-molecular weight, highly isotactic poly(1-hexene). Increases in molecular weight (up to $M_w = 750,000$ where M_w is the weight average molecular weight) and lower molecular weight distributions (down to $M_w/M_n = 1.18$) were found as the polymerization temperature was lowered from ambient to −30 °C.

13.5.6 AMINE-PHENOLATE GROUP 4 TYPE CATALYSTS

Several years ago, Kol and coworkers began investigating the group 4 metal chemistry of amine bis(phenolate) ligands.[46,47] These ligands feature several useful properties, in particular (1) a broad

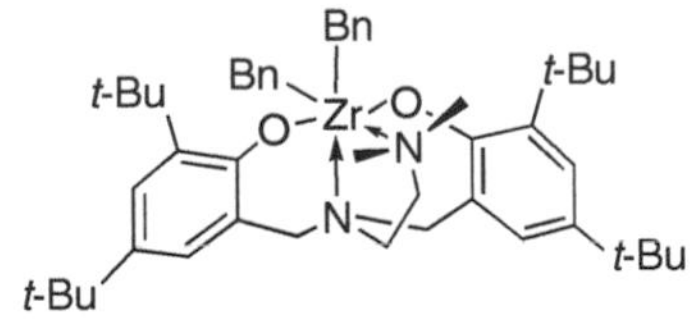

FIGURE 13.6   Representative amine bis(phenolate) ligand complexes (Bn = benzyl).

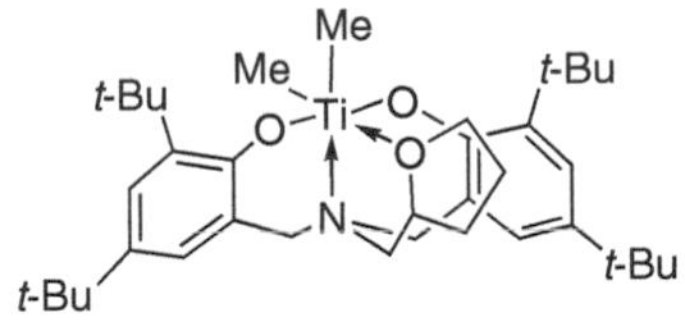

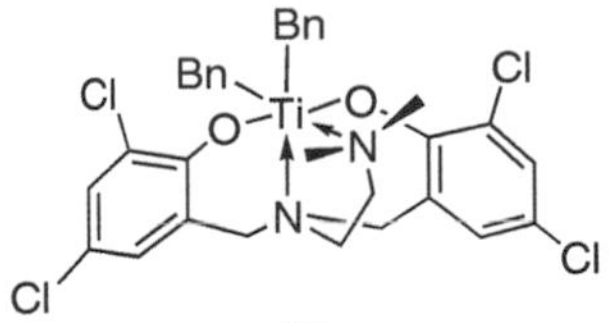

range of ligands featuring different steric and electronic characteristics may be synthesized in a single step from readily available starting materials; (2) the ligands may be attached to group 4 transition metals through several straightforward protocols, the most convenient of which is a one-step reaction between the ligand precursor and an appropriate tetrabenzylmetal precursor.[48] Most importantly, catalysts exhibiting a broad range of activities for propylene and higher α-olefins polymerization may be prepared. Through structural modifications of the ligands, and the use of different group 4 metals, amine bis(phenolate) catalyst structure-activity relationships for 1-hexene polymerization have been revealed. Selected examples of amine bis(phenolate) precatalysts are shown in Figure 13.6.

Complex **21** is a dibenzylzirconium complex of an amine bis(phenolate) ligand bearing *tert*-butyl-substituted phenyl rings and a dimethylamino donor that is bound through an ethylene bridge to the central nitrogen donor. Activation of this complex with $B(C_6F_5)_3$ at room temperature led to a highly active 1-hexene polymerization catalyst (activity $= 21,000$ g PH/mmol Zr·h; PH $=$ poly(1-hexene)). The poly(1-hexene)s formed had M_ws of up to 170,000 and M_w/M_ns of around 2.0, signifying single site catalyst behavior.[49] Analogous zirconium and hafnium complexes exhibited similar activities.[50] Complex **22** is a dimethyltitanium complex featuring a strong tetrahydrofuran (THF) sidearm donor. Upon activation with $B(C_6F_5)_3$, this complex initiated the slow living polymerization of 1-hexene at room temperature (remaining active for up to 6 days), yielding a polymer with an M_w of ca. 1,000,000 and an M_w/M_n of less than 1.10.[51,52] Analogous titanium complexes featuring strong sidearm donors, such as methoxy and *tert*-butylphenolate substituents also catalyzed the living polymerization of 1-hexene or 1-octene, and produced block copolymers of these two monomers at room temperature.[53,54] Complex **23** is a titanium complex of an amine bis(phenolate) ligand featuring a dimethylamino sidearm donor and chloro substituents on the phenolate groups.[55] Activation of this complex with $B(C_6F_5)_3$ led to a highly active 1-hexene polymerization catalyst exhibiting a very high propagation/chain transfer rate ratio, thus enabling the production of ultrahigh molecular weight polymer (M_w ca. 4,500,000) in 1 h. Interestingly, related complexes featuring small phenolate substituents and other sidearm donors (e.g., methoxy or THF) led to substantially less active catalysts, apparently owing to regioerrors (2,1-misinsertions) during α-olefin polymerization that may possibly result from the more open catalytic site.

When ligands with two identical phenolate groups and a nonchiral sidearm donor are used (e.g., complexes of type **21** and **23**), the resulting Ti and Zr complexes are C_s-symmetric. Since the two diastereotopic monomer coordination sites of these catalysts reside on a mirror plane, there is no facial preference for olefin coordination and insertion (assuming that a chain-end stereocontrol mechanism does not operate), and therefore these catalysts produced atactic poly(1-hexene), poly(propylene), and so forth.[56] Even when the catalyst symmetry is reduced to C_1, as in the case of complex **22** (with a chiral THF sidearm donor bound in position 2), or when the two phenolate rings on catalysts of type **21** and **23** have different substituents, no tacticity induction was observed. Apparently, the ligand asymmetry is too remote from the coordination sites to have a significant effect on olefin coordination preference. It thus seemed that tacticity induction would require a more substantial structural change, or, in other words, a ligand skeletal rearrangement.

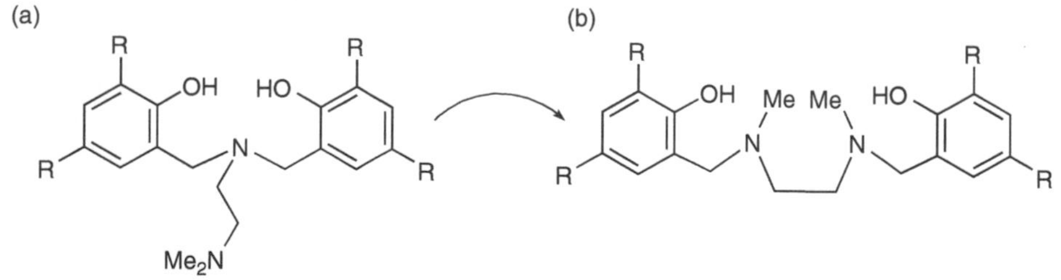

(a)

(b)

FIGURE 13.7 The "divergent" connectivity in amine bis(phenolate) ligands bearing a dimethylamino sidearm donor (a), and their isomeric "sequential" analogues, [ONNO]-type diamine bis(phenolate) (Salan) ligands (b).

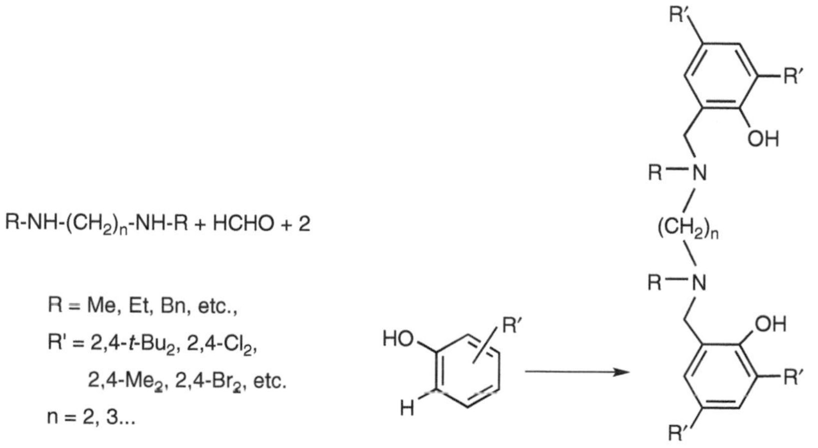

R-NH-(CH$_2$)$_n$-NH-R + HCHO + 2

R = Me, Et, Bn, etc.,

R' = 2,4-*t*-Bu$_2$, 2,4-Cl$_2$,

2,4-Me$_2$, 2,4-Br$_2$, etc.

n = 2, 3...

SCHEME 13.2 A one step Mannich-type synthesis is used to prepare Salan ligand precursors.

The amine bis(phenolate) ligands featuring a sidearm donor may be defined as "divergent" ligands—ligands featuring three donors emerging from a central nitrogen donor. Their typical wrapping mode around an octahedral metal center places the two phenolate oxygens in a trans geometry. In designing the new ligand system, we aimed to keep the donor groups in place but change their connectivity mode, namely, to have the donors connected in a "sequential" mode. These new, sequential diamine bis(phenolate) [ONNO]-type ligands may be regarded as isomers of amine bis(phenolate) ligands that bear a dialkylamino sidearm donor. Such ligands (Salan's, also known as tetrahydro-Salan's) have been employed earlier, but not in group 4 chemistry (Figure 13.7).

Ligands featuring a diaminoethane backbone were initially chosen for complexation studies because they were expected to form a favorable five-membered chelate with the metal center.[57] These ligands are readily accessible through the same single-step Mannich-type chemistry employed for making the amine bis(phenolate) ligands (using a di-secondary amine, formaldehyde, and a substituted phenol as shown in Scheme 13.2). A variety of Salan ligands featuring a broad scope of steric and electronic characteristics are readily accessible.[58]

The sequential arrangement of substituents in the Salan ligand allows for a higher freedom of wrapping modes around an octahedral metal center. The different wrapping modes are conveniently described by use of the *fac/mer* terms for binding of the two O-N-N fragments, namely *fac-fac*, *mer-mer*, and *fac-mer* (Scheme 13.3). The *mer-mer* mode, typical of the intensively studied Salan ligands ([ONNO]-type salicylalkylidene ligands), places the two labile groups (X, typically halide) in a trans geometry, and is therefore not suitable for α-olefin polymerization catalysis. The *fac-mer* mode, while leading to a cis arrangement of the labile groups, results in an overall C_1-symmetry of the octahedral complex, and therefore to less well-defined control over monomer approach. The

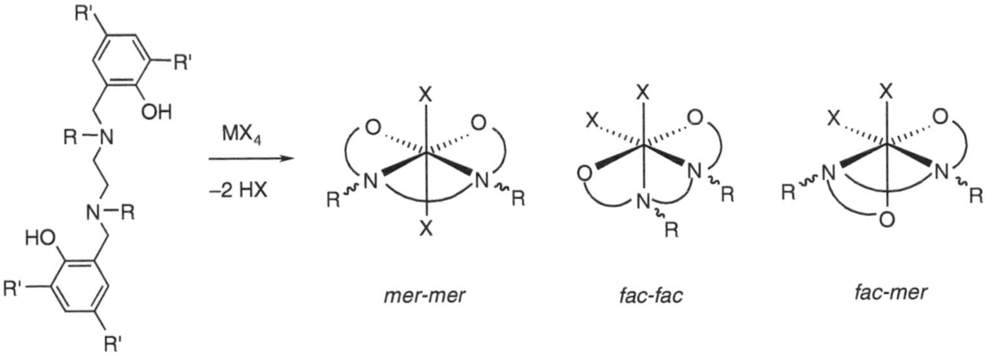

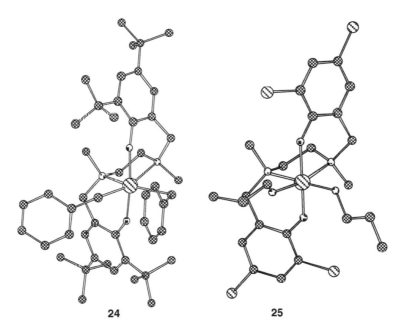

SCHEME 13.3 Synthesis of Salan polymerization precatalysts, and possible wrapping modes of the Salan ligand around octahedral group 4 metal centers.

FIGURE 13.8 Chem-3D representations of the X-ray structures of dibenzylzirconium (**24**) and dipropoxy-titanium (**25**) Salan complexes demonstrating the *fac-fac* Salan wrapping mode (hydrogen atoms are not shown).

preferred wrapping mode is the *fac-fac* mode, which produces C_2-symmetric complexes in which the labile groups have a cis-disposition.

We found that the typical wrapping mode of Salan ligands around Ti and Zr benzyls, alkoxides, and so forth, is indeed the *fac-fac* mode, and in most cases a single stereoisomer is obtained as indicated by nuclear magnetic resonance (NMR) analysis of the crude complexation products.[59] X-ray structures of a zirconium dibenzyl complex of a Salan ligand featuring *tert*-butyl aryl substituents (**24**, Figure 13.8) and of a titanium di-*n*-propoxide complex of a Salan ligand featuring chloro aryl substituents (**25**) confirm the proposed *fac-fac* wrapping modes.[57,63] These, and additional Salan compounds discussed are illustrated in Figure 13.9.

Activation of complex **24** with B(C$_6$F$_5$)$_3$ produced an active 1-hexene polymerization catalyst. The activity of this catalyst in neat 1-hexene at room temperature (ca. 30 g PH/mmol Zr·h), was

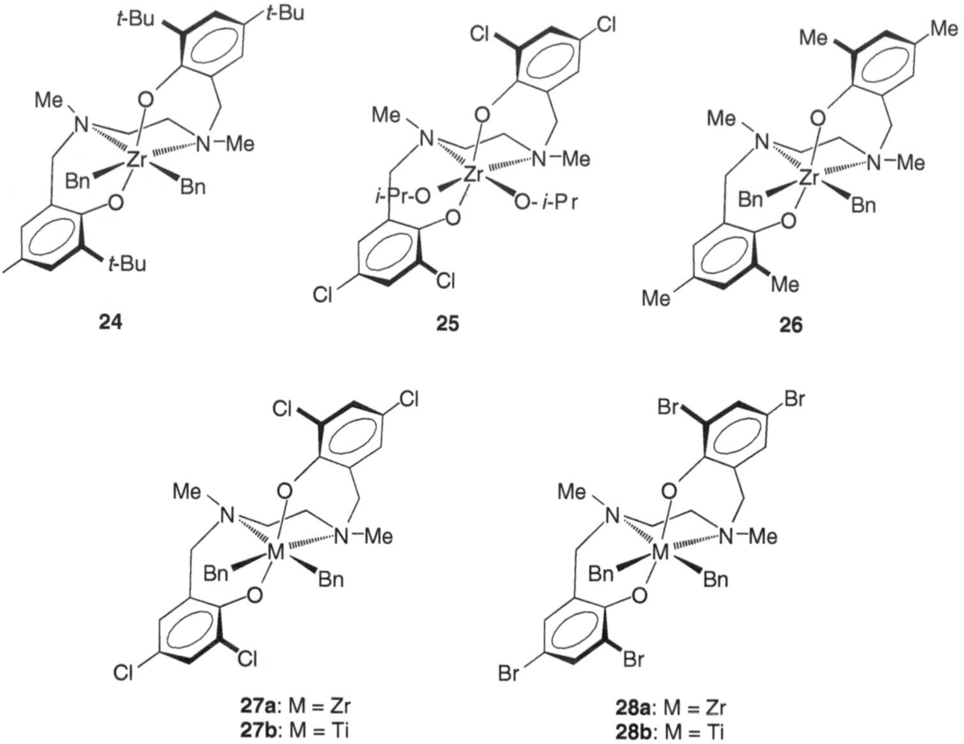

FIGURE 13.9 Representative Salan ligand complexes (Bn = benzyl).

considerably lower than that of its C_s-symmetric amine bis(phenolate) isomeric analogue **21**. Following the progress of the polymerization for 30 min indicated a constant rise in poly(1-hexene) molecular weight and a very narrow molecular weight distribution (1.12–1.15), supporting a living polymerization at room temperature. Most importantly, ^{13}C NMR of the resulting poly(1-hexene) indicated a highly isotactic structure (*mmmm* > 95%). This was the first reported achievement of living, isotactic room temperature α-olefin polymerization.[57] Under the same conditions, a dibenzyl zirconium complex of a Salan ligand featuring less bulky 2,4-methyl phenolate substituents (**26**) acted as a slightly more active catalyst for 1-hexene polymerization. However, the polymer obtained was atactic according to ^{13}C NMR, indicating that the C_2-symmetry of the octahedral zirconium Salan complexes must be supported by bulky aryl substituents at the ortho positions to efficiently direct the approach of the incoming olefin. Busico et al. employed complexes **24** and **26** for the polymerization of propylene, and for the block copolymerization of propylene and ethylene. ^{13}C NMR analysis of the polypropylene homopolymers formed using these catalysts indicated that the catalyst with bulky 2,4-*tert*-butyl phenolate substituents (**24**) produced isotactic polymer formed by an enantiomorphic site control mechanism with 1,2-insertions, whereas the catalyst with the less bulky 2,4-methyl phenolate substituents (**26**) formed a slightly syndiotactic polymer through a chain-end control mechanism with 1,2-insertions.[60,61] A theoretical study by Cavallo and coworkers correlated the structure and stereoselectivity of **24** (nicknamed "New Century Catalyst") to those of traditional heterogeneous Ziegler–Natta catalysts that produce isotactic polymers.[62] Both classes of catalysts polymerize α-olefins by a 1,2-insertion, and their stereoregularity is induced by an enantiomorphic site control mechanism. Transition state calculations indicated that the origin of stereoregularity in both systems is not through a direct interaction between the chiral site and the incoming monomer, but that instead, the chiral site orients the growing chain, and this orientation is responsible for the enantioface selection of the incoming olefin.

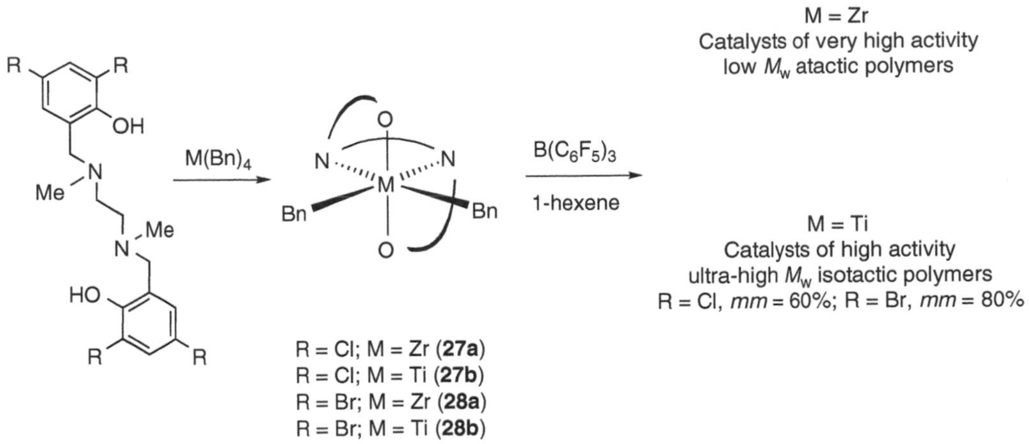

M = Zr
Catalysts of very high activity
low M_w atactic polymers

M = Ti
Catalysts of high activity
ultra-high M_w isotactic polymers
R = Cl, mm = 60%; R = Br, mm = 80%

R = Cl; M = Zr (**27a**)
R = Cl; M = Ti (**27b**)
R = Br; M = Zr (**28a**)
R = Br; M = Ti (**28b**)

SCHEME 13.4 Synthesis and polymerization activity of zirconium and titanium catalysts with electron-deficient Salan ligands.

The lower α-olefin polymerization activities of the C_2-symmetric Salan catalysts **24** and **25**, as compared to those exhibited by related the C_s-symmetrical amine bis(phenolate) zirconium catalysts (e.g., **21**), did not improve upon reducing the size of the Salan group phenolate substituent. This suggested a need to explore the influence of electronic effects.[63] Thus, diamine bis(phenolate) ligand precursors featuring electron-withdrawing *ortho, para*-dichloro-, or dibromo-substituted phenolate rings were synthesized by employing the Mannich synthesis route outlined in Scheme 13.2. The dibenzylzirconium complexes of these ligands, synthesized by reaction of the free ligand precursors with tetrabenzylzirconium (**27a** and **28a**, Scheme 13.4), functioned as highly active 1-hexene polymerization catalysts upon activation with $B(C_6F_5)_3$. The highest activity obtained with these complexes in neat 1-hexene was ca. 5000 g PH/mmol Zr·h. However, gel permeation chromatography (GPC) analysis indicated a low molecular weight for these polymers (M_w ca. 9000 g/mol); more importantly, ^{13}C NMR analysis revealed that they were atactic.

The synthesis of dibenzyltitanium complexes was then investigated, with the hope that the size of these phenolate substituents would be sufficient to induce tacticity with this smaller metal, while still retaining high polymerization activity. The dibenzyltitanium complex **27b** was obtained in high yield by reaction of the ligand precursor with tetrabenzyltitanium (Scheme 13.4). The ligand wrapped around the titanium center in the desired *fac-fac* mode, as evidenced by NMR analysis and by X-ray crystal structure of the analogous di(*n*-propoxytitanium) complex (**25**, Figure 13.8). Even though the dibenzyltitanium complex **27b** was found to be somewhat unstable in solution at room temperature, activation with $B(C_6F_5)_3$ in neat 1-hexene produced a highly active catalyst whose polymerization activity persisted for many hours, giving ultrahigh molecular weight polymers. Polymer samples with very narrow molecular weight distributions and very high-molecular weights could be obtained after a relatively short time (40 min; $M_w = 550,000$; $M_w/M_n = 1.2$). Allowing polymerization to proceed for longer periods resulted in broadening of the molecular weight distribution and the formation of even higher molecular weight polymers (19 h; $M_w = 1,900,000$; $M_w/M_n < 2.0$). Most significantly, these polymers were found to be isotactically enriched (*mm* ca. 60%). From this result, it was predicted that the titanium complex with the bulkier bromo substituents (**28b**) would exert an even higher stereocontrol over the incoming olefin in 1-hexene polymerization. This was indeed found to be the case: Isotactic poly(1-hexene) of *mm* ca. 80% was obtained using this catalyst, with the tacticity arising from an enantiomorphic site control mechanism (as revealed from pentad analysis of polymer ^{13}C NMR spectra). Interestingly, this catalyst was found to be somewhat more active than the previous one, and led to polymers of even higher molecular weights

(75 min polymerization time, $M_w = 1,750,000$, $M_w/M_n = 1.2$; and 18 h, $M_w = 4,000,000$, $M_w/M_n = 2.9$).

13.6 CONCLUSIONS

The search for new catalysts enabling the stereoregular polymerization of α-olefins has yielded several systems diverging from the classical metallocene motif in recent years. These systems afforded some unprecedented reactivities including the living and stereoregular polymerization. Further developments in catalyst design may lead the way to new polymeric materials. These catalysts may also find applications in other processes including asymmetric catalysis.

REFERENCES AND NOTES

1. Natta, G.; Pino, P.; Corradini, P.; Danusso, F.; Mantica, E.; Mazzanti, G.; Moraglio, G. Crystalline high polymers of α-olefins. *J. Am. Chem. Soc.* **1955**, *77*, 1708–1710.

2. Natta, G.; Pino, P.; Mazzanti, G.; Corradini, P.; Giannini, U. Synthesis and properties of highly polymerized crystals of some α-olefins having branched chains. *Rend. Accad. Nazl. Lencei.* **1955**, *19*, 397–403; *Chem. Abstr.* **1956**, *50*, 86332.

3. Natta, G. Stereospecific catalysis and isotactic polymers. *Chim. Ind.* **1956**, *38*, 751–765; *Chem. Abstr.* **1957**, *51*, 11063.

4. Natta, G.; Pino, P.; Mazzanti, G. Stereospecific polymerization of 1-olefins. I. *Gazz. Chim. Ital.* **1957**, *87*, 528–548; *Chem. Abstr.* **1957**, *51*, 88347.

5. Natta, G.; Danusso, F.; Sianesi, D. Stereospecific polymerization and isotactic polymers of vinyl aromatic monomers. *Macromol. Chem.* **1958**, *28*, 253–261.

6. Natta, G. Progress in the stereospecific polymerization. *Macromol. Chem.* **1960**, *35*, 94–131.

7. Natta, G. Kinetic studies of α-olefin polymerization. *J. Polym. Sci.* **1959**, *36*, 21–48.

8. Natta, G. Properties of isotactic, atactic and stereoblock homopolymers, random and block copolymers of α-olefins. *J. Polym. Sci.* **1959**, *36*, 531–549.

9. Montagnoli, G.; Pini, D.; Lucherini, A.; Ciardelli, F.; Pino, P. Quantitative aspects of the stereoselectivity in the stereoregulated polymerization of racemic α-olefins by Ziegler–Natta catalysts. *Macromolecules* **1969**, *2*, 684–686.

10. Ciardelli, F.; Altomare, A.; Carlini, C. Chiral discrimination in the polymerization of α-olefins by Ziegler–Natta initiator systems. *Prog. Polym. Sci.* **1991**, *16*, 259–277.

11. Busico, V.; Corradini, P.; De Biasio, R. 1-Butene polymerization in the presence of $MgCl_2$-supported Ziegler–Natta catalysts: Polymer microstructure in relation to polymerization mechanism. *Macromol. Chem.* **1992**, *193*, 897–907.

12. Kaminsky, W.; Külper, K.; Brintzinger, H. H.; Wild, F. R. W. P. Polymerization of propene and butene with a chiral zirconocene and methylalumoxane as cocatalyst. *Angew. Chem., Int. Ed. Engl.* **1985**, *24*, 507–508.

13. Ewen, J. A.; Jones, R. L.; Razavi, A. Syndiospecific propylene polymerization with group IV metallocene. *J. Am. Chem. Soc.* **1988**, *110*, 6255–6256.

14. Resconi, L.; Abis, L.; Franciscono, G. 1-Olefin polymerization at bis(pentamethylcyclopentadienyl) zirconium and –hafnium centers: Enantioface selectivity. *Macromolecules* **1992**, *25*, 6814–6817.

15. Soga, K.; Yanagihara, H. Extremely highly isospecific polymerization of olefins using Solvay-type $TiCl_3$ and Cp_2TiMe_2 as catalyst. *Macromol. Chem.* **1988**, *189*, 2839–2846.

16. Hoff, M.; Kaminsky, W. Syndiospecific homopolymerisation of higher 1-alkenes with two different bridged [$(RPh)_2C(Cp)(2,7\text{-}tert\text{-}BuFlu)$]$ZrCl_2$ catalysts. *Macromol. Chem. Phys.* **2004**, *205*, 1167–1173.

17. Grumel, V.; Brüll, R.; Pasch, H.; Raubenheimer, H. G.; Sanderson, R.; Wahner, U. M. Poly(pent-1-ene) synthesized with the syndiospecific catalyst $i\text{-}Pr(Cp)(9\text{-}Flu)ZrCl_2$/MAO. *Macromol. Mater. Eng.* **2002**, *287*, 559–564.

18. Coevoet, D.; Cramail, H.; Deffieux, A. Activation of *i*-Pr(CpFluo)ZrCl₂ by methylalumoxane, 3. Kinetic investigation of the syndiospecific hex-1-ene polymerization in hydrocarbon and chlorinated media. *Macromol. Chem. Phys.* **1999**, *200*, 1208–1214.

19. Coevoet, D.; Cramail, H.; Deffieux, A. Activation of *rac*-ethylenebis(indenyl)zirconium dichloride with a low amount of methylaluminoxane (MAO) for olefin polymerization. *Macromol. Chem. Phys.* **1996**, *197*, 855–867.

20. Yamaguchi, Y.; Suzuki, N.; Fries, A.; Mise, T.; Koshino, H.; Ikegami, Y.; Ohmori, H.; Matsumoto, A. Stereospecific polymerization of 1-hexene catalyzed by *ansa*-metallocene/methylaluminoxane systems under high pressures. *J. Polym. Sci., Part A: Polym Chem.* **1999**, *37*, 283–292.

21. Kim, I.; Zhou, J.-M.; Chung, H. Higher α-olefin polymerizations catalyzed by *rac*-Me₂Si(1-C₅H₂-2-CH₃-4-*t*Bu)₂Zr(NMe₂)₂/Al(*i*Bu)₃/[Ph₃C][B(C₆F₅)₄]. *J. Polym. Sci., Part A: Polym Chem.* **2000**, *38*, 1687–1697.

22. Rossi, A.; Odian, G.; Zhang, J. End groups in 1-butene polymerization via methylalumoxane and zirconocene catalyst. *Macromolecules* **1995**, *28*, 1739–1749.

23. Busico, V.; Cipullo, R.; Borriello, A. Regiospecificity of 1-butene polymerization catalyzed by C₂-symmetric group IV metallocenes. *Macromol. Rapid Commun.* **1995**, *16*, 269–274.

24. Borriello, A.; Busico, V.; Cipullo, R.; Fusco, O. Isotactic 1-butene polymerization promoted by C₂-symmetric metallocene catalysts. *Macromol. Chem. Phys.* **1997**, *198*, 1257–1270.

25. Babu, G. N.; Newmark, R. A.; Chien, J. C. W. Microstructure of poly(1-hexene) produced by *ansa*-zirconocenium catalysis. *Macromolecules* **1994**, *27*, 3383–3388.

26. Liu, Z.; Somsook, E.; White, C. B.; Rosaaen, K. A.; Landis, C. R. Kinetics of initiation, propagation, and termination for the [*rac*-(C₂H₄(1-indenyl)₂)ZrMe][MeB(C₆F₅)₃]-catalyzed polymerization of 1-hexene. *J. Am. Chem. Soc.* **2001**, *123*, 11193–11207.

27. Brüll, R.; Pasch, H.; Raubenheimer, H. G.; Sanderson, R.; Wahner, U. M. Polymerization of higher linear α-olefins with (CH₃)₂Si(2-methylbenz[e]indenyl)₂ZrCl₂. *J. Polym. Sci., Part A: Polym Chem.* **2000**, *38*, 2333–2339.

28. Brüll, R.; Kgosane, D.; Neveling, A.; Pasch, H.; Raubenheimer, H. G.; Sanderson, R., Wahner, U. M. Synthesis and properties of poly-1-olefins. *Macromol. Symp.* **2001**, *165*, 11–18.

29. Krentsel, B. A.; Kissin, Y. V.; Kleiner, V. J.; Stotskaya, L. L. *Polymers and Copolymers of Higher Alpha-Olefins: Chemistry, Technology, Applications*; Harsner/Gardner: Cincinnati, 1997.

30. Henschke, O.; Knorr, J.; Arnold, M. Poly-α-olefins from polypropene to poly-1-eicosene made with metallocene catalysts. *J. Macromol. Sci., Pure Appl. Chem.* **1998**, *A35*, 473–481.

31. Fukui, Y.; Murata, M.; Soga, K. Living polymerization of propylene and 1-hexene using bis-Cp type metallocene catalysts. *Macromol. Rapid Commun.* **1999**, *20*, 637–640.

32. Baar, C. R.; Levy, C. J.; Min, E. Y.-J.; Henling, L. M.; Day, M. W.; Bercaw, J. E. Kinetic resolution of chiral α-olefins using optically active *ansa*-zirconocene polymerization catalysts. *J. Am. Chem. Soc.* **2004**, *126*, 8216–8231.

33. van der Linden, A.; Schaverien, C. J.; Mijboom, N.; Ganter, C.; Orpen, A. G. Polymerization of α-olefins and butadiene and catalytic cyclotrimerization of 1-alkynes by a new class of group IV catalysts. Control of molecular weight and polymer microstructure via ligand tuning in sterically hindered chelating phenoxide titanium and zirconium species. *J. Am. Chem. Soc.* **1995**, *117*, 3008–3021.

34. Scollard, J. D.; McConville, D. H.; Payne, N. C.; Vittal, J. J. Polymerization of alpha-olefins by chelating diamide complexes of titanium. *Macromolecules* **1996**, *29*, 5241–5243.

35. Scollard, J. D.; McConville, D. H. Living polymerization of alpha-olefins by chelating diamide complexes of titanium. *J. Am. Chem. Soc.* **1996**, *118*, 10008–10009.

36. Jin, J.; Tsubaki, S.; Uozumi, T.; Sano, T.; Soga, K. Polymerization of propylene with the [ArN(CH₂)₃NAr]TiCl₂ (Ar = 2,6-*i*Pr₂C₆H₃) complex using R₃Al/Ph₃CB(C₆F₅)₄ (R = Me, Et, *i*Bu) as cocatalyst. *Macromol. Rapid Commun.* **1998**, *19*, 597–600.

37. Tsubaki, S.; Jin, J.; Ahn, C.-H.; Sano, T.; Uozumi, T.; Soga, K. Synthesis of isotactic poly(propylene) by titanium based catalysts containing diamide ligand. *Macromol. Chem. Phys.* **2001**, *202*, 482–487.

38. For a recent review on imine-phenolate and related catalysts, see: Mitani, M.; Saito, J.; Ishii, S.-I.; Nakayama, Y.; Makio, H.; Matsukawa, N.; Matsui, S. et al. FI catalysts: New olefin polymerization catalysts for the creation of value-added polymers. *Chem. Record* **2004**, *4*, 137–158.

39. For a recent review on phenoxy based catalysts, see: Suzuki, Y.; Terao, H.; Fujita, T. Recent advances in phenoxy-based catalysts for olefin polymerization. *Bull. Chem. Soc. Jpn.* **2003**, *76*, 1493–1517.

40. Saito, J.; Mitani, M.; Matsui, S.; Kashiwa, N.; Fujita, T. Polymerization of 1-hexene with bis[*N*-(3-*tert*-butylsalicylidene)phenylaminato]titanium(IV) dichloride using ^{i}Bu$_3$Al/Ph$_3$CB(C$_6$F$_5$)$_4$ as a cocatalyst. *Macromol. Rapid Commun.* **2000**, *21*, 1333–1336.

41. Jayaratne, K. C.; Sita, L. R. Stereospecific living Ziegler–Natta polymerization of 1-hexene. *J. Am. Chem. Soc.* **2000**, *122*, 958–959.

42. Jayaratne, K. C.; Keaton, R. J.; Henningsen, D. A.; Sita, L. R. Living Ziegler–Natta cyclo-polymerization of nonconjugated dienes: New classes of microphase-separated polyolefin block copolymers via a tandem polymerization/cyclopolymerization strategy. *J. Am. Chem. Soc.* **2000**, *122*, 10490–10491.

43. Keaton, R. J.; Jayaratne, K. C.; Henningsen, D. A.; Koterwas, L. A.; Sita, L. R. Dramatic enhancement of activities for living Ziegler–Natta polymerizations mediated by "exposed" zirconium acetamidinate initiators: The isospecific living polymerization of vinylcyclohexane. *J. Am. Chem. Soc.* **2001**, *123*, 6197–6198.

44. Keaton, R. J.; Jayaratne, K. C.; Fettinger, J. C.; Sita, L. R. Structural characterization of zir-conium cations derived from a living Ziegler–Natta polymerization system: New insights regarding propagation and termination pathways for homogeneous catalysts. *J. Am. Chem. Soc.* **2000**, *122*, 12909–12910.

45. Ward, B. D.; Bellemin-Laponnaz, S.; Gade, L. H. C_3 chirality in polymerization catalysis: A highly active dicationic scandium(III) catalyst for the isospecific polymerization of 1-hexene. *Angew. Chem., Int. Ed. Engl.* **2005**, *44*, 1668–1671.

46. Tshuva, E. Y.; Versano, M.; Goldberg, I.; Kol, M.; Weitman, H.; Goldschmidt, Z. Titanium complexes of chelating dianionic amine bis(phenolate) ligands: An extra donor makes a big difference. *Inorg. Chem. Commun.* **1999**, *2*, 371–373.

47. Tshuva, E. Y.; Goldberg, I.; Kol, M. Goldschmidt, Z. Coordination chemistry of titanium amine bis(phenolate) complexes: Tuning complex type and structure by ligand modification. *Inorg. Chem.* **2001**, *40*, 4263–4270.

48. Tshuva, E. Y.; Goldberg, I.; Kol, M.; Weitman, H.; Goldschmidt, Z. Novel zirconium complexes of amine bis(phenolate) ligands. Remarkable reactivity in polymerization of 1-hexene due to an extra donor arm. *Chem. Commun.* **2000**, 379–380.

49. Tshuva, E. Y.; Goldberg, I.; Kol, M.; Goldschmidt, Z. Zirconium complexes of amine bis(phenolate) ligands as catalysts for 1-hexene polymerization: Peripheral structural parameters strongly affect reactivity. *Organometallics* **2001**, *20*, 3017–3028.

50. Tshuva, E. Y.; Groysman, S.; Goldberg, I.; Kol, M.; Goldschmidt, Z. [ONXO]-type amine bis(phenolate) zirconium and hafnium complexes as extremely reactive 1-hexene polymerization catalysts: Unusual metal dependent activity pattern. *Organometallics* **2002**, *21*, 662–670.

51. Groysman, S.; Goldberg, I.; Kol, M.; Genizi, E.; Goldschmidt, Z. Group IV complexes of an amine bis(phenolate) ligand featuring a THF sidearm donor: From highly active to living polymerization catalysts of 1-hexene. *Inorg. Chim. Acta* **2003**, *345*, 137–144.

52. Groysman, S.; Goldberg, I.; Kol, M.; Genizi, E.; Goldschmidt, Z. (2003). From THF to furan: Sidearm donor exchange causes a major difference in activity of group IV amine bis(phenolate) polymerization catalysts. *Organometallics* **2003**, *22*, 3013–3015.

53. Tshuva, E. Y.; Goldberg, I.; Kol, M.; Goldschmidt, Z. Living polymerization of 1-hexene due to an extra donor arm on a novel amine bis(phenolate) titanium catalyst. *Inorg. Chem. Commun.* **2000**, *3*, 611–614.

54. Tshuva, E. Y.; Goldberg, I.; Kol, M.; Goldschmidt, Z. Living polymerization and block copoly-merization of α-olefins by an amine bis(phenolate) titanium catalyst. *Chem. Commun.* **2001**, 2120–2121.

55. Groysman, S.; Tshuva, E. Y.; Goldberg, I.; Kol, M.; Goldschmidt, Z.; Shuster, M. Diverse structure-activity trends in amine bis(phenolate) titanium polymerization catalysts. *Organometallics* **2004**, *23*, 5324–5331.

56. For the synthesis of high molecular weight atactic polypropylene with these catalysts, see: Groysman, S.; Tshuva, E. Y.; Reshef, D.; Gendler, S.; Goldberg, I.; Kol, M.; Goldschmidt, Z.; Shuster, M.;

Lidor, G. High-molecular weight atactic polypropylene prepared by zirconium complexes of an amine bis(phenolate) ligand. *Isr. J. Chem.* **2002**, *42*, 373–381.

57. Tshuva, E. Y.; Goldberg, I.; Kol, M. Isospecific living polymerization of 1-hexene by a readily available non-metallocene C_2-symmetrical zirconium catalyst. *J. Am. Chem. Soc.* **2000**, *122*, 10706–10707.

58. Tshuva, E. Y.; Gendeziuk, N.; Kol, M. One-step synthesis of Salans and substituted Salans by Mannich condensation. *Tetrahedron Lett.* **2001**, *42*, 6405–6407.

59. A mixture of stereoisomers may form when the labile groups are halo substituents. Segal, S.; Kol, M. *Unpublished results*, 2003.

60. Busico, V.; Cipullo, R.; Ronca, S.; Budzelaar, P. H. M. Mimicking Ziegler–Natta catalysts in homogeneous phase, 1: C_2-symmetric octahedral Zr(IV) complexes with tetradentate [ONNO]-type ligand. *Macromol. Rapid Commun.* **2001**, *22*, 1405–1410.

61. Busico, V.; Cipullo, R.; Friedrichs, N.; Ronca, S.; Togrou, M. The first molecularly characterized isotactic polypropylene-*block*-polyethylene obtained via "quasi-living" insertion polymerization. *Macromolecules* **2003**, *36*, 3806–3808.

62. Corradini, P.; Guerra, G.; Cavallo, L. Do new century catalysts unravel the mechanism of stereocontrol of old Ziegler–Natta catalysts? *Acc. Chem. Res.* **2004**, *37*, 231–241.

63. Segal, S.; Goldberg, I.; Kol, M. Zirconium and titanium diamine bis(phenolate) catalysts for α-olefin polymerization: From atactic oligomers to ultra-high molecular-weight isotactic polymers. *Organometallics* **2005**, *24*, 200–202.

14 Tactic Polystyrene and Styrene Copolymers

Haiyan Ma and Jiling Huang

CONTENTS

14.1 INTRODUCTION

Atactic polystyrene, as an amorphous material, has been known for centuries. In principle, polystyrene can occur with atactic, isotactic, and syndiotactic configurations. With the extensive studies concerning the stereoselective polymerization of olefins by Ziegler–Natta catalyst systems discovered in the early 1950s, the first stereoregular polystyrene, isotactic polystyrene (iPS), was obtained by Natta using the TiCl$_4$/AlEt$_3$ system in 1955.[1] iPS is a semicrystalline polymer that has a high melting point, $T_m = 240\,°C$.[2] Several companies have tried to commercialize iPS, but its crystallization rate is too slow to be practical in industrial processes.

It was not until 1985, decades after the discovery of isotactic polystyrene, that the initial synthesis of syndiotactic polystyrene (sPS) was reported by Ishihara et al.[3,4] This discovery was enabled, to a large degree, by the introduction of methylaluminoxane (MAO) as a cocatalyst for metallocene-mediated olefin polymerizations by Sinn and Kaminsky at the end of 1970s.[5] Since then, syndiotactic polystyrene has been the subject of intense investigation[6,7] because of its useful properties, which include a high melting point ($T_m = 270\,°C$), a high rate of crystallization, and a high modulus of elasticity. This material also exhibits a low specific gravity and a low dielectric constant, in addition to excellent resistance to water and organic solvents at ambient temperature, thus making sPS a promising material for a large number of applications in many market areas.

Owing to the distinct stereochemical structures, atactic, isotactic, and syndiotactic polystyrene can be distinguished easily from one another by solubility and spectroscopic methods.[3] As shown in Figure 14.1, the ^{13}C NMR spectrum (1,2,4-trichlorobenzene, 130 °C) shows five main peaks in the range of 145.12–146.7 ppm for the phenyl C-1 carbon of atactic polystyrene, corresponding to the various configurational sequences. In contrast, a single and sharp peak is displayed for isotactic polystyrene as well as for syndiotactic polystyrene, because of their highly regular stereochemical structures, albeit distinct chemical shifts (iPS, 146.24 ppm; sPS, 145.13 ppm). Specific characters are also observed in the ^{1}H NMR spectra of these three types of polymers (Figure 14.2).[3]

In this chapter, recent developments and the most important aspects in the stereoregulating catalysis of styrene polymerizations will be reviewed.

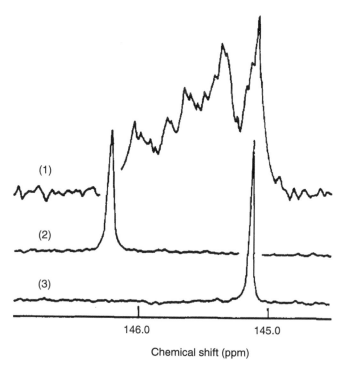

Chemical shift (ppm)

FIGURE 14.1 ^{13}C NMR spectra (1,2,4-trichlorobenzene, 130 °C, 67.8 MHz) of phenyl C-1 carbon of polystyrenes: (1) atactic; (2) isotactic; (3) syndiotactic. (Reprinted with permission from Ishihara, N.; Seimiya, T.; Kuramoto, M.; Uoi, M. *Macromolecules* **1986**, *19*, 2464–2465. Copyright 1986 American Chemical Society.)

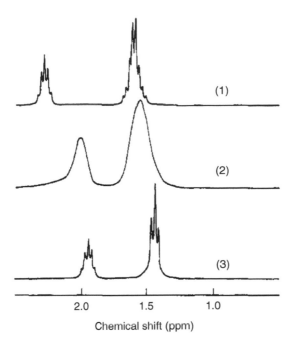

FIGURE 14.2 ^{1}H NMR spectra (1,2,4-trichlorobenzene, 130 °C, 270.1 MHz) of the methine and methylene protons of polystyrenes (1) isotactic; (2) atactic; (3) syndiotactic. (Reprinted with permission from Ishihara, N.; Seimiya, T.; Kuramoto, M.; Uoi, M. *Macromolecules* **1986**, *19*, 2464–2465. Copyright 1986 American Chemical Society.)

14.2 SYNDIOTACTIC POLYSTYRENE

14.2.1 TRANSITION METAL COMPLEXES FOR THE SYNDIOSELECTIVE POLYMERIZATION OF STYRENE

By using homogeneous catalyst systems based on titanium complexes and MAO, Ishihara et al. obtained sPS with an *rrrr* pentad content greater than 98%.[3,4] Based on this pioneering work, a large number of transition metal (mainly titanium) complexes have been synthesized and examined for styrene polymerization, when used in combination with a cocatalyst such as MAO or a tris(pentafluorophenyl)borane derivative.[6,7] Both the catalyst activity and the syndioselectivity are highly dependant on the choice of transition metal complex. Some typical styrene polymerizations using different transition complexes with MAO as the activator are summarized in Table 14.1. From the data, titanium complexes are much preferred as precatalysts for syndioselective styrene polymerization, which may be because of the relatively smaller titanium ionic radius and higher nucleophilicity with respect to some other transition metal compounds. Among the classes of titanium complexes, mono(cyclopentadienyl)titanium complexes have the highest activities and produce polymer with much higher syndiotacticity. Titanium complexes in oxidation states I to IV, but especially III, can behave as catalysts or catalyst precursors.[4,8,9–12]

14.2.1.1 Group 4 Metal Complexes of the Form MX$_n$ (X = Alkyl, Alkoxy, Halide; n = 2–4)

Among group 4 metal complexes of formula MX_n, tetrabenzyltitanium activated with MAO is the most active and syndiospecific catalyst for styrene polymerization ([r] = ~1, sPS% = 93% where sPS% is the percentage of acetone or 2-butanone insoluble fraction in the obtained polymer). The

TABLE 14.1
Polymerization of Styrene in the Presence of Different Transition Metal Complexes Using MAO as the Cocatalyst[a]

Entry #	Precatalyst[b]	S/M[c] (molar)	Al/M[d] (molar)	T_p[e] (°C)	Conv.[f] (wt%)	Activity[g]	M_w	M_w/M_n	sPS[h] (%)	Tacticity[i]	References
1	TiCl$_4$	4,000	800	50	4.1	43				Syndiotactic	4
2	TiBr$_4$	4,000	800	50	2.1	22				Syndiotactic	4
3	Ti(OMe)$_4$	4,000	800	50	3.8	40				Syndiotactic	4
4	Ti(OEt)$_4$	4,000	800	50	9.5	99				Syndiotactic	4
5	Ti(O-n-Bu)$_4$	5,219	100	50	4.2	105	344,000	11.8	91		9
6	Ti(O-n-Bu)$_4$	3,480	1,000	90	4.6	290	49,000	4.21	55	[r] ≈ 1	11
7	Ti(Bn)$_4$	5,219	100	50	6.0	151	153,000	4.6	93		9
8	Ti(Bn)$_4$	3,480	1,000	90	13.9	1,443	60,400	2.54	93	[r] ≈ 1	11
9	Ti(acac)$_2$Cl$_2$	4,000	800	50	0.4	4				Syndiotactic	4
10	Ti(acac)$_3$	5,219	100	50	2.2	55	327,000	13.3	89		9
11	Ti(Ph)$_2$	5,219	100	50	0.8	20	370,000	9.6	88		9
12	Ti(bipy)$_3$	5,219	100	50	Trace					Atactic	9
13	Cp$_2$TiCl$_2$	4,000	600	50	1.0	10				Syndiotactic	4
14	Me$_2$Si(Cp)$_2$TiCl$_2$	3,480	1,000	90	4.1	259	16,400	2.77	75	[r] = 0.94	11
15	CpTiCl$_3$	12,000	600	50	43.9	3,654	54,040	1.93		Syndiotactic	4
16	CpTiCl$_3$	3,480	1,000	90	87.5	45,552	14,700	1.78	89	[r] = 0.94	11
17	Cp*TiCl$_3$	3,480	1,000	90	~100	79,860	198,000	2.34	~100	[r] ≈ 1	11
18	ZrCl$_4$	31,400	800	50	0.4	4				Atactic	4
19	Zr(Bn)$_4$	5,219	100	50	1.8	46	9,000	2.2	56		9
20	Zr(Bn)$_4$	4,702	140	80	13.6	193				[r] = 0.93	11

	Precatalyst[b]	[c]	[d]	[e]	[f]	[g]	[h]	[i]	
21	Cp_2ZrCl_2	31,400	800	50	1.3	13		Atactic	4
22	$Me_2Si(Cp)_2ZrCl_2$	3,480	1,000	50	11.3	59	≈ 0	Atactic	11
23	Cp_2HfCl_2	31,400	800	50	0.7	7		Atactic	4
24	Cp_2VCl_2	31,400	800	50	0.7	7		Atactic	4
25	$V(acac)_2$	17,200	400	50	0.4	8		Atactic	4
26	$Cr(acac)_3$	17,200	400	50	1.4	29		Atactic	4
27	$Co(acac)_3$	17,200	400	50	1.8	37		Atactic	4
28	$Ni(acac)_2$	17,200	400	50	80.8	1,681		Atactic	4

[a] In toluene.
[b] Precatalyst ligand abbreviations: Bn = benzyl; acac = acetylacetonato; bipy = 2,2′-bipyridine; Cp = cyclopentadienyl; Cp* = pentamethylcyclopentadienyl.
[c] Styrene/metal molar ratio.
[d] Aluminum/metal molar ratio.
[e] Polymerization temperature.
[f] wt% styrene monomer converted.
[g] kg PS/(mol M·mol S·h).
[h] wt% of acetone insoluble fraction.
[i] Determined by ^{13}C NMR.

zirconium analogue also shows high syndioselectivity but with less activity for styrene polymerization ([r] = 0.93).[4,9–11] In the presence of MAO, titanium halides, for example, TiCl$_4$ and TiBr$_4$, show high syndioselectivity for the polymerization of styrene, whereas ZrCl$_4$ only gives atactic polystyrene with low activity.[4] Recently, it was reported that in contrast to ZrCl$_4$/MAO, the ZrCl$_4$(THF)$_2$/MAO (THF = tetrahydrofuran) system polymerizes styrene syndioselectively but with low stereoregularity in the polymer chain ([rmr]/[rrr]/[rrm] = 4:82:14).[13] The author suggested that the presence of an organic Lewis base, such as THF, enhanced the syndioselectivity of the catalyst system toward the polymerization of styrene.

The catalyst systems consisting of MAO and titanium alkoxides, such as Ti(OMe)$_4$, Ti(OEt)$_4$, and Ti(O-n-Bu)$_4$, catalyze syndioselective styrene polymerization with moderate activities,[4,9] but are less active than mono(cyclopentadienyl)titanium-based systems. Nevertheless, higher molecular weight syndiotactic polystyrenes are obtained from these catalyst systems (Table 14.1, entries 3–6).

Apart from Ti(IV), titanium complexes of other oxidation states (Ti(III) and Ti(II)), such as Ti(acac)$_3$ (acac = acetylacetonate) and TiPh$_2$, also show syndioselective polymerization activity and produce syndiotactic polystyrene. As for the Ti(0) complex, Ti(bipy)$_3$ (bipy = 2,2′-bipyridine), only atactic polymer is produced (Table 14.1, entries 10–12).[9]

14.2.1.2 Bis(cyclopentadienyl) Complexes of Group 4 Metals

Bis(cyclopentadienyl) titanium complexes activated with MAO can initiate syndioselective styrene polymerization, but with quite low activities as compared to the corresponding monocyclopentadienyl complexes (Table 14.1, entries 13–17).[4,11] This lower activity might be explained by the presence of two cyclopentadienyl rings at the active site of [Cp$_2$TiP]$^+$ (Cp = cyclopentadienyl; P = the polymer chain) versus one for [CpTiP]$^+$, where in the former, the insertion and/or coordination of the monomer may be more sterically hindered and less electronically favored.[4] The polymerization activities of several $ansa$-titanocene complexes (Figure 14.3, **1–6**) as well as their syndioselectivities decrease with increasing bite-angle (Cp$_{centroid}$–Ti–Cp$_{centroid}$ angle).[6,14] The order of decreasing activity for these complexes is **1 > 2 > 3 > 4 > 5 > 6**. Bis(cyclopentadienyl) zirconium and hafnium complexes show no stereoselectivity and quite low activity for styrene polymerization, only producing, in some cases, small amounts of atactic polymers (Table 14.1, entries 21–23).[4,11,15]

14.2.1.3 Mono-Cyclopentadienyl Complexes of Group 4 Metals

Titanium half-sandwich complexes of cyclopentadiene and substituted cyclopentadienes (Cp′) activated with MAO or organic boron compounds especially show a high styrene polymerization activity,

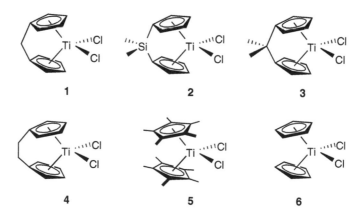

FIGURE 14.3 Structures of titanocene complexes **1–6**.

as well as a high proportion of syndiotactic content in the resulting polymers. The structure of the Cp′ ligand has been widely varied in order to find catalysts with higher activity and syndioselectivity. Pentamethylcyclopentadienyl (Cp*) is one of the most investigated ligands. Under the same experimental conditions, Cp*TiCl$_3$ and its derivatives obtained from substitutions of Cl show higher activities, and produce polystyrenes with higher syndiotacticity and much higher polymer molecular weights as compared to Cp-titanium analogues (Table 14.1, entries 16–17).[4,11,16] It is believed that the addition of electron-donating methyl groups to the Cp ring better stabilizes the electrophilic titanium center; as a result, the concentration of active species present increases that leads to higher catalyst activity. Furthermore, chain transfer through β-H elimination is likely to be suppressed because of the stronger electron-donating ability of the Cp* ligand as compared to Cp, thus leading to higher molecular weight sPS.

Styrene polymerization activities of some mono-cyclopentadienyl titanium complexes increase as the number of electron-releasing substituents on the Cp ring increases: CpTi(OMe)$_3$ < [(Me$_3$Si)$_2$Cp]-Ti(OMe)$_3$ < (Me$_4$Cp)Ti(OMe)$_3$ < [(Me$_3$Si)Me$_4$Cp]Ti(OMe)$_3$ < Cp*Ti(OMe)$_3$ < (EtMe$_4$Cp)-Ti(OMe)$_3$ (10–210 kg sPS/g Ti).[14] An increase in polymer molecular weight and syndioselectivity obtained with these complexes was observed in the same order.

Chien and coworkers observed that the styrene polymerization activities of a set of MAO activated mono-cyclopentadienyl Ti catalysts (Figure 14.4, **7–10**) as well as their syndioselectivity, decreased with increasing steric bulk on the Cp-ring in the following order: **7 > 8 > 10** (sPS% for complexes **7, 8, 10** are 99%, 98%, 75% respectively).[17] In complex **9**, the introduction of diphenylphosphino group caused a drastic decrease of catalyst activity.

The introduction of cyclopentadienyl ligands with a heteroatom-containing substituent capable of bonding with the metal is exemplified by complexes **11–17** (Figure 14.5). When activated with MAO, these complexes form catalyst species with rather low activities and decreased syndioselectivities compared to those without heteroatom substituents.[18–20] The catalyst activities of complexes **11** and

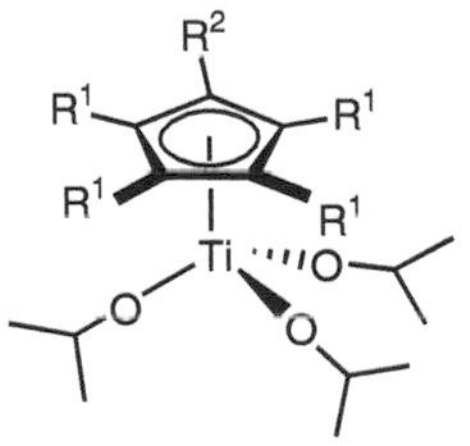

7: R^1 = Me, R^2 = H
8: R^1 = Me, R^2 = SiMe$_3$
9: R^1 = Me, R^2 = PPh$_2$
10: R^1 = Ph, R^2 = H

FIGURE 14.4 Structures of mono-Cp titanium complexes **7–10** with different Cp ring substituents.

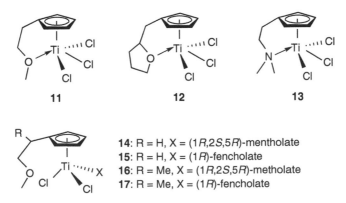

11   **12**   **13**

14: R = H, X = (1*R*,2*S*,5*R*)-mentholate
15: R = H, X = (1*R*)-fencholate
16: R = Me, X = (1*R*,2*S*,5*R*)-metholate
17: R = Me, X = (1*R*)-fencholate

FIGURE 14.5 Structures of mono-Cp titanium complexes **11–17** with a heteroatom-containing substituent on the Cp ring.

12 in the presence of MAO are two orders of magnitude lower as compared to CpTiCl$_3$; the syndiose-lectivities are also as low as 62–75%.[18] It is believed that the pendant heteroatom intramolecularly coordinates to the Lewis acidic metal center reducing its electrophilicity, and sterically hinders the coordination and insertion of the styrene molecule. Nevertheless, an interaction between the heteroatom and the MAO cocatalyst is also possible. The low activity of [(Me$_3$SiO)Me$_4$Cp]TiCl$_3$ most likely arises from such an interaction since the oxygen atom (that is directly bonded to the cyclopentadienyl ring in this complex) is unlikely to act as a unimolecular donor atom to the metal center,[21] although a bimolecular interaction of one CpOR ligand to a different metal center can not be excluded. Not all heteroatom-containing substituents will lead to a dramatic drop in polymerization activity; the activity of complex **18** (Figure 14.6) is still comparable to that of (Me$_3$SiCp)TiCl$_3$.[22] Complex **19**, with a chloride atom in the side arm, shows higher activity and improved syndioselectivity compared to (Me$_3$SiCp)TiCl$_3$. The introduction of an alkenyl substituent into the side arm as a weak donor slightly influences the polymerization behavior of the resulting catalysts (Figure 14.7, **20–23**).[23] In comparison with their alkyl analogues, the styrene polymeriz-ation activities of these complexes increase in the order **20** < **21** < **23** < CpTiCl$_3$ < **22**. Therein, a weak intramolecular coordination of the unsaturated moiety to the titanium center during poly-merization has been proposed to explain the low activity of complex **20**. Whereas in complex **22**, it is thought that the C=C bond of the butenyl group might not coordinate to the active center or coordinate much more weakly to the active center owing to the longer arm, thus leading to higher catalyst activity.

The replacement of a cyclopentadienyl ligand of a mono-Cp Ti complex with an indenyl ligand brings a 50–100% increase in the polymerization activity.[24] The indenyl titanium complex/MAO systems are much less sensitive to polymerization conditions, showing high syndioselectivity for styrene polymerization even at 90 °C. The enhancement in activity and syndioselectivity may be attributed to the greater electron-donating ability of the indenyl ring relative to the Cp moiety. The influence of the indenyl ring substituents on the polymerization has been investigated extens-ively (Figures 14.8 and 14.9, **24–47**).[25–29] The addition of a small substituent, such as methyl or phenyl, at the C(2)-position of the indenyl ring (**27, 28**) significantly increases the activity and syndioselectivity.[25,26,29] When a fused ring was introduced onto the indenyl ligand, as in complex **42**, styrene (S) polymerization activities as high as 1.8×10^8 g PS/(mol Ti·mol S·h) were reported.[25]

18   **19**

FIGURE 14.6 Structures of mono-Cp titanium complexes **18, 19** with a heteroatom-containing substituent on the Cp ring.

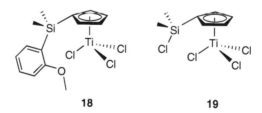

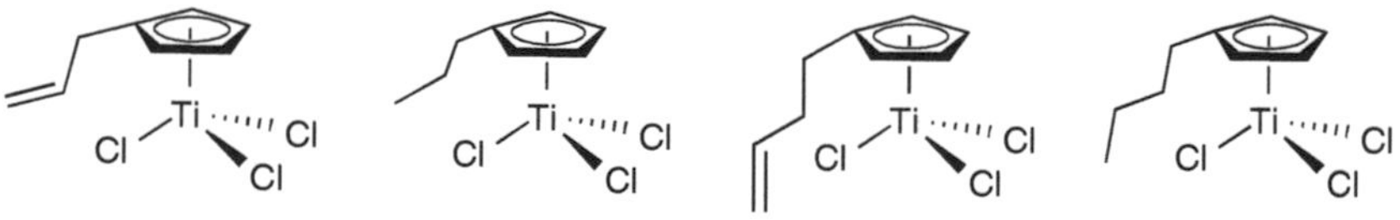

20   **21**   **22**   **23**

FIGURE 14.7 Structures of mono-Cp titanium complexes **20–23** with an alkyl or alkenyl substituent on the Cp ring.

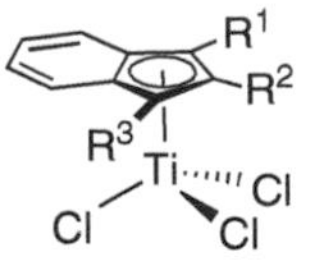

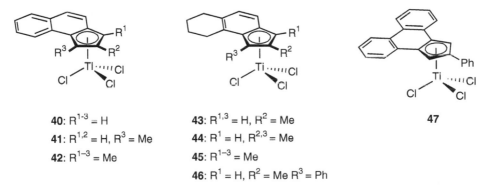

24: R^1 = Me, R^{2,3} = H
25: R^1 = Ph, R^{2,3} = H
26: R^1 = (2-Naph), R^{2,3} = H
27: R^{1,3} = H, R^2 = Me
28: R^{1,3} = H, R^2 = Ph
29: R^{1,3} = H, R^2 = (2-Naph)
30: R^{1,3} = Ph, R^2 = H
31: R^1 = CH$_2$Ph, R^{2,3} = H
32: R^1 = CH$_2$CH$_2$Ph, R^{2,3} = H

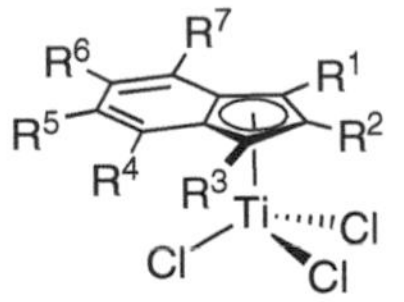

33: R^{1,2} = Me, R$^{3-7}$ = H
34: R^{1,3} = Me, R$^{2,4-7}$ = H
35: R$^{1-3}$ = Me, R$^{4-7}$ = H
36: R^{1,3,4,7} = Me, R^{2,5,6} = H
37: R$^{1-3}$ = H, R$^{4-7}$ = Me
38: R$^{1,2,4-7}$ = Me, R^3 = H
39: R$^{1-7}$ = Me

(Naph = naphthyl)

FIGURE 14.8 Structures of indenyl titanium trichlorides **24–39** with different Ind ring substituents.

40: R$^{1-3}$ = H
41: R^{1,2} = H, R^3 = Me
42: R$^{1-3}$ = Me

43: R^{1,3} = H, R^2 = Me
44: R^1 = H, R^{2,3} = Me
45: R$^{1-3}$ = Me
46: R^1 = H, R^2 = Me R^3 = Ph

47

FIGURE 14.9 Structures of indenyl titanium trichlorides **40–47** with fused rings on the π-ligand.

Complex **47**, having a second fused ring, showed an extremely high catalyst activity of 7.5 × 10^8 g PS/(mol Ti·mol S·h).[30]

The variation of the Ti complex σ-ligands rather than the Cp-substitution pattern, can also influence the polymerization behavior of the catalyst. Kaminsky et al. found[31] that at the low MAO to complex molar ratio (Al:Ti = 300), the polymerization activity of the fluorinated half-sandwich titanium complex Cp′TiF$_3$ (Cp′ = Cp, Cp*, MeCp, EtMe$_4$Cp, n-PrMe$_4$Cp, n-BuMe$_4$Cp, etc.) was about 30–40 times higher than that of the analogous chlorinated system. The highest activity, up to 14,000 kg sPS/(mol Ti·h), was obtained by (MeCp)TiF$_3$ in combination with MAO.[31] The molecular weight, syndiotacticity, and the melting point (T_m = 277 °C) of the polymers produced by the fluorinated complexes were also significantly higher. The same trend was also observed for indenyl titanium complexes.[32] The higher activity and syndioselectivity of the fluoride catalysts as compared to their chloride analogues are attributed to a greater number of more stable Ti(III) active sites, and a higher propagation rate constant for polymerization.[31,32] Qian et al. reported that the treatment of Cp′Ti(OMe)$_3$ with BF$_3$·Et$_2$O resulted in titanium complexes with both fluoride and methoxide σ-ligands, that is, Cp′TiF$_2$(OMe).[33,34] When activated with MAO, these complexes showed an increased styrene polymerization activity as compared to those for Cp′TiCl$_2$(OMe) and CpTiCl$_3$, but were less active than Cp′Ti(OMe)$_3$.

The most commonly applied variation of the σ-ligands in mono-Cp Ti styrene polymerization catalysts is the introduction of alkoxy/aryloxy groups,[7,35,36] which generally leads to catalysts with increased activity and syndioselectivity. Campbell et al.[16] determined the order of deceasing

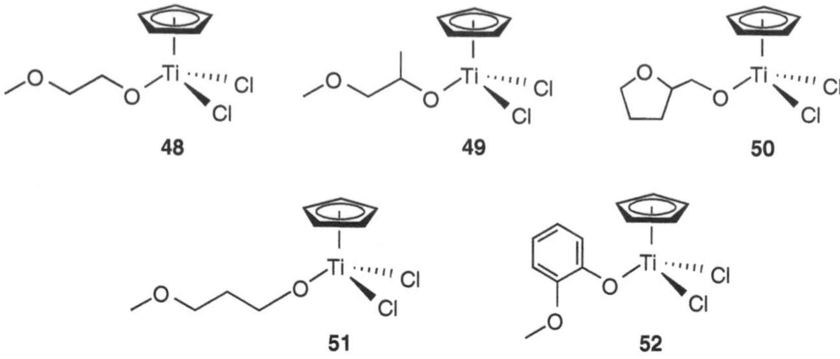

FIGURE 14.10 Structures of mono-Cp titanium complexes **48–52** with an alkoxy σ-ligand.

catalyst activity with various mono-cyclopentadienyl titanium complexes of this type, and found the following trends: CpTi(OEt)$_3$ > CpTi(O-i-Pr)$_3$ > CpTi(OPh)$_3$ ≫ CpTiCl$_3$. Usually, the comparative differences in polymerization activity and syndioselectivity are much larger between the chlorides and the alkoxides than between complexes bearing different alkoxide groups.[7]

MAO activated mono-cyclopentadienyl titanium complexes containing one alkoxide σ-ligand appear to have a superior catalyst performance as compared to the analogous trichlorides, Cp'TiCl$_3$.[18,19,33,37–39] CpTiCl$_2$(OR)/MAO systems have polymerization activities that are double or more than that of CpTiCl$_3$, and exhibit greater syndioselectivities.[37,38] It was reported by Qian and coworkers[19,37–39] that the polymerization activities of a series of titanium complexes with various monoalkoxy group σ-ligands decrease in the order: CpTiCl$_2$(OMe) > CpTiCl$_2$(OEt) > CpTiCl$_2$(O-i-Pr) > CpTiCl$_3$ ≈ CpTiCl$_2$[O-(1R,2S,5R)-menthyl] > CpTiCl$_2$(O-$cyclo$-C$_6$H$_{11}$) > CpTiCl$_2$(O-i-Bu) > CpTiCl$_2$(OBn) (Bn = benzyl) > CpTiCl$_2$(O-n-Bu). For indenyl analogues, the activities decrease as follows: IndTiCl$_2$(OEt) > IndTiCl$_2$(O-i-Pr) ≈ IndTiCl$_2$(OMe) > IndTiCl$_3$ (Ind = indenyl).[40,41] Obviously, both the steric and electronic effects of the alkoxy groups might influence the activities and the syndioselectivities of the catalysts. At a low MAO to titanium complex molar ratio (Al:Ti = 300), the introduction of the alkoxy group significantly promotes the syndioselective polymerization of styrene, affording a higher proportion of syndiotactic polystyrene in the obtained polymer as compared to the polymer from the corresponding trichlorides (sPS% = 70–96% versus 60% for CpTiCl$_3$, 65% for IndTiCl$_3$).[40,42] The presence of an additional oxygen atom in the alkoxy group can further improve the thermal stability of the active catalyst species. The maximum activities for the MAO activated complexes **48–52** (Figure 14.10) were reached at 70 °C, as compared to 50 °C for CpTiCl$_3$ and CpTiCl$_2$(OR) (R = alkoxyl group with one oxygen).[42]

When using boron-based activators, the metallocene chlorides are typically replaced by alkyl groups, such as methyl, benzyl, or trimethylsilylmethyl.[43–46] Cp*TiMe$_3$, Cp*TiBn$_3$, or Cp*Ti(CH$_2$SiMe$_3$)$_3$ when mixed with a boron compound such as B(C$_6$F$_5$)$_3$, [Ph$_3$C]$^+$[B(C$_6$F$_5$)$_4$]$^-$, [Et$_3$NH]$^+$[B(C$_6$F$_5$)$_4$]$^-$ (C$_6$F$_5$ = pentafluorophenyl) in a 1:1 ratio, affords catalytic styrene polymerization systems with high activity and syndioselectivity.[43,44]

14.2.1.4 Group 4 Metal Complexes with Non-Cyclopentadienyl Polydentate Ligands

Titanium complexes with acetylacetonate (acac) ligands were first tested for the polymerization of styrene by Ishihara et al.[4] and Zambelli et al.[9] When activated with MAO, Ti(acac)$_2$Cl$_2$ and Ti(acac)$_3$ can produce syndiotactic polystyrene with low activities, whereas their zirconium analogues only afford atactic polystyrene.[4,9] Hu and coworkers studied a series of MAO activated 1,3-diketonato titanium complexes (Figure 14.11, **53–57**) and found that their activities increased in the following

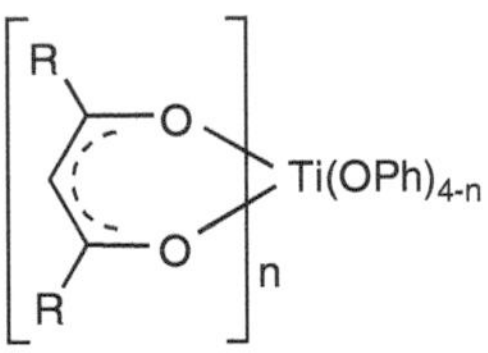

53: R = Me, n = 1
54: R = Me, n = 3
55: R = Ph, n = 1
56: R = Ph, n = 2
57: R = Ph, n = 3

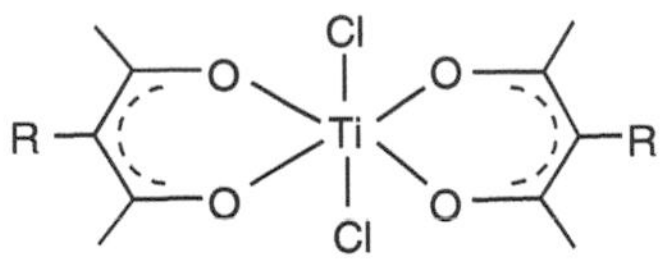

58: R = *n*-Bu
59: R = 1-Ind
60: R = CH_2Ph

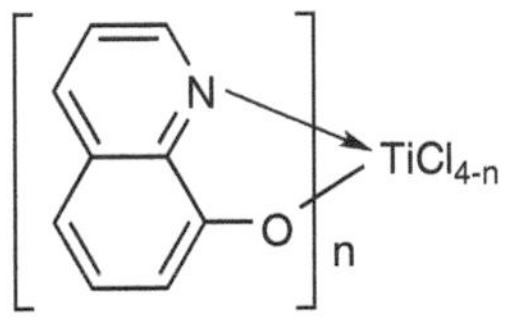

61: n = 1
62: n = 2
63: n = 3

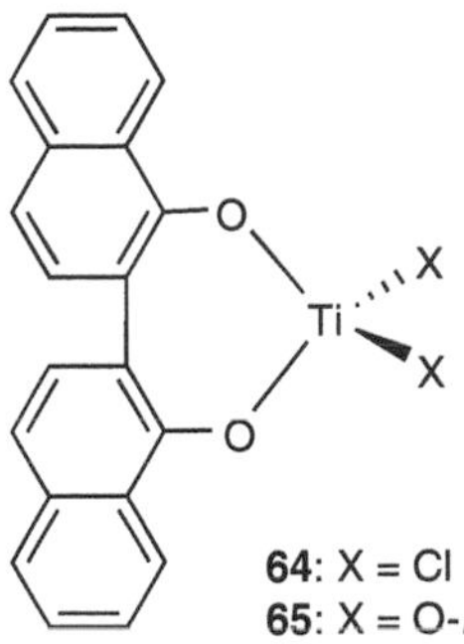

64: X = Cl
65: X = O-*n*-Bu

FIGURE 14.11 Structures of titanium complexes **53–65** with non-Cp polydentate ligands.

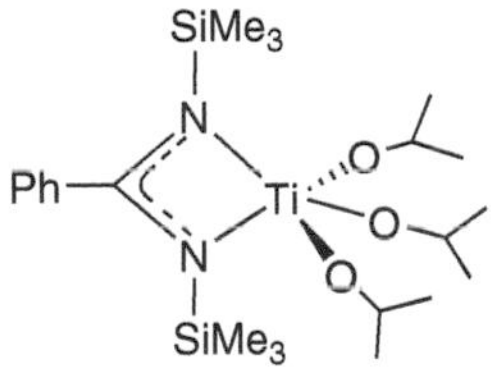

66

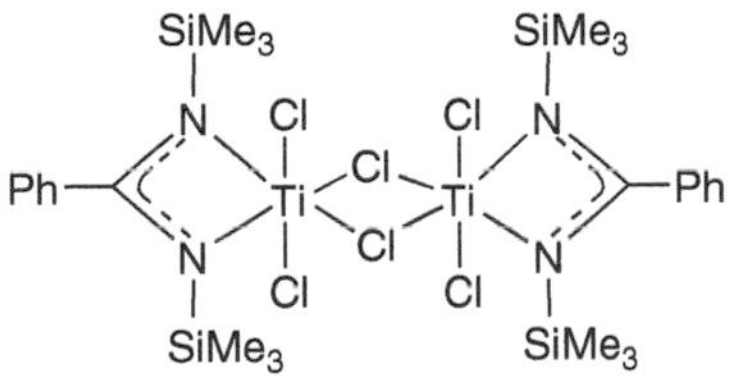

67

FIGURE 14.12 Structures of titanium complexes **66**, **67** with bidentate *N*, *N'*-bis(trimethylsilyl) benzamidate ligands.

order: **53** < **54** < **55** < **56** < **57** (0.43–6.15 × 10^5 g PS/(mol Ti·h)).[47,48] The corresponding series where the phenoxide ligand(s) were replaced by chloride ligand(s) showed lower overall activities that increased in the same order. Substitution at the 2-position increases the activities of the resulting MAO activated complexes (Figure 14.11, **58–60**) by ten times compared to that of the parent unsubstituted complex.[49] Several MAO activated titanium complexes with 8-hydroxyquinolinato (Figure 14.11, **61–63**) or β-dinaphtholate ligands (**64**, **65**) also syndioselectively polymerize styrene to sPS with a high melting point (T_m = 272 °C), but with low catalyst activities (10^5–10^6 g sPS/(mol Ti·mol S·h)).[7,47,50]

Bidentate *N,N'*-bis(trimethylsilyl)amidinate ligands, $[\eta\text{-RC(NSiMe}_3)_2]^-$ are formal three-electron-donating groups that can be regarded as steric equivalents of cyclopentadienyl groups.[7] In comparison with half-sandwich titanium cyclopentadienyl complexes, the titanium complexes $[\eta\text{-PhC(NSiMe}_3)_2\text{TiX}_3]_n$ (n = 1, X = O-*i*-Pr or n = 2, X = Cl) (Figure 14.12, **66**, **67**) show lower polymerization activities in the presence of MAO, but give a high proportion of sPS in the obtained polymers (sPS% = ~95%) with high melting points (T_m = ~273 °C).[46]

Titanium complexes with Schiff-base ligands (Figure 14.13, **68–76**) show low activities for the syndioselective polymerization of styrene when activated with MAO (2.26 × 10^4 g PS/(mol Ti·mol

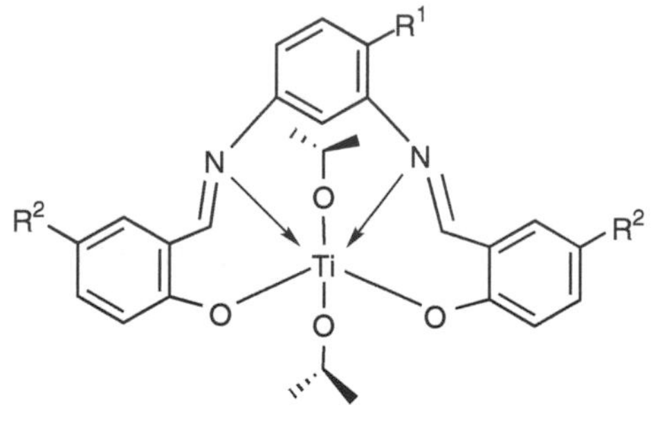

68: R^{1,2} = H   73: R^1 = Me, R^2 = H
69: R^1 = H, R^2 = Me   74: R^1 = Me, R^2 = OMe
70: R^1 = H, R^2 = OMe   75: R^1 = Me, R^2 = Br
71: R^1 = H, R^2 = Cl   76: R^1 = Me, R^2 = t-Bu
72: R^1 = H, R^2 = Br

FIGURE 14.13 Structures of titanium complexes **68–76** with Schiff-base ligands.

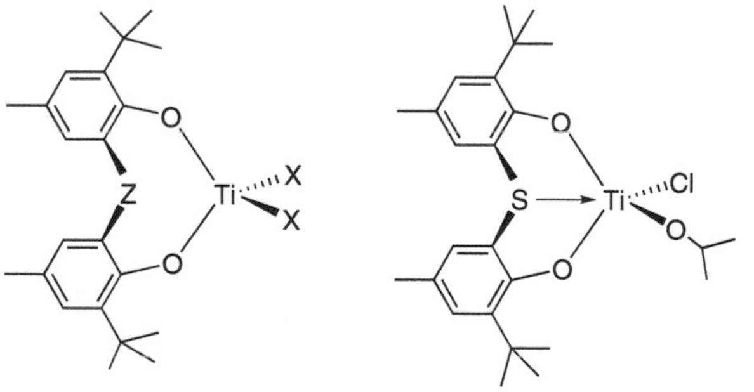

77: Z = CH$_2$, X = Cl   **84**
78: Z = CH$_2$CH$_2$, X =Cl
79: Z = CH$_2$CH$_2$, X = O-i-Pr
80: Z = S, X = Cl
81: Z = S, X = O-i-Pr
82: Z = (S=O), X = Cl
83: Z = (S=O), X = O-i-Pr

FIGURE 14.14 Structures of titanium complexes **77–84** with bridged bis(phenolate) ligands.

S·h)).[51] The variation of ligand substituents in complexes **68–76** slightly influences the polymerization behavior of the catalysts. The introduction of substituents improves the syndioselectivity of the catalyst system as compared to complex **68**. No clear trend is implied.

In bridged bis(phenolato) titanium complexes (Figure 14.14, **77–84**), the catalyst activity for styrene polymerization strongly depends on the nature of the bridging group Z.[52] The sulfur-bridged ligands provide the most active catalysts for the syndioselective polymerization of styrene (**84**, 1.78 × 10^5 g sPS/(mmol Ti·h)). For dichloride complexes, the syndioselective styrene polymerization activity increases in the following order: **77** ≈ **78** < **82** < **80**. The diisopropoxyl complexes show lower activities as compared with the dichloride analogues (**79** < **78**; **83** < **82**).

14.2.1.5 Complexes of Nongroup 4 Transition Metals

In addition to group 4 metal complexes, only a few complexes of other metals are active for the syndio-selective polymerization of styrene. In the presence of $Al(i\text{-}Bu)_3$ and with H_2O, $CHCl_3$ or CCl_4 as the third component, the neodymium complex, $Nd(P_{204})_3$ ($P_{204} = [CH_3(CH_2)_3CH(Et)CH_2O]_2P(O)O\text{-}$), can produce syndiotactic-rich atactic polystyrene.[53] Syndiotactic-rich atactic polystyrene (charac-terized by the major signal of 145.7 ppm in ^{13}C nuclear magnetic resonance (NMR) of the polymer accounting for the *rrrr* pentad, Figure 14.15) can also be obtained from tetra-allyl lanthanides of the formula $[Ln(C_3H_5)_4]Li(dioxane)_{1.5}$ (Ln = Sm, Nd, C_3H_5 = allyl),[54] the neutral nickel σ-acetylide complex, $[Ni(C{\equiv}CPh)_2(P\text{-}n\text{-}Bu_3)_2]$,[55] and indenyl nickel complexes of the formula $(\eta^3\text{-}1\text{-}R\text{-}Ind)Ni(PPh_3)Cl/NaBPh_4$ (R = cyclopentyl, benzyl)[56] without addition of cocatalysts.

Very recently, Hou and coworkers have reported that rare earth half-metallocene complexes (Figure 14.16, **85–88**) activated with $[Ph_3C]^+[B(C_6F_5)_4]^-$ afford highly active systems for syndio-specific styrene polymerization, producing sPS with high syndiotacticities (*rrrr* > 99%) and rather narrow polydispersities ($M_w/M_n = 1.29$–1.55).[57] The activity of scandium complex **85** is compar-able with that for the most active titanium catalysts (1.36×10^7 g sPS/(mol Sc·h)). The neutral allyl lanthanide complexes **89–92** (Figure 14.16) in the absence of a cocatalyst are also active for the syndiospecific polymerization of styrene (*rrrr* ≥ 99%), but with lower activities that are in the order

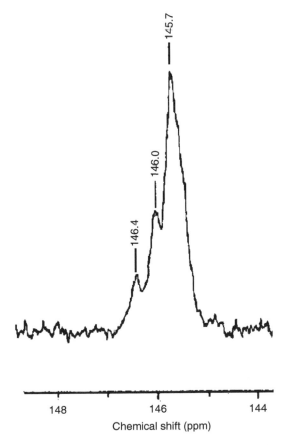

FIGURE 14.15 ^{13}C NMR spectrum (CDCl$_3$, 50 °C, 128 MHz) of the phenyl C-1 carbon of syndiotactic-rich atactic polystyrene prepared from $[Ln(C_3H_5)_4]Li(dioxane)_{1.5}$ (Ln = Sm, Nd, C_3H_5 = allyl). (Reprinted with permission from Baudry-Barbier, D.; Camus, E.; Dormond, A.; Visseaux, M. *Appl. Organomet. Chem.* **1999**, *13*, 813–817. Copyright 1999 John Wiley & Sons, Ltd.)

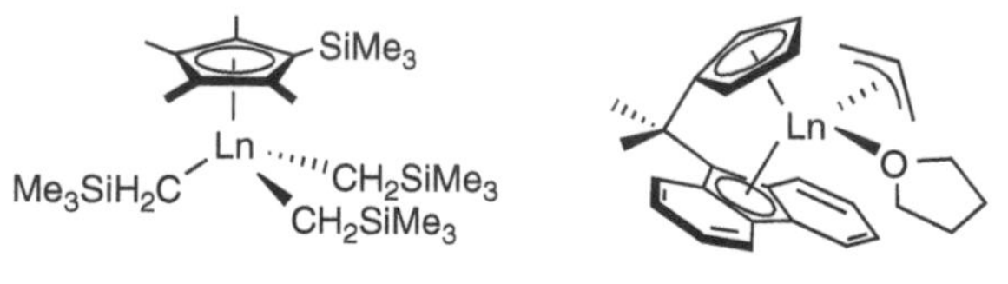

FIGURE 14.16   Structures of rare earth metal complexes **85–92**.

of Nd $\gg$ Sm $>$ La $>$ Y.[58] Among these catalysts, the Nd complex **91** features a remarkably high activity (1.71×10^6 g sPS/(mol Ln·h)).

14.2.2 COCATALYSTS

Transition metal complexes alone typically cannot initiate the syndioselective polymerization of styrene.[4] Like other homogenous alkene polymerizations, the commonly used cocatalysts are MAO and boron-based compounds. For polymerizations, MAO, which is the reaction product of tri-methylaluminum (TMA) and water, is generally used in large excess to the titanium complex. It acts as both a weak reductant and an alkylation agent for the titanium complex to form the active species.[7] For CpTiCl$_3$/MAO catalyzed styrene polymerizations, the polymer yield increases with increasing MAO concentration (Al:Ti = 400–5000).[4] A rather low MAO to titanium complex molar ratio leads to a decrease in the polymerization rate as well as a decrease in the syndiotacticity of the polymer. The amount of the residual TMA in MAO also influences the polymerization behavior of the catalyst system.[6,7] Investigation on Cp*Ti(OMe)$_3$ activated by MAOs with different concentrations of residual TMA showed, that increasing the amount of TMA concentration from 12% to 50% in MAO led to a significant decrease in the activity.[14] The residual TMA is thought to play a deciding role in the reduction of the titanium complex to a Ti(III) species, thus forming the active species for the syndioselective polymerization of styrene. An excess amount of TMA will over reduce the titanium complex to Ti(II), decreasing the polymerization activity.

Boron compounds based on tris(pentafluorophenyl)borane and derivatives (e.g., tetra-kis(pentafluorophenyl)borates (e.g., [Ph$_3$C]$^+$[B(C$_6$F$_5$)$_4$]$^-$ and [Et$_3$NH]$^+$[B(C$_6$F$_5$)$_4$]$^-$) are also used as cocatalysts in syndioselective polymerization of styrene.[43–46] A titanium complex to boron compound 1:1 molar ratio gives the best catalyst performance.[43,59] Higher or lower than a 1:1 molar ratio leads to a significant decrease in both catalyst activity and syndioselectivity. The boron compound reacts with the titanium complex by formation of an active cationic metal species and a noncoordinating borate counter ion. The choice of the tetrakis(pentafluorophenyl)borate countercation (e.g., Ph$_3$C$^+$, Et$_3$NH$^+$) also influences the catalyst activity.[43–46]

14.2.3 ACTIVATORS AND CHAIN TRANSFER AGENTS

Besides the cocatalyst, there are some other additives that are capable of influencing the performance of catalyst systems during styrene polymerization. The addition of a proper amount of triisobutylalu-minum (TIBA) to various titanium complex/MAO systems increases the catalyst activity, whereas the addition of TMA or triethylaluminum (TEA) inhibits polymerization.[4] A relatively high molar ratio of TIBA/MAO will cause a reduction in activity and a sharp decrease in the molecular weights of sPS.[60] When a haloalkylaluminum such as AlEt$_2$Cl is used with the CpTiCl$_3$/MAO system, only atactic polystyrene is produced by a noncoordination polymerization process, that is, a cationic polymerization process.[4] The halo effect is also observed when a halo-containing aluminoxane such as chloroaluminoxane instead of MAO is used with $CpTiCl_3$; the effect is to switch the styrene polymerization dramatically from syndioselective to atactic.[4]

As for borane-activated catalyst systems, such as $Cp^*TiMe_3/B(C_6F_5)_3$, the addition of TIBA is essential to increase the catalyst activity, syndioselectivity, and the molecular weight of the resultant sPS.[43–46,59] Herein, TIBA acts as a scavenger for impurities in the polymerization system.

The binary $Cp^*TiMe_3/B(C_6F_5)_3$ system polymerizes styrene or 4-methylstyrene to an atactic polymer structure at low temperatures from -78 to -20 °C.[61] A carbocationic process initiated from the contact ion pair, $Cp^*Ti^+(IV)Me_2(\mu\text{-Me})B^-(C_6F_5)_3$, is assumed for this nonstereoselective polymerization (Scheme 14.1). The addition of trioctylaluminum (TOA) as a third component to this binary $Cp^*TiMe_3/B(C_6F_5)_3$ system is effective in initiating the syndioselective living homopolymerization of styrene ($M_w/M_n = \sim 1.5$) or 4-methylstyrene ($M_w/M_n = \sim 1.1$), as well as their living random copolymerization at -20 °C.[61] The role of TOA is assumed to shift the equilibrium between contact ion pair $Cp^*Ti^+(IV)Me_2(\mu\text{-Me})B^-(C_6F_5)_3$ and the solvent-separated ion pair $[Cp^*Ti(IV)Me_2]^+[MeB(C_6F_5)_3]^-$ significantly to the latter at -20 °C (Scheme 14.1). Therefore, TOA behaves both as a promoter for the syndioselective polymerization, and as an inhibiter for nonstereoselective polymerization. The use of TOA is also effective in eliminating chain transfer to alkylaluminum that usually occurs when an alkylaluminum compound is used with a catalyst system.[61] The living nature of the polymerization and the high molecular weights of the obtained polymers suggest that no chain transfer to TOA exists in the ternary $Cp^*TiMe_3/B(C_6F_5)_3/TOA$ system.

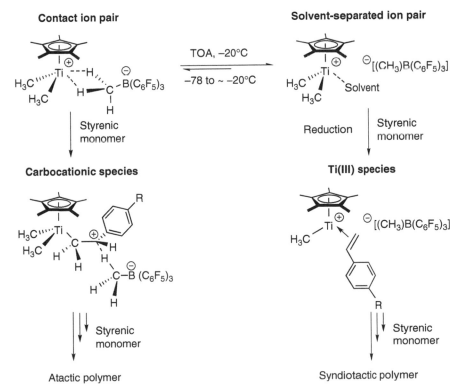

SCHEME 14.1 The role of TOA in the proposed mechanisms for the formation of the active species for atactic and syndioselective polymerizations of styrene with the $Cp^*TiMe_3/B(C_6F_5)_3$ catalyst system at low temperature. Dashed lines represent weak coordination interactions.

14.2.4 Catalyst Immobilization

To be practical in industrial processes, it is desirable to support homogeneous polymerization catalysts on insoluble carriers, such as SiO_2, Al_2O_3, $MgCl_2$, and so forth, in order to carry out the polymerization in slurry or gas phase. When $Ti(O-n-Bu)_4$/MAO is supported on SiO_2, the catalyst system shows higher activity for the syndioselective polymerization of styrene even at 90 °C, which might be due to the higher thermal stability of the active species formed on the support surface.[62,63] Alumina alone is not a good support for $Ti(O-n-Bu)_4$/MAO, but the addition of ethylene glycol to this system leads to a great enhancement in both productivity and syndioselectivity.[63]

Supporting a $CpTiCl_3$/MAO catalyst on silica results in a drop of both activity and syndioselectivity in comparison with the analogous homogeneous system.[64] However, SiO_2-supported $CpTiCl_3$ with MAO as cocatalyst demonstrates better performance in styrene polymerization than MAO/SiO_2-supported $CpTiCl_3$ further activated with additional MAO (MAO/SiO_2 is SiO_2 pretreated with MAO).[65] Higher syndioselectivity and activity are observed for the former. With MAO as cocatalyst, $MgCl_2$ is not a preferred support for $CpTiCl_3$, and leads to lower activity and syndiotacticity of the obtained polystyrene.[66] In the presence of TIBA, MAO/SiO_2-supported $CpTiCl_3$, and $CpTiF_3$ systems behave in a similar manner to each other. Both show the same dependency on $Al_{support}$/Ti molar ratio and have comparable activity, which suggests that the active species of the two systems are essentially the same.[67]

14.2.5 Mechanism of the Syndioselective Polymerization of Styrene

It is generally accepted that the syndioselective polymerization of styrene with titanium complexes in combination with MAO or boron-based activators, proceeds according to a coordination-insertion mechanism (polyinsertion).[4,68] In the presence of MAO that includes residual TMA, half-sandwich titanium complexes of the oxidation state (IV) are converted to $[Cp'Ti^{III}CH_3]^+$ species (Cp' = substituted Cp, Ind, etc.) through the reduction and alkylation by MAO and/or TMA. The addition of styrene significantly accelerates these processes. The precoordination of styrene to the electron deficient Ti^+(III) center leads to the active cationic species for syndioselective styrene polymerization. Competitive coordination of solvent molecule (e.g., toluene) with styrene to the metal center is suggested. The preliminary initiation of the polymerization is realized by a secondary insertion (2,1-insertion)[69] of coordinated styrene into the Ti–CH_3 bond of the active species to form the –$CH(Ph)CH_2CH_3$ end, where a cis-addition to the double bond occurs.[70] After sequential secondary insertions of styrene into the metal–carbon bond of the growing polymer chain, the polymer chain can be terminated through β-hydride transfer to the metal to form a (Ph)C=CH– end group and a Ti-H species that serves as the active species to form a new polymer chain, end-capped with the –$CH(Ph)CH_3$ group. The syndioselective stereochemistry of the polyinsertion process is controlled by the configuration of the tertiary carbon of the last inserted styrene unit (*unlike* 1,3-asymmetric induction),[71,72] that is, the syndiotactic configuration of the polymer chain is determined from the steric repulsive interaction between the phenyl moieties of the last unit of the growing chain end and the incoming monomer (chain-end control).[7,11,43]

In addition, a carbocationic polymerization mechanism has been proposed for the polymerization of styrene with the $Cp*TiMe_3$/B(C_6F_5)$_3$ catalyst system.[73] However, detailed studies confirmed that under normal polymerization conditions (room temperature and above, in toluene), the syndioselective styrene polymerization with the $Cp*TiMe_3$/B(C_6F_5)$_3$ system proceeds through coordination–insertion mechanism instead of a carbocationic mechanism.[74] Using this same catalyst system, a real carbocationic polymerization of styrene becomes significant at low temperature (below 0 °C) or in methylene chloride, and yields atactic polystyrene (Scheme 14.1).[61,73]

14.2.5.1 Initiation and the Active Species for Ti-Based Catalysts

Owing to the complexity of the initiation reaction, the nature and formation of the true active species for syndioselective polymerization have not yet been fully elucidated. The treatment of mono-Cp titanium complexes with MAO leads to trivalent metal species, which have been observed by electron paramagnetic resonance (EPR) spectroscopy.[36] By using redox titration methods, Ti(IV), Ti(III), and Ti(II) species were found to coexist in the catalyst systems.[32,36,75] Cp*-titanium complexes activated with MAO give a higher concentrations of Ti(III) species than MAO activated Cp-titanium complexes. This is consistent with the polymerization activities for Cp*-titanium based catalysts being higher than those for Cp-titanium catalysts.[16] Xu and Ruckenstein also reported that a higher concentration of Ti(III) species was detected in the (1-MeInd)TiF$_3$/MAO system versus the (1-MeInd)TiCl$_3$/MAO system, which corresponds with the higher catalyst activity of the former versus latter.[32] Both EPR investigations and the polymerization results[28,32,36] suggest that Ti(III) species are the major active species for the syndioselective polymerization of styrene, although Ti(IV) and Ti(II) species as active species may not be completely ruled out.

With respect to the activation of Ti-based catalysts with borates, the formation of the ionic complexes [Cp*TiR$_2$]$^+$[B(C$_6$F$_5$)$_3$R]$^-$ or [Cp*TiR$_2$]$^+$[B(C$_6$F$_5$)$_4$]$^-$ (R = Me, Bn) from Cp*TiR$_3$/B(C$_6$F$_5$)$_3$ or Cp*TiR$_3$/[Ph$_3$C]$^+$[B(C$_6$F$_5$)$_4$]$^-$ systems could be characterized spectroscopically.[45,76] Detailed NMR and ESR investigations by Grassi's group and Xu's group on Cp*TiR$_3$/B(C$_6$F$_5$)$_3$ (R = Me, Bn, CH$_2$SiMe$_3$) systems indicated that under the polymerization conditions, the Ti(IV) species [Cp*TiR$_2$]$^+$ slowly decomposed to [Cp*TiR]$^+$, a cationic Ti(III) species, and that the reduction rate was enhanced after the addition of styrene.[44,59,76–78] Furthermore, the concentration of [Cp*TiR]$^+$ as evaluated by ESR measurements was consistent with concentrations of the active species determined by kinetic methods,[77] indicating that Ti(III) species instead of those well-characterized ionic Ti(IV) complexes (i.e., [Cp*TiR$_2$]$^+$[B(C$_6$F$_5$)$_3$R]$^-$ or [Cp*TiR$_2$]$^+$[B(C$_6$F$_5$)$_4$]$^-$) are the active species for the syndioselective polymerization of styrene.[45,77]

When Cp*TiMe$_3$ was premixed with 75% ^{13}C-enriched Al(^{13}CH$_3$)$_3$, a fast methyl exchange between Ti and Al was observed in the ^{1}H NMR spectra with no significant side reaction. Sequential addition of 1 equiv. B(C$_6$F$_5$)$_3$ at 18 °C led to an active catalyst system for the syndioselective polymerization of styrene.[68] From this system, the ^{13}C-enriched end group, -CH(Ph)CH$_2$^{13}CH$_3$, was detected in the NMR spectra of the sPS polymer, suggesting that the syndioselective polymerization of styrene was initiated by the 2,1-insertion of monomer into a Ti-^{13}CH$_3$ bond (Scheme 14.2).[68,73,76,78] Fast methyl exchange between Ti and Al produces Cp*Ti(^{13}CH$_3$)$_3$. After sequential treatment with B(C$_6$F$_5$)$_3$, the Ti(IV) contact ion pair species Cp*Ti$^+$(^{13}CH$_3$)$_2$(μ-^{13}CH$_3$)B$^-$(C$_6$F$_5$)$_3$ and solvent-separated ion pair species [Cp*Ti(^{13}CH$_3$)$_2$(solvent)]$^+$[(^{13}CH$_3$)B(C$_6$F$_5$)$_3$]$^-$ are formed. These cationic Ti(IV) species were observed to be in equilibrium,[73] and were shown to be thermally unstable and to readily undergo reduction to produce Ti(III) species. A doublet and a triplet with the same g value and the same coupling constant were observed in the ESR spectrum of the Cp*Ti(^{13}CH$_3$)$_3$/B(C$_6$F$_5$)$_3$ system at room temperature, and were therefore assigned to the Ti(III) contact ion pair Cp*Ti$^+$(^{13}CH$_3$)(μ-^{13}CH$_3$)B$^-$(C$_6$F$_5$)$_3$ and the solvent-separated ion pair [Cp*Ti(^{13}CH$_3$)(solvent)]$^+$[(^{13}CH$_3$)B(C$_6$F$_5$)$_3$]$^-$, respectively.[76,78] A similar equilibrium between the Ti(III) contact ion pair and the solvent-separated ion pair was anticipated according to the observed spectrum. Syndioselective polymerization of styrene was more likely initiated from the latter.[68,73,76,78]

To further confirm the Ti(III) nature of the active species, several Ti(III) compounds have been investigated for syndioselective styrene polymerization.[6,79] A comparison between Cp*Ti(C$_3$H$_5$)$_2$/[PhMe$_2$NH]$^+$[B(C$_6$F$_5$)$_4$]$^-$ (system **A**, Ti(III)) and Cp*TiBn$_3$/[PhMe$_2$NH]$^+$[B(C$_6$F$_5$)$_4$]$^-$ (system **B**, nominally Ti(IV)) showed that in the dark, only system **A** initiated the syndioselective polymerization of styrene. The syndioselective catalyst activity of system **B** in ambient light arose

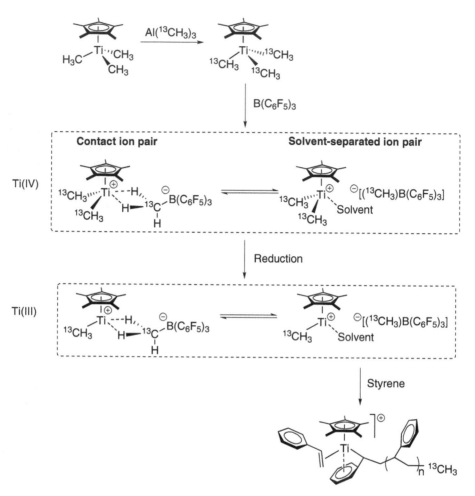

SCHEME 14.2 Proposed mechanism for the formation of active species for syndioselective styrene polymerization using the Cp*TiMe₃/Al(^{13}CH₃)₃/B(C₆H₅)₃ catalyst system. Also shown in the formation of a -CH(Ph)CH$_2^{13}$CH₃ end group. Dashed lines represent weak coordination interactions.

from the reduction of [Cp*TiBn₂]⁺ to the Ti(III) species,[79] leading to the conclusion that Ti(III) species are the active species for the syndioselective polymerization of styrene.

14.2.5.2 Monomer Insertion and Propagation Reactions for Ti-Based Catalysts

The propagation reaction of syndioselective styrene polymerization by titanium catalyst systems involves electrophilic polyinsertion of the monomer. There are two possibilities for the addition of the growing polymer chain to the vinylic double bond of styrene (i.e., cis and trans addition of [Ti]-growing polymer chain bond across the olefin). Through detailed analysis of ^{1}H NMR coupling constants of the random copolymer of perdeuteriostyrene with a small amount of *cis*-β-deuteriostyrene obtained from the TiBn₄/MAO catalyst system,[7,70] Longo and coworkers found that the mode of addition of the growing polymer chain end to the double bond of the monomer is cis (Scheme 14.3).

There are also two other regiochemical pathways for the insertion of styrene into the active metal–carbon bond, primary insertion (1,2-insertion) and secondary insertion (2,1-insertion) as illustrated in Scheme 14.4.

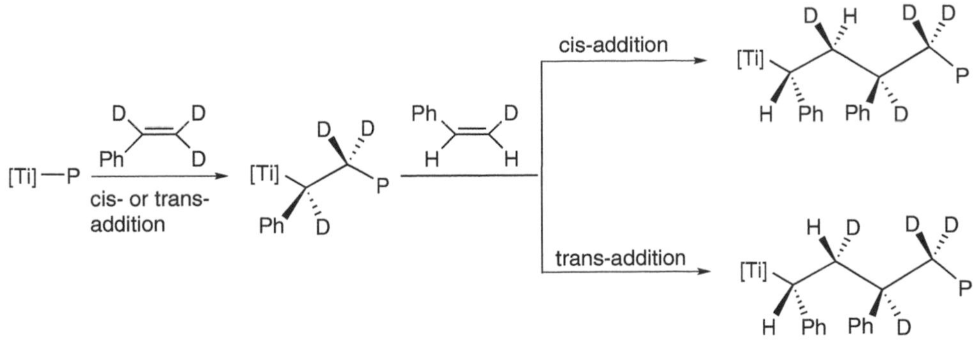

SCHEME 14.3 After the first addition of perdeuteriostyrene, there are two possibilities for the addition of the vinylic double bond of *cis*-β-deuteriostyrene. Only the cis-addition product is observed. [Ti] represents the metal fragment of the active species; P represents the growing polymer chain.

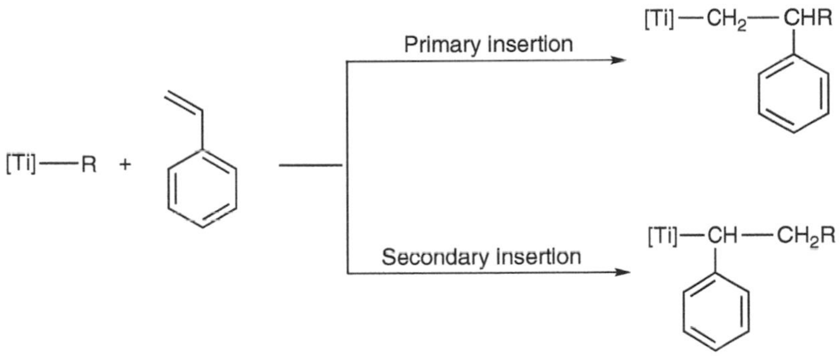

SCHEME 14.4 Two regiochemical pathways of styrene insertion into an active metal–carbon bond. [Ti] represents the metal fragment of the active species; R can be hydride, alkyl groups, and growing polymer chain.

By using the TiBn$_4$/MAO and Al(^{13}CH$_2$CH$_3$)$_3$ (70% ^{13}C-enriched; ^{13}TEA) catalyst system for styrene polymerization, Pellechia et al. observed a sharp methylene resonance at 18.4 ppm in the ^{13}C NMR spectrum of the obtained syndiotactic polystyrene.[69,80] This resonance was assigned to the –CH(Ph)CH$_2^{13}$CH$_2$CH$_3$ end group, which is only accessible through secondary styrene insertion into the active catalyst Ti-^{13}CH$_2$CH$_3$ bond that is formed either via exchange between the catalyst complex and ^{13}TEA, or via chain transfer with ^{13}TEA.[80] The ^{13}C NMR analysis of a sPS sample obtained using the catalyst system Cp*TiMe$_3$/Al(^{13}CH$_3$)$_3$/B(C$_6$F$_5$)$_3$ confirmed the presence of -CH(Ph)CH$_2^{13}$CH$_3$ end groups, which indicates secondary insertion of styrene into the active catalyst Ti-^{13}CH$_3$ bonds as well.[68] In addition, the existence of equal amounts of –CH$_2$CH(Ph)CH$_3$ and (Ph)CH=CH– end groups in low molecular weight sPS prepared with the TiBn$_4$/MAO catalyst system also provided additional evidence for the secondary insertion of styrene,[80] with the former structure being formed by secondary styrene insertion into the active catalyst Ti–H bond (formed by a chain transfer processes), and the latter structure occurring from β-H elimination of the polymer chain.

^{13}C NMR *m* and *r* dyad analysis of syndiotactic polystyrenes prepared with MAO activated CpTiCl$_3$ or Cp*TiCl$_3$ systems was in agreement with the Bernoullian symmetric statistical model for stereoselective propagation.[7,11] The polymer microstructure is a long sequence of syndiotactic dyads with only isolated *m* defects and no consecutive isotactic dyads (Figure 14.17), which is consistent with a chain-end control mechanism directed by the configuration of the tertiary carbon of the last inserted styrene unit (*unlike* 1,3-asymmetric induction).[72,73] Thus, the syndiotactic preference exerted by the chain end arises from the phenyl–phenyl repulsive interaction between the last inserted

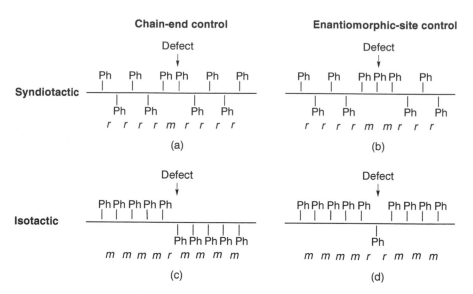

FIGURE 14.17 Possible stereoirregularities of iPS and sPS microstructures by chain-end control and enantiomorphic-site control mechanisms: (a) is usually observed for syndioselective polymerization (through chain-end control); (d) is usually observed for isoselective polymerization (through enantiomorphic-site control).

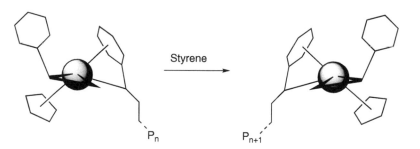

SCHEME 14.5 Incorporation of the coordinated styrene molecule and stereoselective coordination of a new styrene molecule results in the mirror related active species. P_n and P_{n+1} represent the polymer chain with n or $n + 1$ styrene unit. (Reprinted with permission from Longo, P.; Proto, A.; Zambelli, A. *Macromol. Chem. Phys.* **1995**, *196*, 3015–3029. Copyright 1995 Hüthig & Wepf Verlag, Zug.)

styrene unit and the incoming styrene molecule. In Scheme 14.5, the incorporation of a coordinated styrene molecule is fulfilled via two simultaneous steps: cis-migratory insertion, and a nucleophilic substitution of the phenyl moiety of the last unit of the growing polymer chain end by the newly incorporated unit. Sequentially after incorporating this unit, a new styrene coordinates enantioselectively to the metal center, leading to a mirror related active species that implies the syndioselective propagation of the polymer chain.[7,11]

A theoretical study by Cavallo and coworkers[81] with models based on $[CpTi^{III}CH(Ph)CH_3]^+$ species showed that, independent to the chirality of styrene coordination, low energy states are always characterized by the interaction of the aromatic group of the last enchained unit with the metal center. The most stable transition state leads to the formation of a syndiotactic dyad, which is favored by 6 KJ/mol with respect to the transition state leading to an isotactic dyad. The former is favored because the smallest atom on the C_α of the chain, the H atom, can be pointed towards the Cp ligand; whereas for the latter, one of the bigger groups on the growing chain rather than H atom must be oriented toward the Cp ring, leading to higher energy (Figure 14.18). In structure **Syndio**, the shortest interaction distance (3.2 Å) occurs between a C atom of styrene and a C atom of the Cp ring; in structure **Iso**, the shortest interaction distance (~3.0 Å) occurs between styrene

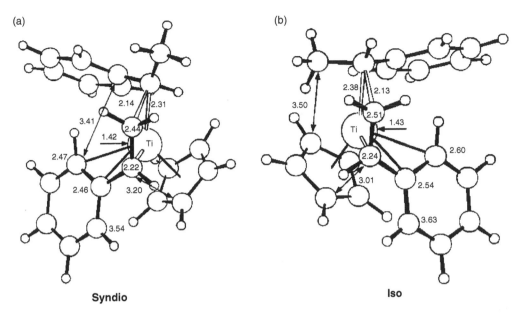

FIGURE 14.18 Geometries of the transition states for the insertion of styrene into the Ti–C bond of the CpTi^{III}CH(Ph)CH$_3^+$ species, in which the chiral CH(Ph)CH$_3$ group is used to simulate a *re*-ending growing chain. The numbers close to the C atoms represent the distance of these atoms from the metal. All distances are reported in Å. (a) Structure **Syndio**, characterized by a *re*-ending chain, a *si*-coordinated monomer and a *R* configuration at the metal atom, is the most stable transition state that leads to the formation of a syndiotactic dyad; (b) Structure **Iso**, characterized by a *re*-ending chain, a *re*-coordinated monomer, and a *S* configuration at the metal atom, is the most stable state that leads to an isotactic dyad. (Reprinted with permission from Minieri, G.; Corradini, P.; Guerra, G.; Zambelli, A.; Cavallo, L. *Macromolecules* **2001**, *34*, 5379–5385. Copyright 2001 American Chemical Society.)

and the Cp ring, as well as the Cp ring and the growing chain. The calculation also showed that [Cp*Ti^{III}CH(Ph)CH$_3$]$^+$ is more stereoselective than [CpTi^{III}CH(Ph)CH$_3$]$^+$, since the most stable transition state leading to a syndiotactic dyad is favored by 16 KJ/mol with respect to the transition stated leading to an isotactic dyad.[81]

14.2.5.3 Chain Transfer and Termination Reactions for Ti-Based Catalysts

The syndioselective polymerization of styrene is typically initiated by insertion of the monomer into a Ti–CH$_3$ bond formed by alkylation of the Ti chloride or alkoxide precatalyst by the MAO. During the polymerization, β-H elimination occurs extensively to form Ti-H active species (Scheme 14.6), which leads to a large amount of sPS end-capped by –CH$_2$CH(Ph)CH$_3$ groups, since the monomer inserts in a secondary fashion into the active Ti–R bond (R = polymer, initiating alkyl group, or hydride). As mentioned in the previous section, the amount of PhCH=CH– end groups formed by β-H elimination is equal to the amount of –CH$_2$CH(Ph)CH$_3$ end groups in the low molecular weight sPS prepared with the TiBn$_4$/MAO catalyst system,[80] which indicates that β-H elimination is the major chain transfer process or termination reaction during the polymerization process.

In addition to chain transfer processes arising from the β-H elimination reaction, the growing polymer chain can also be terminated by transfer reactions to alkylaluminum, MAO and monomer,[7,35] all of which lead to a decrease of the molecular weights of sPS. In the presence of TIBA, the chain transfer reaction to aluminum is the dominant chain transfer reaction instead of β-H elimination. No unsaturated end group could be detected in the obtained polystyrene after deactivation with acidic

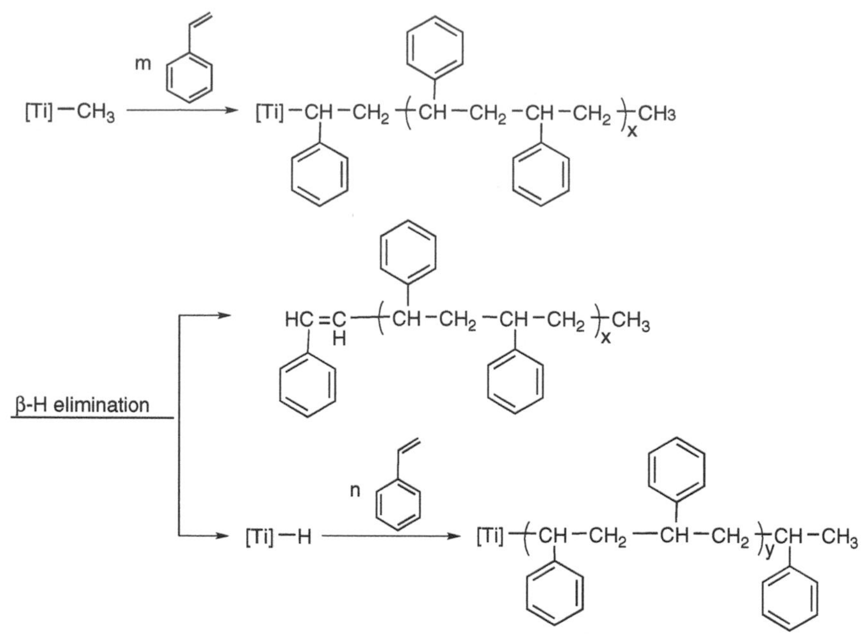

SCHEME 14.6 Illustration for chain termination process through β-H elimination. [Ti] represents the active catalyst site.

methanol, indicating that all growing polymer chains are transferred to aluminum leading to saturated $(Ph)CH_2CH_2$- end groups after deactivation.

A kinetic investigation with the $CpTi(O\text{-}n\text{-}Bu)_3$/MAO system by Chien and coworkers showed that at high styrene concentration ([S] = 1.4 M), the rates of β-H elimination and chain transfer to monomer were comparable. Chain transfer to monomer became slower at lower styrene concentrations ([S] < 1.4 M). Meanwhile, the rate of β-H elimination was independent of styrene concentration and was about 76 times faster than chain transfer to MAO.[35]

14.3 ISOTACTIC POLYSTYRENE

In comparison with the extensive and fruitful investigations in the area of homogeneous syndioselective styrene polymerization, research in the field of isoselective polymerization is still trudging. Isotactic polystyrene, known for almost a half century, is best produced by heterogeneous Ziegler–Natta catalysis.[2] Recently, the homogeneous isoselective polymerization of styrene by soluble metallocenes and nonmetallocene transition metal complexes has been gradually developed.

14.3.1 TRANSITION METAL COMPLEXES FOR ISOSELECTIVE POLYMERIZATION OF STYRENE

14.3.1.1 Group 4 Metal Complexes

Heterogeneous catalyst systems, such as $TiCl_4/SiCl_4/Mg(OEt)_2$/MAO, $TiCl_3/AlEt_3$/α-cyclodextrin, and $TiCl_3/AlEt_3$/butyl ether show very high activities for the isoselective polymerization of styrene (~2000 g PS/(g Ti); 2-butanone insoluble fraction, 14–93%).[82] Solvay-type-$TiCl_3$/Cp_2TiMe_2 (solvay-type-$TiCl_3$:$TiCl_3$ containing coordinated solvent molecules, such as pyridine) and $Ti(O\text{-}n\text{-}Bu)_4/MgCl_2/AlMe_3$ systems give extremely high-isotacticity polystyrene (2-butanone insoluble fraction: 99~100%), but with low activities (0.4 ~ 30 mol PS/(mol Ti·h)).[83] For supported

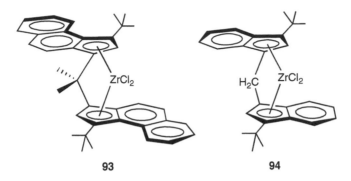

FIGURE 14.19 Structures of *ansa*-zirconocene complexes **93, 94**.

Ti(O-*n*-Bu)$_4$ systems, the tacticity of the obtained polystyrene strongly depends on the support material (i.e., the presence of Cl in the carrier).[84] When MgCl$_2$-supported Ti(O-*n*-Bu)$_4$ is activated with MAO, the soluble part of the catalyst in toluene can afford sPS, whereas the insoluble part of the catalyst gives iPS. This is also true for the Ti(O-*n*-Bu)$_4$/Mg(OH)Cl system when activated with MAO. For Ti(O-*n*-Bu)$_4$/Mg(OH)$_2$/MAO, which has no Cl in the carrier, the insoluble part gives syndiotactic polystyrene, and the soluble part is almost inactive for styrene polymerization.[84] For the TiCl$_3$ (AA)/MAO-system (AA = aluminum-activated), iPS (95% *mmmm* pentad, melting point 221 °C) and sPS are obtained from the insoluble and soluble part of the catalyst respectively.[85]

The great success in the field of stereoselective polymerization of α-olefins promoted the in depth investigation of styrene polymerization using soluble *ansa*-zirconocene complexes and nonmetal-locene complexes. It was reported that isopropylidene-bridged *ansa*-zirconocenes such as complex **93** (Figure 14.19), in the presence of MAO, are capable of producing isotactic polystyrene.[86]

rac-H$_2$C(3-*t*-BuInd)$_2$ZrCl$_2$ (Figure 14.19, **94**) activated by MAO can initiate the isoselective polymerization of styrene.[87] ^{13}C NMR evidence suggests that polymerization proceeds through primary insertion (1,2-) of styrene into a zirconium–hydrogen bond or the zirconium–polymer chain bond. The presence of 3-*tert*-butyl substituents on the indenyl ligands of **94** is thought to be respons-ible for this unusual styrene addition regiochemistry. It should be pointed out that the styrene insertion with *ansa*-zirconocene-based catalysts generally occurs with secondary insertion.[88]

Recently, isoselectivity during styrene polymerization has been achieved using easily access-ible catalyst precursors of the type [MX$_2$(OC$_6$H$_2$-*t*-Bu$_2$-4,6)$_2${S(CH$_2$)$_2$S}] (Figure 14.20, **95–98**) activated with MAO.[89] A remarkable dependence of both the activity and stereoselectivity of the cata-lyst on the ligand backbone is observed, as analogous precursors with 1,5-dithiapentanediyl-linked bis(phenolato) ligands (Figure 14.20, **99, 100**) can only afford syndiotactic polystyrene with low activity. The presence of a C_2-symmetric ligand sphere has been suggested to be critical for isoselect-ive styrene polymerization.[89,90] Isotactic polystyrenes with high melting points (T_m = 208–225 °C) and high molecular weights (M_n = 2.65 × 10^6, M_w/M_n = ~2) can be obtained by using the titanium complexes, **95** and **96**. The narrow molecular weight distribution indicates that polymerization is occurring via a single active catalyst site.[89,90]

14.3.1.2 Complexes of Nongroup 4 Transition Metals

The Ni(acac)$_2$/MAO/Et$_3$N system produces partially isotactic polystyrene with moderate molecular weight (M_w = 9,000–25,000; M_w/M_n = 1.6–4.2; 75–85% *m* diads).[91,92] Ni(α-naph)$_2$ (α-naph = α-nitroacetophenonate, Figure 14.21, **101**) and Ni(hfacac)$_2$ (hfacac = hexafluoroacetylacetonate, Figure 14.21, **102**) in combination with MAO and PCy$_3$ as an ancillary ligand (PCy$_3$ ligates in situ; Cy = cyclohexyl) can also catalyze the polymerization of styrene to produce partially isotactic

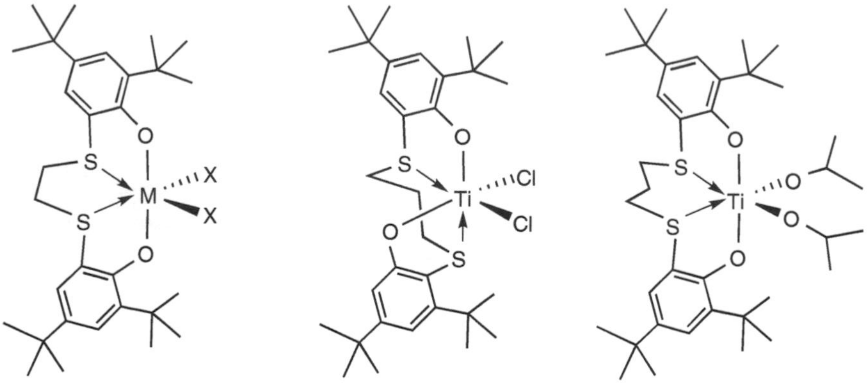

95: M = Ti, X = Cl
96: M = Ti, X = O-*i*-Pr
97: M = Zr, X = CH$_2$Ph
98: M = Hf, X = CH$_2$Ph

99   **100**

FIGURE 14.20 Structures of titanium complexes **95–100** with dithioalkanediyl-bridged bis(phenolate) ligands.

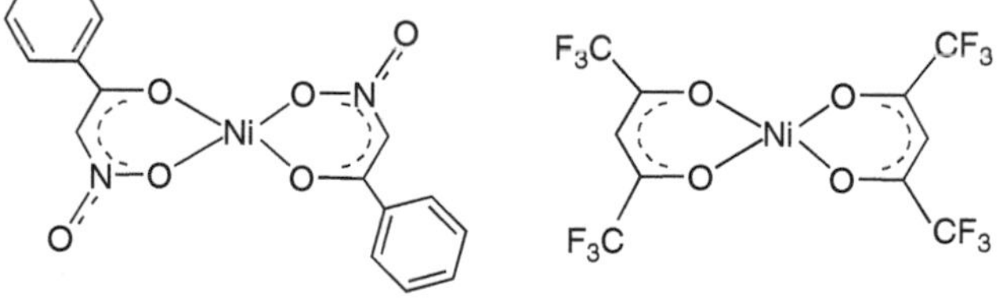

101   **102**

FIGURE 14.21 Structures of Ni(α-naph)$_2$ (**101**) and Ni(hfacac)$_2$ (**102**).

polystyrene.[93] Furthermore, partially isotactic polystyrene is also obtained using the Cp$_2$Ni/MAO system.[94]

The polystyrenes obtained by a heterogeneous vanadium-based catalyst VCl$_3$/AlCl$_3$ in the presence of TEA have a highly isotactic microstructure, and are semicrystalline with a melting point of 220 °C.[95] When activated with TIBA, the MgCl$_2$-supported TiCl$_4$/NdCl$_3$ heterogeneous system catalyzes the isoselective polymerization of styrene with fairly high activity.[82] The addition of NdCl$_3$ significantly improves the catalyst activity, which is nearly four times more active when compared to the corresponding system without NdCl$_3$. The isoselectivity is also greatly improved; iPS with 98% *mmmm* pentads is obtained when using the catalyst system containing NdCl$_3$. Xu and Lin believe that the promoting effect of NdCl$_3$ on the isoselectivity and the catalyst activity may be closely related to the electronegativity of Nd, and the good energy matching and orbital overlap between the f-orbital of Nd and the π-orbital of styrene in the interaction of NdCl$_3$ with TiCl$_4$ during the polymerization process.[82]

Some catalyst systems based on rare earth metal complexes, such as Nd(P$_{507}$)$_3$/H$_2$O/TIBA (P$_{507}$ = RP(OR)(O)O-, R = CH$_3$(CH$_2$)$_3$CH(Et)CH$_2$-), are also active for styrene polymerization and give mixtures of atactic and isotactic polystyrenes (T_m = 220 °C).[96]

14.3.2 MECHANISM OF ISOSELECTIVE POLYMERIZATION OF STYRENE

The isoselective polymerization of styrene can be achieved by different mechanisms depending on the catalyst used. With a Ziegler–Natta catalyst, such as $TiCl_4/TIBA/MgCl_2$, the insertion of the monomer into the metal–carbon bond of the active site is primary (1,2-), and the stereochemistry of the insertion is controlled by the chirality of the active sites (enantiomorphic-site control).[91]

^{13}C NMR analysis of polystyrene samples obtained from the δ-$TiCl_3/Al(^{13}CH_3)_3$ system indicates that the polymerization is highly regioselective. The existence of 2,4-diphenylpentyl polymer end groups also confirms the primary (1,2-) insertion mode of the monomer.[97] From the detailed ^{13}C NMR analysis of the phenyl C-1 carbons of select iPS samples obtained using the $TiCl_4/TIBA/MgCl_2$ catalyst system, a number of minor resonances diagnostic of almost every stereochemical heptad could be assigned. However, resonances accounting for *mmmmrr*, *mmmrrm*, and *mmrrmm* heptads are found to be next intensive after the major signal of the *mmmmmm* heptad. The relatively high intensity of resonances of the *mmmmrr* and *mmmrrm* heptads and their intensity relative to the *mmrrmm* resonance, suggests that the most dominating stereoirregularities in the polymer chain are those with a pair of *r* diads, which can only be formed through an enantiomorphic-site control mechanism. Thus, in the presence of such heterogeneous Ziegler–Natta catalysts, isoselective steric control is due to asymmetric catalytic centers (enantiomorphic-site control, Figure 14.17).[91]

When isoselective polymerization is promoted by nickel catalysts, such as the $Ni(acac)_2/Et_3N/MAO$ system, the insertion of the monomer is mainly secondary (2,1-).[93] The steric control of the polymerization still arises from chiral sites (enantiomorphic-site control), but the configuration of the sites apparently changes rather often during chain propagation since iPS with lower isotacticity is obtained.

Although styrene insertion using δ-$TiCl_3/Al(^{13}CH_3)_3$ seems to follow the generally primary regiochemistry of 1-alkene insertion, experimental evidence supports a secondary regiochemistry of insertion for most C_2- and C_s-symmetric zirconocene systems, that is, formation of a zirconium-benzylidene end group is observed.[98] For *rac*-$H_2C(3$-R-$Ind)_2ZrCl_2/MAO$ systems (Figure 14.22, **103–106**),[99] the regiochemistry of styrene insertion is related to the encumbrance of the substituents in the C(3) position of the indenyl ring. A detailed ^{13}C NMR analysis shows that the regiochemistry of styrene insertion changes from secondary to primary as the bulk of the R substituent in the C(3) position increases: H < Me < Et < *i*-Pr < *t*-Bu. The isopropyl group is the smallest substituent capable of preferentially inducing primary styrene insertion. This regiochemistry becomes strongly prevalent when 3-*tert*-butyl-substituted zirconocene catalysts are used.[87]

14.4 TACTIC STYRENE COPOLYMERS

Although they have high stereoregularity in the polymer backbone, isotactic polystyrene and syndiotactic polystyrene are still not perfect from a material applications point of view. The rather slow crystallization rate of iPS prevents its commercial application; whereas the high syndiotacticity of

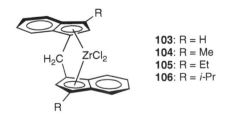

103: R = H
104: R = Me
105: R = Et
106: R = *i*-Pr

FIGURE 14.22 Structures of *rac*-$[H_2C(3$-R-$Ind)_2]ZrCl_2$ **103–106**.

sPS results in high levels of crystallinity that cannot be well controlled, causing problematic brittleness during processing. In order to solve these practical problems in manufacturing polystyrene products, it is desirable to form a copolymer of styrene with other monomers.

Owing to the amount and variety of literatures in this area, only copolymers containing stereoregular styrene–styrene sequences will be discussed here.

14.4.1 Copolymerization with Styrenic Monomers

Semicrystalline styrene-co-4-methylstyrene copolymers produced from TiBn$_4$ or Cp'TiCl$_3$ (Cp' = Cp, Cp*) in combination with MAO are syndiotactic and have melting points ranging from 245 °C at 90 mol% styrene to 220 °C at 56 mol% styrene.[100] The poly(styrene-co-4-methylstyrene) obtained using the CpTiCl$_3$ and Cp*TiCl$_3$ systems possesses, respectively, a random or a slightly blocky structure.[100] Syndiotactic poly(4-methylstyrene)-$block$-polystyrene and poly(4-methylstyrene)-$block$-poly(styrene-co-3-methylstyrene) copolymers can be obtained using the ternary system Cp*TiMe$_3$/B(C$_6$F$_5$)$_3$/TOA at −25 °C, which carries out polymerization in a living manner.[101] The same catalyst system can also be used to carry out the block copolymerization of styrenic monomers bearing functional groups (Figure 14.23).[102]

Lower feed ratios of 4-[($tert$-butyldimethylsilyl)oxy]styrene (**M1**) to styrene (2:98–3:97, mol:mol) lead to useful polymer precursors that can be converted to hydroxyl-functionalized syndiotactic polystyrenes without great loss of physical properties in comparison to those of sPS.[103] The increased incorporation of **M1** leads to decreasing not only the stereoregularity of the copolymer as determined by the melting points, but also the catalyst productivity. Even in the presence of the syndioselective homogeneous catalyst system, IndTiCl$_3$/MAO, at relatively high feed ratios of **M1** to styrene (50:50–70:30, mol:mol), the copolymerization is suggested to proceed through a cationic mechanism, resulting in blocky amorphous copolymers.[103] Auer et al. copolymerized styryl-substituted antioxidant monomers such as **M2** and its trimethylsilylated analogue **M3** with styrene using the IndTiCl$_3$/MAO system to give self-stabilized syndiotactic polystyrene grades.[104] The copolymers possess a highly syndiotactic microstructure as characterized by a very sharp ^{13}C NMR resonance for the phenyl C-1 carbon of styrene at δ = 145.7 ppm,[4,104] but also markedly lower melting points (245–263 °C). These copolymers contain from 2.4 to 6.8 wt% functional units and exhibit enhanced thermo-oxidative stabilities.

When carried out using a syndioselective catalyst system, such as CpTiCl$_3$/MAO, the copolymerization of the styrene-terminated macromonomer, **M4**, with styrene can lead to the formation of graft copolymers with highly syndiotactic polystyrene main chains.[105] The melting point of the graft copolymers decreases with an increasing number of grafts per main polymer chain.

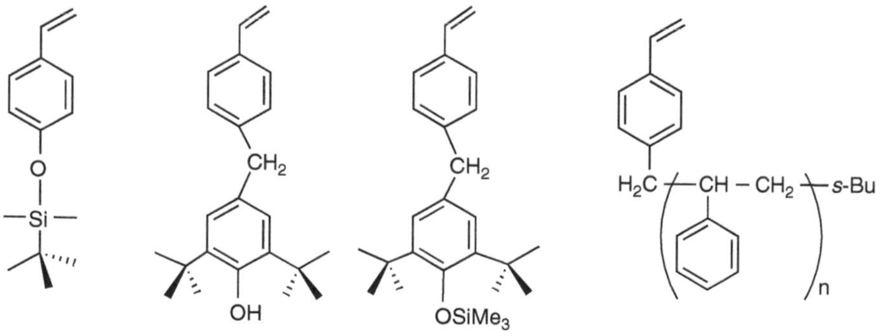

FIGURE 14.23 Structures of styrenic monomers bearing functional groups.

The copolymerization of styrene with a styrene-terminated polyisoprene macromonomer using Ni(acac)$_2$/MAO can easily give a high molecular weight graft copolymer having highly isotactic polystyrene as the main chain and polyisoprene as the side chains.[106]

14.4.2 COPOLYMERIZATION WITH OLEFINS

The structure and composition of copolymers of ethylene and styrene are dependent on the catalyst system used and on the molar ratio of transition metal to cocatalyst (typically MAO). Polymers containing phenylethylene units bridging polyethylene sequences, that is isolated styrene units surrounded by PE segments, are generally formed with titanium-based catalysts.[107]

The constrained geometry catalyst system Me$_2$Si(Cp$'$)(N-t-Bu)TiMe$_2$/[Ph$_3$C]$^+$[B(C$_6$F$_5$)$_4$]$^-$ (Cp$'$ = tetramethylcyclopentadienyl) (Figure 14.24, **107**), shows an alternating tendency for the copolymerizing ethylene with styrene. This system produces relatively high molecular weight copolymers (M_η = 140,000–150,000) even at 80 °C.[108] Ethylene–styrene copolymers obtained with Cp*TiBn$_3$/B(C$_6$F$_5$)$_3$, rac-CH$_2$CH$_2$(Ind)$_2$ZrCl$_2$/MAO (**108**) or MePhC(Cp)(Flu)ZrCl$_2$/ MAO/Al(^{13}CH$_3$)$_3$ (Flu = 9-fluorenyl) (**109**) catalyst systems basically have an alternating structure.[91,109,110] End group analysis shows that in the initiation step, the regioselectivity of styrene insertion into the Zr–^{13}CH$_3$ bonds is prevailingly secondary (2,1-).

The reaction of Me$_2$Si(Flu)(N-t-Bu)TiMe$_2$ and [Ph$_3$C]$^+$[B(C$_6$F$_5$)$_4$]$^-$ nearly quantitatively forms the "cationic" compound [Me$_2$Si(Flu)(N-t-Bu)TiMe]$^+$[B(C$_6$F$_5$)$_4$]$^-$ (Figure 14.25, **110**), which produces a perfectly alternating ethylene and styrene copolymer with the styrene units in a well-defined isotactic arrangement.[111] The structure of the active species (with the bulky fluorenyl ligand) and the following alternating site migratory insertion of comonomer during chain propagation are believed to be responsible for the preferentially alternating, isotactic comonomer incorporation. That is, styrene is coordinated and inserted to the re-face side of the cationic complex **110**, and as a consequence, has the same enantiofacial orientation (isotactic); alternatingly, the ethylene monomer is coordinated and inserted to the si-face side.[111] The copolymerization of ethylene and styrene employing the system consisting of complex Me$_2$C(Ind)$_2$ZrCl$_2$/MAO (Figure 14.25, **111**) affords a copolymer consisting of ethylene–ethylene, styrene–ethylene, head-to-tail styrene–styrene sequences, and highly

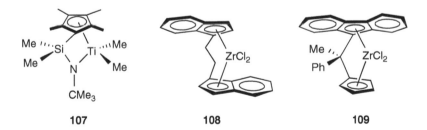

107   **108**   **109**

FIGURE 14.24 Structures of group 4 metal complexes used for ethylene–styrene copolymerizations.

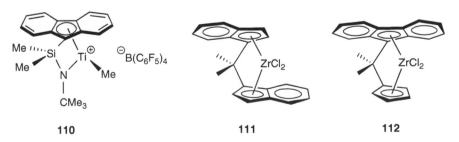

110   **111**   **112**

FIGURE 14.25 Structures of group 4 metal complexes used for copolymerizations of styrene with olefins.

isotactic ethylene–styrene alternating sequences with a melting point of 80–110 °C.[112] MAO activated *rac*-H₂C(3-*t*-BuInd)₂ZrCl₂ (Figure 14.19, **94**), employed as a catalyst for ethylene–styrene copolymerization, produces block copolymers.[87] The polystyrene blocks are isotactic and show crystallinity. The NMR evidences of the existence of the 2,4-diphenyl-1-alkenyl group (formed by β-H elimination, with primary insertion of the last styrene unit) and the 1,3-diphenylalkane group (formed by primary styrene insertion into the catalyst Zr–H bond) suggests a primary (1,2-) styrene insertion into the metal–carbon bond of the growing chain.[87]

Titanium complexes without a π-ligand have also been investigated for the copolymerization of ethylene and styrene. Kakugo et al. reported that an alternating styrene-ethylene copolymer with an isotactic arrangement of styrene units (together with some syndiotactic polystyrene) could be obtained by using a catalyst based on 2,2′-thiobis(4-methyl-6-*tert*-butylphenoxy)titanium dichloride (Figure 14.14, **80**) and MAO.[113] The copolymerization of styrene with small amounts of ethylene using the dithio-linked bis(phenoxy) complex **95**/MAO (Figure 14.20) results in the unprecedented formation of isotactic polystyrene containing isolated ethylene units.[114] The same system can also be used for the copolymerization of propylene and styrene to produce multiblock copolymers, which contain long isotactic styrene sequences interrupted by short isotactic propylene strings,[115] where the opposite regiochemistry of addition of the two monomers (2,1-insertion for styrene and 1,2-insertion for propylene) is retained to give tail-to-tail and head-to-head linkages between the homopolymer blocks.

Very recently, Hou and coworkers reported for the first time that the scandium complex **85** (Figure 14.16) in combination with [Ph₃C]⁺[B(C₆F₅)₄]⁻ efficiently copolymerizes styrene with ethylene, giving random copolymers with syndiotactic styrene–styrene sequences and controllable styrene contents.[57] The narrow, unimodal molecular weight distributions ($M_w/M_n = 1.14$–1.26) of the copolymers indicate the single-site behavior of the catalyst system.

Block copolymerization of propylene and styrene can be achieved with the Cp*Ti(OBn)₃/MAO/TIBA system by means of sequential monomer addition to give a-PP-*block*-s-PS copolymers.[60] In the presence of Me₂C(Cp)(Flu)ZrCl₂ (Figure 14.25, **112**) and MAO, a mixture of propylene and ethylene bubbled into a toluene solution of styrene under ambient conditions leads to an ethylene–propylene–styrene terpolymer, which has an amorphous phase of styrene–ethylene segregated into syndiotactic polypropylene chains.[116] Syndiotactic styrene–butadiene block copolymers can be obtained with the catalyst systems Cp′TiX₃/MAO (Cp′ = Cp, X = Cl, F or Cp′ = Cp*, X = Me) and TiXₙ/MAO (n = 3, X = acac or n = 4, X = O-*t*-Bu).[117] By means of sequential monomer addition, *cis*-(1,4)-polybutadiene-*block*-*syn*-polystyrene copolymers are obtained with the CpTiCl₃/MMAO system (MMAO = modified MAO).[118] By employing the CpTiCl₃/MAO system, styrene can also be copolymerized with (Z)-1,3-pentadiene to afford copolymers mostly containing 1,2-inserted pentadiene units. Both the styrene and the pentadiene units are in a syndiotactic arrangement, but the comonomer sequence distribution is far from the Bernoullian statistical model.[119]

14.4.3 Copolymerization with Polar Monomers

The palladium-catalyzed alternating copolymerization of olefins with carbon monoxide proceeds through alternating insertions of an olefin into a Pd–acyl bond and carbon monoxide into a Pd–alkyl bond. Suitable catalysts for styrene/carbon monoxide copolymerization are mostly cationic Pd(II) complexes with bidentate nitrogen ligands and weakly coordinating anions (Figure 14.26).[120,121] The resulting alternating copolymers are characterized by structural features such as regioisomerism (either monomer 1,2- or 2,1- insertion) and stereoisomerism (iso-, syndio-, or atactic structure). (For more information on copolymers of carbon monoxide, see Chapter 22.)

A true random copolymerization of styrene with a polar monomer was reported by Marks and coworkers.[122] The catalyst for the copolymerization, which is generated in situ through Zn reduction of the Ti(IV) precursor derived from the activation of Cp*TiMe₃ with [Ph₃C]⁺[B(C₆F₅)₄]⁻, can mediate the copolymerization of methyl methacrylate (MMA) and styrene (1:19 molar feed ratio) at

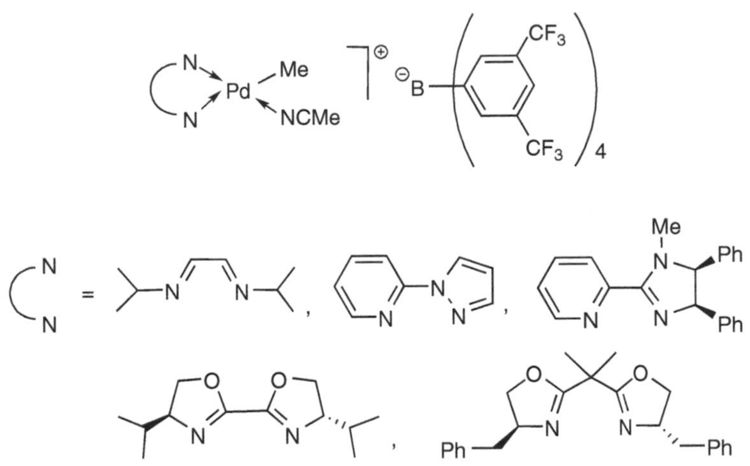

FIGURE 14.26 Cationic Pd(II) complexes with bidentate nitrogen ligands and weakly coordinating anions.

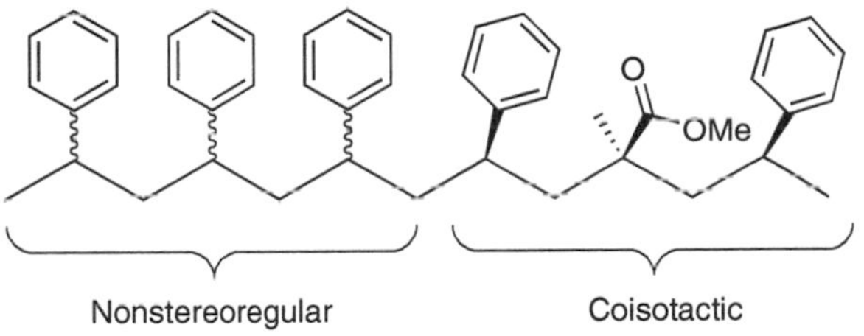

FIGURE 14.27 Illustration for coisotactic structure of poly(S-*co*-MMA). (Reprinted with permission from Jensen, T. R.; Yoon, S. C.; Dash, A. K.; Luo, L.; Marks, T. J. *J. Am. Chem. Soc.* **2003**, *125*, 14482–14494. Copyright 2003 American Chemical Society.)

50 °C to yield random, ~80% coisotactic poly[styrene-*co*-(methylmethacrylate)] containing ~4% MMA (coisotactic, defined as the phenyl group in the styrene unit and the carbomethoxy group in the neighboring MMA unit adopt the same configuration,[123] Figure 14.27). The actual oxidation state of the active species is still not clear, but control experiments suggest that a single-site Ti catalyst is the active species for the copolymerization.[122]

14.5 CONCLUSIONS

The past several decades have witnessed a great breakthrough in the catalysis of stereoregulating polymerizations by homogeneous metallocene catalysts, enabled to a large degree by the introduction of MAO as a cocatalyst by Sinn and Kaminsky at the end of 1970s. Today, we have many efficient and selective catalysts available for stereoregulating polymerization and have gained better control on the fine-tuning of the stereochemical structures of polymer products. Polystyrenes with atactic, isotactic, and syndiotactic stereoregularities can be obtained efficiently through the goal-directed selection of different catalyst systems. The preliminary results and potential possibilities of copolymerizations of styrene with a variety of polar or nonpolar monomers provided by these novel catalyst systems, should further allow the creation of unlimited new polymer architectures, such as alternating, block copolymers and end functional macromolecules. This should allow the opportunity for a thorough investigation of structure and property relationships, and their influence on the mechanical and physical properties of new materials. The main challenge facing this field is the ability to achieve precise control of the polymer stereochemical structure, thus obtaining polymer products with desired structures from the use of the appropriate choice and design of metal complex/cocatalyst systems. It is expected that the development of homogenous metallocene catalysts for commercial processes and commercial polymers will serve as a dominant force in the polymer industry in the near future.

REFERENCES AND NOTES

1. Natta, G.; Pino, P.; Mazzanti, G. Synthesis and structure of some crystalline polyhydrocarbons containing asymmetric carbon atoms in the principle chain. *Chim. Ind. (Milan, Italy)* **1955**, *37*, 927–932; *Chem. Abstr.* **1956**, *50*, 52423.
2. Natta, G.; Danusso, F. Stereospecific polymerization of styrene. II. Notes and problems of crystalline polystyrene. *Chim. Ind. (Milan, Italy)* **1958**, *40*, 445–450; *Chem. Abstr.* **1958**, *52*, 100986.
3. Ishihara, N.; Seimiya, T.; Kuramoto, M.; Uoi, M. Crystalline syndiotactic polystyrene. *Macromolecules* **1986**, *19*, 2464–2465.
4. Ishihara, N.; Kuramoto, M.; Uoi, M. Stereospecific polymerization of styrene giving the syndiotactic polymer. *Macromolecules* **1988**, *21*, 3356–3360.
5. Sinn, H.; Kaminsky, W.; Vollmer, H.-J.; Woldt, R. Living polymers with Ziegler catalysts of high productivity. *Angew. Chem.* **1980**, *92*, 396–402; *Chem. Abstr.* **1980**, *93*, 72366.
6. Tomotsu, N.; Ishihara, N.; Newman, T. H.; Malanga, M. T. Syndiospecific polymerization of styrene. *J. Mol. Catal. A: Chem.* **1998**, *128*, 167–190.
7. Schellenberg, J.; Tomotsu, N. Syndiotactic polystyrene catalysts and polymerization. *Prog. Polym. Sci.* **2002**, *27*, 1925–1982.
8. Kaminsky, W.; Park, Y.-W. Syndiospecific polymerization of styrene with arene titanium(II) complexes as catalyst precursors. *Macromol. Rapid Commun.* **1995**, *16*, 343–346.
9. Zambelli, A.; Oliva, L.; Pellecchia, C. Soluble catalysts for syndiotactic polymerization of styrene. *Macromolecules* **1989**, *22*, 2129–2130.
10. Chien, J. C. W.; Salajka, Z. Syndiospecific polymerization of styrene. I. Tetrabenzyl titanium/methyl aluminoxane catalyst. *J. Polym. Sci., Part A: Polym. Chem.* **1991**, *29*, 1243–1251.
11. Longo, P.; Proto, A.; Zambelli, A. Syndiotactic specific polymerization of styrene: Driving energy of the steric control and reaction mechanism. *Macromol. Chem. Phys.* **1995**, *196*, 3015–3029.
12. Dias, M. L.; Giarrusso, A.; Porri, L. Polymerization of styrene with TiCl$_3$/methylaluminoxane catalyst system. *Macromolecules* **1993**, *26*, 6664–6666.
13. Proto, A.; Luciano, E.; Capacchione, C.; Motta, O. ZrCl$_4$(THF)$_2$/methylaluminoxane as the catalyst for the syndiotactic polymerization of styrene. *Macromol. Rapid Commun.* **2002**, *23*, 183–186.
14. Tomotsu, N.; Ishihara, N. Novel catalysts for syndiospecific polymerization of styrene. *Catal. Sury. Jpn.* **1997**, *1*, 89–110.
15. Green, M. L. H.; Ishihara, N. New *ansa*-metallocene of the group 4 transition metals as homogeneous catalysts for the polymerization of propene and styrene. *J. Chem. Soc., Dalton Trans.* **1994**, 657–665.
16. Campbell, R. E., Jr.; Newman, T. H.; Malanga, M. T. MAO based catalysts for syndiotactic polystyrene (sPS). *Macromol. Symp.* **1995**, *97*, 151–160.
17. Kucht, A.; Kucht, H.; Barry, S.; Chien, J. C. W.; Rausch, M. D. New syndiospecific catalysts for styrene polymerization. *Organometallics* **1993**, *12*, 3075–3078.
18. Liu, J.; Ma, H.; Huang, J.; Qian, Y.; Chan, A. S. C. Syndiotactic polymerization of styrene with half-sandwich titanocenes. *Eur. Polym. J.* **1999**, *35*, 543–545.
19. Qian, Y.; Zhang, H.; Zhou, J.; Zhao, W.; Sun, X.; Huang, J. Synthesis and polymerization behavior of various substituted half-sandwich titanium complexes Cp′TiCl$_2$(OR*) as catalysts for syndiotactic polystyrene. *J. Mol. Catal. A: Chem.* **2004**, *208*, 45–54.
20. Flores, J. C.; Chien, J. C. W.; Rausch, M. D. {[2-(Dimethylamino)ethyl]cyclopentadienyl} trichlorotitanium: A new type of olefin polymerization catalyst. *Organometallics* **1994**, *13*, 4140–4142.
21. Tian, G.; Xu, S.; Zhang, Y.; Wang, B.; Zhou, X. Siloxy-substituted tetramethylcyclopentadienyl and indenyl trichlorotitanium complexes for syndiospecific polymerization of styrene. *J. Organomet. Chem.* **1998**, *558*, 231–233.
22. Sun, X.; Xie, J.; Zhang, H.; Huang, J. Various silyl-substituted cyclopentadienyl titanium complexes [CpSi(CH$_3$)$_2$X]TiCl$_3$ as catalysts for syndiotactic polystyrene. *Eur. Polym. J.* **2004**, *40*, 1903–1908.

23. Qian, X.; Huang, J.; Qian, Y.; Wang, C. Syndiotactic polymerization of styrene catalyzed by alkenyl-substituted cyclopentadienyltitanium trichlorides. *Appl. Organomet. Chem.* **2003**, *17*, 277–281.

24. Ready, T. E.; Day, R. O.; Chien, J. C. W.; Rausch, M. D. (η^5-Indenyl)trichlorotitanium. An improved syndiotactic polymerization catalyst for styrene. *Macromolecules* **1993**, *26*, 5822–5823.

25. Foster, P.; Chien, J. C. W.; Rausch, M. D. Highly stable catalysts for the stereospecific polymerization of styrene. *Organometallics* **1996**, *15*, 2404–2409.

26. Ready, T. E.; Chien, J. C. W.; Rausch, M. D. Alkyl-substituted indenyl titanium precursors for syndiospecific Ziegler–Natta polymerization of styrene. *J. Organomet. Chem.* **1996**, *519*, 21–28.

27. Ready, T. E.; Chien, J. C. W.; Rausch, M. D. New indenyl titanium catalysts for syndiospecific styrene polymerizations. *J. Organomet. Chem.* **1999**, *583*, 11–27.

28. Xu, G.; Cheng, D. Syndiospecific polymerization of styrene with half-sandwich titanocene catalysts. Influence of ligand pattern on polymerization behavior. *Macromolecules* **2000**, *33*, 2825–2831.

29. Stojkovic, O.; Kaminsky, W. Syndiospecific polymerization of styrene with aryl substituted indenyl half-sandwich titanocenes. *Macromol. Chem. Phys.* **2004**, *205*, 357–362.

30. Schneider, N.; Prosenc, M.-H.; Brintzinger, H.-H. Cyclopenta[1]phenanthrene titanium trichloride derivatives: Syntheses, crystal structure and properties as catalysts for styrene polymerization. *J. Organomet. Chem.* **1997**, *545–546*, 291–295.

31. Kaminsky, W.; Lenk, S.; Scholz, V.; Roesky, H. W.; Herzog, A. Fluorinated half-sandwich complexes as catalysts in syndiospecific styrene polymerization. *Macromolecules* **1997**, *30*, 7647–7650.

32. Xu, G.; Ruckenstein, E. Syndiospecific polymerization of styrene using fluorinated indenyltitanium complexes. *J. Polym. Sci., Part A: Polym. Chem.* **1999**, *37*, 2481–2488.

33. Qian, Y.; Zhang, H.; Qian, X.; Huang, J.; Shen, C. Syndiospecific polymerization of styrene catalyzed in situ by alkoxyl substituted half-sandwich titanocene and $BF_3 \cdot Et_2O$. *J. Mol. Catal. A: Chem.* **2003**, *192*, 25–33.

34. Qian, X.; Huang, J.; Qian, Y. Synthesis of new substituted cyclopentadienyl titanium monomethoxydifluorides with $BF_3 \cdot Et_2O$ as fluorinating reagent and their use in syndiotactic polymerization of styrene. *J. Organomet. Chem.* **2004**, *689*, 1503–1510.

35. Chien, J. C. W.; Salajka, Z. Syndiospecific polymerization of styrene. II Monocyclopentadienyl-tributoxy titanium/methylaluminoxane catalyst. *J. Polym. Sci., Part A: Polym. Chem.* **1991**, *29*, 1253–1263.

36. Chien, J. C. W.; Salajka, Z.; Dong, S. Syndiospecific polymerization of styrene. 3. Catalyst structure. *Macromolecules* **1992**, *25*, 3199–3203.

37. Liu, J.; Huang, J.; Qian, Y.; Wang, F.; Chan, A. S. C. A catalyst for syndiotactic polymerization of styrene. *Polym. J.* **1997**, *29*, 182–183.

38. Liu, J.; Huang, J.; Qian, Y.; Chan, A. S. C. Syndiotactic polymerization of styrene with $CpTiCl_2(O^iPr)$/methylaluminoxane. *Polym. Bull.* **1996**, *37*, 719–721.

39. Liu, J.; Ma, H.; Huang, J.; Qian, Y. Syndiotactic polymerization of styrene with $CpTiCl_2(OR)$/MAO system. *Eur. Polym. J.* **2000**, *36*, 2055–2058.

40. Ma, H.; Chen, B.; Huang, J.; Qian, Y. Steric and electronic effects of the R in $IndTiCl_2(OR)$ catalysts on the syndiospecific polymerization of styrene. *J. Mol. Catal. A: Chem.* **2001**, *170*, 67–73.

41. Qian, Y.; Zhang, H.; Qian, X.; Chen, B.; Huang, J. Syndiotactic polymerization of styrene by 1-(Me)$IndTiCl_2(OR)$/MAO system. *Eur. Polym. J.* **2002**, *38*, 1613–1618.

42. Ma, H.; Zhang, Y.; Chen, B.; Huang, J.; Qian, Y. Syndiospecific polymerization of styrene catalyzed by $CpTiCl_2(OR)$ complexes. *J. Polym. Sci., Part A: Polym. Chem.* **2001**, *39*, 1817–1824.

43. Pellecchia, C.; Longo, P.; Proto, A.; Zambelli, A. Novel aluminoxane-free catalysts for syndiotactic-specific polymerization of styrene. *Makromol. Chem., Rapid Commun.* **1992**, *13*, 265–268.

44. Xu, G. Cationic titanocene catalysts for syndiospecific polymerization of styrene. *Macromolecules* **1998**, *31*, 586–591.

45. Grassi, A.; Pellecchia, C.; Oliva, L.; Laschi, F. A combined NMR and electron spin resonance investigation of the $(C_5(CH_3)_5)Ti(CH_2C_6H_5)_3$/$B(C_6F_5)_3$ catalyst system active in the syndiospecific styrene polymerization. *Macromol. Chem. Phys.* **1995**, *196*, 1093–1100.

46. Flores, J. C.; Chien, J. C. W.; Rausch, M. D. [N,N'-Bis(trimethylsilyl)benzamidinato]titanium and -zirconium compounds. Synthesis and application as precursors for the syndiospecific polymerization of styrene. *Organometallics* **1995**, *14*, 1827–1833.

47. Xie, G.; Xu, X.; Zhou, N.; Yan, W.; Hu, Y.; Li, Y.; Chen, X.; Chang, W. Studies and developments of non-metallocene based catalysts for syndiotactic polymerization of styrene. *Hecheng Shuzhi Ji Shuliao* **1997**, *14(2)*, 8–11; *Chem. Abstr.* **1998**, *129*, 95773.

48. Yan, W.; Liu, L.; Jiang, T.; Li, D.; Zhou, N.; Hu, Y. Syndiotactic polystyrene catalytically prepared over β-diketonate titanium complexes/MAO and its thermal properties. *Yingyong Huaxue* **2001**, *18(2)*, 116–119; *Chem. Abstr.* **2001**, *134*, 326830.

49. Wang, J.; Liu, Z.; Wang, D.; Guo, D. Syndiospecific polymerization of styrene based on dichlorobis(1,3-diketonato)titanium/methylaluminoxane catalyst system: Effects of substituents on catalyst activity. *Polym. Int.* **2000**, *49*, 1665–1669.

50. Xu, X.; Zhou, N.; Xie, G. Syndiotactic polymerization of styrene using titanium-8-hydroxyquinolinate complexes/methylaluminoxane systems. *Gaofenzi Xuebao* **1997**, 6, 746–748; *Chem. Abstr.* **1998**, *128*, 128339.

51. Yong, L.; Huang, J.; Lian, B.; Qian, Y. Synthesis of titanium(IV) complexes with Schiff-base ligand and their catalytic activities for polymerization of ethylene and styrene. *Chin. J. Chem.* **2001**, *19*, 429–432.

52. Okuda, J.; Masoud, E. Syndiospecific polymerization of styrene using methylaluminoxane-activated bis(phenolato)titanium complexes. *Macromol. Chem. Phys.* **1998**, *199*, 543–545.

53. Yang, M., Cha, C., Shen, Z. Polymerization of styrene by rare earth coordination catalysts. *Polym. J.* **1990**, *22(10)*, 919–923.

54. Baudry-Barbier, D.; Camus, E.; Dormond, A.; Visseaux, M. Homogeneous organolanthanide catalysts for the selective polymerization of styrene without aluminium cocatalysts. *Appl. Organomet. Chem.* **1999**, *13*, 813–817.

55. Sun, H.; Shen, Q.; Yang, M. New neutral Ni(II)- and Pd(II)-based initiators for polymerization of styrene. *Eur. Polym. J.* **2002**, *38*, 2045–2049.

56. Sun, H.; Li, W.; Han, X.; Shen, Q.; Zhang, Y. Indenyl nickel complexes: Synthesis, characterization and styrene polymerization catalysis. *J. Organomet. Chem.* **2003**, *688*, 132–137.

57. Luo, Y.; Baldamus, J.; Hou, Z. Scandium half-metallocene-catalyzed syndiospecific polymerization and styrene-ethylene copolymerization: Unprecedented incorporation of syndiotactic styrene-styrene sequences in styrene-ethylene copolymers. *J. Am. Chem. Soc.* **2004**, *126*, 13910–13911.

58. Kirillov, E.; Lehmann, C. W.; Razavi, A.; Carpentier, J.-F. Highly syndiospecific polymerization of styrene catalyzed by allyl lanthanide complexes. *J. Am. Chem. Soc.* **2004**, *126*, 12240–12241.

59. Xu, G.; Liu, S. Syndiospecific polymerization of styrene catalyzed by metallocene catalysts I. $Cp*Ti(CH_2SiMe_3)_3/B(C_6F_5)_3$ catalyst. *Cuihua Xuebao* **1998**, *19(3)*, 251–254; *Chem. Abstr.* **1998**, *129*, 41440.

60. Lin, S.; Wu, Q.; Chen, R.; Zhu, F. Syntheses and characterization of polypropene, polystyrene and their block copolymer with titanium catalysts. *Macromol. Symp.* **2003**, *195*, 63–68.

61. Kawabe, M.; Murata, M. Syndiospecific living polymerization of 4-methylstyrene and styrene with (trimethyl)pentamethylcyclopentadienyltitanium/ tri(pentafluorophenylborane/trioctylaluminum catalytic system. *J. Polym. Sci., Part A: Polym. Chem.* **2001**, *39*, 3692–3706.

62. Soga, K.; Nakatani, H. Syndiotactic polymerization of styrene with supported Kaminsky-Sinn catalysts. *Macromolecules* **1990**, *23*, 957–959.

63. Pasquet, V.; Spitz, R. Improvement of supported catalysts for syndiotactic polymerization of styrene. *Macromol. Chem. Phys.* **1999**, *200*, 1453–1457.

64. Xu, J.; Zhao, J.; Fan, Z.; Feng, L. ESR study on SiO_2-supported half-titanocene catalyst for syndiospecific polymerization of styrene. *Macromol. Rapid Commun.* **1997**, *18*, 875–882.

65. Yim, J.-H.; Chu, K.-J.; Choi, K.-W.; Ihm, S.-K. Syndiospecific polymerization of styrene over silica supported $CpTiCl_3$ catalysts. *Eur. Polym. J.* **1996**, *32*, 1381–1385.

66. Xu, J.; Zhao, J.; Fan, Z.; Feng, L. ESR study on $MgCl_2$-supported half-titanocene catalyst for syndiospecific polymerization of styrene. *Eur. Polym. J.* **1999**, *35*, 127–132.

67. Kamisky, W.; Arrowsmith, D.; Strübel, C. Polymerization of styrene with supported half-sandwich complexes. *J. Polym. Sci., Part A: Polym. Chem.* **1999**, *37*, 2959–2968.

68. Pellecchia, C.; Pappalardo, D.; Oliva, L.; Zambelli, A. η^5-$C_5Me_5TiMe_3$-$B(C_6F_5)_3$: A true Ziegler–Natta catalyst for the syndiotactic-specific polymerization of styrene. *J. Am. Chem. Soc.* **1995**, *117*, 6593–6594.

69. Pellecchia, C.; Longo, P.; Grassi, A.; Ammendola, P.; Zambelli, A. Synthesis of highly syndiotactic polystyrene with organometallic catalysts and monomer insertion. *Makromol. Chem., Rapid Commun.* **1987**, *8*, 277–279.

70. Longo, P.; Grassi, A.; Proto, A.; Ammendola, P. Syndiotactic polymerization of styrene: Mode of addition to the double bond. *Macromolecules* **1988**, *21*, 24–25.

71. Ammendola, P.; Pellecchia, C.; Longo, P.; Zambelli, A. Polymerization of vinyl monomers with achiral titanocene catalysts: From "like" to "unlike" 1, 3-asymmetric induction. *Gazz. Chim. Ital.* **1987**, *117*, 65–66.

72. Ewen, J. A. Mechanisms of stereochemical control in propylene polymerization with soluble group 4B metallocene/methylaluminoxane catalysts. *J. Am. Chem. Soc.* **1984**, *106*, 6355–6364.

73. Wang, Q.; Quyoum, R.; Gillis, D. J.; Tudoret, M.-J.; Jeremic, D.; Hunter, B. K.; Baird, M. C. Ethylene, styrene, and α-methylstyrene polymerization by mono(pentamethylcyclopentadienyl) (Cp*) complexes of titanium, zirconium, and hafnium: Roles of cationic complexes of the type $[Cp^*MR_2]^+$ (R = alkyl) as both coordination polymerization catalysts and carbocationic polymerization initiators. *Organometallics* **1996**, *15*, 693–703.

74. Pellecchia, C.; Grassi, A. Syndiotactic-specific polymerization of styrene: Catalyst structure and polymerization mechanism. *Top. Catal.* **1999**, *7*, 125–132.

75. Ready, T. E.; Gurge, R.; Chien, J. C. W.; Rausch, M. D. Oxidation states of active species for syndiotactic-specific polymerization of styrene. *Organometallics* **1998**, *17*, 5236–5239.

76. Grassi, A.; Zambelli, A. Reductive decomposition of cationic half-titanocene(IV) complexes, precursors of the active species in syndiospecific styrene polymerization. *Organometallics* **1996**, *15*, 480–482.

77. Grassi, A.; Lamberti, C.; Zambelli, A.; Mingozzi, I. Syndiospecific styrene polymerization promoted by half-titanocene catalysts: A kinetic investigation providing a closer insight to the active species. *Macromolecules* **1997**, *30*, 1884–1889.

78. Grassi, A.; Saccheo, S.; Zambelli, A. Reactivity of the $[(\eta^5\text{-}C_5Me_5)TiCH_3][RB(C_6F_5)_3]$ complexes identified as active species in syndiospecific styrene polymerization. *Macromolecules* **1998**, *31*, 5588–5891.

79. Mahanthappa, M. K.; Waymouth, R. M. Titanium-mediated syndiospecific styrene polymerizations: Role of oxidation state. *J. Am. Chem. Soc.* **2001**, *123*, 12093–12094.

80. Zambelli, A.; Longo, P.; Pellecchia, C.; Grasi, A. β-Hydrogen abstraction and regiospecific insertion in syndiotactic polymerization of styrene. *Macromolecules* **1987**, *20*, 2035–2037.

81. Minieri, G.; Corradini, P.; Guerra, G.; Zambelli, A.; Cavallo, L. A theoretical study of syndiospecific styrene polymerization with Cp-based and Cp-free titanium catalysts. 2. Mechanism of chain-end stereocontrol. *Macromolecules* **2001**, *34*, 5379–5385.

82. Xu, G.; Lin, S. A novel NdCl$_3$-modified Ziegler–Natta catalyst for the isotactic-specific polymerization of styrene. *Macromol. Rapid Commun.* **1994**, *15*, 873–877.

83. Soga, K.; Uozumi, T.; Yanagihara, H.; Siono, T. Polymerization of styrene with heterogeneous Ziegler–Natta catalysts. *Makromol. Chem., Rapid Commun.* **1990**, *11*, 229–234.

84. Soga, K.; Monoi, T. Polymerization of styrene with Mg(OH)$_x$Cl$_{2-x}$-supported (x = 0–2) Ti(O^{n}Bu)$_4$ catalysts combined with MAO. *Macromolecules* **1990**, *23*, 1558–1560.

85. Mani, R.; Burns, C. M. Homo- and copolymerization of ethylene and styrene using TiCl$_3$ (AA)/methylaluminoxane. *Macromolecules* **1991**, *24*, 5476–5477.

86. Arai, T.; Ohtsu, T.; Suzuki, S. Homo- and copolymerization of styrene by bridged zirconocene complexes with benzindenyl ligands. *Polym. Prepr. (Am. Chem. Soc., Div. Polym. Chem.)* **1998**, *39(1)*, 220–221.

87. Caporaso, L.; Izzo, L.; Sisti, I.; Oliva, L. Stereospecific ethylene-styrene block copolymerization with *ansa*-zirconocene-based catalyst. *Macromolecules* **2002**, *35*, 4866–4870.

88. Caporaso, L.; Izzo, L.; Oliva, L. Ethylene as catalyst reactivator in the propene-styrene copolymerization. *Macromolecules* **1999**, *32*, 7329–7331.

89. Capacchione, C.; Proto, A.; Ebeling, H.; Mülhaupt, R.; Möller, K.; Spaniol, T. P.; Okuda, J. Ancillary ligand effect on single-site styrene polymerization: Isospecificity of group 4 metal bis(phenolate) catalysts. *J. Am. Chem. Soc.* **2003**, *125*, 4964–4965.

90. Capacchione, C.; Proto, A.; Ebeling, H.; Mülhaupt, R.; Möller, K.; Manivannan, R.; Spaniol, T. P.; Okuda, J. Non-metallocene catalysts for the styrene polymerization: Isospecific group 4 metal bis(phenolate) catalysts. *J. Mol. Catal. A: Chem.* **2004**, *213*, 137–140.

91. Long, P.; Grassi, A.; Oliva, L.; Ammendola, P. Some ^{13}C NMR evidence on isotactic polymerization of styrene. *Makromol. Chem.* **1990**, *191*, 237–242.

92. Po, R.; Cardi, N.; Santi, R.; Romano, A. M.; Zannoni, C.; Spera, S. Polymerization of styrene with nickel complex/methylaluminoxane catalytic systems. *J. Polym. Sci., Part A: Polym. Chem.* **1998**, *36*, 2119–2126.

93. Carlini, C.; Galletti, A. M. R.; Sbrana, G.; Caretti, D. Homo- and co-polymerization of styrene with ethylene by novel nickel catalysts. *Polymer* **2001**, *42*, 5069–5078.

94. Longo, P.; Grisi, F.; Proto, A.; Zambelli, A. New Ni(II) based catalysts active in the polymerization of olefins. *Macromol. Rapid Commun.* **1998**, *19*, 31–34.

95. Ribeiro, M. R.; Portela, M. F.; Deffieux, A.; Cramail, H.; Rocha, M. F. Isospecific homo- and copolymerization of styrene with ethylene in the presence of VCl$_3$, AlCl$_3$ as catalyst. *Macromol. Rapid Commun.* **1996**, *17*, 461–469.

96. Liu, L.; Gong, Z.; Zheng, Y.; Jing, X. Stereospecfic polymerization of styrene using homogeneous lanthanide catalyst. *Macromol. Rapid Commun.* **1997**, *18*, 859–864.

97. Ammendola, P.; Tancredi, T.; Zambelli, A. Isotactic polymerization of styrene and vinylcyclohexane in the presence of a ^{13}C-enriched Ziegler–Natta catalyst: Regioselectivity and enantioselectivity of the insertion into metal-methyl bonds. *Macromolecules* **1986**, *19*, 307–310.

98. Oliva, L.; Caporaso, L.; Pellecchia, C.; Zambelli, A. Regiospecificity of ethylene-styrene copolymerization with a homogeneous zirconocene catalyst. *Macromolecules* **1995**, *28*, 4665–4667.

99. Izzo, L.; Napoli, M.; Oliva, L. Regiochemistry of the styrene insertion with CH$_2$-bridged *ansa*-zirconocene-based catalysts. *Macromolecules* **2003**, *36*, 9340–9345.

100. Nakatani, H.; Nitta, K.-H.; Soga, K.; Takata, T. Synthesis and thermal properties of poly(styrene-*co*-4-methylstyrene) produced with syndiospecific metallocene catalysts. *Polymer* **1997**, *38*, 4751–4756.

101. Kawabe, M.; Murata M. Synthesis of block graft copolymer with syndiospecific living polymerization of styrene derivatives by (trimethyl)pentamethylcyclopentadienyl titanium/tris(pentafluorophenyl)borane/trioctylaluminium catalytic system. *Macromol. Chem. Phys.* **2001**, *202*, 1799–1805.

102. Kawabe, M.; Murata M. Syndiospecific living block copolymerization of styrenic monomers containing functional groups, and preparation of syndiotactic poly{(4-hydroxystyrene)-*block*-[(4-methylstyrene)-*co*-(4-hydroxystyrene)]}. *Macromol. Chem. Phys.* **2002**, *203*, 24–30.

103. Kim, K. H.; Jo, W. H.; Kwak, S.; Kim, K. U.; Hwang, S. S.; Kim, J. Synthesis and characterization of poly(styrene-*co*-4-[(*tert*-butyldimethylsilyl)oxy]styrene) as a precursor of hydroxyl-functionalized syndiotactic polystyrene. *Macromolecules* **1999**, *32*, 8703–8710.

104. Auer, M.; Nicolas, R.; Rosling, A.; Wilén, C.-E. Synthesis of novel-*dl*-α-tocopherol-based and sterically-hindered-phenol-based monomers and their utilization in copolymerizations over metallocene/MAO catalyst systems. A strategy to remove concerns about additive compatibility and migration. *Macromolecules* **2003**, *36*, 8346–8352.

105. Endo, K.; Senoo, K. Syndiospecific copolymerization of styrene with styrene macromonomer bearing terminal styryl group by CpTiCl$_3$-methylaluminoxane catalyst. *Polymer* **1999**, *40*, 5977–5980.

106. Senoo, K.; Endo, K. Isospecific copolymerization of styrene and a styrene-terminated polyisoprene macromonomer with the Ni(acac)$_2$/MAO catalyst. *J. Polym. Sci., Part A: Polym. Chem.* **2000**, *38*, 1241–1246.

107. Longo, P.; Grassi, A.; Oliva, L. Copolymerization of styrene and ethylene in the presence of different syndiospecific catalysts. *Makromol. Chem.* **1990**, *191*, 2387–2396.

108. Sukhova, T. A.; Panin, A. N.; Babkina, O. N.; Bravaya, N. M. Catalytic systems Me$_2$SiCp*N^{t}BuMX$_2$/(CPh$_3$)(B(C$_6$F$_5$)$_4$) (M = Ti, X = CH$_3$, M = Zr, X = ^{i}Bu) in copolymerization of ethylene with styrene. *J. Polym. Sci., Part A: Polym. Chem.* **1999**, *37*, 1083–1093.

109. Pellecchia, C.; Pappalardo, D.; D'Arco, M.; Zambelli, A. Alternating ethylene-styrene copolymerization with a methylaluminoxane-free half-titanocene catalyst. *Macromolecules* **1996**, *29*, 1158–1162.

110. Oliva, L.; Longo, P.; Izzo, L.; Serio, M. D. Zirconocene-based catalysts for the ethylene-styrene copolymerization: Reactivity ratios and reaction mechanism. *Macromolecules* **1997**, *30*, 5616–5619.

111. Xu, G. Copolymerization of ethylene with styrene catalyzed by the [η^1: η^5-*tert*-butyl(dimethyl-fluorenylsilyl)amido]methyltitanium "cation". *Macromolecules* **1998**, *31*, 2395–2402.

112. Arai, T.; Ohtsu, T.; Suzuki, S. Stereoregular and Bernoullian copolymerization of styrene and ethylene by bridged metallocene catalysts. *Macromol. Chem. Phys.* **1998**, *19*, 327–331.

113. Kakugo, M.; Miyatake, T.; Mizunuma, K. Polymerization of styrene and copolymerization of styrene with olefin in the presence of soluble Ziegler–Natta catalysts. *Stud. Surf. Sci. Catal.* **1990**, *56*, 517–529.

114. Capacchione, C.; D'Acunzi, M.; Motta, O.; Oliva, L.; Proto, A.; Okuda, J. Isolated ethylene units in isotactic polystyrene chain: Stereocontrol of an isospecific post-metallocene titanium catalyst. *Macromol. Chem. Phys.* **2004**, *205*, 370–373.

115. Capacchione, C.; Carlo, F. D.; Zannoni, C.; Okuda, J.; Proto, A. Propylene-styrene multiblock copolymers: Evidence for monomer enchainment via opposite insertion regiochemistry by a single-site catalyst. *Macromolecules* **2004**, *37*, 8918–8922.

116. Rosa, C. D.; Buono, A.; Caporaso, L.; Oliva, L. Structural characterization of syndiotactic propylene-styrene-ethylene terpolymers. *Macromolecules* **2003**, *36*, 7119–7125.

117. Zambelli, A.; Caprio, M.; Grassi, A.; Bowen, D. E. Syndiotactic styrene-butadiene block copolymers synthesized with CpTiX$_3$/MAO (Cp = C$_5$H$_5$, X = Cl, F; Cp = C$_5$Me$_5$, X = Me) and TiX$_n$/MAO (n = 3, X = acac; n = 4, X = O-*tert*-Bu). *Macromol. Chem. Phys.* **2000**, *201*, 393–400.

118. Ban, H. T.; Tsunogae, Y.; Shiono, T. Synthesis and characterization of *cis*-polybutadiene-*block-syn*-polystyrene copolymers with a cyclopentadienyl titanium trichloride/modified methylaluminoxane catalyst. *J. Polym. Sci., Part A: Polym. Chem.* **2004**, *42*, 2698–2704.

119. Longo, P.; Proto, A.; Oliva, L.; Sessa, I.; Zambelli, A. Copolymerization of styrene with (Z)-1,3-pentadiene in the presence of a syndiotactic-specific catalyst. *J. Polym. Sci., Part A: Polym. Chem.* **1997**, *35*, 2697–2702.

120. Binotti, B.; Carfagna, C.; Gatti, G.; Martini, D.; Mosca, L.; Pettinari, C. Mechanistic aspects of isotactic CO/styrene copolymerization catalyzed by oxazoline palladium(II) complexes. *Organometallics* **2003**, *22*, 1115–1123.

121. Bastero, A.; Claver, C.; Ruiz, A.; Castillón, S.; Daura, E.; Bo, C.; Zangrando, E. Insights into CO/styrene copolymerization by using Pd^{II} catalysts containing modular pyridine-imidazoline ligands. *Chem. Eur. J.* **2004**, *10*, 3747–3760.

122. Jensen, T. R.; Yoon, S. C.; Dash, A. K.; Luo, L.; Marks, T. J. Organotitanium-mediated stereoselective coordinative/insertive homopolymerizations and copolymerizations of styrene and methyl methacrylate. *J. Am. Chem. Soc.* **2003**, *125*, 14482–14494.

123. Yokota, K.; Hirabayashi, T. Determination of coisotacticities of some alternating styrene-acrylic copolymers by NMR spectra. *J. Polym. Sci.* **1976**, *14*, 57–71.

15 Cyclopentene Homo- and Copolymers Made with Early and Late Transition Metal Catalysts

Naofumi Naga

CONTENTS

15.1 INTRODUCTION

Cyclopentene (CPE) is obtained from one of the C_5 fractions of cyclopentadiene, which is a sub-product of cracking naphtha. CPE is prepared by partial hydrogenation of cyclopentadiene or reduction of cyclopentadiene by sodium in liquid nitrogen. CPE is a unique olefin that, owing to its diverse insertion modes, can be polymerized with various transition metal catalysts. Figure 15.1 shows the four types of CPE polymerization insertion and enchainment modes: 1,2-cis-enchainment, 1,2-trans-enchainment, 1,3-cis-enchainment, and 1,3-trans-enchainment.

The homopolymerization of CPE and the characterization of the resulting poly(cyclopentene) (PCPE) have been widely investigated. Crystalline homo-PCPE is a largely insoluble material having a melting temperature (T_m) of 395 °C (under vacuum), which is much higher than its decomposition temperature. Copolymerizations of CPE with the common olefins ethylene and propylene have also been investigated in order to control the properties of the resulting copolymers and to clarify the insertion mechanisms of CPE homopolymerization. When CPE units are incorporated as comonomers into polyethylene or isotactic polypropylene (iPP), the T_ms of the polymers are reduced, and the glass transition temperatures (T_gs) are increased to room temperature. This chapter reviews recent developments in the polymerization and copolymerization of CPE with early and late transition metal catalysts.

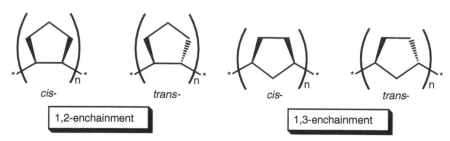

FIGURE 15.1   Enchainment modes of CPE.

15.2 HOMOPOLYMERIZATION OF CYCLOPENTENE WITH TRANSITION METAL CATALYSTS

15.2.1 HOMOPOLYMERIZATION WITH EARLY TRANSITION METAL CATALYSTS

The polymerization of cycloolefins using traditional heterogeneous Ziegler–Natta catalysts yields only products generated by ring-opening polymerization.[1–3] Kaminsky et al. first achieved the vinyl-type polymerization of CPE using early transition metal catalysts in 1988.[4–6] They conducted polymerization with the isoselective zirconocene catalysts rac-ethylenebis(indenyl)zirconium dichloride (rac-Et(Ind)$_2$ZrCl$_2$) and rac-ethylenebis(tetrahydroindenyl)zirconium dichloride (rac-Et(H$_4$Ind)$_2$ZrCl$_2$), using methylaluminoxane (MAO) as a cocatalyst at ambient temperature. The resulting PCPE was a crystalline polymer with a high melting temperature of 395 °C. The wide-angle X-ray diffraction (WAXD) pattern of this material showed three diffraction peaks at $2\theta = 16$ °, 19.5 °, and 24 °. PCPE exhibiting this WAXD pattern is defined as having crystal Type I. The WAXD pattern of Type I PCPE exhibits a small halo region derived from the amorphous phase, indicating a high degree of crystallinity.

Since Type I crystalline PCPE has low solubility in typical organic solvents, it is difficult to study its microstructure by solution nuclear magnetic resonance (NMR) spectroscopy. Kaminsky et al. thus synthesized soluble oligomeric CPEs for ^{13}C NMR spectroscopic studies by employing high temperatures, a high catalyst concentration, and a low CPE concentration during polymerization. The oligomeric CPEs obtained were soluble in 1,2,4-trichlorobenzene at ca. 100 °C for ^{13}C NMR studies and exhibited three strong peaks at 31, 39, and 47 ppm. From these studies, it was concluded that most of the polymer had cis-1,2- or trans-1,2-enchainment.

Hydrooligomerization of CPE (carrying out polymerization in the presence of added H$_2$ as a chain transfer agent) is another effective method for decreasing PCPE molecular weight and improving the solubility of the product for characterization purposes. Collins and Kelly have hydrooligomerized CPE using rac-Et(Ind)$_2$ZrCl$_2$ and bis(cyclopentadienyl)zirconium dichloride (Cp$_2$ZrCl$_2$) catalysts activated with MAO. From the mixture of oligomers produced, pure CPE trimer and tetramer were isolated and studied in order to better understand how CPE polymerization proceeds with these catalysts.[7] Model samples of CPE trimers having known regio- and stereochemistries were separately produced by means of total synthesis as comparatives. ^{13}C NMR spectroscopy of the model trimers and hydrooligomers clearly showed that the enchainment of CPE during polymerization with rac-Et(Ind)$_2$ZrCl$_2$/MAO occurs predominantly through cis-1,3-insertion.

Two possible stereoisomeric tetramers can result from the cis-1,3-enchainment of CPE, based on the structure of the CPE trimers studied: erythrodisyndiotactic (cis–cis syndiotactic) and erythrodiisotactic (cis–cis isotactic) (Figure 15.2). Chiral column gas chromatography (GC) analysis of CPE tetramers obtained using rac-Et(Ind)$_2$ZrCl$_2$/MAO showed that the tetramers consisted of a single stereocomponent (not a mixture of diastereomers). It was impossible to determine whether the 1,3-PCPE was diisotactic or disyndiotactic. Because CPE is not prochiral, it is therefore not clear that the chiral rac-Et(Ind)$_2$ZrCl$_2$ catalyst would intrinsically favor an isotactic polymer structure (as it

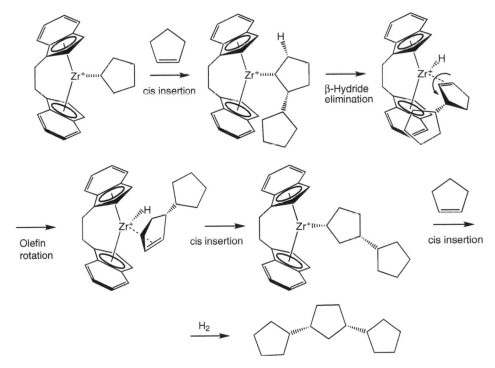

Erythrodisyndiotactic, cis–cis syndiotactic

Erythrodiisotactic, cis–cis isotactic

FIGURE 15.2 Possible stereoisomeric tetramers for cis-1,3 CPE enchainment. (Reprinted with permission from Kelly, W. M.; Taylor, N. J.; Collins, S. *Macromolecules* **1994**, *27*, 4477–4485. Copyright 1994 American Chemical Society.)

SCHEME 15.1 cis-1,3-Insertion mechanism for CPE during hydrotrimerization. (Reprinted with permission from Collins, S.; Kelly, W. M. *Macromolecules* **1992**, *25*, 233–237. Copyright 1992 American Chemical Society.)

does for α-olefins). Scheme 15.1 outlines a mechanism proposed by Collins for the 1,3-enchainment of CPE during hydrotrimerization (and polymerization) with *rac*-Et(Ind)$_2$ZrCl$_2$/MAO: (1) insertion of CPE into a Zr–H bond (obtained following a β-hydride or hydrogen chain transfer event, not shown), (2) cis insertion of a second CPE monomer unit (forming a bis(cyclopentyl) intermediate), (3) β-hydride elimination to form a cycloolefin hydride complex, (4) rotation of the resultant coordinated cycloolefin about the Zr center, and finally, (5) *cis* reinsertion of this cycloolefin at the olefinic carbon farthest from the polymer chain.

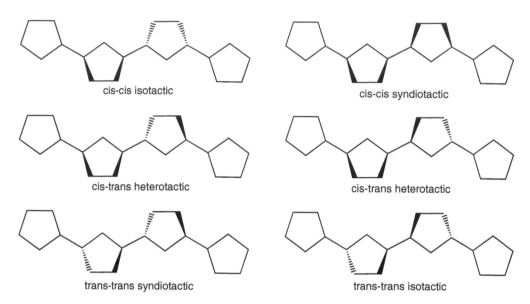

cis-cis isotactic

cis-cis syndiotactic

cis-trans heterotactic

cis-trans heterotactic

trans-trans syndiotactic

trans-trans isotactic

FIGURE 15.3 Possible stereoisomeric tetramers for cis- and trans-1,3 CPE enchainments. (Reprinted with permission from Kelly, W. M.; Taylor, N. J.; Collins, S. *Macromolecules* **1994**, *27*, 4477–4485. Copyright 1994 American Chemical Society.)

Collins et al. also investigated the hydrooligomerization and polymerization of CPE with *rac*-Et(H$_4$Ind)$_2$ZrCl$_2$/MAO. They found not only *cis*-1,3-enchainment of CPE in the polymer, but also *trans*-1,3-enchainment.[8,9] This determination was accomplished by ^{13}C NMR comparison of the actual CPE tetramers formed through hydrooligomerization to independently synthesize authentic samples of the six possible stereoisomeric tetramers (Figure 15.3).

Collins and coworkers have proposed two mechanisms for the formation of trans-1,3-enchained CPE units with *rac*-Et(H$_4$Ind)$_2$ZrCl$_2$.[9] In the first of these (reversible chain transfer, Scheme 15.2), CPE monomer displaces coordinated, β-hydride-eliminated polymer chains having 2-cyclopentenyl end groups (i). Subsequent recoordination/reinsertion of the polymer chains with the opposite orientation (ii) would lead to trans-1,3-insertion (iii). An alternative mechanism involves indirect isomerization through an intermediate σ-CH complex (Scheme 15.3). Collins et al. also investigated the deuterium isotope effect on the stereochemistry of oligomerization using cyclopentene-d_8. The observations agreed with the alternative process involving the σ-CH complex intermediate.

Kaminsky and Arndt also investigated the hydrooligomerization and polymerization of CPE with Cp$_2$ZrCl$_2$, *rac*-dimethylsilylenebis(indenyl)zirconium dichloride (*rac*-Me$_2$Si(Ind)$_2$ZrCl$_2$), and diphenylmethylene(cyclopentadienyl)(9-fluorenyl)zirconium dichloride (Ph$_2$C(Cp)(Flu)ZrCl$_2$). The microstructures of the resulting oligomers and polymers were studied using ^{13}C, HH COSY, DEPT, and heteronuclear correlation NMR spectroscopy.[10,11] The C$_2$-symmetric *rac*-Me$_2$Si(Ind)$_2$ZrCl$_2$ catalyst produces crystalline polymer and an erythrodiisotactic hydrotetramer (100% *m* dyads), indicating highly stereoselective insertion. On the other hand, the C$_{2v}$-symmetric Cp$_2$ZrCl$_2$ and C$_s$-symmetric Ph$_2$C(Cp)(Flu)ZrCl$_2$ catalysts show little or no stereoselectivity in the polymerization and yield amorphous polymers. The ^{13}C NMR spectra of the amorphous PCPE polymers showed three strong peaks at 31, 39, and 47 ppm, as observed in the spectra of oligomers obtained with C$_2$-symmetric zirconocene catalysts.[4–6,10] Oligomerization with Cp$_2$ZrCl$_2$ and Ph$_2$C(Cp)(Flu)ZrCl$_2$ yielded a mixture of the two diastereomeric cis,cis-hydrotetramers in each case.[10] The low stereoselectivity seen in polymerization with Ph$_2$C(Cp)(Flu)ZrCl$_2$ may be the result of unfavorable ligand-monomer interactions; other possible causes are site isomerization errors or two consecutive chain elimination/readdition steps.

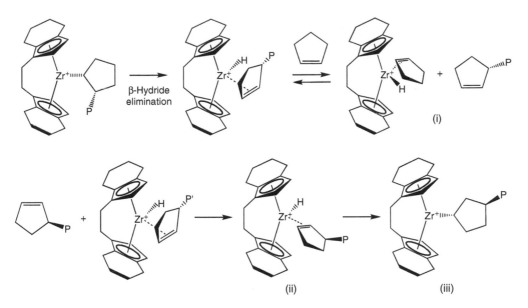

SCHEME 15.2 Proposed mechanism for trans-1,3-insertion of CPE through reversible chain transfer (P and P' = polymer chain). (Reprinted with permission from Kelly, W. M.; Wang, S.; Collins, S. *Macromolecules* **1997**, *30*, 3151–3158. Copyright 1997 American Chemical Society.)

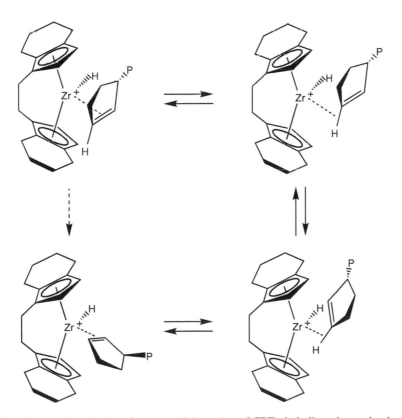

SCHEME 15.3 Proposed mechanism for trans-1,3-insertion of CPE via indirect isomerization through a σ-CH complex (P = polymer chain). (Reprinted with permission from Kelly, W. M.; Wang, S.; Collins, S. *Macromolecules* **1997**, *30*, 3151–3158. Copyright 1997 American Chemical Society.)

Asanuma et al. investigated CPE polymerization with the C_s-symmetric catalyst isopropylidene(cyclopentadienyl)(9-fluorenyl)zirconium dichloride (iPr(Cp)(Flu)ZrCl$_2$), using MAO as a cocatalyst, and produced PCPE with a new type of crystalline structure.[12] The WAXD pattern of this PCPE shows three diffraction peaks at $2\theta = 17\,°$, $19\,°$, and $21\,°$ with a wide-based reflection peak. PCPE exhibiting this WAXD pattern is defined as having crystal Type II. The WAXD pattern of Type II PCPE has a large halo region derived from the amorphous phase, indicating a lower crystallinity than Type I PCPE. The ^{13}CNMR spectrum of this material showed three strong peaks at 31, 39, and 47 ppm, as observed in the spectra of amorphous PCPEs prepared using C_s-symmetric Ph$_2$C(Cp)(Flu)ZrCl$_2$.[10] The nature of the ligand ring bridge in C_s-symmetric polymerization catalysts affects the morphology of the resulting PCPEs (crystalline for iPr(Cp)(Flu)ZrCl$_2$ versus amorphous for Ph$_2$C(Cp)(Flu)ZrCl$_2$).

15.2.2 HOMOPOLYMERIZATION WITH LATE TRANSITION METAL CATALYSTS

CPE is polymerized by some late transition metal catalysts. McLain et al. investigated polymerization and hydrooligomerization with MAO- and borate-activated Ni and Pd catalysts having diimine ligands.[13,14] These catalysts produce crystalline PCPE of Type II (similar to PCPE obtained with iPr(Cp)(Flu)ZrCl$_2$).[12] ^{13}C, DEPT, and 2D INADEQUATE NMR spectroscopy of this PCPE shows a cis-1,3-enchainment of the CPE units. The tacticity of polymerization was investigated by hydrooligomerizing CPE using a Ni diimine/MAO catalyst and performing a GC separation of the resultant pentamers (Scheme 15.4) and other oligomers. The tacticity of the oligomers was essentially atactic (52% m dyads) to moderately isotactic (66% m dyads). It was concluded that the crystalline structure of Type II PCPE differs in tacticity from the Type I structure. The T_ms of the Type II PCPEs obtained with Ni and Pd diimine catalysts ranged from 240 to 330 °C, lower than T_ms for Type I PCPEs. These late transition metal catalysts also polymerize substituted cyclopentenes such as 3-methylcyclopentene, 3-ethylcyclopentene, 4-methylcyclopentene, and 3-cyclopentylcyclopentene.[15,16]

The relationship between the structures of the zirconocene (and nickel) catalysts used for CPE polymerization and the features of the resulting PCPE are summarized in Table 15.1. While the

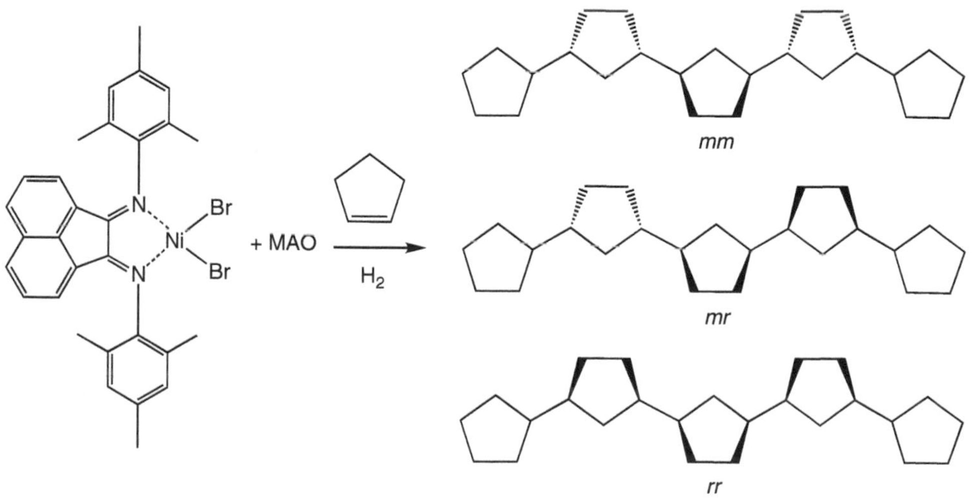

SCHEME 15.4 Hydrooligomerization of CPE with a Ni diimine catalyst to give pentamers. (Reprinted with permission from McLain, S. J.; Feldman, J.; McCord, E. F.; Gardner, K. H.; Teasley, M. F.; Coughlin, E. B.; Sweetman, K. J.; Johnson, L. K.; Brookhart, M. *Macromolecules* **1998**, *31*, 6705–6707. Copyright 1998 American Chemical Society.)

TABLE 15.1

Relationship between Catalyst Structure and Morphology of PCPE

Catalyst	Symmetry of Catalyst	CPE Enchainment	PCPE Tacticity	PCPE Morphology	References
Cp_2ZrCl_2	C_{2v}	cis	Atactic	Amorphous	10,11
rac-Et(Ind)$_2$ZrCl$_2$	C_2	cis	Isotactic	Crystal Type I	4
rac-Et(H$_4$Ind)$_2$ZrCl$_2$	C_2	cis, trans	Not determined	Not determined	8
rac-Me$_2$Si(Ind)$_2$ZrCl$_2$	C_2	cis	Isotactic	Crystal	10,11
iPr(Cp)(Flu)ZrCl$_2$	C_s	Not determined	Not determined	Crystal Type II	12
Ph$_2$C(Cp)(Flu)ZrCl$_2$	C_s	cis	Atactic	Amorphous	10,11
Ni diimine	—	cis	Atactic	Crystal Type II	13

catalyst nature affects the morphology of the resulting polymers, no clear correlation is observed between the symmetry of the catalyst and the morphology of the polymer. Relationships between catalyst nature and PCPE tacticity have been reported in fragments. Some C_2-symmetric catalysts, that is, rac-Et(Ind)$_2$ZrCl$_2$ and rac-Me$_2$Si(Ind)$_2$ZrCl$_2$, produce isotactic polymer. On the other hand, C_{2v}-symmetric Cp$_2$ZrCl$_2$, C_s-symmetric Ph$_2$C(Cp)(Flu)ZrCl$_2$, and nonmetallocene Ni diimine catalysts yield atactic polymer. Further consideration is needed to make general comments about the correlation between the catalyst nature, tacticity, and morphology of PCPE.

15.3 COPOLYMERIZATION OF ETHYLENE AND CYCLOPENTENE WITH TRANSITION METAL CATALYSTS

15.3.1 COPOLYMERIZATION WITH ZIRCONOCENE CATALYSTS

The copolymerization of ethylene and CPE was investigated with classical heterogeneous and homogeneous Ziegler–Natta catalysts by Natta. CPE was polymerized through both ring-opening and vinyl-type polymerization mechanisms, and yielded crystalline and elastomeric copolymers.[17–19] Kaminsky and Spiehl were the first to investigate the copolymerization of ethylene and CPE with homogeneous metallocene catalysts.[20] They achieved copolymerization under various polymerization conditions (−10 to 20 °C, 0:1–23:1 CPE:ethylene molar ratio, 0.296 mol/L ethylene concentration) using the MAO-activated isoselective zirconocene catalyst rac-Et(Ind)$_2$ZrCl$_2$ (0.1–6.4 μmol/L catalyst concentration, 220 mmol/L MAO concentration). The consumption of ethylene was used to monitor the copolymerization rate over time, and the maximum polymerization rate was observed at 3 min after initiation at the following conditions: 20 or 30 °C, 7.7:1 molar ratio of CPE:ethylene, and 0.64 μmol/L catalyst concentration. The highest overall polymerization rate was attained when the molar ratio of CPE:ethylene in the feed was 15.3:1. The CPE content in the copolymer increased linearly as the amount of CPE in the feed was increased, but decreased when the polymerization temperature was increased or catalyst concentration was decreased.

^{13}C NMR spectroscopy showed that CPE was preferentially incorporated in the copolymers through the 1,2-insertion mode, without ring-opening metathesis. The 1,2-enchainment pattern is a result of the facile coordination of ethylene to the active metal center: β-hydride elimination of a CPE-ended chain, which is needed to form a 1,3-enchained CPE unit, is relatively slow compared to ethylene insertion. In the NMR spectrum of the copolymer containing 28 mol% CPE, short blocks of PCPE units were detected.

Kaminsky and Spiehl also investigated the copolymerization of ethylene with cyclohexene, cycloheptene, and cyclooctene using the isoselective zirconocene rac-Et(Ind)$_2$ZrCl$_2$ under similar conditions to those used for ethylene/CPE copolymerization.[20] Ethylene copolymerization with

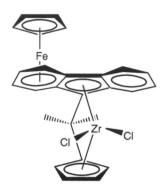

FIGURE 15.4 *rac*-Dimethylsilandiyl(ferroceno[2,3]inden-1-yl)(cyclopentadienyl)zirconium dichloride. (Reprinted with permission from Jerschow, A.; Ernst, E.; Hermann, W.; Müller, N. *Macromolecules* **1995**, *28*, 7095–7099. Copyright 1995 American Chemical Society.)

cycloheptene and cyclooctene yielded the corresponding copolymers (containing 3.2 mol% cycloheptene and 1.2 mol% cyclooctene, respectively) with vinyl-type polymerization of the cycloolefins. The activity for copolymerization with cycloheptene was four times smaller than that for copolymerization with CPE. In copolymerization with cyclooctene, the incorporation of cyclooctene caused a considerable decrease of polymer molecular weight with increasing comonomer content. Cyclohexene did not copolymerize, owing to the stability and the low ring strain of its six-membered ring structure.

Müller and coworkers investigated the copolymerization of ethylene and CPE using the C_1-symmetric zirconocene *rac*-dimethylsilandiyl(ferroceno[2,3]inden-1-yl)(cyclopentadienyl) zirconium dichloride (Figure 15.4) at 50–70 °C with MAO cocatalyst.[21] The microstructure of the copolymer was studied using one- and two-dimensional heteronuclear correlated NMR. Both 1,2-enchained and 1,3-enchained CPE units were detected. The ratio of 1,3- to 1,2-units increased as the CPE content in the copolymer increased, with up to 64% of the CPE units having a 1,3-structure. This catalyst also produced PCPE homopolymers having both 1,3- and 1,2-units. The cis/trans ratio was quantified for homo-PCPE sequences in these polymers; 10.0% of the 1,3-units and 2.9% of 1,2-units are trans. Owing to its different structure, *rac*-dimethylsilandiyl(ferroceno[2,3]inden-1-yl)(cyclopentadienyl)zirconium dichloride should produce different PCPE homopolymer and ethylene/CPE copolymer microstructures than the *rac*-Et(Ind)$_2$ZrCl$_2$ catalyst.

Naga and Imanishi have studied the effects of zirconocene catalyst ligand structure on the copolymerization of ethylene and CPE.[22] The catalysts used in the investigation were the nonbridged (aspecific) catalysts Cp$_2$ZrCl$_2$, bis(pentamethylcyclopentadienyl)zirconium dichloride (Cp*$_2$ZrCl$_2$), and bis(indenyl)zirconium dichloride (Ind$_2$ZrCl$_2$); the bridged aspecific catalyst dimethylsilylenebis(cyclopentadienyl)zirconium dichloride (Me$_2$SiCp$_2$ZrCl$_2$); the isoselective catalysts *rac*-Et(Ind)$_2$ZrCl$_2$ and *rac*-Me$_2$Si(Ind)$_2$ZrCl$_2$; and the syndioselective catalyst Ph$_2$C(Cp)(Flu)ZrCl$_2$ (conditions: toluene solvent, 40 °C, 1.0 atm ethylene, MAO activation). All of the nonbridged catalysts produced polyethylene (homopolymer) without incorporation of CPE. On the other hand, all of the bridged catalysts yielded copolymer. Amongst the catalysts used, *rac*-Et(Ind)$_2$ZrCl$_2$ showed the highest reactivity toward CPE, producing a copolymer with 38 mol% CPE. The CPE units in the copolymers obtained with all of the bridged catalysts except for *rac*-Et(Ind)$_2$ZrCl$_2$ were mainly enchained as cis-1,2 units as determined by ^{13}C NMR, independent of the symmetry of the catalyst.

The copolymers obtained with *rac*-Et(Ind)$_2$ZrCl$_2$ contained not only cis-1,2 CPE units but also cis-1,3 CPE units (20%–30% of total CPE units). These 1,3-inserted CPE units were also detected in high-CPE content copolymers (>20 mol%) obtained with the other bridged catalysts. The ^{13}C NMR CCC triad sequence (assigned by Müller[21]), where C represents the CPE unit, was not observed in any of the copolymers. Small amounts of ^{13}C NMR CC dyad sequences[21] were detected in the copolymers obtained with $Me_2SiCp_2ZrCl_2$ and rac-$Me_2Si(Ind)_2ZrCl_2$, and no [CC] dyads were detected in the copolymer obtained with $Ph_2C(Cp)(Flu)ZrCl_2$. These results indicate the difficultly of reacting CPE monomer with a growing polymer chain having a CPE unit as its last-inserted monomer. Moreover, there is the possibility of the highly alternating copolymerization of ethylene and CPE.

Differential scanning calorimetry (DSC) analysis of the copolymers obtained with $MeSiCp_2ZrCl_2$ and rac-$Et(Ind)_2ZrCl_2$ showed multiple melting endotherms. Temperature rising elution fractionation (TREF) of the copolymers showed a broad composition distribution, which should arise from multiple active sites during copolymerization. Kaminsky and Spiehl described a broad molecular weight distribution for ethylene/CPE copolymers obtained with rac-$Et(Ind)_2ZrCl_2$,[20] a finding that suggests the same broad distribution of copolymer composition. On the other hand, a narrow copolymer composition distribution was confirmed by TREF for ethylene/CPE copolymers obtained with rac-$Me_2Si(Ind)_2ZrCl_2$. This result suggests that the nature of the ring bridge in the catalyst affects the composition distribution of the copolymer. A possible explanation is a distortion of indenyl ligands of rac-$Et(Ind)_2ZrCl_2$ owing to internal rotation of the ethylidene bridge. This catalyst forms either a right-handed or a left-handed ring helix as reported by Kaminsky et al. The internal rotation between its right- and left-handed forms changes the geometry of the coordination sites, and therefore the monomer reactivity of the active sites.[23]

15.3.2 SYNTHESIS OF ETHYLENE/CYCLOPENTENE COPOLYMERS HAVING ALTERNATING AND BLOCK MONOMER SEQUENCES

Alternating copolymerizations of ethylene and CPE have been achieved using some titanium catalysts. Coates and Fujita reported the synthesis of highly alternating ethylene-CPE copolymers (no CC dyad sequences) using a MAO-activated bis(phenoxyimine)titanium dichloride catalyst (Scheme 15.5a).[24] The T_g of the copolymers increased as the CPE content in the copolymer increased, and ranged from −27.3 (27 mol% CPE) to 10.1 °C (47 mol% CPE). Coates et al. also synthesized comparative atactic and isotactic poly(ethylene-alt-(cis-1,2-cyclopentene))s, using the ring-opening metathesis polymerization of bicyclo[3.2.0]hept-6-ene and subsequent hydrogenation of the unsaturated main chain. The ^{13}C NMR spectra of these authentic samples clearly show that the bis(phenoxyimine)titanium dichloride catalyst produces atactic, highly alternating ethylene-CPE copolymers. Block copolymers of ethylene and ethylene-alt-(cis-1,2-cyclopentene) were also synthesized with this catalyst by repeatedly raising and lowering the ethylene pressure of the ethylene/CPE copolymerization (Scheme 15.5b). This reaction produced several kinds of diblock and multiblock copolymers.

Waymouth et al. investigated highly alternating CPE/ethylene copolymerization using constrained geometry catalysts (CGCs).[25] Copolymerization with a dimethylsilylene(tetramethyl cyclopentadienyl)(N-$tert$-butyl)titanium dichloride catalyst activated with a modified methyl-isobutyl aluminoxane (MMAO) yielded highly alternating, atactic copolymers with cis-1,2-enchained CPE units (Scheme 15.6a). On the other hand, a chiral indenyl CGC, dimethylsilylene(indenyl)(N-$tert$-butyl)titanium dichloride, was able to induce the highly isoselective and perfectly alternating copolymerization of ethylene and CPE (Scheme 15.6b). During copolymerization with these CGCs, CPE was incorporated preferentially through cis-1,2-insertion. The isotactic copolymer had a higher T_g and T_m (50 mol% CPE, $T_g = 16.3\,°C$, $T_m = 182.5\,°C$) than the atactic copolymer (40 mol% CPE, $T_g = -21.8\,°C$, $T_m = 127.2\,°C$).

Waymouth and Lavoie also recently reported the highly alternating stereoselective copolymerization of ethylene with CPE, cycloheptene, and cyclooctene using a series of CGCs.[26] Highly alternating CPE/ethylene copolymerization was also achieved using CGCs having benz[6,7]indenyl ligands.

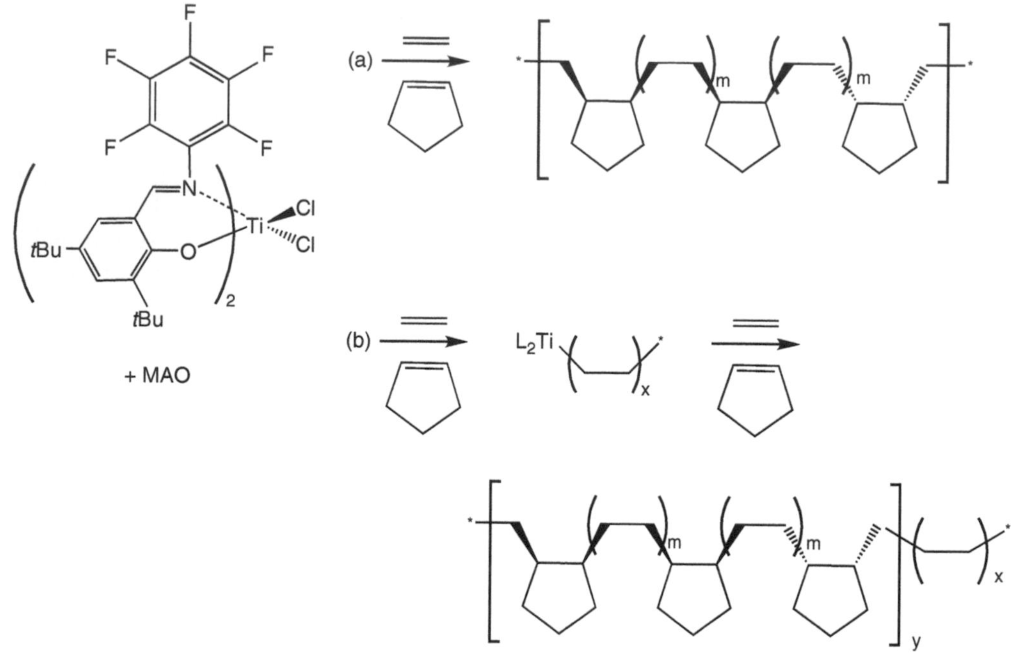

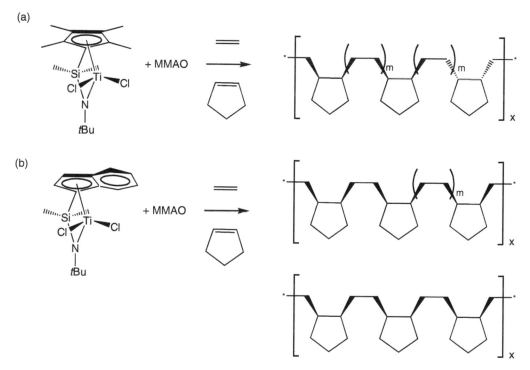

SCHEME 15.5 Copolymerization of ethylene and CPE with a bis(phenoxyimine)titanium dichloride catalyst: (a) alternating copolymerization; (b) block copolymerization producing blocks of polyethylene and poly(ethylene-*alt*-(*cis*-1,2-cyclopentene) (L = phenoxyimine ligand). (Reprinted with permission from Fujita, M.; Coates, G. W. *Macromolecules* **2002**, *35*, 9640–9647. Copyright 2002 American Chemical Society.)

SCHEME 15.6 Copolymerization of ethylene and CPE with (a) pentamethylcyclopentadienyl and (b) indenyl constrained geometry catalysts. (Lavoie, A. R.; Ho, M. H.; Waymouth, R. M. *Chem. Commun.* **2003**, 864–865. Reproduced by permission of The Royal Society of Chemistry.)

15.4 COPOLYMERIZATION OF PROPYLENE AND CYCLOPENTENE WITH TRANSITION METAL CATALYSTS

CPE may be incorporated into polypropylenes in order to lower their T_ms and raise their T_gs. Kaminsky et al. reported the copolymerization of propylene and CPE using the isoselective zirconocenes rac-Et(Ind)$_2$ZrCl$_2$ and rac-Me$_2$Si(Ind)$_2$ZrCl$_2$.[5,27] ^{13}C NMR spectroscopy of the resulting isotactic copolymers (containing 2–5 mol% CPE) revealed a 1,2-enchainment mode for the CPE units; no PCPE blocks were detected (Figure 15.5).

Köller and coworkers investigated propylene/CPE copolymerization with rac-Me$_2$Si(Ind)$_2$ZrCl$_2$ and prepared copolymers with various CPE contents ranging from 0.5 to 30.5 mol%.[28] The T_gs of the copolymers increased as the CPE content increased, and reached 19 °C at 30.5 mol% CPE. The calculated reactivity ratios were $r_p = 40$ and $r_c = 0.001$, where $r_p = k_{pp}/k_{pc}$, $r_c = k_{cc}/k_{cp}$, and k_{pp}, k_{pc}, k_{cp}, and k_{cc} are the rate constants of propylene addition to a propylene terminal chain, CPE addition to a propylene terminal chain, propylene addition to a CPE terminal chain, and CPE addition to a CPE terminal chain, respectively.

In order to study the precise insertion mode for CPE in its copolymerization with propylene, Naga and Imanishi investigated copolymerization using the isoselective catalysts rac-Et(Ind)$_2$ZrCl$_2$, rac-Me$_2$Si(Ind)$_2$ZrCl$_2$, and rac-dimethylsilylenebis(2-methylindenyl)zirconium dichloride (rac-Me$_2$Si(2-Me-Ind)$_2$ZrCl$_2$) and the syndioselective catalyst Ph$_2$C(Cp)(Flu)ZrCl$_2$.[29] The isoselective zirconocenes produced isotactic propylene-CPE copolymers with narrow molecular weight distributions, whereas the syndioselective zirconocene produced syndiotactic polypropylene (homopolymer). The catalysts rac-Et(Ind)$_2$ZrCl$_2$ and rac-Me$_2$Si(Ind)$_2$ZrCl$_2$, which show regioirregularity and relatively low isoselectivity in propylene polymerization,[30] produced isotactic copolymers containing 1.4–10.1 mol% CPE enchained as both cis-1,2- and cis-1,3-units (Scheme 15.7). The 1,3-enchained CPE units comprised 52%–62% of the total CPE units present in the copolymers obtained with rac-Et(Ind)$_2$ZrCl$_2$ and 23%–32% for the copolymers obtained with rac-Me$_2$Si(Ind)$_2$ZrCl$_2$. On the other hand, the highly isoselective catalyst rac-Me$_2$Si(2-Me-Ind)$_2$ZrCl$_2$ produced an isotactic copolymer incorporating 6.1 mol% CPE as preferentially cis-1,2-inserted CPE units.

The kinetic parameters for copolymerization, that is, the relative rates of CPE 1,2-insertion and 1,3-insertion (through isomerization), were estimated from the relative amounts of 1,2- and 1,3-inserted units in the copolymers. The proportion of 1,3-inserted units increased as the CPE content in the copolymers increased. The rate constant ratio k_i/k_{cp} for copolymerization with rac-Et(Ind)$_2$ZrCl$_2$, where k_i and k_{cp} are the rate constants of, respectively, isomerization of a 1,2-inserted CPE unit to form a 1,3-inserted CPE unit and propylene addition to a 1,2-inserted CPE propagating chain end, is 3.6 times greater than k_i/k_{cp} for rac-Me$_2$Si(Ind)$_2$ZrCl$_2$ (Scheme 15.7). This indicates that propylene is more reactive with the 1,2-inserted CPE propagating chain end of the rac-Me$_2$Si(Ind)$_2$Cl$_2$ catalyst than with that of the rac-Et(Ind)$_2$ZrCl$_2$ catalyst. This trend agrees with the order of the catalysts' r_p values; $r_p = 45.5$ for rac-Et(Ind)$_2$ZrCl$_2$ and 72.0 for rac-Me$_2$Si(Ind)$_2$ZrCl$_2$. The structural differences between these catalysts, that is, the somewhat narrower coordination space of rac-Me$_2$Si(Ind)$_2$ZrCl$_2$ as compared to rac-Et(Ind)$_2$ZrCl$_2$, would cause the lower CPE reactivity and increased r_p value seen in copolymerization using the former catalyst.[31,32]

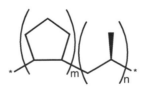

FIGURE 15.5 Enchainment modes of CPE and propylene units for propylene/CPE copolymerization with isospecific zirconocene catalysts (a block copolymer structure is not implied).

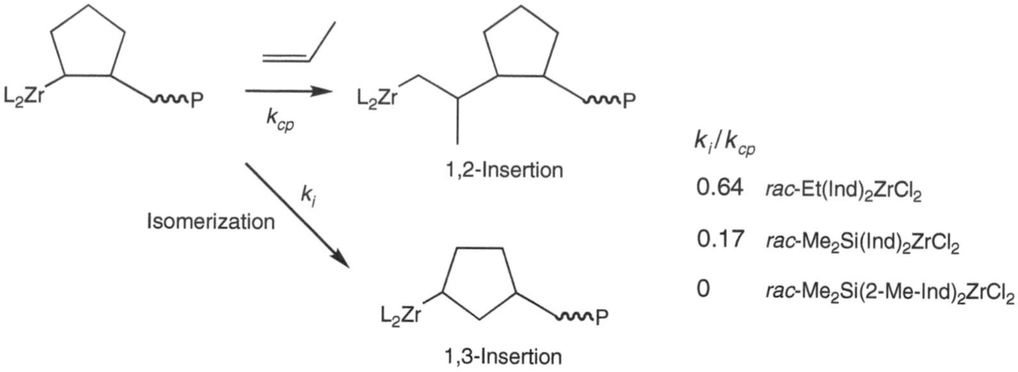

SCHEME 15.7   Insertion modes of CPE (1,2 versus 1,3) in the copolymerization of propylene and CPE with isospecific zirconocene catalysts ($L_2 = $ *ansa*-indenyl ligand and P = polymer chain). (Reprinted from Naga, M.; Imanishi, Y. *Polymer* **2002**, *43*, 2133–2139. Copyright 2002, with permission from Elsevier.)

15.5 CONCLUSIONS

The development of early and late transition metal catalysts has advanced olefin polymerization and the synthesis of new polyolefins. Cyclopentene (co)polymers are interesting polyolefins owing to the diverse insertion modes and controllable monomer sequence distributions possible with this monomer. However, CPE polymerization and copolymerization require further research into achieving precise control of the polymerization process in order to best tailor polymer properties for commercial applications. For example, while PCPE is an excellent nucleating agent for isotactic polypropylene,[33] it is not suitable for construction materials because its high melting temperature makes it difficult to process. New polymerization technologies, particularly the further development of transition metal catalysts, will hopefully solve these problems and make the tailored synthesis and commercialization of CPE (co)polymers possible for special uses, such as heat-resistant polyolefins or modifiers for other polyolefin materials.

REFERENCES AND NOTES

1. Boor, J. E. A.; Youngman, E. A.; Dimbat, M. Polymerization of cyclopentene, 3-methylcyclopentene, and 3-methylcyclohexene. *Makromol. Chem.* **1966**, *90*, 26–37.
2. Natta, G.; Dall'Asta, G.; Bassi, I. W.; Carella, G. Stereospecific ring cleavage homopolymerization of cycloolefins and structural examination of the resulting homologous series of linear crystalline *trans*-polyalkenamers. *Makromol. Chem.* **1966**, *91*, 87–106.
3. Boor, J. *Ziegler–Natta Catalysts and Polymerization*; Academic Press: New York, 1979; p 526.
4. Kaminsky, W.; Bark, A.; Spiehl, R.; Moller-Lindenhof, N.; Niedoba, S. Isotactic polymerization of olefins with homogeneous zirconium catalysts. In *Transition Metals and Organometallics as Catalysts for Olefin Polymerization*; Kaminsky, W., Sinn, H., Eds.; Springer-Verlag: Berlin, 1988; pp 291–301.
5. Kaminsky, W.; Arndt, M.; Bark, A. New results of the polymerization of olefins with metallocene/aluminoxane catalysts. *Polym. Prepr. (Am. Chem. Soc. Div. Polym. Chem.)* **1991**, *32(1)*, 467–468.
6. Kaminsky, W.; Bark, A.; Däke, I. Polymerization of cyclic olefins with homogeneous catalysts. In *Catalytic Olefin Polymerization*; Keii, T., Soga, K., Eds.; Kodansha Ltd.: Tokyo, 1990; pp 425–438.
7. Collins, S.; Kelly, W. M. The microstructure of poly(cyclopentene) produced by polymerization of cyclopentene with homogeneous Ziegler–Natta catalysts. *Macromolecules* **1992**, *25*, 233–237.
8. Kelly, W. M.; Taylor, N. J.; Collins, S. Polymerization of cyclopentene using metallocene catalysts: Polymer tacticity and properties. *Macromolecules* **1994**, *27*, 4477–4485.
9. Kelly, W. M.; Wang, S.; Collins, S. Polymerization of cyclopentene using metallocene catalysts: Competitive cis- and trans-1,3 insertion mechanisms. *Macromolecules* **1997**, *30*, 3151–3158.

10. Arndt, M.; Kaminsky, W. Polymerization with metallocene catalysts. Hydrooligomerization and NMR investigations concerning the microstructure of poly(cyclopentenes). *Macromol. Symp.* **1995**, *95*, 167–183.

11. Arndt, M.; Kaminsky, W. Microstructure of poly(cycloolefins) produced by metallocene/methylaluminoxane (mao) catalysts. *Macromol. Symp.* **1995**, *97*, 225–246.

12. Asanuma, T.; Tamai, Y. Stereoregular polycyclopentene and its manufacture. Jpn. Kokai Tokkyo Koho JP 03139506 A2 (Mitsui Toatsu), June 13, 1991; *Chem. Abstr.* **1991**, *115*, 208858.

13. McLain, S. J.; Feldman, J.; McCord, E. F.; Gardner, K. H.; Teasley, M. F.; Coughlin, E. B.; Sweetman, K. J.; Johnson, L. K.; Brookhart, M. Addition polymerization of cyclopentene with nickel and palladium catalysts. *Macromolecules* **1998**, *31*, 6705–6707.

14. McLain, S. J.; Feldman, J.; McCord, E. F.; Gardner, K. H.; Teasley, M. F.; Coughlin, E. B.; Sweetman, K. J.; Johnson, L. K.; Brookhart, M. Poly(cyclopentene): A new processible high-melting polyolefin made from Ni and Pd catalysts. *Polym. Mat. Sci. Eng.* **1997**, *76*, 20–21.

15. McLain, S. J.; McCord, E. F.; Bennett, A. M. A.; Ittel, S. D.; Sweetman, K. J.; Teasley, M. F. Polymers of substituted cyclopentenes. PCT Int. Pat. Appl. WO 99/50320 A2 (du Pont), October 7, 1999.

16. Ittel, S. D.; Johnson, L. K.; Brookhart, M. Late-metal catalysts for ethylene homo- and copolymerization. *Chem. Rev.* **2000**, *100*, 1169–1203.

17. Natta, G.; Dell'Asta, G.; Mazzanti, G.; Pasqunn, I.; Valvassori, A.; Zambelli, A. Crystalline, alternating ethylene-cyclopentene copolymers and other ethylene-cycloolefin copolymers. *Makromol. Chem.* **1962**, *54*, 95–101.

18. Natta, G.; Allegra, G.; Bassi, I. W.; Corradini, P.; Ganis, P. *Makromol. Chem.* **1962**, *58*, 242–243.

19. Dall'Asta, G.; Mazzanti, G. Reactivity of various cycloolefins in copolymerization with ethylene by using anionic coordination catalysts. *Makromol. Chem.* **1963**, *61*, 178–197.

20. Kaminsky, W.; Spiehl, R. Copolymerization of cycloalkenes with ethylene in presence of chiral zirconocene catalysts. *Makromol. Chem.* **1989**, *190*, 515–526.

21. Jerschow, A.; Ernst, E.; Hermann, W.; Müller, N. Nuclear magnetic resonance evidence for a new microstructure in ethene-cyclopentene copolymers. *Macromolecules* **1995**, *28*, 7095–7099.

22. Naga, N.; Imanishi, Y. Copolymerization of ethylene and cyclopentene with zirconocene catalysts: Effect of ligand structure of zirconocenes. *Macromol. Chem. Phys.* **2002**, *203*, 159–165.

23. Kaminsky, W.; Rabe, O.; Schauwienold, A. M.; Schupfner, G. U.; Hanss, J.; Kopf, J. Crystal structure and propylene polymerization characteristics of bridged zirconocene catalysts. *J. Organomet. Chem.* **1995**, *497*, 181–193.

24. Fujita, M.; Coates, G. W. Synthesis and characterization of alternating and multiblock copolymers from ethylene and cyclopentene. *Macromolecules* **2002**, *35*, 9640–9647.

25. Lavoie, A. R.; Ho, M. H.; Waymouth, R. M. Alternating stereospecific copolymerization of cyclopentene and ethylene with constrained geometry catalysts. *Chem. Commun.* **2003**, 864–865.

26. Lavoie, A. R.; Waymouth, R. M. Catalytic syntheses of alternating, stereoregular ethylene/cycloolefin copolymers. *Tetrahedron* **2004**, *60*, 7147–7155.

27. Kaminsky, W.; Bark, A.; Arndt, M. New polymers by homogenous zirconocene/aluminoxane catalysts. *Makromol. Chem., Macromol. Symp.* **1991**, *47*, 83–93.

28. Arnold, M.; Henschke, O.; Köller, F. Copolymerization of propene and cyclopentene with the metallocene catalyst $(CH_3)_2Si[Ind]_2ZrCl_2$/methyl aluminoxane. *Macromol. Rep.* **1996**, *A33 (Suppl. 3 and 4)*, 219–227.

29. Naga, N.; Imanishi, Y. Structure of cyclopentene unit in the copolymer with propylene obtained by stereospecific zirconocene catalysts. *Polymer* **2002**, *43*, 2133–2139.

30. Resconi, L.; Piemontesi, F.; Camurati, I.; Sudmeijer, O.; Nifant'ev, I. E.; Ivchenko, P. V.; Kuz'mina, L. G. A new class of isospecific, highly regiospecific zirconocene catalysts for the polymerization of propene. *J. Am. Chem. Soc.* **1998**, *120*, 2308–2321.

31. Wild, F. R. W. P.; Wasiucionek, M.; Huttner, G.; Brintzinger, H. H. *ansa*-Mettallocene derivatives. VII. Synthesis and crystal structure of a chiral *ansa*-zirconocene derivative with ethylene-bridged tetrahydroindenyl ligands. *J. Organomet. Chem.* **1985**, *288*, 63–67.

32. Spaleck, W.; Antberg, M.; Rohrmann, J.; Winter, A.; Bachmann, B.; Kiprof, P.; Behm, J.; Hermann, W. High-molecular-weight polypropylene via mass-tailored zirconocene-type catalysts. *Angew. Chem., Int. Ed. Engl.* **1992**, *31*, 1347–1350.

33. Lee, D. H.; Yoon, K. B. Effect of polycyclopentene on crystallization of isotactic polypropylene. *J. Appl. Polym. Sci.* **1994**, *54*, 1507–1511.

16 Tactic Norbornene Homo- and Copolymers Made with Early and Late Transition Metal Catalysts

Walter Kaminsky and Michael Arndt-Rosenau

CONTENTS

16.1 INTRODUCTION

Metallocene/methylaluminoxane (MAO) and other single site catalysts for olefin polymerization have opened a new field of synthesis in polymer chemistry.[1–3] Strained cyclic olefins such as cyclobutene, cyclopentene, norbornene (NB), and their substituted compounds can be used as monomers and comonomers in a wide variety of polymers.[4] Much interest is focused on norbornene homo- and copolymers because of the easy availability of norbornene and the special properties of their polymers. Norbornene can be polymerized by ring opening metathesis polymerization (ROMP), giving elastomeric materials,[5] or by double bond opening (addition polymerization). Homopolymerization of norbornene by double bond opening can be achieved by early and late transition metal catalysts, namely Ti, Zr, Hf,[6–8] Ni,[9,10] and Pd[11–13] (Scheme 16.1).

Polymerization of norbornene with heterogeneous Ziegler–Natta catalysts is accompanied by ROMP,[14] whereas homogeneous metallocene, Ni, and Pd catalysts promote addition polymerization. The polymers feature two chiral centers per monomer unit and therefore are ditactic.

Cyclic monomers can be divided into achiral and prochiral types. Polymerization of both types by chiral metallocenes may yield tactic, crystalline homopolymers.[15] The melting points (T_ms) of

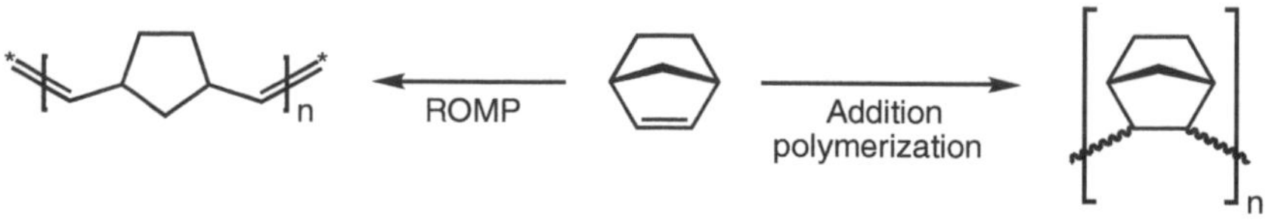

SCHEME 16.1   Mechanisms for the transition metal-catalyzed polymerization of norbornene (NB).

these homopolymers are extremely high, about 600 °C, and decomposition occurs before melting. Whereas the atactic polymers can be dissolved in hydrocarbon solvents at least to some extent, the tactic polymers are hardly soluble.

Polynorbornene (PNB) homopolymers of industrial interest are not processible owing to their high T_ms and insolubility in common organic solvents. By copolymerization of norbornene with ethylene, a cycloolefin copolymer (COC) can be produced. These new materials have been the focus of academic and industrial research. Ethylene/norbornene (E/NB) copolymers are usually amorphous and show an excellent transparency and high refractive index, making them suitable for optical applications.[16]

16.2 POLYNORBORNENE

16.2.1 Early Transition Metal Catalysts

The use of early transition metal catalysts for the homopolymerization of norbornene dates back to 1963. Sartori et al.[17] reported on the polymerization of norbornene by $TiCl_4/(i\text{-}Bu)_3Al$ (Ti:Al=1:2). Further reports refer to the analogous $TiCl_4/Et_3Al$ system, which produces a mixture of ROMP and addition polymer.[18]

While the copolymerization of norbornene and ethylene by vanadium and titanium catalysts had been investigated in the 1970s, it was not until the invention of metallocene/MAO catalysts and their application to the homopolymerization of norbornene and other cycloolefins by Kaminsky et al.[8,15,19] that the use of early transition metal catalysts for the addition polymerization of norbornene drew new attention. Nevertheless, the industrial relevance of E/NB copolymers, and the nature of the homopolymers, described in the first reports as insoluble in organic solvents, crystalline, and extremely high melting, focused further investigations on copolymers rather than on norbornene homopolymers.

Thus, it was not until 1990 that the group of Kaminsky and Arndt-Rosenau took a more detailed look at the homopolymerization of norbornene and the structures of the resulting polymers. Driven by the growing interest in copolymers with high norbornene contents and high glass transition (T_g) temperatures, as well as the unusual properties of PNBs, Arndt-Rosenau et al.[20,21] used the hydrooligomerization technique to produce saturated model norbornene dimers and trimers with metallocene catalysts known to produce atactic, isotactic, and syndiotactic poly(α-olefins) (**1–3**, Table 16.1).

The authors analyzed the structures and distributions of the oligomers and tried to correlate results with the metallocene structure and the polymerization mechanism, in order to extrapolate the structural findings to the polymer microstructure. It was shown that oligomers (and polymers) of different stereochemistries (tacticities) can be produced using metallocene catalysts, and that while some of the PNBs were insoluble in common organic solvents, other catalysts gave polymers soluble in, for example, toluene.

Their investigations showed that norbornene was inserted into the M–C bonds of the growing polymer chain (or into M–H bonds formed by chain transfer) in a cis-exo fashion. Owing to the structure of norbornene, each insertion generates two chiral centers of opposite stereochemistry and

TABLE 16.1
Characteristics of Metallocene/MAO Catalysts Used for Norbornene Hydrooligomerization by Arndt-Rosenau[20]

Structure

Name[a]	Cp$_2$ZrCl$_2$ (1)	rac-[Me$_2$Si(Ind)$_2$]ZrCl$_2$ (2)	[Ph$_2$C(Cp)(Flu)]ZrCl$_2$ (3)
Symmetry	C_{2v}	C_2	C_s
Topology of sites for monomer coordination at [L$_2$Zr(P)]$^+$[b]	Homotopic	Homotopic	Enantiotopic
Topology of sites for [L$_2$Zr(NB)(P)]$^+$[b]	Nonchirotopic, nonstereogenic	Chirotopic, nonstereogenic	Chirotopic, stereogenic
Tacticity in α-olefin polymerization	Atactic	Isotactic	Syndiotactic

[a] Cp = cyclopentadienyl; Ind = 1-indenyl; Flu = 9-fluorenyl.
[b] P = polymer chain; L = ligand (Cp, Ind, Flu).

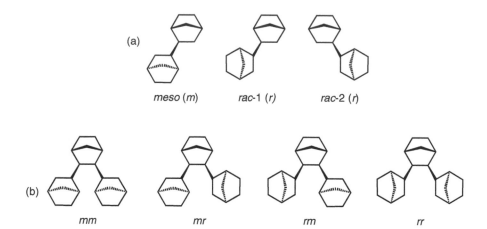

FIGURE 16.1 Structures of the hydrodimers (a) and hydrotrimers (b) of norbornene according to Arndt-Rosenau.

therefore the resulting polymers are erythroditactic. The possible configurational base units (the erythroditacticities) of homo-PNB are represented by the following configurations:

erythrodiisotactic	-(RS)(RS)(RS)(RS)(RS)(RS)(RS)(RS)-
	m m m m m m m
erythrodisyndiotactic	-(RS)(SR)(RS)(SR)(RS)(SR)(RS)(SR)-
	r r r r r r r
erythroatactic	-(RS)(SR)(SR)(RS)(SR)(SR)(RS)(RS)-
	r m r r m r m

Detailed 2D nuclear magnetic resonance (NMR) investigations on the hydrodimers formed by 1–3 showed that all three catalyst precursors (when combined with MAO) yielded two diastereomeric hydrodimers, which were found to be the *meso* and racemic isomers shown in Figure 16.1a.

Despite the catalysts' different stereoselectivity in α-olefin polymerization, all three favored the formation of the *meso*-hydrodimer. In case of the achiral catalyst Cp_2ZrCl_2 (**1**, Cp = cyclopentadienyl), this has to be attributed to chain-end control since the Cp ligands do not induce stereopreference. From the *meso/rac* ratio, a difference in the free energies of activation ($\Delta G^{\#}_{meso} - \Delta G^{\#}_{rac}$) of 1.5 kJ/mol at 30 °C has been calculated.

Of the three hydrotrimers that may result from a cis-exo insertion (Figure 16.1b), **1** was found to produce only two: the *mm* and *mr* isomers (Table 16.2). No *rr*-hydrotrimer is produced, showing that a racemic enchainment (*r* dyad) formed by the first two insertions preferably is followed by a *meso* linkage (*m* dyad). In addition, the high amount of *mr* isomer found in combination with the distribution of hydrodimers indicates that a *meso* enchainment formed by the first two insertions is preferably followed by a racemic one. Thus, it can be concluded that the penultimate unit has a rather strong influence on the monomer insertion (one has to bear in mind that every main chain atom of the growing polymer is a chiral center and that norbornene has to be considered as a bulky and rigid monomer). Based on the distribution of hydrodimers and trimers, a polymer formed by the same mechanisms should have a heterotactic (*mrmrmrm*) structure disturbed by a significant amount of *mm* sequences.

Based on the mechanisms known from α-olefin polymerization, C_s-symmetric [Ph_2C(Cp)(Flu)]$ZrCl_2$ (**3**, Flu = 9-fluorenyl) should produce an erythrodisyndiotactic polymer and therefore favor the formation of the *rac*-hydrodimer, whereas C_2-symmetric *rac*-[Me_2Si(Ind)_2]ZrCl_2

TABLE 16.2

Mol% Distribution of the Hydrodimers and -Trimers of Norbornene Produced at 30 °C by Metallocene/MAO Catalysts 1–3[20]

	Cp_2ZrCl_2 (1) C_{2v}	rac-$[Me_2Si(Ind)_2]ZrCl_2$ (2) C_2	$[Ph_2C(Cp)(Flu)]ZrCl_2$ (3) C_s
meso-hydrodimer	65	58	53
rac-hydrodimer	35	42	47
mm hydrotrimer	23	62	8
(*mr* + *rm*) hydrotrimer	77	38	15
rr hydrotrimer	0	0	77

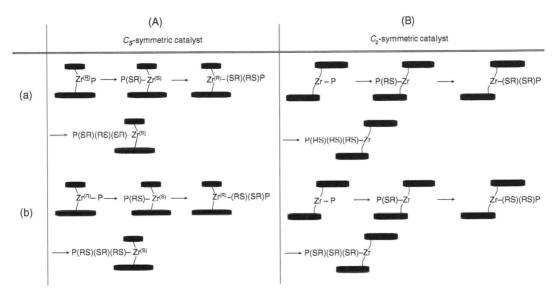

SCHEME 16.2   Stereochemistry during the polymerization of a prochiral cycloolefin such as norbornene using C_s- (A) and C_2-symmetric (B) catalysts. Independent of the relative topicities of the single insertion events (contrasted by pathways a and b), application of the mechanisms known from α-olefin polymerization predict an erythrodisyndiotactic polymer to be formed by C_s-symmetric catalysts and an erythrodiisotactic polymer to be formed by C_2-symmetric catalysts.

(**2**, Ind = 1-indenyl) should produce an erythrodiisotactic polymer and therefore preferably yield the *meso*-hydrodimer.

As shown in Scheme 16.2 and Table 16.2, both catalysts yield mixtures of *meso*- and *rac*-hydrodimers with a *meso/rac* ratio greater than one. This can be explained by a change of the relative topicities of insertion from the first to the second insertion, and is in accordance with observations by Pino[22] and Corradini[23] showing that the insertion into an M–H bond of an active metallocene catalyst species is governed directly by steric interactions of the monomer with the ligand framework of the catalyst (direct stereocontrol), while further insertions are controlled by the orientation of the (bulky) growing polymer chain and its interactions with the ligand framework (indirect stereocontrol).

In accordance with this mechanism, **2** produces no *rr*-hydrotrimer; thus a low stereoselectivity of the first insertion (forming both *m* and *r* dimers) is followed by highly stereoselective consecutive insertions producing *meso* enchainments only (*mm* and *rm* trimers). As expected, **3**

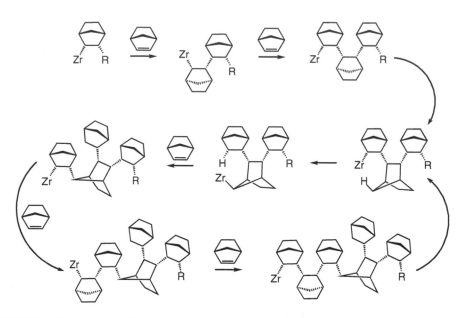

SCHEME 16.3 Mechanism of the formation of 2-*exo*,7'-*syn*-enchained units during norbornene polymerization by *rac*-[Me$_2$C(Ind)$_2$]ZrCl$_2$/MAO catalysts according to Fink[25] (R = H, Me; bis(indenyl) ligand and charge on Zr center are not shown).

produces preferably the *rr*-hydrotrimer plus some *mr*-hydrotrimer; nevertheless, a small amount of *mm*-hydrotrimer was also found, indicating the lower stereoselectivity of this catalyst compared to **2**.

In a later paper, Arndt-Rosenau and Gosmann[24] reported on the crystal structure of a hydropentamer of norbornene produced using *rac*-[C$_2$H$_4$(H$_4$Ind)$_2$]ZrCl$_2$ (H$_4$Ind = 4,5,6,7-tetrahydroindenyl; **4**, *vide infra*) that shows a trisubstituted central norbornene linking unit and consists of a stereoregular (erythrodiisotactic) trimer to which a stereoregular (*meso*) dimer is attached. More recently, Fink et al.[25] have extended the work of Arndt-Rosenau and characterized tetramers and pentamers obtained by the hydrooligomerization of norbornene using *rac*-[Me$_2$C(Ind)$_2$]ZrCl$_2$ (**5**, *vide infra*, Scheme 16.3). They propose C–H activation at the C7 carbon of the most recently inserted norbornene unit to be the root cause of the formation of trisubstituted units, and based on the results of deuterium labeling experiments and molecular modeling, assume that PNBs produced by some metallocene catalysts may have a regular structure involving such trisubstituted norbornene units.

By 1994, Arndt[20] had pointed out that norbornene behaves differently from other olefins in addition polymerization owing to its rigid structure, which prevents the formation of agostic interactions with hydrogen atoms attached to main chain carbon centers, and therefore leaves the active metallocene center highly unsaturated. The calculations by Fink et al.[25] indicate that this is the root cause of the C7-linkage formation. If this mechanism involving C–H σ-bond metathesis and subsequent monomer insertion at the activated C7 site is regular, a PNB with alternating 2-*exo*,7'-*syn* and 2-*exo*,2'-*exo*-enchainments in the main chain and 2-*exo*,2'-*exo*-linked norbornyl branches would be the result (Figure 16.2). Differences in solubilities and ^{13}C-CPMAS (cross-polarization/magic-angle spinning) NMR spectra of polymers produced by different catalysts indicate that not all metallocenes produce PNBs of this special structure.

Half-sandwich complexes of titanium, when activated by a cocatalyst such as MAO, are even more active for the homopolymerization of norbornene than metallocenes.[26,27] Heitz and Peucker[28] found that chromium-based half-sandwich complexes can also be activated to produce PNBs and E/NB copolymers. Nevertheless, detailed investigations on half-sandwich complexes for the addition polymerization of norbornene do not appear in the literature.

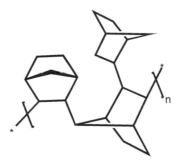

FIGURE 16.2 Structure of a polynorbornene containing alternating 2-*exo*,7′-*syn*- and 2-*exo*,2′-*exo*-enchainments in the main chain and 2-*exo*,2′-*exo*-linked norbornyl branches.

16.2.2 LATE TRANSITION METAL CATALYSTS

The first addition polymerization of norbornene catalyzed by a late transition metal compound dates back to work of Schultz in 1966 using $PdCl_2$ as a catalyst.[29] In the late 1970s, palladium complexes of the type L_2PdCl_2 (L = C_6H_5CN, Ph_3P) were reported to promote the addition polymerization of norbornene,[30–32] and based on the work of Sen and Lai,[33,34] the dicationic system $[(CH_3CN)_4Pd]^{2+}[BF_4]_2^-$ was found to be a very active catalyst.

As with early transition metal catalysts, renewed interest in the Pd-catalyzed polymerization of norbornene and its derivatives grew in the early 1990s when Heitz and Risse[12,13,35–38] began to investigate the Pd-catalyzed homo- and copolymerization of norbornene and its derivatives. Also in the early 1990s, the first Ni-based catalysts to homopolymerize norbornene to low molecular weight polymers with narrow molecular weight distributions were reported by Novak and Deming.[9] Since that time, Ni-based catalysts activated by MAO have been reported to catalyze the formation of high molecular weight amorphous addition PNBs.

Goodall and researchers at BF Goodrich[39] recognized the potential of polymers based on norbornene derivatives, and devised a toolbox of Pd- and Ni-based catalysts for their homo- and copolymerization. They used these newly developed, highly active catalysts to design several product groups based on addition polymers of norbornene derivatives.

The catalysts developed at BF Goodrich can be divided into three distinct groups: (1) catalysts based on cationic "naked"-type nickel and palladium complexes, or multicomponent systems able to generate similar active species; (2) catalysts based on neutral Ni and Pd compounds bearing highly electrophilic groups; and (3) catalysts based on cationic palladium complexes.

The "naked" nickel and palladium catalysts[10] are based on the concept of "naked" nickel catalysts as used in butadiene polymerization. A catalyst precursor is selected, which in the presence of the monomer, forms a catalytically active center bearing only monomer and the growing polymer chain as ligands. Among the simple olefins, norbornene can be considered to be a strong π-donor; therefore, cationic π-allyl complexes of nickel and palladium stabilized by cyclooctadiene (COD) ligands and noncoordinating counterions such as PF_6^- are prototypes of this class of catalyst precursors (Figure 16.3). As depicted in Figure 16.4 for Ni, the active species is assumed to consist of a metal center bearing two norbornene monomer ligands and the growing polymer chain.

Similar active species are most probably formed in situ from several of the multicomponent systems. Thus, the BF Goodrich group found that a "naked" nickel catalyst system based on di(2-ethylhexanoate)nickel, $BF_3 \cdot Et_2O$, and $AlEt_3$ in a 1:9:10 ratio forms, in the presence of butadiene, a catalyst precursor that may be activated by $HSbF_6$ to yield a cationic active species for norbornene polymerization. In a very similar mode, bis(1,4-cyclooctadiene)nickel(0)/butadiene/$HSbF_6$ and di(2-ethylhexanoate)nickel/butadiene/$HSbF_6$ give highly active catalysts for norbornene polymerization.

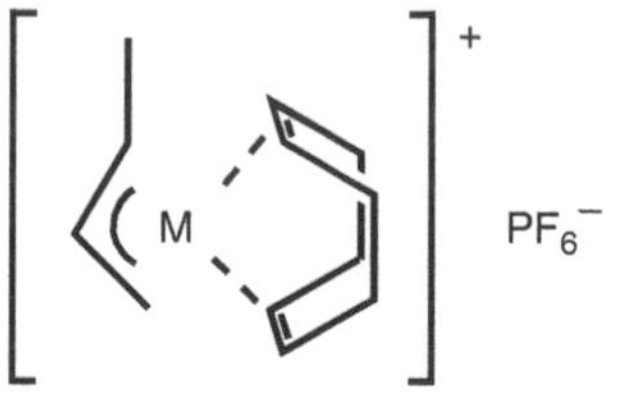

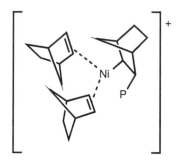

FIGURE 16.3 "Naked"-type nickel and palladium catalysts for norbornene polymerization developed by the group at BF Goodrich (M = Ni, Pd).

FIGURE 16.4 Active species for norbornene polymerization by "naked"-type nickel catalysts according to Goodall (P = polymer chain).

The PNBs from all of these Ni-based catalysts are soluble in common organic solvents such as toluene or hexane, and resemble PNBs produced using Ni(acac)$_2$ (acac = acetylacetonate) or di(2-ethylhexanoate)nickel in combination with MAO. Goodall and coworkers[40] describe the structure of PNB formed using $[(\eta^3\text{-crotyl})Ni(1,4\text{-COD})]^+[PF_6]^-$ as containing almost equal amounts of *mm* and *mr* triads.

By introducing α-olefins into the polymerization system as chain transfer agents (CTAs), the molecular weight of the PNBs from the Ni cation-based catalysts can be controlled; the insertion of an α-olefin is immediately followed by β-hydrogen elimination, and a vinyl-terminated PNB and a Ni hydride are formed (Scheme 16.4). The Ni hydride species serves as a starting point for another chain growth reaction.

Compared to its nickel analogue, the palladium catalyst $[(\eta^3\text{-crotyl})Pd(1,4\text{-COD})]^+[PF_6]^-$ according to Goodall et al. is substantially less active and results in insoluble or hardly soluble polymers that are proposed to be erythrodiisotactic. Similar insoluble PNBs were produced by Arndt-Rosenau and Gosmann using Pd(acac)$_2$ in combination with MAO.[24]

In the literature, many Pd and Ni complexes are described as yielding PNBs after activation with MAO. Nevertheless, in most cases the catalysts produce products identical to those obtained from simple Pd or Ni salts combined with MAO, and it has to be assumed that the cocatalyst in the former cases simply removes the ligand to generate a catalytically active species. Therefore, in contrast to what has been shown for metallocene catalysts, in most cases there is no influence of the ligand structure on the polymer microstructure and properties, although there is an influence on kinetic profiles and conversion during polymer formation.

The second class of active Ni and Pd catalysts described by the BF Goodrich group[40] is based on neutral Ni or Pd compounds bearing highly electrophilic groups. It was found during investigations of the activation of Ni complexes with B(C$_6$F$_5$)$_3$ that simple Ni salts such as Ni(dpm)$_2$ (dpm = 2,2,6,6-tetramethyl-3,5-heptanedionate) or Ni(2-ethylhexanoate)$_2$ may be activated by B(C$_6$F$_5$)$_3$ and other boranes bearing highly electrophilic aryl groups, in the absence of any aluminum alkyl, to yield active catalysts for norbornene polymerization (Scheme 16.5a). These multicomponent catalysts

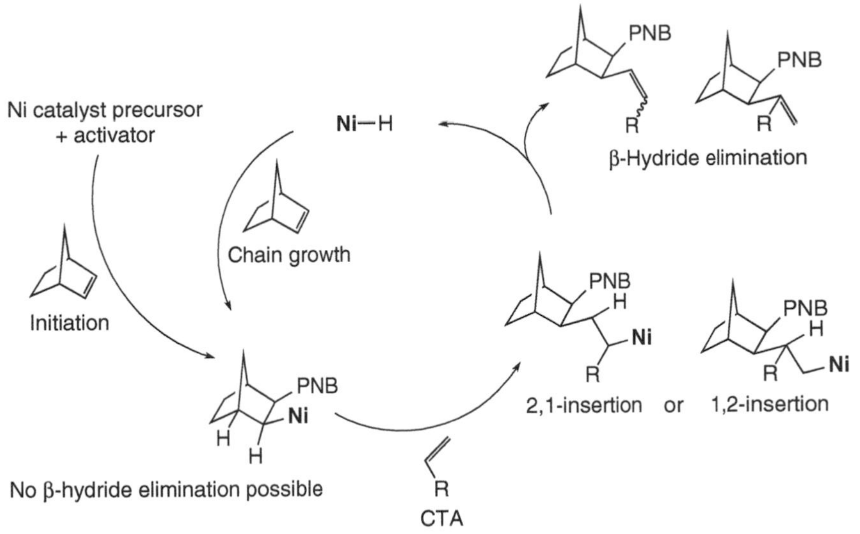

SCHEME 16.4 Mechanism of molecular weight regulation by the use of α-olefins as chain transfer agents (CTAs) in the addition polymerization of norbornene by Ni-based catalysts.

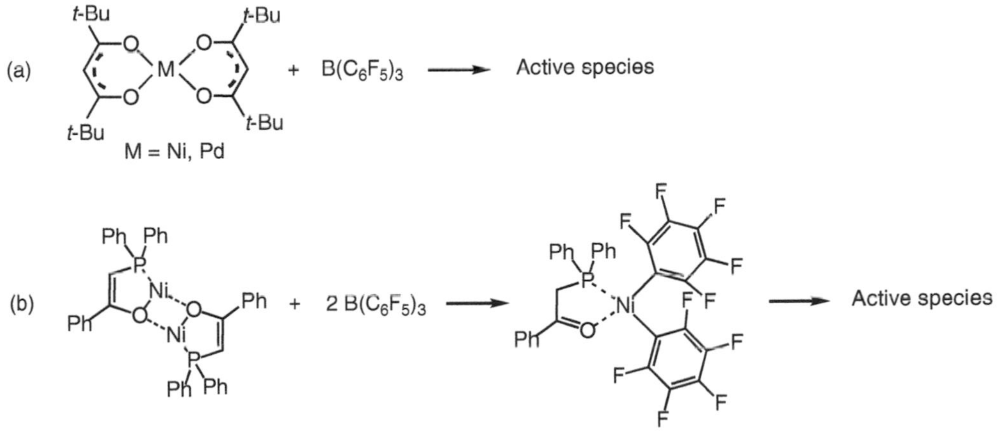

SCHEME 16.5 Multicomponent Ni and Pd (a) acac-type and (b) SHOP-type catalysts/initiators for the polymerization of norbornene.

were shown to be based upon neutral active species generated by the transfer of two pentafluorophenyl groups to the Ni center, as shown for a SHOP-type (SHOP = Shell Higher Olefins Process) catalyst in Scheme 16.5b.

Single component catalyst precursors are also based on nickel bis(pentafluorophenyl) complexes (Figure 16.5a and b). It can be assumed that in the active species (Figure 16.5c), one of the pentafluorophenyl groups acts as a starting point for the growing polymer chain, whereas the other reduces electron density at the transition metal center.

The polymers generated from the Ni catalyst according to Goodall et al. show a slightly erythrodiisotactic structure and feature pentafluorophenyl end groups, while polymers from the analogous palladium systems could not be investigated owing to their insolubility. Molecular weight regulation by α-olefins is possible, but is accompanied by catalyst deactivation due to the reductive elimination of pentafluorobenzene. Another method for molecular weight regulation is the addition of small

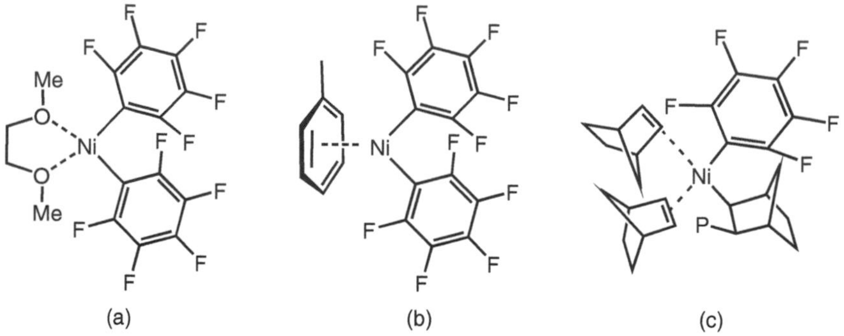

(a)   (b)   (c)

FIGURE 16.5 (a,b) Neutral bis(pentafluorophenyl)nickel initiators for the polymerization of norbornene and (c) proposed structure of the active species (P = polymer chain).

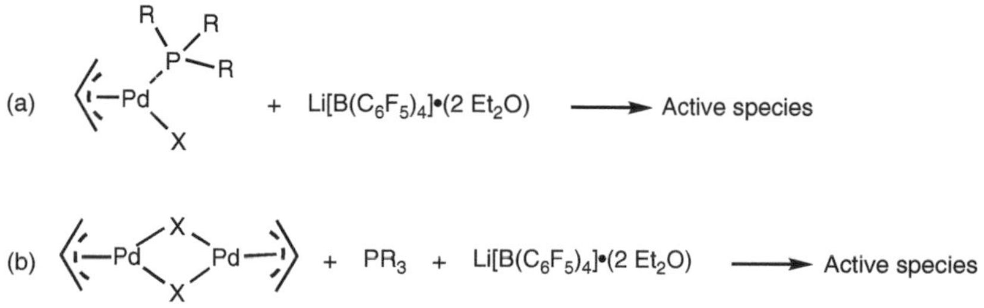

SCHEME 16.6 Multicomponent catalysts for norbornene polymerization based on cationic palladium phosphino-complexes (X = Cl, NO$_3$, O$_2$CCH$_3$, O$_2$CCF$_3$; R = Ph, n-Bu, t-Bu, o-tolyl, benzyl, C$_6$F$_5$, cyclohexyl).

amounts of water or alcohols to induce protolysis of the growing chain. Therefore, these catalysts can be regarded as initiators rather than as true catalysts, forming multiple polymer chains.

Finally, palladium allyl compounds[41] bearing a phosphine ligand and a leaving group were found by the BF Goodrich group to be single component catalyst precursors, yielding palladium cations that are highly active for norbornene polymerization (Scheme 16.6). Multicomponent catalyst systems based on the combination of a palladium allyl compound bearing a phosphine ligand and an activator (to replace a leaving group with a noncoordinating anion) can be used, as can the combination of allyl palladium chloride dimer with an activator in the presence of norbornene and a suitable phosphine.

The molecular weight of the polymers obtained depends upon the phosphine used, and the addition of α-olefins can be used to induce chain transfer; nevertheless, the system is less susceptible to the α-olefin chain transfer method than the aforementioned Ni catalysts. A major advantage of this catalyst system is its tolerance toward polar groups, enabling its use in the copolymerization of functional derivatives of norbornene and for polymerization in suspension or emulsion.

Iron- and cobalt-based catalysts[26,42,43] have been reported to be active for the polymerization of norbornene, especially when activated with MAO. Similar to the case of many nickel and palladium catalysts, PNBs generated with simple salts such as cobalt neodecanoate are similar to PNBs prepared with ligand-bearing catalysts (e.g., Brookhart–Gibson type pyridyldiimine cobalt catalysts).

16.2.3 Structure

Generally, PNBs may be grouped into two classes: those that are soluble in toluene and those that precipitate during polymerization. Wide-angle X-ray scattering (WAXS) and ^{13}C-CPMAS NMR

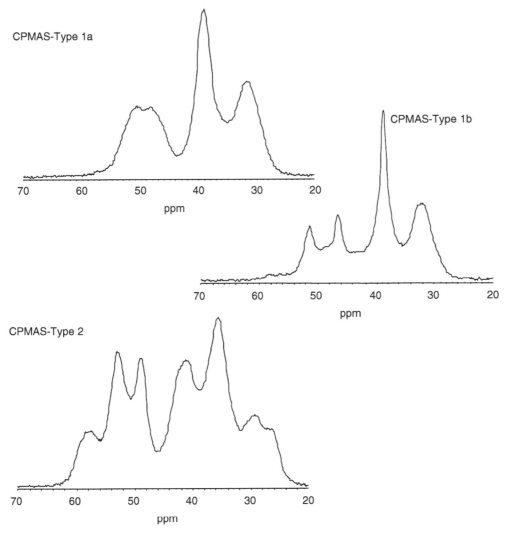

FIGURE 16.6 ^{13}C NMR CPMAS spectra of the three basic types of PNBs described by Arndt-Rosenau and Gosmann, prepared using Ni(acac)$_2$/MAO (Type 1a), **2**/MAO (Type 1b), and Pd(acac)$_2$/MAO (Type 2).[24,26]

investigations, as well as high temperature high-resolution ^{13}C NMR investigations of the soluble polymers, confirm the different structures of the polymers and enable a further classification (Figure 16.6).[24,26]

Although their ^{13}C CPMAS NMR spectra are similar, the polymers generated by Ni(acac)$_2$/MAO and other "naked" nickel-type catalysts (Type 1a, Figure 16.6) differ significantly in their solubility from those derived from metallocene **2** (Type 1b, Figure 16.6). The PNB generated from the Ni catalysts is totally soluble in toluene, whereas the polymer generated from the zirconocene is insoluble. Taking WAXS data into account, this difference must be attributed to a different tacticity of the polymers; while the Type 1a product from the Ni catalysts only shows two amorphous halos, the Type 1b polymer generated by **2** shows three amorphous halos superimposed by sharp reflections showing crystallinity. Nevertheless, not all metallocenes produce this type of PNB. PNBs generated by **1**/MAO are soluble in toluene and their spectra are similar to those of the "naked"-type nickel catalysts.

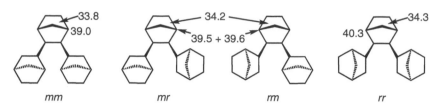

FIGURE 16.7 ^{13}C NMR chemical shifts of bridge and bridgehead carbon atoms in norbornene hydrotrimers according to Arndt-Rosenau.

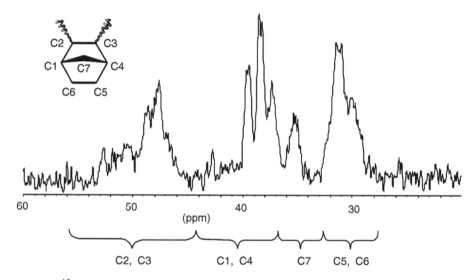

FIGURE 16.8 ^{13}C NMR spectrum of a soluble Type 1a PNB generated with "naked" nickel-type catalysts/MAO and peak assignments according to Goodall et al.

Based on the assignments made by Arndt-Rosenau for norbornene hydrotrimers (Figure 16.7),[21] Goodall and coworkers assigned the ^{13}C NMR resonances of the soluble PNBs made with "naked"-type nickel catalysts as shown in Figure 16.8.[40] Using ^{13}C NMR DEPT (distortionless enhancement by polarization transfer) spectra, they were able to assign two resonances to C7 (Figure 16.9(a)). Owing to the fact that both appear below 40 ppm, comparison to the hydrotrimers indicates that they belong to *mm* (33.8 ppm) and *mr* + *rm* or *rr* (34.2–34.3 ppm) triads. The proximity of the resonances from *mr* + *rm* and *rr* triads prevents an unambiguous assignment of the latter peak.

Their investigations were aided by comparison of the PNBs from "naked"-type nickel catalysts to those from neutral nickel catalysts bearing pentafluorophenyl (highly electrophilic) groups, which produced a substantially different material (Figure 16.9b). The DEPT spectra for this latter polymer showed only one resonance for C7 at about 35 ppm, which indicated no substantial content of *mm* triads. Again, the overlapping of the *mr* + *rm* and *rr* triad signals prevents a clear assignment of tacticity.

The results of Goodall et al. and Arndt-Rosenau shed some light on the microstructure of PNB. Nevertheless, the recent observations by Fink and Arndt-Rosenau indicate that further investigations are needed to assign clearly the microstructures of PNBs, and that extrapolation of results from trimers to polymers is not always possible.

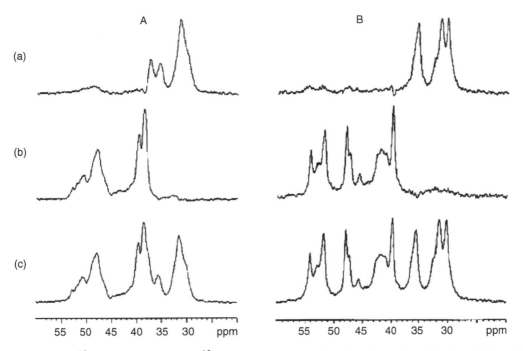

FIGURE 16.9 ^{13}C NMR spectra (c) and ^{13}C NMR DEPT spectra showing the methine (b) and methylene (a) carbon atoms of PNBs generated by the "naked"-type nickel catalyst [(η^3-crotyl)Ni(1,4-COD)]$^+$[PF$_6$]$^-$ (A) and Ni(dpm)$_2$/AlEt$_3$/B$_6$F$_5$ (B).[40]

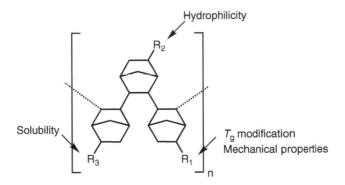

FIGURE 16.10 Basic concept of Promerus product lines based on copolymers of norbornene derivatives. Different substituents are used to tailor specific polymer properties (R$_1$ = alkyl group, oxygen-containing group; R$_2$ = trialkylsiloxy group; R$_3$ = photosensitive ester group).

16.2.4 INDUSTRIAL APPLICATIONS

Copolymers based on norbornene derivatives bearing different functional groups (Figure 16.10) are commercialized by Promerus,[44] a subsidiary of Sumitomo Bakelite Co. who took over this business from BF Goodrich. Using Ni- and Pd-based catalysts, they produce a range of tailor-made homo-, co-, and terpolymers based on substituted norbornenes for applications in electronic materials. Three basic products have been developed at BF Goodrich:[39] Avatrel,™ Appear,™ and DUVCOR.™

Avatrel™ is a group of dielectric polymers based on copolymers of alkyl norbornenes (>90 wt%) and 5-norbornene-2-triethoxysilane (2–10 wt%), in which the alkyl group (R$_1$ in Figure 16.10) is used to tailor the polymer's T_g and toughness, and the trialkylsiloxy group (R_2 in Figure 16.10) is used to impart good adhesion to metals.[45] Dielectric polymers[46] are used for electronic packaging, for example, in multichip modules, where their low dielectric constants (2.4–2.6) enable close packing of conducting lines and thereby a high interconnect density. Their hydrocarbon nature causes a low water uptake and therefore a stable dielectric character, which gives them an advantage over polyimides, which are also used for these applications.

Appear™ polymers have a similar composition and consist of >90 wt% of an alkyl norbornene and <10 wt% of an oxygen-containing norbornene derivative, which is used to increase chain–chain interactions and thereby the overall polymer properties (also represented by R_1 in Figure 16.10). Applications for these polymers are flat panel displays[47–49] and optical wave guides,[50] both of which are accessible owing to the high optical transmission and low birefringence of the polymers, combined with their excellent moisture resistance and ability to be used at high temperature.

DUVCOR™ polymers are used in photolithographic applications, primarily in deep UV (197 and 153 nm) positive photoresists.[51] Whereas the cycloaliphatic backbone of PNB ensures a good transparency and a high reactive ion etch resistance, norbornene comonomers bearing functional substituents such as esters or ethers are used to tailor the adhesive properties of the material (R_3 in Figure 16.10). A high amount (10–40%) of norbornene comonomers bearing acid-sensitive groups (R_3 in Figure 16.10), for example *tert*-butyl carboxylic ester, is used to enable subsequent acid-catalyzed deprotection to change the solubility of the copolymer. Typically, photosensitive acid generators such as triarylsulfonium hexafluorophosphate are used to induce cleavage of the ester moiety.

Further products based on PNB copolymers have been developed for adhesive encapsulation and cover coatings, as well as sacrificial materials for microelectromechanical applications.[52,53]

16.3 COPOLYMERS OF NORBORNENE

Norbornene can be copolymerized with olefins such as ethylene, propylene, 1-butene, and longer-chain α-olefins using early and late transition metal catalysts. The resultant copolymer properties depend on different parameters, such as comonomer content and distribution throughout the polymer chain, as well as the conformational orientation of the comonomer units. The microstructure of the copolymer can be controlled by the appropriate choice of reaction conditions and catalyst structure. The most powerful method to determine copolymer microstructure is ^{13}C NMR spectroscopy. In the past years, much progress has been achieved in making peak assignments for olefin–norbornene copolymers.[54–58]

Metallocene catalysts are very active for the copolymerization of norbornene with ethylene, but are also very sensitive to polar impurities such as aldehydes, esters, and ketones. The incorporation of norbornene into the polymer chain is slower than the incorporation of ethylene, and it is necessary to use high molar fractions of norbornene in the monomer feed to achieve all technically interesting norbornene incorporation levels. Since Brookhart and coworkers discovered that cationic nickel and palladium complexes possessing bulky α-diimine ligands are able to produce high molar mass polymers,[59,60] catalysts based on late transition metals have become an interesting alternative to metallocene catalysts.[61,62] In 1998, Goodall et al. reported ethylene/norbornene copolymerization using a variety of late transition metal-based catalysts.[10] α-Diimine-ligated palladium(II) catalyst systems were able to copolymerize ethylene with norbornene,[63] in contrast to their nickel analogues, which did not incorporate norbornene. These palladium catalysts are relatively insensitive to polar impurities.

Another interesting feature of these catalyst systems is that they generate highly branched or even hyper-branched (i.e., containing branches on branches) products when used for ethylene polymerization.[64] E/NB copolymers with low norbornene contents also show these hyper-branches.

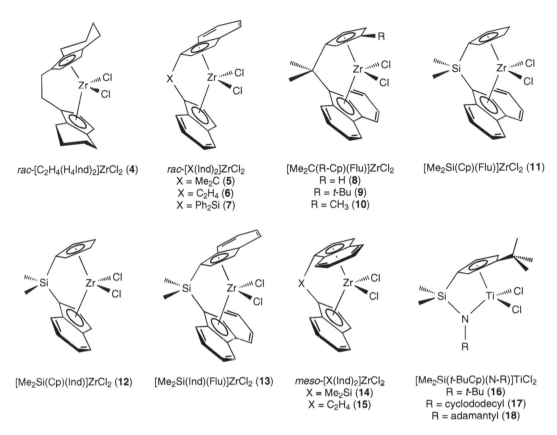

rac-[C$_2$H$_4$(H$_4$Ind)$_2$]ZrCl$_2$ (4)

rac-[X(Ind)$_2$]ZrCl$_2$
X = Me$_2$C (5)
X = C$_2$H$_4$ (6)
X = Ph$_2$Si (7)

[Me$_2$C(R-Cp)(Flu)]ZrCl$_2$
R = H (8)
R = t-Bu (9)
R = CH$_3$ (10)

[Me$_2$Si(Cp)(Flu)]ZrCl$_2$ (11)

[Me$_2$Si(Cp)(Ind)]ZrCl$_2$ (12)

[Me$_2$Si(Ind)(Flu)]ZrCl$_2$ (13)

$meso$-[X(Ind)$_2$]ZrCl$_2$
X = Me$_2$Si (14)
X = C$_2$H$_4$ (15)

[Me$_2$Si(t-BuCp)(N-R)]TiCl$_2$
R = t-Bu (16)
R = cyclododecyl (17)
R = adamantyl (18)

FIGURE 16.11 Molecular structures of ethylene/norbornene copolymerization catalysts used for experiments presented in Tables 16.3–16.5.

The mechanical properties of these branched polymers are assumed to differ from those of metallocene/MAO-based COCs.[64] Other amorphous copolymers of norbornene and ethylene are also available using less active vanadium catalysts.[65]

16.3.1 EARLY TRANSITION METAL CATALYSTS

Early attempts to produce E/NB copolymers utilized heterogeneous TiCl$_4$/AlEt$_2$Cl or vanadium catalysts, but real progress was achieved utilizing metallocene catalysts for this purpose. Metallocenes are about ten times more active than vanadium systems, and by choosing the metallocene, the norbornene/ethylene comonomer sequence distribution in the copolymer may be varied from statistical (random) to alternating.

For the copolymerization of norbornene and ethylene, different metallocene catalysts having C_1, C_2, C_{2v}, and C_s symmetry were studied by the authors (**1–3**, Table 16.1, and **4–15**, Figure 16.11). The copolymers obtained differ not only regarding the incorporation of norbornene, but also in microstructure and stereostructure, depending on the symmetry of the catalyst used.

The copolymerizations were carried out under argon using a 1 L Büchi A6 Type I autoclave equipped with an additional external cooling system.[66] For the standard experiments, the reactor was evacuated at 95 °C for 1 h and subsequently charged with a solution of norbornene in toluene, 190 mL toluene solvent, 500 mg MAO in 10 mL toluene (from Witco/Crompton), and ethylene at different pressures. Norbornene was dried over triisobutylaluminum and subsequently distilled before use. Polymerizations were initiated by injection of a toluenic metallocene solution into the reaction vessel

(5 μmol/L = 1 μmol zirconocene). The MAO Al:Zr ratio was 8600; the time of runs was 2 h if not otherwise specified. The ethylene concentration was 0.237 mol/L = 2 bar ethylene pressure (during the reactions, the ethylene pressure was kept constant at the noted pressure). Polymerizations were quenched by addition of 5 mL ethanol. Workup proceeded by adding the quenched polymer solution to 50 mL diluted aqueous HCl (0.2 mol/L) and stirring overnight, followed by neutralization of the toluene solution with aqueous $NaHCO_3$ and washing with water. After phase separation, the copolymer was precipitated from the organic (toluene) phase, if possible. Otherwise, the organic solvent was removed from the organic phase under reduced pressure and the polymer obtained was dried in vacuum.

The polymers were characterized with the following protocols: ^{13}C NMR spectra were recorded on a Bruker Avance 400 Ultrashield spectrometer at 100.62 MHz (pulse program: waltz-16, pulse angle: 60°, delay time: 5 s, number of scans: 1000–4000) and 100 °C using 200–300 mg of polymer in 2.7 mL of 1,2,4-trichlorobenzene and 0.3 mL of 1,1,2,2-tetrachloroethane-d_2. Chemical shifts are reported relative to $C_2D_2Cl_4$ ($\delta = 74.24$ for ^{13}C). High temperature gel permeation chromatography (GPC) measurements were performed in 1,2,4-trichlorobenzene at 135 °C using a Waters GPCV 2000 instrument with Ultrastyragel columns. Calibration was applied using polystyrene standards. Differential scanning calorimetry (DSC) curves were recorded on a Mettler Toledo DSC 821e instrument calibrated with n-heptane ($T_m = -90.6$ °C), mercury ($T_m = -38.9$ °C), and gallium ($T_m = 419.5$ °C). Results of the second thermal cycle are presented exclusively. Viscosimetric measurements were performed in decahydronaphthalene at 135 °C using an Ubbelohde viscosimeter (Oa capillary, K = 0.005 mm²/s²). In the case of copolymers with norbornene incorporation levels of less than 5%, the Mark–Houwink constants of polyethylene (K = 4.34×10^{-2} mL/g; a = 0.724) were applied. For all other copolymers, molar masses were calculated using Mark–Houwink constants for a highly alternating E/NB copolymer (K = 4.93×10^{-2} mL/g; a = 0.589).[67]

The relative reactivities of a comonomer pair (monomers 1 and 2) in a copolymerization are typically expressed using the reactivity ratios r_1 and r_2 and their product r_1r_2. For the comonomer pair ethylene/norbornene (monomer 1 = ethylene, monomer 2 = norbornene; that is, $r_1 = r_E$ and $r_2 = r_N$), the copolymerization parameter r_1 quantifies how much faster ethylene is inserted than norbornene when the previous insertion was ethylene (e.g., an r_1 value of 2 indicates that ethylene is inserted two times faster than norbornene). The parameter r_2 quantifies how much slower norbornene is inserted than ethylene when the previous insertion was norbornene. If $r_2 = 0$, it is impossible to insert a norbornene when the previous insertion was norbornene; in this case it is impossible to synthesize dyads or longer blocks of norbornene units in the polymer chain. The product r_1r_2 indicates the nature of the copolymerization. If $r_1r_2 = 1$, the copolymerization is statistical (random). If $r_1r_2 = 0$, no blocks of monomer 2 are possible; the result is an alternating structure of both monomers even with a high excess of monomer 2 in the starting phase. If the product $r_1r_2 > 1$, there are longer blocks of monomer 2 formed in the copolymer (see Chapter 12 for a further discussion of reactivity ratios).

Table 16.3 gives r_1, r_2, and r_1r_2 values for norbornene/ethylene copolymerization with metallocenes of various symmetries (1–4, 6–10, 15). As compared to other sterically hindered olefins such as 1-hexene or 1-octene, for which r_1 values are >5, the insertion of norbornene units into the growing polymer chain is more facile. The r_1 value for the *ansa*-zirconocene 4, which indicates facile norbornene insertion at 0 °C, increases with polymerization temperature indicating that NB addition becomes less likely as temperature increases. The incorporation as measured by r_1 is also high for the C_s-symmetric complex 3. In general, metallocene catalysts are more likely to form statistical (random) to alternating copolymers than blocky E/NB copolymers. This can be seen by the products of r_1r_2 in Table 16.3, which are <1. C_1-symmetric metallocenes should produce stereoregular alternating copolymers ($r_1r_2 \approx 0$), depending on the symmetry of the metallocene, whereas *meso* compounds (C_s-symmetric diastereotopes) should produce atactic alternating copolymers (for polyolefins in general, C_{2v}-symmetric and *meso* metallocenes make atactic materials, C_2- and C_1-symmetric metallocenes make isotactic materials, and C_s-symmetric (homotope)

TABLE 16.3

Parameters r_1, r_2, and $r_1 r_2$ for Ethylene/Norbornene Copolymerization with Different Metallocene/MAO Catalysts (Monomer 1 = Ethylene; Monomer 2 = Norbornene)

Metallocene	Symmetry	T_p (°C)[a]	r_1	r_2	$r_1 r_2$
rac-[C$_2$H$_4$(H$_4$Ind)$_2$]ZrCl$_2$ (**4**)	C_2	0	1.9	<1	~1
		25	2.2	<1	~1
		50	3.2	<1	~1
rac-[C$_2$H$_4$(Ind)$_2$]ZrCl$_2$ (**6**)	C_2	25	6.6	0.10	0.7
rac-[Ph$_2$Si(Ind)$_2$]ZrCl$_2$ (**7**)	C_2	30	3.44	<1	~1
rac-[Me$_2$Si(Ind)$_2$]ZrCl$_2$ (**2**)	C_2	30	2.6	<2	~1
[Me$_2$C(Cp)(Flu)]ZrCl$_2$ (**8**)	C_s	30	3.4	0.06	0.2
[Ph$_2$C(Cp)(Flu)]ZrCl$_2$ (**3**)	C_s	0	2.0	0.05	0.1
		30	3.0	0.05	0.15
[Me$_2$C(3-t-BuCp)(Flu)]ZrCl$_2$ (**9**)	C_1	30	3.1	0	0
[Me$_2$C(3-MeCp)(Flu)]ZrCl$_2$ (**10**)	C_1	30	3.3	0.001	0.03
meso-[C$_2$H$_4$(Ind)$_2$]ZrCl$_2$ (**15**)	C_s diastereotope	30	18.1	0.007	0.13
Cp$_2$ZrCl$_2$ (**1**)	C_{2v}	23	20	<0.1	~1

[a] Polymerization temperature; see text for other conditions.

metallocenes make syndiotactic materials; C_2- and C_{2v}-symmetric catalysts make more random copolymers, C_s-symmetric catalysts make random to alternating copolymers, and C_1-symmetric catalysts make more alternating copolymers).

The differing steric demands of the two inequivalent coordination sides in the C_1-symmetric catalyst Me$_2$C(3-t-BuCp)(Flu)]ZrCl$_2$ (**9**) were exploited to control the distribution of monomers along the polymer chain, and to promote a specific copolymerization mechanism. The more open coordination site of this catalyst can better accommodate norbornene (*vide infra*); in this manner, the formation of alternating copolymers was accomplished (i.e., $r_1 r_2 = 0$).

Table 16.4 shows activity, mol% norbornene incorporation, and copolymer molecular weight and T_g/T_m for ethylene/norbornene copolymerization using MAO-activated zirconocenes **8–15**. Of the catalysts used in this investigation, [Me$_2$Si(Cp)(Flu)]ZrCl$_2$ (**11**) was the most active, providing high molecular weights along with good comonomer incorporations in the polymer. The observed T_gs of polymers formed with **11** reach 133 °C; the highest T_gs (200 °C) are achieved using [Me$_2$C(Cp)(Flu)]ZrCl$_2$ (**8**).

Surprisingly, in some cases copolymer molecular weight increases with increasing norbornene incorporation, which is remarkably illustrated by the catalyst [Me$_2$C(3-MeCp)(Flu)]ZrCl$_2$ (**10**). The lowest weight average molecular weight (M_w) of 87,000 g/mol is found for a feed composition of $x_N = 0.37$ (37 mol% norbornene), reaching 431,000 g/mol at $x_N = 0.94$.

As might be expected, the T_gs of the copolymers depend upon microstructure and norbornene content. At about 14 mol% norbornene content in the copolymer ($X_N = 0.14$), a T_g of 0 °C is found for all of the catalysts tested and increases for statistically distributed (random) copolymers to 200 °C. A random copolymer with $X_N = 0.50$ has a T_g of about 145 °C.

The activities of [Me$_2$Si(Ind)(Flu)]ZrCl$_2$ (**13**), meso-[Me$_2$Si(Ind)$_2$]ZrCl$_2$ (**14**), and meso-[C$_2$H$_4$(Ind)$_2$]ZrCl$_2$ (**15**) are low, as are as the molar masses of the obtained polymers. Norbornene block sequences were observed in these copolymers, which is remarkable, considering the generally poor incorporation of norbornene by these systems. Multimodal molar mass distributions are found for the products of **13**/MAO (e.g., $M_w/M_n = 8.1$, $x_N = 0.59$; M_n = number average molecular weight) and **14**/MAO (e.g., $M_w/M_n = 8.8$, $x_N = 0.31$), whereas the M_w/M_n distribution for copolymers from the other catalysts is about 2.

TABLE 16.4

Ethylene/Norbornene Copolymerization with Various Zirconocene/MAO Catalysts

Zirconocene[a]	x_N[b]	X_N[c]	Activity (kg$_{Pol}$/mol Zr·h)	M_w[d,e] (g/mol)	T_g $[T_m]$[f,e] (°C)
[Me$_2$C(Cp)(Flu)]ZrCl$_2$ (**8**)	0	0	1,540	558,000	[136]
	0.21	0.090	5,880	178,000	[7]
	0.40	0.145	5,550	230,000	0
	0.56	0.260	5,080	114,000	46
	0.80	0.371	3,310	129,000	92
	0.95	0.495	1,780	158,000	135
	0.99	0.626	2,080	143,000	200
[Me$_2$C(3-MeCp)(Flu)]ZrCl$_2$ (**10**)	0.19	0.073	3,430	145,000	n.d.
	0.37	0.150	3,270	87,000	3
	0.55	0.218	3,920	99,000	40
	0.77	0.324	4,420	171,000	86
	0.89	0.407	5,720	252,000	106 [215]
	0.94	0.438	1,600	431,000	118 [243]
	0.96	0.456	20	n.d.	124 [276]
[Me$_2$C(3-t-BuCp)(Flu)]ZrCl$_2$ (**9**)	0.23	0.055	1,850	230,000	−96
	0.43	0.080	1,120	159,000	−66
	0.58	0.160	296	101,000	14
	0.68	0.220	355	90,000	37
	0.83	0.313	437	82,000	74
	0.91	0.373	205	7,300	97 [242]
	0.93	0.388	145	8,900	103 [255]
[Me$_2$Si(Cp)(Flu)]ZrCl$_2$ (**11**)	0.21	0.054	38,900	842,000	[88]
	0.40	0.123	27,100	705,000	−4 [44]
	0.57	0.184	30,200	808,000	15
	0.79	0.323	27,400	1,011,000	70
	0.90	0.407	14,300	1,123,000	65
	0.95	0.467	11,000	431,000	129
	0.99	0.644	432	190,000	133
[Me$_2$Si(Cp)(Ind)]ZrCl$_2$ (**12**)	0.20	0.090	6,900	29,600	[8]
	0.45	0.170	3,800	28,000	15
	0.80	0.440	460	13,800	120
[Me$_2$Si(Ind)(Flu)]ZrCl$_2$ (**13**)	0.21	0.028	332	360,000	[124]
	0.42	0.088	370	n.d.	[127]
	0.59	0.107	170	182,000	40 [124]
	0.80	0.257	70	135,000	21 [126]
	0.91	0.216	86	57,000	82 [129]
$meso$-[Me$_2$Si(Ind)$_2$]ZrCl$_2$ (**14**)	0	0	2,780	4,400	[108]
	0.31	0.096	654	12,600	[86]
	0.49	0.170	260	200,000	2 [118]
	0.70	0.299	27	222,000	77
$meso$-[C$_2$H$_4$(Ind)$_2$]ZrCl$_2$ (**15**)	0.40	0.41	1,970	221,000	[102]
	0.49	0.57	688	210,000	2 [65]
	0.90	0.253	180	245,000	51
	0.98	0.387	7	81,400	n.d.

[a] Conditions: [ethylene] = 0.237 mol/L = 2 bar, [Zr] = 5 μmol/L, [MAO] = 2.5 g/L (Al:Zr = 8600), 200 mL toluene, 2 h, 30 °C, batch process; see text for other conditions and analytical procedures. [b] Norbornene molar fraction in feed. [c] Norbornene molar fraction in copolymer (^{13}C NMR, 1,2,4-trichlorobenzene, 100 °C). [d] By GPC in 1,2,4-trichlorobenzene at 135 °C, versus polystyrene. [e] n.d. = Not detected. [f] By DSC.

TABLE 16.5
Ethylene/Norbornene Copolymerization with Constrained Geometry Catalysts/MAO

CGC Catalyst[a]	x_N[b]	X_N[c]	Activity (kg$_{Pol}$/mol Ti/Zr·h)	M_η[d,f] (g/mol)	T_g [T_m][e,f] (°C)
[Me$_2$Si(3-t-BuCp)(N-t-Bu)]TiCl$_2$ (**16**)	0	0	12,000	n.d.	[135]
	0.20	0.01	20,000	45,000	n.d.
	0.40	0.03	27,000	48,000	n.d.
	0.60	0.09	30,000	46,000	−3
	0.80	0.18	18,000	37,000	17
	0.95	0.35	4,000	20,000	78
[Me$_2$Si(3-t-BuCp)(NC$_{12}$H$_{23}$)]TiCl$_2$ (**17**)	0.40	0.17	350	54,000	16
	0.60	0.25	250	43,000	40
	0.80	0.35	200	40,000	79
	0.95	0.49	70	22,000	140
[Me$_2$Si(3-t-BuCp)(NAdam)]ZrCl$_2$ (**18**)	0.20	0.10	400	n.d.	−1
	0.40	0.22	590	6,000	35
	0.60	0.31	600	5,300	51
	0.80	0.42	1,200	6,000	77
	0.95	0.51	380	4,400	108

[a] Conditions: [ethylene]$_0$ = 2 bar, [Zr] = 5–10 μmol/L, [MAO] = 2.5 g/L (Al:Zr ratio = 4300 – 8600), 200 mL toluene, 1 h, 90 °C; see text for other conditions and analytical procedures.
[b] Norbornene molar fraction in feed.
[c] Norbornene molar fraction in copolymer (^{13}C NMR, 1,2,4-trichlorobenzene, 100 °C).
[d] Copolymer molar mass determined by viscosimetry in 1,2,4-trichlorobenzene at 135 °C.
[e] n.d. = Not detected.
[f] By DSC.

Constrained geometry catalysts (CGCs) are also able to copolymerize ethylene and norbornene.[68,69–72] The MAO-activated CGCs Me$_2$Si[(3-t-BuCp)(N-t-Bu)]TiCl$_2$ (**16**), [Me$_2$Si(3-t-BuCp)(NC$_{12}$H$_{23}$)]TiCl$_2$ (**17**), and Me$_2$Si(3-t-BuCp)(NAdam)]ZrCl$_2$ (**18**, Adam = adamantyl) (Figure 16.11) were studied as copolymerization catalysts. Table 16.5 shows the results of these copolymerizations.

The CGC catalysts are less active than C_1-symmetric zirconocene/MAO systems, but produce nearly perfectly alternating copolymers. For all three CGC catalysts, the product $r_E r_N (r_1 r_2)$ is <0.03. The 90 °C copolymerization parameters for the three catalysts are: **16**, r_E = 14.6, r_N = 5 × 10^{-8}; **17**, r_E = 3.5, r_N = 0.0006; **18**, r_E = 1.63, r_N = 0.014. As seen for Cp-Flu zirconocenes, the *tert*-butyl Cp substituent is a necessary condition to produce alternating structures. The norbornene incorporation decreases with the following substitution at the ligand N-atom: adamantyl (**18**) > cyclododecyl (**17**) > *tert*-butyl (**16**).

The mechanism by which ethylene/norbornene copolymerization proceeds depends on the exact nature of the metallocene or CGC catalyst employed. The insertion rate depends not only on the last inserted monomer unit, but can also be influenced by the second last inserted monomer unit. The copolymerization parameters described above (Markov 1 model) are derived from the following insertion events:

$$Cat–E–P + E \xrightarrow{k_{EE}} Cat–E–E–P$$

$$Cat–E–P + N \xrightarrow{k_{EN}} Cat–N–E–P$$

$$\text{Cat–N–P} + \text{E} \xrightarrow{k_{NE}} \text{Cat–E–N–P}$$

$$\text{Cat–N–P} + \text{N} \xrightarrow{k_{NN}} \text{Cat–N–N–P}$$

where E = ethylene unit (or enchained ethylene unit), N = norbornene unit (or enchained norbornene unit), P = growing polymer chain, and Cat = active metal center of catalyst, and

$$r_E = \frac{k_{EE}}{k_{EN}}$$

$$r_N = \frac{k_{NN}}{k_{NE}}$$

If the second last unit plays a role, eight equations are given (Markov 2 model); for example,

$$\text{Cat–E–E–P} + \text{N} \xrightarrow{k_{EEN}} \text{Cat–N–E–E–P}$$

and

$$r_{EE} = \frac{k_{EEE}}{k_{EEN}}$$

$$r_{EN} = \frac{k_{ENN}}{k_{ENE}}$$

$$r_{NE} = \frac{k_{NEE}}{k_{NEN}}$$

$$r_{NN} = \frac{k_{NNN}}{k_{NNE}}$$

C_1-symmetric catalysts such as **9** and **10** bear a Cp ring substituent pointing in the same direction as one of the chlorine atoms on the Zr center. In the active catalyst species, the chlorine is replaced by the growing polymer chain or a π-complex-bonded olefin. The coordination site on the same side as the Cp substituent (site B) is more sterically hindered than the other coordination site (site A). After insertion of a monomer coordinating at site B, the polymer chain is "walked" back to less sterically hindered site A by the next monomer insertion (alternating/chain migratory mechanism). But in some cases, the polymer chain can "back skip" to site A before the next monomer insertion (site epimerization). Norbornene, as a bulky olefin, can be π-bonded only at site A[73] (see Chapter 2 for a further discussion of the alternating and site epimerization mechanisms with catalyst **9**).

A chain migratory insertion mechanism for ethylene/norbornene copolymerization utilizing both A and B coordination sites in an alternating fashion[66,73] is implied by the derived parameter set $r_E^A = 3.08$, $r_N^A = 0$, $r_E^B = 500$, and $r_N^B = 0$ for the C_1-symmetric *tert*-butyl-substituted Cp-Flu metallocene **9** at 30 °C. However, for the analogous methyl-substituted catalyst **10**, at relatively low norbornene feed concentrations (i.e., $x_N < 0.93$ and $X_N < 0.46$) only one of the two coordination sites is used, and this site inserts both monomers according to the derived parameters $r_E = 3.3$ and $r_N = 0.001$. At higher norbornene feed concentrations (i.e., $0.93 < x_N < 0.98$) with **10**, norbornene blocks are formed by this chain back-skip mechanism.

In the case of the unsubstituted C_s-symmetric metallocenes **8** and **11**, copolymerization proceeds under control of the last inserted monomer unit (chain-end control), that is, it can be described by a second order Markov model. Ethylene is inserted with these zirconocenes[66] three times faster than norbornene. No norbornene block sequences longer than two (NN units) are formed, in agreement with parameters calculated for **8** ($r_{EE} = 2.40$, $r_{NE} = 4.34$, $r_{EN} = 0.03$, and $r_{NN} = 0.00$).[66] This result easily explains the maximum observed $X_N = 0.66$.

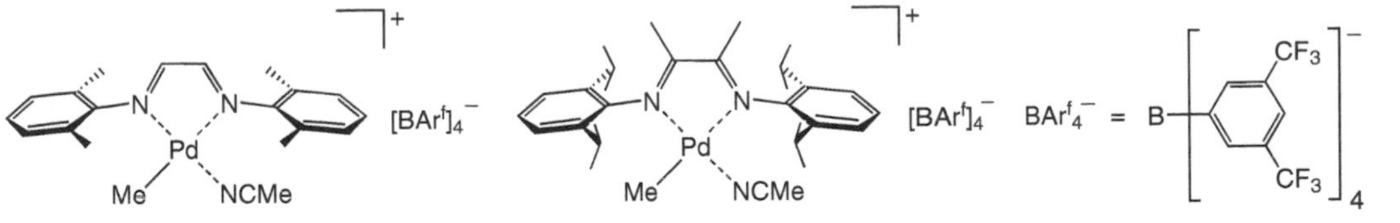

[(2,6-Me$_2$C$_6$H$_3$)$_2$GLY]Pd (**19**)   [(2,6-i-Pr$_2$C$_6$H$_3$)$_2$BUD]Pd (**20**)

FIGURE 16.12 Palladium α-diimine catalysts used for ethylene/norbornene copolymerization (GLY = 1,2-ethanediimine = glycine-derivate; BUD = 2,3-butanediimine).

Ethylene/norbornene copolymerization with the CGC catalysts could be described by a Markov first order model. No temperature effects on incorporation and microstructure were observed.

16.3.2 LATE TRANSITION METAL CATALYSTS

Ethylene/norbornene copolymerzations were carried out by the authors using palladium catalysts incorporating symmetric α-diimine ligands.[26,67] The synthesis of these catalysts is comparatively easy and fast, making them a candidate for combinatorial chemistry. Symyx Technologies has reported the combinatorial syntheses of large Ni[II] and Pd[II] α-diimine complex libraries and the investigation of their ethylene polymerization behavior in high-throughput screenings.[74] Two α-diimine catalysts with interesting performance, [(2,6-Me$_2$C$_6$H$_3$)$_2$GLY]Pd (**19**, GLY = 1,2-ethanediimine = glycine-derivate) and [(2,6-i-Pr$_2$C$_6$H$_3$)$_2$BUD]Pd (**20**, BUD = 2,3-butanediimine) (Figure 16.12) were synthesized as discrete cationic species according to the literature,[60] and their behavior in ethylene/norbornene copolymerization was investigated in detail by the authors (Table 16.6). The activities of these catalysts range from 8 to 243 kg polymer/mol Pd·h. Considering the high incorporation rates of norbornene achieved, these activities are about one order of magnitude lower than for typical metallocene/MAO catalyst systems.

Catalysts **19** and **20** show notably different polymerization behaviors. Catalyst **19** incorporates norbornene much better than ethylene (Figure 16.13). Even at very low molar fractions of norbornene in the feed (e.g., $x_N = 0.05$), the incorporation of norbornene is at about $X_N = 0.44$ (Table 16.6, Run 2). When metallocene/MAO catalyst systems are used similarly, it is necessary to carry out copolymerization at feed compositions of up to $x_N = 0.90$ to incorporate norbornene in the same amount. At higher x_Ns with **19**, incorporation is nearly independent of feed composition and reaches a plateau of about $X_N = 0.60$. Polymerizations carried out at $x_N > 0.60$ yielded partially insoluble polymers that were not evaluable by standard ^{13}C NMR characterization protocols.

In contrast, **20** shows almost ideal copolymerization behavior at low to moderate values of x_N, that is, the norbornene content of the polymer reflects the feed composition. The norbornene incorporation levels are low compared to those seen with **19**, although still higher than for most metallocene/MAO catalysts. The norbornene content does not surpass $X_N = 0.40$. No norbornene block structures could be observed in the ^{13}C NMR spectra; even at higher x_N, sequences with isolated norbornene units are formed exclusively as determined by ^{13}C NMR.[67] The coordination sites of **20** are blocked by its isopropyl ligand aryl substituents and the ligand is inflexible because of its 2,3-butanediimine bridge system. The steric bulk of the growing polymer chain, if norbornene is the last inserted unit, disfavors the coordination of norbornene. The formation of norbornene block sequences is therefore improbable. The coordination of ethylene is much more likely, leading to copolymers with isolated norbornene units.

In contrast, the steric bulk of the methyl ligand aryl substituent groups in **19** is much lower, and its 1,2-ethanediimine bridge is more flexible than the 2,3-butanediimine backbone of **20**. The

TABLE 16.6

Ethylene/Norbornene Copolymerization with Pd Diimine Catalysts (2,6-Me$_2$C$_6$H$_3$)$_2$GLYPd (19) and (2,6-i-Pr$_2$C$_6$H$_3$)$_2$BUDPd (20)

Run	x_N[b]	Reaction conditions[a] $[E]_0$[c] (mol/L)	$[N]_0$[c] (mol/L)	Pd (μmol)	A[d]	x_N[e]	Results T_g[f] (°C)	M_η[g] (g/mol)	M_w/M_n[h]
[(2,6-Me$_2$C$_6$H$_3$)$_2$GLY]Pd (19)									
1	0.00	0.70	—	16.76	36	0.00	−75	1,500[i]	1.4
2	0.05	0.72	0.04	4.19	161	0.44	98	13,000	1.4
3	0.10	0.72	0.08	4.19	243	0.48	126	36,000	1.5
4	0.20	0.57	0.14	4.19	120	0.54	146	40,000	1.7
5	0.34	0.47	0.24	8.38	67	0.59	169	55,000	1.9
6	0.40	0.42	0.28	16.76	57	0.60	177	54,000	1.8
7	0.50	0.35	0.35	16.76	50	0.62	189	36,000	1.7
8	0.59	0.29	0.41	16.76	34	0.62	216	18,000	1.7
9	0.80	0.14	0.56	16.76	11	[j]	[k]	12,000	1.7
10	0.90	0.07	0.63	33.54	10	[j]	[k]	7,000	2.7[l]
11	1.00	—	0.88	33.54	<1	1.00	[k]	[m]	[m]
[(2,6-i-Pr$_2$C$_6$H$_3$)$_2$BUD]Pd (20)									
1	0.00	0.69	—	7.77	185	0.00	−67	73,000	1.1
2	0.10	0.63	0.07	7.77	75	0.09	−28	502,000	1.7
3	0.20	0.55	0.14	7.77	63	0.16	10	358,000	1.7
4	0.34	0.47	0.24	7.77	53	0.26	48	287,000	1.7
5	0.40	0.42	0.28	7.77	57	0.29	63	248,000	1.6
6	0.49	0.36	0.35	7.82	48	0.34	83	231,000	1.5
7	0.59	0.29	0.42	7.82	37	0.40	97	157,000	2.7
8	0.81	0.17	0.70	7.82	8	0.40	120	22,000	6.3
9	0.90	0.17	1.52	15.65	4	[j]	[k]	[m]	[m]
10	1.00	—	0.79	26.22	<1	1.00	[j]	[l]	[l]

[a]Conditions: 200 mL toluene, 30 °C (see Reference 67).

[b]Norbornene molar fraction in feed.

[c]E = ethylene; N = norbornene.

[d]Activity in kg$_{Pol}$/mol Pd·h.

[e]Norbornene molar fraction in copolymer (^{13}C NMR, 1,2,4-trichlorobenzene, 100 °C).

[f]By DSC.

[g]Copolymer molar mass determined by viscosimetry in 1,2,4-trichlorobenzene at 135 °C.

[h]By GPC in 1,2,4-trichlorobenzene at 135 °C, versus polystyrene.

[i]M_w by GPC.

[j]Not evaluable by ^{13}C NMR spectroscopy due to insolubility.

[k]Polymer decomposes.

[l]Polymer molecular weight distribution is bi- or multimodal.

[m]Polymer is not soluble.

coordination sites are therefore less blocked in **19**, making norbornene block sequences more likely. At x_Ns of less than 10%, alternating structures are dominant.[67]

There is a comonomer feed composition effect seen on the polymerization activity of **19**. Activity reaches a maximum at $x_N = 0.1$ (Table 16.6, Run 3) and is seven times higher than that for the homopolymerization of ethylene. At higher x_Ns, activity decreases. Catalyst **20** does not show such an effect. The E/NB copolymer molecular weights obtained with the Pd catalysts range between 7,000 and 502,000 g/mol. For both catalysts, a comonomer feed composition effect on molecular

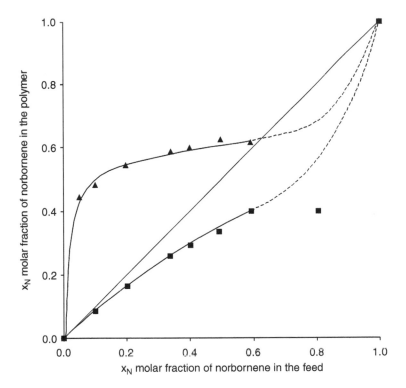

FIGURE 16.13 Diagram for the copolymerization of ethylene and norbornene with palladium α-diimine catalysts (triangles = **19**; squares = **20**).

weight is seen. The molecular weights increase with higher x_Ns, and reach a maximum for **19** by $x_N = 0.34$ and for **20** by $x_N = 0.1$. The polydispersities (M_w/M_n) of the copolymers formed with the Pd catalysts are generally lower than 2, indicating that the polymerization mechanism is intermediate between single site ($M_w/M_n = 2$) and living ($M_w/M_n = 1$) character. Polymers produced at high x_Ns are bi- or multimodal.

The amount of norbornene incorporated into the copolymer also influences T_g. The T_gs of E/NB copolymers produced with **19** are very high and range from 98 to 217 °C. Copolymers produced at $x_N = 0.80$ and higher, as well as homo-PNB, show no T_gs or T_ms under 350 °C, and decompose at >350 °C. The T_gs of copolymers produced by **20** range from −28 to 120 °C, owing to their lower norbornene content.

16.3.3 ANALYSIS AND STRUCTURE

The best method for E/NB copolymer micro- and stereostructural analysis is ^{13}C NMR spectroscopy,[60,66,73,75] which can be used to determine if the polymer chain contains dyads and triads (short norbornene blocks) or only isolated norbornene units. This section compares the different microstructures of E/NB copolymers made with the different types of catalysts discussed in this chapter. Details can be found in the literature.[56,66,75,76]

To begin a discussion of the micro- and stereostructures of E/NB copolymers produced with different types of catalysts, trends for C_1-symmetric metallocenes will first be considered. While C_2-symmetric catalysts (e.g., **2**, **4**, **6**, **7**) produce random copolymers with small norbornene blocks if the norbornene concentration in the feed is high, C_1-symmetric catalysts produce more alternating (**10**) or purely alternating (**9**) polymers. The catalyst system **10**/MAO was found to incorporate norbornene slightly better than **9**/MAO (Table 16.4); whereas exclusively isolated and alternating norbornene

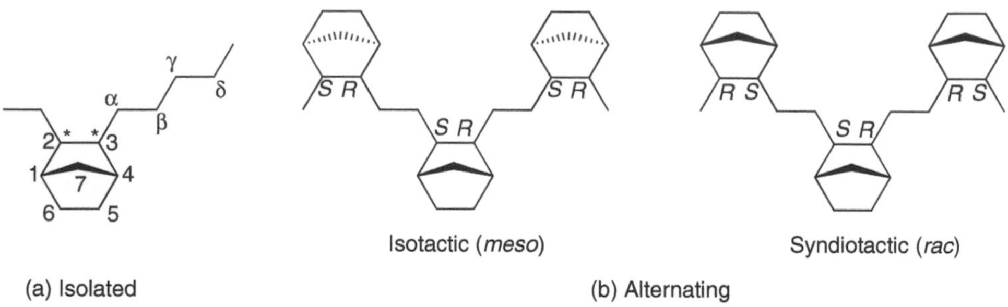

(a) Isolated   Isotactic (*meso*)   Syndiotactic (*rac*)

(b) Alternating

FIGURE 16.14 Structural units in E/NB copolymers with (a) isolated norbornene units and (b) alternating norbornene/ethylene units.

TABLE 16.7

Assignment of the Norbornene C_1–C_7 and Ethylene C_α–C_δ Carbon Atoms in E/NB Copolymers with Isolated or Alternating Norbornene Units to Different Groups of ^{13}C NMR Peaks as Depicted in Figure 16.14

Signal Area	δ ^{13}C (ppm)	Assignments
A	56–44.8	C2, C3
B	44.8–36.8	C1, C4
C	36.8–32.8	C7
D	32.8–10	C5, C6, Cα, Cβ, Cγ, Cδ

sequences (but no norbornene blocks) are formed by **9**/MAO at 30 °C, polymers containing more than 47 mol% of norbornene produced by **10**/MAO feature norbornene blocks (dyads).[66] At a norbornene feed concentration of $x_N = 0.998$, copolymers made with **10** reach X_Ns of 0.60. The molar masses of the copolymers produced by **10**/MAO are significantly lower than average values in this study. Their alternating microstructures lead to T_ms in the range of 215–280 °C when $0.40 < X_N < 0.46$ (at even higher values for X_N, the crystallinity from the alternating microstructure is disturbed by norbornene blocks). In comparison, copolymers produced by **9**/MAO at 30 °C, which have only alternating and isolated norbornene units, show T_ms of up to 320 °C, peaking at 50 mol% norbornene in the polymer.[73]

The incorporation level of norbornene, as well as its distribution and enchainment orientation, have a great influence on the properties of E/NB copolymers. Norbornene copolymerization with early and late transition metal complexes happens exclusively by double bond addition (2,3-insertion). The norbornene unit has a 2,3-cis-exo-orientation (Figure 16.14a), and two stereocenters are formed in the polymer chain. Copolymers with isolated norbornene units therefore have different possible tacticities than alternating copolymers (Figure 16.14b).

Table 16.7 gives correlations of ^{13}C NMR peaks to the carbon atoms in the E/NB copolymer chain as shown in Figure 16.14a (signal areas A, B, C, D stand for peak regions of the ^{13}C NMR spectrum). The signals for general E/NB copolymer ^{13}C NMR spectra (i.e., for both isolated norbornene units and norbornene–norbornene linkages) were assigned to the different compositional and stereochemical pentads by different authors (Arndt,[66,73] Fink,[75,76] Tritto[77]). Table 16.8 shows a correlation of the ^{13}C NMR peak assignments with the carbon atoms C2/C3. The assignments of Fink were carried out using ^{13}C-enriched monomers.[75,76]

The major differences in the pentads are reflections of the differing abilities of various metallocenes to form blocks of norbornene units and different stereoregularities within the norbornene

TABLE 16.8

Pentad and Tetrad Assignments of ^{13}C NMR Peaks for E/NB Copolymer Carbon Atoms C2/C3[66,73,75,76]

Peak Number	δ ^{13}C (ppm)	Pentad[a]	C-atom
1	55.6	*r,m*-EN<u>N</u>NE	C2/C3
2	55.0	*m,m*-EN<u>N</u>NE	C2/C3
3	52.0	*m,m*-NN<u>NE</u>	C2/C3
4	51.5	*m,r*-EN<u>N</u>NE	C2/C3
5	50.4	*m,m*-E<u>N</u>NN	C2/C3
6	50.3	*r*-ENNE	C2
7	49.4	*m*-ENNE	C2
8	48.3	*m*-ENNEN/*m*-NENNE	C3
9	48.23	*m*-ENNE	C3
10	48.05	*m,m*-NENEN + 0.5 *m*-EENEN/*m*-NENEE = 0.5 *m*-ENEN/*m*-NENE	C2/C3
11	47.8	*r*-ENNE	C3
12	47.7	*m,m*-EE<u>N</u>NN	C2/C3
13	47.55	*r,r*-NENEN + 0.5 *r*-EENEN/*r*-NENEE = 0.5 *r*-ENEN/*r*-NENE	C2/C3
14	47.35	EENEE + 0.5 EENEN/NENEE = 0.5 EENE/ENEE	C2/C3
15	45.8	*m,r*-NN<u>NE</u>	C2/C3
16	44.8	*m,r*-E<u>N</u>NN	C2/C3

[a]The norbornene unit underlined in the pentad is considered for peaks assigned to C2/C3.

blocks.[66] The C_1-symmetric metallocene system **10**/MAO (which gives highly alternating copolymers at $X_N < 0.50$ but can make blocks at higher levels) produces a copolymer consisting mainly of alternating NENEN sequences, as characterized by a sharp ^{13}C NMR resonance at 48 ppm, and only a few (E)NN(E) blocks, characterized by small resonances at 51–49 ppm and at 47.5–47 and 29.5–28 ppm. In contrast, at an identical incorporation level of norbornene (49 mol%), the copolymer produced by the unsubstituted **8**/MAO catalyst system features a significantly higher amount of (E)NN(E) blocks. Thus, the trend is a correlation of greater Cp ligand substitution to less blockiness, that is, unsubstituted (**8**) > 3-Me (**10**) > 3-*t*-Bu (**9**). The 3-*tert*-butyl-Cp ligand substitution pattern is so selective that no more than 50 mol% norbornene can be incorporated. The stereochemistry of the NN linkage in the copolymers prepared with **8**/MAO is most probably predominantly racemic (syndiotactic), in accordance with the stereoselectivity of this catalyst system in the hydrooligomerization of norbornene. The polymer produced by the C_2-symmetric *ansa*-metallocene system **2**/MAO features *meso* (isotactic) (E)NN(E) blocks and longer N(N)N sequences, which increase significantly as the norbornene incorporation rate increases, and are indicated by the appearance of ^{13}C NMR resonances between 35 and 40 ppm.

In contrast to alternating polymers made using other catalyst systems, the alternating copolymer structures obtained with **9** are not due to a chain-end control mechanism (that produces alternating structures because sterically demanding monomers cannot insert twice in a row). Instead, this is due to the structure of the catalyst in combination with a chain migratory insertion mechanism. In this process, norbornene can insert at only one site of the catalyst, whereas the smaller monomer ethylene can insert at either site, which results in copolymers containing only odd-numbered ethylene sequence lengths.[73] In the case of high norbornene:ethylene feed ratios, an alternating structure results.

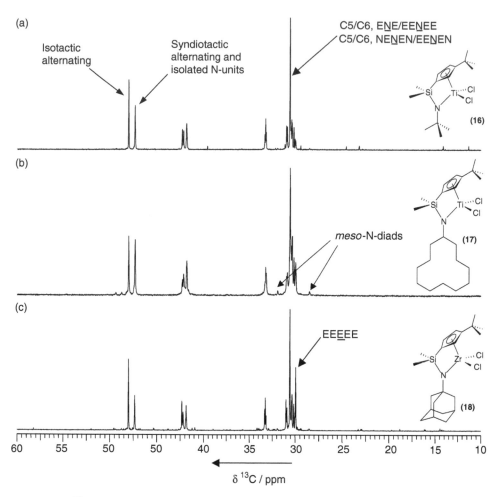

FIGURE 16.15 ^{13}C NMR spectra of E/NB copolymers synthesized using constrained geometry catalysts **16** ($X_N = 0.42$), **17** ($X_N = 0.40$), and **18** ($X_N = 0.41$).

Broadening this comparison to include copolymers prepared by both early and late transition metal catalysts, the results discussed immediately above show that C_1-symmetric zirconocenes such as **9**/MAO produce only copolymers with isolated norbornene units or alternating structures (at 30 °C), mainly with isotactic (*meso*) configurations. C_2-symmetric zirconocenes such as **2**/MAO readily produce norbornene dyads that are exclusively *meso*-linked (isotactic). In accordance with their catalyst structures, C_s-symmetric zirconocenes such as **8**/MAO produce norbornene dyads with a *rac*-linkage (syndiotactic), although with a generally lower stereoselectivity. Palladium α-diimine catalysts, despite the homotopic nature of their coordination sides (that would be expected to give a mixture of *meso* and racemic blocks), produce norbornene dyads that are solely *rac*-connected. This behavior can be attributed to a chain-end control type polymerization mechanism.

Alternating E/NB copolymers synthesized by CGC catalysts show a range of different micro-structures and tacticities (Figure 16.15). Catalysts **16** and **18** are not able to make copolymers with norbornene–norbornene dyads, even with a high molar excess of norbornene in the feed. Small amounts of *meso* norbornene dyads can be observed if catalyst **17**, which contains a relatively flexible cyclododecyl ligand N-substituent, is used.

Alternating copolymers prepared with **16**/MAO and **17**/MAO are predominantly isotactic. However, the low amounts of syndiotactic alternating structures present are sufficient to prevent

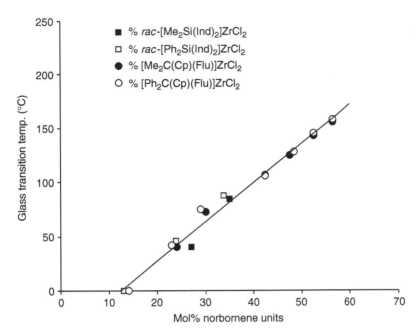

FIGURE 16.16 T_gs of E/NB copolymers made with different zirconocenes (filled squares = **2**; open squares = **7**; filled circles = **8**; open circles = **3**.

crystallization, and these materials are amorphous. Copolymers prepared with **18**/MAO do not show any syndiotactic alternating sequences, and are semicrystalline with a melting point of 280 °C if the norbornene content is more than 40%. The melting point of the alternating sequences is not sensitive to composition; only the amount of crystallinity depends on the composition, and increases with the norbornene content at compositions of over 40 mol% norbornene. Longer ethylene blocks are formed with the N-adamantyl catalyst **18**, which has the bulkiest N-substituent, as compared to CGCs **16** and **17**.

16.3.4 PROPERTIES AND INDUSTRIAL APPLICATIONS

E/NB copolymers are characterized by excellent transparency and high, long-life service temperatures. They are solvent- and chemical-resistant and can be melt processed. Owing to their high carbon/hydrogen ratio, these polymers have a high refractive index (1.53 for a 50:50 ethylene/norbornene copolymer). Their stability against hydrolysis and chemical degradation, in combination with their stiffness, makes them interesting materials for optical applications, for example in compact discs, lenses, or films. The T_gs of E/NB copolymers (both statistically random and alternating types) are correlated with norbornene incorporation (Figure 16.16).

At a norbornene content of 50 mol%, the T_g of a statistically random or alternating E/NB copolymer is about 150 °C. A material with 75 mol% norbornene has a T_g of about 200 °C. Alternating copolymers are semicrystalline if the norbornene content is above 37 mol%. They feature T_gs of 100–130 °C and T_ms of 270–320 °C. Alternating copolymers have an even greater solvent resistance than statistical copolymers, although they are still transparent owing to the small size of their crystalline regions (5 nm).

The T_g of COC copolymers can be increased if higher condensed cyclic olefin comonomers are used, such as dimethanooctahydronaphthalene (DMON) or trimethanododecahydroanthracene (TMDA) (Figure 16.17).[78,79] With increasing monomer bulk, incorporation of the cycloolefin

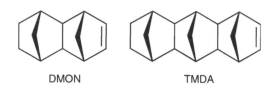

DMON   TMDA

FIGURE 16.17 Structures of dimethanooctahydronaphthalene (DMON) and trimethanododecahydro-anthracene (TMDA).

becomes more and more difficult. While r_1 for norbornene is similar to that of propylene, r_1 values for DMON and TMDA are comparable to those of 1-butene and 1-hexene.

Because polypropylene has a much higher T_g than polyethylene, the copolymerization of propylene and norbornene yields copolymers featuring higher T_gs than E/NB copolymers at similar norbornene contents. The rate of propylene/norbornene copolymerization is significantly lower than that observed for ethylene/norbornene copolymerization, most reasonably explained by the difficulty of inserting propylene after the cycloolefin owing to steric interactions of the propylene methyl group and the norbornene unit. Therefore, the rate of insertion is lowered relative to the rate of chain termination, and the molecular weights of these copolymers decrease dramatically with increasing norbornene:propylene feed ratios.

A commercial plant for the production of COC material (E/NB copolymer) was built in 2000 by Ticona in Oberhausen, Germany, with a capacity of 30,000 t/a (tons per annum). Mitsui produces E/NB copolymers using vanadium-based catalysts. The industrially produced copolymers have norbornene contents between 30 and 60 mol% and T_gs of 120–180 °C. The copolymer densities are low and near 1. For many applications, these COC materials show better mechanical properties than comparable amorphous thermoplastics, and are processible by all conventional methods. E/NB copolymers are proving valuable as materials for high capacity CDs and DVDs, lenses, blister foils, medical equipment, capacitors, and packaging.[16]

16.4 CONCLUSIONS

Single site catalysts, such as metallocene compounds, CGCs, and nickel or palladium diimine complexes, used in combination with MAO or borate cocatalysts, are highly active for the homopolymerization of norbornene and its copolymerization with ethylene. The structure of the norbornene homo- and copolymers can be widely influenced by the symmetry and structure of the ligands on the transition metal complexes.

E/NB copolymers can be obtained as random amorphous materials with high T_gs or as alternating, partially crystalline materials with high T_ms. Amorphous copolymers have short blocks of norbornene units (dyads or triads), which account for their high T_gs and excellent optical properties. All norbornene homo- and copolymers made by single site catalysts are characterized by narrow molecular weight distributions, which make technical processing easier. The first commercial norbornene copolymer products are already available.

REFERENCES AND NOTES

1. *Metallocene-Based Polyolefins, Preparation, Properties and Technology*; Scheirs, J., Kaminsky, W., Eds.; Wiley: Chichester, 2000; Vol. 1 and 2.
2. Resconi, L.; Cavallo, L.; Fait, A.; Piemontesi, F. Selectivity in propene polymerization with metallocene catalysts. *Chem. Rev.* **2000**, *100*, 1253–1345.
3. Gibson, V. C.; Spitzmesser, S. K. Advances in non-metallocene olefin polymerization catalysis. *Chem. Rev.* **2003**, *103*, 283–316.

4. Kaminsky, W.; Spiehl, R. Copolymerization of cycloalkenes with ethylene in presence of chiral zirconocene catalysts. *Makromol. Chem.* **1989**, *19*, 515–526.

5. Ivin, K. J. *Olefin Metathesis*; Academic Press: London, 1983.

6. Kaminsky, W.; Arndt, M. Metallocenes for polymer catalysis. *Adv. Polym. Sci.* **1997**, *127*, 143–187.

7. McKnight, A. L.; Waymouth, R. M. Ethylene/norbornene copolymerizations with titanium CpA catalysts. *Macromolecules* **1999**, *32*, 2816–2825.

8. Kaminsky, W.; Bark, A.; Arndt, M. New polymers by homogeneous zirconocene/aluminoxane catalysts. *Makromol. Chem., Macromol. Symp.* **1991**, *47*, 83–93.

9. Deming, T. J.; Novak, B. M. Preparation and reactivity studies of highly versatile, nickel based polymerization catalysts systems. *Macromolecules* **1993**, *26*, 7089–7091.

10. Goodall, B. L.; Barnes, D. A.; Benedikt, G. M.; McIntosh, L. H., III; Rhodes, L. F. Novel heat-resistant cyclic olefin polymers made using single components nickel and palladium catalysts. *Polym. Mater. Sci. Eng.* **1997**, *76*, 56–57.

11. Sen, A.; Lai, T. W.; Thomas, R. R. Reactions of electrophilic transition metal cations with olefins and small ring compounds. Rearrangements and polymerizations. *J. Organomet. Chem.* **1988**, *358*, 567–588.

12. Heitz, W.; Haselwander, T. F. A.; Maskos, M. Vinylic polymerization of norbornene by Pd(II) catalysis in the presence of ethylene. *Macromol. Rapid Commun.* **1997**, *18*, 689–697.

13. Mehler, C.; Risse, W. Addition polymerization of norbornene catalyzed by palladium(2+) compounds. A polymerization reaction with rare chain transfer and chain termination. *Macromolecules* **1992**, *25*, 4226–4228.

14. Saegusa, T.; Tsujino, T.; Furukawa, J. Polymerization of norbornene by modified Ziegler catalyst. *Makromol. Chem.* **1964**, *78*, 231–233.

15. Kaminsky, W.; Bark, A.; Däke, I. Polymerization of cyclic olefins with homogeneous catalysts. *Stud. Surf. Sci. Catal.* **1990**, *56 (Catal. Olefin Polym.)*, 425–438.

16. Lamonte, R. R.; McNally, D. Cyclic olefin copolymers. *Adv. Mater. Processes* **2001**, *159*, 33–36.

17. Sartori, G.; Ciampelli, F. C.; Cameli, N. Polymerization of norbornene. *Chim. Ind. (Milan)* **1963**, *45*, 1478–1482.

18. Tsujino, T.; Saegusa, T.; Furukawa, J. Polymerization of norbornene by modified Ziegler catalysts. *Makromol. Chem.* **1965**, *85*, 71–79.

19. Kaminsky, W.; Arndt, M.; Bark, A. New results of the polymerization of olefins with metallocene/aluminoxane catalysts. *Polym. Prep. (Am. Chem. Soc. Div. Polym. Chem.)* **1989**, *32(1)*, 467–468.

20. Arndt, M. *Grundlagen und Mechanismen der Polymerisation von Cycloolefinen unter Verwendung homogener Ziegler–Natta Katalysatoren;* Verlag Shaker: Aachen, 1994.

21. Arndt, M.; Engehausen, R.; Kaminsky, W; Zoumis, K. Hydrooligomerization of cycloolefins—a view of the microstructure of polynorbornene. *J. Mol. Catal. A: Chem.* **1995**, *101*, 171–178.

22. Pino, M.; Galimberti, M. Asymmetric deuteration and deuteriooligomerization of 1-pentene. *J. Organomet. Chem.* **1989**, *370*, 1–7.

23. Corradini, P.; Guerra, G. Models for stereospecificity in homogeneous and heterogeneous Ziegler–Natta polymerizations. *Progr. Polym. Sci.* **1991**, *16*, 239–257.

24. Arndt, M.; Gosmann, M. Transition metal-catalyzed polymerization of norbornene. *Polym. Bull.* **1998**, *41*, 433–440.

25. Karafilidis, C.; Hermann, H.; Rufinska, A.; Gabor, B.; Mynott, R. J.; Breitenbruch, G; Weidenthaler, C.; Rust, J.; Joppek, W.; Brookhart, M. S.; Thiel, W.; Fink, G. Metallocene-catalyzed C7-linkage in the hydrooligomerization of norbornene by sigma-bond metathesis: Insight into the microstructure of polynorbornene. *Angew. Chem., Int. Ed. Engl.* **2004**, *43*, 2444–2446.

26. Gosmann, M. *Norbornen Homo- und Copolymerisation durch Katalysatoren auf Basis später Übergangsmetalle.* Ph.D. Dissertation, University of Hamburg, Germany, 2000.

27. Wu, Q.; Lu; Y. Synthesis of a soluble vinyl-type polynorbornene with a half-titanocene/methylaluminoxane catalyst. *J. Polym. Sci., Part A: Polym. Chem.* **2002**, *40*, 1421–1425.

28. Peucker, U.; Heitz, W. Vinylic polymerization and copolymerization of norbornene and ethene by homogeneous chromium(III) catalysts. *Macromol. Chem.* **2001**, *202*, 1289–1297.

29. Schultz; R. G. Chemistry of palladium complexes. III. Polymerization of norbornene systems catalyzed by palladium chloride. *J. Polym. Sci., Part B: Polym. Lett.* **1966**, *4*, 541–546.

30. Gaylord, N. G.; Deshpande, A. B. Structure of vinyl-type polynorbornenes prepared with Ziegler–Natta catalysts. *J. Polym. Sci., Part B: Polym. Lett.* **1976**, *14*, 613–617.

31. Gaylord, N. G.; Deshpande, A. B.; Mandal, B. M.; Martan, M. 2,3- and 2,7-Bicyclo[2.2.1]hept-2-enes. Preparation and structures of polynorbornenes. *J. Macromol. Sci., Chem.* **1977**, *A11*, 1053–1070.

32. Tanielian, C.; Kiennemann, A.; Osparpucu, T. Effect of different catalysts on the basis of group VIII transition elements for the polymerization of norbornene. *Can. J. Chem.* **1979**, *57*, 2022–2027.

33. Sen, A.; Lai, T.-W. Catalysis by solvated transition-metal cations. Novel catalytic transformations of alkenes by tetrakis(acetonitrile)palladium ditetrafluoroborate. Evidence for the formation of incipient carbonium ions as intermediates. *J. Am. Chem. Soc.* **1981**, *103*, 4627–4629.

34. Sen, A.; Lai, T. W. Catalytic polymerization of acetylenes and olefins by tetrakis(acetonitrile)palladium (II) ditetrafluoroborate. *Organometallics* **1982**, *1*, 415–417.

35. Mehler, C.; Risse, W. The palladium(II)-catalyzed polymerization of norbornene. *Makromol. Chem., Rapid Commun.* **1991**, *12*, 255–259.

36. Breunig, S.; Risse, W. Transition metal-catalyzed vinyl addition polymerizations of norbornene derivatives with ester groups. *Makromol. Chem.* **1992**, *193*, 2915–2927.

37. Melia, J.; Connor, E.; Rush, S.; Breunig, S.; Mehler, C; Risse, W. Pd(II)-catalyzed addition polymerizations of strained polycyclic olefins. *Macromol. Symp.* **1995**, *89*, 433–442.

38. Haselwander, T. F. A.; Heitz, W.; Krügel, A. S. A.; Wendorff, J. H. Polynorbornene. Synthesis, properties, and simulations. *Macromol. Chem. Phys.* **1996**, *197*, 3435–3453.

39. Goodall, B. L. Cycloaliphatic polymers via late transition metal catalysis. In *Late Transition Metal Polymerization Catalyis;* Rieger, B., Baugh, L. S., Kacker, S., Striegler, S., Eds.; Wiley-VCH: Weinheim, 2003; pp 101–154.

40. Barnes, D. A.; Benedikt, G. M.; Goodall, B. L.; Huang, S. S.; Kalamarides, H. A.; Lenhard, S.; McIntosh, L. H. III; Selvy, K. T.; Shick, R. A.; Rhodes, L. F. Addition polymerization of norbornene-type monomers using neutral nickel complexes containing fluorinated aryl ligands. *Macromolecules* **2003**, *36*, 2623–2632.

41. Lipian, J.; Mimna, R. A.; Fondran, J. C.; Yandulov, D.; Shick, R. A.; Goodall, B. L.; Rhodes, L. F.; Huffman, J. C. Addition polymerization of norbornene-type monomers. High activity cationic allyl palladium catalysts. *Macromolecules* **2002**, *35*, 8969–8977.

42. Goodall, B. L.; McIntosh, L. H. III; Rhodes, L. F. New catalysts for the polymerization of cyclic olefins. *Makromol. Symp.* **1995**, *89*, 421–432.

43. Alt, F. P.; Heitz, W. Vinylic polymerization of bicyclo[2.2.1]hept-2-ene by Co(II) catalysis. *Macromol. Chem. Phys.* **1998**, *199*, 1951–1956.

44. Promerus Electronic Materials Home Page. http://www.promerus.com (accessed Sept 2006).

45. Grove, N. R.; Kohl, P. A.; Bidstrup-Allen, S. A.; Jayaraman, S.; Shick, R. Functionalized polynorbornene dielectric polymers: Adhesion and mechanical properties. *J. Polym. Sci., Part B: Polym. Phys.* **1999**, *37*, 3003–3010.

46. Shick, R. A.; Jayaraman, S. K.; Goodall, B. L.; Rhodes, L. F.; McDougall, W. C.; Kohl, P.; Bidstrup-Allen, S. A.; Chiniwalla, P. Avatrel dielectric polymers for electronic packaging. *Adv. Microelectron.* **1998**, *25*, 13–14.

47. Angiolini, S.; Avidano, M.; Barlocco, C.; Bracco, R.; Bacskay, J. J.; Lipian, J.-H.; Neal, P. S.; Rhodes, L. F.; Shick, R. A.; Zhao, X.-M.; Freeman, G. (Title unavailable). In *Proc. Eurodisplay*, Presented at The 22nd International Display Research Conference, Nice, France, October 2002; paper LN-19b, 907–910.

48. Angiolini, S.; Avidano, M.; Bracco, R.; Barlocco, C.; Young, N. D.; Trainor, M.; Zhao, X.-M. (Title unavailable). *Proc. SID 2003 (Digest of Technical Papers)*, May 2003, Baltimore, MD, USA, *34*, 1325–1328.

49. Jayachandran, P. J.; Kelleher, H. A.; Bidstrup-Allen, S. A.; Kohl, P. A. Improved fabrication of micro air-channels by incorporation of a structural barrier. *J. Micromech. Microeng.* **2005**, *15*, 35–42.

50. Glukh, K.; Lipian, J.-H.; Mimna, R.; Neal, P. S.; Ravikiran, R.; Rhodes, L. F.; Shick, R. A.; Zhao, X.-M. High-performance polymeric materials for waveguide applications. *Proc. SPIE (Int. Soc. Opt. Eng.)* **2000**, *4106 (Linear, Nonlinear, and Power-Limiting Organics)*, 43–53.

51. Li, W., Rhodes, L., Langsdorf, L. (Title unavailable). SPIE Microlithography 2003 presentation, February 2003, Santa Clara, CA.

52. Bhusari, D.; Reed, H. A.; Wedlake, M.; Padovani, A. M.; Allen, S. A. B.; Kohl, P. A. Fabrication of air-channel structures for microfluidic, microelectromechanical, and microelectronic applications. *J. Microelectromech. Syst.* **2001**, *10*, 400–408.

53. Kohl, P. A.; Zhao, Q.; Patel, K.; Schmidt, D.; Bidstrup-Allen, S. A.; Shick, R.; Jayaraman, S. Air-gaps for electrical interconnections. *Electrochem. Solid-State Lett.* **1998**, *1*, 49–51.

54. Cherdron, H.; Brekner, M.-J.; Osan, F. Cycloolefin-Copolymere: Eine neue Klasse transparenter Thermoplaste. *Angew. Makromol. Chem.* **1994**, *22*, 121–133.

55. Rische, T.; Waddon, A. J; Dickinson, L. C.; MacKnight, W. J. Microstructure and morphology of cycloolefin copolymers. *Macromolecules* **1998**, *31*, 1871–1874.

56. Tritto, I.; Marestin, C.; Boggioni, L.; Sacchi, M. C.; Brintzinger, H.-H.; Ferro, D. R. Stereoregular and stereoirregular alternating ethylene-norbornene copolymers. *Macromolecules* **2001**, *34*, 5770–5777.

57. Bergström, C. H.; Sperlich, B. R.; Ruotoistenmäki, J.; Seppälä, J. V. Investigation of the microstructure of metallocene-catalyzed norbornene-ethylene copolymers using NMR spectroscopy. *J. Polym. Sci., Part A: Polym. Chem.* **1998**, *36*, 1633–1638.

58. Hasan, T.; Ikeda, T.; Shiono, T. Ethene-norbornene copolymers with high norbornene content produced by *ansa*-fluorenylamidodimethyltitanium complex using a suitable activator. *Macromolecules* **2004**, *3*, 8503–8509.

59. Johnson, L. K.; Killian, C. M.; Brookhart, M. New Pd(II)- and Ni(II)-based catalysts for polymerization of ethylene and α-olefins. *J. Am. Chem. Soc.* **1995**, *117*, 6414–6415.

60. Johnson, L. K.; Kilian, C. M.; Arthur, S. D.; Feldman, J.; McCord, E. F.; McLain, S. J.; Kreutzer, K. A.; Bennett, M. A.; Coughlin, E. B.; Ittel, S. D.; Parthasarathy, A.; Tempel, D. J.; Brookhart, M. α-Olefins and olefin polymers and processes therefor. PCT Int. Pat. Appl. WO 96/23010 (University of North Carolina, Chapel Hill/du Pont de Nemours), August 1, 1996.

61. Britovsek, G. J. P.; Gibson, V. C.; Wass, D. F. The Search for new-generation olefin polymerization catalysts: Life beyond metallocenes. *Angew. Chem., Int. Ed Engl.* **1999**, *38*, 428–447.

62. Ittel, S. D.; Johnson, L. K.; Brookhart, M. Late-metal catalysts for ethylene homo- and copolymerization. *Chem. Rev.* **2000**, *100*, 1169–1203.

63. Goodall, B. L.; McIntosh, L. F. III. Method for the preparation of copolymers of ethylene and norbornene-type monomers with cationic palladium catalysts. PCT Int. Pat. Appl. WO98/56839 A1 (B.F. Goodrich), December 17,

64. Johnson, L. K.; Mecking, S.; Brookhart, M. Copolymerization of ethylene and propylene with functionalized vinyl monomers by palladium(II) catalysts. *J. Am. Chem. Soc.* **1996**, *118*, 267–268.

65. Benedikt, G. M.; Goodall, B. L.; Marchant, N. S.; Rhodes, L. F. Polymerization of multicyclic monomers using zirconocene catalysts. Effect of polymer microstructure on thermal properties. *New J. Chem.* **1994**, *18*, 105–114.

66. Arndt-Rosenau, M.; Beulich, I. Microstructure of ethene/norbornene copolymers. *Macromolecules* **1999**, *32*, 7335–7343.

67. Kiesewetter, J.; Kaminsky, W. Ethene/norbornene copolymerization with palladium(II) α-diimine catalysts. From ligand screening to discrete catalyst species. *Chem. Eur. J.* **2003**, *9*, 1750–1758.

68. Stevens, J. C. INSITE catalysts structure/activity relationship for olefin polymerization. *Stud. Surf. Sci. Catal.* **1994**, *89 (Catalyst Design for Tailor-Made Polyolefins)*, 277–284.

69. Canich, J. M. Olefin polymerization catalysts. European Patent Application EP 420,436 A1 (Exxon Chemical Patents, Inc., USA), September 10, 1990.

70. Stevens, J. C.; Timmers, F. J.; Wilson, D. R.; Schmidt, G. F.; Nickias, P. N.; Rosen, R. K.; Knight, G. W.; Lai, S. Y. Constrained geometry addition polymerization catalysts, processes for their preparation, precursors therefore, methods of use, and novel polymers formed therewith. European Patent Application EP 416,815 A2 (Dow Chemical Company), August 30, 1990.

71. Harrington, B. A.; Crowther, D. J. Stereoregular, alternating ethylene-norbornene copolymers from monocyclopentadienyl catalysts activated with non-coordinating discrete anions. *J. Mol. Catal. A: Chem.* **1998**, *128*, 79–84.

72. Tran, P. D.; Kaminsky, W. Ethene/norbornene copolymerization by [Me$_2$Si(3-*t*-BuCp)(N-*t*-Bu)]TiCl$_2$/MAO-catalyst. *J. Zhejiang University, Science* **2003**, *4*, 121–130.

73. Arndt, M.; Beulich, I. C$_1$-symmetric metallocenes for olefin polymerization. Part 1: Catalytic performance of [Me$_2$C(3-^{tert}BuCp)(Flu)]ZrCl$_2$ in ethene/norbornene copolymerization. *Macromol. Chem. Phys.* **1998**, *199*, 1221–1232.

74. Boussie, T. R.; Coutard, C.; Turner, H.; Murphy, V.; Powers, T. S. Solid-phase synthesis and encoding strategies for olefin polymerization catalyst libraries. *Angew. Chem., Int. Ed. Engl.* **1998**, *37*, 3272–3275.

75. Wendt, R. A.; Mynott, R.; Hauschild, K.; Ruchatz, D.; Fink, G. ^{13}C NMR studies of ethene-norbornene copolymers. Assignment of sequence distribution using ^{13}C-enriched monomers and determination of the copolymerization parameters. *Macromol. Chem. Phys.* **1999**, *200*, 1340–1350.

76. Wendt, R. A.; Fink, G. ^{13}C NMR studies of ethene/norbornene copolymers using ^{13}C-enriched monomers: Signal assignments of copolymers containing norbornene microblocks of up to a length of three norbornene units. *Macromol. Chem. Phys.* **2001**, *202*, 3490–3501.

77. Provasoli, A.; Ferro, D. R.; Tritto, I.; Boggioni, L. The conformational characteristics of ethylene-norbornene copolymers and their influence on the ^{13}C NMR spectra. *Macromolecules* **1999**, *32*, 6697–6706.

78. Kaminsky, W.; Bark, A. Copolymerization of ethene and dimethanooctahydronaphthalene with aluminoxane containing catalysts. *Polym. Int.* **1992**, *2*, 251–253.

79. Kaminsky, W.; Engehausen, R.; Kopf, J. A tailor-made metallocene for the copolymerization of ethene with bulky cycloalkenes. *Angew. Chem., Int. Ed. Engl.* **1995**, *34*, 2273–2275.

Part IV

Diene Monomers and
Olefin-Containing Polymers

17 The Stereoselective Polymerization of Linear Conjugated Dienes

Antonio Proto and Carmine Capacchione

CONTENTS

17.1 INTRODUCTION

Polydienes are amongst the largest worldwide produced and manufactured classes of polymers. This is because of their elastomeric properties, which make them suitable for many applications as synthetic rubbers.[1]

The peculiar features of polydienes are due not only to the presence of unsaturated double bonds in the polymer chain, but also to their particular microstructural characteristics (chemo-, regio-, and stereoselectivity). Owing to the complexity of polydiene structures, before going into detail concerning the different stereoregular polymers that can be obtained from a given monomer, it is worth clarifying the meaning of the terms used to describe the selectivity of the polymerization reactions, because these reactions determine the microstructures of the resulting polydienes.

1. *Chemoselectivity* refers, in a general sense, to the preferential reactivity of a chemical functionality with respect to another functionality present in the same molecule. In the case of diene polymerization, this term is used to depict the preferential reactivity of one double bond with respect to another double bond, leading (in the case of complete selectivity) to the formation of a polymer constituted of only one repeat unit (i.e., a 1,2-, 1,4-, or 3,4-polydiene).
2. *Regioselectivity* refers, in analogy to α-olefin polymerization, to the mode of insertion of the double bond into the bond between the metal and the growing polymer chain (i.e., Markovnikov or anti-Markovnikov).
3. *Stereoselectivity* describes the preferential formation of one stereoisomer when two or more stereoisomers are possible. In the case of polydienes, in addition to the iso–syndio isomerism possible due to the presence of a prochiral carbon atom (as observed for α-olefin monomers), an additional source of stereoisomerism is present. This is the presence of the double bond in the polymer backbone, which can assume two possible configurations (cis–trans).

As a consequence, even for the simplest diene monomer, butadiene (BD), one can obtain four different kinds of stereoregular polybutadienes (PBDs), as shown in Figure 17.1. Isomerism becomes even more complex in polymers of substituted (isoprene-like) monomers having the general formula $H_2C{=}CRCH{=}CH_2$ (R = alkyl or aryl group), because in this case it is possible to obtain additional enchainment structures, as shown in Figure 17.2. Moreover, for monomers having the general formula $H_2C{=}CHCH{=}CHR$, the presence of the additional asymmetric (R group-bearing) carbon in combination with the stereoisomerism arising from the presence of the double bond causes an even higher number of possible stereoregular polymer microstructures.

Dienes can be efficiently polymerized by anionic polymerization, but this polymerization process only yields polymers with a low degree of stereoregularity.[2] In analogy to α-olefin polymerization, the tool that has paved the way for the synthesis of highly stereoregular diene polymers is the use of transition metal catalysts. Using suitable catalytic systems, it is possible, for example, to obtain BD polymers of all possible microstructures with a high degree of stereoregularity. As with the polymerization of α-olefins, the mechanism commonly accepted for this polymerization is an insertion mechanism. However, as compared to stereoregular α-olefin polymerizations, the polydiene microstructure is less easily predictable from catalyst design. In many cases, the nature of the metal center and reaction conditions play a decisive role in stereocontrol. This behavior is most likely due to the more remarkable role of electronic factors in diene polymerization as compared to steric factors. The stronger influence of electronic factors is a result of the additional double bond in diene monomers, which can interact with the metal center to give allylic-type interactions that can have a strong influence on stereocontrol.

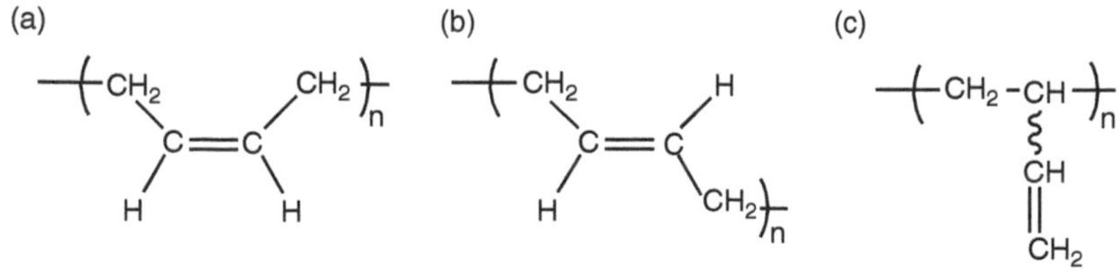

FIGURE 17.1 Structures of stereoregular polybutadienes: (a) *cis*-1,4-polybutadiene; (b) *trans*-1,4-polybutadiene; (c) isotactic or syndiotactic 1,2-polybutadiene.

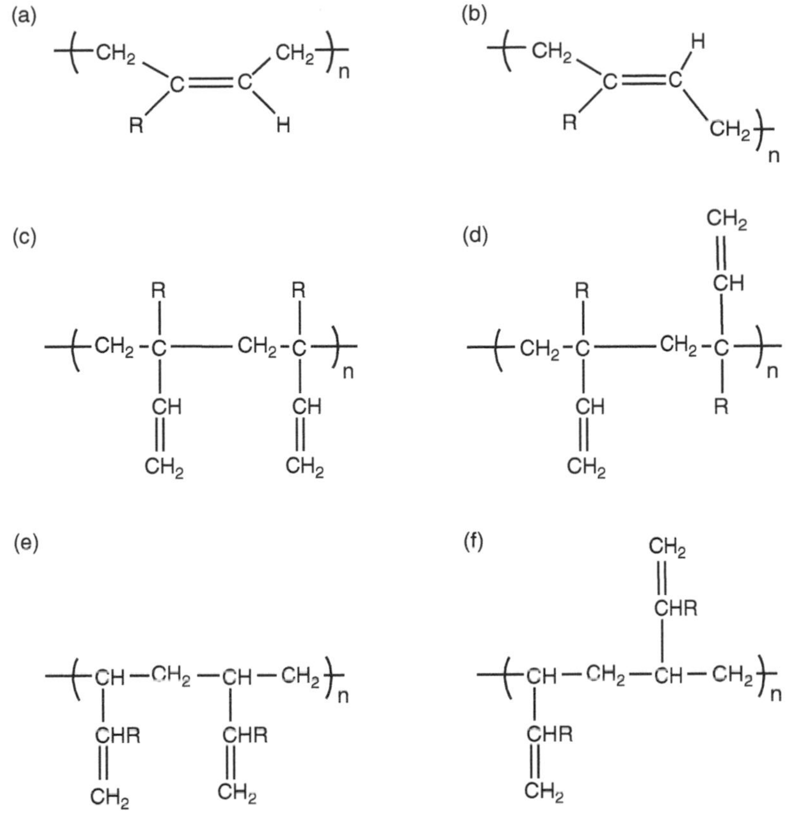

(a) (b) (c) (d) (e) (f)

FIGURE 17.2 Structures of stereoregular 2-substituted (isoprene-like) polybutadienes (R = alkyl or aryl group): (a) *cis*-1,4-polymer; (b) *trans*-1,4-polymer; (c) isotactic 1,2-polymer; (d) syndiotactic 1,2-polymer; (e) isotactic 3,4-polymer; (f) syndiotactic 3,4-polymer.

In recent years, however, many endeavors have been devoted to rationalize the polymerization behavior of conjugated dienes from both the experimental and theoretical points of view. This chapter will cover the latest developments in the field with particular attention given to stereoselective polymerizations performed in the presence of homogeneous catalytic systems.

17.2 BUTADIENE POLYMERS

cis-1,4-Polybutadiene is one of the most important rubbers used for technical purposes and is produced with a high degree of stereoregularity using conventional Ziegler–Natta catalysts.[1] Moreover, the other possible stereoregular microstructures are also known for PBD (*trans*-1,4-polybutadiene, isotactic 1,2-polybutadiene, and syndiotactic 1,2-polybutadiene).

17.2.1 *CIS*-1,4-POLYBUTADIENE

The first and most extensively studied catalytic systems for the polymerization of BD were the conventional Ziegler–Natta systems comprising titanium salts and aluminum alkyls: AlR$_3$/TiCl$_4$, AlR$_3$/TiBr$_4$, and AlR$_3$/TiI$_4$ (R = Me, Et, *i*-Bu).[3] These systems produce 1,4-PBD with a cis content ranging from 65% in the case of the chloride derivatives of titanium to 94% in the case of the iodide derivatives, with the concomitant production of crystalline trans polymer (~2%) and a low content of 1,2 units (~4%). The most commonly used aluminum alkyl is Al(*i*-Bu)$_3$, but other types

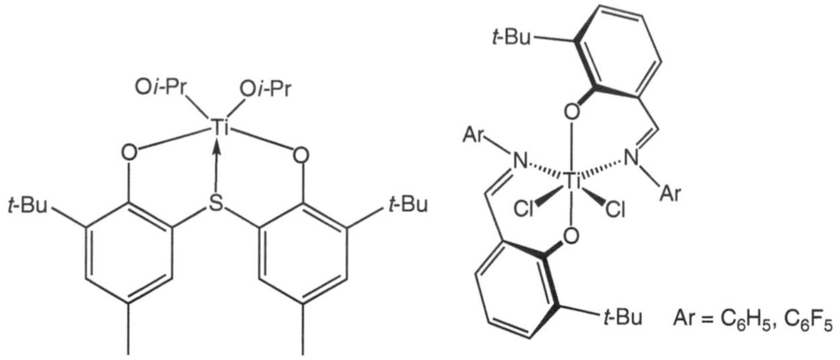

FIGURE 17.3 Bis(aryloxo)-based titanium catalysts: 2,2′-thiobis(6-*tert*-butyl-4-methylphenoxy)titanium diisopropoxide (**1**) and bis(phenoxyimino)titanium dichlorides (**2**). (Reprinted with permission from Lopez-Sanchez, J. A.; Lamberti, M.; Pappalardo, D.; Pellecchia, C. *Macromolecules* **2003**, *36*, 9260–9263. Copyright 2003 American Chemical Society.)

of aluminum alkyls have also been used. These catalysts were the basis for the first commercial production of PBD; however, it is not possible to use them to obtain a polymer with a cis content of higher then 95%. In analogy to α-olefin polymerization, these catalysts produce polymers with broad molecular weight distributions owing to the presence of multiple active sites in the catalyst system.

Recently, homogeneus titanium half-sandwich catalysts of the general formula Cp′TiX$_3$ (Cp′ = cyclopentadienyl (C$_5$H$_5$) or alkyl-substituted cyclopentadienyl; X = F, Cl, Br, OR (R = alkyl group)), when activated by methylaluminoxane (MAO), have demonstrated high activities for the production of 1,4-PBD with a cis content of 75%–85%.[4] In contrast to heterogeneous catalytic systems, the polymers obtained in this case have a very low content of trans units (1%–5%) but show a higher content of vinyl (1,2-enchained) units (15%–20%) that are useful for subsequent vulcanization. In addition, Miyazawa et al. have shown that it is possible to obtain polymerization with living behavior by performing the polymerization at subambient temperatures.[5] Both Ti(CH$_2$Ph)$_4$ and bis(aryloxo)-based titanium compounds, such as 2,2′-thiobis(6-*tert*-butyl-4-methylphenoxy)titanium diisopropoxide (**1**) and bis(phenoxyimino)titanium dichlorides (**2**) (Figure 17.3), when activated with MAO, show analogous behavior to that of half-sandwich titanocenes, leading to the hypothesis that the same catalytic species is formed during the polymerization regardless of the catalyst precursor structure.[6]

Vanadium(III) complexes, namely Cp$_2$VCl and CpVCl$_2$(PEt$_3$)$_2$ (Cp = C$_5$H$_5$), produce PBD with a predominantly cis-1,4 structure (87%) containing some 1,2 units (12%) and very low trans content (1%) when activated by MAO.[7] Compounds **3** and **4** (Figure 17.4), when activated by MAO, produce PBD with similar contents of cis-1,4 (91%) and 1,2 units (9%), but with no detectable trans content.[8]

Homogeneous catalytic systems based on cobalt salts of organic acids (typically Co octanoate) and aluminum alkyls containing chlorine (i.e., AlEt$_2$Cl, Al$_2$Et$_3$Cl$_3$, or AlEtCl$_2$) display high activity and produce PBD with cis contents of up to 95% depending on the polymerization conditions (temperature, Cl:Al ratio in the alkylaluminum chloride compound, and the nature of the solvent).[9] The systems Co(stearate)$_3$/MAO/*tert*-butyl chloride and Co(acac)$_3$/MAO (acac = acetylacetonate) have also been shown to produce PBD with a high cis content (>97%) and a narrow molecular weight distribution.[10]

Several catalytic systems based on nickel give high cis content PBDs with activities and stereoselectivities similar to those observed for cobalt compounds.[11] One such catalytic system is either a soluble nickel salt (Ni(acac)$_2$ or Ni(octanoate)$_2$) plus an alkylaluminum chloride,

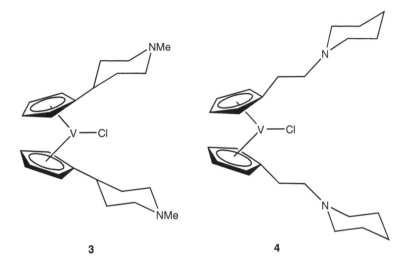

3   **4**

FIGURE 17.4 Bis(cyclopentadienyl) vanadium(III) complexes active for the cis-1,4 polymerization of butadiene. (Reprinted with permission from Bradley, S.; Camm, K. D.; Furtado, S. J.; Gott, A. L.; McGowan, P. C.; Podesta, T. J.; Thornton-Pett, M. *Organometallics* **2002**, *21*, 3443–3453. Copyright 2002 American Chemical Society.)

or a three-component $AlEt_3/Ni(carboxylate)_2/(BF_3 \cdot OEt_2)$ system. These systems produce PBD with cis contents of up to 98%. In addition, the aluminum alkyl-free $Ni(cod)_2/B(C_6F_5)_3$ (cod = 1,5-cyclooctadiene) system also produces PBD with a predominantly (up to 91%) cis-1,4 structure.[12]

The best results in terms of activity, cis selectivity, and polymer molecular weight have so far been obtained using catalytic systems based on lanthanides, in particular on neodymium. Amongst these we can distinguish two class of catalytic systems: $AlEt_2Cl/Nd(carboxylate)_3/Al(i-Bu)_3$ and $Al(i-Bu)_3/NdCl_3(L)_n$ (where L is a donor such as tetrahydrofuran, dimethylsulfoxide, or pyridine).[13] Both systems are able to produce PBD with cis contents of up to 99% and negligible trans content. Notably, the PBD molecular weight increases continuously with the polymerization time, which is the most distinctive feature of these catalytic systems. BD polymerization using these systems cannot be considered to be a true living polymerization because of the relatively broad molecular weight distributions obtained ($M_w/M_n > 2$). Recently, Taube and coworkers have shown that neodymium tris-allyl complexes of the general formula $Nd(C_3H_4R)_3$ (R = H or $-(C_6H_4)_xC_3H_5$, with x = 0–15), when activated by various aluminum alkyls or aluminoxanes, are able to produce PBD with high cis contents and narrow molecular weight distributions ($M_w/M_n < 1.4$) at high temperatures (50 °C).[14] A very high degree of cis-1,4 selectivity (>99.5%) and "living" character ($M_w/M_n = 1.4–1.8$) has been obtained by Hou and coworkers using a samarocene catalyst (**5**, Figure 17.5) in combination with modified methylaluminoxane (MMAO).[15]

17.2.2 *TRANS*-1,4-POLYBUTADIENE

Despite the low commercial interest in *trans*-1,4-polybutadiene (due to its crystalline nature, which prevents its use in applications as a synthetic rubber), many catalytic systems are able to produce PBD with a high trans-1,4 selectivity.

Heterogeneous titanium systems such as $TiCl_3/AlEt_3$ produce mixtures of PBDs containing various amounts of trans-1,4 polymer, which can be recovered after extraction with a suitable solvent.[16] A polymer consisting of 93%–94% *trans*-1,4-polybutadiene can be obtained using the catalytic system $AlEtCl_2/Ti(On-Bu)_4$.[17]

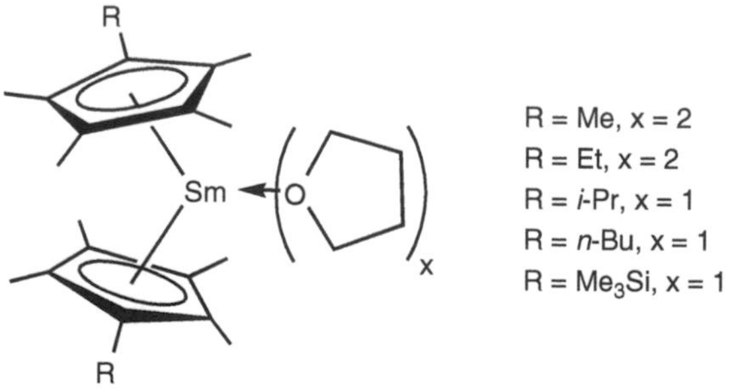

| R = Me, x = 2 |
| R = Et, x = 2 |
| R = i-Pr, x = 1 |
| R = n-Bu, x = 1 |
| R = Me$_3$Si, x = 1 |

5

FIGURE 17.5 Samarium(II) complexes active for the cis-1,4 polymerization of butadiene. (Reprinted with permission from Kaita, S.; Takeguchi, Y.; Hou, Z.; Nishiura, M.; Doi, Y.; Wakatsuki, Y. *Macromolecules* **2003**, *36*, 7923–7926. Copyright 2003 American Chemical Society.)

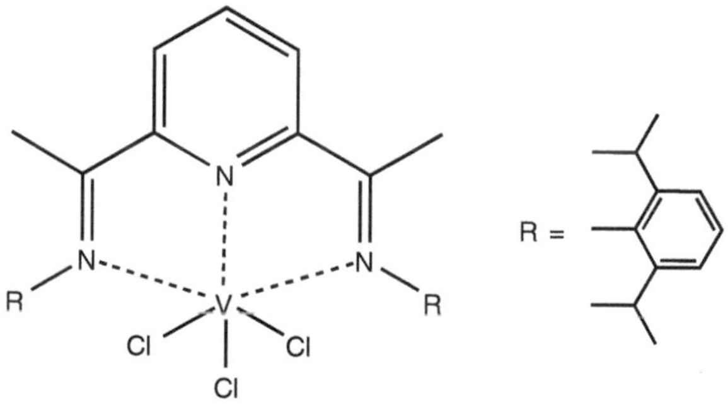

6

FIGURE 17.6 A bis(imino)pyridyl vanadium(III) complex, {2,6-[(2,6-i-Pr$_2$C$_6$N$_3$)N=C(Me)]$_2$ (C$_5$H$_3$N)}VCl$_3$] (**6**), active for the trans-1,4 polymerization of butadiene. (From Colamarco, E.; Milione, S.; Cuomo, C.; Grassi, A. *Macromol. Rapid Commun.* **2004**, *25*, 450–454. With permission.)

Various heterogeneous and homogeneous systems consisting of vanadium halides (VCl$_3$, VCl$_4$, VOCl$_3$) and aluminum alkyls produce high molecular weight PBD with trans selectivities of up to 99%.[18] Other MAO-activated soluble vanadium compounds, such as bis(imino)pyridyl vanadium complexes (i.e., **6**, Figure 17.6), have been shown to produce PBD with high trans selectivity. However, the PBDs obtained with both the vanadium halides and other catalyst systems exhibit broad molecular weight distributions.[19]

Recently, Yasuda and coworkers have obtained PBD with high trans-1,4 selectivity by using MMAO-activated iron complexes bearing tridentate terpyridyl *N*,*N*,*N*-donor ligands (**7**, Figure 17.7) as catalysts.[20] By adding a neutral donor such as a tertiary amine (e.g., NEt$_3$) to the cobalt salt-based catalytic systems described earlier for the synthesis of *cis*-1,4-polybutadiene, it is also possible to prepare PBD with high trans selectivity (up to 95%).[21] Simple rhodium salts such as Rh(NO$_3$)$_3$·2H$_2$O and RhCl$_3$·3H$_2$O also produce PBD with high trans selectivity in aqueous or alcoholic solution; notably, no alkylating agents are required.[22] Finally, single-component lanthanide tris-allyl complexes and binary systems consisting of neodymium alkoxides or aryloxides and dialkylmagnesiums (in place of the alkylaluminum reagents used to provide a cis-1,4-specific catalyst as described previously) also produce PBD with high trans selectivity (95%) and narrow molecular weight distributions (M_w/M_n = 1.1–1.8).[23]

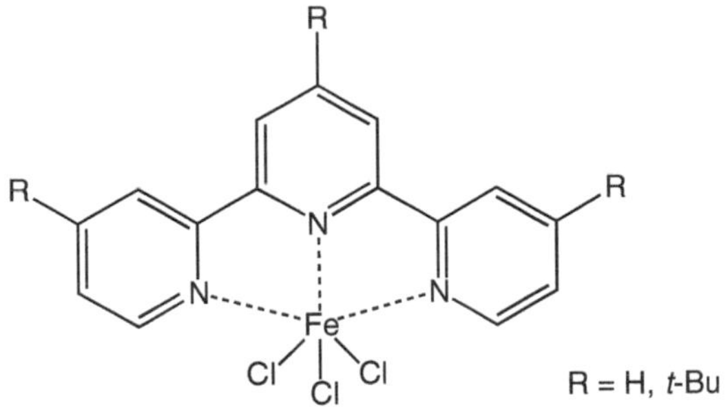

7

FIGURE 17.7 Terpyridyl iron(III) complexes active for the trans-1,4 polymerization of butadiene and the 3,4-polymerization of isoprene. (Reprinted with permission from Nakayama, Y.; Baba, Y.; Yasuda H.; Kawakita, K.; Ueyama, N. *Macromolecules* **2003**, *36*, 7953–7958. Copyright 2003 American Chemical Society.)

17.2.3 SYNDIOTACTIC AND ISOTACTIC 1,2-POLYBUTADIENES

While these crystalline stereoregular polymers are not of commercial interest, their synthesis has been accomplished using a variety of catalytic systems. Titanium alkoxides in combination with aluminum alkyls produce syndiotactic 1,2-polybutadiene with the concomitant production of an amorphous polymer.[24] V(acac)$_3$, Co(acac)$_3$, Mo(acac)$_3$, and other cobalt and molybdenum compounds such as CoBr$_2$, Co(SCN)$_2$, MoCl$_5$, and MoO$_2$(acac)$_2$, when used in combination with aluminum alkyls, produce mixtures of PBDs in which the amount of 1,2-syndiotactic polymer can rise to above 80%.[25] The best results in terms of 1,2 selectivity (>99%) and polymer crystallinity (>79%) are obtained using the ternary system Co(acac)$_3$/organoaluminum/CS$_2$ (organoaluminum = aluminoxane, aluminum alkyl, or alkylaluminum halide).[26] Porri and coworkers have recently obtained PBD with a prevalently 1,2-syndiotactic microstructure (up to 70%) using various MAO-activated iron complexes such as Fe(bipy)$_2$Et$_2$ (**8**; bipy = 2,2'-bipyridine) and Fe(phen)$_2$Cl$_2$ (**9**; phen = 1,10-phenanthroline) (Figure 17.8).[27]

Isotactic 1,2-polybutadiene has so far only been obtained using catalyst systems composed of aluminum alkyls and soluble chromium compounds such as Cr(acac)$_3$, Cr(C=NPh)$_6$, Cr(CO)$_6$, and Cr(CO)$_3$Py$_3$.[28] In all of these cases, high Al:Cr ratios and ageing of the catalytic system are crucial to obtain prevalently isotactic polymer; otherwise, syndiotactic polymer is produced. This behavior indicates that the catalytic species actually responsible for the isotactic polymerization are formed by reduction of the initial Cr complex by the alkylaluminum reagent.

17.3 ISOPRENE POLYMERS

The presence of an additional methyl group on the C2 carbon of the BD skeleton gives rise to a more complicated situation regarding the possible number of polymer stereoisomers for isoprene (Figure 17.2; R = CH$_3$). However, owing to the lower reactivity of this monomer as compared to BD, only a few stereoregular polyisoprenes (PIs) have been obtained so far. Synthetic cis-1,4-polyisoprene obtained by transition metal catalysis is practically identical to natural rubber, with a cis content of ca. 97%. Highly 1,4-trans polymer (>99%) with a structure equivalent to that of natural balata or gutta-percha rubber can be also obtained by judicious choice of the catalytic polymerization system. The industrial production of these polymers is less profitable than that of PBD owing to the high cost of isoprene monomer and the large availability of the natural products.[1]

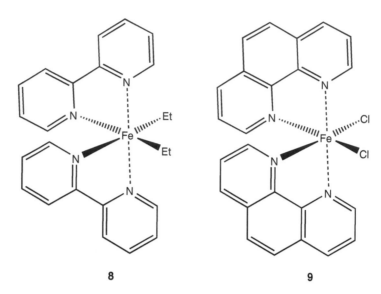

8   **9**

FIGURE 17.8 Iron(II) complexes active for the 1,2-syndiotactic polymerization of butadiene and the 3,4-polymerization of isoprene: Fe(bipy)$_2$Et$_2$ (**8**; bipy = 2,2′-bipyridine) and Fe(phen)$_2$Cl$_2$ (**9**; phen = 1,10-phenanthroline).

17.3.1 *CIS*-1,4-POLYISOPRENE

Classical heterogeneous Ziegler–Natta catalytic systems based on TiCl$_4$/AlR$_3$ (R = Me, Et, *i*-Bu) were initially used to obtain 1,4-polyisoprene with high cis content. These systems have been thoroughly studied. Notably, several parameters such as Al:Ti ratio, reaction temperature, the nature of the alkyl groups in the organoaluminum compounds, catalyst ageing, and the presence of additional atom donors such as Ph$_2$O and CS$_2$ can influence both catalyst activity and stereoselectivity.[29]

In spite of the high activities shown for BD polymerization by homogeneous catalyst systems, amongst these only a few have been found to be active for isoprene polymerization. MAO-activated titanium complexes produce 1,4-polyisoprene polymer with a prevalently cis microstructure (>94%).[6a,c] Recently, Miyazawa et al. have shown that monocyclopentadienyl titanium complexes activated by MAO can promote isoprene polymerization, giving polymers with narrow molecular weight distributions (M_w/M_n < 2) and cis contents of up to 92% with small amounts of 1,2- and 3,4-enchained structures also present.[30]

Neodymium-based complexes, analogous to those previously discussed for high *cis*-1,4-polybutadiene, also polymerize isoprene with high cis selectivity (>95%).[31]

17.3.2 *TRANS*-1,4-POLYISOPRENE

Titanium-based TiCl$_3$/AlR$_3$ (R = Me, Et, *i*-Bu) catalysts produce PI with high trans contents (>85%).[16] However, the most active and stereoselective systems for the synthesis of *trans*-1,4-polyisoprene are those based on vanadium compounds. Vanadium halides and mixtures of vanadium and titanium halides, in combination with aluminum alkyls, give high *trans*-1,4-polyisoprene (~90%).[18]

17.3.3 3,4-POLYISOPRENE

Recently, PIs with high 3,4 contents (up to 99%) have been obtained using various catalytic systems based on iron complexes and MAO or MMAO. The highest selectivity is obtained in the presence of the previously mentioned terpyridyl iron complexes **7** (>99%),[20] while Fe(bipy)$_2$Cl$_2$ complexes activated by MAO give PI with a lower degree of 3,4 content (93% at -78 °C).[27] In spite of the high crystallinity shown by these polymers, they seem to possess a low degree of stereoregularity, displaying a comparable amount of isotactic and syndiotactic dyads in their ^{13}C nuclear magnetic resonance (NMR) spectra.

17.4   1,3-PENTADIENE POLYMERS

1,3-Pentadiene (PD) exists in two isomeric forms, (E) and (Z), having different reactivities and chemoselectivies depending on the type of catalyst used. The presence of an additional source of stereoisomerism owing to the presence of a terminal methyl group in the monomer causes a multiplication of the possible stereoregular polymers (Figure 17.9). Amongst the 11 possible stereoregular polymers, however, only five have been prepared so far: *trans*-1,4-isotactic, *cis*-1,4-isotactic, *cis*-1,4-syndiotactic, *cis*-1,2-syndiotactic, and *trans*-1,2-syndiotactic poly(1,3-pentadiene) (PPD).

17.4.1   Isotactic *trans*-1,4-Polypentadiene

Highly crystalline *trans*-1,4-isotactic PPD has been prepared using the heterogeneous system AlEt$_3$/VCl$_3$[32] starting from either the (E) or (Z) isomers; however, at present, no homogeneous catalyst is known that gives *trans*-1,4-isotactic polymers from PD.

17.4.2   Isotactic *cis*-1,4-Polypentadiene

cis-1,4-Isotactic polypentadiene was first synthesized with the homogeneous catalytic system AlR$_3$/Ti(On-Bu)$_4$ (R = Et, i-Bu).[33] The crude polymer also contains 1,2 and 3,4 polymers in the amount of about 30%, which can be removed by extraction to give the crystalline product polymer. It is worth noting that both the (E) and (Z) isomers of PD can be polymerized with this system. Interestingly, when the pure (Z) isomer is used, a mixture of the two isomers is recovered in the presence of the catalyst at about 10% conversion, showing that the (Z) isomer is transformed into the (E) isomer during the polymerization process. A more active and stereoselective system for the synthesis of *cis*-1,4-isotactic PPD is the ternary system AlEt$_2$Cl/Nd(octanoate)$_3$/Al(i-Bu)$_3$, which produces highly crystalline polymers containing up to 95% cis-isotactic units.[34] The non-cis units are predominantly 1,2 with no traces of 3,4 units. In this case, only the (E) isomer is efficiently polymerized.

Various MAO-activated catalytic systems based on chromium(II) compounds, such as CrCl$_2$(dmpe)$_2$ and Cr(CH$_3$)$_2$(dmpe)$_2$ (dmpe = 1,2-bis(dimethylphosphino)ethane), produce *cis*-1,4-isotactic PPD from (E)-1,3-pentadiene and atactic *cis*-1,4-polypentadiene from (Z)-1,3-pentadiene.[35]

CpTiCl$_3$ activated by MAO polymerizes (Z)-1,3-pentadiene, affording a polymer with high cis content ($\geq$99%) at room temperature. In this case, the polymer does not show crystallinity, suggesting an atactic arrangement of the methyl groups.[4a] Notably, in this case the (E) isomer is more reactive, but its polymerization results in a mixture of 1,4-cis (50%), 1,4-trans (40%), and 1,2 (10%) units. Similar results have also been obtained using bis(phenoxyimino) titanium catalysts **2** activated by MAO.[6c]

17.4.3   Syndiotactic *cis*-1,4-Polypentadiene

This polymer is produced by cobalt-based catalyst systems such as Co(acac)$_3$/AlEt$_2$Cl/H$_2$O[36] and Co(acac)$_3$/MAO.[37] The presence of water in the former catalytic system is needed to produce an aluminoxane, which is the true activator. The polymer produced is highly crystalline, with a 1,4-cis

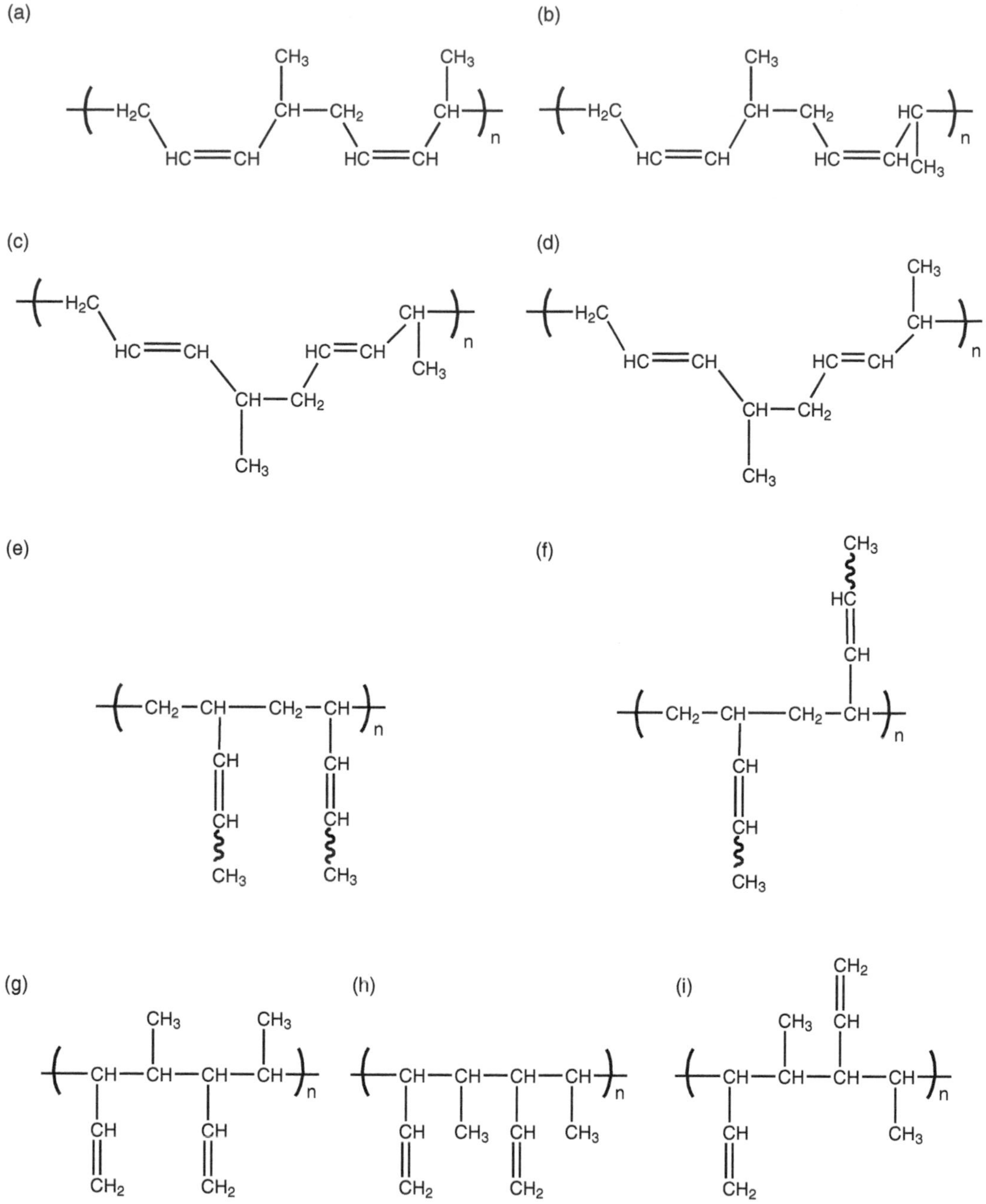

FIGURE 17.9 Structures of the possible stereoregular polymers of 1,3-pentadiene: (a) *cis*-1,4-isotactic; (b) *cis*-1,4-syndiotactic; (c) *trans*-1,4-isotactic; (d) *trans*-1,4-syndiotactic; (e) 1,2-isotactic (*cis* and *trans*); (f) 1,2-syndiotactic (*cis* and *trans*); (g) 3,4-erythrodiisotactic; (h) 3,4-threodiisotactic; (i) 3,4-disyndiotactic.

content of up to 94% and minor amounts of 1,2 (5%) and 1,4-trans units (<1%). Crystalline *cis*-1,4-syndiotactic PPD is also obtained in presence of AlEt$_3$/Ni(octanoate)$_2$/BF$_3$ with lower 1,4-cis content (88%).[38] Ni(acac)$_2$[4a] and Cp$_2$Ni[39] catalysts activated with MAO also produce a polymer with a predominantly *cis*-1,4-syndiotactic structure (85%) and lower amounts of 1,4-trans (12%)

and 1,2 units (3%). It is worth noting that in the presence of these catalytic systems only the (E)-1,3-pentadiene isomer is polymerized; even in mixtures of the two isomers, the (Z)-1,3-pentadiene isomer is completely unreactive.

17.4.4 SYNDIOTACTIC *CIS*-1,2-POLYPENTADIENE

The catalytic system CpTiCl$_3$, which at room temperature produces PPD with an atactic 1,4-cis microstructure, instead gives at polymerization temperatures of $\leq -30\,°C$ highly crystalline, stereoregular *cis*-1,2-syndiotactic ($\geq 99\%$) PPD having a melting point (T_m) of ca. $100\,°C$.[40] This polymer was subsequently fully characterized by ^{1}H and ^{13}C NMR, infrared (IR) spectroscopy, and X-ray diffraction.[41]

17.4.5 SYNDIOTACTIC *TRANS*-1,2-POLYPENTADIENE

Highly crystalline PPD (T_m 132 °C) consisting of *trans*-1,2-syndiotactic units ($\geq 99\%$) was obtained for the first time using the catalytic system CoCl$_2$(Pi-PrPh$_2$)$_2$/MAO.[42] In this case, only the (E) isomer is reactive, whereas the (Z) isomer is completely unreactive. The C_2-symmetric zirconocene compound *rac*-[CH$_2$(3-*tert*-butyl-1-indenyl)$_2$]ZrCl$_2$, when activated by MAO, produces a prevalently 1,2-syndiotactic (80%–95%) polymer from (E)-1,3-pentadiene with low activity.[43] In this case, the 1,4-trans units present in the polymer (5%–20%) lower the crystallinity, resulting in an amorphous polymer. This system also polymerizes (Z)-1,3-pentadiene, producing atactic 1,4-*trans* polymer. PPDs with prevalently 1,2-syndiotactic structures were also obtained in the presence of AlEt$_2$Cl/Co(acac)$_3$ in heptane[44] and AlEt$_3$/Fe(octanoate)/BuSCN,[38] in both cases with a low degree of stereoregularity.

17.5 OTHER CONJUGATED DIENE POLYMERS

1,3-Dialkenes having different structures than butadiene, isoprene, and 1,3-pentadiene can be described as substituted butadienes bearing alkyl groups at the C1 or C2 carbon atoms, as shown in Figure 17.10.

The interest in the synthesis of stereoregular polymers based on these 1,3-diene monomers is rather limited owing to their scarce availability and consequent high cost. However, some polymerizations have been performed, mainly to gain (from an analysis of polymer microstructure) more information about the mechanism of polymerization in the presence of a given catalyst. In this area, most studies date back to the infancy of Ziegler–Natta catalysis and therefore deal with heterogeneous catalysts based on titanium or vanadium salts.[1] Only in a few cases have such monomers been used more recently with homogeneous catalysts. Longo and coworkers have reported the polymerization of phenyl-1,3-butadienes, namely, 2-phenyl-1,3-butadiene and (E)-1-phenyl-1,3-butadiene, in the presence of CpTiCl$_3$ and Ni(acac)$_2$ activated by MAO.[45] 2-Phenyl-1,3-butadiene behaves like other dienes (e.g., butadiene, isoprene), giving a polymer with high cis content ($>99\%$) with titanium catalysts, while affording polymer with a low degree of stereoregularity (63% 1,4-cis; 26%

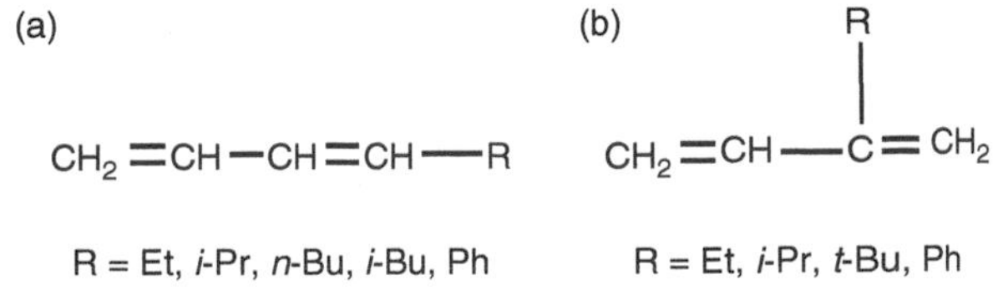

(a)　　　　　　　　　　　　　(b)

$$CH_2\!=\!CH\!-\!CH\!=\!CH\!-\!R \qquad CH_2\!=\!CH\!-\!\underset{\underset{R}{|}}{C}\!=\!CH_2$$

R = Et, *i*-Pr, *n*-Bu, *i*-Bu, Ph　　　R = Et, *i*-Pr, *t*-Bu, Ph

FIGURE 17.10 Other conjugated diene monomers: (a) 1-substituted butadienes; (b) 2-substituted butadienes.

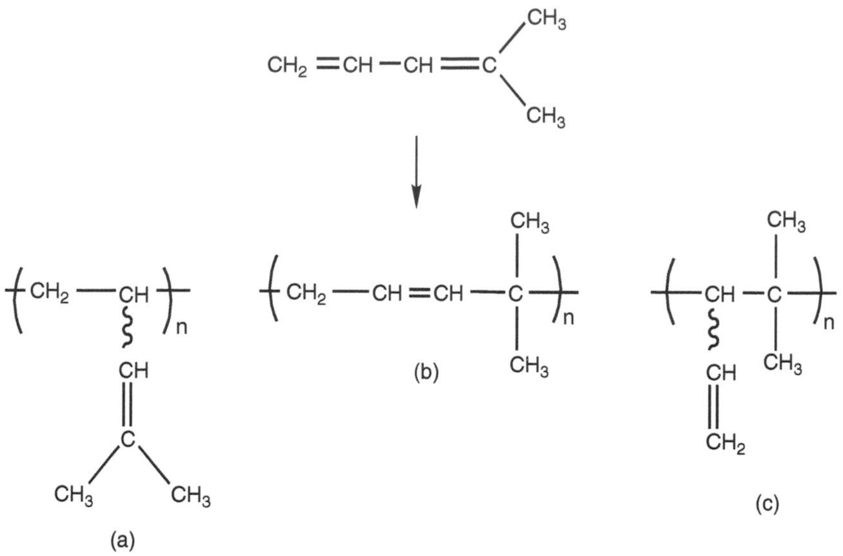

FIGURE 17.11 The possible stereoregular polymers of 4-methyl-1,3-pentadiene: (a) 1,2-isotactic or -syndiotactic; (b) 1,4-cis or -trans; (c) 3,4-isotactic or -syndiotactic.

1,4-trans; 1% 3,4) with the nickel catalyst. In contrast, with both catalytic systems, (E)-1-phenyl-1,3-butadiene gives a polymer with a predominantly 3,4 structure (76%–84%) and minor amounts of 1,4-cis (11%–16%) and 1,4-trans (5%–8%) units. These results have been explained by considering the positive electronic influence of the phenyl ring on the stability of the allylic intermediate involved in the polymerization process for 2-phenyl-1,3-butadiene, and in terms of steric effects that prevent the formation of such an allylic intermediate for (E)-1-phenyl-1,3-butadiene, allowing only its less congested (3,4) double bond to react. Several catalytic systems have been used to polymerize 3-methyl-1,3-pentadiene (3-MPD). Ni(acac)$_2$ activated by MAO affords a polymer with a *cis*-1,4-syndiotactic structure,[4a] while the ternary system AlEt$_2$Cl/Nd(octanoate)$_3$/Al(i-Bu)$_3$ gives a polymer with a *cis*-1,4-isotactic structure.[46] More recently, Porri et al. and Ricci et al. have reported the polymerization of 1,3-dienes such as 2,3-dimethyl-1,3-butadiene (DMB) and 3-MPD using (bipy)$_2$FeEt$_2$ activated by MAO.[27a,b] In the case of DMB, a 1,4-cis microstructure was attributed to the highly crystalline polymer obtained (T_m 200 °C) on the basis of X-ray analysis. Poly(3-MPD) presents some degree of crystallinity but its microstructure has not been completely elucidated.

17.6　4-METHYL-1,3-PENTADIENE POLYMERS

The presence of a quaternary carbon atom bearing two methyl groups in 4-methyl-1,3-pentadiene (4-MPD) renders the polymerization behavior of this monomer quite different from that of other dienes. In fact, amongst the six possible stereoregular polymers for this monomer (Figure 17.11), only the two stereoregular 1,2 structures are known so far. Therefore, this monomer can be regarded as being more similar to styrene than to other conjugated 1,3-diolefins.

17.6.1　ISOTACTIC 1,2-POLY(4-METHYL-1,3-PENTADIENE)

The polymerization of 4-MPD to give 1,2-isotactic polymer has been performed with various heterogeneous titanium catalysts.[47] The product in this case contains a portion (20–30%) of soluble, amorphous 1,4 polymer. The insoluble polymer material is highly crystalline with a T_m of

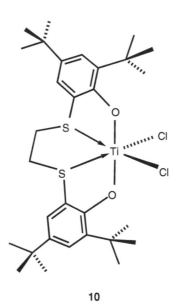

10

FIGURE 17.12 Dichloro[1,4-dithiabutanediyl-2,2′-bis(4,6-di-*tert*-butyl-phenoxy)]titanium (**10**), an active catalyst for the isospecific 1,2-polymerization of 4-methyl-1,3-pentadiene. (Reprinted with permission from Proto, A.; Capacchione, C.; Venditto, V.; Okuda, J. *Macromolecules* **2003**, *36*, 9249–9251. Copyright 2003 American Chemical Society.)

166 °C. The only homogeneous catalyst system able to produce highly 1,2-isotactic poly(4-MPD) hitherto reported is the MAO-activated post-metallocene catalyst dichloro[1,4-dithiabutanediyl-2,2′-bis(4,6-di-*tert*-butyl-phenoxy)]titanium (**10**, Figure 17.12).[48] This catalyst was first used to promote the isospecific polymerization of styrene;[49] its successful use to produce highly isotactic 1,2-poly(4-MPD) confirms the known similar polymerization behavior of these two monomers.

As judged by ^{1}H and ^{13}C NMR, this polymer displays a high degree of stereoregularity with a very small amount of 1,4 units ($\leq$3%). X-ray and IR analyses are in agreement with those reported for poly(4-MPD) obtained in the presence of the heterogeneous titanium catalysts; the slightly lower T_m observed (146 °C) was attributed to the presence of the 1,4 units, which can disturb the crystallinity of the polymer even in small amounts. It is worth noting that, in this case, ^{13}C NMR analysis is ineffective for distinguishing between isotactic and syndiotactic poly(4-MPD) because the difference between their carbon chemical shifts is too small (Table 17.1).

Conversely, the poly(4-MPD) ^{1}H NMR spectrum allows one to discriminate elegantly between the two possible microstructures. As shown in Figure 17.13a, two distinct multiplets between 0.6 and 1.3 ppm are clearly visible for isotactic 1,2-poly(4-MPD). Such a spectrum is expected for two magnetically inequivalent geminal protons (H_a and H_b), as in the case of the isotactic structure, while a single resonance (deceptive triplet) due to the equivalence of these protons is expected in the case of the syndiotactic structure (Figure 17.13b).

17.6.2 SYNDIOTACTIC 1,2-POLY(4-METHYL-1,3-PENTADIENE)

The discovery of homogeneous catalysts for the synthesis of syndiotactic polystyrene has generated an interest in the stereocontrolled polymerization of 4-MPD.[50] Highly syndiotactic 1,2-poly(4-MPD) was first synthesized and structurally characterized through ^{1}H NMR and ^{13}C NMR by Zambelli's group in 1988 using a Ti(CH$_2$Ph)$_4$/MAO catalytic system.[6a] The ^{13}C NMR spectrum of this amorphous polymer consists of six sharp resonances, which are diagnostic of a highly regioregular and

TABLE 17.1

Assignments of the ¹³C NMR Resonances for Isotactic and Syndiotactic 1,2-Poly(4-methyl-1,3-pentadiene) (poly(4-MPD))

Carbon	Isotactic 1,2-poly(4-MPD) (ppm)	Syndiotactic 1,2-poly(4-MPD) (ppm)
C(1)	42.00	42.63
C(2)	33.69	33.52
C(3)	131.75	131.54
C(4)	128.86	129.54
C(5)	25.81	25.86
C(5')	18.11	17.95

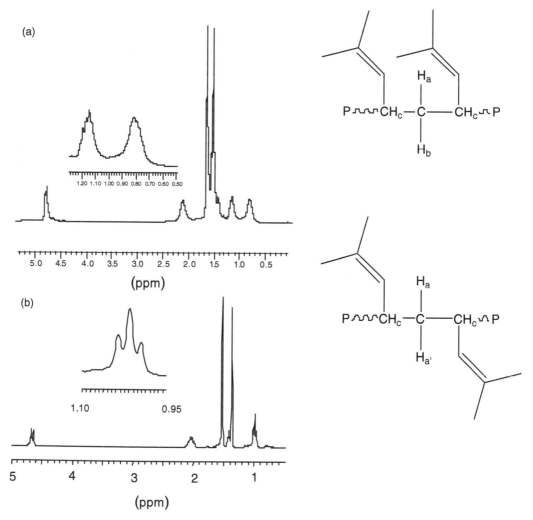

Source: Reprinted with permission from Proto, A.; Capacchione, C.; Venditto, V.; Okuda, J. *Macromolecules* **2003**, *36*, 9249–9251. Copyright 2003 American Chemical Society.

FIGURE 17.13   ¹H NMR spectra of (a) isotactic and (b) syndiotactic 1,2-poly(4-methyl-1,3-pentadiene) (P = polymer chain). (Reprinted with permission from Zambelli, A.; Ammendola, P.; Proto A. *Macromolecules* **1989**, *22*, 2126–2128; Proto, A.; Capacchione, C.; Venditto, V.; Okuda, J. *Macromolecules* **2003**, *36*, 9249–9251. Copyright 1989 and 2003 American Chemical Society.)

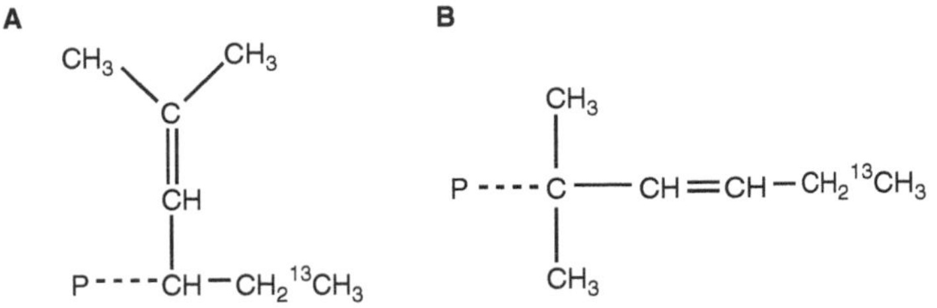

FIGURE 17.14 End groups detected in poly(4-methyl-1,3 pentadiene) polymerized with $CpTiCl_3$/MAO in the presence of $Al(^{13}CH_3)_3$: 2,1-insertion (**A**); 1,4-insertion (**B**) (P = polymer chain).

stereoregular structure (less than 12% of 1,4 units); a syndiotactic arrangement was inferred on the basis of the polymer ^1H NMR spectrum. As shown in Figure 17.13b, the methylene region of syndiotactic 1,2-poly(4-MPD) consists of a triplet centered at 1.36 ppm, which arises from the two magnetically equivalent methylene protons H_a and H'_a, as predicted for a polymer chain with syndiotactic structure. Furthermore, catalytic hydrogenation of this polymer and comparison of the ^{13}C NMR spectra of the resulting material with a genuine sample of isotactic poly(4-methyl-1-pentene) confirms the syndiotactic structure. Subsequently, Porri and coworkers also obtained such polymer using catalytic systems composed of MAO-activated titanocene and half-titanocene catalysts (Cp_2TiCl_2, $CpTiCl_3$, $CpTiCl_2$).[4d,51] A mixture of isotactic and syndiotactic polymers is obtained in presence of the catalytic system $Ti(On\text{-}Bu)_4$/MAO.[52] All of the abovementioned catalysts display the same stereospecificity for the polymerization of styrene and 4-MPD, producing syndiotactic polymer in both cases. Surprisingly, the MAO-activated catalyst rac-[CH_2(3-tert-butyl-1-indenyl)$_2$]$ZrCl_2$, which polymerizes styrene isospecifically, produces highly syndiotactic poly(4-MPD).[43]

The regiochemistry of 4-MPD polymerization was also studied using the catalytic system $CpTiCl_3$/MAO in presence of isotopically ^{13}C-enriched $Al(^{13}CH_3)_3$.[53] In addition to the signals due to the 1,2-syndiotactic polymer, the ^{13}C NMR spectrum of the polymer obtained under these conditions displayed two additional signals at 11.9 and 13.9 ppm in a 7:3 intensity ratio. These signals were attributed to, respectively, the enriched methyl groups in the polymer end groups **A** and **B** shown in Figure 17.14.

End group **A** is similar to that observed in styrene polymerization[49a] and is due to the secondary (Markovnikov or 2,1) insertion of the monomer into the Ti–$^{13}CH_3$ bond. In the case of styrene polymerization, the further absence of detectable regioinversions in the polymer structure suggests that the polymerization is highly regiospecific either in the initiation step or the propagation step. The additional presence of end group **B**, due to 1,4 insertion, is quite unexpected, because of the absence of 1,4 units in the polymer prepared with the same catalytic system but without added isotopically enriched $Al(^{13}CH_3)_3$. The difference in chemoselectivity observed between the initiation and propagation steps can be explained by considering that after a 2,1 insertion of the first monomer unit into the Ti–$^{13}CH_3$ bond, an allylic intermediate is formed; thus, the next inserting monomer unit can add at either carbon C_a or C_b, resulting in the formation of end groups **A** and **B**, respectively (Scheme 17.1). The subsequent incoming monomers (i.e., propagation steps) are always attacked at their terminal C1 carbon by the carbon labeled C_c or $C_{c'}$ of the growing polymer chain, because the double bond in the growing chain interacts with the metal center (back-biting coordination), favoring this mode of insertion and resulting in a highly regioregular polymer.

The role of back-biting coordination in governing 4-MPD polymerization chemoselectivity was subsequently confirmed using ethylene/4-MPD copolymerization.[54] In this case, a considerable amount of 1,4-enchained 4-MPD units were detected adjacent to ethylene units along the polymer chain. This suggests that the absence of a double bond in the penultimate unit of the growing chain (i.e., after ethylene insertion), and thus the impossibility of back-biting coordination, causes a decrease in chemoselectivity for 4-MPD incorporation.

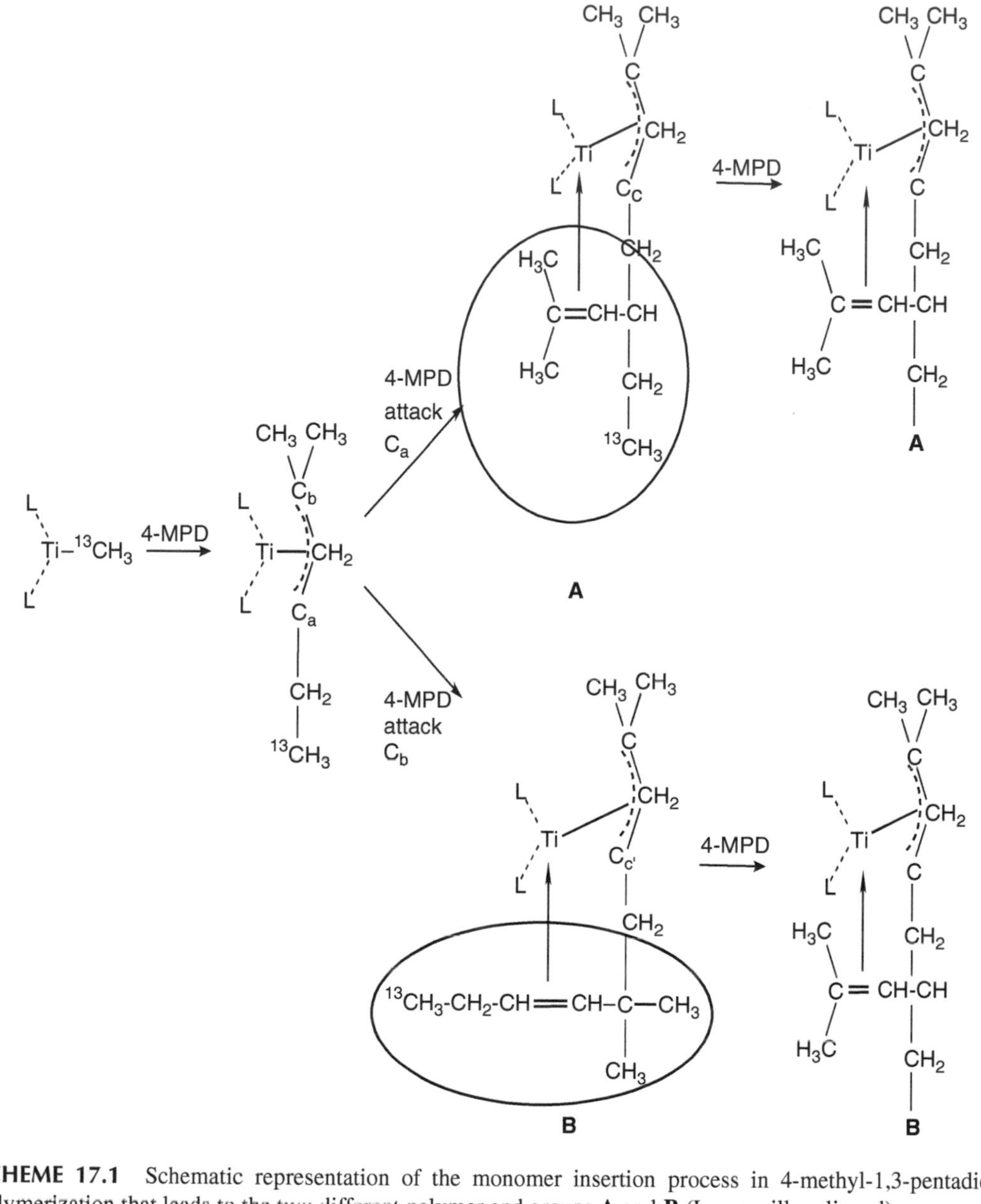

SCHEME 17.1 Schematic representation of the monomer insertion process in 4-methyl-1,3-pentadiene polymerization that leads to the two different polymer end groups **A** and **B** (L = ancillary ligand).

17.7 MECHANISM OF CHEMOSELECTIVITY AND STEREOSELECTIVITY IN DIENE POLYMERIZATION

As discussed in the earlier sections, many catalysts can be used to polymerize linear conjugated dienes, making it possible to achieve a wide variety of structurally different materials characterized by high chemo- and stereoregularity. As is known for monoalkenes, the most accredited mechanism for conjugated diene polymerization by homogeneous single site catalysts involves two steps, namely, coordination of the incoming monomer to the catalyst active site and subsequent monomer insertion into a metal–carbon bond. However, the polymerization mechanism for conjugated dienes presents several peculiar aspects, mainly related to the type of bond between the transition metal of the catalyst and the growing chain. This bond is of a σ type in monoalkene polymerizations, but is

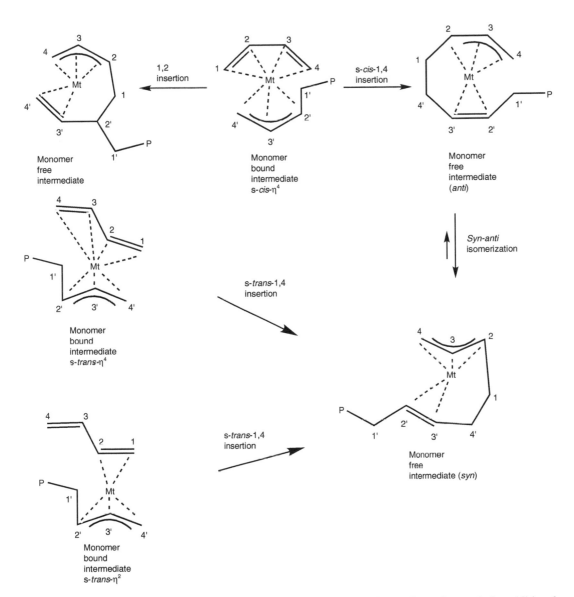

SCHEME 17.2 Mechanism of formation of 1,2 and 1,4-*cis/trans* polydienes (P = polymer chain; additional ligands eventually present on the metal center omitted for clarity). (Modified with permission from Costabile, C.; Milano, G.; Cavallo, L.; Guerra, G. *Macromolecules* **2001**, *34*, 7952–7960. Copyright 2001 American Chemical Society.)

of the allylic type (η^3) in conjugated diene polymerizations.[1] Furthermore, the conjugated diene monomer can coordinate to the metal center assuming different conformations and hapticities: s-cis-η^4 (cis conformation around the single bond and coordination of the two double bonds), s-trans-η^4 (trans conformation around the single bond and coordination of the two double bonds), or s-trans-η^2 (trans conformation around the single bond and coordination of only one double bond) (Scheme 17.2). For simple dienes such as butadiene, isoprene, and (*E*)-1,3-pentadiene, cis-η^4 coordination is by far the most energetically favored mode of coordination. The trans-η^4 coordination, although less common, has been observed in some complexes of Zr[55] and Mo;[56] therefore, one cannot exclude its intermediacy in polymerization with some transition metal catalysts. Finally, trans-η^2 coordination likely occurs when only one coordination site is available. The most accepted scheme to explain chemoselectivity and cis–trans stereoisomerism is depicted in Scheme 17.2.[1a]

Either 1,4-cis or 1,2 units can arise from the intermediate involving an s-cis-η^4 coordinated diene monomer and the η^3-coordinated allyl terminus of the growing chain presenting an *anti* structure (Scheme 17.2, top center). This kind of ligand arrangement gives rise to either 1,4-cis or 1,2 units, depending on whether the incoming monomer reacts at, respectively, the terminal C4′ or internal C2′ allyl carbon of the growing chain. Conversely, 1,4-trans units can derive from an intermediate involving an s-trans-η^4 or s-trans-η^2 coordinated diene monomer and the η^3-coordinated allyl terminus of the growing chain presenting a *syn* structure (Scheme 17.2, middle left and bottom left). In this case, insertion at the terminal allyl carbon C4′ affords a monomer-free intermediate with a *syn* structure (Scheme 17.2, bottom right). An alternative route that also gives rise to such an intermediate is isomerization of the corresponding *anti* form arising from the intermediate involved in the *cis*-1,4-polymerization mechanism (Scheme 17.2, top right). In the absence of steric effects due to the presence of substituents at the monomer C2 carbon, this equilibrium is completely shifted towards the *syn* form, which is thermodynamically more stable.[57]

It is thus evident that the cis–trans selectivity of a given catalytic system depends essentially on two factors: (1) the insertion process of a new monomer unit at the allylic group of the growing polymer chain coordinated to the metal center, and (2) the relative rate of the insertion process with respect to the rate of the *anti*→*syn* isomerization.

In addition to cis–trans stereoisomerism, if an asymmetric carbon is formed during the polymerization, as in the 1,2-polymerization of a generic diene monomer, or the cis-1,4-polymerization of a 4-monosubstituted or 1,4-disubstituted diene monomer, the resulting polymer can have an isotactic or syndiotactic microstructure.[58] As depicted in Scheme 17.2, 1,4-*cis* and 1,2 polymers arise from the same intermediate; the growing polymer chain is *anti*-η^3 bonded to the metal center, and the new incoming monomer is *cis*-η^4 coordinated. The tacticity of the resulting polymer is thought to be imposed by the mutual orientation of the incoming monomer and the last-inserted monomer unit as shown in Figure 17.15. The incoming monomer can adopt two orientations with respect to the last-inserted allylic (butenyl) unit, and in each orientation it can react either at the last-inserted unit's terminal C1 carbon to give a cis-1,4-enchained unit, or at its internal C3 carbon to give a 1,2-enchained unit. The combination of these two factors results in the formation of a new butenyl group having either the same or the opposite chirality. Reaction of the incoming monomer at the last-inserted unit's C3 carbon with an orientation as shown in Figure 17.15a gives rise to a 1,2-enchained unit and a new butenyl group having opposite chirality with respect to the previous one, whereas an orientation as shown in Figure 17.15c results in a butenyl group having the same chirality; hence, 1,2-syndiotactic and 1,2-isotactic polymers are formed, respectively. On the other hand, insertion of the new monomer at the last-inserted unit's C1 carbon will give a situation as shown in Figure 17.15b, producing a 1,4-cis-enchained unit and a butenyl group with the same chirality as the previous one; in the case of the path shown in Figure 17.15d, the newly formed butenyl group will have an opposite chirality to the previous one. These pathways result in, respectively, *cis*-1,4-isotactic and *cis*-1,4-syndiotactic polymer.

The scenario depicted above is of general validity, and the way in which the different mechanisms are achieved depends on the balance between the energy required for structural variation and the energy gained from stronger interactions between the reactants in the coordination sphere of the metal atom. Therefore, mechanistic studies should take into account the nature of the metal center, the ancillary ligands eventually present, and the steric and electronic features of the diene monomer.

From both the experimental and theoretical points of view, the most thoroughly studied catalytic systems are undoubtedly allylnickel(II) systems[59] and monocyclopentadienyl titanium complexes.[60] In the case of the nickel systems, chain growth proceeding by BD insertion into the allyl-transition metal bond was proven directly by NMR spectroscopy for both 1,4-trans- and 1,4-cis-regulating catalysts.[61] In this case, the proposed mechanism for stereoregulation suggests that the cis–trans

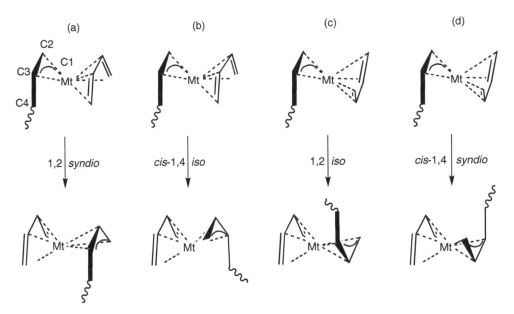

FIGURE 17.15 Possible mutual orientation of the monomer and growing polymer chain around the metal center and mechanism of iso-syndio selectivity in diene polymerization (note that carbon atoms for the last-inserted monomer unit are numbered differently than in Scheme 17.2). (From Porri, L.; Giarrusso, A.; Ricci, G. *Macromol. Symp.* **2002**, *178*, 55–68. With permission.)

selectivity is not determined by the rate of *anti-syn* isomerization, but by the different reactivities of the *anti-* and *syn*-butenylnickel(II) complexes with respect to the mode of BD coordination. Theoretical studies have confirmed this mechanism through density functional theory methods[59] and have pointed out that the two crucial steps for the cis–trans regulation are monomer insertion and *syn-anti* isomerization. They have shown that chain propagation takes place by *cis*-butadiene insertion into the π-butenylnickel(II) bond (π-allyl-insertion mechanism) through a quasi-planar four-membered transition state. The insertion most likely occurs from a prone orientation of *cis*-butadiene (*vide infra*).

Furthermore, the two isomeric butenylnickel(II) forms of the active catalyst complex, the *anti* and *syn* forms, can differ in their reactivity depending on the catalyst structure. The key to understanding the cis–trans regulation lies in the different reactivities of the *anti-* and *syn*-butenylnickel(II) forms of the active catalyst complex, both in relation to their interconversion and together with the associated *anti-syn* equilibrium.

As discussed earlier, the catalytic system CpTiCl3/MAO promotes the polymerization of various diene monomers, yielding a polymer with a prevalently 1,4-cis microstructure in the case of BD.[4a] Assuming that the catalytic species is the organotitanium cation [CpTi-P]$^+$ (P = growing polymer chain), Peluso et al.[60b,c] have proposed that the polymer chain P in the active species is coordinated to the metal center through both the π-allyl group of its terminal unit and the π bond of its penultimate diene unit (backbiting), and that *cis*-η^4 coordination of an incoming monomer to the active species requires the breakage of the latter (backbiting) interaction and a change of the terminal allyl coordination mode from η^3 to η^1. The rearrangement of the growing polymer chain is predicted to be the rate-determining step of the whole propagation reaction, and should be much easier when the ending unit of the growing polymer chain is a butenyl rather than a 2-methylbutenyl group (explaining the large difference observed between the polymerization rates of BD and isoprene). The stereo- and chemoselectivity for this catalytic system was also explained, considering an intermediate bearing a Cp ring as an ancillary ligand and considering that the back-biting interaction is removed during the insertion step. In this case, assuming an *anti*-allyl coordination

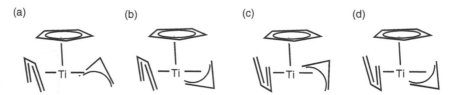

FIGURE 17.16 Possible orientations of the coordinated monomer and the allylic end group of the growing polymer chain around the metal center in diene polymerization with monocyclopentadienyl titanium catalyst systems: (a) endo–endo (prone–prone); (b) endo–exo (prone–supine); (c) exo–endo (supine–prone); (d) exo–exo (supine–supine). (Reprinted with permission from Peluso, A.; Improta, R.; Zambelli, A. *Organometallics* **2000**, *19*, 411–419. Copyright 2000 American Chemical Society.)

of the growing chain, most insertions would occur starting from the endo–endo (prone–prone) monomer-coordinated intermediate (Figure 17.16a), that is, from a s-*cis*-η^4-coordinated monomer having its concavity oriented toward the terminal allyl group of the growing chain.

This insertion would lead, through least nuclear motions, essentially only to *1,4-like* (two consecutive units enchained with the same enantiofaces) and *1,2-unlike* (two consecutive units enchained with opposite enantiofaces) monomer-free intermediates. This situation is assumed to be due to the presence of an ancillary ligand in the coordination sphere that allows only one back-biting interaction, and it is not dependent on the nature of both the metal and the ancillary ligand.[60e] In contrast, for monomers with high-energy s-cis-η^4 configurations, such as (Z)-pentadiene and 4-methyl-pentadiene, the favored insertion reaction involves a s-trans-η^2 monomer coordination and a backbiting *syn*-allyl (η^3-η^2) coordinated growing polymer chain. This mechanism is able to account for the high stereoselectivity in favor of 1,2-syndiotactic polymerization observed for these dienes. In addition, the switching to a s-cis-η^4 coordination of the monomer when backbiting of the penultimate unit of the growing chain is unfeasible, both for the initiation step of polymerization and for diene insertion steps following an ethylene insertion during copolymerization, explains the loss of chemoselectivity observed in these cases.[60f]

17.8 CONCLUSIONS

The development of homogeneous catalysis, similar to the case for monoalkenes, has engendered new possibilities for the stereoselective polymerization of dienes. As discussed in this chapter, new polymers with unprecedented microstructural features have been synthesized in recent years, and through parallel experimental and theoretical studies, a deeper knowledge of the factors governing chemo- and stereoregulation has been reached. Nevertheless, complete comprehension of the factors regulating the chemo- and stereoselectivity of diene polymerizations promoted by homogeneous catalytic systems still remains a challenge. The most relevant difference with respect to α-olefin polymerization is, in our opinion, the lack of a precise understanding of the relationship between the structure of the ancillary ligand of the precatalyst complex and the activity and selectivity of the resulting catalytic system, which renders it difficult to design a catalyst for the purpose of obtaining a polymer with a given set of microstructural features. This gap will probably never be completely overcome, owing to the intrinsically different electronic properties of different diene monomers, but presents a stimulating objective for future research in this area.

REFERENCES AND NOTES

1. For reviews, see: (a) Porri, L.; Giarrusso, A. Conjugated diene polymerization. In *Comprehensive Polymer Science*, Eastmond, G. C., Ledwith, A., Russo, S., Sigwalt, P., Eds.; Pergamon Press: Oxford, 1989; Vol. 4, pp 53–108. (b) Porri, L.; Giarrusso, A.; Ricci, G. Recent views on the mechanism of

diolefin polymerization with transition metal initiator systems. *Prog. Polym. Sci.* **1991**, *16*, 405–441. (c) Taube, R.; Sylvester, G. Stereospecific polymerization of butadiene or isoprene. In *Applied Homogeneous Catalysis with Organometallic Compounds*, Cornils, B., Hermann, W. A., Eds.; Wiley-VCH: Weinheim, 1996; Vol. 1, pp 280–318. (d) Taube, R.; Windisch, H.; Maiwald, S. The catalysis of the stereospecific butadiene polymerization by allyl nickel and allyl lanthanide complexes—a mechanistic comparison. *Macromol. Symp.* **1995**, *89*, 393–409.

2. Bywater, S. Carbanionic polymerization: Polymer configuration and the stereoregulation process. In *Comprehensive Polymer Science*, 1st ed.; Allen, G., Bevington, J. C., Eds.; Pergamon Press: Oxford, 1990; Vol. 3, pp 443–451.

3. (a) Natta, G.; Porri, L. Elastomers by coordinated anionic mechanism. A. Diene elastomers. In *Polymer Chemistry of Synthetic Elastomers*, Kennedy, J. P., Tornqvist, E., Eds.; Wiley-Interscience: New York, 1969; Vol. 2, pp 597–678. (b) Natta, G.; Porri, L.; Mazzei, A. Stereospecific polymerizations of conjugated diolefins. II. Polymerization of butadiene with catalysts prepared from aluminum alkyls and soluble vanadium chlorides. *Chim. Ind. (Milan)*, **1959**, *41*, 116–122; *Chem. Abstr.* **1959**, *53*, 86803. (c) Henderson, J. F. Polymerization of butadiene by triisobutylaluminum diisopropyl etherate and titanium tetraiodide. *J. Polym. Sci., Part C: Polym. Symp.* **1964**, *4*, 233–247.

4. (a) Oliva, L.; Longo, P.; Grassi, A.; Ammendola, P.; Pellecchia, C. Polymerization of 1,3-alkadienes in presence of Ni- and Ti-based catalytic systems containing methylalumoxane. *Makromol. Chem., Rapid Commun.* **1990**, *11*, 519–524. (b) Ricci, G.; Italia, S.; Giarrusso, A.; Porri, L. Polymerization of 1,3-dienes with the soluble catalyst system methylaluminoxanes-[CpTiCl₃]. Influence of monomer structure on polymerization stereospecificity. *J. Organomet. Chem.* **1993**, *451*, 67–72. (c) Pellecchia, C.; Zambelli, A. Copolymerization of hydrocarbon monomers in the presence of CpTiCl₃-MAO: Some information on the reaction mechanism from kinetic data and model compounds. In *Catalyst Design for Tailor-Made Polyolefins*, Soga, K., Terano, M., Eds.; Elsevier: Amsterdam, 1994; pp 209–219. (d) Ricci, G.; Bosisio, C. P.; Porri, L. Polymerization of 1,3-butadiene, 4-methyl-1,3-pentadiene, and styrene with catalyst systems based on bis-cyclopentadienyl derivatives of titanium. *Macromol. Rapid Commun.* **1996**, *17*, 781–785. (e) Foster, P.; Rausch, M. D.; Chien, J. C. W. The synthesis and polymerization behavior of methoxy-substituted (indenyl)trichlorotitanium complexes. *J. Organomet. Chem.* **1997**, *527*, 71–74. (f) Ikai, S.; Yamashita, J.; Kai, Y.; Murakami, M.; Yano, T.; Qian, Y.; Huang, J. Butadiene polymerization with various half-titanocenes. *J. Mol. Catal. A: Chem.* **1999**, *140*, 115–119. (g) Kaminsky, W. New elastomers by metallocene catalysis. *Macromol. Symp.* **2001**, *174*, 269–276.

5. (a) Miyazawa, A.; Kase, T.; Soga, K. Living polymerization of 1,3-butadiene catalyzed by some cyclopentadienyltitanium trichlorides with MAO. *Polym. Prepr. (Am. Chem. Soc. Div. Polym. Chem.)* **1999**, *40(1)*, 109–110. (b) Miyzawa, A.; Kase, T.; Soga, K. *cis*-Specific living polymerization of 1,3-butadiene catalyzed by alkyl and alkylsilyl substituted cyclopentadienyltitanium trichlorides with MAO. *Macromolecules* **2000**, *33*, 2796–2800.

6. (a) Zambelli, A.; Ammendola, P.; Proto, A. Synthesis of syndiotactic poly-1,2-(4-methyl-1,3-pentadiene). *Macromolecules* **1989**, *22*, 2126–2128. (b) Miyatake, T.; Mizunuma, K.; Kakugo, M. Ti complex catalysts including thiobisphenoxy group as a ligand for olefin polymerization. *Makromol. Chem., Macromol. Symp.* **1993**, *66*, 203–214. (c) Lopez-Sanchez, J. A.; Lamberti, M.; Pappalardo, D.; Pellecchia, C. Polymerization of conjugated dienes promoted by bis(phenoxyimino)titanium catalysts. *Macromolecules* **2003**, *36*, 9260–9263.

7. Ricci, G.; Panagia, A. P.; Porri, L. Polymerization of 1,3-dienes with catalysts based on mono- and bis-cyclopentadienyl derivatives of vanadium. *Polymer* **1996**, *37*, 363–365.

8. Bradley, S.; Camm, K. D.; Furtado, S. J.; Gott, A. L.; McGowan, P. C.; Podesta, T. J.; Thornton-Pett, M. Synthesis and structure of amino-functionalized cyclopentadienyl vanadium complexes and evaluation of their butadiene polymerization behavior. *Organometallics* **2002**, *21*, 3443–3453.

9. (a) Zgonnik, V. N.; Dolgoplosk, B. A.; Nikolaiev, N. I.; Kropachev, V. A. The effect of water on the polymerization of butadiene on a cobalt catalyst. *Vysokomol. Soedin.* **1965**, *7*, 308–311; *Chem. Abstr.* **1965**, *62*, 82972. (b) Timofeyeva, G. V.; Kokorina, N. A.; Medvedev, S. S. Butadiene polymerization on the catalytic system dichlorobis(pyridine)cobalt-diethylaluminum chloride in the presence of electron-acceptor additives. *Vysokomol. Soedin., Ser. A* **1969**, *11*, 596–600; *Chem. Abstr.* **1969**, *70*, 115606. (c) Racanelli, P.; Porri, L. *cis*-1,4-Polybutadiene by cobalt catalysts. Features of the catalysts prepared from alkyl aluminum compounds containing aluminum-oxygen-aluminum. *Eur. Polym. J.* **1970**, *6*,

751–761. (d) Medvedev, S. S.; Volkov, L. A.; Byrikhin, V. S.; Timofeyeva, G. V. Polymerization kinetics of 1,3-butadiene in the presence of cobalt catalyst systems. *Vysokomol. Soedin., Ser. A.* **1971**, *13*, 1388–1396; *Chem. Abstr.* **1971**, *75*, 89216. (e) Ricci, G.; Italia, S.; Comitani, C.; Porri, L. Polymerization of conjugated dialkenes with transition metal catalysts. Influence of methyl aluminoxane on catalyst activity and stereospecificity. *Polym. Commun.* **1991**, *32*, 514–517. (f) Endo, K.; Uchida, Y.; Matsuda, Y. Polymerizations of butadiene with Ni(acac)$_2$-methylaluminoxane catalysts. *Macromol. Chem. Phys.* **1996**, *197*, 3515–3521.

10. (a) Cass, P.; Pratt, K.; Mann, T.; Laslett, B.; Rizzardo, E.; Burford, R. Investigation of methylaluminoxane as a cocatalyst for the polymerization of 1,3-butadiene to high *cis*-1,4-polybutadiene. *J. Polym. Sci., Part A: Polym. Chem.* **1999**, *37*, 3277–3284. (b) Endo, K.; Hatakeyama, N. Stereospecific and molecular weight-controlled polymerization of 1,3-butadiene with Co(acac)$_3$-MAO catalyst. *J. Polym. Sci., Part A: Polym. Chem.* **2001**, *39*, 2793–2798.

11. (a) Schleimer, B.; Weber, H. Short-chain, linear polybutadienes. *Angew. Makromol. Chem.* **1971**, *16/17*, 253–269. (b) Dawans, F.; Teyssie, P. Polymerization by transition metal derivatives. VI. *cis*-Polymerization of 1,3-butadiene by bis(cyclooctadiene)nickel(0) and acidic metal salts. *J. Polym. Sci., Part B: Polym. Lett.* **1965**, *3*, 1045–1048. (c) Durand, J. P.; Dawans, F.; Teyssie, P. Polymerization by transition metal derivatives. XII. Factors controlling activity and stereospecificity in the 1,4 polymerization of butadiene by monometallic nickel catalysts. *J. Polym. Sci., Part A-1: Polym. Chem.* **1970**, *8*, 979–990. (d) Yoshimoto, T.; Komatsu, K.; Sakada, R.; Yamamoto, K.; Takeuchi, Y.; Onishi, A.; Ueda, K. Kinetic study of *cis*-1,4 polymerization of butadiene with nickel carboxylate/boron trifluoride etherate/triethylaluminum catalyst. *Makromol. Chem.* **1970**, *139*, 61–72. (e) Throckmorton, M. C.; Farson, F. S. Hydrogen fluoride-nickel-aluminum trialkyl catalyst system for producing high *cis*-1,4-polybutadiene. *Rubber Chem. Technol.* **1971**, *44*, 268–277. (f) Sakata, R.; Hosono, J.; Onishi, A.; Ueda, K. Effect of unsaturated hydrocarbons on the polymerization of butadiene with nickel catalyst. *Makromol. Chem.* **1970**, *139*, 73–81. (g) Hadjiandreou, P.; Julemont, M.; Teyssie, P. Butadiene 1,4-polymerization initiated by bis[(η^3-allyl)(trifluoroacetato)nickel]: A perfectly "living" coordination system. *Macromolecules* **1984**, *17*, 2455–2456. (h) Oehme, A.; Gebauer, U.; Gehrke, K.; Lechner, M. D. The influence of ageing and polymerization conditions on the polymerization of butadiene using a neodymium catalyst system. *Angew. Makromol. Chem.* **1996**, *235*, 121–130.

12. Jang, Y.; Choi, D. S.; Han, S. Effects of tris(pentafluorophenyl)borane on the activation of a metal alkyl-free Ni-based catalyst in the polymerization of 1,3-butadiene. *J. Polym. Sci., Part A: Polym. Chem.* **2004**, *42*, 1164–1173.

13. (a) Marina, N. G.; Monakov, Y. B.; Rafikov, S. R.; Gadaleva, K. K. Polymerization of dienes in the presence of lanthanide-containing catalytic systems. *Vysokomol. Soedin., Ser. A.* **1984**, *26*, 1123–1138 and references therein. (b) Wilson, D. J. Recent advances in the neodymium-catalysed polymerization of 1,3-dienes. *Makromol. Chem., Macromol. Symp.* **1993**, *66*, 273–288. (c) Zhiquan, S.; Jun, O.; Fusong, W.; Zhenya, H.; Baogong, Z. H. The characteristics of lanthanide coordination catalysts and the *cis*-polydienes prepared therewith. *J. Polym. Sci., Polym. Chem. Ed.* **1980**, *18*, 3345–3357. (d) Ricci, G.; Italia, S.; Cabassi, F.; Porri, L. Neodymium catalysts for 1,3-diene polymerization: Influence of the preparation conditions on activity. *Polym. Commun.* **1987**, *28*, 223–226. (e) Pross, A.; Marquardt, P.; Reichert, K. H.; Nentwig, W.; Knauf, T. Modelling the polymerization of 1,3-butadiene in solution with a neodymium catalyst. *Angew. Makromol. Chem.* **1993**, *211*, 89–101. (f) Quirk, R. P.; Kells, A. M.; Yunlu, K.; Cuif, J. P. Butadiene polymerization using neodymium versatate-based catalysts: Catalyst optimization and effects of water and excess versatic acid. *Polymer* **2000**, *41*, 5903–5908. (g) Barbotin, F.; Spitz, R.; Boisson, C. Heterogeneous Ziegler–Natta catalyst based on neodymium for the stereospecific polymerization of butadiene. *Macromol. Rapid Commun.* **2001**, *22*, 1411–1414.

14. (a) Maiwald, S.; Weißenborn, H.; Windisch, H.; Sommer, C.; Müller, G; Taube, R. On the catalysis of stereospecific polymerization of butadiene with the catalyst systems from Nd(η^3-C$_3$H$_5$)$_3$·dioxane and methyl aluminoxane as well as hexaisobutyl aluminoxane. *Macromol. Chem. Phys.* **1997**, *198*, 3305–3315. (b) Maiwald, S.; Sommer, C.; Müller, G; Taube, R. On the 1,4-cis-polymerization of butadiene with the highly active catalyst systems Nd(C$_3$H$_5$)$_2$Cl·1.5 THF/hexaisobutylaluminoxane (HIBAO), Nd(C$_3$H$_5$)Cl$_2$·2 THF/HIBAO and Nd(C$_3$H$_5$)Cl$_2$·THF/methylaluminoxane (MAO)—degree of polymerization, polydispersity, kinetics and catalyst formation. *Macromol. Chem. Phys.* **2001**, *202*, 1446–1456. (c) Maiwald, S.; Sommer, C.; Müller, G; Taube, R. Highly active single-site catalysts for

the 1,4-cis polymerization of butadiene from allylneodymium(III) chlorides and trialkylaluminums—a contribution to the activation of tris(allyl)neodymium(III) and the further elucidation of the structure-activity relationship. *Macromol. Chem. Phys.* **2002**, *203*, 1029–1039.

15. (a) Kaita, S.; Hou, Z.; Wakatsuki, Y. Stereospecific polymerization of 1,3-butadiene with samarocene-based catalysts. *Macromolecules* **1999**, *32*, 9078–9079. (b) Kaita, S.; Takeguchi, Y.; Hou, Z.; Nishiura, M.; Doi, Y.; Wakatsuki, Y. Pronounced enhancement brought in by substituents on the cyclopentadienyl ligand: Catalyst system $(C_5Me_4R)_2Sm(THF)_x$/MMAO (R = Et, *i*-Pr, *n*-Bu, TMS; MMAO = modified methylaluminoxane) for 1,4-cis stereospecific polymerization of 1,3-butadiene in cyclohexane solvent. *Macromolecules* **2003**, *36*, 7923–7926.

16. Natta, G.; Porri, L.; Fiore, L. Stereospecific polymerization of conjugated diolefins with catalysts containing forms of $TiCl_3$ with various lattice structures. *Gazz. Chim. Ital.* **1959**, *89*, 761–774; *Chem. Abstr.* **1960**, *54*, 117500.

17. Cuccinella, S.; Mazzei, A.; Marconi, W.; Busetto, C. Reactions between $AlRCl_2$ and Ti(OR')$_4$ and activity in diolefin polymerization *J. Macromol. Sci., Chem.* **1970**, *4*, 1549–1561.

18. Natta, G.; Porri, L.; Corradini, P.; Morero, D. Stereospecific polymerization of conjugated diolefins. I. Synthesis and structure of compounds with 1,4-*trans* linkage. *Chim. Ind. (Milan)* **1958**, *40*, 362–371; *Chem. Abstr.* **1959**, *52*, 86086.

19. (a) Ricci, G.; Italia, S.; Comitani, C.; Porri L. Polymerization of conjugated dialkenes with transition metal catalysts. Influence of methylaluminoxane on catalyst activity and stereospecificity. *Polym. Commun.* **1991**, *32*, 513–517. (b) Ricci, G.; Italia, S.; Porri, L. Polymerization of 1,3-dienes with methylaluminoxane-triacetylacetonatovanadium. *Macromol. Chem. Phys.* **1994**, *194*, 1389–1397. (c) Colamarco, E.; Milione, S.; Cuomo, C.; Grassi, A. Homo- and copolymerization of butadiene catalyzed by an bis(imino)pyridyl vanadium complex. *Macromol. Rapid Commun.* **2004**, *25*, 450–454.

20. Nakayama, Y.; Baba, Y.; Yasuda H.; Kawakita, K.; Ueyama, N. Stereospecific polymerizations of conjugated dienes by single site iron complexes having chelating *N,N,N*-donor ligands. *Macromolecules* **2003**, *36*, 7953–7958.

21. Cooper, W.; Eaves, D. E.; Vaughan, G. Electron donors in diene polymerization. *Adv. Chem. Ser.* **1966**, *52*, 46–66.

22. (a) Rinehart, R. E.; Smith, H. P.; Witt, H.; Romeyn, H. Jr. The preparation of *trans* 1,4-polybutadiene by rhodium salts in solution. *J. Am. Chem. Soc.* **1961**, *83*, 4864–4865. (b) Rinehart, R. E.; Smith, H. P.; Witt, H.; Romeyn, H. Jr. Rhodium salts as catalysts for the polymerization of butadiene. *J. Am. Chem. Soc.* **1962**, *84*, 4145–4147.

23. (a) Maiwald, S.; Weißenborn, H.; Sommer, C.; Müller, G.; Taube, R. Complex catalysis. LIX. Catalysis of 1,4-*trans*-polymerization of butadiene with tris(allyl)neodymium(III) Nd(η^3-C$_3$H$_5$)$_3$ as single-component catalyst - kinetics and reaction mechanism. *J. Organomet. Chem.* **2001**, *640*, 1–9. (b) Taube, R.; Maiwald, S.; Sieler, J. Simplified synthesis of Nd(π-C$_3$H$_5$)$_3$·C$_4$H$_8$O$_2$ via the Grignard method and preparation of the novel allylneodymium(III) complexes [Nd(π-C$_5$Me$_5$)(π-C$_3$H$_5$)$_2$·C$_4$H$_8$O$_2$] and [Nd(π-C$_3$H$_5$)Cl(THF)$_5$]B(C$_6$H$_5$)$_4$·THF as precatalysts for stereospecific butadiene polymerization. *J. Organomet. Chem.* **2001**, *621*, 327–336. (c) Jenkins, D. K. Butadiene polymerization with a rare earth compound using a magnesium alkyl cocatalyst. 1. *Polymer* **1985**, *26*, 147–151. (d) Jenkins, D. K. Butadiene polymerization with a rare earth compound using a magnesium alkyl cocatalyst. 2. *Polymer* **1985**, *26*, 152–158.

24. Natta, G.; Porri, L.; Carbonaro, A. Polymerization of conjugated diolefins by homogeneous aluminum alkyl-titanium alkoxide catalyst systems. II. 1,2-Polybutadiene and 3,4-polyisoprene. *Makromol. Chem.* **1964**, *77*, 126–138.

25. (a) Natta, G.; Porri, L.; Zanini, G.; Fiore, L. Stereospecific polymerization of conjugated diolefins. IV. Preparation of syndiotactic 1,2-polybutadiene. *Chim. Ind. (Milan)* **1959**, *41*, 526–533; *Chem. Abstr.* **1960**, *54*, 6376. (b) Natta, G. Stereospecific catalysis and isotactic polymers. *Chim. Ind. (Milan)* **1956**, *38*, 751–765; *Chem. Abstr.* **1957**, *51*, 11063. (c) Susa, E. Cobalt catalysts for preparing syndiotactic 1,2-polybutadiene. *J. Polym. Sci., Part C: Polym. Symp.* **1963**, *4*, 399–410.

26. (a) Ashitaka, H.; Ishikawa, H.; Ueno, H.; Nagasaka, A. Syndiotactic 1,2-polybutadiene with cobalt-carbon disulfide catalyst system. I. Preparation, properties, and application of highly crystalline syndiotactic 1,2-polybutadiene. *J. Polym. Sci., Polym. Chem. Ed.* **1983**, *21*, 1853–1860. (b) Ashitaka, H.; Jinda, K.; Ueno, H. Syndiotactic 1,2-polybutadiene with cobalt-carbon disulfide catalyst

system. II. Catalysts for stereospecific polymerization of butadiene to syndiotactic 1,2-polybutadiene. *J. Polym. Sci., Polym. Chem. Ed.* **1983**, *21*, 1951–1972. (c) Ashitaka, H.; Jinda, K.; Ueno, H. Syndiotactic 1,2-polybutadiene with cobalt-carbon disulfide catalyst system. III. ^{1}H- and ^{13}C-NMR study of highly syndiotactic 1,2-polybutadiene. *J. Polym. Sci., Polym. Chem. Ed.* **1983**, *21*, 1973–1988.

27. (a) Bazzini, C.; Giarrusso, A.; Porri, L. Diethylbis(2,2'-bipyridine)iron/MAO. A very active and stereospecific catalyst for 1,3-diene polymerization. *Macromol. Rapid Commun.* **2002**, *23*, 922–927. (b) Ricci, G.; Morganti, D.; Sommazzi, A.; Santi, R.; Masi, F. Polymerization of 1,3-dienes with iron complexes based catalysts. Influence of the ligand on catalyst activity and stereospecificity. *J. Mol. Catal. A: Chem.* **2003**, *204*, 287–293. (c) Bazzini, C.; Giarrusso, A.; Porri, L.; Pirozzi, B.; Napolitano, R. Synthesis and characterization of syndiotactic 3,4-polyisoprene prepared with diethylbis(2,2'-bipyridine)iron-MAO. *Polymer* **2004**, *45*, 2871–2875.

28. (a) Natta, G.; Porri, L.; Zanini, G.; Palvarini, A. Stereospecific polymerization of conjugated dienes. V. Preparation and properties of isotactic 1,2-polybutadiene. *Chim. Ind. (Milan)* **1959**, *41*, 1165–1169; *Chem. Abstr.* **1961**, *55*, 111532. (b) Natta, G.; Porri, L.; Corradini, P.; Morero, D.; Borghi, I. Hydrogenation of crystalline isotactic and syndiotactic 1,2-polybutadienes. *Atti Accad. Nazl. Lincei, Rend., Classe Sci. Fis., Mat. e Nat.* **1960**, *28*, 452–460; *Chem. Abstr.* **1961**, *55*, 17467.

29. (a) Horne, S. E. Jr.; Kichl, J. P.; Shipman, J. J.; Folt, V. L.; Gibbs, C. T. Ameripol SN—a synthetic *cis*-1,4-polyisoprene. *Ind. Eng. Chem.* **1956**, *48*, 784–791. (b) Schoenberg, E.; Chalfant, D. L.; Mayor, R. H. Preformed alkylaluminum-titanium tetrachloride catalysts for isoprene. Polymerization effect of groups attached to the aluminum on catalytic performance. *Rubber. Chem. Technol.* **1964**, *5*, 1096–1101. (c) Iwamoto, M.; Yuguchi, S. New catalyst for vinyl polymerization of butadiene and isoprene. *J. Polym. Sci., Part B: Polym. Lett.* **1967**, *5*, 1007–1011.

30. Miyazawa, A.; Kase, T.; Shibuya, T. Polymerization of isoprene with η^5-C_5H_4(*tert*-Bu)TiCl$_3$/MAO catalyst. *J. Polym. Sci., Part A: Polym. Chem.* **2004**, *42*, 1841–1844.

31. (a) Shen, Z.; Ouyang, J.; Wang, F.; Hu, Z.; Yu, F., Qian, B. The characteristics of lanthanide coordination catalysts and the *cis*-polydienes prepared therewith. *J. Polym. Sci., Polym. Chem. Ed.* **1980**, *18*, 3345–3357. (b) Yang, J. H.; Tsutsui, M.; Chen, Z.; Bergbreiter, E. New binary lanthanide catalysts for stereospecific diene polymerization. *Macromolecules* **1982**, *15*, 230–233. (c) Hsieh, H. L.; Yeh, H. C. Polymerization of butadiene and isoprene with lanthanide catalysts; characterization and properties of homopolymers and copolymers. *Rubber. Chem. Technol.* **1985**, *58*, 117–145.

32. (a) Natta, L.; Porri, L.; Corradini, P.; Morero, D. Stereospecific polymerization of conjugated diolefins. I. Synthesis and structure of compounds with 1,4-trans linkage. *Chim. Ind. (Milan)* **1958**, *40*, 362–371; *Chem. Abstr.* **1959**, *53*, 1614. (b) Natta, L.; Porri, L.; Corradini, P.; Zanini, G.; Ciampelli, F. Isotactic trans-1,4 polymers of 1,3 pentadiene. *J. Polym. Sci.* **1961**, *51*, 463–474.

33. (a) Natta, G.; Porri, L.; Stoppa, G.; Allegra, G.; Ciampelli, F. Isotactic *cis*-1,4-poly-1,3-pentadiene. *J. Polym. Sci., Part B: Polym. Lett.* **1963**, *1*, 67–71. (b) Natta, G.; Porri, L.; Carbonaro, A.; Stoppa, G. Polymerization of conjugated diolefins by homogeneous aluminum alkyl-titanium alkoxide catalyst systems. I. Cis 1,4 isotactic poly-1,3-pentadiene. *Makromol. Chem.* **1964**, *77*, 114–125.

34. (a) Ricci, G.; Boffa, G.; Porri, L. Polymerization of 1,3-dialkenes with neodymium catalysts. Some remarks on the influence of the solvent. *Makromol. Chem., Rapid Commun.* **1986**, *7*, 355–359. (b) Purevsuren, B.; Allegra, G.; Meille, S. V.; Farina, A.; Porri, L.; Ricci, G. *cis*-Isotactic 1,4-polypentadiene. NMR solution characterization and crystal structure of polymers prepared with neodymium-catalytic systems. *Polym. J.* **1998**, *30*, 431–434. (c) Porri, L.; Ricci, G.; Shubin, N. Polymerization of 1,3-dienes with neodymium catalysts. *Macromol. Symp.* **1998**, *128*, 53–61.

35. Ricci, G.; Battistella, M.; Porri, L. Chemoselectivity and stereospecificity of chromium (II) catalysts for 1,3-diene polymerization. *Macromolecules* **2001**, *34*, 5766–5769.

36. Natta, G.; Porri, L.; Carbonaro, A.; Ciampelli, F.; Allegra, G. Polymer of 1,3-pentadiene with a *cis*-1,4 syndiotactic structure. *Makromol. Chem.* **1962**, *51*, 229–232.

37. Ricci, G.; Italia, S.; Comitani, C.; Porri, L. Polymerization of conjugated dialkenes with transition metal catalysts. Influence of methylaluminoxane on catalyst activity and stereospecificity. *Polym. Commun.* **1991**, *32*, 514–517.

38. Beebe, D. H.; Gordon, C. E.; Thudium, R. N.; Throckmorton, M. C.; Hanlon, T. L. Microstructure determination of poly(1,3-pentadiene) by a combination of infrared, 60-MHz NMR, 300-MHz NMR, and X-ray diffraction spectroscopy. *J.Polym. Sci, Part A-1: Polym. Chem.* **1978**, *16*, 2285–2301.

39. Longo, P.; Amendola, A. G.; Grisi, F.; Proto, A. Polymerization of styrene and conjugated diolefins in presence of nickelocenes-based catalysts. *Macromol. Chem. Phys.* **1999**, *200*, 2461–2466.

40. Ricci, G.; Italia, S.; Porri, L. Polymerization of (Z)-1,3-pentadiene with $CpTiCl_3/MAO$. Effect of temperature on polymer structure and mechanistic implications. *Macromolecules* **1994**, *27*, 868–869.

41. Ricci, G.; Alberti, E.; Zetta, L.; Motta, T.; Bertini, F.; Mendichi, R.; Arosio, P.; Famulari, A.; Meille, S. V. Synthesis, characterization and molecular conformation of syndiotactic 1,2 polypentadiene: The *cis* polymer. *Macromolecules* **2005**, *38*, 8353–8361.

42. (a) Ricci, G.; Forni, A.; Boglia, A.; Motta, T.; Zannoni, G.; Canetti, M.; Bertini, F. Synthesis and X-ray structure of $CoCl_2(Pi\text{-}PrPh_2)_2$. A new highly active and stereospecific catalyst for 1,2 polymerization of conjugated dienes when used in association with MAO. *Macromolecules* **2005**, *38*, 1064–1070. (b) Ricci, G.; Forni, A.; Motta, T.; Boglia, A.; Alberti, E.; Zetta, L.; Bertini, F.; Arosio, P.; Famulari, A.; Mille, S. V. Synthesis, characterization, and crystalline structure of syndiotactic 1,2-polypentadiene: The trans polymer. *Macromolecules* **2005**, *38*, 8345–8352.

43. Pragliola, S.; Forlenza, E.; Longo, P. C_2-Symmetric zirconocenes in the polymerization of conjugated diolefins. *Macromol. Rapid Commun.* **2001**, *22*, 783–786.

44. Porri, L.; Di Corato, A.; Natta, G. Polymerization of 1,3-pentadiene by cobalt catalysts. Synthesis of 1,2 and *cis*-1,4 syndiotactic polypentadienes. *Eur. Polym. J.* **1969**, *5*, 1–13.

45. Pragliola, S.; Cipriano, M.; Boccia, A. C.; Longo, P. Polymerization of phenyl-1,3-butadienes in the presence of Ziegler–Natta catalysts. *Macromol. Rapid Commun.* **2002**, *23*, 356–361.

46. (a) Ricci, G.; Zetta, L.; Porri, L.; Meille, S. V. Synthesis and characterization of isotactic *cis*-1,4-poly(3-methyl-1,3-pentadiene). *Macromol. Chem. Phys.* **1995**, *196*, 2785–2793. (b) Meille, S. V.; Capelli, S.; Allegra, G.; Ricci, G. Isotactic *cis*-1,4-poly(3-methyl-1,3-pentadiene): A new conformation for isotactic *cis*-1,4-polydienes. *Macromol. Rapid Commun.* **1995**, *16*, 329–335.

47. (a) Porri, L.; Gallazzi, M. C. Effect of substituents at position 4 on the stereospecific polymerization of 1,3-diolefins. Polymers with a 1,2 isotactic structure from 4-methyl-1,3-pentadiene. *Eur. Polym. J.* **1966**, *2*, 189–198. (b) Natta, G.; Corradini, P.; Bassi, I. W.; Fagherazzi, G. The crystal structure of 1,2-isotactic poly(4-methyl-1,3-pentadiene). *Eur. Polym. J.* **1968**, *4*, 297–315.

48. Proto, A.; Capacchione, C.; Venditto, V.; Okuda, J. Synthesis of isotactic poly-1,2-(4-methyl-1,3-pentadiene) by a homogeneous titanium catalyst. *Macromolecules* **2003**, *36*, 9249–9251.

49. (a) Capacchione, C.; Proto, A.; Ebeling, H.; Mülhaupt, R.; Spaniol, T. P.; Möller, K.; Okuda, J. Ancillary ligand effect on single-site styrene polymerization: Isospecificity of group 4 metal bis(phenolate) catalysts. *J. Am. Chem. Soc.* **2003**, *125*, 4964–4965. (b) Capacchione, C.; Proto, A.; Ebeling, H.; Mülhaupt, R.; Möller, K.; Manivannan, R.; Spaniol, T. P.; Okuda, J. Non-metallocene catalysts for the styrene polymerization: Isospecific group 4 metal bis(phenolate) catalysts. *J. Mol. Catal. A: Chem.* **2004**, *213*, 137–140. (c) Beckerle, K.; Capacchione, C.; Ebeling, H.; Manivannan, R.; Mülhaupt, R.; Proto, A.; Spaniol, T. P.; Okuda, J. Stereospecific post-metallocene polymerization catalysts: The example of isospecific styrene polymerization. *J. Organomet. Chem.* **2004**, *689*, 4636–4641. (d) Capacchione, C.; Manivannan, R.; Barone, M.; Beckerle, K.; Centore, R.; Oliva, L.; Proto, A.; Tuzi, A.; Spaniol, T. P.; Okuda, J. Isospecific styrene polymerization by chiral titanium complexes that contain a tetradentate [OSSO]-type bis(phenolato) ligand. *Organometallics* **2005**, *24*, 2971–2982.

50. For reviews, see: (a) Po, R.; Cardi, N. Synthesis of syndiotactic polystyrene: Reaction mechanisms and catalysis. *Prog. Polym. Sci.* **1996**, *21*, 47–88. (b) Tomotsu, N.; Ishihara, N.; Newmann, T. H.; Malanga, M. Syndiospecific polymerization of styrene. *J. Mol. Catal. A: Chem.* **1998**, *128*, 167–190. For articles, see: (c) Ishihara, N.; Seimiya, T.; Kuramoto, M.; Uoi, M. Crystalline syndiotactic polystyrene. *Macromolecules* **1986**, *19*, 2464–2465. (d) Pellecchia, C.; Longo, P.; Grassi, A.; Ammendola, P.; Zambelli, A. Synthesis of highly syndiotactic polystyrene with organometallic catalysts and monomer insertion. *Makromol. Chem., Rapid Commun.* **1987**, *8*, 277–279. (e) Longo, P.; Proto, A.; Oliva, L. Zirconium catalysts for the syndiotactic polymerization of styrene. *Makromol. Chem., Rapid Commun.* **1994**, *15*, 151–154. (f) Pellecchia, C.; Longo, P.; Proto, A.; Zambelli, A. Novel aluminoxane-free catalysts for syndiotactic-specific polymerization of styrene. *Makromol. Chem.,*

Rapid Commun. **1992**, *13*, 265–268. (g) Pellecchia, C.; Pappalardo, D.; Oliva, L.; Zambelli, A. η^5-$C_5Me_5TiMe_3$-$B(C_6F_5)_3$: A true Ziegler–Natta catalyst for the syndiotactic-specific polymerization of styrene. *J. Am. Chem. Soc.* **1995**, *117*, 6593–6594. (h) Proto, A.; Luciano, E.; Capacchione, C.; Motta O. $ZrCl_4(THF)_2$/methylaluminoxane as the catalyst for the syndiotatic polymerization of styrene. *Macromol. Rapid Commun.* **2002**, *23*, 183–186.

51. Cugini, C.; Rombolà, O. A.; Giarrusso, A.; Porri, L.; Ricci, G. Polymerization of 4-methyl-1,3-pentadiene with catalysts based on cyclopentadienyl titanium chlorides: Effect of *anti/syn* isomerism of the allylic group on the chemoselectivity and the role of backbiting coordination in 1,3-diene polymerization. *Macromol. Chem. Phys.* **2005**, *206*, 1684–1690.

52. Ricci, G.; Porri, L. Polymerization of 4-methyl-1,3-pentadiene with MAO/Ti(OBu)$_4$. The influence of preparation/aging temperature upon the stereospecificity of the catalyst. *Polymer* **1997**, *17*, 4499–4503.

53. Longo, P.; Proto, A.; Oliva, P.; Zambelli, A. Syndiotactic-specific polymerization of 4-methyl-1,3-pentadiene: Insertion on a Mt-CH$_3$ bond. *Macromolecules* **1996**, *29*, 5500–5501.

54. Longo, P.; Grisi, F.; Proto, A.; Zambelli, A. Chemoselectivity in 4-methyl-1,3-pentadiene polymerization in the presence of homogeneous Ti-based catalysts. *Macromol. Rapid Commun.* **1997**, *18*, 193–190.

55. (a) Erker, G.; Wickler, J.; Engel, K.; Kruger, C. (s-*trans*-η^4-Diene)zirconocene complexes. *Chem. Ber.* **1982**, *115*, 3300–3310. (b) Benn, R.; Schroth, G. The structure and fluxional behaviour of butadiene in butadiene-transition metal complexes of Ti, Zr, Hf, Mo, W, and Co. *J. Organomet. Chem.* **1982**, *228*, 71–85.

56. (a) Hunter, A. D.; Legzdnis, P.; Nurse, C. R.; Einstein, F. W. B.; Willis, A. C. Very twisted η^4-*trans*-diene complexes. *J. Am. Chem. Soc.* **1985**, *107*, 1791–1792. (b) Hunter, A. D.; Legzdnis, P.; Einstein, F. W. B.; Willis, A. C.; Bursten, B. E.; Gatter, M. G. Organometallic nitrosyl chemistry. 30. Structural and electronic consequences of coordinating butadienes to (η^5-C_5H_5)Mo(NO). *J. Am. Chem. Soc.* **1986**, *108*, 3843–3844.

57. (a) Van Leeuwen, P. W. N. M.; Praat, A. NMR studies of π-methallypalladium compounds. *J. Organomet. Chem.* **1970**, *21*, 501–515. (b) Lukas, J.; Van Leeuwen, P. W. N. M.; Volger, H. C.; Kouwenhoven, A. P. The stereospecific addition of dienes to palladium chloride. *J. Organomet. Chem.* **1973**, *47*, 153–163.

58. Porri, L.; Giarrusso, A.; Ricci, G. On the mechanism of formation of isotactic and syndiotactic polydiolefins. *Macromol. Symp.* **2002**, *178*, 55–68.

59. (a) Tobisch, S. Theoretical investigation of the mechanism of cis-trans regulation for the allylnickel(II)-catalyzed 1,4 polymerization of butadiene. *Acc. Chem. Res.* **2002**, *35*, 96–104. (b) Tobisch, S.; Taube, R. Mechanistic studies of the 1,4-polymerization of butadiene according to the π-allyl-insertion mechanism. 3. Density functional study of the C-C bond formation reaction in cationic "ligand-free" (η^2:η^3-heptadienyl)(η^2-/η^4-butadiene)nickel(II) complexes [Ni(C$_7$H$_{11}$)(C$_4$H$_6$)]$^+$. *Organometallics* **1999**, *18*, 5204–5218. (c) Tobisch, S.; Taube, R. Density functional (DFT) study of the *anti-syn* isomerization of the butenyl group in cationic and neutral (butenyl)(butadiene)(monoligand)nickel(II) complexes. *Organometallics* **1999**, *18*, 3045–3060. (d) Tobisch, S.; Bögel, H.; Taube, R. Mechanistic studies of the 1,4-polymerization of butadiene according to the π-allyl-insertion mechanism. 2. Density functional study of the C-C bond formation reaction in cationic and neutral (η^3-crotyl)(η^2-/η^4-butadiene)nickel(II) complexes [Ni(C$_4$H$_7$)(C$_4$H$_6$)]$^+$, [Ni(C$_4$H$_7$)(C$_4$H$_6$)L]$^+$ (L = C$_2$H$_4$, PH$_3$), and [Ni(C$_4$H$_7$)(C$_4$H$_6$)X] (X$^-$ = I$^-$). *Organometallics* **1998**, *17*, 1177–1196. (e) Tobisch, S.; Bogel, H.; Taube, R. Mechanistic studies of the 1,4-cis polymerization of butadiene according to the π-allyl insertion mechanism. 1. Density functional study of the C-C bond formation reaction in cationic (η^3-allyl)(η^2-/η^4-butadiene)nickel(II) complexes [Ni(C$_3$H$_5$)(C$_4$H$_6$)]$^+$ and [Ni(C$_3$H$_5$)(C$_4$H$_6$)(C$_2$H$_4$)]$^+$. *Organometallics* **1996**, *15*, 3563–3561.

60. (a) Guerra, G.; Cavallo, L.; Corradini, P.; Fusco, R. Molecular mechanics and stereospecificity in Ziegler–Natta 1,2 and cis-1,4 polymerizations of conjugated dienes. *Macromolecules* **1997**, *30*, 677–684. (b) Peluso, A.; Improta, R.; Zambelli, A. Mechanism of isoprene and butadiene polymerization in the presence of CpTiCl$_3$-MAO initiator: A theoretical study. *Macromolecules* **1997**, *30*, 2219–2227. (c) Improta, R.; Peluso, A. The influence of back-biting interaction on the polymerization of conjugated dienes in the presence of Ziegler–Natta catalysts. *Macromolecules* **1999**, *32*, 6852–6855. (d) Peluso, A.; Improta, R.; Zambelli, A. Polymerization mechanism of conjugated dienes in the presence of Ziegler–Natta type catalysts: Theoretical study of butadiene and isoprene polymerization with

CpTiCl$_3$-MAO initiator. *Organometallics* **2000**, *19*, 411–419. (e) Costabile, C.; Milano, G.; Cavallo, L.; Guerra, G. Stereoselectivity and chemoselectivity in Ziegler–Natta polymerization of conjugated dienes. 1. Monomers with low energy s-*cis* η^4 coordination. *Macromolecules* **2001**, *34*, 7952–7960. (f) Costabile, C.; Milano, G.; Cavallo, L.; Longo, P.; Guerra, G.; Zambelli, A. Stereoselectivity and chemoselectivity in Ziegler–Natta polymerization of conjugated dienes. 2. Monomers with high energy s-*cis* η^4 coordination. *Polymer* **2004**, *45*, 467–485.

61. (a) Harrod, J. F.; Wallace, L. R. Kinetics of π-crotylnickel iodide catalyzed butadiene polymerization. *Macromolecules* **1969**, *2*, 449–452. (b) Hughes, R. P.; Powell, J. Transition metal promoted reactions of unsaturated hydrocarbons. I. Mechanism of 1,3-diene insertion into allyl-palladium bonds. *J. Am. Chem. Soc.* **1972**, *94*, 7723–7732. (c) Druz, N. N.; Zak, A. V.; Lobach, M. I.; Vasiliev, V. A.; Kormer, V. A. Investigation of the individual stages of 2-alkylbutadiene polymerization with bis-(π-crotylnickel iodide) propagation reactions. *Eur. Polym. J.* **1978**, *14*, 21–24. (d) Warin, R.; Julemont, M.; Teyssie, P. Butadiene 1,4-polymerization catalysis: II. An NMR study of the structure and dynamics of bis(η^3-allylnickeltrifluoracetate) and its significance for the mechanism of the polymerization. *J. Organomet. Chem.* **1980**, *185*, 413–425. (e) Taube, R.; Gehrke, J.-P.; Radeglia, R. ^{31}P-NMR spectroscopic characterization of *anti-* and *syn*-structure of C$_3$-substituted η^3-allylbis(triarylphosphite)-nickel(II)-hexafluorophosphate complexes and mechanism of [C$_3$H$_5$Ni(P(OPh)$_3$)$_2$]PF$_6$ catalyzed 1,4-trans polymerization of butadiene. *J. Organomet. Chem.* **1985**, *291*, 101–115. (f) Taube, R.; Wache, S. η^2, η^3, η^2-Dodeca-2(*E*),6(*E*),10(*Z*)-trien-1-yl-nickel(II)-tetrakis[3,5-bis(trifluoromethyl)phenyl]borate as catalyst for the 1,4-cis polymerization of butadiene. Activity and selectivity of the C$_{12}$-allylnickel(II)-cation. *J. Organomet. Chem.* **1992**, *428*, 431–442.

18 The Stereoselective Polymerization of Cyclic Conjugated Dienes

Mitsuru Nakano

CONTENTS

18.1 INTRODUCTION

Interest in cyclic olefin polymers has been growing over the past decade owing to their thermal and chemical stability, high mechanical strength, and modulus, which originate from the restricted molecular motion of cyclic groups. In addition, some cyclic olefin polymers are recognized as promising optical materials because of their low birefringence and high transparency. Among the various cyclic unsaturated compounds, cyclic conjugated dienes are quite attractive monomer candidates because some of them can be derived from desirable raw materials. For instance, 1,3-cyclohexadiene (CHD) is readily prepared from cyclohexene, a known inexpensive byproduct from ε-caprolactam synthesis. Another example is *cis*-5,6-dihydroxy-1,3-cyclohexadiene, which is a microbial oxidization product of benzene utilizing *Pseudomonas putida*.[1] If the cyclic diene compound is polymerized with controlled microstructure, the obtained polymer is expected to form a soluble precursor of an electron-conductive poly(*p*-phenylene) (PPP) (Scheme 18.1).[1]

Although cyclic conjugated diene polymers are potentially fascinating materials, examples of these materials have been so far limited because of the following reasons:

1. Few active catalysts have been found for cyclic conjugated diene polymerization, whereas a wide variety of catalysts for linear conjugated diene polymerization are known (see Chapter 17 and other reviews[2]). Only two types of polymerization systems are considered to be effective. They are Li complex-catalyzed anionic polymerization and Ni-mediated coordination polymerization.
2. The monomer variety is quite limited. In most cases, only six-membered-ring monomers, for example, CHD and its derivatives, could provide a polymer in relatively high yield.
3. In the case of regio- and stereoselective polymerization of CHD, the obtained polymer is hard to characterize fully because it is insoluble in organic solvents.

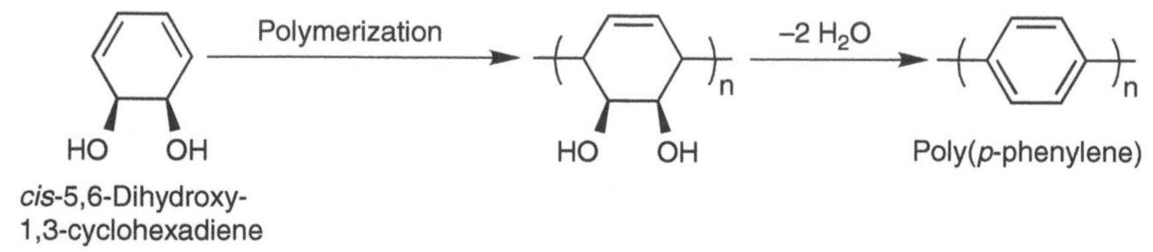

SCHEME 18.1 Synthesis of poly(*cis*-5,6-dihydroxy-1,3-cyclohexadiene) as a poly(*p*-phenylene) precursor.

This chapter begins with a brief discussion of the regio- and stereostructures of poly(cyclic conjugated dienes). Second, the anionic polymerization of CHD will be mentioned. Although ionic polymerization is not a central topic of this book, it is beneficial to compare anionically synthesized polymers with transition metal-derived ones. Recent breakthroughs concerning the living anionic polymerization of CHD provide another reason for including this topic. Subsequently, coordination polymerizations of CHD and its derivatives using several transition metal complexes are introduced. Nickel-catalyzed polymerization is the central theme of this chapter from the viewpoint of both regio- and stereocontrol and catalyst activity. Finally, a brief overall perspective of cyclic conjugated diene polymerization is given.

18.2 REGIO- AND STEREOSTRUCTURE OF POLY(CYCLIC CONJUGATED DIENES)

This section focuses on the regio- and stereostructure of CHD polymers, because very few cyclic conjugated diene monomers besides CHD can be polymerized. Although poly(*cis,cis*-1,3-cycloheptadiene) and poly(*cis,cis*-1,3-cyclooctadiene) were synthesized by cationic polymerization,[3] the obtained products were found to be oligomers and their detailed structures were not clear.

The regiochemistry of poly(CHD) is similar to that of poly(1,3-butadiene) (PBD); 1,4- and 1,2-linked structures are shown in Scheme 18.2. For reference, the cis and trans possibilities for the 1,4- and 1,2-structures are also depicted. In this context, cis (or trans) denotes that the bridging C–C bonds between monomer units are in a cis (or trans) relationship on the face of the cyclohexenyl ring. In contrast to PBD, it is difficult to produce high molecular weight poly(CHD) solely through 1,2-addition because of monomer steric hindrance.

Concerning the stereochemistry of 1,4-linked poly(CHD), there are four stereoisomers: *cis*-diisotactic (erythrodiisotactic), *cis*-disyndiotactic (erythrodisyndiotactic), *trans*-diisotactic (threodiisotactic), and *trans*-disyndiotactic (threodisyndiotactic), shown in Figure 18.1. Both the regiostructure and the stereostructure of poly(CHD) affect physical properties such as solubility and crystallinity (*vide infra*).

18.3 CHARACTERIZATION AND PROPERTIES OF POLY(CYCLIC CONJUGATED DIENES)

The relationship between physical properties and the regio-/stereostructure of poly(CHD) homopolymer has not yet been fully elucidated, but the general tendencies are as follows:

1. Poly(CHD)s with 1,2-linked and 1,2-/1,4-linked mixed regiostructures are amorphous and soluble in organic solvents.
2. 1,4-linked regioregular poly(CHD) with low stereoregularity is also amorphous and soluble in organic solvents. Crystallinity is occasionally observed in partly soluble 1,4-linked poly(CHD), but the melting temperature (T_m) of the polymer is below 200 °C.
3. Insoluble and crystalline poly(CHD) is considered to be 1,4-linked poly(CHD) with high stereoregularity. The T_m range of these poly(CHD)s is from ca. 270 to 325 °C.

Polymerization
and/or
1,4-linked structure
1,2-linked structure
cis
trans
cis
trans

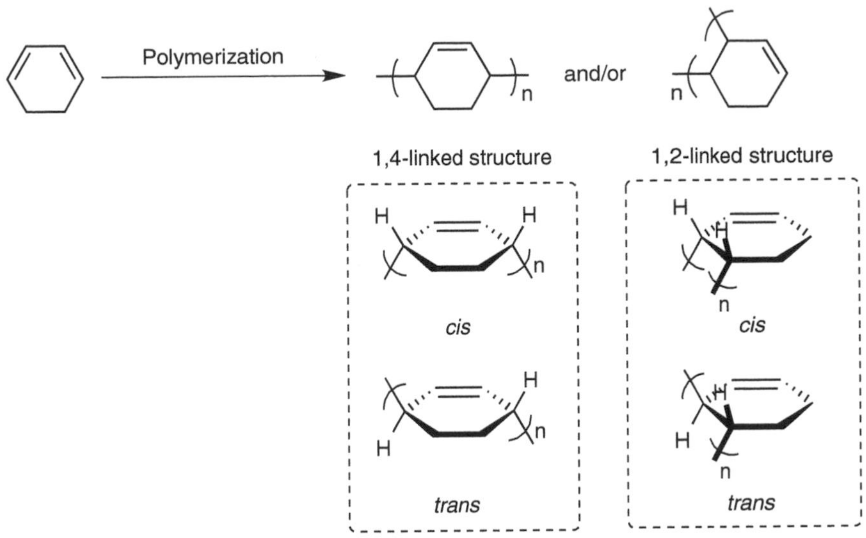

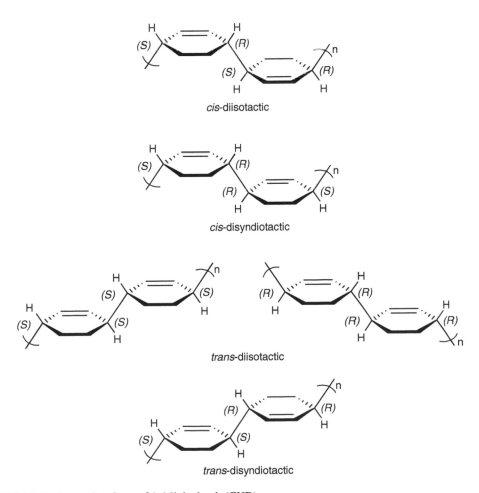

SCHEME 18.2 Regiochemistry of poly(CHD).

cis-diisotactic
cis-disyndiotactic
trans-diisotactic
trans-disyndiotactic

FIGURE 18.1 Stereochemistry of 1,4-linked poly(CHD).

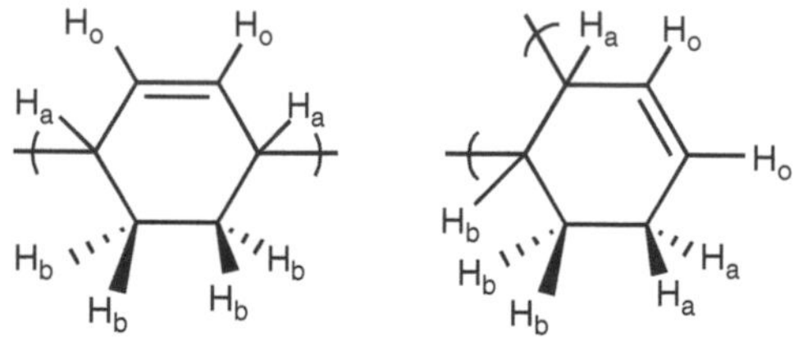

1,4-linked unit   1,2-linked unit

FIGURE 18.2 ^{1}H NMR proton assignments for 1,4- and 1,2-linked poly(CHD) units.

Regiostructures for soluble poly(CHD) and its copolymers are determined by solution ^{1}H nuclear magnetic resonance (NMR) spectroscopy. For the poly(CHD) unit, the H_o, H_a, and H_b protons (Figure 18.2) are identified with signals at, respectively, 5.73, 2.01, and 1.56 ppm. There are equal numbers of H_o and H_a protons in the 1,4-unit of poly(CHD), while the ratio of H_o to H_a protons in the 1,2-unit is 2:3.[3] Therefore, regioregularity can be estimated by the H_o/H_a integral value ratio.

Some insoluble poly(CHD)s are characterized using solid-state cross-polarization/magic-angle spinning (CPMAS) ^{13}C NMR. As shown in Scheme 18.2, 1,4-linked poly(CHD) should have only three types of carbon atoms in its ^{13}C NMR spectrum, whereas 1,2-linked poly(CHD) possesses six different types of carbon atoms. In the case of insoluble poly(CHD) synthesized using Ni complexes, carbon signals are observed at around 24, 40, and 131 ppm, which are attributed to, respectively, methylene, saturated methine, and unsaturated methine groups.[4,5] Concerning the stereostructure of poly(CHD), virtually no useful characterization method has so far been established because highly stereoregular poly(CHD) is considered to be insoluble.

The crystallinity of poly(CHD) is estimated by X-ray diffraction (XRD). Three major peaks (d = 5.28, 4.51, and 3.91 Å) are observed in its XRD spectrum in most cases.[5,6] Although differential scanning calorimetry (DSC) has been adopted to investigate T_m, the degree of polymer crystallinity cannot be quantitatively determined because the heat of fusion (ΔH_f) value (in J/g) of a poly(CHD) single crystal has not yet been determined.

18.4 ANIONIC POLYMERIZATION OF CYCLIC CONJUGATED DIENES

Although various alkyllithiums are well-known anionic polymerization initiators,[7,8] a CHD polymer with high molecular weight had not been obtained anionically until the mid-1990s. Owing to allylic hydrogen abstraction from CHD by the organolithium species, 1,4-cyclohexadiene and benzene are formed during the polymerization.[9,10] The breakthrough in this field was the discovery of alkyllithium/chelate compound systems, which induce the living polymerization of CHD.[11−13] By the use of two-component catalysts, for example, n-BuLi/N,N,N',N'-tetramethylethylenediamine (TMEDA), the molecular weight and the molecular weight distribution of poly(CHD) become well-controlled. The ratio of 1,2- and 1,4-linked structures in the obtained poly(CHD) depends mainly on the nature of the nitrogen-containing chelate compound used. When TMEDA is used as the additive, the obtained polymer has a relatively high 1,2-unit content, whereas 1,4-units were found to predominate for the alkyllithium/1,4-diazabicyclo[2.2.2]octane (DABCO) initiating system. The stereoregularity of the 1,4-linked poly(CHD) mentioned here must be relatively low because the obtained polymers are soluble in organic solvents. In addition, Long and coworkers elucidated that the 1,4-units in the polymer main chain have mixed cis/trans stereostructures by analysis of functionalized end groups.[14] Quirk et al. revealed that sec-BuLi combined with DABCO provided partly insoluble poly(CHD) with crystallinity,[15] which indicates a more stereoregular structure. The obtained polymer, however, must still only have moderate stereoregularity because its T_m (178 °C) is much lower than those of the Ni-derived poly(CHD)s discussed in the following section ($T_m > 300$ °C).

18.5 TRANSITION METAL-CATALYZED COORDINATION POLYMERIZATION OF CYCLIC CONJUGATED DIENES

A wide variety of transition metal catalysts, such as Ti-, Mo-, W-, Ni-, and Pd-based complexes, have been investigated since the late 1950s for cyclic conjugated diene polymerization. Because this chapter focuses on the "coordination" polymerization of CHD and its derivatives, cationic polymerization initiated by transition metal compounds[3] will be excluded in this section.

To our knowledge, titanium complexes were the first catalysts used for cyclic conjugated diene polymerization. Marvel, Hartzell, and Dolgoplosk et al. reported $TiCl_4/Al(i\text{-}Bu)_3$-catalyzed CHD polymerization, but very little information about microstructure was given.[16,17] Because the poly(CHD) obtained is soluble in aromatic solvents, its regio- and/or stereostructure is likely not well-controlled. Lefebvre and Dawans also tried to polymerize CHD using $TiCl_4$-based catalysts, but the obtained polymer was considered to be a mixture with a cationically polymerized product.[7]

Sen's research group found that $[Pd(CH_3CN)_4]^{2+}[BF_4]_2^-$ catalyzes the polymerization of CHD in addition to the polymerization of other unsaturated compounds such as styrene, phenylacetylene, and norbornene.[18] Unfortunately, the Pd complex used was not very active, giving poly(CHD) in 40% yield with a number average molecular weight (M_n) of 2000. On the basis of the ^{1}H NMR spectrum of the polymer, the 1,2-linkage was predominant along the main chain. This regioselectivity might be the reason for the low molecular weight of the polymer obtained. The complexes $[M(NO)_2(CH_3CN)_4]^{2+}[BF_4]_2^-$ (M = Mo, W) are other CHD polymerization catalysts developed by Sen and Thomas.[19] Again, the polymers obtained with these catalysts were oligomers. Judging from their ^{1}H NMR spectra, the regiostructure of these poly(CHD)s was found to be a mixture of 1,2- and 1,4-linkages, with the latter predominating.

In contrast to the transition metal catalysts above, nickel complexes have been studied intensely for the polymerization of CHD and its derivatives. π-Allylnickel-based complexes were employed by Dolgoplosk et al. for CHD polymerization.[17,20] They used π-alkenylnickel halides, for example, π-metallyl nickel dichloride and π-allyl nickel dibromide, combined with electron acceptors such as chloranil (tetrachloro-p-quinone) or nickel trichloroacetate. Unfortunately, the true propagating species of the above catalytic systems are not clear, but they were moderately active for CHD polymerization and the polymer obtained appeared to have a predominantly 1,4-linked structure. However, the stereoregularity of the polymer was not very high based on its T_m (270 °C).

Hampton et al. synthesized a series of π-allylnickel complexes of the formula $[Ni(\eta^3\text{-allyl})(\mu\text{-}X)]_2$ (X = OAr, SAr; Ar = aryl) (**1–8**, Figure 18.3) for the polymerization of 1,3-butadiene (BD) and CHD.[4] Complexes **5–7** were active enough to provide poly(CHD) quantitatively, whereas complexes **1–4** and **8** were inactive as polymerization initiators. Although an electron-deficient X ligand is a necessary prerequisite, it is not sufficient for the $[Ni(\eta^3\text{-allyl})(\mu\text{-}X)]_2$ complexes to exhibit polymerization activity. In the case of complexes **3** and **4**, inactivity could be due to the coordination of the o-chlorine or -fluorine atoms to the nickel center. The inactivity of complex **8** could be due to the better π-donor ability of arenethiolates as compared to phenols, which could stabilize its dimeric structure compared with complex **5**. In general, reduction of coordinative unsaturation makes the nickel complexes less active for diene polymerization. The regiostructure of insoluble poly(CHD)s prepared using complexes **5–7** was characterized using solid-state CPMAS ^{13}C NMR. These polymers should have >95% 1,4-linked structures because there are only three types of carbon atoms in their NMR spectra.[4] The stereochemistry of polymerization is believed to be *syn*-coordinative addition to CHD, resulting in the formation of a cis structure. The *syn*-addition mechanism was originally proposed by Porri and Aglietto for linear conjugated diene polymerization (see Chapter 17).[21]

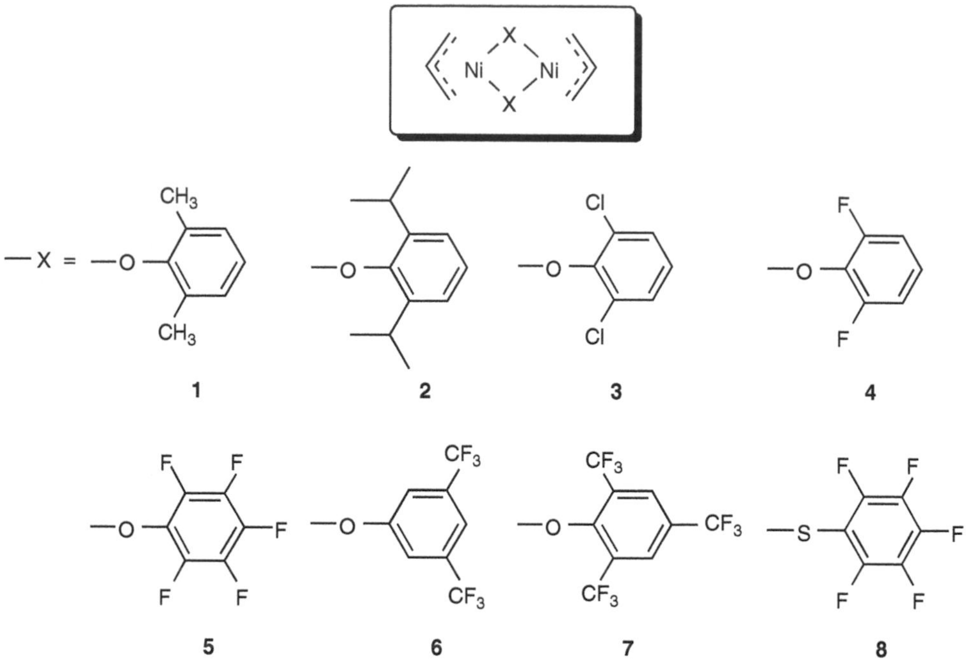

FIGURE 18.3 π-Allylnickel complexes [Ni(η³-allyl)(μ-X)]₂ (X = OAr, SAr; Ar = aryl) used for CHD polymerization.

Grubbs and coworkers attempted to polymerize CHD and its functionalized derivatives (Figure 18.4) using bis[(η³-allyl)(trifluoroacetato)nickel(II)] (ANiTFA).[22,23] Their ultimate goal was to make a PPP without structural defects via soluble precursor methods. In order to achieve this goal, regio- and stereocontrol are important: 1,4-regioregular poly(CHD) derivatives yield perfectly *para*-linked poly(phenylene) after elimination of the functional substituents. For facile substituent elimination (aromatization) to PPP, the 1,4-*SSRR* repeat unit shown in Scheme 18.3 is desirable. In addition, ANiTFA was expected to work as a catalyst because it is potentially tolerant to monomer functionalities.[24,25] Among the CHD derivatives studied, only *cis*-5,6-bis(trimethylsiloxy)-1,3-cyclohexadiene (**TMSO-CHD**) was polymerized by ANiTFA in high yield (Scheme 18.4). The authors concluded that the obtained poly(**TMSO-CHD**) has a 1,4-linked regioregular structure because no ¹H NMR proton resonance in the 1.8–2.1 ppm range was observed (proton signals around 2.0 ppm are characteristic for the 1,2-unit of poly(**TMSO-CHD**) synthesized using a radical initiator).[19] The obtained poly(**TMSO-CHD**) was thus considered to have either a 1,4-linked *cis*-diisotactic or a 1,4-linked *cis*-disyndiotactic structure, because of the following reasons: (1) its XRD spectrum revealed a partially ordered structure (a single diffraction peak was observed at 9.73 Å), (2) it had a relatively high Mark–Houwink–Sakurada coefficient *a* of 0.9–1.1 in tetrahydrofuran (THF), indicating a rodlike conformation. Although the detailed stereostructure of poly(**TMSO-CHD**) is not yet elucidated, the authors succeeded in converting poly(**TMSO-CHD**) into PPP without *ortho*-linkages[26] (note that the NMR spectra and XRD patterns of poly(**TMSO-CHD**) are different from those discussed earlier for poly(CHD), because of the trimethylsiloxy substituents).

Recent achievements in nickel-mediated CHD polymerization involve the development of methylaluminoxane (MAO) activated catalytic systems, which have been studied independently by a few research groups. Although ANiTFA is a good catalyst for CHD polymerization, the complexes discussed below are more active.

Our group at Toyota Central R&D Labs has discovered a remarkably active bis(allylnickel bromide)/MAO (ANiBr/MAO) catalyst for CHD polymerization.[6,27] As shown in Table 18.1, CHD

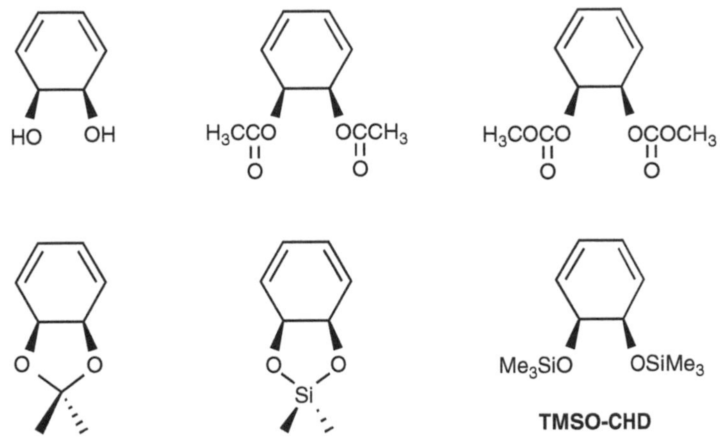

FIGURE 18.4 5,6-Disubstituted 1,3-cyclohexadiene monomers.

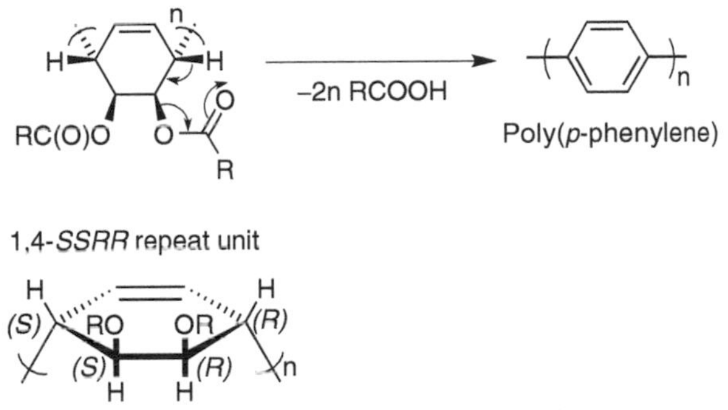

SCHEME 18.3 1,4-*SSRR* repeat unit of a poly(5,6-disubstituted 1,3-cyclohexadiene).

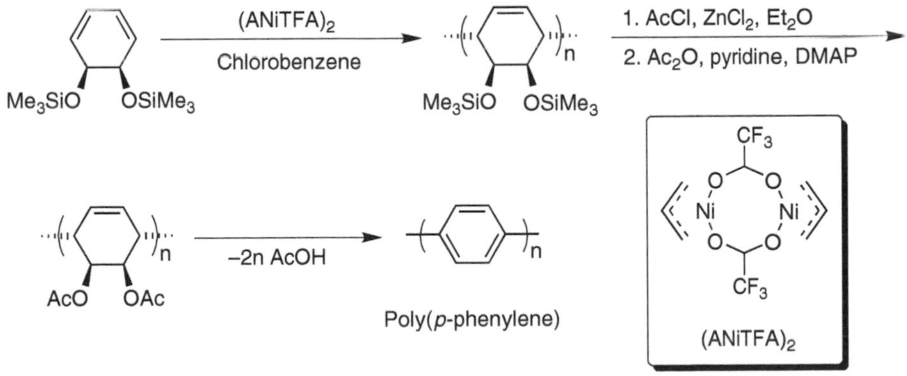

SCHEME 18.4 Regio- and stereospecific polymerization of a functionalized CHD monomer, **TMSO-CHD**, in the presence of ANiTFA (Ac = acetyl, DMAP = 4-(dimethylamino)pyridine).

polymerization in the presence of ANiBr/MAO is complete immediately (under 5 min) at room temperature to afford a polymer in high yield (78–96%). On the other hand, Zr, Ti-, and Pd-based complexes were found to be much less active for CHD polymerization, and ANiTFA provided a polymer in 40% yield after a 34 h polymerization time.

TABLE 18.1

Polymerization of 1,3-Cyclohexadiene (CHD) Using Various Transition Metal Catalysts

Entry[a]	Catalyst	Cocatalyst	[Al]/[cat]	Solvent	Time	Polymer Yield (%)
1	ANiBr	MAO	100	Toluene	5 min	78
2	ANiBr	MAO	100	Chlorobenzene	<1 min	88
3	ANiBr	MAO	100	o-Dichlorobenzene	<1 min	96
4	ANiBr	MAO	100	Cyclohexane	<1 min	92
5	rac-Et(H$_4$Ind)$_2$ZrCl$_2$	MAO	1000	Toluene	5 days	10
6	CpTiCl$_3$	MAO	1000	Toluene	4 days	30
7	[(η^3-C$_3$H$_5$)Ni(OC(O)CF$_3$)]$_2$	None	—	Toluene	34 h[b]	40
8	[(η^3-C$_3$H$_5$)Pd]$^+$[SbF$_6$]$^-$	None	—	Dichloromethane	7 days	10
9	[(η^3-C$_3$H$_5$)Ni(cod)]$^+$[B(C$_6$F$_5$)$_4$]$^-$	None	—	o-Dichlorobenzene	20 min	72

[a] Polymerization conditions: [CHD]$_0$ = 1.5 mol/L, [CHD]$_0$/[catalyst] = 500:1, room temperature; MAO = methylaluminoxane; H$_4$Ind = 4,5,6,7-tetrahydro-1-indenyl; Cp = C$_5$H$_5$; cod = 1,5-cyclooctadiene.

[b] Polymerization was carried out at room temperature for 24 h followed by 10 h at 50 °C.

Source: Nakano, M.; Yao, Q.; Usuki, A.; Tanimura, S.; Matsuoka, T. *Chem. Commun.* **2000**, 2207–2208. Reproduced by permission of The Royal Society of Chemistry.

The catalytic activity of the ANiBr/MAO system depends on the amount of MAO present. Adding excess MAO to ANiBr generally enhances the rate and the yield of CHD polymerization, although 100–200 equivalents of Al per Ni are sufficient. The solvent also plays an important role for controlling polymerization behavior. Halogenated aromatic solvents (Table 18.1, entries 2 and 3) and cyclohexane were found to enhance catalyst activity. In contrast, polar solvents such as diethylether or THF deactivated the catalyst, resulting in polymer formation in low yield. It is interesting, however, that ANiBr/MAO exhibited extremely high activity in toluene containing a small amount of THF. Presumably, the THF molecule would cause the dimeric nickel complex to dissociate to a monomeric state, and would moderately stabilize the metal center of the propagating species.

The ANiBr/MAO catalyst showed high activity for the following reasons: (1) π-allylnickel should have an identical structure to the propagating species of polymerization, allowing for immediate initiation; (2) MAO activates ANiBr to form a highly electron-deficient species favorable for binding unsaturated compounds; (3) ANiBr has a very high solubility in various organic solvents; (4) for sterically demanding monomers such as CHD, the π-allylnickel species should have enough space around the metal center for monomer insertion.

Borate-based cocatalysts have also been investigated. Most borate-based activators were quite effective for CHD polymerization using ANiBr.[28] In particular, [Ph$_3$C]$^+$[B(C$_6$F$_5$)$_4$]$^-$ and Li$^+$[B(C$_6$F$_5$)$_4$]$^-$ effectively activated ANiBr to afford quantitative yields of poly(CHD)s, even with equimolar or smaller amounts of borate-based cocatalysts per ANiBr.

In order to explore the CHD polymerization mechanism catalyzed by ANiBr combined with MAO or borate-based cocatalysts, single-component Ni complexes[29] were synthesized (Figure 18.5). Both [(η^3-allyl)Ni(1,5-cyclooctadiene)]$^+$[PF$_6$]$^-$ and [(η^3-allyl)Ni(1,5-cyclooctadiene)]$^+$[B(C$_6$F$_5$)$_4$]$^-$ provided poly(CHD) in high yield (>70%). In addition, the polymer obtained showed the same regio- and stereostructure as polymer produced using ANiBr/MAO (*vide infra*). Furthermore, the complex with [B(C$_6$F$_5$)$_4$]$^-$ counterion showed higher activity than the complex with [PF$_6$]$^-$, implying that the cationic allylnickel center (and not the anion) is presumably the true propagating species in the CHD polymerization. In contrast, the well-defined Pd analogues [(η^3-allyl)Pd(1,5-cyclooctadiene)]$^+$[PF$_6$]$^-$ and [(η^3-allyl)Pd(1,5-cyclooctadiene)]$^+$[B(C$_6$F$_5$)$_4$]$^-$ showed no catalytic activity for CHD polymerization.[6]

Ni
PF6
Ni
B
F
F
F
F
F
4

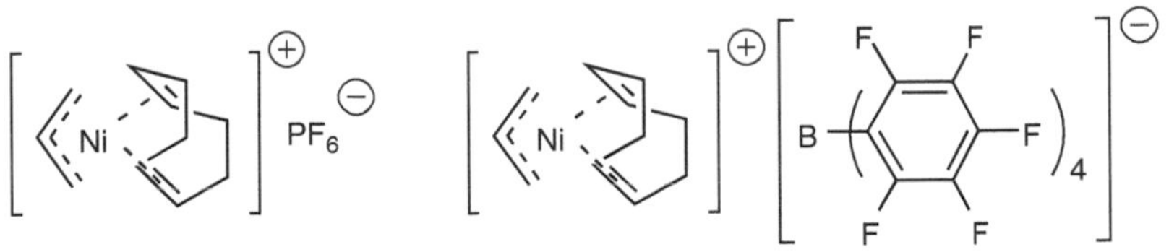

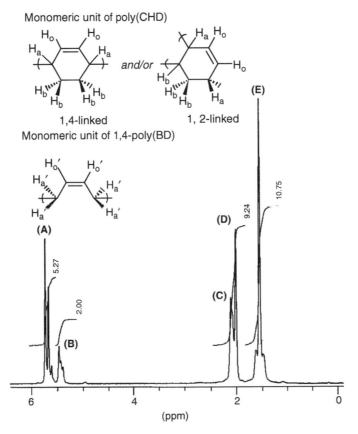

FIGURE 18.5 Well-defined nickel complexes used for CHD polymerization.

Monomeric unit of poly(CHD)
and/or
1,4-linked
1, 2-linked
Monomeric unit of 1,4-poly(BD)
(A)
(B)
(C)
(D)
(E)
5.27
2.00
9.24
10.75
6
4
2
0
(ppm)

FIGURE 18.6 ^{1}H NMR spectrum of a CHD/1,3-butadiene copolymer prepared using ANiBr/MAO. Peaks (B) and (C) are due to the H_o' and H_a' protons of the 1,4-PBD units. (Nakano, M.; Yao, Q.; Usuki, A.; Tanimura, S.; Matsuoka, T. *Chem. Commun.* **2000**, 2207–2208. Reproduced by permission of The Royal Society of Chemistry.)

Unfortunately, the CHD homopolymer obtained with ANiBr/MAO was insoluble in organic solvents. Therefore, 30 mol% of BD was randomly copolymerized with CHD in the presence of ANiBr/MAO in order to produce a soluble polymer.[6] On the basis of its ^{1}H NMR spectra, the observed H_o/H_a integral value ratio of the poly(CHD) unit agreed well with the calculated one, leading to the postulation that the 1,4-linkage is dominant in the poly(CHD) main chain (Figure 18.6).

Another experiment performed to confirm the polymer regiostructure was the synthesis of a poly(CHD) oligomer anchored by a soluble poly(norbornene) with a high molecular weight. The propagating species of ANiBr/MAO-catalyzed norbornene polymerization was not perfectly living, but was long-lived.[6,28] After norbornene polymerization by ANiBr/MAO, CHD was added to give a soluble norbornene-CHD diblock copolymer. The ^{1}H NMR spectrum of this material showed a sharp

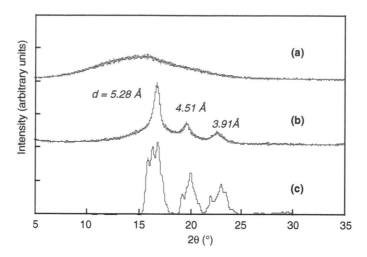

FIGURE 18.7   X-ray diffraction patterns for: (a) poly(CHD) obtained using *n*-BuLi/TMEDA; (b) poly(CHD) obtained using ANiBr/MAO; (c) *cis*-syndiotactic 1,4-poly(CHD) simulated using the COMPASS force field. (Nakano, M.; Yao, Q.; Usuki, A.; Tanimura, S.; Matsuoka, T. *Chem. Commun.* **2000**, 2207–2208. Reproduced by permission of The Royal Society of Chemistry.)

single peak for olefinic protons at 5.75 ppm, which also supports a regioregular 1,4-structure for the oligo(CHD) chain attached to the poly(norbornene). Taken together, these observations suggest that a 1,4-linked structure is predominant in the poly(CHD) prepared using ANiBr/MAO.

As shown in Figure 18.7a, the poly(CHD) prepared using a *n*-BuLi-based initiator gave a typical amorphous powder XRD pattern, even when the polymer main chain consisted of >90% 1,4-units. On the other hand, 1,4-poly(CHD) obtained with the ANiBr-based catalysts or with [(η^3-allyl)Ni(1,5-cyclooctadiene)]$^+$[B(C$_6$F$_5$)$_4$]$^-$ was found to be highly crystalline, with three major peaks in its XRD spectrum (Figure 18.7b).[6] Therefore, the crystallinity of poly(CHD) initiated with ANiBr/MAO catalyst originates from the stereoregularity, rather than the regioregularity, of its structure. Since the obtained poly(CHD) is insoluble, constant-temperature and constant-pressure molecular dynamics simulations of the crystalline poly(CHD) were performed at 300 K and under zero pressure using the COMPASS force field.[30] By use of the simulated atomic coordinates of the crystalline poly(CHD), time-averaged powder XRD spectra were calculated for the four stereoisomers shown in Figure 18.1. Interestingly, it was confirmed that only the simulated crystalline structure of the *cis*-syndiotactic 1,4-linked poly(CHD) shown in Figure 18.7c could reproduce the experimental XRD pattern. Ni-catalyzed poly(CHD) is believed to be a cis-rich structure arising from a *syn*-coordinative mechanism.[21] This simulation reinforces the supposed mechanism and provides additional information about the tacticity of 1,4-linked *cis*-poly(CHD).

An additional series of nickel complexes was synthesized to investigate ligand effects on polymerization behavior and polymer microstructure.[28,31] MAO-activated nickel complexes with bulky diimine ligands[27] (Figure 18.8) were found to give reduced polymerization activities as compared to the ANiBr/MAO catalyst. Interestingly, the polymerization rate of CHD with these complexes strongly depends on the R substituents attached to the backbone carbons of the diimine ligand, with less bulky R groups providing higher catalytic activities. Bulkier backbones such as acenaphthene tend to suppress the rotation of the 2,6-diisopropylphenyl substituents, which can block the apical positions of the Ni center more strictly than when these substituents can freely rotate.[32] Because CHD is a bulky monomer, its coordination and insertion processes in polymerization are considered to be sensitive to the steric effects around the metal center. All of the polymers obtained using the above catalysts were insoluble in organic solvents and were crystalline with T_ms of 310–320 °C. The XRD patterns of the polymers are basically the same as those of polymers prepared with ANiBr/MAO.

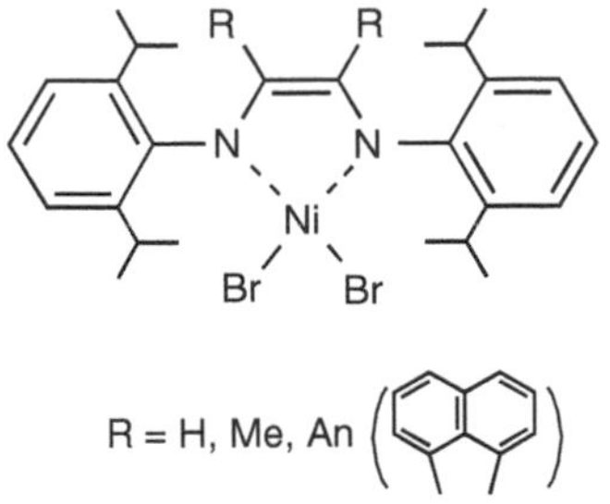

FIGURE 18.8 Diimine nickel complexes used for CHD polymerization (An = acenaphthyl).

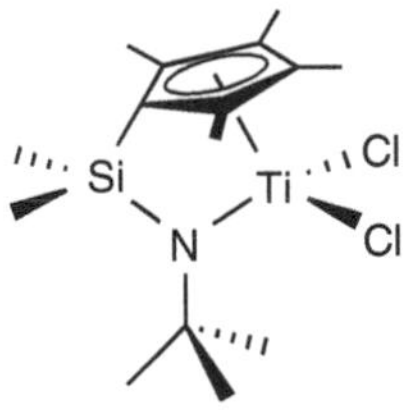

FIGURE 18.9 Constrained geometry catalyst used for CHD polymerization.

Related catalyst systems have been proposed by Po et al.[33] and Longo et al.[5] The former group employed $Ni(acac)_2/MAO$ (acac = acetylacetonate) for CHD polymerization while the latter adopted Cp_2Ni/MAO (Cp = C_5H_5). $Ni(acac)_2/MAO$ ([Al]/[Ni] = 100:1) was so active that it provided poly(CHD) almost quantitatively at 50 °C; Cp_2Ni/MAO was less active. The polymerization yield using Cp_2Ni/MAO was moderate even at a higher [Al]/[Ni] ratio (2000:1) and a higher polymerization temperature (75 °C). As is the case with ANiBr/MAO, the above catalytic systems should generate a cationic π-allyl nickel species in situ in the polymerization as a propagating species. Judging from the polymer solid-state ^{13}C NMR spectra, XRD patterns, and T_ms (310–320 °C), the regio- and stereostructures of the obtained poly(CHD)s are almost identical to polymers made in our group at Toyota using ANiBr-based catalysts or single-component Ni complexes such as $[(\eta^3\text{-allyl})Ni(1,5\text{-cyclooctadiene})]^+[B(C_6F_5)_4]^-$.

Recently, a constrained geometry catalyst, $[Me_2Si(Nt\text{-}Bu)(Me_4Cp)]TiCl_2$ (Figure 18.9), was proposed as a CHD polymerization catalyst.[34] When activated by MAO, this catalyst copolymerized ethylene with CHD, although CHD incorporation into the copolymer was limited to ≤ 12 mol%. The CHD units in the obtained copolymer were 1,4-linked, but information concerning stereoregularity was not given.

18.6 CONCLUSIONS

This chapter has reviewed cyclic conjugated diene polymerization using a wide variety of catalysts. Ni complexes have been primarily discussed because they are quite active for CHD polymerization and because the obtained polymer has a regio- and stereoregular structure, which induces crystallization. However, stereoregular poly(CHD) as a practical material has some problems that require improvement. For example, the decomposition temperature of poly(CHD) (320–330 °C) is so close to its melting point that the polymer is not melt processable. Although we and other groups have tried to control microstructure by changing polymerization conditions or the ligands of catalyst complexes, the regio- and stereostructure of poly(CHD) has not varied much upon changing these parameters.

One of the key technologies needed to make cyclic conjugated diene polymers useful is an expansion of monomer availability. Presently, neither 1,3-cycloheptadiene nor 1,3-cyclooctadiene has been coordinatively polymerized, even with highly active cationic Ni complexes. The polymerization of functionalized CHDs is, so far, limited to ANiTFA. In order to provide processability and functionality to cyclic conjugated diene polymers, these problems must be overcome. The progress of transition metal-catalyzed polymerization may make this possible in the near future.

ACKNOWLEDGMENTS

Dr. A. Usuki at Toyota Central R&D Laboratories, Inc. is thanked for helpful discussions.

REFERENCES AND NOTES

1. Ballard, D. G. H.; Courtis, A.; Shirley, I. M.; Taylor, S. C. Synthesis of polyphenylene from a *cis*-dihydrocatechol biologically produced monomer. *Macromolecules* **1988**, *21*, 294–304.
2. For example, see: Taube, R.; Sylvester, G. Stereospecific polymerization of butadiene or isoprene. In *Applied Homogeneous Catalysis with Organometallic Compounds*; Cornils, B.; Herrmann, W. A., Eds.; Wiley-VCH: Weinheim, 1996; Vol. 1, pp 280–318.
3. For example, see: Imanishi, Y.; Matsuzaki, K.; Yamane, T.; Kohjiya, S.; Okamura, S. Cationic polymerization of cyclic dienes. IX. The structure of poly(1,3-cyclohexadiene) and poly(*cis,cis*-1,3-cyclooctadiene). *J. Macromol. Sci., Chem.* **1969**, *A3*, 249–259.
4. Hampton, P. D.; Wu, S.; Alam, T. M.; Claverie, J. P. Synthesis of allylnickel aryloxides and arenethiolates: Study of their dynamic isomerization and 1,3-diene polymerization activity. *Organometallics* **1994**, *13*, 2066–2074.
5. Longo, P.; Freda, C.; de Ballesteros, O. R.; Grisi, F. Highly stereoregular polymerization of 1,3-cyclohexadiene in the presence of Cp$_2$Ni-MAO catalyst. *Macromol. Chem. Phys.* **2001**, *202*, 409–412.
6. Nakano, M.; Yao, Q.; Usuki, A.; Tanimura, S.; Matsuoka, T. Stereo- and regiospecific polymerization of cyclic conjugated dienes using highly active nickel catalysts. *Chem. Commun.* **2000**, 2207–2208
7. Lefebvre, G.; Dawans, F. 1,3-Cyclohexadiene polymers. Part I. Preparation and aromatization of poly-1,3-cyclohexadiene. *J. Polym. Sci., Part A* **1964**, *2*, 3277–3295.
8. Cassidy, P. E.; Marvel, C. S.; Ray, S. Preparation and aromatization of poly-1,3- cyclohexadiene and subsequent crosslinking. *J. Polym. Sci., Part A* **1965**, *3*, 1553–1564.
9. Zhong, X. F.; Francois, B. Kinetics of 1,3-cyclohexadiene polymerization initiated by organolithium compounds in a non-polar medium, 1. Pure propagation step. *Makromol. Chem.* **1990**, *191*, 2735–2741.
10. Francois, B.; Zhong, X. F. Kinetics of 1,3-cyclohexadiene polymerization initiated by organolithium compounds in a non-polar medium, 2. Secondary reactions in the propagation step. *Makromol. Chem.* **1990**, *191*, 2743–2753.
11. Natori, I. Synthesis of polymers with an alicyclic structure in the main chain. Living anionic polymerization of 1,3-cyclohexadiene with the *n*-butyllithium/*N*,*N*,*N'*,*N'*-tetramethylethylenediamine system. *Macromolecules* **1997**, *30*, 3696–3697.
12. Natori, I.; Inoue, S. Anionic polymerization of 1,3-cyclohexadiene with alkyllithium/amine systems. Characteristics of *n*-butyllithium/*N*,*N*,*N'*,*N'*-tetramethylethylenediamine system for living anionic polymerization. *Macromolecules* **1998**, *31*, 4687–4694.
13. Hong, K.; Mays, J. W. 1,3-Cyclohexadiene Polymers. 1. Anionic Polymerization. *Macromolecules* **2001**, *34*, 782–786.
14. Williamson, D. T.; Glass, T. E.; Long, T. E. Determination of the stereochemistry of poly(1,3-cyclohexadiene) via end group functionalization. *Macromolecules* **2001**, *34*, 6144–6146.
15. Quirk, R. P.; You, F. X.; Zhu, L.; Cheng, S. Z. D. Anionic polymerization of cyclohexa-1,3-diene in cyclohexane with high stereoregularity and the formation of crystalline poly(cyclohexa-1,3-diene). *Macromol. Chem. Phys.* **2003**, *204*, 755–761.
16. Marvel, C. S.; Hartzell, G. E. Preparation and aromatization of poly-1,3-cyclohexadiene. *J. Am. Chem. Soc.* **1959**, *81*, 448–452.

17. Dolgoplosk, B. A.; Beilin, S. I.; Korshak, Yu. V.; Makovetsky, K. L.; Tinyakova, E. I. Regularities of diene copolymerization by coordination catalytic systems and the mechanism of stereoregulation. *J. Polym. Sci., Part A-1: Polymer Chem. Ed.* **1973**, *11*, 2569–2590.

18. Sen, A.; Lai, T.-W. Catalytic polymerization of acetylenes and olefins by tetrakis(acetonitrile) palladium(II) bis(tetrafluoroborate). *Organometallics* **1982**, *1*, 415–417.

19. Sen, A.; Thomas, R. R. Catalysis by solvated transition-metal cations. 3. Novel catalytic transformations of alkenes by cationic compounds of molybdenum and tungsten. *Organometallics* **1982**, *1*, 1251–1254.

20. Dolgoplosk, B. A.; Beilin, S. I.; Korshak, Y. V.; Chernenko, G. M.; Vardanyan, L. M.; Teterina. M. P. Copolymerization of dienes under influence of π-allyl complexes of nickel. *Eur. Polym. J.* **1973**, *9*, 895–908.

21. Porri, L.; Aglietto, M. Stereospecific polymerization of *cis,cis*-1,4-dideuterio-1,3-butadiene to trans-1,4 or cis-1,4 polymers. *Makromol. Chem.* **1976**, *177*, 1465–1476.

22. Gin, D. L.; Conticello, V. P.; Grubbs, R. H. Transition-metal-catalyzed polymerization of heteroatom-functionalized cyclohexadienes: Stereoregular precursors to poly(*p*-phenylene). *J. Am. Chem. Soc.* **1992**, *114*, 3167–3169.

23. Gin, D. L.; Conticello, V. P.; Grubbs, R. H. Stereoregular precursors to poly(*p*-phenylene) via transition-metal-catalyzed polymerization. 1. Precursor design and synthesis. *J. Am. Chem. Soc.* **1994**, *116*, 10507–10519.

24. Deming, T. J.; Novak, B. M. Polyisocyanides using [(η^3-C$_3$H$_5$)Ni(OC(O)CF$_3$)]$_2$: rational design and implementation of a living polymerization catalyst. *Macromolecules* **1991**, *24*, 6043–6045.

25. Deming, T. J.; Novak, B. M. Mechanistic studies on the nickel-catalyzed polymerization of isocyanides. *J. Am. Chem. Soc.* **1993**, *115*, 9101–9111.

26. Gin, D. L.; Conticello, V. P.; Grubbs, R. H. Stereoregular precursors to poly(*p*-phenylene) via transition-metal-catalyzed polymerization. 2. The effects of polymer stereochemistry and acid catalysts on precursor aromatization: A characterization study. *J. Am. Chem. Soc.* **1994**, *116*, 10934–10947.

27. Nakano, M. Synthesis of novel crystalline cyclic olefin polymer catalyzed by highly active nickel complexes. *R&D Review of Toyota CRDL* **2000**, *35*, 82; *Chem. Abstr.* **2000**, *133*, 238344.

28. Nakano, M.; Usuki, A. Stereo- and regiospecific polymerization of cyclic conjugated dienes using highly active nickel catalysts. *Kobunshi Ronbunshu (Jpn. J. Polym. Sci. Tech.)* 2002, 59, 356–363; *Chem. Abstr.* **2002**, *137*, 217259.

29. Ascenso, J. R.; Dias, A. R.; Gomes, P. T.; Romão, C. C.; Tkatchenko, I.; Revillon, A.; Pham, Q.-T. Isospecific oligo-/polymerization of styrene with soluble cationic nickel complexes. The influence of phosphorus(III) ligands. *Macromolecules* **1996**, *29*, 4172–4179.

30. Tanimura, S.; Matsuoka, T.; Nakano, M.; Usuki, A. Molecular dynamics study of the stereostructure of 1,4-linked poly(cyclohexa-1,3-diene) obtained with π-allylnickel-based catalysts. *J. Polym. Sci., Part B: Polym. Phys.* **2001**, *39*, 973–978.

31. Nakano, M.; Yao, Q.; Usuki, A. Control of cyclic conjugated diene polymerization using highly active nickel catalysts. *Polym. Prep. Jpn. (Engl. Ed.)* **2000**, *49*, E73–E73.

32. Johnson, L. K.; Killian, C. M.; Brookhart, M. New Pd(II)- and Ni(II)-based catalysts for polymerization of ethylene and α-olefins. *J. Am. Chem. Soc.* **1995**, *117*, 6414–6415.

33. Po, R.; Santi R.; Cardaci, M. A. Polymerization of 1,3-cyclohexadiene with nickel/MAO catalytic systems. *J. Polym. Sci., Part A: Polym. Chem.* **2000**, *38*, 3004–3009.

34. Heiser, D.; Mülhaupt, R. Synthesis of new cycloolefin copolymers based upon 1,3-cyclohexadiene. *Polym. Mat. Sci. Eng.* **2004**, *91*, 554–555.

19 Tactic Nonconjugated Diene Cyclopolymerization and Cyclocopolymerization

Il Kim

CONTENTS

19.1 INTRODUCTION

The polymerization of nonconjugated dienes to form polymers and copolymers that incorporate recurring units of cyclic moieties in the polymer or copolymer backbone is well known.[1,2] While it had been initially presumed that the polymerization of nonconjugated dienes produces only crosslinked polymers, the cyclopolymerization of diallyl quaternary ammonium salts to noncrosslinked polymers was first reported by G. B. Butler in 1947.[3] A polymerization mechanism involving alternating intramolecular–intermolecular chain propagation was proposed to account for the linear cyclopolymers. The proposed cyclic structures in the polymer backbone were substantiated by degradation of representative polymers. Since then, a large variety of nonconjugated dienes have been reported to undergo cyclopolymerization. Later studies have shown that in numerous cases, cyclic structures are derived by propagation through the less stable intermediate, that is, the reactions are under kinetic rather than thermodynamic control.[3]

The copolymerization of nonconjugated dienes with monoolefinic vinyl monomers is also known.[1,2] Depending upon the comonomer pair and the copolymerization conditions, the product copolymer may contain (1) cyclized units made from sequential addition of both ends of the diene monomer; (2) cyclized units made from addition of one end of the diene followed by the monoolefin comonomer or one side of a second diene, followed by the other end of the first diene; (3) non-cyclized diene units bearing pendant olefins, and (4) crosslinked dienes formed by incorporation of these pendant olefins into another polymer chain (Scheme 19.1). For symmetrical, nonconjugated dienes, it has been shown that all of the well-known methods of polymerization can be employed to

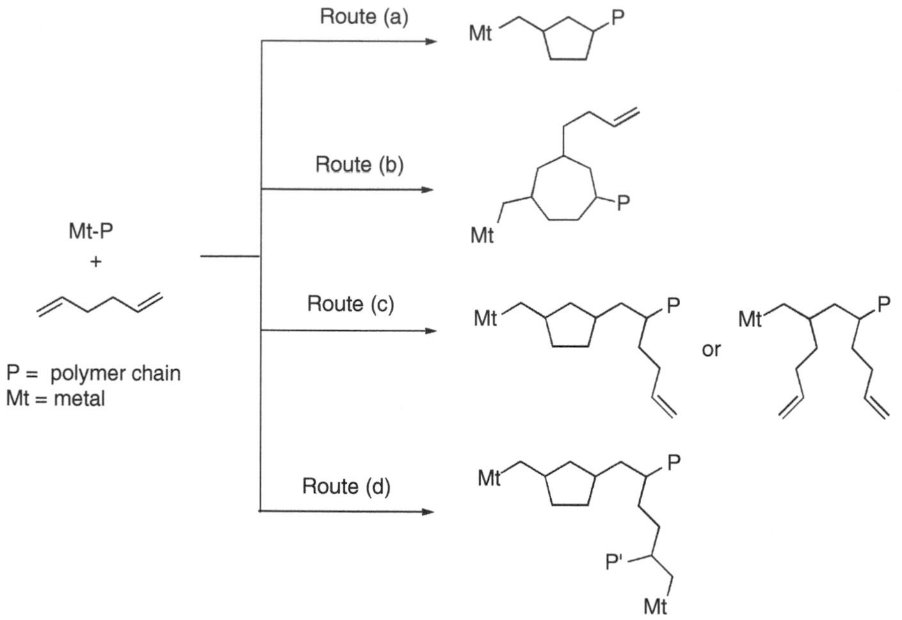

SCHEME 19.1 The possible structures of polymers synthesized by 1,5-hexadiene (1,5-HD) polymerization.

initiate cyclopolymerization. Polymerizations by free-radical, cationic, and anionic initiation have been demonstrated.[3]

The stereocontrol capabilities of homogeneous metallocene polymerization catalysts present an opportunity for the microstructural design of polymers that are difficult or impossible to prepare with conventional heterogeneous catalysts. The availability of various chiral metallocene compounds provides an opportunity to design new chiral polymers. For example, the homogeneous cyclopolymerization of nonconjugated dienes yields an array of structurally diverse cyclopolymers. The Waymouth research group reported the diastereo- and enantioselective cyclopolymerization of 1,5-hexadiene (1,5-HD) to give stereoregular poly(methylene-1,3-cyclopentane) (PMCP).[1,4–8]

This chapter will discuss various homogeneous catalyst systems for the cyclopolymerization of nonconjugated diolefins, particularly 1,5-HD, which control the stereochemistry and microstructure and therefore the physical properties of the polymers synthesized. In addition, the effect of metallocene structure on ethylene/1,5-HD and propylene/1,5-HD copolymerizations will be described.

19.2 STEREOREGULAR CYCLOPOLYMERIZATION OF NONCONJUGATED DIENES

Some nonconjugated dienes, such as 1,5-HD, 2-methyl-1,5-hexadiene, 1,6-heptadiene, and 1,7-octadiene (1,7-OD), polymerize in the presence of metallocene catalysts to give polymers with cyclic structures. Whereas poly(α-olefin)s have only two microstructures of maximum order (isotactic and syndiotactic),[9] cyclopolymers have four such microstructures (Figure 19.1), since the cyclization step introduces an additional type of selectivity (diastereoselectivity) that concerns the formation of cis and trans rings.[1] Polymers with vinyl-ended branches resulting from 1,2-addition (Scheme 19.1c) are also possible. Even though metallocene-based catalysts exhibit high cyclo- and regioselectivities for the cyclopolymerization of α,ω-dienes, the selectivities depend on the catalysts, monomers, and

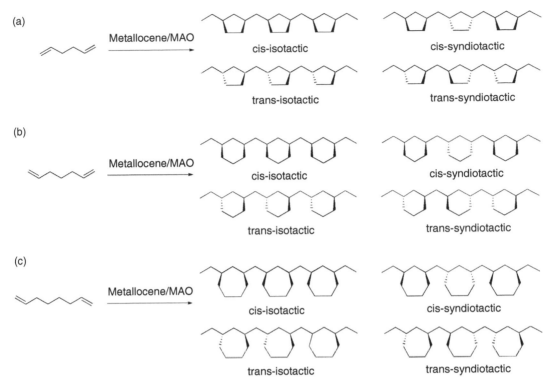

FIGURE 19.1 Maximum order structures of the cyclopolymers produced by the cyclopolymerization of 1,5-HD (a), 1,6-heptadiene (b), and 1,7-OD (c) with metallocene catalysts. These structures are also referred to as cis-diisotactic (*meso*-diisotactic), cis-disyndiotactic (*meso*-disyndiotactic), trans-diisotactic (*racemo*-diisotactic), and trans-disyndiotactic (*racemo*-disyndiotactic).

reaction conditions. In the following sections, the factors that govern the diastereoselectivity of diene cyclopolymerization will be presented according to the monomer used.

19.2.1 STEREOREGULAR CYCLOPOLYMERIZATION OF 1,5-HEXADIENE WITH METALLOCENE CATALYSTS

The Waymouth research group extensively investigated group 4 metallocene/methylaluminoxane (MAO) catalysts for the polymerization of 1,5-HD, finding that cyclization is complete in the case of these homogeneous catalysts.[4–8] Four microstructures of maximum order are possible for the cyclopolymers (Figure 19.1).[1] In addition, the cyclopolymerization can be diastereoselective depending upon the type of ligands present in the metallocene complexes. For example, the cyclopolymerization of 1,5-HD by optically active metallocene catalysts led to optically active *trans*-isotactic PMCP as a major product.[4–8] The trans-isotactic microstructure is a novel example of a polymer that is chiral by virtue of configurational main chain stereochemistry.[10,11]

Marvel and Garrison proposed the widely accepted cyclopolymerzation mechanism through their initial work on the cyclopolymerization of various α, ω-dienes ($H_2C=CH(CH_2)_nCH=CH_2$, where $n = 4$–12, 14, 18) using $TiCl_4/Al(iBu)_3$ catalyst.[3,12,13] There are two distinct stereochemical events, olefin 1,2-addition and olefin cyclization. The enantioselectivity of the first olefin 1,2-addition of the diene monomer determines the tacticity of the polymer, and the diastereoselectivity of the cyclization step determines whether cis or trans rings are formed.[1,4–8] As discovered by the Waymouth research group, MAO-activated Cp_2ZrCl_2 (Cp = C_5H_5; **1a**, Figure 19.2) gives *trans*-atactic PMCP, and the more sterically hindered MAO-activated $Cp_2^*ZrCl_2$ (Cp* = C_5Me_5; **2a**, Figure 19.2) gives

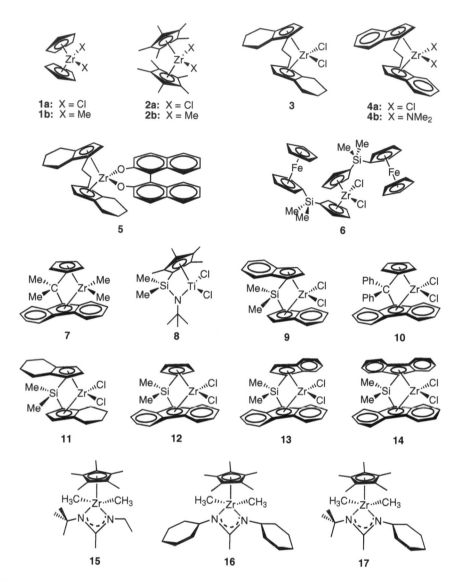

FIGURE 19.2 Various metallocene and half-metallocene catalyst precursors for the cyclo(co)polymerization of nonconjugated dienes.

cis-atactic PMCP.[4–8,10,11] The same research group also reported that isotactic polymers with trans rings could be obtained by using chiral metallocenes of the Brintzinger type, *rac*-Et(H₄Ind)₂ZrCl₂ (H₄Ind = 4,5,6,7-tetrahydro-1-indenyl; **3**, Figure 19.2) and *rac*-Et(Ind)₂ZrCl₂ (Ind = 1-indenyl; **4a**, Figure 19.2).[4–8]

As expected, the cis/trans diastereoselectivity is influenced by the structure of the catalyst precursor, and is controllable by choosing a proper catalyst and polymerization conditions. The enantioselectivity (the relative stereochemistry between the rings) of PMCP is also affected by the catalyst structure. Complexes **1a**, **1b** (Figure 19.2), and **2a**, which give atactic poly(α-olefin)s, produce atactic PMCP, and the isoselective catalysts **3** and **4a** yield isotactic PMCPs. These differences in enantioselectivity versus catalyst type are consistent with those for the polymerization of α-olefins.[1,9] *trans*-Isotactic polymers can be optically active (chiral) if homochiral catalysts are used. The Waymouth research group showed that the MAO-activated homochiral *ansa*-zirconocene BINOL complex **5** (BINOL = 1,1′-bi-2-naphtholate; Figure 19.2) gave optically active *trans*-polymer.

A recent paper by Mukaiyama has reported a bis(ferrocenyl) zirconocene complex (**6**, Figure 19.2) that exhibits an extremely high trans selectivity for ring formation in PMCP.[14]

Cavallo et al. carried out a conformational modeling study for 1,5-HD polymerization using bis-Cp (**1**) and bis-Cp* (**2**) zirconocenes in order to investigate the origin of diastereoselectivity.[7] Active catalyst **1** formed a chair structure with the growing polymer in a pseudoequatorial position as the lowest energy conformation, which yields a trans ring, while the bulkier active catalyst **2** formed a twisted-boat arrangement in a pseudoequatorial placement as a minimum energy conformation, which results in the formation of a cis ring. According to these results, it might be expected that highly stereoselective catalyst **5** would effect a homofacial 1,2-addition/cyclization process, yielding a cyclopolymer with predominantly cis rings. However, the polymer produced by catalyst **5** contains predominantly trans rings. The trans selectivity of both less sterically hindered active catalyst **1** and more sterically hindered catalysts **4** and **5** demonstrates that these catalysts prefer a heterofacial 1,2-addition/cyclization sequence, rather than a homofacial 1,2-addition/cyclization that leads to cis selectivity. In the heterofacial 1,2-addition/cyclization sequence, the cyclization step occurs on a diastereoface of opposite topicity to the enantioface selected for the initial 1,2-addition step.

In our group, cyclopolymerizations of 1,5-HD were investigated in the presence of different metallocene catalysts (**2b**, **4b**, **7**, Figure 19.2).[15] In the presence of MAO or noncoordinating anion activators such as $[Ph_3C]^+[B(C_6F_5)_4]^-$, precatalysts having the spectator ligand structure present in **2b** yield atactic polymers in α-olefin polymerization, while those with the ligand structure of **4b** yield isotactic polymers and those with the ligand structure of **7** yield syndiotactic polymers.[9] Table 19.1 shows the results of 1,5-HD polymerizations with **2b**, **4b**, and **7** in conjunction with $[Ph_3C]^+[B(C_6F_5)_4]^-/Al(i\text{-}Bu)_3$ cocatalyst. Polymerization of 1,5-HD using **4b** proceeded much more rapidly than polymerization using **7** and **2b**, producing insoluble, rubbery products. With catalyst **4b**, polymerization reactivity (as measured by yield) increased as the polymerization temperature was increased under conditions of $[Zr] = 39\mu M$ and $[1,5\text{-}HD]_0 = 0.88 M$ (runs 1, 2, 5, 8, 10). Most of the products were insoluble in toluene, suggesting that crosslinking occurred. The toluene-soluble percentage of the polymer product, which was determined by extraction in toluene and represents the noncrosslinked and cyclized product percentage, was highest at $-25\,^\circ C$ and lowest at $50\,^\circ C$. Under more dilute conditions, the content of toluene-soluble polymer increased, but yield decreased (runs 3, 4, 6, 7, 9). It is interesting to note that resonances corresponding to uncyclized monomer units were barely detectable in ^1H and ^{13}C nuclear magnetic resonance (NMR) analyses of the toluene-soluble polymers; thus, under these conditions, cyclization had predominantly taken place.

With catalysts **2b**, **7**, and **1a** (using comparative data from the literature[6]), conversion of monomer was not so high; however, all of the polymer products were soluble in toluene. The dependence of the reactivities of **2b** and **7** on the polymerization temperature was in contrast to that seen with **4b**. As shown in the Table 19.1 runs with **4b**, when the conversion of monomer (as measured by yield) is more than about 36% under the polymerization conditions used, much of the product is insoluble in toluene. Thus, it is assumed that the percentage of toluene-soluble polymer is proportional to the degree of cyclization. The degree of cyclization is higher under more dilute conditions.[15]

By using the two-step mechanism of olefin 1,2-addition and cyclization proposed by Marvel and Garrison,[12,13] the degree of olefin cyclization can be interpreted as the point of competitive intramolecular cyclization and intermolecular propagation (Scheme 19.2).[4–8,15,16] Because the intramolecular cyclization is unimolecular (Step 2, Scheme 19.2) while the intermolecular propagation is bimolecular (Step 3, Scheme 19.2), the rate of intramolecular cyclization/intermolecular propagation increases with more dilute conditions. In other words, the concentration of monomer does not affect the cyclization rate, but affects the intermolecular propagation rate.[17] Thus, the relative amount of cyclized product increases under more dilute conditions compared to product formed by intermolecular propagation. Some unreacted vinyl groups remaining as pendant groups (intermediate **D** in Scheme 19.2), which come from intermolecular propagation (Step 3), may react continuously with other active species (Step 4, Scheme 19.2), and produce insoluble material (intermediate **E**).

TABLE 19.1
Results of 1,5-Hexadiene Polymerization with Metallocene Catalysts 2b, 4b, and 7

Run Number[a]	Catalyst	[Zr] (×10^4 M)	[1,5-HD]$_0$ (M)	T_p (°C)	Time (min)	Yield (%)	Toluene Soluble[b] (%)	Trans[c] (%)	Cyclization[d] (%)	T_m (°C)	ΔH_f (J/g)	[η][e] (dL/g)
1	4b	0.39	0.88	−25	120	2.9	100	62.3	93.3	ND[f]	ND[f]	0.105
2	4b	0.39	0.88	0	120	76.8	79.89	65.0	>99	70.5	2.2	0.133
3	4b	0.21	0.88	25	120	11.7	100	NA[i]	NA[i]	72.6	4.4	0.478
4	4b	0.30	0.88	25	60	31.4	8.77	NA[i]	NA[i]	66.5	5.6	0.434
5	4b	0.39	0.88	25	60	86.1	7.61	62.9	98.0	85.0	11.5	0.365
6	4b	0.21	0.44	50	30	61.1	11.77	NA[i]	NA[i]	82.6	13.8	0.462
7	4b	0.30	0.88	50	5	43.8	9.48	NA[i]	NA[i]	77.4	12.6	0.448
8	4b	0.39	0.88	50	30	100	7.36	55.0	98.5	98.2	12.2	0.391
9	4b	0.21	0.44	70	60	34.5	100	NA[i]	NA[i]	77.4	14.5	0.551
10	4b	0.39	0.88	70	5	70.9	19.27	56.4	98.9	84.9	9.0	0.337
11	7	0.83	0.88	−25	1440	34.4	100	67.5	90.7	ND[f]	ND[f]	0.396
12	7	0.83	0.88	0	1440	21.0	100	67.2	95.8	61.4	0.3	0.378
13	7	0.83	0.88	25	120	35.5	100	NA[i]	NA[i]	70.0	16.0	0.311
14	7	0.83	0.88	25	1440	36.2	100	67.6	97.7	76.7	12.4	0.352
15	7	0.83	0.88	50	1440	21.6	100	65.2	98.0	82.7	17.4	0.205
16	7	0.83	0.88	70	1440	2.8	100	64.6	98.6	90.1	20.3	0.156
17	2b	0.83	0.88	−25	120	3.2	100	6.5	>99	190.5	56.9	0.516
18	2b	0.83	0.88	0	120	8.4	100	10.3	>99	154.5	55.8	0.143
19	2b	0.83	0.88	25	120	4.3	100	16.3	98.0	119.9	37.6	0.086
20	2b	0.83	0.88	50	120	3.0	100	38.5	NA[i]	ND[f]	ND[f]	0.036
21	2b	0.83	0.88	70	120	2.3	100	42.7	NA[i]	ND[f]	ND[f]	0.021
22[g]	1a	0.5	0.21	21	60	11.1	NA[i]	80.0	>99	NA[i]	NA[i]	27,000[h]

[a]Conditions: Al(i-Bu)$_3$/[Ph$_3$C]$^+$[B(C$_6$F$_5$)$_4$]$^-$ cocatalyst; [Al(i-Bu)$_3$]/[Zr] = 80:1, [Ph$_3$C]$^+$[B(C$_6$F$_5$)$_4$]$^-$/[Zr] = 1:1, in 80 mL toluene solvent. [b]Of polymer; extracted in boiling toluene for 24 h. [c]Determined by ^{13}C NMR. [d]Degree of cyclization determined by ^{1}H NMR. [e]Measured in 1,2,4-trichlorobenzene at 135 °C. [f]Not detected. [g]Data from Reference 6. [h]M_W value determined by GPC. [i]Not analyzed.

Source: From Kim, I.; Shin, Y.-S.; Lee, J. K.; Won, M.-S. *J. Polym. Sci., Part A: Polym. Chem.* **2000**, *38*, 1520–1527. With permission.

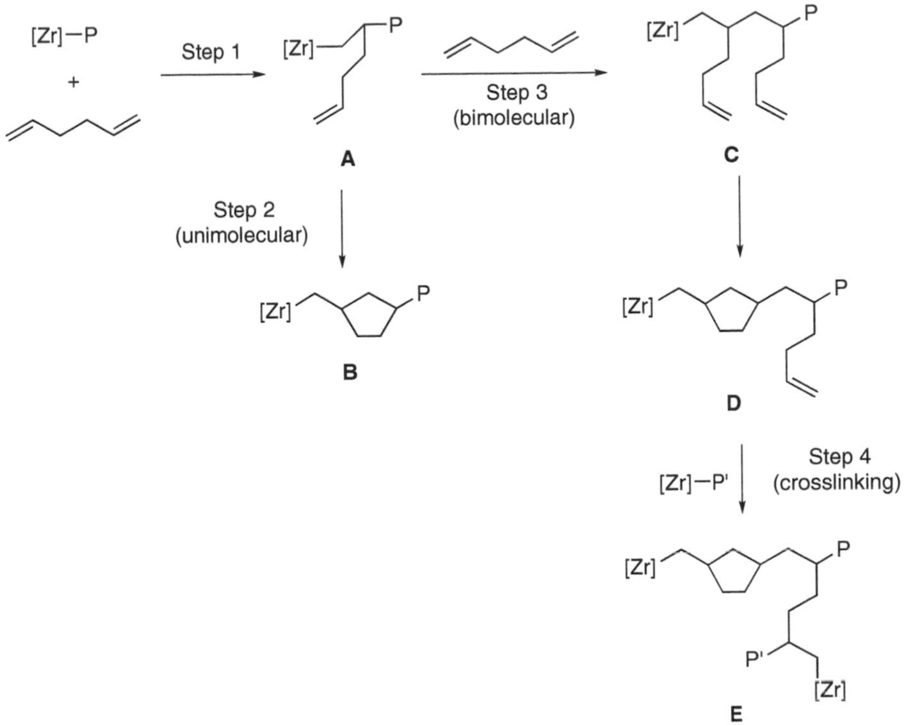

SCHEME 19.2 Proposed unimolecular and bimolecular mechanisms forming cyclopolymer and network polymer in 1,5-HD polymerization (P = polymer chain). (From Kim, I.; Shin, Y.-S.; Lee, J. K.; Won, M.-S. *J. Polym. Sci., Part A: Polym. Chem.* **2000**, *38*, 1520–1527. With permission.)

The ^{1}H NMR spectra of toluene-soluble PMCP fractions showed resonances in the olefin region resulting from 1,2-addition,[4–8,18–20] indicating the presence of pendant olefins. The degree of cyclization of 1,5-HD showed a dependence on polymerization temperature;[4–8,16] however, it was not always proportional to temperature. The degree of cyclization by catalysts **4b** and **7** increased as the polymerization temperature increased, but that for catalyst **2b** decreased under these conditions. Thus, it can be said that more dilute conditions enhance the cyclization of 1,5-HD, and the effect of polymerization temperature on cyclization is dependent on the structure of the catalyst.

In the polymerization of 1,5-HD, 1,2-addition and cyclization reactions can be described by Equations 1.1 through 1.3 (where r = the rate of reaction), on the condition that the 1,2-addition obeys a first-order Markovian process and the cyclization follows a Bernoullian process:[15]

$$r_{vv} = k_{vv}[M_v^*][M] \tag{19.1}$$

$$r_{cv} = k_{cv}[M_c^*][M] \tag{19.2}$$

$$r_c = k_c[M_v^*] \tag{19.3}$$

where $[M_v^*]$, $[M_c^*]$, $[M]$, k_{vv}, k_{cv}, and k_c are, respectively, the number of propagating chain ends of 1,2-vinyl insertion units (V), the number of propagating chain ends of cyclized units (C), monomer concentration, the rate constant of 1,2-vinyl insertion of a second monomer after 1,2-vinyl addition, the rate constant of 1,2-vinyl insertion of a second monomer after cyclization, and the rate constant of cyclization. The ratio of 1,2-addition units (d[M_v]) to cyclization units (d[M_c]) in the polymer can be expressed by Equation 19.4

$$\frac{d[M_v]}{d[M_c]} = \frac{k_{vv}[M_v^*][M]}{k_c[M_v^*]} = \left(\frac{k_{vv}}{k_c}\right)[M] \tag{19.4}$$

If the 1,2-addition and cyclization reactions follow this equation, the values of $[V]/[C]$ should increase almost proportionally to an increase of $[M]$.[15]

Cyclopolymerization of 1,5-HD using the MAO-activated constrained geometry catalyst Me$_2$Si(Me$_4$Cp)(N-t-Bu)TiCl$_2$ (8, Figure 19.2) was also investigated by Mülhaupt and Waymouth and coworkers.[21] This catalyst system afforded randomly distributed *cis*- and *trans*-cyclopentane rings in the backbone. 1,5-HD incorporation reached 52 mol% and the ratio of vinyl side chains (noncyclized units) to cyclopentane rings was controlled by the 1,5-HD concentration, where low 1,5-HD concentrations promoted cyclopolymerization.

19.2.2 STEREOREGULAR CYCLOPOLYMERIZATION OF 1,7-OCTADIENE WITH METALLOCENE CATALYSTS

Polymerization of 1,7-OD was conducted by Naga et al. with MAO-activated aselective (1a), isoselective (9, Figure 19.2),and syndioselective (10, Figure 19.2) metallocene catalysts.[22] The polymerization activity decreased in the order 9 > 10 > 1a. The molecular weight distributions (M_w/M_n)s of the polymers obtained with 9 and 10 were 1.7–2.5 at low conversion (with conversion measured by yield). The results indicated that polymerization of 1,7-OD proceeds with a single active species.[1,9] Three kinds of incorporated 1,7-OD units are proposed on the assumption that vinyl addition proceeds preferentially with 1,2-regiochemistry; the structures of the units are shown in Figure 19.3.

The first structure is a 1,2-inserted unit with pendant vinyl group (F). The second is a cycloaddition unit formed through intramolecular 1,2-addition of the pendant vinyl group (G). The third is a graft addition unit formed through intermolecular 1,2-addition of the pendant vinyl group (H). The structures of the polymers obtained were confirmed using ^{13}C NMR and DEPT (distortionless enhancement by polarization transfer) spectroscopy.[22,23] The resonances at 26.7 (d), 30.0 (e) 33.9 (b), 34.0 (c), 35.5 (f), 41.8 and 44.0–45.0 (a), and 114.5 and 139.5 (g,h) ppm are assigned to the pendant vinyl group structure (F). The resonances at 26.7 (d′,e′), 36.0 (c′,f′), 37.2 (b′,g′), 43.2 (a′), and 44.0–45.0, 47.8, and 48.5 (h′) ppm are assigned to the cyclic polymer structure (G). Even

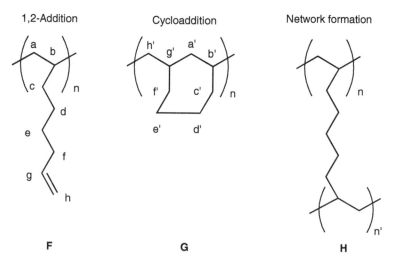

FIGURE 19.3   The possible structures of enchained units in 1,7-OD polymerization.

though the structure of the network polymer (**H**) could not be distinguished from **F** by NMR owing to structural similarity, solvent (toluene) extraction can be utilized for the separation of each polymer. The high glass transition temperatures (T_gs) of poly(1,7-OD) obtained with catalyst **9** also indicate the presence of cyclic units in these polymers. Investigation of the effect of monomer conversion (as measured by yield) on the microstructures of the resultant polymers showed that the fraction of saturated units in the poly(1,7-OD)s, that is, the cyclization selectivity, decreased in the order **9** > **1a** > **10**.[22]

The cyclization selectivity decreased with lower polymerization temperatures. Waymouth and Coates reported the same tendency in the cyclopolymerization of 1,5-HD.[8] The monomer concentration affects the microstructure of poly(1,7-OD), and the cyclization selectivity is drastically decreased with increasing monomer concentration. Similar results with 1,5-HD polymerizations were also reported.[15,21] Poly(1,7-OD) cyclopolymers have four possible microstructures, since the cyclization step introduces diastereoselectivity that concerns the formation of cis and trans rings (Figure 19.1).[1] As compared to 1,5-HD cyclopolymerization, the longer diene monomer structure of 1,7-OD results in a decrease in the conformational rigidity of the incipient ring during cyclization.[4–8,15,22] Since a homofacial insertion/cyclization sequence results in a cis ring, an increase in diene length will result in an increase in the cis ring content of the polymer. Waymouth and Coates demonstrated that polymerizations of 1,5-HD, 1,6-heptadiene, and 1,7-OD with catalyst **5** yielded polymers with cis ring contents of 28%, 50%, and 78%, respectively.[24]

19.2.3 Cyclopolymerization of Nonconjugated Dienes with Nonmetallocene Catalysts

The polymerization of nonconjugated dienes with various homogeneous Ziegler–Natta catalysts has been investigated for the synthesis of polyolefins having cyclic units and pendant vinyl groups. Cyclopolymerization of 1,5-HD to PMCP was reported initially by Marvel et al.,[12,13] and further investigated by Makowski.[17] Ziegler–Natta catalysts formed from $TiCl_4/Al(i\text{-}Bu)_3$ and $TiCl_4/AlEt_3$ showed low activities and incomplete cyclization of the diene. Cheng reported the cyclopolymerization of 1,5-HD using $TiCl_3/AlEt_2Cl$ catalyst.[20] ^{13}C NMR analysis of the resulting polymer indicated complete cyclization. Doi et al. reported that the living polymerization of 1,5-HD with $V(acac)_3/AlEt_2Cl$ in toluene at −78 °C gave a polymer composed of alternating methylene-1,3-cyclopentane (MCP) and 1-vinyl tetramethylene (VTM) units.[25]

Other polycyclic systems have been introduced into polymers using nonmetallocene catalysts.[3] For example, by using a Ziegler–Natta catalyst, the triolefinic monomer 3-vinyl-1,5-hexadiene has been converted into a soluble polymer having little unsaturation. An extensive linear chain structure containing 2,6-methylene-linked bicyclo[2.2.1]-heptyl rings is indicated. Triallylmethylsilane has also been polymerized by Ziegler–Natta catalysts to give soluble, solid polymers showing little residual unsaturation, properties consistent with a large bicyclic [3.3.1] ring content.

Cyclopolymers were also produced by the cyclopolymerization of 1,5-HD employing a reduced valence state group VIB metal oxide catalyst on a porous support by Ho and Wu.[26] The preferred catalyst comprises carbon monoxide-reduced chromium on silica. The ^{13}C NMR analysis of the resulting polymer showed that the chromium catalyst yielded predominantly cyclic polymer with a small amount of unsaturated double bonds at the end of each polymer chain and recurring units of irregular enchainment, that is, head-to-head, head-to-tail, and so forth. The polymers did not exhibit stereochemical regularity or block structures. The recurring units of poly(1,5-HD) polymerized according to this process are those formed by the routes shown in Scheme 19.1a and c.[26] These products with unique chemical structures have unusual properties for application as lubricants, additives, or chemical intermediates.

Recently, Hustad and Coates reported cyclopolymerization of 1,5-HD[27] and 1,6-heptadiene[28] using a MAO-activated fluorinated bis(phenoxyimine)-based titanium complex that yields syndiotactic polymer in propylene polymerization (Scheme 19.3). Detailed analysis of the resulting

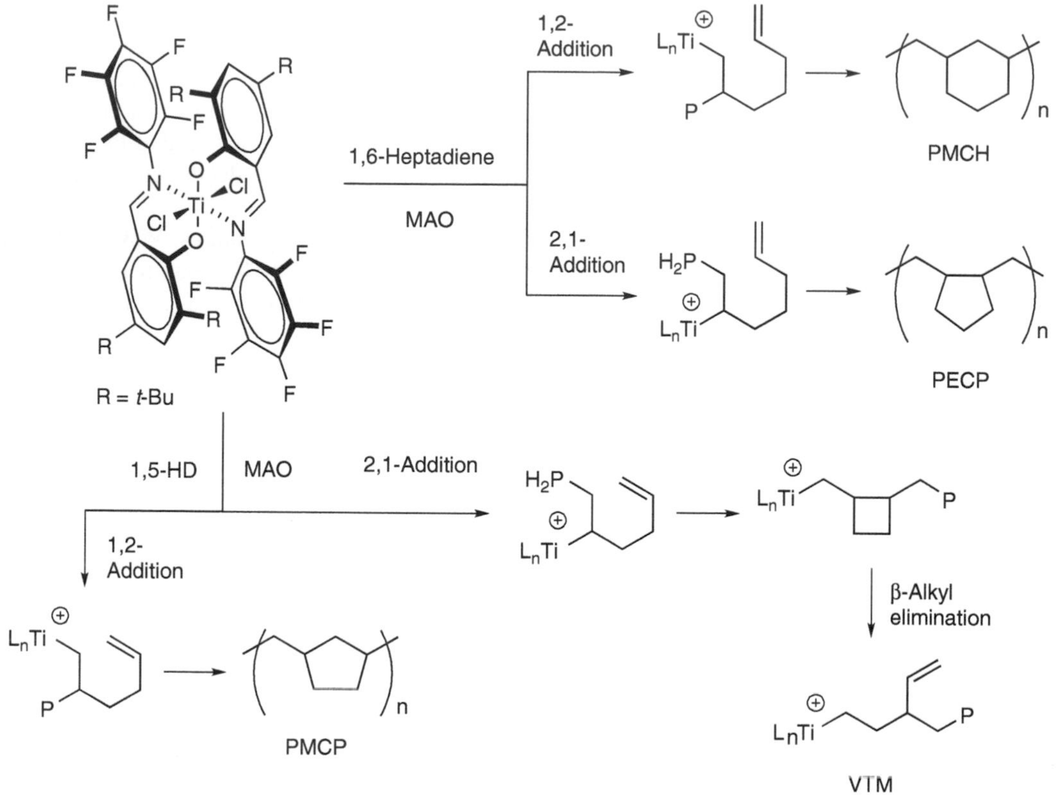

SCHEME 19.3   Primary (1,2-) and secondary (2,1-) insertion/cyclization mechanisms in 1,5-HD and 1,7-OD cyclopolymerizations with a fluorinated bis(phenoxyimine) titanium catalyst (P = polymer chain). (Adapted with permission from Hustad, P. D.; Coates, G. W. *J. Am. Chem. Soc.* **2002**, *124*, 11578–11579; Hustad, P. D.; Tian, J.; Coates, G. W. *J. Am. Chem. Soc.* **2002**, *124*, 3614–3621. Copyright 2002 American Chemical Society.)

poly(1,5-HD) microstructures revealed that the polymers contained MCP units (63%) as well as 3-vinyl tetramethylene units (37%). Owing to the propensity for 2,1-addition, this catalyst system resulted in the insertion/isomerization polymerization of 1,5-HD, giving a material composed of VTM and MCP repeat units (Scheme 19.3).[27,28] Cyclopolymerization of 1,6-heptadiene with the same catalyst produced a polymer with no observable unsaturation, indicating quantitative cyclization. The microstructure of the poly(1,6-heptadiene) displayed resonances consistent with poly(methylene-1,3-cyclohexane) (PMCH) having 80% cis rings; however, the polymer ^{13}C NMR spectrum also contained new peaks at 44–43, 31, 29, and 23 ppm which matched those of poly(ethylene-1,2-cyclopentane) (PECP; prepared independently by the ring opening metathesis polymerization (ROMP) of bicyclo[3.2.0]-hept-6-ene and subsequent hydrogenation).[28] The presence of both methylene-1,3-cyclohexane (MCH) and ethylene-1,2-cyclopentane (ECP) units in the polymer was caused by the high occurrence of secondary insertions in this catalytic system, since the ECP units can only form following monomer insertion with 2,1-regiochemistry.

19.3   COPOLYMERIZATION OF NONCONJUGATED DIENES AND MONOOLEFINS

Copolymerizations of ethylene and/or propylene with nonconjugated dienes using various metallocene catalysts are a useful method to synthesize polyolefins with cyclic backbones.

Olefin/nonconjugated diene copolymerizations can be an attractive alternative synthetic method to olefin/cycloolefin copolymerization, because homo- and copolymerizations of cycloolefins typically show low productivity. Bergemann et al. studied ethylene/1,5-HD copolymerization with the MAO-activated catalyst **11** (Figure 19.2) under a high ethylene pressure of 1500 bar.[29] According to ^{13}C NMR analysis, that is, resonances at 33.1 and 32.0 ppm assigned to cis- and trans-cyclopentane rings in the polymer backbone, respectively, and a resonance at 29.8 ppm assigned to the methylene group, the copolymer contained 4.2 mol% of cyclic 1,5-HD units with predominantly trans rings (feed concentration of 1,5-HD during polymerization = 72 mol%). Naga and Imanishi investigated ethylene/1,5-HD copolymerizations with various MAO-activated nonbridged (e.g., **1a**, **2a**) and bridged (e.g., **4a**, **9**, **10**) zirconocene catalysts.[30] They found that the ligand structure of the zirconocene, the amount of 1,5-HD in the feed, and the catalyst concentration strongly affected the resulting copolymer composition. The bridged catalysts showed greater reactivity toward 1,5-HD than the nonbridged catalysts.

Mülhaupt and coworkers studied homo- and copolymerizations of 1,5-HD with ethylene and styrene using the MAO-activated constrained geometry catalyst **8**.[21] This catalyst system afforded very high 1,5-HD incorporation (reaching 52 mol%) with randomly distributed cis- and trans-cyclopentane rings in the homo- and copolymer backbones. The ratio of vinyl side chains to cyclic rings was controlled by the 1,5-HD concentration, where low concentrations of 1,5-HD promoted cyclopolymerization.

We have investigated the effects of metallocene stereoselectivity on ethylene/1,5-HD[31] and propylene/1,5-HD[32] copolymerizations with two metallocene catalysts, isoselective **4b** and syndioselective **7**, in combination with $Al(i\text{-}Bu)_3/[Ph_3C]^+[B(C_6F_5)_4]^-$ cocatalyst. In ethylene/1,5-HD copolymerizations, a consistently higher incorporation of 1,5-HD was observed with syndioselective **7** (Figure 19.4). This tendency is in good agreement with the results of propylene/1,5-HD and ethylene/α-olefin copolymerizations with the same catalysts, in which the syndioselective catalyst shows higher comonomer reactivity than the isoselective catalyst.[33] In the simplest case, a copolymerization reaction of ethylene (E) and 1,5-HD, four propagation reactions are possible. The two participants in each chain propagation step are a monomer molecule and an active center (C*) that carries a polymer chain ending with a monomer unit derived either from E or from 1,5-HD (Equations 19.5 through 19.8)

$$C^*\text{--}E\text{--polymer} + E \xrightarrow{k_{EE}} C^*\text{--}E\text{--}E\text{--polymer} \tag{19.5}$$

$$C^*\text{--}E\text{--polymer} + 1,5\text{-HD} \xrightarrow{k_{EHD}} C^*\text{--}1,5\text{-HD}\text{--}E\text{--polymer} \tag{19.6}$$

$$C^*\text{--}1,5\text{-HD}\text{--polymer} + E \xrightarrow{k_{HDE}} C^*\text{--}E\text{--}1,5\text{-HD}\text{--polymer} \tag{19.7}$$

$$C^*\text{--}1,5\text{-HD}\text{--polymer} + 1,5\text{-HD} \xrightarrow{k_{HDHD}} C^*\text{--}1,5\text{-HD}\text{--}1,5\text{-HD}\text{--polymer} \tag{19.8}$$

where k_{EE} is the rate constant for a propagating chain ending in E adding to monomer E, k_{EHD} that for a propagating chain ending in E adding to 1,5-HD, and so on. The four rate constants in the copolymerization reactions are traditionally grouped into two reactivity ratios (Equations 19.9 and 19.10)

$$r_E = k_{EE}/k_{EHD} \tag{19.9}$$

$$r_{HD} = k_{HDHD}/k_{HDE} \tag{19.10}$$

Most procedures for evaluating r_E and r_{HD} involve the experimental determination of the copolymer composition for several different comonomer feed compositions in conjunction with a differential form of the copolymerization equation.

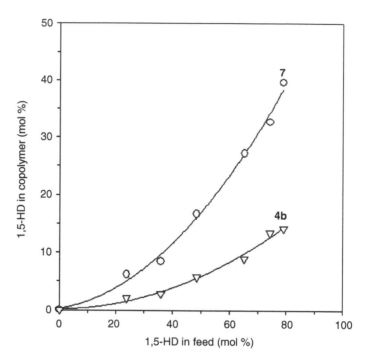

FIGURE 19.4 Plot of mol% 1,5-HD incorporated into the copolymer versus mol% 1,5-HD in the monomer feed for ethylene/1,5-HD cyclocopolymerization using **4b** and **7** (Al(*i*-Bu)$_3$/[Ph$_3$C]$^+$[B(C$_6$F$_6$)$_4$]$^-$ cocatalyst). (Reprinted from Kim, I.; Shin, Y. S.; Lee, J.-K.; Cho, N. J.; Lee, J.-O.; Won, M.-S. *Polymer* **2001**, *42*, 9393–9403. With permission from Elsevier.)

The calculation of copolymerization parameters according to a Kelen–Tüdös plot[34,35] using the data in Figure 19.4 resulted in r_E = 17.44 and r_{HD} = 0.02 ($r_E r_{HD}$ = 0.35) with **4b**/Al(*i*-Bu)$_3$/[Ph$_3$C]$^+$[B(C$_6$F$_6$)$_4$]$^-$ catalyst, and r_E = 4.48 and r_{HD} = 0.12 ($r_E r_{HD}$ = 0.54) with **7**/Al(*i*-Bu)$_3$/[Ph$_3$C]$^+$[B(C$_6$F$_6$)$_4$]$^-$ catalyst.

Since ethylene is much more reactive than 1,5-HD, the r_E values are higher than 1 and the r_{HD} values are lower than 1. Different types of copolymerization behavior are observed depending on the values of the monomer reactivity ratios. Copolymerizations can be classified into three types based on whether the product of the two monomer reactivity ratios $r_1 r_2$ is unity, less than unity, or greater than unity. It is useful to recall that when $r_1 r_2 \approx 1$ the resultant copolymer shows a random structure; when $r_1 r_2 > 1$ a blocky structure is evident, and when $r_1 r_2 < 1$ the copolymer has an alternating structure.[36] In this sense, poly(ethylene-*co*-1,5-HD) copolymers produced using **4b** and **7** appear to show alternating structures to some degree. However, detailed analysis of the copolymer microstructure by ^{13}C NMR spectroscopy[31] shows that **4b** produces copolymers having a random distribution of 1,5-HD units and **7** produces copolymers with an alternating distribution of 1,5-HD and ethylene units. The 1,5-HD units incorporated in the copolymers produced by both catalysts were completely cyclized to MCP units derived from the intramolecular cyclization of 1,2-added 1,5-HD.

The microstructure of ethylene/1,5-HD copolymers is considerably more complicated than that of simple ethylene/α-olefin copolymers, since it includes structures resulting from cyclization of 1,2-added 1,5-HD. Cyclopolymerization of 1,5-HD giving PMCP is a chain growth reaction during which a conventional 1,2-addition of a vinylic function into the metal–carbon bond is followed by an intramolecular 1,2-addition, resulting in the formation of alicyclic rings connected by methylene groups (Figure 19.5). The cyclic units formed from 1,5-HD are described by both the cis/trans stereochemistry of the rings and the relative stereochemistry between the rings.[4–8,10,11,15] In addition,

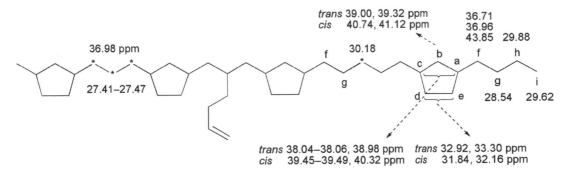

FIGURE 19.5 The plausible microstructures of poly(ethylene-*co*-1,5-HD) and its characterization by ^{13}C NMR spectroscopy. (Reprinted from Kim, I.; Shin, Y. S.; Lee, J.-K.; Cho, N. J.; Lee, J.-O.; Won, M.-S. *Polymer* **2001**, *42*, 9393–9403. With permission from Elsevier.)

intermolecular propagation of 1,2-added 1,5-HD yields olefin-bearing side chains in the copolymer. The degree of cyclization of 1,2-added 1,5-HD is dependent upon the concentration of 1,5-HD in the feed, owing to the competition between the intramolecular cyclization and intermolecular propagation reactions. However, the degree of cyclization of 1,2-inserted 1,5-HD, as determined by ^{1}H NMR spectroscopy of the copolymers obtained from catalysts **4b** and **7**, did not vary much with variation in 1,5-HD concentration in the feed within the experimental range used in our experiments; that is, only a small amount of side chain vinylic double bonds were detected (4.85–5.02 ppm, =CH$_2$, and 5.5–5.9 ppm, =CH-) at high concentrations of 1,5-HD in the copolymerization feed.

Copolymerization of propylene (P) and 1,5-HD was also carried out at 30 °C with the **4b**/Al(*i*-Bu)$_3$/[Ph$_3$C]$^+$[B(C$_6$F$_6$)$_4$]$^-$ and **7** /Al(*i*-Bu)$_3$/[Ph$_3$C]$^+$[B(C$_6$F$_6$)$_4$]$^-$ catalyst systems.[32] Figure 19.6 shows plots of 1,5-HD percentage in the feed versus 1,5-HD percentage in the resultant copolymers. The calculation of copolymerization parameters according to a Kelen–Tüdös plot resulted in $r_P = 16.25$ and $r_{HD} = 0.34$ ($r_P r_{HD} = 5.53$) for **4b** and $r_P = 8.85$ and $r_{HD} = 0.274$ ($r_P r_{HD} = 2.74$) for **7**. Catalyst **7** gives higher incorporations of 1,5-HD than catalyst **4b**. These results indicate that a syndioselective catalyst shows a higher reactivity toward 1,5-HD than an isoselective catalyst. Similar results for relative comonomer reactivities were reported for the copolymerization of propylene and 1-hexene.[37–39] Considering the products of the monomer reactivity ratios, the poly(propylene-*co*-1,5-HD) copolymers appear to be blocky, the more blocky the more isoselective the catalyst.

As in the case of ethylene/1,5-HD copolymerization, the degree of cyclization of 1,2-inserted 1,5-HD in the poly(propylene-*co*-1,5-HD) copolymers (determined by ^{1}H NMR spectroscopy) was dependent upon the 1,5-HD concentration in the feed.[32] At low 1,5-HD feed concentration, the degree of cyclization approaches almost 100%; however, it decreases to some degree as the concentration of 1,5-HD increases. The selectivity of 1,2-added 1,5-HD cyclization can be interpreted as the point of competition between the intramolecular cyclization and intermolecular propagation reactions.[31,32]

Shiono and coworkers studied copolymerizations of propene with 1,5-HD and 1,7-OD with MAO-activated isoselective catalyst **9** and syndioselective catalyst **10**.[40] The incorporation of the nonconjugated dienes in the copolymers was higher with **10** than with **9**. The microstructures of the copolymers were studied by ^{13}C NMR and DEPT spectroscopy; detailed assignments of the resonances are shown in Figure 19.7.[40] The stereoregulations of the cyclic carbons in the isolated MCP units next to propene units were defined as cis-*m*, cis-*r*, trans-*m*, and trans-*r* and were assigned the ^{13}C NMR resonances shown in Figure 19.7. The stereoselectivity in the cycloaddition of 1,5-HD was investigated based on the structures of isolated MCP units. It was found that 1,5-HD was inserted stereoselectively by enantiomorphic site control with both catalysts. The cyclization selectivity for 1,5-HD copolymerization was higher than that for 1,7-OD copolymerization. Catalyst **9** gave copolymers with higher cyclization selectivity than catalyst **10** in the propene/1,5-HD copolymerization.

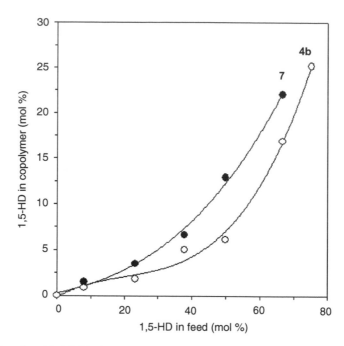

FIGURE 19.6 Plot of mol% 1,5-HD incorporated into the copolymer versus mol% 1,5-HD in the monomer feed for propylene/1,5-HD cyclocopolymerization using **4b** and **7** (Al(i-Bu)$_3$/[Ph$_3$C]$^+$[B(C$_6$F$_6$)$_4$]$^-$ cocatalyst). (From Kim, I.; Shin, Y. S.; Lee, J.-K. *J. Polym. Sci., Part A: Polym. Chem.* **2000**, *38*, 1590–1598. With permission.)

Recently, Choo and Waymouth performed the copolymerization of ethylene with 1,5-HD using various metallocene catalysts (**12**, **13**, **14**, Figure 19.2).[41] 1,5-HD cyclopolymerized exclusively to give MCP units in the copolymers, with only traces of uncyclized 1,2-inserted 1,5-HD. The diastereoselectivity of the cyclocopolymerization favored the formation of *trans*-1,3-cyclopentane rings for metallocenes (74% trans for **12**, 81% trans for **13**, and 66% trans for **14**). For metallocenes **12** and **14**, the ethylene/1,5-HD copolymerization yielded copolymers with similar comonomer compositions and sequence distributions to those observed for ethylene/1-hexene copolymerization with these catalysts. On the other hand, the copolymers derived from metallocene **13** showed very different compositions and sequence distributions. At comparable comonomer feed ratios, the poly(ethylene-*co*-1,5-HD)s were enriched in the 1,5-HD comonomer and deficient in ethylene as compared to the analogous polymers prepared from ethylene and 1-hexene. The copolymerization behavior of **13** provided support for a dual-site alternating mechanism for 1,5-HD incorporation, wherein one coordination site of the active catalyst center is highly selective for the initial 1,2-inserion of 1,5-HD and the other site is selective for cyclization.

Sita and Jayaratne[42–44] invented very effective half-metallocene catalysts (**15**, **16**, **17**, Figure 19.2) that can polymerize α-olefins upon activation with borate cocatalysts. These transition metal complexes possess the ability to polymerize α-olefins and dienes in both a stereoselective and living fashion. As a result, these catalysts can be utilized to design new block copolymers. Cyclopolymerization of 1,5-HD using **15–17** in combination with [PhNMe$_2$H]$^+$[B(C$_6$F$_5$)$_4$]$^-$ cocatalyst in chlorobenzene at −10 °C gave high molecular weight PMCP materials possessing extremely narrow polydispersities (M_w/M_n < 1.1) in a living fashion.[44] Polymerization activity was attenuated as the steric bulk of the amidinate substituents in compounds **15–17** was increased. All of the catalysts were found to be at least 98% selective for cyclopolymerization versus linear 1,2-addition of 1,5-HD. Additional support that cyclopolymerizations of 1,5-HD were occurring in a living fashion

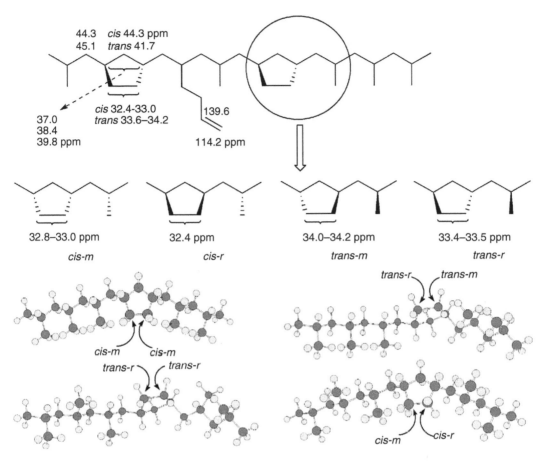

FIGURE 19.7 The microstructures of isolated MCP units next to propylene units in propylene/1,5-HD cyclo-copolymerization and their ^{13}C NMR characterization. (Adapted with permission from Naga, N.; Shiono, T.; Ikeda, T. *Macromolecules* **1999**, *32*, 1348–1355. Copyright 1999 American Chemical Society.)

with these catalysts was provided by the successful synthesis of di- and triblock copolymers of isotactic poly(1-hexene) and PMCP with narrow polydispersity.

19.4 PROPERTIES OF POLY(METHYLENE-1,3-CYCLOPENTANE) AND MONOOLEFIN/1,5-HEXADIENE COPOLYMERS

Polyolefins with cyclic units in the backbone show high T_gs and high transparencies, and thus are suitable for optical and medical applications.[45,46] Even though they can be prepared by the copolymerization of ethylene and cyclic olefins (such as norbornene) using metallocene catalysts, the cyclopolymerization of nonconjugated dienes offers another access route into cyclopolymer materials.

According to a recent X-ray diffraction (XRD) study of PMCP materials by Auriemma and coworkers,[10,11] a common disordered crystalline form presenting a long range positional order (pseudo-hexagonal) for the axes of configurationally and conformationally disordered chains was observed. A conformational analysis for PMCP of different microstructures revealed that, independent of the PMCP microstructure, extended chain conformations suitable for the disordered crystalline phase were proven to be geometrically and energetically feasible. Configurational order (i.e., cis, trans, or isotactic structures) resulted in large increases in the entropy of melting for the polymers, owing to the related higher conformational order in the disordered crystalline phase.

The melting temperatures (T_ms) of the PMCPs studied ranged from 86 to 171 °C and were dependent upon the cis content of the different samples, but were scarcely dependent upon the relative stereochemistry between the rings (e.g., a trans-atactic PMCP and a trans-isotactic PMCP showed the same T_m value). Thus, PMCP with 81% trans rings had a T_m of 86 °C; PMCPs with 63% and 66% trans rings had the same T_m value of 86 °C; random-isotactic PMCP with 52% cis rings had a T_m of 120 °C; and cis-atactic PMCP with 86% cis rings had a T_m of 171 °C.[10,11] A cis-atactic PMCP with a higher cis ring content (93.5%), produced by 7 at −25 °C, had an even higher T_m of 190.5 °C.[15] These results suggest that the strong increase of the melting temperature of PMCP with increasing cis content is related to the more efficient lateral packing of the chains, pointed out by the shorter interchain distances.

The T_m values of homo-polyethylene and homo-PMCP (67.6% trans rings) produced using catalyst 7 were 138.0 and 76.7 °C, respectively. The poly(ethylene-co-1,5-HD)s produced by the same catalysts showed multiple melt transitions.[31] As an example, a copolymer containing 16.7 mol% of 1,5-HD units in the main chain showed T_ms at 88.6 and 116.4 °C, and a copolymer containing 39.6 mol% of 1,5-HD units in the main chain showed T_ms at 73.8 and 96.9 °C. The crystallinity of the copolymers, as measured by XRD analysis, decreased monotonously as the content of 1,5-HD units in the main chain increased.[31] Poly(propylene-co-1,5-HD)s produced by 7 lost their crystallinity, as evidenced by both XRD and differential scanning calorimetry (DSC) analysis, as the content of 1,5-HD units in the main chain was increased.[32] Thus, when the content of enchained 1,5-HD units reached 6.7%, the copolymer showed no T_m and no characteristic XRD peaks.

As a means of developing applications, we have investigated the thermal-responsive shape memory properties of trans-isotactic PMCP prepared using catalyst 4b.[47] The PMCP polymer resembles a ROMP poly(norbornene), which is an example of a commercially available shape memory polymer.[48] The PMCP polymer of this study was partially crystalline and had an elongation at break of more than 400% at 25–85 °C.[48] The shape memory effect of PMCP of moderate molecular weight was enhanced by polyethylene segments introduced by sequential copolymerization; that is, after polymerization of 1,5-HD for 24 h using 4b, ethylene was continuously polymerized, leading to poly(ethylene-block-1,5-HD) copolymers. The crystalline phase of the polyethylene segments seems to strengthen the fixed structure that stores (memorizes) the shape. In this system, the T_g or T_m of the PMCP was selectively used as the shape recovery temperature when an appropriate deformation temperature was chosen.

19.5 CONCLUSIONS

Recent advances in tactic diene cyclopolymerization and cyclocopolymerization with various metal-locene catalysts demonstrate the value of metallocene compounds in new cyclopolymer syntheses. The use of metallocene catalysts provides better control of enantioselectivity in nonconjugated diene cyclopolymerizations than the use of conventional Ziegler–Natta catalysts. In addition, metallocene-based catalysts exhibit extremely high cyclo- and regioselectivities for the cyclopolymerization of α,ω-dienes. The cyclization step introduces an additional type of selectivity (diastereoselectivity) that concerns the formation of cis and trans rings. Thus, by controlling the ligand structure of the metallocene catalyst used for polymerization, PMCPs having several of the different potential kinds of microstructures have been obtained: random-atactic, random-isotactic, cis-atactic, trans-atactic, and trans-isotactic. Longer aliphatic dienes such as 1,7-OD result in a decrease in the conformational rigidity of the incipient ring during cyclization. As a result, an increase in diene monomer length results in an increase in the cis ring content of the polymer.

In our work, the isoselective catalyst 4b was found to show higher activity but lower reactivity toward 1,5-HD than the syndioselective catalyst 7 in the copolymerization of 1,5-HD with ethylene or propylene. According to the values of $r_{E}r_{HD}$ and $r_{P}r_{HD}$, poly(ethylene-co-1,5-HD)s produced by **4b** and **7** show alternating structures to some degree, and poly(propylene-co-1,5-HD)s show blocky structures to some degree. In copolymerizations, the insertion of 1,5-HD proceeds by enantiomorphic site control. Poly(ethylene-co-1,5-HD) shows multiple T_ms even at high 1,5-HD contents (e.g., 39.6 mol%) and a characteristic XRD pattern; however, poly(propylene-co-1,5-HD)s lose their crystallinity at a moderate 1,5-HD content (i.e., ≥ 6.7 mol%).

It is expected that, in the coming decade, polymer chemists will continue to find that transition metal complexes, including metallocenes, present advantages for the synthesis of novel cyclopolymers with controlled structures using nonconjugated diene monomers. The controlled nature of such polymerizations also provides a route into new functional polymers, such as polyolefin-based block copolymers containing functional domains. In addition, suitable applications for the new cyclopolymers are expected to be developed, along with more detailed mechanistic investigations of cyclopolymerizations.

REFERENCES AND NOTES

1. Coates, G. W. Precise control of polyolefin stereochemistry using single-site metal catalysts. *Chem. Rev.* **2000**, *100*, 1223–1252.
2. Suzuki, N. Stereospecific olefin polymerization catalyzed by metallocene complexes. *Top. Organomet. Chem.* **2004**, *8*, 177–215.
3. Butler G. B. Cyclopolymerization. In *Encyclopedia of Polymer Science and Engineering*; 2nd ed.; Mark H. F., Bikales, N. M., Overberger, C. G., Menges, G., Kroschwitz, J. I., Eds.; John Wiley and Sons: New York, 1986; Vol. 4, pp 543–598.
4. Resconi, L.; Waymouth, R. M. Diastereoselectivity in the homogeneous cyclopolymerization of 1,5-hexadiene. *J. Am. Chem. Soc.* **1990**, *112*, 4953–4954.
5. Coates, G. W.; Waymouth, R. M. Enantioselective cyclopolymerization: optically active poly(methylene-1,3-cyclopentane). *J. Am. Chem. Soc.* **1991**, *113*, 6270–6271.
6. Resconi, L.; Coates, G. W.; Mogstad, A.; Waymouth, R. M. Stereospecific cyclopolymerization with group-4 metallocenes. *J. Macromol. Sci., Chem.* **1991**, *A28*, 1225–1234.
7. Cavallo, L.; Guerra, G.; Corradini, P.; Resconi, L.; Waymouth, R. M. Model catalytic sites for olefin polymerization and diastereoselectivity in the cyclopolymerization of 1,5-hexadiene. *Macromolecules* **1993**, *26*, 260–267.
8. Coates, G. W.; Waymouth, R. M. Enantioselective cyclopolymerization of 1,5-hexadiene catalyzed by chiral zirconocenes: a novel strategy for the synthesis of optically active polymers with chirality in the main chain. *J. Am. Chem. Soc.* **1993**, *115*, 91–98.
9. Resconi, L.; Cavallo, L.; Fait, A.; Piemontesi, F. Selectivity in propene polymerization with metallocene catalysts. *Chem. Rev.* **2000**, *100*, 1253–1346.
10. Ruiz de Ballesteros, O.; Venditto, V.; Auriemma, F.; Guerra, G.; Resconi, L.; Waymouth, R.; Mogstad, A. Thermal and structural characterization of poly(methylene-1,3-cyclopentane) samples of different microstructures. *Macromolecules* **1995**, *28*, 2383–2388.
11. Ruiz de Ballesteros, O.; Cavallo, L.; Auriemma, F.; Guerra G. Conformational analysis of poly(methylene-1,3-cyclopentylene) and chain conformation in the crystalline phase. *Macromolecules* **1995**, *28*, 7355–7362.
12. Marvel, C. S.; Stille, J. K. Intermolecular-intramolecular polymerization of α-diolefins by metal alkyl coördination catalysts. *J. Am. Chem. Soc.* **1958**, *80*, 1740–1744.
13. Marvel, C. S.; Garrison, W. E. Jr. Polymerization of higher α-diolefins with metal alkyl coordination catalysts. *J. Am. Chem. Soc.* **1959**, *81*, 4737–4744.
14. Mitani, M.; Oouchi, K.; Hayakawa, M.; Yamada, T.; Mukaiyama, T. Stereoselective cyclopolymerization of 1,5-hexadiene using novel bis(ferrocenyl)zirconocene catalyst. *Chem. Lett.* **1995**, 905–906.
15. Kim, I.; Shin, Y.-S.; Lee, J. K.; Won, M.-S. Cyclopolymerization of 1,5-hexadiene catalyzed by various stereospecific metallocene compounds. *J. Polym. Sci., Part A: Polym. Chem.* **2000**, *38*, 1520–1527.

16. Guaita, M. Temperature dependence of cyclization ratios in cyclopolymerization. In *Cyclopolymerization and Polymers with Chain-Ring Structures*; Butler, G. B., Kresta, J. E., Eds.; ACS Symposium Series 195; American Chemical Society: Washington, DC, 1982; pp 11–28.

17. Makowski, H. S.; Shim, B. K. C.; Wilchinsky, Z. W. 1,5-Hexadiene polymers. I. Structure and properties of poly-1,5-hexadiene. *J. Polym. Sci., Part A* **1964**, *2*, 1549–1566.

18. Kaminsky, W.; Drögemüller, H. Terpolymers of ethylene, propene and 1,5-hexadiene synthesized with zirconocene/methylaluminoxane. *Makromol. Chem., Rapid Commun.* **1990**, *11*, 89–94.

19. Farina, M. The stereochemistry of linear macromolecules. *Top. Stereochem.* **1987**, *17*, 1–111.

20. Cheng, H. N.; Khasat, N. P. Carbon-13 NMR characterization of poly(1,5-hexadiene). *J. Appl. Polym. Sci.* **1988**, *35*, 825–829.

21. Sernetz, F. G.; Mülhaupt, R.; Waymouth, R. M. Homo-, co- and terpolymerization of 1,5-hexadiene using a methylalumoxane activated mono-Cp-amido-complex. *Polym. Bull.* **1997**, *38*, 141–148.

22. Naga, N.; Shiono, T.; Ikeda, T. Cyclopolymerization of 1,7-octadiene with metallocene/methylaluminoxane. *Macromol. Chem. Phys.* **1999**, *200*, 1466–1472.

23. Doi, Y.; Tokuhiro, N.; Soga, K. Polymerization of diolefins by a soluble vanadium-based catalyst. *Kobunshi Ronbunshu* **1989**, *46*, 215–222; *Chem. Abstr.* **1989**, *111*, 78655.

24. Coates, G. W.; Waymouth, R. M. Chiral polymers via cyclopolymerization. *J. Mol. Catal.* **1992**, *76*, 189–194.

25. Doi, Y.; Tokuhiro, N.; Soga, K. Synthesis and structure of a "living" copolymer of propylene and 1,5-hexadiene. *Makromol. Chem.* **1989**, *190*, 643–651.

26. Ho, S. C. H.; Wu, M. M.; Xiong, Y. Novel cyclopolymerization polymers from nonconjugated dienes and 1-alkenes. PCT International Patent Application WO 95/06669 (Mobil Oil Corp.), March 9, 1995.

27. Hustad, P. D.; Coates, G. W. Insertion/isomerization polymerization of 1,5-hexadiene: synthesis of functional propylene copolymers and block copolymers. *J. Am. Chem. Soc.* **2002**, *124*, 11578–11579.

28. Hustad, P. D.; Tian, J.; Coates, G. W. Mechanism of propylene insertion using bis(phenoxyimine)-based titanium catalysts: an unusual secondary insertion of propylene in a group IV catalyst system. *J. Am. Chem. Soc.* **2002**, *124*, 3614–3621.

29. Bergemann, C.; Cropp, R.; Luft, G. Copolymerization of ethylene and 1,5-hexadiene under high pressure catalyzed by a metallocene. *J. Mol. Catal. A: Chem.* **1997**, *116*, 317–322.

30. Naga, N.; Imanishi, Y. Copolymerization of ethylene and 1,5-hexadiene with zirconocene catalysts. *Macromol. Chem. Phys.* **2002**, *203*, 771–777.

31. Kim, I.; Shin, Y. S.; Lee, J.-K.; Cho, N. J.; Lee, J.-O.; Won, M.-S. Copolymerization of ethylene and 1,5-hexadiene by stereospecific metallocenes in the presence of $Al(iBu)_3/[Ph_3C][B(C_6F_5)_4]$. *Polymer* **2001**, *42*, 9393–9403.

32. Kim, I.; Shin, Y. S.; Lee, J.-K. Copolymerization of propylene and 1,5-hexadiene with stereospecific metallocene/$Al(iBu)_3/[Ph_3C][B(C_6F_5)_4]$. *J. Polym. Sci., Part A: Polym. Chem.* **2000**, *38*, 1590–1598.

33. Hlatky, G. G. Single-site catalysts for olefin polymerization: Annual review for 1997. *Coord. Chem. Rev.* **2000**, *199*, 235–329.

34. Kelen, T.; Tüdös, F. Analysis of the linear methods for determining copolymerization reactivity ratios. I. New improved linear graphic method. *J. Macromol. Sci., Chem.* **1975**, *A9*, 1–27.

35. Kelen, T.; Tüdös, F. Analysis of the linear methods of determining copolymerization reactivity ratios. 10. Estimation of errors and planning of experiments in penultimate systems. *Makromol. Chem.* **1990**, *191*, 1863–1869.

36. Odian, G. *Principles of Polymerization*, 4th ed.; John Wiley and Sons: New York, **2004**; p 464.

37. Van Reenen, A. J.; Brull, R.; Wahner, U. M.; Raubenheimer, H. G.; Sanderson, R. D.; Pasch, H. The copolymerization of propylene with higher linear α-olefins. *J. Polym. Sci., Part A: Polym. Chem.* **2000**, *38*, 4110–4118.

38. Xu, J.-T.; Zhu, Y.-B.; Fan, Z.-Q.; Feng, L.-X. Copolymerization of propylene with various higher α-olefins using silica-supported *rac*-$Me_2Si(Ind)_2ZrCl_2$. *J. Polym. Sci., Part A: Polym. Chem.* **2001**, *39*, 3294–3303.

39. Graef, S. M.; Wahner, U. M.; Van Reenen, A. J.; Brull, R.; Sanderson, R. D.; Pasch, H. Copolymerization of propylene with higher α-olefins in the presence of the syndiospecific catalyst *i*-Pr(Cp)(9-Flu)ZrCl$_2$/MAO. *J. Polym. Sci., Part A: Polym. Chem.* **2002**, *40*, 128–140.

40. Naga, N.; Shiono, T.; Ikeda, T. Copolymerization of propene and nonconjugated diene involving intramolecular cyclization with metallocene/methyl aluminoxane. *Macromolecules* **1999**, *32*, 1348–1355.

41. Choo, T. N.; Waymouth, R. M. Cyclocopolymerization: a mechanistic probe for dual-site alternating copolymerization of ethylene and α-olefins. *J. Am. Chem. Soc.* **2002**, *124*, 4188–4189.
42. Jayaratne, K. C.; Sita, L. R. Stereospecific living Ziegler–Natta polymerization of 1-hexene. *J. Am. Chem. Soc.* **2000**, *122*, 958–959.
43. Jayaratne, K. C.; Sita, L. R. Direct methyl group exchange between cationic zirconium Ziegler–Natta initiators and their living polymers: Ramifications for the production of stereoblock polyolefins. *J. Am. Chem. Soc.* **2001**, *123*, 10754–10755.
44. Sita, L. R.; Jayaratne, K. C. Stereospecific living polymerization of olefins by a novel Ziegler–Natta catalyst composition. U.S. Patent 6,579,998 B2 (University of Maryland), June 17, 2003.
45. Kaminsky, W. New polymers by metallocene catalysis. *Macromol. Chem. Phys.* **1996**, *197*, 3907–3945.
46. Olabisi, O.; Atiqullah, M.; Kaminsky, W. Group 4 metallocenes: supported and unsupported. *J. Macromol. Sci., Rev. Macromol. Chem. Phys.* **1997**, *C37*, 519–554.
47. Jeong, H. M.; Song, J. H.; Chi, K. W.; Kim, I.; Kim, K. T. Shape memory effect of poly(methylene-1,3-cyclopentane) and its copolymer with polyethylene. *Polym. Int.* **2002**, *51*, 275–280.
48. Nagata, N. Development of polynorbornene-based shape-memory resins. *Kagaku (Kyoto)* **1990**, *45*, 554–557; *Chem. Abstr.* **1990**, *113*, 213109.

20 Stereochemistry of Polymers Formed by Metathesis Polymerization of Bicyclic and Polycyclic Olefins

Darragh Breen, Katherine Curran, and Wilhelm Risse

CONTENTS

20.1 INTRODUCTION TO THE OLEFIN METATHESIS REACTION

Over the last few decades, the olefin metathesis reaction has become a very important reaction in organic synthesis and polymer synthesis.[1-7] It involves the transition metal-catalyzed redistribution of carbon–carbon double bonds. It can be understood as a reaction, in which the σ- and π-bonds of the C=C units are cleaved, and double bonds are reformed with the alkylidene groups exchanged (Scheme 20.1).

Detailed investigations of the catalytic mechanism of metathesis revealed that transition metal carbenes and metallacyclobutanes represent key intermediates.[8-12] The metathesis reaction (Scheme 20.2) generally proceeds through a [2 + 2] addition of an alkene to a transition metal carbene (**A**) to give a metallacyclobutane (**B**). The metallacyclobutane subsequently cleaves to reform a transition metal carbene (**C**), which can continue the catalytic cycle. Transition metal carbenes with metals in a high oxidation state are also called transition metal alkylidenes.

Important classes of olefin metathesis reactions include cross metathesis (CM),[1-7,13] ring-closing metathesis (RCM),[1-7,14-16] acyclic diene metathesis polymerization (ADMET),[1-7,17,18] and ring-opening olefin metathesis polymerization (ROMP).[1-7,18-20] This chapter focuses on the last class of metathesis reactions, ROMP, and in particular on the stereochemistry of the resulting macromolecular products.

When cyclic olefins are subjected to olefin metathesis, breaking of the double bonds proceeds with opening of the unsaturated cyclic unit, and poly(alkenylenes), that is, macromolecular products that are also known as polyalkenamers,[21] can be obtained (Scheme 20.3). Common examples of cyclic

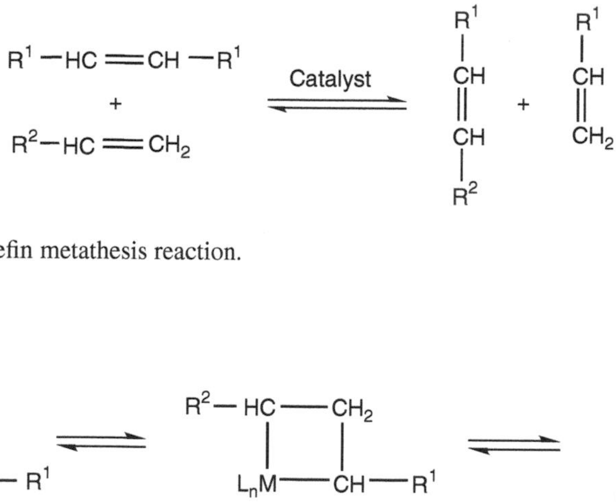

SCHEME 20.1   The olefin metathesis reaction.

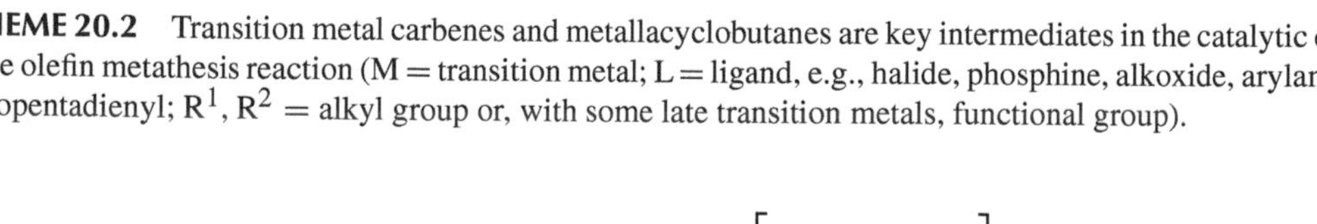

SCHEME 20.2   Transition metal carbenes and metallacyclobutanes are key intermediates in the catalytic cycle of the olefin metathesis reaction (M = transition metal; L = ligand, e.g., halide, phosphine, alkoxide, arylamido, cyclopentadienyl; R^1, R^2 = alkyl group or, with some late transition metals, functional group).

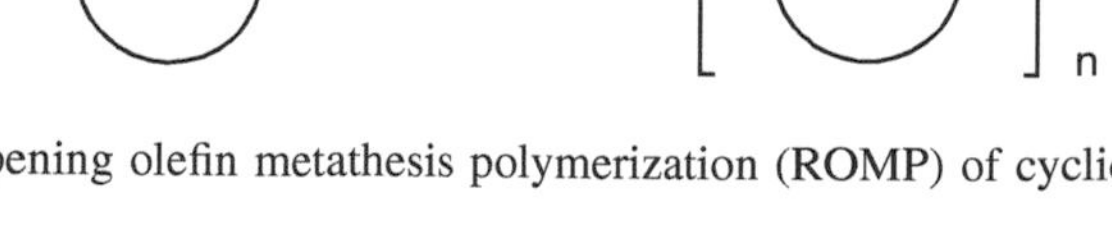

SCHEME 20.3   Ring-opening olefin metathesis polymerization (ROMP) of cyclic olefins affording macromolecular structures.

olefin ROMP monomers are cyclobutene, cyclopentene, cyclooctene, norbornene, norbornadiene, and dicyclopentadiene.[1-7]

The release of ring strain contributes to the driving force for this reaction.[22] Cyclohexene displays very little ring strain and accordingly is difficult to polymerize.[23] By comparison, norbornene and norbornene derivatives are significantly strained and undergo reaction readily with a large number of metathesis catalysts.

A wide range of catalyst compositions based on the transition elements Ti, Nb, Ta, Cr, Mo, W, Re, Co, Ir, Ru, and Os are active for ROMP.[1,2] These can be categorized as either conventional catalysts or recently developed well-defined catalysts. The active metathesis catalyst is generally a transition metal carbene (the metallacyclobutane form represents the actual resting state of the catalyst in the case of several Ti- and Ta-based catalysts).

Conventional catalysts are often multicomponent systems, in which the metathesis-active carbene is generated from a precursor such as a transition metal halide, oxohalide, or oxide compound through reaction with a main group organometallic compound. For example, WCl_6 and $MoCl_5$ form highly active metathesis catalysts upon reaction with Me_4Sn or Et_3Al,[24,25] whereby reaction of the Sn or Al compound with the halide produces the transition metal carbene in situ.[1,2] Late metal halides such as $RuCl_3 \cdot (H_2O)_3$, $OsCl_3$, and $OsCl_3 \cdot (H_2O)_n$ become active catalysts upon addition of a strained cyclic olefin monomer in a polar solvent.[26-30] $ReCl_5$ can act as an effective polymerization catalyst for bicyclic olefins and is generally used in either benzene or chlorobenzene solvent.

In recent years, well-defined catalysts have found increasing use in metathesis polymerizations.[1-7] These are single-component transition metal complexes, which can be isolated and stored and which do not require reaction with a second component for catalyst activation.

20.2 STEREOCHEMICAL CONFIGURATIONS OF POLYMERS PREPARED BY ROMP

Metathesis polymers obtained from unsubstituted monocyclic olefins, such as cyclopentene or cyclooctene, can contain two types of repeating units. These are units in which the carbon–carbon double bond has the cis or trans configuration.[2,24,31-33] Sequences of cis- and trans-enchained poly(1-octenylene) are shown in Figure 20.1.

ROMP of 4-methylcyclopentene leads to the formation of a polymer that can adopt different stereochemical configurations owing to the relative positioning of the methyl substituents in neighboring repeating units (dyads).[2,34] The four possible dyad structures are displayed in Figure 20.2.

The stereochemistry of ROMP products derived from norbornene (NB) and norbornadiene (NBD) and their derivatives has been extensively investigated and is described in a similar manner to the stereochemistry of addition polymers (isotactic, syndiotactic, etc.).[1-7] These polymers are comprised of an alternating sequence of olefin units and cis-1,3-enchained five-membered rings.[35] The tertiary carbon atoms attached to the olefin units represent chiral centers. Accordingly, four different isomeric dyad structures are possible for metathesis polymers of NB, NBD, and symmetrically substituted norbornene and norbornadiene derivatives.[1,2,20,36-38] The four dyad structures for polynorbornene

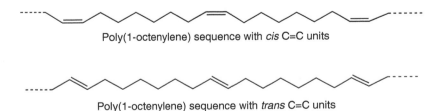

Poly(1-octenylene) sequence with *cis* C=C units

Poly(1-octenylene) sequence with *trans* C=C units

FIGURE 20.1 Cis and trans isomerism in polymers obtained by ROMP of cyclooctene.

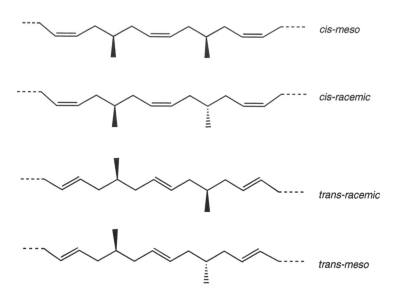

FIGURE 20.2   Stereochemical configurations of polymers prepared by ROMP of 4-methylcyclopentene.

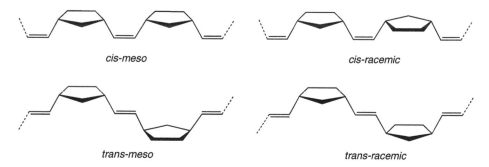

FIGURE 20.3   Dyad structures for polynorbornene prepared by ROMP.

(poly(NB)) are depicted in Figure 20.3. Isotactic polymers are composed primarily of *meso* (*m*) units; syndiotactic polymers are composed primarily of racemic (*r*) units. Atactic polymers contain an approximately equal amount of *meso* and racemic units. It is possible to prepare polymers of NB and related polycyclic olefins that display a wide range of cis/trans olefin contents and significantly different tacticities.

In the case of polymers derived from unsymmetrically substituted norbornene derivatives, such as 1- and 5-alkylnorbornenes, the possibility of regioisomerism exists.[39–45] The substituents of neighboring units may be oriented in the same direction or in opposite directions, giving rise to head–tail (tail–head), head–head, and tail–tail structures (Figure 20.4). A head–tail dyad is structurally identical with a tail–head dyad. A particular carbon atom, which is part of a head unit that is neighbored by a tail unit, is labeled HT; a carbon atom belonging to a tail unit neighbored by a head unit is labeled TH.

20.3   STEREOCHEMISTRY OF POLYMERS PREPARED FROM BICYCLIC OLEFIN MONOMERS EMPLOYING CONVENTIONAL METATHESIS CATALYSTS

Ivin and Rooney et al. pioneered ^{13}C nuclear magnetic resonance (NMR) polymer microstructural analysis of ROMP polynorbornenes, using conventional metathesis catalysts in their early

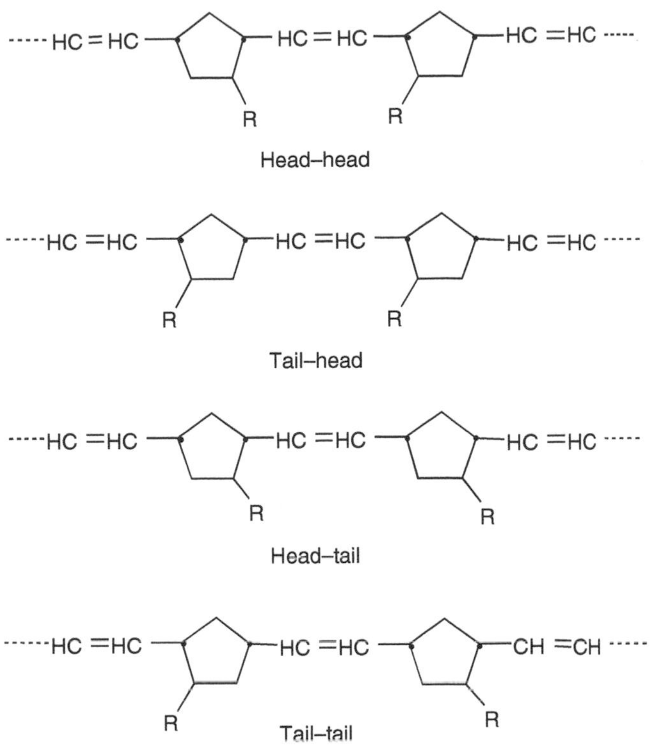

FIGURE 20.4 Regioisomerism in a metathesis polymer prepared from a 5-substituted norbornene derivative (head–tail dyads are structurally identical with tail–head dyads).

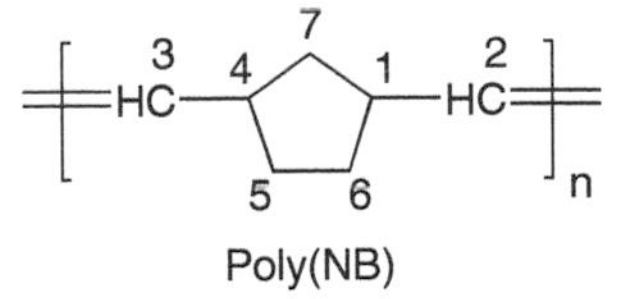

Poly(NB)

FIGURE 20.5 Carbon numbering scheme for ROMP polynorbornene.

studies.[2,37,39–45] They found that the polymer stereochemistry depends on various factors involving the catalyst, the monomer, and the polymerization solvent. The ^{13}C NMR spectrum of the parent polymer poly(NB) displays significantly different chemical shifts for cis- and trans-enchained repeating units.[37] As for polymers prepared from monocyclic olefins, there is a characteristic 4–5 ppm shift difference between the resonances of the carbon atoms α to trans and α to cis C=C units. The allylic carbon atoms attached to cis C=C units resonate at higher field. It was possible to assign signals to the cis/trans isomerism at the triad level (three repeating units). However, the shift differences for structures having the same C=C enchainment (cis or trans) but different tacticities (*meso* or *racemic*) are very small for poly(NB), and have not been resolved so far (with spectrometers of MHz $\leq$ 500). The signals corresponding to the allylic carbons ($C^{1,4}$; numbering system as shown in Figure 20.5)[2] display a small degree of line broadening as compared to the signals of the C_2H_4 segment of the five-membered rings ($C^{5,6}$).[46]

20.3.1 POLYMERIZATION OF 5,5-DIMETHYLNORBORNENE

Ivin and Rooney recognized that an indirect method could be used to distinguish between isotactic and syndiotactic structures in ROMP polynorbornenes.[37,39–45] This method exploits the

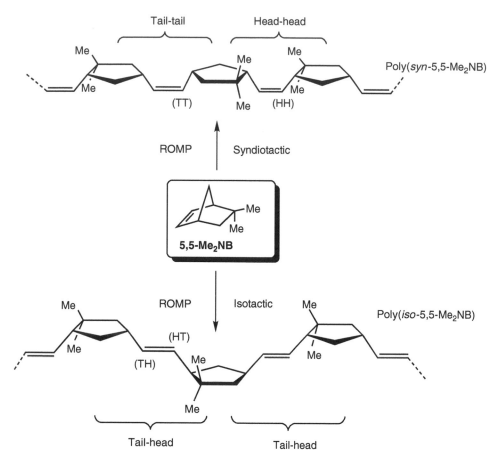

SCHEME 20.4 Use of a single enantiomer of 5,5-dimethylnorbornene in the determination of polymer tacticity.

observable olefin carbon NMR shift differences seen for the various head/tail enchainments in poly(5-substituted)norbornenes, and relates them to structures with different tacticities. A perfectly isotactic polymer, poly(*iso*-5,5-Me$_2$NB), derived from a pure enantiomer of 5,5-dimethylnorbornene (5,5-Me$_2$NB), would by nature be exclusively composed of tail–head (head–tail) dyads (with the corresponding olefin carbons denoted as TH and HT in Scheme 20.4), whereas the corresponding perfectly syndiotactic polymer, (poly(*syn*-5,5-Me$_2$NB)), would contain exclusively head–head and tail–tail dyads (with the corresponding olefin carbons labeled HH and TT).

Enantiomerically enriched 5,5-Me$_2$NB was subjected to metathesis polymerization. ReCl$_5$ catalyst (in chlorobenzene solvent) was found to produce a highly stereoregular cis-syndiotactic polymer.[41–44] The syndiotactic structure was deduced from its ^{13}C NMR spectrum, which displays two prominent signals corresponding to the HH and TT olefin carbons of poly(*syn*-5,5-Me$_2$NB) (two low-intensity signals are also visible that correspond to the HT and TH olefin carbons of the incorporated minor stereoisomer of 5,5-Me$_2$NB). The cis-vinylene structure in this polymer, and in all subsequently discussed metathesis polymers of bicyclic and polycyclic monomers, is identified by the characteristic upfield shift of the ^{13}C NMR signal for the cis-allylic carbon nuclei (as compared to trans-allylic carbons).

Subsequently, a high-cis syndiotactic (90%) polymer of 5,5-Me$_2$NB was also obtained using the catalyst mixture OsCl$_3$/phenylacetylene in tetrahydrofuran (THF) solvent.[44] The tungsten complexes **1a–c** (Figure 20.6), upon activation with Et$_2$AlCl, produce high-cis poly(5,5-Me$_2$NB) samples

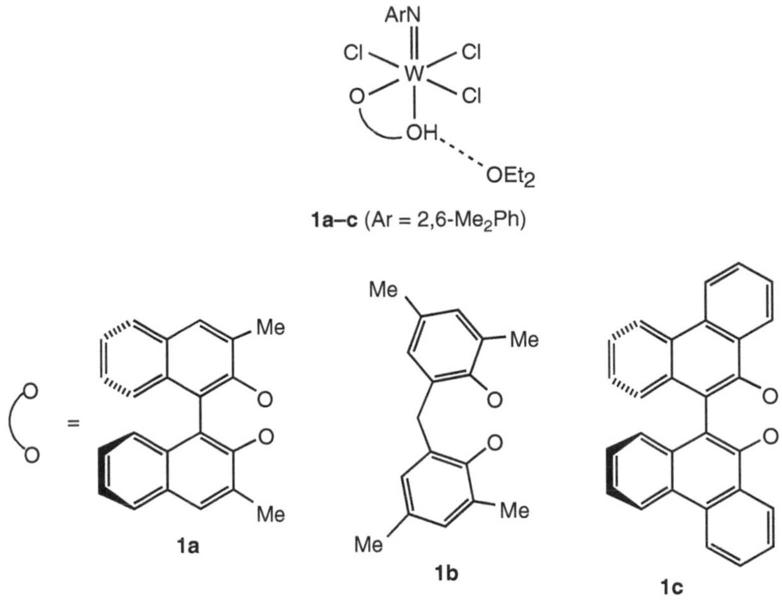

1a–c (Ar = 2,6-Me$_2$Ph)

1a   **1b**   **1c**

FIGURE 20.6 Tungsten diolate complexes for the preparation of poly(5,5-Me$_2$NB) enriched in cis-syndiotactic structures.

reported to be syndiotactic.[47] The high cis content (>95%) is suggested to arise from minimized steric repulsion in the transition state for chain propagation.

The catalyst (mesitylene)W(CO)$_3$/EtAlCl$_2$/exo-2,3-epoxynorbornane produces predominantly trans-isotactic poly(iso-5,5-Me$_2$NB) (85:15 trans:cis) at a monomer concentration of 0.4 mol/L.[43] The ^{13}C NMR spectrum of the polymer obtained from enantiomerically enriched monomer displays two prominent signals in the region of δ 128–138 ppm that correspond to the trans-HT and trans-TH olefin carbon nuclei. At higher monomer concentrations, that is, 2.7 mol/L, the proportion of cis units increases to 40% (it is noted that the cis/trans and m/r dyad ratios of metathesis polymers are often measured by the intensity of the relevant ^{13}C NMR signals, and small errors in these ratios can occur when spin relaxation is not complete).

The amount of trans olefin units in poly(5,5-Me$_2$NB) can be increased to 95% by using RuCl$_3$·(H$_2$O)$_3$ catalyst (solvent = ethanol/chlorobenzene). However, the ^{13}C NMR spectrum of the polymer made from enantiomerically enriched monomer shows four olefin signals of nearly equal intensity, which indicates that the polymer is predominantly atactic (57% m and 43% r).[42] The ^{13}C NMR spectrum of poly(5,5-Me$_2$NB) prepared from racemic monomer also reveals an atactic structure, as it displays splitting of the HH olefin signal into two peaks owing to the presence of syndiotactic and isotactic head–head units. (The other three olefin signals (HT, TH, TT) assigned to the head–tail and tail–tail units are not further split.)

The choice of a solvent can be important for the stereochemistry of products obtained by metathesis polymerization. Poly(5,5-Me$_2$NB) prepared with WCl$_6$/Me$_4$Sn in chlorobenzene is mainly atactic and contains 69% cis olefin units. In comparison, poly(5,5-Me$_2$NB) obtained with WCl$_6$/Me$_4$Sn in dioxane is an all-cis, slightly syndiotactic polymer (r:m = 70:30).[46] The same stereochemistry is achieved with MoCl$_5$/Me$_4$Sn in dioxane.

Polymers with different tacticities and cis/trans olefin contents were also synthesized from racemic and enantiomerically enriched exo- and endo-5-methylnorbornene.[39,40,44] High-cis syndiotactic polymers were prepared with ReCl$_5$ (in benzene or chlorobenzene) and high-trans atactic polymers were obtained with two Ru-based catalysts, RuCl$_3$·(H$_2$O)$_3$ in PhCl/EtOH and an in

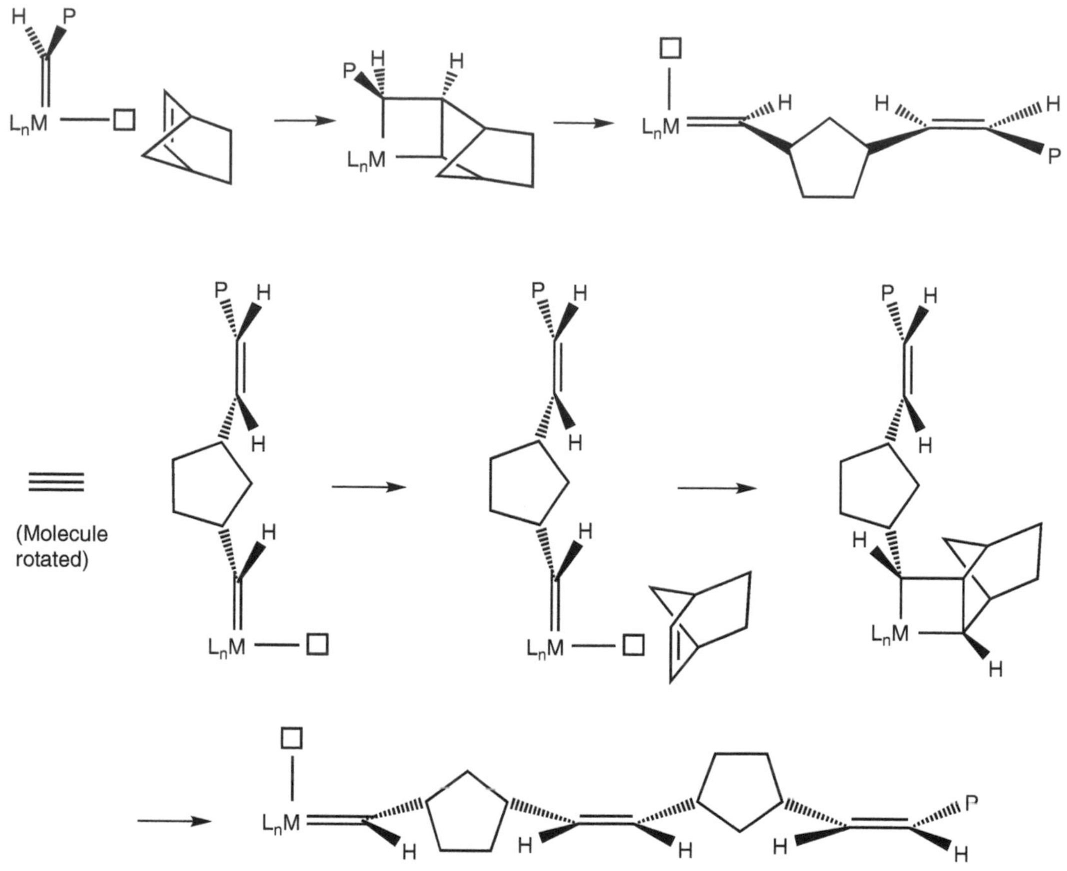

SCHEME 20.5 An early model for the formation of syndiotactic *cis*-polynorbornene (M = transition metal, L = ligand, P = polymer chain, □ = vacant coordination site).

situ-prepared Ru cyclooctadiene complex. Tacticity was further confirmed by ^{13}C NMR analysis of the hydrogenated versions of 5-methyl- and 5,5-dimethyl-substituted norbornene polymers.[44]

An early model was suggested by Calderon and Kelly[38] and Ivin et al.[37] to account for a predominantly syndiotactic microstructure in high-cis metathesis polymers (Scheme 20.5) and a predominantly isotactic structure in high-trans polymers (Scheme 20.6) when conventional metathesis catalysts are employed. This is a theoretical model, which assumes chain propagation to occur at chiral transition metal alkylidene species (L$_n$M=CH-R). It was proposed that the rate of rotation about the metal–carbon double bond is significantly lower than the rate of chain propagation, and that scrambling of the vacant coordination site with any of the ligand sites does not occur. Accordingly, a syndiotactic sequence is formed in high-cis poly(NB). In the top row of Scheme 20.5, both the methylene bridge of the monomer and the polymer chain P point in the same direction, that is, out of the plane of projection, as the monomer approaches the metal alkylidene, and a *cis*-metallacyclobutane structure is formed. The metallacyclobutane subsequently cleaves to reform a metal alkylidene, and a *cis*-vinylene structure is obtained. In the middle row of Scheme 20.5, the metal alkylidene (complete molecule rotated around two axes) is approached by a second monomer molecule, whereby the CH$_2$ bridge and polymer chain P are pointing behind the plane of projection. A second metallacycle is formed and cleaved to produce a syndiotactic dyad with a cis double bond (bottom row of Scheme 20.5). The formation of trans-isotactic poly(NB) occurs in a similar fashion, and part of the relevant sequence is displayed in Scheme 20.6.

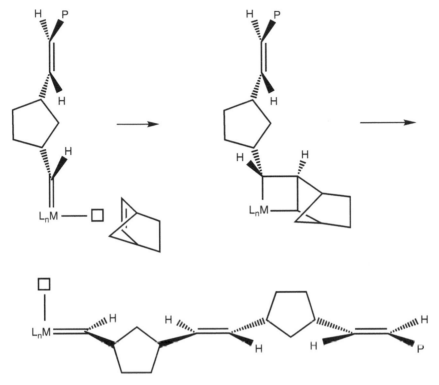

SCHEME 20.6 An early model for the formation of isotactic *trans*-polynorbornene (M = transition metal, L = ligand, P = polymer chain, □ = vacant coordination site).

It was further suggested that the cis:trans ratio in the polymer is determined by the extent that the last reacted monomer unit remains π-bonded to the transition metal center when the next monomer molecule is approaching. It must be emphasized here that these models are based on speculation, as the transition metal carbene and metallacyclobutane intermediates responsible for polymer chain growth in these conventional polymerization systems have not yet been identified and characterized.

20.3.2 POLYMERIZATION OF 1-METHYLNORBORNENE

Several catalysts convert 1-methylnorbornene (1-MeNB) into poly(1-MeNB) samples that contain exclusively or predominantly head–tail linkages.[45] The catalyst $ReCl_5$ (in chlorobenzene) produces an all head–tail cis-syndiotactic polymer. This means that poly(1-MeNB) consists of a strictly alternating sequence of the two enantiomers of 1-MeNB (Scheme 20.7). So far, reactions of $ReCl_5$ with enantiomerically pure 1-MeNB have not produced any macromolecular products.

The catalyst $OsCl_3$ (in a 1:1 by volume mixture of ethanol/chlorobenzene) converts racemic 1-MeNB into an atactic, all-trans polymer with predominantly head–tail structures at low monomer concentrations (0.2 mol/L). The ratio of (HT + TH)/(HH + TT) signals is 9:1. By comparison, polymerization of a single isomer of 1-MeNB gives isotactic poly(1-MeNB), as the selective head–tail enchainment necessarily leads to the *meso* stereochemistry when enantiomerically pure monomer is used (Scheme 20.8). At an increased monomer concentration (1.5 mol/L), a polymer with 16% cis units is formed. This polymer does not contain any cis-head–head sequences.

Polymerization of 1-MeNB with $RuCl_3 \cdot (H_2O)_3$ (in 1:1 EtOH/PhCl) produces poly(1-MeNB) exclusively composed of repeating units that contain trans olefin structures. This polymer contains a completely random sequence of HT, HH, and TT units and is atactic, both when prepared from a racemic mixture of 1-MeNB and when synthesized from a single enantiomer of 1-MeNB.[45]

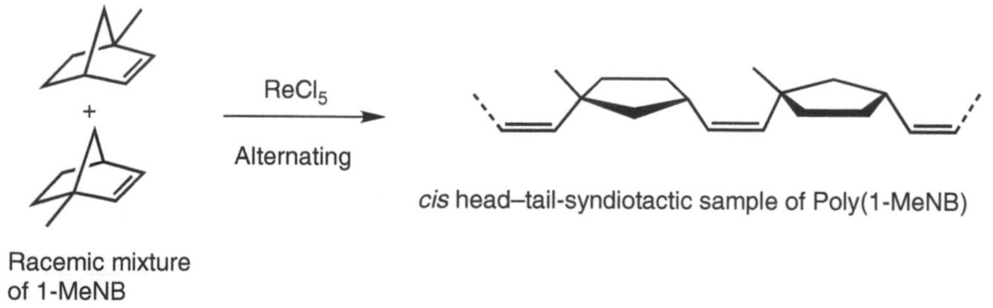

SCHEME 20.7 ROMP of 1-methylnorbornene catalyzed by ReCl₅.

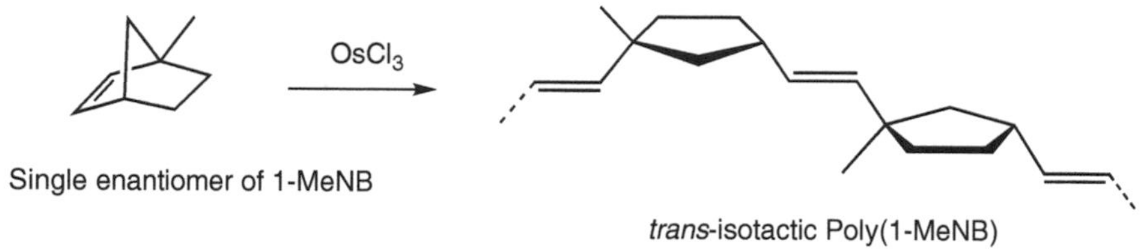

SCHEME 20.8 ROMP of 1-methylnorbornene with OsCl₃.

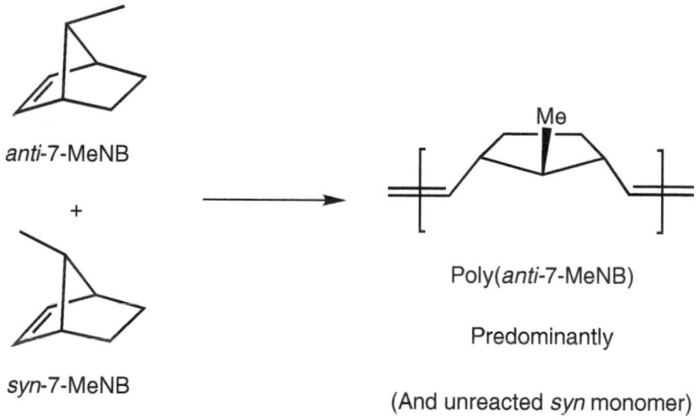

SCHEME 20.9 ROMP of a 1:1 mixture of *syn*- and *anti*-7-methylnorbornene is strongly selective for polymerization of the *anti* isomer.

20.3.3 POLYMERIZATION OF 7-METHYLNORBORNENE AND 7-METHYLNORBORNADIENE

Both ReCl₅ and (mesitylene)W(CO)₃/EtAlCl₂/*exo*-2,3-epoxynorbornane produce polymers of *anti*-7-methylnorbornene (*anti*-7-MeNB) with highly tactic sequences (Scheme 20.9).[48] The stereochemistry of poly(*anti*-7-MeNB) is assigned assuming that the tacticity is the same as for poly(5,5-Me₂NB): cis-syndiotactic when polymerization of *anti*-7-MeNB is carried out with ReCl₅ (90% cis), and mixed trans-isotactic and cis-syndiotactic when (mesitylene)W(CO)₃/EtAlCl₂/*exo*-2,3-epoxynorbornane is used. In the case of the W-based catalyst, the trans/cis ratio for poly(*anti*-7-MeNB) is 55:45, which is lower than that for poly(5,5-Me₂NB) prepared with the same catalyst.

Poly(*anti*-7-MeNB) prepared with RuCl₃·(H₂O)₃ (in 1:1 EtOH/PhCl) is all trans and atactic. Its ¹³C NMR spectrum displays splitting owing to tacticity for all carbon nuclei with the exception of C⁷ (numbering scheme according to Figure 20.5). Polymerization with OsCl₃ in EtOH/PhCl produces poly(*anti*-7-MeNB) with 87% trans-atactic structures.[48] The trans olefin content can be further decreased to 48% by employing WCl_6/Bu_4Sn in chlorobenzene as the catalyst. Both the cis and trans olefin sequences are atactic. Use of a ruthenium trifluoroacetate catalyst (prepared from $RuCl_3 \cdot (H_2O)_3$ and CF_3COOAg) gives poly(*anti*-7-MeNB) that contains 43% trans-atactic and 57% cis-syndiotactic dyads. This stereochemistry differs from that of poly(*anti*-7-MeNB) made with $RuCl_3 \cdot (H_2O)_3$.

Technically, these polymerizations were carried out with a 1:1 mixture of *syn*- and *anti*-7-methylnorbornene (Scheme 20.9) rather than with pure *anti*-7-MeNB.[48] The *anti* isomer selectively undergoes polymerization, and the *syn* isomer remains largely unreacted, usually to an extent of 90% or more.

This result suggests that the polymerization of norbornene-type monomers occurs selectively on the exo face. Samples of poly(7-MeNB) prepared with (mesitylene)$W(CO)_3/EtAlCl_2/exo$-2,3-epoxynorbornane contain a small amount of the *syn* monomer incorporated as measured by ^{13}C NMR. This catalyst is also capable of polymerizing the *syn* monomer in the presence of a small amount of the *anti* monomer or norbornene; the resulting product is mainly composed of repeating units of the *syn* monomer.[48] However, the polymerization of pure *syn*-7-MeNB with (mesitylene)$W(CO)_3/EtAlCl_2/exo$-2,3-epoxynorbornane was reported to have failed.

The monomer 7-methylnorbornadiene (7-MeNBD) also undergoes polymerization in the presence of several conventional metathesis catalysts to form poly(7-MeNBD). It is noteworthy that the catalyst $OsCl_3$ (in 1:1 EtOH/PhCl) leads to polymer stereochemistry that is significantly different from that of poly(*anti*-7-MeNB). The $OsCl_3$-derived polymer of the monoolefin contains predominantly atactic trans olefin structures, whereas the analogous diolefin polymer is composed of nearly exclusively cis alkene structures (97%) and predominantly syndiotactic dyads ($r:m = 75:25$ through the ^{13}C NMR signals of the methyl substituent at δ 16.2–17.0 ppm).[49] The catalysts $ReCl_5$ and WCl_6/Me_4Sn produce poly(7-MeNBD) with a similar stereochemical configuration to that of poly(*anti*-7-MeNB) made with these catalysts. Both of the $ReCl_5$-based products are cis-syndiotactic, but the cis olefin content is slightly lower in the case of the diene-based polymer, that is, 80% versus 90%.

The WCl_6/Me_4Sn-derived poly(7-MeNBD) contains 42% cis and 58% trans units, both of which are atactic. A catalyst based on (mesitylene)$W(CO)_3/EtAlCl_2$ produces poly(7-MeNBD) with 76% cis C=C units that are highly syndiotactic and 24% trans C=C structures that are highly isotactic. A $MoCl_5/Me_4Sn/Et_2O$ catalyst gives an atactic, predominantly trans polymer (80% trans C=C). With all of these catalysts, polymerization occurs selectively at the *anti* face of the monomer, and the resulting products generally contain only between 6% and 8% repeating units that are derived from the reaction at the *syn* face (Scheme 20.10).

The stereochemistry of the *syn* units of poly(7-MeNBD) was investigated for products made with $OsCl_3$ (in EtOH/PhCl) and WCl_6/Me_4Sn (in PhCl) and found to be very similar to the microstructure of the *anti* units. The tacticity results for poly(7-MeNBD) were confirmed by ^{13}C NMR analysis of the hydrogenated polymer.

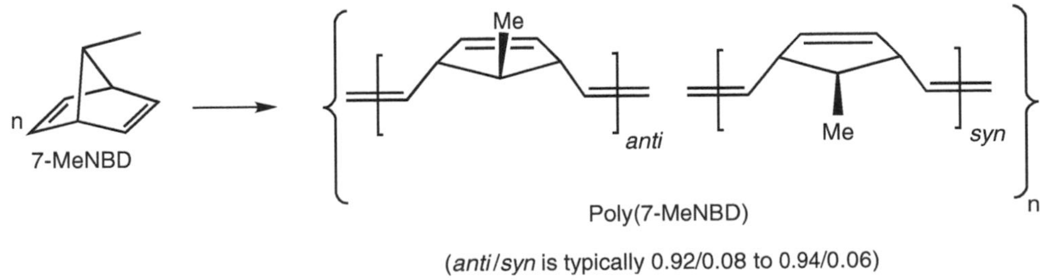

(*anti/syn* is typically 0.92/0.08 to 0.94/0.06)

SCHEME 20.10 ROMP of 7-methylnorbornadiene can occur at either the *syn* or *anti* face, with predominant reaction at the *anti* face.

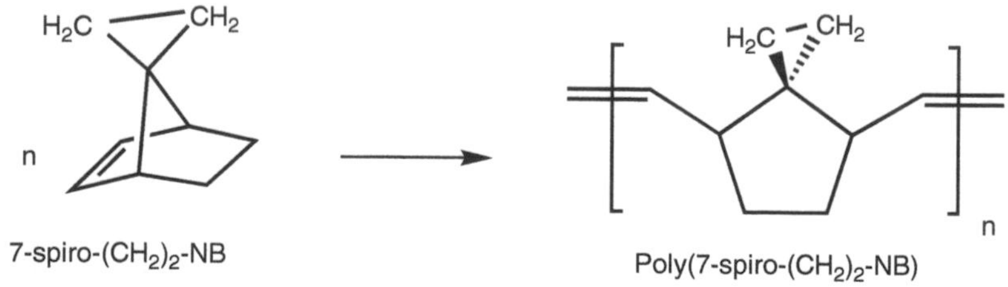

SCHEME 20.11 Polymerization of spiro(bicyclo[2.2.1]hept-2-ene-7,1'-cyclopropane).

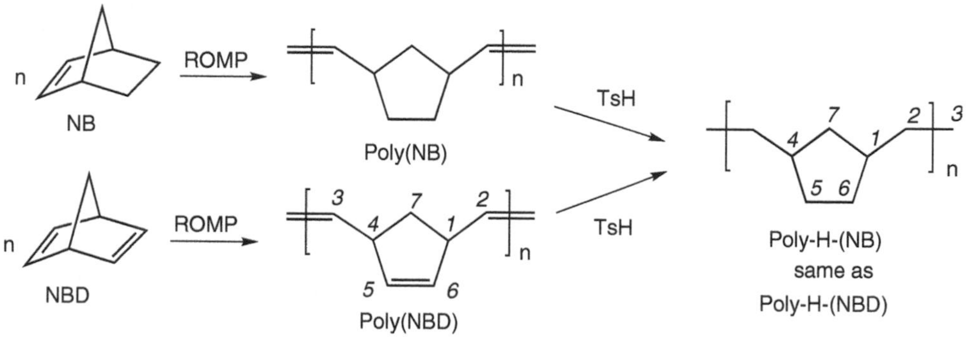

SCHEME 20.12 ROMP of norbornene and norbornadiene with subsequent hydrogenation for ^{13}C NMR analysis (TsH = p-toluenesulfonhydrazide).

20.3.4 POLYMERS OF SPIRO(BICYCLO[2.2.1]HEPT-2-ENE-7,1'-CYCLOPROPANE)

The spiro bridge of spiro(bicyclo[2.2.1]hept-2-ene-7,1'-cyclopropane) (7-spiro-(CH$_2$)$_2$-NB) is less sterically demanding than the syn-methyl substituent of syn-7-MeNB, and yields greater than 90% of poly(7-spiro-(CH$_2$)$_2$-NB) (Scheme 20.11) are obtained with RuCl$_3$·(H$_2$O)$_3$ (12:1 toluene/EtOH, 18 h, 75 °C) and WCl$_6$/Ph$_4$Sn (toluene, 15 h, 20 °C). Samples of poly(7-spiro-(CH$_2$)$_2$-NB) made using both catalysts are atactic and contain predominantly trans olefin linkages (85% for the W-based catalyst and 95% for the Ru catalyst).[50]

20.3.5 POLYMERS OF NORBORNENE AND NORBORNADIENE

It became possible to estimate the tacticity of the parent polymers, poly(NB) and polynorbornadiene (poly(NBD)), after hydrogenation of the olefin double bonds (Scheme 20.12), employing a high-field NMR spectrometer. The carbon nuclei C^7 and C^{5,6} of the hydrogenated products poly-H-(NB)/poly-H-(NBD) display splitting with partial signal resolution owing to r/m stereochemistry.[51]

The polymerization of norbornene with ReCl$_5$ (in PhCl) affords a high-cis, predominantly syndiotactic poly(NB) (Table 20.1).[51] The ^{13}C NMR spectrum of the hydrogenated polymer qualitatively reveals that the stereoregularity is not quite as high as in poly(5,5-Me$_2$NB) prepared with ReCl$_5$. Poly(NB) with higher stereoregularity can be obtained by employing a modified OsCl$_3$ catalyst. In the case of the polymerization of $anti$-7-MeNB, it was shown that OsCl$_3$ (in 1:1 EtOH/PhCl) gave a high-trans atactic polymer.[48] However, through the addition of p-benzoquinone or by employing phenylacetylene in THF, the Os catalyst becomes highly cis-selective, and poly(NB) samples with 95% and 98% cis double bonds, respectively, are obtained.[52] These polymers are more than 90% syndiotactic (Table 20.1).[46,52] In the case of the p-benzoquinone-modified catalyst, it is essential to employ a sufficiently high monomer concentration, such as 0.8 mol/L, to retain selectivity. The cis

TABLE 20.1

Stereochemical Configuration of Polynorbornene (poly(NB)) and Polynorbornadiene (poly(NBD)) Samples Obtained with Conventional ROMP Catalysts

Monomer	Catalyst/Solvent	Olefin Stereochemistry	Tacticity
NB	ReCl$_5$/PhCl	High-cis	Predominantly syndiotactic
NB	OsCl$_3$/p-benzoquinone/EtOH/PhCl	95% cis	>90% syndiotactic
NB	OsCl$_3$/phenylacetylene/THF	98% cis	>90% syndiotactic
NB	WCl$_6$/Me$_4$Sn/dioxane	>90% cis	Atactic
NB	MoCl$_5$/Me$_4$Sn/dioxane	>90% cis	Atactic
NB	NbCl$_5$/Me$_4$Sn/dioxane	>90% cis	Atactic
NB	TaCl$_5$/Me$_4$Sn/dioxane	>90% cis	Atactic
NB	(mesitylene)W(CO)$_3$/EtAlCl$_2$/PhCl	16% cis	Predominantly atactic
NB	(mesitylene)W(CO)$_3$/EtAlCl$_2$/dioxane	95% cis	Predominantly atactic
NB	RuCl$_3$·(H$_2$O)$_3$/EtOH/PhCl	High-trans	Atactic
NBD	OsCl$_3$/phenylacetylene/THF	>95% cis	>90% syndiotactic
NBD	OsCl$_3$/EtOH/PhCl	93% cis	65–80% syndiotactic
NBD	WCl$_6$/Me$_4$Sn/dioxane	High-cis	Atactic
NBD	MoCl$_5$/Me$_4$Sn/dioxane	High-cis	Atactic
NBD	NbCl$_5$/Me$_4$Sn/dioxane	High-cis	Atactic
NBD	TaCl$_5$/Me$_1$Sn/dioxane	High-cis	Atactic

olefin content drops to 31% when a monomer concentration of 0.3 mol/L is used in a 1:1 mixture of chlorobenzene/ethanol.[52]

The catalysts WCl$_6$/Me$_4$Sn, MoCl$_5$/Me$_4$Sn, NbCl$_5$/Me$_4$Sn, and TaCl$_5$/Me$_4$Sn, in the solvent dioxane, also produce poly(NB) with a cis double bond content of more than 90% (Table 20.1).[46] These polymers are atactic, which slightly contrasts with the 70% r structure of poly(5,5-Me$_2$NB) prepared with WCl$_6$/Me$_4$Sn/dioxane or MoCl$_5$/Me$_4$Sn/dioxane. The cis content of poly(NB) prepared with these catalysts drops to 50% or less when the polymerization solvent is changed from dioxane to chlorobenzene. (Mesitylene)W(CO)$_3$/EtAlCl$_2$ also produces poly(NB) with a different stereochemical structure when the solvent system is changed.[51] A polymer containing 84% *trans*-vinylene units is obtained in chlorobenzene, but the trans content decreases to 25% when a 0.7:3.7 volume ratio of dioxane/chlorobenzene is used as solvent. The *trans*-vinylene content is further decreased to 5% when the polymerization is conducted in pure dioxane (Table 20.1).

High-trans atactic poly(NB) can be prepared with RuCl$_3$·(H$_2$O)$_3$ (in 1:1 EtOH/PhCl) and [(p-cymene)RuCl$_2$]$_2$(in 1:10 EtOH/PhCl). In the case of RuCl$_3$·(H$_2$O)$_3$, the addition of phenylacetylene as an additive to the polymerization system does not lead to a different polymer microstructure,[51] in contrast to OsCl$_3$. *cis*-Syndiotactic poly(NB) obtained with OsCl$_3$/phenylacetylene/THF (Table 20.1) is compared with trans-atactic poly(1-MeNB) and poly(*anti*-7-MeNB) prepared with OsCl$_3$/EtOH/PhCl, assuming that polymer stereochemistry is not significantly changed in the presence of a 1- or 7-Me substituent.

Differential scanning calorimetry (DSC) analysis of an atactic hydrogenated poly(NB) sample reveals a melt transition (T_m) with a peak maximum at 146 °C ($\Delta H = 49$ J/g).[53,54] This demonstrates that atactic hydrogenated poly(NB) is moderately crystalline despite its irregular microstructure.

When norbornene is replaced by norbornadiene, very similar stereochemistry results are obtained using the catalysts WCl$_6$/Me$_4$Sn, MoCl$_5$/Me$_4$Sn, NbCl$_5$/Me$_4$Sn, and TaCl$_5$/Me$_4$Sn in the solvent dioxane.[46] Norbornadiene polymerizations are often carried out in the presence of a small amount of a linear olefin to prevent crosslinking or the formation of very high molar mass polymers.[55] Again, a highly stereoregular polymer is isolated when OsCl$_3$/phenylacetylene (in THF) is used as the catalyst (>95% cis double bonds and >90% r dyads) (Table 20.1).[46] This tacticity is higher than in poly(NBD) prepared with unmodified OsCl$_3$(in EtOH/PhCl) (high-cis content and about 65%–80% r).[51] The latter polymer has a very similar stereochemistry to that of poly(7-MeNBD) obtained with OsCl$_3$ but differs from the high-trans-atactic structure of poly(*anti*-7-MeNB) obtained with OsCl$_3$.

In many earlier studies, the stereochemistry of poly(NBD) was determined only after hydrogenation of the polymer products.[51] Subsequently, it was found that poly(NBD) microstructure can be directly assessed by examining the ^{13}C NMR signal corresponding to the olefin carbons C^{5,6} of the five-membered rings (Scheme 20.12). This signal is split when both m and r dyads are present.[46]

20.3.6 POLYMERS OF *ENDO,EXO*-5,6-DIMETHYLNORBORNENE

The conventional catalyst system OsCl$_3$/phenylacetylene (in THF) produces moderately regular metathesis polymers of enantiomerically enriched (90% ee) (Scheme 20.13) and racemic *endo,exo*-5,6-dimethylnorbornene (*endo,exo*-5,6-Me$_2$NB).[56] The stereoregularity is not as pronounced as for the highly syndiotactic poly(NB) and poly(NBD) samples prepared with this same catalyst system. The product formed from (+)-*endo,exo*-5,6-Me$_2$NB is reported to contain approximately 61% cis C=C units, and the polymer made from the racemic monomer contains about 85% cis units. The r/m ratio is established after hydrogenation of the C=C double bonds (Figure 20.7). The intensity of the ^{13}C NMR signals assigned to the rr triad of hydrogenated poly(+)-*endo,exo*-5,6-Me$_2$NB (poly-H-((+)-*endo,exo*-5,6-Me$_2$NB)) is approximately twice the intensity of the corresponding mm triad resonances.

It is noteworthy that the fraction of rm and mr triads is significantly smaller than the fraction of mm triads, which indicates that the polymer is predominantly composed of block sequences of syndiotactic and isotactic structures (Figure 20.7). This is a syndiotactic-biased stereoblock polymer. The authors suggested that the Os-based catalytic sites responsible for the formation of cis C=C structures produce syndiotactic sequences, and the trans-specific sites form isotactic structures. By comparison, poly((+)-*endo,exo*-5,6-Me$_2$NB) prepared with RuCl$_3$·(H$_2$O)$_3$ (in EtOH/PhCl) is all-trans and atactic.

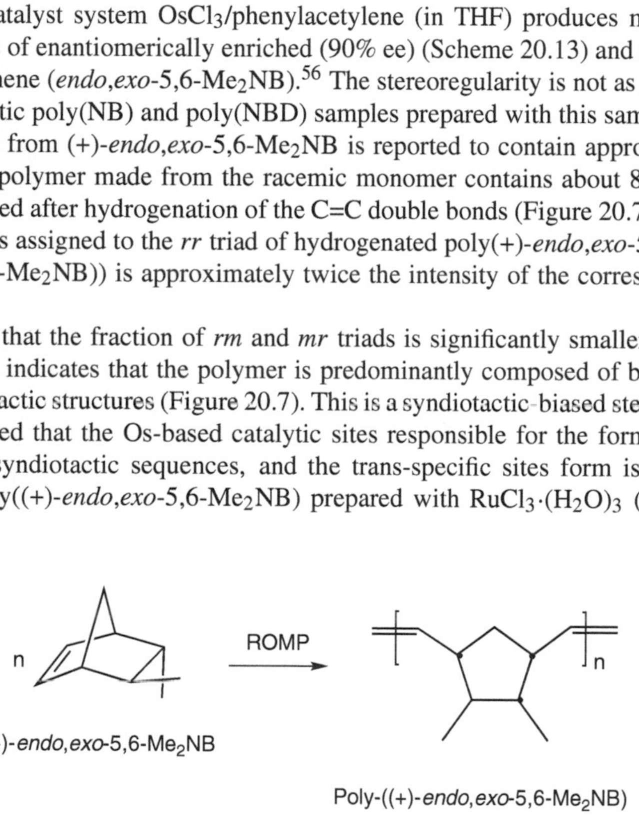

SCHEME 20.13   ROMP of (+)-*endo,exo*-5,6-dimethylnorbornene.

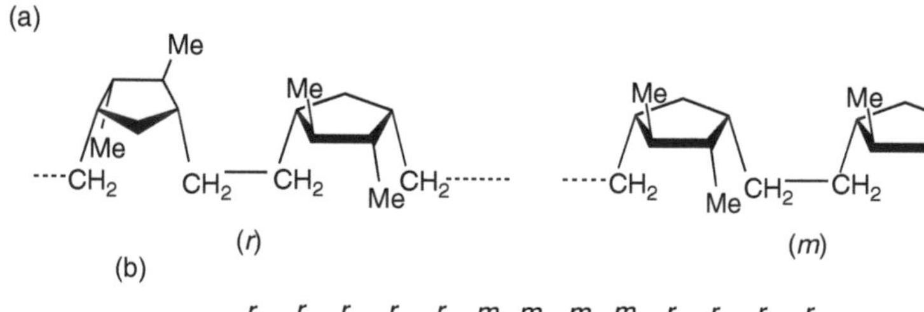

FIGURE 20.7   Dyad structures of hydrogenated poly((+)-*endo,exo*-5,6-dimethylnorbornene) (a) and stereoblock polymer composed of repeating monomer (M) units linked in a racemic (*r*) and *meso* (*m*) fashion (b).

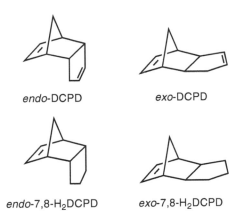

endo-DCPD   exo-DCPD

endo-7,8-H$_2$DCPD   exo-7,8-H$_2$DCPD

FIGURE 20.8 Tricyclic ROMP monomers.

20.3.7 POLYMERIZATION OF *ENDO*- AND *EXO*-DICYCLOPENTADIENE

Metathesis polymers of *endo*- and *exo*-dicyclopentadiene (*endo*- and *exo*-DCPD, Figure 20.8) prepared with WCl$_6$/Me$_4$Sn (in PhCl) are atactic and contain both cis and trans olefin structures. Polymer tacticity was established after complete hydrogenation of the polymers.[57] Products prepared with ReCl$_5$ (in PhCl) have essentially all-cis olefin linkages. It is noteworthy that poly(*endo*-DCPD) synthesized with RuCl$_3$·(H$_2$O)$_3$ (in EtOH/PhCl) is an all-cis polymer. This contrasts with polymers prepared from *exo*-DCPD, *endo*- and *exo*-7,8-dihydrodicyclopentadiene (*endo*-7,8-H$_2$DCPD and *exo*-7,8-H$_2$DCPD, Figure 20.8), and other norbornene-type monomers prepared using RuCl$_3$·(H$_2$O)$_3$, which contain 90% or more trans olefin linkages. The high-trans poly(*endo*-7,8-H$_2$DCPD) is atactic.

The authors tentatively proposed that the steric bulk of the catalytic site caused by coordination (in a chelating fashion) of an additional *endo*-DCPD molecule, which is not involved in chain propagation, is responsible for the high-cis structure of poly(*endo*-DCPD) that is obtained with RuCl$_3$·(H$_2$O)$_3$.

20.3.8 POLYMERS OF 2,3-DICARBOALKOXYNORBORNADIENES AND 2,3-DICARBOALKOXY-7-OXANORBORNADIENES

As outlined in the sections above, the macromolecular stereochemistry of ROMP polymers depends on factors involving the catalyst, cocatalyst, solvent, and monomer employed. An additional factor affecting microstructure—the presence or absence of phenylacetylene as an additive in the polymerization system—is also found when 2,3-dicarbomethoxynorbornadiene (2,3-(CO$_2$Me)$_2$NBD, Figure 20.9) is polymerized by OsCl$_3$.[58] Polymerization without added phenylacetylene gives atactic, high-trans (90%) poly(2,3-dicarbomethoxynorbornadiene) (poly(2,3-(CO$_2$Me)$_2$NBD)), whereas polymerization with OsCl$_3$/phenylacetylene produces a high-cis (83%) polymer enriched in syndiotactic dyads (ca. 80% *r*). The high-trans composition of the former polymer contrasts with the high-cis structure of poly(NBD) prepared with unmodified OsCl$_3$, but is similar to that of trans-atactic, OsCl$_3$-derived poly(*anti*-7-MeNB) (i.e., that obtained when no phenylacetylene is added to the polymerization system).

The trans-atactic stereochemistry of the OsCl$_3$-derived poly(2,3-(CO$_2$Me)$_2$NBD) also differs significantly from the high-cis (76%–86% cis), syndiotactic (ca. 80% *r*) microstructure of the polymer obtained from the OsCl$_3$-catalyzed polymerization of the analogous monoolefin, *exo,cis*-2,3-dicarbomethoxynorborn-5-ene (*exo,cis*-2,3-(CO$_2$Me)$_2$NB, Figure 20.9). By comparison, RuCl$_3$·(H$_2$O)$_3$-catalyzed polymerizations of 2,3-(CO$_2$Me)$_2$NBD and 2,3-*exo,cis*-(CO$_2$Me)$_2$NB give

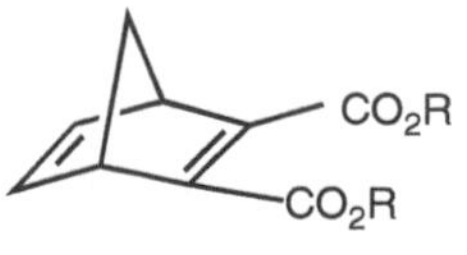

2,3-$(CO_2R)_2$NBD

R = Me, Et, i-Pr

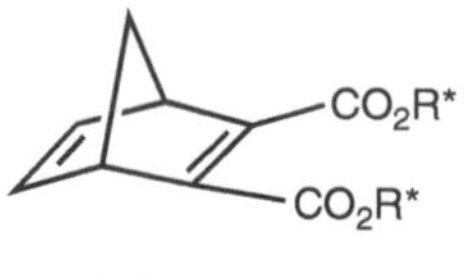

2,3-$(CO_2R^*)_2$NBD

Chiral R* = C*H(Me)C_2H_5,
CH$_2$C*H(Me)C_2H_5

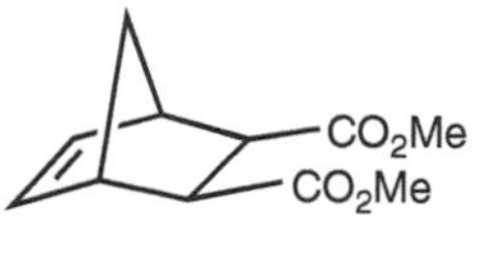

exo,cis-2,3-$(CO_2Me)_2$NB

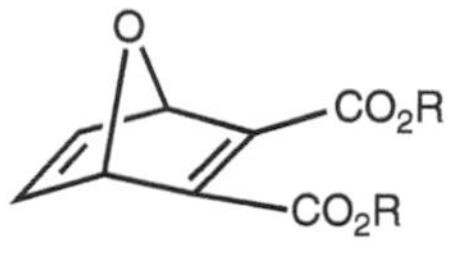

2,3-$(CO_2R)_2$ONBD

R = Me, Et, i-Pr

FIGURE 20.9 Carboxylic ester-substituted bicyclic olefin monomers bearing two (identical) ester substituents R or R*.

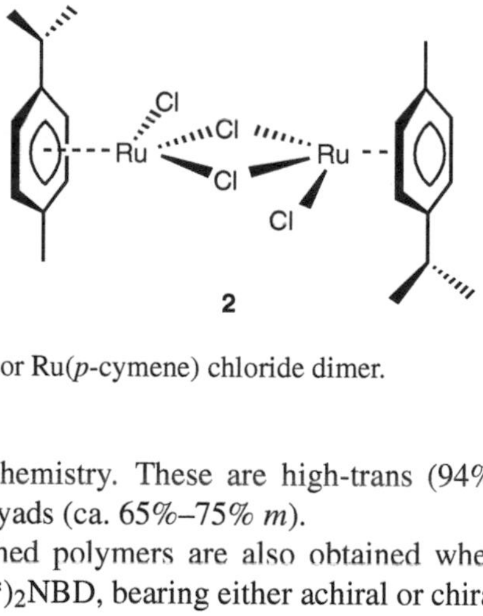

2

FIGURE 20.10 Catalyst precursor Ru(p-cymene) chloride dimer.

polymers with similar stereochemistry. These are high-trans (94% and 99%, respectively) and moderately enriched in *meso* dyads (ca. 65%–75% *m*).

High-trans, isotactic-enriched polymers are also obtained when norbornadiene diesters 2,3-$(CO_2R)_2$NBD and 2,3-$(CO_2R^*)_2$NBD, bearing either achiral or chiral alkoxy groups (Figure 20.9), are reacted with [RuCl$_2$(p-cymene)]$_2$ (**2**, Figure 20.10) activated by trimethylsilyldiazomethane.[59,60] In addition, RuCl$_3$·(H$_2$O)$_3$ and **2**/trimethylsilyldiazomethane both polymerize 2,3-dicarbomethoxy-7-oxanorbornadiene (2,3-$(CO_2Me)_2$ONBD, Figure 20.9) to give a high-trans material;[60,61] the latter catalyst system has also been reported to form high-trans poly(2,3-$(CO_2Et)_2$ONBD) and poly(2,3-$(CO_2i$-Pr)$_2$ONBD) from the higher 7-oxanorbornadiene esters 2,3-dicarboethoxy-7-oxanorbornadiene and 2,3-dicarboisopropoxynorbornadiene.[60]

To this date, poly(($CO_2Me)_2$NBD) prepared with the catalyst system MoCl$_5$/Me$_4$Sn/dioxane stands out as a rare example of an isotactic, all-cis polymer obtained with a conventional transition metal halide catalyst.[58] This microstructure contrasts with the atactic and moderately syndiotactic structures of poly(NB) and poly(5,5-Me$_2$NB), respectively, that are obtained with the MoCl$_5$/Me$_4$Sn/dioxane system.

20.3.9 ALTERNATING ETHYLENE/NORBORNENE COPOLYMERS

A polymer composed of a strictly alternating sequence of norbornene and ethylene repeating units can be obtained through a reaction sequence that involves ROMP of *exo*-tricyclo[4.2.1.0^{2,5}]non-3-ene (*exo*-TCN) and subsequent hydrogenation of the polymer double bonds (Scheme 20.14).[62]

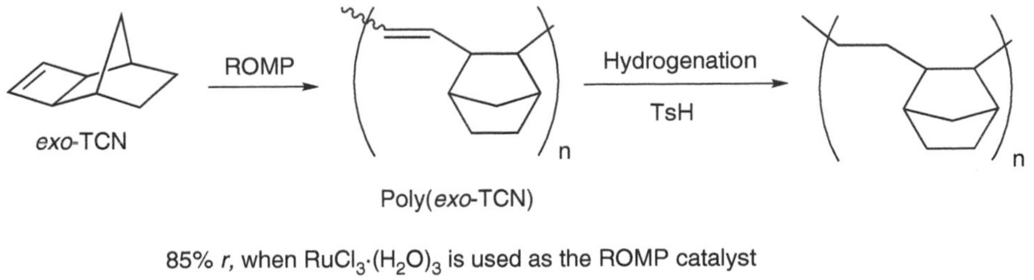

Poly(exo-TCN)

85% *r*, when RuCl$_3$·(H$_2$O)$_3$ is used as the ROMP catalyst

SCHEME 20.14 Strictly alternating ethylene-norbornene copolymers enriched in *r* dyads prepared through ROMP and subsequent hydrogenation (TsH = *p*-toluenesulfonhydrazide).

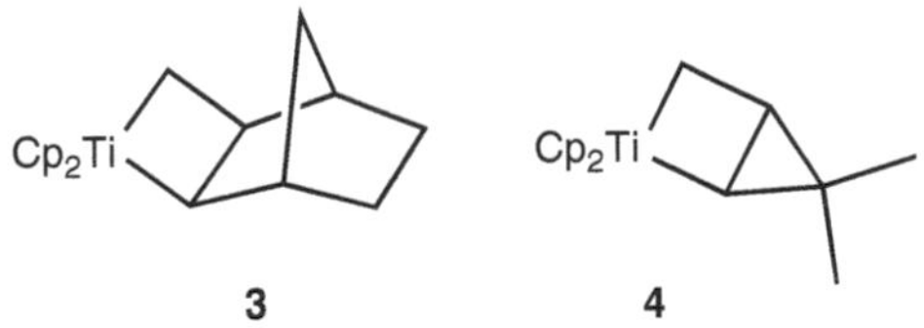

FIGURE 20.11 Titanacyclobutane compounds that initiate "living" ROMP (Cp = cyclopentadienyl).

This polymer is enriched in syndiotactic sequences (85% *r*) when RuCl$_3$·(H$_2$O)$_3$ is employed as the metathesis catalyst.

20.4 POLYMERIZATIONS WITH WELL-DEFINED METATHESIS CATALYSTS

Over the last two decades, several well-defined metallacyclobutanes and transition metal alkylidenes have been developed. They often catalyze metathesis polymerizations of norbornene-type monomers in a "living" fashion.[1–7] "Living" polymerizations are those that proceed with fast initiation, and generally, no chain termination and chain transfer occur, thus providing control of molar mass and molar mass distribution.[93] Some of these well-defined catalysts can be employed for stereoselective polymerizations, and they provide further insight into the mechanisms of stereocontrol in ROMP. The following sections describe the use of catalyst/monomer combinations in which the catalyst is a metallacyclobutane, a transition metal alkylidene compound, or a structurally well-defined transition metal complex that forms a ROMP-active transition metal alkylidene compound upon thermal activation. In the case of the well-defined ROMP catalyst precursors, polymerization often proceeds in a nonliving fashion when initiation is slow with regard to chain propagation or when the chain propagating alkylidene is either not sufficiently stable or not highly ROMP-selective.

20.4.1 TITANACYCLOBUTANE CATALYSTS

The first "living" metathesis polymerizations were established using metallacyclobutane complexes **3** and **4** as catalysts (Figure 20.11) and NB as the monomer.[63] The latter complex has been employed for the synthesis of moderately regular (80:20 trans:cis) poly(*anti*-7-methylnorbornene) (poly(*anti*-7-MeNB)) that contains a small excess of syndiotactic structures.[64] Trans double bonds are primarily associated with *r* dyads (*r:m* = 75:25), whereas cis double bonds are predominantly associated with *m* dyads. The overall *r:m* ratio of the hydrogenated derivative of the polymer (poly-H-(*anti*-7-MeNB), Figure 20.12) is 64:36.

These relationships of trans-racemic and cis-*meso* structures differ from those observed for ROMP products obtained with most conventional catalyst systems (described in the previous sections). For

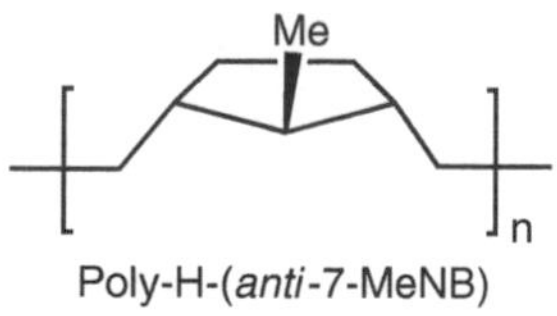

Poly-H-(*anti*-7-MeNB)

FIGURE 20.12   Hydrogenated poly(*anti*-7-methylnorbornene).

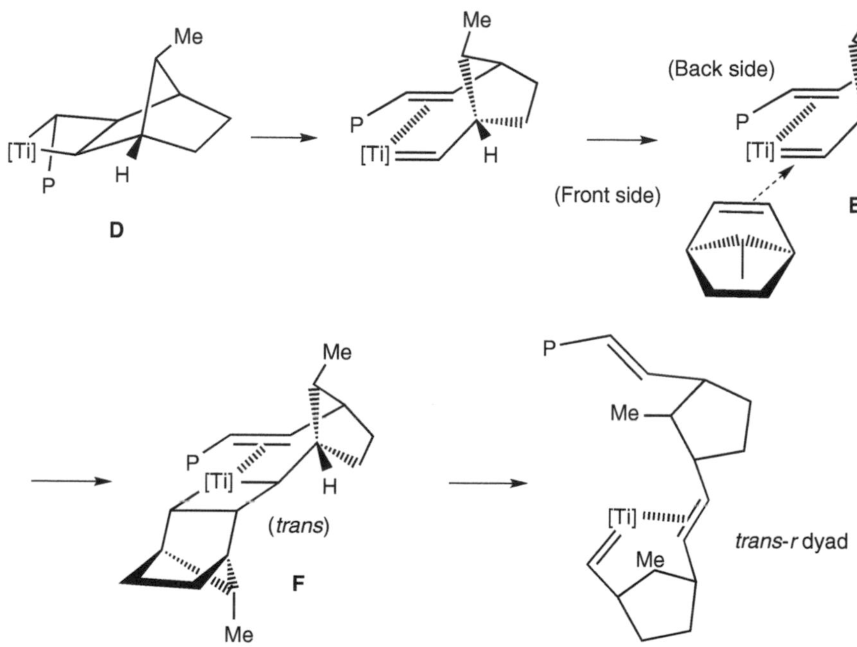

SCHEME 20.15   Coordination of the previously formed double bond to Ti gives a trans-racemic dyad when *anti*-7-methylnorbornene approaches the Ti=C face of **E** from the front side with its cyclic structure pointing down ([Ti] = Cp$_2$Ti, P = polymer chain, arrow used to indicate monomer approach).

conventional catalysts, cis C=C structures of ROMP products are often associated with syndiotactic sequences, and trans structures are present as either atactic or isotactic sequences. It is believed that in titanacycle-catalyzed ROMP (Scheme 20.15), the titanacycle **D** cleaves to give a titanium carbene intermediate, and the olefin bond formed can remain coordinated to the transition metal when the next monomer molecule is approaching (structure **E**).[64] The monomer advances toward the face of the titanium alkylidene that is opposite to the coordinated olefin unit (approach from the front side of **E**). If the monomer approaches the alkylidene in such a fashion that a trans double bond is formed, that is, the cyclic structure pointing down, away from the alkylidene α-substituent (i.e., the five-membered ring) which is pointing up, *trans*-titanacycle **F** is formed. This titanacycle subsequently cleaves to produce a trans-racemic dyad. When, sporadically, the 6-membered-ring of the monomer is pointing up instead (cis with respect to the five-membered ring linked to the alkylidene carbon), a cis-*meso* dyad is formed. It is assumed that occasionally the previously formed C=C double bond is not bound to Ti, and then the monomer can approach both faces of the alkylidene (front side and back side). Under those circumstances, cis and trans olefin units with both *meso* and racemic stereochemistry can be formed.

Trans/cis olefin ratios were also determined for poly(1-MeNB) and poly(NB) prepared with **3** and **4**, respectively.[64] In the first case, the trans/cis ratio is between 90:10 and 95:5, and in the latter

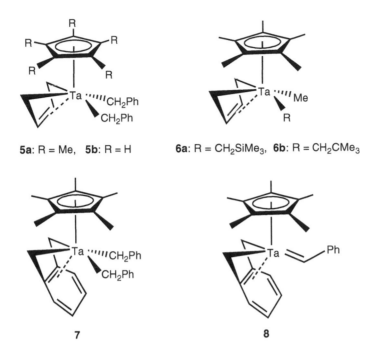

5a: R = Me, 5b: R = H   6a: R = CH$_2$SiMe$_3$, 6b: R = CH$_2$CMe$_3$

7   8

FIGURE 20.13   Ta catalysts for cis and trans selective ROMP of norbornene.

case, the trans/cis ratio is 64:36. Poly(1-MeNB) obtained with **3** contains a significant amount of head–head and tail–tail structures, with (HT + TH)/(HH + TT) = 2.6:1.

20.4.2 TANTALUM-BASED CATALYSTS

Tantalum complexes **5–7** (Figure 20.13) catalyze the polymerization of NB and *anti*-7-MeNB when they are heated above ambient temperature.[66] It is believed that thermolysis leads to the formation of metathesis-active tantalum alkylidene compounds. Dibenzyltantalum complex **5a** acts as a cis-isospecific ROMP catalyst for *anti*-7-MeNB, producing poly(*anti*-7-MeNB) with 99% cis olefins and 88% *m* dyads when polymerization is conducted at ≥45 °C.[65] By comparison, poly(NB) prepared with **5a** is high-cis (97%–98%) but atactic (55%–57% *m*).[66] Ta complexes **6a** and **6b** also produce poly(NB) with a cis content of 97%–98%, whereas *o*-xylylene complexes **7** and **8** (Figure 20.13) produce high-trans (90%–95%) poly(NB). Ta complex **5b** (bearing a cyclopentadienyl ligand instead of pentamethylcyclopentadienyl as in **5a**) catalyzes the polymerization of norbornene in a nonselective manner. The resulting polymer contains 50% cis double bonds, which contrasts with the formation of high-trans polymer when *anti*-7-MeNB is polymerized using **5b**. All of the polymers synthesized with Ta complexes **5–8** display moderately broad molar mass distributions, and are therefore not the result of "living" polymerizations.

20.4.3 RUTHENIUM ALKYLIDENE CATALYSTS

The ruthenium alkylidene complexes **9–11** (Figure 20.14) combine high metathesis activity with remarkable tolerance of functional groups, and have found significant use in a wide range of olefin metathesis reactions.[1,5] The polymerization of NB and NBD with Grubbs' Ru catalyst **9** is not very selective with regard to tacticity. While both polymers display only a very small excess of *m* dyads and can be classified as atactic,[67] they contain 83% and 86% trans double bonds, respectively.

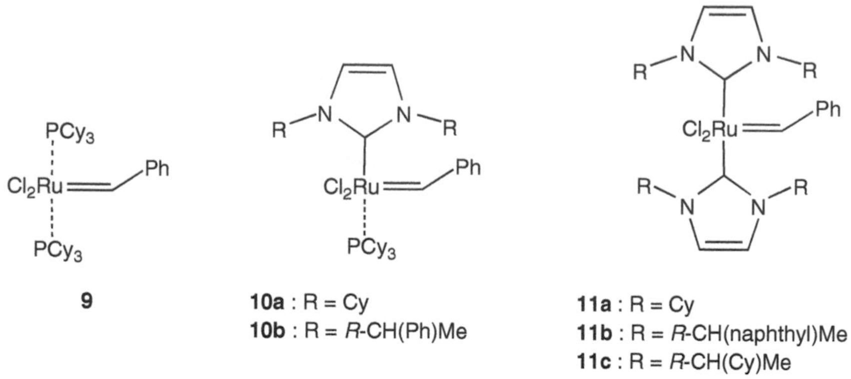

FIGURE 20.14   Ru benzylidene complexes that are metathesis catalysts.

By comparison, the polymerization of *anti*-7-MeNB with **9** gives moderately isotactic poly(*anti*-7-MeNB). This polymer contains between 65% and 80% *meso* structures for both cis- and trans-enchained dyads, the trans/cis ratio being 80:20.[67]

The trans C=C content of poly(7-MeNBD) synthesized with **9** is similar to that for poly(*anti*-7-MeNB), and is reported to range from 71% to 83%.[67–69] However, this polymer is atactic overall with 55% of the trans- and 62% of the cis-enchained dyads displaying racemic geometry.[68] This contrasts with the moderately isotactic structure of poly(*anti*-7-MeNB) obtained with this catalyst.

In Ru benzylidene catalysts **10–11**, one or both of the PCy₃ ligands of **9** have been exchanged for *N*-heterocyclic carbene ligands, respectively. The trans selectivity of **10a,b** and **11a-c** in the ROMP of *anti*-7-MeNB ranges from 44% to 66%, which is lower than when **9** is used. Catalyst **11a** produces poly(*anti*-7-MeNB) with 66% trans C=C enchained dyads that are predominantly syndiotactic,[69] which differs from the microstructure of poly(*anti*-7-MeNB) prepared with **9**. Poly(*anti*-7-MeNB) prepared with **10a,b** and **11b,c** is also reported to contain *trans* C=C linked units that are predominantly syndiotactic. A moderate excess of isotactic dyads is detected for the cis C=C-enchained units of poly(*anti*-7-MeNB) produced with all of **9–11**.[69]

Poly(7-MeNBD) prepared with **10b** and **11a** contains between 47% and 52% trans C=C units, which are atactic.[69] It is reported that the cis sequences in these samples comprise more *r* than *m* dyads. A model has been proposed suggesting that the change in polymer tacticity from syndiotactic to atactic is due to switching from a mode of coordination of the previously formed C=C double bond (similar to structure **E** in Scheme 20.15) to a mode of noncoordination.

20.4.4   TUNGSTEN-BASED CATALYSTS

Tungsten alkylidene complex **12** (Figure 20.15) bears the same binaphtholate ligand as tungsten complex **1a**. However, the polymerization of 5,5-Me₂NB by **12**/GaBr₃ produces an atactic polymer (48% *r*) that contains 41% cis double bonds.[47] This contrasts with the high-cis, syndiotactic poly(5,5-Me₂NB) formed with **1a**/Et₂AlCl. The cis content of poly(NB) prepared with **12**/GaBr₃ is lower than that in similarly prepared poly(5,5-Me₂NB) and is reported to be 24%.

Similar to most conventional catalysts, tungsten alkylidene complex **12** (without GaBr₃) selectively polymerizes the anti isomer of 7-MeNB from a 1:1 mixture of the syn and anti isomers. The resulting poly(*anti*-7-MeNB) contains less than 5% 7-*syn*-MeNB incorporated. This polymer is atactic and contains 41% trans C=C structures.[70] The hydrogenated polymer (poly-H-(*anti*-7-MeNB)) is semicrystalline despite its atactic stereochemistry. This is similar to the partial crystallinity of the atactic hydrogenated polynorbornene previously described. The DSC curve for poly-H-(*anti*-7-MeNB) displays an endothermic melt transition with a peak maximum at $T_m = 176\ °C$ ($\Delta H = 26\ J/g$) (there is an additional minor peak at 137 °C). This result shows that partial

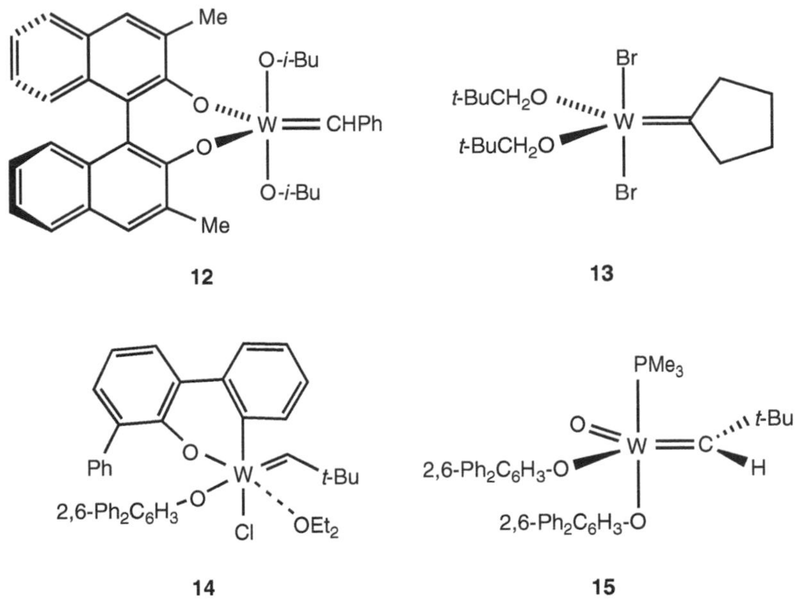

FIGURE 20.15 Tungsten alkylidene ROMP catalysts.

crystallinity is not necessarily associated with high tacticity in hydrogenated metathesis polymers prepared from norbornene-type monomers.

Tungsten catalyst **13** (Figure 20.15) becomes highly ROMP active upon addition of $GaBr_3$, and syn-7-MeNB also undergoes polymerization at temperatures above $-38\,°C$. Poly(syn-7-MeNB) is formed from isomerically pure syn-7-MeNB and **13**/$GaBr_3$ and contains 90% trans C=C units.[70] The ^{13}C NMR spectrum does not show fine structure beyond trans/cis isomerism that indicates that this polymer is highly tactic. Ivin et al. suggested that it is isotactic on the basis of the mechanistic model presented in Scheme 20.6, but this has not yet been confirmed. The DSC curve (second heat cycle) of hydrogenated poly-H-(syn-7-MeNB) shows two endotherms, with peak maxima at $T_1 = 179\,°C$ ($\Delta H = 13\,J/g$) and $T_2 = 230\,°C$ ($\Delta H = 26\,J/g$). Synthesis of a block copolymer, poly(anti-7-MeNB)-block-poly(syn-7-MeNB), is achieved by polymerization of a 1:1 mixture of the anti and syn isomers of 7-MeNB by **13**, initially with no $GaBr_3$ present and subsequently with $GaBr_3$ added to the polymerization mixture after all of the anti-7-MeNB has been consumed.[70]

A moderately stereoregular all-head-tail poly(1-MeNB) is obtained with tungsten alkylidene **14** (Figure 20.15). This metathesis polymer is high-cis and slightly syndiotactic (ca. 70% r).[53] This microstructure contrasts with the high-trans, predominantly head–tail atactic stereochemistry of poly(1-MeNB) prepared using $OsCl_3$.[45]

Tungsten oxo alkylidene complex **15** (Figure 20.15) is reported to polymerize both 2,3-bis(trifluoromethyl)norbornadiene (2,3-$(CF_3)_2$NBD, Figure 20.16) and 2,3-dicarbomethoxynorbornadiene (2,3-$(CO_2Me)_2$NBD, Figure 20.9) in a very stereoselective manner.[71] The resulting poly(2,3-$CO_2Me)_2$NBD) and poly(2,3-$(CF_3)_2$NBD) are high-cis and isotactic (>95% cis; >95% m). This is suggested to result from a chain-end control mechanism (vide infra, Section 20.4.5.2). Further, the molar mass distributions of these products obtained with **15** are narrow, as the polymerizations proceed in a living fashion.

20.4.5 Imido Molybdenum Alkylidene Complexes

Schrock et al. developed and explored a wide range of achiral and chiral molybdenum alkylidene compounds that can be employed as well-defined ROMP catalysts.[1,2,4,7,20,72–74] Some examples of

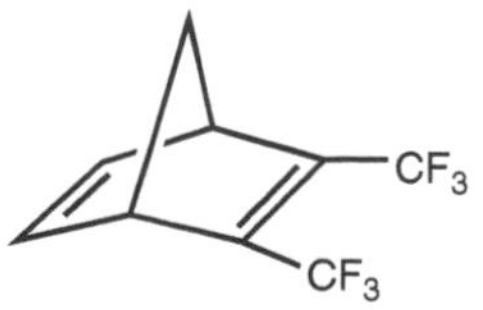

2,3-(CF$_3$)$_2$NBD

FIGURE 20.16 2,3-Bis(trifluoromethyl)norbornadiene.

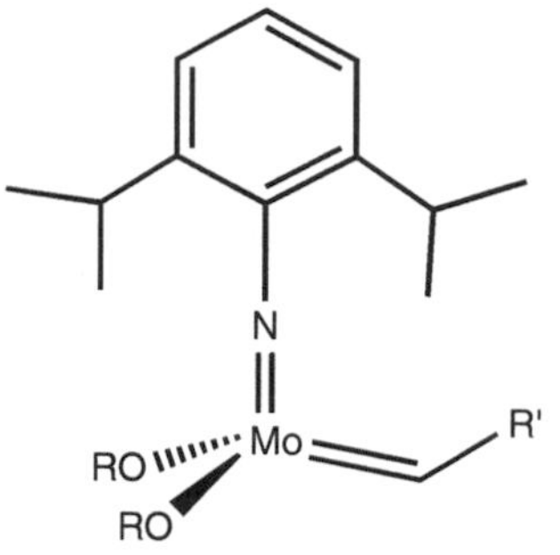

16a: R = R' = t-Bu

16b: R = t-Bu, R' = CMe$_2$Ph

16c: R = CMe$_2$(CF$_3$), R' = t-Bu

16d: R = CMe$_2$(CF$_3$), R' = CMe$_2$Ph

16e: R = CMe(CF$_3$)$_2$, R' = t-Bu

16f: R = CMe(CF$_3$)$_2$, R' = CMe$_2$Ph

16a–f

FIGURE 20.17 Well-defined imido molybdenum alkylidene ROMP catalysts.

achiral Mo alkylidenes are compounds **16a–f** (Figure 20.17). These are pseudotetrahedral complexes that are stabilized by bulky arylimido and alkoxide ligands. The alkylidene carbon generally bears a relatively large α-substituent (R'). The transition metal center becomes more electrophilic upon increasing the electron withdrawing character of the alkoxide ligand, and the metathesis activity increases substantially when changing from the *tert*-butoxide ligand in **16a,b** to the hexafluoro-*tert*-butoxide ligand in **16e,f**. Changes in the size and electron withdrawing nature of the ligands coordinated to the transition metal are also often accompanied by a significant change in the stereochemistry of the ROMP products. Complexes **16a–f** have been employed in numerous metathesis polymerization systems. They provide living metathesis polymerizations of strained bicyclic olefins, and they display reasonably good functional group tolerance (even though they react with oxygen and water).

20.4.5.1 Cis/Trans Selectivity in Polymerizations Catalyzed by Imido Molybdenum Alkylidene Complexes

Feast and Gibson et al. joined Schrock et al. in extensive studies of the stereocontrol that can be exerted with Mo-based ROMP catalysts/initiators.[75,76] 2,3-(CF$_3$)$_2$NBD and 2,3-dicarboalkoxynorbornadienes (e.g., 2,3-(CO$_2$Me)$_2$NBD) are moderately reactive monomers that were found to be well suited for investigations of the stereoselectivity of Mo alkylidene catalysts/initiators. Polymerization of 2,3-(CF$_3$)$_2$NBD by *tert*-butoxide-based Mo alkylidene **16a** produces high-trans poly(2,3-(CF$_3$)$_2$NBD) (≥98% trans),[75] and ROMP of 2,3-(CF$_3$)$_2$NBD with the hexafluoro-*tert*-butoxide based Mo alkylidene **16e** yields a high-cis polymer (≥97% cis).[76] A mechanism of cis/trans selectivity in these Mo alkylidene-catalyzed metathesis polymerizations has been proposed that is based on detailed kinetic investigations.[77] In contrast to most conventional

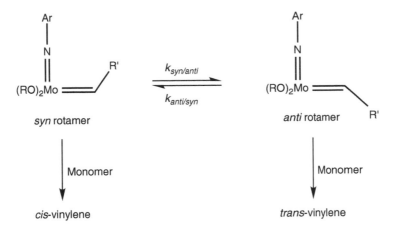

SCHEME 20.16 *Syn* and *anti* rotamers of Mo alkylidenes **16a–f** in chain propagation of ROMP (Ar = 2,6-*i*-Pr$_2$C$_6$H$_3$).

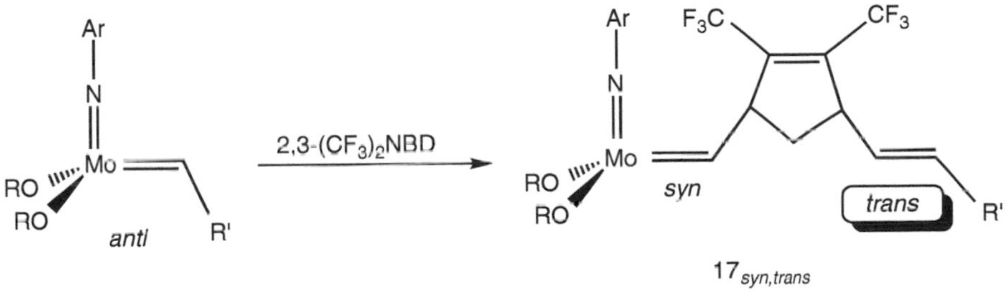

SCHEME 20.17 Reaction of the *anti* rotamer of hexafluoro-*tert*-butoxide-based Mo alkylidenes with the bicyclic olefin 2,3-(CF$_3$)$_2$NBD produces a trans C=C linkage (R = CMe(CF$_3$)$_2$, R$'$ = *t*-Bu or CMe$_2$Ph, Ar = 2,6-*i*-Pr$_2$C$_6$H$_3$).

metathesis catalysts, Mo alkylidenes **16a–f** are well-defined, and thus are well suited for mechanistic studies through NMR analysis.

Two rotameric forms exist of the initiating species **16a–f** and the subsequently formed propagating alkylidenes, which produce characteristic ^{1}H NMR signals in the range of δ 10–14 ppm.[75,78,79] The alkylidene substituent R$'$ is *syn* with regard to the imido ligand NAr in one of these isomers, and *anti* in the other isomer (Scheme 20.16). It was suggested by Schrock and Oskam that the *syn* rotamer is responsible for producing cis double bonds, and that the *anti* form leads to the formation of trans olefin structures, upon reaction with cyclic olefin monomers.[77] A mechanism for the overall cis/trans stereocontrol in ROMP with **16a–f** was proposed, which is based on kinetic data of *syn/anti* isomerization compared with the metathesis activity of each of the rotameric forms as outlined in the following discussion.

The *anti/syn* ratio can vary over several orders of magnitude depending on the alkylidene substituent R$'$ (i.e., on the monomer used). The *syn* rotamer is usually the predominant isomer, and in one of the prominent polymerization systems, that is, the polymerization of 2,3-(CF$_3$)$_2$NBD, the *anti/syn* ratio is smaller than 10^{-2}:1. This means that it can be difficult to detect the *anti* rotamer when employing common NMR methods; however, in oligomers obtained by the reaction of 2,3-(CO$_2$Me)$_2$NBD with catalyst **16a**[75] the *anti/syn* ratio is approximately 0.2:1, and both rotamers are observed by NMR. In complexes **16a–f** and in the reaction products of **16a–f** with 2,3-(CF$_3$)$_2$NBD, for example, **17**$_{syn,trans}$ and **17**$_{syn,cis}$ (Schemes 20.17 and 20.18), the concentration of the energetically less favorable *anti* rotamer is very small. The *anti/syn* rotamer ratio is substantially increased

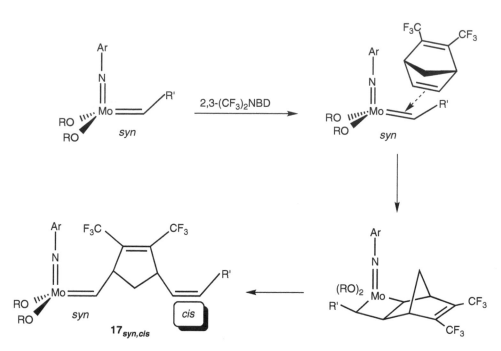

SCHEME 20.18 Suggested mechanism for the addition of 2,3-(CF$_3$)$_2$NBD to the *syn* rotamer of hexafluoro-*tert*-butoxide-based Mo alkylidenes producing a cis C=C linkage (R = CMe(CF$_3$)$_2$, R' = *t*-Bu or CMe$_2$Ph, Ar = 2,6-*i*-Pr$_2$C$_6$H$_3$, arrow used to indicate monomer approach from the back side).

(to approximately 0.5:1) upon low temperature photolysis.[77] Afterwards, NMR signals of the *anti* rotamer are clearly visible, and both the rate of the *anti* to *syn* rotamer transformation and the rate of metathesis reaction can be determined by low temperature NMR studies. The *anti* rotamer **16f**$_{anti}$ is able to react with 2,3-(CF$_3$)$_2$NBD at −80 °C in toluene as the solvent. The insertion product **17**$_{syn,trans}$ contains a trans C=C structure (Scheme 20.17), and is formed at the same rate that **16f**$_{anti}$ is consumed. The *syn* rotamer **16f**$_{syn}$ and the newly formed **17**$_{syn,trans}$ do not undergo reaction at this temperature; however, they do slowly react with excess 2,3-(CF$_3$)$_2$NBD monomer at temperatures above −40 °C. This indicates that the *anti* species react with 2,3-(CF$_3$)$_2$NBD at least two orders of magnitude faster than the *syn* rotameric forms of the initiator and chain propagation species.

The first insertion product obtained from the *syn* rotamer **16f**$_{syn}$ (above −40 °C) contains exclusively a cis olefin structure (**17**$_{syn,cis}$ in Scheme 20.18). The interconversion of *syn* to *anti* rotamer for **16f** in the temperature range of −40 to 25 °C is very slow. This means that the more stable *syn*-Mo=CHR' structure dominates the chain propagation of poly(2,3-(CF$_3$)$_2$NBD) polymerized by **16f**, even though the *anti* form is more reactive. Propagation by *syn*-Mo=CHR' is faster than rotamer interconversion, and the second and subsequent insertions of 2,3-(CF$_3$)$_2$NBD predominantly involve the *syn* rotamer, and high-cis poly(2,3-(CF$_3$)$_2$NBD) is formed. A mechanism has been suggested by Schrock and Oskam in which the bicyclic monomer approaches a CNO face of the chain propagating alkylidene, whereby C, N, and O represent the alkylidene carbon, the imido nitrogen, and one of the two alkoxide oxygen atoms, respectively.[77] The monomer C=C and catalyst Mo=C bonds are approximately parallel to one another and the methylene bridge points toward the aryl ring of the imido ligand (Scheme 20.18). It is proposed that a cis-metallacyclobutane intermediate is formed and is subsequently cleaved, extending the growing polymer chain by one repeat unit. In this fashion, a *cis*-vinylene and a *syn* alkylidene structure are produced.

Depending on the electron withdrawing character of the alkoxide ligand present in the Mo catalyst, the rate of rotamer interconversion can vary by up to six orders of magnitude. It is very

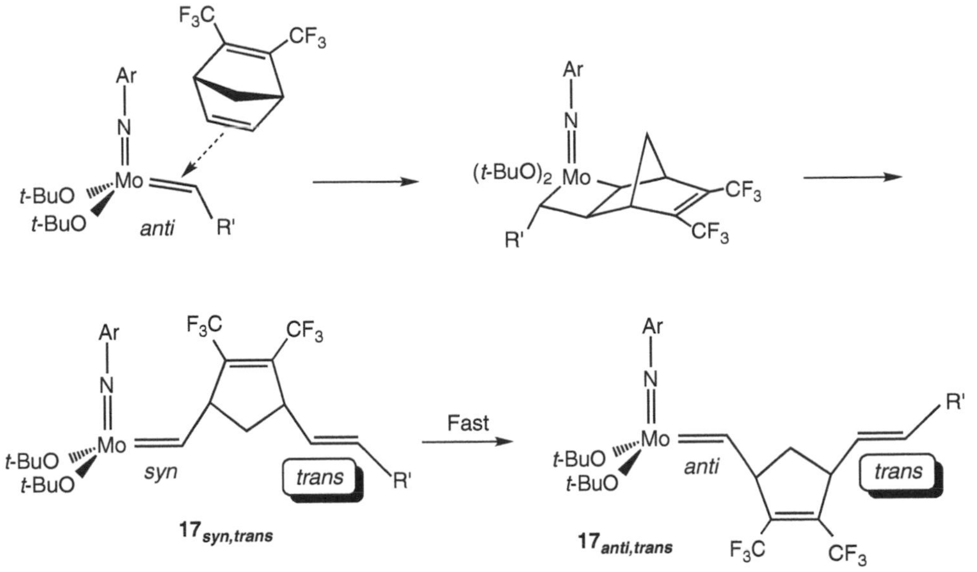

SCHEME 20.19 Suggested mechanism for the addition of 2,3-(CF$_3$)$_2$NBD to the *anti* form of *tert*-butoxide-based Mo alkylidenes producing a trans C=C linkage (R′ = *t*-Bu or CMe$_2$Ph, Ar = 1 + [0 − 9]2,6-*i*-Pr$_2$C$_6$H$_3$, arrow used to indicate monomer approach from the back side).

slow for the hexafluoro-*tert*-butoxide-based catalysts **16e,f** with $k_{syn/anti} = 7 \times 10^{-5}$/s at 25 °C (in toluene), but it is very fast for alkylidenes **16a,b** which bear nonfluorinated *tert*-butoxide ligands ($k_{syn/anti}$ represents the rate constant for the conversion of the *syn* rotamer into the *anti* rotamer as shown in Scheme 20.16; $k_{anti/syn}$ represents the opposite process).[77] For the latter catalysts (**16a,b**), $k_{syn/anti}$ is approximately 1/s at 25 °C. It is suggested that in polymerizations of 2,3-(CF$_3$)$_2$NBD catalyzed by **16a,b**, the *anti* rotamer is dominating in the chain propagation process, even though its concentration is very small (for **16b**, the ratio of $k_{syn/anti}/k_{anti/syn}$ in Scheme 20.16 is 1400:1, and the ratio of $k_{syn/anti}/k_{anti/syn}$ for the active Mo alkylidene end groups of growing oligo- and poly(2,3-(CF$_3$)$_2$)NBD) is larger than 100:1). In these polymerizations catalyzed by **16a,b**, conversion of the *syn* to the *anti* rotamer is faster than propagation (Scheme 20.19), whereas in the previously described polymerizations catalyzed by **16e,f**, propagation is faster than rotamer isomerization. It is assumed (Scheme 20.19) that 2,3-(CF$_3$)$_2$NBD approaches the chain propagating *anti* form of the alkylidene with the CH$_2$ bridge directed toward the arylimido ligand (as in the case of **16e,f** presented in Scheme 20.18). A trans-metallacyclobutane structure is formed (Scheme 20.19), which subsequently cleaves to produce a trans-vinylene structure and a new Mo alkylidene, which has the *syn* form **17**$_{syn,trans}$. Conversion of the *syn* to the *anti* rotamer **17**$_{anti,trans}$ (Scheme 20.19) is rapid relative to chain propagation, and the next molecule of 2,3-(CF$_3$)$_2$NBD is very likely to react with the *anti* form **17**$_{anti,trans}$ to give again a *trans*-vinylene structure.

The cis/trans ratio of poly(2,3-(CF$_3$)$_2$NBD) prepared with Mo alkylidenes bearing the hexafluoro-*tert*-butoxide ligand (**16e,f**) depends on the reaction temperature. When the reaction temperature is increased from 25 °C to 65 °C, the cis content of poly(2,3-(CF$_3$)$_2$NBD) prepared with **16f** in toluene drops from 97% to 88%.[80] This is consistent with a greater participation of the *anti* rotamer at the higher temperature. The change in cis/trans selectivity is more pronounced when Mo(=N-2-*t*-BuC$_6$H$_4$)(=CHCMe$_2$Ph)[OCMe(CF$_3$)$_2$]$_2$ (**18**, Figure 20.18) is used as the catalyst/initiator. In this complex (bearing one *tert*-butyl N-aryl ortho-substituent), the *syn/anti* rotamer interconversion is approximately three orders of magnitude faster than for complexes **16e,f** (bearing two isopropyl N-aryl ortho-substituents). The metathesis polymerization of 2,3-(CF$_3$)$_2$NBD with **18** must be carried

FIGURE 20.18 *tert*-Butylphenylimido molybdenum alkylidene catalyst ($R = CMe(CF_3)_2$).

TABLE 20.2

Cis Olefin Content in Poly(2,3-bis(trifluoromethyl)norbornadiene) (poly(2,3-(CF$_3$)$_2$NBD)) Samples Prepared at Different Polymerization Temperatures Employing Mo Alkylidenes with Different Arylimido Ligands[a]

Catalyst	Arylimido Ligand[b]	Alkoxide Ligand	$T_p(°C)$[c]	cis Content (%)
16f	NAr	$(CF_3)_2MeC-O$	25	97
16f	NAr	$(CF_3)_2MeC-O$	65	88
18	NAr*	$(CF_3)_2MeC-O$	−35	97
18	NAr*	$(CF_3)_2MeC-O$	23	78
18	NAr*	$(CF_3)_2MeC-O$	65	29
16d	NAr	$(CF_3)Me_2C-O$	−35	56
16d	NAr	$(CF_3)Me_2C-O$	25	36
16d	NAr	$(CF_3)Me_2C-O$	65	9

[a] Reference 80.
[b] $Ar = 2,6$-i-$Pr_2C_6H_3$, $Ar^* = 2$-t-BuC_6H_4.
[c] Solvent = toluene.

out at low temperature (−35 °C) in order to obtain poly(2,3-(CF$_3$)$_2$NBD) with a high cis olefin content (97%). The cis content drops to 78% at 25 °C and to 29% at 65 °C in toluene (Table 20.2).[80]

The cis-vinylene content is also significantly reduced in poly(2,3-(CF$_3$)$_2$NBD) prepared with the trifluoro-*tert*-butoxide-based catalyst **16d**. The *syn/anti* rotamer interconversion in **16d** is approximately two orders of magnitude faster than that for hexafluoro-*tert*-butoxide-based catalyst **16f**. This implies that the *anti* rotamer is more accessible for **16d**, and the cis content of poly(2,3-(CF$_3$)$_2$NBD) prepared with **16d** decreases from 56% at −35 °C to 36% at 23 °C and to 9% at 65 °C (Table 20.2).

The *cis*-vinylene content of poly(2,3-(CF$_3$)$_2$NBD) can be varied between 2% and 98% by employing mixtures of **16a** and **16e** at different ratios.[76] Exchange of the alkoxide ligands between the two different catalyst species is fast compared to chain propagation, and polymer chains of uniform cis/trans composition are formed. For example, the cis content of poly(2,3-(CF$_3$)$_2$NBD) is approximately 30% when an equimolar ratio of **16a** and **16e** is used for polymerization at 20 °C.

Stereoblock copolymers comprised of a sequence of high-trans poly(2,3-(CF$_3$)$_2$NBD) followed by a sequence of high-cis poly(2,3-(CF$_3$)$_2$NBD) are obtained when ROMP is carried out initially with **16a** and the *tert*-butoxide ligands are subsequently replaced by hexafluoro-*tert*-butoxide before

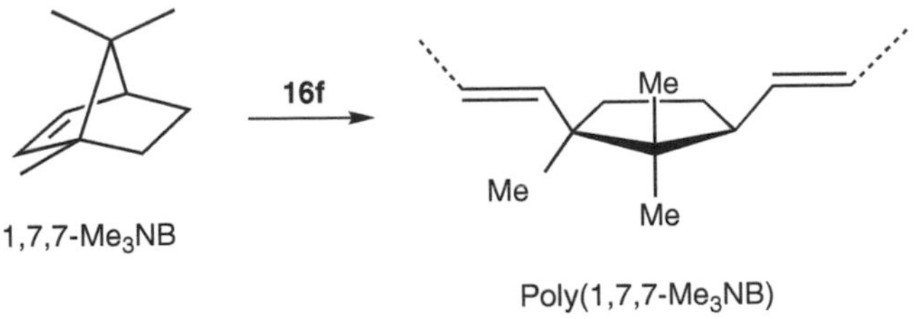

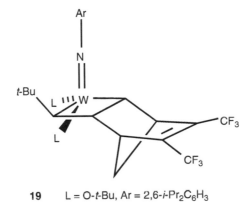

SCHEME 20.20 Polymerization of 1,7,7-trimethylnorbornene.

FIGURE 20.19 Tungstacyclobutane derived from 2,3-bis(trifluoro)norbornadiene.

the addition of further monomer.[81] This AB-type block copolymer displays phase separation. Thus, two glass transition temperatures ($T_{g,1}$ and $T_{g,2}$) are recorded at 95 and 145 °C by DSC. Conformational mobility in the high-cis polymer is more restricted than in the high-trans polymer, and accordingly, the T_g of cis-enchained polymer is higher.

The same trend, an increased T_g for a high-cis polymer as compared to the corresponding trans polymer, is observed for poly(*endo,exo*-5,6-Me$_2$NB) (Scheme 20.13): T_g = 85 or 79 °C for poly(*endo,exo*-5,6-Me$_2$NB) with 85% cis (made using **16e**), versus T_g = 55 °C for poly(*endo,exo*-5,6-Me$_2$NB) containing 5% cis units (made using **16a**).[82]

The cis/trans selectivity also depends on the steric bulk of the monomer. Racemic and enantiomeric 1,7,7-trimethylnorbornene (1,7,7-Me$_3$NB) monomers are very bulky and polymerize only very slowly with the highly active hexafluoro-*tert*-butoxide-based catalyst **16f** (Scheme 20.20). Polymerization using 34 equivalents of monomer requires a reaction time of several weeks at 40 °C in order to proceed to completion. The resulting poly(1,7,7,-Me$_3$NB) is an all-head-tail, all-trans polymer,[83] which contrasts with the high-cis structures of poly(2,3-(CF$_3$)$_2$NBD) and poly(*exo,endo*-5,6-Me$_2$NB) prepared with **16f**. At high monomer concentrations [M], the rate of polymerization is essentially independent of [M] with a rate constant of k = 6.1 × 10^{-5}/s at 20 °C. This value is very similar to $k_{syn/anti}$ = 7 × 10^{-5}/s determined for the *syn/anti* rotamer interconversion of **16f** at 25 °C. This means that the rate-determining step in the chain propagation of 1,7,7,-Me$_3$NB polymerization is the transformation of the *syn* alkylidene into the *anti* form, and thus confirms that the mechanism of cis/trans selectivity outlined in Schemes 20.16 through 20.19 applies to several ROMP monomer/catalyst combinations.

It is to be noted, however, that the existence of monomer/catalyst combinations in which the *syn* rotamer can also contribute to the formation of trans C=C structures, and the *anti* rotamer can produce some cis-vinylene units, can currently not be excluded. For example, tungstacyclobutane compound **19** (Figure 20.19) was obtained from the reaction of monomer 2,3-(CF$_3$)$_2$NBD with

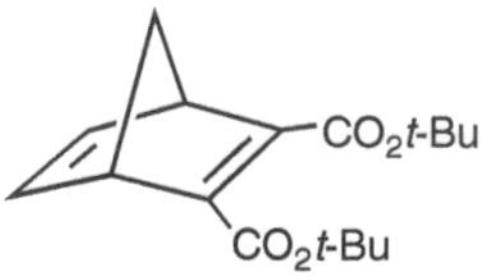

(CO$_2$t-Bu)$_2$NBD

FIGURE 20.20 2,3-Dicarbo-*tert*-butoxynorbornadiene.

W(=N-2,6-*i*-Pr$_2$C$_6$H$_3$)(=CHC-*t*-Bu)(O-*t*-Bu)$_2$ (which is the W-based equivalent of Mo alkylidene **16a**).[75] The *tert*-butyl substituent, which is cis with regard to the imido ligand, is trans with regard to the bicyclic ring system in **19**. Thus, 2,3-(CF$_3$)$_2$NBD approaches the *syn* rotamer of W=CH-*t*-Bu in a trans fashion with the methylene bridge pointing toward the alkoxide ligands rather than the imido ligand.

The cis/trans selectivity in Mo alkylidene-catalyzed polymerizations of 2,3-(CO$_2$Me)$_2$NBD and 2,3-(CF$_3$)$_2$NBD is similar. Poly(2,3-(CO$_2$Me)$_2$NBD) prepared with the *tert*-butoxide-based catalyst/initiator **16a** contains between 90% and 95% trans-vinylene units,[75] and poly(2,3-(CO$_2$Me)$_2$NBD) synthesized with the hexafluoro-*tert*-butoxide-based catalyst **16f** is high-cis (98% *cis*-vinylene units).[80,84] Similarly, the metathesis polymers of 2,3-dicarboethoxy-, 2,3-dicarboisopropoxy-, and 2,3-dicarbo-*tert*-butoxynorbornadiene (2,3-(CO$_2$Et)$_2$NBD, 2,3-(CO$_2$*i*-Pr)$_2$NBD, and 2,3-(CO$_2$*t*-Bu)$_2$NBD, Figures 20.9 and 20.20) contain between 97% and 99% cis C=C structures when **16f** and related perfluoroalkoxy-based Mo alkylidenes are employed as catalysts.[80]

Furthermore, poly(*endo,exo*-2,3-dicarbomethoxynorborn-5-ene) (poly(*endo,exo*-2,3-(CO$_2$Me)$_2$ NB)), poly(*exo-cis*-2,3-dicarbomethoxynorborn-5-ene) (poly(*exo-cis*-2,3-(CO$_2$Me)$_2$NB)), and poly(*endo,exo*-2,3-dimethoxymethylnorborn-5-ene) (poly(*endo,exo*-2,3-(MeOCH$_2$)$_2$NB)) prepared with the *tert*-butoxide-based Mo complex **16b** are highly trans, that is, 90%, 95%, and 92% trans, respectively (Table 20.3),[85] which is similar to the trans content of poly(*endo,exo*-5,6-Me$_2$NB) prepared with **16b**. By comparison, the trans-vinylene content is lower in similarly prepared polymers of norbornene, benzonorbornadiene (BNBD), *exo-cis*-2,3-norbornenediol diacetate (*exo-cis*-2,3-(AcO)$_2$NB), and *exo-cis*-O,O′-isopropylidene-2,3-norbornenediol (*exo-cis*-2,3-(ketal)NB) (Table 20.3)[75,86] and ranges from 60% to 79%.

The *exo*- and *endo*-5-norbornene-2,3-dicarboximide compounds shown in Figure 20.21, which are derived from methyl esters of natural α-amino acids, undergo ROMP in the presence of Mo alkylidenes **16b**, **16d**, and **16f**.[87] In the case of the *exo*-dicarboximides *exo*-NB-2,3-(CO)$_2$N-Gly, *exo*-NB-2,3-(CO)$_2$N-Ala, and *exo*-NB-2,3-(CO)$_2$N-Ile, the *tert*-butoxide-derived Mo catalyst **16b** gives high-trans polymers (>90%), whereas the electron deficient hexafluoro-*tert*-butoxide catalyst **16f** produces polymers enriched in *cis*-vinylene units (70–80% cis). An intermediate *cis* olefin content is achieved with the Mo alkylidene **16d**, which contains only one trifluoromethyl substituent per alkoxy group. The *endo*-dicarboximides *endo*-NB-2,3-(CO)$_2$N-Gly, *endo*-NB-2,3-(CO)$_2$N-Ala, and *endo*-NB-2,3-(CO)$_2$N-Ile are generally less reactive than the exo derivatives and give high-trans polymers with all three Mo alkylidenes **16b**, **16d**, and **16f**.

In addition, *exo*-5-norbornene-2,3-dicarboximides with *n*-alkyl substituents form high-trans ROMP products (97% trans) with Mo compound **16b**, and form polymers with a lower trans content (approximately 30%) with Mo catalyst **16f**.[88] In the case of the corresponding *endo* isomers, a 93% *trans*-vinylene content is obtained with **16b** and a 57% *trans*-vinylene content is achieved with **16f**.

20.4.5.2 Tacticity in Polymerizations Catalyzed by Imido Molybdenum Alkylidene Complexes

High-trans poly(2,3-(CF$_3$)$_2$NBD) obtained with the *tert*-butoxide-based Mo catalysts **16a** and **16b** is highly syndiotactic (>90% *r*).[75,85] This was initially deduced from the high relaxed dielectric

TABLE 20.3

***trans*-Vinylene Content of ROMP Products of Norbornene and Norbornadiene Derivatives Obtained with *tert*-Butoxide-Based Mo Alkylidene Catalysts 16a,b[a]**

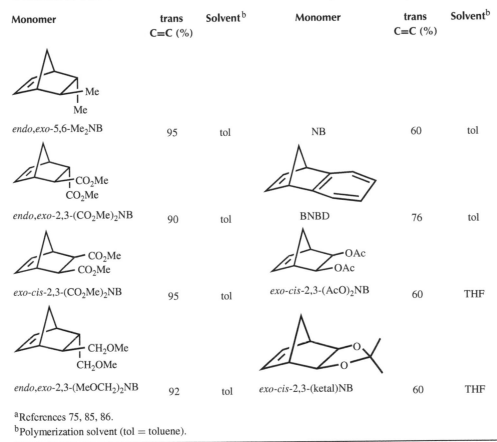

Monomer	trans C=C (%)	Solvent[b]	Monomer	trans C=C (%)	Solvent[b]
endo,exo-5,6-Me₂NB	95	tol	NB	60	tol
endo,exo-2,3-(CO₂Me)₂NB	90	tol	BNBD	76	tol
exo-cis-2,3-(CO₂Me)₂NB	95	tol	*exo-cis*-2,3-(AcO)₂NB	60	THF
endo,exo-2,3-(MeOCH₂)₂NB	92	tol	*exo-cis*-2,3-(ketal)NB	60	THF

[a]References 75, 85, 86.
[b]Polymerization solvent (tol = toluene).

exo-and endo-NB-2,3-(CO)₂N-Gly: R = H

exo- and endo-NB-2,3-(CO)₂N-Ala: R = Me

exo- and endo-NB-2,3-(CO)₂N-Ile: R = CHMeEt

exo-NB-2,3-(CO)₂N-R

endo-NB-2,3-(CO)₂N-R

FIGURE 20.21 *Exo-* and *endo*-5-Norbornene-2,3-dicarboximide monomers.

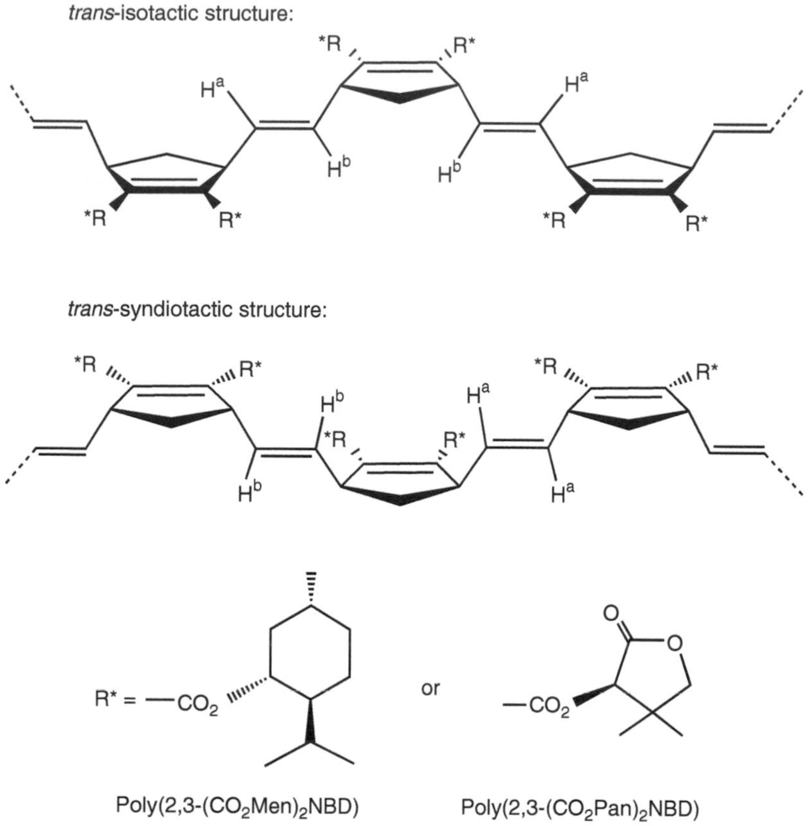

trans-isotactic structure:

trans-syndiotactic structure:

R* = —CO$_2$ or   —CO$_2$

Poly(2,3-(CO$_2$Men)$_2$NBD)   Poly(2,3-(CO$_2$Pan)$_2$NBD)

FIGURE 20.22 Olefin protons in trans-enchained ROMP poly(norbornadiene-2,3-dicarboxylate)s bearing chiral substituents R*.

constant of the polymer, which reflects a high dipole moment. The predominantly syndiotactic stereochemistry was subsequently confirmed by ^{1}H,^{1}H COSY NMR analysis of related metathesis polymers that contain chiral substituents.[85] These polymers, poly(2,3-dicarboalkoxynorbornadienes) poly(2,3-(CO$_2$Men)$_2$NBD) and poly(2,3-(CO$_2$Pan)$_2$NBD) (Figure 20.22), were prepared by ROMP of enantiomerically pure norbornadiene-2,3-dicarboxylic ester monomers bearing two chiral menthyl (2,3-(CO$_2$Men)$_2$NBD) and pantalactonyl ester substituents (2,3-(CO$_2$Pan)$_2$NBD), respectively. In the trans-isotactic sequence displayed in Figure 20.22, the two protons H^a and H^b within each vinylene unit are not equivalent owing to the chirality of substituent R*. Accordingly, two different olefin proton resonances would be recorded that would display correlation in the ^{1}H,^{1}H COSY NMR spectrum. However, ^{1}H,^{1}H COSY NMR spectra of the highly stereoregular trans polymers show that the corresponding two olefin protons are not coupled and therefore reside in the equivalent environment provided by the syndiotactic structure. These NMR results demonstrate that the metathesis polymers obtained from monomers 2,3-(CO$_2$Men)$_2$NBD and 2,3-(CO$_2$Pan)$_2$NBD using catalyst **16b** are syndiotactic.[85] The two protons H^a attached to the same vinylene unit in the trans-syndiotactic structure are equivalent, as shown in Figure 20.22. Protons H^b of the neighboring vinylene unit produce the second olefin proton resonance, and no coupling with H^a is detected.

High-cis (≥97% cis) poly(2,3-(CF$_3$)$_2$NBD) and poly(2,3-(CO$_2$Me)$_2$NBD) obtained with catalysts **16e** and **16f** are slightly isotactic (containing approximately 75% *meso* units).[85] This stereochemistry was again elucidated with the aid of 2D NMR spectroscopy, employing related highly stereoregular polymers of 2,3-(CO$_2$Men)$_2$NBD and 2,3-(CO$_2$Pan)$_2$NBD (>99% cis and

FIGURE 20.23 cis-Isotactic polymers derived from the chiral monomers 2,3-(CO$_2$Men)$_2$NBD and 2,3-(CO$_2$Pan)$_2$NBD (R* = CO$_2$Men or CO$_2$Pan).

>90% tactic). ^{1}H,^{1}H COSY NMR spectra reveal significant coupling between the relevant olefin protons H^a and H^b and confirm the cis-isotactic microstructure displayed in Figure 20.23.

A chain-end control mechanism has been proposed that can explain the stereoselectivity in metathesis polymerizations of trifluoromethyl- and carboalkoxy-susbstituted norbornadienes 2,3-(CF$_3$)$_2$NBD, 2,3-(CO$_2$Me)$_2$NBD, 2,3-(CO$_2$Men)$_2$NBD, and 2,3-(CO$_2$Pan)$_2$NBD catalyzed by achiral complexes **16a,b** and **16e,f**.[85] These pseudotetrahedral catalysts are characterized by two different CNO faces (whereby C, N, and O represent the alkylidene carbon, the imido nitrogen, and one of the two alkoxy oxygen atoms, respectively). If successive monomer additions/insertions occur from opposite CNO faces of the catalyst, back side followed by front side approach of the monomer (i.e., the monomer approaches the two CNO faces in an alternating fashion), a syndiotactic polymer is formed (Scheme 20.21). Polymers synthesized with catalysts **16a** and **16b** are found to be high-trans and highly syndiotactic.[85] After the first addition/insertion involving the *anti* rotamer (Scheme 20.21), the *syn* form of the Mo alkylidene is obtained, which rapidly interconverts into the *anti* rotamer, as previously established for *tert*-butoxide-based Mo alkylidenes **16a,b**. Subsequently, the monomer approaches the front side. In this fashion, it is always the −(R)C=C(R)− part of the previously inserted monomer unit that is presented to the incoming monomer. This means that the polymer chain end attached to the transition metal controls the polymer stereochemistry.

This chain-end control mechanism was found to be ineffective in the polymerizations of *endo,exo*-5,6-Me$_2$NB, *trans*-2,3-(CO$_2$Me)$_2$NB, and *trans*-2,3-(MeOCH$_2$)$_2$NB catalyzed by **16a,b**. The resulting polymers are highly trans but atactic.[56,82,85] In the case of 7-MeNBD, Mo compounds **16a,b** produce polymers that contain 68% cis-vinylene structures. It is reported that the cis-enchained units are highly syndiotactic and that dyads containing trans C=C structures are predominantly isotactic (82% *m*).[68] This result is markedly different from the stereochemistry results obtained for 2,3-(CF$_3$)$_2$NBD, 2,3-(CO$_2$Me)$_2$NBD, 2,3-(CO$_2$Men)$_2$NBD, and 2,3-(CO$_2$Pan)$_2$NBD using **16a,b**.

In contrast to stereoselective ROMP by *tert*-butoxide-based Mo alkylidenes **16a** and **16b**, the overall cis-selective hexafluoro-*tert*-butoxide-based Mo complexes **16e** and **16f** exert only moderate chain-end control in polymerizations of 2,3-(CF$_3$)$_2$NBD, 2,3-(CO$_2$Me)$_2$NBD, and *endo,exo*-5,6-Me$_2$NB. High-cis polymers are formed that are isotactic to a small degree (*mm* triad content ranges between 60% and 78%).[80,82,85] The *mm* triad content increases to 88% when (CO$_2$*t*-Bu)$_2$NBD, with bulky carbo-*tert*-butoxy substituents, is employed as the monomer.[80] As previously discussed, it is assumed that chain propagation catalyzed by **16e** and **16f** involves predominantly the *syn* rotamer, since *syn/anti* interconversions are very slow relative to chain propagation.[77] Schrock and coworkers proposed that the monomer approaches the same CNO face in each propagation step during the formation of the short to moderately long isotactic sequences (Scheme 20.22).[85]

However, a moderately syndiotactic poly(2,3-(CO$_2$Me)$_2$NBD) is obtained with the related catalyst Mo(=N-2,6-*i*-Pr$_2$C$_6$H$_3$)(=CHCMe$_2$Ph)[OC(CF$_3$)$_2$CF$_2$CF$_2$CF$_3$]$_2$ (**16g**, Figure 20.24).[80] The 77% *rr* triad content of this material contrasts with the isotactic structure of poly(2,3-(CO$_2$*t*-Bu)$_2$NBD) that is produced with the same catalyst. The latter polymer, bearing more bulky *tert*-butyl carboxylate substituents, contains 90% *mm* triads. This difference in tacticity cannot yet be explained. Poly(7-MeNBD) prepared with Mo compound **16f** contains 88% cis double bonds and is atactic.[68]

As noted previously, the commonly cis-selective catalyst **16f** produces an all-trans polymer of 1,7,7-Me$_3$NB (Scheme 20.20), because the steric bulk of the monomer causes the rate

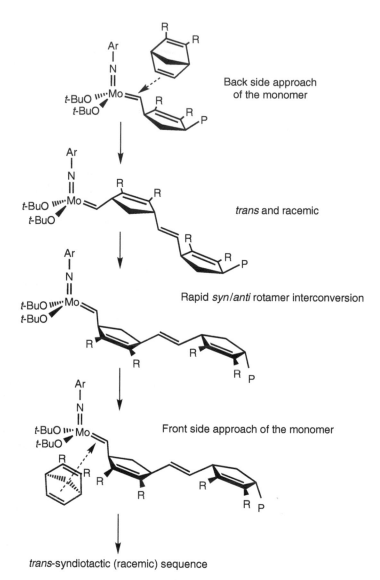

SCHEME 20.21   Proposed chain end control mechanism producing trans-syndiotactic polymers in ROMP of 2,3-disubstituted norbornadienes (Ar = 2,6-i-Pr$_2$C$_6$H$_3$, P = polymer chain; R = CF$_3$, CO$_2$Me, or chiral substituent R*; arrow indicates monomer approach) (Reference 85).

of chain propagation to be significantly reduced in comparison to the rate of *syn/anti* rotamer interconversion.[83] Chain propagation is proposed by Feast et al. to involve exclusively the *anti* rotamer and accordingly gives an all-trans polymer, which contrasts with the polymerizations of (CO$_2$Me)$_2$NBD and (CF$_3$)$_2$NBD using **16f**. Polymerization of 1,7,7-Me$_3$NB is head–tail selective and accordingly gives an isotactic polymer when a single enantiomer of 1,7,7-Me$_3$NB is employed. ROMP of a racemic mixture of 1,7,7-Me$_3$NB by **16f** leads to the formation of atactic polymer.

Polymerization of a single enantiomer of methyl-*N*-(1-phenylethyl)-2-azabicyclo[2.2.1]hept-5-ene-3-carboxylate (Scheme 20.23) by the hexafluoro-*tert*-butoxide-based Mo complex **16e** produces high-cis polymer (>98% cis) with an all-head-tail structure.[89] This product is isotactic as a result of the head–tail enchainment of enantiomerically pure monomer. By comparison, the all-head-tail polymer formed from a racemic mixture of this azabicyclic olefin monomer is atactic.

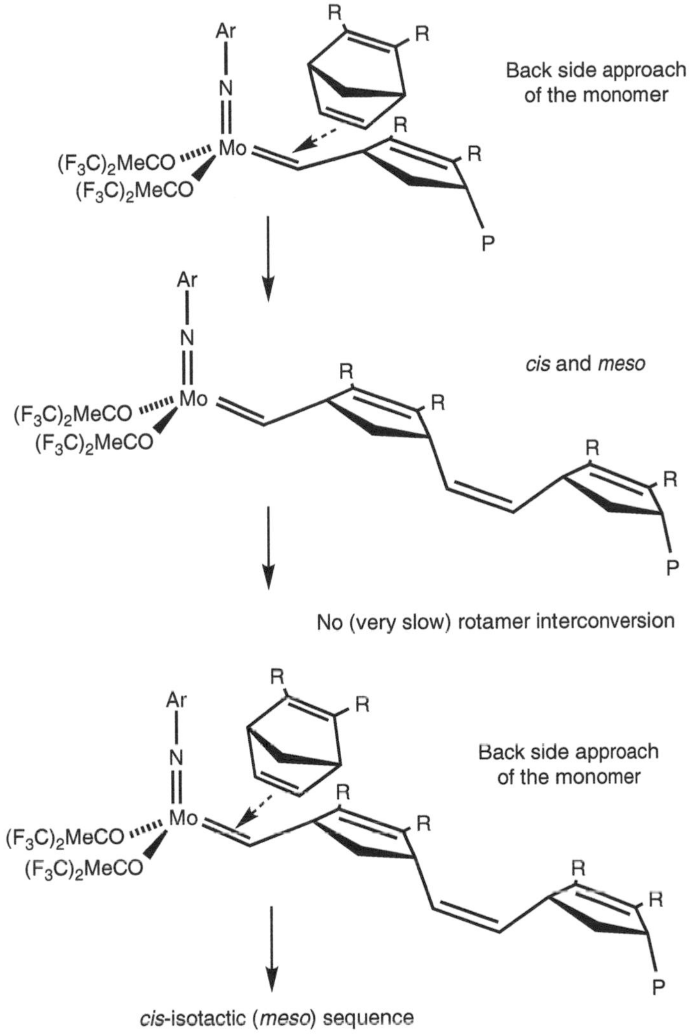

SCHEME 20.22 Partly stereoregular high-cis polymers obtained with an achiral Mo alkylidene complex bearing electron deficient alkoxide ligands (Ar = 2,6-i-Pr$_2$C$_6$H$_3$, P = polymer chain; R = CF$_3$, CO$_2$Me, or chiral R*; arrow indicates back side approach of the monomer). When successive monomers approach the same CNO face, isotactic sequences are obtained (Reference 85).

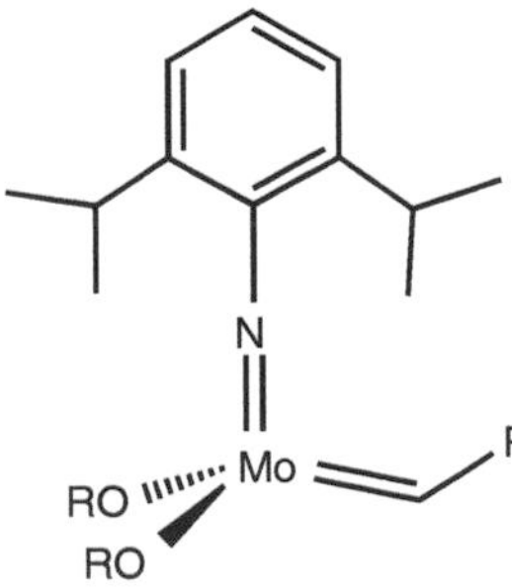

16g R = C(CF$_3$)$_2$CF$_2$CF$_2$CF$_3$, R' = CMe$_2$Ph

FIGURE 20.24 Imido molybdenum alkylidene ROMP catalyst with perfluoroalkoxide ligands.

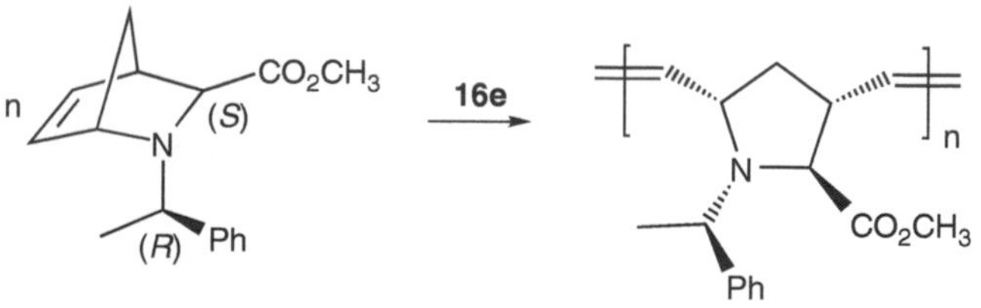
CO2CH3
n
(S)
N
(R)
Ph
16e
n
N
CO2CH3
Ph

SCHEME 20.23 ROMP of methyl-*N*-(1-phenylethyl)-2-azabicyclo[2.2.1]hept-5-ene-3-carboxylate.

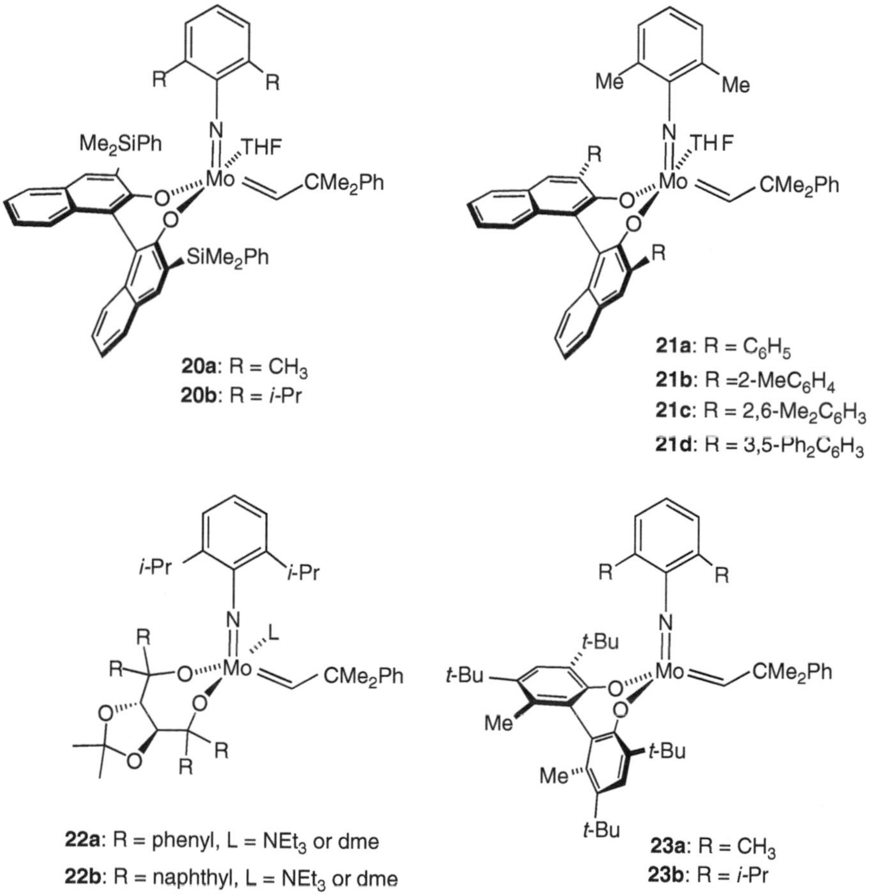
R
R
Me2SiPh
N
THF
Mo
CMe2Ph
O
O
SiMe2Ph
20a: R = CH3
20b: R = i-Pr
Me
Me
R
N
THF
Mo
CMe2Ph
O
O
R
21a: R = C6H5
21b: R =2-MeC6H4
21c: R = 2,6-Me2C6H3
21d: R = 3,5-Ph2C6H3
i-Pr
i-Pr
N
L
R
R
O
Mo
CMe2Ph
O
O
O
R
R
22a: R = phenyl, L = NEt3 or dme
22b: R = naphthyl, L = NEt3 or dme
R
R
N
t-Bu
t-Bu
O
Mo
CMe2Ph
Me
O
t-Bu
Me
t-Bu
23a: R = CH3
23b: R = i-Pr

FIGURE 20.25 Mo alkylidene complexes with chelating diolate ligands (dme = dimethoxyethane).

High-cis polymers with a consistently high degree of isotacticity can be obtained from a wide range of monomers by replacing the alkoxide ligands of **16e,f** with C_2-symmetric diolate ligands. Mo alkylidene complex **20a** (Figure 20.25), bearing the silyl-substituted chelating binaphtholate ligand (±)-BINO$(SiMe_2Ph)_2$, stands out as a catalyst/initiator that produces essentially all-cis (>99% cis) and highly isotactic poly(2,3-$(CF_3)_2$NBD) and poly(2,3-$(CO_2Me)_2$NBD) with >99% *meso* units (Table 20.4).[84,85] It has been proposed by Schrock and coworkers that the asymmetric configuration of the diolate ligand is responsible for the enhanced stereocontrol in these polymerizations. This is an enantiomorphic site control mechanism (Scheme 20.24).[85]

TABLE 20.4

Proposed Enantiomorphic Site Control Mechanism in the Polymerization of Norbornene and Norbornadiene Derivatives with Mo Alkylidene Complexes Mo(=CHCMe$_2$Ph)(=NAr)[(±)-BINO(SiMe$_2$Ph)$_2$](THF)[a]

Monomer	=NAr (catalyst)	cis Content (%)	m Content (%)
R = CF$_3$ (2,3-(CF$_3$)$_2$NBD)	=N-2,6-Me$_2$C$_6$H$_3$ (**20a**)	>99	>99
2,3-(CF$_3$)$_2$NBD	=N-2,6-i-Pr$_2$C$_6$H$_3$ (**20b**)	71	86
R = CO$_2$Me (2,3-(CO$_2$Me)$_2$NBD)	=N-2,6-Me$_2$C$_6$H$_3$ (**20a**)	>99	>99
2,3-(CO$_2$Me)$_2$NBD	=N-2,6-i-Pr$_2$C$_6$H$_3$ (**20b**)	93	97
2,3-(CO$_2$Men)$_2$NBD	=N-2,6-Me$_2$C$_6$H$_3$ (**20a**)	99	>90
2,3-(CO$_2$Pan)$_2$NBD	=N-2,6-Me$_2$C$_6$H$_3$ (**20a**)	99	>90
R^1 = Me (*endo,exo*-5,6-Me$_2$NB)	=N-2,6-Me$_2$C$_6$H$_3$ (**20a**)	ca. 95	ca. 95
R^1 = CO$_2$Me (*endo,exo*-2,3-(CO$_2$Me)$_2$NB)	=N-2,6-Me$_2$C$_6$H$_3$ (**20a**)	99	93
R^1 = MeOCH$_2$ (*endo,exo*-2,3-(MeOCH$_2$)$_2$NB)	=N-2,6-Me$_2$C$_6$H$_3$ (**20a**)	99	91

[a]References 56, 84, 85.

The comparatively small methyl substituents of the arylimido ligand in **20a** have the right size for excellent stereocontrol in the polymerization of 2,3-(CF$_3$)$_2$NBD. Poly(2,3-(CF$_3$)$_2$NBD) prepared with **20b** (Figure 20.25), bearing the bulkier 2,6-diisopropylphenylimido ligand, is less regular. It contains a smaller proportion of cis-vinylene units (71% cis) than poly(2,3-(CF$_3$)$_2$NBD) synthesized with **20a**; 86% of these cis units have the *meso* configuration.[84] By comparison, stereocontrol is still good in the polymerization of monomer (CO$_2$Me)$_2$NBD with catalyst **20b** (93% cis; 97% of cis units are *meso*) (Table 20.4).

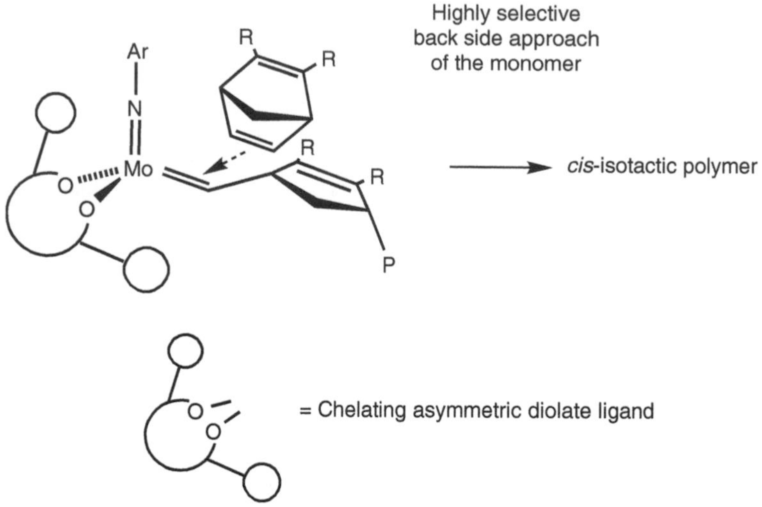

SCHEME 20.24 Enantiomorphic site control with Mo alkylidene catalysts bearing chelating asymmetric diolate ligands (Ar = 2,6-Me$_2$Ar, full circle representing either one or two bulky substituents, R = CF$_3$, CO$_2$Me, CO$_2$Men or CO$_2$Pan).

Chiral ester-substituted 2,3-(CO$_2$Men)$_2$NBD and 2,3-(CO$_2$Pan)$_2$NBD polymers obtained with **20a** are also very stereoregular. They contain 99% cis-vinylene structures and more than 90% m dyads.[85] Poly(endo,exo-2,3-(CO$_2$Me)$_2$NB) and poly(endo,exo-2,3-(MeOCH$_2$)$_2$NB) prepared with the same Mo complex are comprised of more than 99% cis-vinylene structures and 93% and 91% m dyads, respectively. Poly(endo,exo-5,6-Me$_2$NB) prepared with **20a** is highly cis, with 95% cis and 95% m units (Table 20.4).[56] Other binaptholate-based Mo alkylidene complexes for the preparation of stereoregular metathesis polymers are compounds **21a–d** (Figure 20.25), which contain aryl substituents in the 3 and 3′ positions of the chelating ligand instead of the 3,3′-silyl substitutents in **20a**.[90]

Overall, very good stereocontrol has also been reported for the Mo alkylidene complexes **22a,b** (Figure 20.25), which bear C_2-symmetric R$_4$tartH$_2$ ligands ("tart" refers to tartaric acid-derived) with R = phenyl or naphthyl.[84] In polymerizations of 2,3-(CF$_3$)$_2$NBD with the biphenolate-based Mo complexes **23a** and **23b** (Figure 20.25), the size of the imido ligand is again critical for good stereocontrol. In the case of compound **23a**, which bears the smaller 2,6-dimethylphenylimido ligand, the cis selectivity is 96%, and more than 99% of these cis-vinylene structures are part of isotactic units.[90] The cis selectivity is significantly reduced to 44% when complex **23b**, bearing a larger 2,6-diisopropylphenylimido ligand, is used for polymerization of 2,3-(CF$_3$)$_2$NBD instead of **23a**.

20.4.6 ALTERNATING ETHYLENE/CYCLOPENTENE COPOLYMERS

Metathesis polymerization of bicyclo[3.2.0]hept-6-ene ([3.2.0]-6), followed by hydrogenation with p-toluenesulfonhydrazide (TsH), affords the strictly alternating ethylene/cyclopentene copolymer poly-H-([3.2.0]-6) (Scheme 20.25).[91] A polymer sample of poly-H-([3.2.0]-6) obtained with the Grubbs catalyst Ru(=CHPh)(PCy$_3$)$_2$Cl$_2$ (**9**) is slightly syndiotactic (66% r dyads). This product is semicrystalline with a T_g of 19 °C and a T_m with a peak maximum at 123 °C. A highly stereoregular poly([3.2.0]-6) is obtained with Schrock–Hoveyda catalyst **24** (Scheme 20.25; commercially available). This polymer is isotactic and contains >99% cis-vinylene units; poly-H-([3.2.0]-6) made from this material displays the thermal transitions $T_g = 17$ °C and $T_m = 182$ °C.

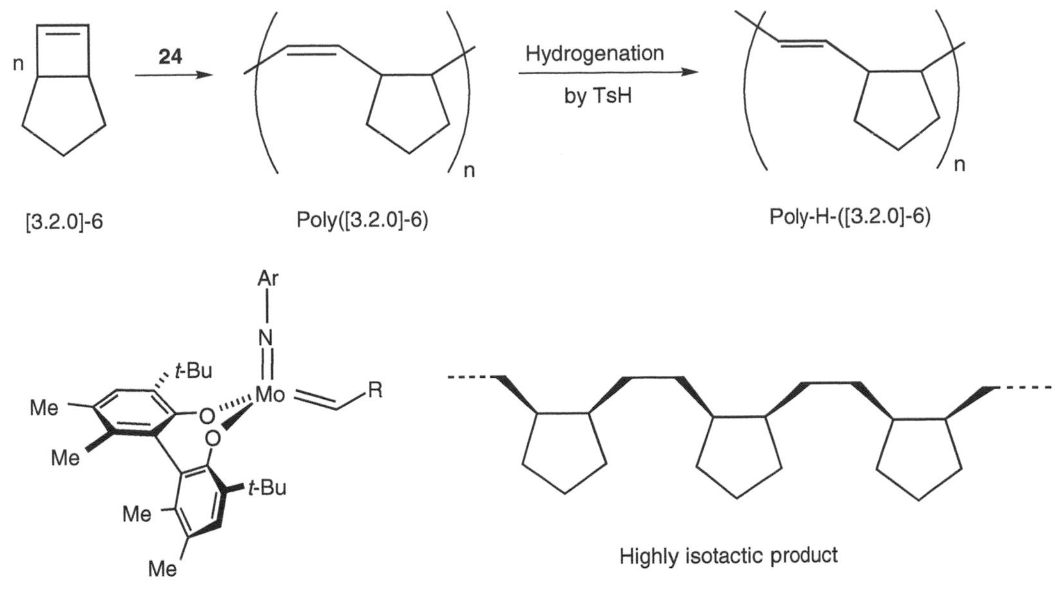

[3.2.0]-6   Poly([3.2.0]-6)   Poly-H-([3.2.0]-6)

Highly isotactic product

24: R = CMe$_2$Ph, Ar = 2,6-i-Pr$_2$C$_6$H$_3$

SCHEME 20.25 Strictly alternating ethylene/cyclopentene copolymers via ROMP (TsH = p-toluenesulfonhydrazide).

20.4.7 CRYSTALLINE POLY(*ENDO*-DICYCLOPENTADIENE) AND HYDROGENATED POLY(*ENDO*-DICYCLOPENTADIENE)

The Schrock–Hoveyda catalyst **24** was also successfully employed in the preparation of highly stereoregular poly(*endo*-dicyclopentadiene) (poly(*endo*-DCPD)).[92] This product contains 95% *cis*-vinylene structures and displays a T_m of 243 °C ($\Delta H = 14$ J/g). The corresponding hydrogenated poly-H-(*endo*-DCPD) (Figure 20.26) is crystalline, with $T_m = 290$ °C ($\Delta H = 42$ J/g). This product is believed to be isotactic, although the tacticity was not determined owing to the low solubility. Crystalline samples of poly(*endo*-DCPD) and poly-H-(*endo*-DCPD) with similar T_ms were also obtained with the two-component catalyst systems **25a**/n-BuLi and **25b**/n-BuLi (Figure 20.26). The *cis*-vinylene contents of these products are 85% and 91%, respectively, and the corresponding poly-H-(*endo*-DCPD) samples obtained after hydrogenation are 95% and 96% isotactic. By comparison, poly(*endo*-DCPD) obtained with Schrock catalyst **16b** (bearing *tert*-butoxide ligands) contains 55% trans double bonds and is amorphous, with $T_g = 145$ °C. The hydrogenated product derived from this poly(*endo*-DCPD) is predominantly atactic (58% m dyads) and displays a T_g of 100 °C. The M_w/M_n ratio of these polymers (prepared with both the single-component and two-component catalysts) is in the range of 1.9 and 2.8. NMR data of the poly(*endo*-DCPD) samples suggest that all of these products are noncrosslinked.

20.5 CONCLUSIONS

A wide range of multi- and single-component catalysts can be employed in the ROMP of bicyclic and polycyclic olefin monomers. Many catalysts give polymers with low or moderate regio- and/or stereoregularity. Selected monomer/catalyst combinations provide access to highly stereoregular polymers. Noteworthy among the conventional catalyst systems are ReCl$_5$ and OsCl$_3$/phenylacetylene, which give highly regular cis-syndiotactic poly(5,5-dimethylnorbornene).

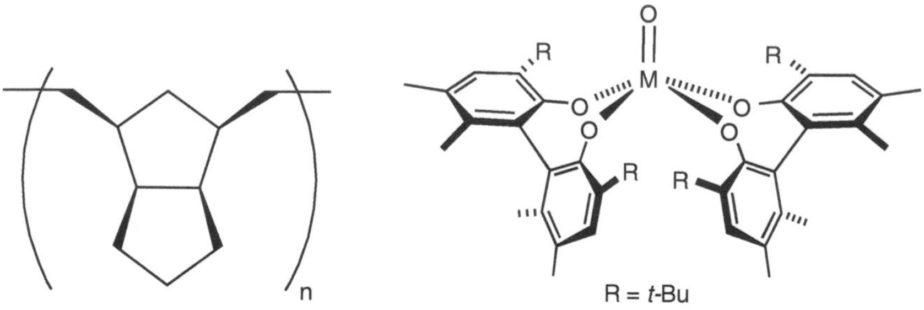

Poly-H-(*endo*-DCPD)

25a: M = Mo; **25b**: M = W

FIGURE 20.26   Hydrogenated poly(*endo*-dicyclopentadiene) prepared through ROMP with **16b**, **24**, or **25a,b** and subsequent hydrogenation.

In the case of the OsCl₃/phenylacetylene catalyst, norbornene and norbornadiene are also polymerized in a highly regular manner, and products with more than 90% cis-syndiotactic units are obtained. Most high-trans polymers produced with conventional catalysts are atactic. One of the few exceptions is poly(5,5-dimethylnorbornene) prepared with (mesitylene)W(CO)₃/EtAlCl₂/*exo*-2,3-epoxynorbornane. This metathesis polymer contains 85% *trans*-vinylene enchained units that are predominantly isotactic.

Imido molybdenum alkylidene complexes bearing two *tert*-butoxide ligands produce highly stereoregular poly(2,3-carboalkoxynorbornadiene) and poly(2,3-bis(trifluoromethyl)norbornadiene) that are trans-syndiotactic. By comparison, a tungsten oxo alkylidene complex bearing two phenoxide ligands converts these carboalkoxy- and trifluoromethyl-substituted norbornadienes into high-cis polymers that are nearly exclusively isotactic. It has been proposed that chain-end control mechanisms are responsible for the formation of the cis-isotactic and trans-syndiotactic structures in these polymers.

Several molybdenum alkylidene complexes bearing bulky C_2-symmetric biphenolate and bisnaphtholate ligands exert excellent stereocontrol in a wider range of metathesis polymerizations involving several bi- and tricyclic olefin monomers. The reaction products are cis-isotactic. These polymer structures are highly regular, and it has been suggested that an enantiomorphic site control mechanism controls the stereochemistry.

REFERENCES AND NOTES

1. Grubbs, R. H. *Handbook of Metathesis,* Vol. 1–3; Wiley-VCH: Weinheim, Germany, 2003.
2. Ivin, K. J., Mol, J. C. *Olefin Metathesis and Metathesis Polymerization*; Academic Press: San Diego, CA, 1997.
3. Grubbs, R. H. Olefin metathesis. *Tetrahedron* **2004**, *60*, 7117–7140.
4. Schrock, R. R.; Hoveyda, A. H. Molybdenum and tungsten imido alkylidene complexes as efficient olefin-metathesis catalysts. *Angew. Chem., Int. Ed. Engl.* **2003**, *42*, 4592–4633.
5. Trnka, T. M.; Grubbs, R. H. The development of L₂X₂Ru=CHR olefin metathesis catalysts: An organometallic success story. *Acc. Chem. Res.* **2001**, *34*, 18–29.
6. Fuerstner, A. Olefin metathesis and beyond. *Angew. Chem., Int. Ed. Engl.* **2000**, *39*, 3012–3043.
7. Schrock, R. R. Olefin metathesis by molybdenum imido alkylidene catalysts. *Tetrahedron* **1999**, *55*, 8141–8153.
8. Herrison, J.-L.; Chauvin, Y. Catalysis of olefin transformations by tungsten complexes. II. Telomerization of cyclic olefins in the presence of acyclic olefins. *Makromol. Chem.* **1971**, *141*, 161–176.
9. Grubbs, R. H. Alkene and alkyne metathesis reactions. In *Comprehensive Organometallic Chemistry*; Wilkinson, G., Ed.; Pergamon: Oxford (U.K.), 1982; Vol. 8, pp 499–551.

10. Howard, T. R.; Lee, J. B.; Grubbs, R. H. Titanium metallacarbene-metallacyclobutane reactions: Stepwise metathesis. *J. Am. Chem. Soc.* **1980**, *102*, 6876–6878.

11. Lee, J. B.; Ott, K. C.; Grubbs, R. H. Kinetics and stereochemistry of the titanacyclobutane-titanaethylene interconversion. Investigation of a degenerate olefin metathesis reaction. *J. Am. Chem. Soc.* **1982**, *104*, 7491–7496.

12. Casey, C. P.; Tuinstra, H. E.; Saeman, M. C. Reactions of (CO)$_5$WC(Tol)$_2$ with alkenes. A model for structural selectivity in the olefin metathesis reaction. *J. Am. Chem. Soc.* **1976**, *98*, 608–609.

13. Connon, S. J.; Blechert, S. Recent developments in olefin cross metathesis. *Angew. Chem., Int. Ed. Engl.* **2003**, *42*, 1900–1923.

14. Grubbs, R. H.; Miller, S. J.; Fu, G. C. Ring-closing metathesis and related processes in organic synthesis. *Acc. Chem. Res.* **1995**, *28*, 446–452.

15. Deiters, A.; Martin, S. F. Synthesis of oxygen- and nitrogen-containing heterocycles by ring-closing metathesis. *Chem. Rev.* **2004**, *104*, 2199–2238.

16. McReynolds, M. D.; Dougherty, J. M.; Hanson, P. R. Synthesis of phosphorus and sulfur heterocycles via ring-closing olefin metathesis. *Chem. Rev.* **2004**, *104*, 2239–2258.

17. Schwendeman, J. E.; Church, A. C.; Wagener, K. B. Synthesis and catalyst issues associated with ADMET polymerization. *Adv. Synth. Catal.* **2002**, *344*, 597–613.

18. Buchmeiser, M. R. Homogeneous metathesis polymerization by well-defined group VI and group VIII transition metal alkylidenes: Fundamentals and applications in the preparation of advanced materials. *Chem. Rev.* **2000**, *100*, 1565–1604.

19. Slugovc, C. The ring opening metathesis polymerization toolbox. *Macromol. Rapid Commun.* **2004**, *25*, 1283–1297.

20. Schrock, R. R. Living ring-opening metathesis polymerization catalyzed by well-characterized transition metal alkylidene complexes. *Acc. Chem. Res.* **1990**, *23*, 158–165.

21. Ofstead, E. A. Polyalkenamers. In *Encyclopedia of Polymer Science and Engineering*, 2nd ed.; Kroschwitz, J. I., Ed.; Wiley: New York, 1987; Vol. 11, pp 287–315.

22. Schleyer, P. v. R.; Williams, J. E., Jr.; Blanchard, K. R. Evaluation of strain in hydrocarbons. The strain in adamantane and its origin. *J. Am. Chem. Soc.* **1970**, *92*, 2377–2386.

23. Patton, P. A.; Lillya, C. P.; McCarthy, T. J. Olefin metathesis of cyclohexene. *Macromolecules* **1986**, *19*, 1266–1268.

24. Natta, G.; Dall'Asta, G.; Mazzanti, G. Stereospecific homopolymerization of cyclopentene. *Angew. Chem., Int. Ed. Engl.* **1964**, *3*, 723–730.

25. Natta, G.; Dall'Asta, G.; Mazzanti, G.; Motroni, G. Stereospecific polymerization of cyclobutene. *Makromol. Chem.* **1963**, *69*, 163–179.

26. Novak, B. M.; Grubbs, R. H. Catalytic organometallic chemistry in water: The aqueous ring-opening metathesis polymerization of 7-oxanorbornene derivatives. *J. Am. Chem. Soc.* **1988**, *110*, 7542–7543.

27. Novak, B. M.; Grubbs, R. H. The ring opening metathesis polymerization of 7-oxabicyclo[2.2.1]hept-5-ene derivatives: A new acyclic polymeric ionophore. *J. Am. Chem. Soc.* **1988**, *110*, 960–961.

28. Rinehart, R. E.; Smith, H. P. Emulsion polymerization of the norbornene ring system catalyzed by noble metal compounds. *J. Polym. Sci., Part B: Polym. Lett.* **1965**, *3*, 1049–1052.

29. Michelotti, F. W.; Keaveney, W. P. Coordinated polymerization of the bicyclo[2.2.1]hept-2-ene ring system (norbornene) in polar media. *J. Polym. Sci., Part A* **1965**, *3*, 895–905.

30. Natta, G.; Dall'Asta, G.; Porri, L. Polymerization of cyclobutene and 3-methylcyclobutene by RuCl$_3$ in polar protic solvents. *Makromol. Chem.* **1965**, *81*, 253–257.

31. Ofstead, E. A.; Lawrence, J. P.; Senyek, M. L.; Calderon, N. Stereochemistry of olefin metathesis—steric control and molecular weight regulation in polypentenamer synthesis. *J. Mol. Catal.* **1980**, *8*, 227–242.

32. Guenther, P.; Haas, F.; Marwede, G.; Nuetzel, K.; Oberkirch, W.; Pampus, G.; Schoen, N.; Witte, J. Ring-opening polymerization of cyclopentene (catalysts, polymerization, and product properties). *Angew. Makromol. Chem.* **1970**, *14*, 87–109.

33. Natta, G.; Dall'Asta, G.; Bassi, I. W.; Carella, G. Stereospecific ring cleavage homopolymerization of cycloolefins and structural examination of the resulting homologous series of linear crystalline *trans*-polyalkenamers. *Makromol. Chem.* **1966**, *91*, 87–106.

34. Kenwright, A. M. High-resolution NMR and ROMP. In *Ring Opening Metathesis Polymerization and Related Chemistry*; Khosravi, E., Szymanska-Buzar, T., Eds.; NATO Sci. Ser. II., Math. Phys. Chem. 56; Kluwer Academic Publishers: Dordrecht, Netherlands, 2002; pp 57–67.

35. Truett, W. L.; Johnson, D. R.; Robinson, I. M.; Montague, B. A. Polynorbornene by coördination polymerization. *J. Am. Chem. Soc.* **1960**, *82*, 2337–2340.

36. Ivin, K. J. ROMP and related chemistry: Past, present and future. In *Ring Opening Metathesis Polymerization and Related Chemistry*; Khosravi, E., Szymanska-Buzar, T., Eds.; NATO Sci. Ser. II, Math. Phys. Chem. 56; Kluwer Academic Publishers: Dordrecht, Netherlands, **2002**; pp 1–15.

37. Ivin, K. J.; Laverty, T.; Rooney, J. J. The ^{13}C NMR spectra of poly(1-pentenylene) and poly(1,3-cyclopentylenevinylene). *Makromol. Chem.* **1977**, *178*, 1545–1560.

38. Kelly, W. J.; Calderon, N. Selectivity aspects in cross metathesis reactions; *J. Macromol. Sci., Chem.* **1975**, *9*, 911–929.

39. Ivin, K. J.; Lam, L.-M.; Rooney, J. J. Carbon-13 NMR spectra of polymers made by ring-opening polymerization of (±)-*endo*-5-methylbicyclo[2.2.1]hept-2-ene using metathesis catalysts. *Makromol. Chem.* **1981**, *182*, 1847–1854.

40. Ivin, K. J.; Lapienis, G.; Rooney, J. J. ^{13}C NMR spectra of polymers made by ring-opening polymerization of (±)- and (+)-*exo*-5-methylbicyclo[2.2.1]hept-2-ene using metathesis catalysts. *Polymer* **1980**, *21*, 436–443.

41. Huu Thoi Ho; Ivin, K. J.; Rooney, J. J. Carbon-13 NMR spectra of polymers made by ring-opening polymerization of optically active and racemic 5,5-dimethylbicyclo[2.2.1]hept-2-ene using metathesis catalysts. *Makromol. Chem.* **1982**, *183*, 1629–1646.

42. Ho Huu Thoi; Ivin, K. J.; Rooney, J. J. Tacticity and stereochemistry in the ring-opening polymerization of 5,5-dimethylbicyclo[2.2.1]hept-2-ene initiated by metathesis catalysts. *J. Mol. Catal.* **1982**, *15*, 245–270.

43. Devine, G. I.; Huu Thoi Ho; Ivin, K. J.; Mohamed, M. A.; Rooney, J. J. Production of high-trans isotactic polymer of 5,5-dimethylbicyclo[2.2.1]hept-2-ene using a metathesis catalyst. *Chem. Commun.* **1982**, 1229–1231.

44. Carvill, A. G.; Greene, R. M. E.; Hamilton, J. G.; Ivin, K. J.; Kenwright, A. M.; Rooney, J. J. ^{13}C NMR spectra of hydrogenated polymers of *exo*-5-methyl-, *endo*-5-methyl-, and 5,5-dimethyl-derivatives of bicyclo[2.2.1]hept-2-ene prepared by ring-opening metathesis polymerization. *Macromol. Chem. Phys.* **1998**, *199*, 687–693.

45. Hamilton, J. G.; Ivin, K. J.; Rooney, J. J. ^{13}C NMR spectra of ring-opened polymers of 1-methylbicyclo[2.2.1]hept-2-ene and their hydrogenated products. *Brit. Polym. J.* **1984**, *16*, 21–33.

46. Al-Samak, B.; Amir-Ebrahimi, V.; Corry, D. G.; Hamilton, J. G.; Rigby, S.; Rooney, J. J.; Thompson, J. M. Dramatic solvent effects on ring-opening metathesis polymerization of cycloalkenes. *J. Mol. Catal. A: Chem.* **2000**, *160*, 13–21.

47. Eilerts, N. W.; Heppert, J. A. Bisphenol ligands in the stereocontrol of transformations catalyzed by titanium and tungsten. *Polyhedron* **1995**, *14*, 3255–3271.

48. Hamilton, J. G.; Ivin, K. J.; Rooney, J. J. Microstructure and mechanism of formation of the ring-opened polymers of *syn*- and *anti*-7-methylbicyclo[2.2.1]hept-2-ene initiated with metathesis catalysts. *J. Mol. Catal.* **1985**, *28*, 255–278.

49. Hamilton, J. G.; Rooney, J. J.; Snowden, D. G. Ring-opening metathesis polymerization of 7-methylnorbornadiene. *Makromol. Chem.* **1993**, *194*, 2907–2922.

50. Seehof, N.; Risse, W. Ring-opening olefin metathesis polymerization of spiro(bicyclo[2.2.1]hept-2-ene-7,1'-cyclopropane). *Macromolecules* **1993**, *26*, 5971–5975.

51. Al-Samak, B.; Amir-Ebrahimi, V.; Carvill, A. G.; Hamilton, J. G.; Rooney, J. J.. Determination of the tacticity of ring-opened metathesis polymers of norbornene and norbornadiene by ^{13}C NMR spectroscopy of their hydrogenated derivatives. *Polym. Int.* **1996**, *41*, 85–92.

52. Amir-Ebrahimi, V.; Carvill, A. G.; Hamilton, J. G.; Rooney, J. J.; Tuffy, C. Copolymerization of cycloalkenes as a probe of the propagation steps in olefin metathesis. *J. Mol. Catal. A: Chem.* **1997**, *115*, 85–94.

53. Bassett, J.-M.; Leconte, M.; Lefebvre, F.; Hamilton, J. G.; Rooney, J. J. Stereoselectivity in cyclic and acyclic metathesis reactions. *Macromol. Chem. Phys.* **1997**, *198*, 3499–3506.

54. Cataldo, F. FT-IR spectroscopic characterization of hydrogenated polyoctenamer and polynorbornene and DSC study of their thermal properties. *Polym. Int.* **1994**, *34*, 49–57.

55. Bell, B.; Hamilton, J. G.; Mackey, O. N. D.; Rooney, J. J. Microstructure of ring-opened polymers and copolymers of norbornadiene. *J. Mol. Catal.* **1992**, *77*, 61–73.

56. Ivin, K. J.; Kenwright, A. M.; Hofmeister, G. E.; McConville, D. H.; Schrock, R. R.; Amir-Ebrahimi, V.; Carvill, A. G.; Hamilton, J. G.; Rooney, J. J. ^{13}C NMR spectra of tactic and atactic hydrogenated ring-opened polymers of enantiomeric and racemic *endo,exo*-5,6-dimethylnorbornene. *Macromol. Chem. Phys.* **1998**, *199*, 547–553.

57. Hamilton, J. G.; Ivin, K. J.; Rooney, J. J. Ring-opening polymerization of *endo*- and *exo*-dicyclopentadiene and their 7,8-dihydro derivatives. *J. Mol. Catal.* **1986**, *36*, 115–125.

58. Amir-Ebrahimi, V.; Corry, D. A. K.; Hamilton, J. G.; Rooney, J. J. Determination of the tacticities of ring-opened metathesis polymers of symmetrical 5,6-disubstituted derivatives of norbornene and norbornadiene from the ^{13}C NMR spectra of their hydrogenated derivatives. *J. Mol. Catal. A: Chem.* **1998**, *133*, 115–122.

59. Delaude, L.; Demonceau, A.; Noels, A. F. Probing the stereoselectivity of the ruthenium-catalyzed ring-opening metathesis polymerization of norbornene and norbonadiene diesters. *Macromolecules* **2003**, *36*, 1446–1456.

60. Delaude, L.; Demonceau, A.; Noels, A. F. Highly stereoselective ruthenium-catalyzed ring-opening metathesis polymerization of 2,3-difunctionalized norbornadienes and their 7-oxa analogues. *Macromolecules* **1999**, *32*, 2091–2103.

61. Amir-Ebrahimi, V.; Byrne, D.; Hamilton, J. G.; Rooney, J. J. A UV/visible spectroscopic study of water-soluble conjugated polyenes prepared via the ring-opening metathesis polymerization reaction. *Macromol. Chem. Phys.* **1995**, *196*, 327–342.

62. Risse, W.; Destro, M. Alternating ethylene-norbornene copolymers via ring-opening olefin metathesis polymerization. Presented at the 224th National Meeting of the American Chemical Society, Boston, MA, August 18-22, 2002; ORGN 306.

63. Gilliom, L. R.; Grubbs, R. H. Titanacyclobutanes from strained cyclic olefins: The living polymerization of norbornene. *J. Am. Chem. Soc.* **1986**, *108*, 733–742.

64. Gilliom, L.; Grubbs, R. H. The stereochemistry of norbornene polymerization by titanacyclobutanes. *J. Mol. Catal.* **1988**, *46*, 255–266.

65. Mashima, K.; Tanaka, Y.; Kaidzu, M.; Nakamura, A. cis-Iso-specific polymerization of norbornenes by a unique combination of Cp* and 1,3-butadiene ligands on tantalum: Crystal structures of Cp*(η^4-C$_4$H$_6$)Ta(CH$_2$Ph)$_2$ and Cp*(η^4-C$_4$H$_6$)Ta(=CHPh)(PMe$_3$). *Organometallics* **1996**, *15*, 2431–2433.

66. Mashima, K.; Kaidzu, M.; Tanaka, Y.; Nakayama, Y.; Nakamura, A.; Hamilton, J. G.; Rooney, J. J. Control of stereoselectivity in the ring-opening metathesis polymerization of norbornene by the auxiliary ligands butadiene and *o*-xylylene in well-defined pentamethylcyclopentadiene tantalum carbene complexes. *Organometallics* **1998**, *17*, 4183–4195.

67. Amir-Ebrahimi, V.; Corry, D. A.; Hamilton, J. G.; Thompson, J. M.; Rooney, J. J. Characteristics of RuCl$_2$(CHPh)(PCy$_3$)$_2$ as a catalyst for ring-opening metathesis polymerization. *Macromolecules* **2002**, *33*, 717–724.

68. Ivin, K. J.; Kenwright, A. M.; Khosravi, E.; Hamilton, J. G. Ring-opening metathesis polymerization of 7-methylbicyclo[2.2.1]hepta-2,5-diene initiated by well-defined molybdenum and ruthenium carbene complexes. *J. Organomet. Chem.* **2000**, *606*, 37–48.

69. Hamilton, J. G.; Frenzel, U.; Kohl, F. J.; Weskamp, T.; Rooney, J. J.; Herrmann, W. A.; Nuyken, O. *N*-Heterocyclic carbenes (NHC) in olefin metathesis: Influence of the NHC-ligands on polymer tacticity. *J. Organomet. Chem.* **2000**, *606*, 8–12.

70. Kress, J.; Ivin, K. J.; Amir-Ebrahimi, V.; Weber, P. Studies of the metathesis polymerization and copolymerization of *syn*- and *anti*-7-methylnorbornene initiated by the tungsten-carbene complex W[(=C<(CH$_2$)$_4$](OCH$_2$CMe$_3$)$_2$Br$_2$·GaBr$_3$. *Makromol. Chem.* **1990**, *191*, 2237–2251.

71. O'Donoghue, M. B.; Schrock, R. R.; LaPointe, A. M.; Davis, W. M. Preparation of well-defined, metathetically active oxo alkylidene complexes of tungsten. *Organometallics* **1996**, *15*, 1334–1336.

72. Schrock, R. R.; Murdzek, J. S.; Bazan, G. C.; Robbins, J.; DiMare, M.; O'Regan, M. Synthesis of molybdenum imido alkylidene complexes and some reactions involving acyclic olefins. *J. Am. Chem. Soc.* **1990**, *112*, 3875–3886.

73. Murdzek, J. S.; Schrock, R. R. Low polydispersity homopolymers and block copolymers by ring-opening of 5,6-dicarbomethoxynorbornene. *Macromolecules* **1987**, *20*, 2640–2642.

74. Murdzek, J. S., Schrock, R. R. Well-characterized olefin metathesis catalysts that contain molybdenum. *Organometallics* **1987**, *6*, 1373–1374.

75. Bazan, G. C.; Khosravi, E.; Schrock, R. R.; Feast, W. J.; Gibson, V. C.; O'Regan, M. B.; Thomas, J. K.; Davis, W. M. Living ring-opening metathesis polymerization of 2,3-difunctional norbornadienes by Mo(CH-*t*-Bu)(N-2,6-C$_6$H$_3$-*i*-Pr$_2$)(O-*t*-Bu)$_2$. *J. Am. Chem. Soc.* **1990**, *112*, 8378–8387.

76. Feast, W. J.; Gibson, V. C.; Marshall, E. L. A remarkable ancillary ligand effect in living ring-opening metathesis polymerisation. *Chem. Commun.* **1992**, 1157–1158.

77. Oskam, J. H.; Schrock, R. R. Rotational isomers of Mo(VI) alkylidene complexes and cis/trans polymer structure: Investigations in ring-opening metathesis polymerization. *J. Am. Chem. Soc.* **1993**, *115*, 11831–11845.

78. Schrock, R. R.; Crowe, W. E.; Bazan, G. C.; DiMare, M.; O'Regan, M. B.; Schofield, M. H. Monoadducts of imido alkylidene complexes, *syn* and *anti* rotamers, and alkylidene ligand rotation. *Organometallics* **1991**, *10*, 1832–1843.

79. Feldman, J.; Schrock, R. R. Recent advances in the chemistry of "d0" alkylidene and metallacyclobutane complexes. *Prog. Inorg. Chem.* **1991**, *39*, 1–81.

80. Schrock, R. R.; Lee, J.-K.; O'Dell, R.; Oskam, J. H. Exploring factors that determine cis/trans structure and tacticity in polymers prepared by ring-opening metathesis polymerization with initiators of the type *syn*- and *anti*-Mo(NAr)(CHCMe$_2$Ph)(OR)$_2$. Observation of a temperature-dependent cis/trans ratio. *Macromolecules* **1995**, *28*, 5933–5940.

81. Broeders, J.; Feast, W. J.; Gibson, V. C.; Khosravi, E. Synthesis of stereoblock copolymers via ligand exchange in living stereoselective ring opening metathesis polymerisation of 2,3-bis(trifluoromethyl)bicyclo[2.2.1]hepta-2,5-diene. *Chem. Commun.* **1996**, 343–344.

82. Sunaga, T.; Ivin, K. J.; Hofmeister, G. E.; Oskam, J. H.; Schrock, R. R. Ring-opening metathesis polymerization of (+) and (±)-*endo,exo*-5,6-dimethylbicyclo[2.2.1]hept-2-ene by Mo(CH-*t*-Bu)(N-2,6-C$_6$H$_3$-*i*-Pr$_2$)(OR)$_2$. *Macromolecules* **1994**, *27*, 4043–4050.

83. Feast, W. J.; Gibson, V. C.; Ivin, K. J.; Kenwright, A. M.; Khosravi, E. Metathesis polymerisation of 1,7,7-trimethylbicyclo[2.2.1]hept-2-ene using a well defined molybdenum initiator. *J. Mol. Catal.* **1994**, *90*, 87–99.

84. McConville, D. H.; Wolf, J. R.; Schrock, R. R. Synthesis of chiral molybdenum ROMP initiators and all-cis highly tactic poly(2,3-(R)$_2$norbornadiene) (R = CF$_3$ or CO$_2$Me). *J. Am. Chem. Soc.* **1993**, *115*, 4413–4414.

85. O'Dell, R.; McConville, D. H.; Hofmeister, G. E.; Schrock, R. R. Polymerization of enantiomerically pure 2,3-dicarboalkoxynorbornadienes and 5,6-disubstituted norbornenes by well-characterized molybdenum ring-opening metathesis polymerization initiators. Direct determination of tacticity in cis, highly tactic and trans, highly tactic polymers. *J. Am. Chem. Soc.* **1994**, *116*, 3414–3423.

86. Bazan, G. C.; Schrock, R. R.; Cho, H.-N.; Gibson, V. C. Polymerization of functionalized norbornenes employing Mo(CH-*t*-Bu)(NAr)(O-*t*-Bu)$_2$ as the initiator. *Macromolecules* **1991**, *24*, 4495–4502.

87. Biagini, S. C. G.; Coles, M. P.; Gibson, V. C.; Giles, M. R.; Marshall, E. L.; North, M. Living ring-opening metathesis polymerization of amino ester functionalized norbornenes. *Polymer* **1998**, *39*, 1007–1014.

88. Khosravi, E.; Feast, W. J.; Al-Hajaji, A. A.; Leejarkpai, T. ROMP of *n*-alkyl norbornene dicarboxyimides: From classical to well-defined initiators, an overview. *J. Mol. Catal. A: Chem.* **2000**, *160*, 1–11.

89. Schitter, R. M. E.; Steinhäusler, T.; Stelzer, F. Ring opening metathesis polymerization of methyl-*N*-(1-phenylethyl)-2-azabicyclo[2.2.1]hept-5-ene-3-carboxylate. *J. Mol. Catal.* **1997**, *115*, 11–20.

90. Totland, K. M.; Boyd, T. J.; Lavoie, G. G.; Davis, W. M.; Schrock, R. R. Ring-opening metathesis polymerization with binaphtholate or biphenolate complexes of molybdenum. *Macromolecules* **1996**, *29*, 6114–6125.

91. Fujita, M.; Coates, G. W. Synthesis and characterization of alternating and multiblock copolymers from ethylene and cyclopentene. *Macromolecules* **2002**, *36*, 9640–9647.

92. Hayano, S.; Kurakata, H.; Uchida, D.; Sakamoto, M.; Kishi, N.; Matsumoto, H.; Tsunogae, Y.; Igarashi, I. Stereospecific ring-opening metathesis polymerization of *endo*-dicyclopentadiene by Schrock-Hoveyda catalyst and novel Mo- and W-based complexes. Development of crystalline hydrogenated poly(*endo*-dicyclopentadiene). *Chem. Lett.* **2003**, *32*, 670–671.

93. Jenkins, A. D.; Kratochvil, P.; Stepto, R. F. T.; Suter, U. W. Glossary of basic terms in polymer science. *Pure Appl. Chem.* **1996**, *68*, 2287–2311.

21 Stereoregularity of Polyacetylene and Its Derivatives

Toshio Masuda, Masashi Shiotsuki, and Junichi Tabei

CONTENTS

21.1 INTRODUCTION

Acetylene and its mono- and disubstituted derivatives polymerize in the presence of suitable catalysts to provide linear polymers (Scheme 21.1).[1–6] Their polymerization behavior greatly varies, depending on the number and kind of substituents on the monomer. Unlike vinyl polymers, the formed polyacetylenes possess carbon–carbon alternating double bonds along the main chain, and hence they belong to the group of conjugated polymers. The alternating double bond structure endows these polymers with unique properties such as electrical conductivity, paramagnetism, color, photoconductivity, electroluminescence, and electrochromism.[7–10] Although various catalysts, including radical and ionic initiators and metal complexes, have been examined for the polymerization of acetylenes, only a restricted number of transition metal catalysts are able to selectively polymerize them. Polymerization of acetylenes is often accompanied by cyclotrimerization and linear oligomerization; hence, judicious choice of catalysts is essential. Furthermore, the type of monomers polymerizable with a particular catalyst is typically rather restricted, and therefore it is important to recognize the characteristics of each catalyst. Figure 21.1 shows typical catalysts used for the polymerization of

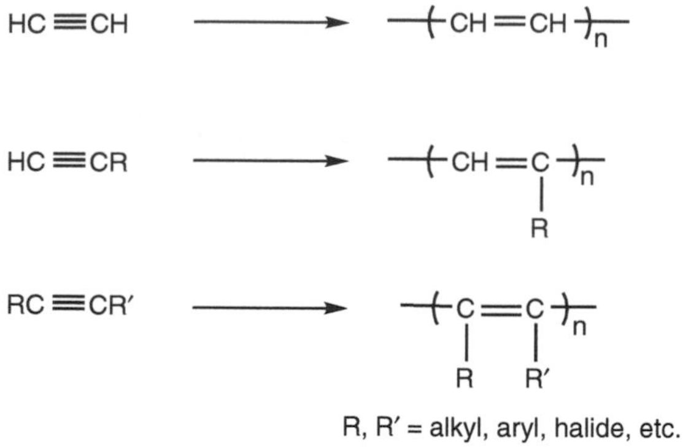
HC≡CH
CH=CH
n
HC≡CR
CH=C
R
RC≡CR′
C=C
R R′
R, R′ = alkyl, aryl, halide, etc.

SCHEME 21.1 Polymerization of acetylenes.

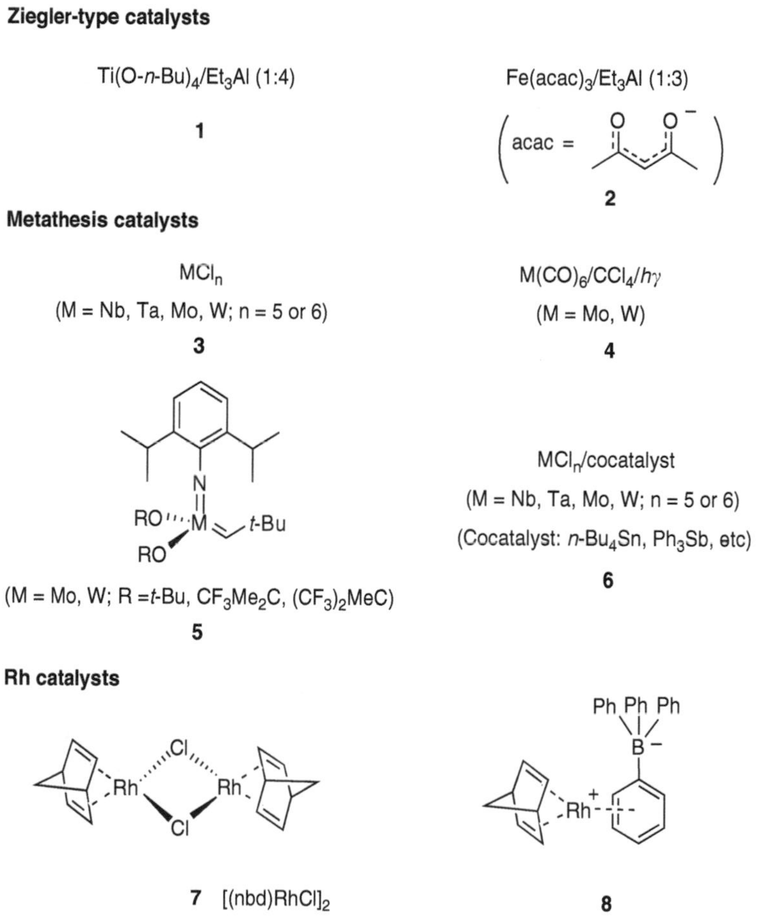
Ziegler-type catalysts
Ti(O-n-Bu)4/Et3Al (1:4)
1
Fe(acac)3/Et3Al (1:3)
acac =
2
Metathesis catalysts
MCln
(M = Nb, Ta, Mo, W; n = 5 or 6)
3
M(CO)6/CCl4/hγ
(M = Mo, W)
4
N
RO
M
t-Bu
RO
(M = Mo, W; R = t-Bu, CF3Me2C, (CF3)2MeC)
5
MCln/cocatalyst
(M = Nb, Ta, Mo, W; n = 5 or 6)
(Cocatalyst: n-Bu4Sn, Ph3Sb, etc)
6
Rh catalysts
Cl
Rh
Rh
Cl
7 [(nbd)RhCl]2
Ph Ph Ph
B
Rh
8

FIGURE 21.1 Catalysts for the polymerization of acetylene include Ziegler-type catalysts, metathesis catalysts, and Rh-based catalysts.

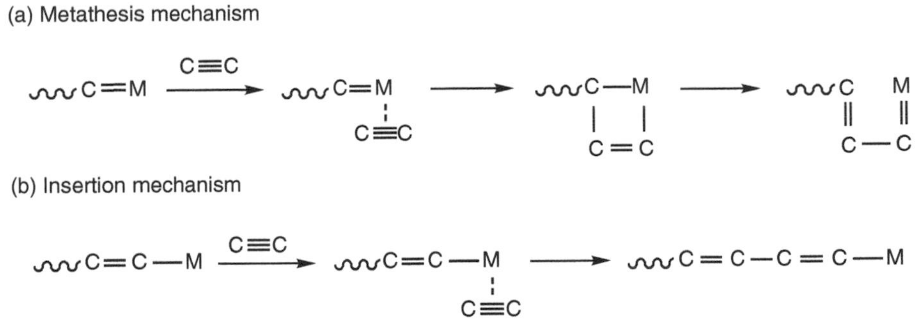

(a) Metathesis mechanism

$$\sim\!\!\sim\!\!C\!=\!M \xrightarrow{C\equiv C} \sim\!\!\sim\!\!C\!=\!M \longrightarrow \sim\!\!\sim\!\!C\!-\!M \longrightarrow \sim\!\!\sim\!\!C \quad M$$

(b) Insertion mechanism

$$\sim\!\!\sim\!\!C\!=\!C\!-\!M \xrightarrow{C\equiv C} \sim\!\!\sim\!\!C\!=\!C\!-\!M \longrightarrow \sim\!\!\sim\!\!C\!=\!C\!-\!C\!=\!C\!-\!M$$

SCHEME 21.2　Polymerization mechanisms for acetylenes where M represents the active catalyst site.

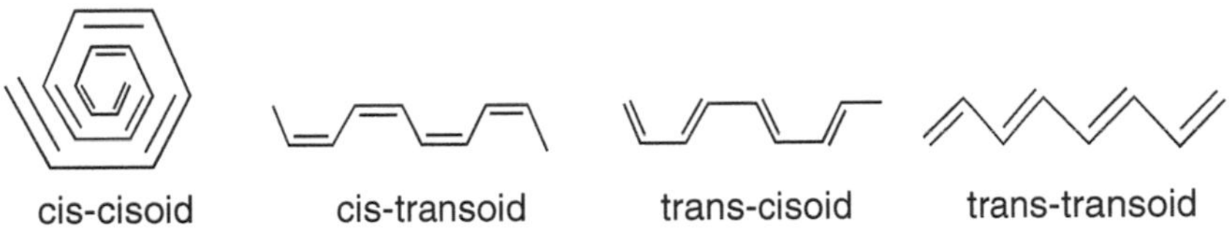

cis-cisoid　　　　cis-transoid　　　　trans-cisoid　　　　trans-transoid

FIGURE 21.2　Possible steric structures of polyacetylenes.

acetylenes. They are divided into three types: Ziegler-type (**1, 2**), metathesis (**3–6**), and Rh (**7, 8**) catalysts. The polymerization induced by the Ziegler-type and Rh catalysts proceeds through an alkyne insertion mechanism, where the propagating species are alkenyl metals (Scheme 21.2). On the other hand, the polymerization using metathesis catalysts involves metal carbenes as the propagating species. Although the two mechanisms lead to the same alternating double bond structure, they can be distinguished from each other by detailed studies on the polymer chain ends and propagating species.

As shown in Figure 21.2, four steric (geometric) structures are theoretically possible for polyacetylenes, that is, cis-cisoid, cis-transoid, trans-cisoid, and trans-transoid, because the rotation of the single bond between two main chain double bonds in the main chain is more or less restricted. Polyacetylene can be obtained in the membrane form by use of a mixed catalyst composed of $Ti(O\text{-}n\text{-}Bu)_4$ and Et_3Al, the so-called Shirakawa catalyst (**1**); both the cis- and trans-isomers are known, which are thought to have cis-transoidal and trans-transoidal structures, respectively (Table 21.1).[11–13] Phenylacetylene can be polymerized with a Ziegler-type catalyst, $Fe(acac)_3/Et_3Al$ (**2**) (acac = acetylacetonate), Rh catalysts (**7**), and metathesis catalysts (**3–5**) that contain Mo and W as the central metals, to provide cis-cisoidal, cis-transoidal, cis-rich, or trans-rich polymers, respectively.

tert-Butylacetylene readily polymerizes with metathesis catalysts, and all-cis polymer is obtained with Mo catalysts under suitable conditions. Many disubstituted acetylenes can be polymerized with the MCl_n/cocatalyst systems (**6**), but the geometric structure of the formed polymers are not well known, because they have a tetrasubstituted-olefin main chain structure, and their geometric structure is difficult to characterize by ^1H and ^{13}C NMR. Rh catalysts such as **7** and **8** produce polymers from specific monosubstituted acetylene monomers including phenylacetylenes ($HC\equiv CPh$), alkyl propiolates ($HC\equiv C\text{-}C(=O)OR$), and N-propargylamides ($HC\equiv C\text{-}CH_2NHC(=O)R$), to yield cis-transoidal polymers. In most cases, the cis contents of the polymers are quantitative, which is easily verified by the presence of the sharp cis-olefinic signal in the ^1H NMR spectra.

In this chapter, the stereoregularity of polyacetylene and its derivatives are discussed in correlation with monomer, catalyst type, polymerization mechanism, polymer properties, and so forth.

TABLE 21.1

Relationships Between Catalysts and Monomers in Acetylene Polymerization

Catalyst	Monomer[a]	Stereoregularity	Mechanism
1	$HC \equiv CH$	all-cis, all-trans	Insertion
2	$HC \equiv C\text{-}R$		Insertion
	$HC \equiv C\text{-}Ar$	cis-cisoidal	Insertion
3–5	$HC \equiv C\text{-}R$		Metathesis
	$HC \equiv C\text{-}t\text{-}Bu$	all-cis, trans-rich	Metathesis
	$HC \equiv C\text{-}Ar$	cis-rich, trans-rich	Metathesis
6	$R\text{-}C \equiv C\text{-}R'$		Metathesis
	$R\text{-}C \equiv C\text{-}Ar$		Metathesis
	$Ar\text{-}C \equiv C\text{-}Ar'$		Metathesis
7, 8	$HC \equiv C\text{-}Ar$	cis-transoidal	Insertion
	$HC \equiv C\text{-}C(=O)OR$	cis-transoidal	Insertion
	$HC \equiv C\text{-}CH_2NHC(=O)\text{-}R$	cis-transoidal	Insertion

[a] R, R' alkyl; Ar, Ar' aryl.

21.2 STEREOREGULARITY OF POLYACETYLENE

21.2.1 POLYMERIZATION CATALYSTS AND OTHER CONDITIONS

According to a broad definition, Ziegler-type catalysts are defined as catalysts comprising group 4–8 transition metal compounds in combination with group 1–3 organometals (including hydrides). Figure 21.3 shows typical Ziegler-type catalysts employed for the polymerization of acetylene. The Shirakawa catalyst, $Ti(O\text{-}n\text{-}Bu)_4/Et_3Al$, (1:4 ratio), is a homogeneous catalyst used more frequently for acetylene polymerization than for substituted acetylenes polymerization.[14,15] After this catalyst is aged in toluene solution at room temperature for a certain time, acetylene gas is introduced to the system. Polyacetylene membranes can be directly obtained with the Shirakawa catalyst when high catalyst concentrations are used. If the catalyst concentration is low, a polymer membrane cannot be obtained; instead, only powder or gel is formed. Typical polymerization conditions are: $[Ti(O\text{-}n\text{-}Bu)_4] = 0.25$ M, $[Et_3Al] = 1.0$ M in toluene, $-78\ °C$, and 500–600 mmHg acetylene pressure. Since polyacetylene is insoluble in all solvents, its molecular weight cannot be determined directly. Its molecular weight has been estimated to be typically around 10,000 (M_n) as determined from the molecular weights of polyethylenes obtained by the hydrogenation of polyacetylene. Naarmann and Theophilou have reported the use of a $Ti(O\text{-}n\text{-}Bu)_4/Et_3Al$ (1:2 ratio) catalyst aged at high temperatures (e.g., 120 °C in silicone oil), which affords a highly conducting polyacetylene membrane.[16,17] The Luttinger catalyst is less active than the Shirakawa catalyst, but has the advantage that flammable Et_3Al is unnecessary.[18,19] Synthesis of polyacetylene was performed with the Luttinger catalyst without the need of any solvents during catalyst preparation, or during the polymerization.[20] By this method, a ductile polyacetylene with large Young's modulus and tensile strength is obtained, which provides a highly oriented membrane. Furthermore, helical polyacetylene has been prepared from the Luttinger catalyst by performing the polymerization in chiral nematic liquid crystalline solvents.[21]

21.2.2 GEOMETRIC STRUCTURE OF POLYACETYLENE

The infrared (IR) spectra of *cis*- and *trans*-polyacetylenes are fairly simple because of their symmetrical structures.[14] *cis*-Polyacetylene exhibits absorptions at 1329 and 1249 cm^{-1} (C-H in-plain

Ti(O-*n*-Bu)$_4$/Et$_3$Al (1:4)   (Shirakawa catalyst)
(aged at r.t.)

Ti(O-*n*-Bu)$_4$/Et$_3$Al (1:2)   (Naarmann catalyst)
(aged at 120 °C)

Co(NO$_3$)$_2$/NaBH$_4$ (Luttinger catalyst)

FIGURE 21.3   Ziegler-type catalysts for acetylene polymerization.

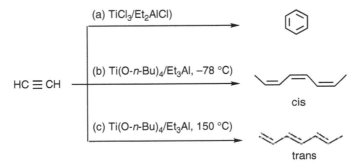

SCHEME 21.3   The cyclotrimerization of acetylene to form benzene (a), and the oligomerization of acetylene forming a cis structure (b), or a trans structure (c).

deformation) and at 740 cm^{-1} (C-H out-of-plain deformation), whereas the trans isomer displays a large absorption at 1015 cm^{-1} (C-H out-of-plain deformation). The geometric structure of polyacetylene can be readily determined by IR spectroscopy by use of Equation 21.1.

$$\text{cis content(\%)} = \frac{1.30\,A_{740}}{1.30\,A_{740} + A_{1015}} \times 100 \tag{21.1}$$

where A_{740} and A_{1015} are the absorbances at 740 and 1015 cm^{-1} of a sample, respectively. The geometric structure of polyacetylene can also be evaluated by ^{13}C NMR[22] and Raman spectroscopy[23]. Polyacetylene, however, tends to undergo isomerization from laser light during Raman spectrum measurement, making this a less valuable characterization technique.

In the polymerization of acetylene, the proportion of benzene (cyclotrimer) in the product strongly depends on the Lewis acidity of the catalyst components;[24] for example, benzene yield 100% (TiCl$_3$/Et$_2$AlCl); 70% (TiCl$_4$/Et$_3$Al); 80%–90% (Ti(acac)$_3$/Et$_2$AlCl); <1% (Ti(acac)$_3$/Et$_3$Al); <1% (Ti(O-*n*-Bu)$_4$/Et$_3$Al). The cis content of polyacetylene can be controlled by varying the polymerization temperature.[14,15] For example, when TiCl$_4$/Et$_3$Al (1:4 molar ratio) is used, the cis content is 98% at −78 °C, while it is 0% at 150 °C (Scheme 21.3). The cis content also depends on the Al/Ti molar ratio; that is, when TiCl$_4$/Et$_3$Al is used at 0 °C, the cis content is 2% at Al/Ti = 1, while it is 70% at Al/Ti = 3. Furthermore, *cis*-polyacetylene (95% cis) can be isomerized into *trans*-polyacetylene (95% trans) by heat treatment at 180 °C for 1 h.

TABLE 21.2

Polymerization of Substituted Acetylenes by Metathesis Catalysts

Monomer	Catalyst	$M_w \times 10^{-3}$ or $[\eta]^a$	
a) Monosubstituted hydrocarbon acetylenes			
$HC{\equiv}C\text{-}n\text{-}Bu$	$WCl_2(OC_6H_4\text{-}2,6\text{-}Me_2)_4$	170	(M_n)
$HC{\equiv}C\text{-}t\text{-}Bu$	$MoCl_5$	33	(M_n)
$HC{\equiv}CPh$	WCl_6/Ph_4Sn	15	(M_n)
$HC{\equiv}CC_6H_2\text{-}2,6\text{-}Me_2\text{-}4\text{-}t\text{-}Bu$	$W(CO)_6/CCl_4/h\nu$	600	(M_w)
b) Monosubstituted heteroatom-containing acetylenes			
$HC{\equiv}CCH(SiMe_3)\text{-}n\text{-}C_5H_{11}$	$MoCl_5/Et_3SiH$	4500	(M_w)
$HC{\equiv}CC_6H_4\text{-}o\text{-}SiMe_3$	$W(CO)_6/CCl_4/h\nu$	3400	(M_w)
$HC{\equiv}CC_6H_4\text{-}o\text{-}CF_3$	$W(CO)_6/CCl_4/h\nu$	1600	(M_w)
$HC{\equiv}CC_6F_5$	WCl_6/Ph_4Sn	0.61	$([\eta])$
c) Disubstituted hydrocarbon acetylenes			
$MeC{\equiv}C\text{-}n\text{-}Pr$	$MoCl_5/n\text{-}Bu_4Sn$	1100	(M_w)
$PhC{\equiv}CMe$	$TaCl_5/n\text{-}Bu_4Sn$	1500	(M_w)
$PhC{\equiv}CPh$	$TaCl_5/n\text{-}Bu_4Sn$	insoluble	(M_w)
$PhC{\equiv}CC_6H_4\text{-}p\text{-}t\text{-}Bu$	$TaCl_5/n\text{-}Bu_4Sn$	3600	(M_w)
d) Disubstituted heteroatom-containing acetylenes			
$ClC{\equiv}C\text{-}n\text{-}C_6H_{13}$	$MoCl_5/n\text{-}Bu_4Sn$	1100	(M_w)
$ClC{\equiv}CPh$	$MoCl_5/n\text{-}Bu_4Sn$	690	(M_w)
$MeC{\equiv}CSiMe_3$	$TaCl_5$	730	(M_w)
$PhC{\equiv}CC_6H_4\text{-}p\text{-}SiMe_3$	$TaCl_5/n\text{-}Bu_4Sn$	2200	(M_w)

^{a} Notation in parenthesis indicates the type of molecular weight (M_w) where M_w is the weight average molecular weight in g/mol; M_n is the number average molecular weight in g/mol), and [η] is the polymer viscosity in dL/g.

21.3 METATHESIS POLYMERIZATION AND STEREOREGULARITY OF THE FORMED POLYMERS

21.3.1 RELATIONSHIP BETWEEN MONOMERS AND CATALYSTS

Table 21.2 presents examples of the monomers and catalysts used in the metathesis polymerization of mono- and disubstituted acetylenes. See other review articles for detailed information including references therein.[1,2,5,6] In general, when suitable Mo and W catalysts are chosen, they are capable of polymerizing monosubstituted acetylenes including hydrocarbon-based monomers, as well as heteroatom-containing monomers. Furthermore, sterically unhindered monomers as well as very crowded monomers can be polymerized. In contrast, Nb and Ta catalysts induce merely cyclotrimerization for monosubstituted acetylenes. Typical monosubstituted hydrocarbon monomers are *tert*-butylacetylene and phenylacetylene. Examples of common heteroatoms in monosubstituted acetylene substituents include Si, Ge, and halogens. In particular, Si- and F-containing substituents are unlikely to deactivate polymerization catalysts, and tend to provide relatively high molecular weight polymers. An interesting point in the polymerization of phenylacetylenes by W and Mo catalysts is that phenylacetylene derivatives with bulky *ortho*-substituents (e.g., SiMe_3 and CF_3) yield polymers whose molecular weights reach about one million. This ortho-substituent effect is quite remarkable since the molecular weight of unsubstituted poly(phenylacetylene)s (PPAs) prepared using the same catalysts are about two orders of magnitude smaller. This demonstrates that W and Mo catalysts are especially effective in polymerizing sterically crowded monosubstituted acetylenes.

In general, disubstituted acetylenes are sterically more hindered than monosubstituted ones. Consequently, they do not polymerize with either Ziegler-type or Rh catalysts, but are polymerized only by metathesis catalysts based on group 5 and 6 transition metals. Disubstituted monomers with less steric hindrance tend to polymerize with Mo and W catalysts, whereas their more sterically crowded counterparts polymerize only with Nb and Ta catalysts. For instance, n-dialkylacetylenes selectively yield polymers with $MoCl_5$/cocatalyst and WCl_6/cocatalyst systems, whereas the polymerization of these monomers produces cyclotrimers as byproducts when Nb and Ta catalyst systems are used. In the polymerization of 1-phenyl-1-alkynes, in contrast, W-based catalysts are somewhat active, with Nb and Ta catalysts being much more effective. Furthermore, Mo-, W-, and Nb-based catalysts only minorly induce the polymerization of diphenylacetylene, whereas $TaCl_5$/cocatalyst systems produce polymer. Poly(diphenylacetylene) is insoluble in all solvents, whereas its derivatives having bulky *para*-substituents are soluble in toluene and chloroform. One of the most reactive and interesting heteroatom-containing disubstituted acetylenes is 1-(trimethylsilyl)-1-propyne, which is quantitatively polymerized by Nb and Ta catalysts. The M_w of the polymer obtained with the $TaCl_5$/Ph_3Bi system reaches four million. The Cl-containing monomers in Table 21.2 are polymerizable only by Mo-based catalysts, which appears to be due to both steric and electronic reasons.

21.3.2 GEOMETRIC STRUCTURE OF THE FORMED POLYMERS

As described above, a variety of polymers are obtained from both mono- and disubstituted acetylene monomers using metathesis catalysts. Among these monomers, the geometric structures of the formed polymers have been studied in detail for phenylacetylene, *tert*-butylacetylene, and a few disubstituted monomers. Nuclear magnetic resonance (NMR) and IR spectroscopies are often used to obtain information about the geometric structure of the polymers. Generally, it is difficult to characterize the geometric structure of disubstituted acetylene polymers because their main chain double bonds have tetrasubstituted ethylene structural units. Discussed below are the geometric structures of the polymers obtained from phenylacetylene, *tert*-butylacetylene, and a few disubstituted acetylenes.

21.3.2.1 Poly(phenylacetylene) and Other Poly(arylacetylenes)

The steric structure of poly(phenylacetylene) has been studied extensively, but only qualitatively in most cases.[25] The poly(phenylacetylene) prepared with $Fe(acac)_3$/Et_3Al (2), a Ziegler-type catalyst, possesses mainly the cis-cisoidal structure (as evidenced by the C–H out-of-plane deformation at 740 cm^{-1} in the IR spectrum); this polymer is crimson in color and insoluble in all solvents owing to its high crystallinity.[25,26] On the other hand, the polymerization of phenylacetylene by WCl_6 or $MoCl_5$ provides completely soluble polymers, having relatively high molecular weights (M_n) of ca. 5,000–15,000.[27] The W-derived polymer, which possesses a weak absorption at 870 cm^{-1} in the IR spectrum and an emission at 430 nm upon excitation at 250 nm, is auburn in color, and has a high softening point (~226 °C). In contrast, the Mo-derived polymer, which has a strong absorption at 870 cm^{-1} and an emission at 360 nm upon excitation at 250 nm, is dull yellow in color, and has a high softening point (~215 °C). These results suggest that the W- and Mo-derived polymers have, respectively, trans-rich and cis-rich structures.

Poly(β-naphthylacetylene) (M_n = 92,000) obtained from WCl_6 is fully soluble in toluene and is dark brown in color, whereas polymer made from $MoCl_5$ is only marginally soluble and is red in color, suggesting a difference in their geometric structures.[28] On the basis of IR spectroscopy, differential thermal analysis, and X-ray diffraction, it is inferred that the former polymer is trans-rich while the latter is cis-rich and more crystalline. W-derived poly(α-naphthylacetylene) (M_n = 140,000) is completely soluble in toluene, is dark purple in color, and has an absorption at 510 nm, whereas the Mo-derived polymer is insoluble and red in color.[29] Thus the W-derived polymer has a wider conjugation that is attributable to a more trans-rich structure. Poly(9-anthrylacetylene) can be synthesized using both WCl_6 and $MoCl_5$, and is a black polymer insoluble in all solvents irrespective of the identity of the catalyst used.[30] Poly(1-pyrenylacetylene) has also been prepared with both W and Mo catalysts systems.[31] The W-derived polymer (M_n = 140,000) is a dark purple

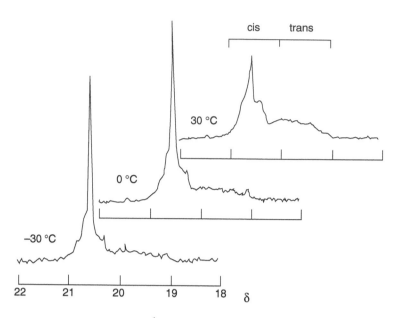

FIGURE 21.4 The methyl signal in the ^{1}H NMR spectra of poly(o-methylphenylacetylene)s obtained with MoOCl$_4$/n-Bu$_4$Sn/EtOH (1:1:1) in toluene at different temperatures. (Reprinted with permission from Kaneshiro, H.; Masuda, T.; Higashimura, T. *Polym. Bull.* **1995**, *35*, 17. Copyright 1995 Springer-Verlag GmbH.)

polymer soluble in o-dichlorobenzene, whereas the Mo-derived polymer is insoluble in any organic solvents; this difference is also explainable in terms of a difference in geometric structure.

The cis content of poly(o-methylphenylacetylene) can be determined by NMR because the methyl signal shows different chemical shifts, depending on the geometric structure of the backbone olefin units (cis or trans).[32] The MoOCl$_4$/n-Bu$_4$Sn/EtOH catalyst provides a cis-rich, "living" poly(o-methylphenylacetylene) (cis content 77%, M_w/M_n = 1.21) at 0 and -30 °C (Figure 21.4). Whereas poly(phenylacetylene) made using Fe catalysts is crystalline and insoluble owing to its cis-cisoidal structure, poly(phenylacetylene) derivatives having p-adamantyl, p-$tert$-butyl, and p-n-butyl substituents are soluble, irrespective of the catalyst used in their preparation.[33] According to ^{1}H and ^{13}C NMR, substituted poly(phenylacetylene) polymers obtained from either Fe or Rh catalysts have virtually an all-cis structure, although it is difficult to distinguish between the cis-cisoidal and cis-transoidal structures. On the other hand, Mo- and W-derived substituted poly(phenylacetylene)s have, respectively, cis-rich and trans-rich structures.

21.3.2.2 Poly($tert$-butylacetylene)

The geometric structure of poly($tert$-butylacetylene) can easily be determined by ^{13}C NMR[34] and ^{1}H NMR.[35] MoCl$_5$ produces polymer with a higher cis content than WCl$_6$. The cis content is higher when oxygen-containing polymerization solvents like anisole are used, as compared to hydrocarbon solvents such as toluene.[34] In the extreme case, all-cis poly($tert$-butylacetylene) can be obtained by utilizing MoCl$_5$ in anisole, whereas the cis content for the WCl$_6$/toluene system is the lowest and is about 50%. Poly($tert$-butylacetylene), obtained with the MoOCl$_4$/n-Bu$_4$Sn system by terminating polymerization just after complete monomer consumption, possesses an all-cis structure.[36] However, stereoregularity decreases when the polymerization is allowed to stand for a further period after complete monomer consumption. This is because acid-catalyzed geometric isomerization of the polymer occurs, even though the propagation reaction itself proceeds stereoselectively.

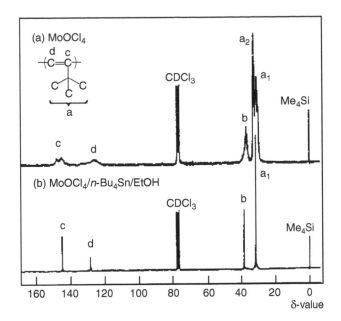

FIGURE 21.5 ^{13}C NMR spectra of poly(*tert*-butylacetylene)s obtained with (a) MoOCl$_4$ and (b) MoOCl$_4$ /*n*-Bu$_4$ Sn/EtOH (1:1:1) in toluene at 0 °C. (Reprinted with permission from Nakano, M.; Masuda, T.; Higashimura, T. *Macromolecules* **1994**, *27*, 1344. Copyright 1994 American Chemical Society.)

tert-Butylacetylene undergoes stereoselective living polymerization with MoOCl$_4$/*n*-Bu$_4$Sn/EtOH.[37] The methyl carbon signal in the ^{13}C NMR of poly(*tert*-butylacetylene) obtained from a single-component MoOCl$_4$ catalyst splits into two peaks, indicating the presence of both cis and trans structures (Figures 21.5a and 21.6). In contrast, polymer obtained from the MoOCl$_4$/*n*-Bu$_4$Sn/EtOH (1:1:1) catalyst system shows virtually only one peak for the methyl carbons (Figure 21.5b), and hence is thought to exclusively consist of the cis structure (cis content 97%, $M_w/M_n = 1.12$). Similar, but somewhat inferior results are obtained with the MoCl$_5$/*n*-Bu$_4$Sn/EtOH (1:1:1) system (cis content 90%, $M_w/M_n = 1.24$).

21.3.2.3   Polymers from Disubstituted Acetylenes

Although it is difficult to gain knowledge about the geometric structure of the polymers derived from disubstituted acetylenes, some information is available regarding poly(diphenylacetylene) and poly[1-(trimethylsilyl)-1-propyne]. The relationship between the catalyst used and the resultant polymer geometric structure for poly(diphenylacetylene) is similar to that for poly(phenylacetylene): the W-derived polymer is trans-rich while the Mo-derived polymer is cis-rich. These assignments were made according to IR spectroscopy and differential thermal analysis; both polymers are completely insoluble.[38]

Poly[1-(trimethylsilyl)-1-propyne] is a unique polymer that shows very high gas permeability.[39] This polymer can be obtained by using either TaCl$_5$ or NbCl$_5$ as polymerization catalysts. All four carbon atoms in this polymer display splitting of their ^{13}C NMR signal into two peaks (Figure 21.7).[40] From the peak area ratios, it can be concluded that the Nb-derived polymer has an approximately 60% cis structure, whereas its Ta-derived counterpart has a 38% cis structure. It is postulated that NbCl$_5$ provides poly[1-(trimethylsilyl)-1-propyne]s with higher cis content as compared to TaCl$_5$ in a similar way to that in which Mo-based catalysts produce higher cis contents as compared to W-based catalysts. Provided that the same catalyst is used, the geometric structure of this polymer varies little with the polymerization temperature over a range of 30–100 °C, or with the polymerization solvent (e.g., cyclohexane, toluene, 1,2-dichloroethane, anisole).

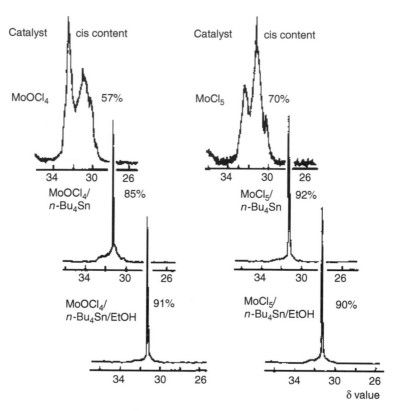

FIGURE 21.6 The methyl signal in the ^{13}C NMR spectra of poly(*tert*-butylacetylene)s obtained with various catalyst systems in toluene at 0 °C. (Reprinted with permission from Nakano, M.; Masuda, T.; Higashimura, T. *Macromolecules* **1994**, *27*, 1344. Copyright 1994 American Chemical Society.)

21.4 POLYMERIZATION WITH RHODIUM CATALYSTS AND STEREOREGULARITY OF THE FORMED POLYMERS

In 1969, Kern first reported that the Wilkinson catalyst, RhCl(PPh$_3$)$_3$, could be used to polymerize phenylacetylene monomer, and that the color of the resulting poly(phenylacetylene) was dependent on the polymerization conditions used.[41] Recently, analogues of the Wilkinson catalyst such as [(2,5-norbornadiene)RhCl]$_2$ (**7**, Figure 21.1) have been developed that selectively produce *cis*-PPA even at room temperature.[42–46] This is in contrast to the low temperature (−78 °C) required when using Ziegler–Natta catalysts to produce perfect *cis*-polyacetylene.[47]

Rh-catalyzed polymerizations of substituted acetylenes proceed through the insertion mechanism (Scheme 21.2b). Furthermore, the excellent ability of Rh catalysts to tolerate various functional groups allows for the synthesis of highly functionalized polymers. Thus, the monomers utilizable include phenylacetylenes bearing amino, hydroxyl, azo, and precursors of radical groups (such as a nitroxide group), in addition to propiolic acid esters, *N*-propargylamides, and cyclooctyne (Figure 21.8). The polymerization of electron-donor-substituted monomers, such as *p*-methoxyethynylbenzene, gives a lower polymer yield as compared to the polymerization of electron-accepting monomers such as *p*-chloroethynylbenzene. The opposite relationship is observed when metathesis polymerization catalysts are used, demonstrating that the mechanisms of polymerizations involving these two kinds of catalysts differ from each other.[48]

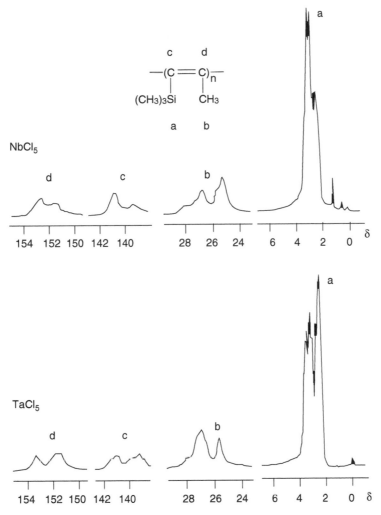

FIGURE 21.7 ^{13}C NMR spectra of poly[1-(trimethylsilyl)-1-propyne]s obtained with NbCl$_5$ and TaCl$_5$ catalysts in toluene at 80 °C. (Reprinted with permission from Izumikawa, H.; Masuda, T.; Higashimura, T. *Polym. Bull.* **1991**, *27*, 193. Copyright 1991 Springer-Verlag GmbH.)

21.4.1 Ligand and Solvent Effects in the Polymerization

Dimeric Rh complexes of the formula [(L-L)RhCl]$_2$ (L-L = 2,5-norbornadiene (nbd) (**7**); 1,5-cyclooctadiene (cod) (**7a**)) and [(L)$_2$RhCl]$_2$ (L = cyclooctene (coe) (**7b**)) have frequently been employed for the polymerization of phenylacetylene (Figure 21.9).[42,43,49–51] Of these Rh complex catalysts, [(nbd)RhCl]$_2$ gives the highest polymerization yield (almost quantitative), whereas [(coe)$_2$RhCl]$_2$ does not yield any polymer.[43] This suggests that the nbd ligand is strongly coordinated to the Rh atom during the polymerization; that is, no replacement of this ligand with electron-donating molecules including monomer takes place to decrease the polymerization yield. The order of catalytic activity of the dimeric Rh complexes from highest activity to lowest is as follows: [(nbd)RhCl]$_2$ > [(cod)RhCl]$_2$ ≫ [(coe)$_2$RhCl]$_2$.[43,51]

The Rh-catalyzed polymerization of monosubstituted acetylenes proceeds in various solvents including benzene, tetrahydrofuran (THF), ethanol, and triethylamine (Et$_3$N). Among these solvents, ethanol and Et$_3$N are most favorable for phenylacetylenes from the viewpoint of both

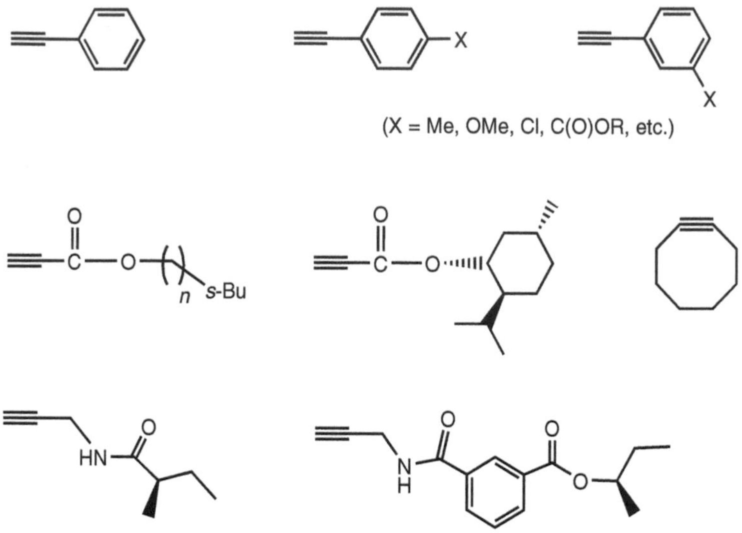

(X = Me, OMe, Cl, C(O)OR, etc.)

FIGURE 21.8 Substituted acetylene monomers polymerizable with Rh-based catalysts.

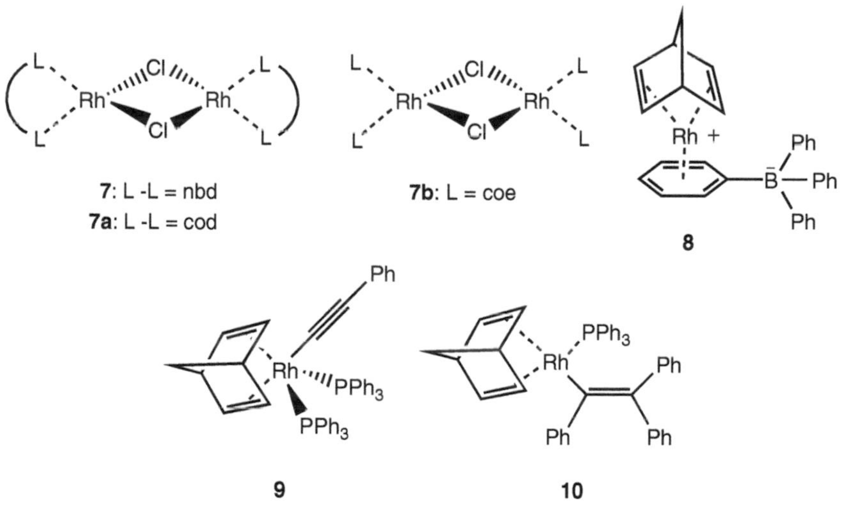

7: L -L = nbd
7a: L -L = cod

7b: L = coe

8

9   **10**

FIGURE 21.9 Rh catalysts for the polymerization of acetylenes.

polymerization rate and polymer molecular weight. This polymerization feature is different from that with metathesis catalysts, such as WCl_6, which form stable complexes with amines, and react with alcohols to give alkoxy complexes. The most widely employed Rh catalyst system is [(nbd)RhCl]$_2$/Et$_3$N. This catalyst gives excellent yields of poly(phenylacetylenes) having high molecular weights ($M_n > 10^5$). Combinations of [(nbd)RhCl]$_2$ with suitable organometallics such as n-BuLi and Et$_3$Al greatly accelerate the polymerization of phenylacetylene.[52]

As shown in Figure 21.10, the ^{1}H NMR spectra of various Rh-derived monosubstituted acetylenes polymers display sharp singlet signals at about δ 7–6 ppm, which can be assigned to the backbone hydrogen resonances. In particular, poly(phenylacetylene) shows a sharp peak at δ 5.8 ppm characteristic of the cis-olefinic protons of the main chain. The highly stereoregular polymers shown in Figure 21.10 possess a completely head-to-tail structure, as clearly evidenced by the

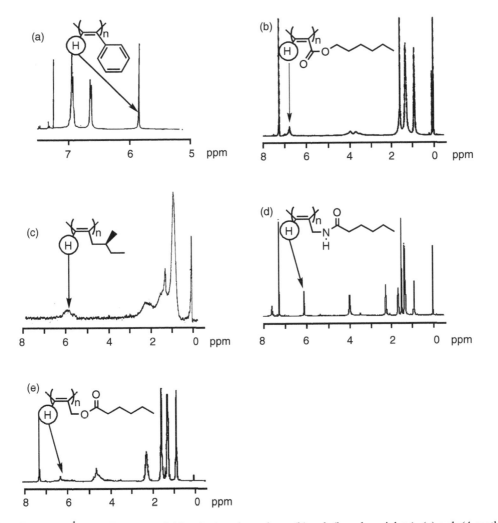

FIGURE 21.10 ^{1}H NMR spectra of (a) polyphenylacetylene, (b) poly(hexylpropiolate), (c) poly(4-methyl-1-hexyne), (d) poly(N-propargylhexanamide), and (e) poly(propargylhexanoate). (Reprinted with permission from Ciardelli, F.; Lanzillo, S.; Pieroni, O. *Macromolecules* **1974**, 7, 174. Copyright 1974 American Chemical Society.)

presence of a singlet vinyl proton signal in the ^{1}H NMR. The ^{13}C NMR spectra of two of these polymers (Figure 21.11) display backbone resonances at about δ 130–140 ppm. Furthermore, their Raman spectra display two large peaks at 1550 and 1330 cm^{-1} assigned to the cis-C=C bond in the cis-transoidal structure, and to the C–C bond coupled with the C–H bond in the cis-polymer, respectively.[53,54] From these results, it is concluded that the Rh-based PPAs selectively possess the cis-transoidal structure.

The UV-VIS spectrum of the [(nbd)RhCl]$_2$ complex has been measured in various solvents to gain insight into the active species of polymerization.[43] An absorption shoulder at 360 nm in THF or ethanol, which is also observed in the solid state, is assignable to the dimeric Rh complex. On the other hand, when the catalyst is dissolved in neat Et$_3$N, the UV-VIS peak observed at 380 nm can be ascribed to a typical monomeric species of the Rh complex. The appearance of the monomeric species of the Rh complex indicates that Et$_3$N promotes the dissociation of the dimeric complex, [(nbd)RhCl]$_2$, to produce the Et$_3$N coordinated monomeric one, which in turn must lose the Et$_3$N ligand to form the propagating species for the polymerization. The dissociation of the dimeric Rh

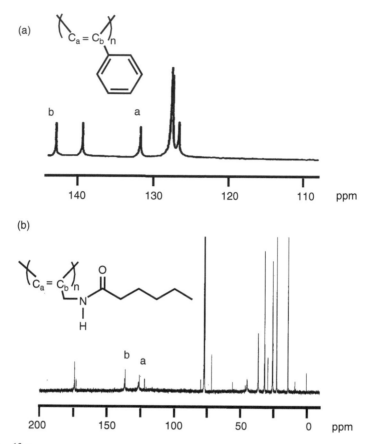

FIGURE 21.11 ^{13}C NMR spectra of (a) polyphenylacetylene and (b) poly(N-propargylhexanamide) in CDCl$_3$. (Reprinted with permission from Furlani, A.; Napoletano, C.; Russo, M. V.; Feast, W. *J. Polym. Bull.* **1986**, *16*, 311. Copyright 1986 Springer-Verlag GmbH.)

complex into the monomeric one is not fully complete in Et$_3$N; the extent of this dissociation varies depending on the solvent used.

As shown in Figure 21.12, the ^1H NMR spectrum of [(nbd)RhCl]$_2$ shows peaks at 1.2, 3.8, and 3.9 ppm that can be assigned to, respectively, the bridging methylene protons, and the olefinic protons in the ligating and free nbd molecules. This indicates that the complex has a planar square structure. A large shift of the olefinic proton (H^a) in the ligand molecule from that of the free species indicates strong coordination of the molecules to the Rh atom.[43,55] Among the bis-coe, cod, and nbd ligands, the largest upfield shift of the olefinic proton (from 6.63 ppm in the free molecule to 3.9 ppm in the coordinated molecule) is observed for nbd. This remarkable shift reflects a much stronger coordination of the nbd molecules to the Rh atom, which should generate a more stable propagating species. This is consistent with the fact that no polymer is formed with [(coe)$_2$RhCl]$_2$ catalyst, owing to weaker coordination of coe to the Rh atom.

The ^1H NMR spectrum of [(nbd)RhCl]$_2$ was also measured in CDCl$_3$ solution after the addition of Et$_3$N. The signals observed at 1.2 and 3.7 ppm are assigned to the ligand methylene bridge protons (H^d) and methine protons (H^b), respectively. The two signals observed at 3.3 and 4.1 ppm after the addition of Et$_3$N can be assigned to the olefinic protons (H^a and H^c) in the monomeric species, (nbd)RhCl/Et$_3$N. Therefore, the spectral change induced by the addition of Et$_3$N can be explained in terms of dissociation of the dinuclear complex into the monomeric species as previously mentioned: the two sets of protons (H^a and H^c) never become magnetically equivalent owing to the lower

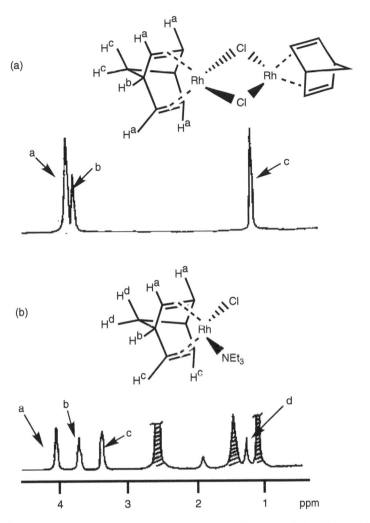

FIGURE 21.12 ^{1}H NMR spectra of (a) [(nbd)RhCl]$_2$ and (b) a Et$_3$N containing [(nbd)RhCl]$_2$ measured in CDCl$_3$ at 30 °C. (Reprinted with permission from Tabata, M.; Yang, W.; Yokota, K. *J. Polym. Sci., Part A: Polym. Chem.* **1994**, *32*, 1113. Copyright 1994 John Wiley & Sons, Inc.)

symmetry in the monomeric form. The signals marked with a hatch at 1.2 and 2.6 ppm may be assigned to the methyl and methylene protons of free Et$_3$N molecules, respectively. Thus, the ^{1}H NMR spectrum manifests that the Rh complex is stable even at room temperature, which agrees with the fact that the Rh-catalyzed polymerization of monosubstituted acetylenes proceeds at room temperature with fairly high polymer yields. The ^{1}H NMR data regarding the dissociation of the Rh catalyst is consistent with the of the UV-VIS data. Therefore, it is concluded that both Et$_3$N and methanol solvents, which are frequently used in this polymerization, are coordinated to the monomeric Rh species to stabilize it, and that the order of the ability of the solvents to coordinate to the Rh atom is as follows: Et$_3$N > MeOH $\gg$ THF.

The polymerization of phenylacetylenes is feasible even in aqueous media when water-soluble catalysts are used. For example, [(cod)Rh(mid)$_2$]$^+$[PF$_6$]$^-$ (mid = N-methylimidazole) provides cis-transoidal poly(phenylacetylene) (cis content 98%) in high yield (98%).[56] Other complexes such as (cod)Rh(SO$_3$C$_6$H$_4$-p-CH$_3$)(H$_2$O) and (nbd)Rh(SO$_3$C$_6$H$_4$-p-CH$_3$)(H$_2$O) also work as water-soluble catalysts. The polymerization of phenylacetylene in compressed (liquid or supercritical) CO$_2$ has been studied using a the Rh catalyst, $[(nbd)Rh(acac)]_2$.[57] Polymerization is accelerated in CO_2 compared to the case of conventional organic solvents such as THF and hexane. Quite recently, ionic liquids have been examined as media for Rh-catalyzed polymerization of phenylacetylene.[58]

21.4.2 POLYMERIZATION OF VARIOUS ACETYLENES

The excellent ability of Rh catalysts to tolerate functional groups enables the polymerization of many functional acetylene monomers. For example, a substituted phenylacetylene bearing an N,N-dimethylamino group on the phenyl ring[48] has been used to prepare radical-carrying polymers. Phenylacetylenes with an OH group on the phenyl ring[59,60] or an NH moiety on the phenyl ring have been used as precursors for oxy radicals, and nitroxide groups[61] for the formation of polymeric magnets. Several alkyl propiolates are polymerized in moderate yields using Rh catalysts ($[(nbd)RhCl]_2$). Relatively high yields of poly(propiolate)s with high molecular weights are accessible when the polymerization is conducted in alcoholic solvents or acetonitrile at high monomer and catalyst concentrations.[62,63] In general, the cis olefinic proton of poly(propiolates) appears at around 6.7 ppm in the ^{1}H NMR spectrum (Figure 21.10), and according to this peak's relative magnitude, poly(propiolates) with alkylene spacers between the ester group and the chiral carbon possess a quantitative cis content (100%). The cis contents of the polymers without alkylene spacers are lower than those of the other polymers (~80%).

Although carboxylic acids are known to work as terminating agents for acetylene polymerization, sodium p-ethynylphenylcarboxylate and sodium propiolate polymerize in the presence of various water-soluble Rh complexes, including $[Rh(cod)_2]^+[BF_4]^-$, $Rh(cod)(tosylate)(H_2O)$, and $[Rh(nbd)_2]^+[ClO_4]^-$.[64,65] These carboxylate- and propiolate-bearing polyacetylenes can assume predominantly one-handed helical conformations upon complexation with chiral amines. These polymers can be used as probes for the chirality assignment of various chiral compounds.

Recently, it has been found that N-propargylamides can be polymerized by the cationic Rh complex, $[(nbd)Rh]^+[\eta^6\text{-}C_6H_5B(C_6H_5)_3]^-$ (**8**, Figure 21.9), to produce helical polymers.[66,67] The helicity of these polymers is induced by intramolecular hydrogen bonding. The ^{1}H NMR spectra of poly(N-propargylalkylamides) show very broad signals for the olefinic protons of the main chain. This peak broadening implies that the helical structure restrains the mobility of the main chain. When a few drops of CD_3OD are added to the $CDCl_3$ solution of the polymer sample, the signal of the main chain proton sharpens in the ^{1}H NMR at 60 °C (6.0 ppm), and verifies that the polymers possess almost perfect cis-stereoregularity (Figure 21.10). The cis contents (~80%) of poly(N-propargylbenzamides) tend to be lower than those of the other poly(N-propargylalkylamides).

In general, Rh catalysts are not very effective in stereoselectively polymerizing sterically crowded monosubstituted acetylenes such as *tert*-butylacetylene and ortho-substituted phenylacetylenes. Rh catalysts are also not capable of polymerizing disubstituted acetylenes. One exception is cyclooctyne, whose very large ring strain (~38 kJ/mol) enables fast polymerization with $[(nbd)RhCl]_2$, giving an insoluble polymer in good yield.[68]

21.4.3 STEREOSELECTIVE LIVING POLYMERIZATION

Rh-catalyzed living polymerization was first accomplished in 1994.[69] The excellent ability of pre-isolated, well-defined catalyst to produce quantitative yields of poly(phenylacetylenes) with narrow polydispersities was demonstrated. The catalyst used, $(nbd)(PPh_3)_2Rh\text{-}C\equiv C\text{-}Ph$ (**9**, Figure 21.9), has been completely characterized by single-crystal X-ray analysis. It has been disclosed that the initiation reaction of acetylene polymerization with this catalyst proceeds not through direct insertion of monomer into the Rh–C bond of (nbd)(PPh$_3$)$_2$Rh–C≡C–Ph, but through the initial formation of a Rh–H complex by the reaction of (nbd)(PPh$_3$)$_2$Rh–C≡C–Ph with the acetylene monomer, followed by monomer insertion.[70] Polymerization of phenylacetylene with (nbd)(PPh$_3$)$_2$Rh–C≡C–Ph in the presence of 4-(N,N-dimethylamino)pyridine (DMAP) provides a well-defined polymer having a long-lived active site at the propagating chain end. Without added DMAP, the polydispersity increases from about 1.0 to 1.3. The initiation efficiency of this Rh catalyst system, however, remains low (no more than 35%). The high stability of the propagating centers of this system allows for the isolation of poly(phenylacetylene) having active propagating sites, which can then be used to sequentially polymerize different monomers, giving block copolymers with precisely controlled structures.

One striking feature of the stereoregular poly(phenylacetylenes) is their simple NMR spectra, which facilitate the investigation of the polymerization mechanism as well as the polymer structure. A copolymer of phenylacetylene with partially ^{13}C-labeled phenylacetylene (Ph^{13}C=^{13}C) shows two doublet carbon signals with a $J_{13C-13C}$ of 72 Hz, indicating that the presence of a ^{13}C=^{13}C bond in the polymer backbone.[70] This is a clear indication of the presence of the insertion mechanism instead of a metathesis pathway.

Following the discovery of the well-defined Rh-based living polymerization catalyst **9**, a new living polymerization catalyst, [(nbd)RhOMe]$_2$/PPh$_3$/DMAP, was further developed, which increased initiation efficiency from 35% to 70%.[71] Polymerization of acetylene monomers with [(nbd)RhOMe]$_2$/PPh$_3$/DMAP is three to four times faster than with **9**. Isolation of [(nbd)RhOMe]$_2$ is unnecessary; a simple mixture of commercially available [(nbd)RhCl]$_2$, Ph$_3$P, NaOMe, and DMAP induces the living polymerization of phenylacetylene without broadening polydispersity.

A ternary catalyst, [(nbd)RhCl]$_2$/LiCPh=CPh$_2$/PPh$_3$, polymerizes phenylacetylene and its para-substituted derivatives to give living polymers.[72] Living polymerization is also possible in the presence of water. The isolation of the active species from the [(nbd)RhCl]$_2$/LiCPh=CPh$_2$/PPh$_3$ mixture is not an essential requirement; when the complex is formed in situ by reaction of [(nbd)RhCl]$_2$ with LiCPh=CPh$_2$ and Ph$_3$P, living polymerization is induced with quantitative initiation efficiency. A new vinylrhodium complex, **10** (Figure 21.9) has also been prepared, isolated, and fully characterized by X-ray analysis. This complex proved to be effective in the living polymerization of phenylacetylenes.[73] A remarkable feature of this polymerization system is the ability to introduce functional groups at the initiating chain end. For example, living poly(phenylacetylene) bearing a terminal hydroxyl group is readily obtained by the polymerization with a ternary catalyst comprising [(nbd)RhCl]$_2$, LiCPh=C(Ph)(C$_6$H$_4$-p-OSi(CH$_3$)$_2$-t-Bu), and Ph$_3$P, followed by deprotection (desilylation) of the formed polymer. Polymerization of β-propiolactone with the terminal phenoxide anion of this polymer gives a new block copolymer of phenylacetylene with β-propiolactone.[74]

21.5 CONCLUSIONS

Polyacetylenes are the most typical and basic π-conjugated polymers, and can ideally take four geometrical structures (trans-transoid, trans-cisoid, cis-transoid, cis-cisoid). At present, not only early transition metals, but also many late transition metals are used as catalysts for the polymerization of substituted acetylenes. However, the effective catalysts are restricted to some extent, and Ta, Nd, Mo, and W of transition metal groups 5 and 6, and Fe and Rh of transition metal groups 8 and 9 are mainly used. The polymerization mechanism of Ta, Nd, W, and Mo based catalysts is a metathesis mechanism, and that of Ti, Fe, and Rh based catalysts is an insertion mechanism. Most of the substituted polyacetylenes prepared with W and Mo catalysts provide trans-rich and cis-rich geometries respectively. Polymers formed with Fe and Rh catalysts selectively possess stereoregular cis main chains.

REFERENCES AND NOTES

1. Masuda, T.; Sanda, F. Polymerization of substituted acetylenes. In *Handbook of Metathesis*; Grubbs, R. H., Ed.; Wiley-VCH: Weinheim, 2003; Vol. 3, pp 375–406.

2. Nomura, R.; Masuda T. Acetylenic polymers, substituted. In: *Encyclopedia of Polymer Science and Technology*, 3rd ed.; Kroshwitz, J. I., Ed.; Wiley: New York 2003; Vol. IA, p 1.

3. Tabata, M.; Sone, T.; Mawatari, Y.; Yonemoto, D.; Miyasaka, A.; Fukushima, T.; Sadahiro, Y. π-Conjugated columnar polyacetylenes prepared with Rh complex catalyst. *Macromol. Symp.* **2003**, *192*, 75–97.

4. Tabata, M.; Sone, T.; Sadahiro, Y. Precise synthesis of monosubstituted polyacetylenes using Rh complex catalysts. Control of solid structure and π-conjugation length. *Macromol. Chem. Phys.* **1999**, *200*, 265–282.

5. Masuda, T. Acetylene polymerization. In *Catalysis in Precision Polymerization*; Kobayashi, S., Ed.; Wiley: Chichester, 1997; pp 67–97.

6. Masuda, T. Acetylenic polymer. In *Polymeric Material Encyclopedia;* Salamone, J. C., Ed.; CRC Press: New York, 1996; Vol. 1, pp 32–40.

7. Lam, J. W. Y.; Tang, B. Z. Liquid-crystalline and light-emitting polyacetylenes. *J. Polym. Sci., Part A: Polym. Chem.* **2003**, *41*, 2607–2629.

8. Yashima, E.; Maeda, K. Helicity induction on optically inactive polyacetylenes and polyphosphazenes. In *Synthetic Macromolecules with Higher Structural Order*; Kahn, T. I. M., Ed.; ACS Symposium Series 812; American Chemical Society: Washington, DC, 2002; pp 41–53.

9. McQuade, D. T.; Pullen, A. E.; Swager, T. M. Conjugated polymer-based chemical sensors. *Chem. Rev.* **2000**, *100*, 2537–2574.

10. Kraft, A.; Grimsdale, A. C.; Holmes, A. B. Electroluminescent conjugated polymers-seeing polymers in a new light. *Angew. Chem. Int. Ed. Engl.* **1998**, *37*, 403–428.

11. Curran, S.; Stark-Hauser, A.; Roth, S. Polyacetylene. In *Handbook of Organic Conductive Molecules and Polymers*; Nalwa, H. S., Ed.; Wiley: Chichester 1997; Vol. 2, Chapter 1.

12. Saxman, A. M.; Liepens, R.; Aldissi, M. Polyacetylene: Its synthesis, doping and structure. *Prog. Polym. Sci.* **1985**, *11*, 57–89.

13. Chien, J. C. W. *Polyacetylene*; Academic Press: New York, 1984.

14. Shirakawa, H.; Ikeda, S. Infrared spectra of polyacetylene. *Polym. J.* **1971**, *2*, 231–244.

15. Ito, T.; Shirakawa, H.; Ikeda, S. J. Simultaneous polymerization and formation of polyacetylene film on the surface of a concentrated soluble Ziegler-type catalyst solution. *Polym. Sci., Polym. Chem. Ed.* **1974**, *12*, 11–20.

16. Naarmann, H.; Theophilou, N. New process for the production of metal-like, stable polyacetylene. *Synth. Met.* **1987**, *22*, 1–8.

17. Naarmann, H. Synthesis of new conductive polymers. *Synth. Met.* **1987**, *17*, 223–228.

18. Luttinger, L. B. *Chem. Ind.* (London) **1960**, 1135.

19. Terlemezyan, L.; Mihailov, M. On the configuration of polyacetylene obtained by Luttinger's catalyst. *Makromol. Chem., Rapid Commun.* **1982**, *3*, 613–161.

20. Akagi, K.; Sakamaki, K.; Shirakawa, H. Intrinsic nonsolvent polymerization method for synthesis of highly stretchable and highly conductive polyacetylene films. *Macromolecules* **1992**, *25*, 6725–6726.

21. Akagi, K.; Piao, G.; Kaneko, S.; Sakamaki, K.; Shirakawa, H.; Kyotani, M. Helical polyacetylene synthesized with a chiral nematic reaction field. *Science* **1998**, *282*, 1683–1686.

22. Maricq, M. M.; Waugh, J. S.; MacDiarmid, A. G.; Shirakawa, H.; Heeger, A. J. Carbon-13 nuclear magnetic resonance of *cis*- and *trans*-polyacetylenes. *J. Am. Chem. Soc.* **1978**, *100*, 7729–7730.

23. Shirakawa, H.; Ito, T.; Ikeda, S. Raman scattering of electronic spectra of polyacetylene. *Polym. J.* **1973**, *4*, 460–462.

24. Shirakawa, H.; Ikeda, S. Cyclotrimerization of acetylene by the tris(acetylacetonato)titanium(III)-diethylaluminum chloride system. *J. Polym. Sci., Polym. Chem. Ed.* **1974**, *12*, 929–937.

25. Simionescu, C. I.; Percec, V. Progress in polyacetylene chemistry. *Prog. Polym. Sci.* **1982**, *8*, 133–214.

26. Biyani, B.; Campagna, A. J.; Daruwalla, D.; Srivastava, C. M.; Ehrlich, P. Crystallizable polyphenylacetylene. Preparation and solution properties. *J. Macromol. Sci., Chem.* **1975**, *A9*, 327–339.

27. Masuda, T.; Sasaki, N.; Higashimura, T. Polymerization of phenylacetylenes. III. Structure and properties of poly(phenylacetylene)s obtained by WCl_6 and $MoCl_5$. *Macromolecules* **1975**, *8*, 717–721.

28. Ohtori, T.; Masuda, T.; Higashimura, T. Polymerization of phenylacetylenes. IX. Polymerization of β-naphthyl-acetylene by WCl_6 and $MoCl_5$. *Polym. J.* **1979**, *11*, 805–811.

29. Nanjo, K.; Karim, S. M. A.; Nomura, R.; Wada, T.; Sasabe, H.; Masuda, T. Synthesis and properties of poly(1-naphthylacetylene) and poly(9-anthrylacetylene). *J. Polym. Sci., Part A: Polym. Chem.* **1999**, *37*, 277–282.

30. Musikabhumma, K.; Masuda. T. Synthesis and properties of widely conjugated polyacetylenes having anthryl and phenanthryl pendant groups. *J. Polym. Sci., Part A: Polym. Chem.* **1998**, *36*, 3131–3137.

31. Karim, S. M. A.; Musikabhumma, K.; Nomura, R.; Masuda. T. Synthesis and properties of poly(9-phenanthrylacetylene) and poly(1-pyrenylacetylene); *Proc. Jpn. Acad., Ser. B: Phys. Biol. Sci.* **1999**, *75B*, 97–100.

32. Kaneshiro, H.; Masuda, T.; Higashimura, T. Synthesis of a cis-rich, living poly[(*o*-methylphenyl)acetylene] by use of the $MoOCl_4$–*n*-Bu_4Sn–EtOH catalyst. *Polym. Bull.* **1995**, *35*, 17–23.

33. Fujita, Y.; Misumi, Y.; Tabata, M.; Masuda. T. Synthesis, geometric structure, and properties of poly(phenylacetylenes) with bulky *para*-substituents. *J. Polym. Sci., Part A: Polym. Chem.* **1998**, *36*, 3157–3163.

34. Okano, Y.; Masuda, T.; Higashimura, T. Polymerization of *t*-butylacetylene by group 6 transition metal catalysts: Geometric structure control by reaction conditions. *Polym. J.* **1982**, *14*, 477–483.

35. Katz, T. J.; Ho, T. H.; Shih, N.-Y.; Ying, Y.-C.; Stuart, V. I. W. Polymerization of acetylenes and cyclic olefins induced by metal carbynes. *J. Am. Chem. Soc.* **1984**, *106*, 2659–2668.

36. Masuda, T.; Izumikawa, H.; Misumi, Y.; Higashimura, T. Stereospecific polymerization of *tert*-butylacetylene by molybdenum catalysts. Effect of acid-catalyzed geometric isomerization. *Macromolecules* **1996**, *29*, 1167–1171.

37. Nakano, M.; Masuda, T.; Higashimura, T. Stereospecific living polymerization of *tert*-butylacetylene by molybdenum-based ternary catalyst systems. *Macromolecules* **1994**, *27*, 1344–1348.

38. Masuda, T.; Kawai, H.; Ohtori, T.; Higashimura, T. Polymerization of phenylacetylenes. X. Polymerization of diphenylacletylene by WCl_6- and $MoCl_5$-based catalysts. *Polym. J.* **1979**, *11*, 813–818.

39. Nagai, K.; Masuda, T.; Nakagawa, T.; Freeman, B. D.; Pinnau, I. Poly[1-(trimethylsilyl)-1-propyne] and related polymers: Synthesis, properties and functions. *Prog. Polym. Sci.* **2001**, *26*, 721–798.

40. Izumikawa, H.; Masuda, T.; Higashimura, T. Study on the geometric structure of poly[1-(trimethylsilyl)-1-propyne] by ^{13}C and ^{29}Si NMR spectroscopies. *Polym. Bull.* **1991**, *27*, 193–199.

41. Kern, R. J. Preparation and properties of isomeric poly(phenylacetylenes). *J. Polym. Sci., Polym. Chem. Ed.* **1969**, *7*, 621–631.

42. Tabata, M.; Yang, W.; Yokota, K. Polymerization of *m*-chlorophenylacetylene initiated by [Rh(norbornadiene)Cl]$_2$-triethylamine catalyst containing long-lived propagation species. *Polym. J.* **1990**, *22*, 1105–1107.

43. Tabata, M.; Yang, W.; Yokota, K. ^{1}H-NMR and UV studies of Rh complexes as a stereoregular polymerization catalysts for phenylacetylenes: Effects of ligands and solvents on its catalyst activity. *J. Polym. Sci., Part A: Polym. Chem.* **1994**, *32*, 1113–1120.

44. Tabata, M.; Sadahiro, Y.; Yokota, K.; Kobayashi, S. Formation of columnar from poly(*p*-methylethynylbenzene) polymerized using [Rh(norbornadiene)Cl]$_2$ as a catalyst. *Polym. J.* **1996**, *35*, 5411–5415.

45. Furlani, A.; Licoccia, S.; Russo, A. M.; Camius, A. M.; Marsich, N. Rhodium and platinum complexes as catalysts for the polymerization of phenylacetylene. *J. Polym. Sci., Part A: Polym. Chem.* **1986**, *24*, 991–1005.

46. Furlani, A.; Napoletano, C.; Russo, M. V.; Feast, W. J. Stereoregular polyphenylacetylene. *Polym. Bull.* **1986**, *16*, 311–317.

47. Ferraro, J. R.; Williams, J. M. *Introduction to Synthetic Electrical Conductors*; Academic Press: New York, 1987.

48. Tabata, M. Personal communication, 1997.

49. Lindgren, M.; Lee, H-S.; Tabata, M.; Yang, W.; Yokota, K. Synthesis of soluble poly(phenylacetylenes) containing a strong donor function. *Polymer* **1991**, *32*, 1531–1534.

50. Tabata, M.; Tanaka, Y.; Sadahiro, Y.; Sone, T.; Yokota, K.; Miura, I. Pressure-induced cis to trans isomerization of aromatic polyacetylenes. 2. Poly((*o*-ethoxyphenyl)acetylene) stereoregularly polymerized using a Rh complex catalyst. *Macromolecules* **1997**, *30*, 5200–5204.

51. Tabata, M.; Sone, T.; Sadahiro, Y.; Yokota, K.; Nozaki, Y. Pressure-induced cis to trans isomerization of aromatic polyacetylenes prepared using a Rh complex catalyst: A control of π-conjugation length. *J. Polym. Sci., Part A: Polym. Chem.* **1998**, *36*, 217–223.

52. Kanki, K.; Misumi, Y.; Masuda, T. Remarkable cocatalytic effect of organometallics and rate control by triphenylphosphine in the Rh-catalyzed polymerization of phenylacetylene. *Macromolecules* **1999**, *32*, 2384–2386.

53. Tabata, M.; Takamura, H.; Yokoto, K.; Nozaki, Y.; Hoshina, H.; Minakawa, H.; Kodaira, K. Pressure-induced cis to trans isomerization of poly(*o*-methoxyphenylacetylene) polymerized by Rh complex catalyst. A Raman, X-ray, and ESR study. *Macromolecules* **1994**, *27*, 6234–6236.

54. Tabata, M.; Nozaki, M.; Yokota, K.; Minakawa, H. Structural differences of poly(α-ethynylnaphthalene)s obtained with [Rh(norbornadiene)Cl]$_2$ and WCl$_6$ catalysts: An electron spin resonance and Raman study. *Polymer* **1996**, *37*, 1959–1963.

55. Crabtree, R. H. *The Organometallic Chemistry of the Transition Metals*; Wiley: New York, 1988; p 102.

56. Tang, B. Z.; Poon, W. H.; Leung, S. M.; Leung, W. H.; Peng, H. Synthesis of stereoregular poly(phenylacetylene)s by organorhodium complexes in aqueous media. *Macromolecules* **1997**, *30*, 2209–2212.

57. Hori, H.; Six, C.; Leitner, W. Rhodium-catalyzed phenylacetylene polymerization in compressed carbon dioxide. *Macromolecules* **1999**, *32*, 3178–3182.

58. Mastrorilli, P.; Nobile, C. F.; Gallo, V.; Suranna, G. P.; Farinola, G. Rhodium(I) catalyzed polymerization of phenylacetylene in ionic liquids. *J. Mol. Catal. A.* **2002**, *184*, 73–78.

59. Togo, Y.; Nakamura, N.; Iwamura, H. Synthesis and photolysis of *N*-(phenoxycarbonyloxy)-2-thiopyridone derivatives. A new unimolecular route to quantitative generation of phenoxyl radicals. *Chem. Lett.* **1991**, *7*, 1201–1204.

60. Yoshioka, N.; Nishide, H.; Kaneko, T.; Yoshiki, H.; Tsuchida, E. Poly[(*p*-ethynylphenyl) hydrogalvinoxy] and its polyradical derivative with high spin concentration. *Macromolecules* **1992**, *25*, 3838–3842.

61. Iwamura, H. Magnetic coupling of two triplet phenylnitrene units joined through an acetylenic or a diacetylenic linkage. *Mol. Cryst. Liq. Cryst.* **1989**, *176*, 33–48.

62. Nakako, H.; Nomura, R.; Tabata, M.; Masuda, T. Synthesis and structure in solution of poly[(-)-menthyl propiolate as a new class of helical polyacetylene. *Macromolecules* **1999**, *32*, 2861–2864.

63. Nakako, H.; Mayahara, Y.; Nomura, R.; Tabata, M.; Masuda, T. Effect of chiral substituents on the helical conformation of poly(propiolic esters). *Macromolecules* **2000**, *33*, 3978–3982.

64. Sato, M. A.; Maeda, K.; Onouchi, H.; Yashima, E. Synthesis and macromolecular helicity induction of a stereoregular polyacetylene bearing a carboxy group with natural amino acids in water. *Macromolecules* **2000**, *33*, 4616–4618.

65. Maeda, K.; Goto, H.; Yashima, E. Stereospecific polymerization of propiolic acid with rhodium complexes in the presence of bases and helix induction on the polymer in water. *Macromolecules* **2001**, *34*, 1160–1164.

66. Nomura, R.; Tabei, J.; Masuda, T. Biomimetic stabilization of helical structure in a synthetic polymer by means of intramolecular hydrogen bonds. *J. Am. Chem. Soc.* **2001**, *123*, 8430–8431.

67. Nomura, R.; Tabei, J.; Masuda, T. Effect of side chain structure on the conformation of poly(*N*-propargylalkylamide). *Macromolecules* **2002**, *35*, 2955–2961.

68. Yamada, K.; Nomura, R.; Masuda, T. Polymerization of cyclooctyne with late transition metal catalysts. *Macromolecules* **2000**, *33*, 9179–9181.

69. Kishimoto, Y.; Eckerle, P.; Miyatake, T.; Ikariya, T.; Noyori, R. Living polymerization of phenylacetylenes initiated by Rh(C≡CC$_6$H$_5$)(2,5-norbornadiene)[P(C$_6$H$_5$)$_3$]$_2$. *J. Am. Chem. Soc.* **1994**, *116*, 12131–12132.

70. Kishimoto, Y.; Eckerle, P.; Miyatake, T.; Kainosho, M.; Ono, A.; Ikariya, T.; Noyori, R. Well-controlled polymerization of phenylacetylenes with organorhodium(I) complexes: Mechanism and structure of the polyenes. *J. Am. Chem. Soc.* **1999**, *121*, 12035–12044.

71. Kishimoto, Y.; Miyatake, T.; Ikariya, T.; Noyori, R. An efficient rhodium(I) initiator for stereospecific living polymerization of phenylacetylenes. *Macromolecules* **1996**, *29*, 5054–5055.

72. Misumi, Y.; Masuda, T. Living polymerization of phenylacetylene by novel rhodium catalysts. Quantitative initiation and introduction of functional groups at the initiating chain end. *Macromolecules* **1998**, *31*, 7572–7573.

73. Miyake, M.; Misumi, Y.; Masuda, T. Living polymerization of phenylacetylene by isolated rhodium complexes, $Rh[C(C_6H_5)=C(C_6H_5)_2](nbd)(4-XC_6H_4)_3P$ (X = F, Cl). *Macromolecules* **2000**, *33*, 6636–6639.

74. Kanki, K.; Misumi, Y.; Masuda, T. Synthesis of poly(phenylacetylene)-*block*-poly(β-propiolactone) by use of Rh-catalyzed living polymerization of phenylacetylene. *Inorg. Chim. Acta* **2002**, *336*, 101–104.

75. Ciardelli, F.; Lanzillo, S.; Pieroni, O. Optically active polymer of 1-alkynes. *Macromolecules* **1974**, *7*, 174–179.

Part V

Functional and Non-Olefinic Monomers

22 Tacticity in Ethylene/Carbon Monoxide/Vinyl Co- and Terpolymerizations

Kyoko Nozaki

CONTENTS

22.1 INTRODUCTION

The study of the transition metal-catalyzed alternating copolymerization of ethylene with CO dates back to the 1950s.[1] While the polymer ethylene/CO ratio is higher than one when the copolymer is produced via a radical process, a completely alternating copolymer is produced when transition metal complexes, particularly palladium species, are employed as precatalysts.[2] Because the melting point of ethylene/CO alternating copolymer is almost as high as its decomposition temperature, decreasing its melting point is essential to obtain a melt-processable material. Thus, for practical usage, a termonomer like propylene or a higher 1-alkene is typically added. The catalytic systems most frequently used for ethylene/propylene/CO terpolymerization and propylene/CO copolymerization are palladium complexes of the type $[Pd(L^{\wedge}L')(S)_2][X]_2$, where $L^{\wedge}L'$ (L = or $\neq$ L') is a cis-chelating bis(phosphine) ligand such as 1,3-propanediyl-bis(diphenylphosphine) (dppp), S is a solvent molecule, and X is an anion with low coordination capability. In addition, an oxidant is often added to the dicationic palladium complexes in order to minimize the formation of inactive reduced palladium species.[3,4] On the other hand, for styrene/CO copolymerization or for ethylene/styrene/CO terpolymerization, bis-sp^2-nitrogen ligands, such as bipyridine or diimines, are suitable as the $L^{\wedge}L'$ in $[Pd(L^{\wedge}L')(S)_2][X]_2$.

Olefin/CO copolymerization using dicationic $[Pd(L^{\wedge}L')(S)_2][X]_2$ compounds is often carried out in methanol. The methanol reacts with the Pd dication to form a Pd–OMe bond generating a proton. The initiation step is reported to be carbonyl insertion into a Pd–OMe bond, followed by the subsequent alternating insertion of olefins and CO (Scheme 22.1a). Chain transfer occurs by methanolysis of the Pd–C(=O)–polymer species, generating a Pd–H species and a polymer with an ester chain end (MeOC(=O)–polymer). Olefin insertion into the resulting Pd–H bond gives an alkylpalladium species to which CO and olefin insert in an alternating manner to propagate a new polymer chain. Oxidant, such as quinone, is often added to prevent the active Pd^{2+} species from being reduced to inactive Pd^0. The presence of the oxidant can be avoided by using fluorinated alcohols as polymerization solvents instead of methanol, although the reason is still not clear.[5] Alternatively, alkylpalladium complexes of the general formula, [Pd(L^L′)(CH$_3$)(S)][X], are employed in some cases, especially to initiate polymerizations carried out in aprotic solvents (Scheme 22.1b). In this case, carbonyl insertion to the Pd–Me bond initiates the reaction. The propagation is the same as in the other case. In this system, chain-transfer results from β-hydride elimination, generating enone terminated polymer and a Pd–H species. Olefin insertion into the Pd–H bond regenerates the Pd–alkyl species, which corresponds to the original Pd–Me. The methanolysis chain elimination process is operative only when polymerizations are carried out in methanol, although β-hydride elimination takes place in both mechanisms. Thus, one can expect the polymer products with higher molecular weight when the polymerization is initiated by an alkyl–Pd species in an aprotic solvent.

Recently, several excellent review articles on the catalytic synthesis of olefin/CO copolymers have been published, and include information on catalysts, substrates, and polymer

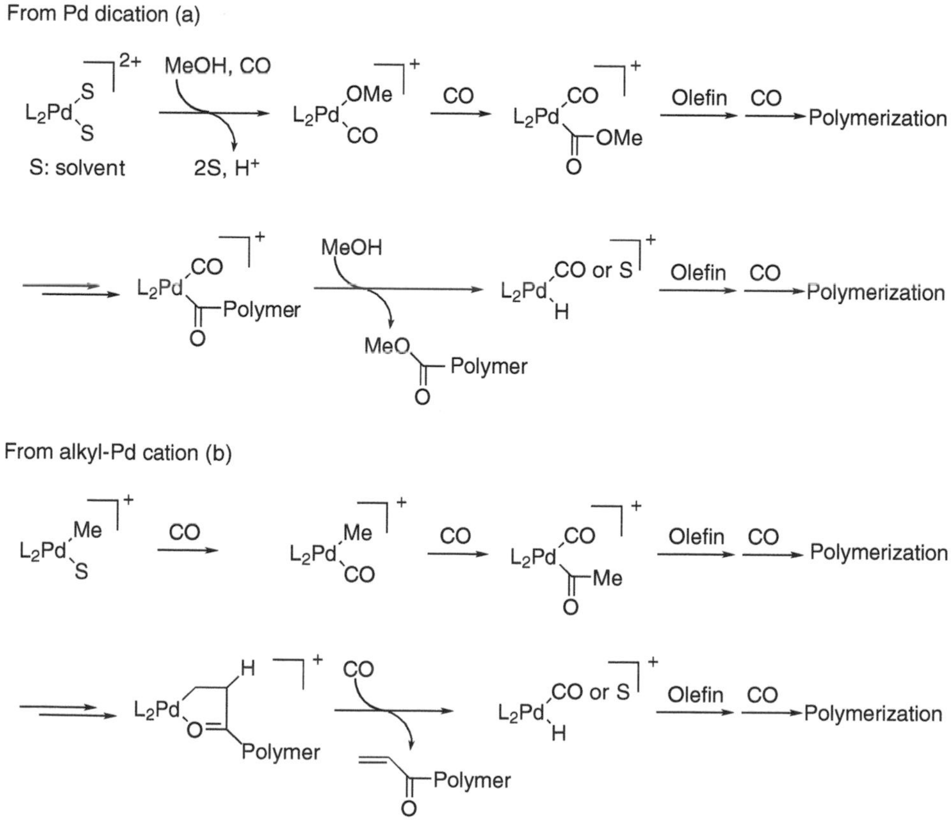

SCHEME 22.1   Initiation, propagation, and chain transfer for olefin/CO alternating copolymerization initiated by (a) a palladium dication in the presence of methanol, or (b) an alkylpalladium cation in an aprotic solvent. (L$_2$ is a cis-chelating bidentate ligand L^L′ and S represents solvent(s).)

performance.[3,4,6–12] This chapter focuses on factors that control polyketone tacticities in palladium-catalyzed alternating copolymerizations of olefins with carbon monoxide.

22.2 TACTICITIES IN ALTERNATING OLEFIN/CO CO- AND TERPOLYMERS

Propylene/CO and styrene/CO copolymers possess side chains (methyl and phenyl groups, respectively); thus, multiple possibilities for regioisomers and stereoisomers exist. Several mechanistic studies have proposed that the key step which determines the regio- and stereochemistry of olefin/CO copolymers is olefin insertion into the acylpalladium species.[6,7] Polymers with high regioregularity are produced if the olefin insertion reaction is regioselective to either 1,2- or 2,1-addition. As shown in Scheme 22.2, propylene insertion is mostly 1,2- whereas styrene insertion is usually 2,1-.

There are two types of tacticity control; namely, chain-end control and enantiomorphic-site control. For styrene/CO copolymerization, syndiotactic copolymers are obtained if efficient chain-end control to the *unlike*[13] diad controls chain propagation (Scheme 22.2b; see Reference 13 for a definition of *like* and *unlike* diads). This is the case when achiral bis(nitrogen) ligands, such as bipyridine or 1,10-phenanthroline, are employed. On the other hand, isotactic copolymers arise from enantiofacial selection by the catalyst (enantiomorphic-site control). Using chiral catalysts that differentiate between the two-olefin enantiofaces, isotactic copolymers are produced both for propylene/CO and styrene/CO (Scheme 22.2a and c). Isotactic copolymers can also be produced using an achiral catalyst if efficient chain-end control prefers the *like* diad to the *unlike* diad. In fact, propylene/CO copolymerization catalyzed by Pd-complexes containing 1,3-bis(diethylphosphino)propane[14] or bis(diarylphosphinomethyl)-1,2-phenylene ligands gives stereoregular, isotactic polyketones.[15] It should be noted, however, this *like* selectivity might be attributed to the enantiomorphic-site control if the achiral ligands create chiral complexes upon their coordination.[16] In any case, for propylene or styrene, the stereoregular olefin/CO copolymers reported to date are: isotactic poly(propylene-*ALT*-CO), syndiotactic poly(styrene-*ALT*-CO), and isotactic poly(styrene-*ALT*-CO) (Scheme 22.2a, b, and c, respectively). Terpolymers of propylene/ethylene/CO and styrene/ethylene/CO have also been produced in which the α-olefin side-chains are either atactic or isotactic (*vide infra*).

It should be noted that, unlike for polypropylene or polystyrene, there exist asymmetric centers in the main chain of the propylene/CO and styrene/CO copolymers. Thus, one enantiomer of a

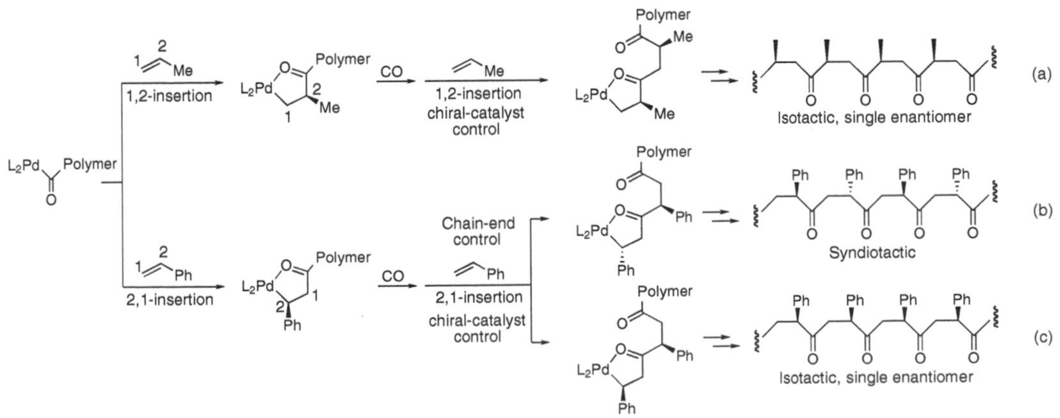

SCHEME 22.2 Tactic polyketones synthesized by the alternating copolymerization of olefins with CO. (L$_2$ is a cis-chelating bidentate ligand L^L'.)

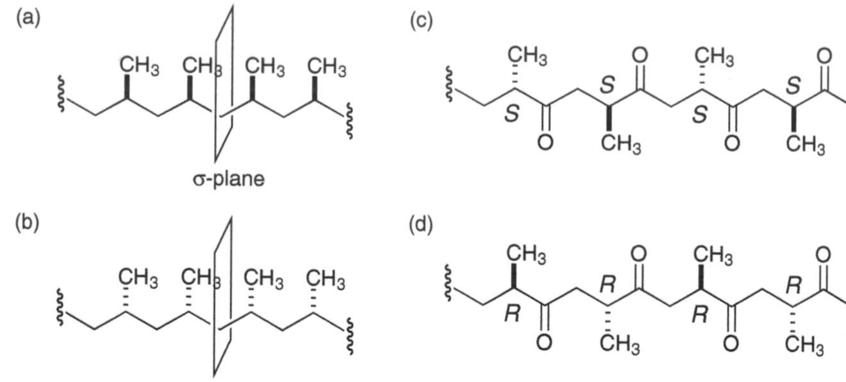

FIGURE 22.1 Isotactic polypropylene (a and b) and isotactic poly(propylene-*ALT*-CO) (c and d). A σ-plane exists in polypropylene; in other words, (a) and (b) are essentially the same. On the other hand, (c) and (d) are enantiomers.

chiral catalyst should produce the corresponding enantiomer of the isotactic polyketone as long as stereoregularity arises from catalyst control (enantiomorphic-site control). In other words, optically active catalysts potentially produce optically active polyketones.

Asymmetric synthesis polymerization of prochiral monomers is a reaction that produces polymers with configurational chirality in the main chain.[17,18] Even if effective chiral induction takes place during the polymerization of an α-olefin monomer (CH_2=CHR), the resulting stereoregular (e.g., isotactic) polymer is hardly optically active because the polymer chain has a plane of symmetry if one ignores the trifling difference of the chain ends (Figure 22.1a and b). On the other hand, alternating copolymers of α-olefin monomers with carbon monoxide possess true chiral centers in the polymer main chain, owing to the absence of a plane of symmetry. Accordingly, there exist two enantiomers, namely, *RRRR*··· and *SSSS*···, for isotactic polyketones (Figure 22.1c and d).

22.3 ISOTACTIC POLY(PROPYLENE-*ALT*-CO)

22.3.1 Synthesis of Isotactic Poly(propylene-*ALT*-CO) Using Achiral Ligands

The first propylene/CO alternating copolymer was obtained using Pd dication complexes having ligands with the structure $Ar_2P(CH_2)_3PAr_2$ (**1a**, Ar = phenyl; **1b**, Ar = 2-methoxyphenyl; Figure 22.2), however, the regioregularity and stereoregularity of this material were low as determined by ^{13}C NMR spectroscopy in the region of the carbonyl resonances.[19] The head-to-head (h–h) and the tail-to-tail (t–t) propylene enchainments exhibit peaks at 223 and 214 ppm, respectively, whereas the peaks around 218–219 ppm are attributed to the head-to-tail (h–t) structure. The sharpness of the δ 219 peak, which corresponds to the *like–like* triad, provides an indication of the degree of isotacticity present in the polymer.

In the 1990s, investigations into bis(phosphine) ligand design opened a new area for alternating propylene/CO copolymerizations, namely, the ability to synthesize highly isotactic polyketones. Representative ligands on Pd dication employed for these copolymerizations are shown in Figure 22.2. While having the same 1,3-propanediyl backbone as previously used complexes (**1a–b**), simple replacement of the aryl groups in $Ar_2P(CH_2)_3PAr_2$ by ethyl groups to give $Et_2P(CH_2)_3PEt_2$ (**1c**) improved the regioselectivity of polymerization, and thus completely regioregular, isotactic-rich copolymer was obtained although quantitative evaluation of the tacticities was not included in literature.[20] Another achiral ligand, bis(diarylphosphinomethyl)-1,2-phenylene (**2**), also provided

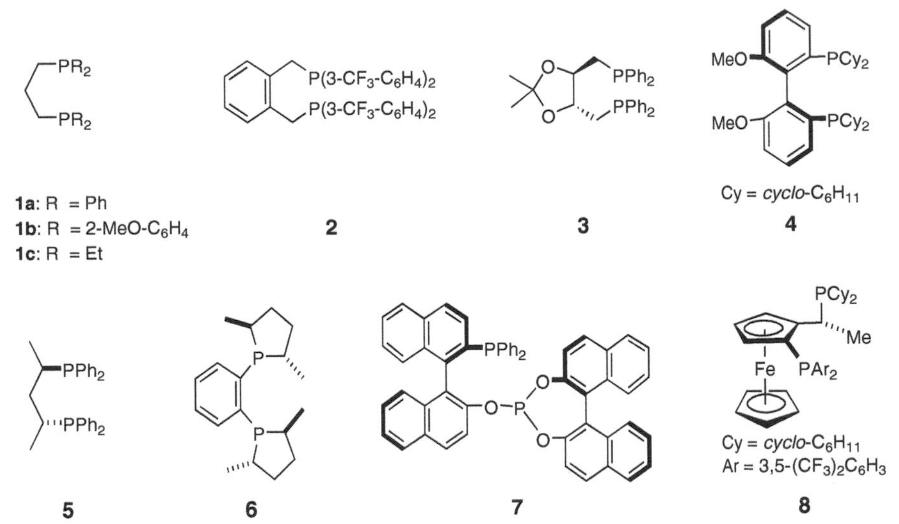
PR2
PR2
1a: R = Ph
1b: R = 2-MeO-C6H4
1c: R = Et
P(3-CF3-C6H4)2
P(3-CF3-C6H4)2
2
PPh2
PPh2
3
MeO
PCy2
MeO
PCy2
Cy = cyclo-C6H11
4
PPh2
PPh2
5
6
PPh2
7
PCy2
Me
Fe
PAr2
Cy = cyclo-C6H11
Ar = 3,5-(CF3)2C6H3
8

FIGURE 22.2 Ligands for palladium dication or alkylpalladium complexes employed for the synthesis of isotactic poly(propylene-*ALT*-CO).

the regioregular highly isotactic CO/propylene polyketone (82% for the major isotactic head-to-tail carbonyl peak at 219 ppm).

22.3.2 Synthesis of Isotactic Poly(propylene-*ALT*-CO) Using Chiral Ligands

Chiral ligands were first employed to improve isotacticity in propylene/CO alternating copolymerization, by exploiting catalyst control over the enantiofacial selection of propylene. The first attempted synthesis of isotactic poly(propylene-*ALT*-CO) using an optically active ligand involved DIOP (**3**); however, the product was both regioirregular and stereoirregular.[21] In 1992, Consiglio succeeded in synthesizing of isotactic poly(propylene-*ALT*-CO) using 6,6′-disubstituted-2,2′-bis(dialkylphosphino)-1,1′-biphenyl ligands such as **4** (100% h–t polymer, with 91% for the major peak at δ 219).[20,22] Unlike dppp (**1a**), a chiral dimethyl-substituted dppp, 2,4-bis(diphenylphosphino)pentane (**5**), provided highly isotactic copolymer.[23] Highly isotactic poly(propylene-*ALT*-CO) was also produced using Pd complexes of chiral ligands **6**,[24] **7**,[25,26] and **8**.[27,28] Notably, not only the C_2-symmetric ligands **1–6**, but also C_1-symmetric **7** and **8**, efficiently produced the corresponding isotactic polyketone. The highest productivity for propylene/CO alternating copolymerization, 1797 g polyketone/g Pdh has been reported using the catalyst system $Pd(OAc)_2$/ligand **7**/$BF_3{\cdot}OEt_2$ (OAc = acetate) in CH_2Cl_2-MeOH at 50 °C under 7.5 MPa of CO pressure (polymer $M_n = 14,000$; >99% h–t; >97.5% for the major peak at δ 219).

22.3.3 Enantioselectivity in Isotactic Copolymers

As mentioned before, enantiomers exist for isotactic polyketones. It should be noted that high isotacticity does not necessarily mean high polymer enantiopurity, since both *RRRR*··· and *SSSS*··· chain segments can exist in the same polymer in amounts determined by the nature of its stereoerrors. For the asymmetric centers in the main chain of poly(propylene-*ALT*-CO) prepared using a catalyst system containing ligand **6**, almost complete enantioselectivity (preferential existence of one chain stereoisomer) was confirmed by ^{13}C NMR analysis utilizing a chiral shift reagent.[24] Shorter units of the copolymer, that contains one propylene unit or two propylene units were prepared by **4**[29] and by **7**,[24] respectively, the results also supporting the exclusive formation of a single poly(propylene-*ALT*-CO) enantiomer.

More practically, the molar absorption for the carbonyl group by circular dichroism (CD) can be used as an indicator for the enantioselectivity of the poly(propylene-*ALT*-CO) chain. High values, such as $\Delta\varepsilon$ +1.73 with a catalyst system containing **4**[24] and $\Delta\varepsilon$ −1.66 with a catalyst system containing **7**,[26] were reported using $(CF_3)_2CHOH$ as a solvent. Although the presence of higher-order polymer structures, for example, helical structures, would affect CD absorption, the absence of such structures was indicated by comparing the CD spectrum of the polyketone to that of (*S*)-3-methyl-2,5-hexanedione, a single configurational unit analogue.[30] Accordingly, the molar absorption of poly(propylene-*ALT*-CO) by CD can be taken as proportional to the enantiomeric excess of the asymmetric center in the main chain.

22.3.4 Spiroketal Formation

Depending on the reaction conditions, poly(propylene-*ALT*-CO)s can be isolated as either the true polyketone, poly(1-methyl-2-oxo-propanediyl), or as a polyspiroketal, poly[spiro-2,5-(3-methyltetrahydrofuran)] (Scheme 22.3).[31,32] The latter polymer can be transformed into the

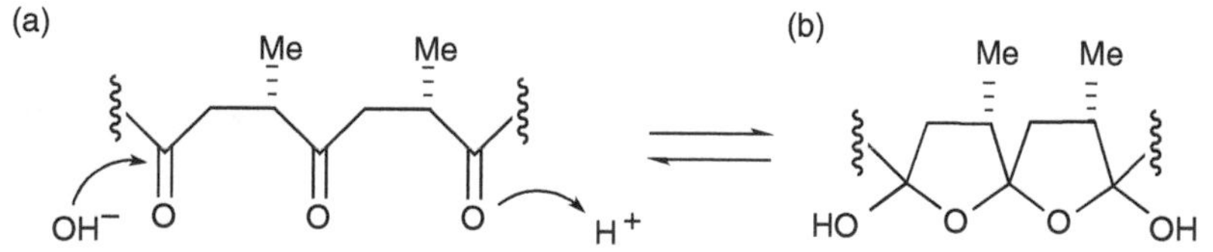

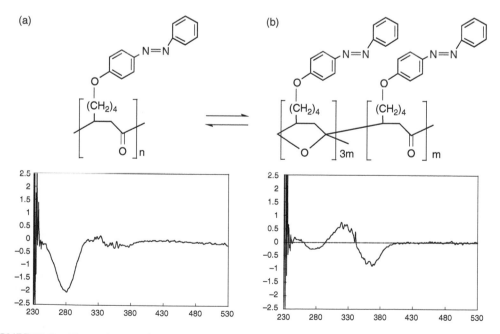

SCHEME 22.3 For highly isotactic poly(1-methyl-2-oxo-propanediyl) (poly(propylene-*ALT*-CO)), a solution equilibrium exists between polyketone (a) and polyspiroketal (b) forms. The polyketone form is preferred in $(CF_3)_2CHOH$.

FIGURE 22.3 Alternating copolymer of α-olefin having azobenzene side chain with CO in a pure polyketone form (a) and a 3:1 mixture of polyspiroketal and polyketone (b). Strong Cotton effect around 340 nm in CD spectrum is attributed to the helical orientation of the azobenzene moiety in the polyspiroketal structure.

polyketone either thermally in the absence of solvent, or by dissolution in $(CF_3)_2CHOH$ (abbreviated as HFIP).[14,24] There are no solvents other than HFIP that stabilize the polyketone form and the reason is still not clear. In solvents such as $CHCl_3$-MeOH, an equilibrium exists between the polyketone and polyspiroketal. The ketal formation is unique for highly isotactic copolymers generated from propylene or other higher aliphatic 1-alkenes. A helical structure was suggested by CD spectrum for a spiroketal having azobenzene side chains. Asymmetric alternating copolymerization of $CH_2=CH-(CH_2)_4-O-C_6H_4-N=N-C_6H_5$ with CO provided a 3:1 mixture of polyspiroketal and polyketone upon reprecipitation from methanol and $CHCl_3$. Strong Cotton effect of the azobenzene moiety around 340 nm was indicative of the helical orientation of the azobenzene moiety in the polyspiroketal structure (Figure 22.3).[33]

22.3.5 MECHANISTIC STUDIES

The key intermediates for the copolymerization of propylene and CO were successfully characterized by ^1H, ^{13}C, and ^{31}P NMR spectroscopy for copolymerization using a catalyst system with ligand **7** (Scheme 22.4).[26] By treatment of the cationic acylpalladium complex $[(7)Pd(C(=O)Me)(MeCN)]^+[BAr_4]^-$ (Ar = $3,5-(CF_3)_2C_6H_3$) with propylene, alkylpalladium species were formed as a result of propylene 1,2-insertion (Scheme 22.4, top right). The major

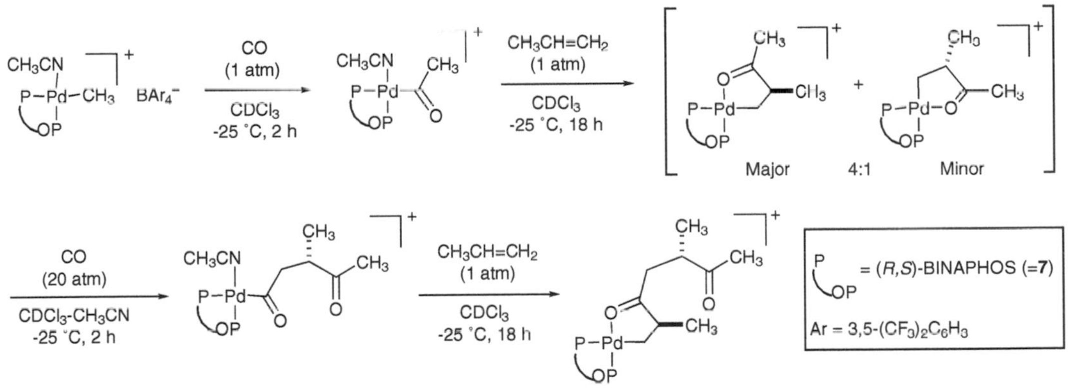

SCHEME 22.4 Stepwise observation of propylene/CO alternating copolymerization. All structures were spectroscopically confirmed.

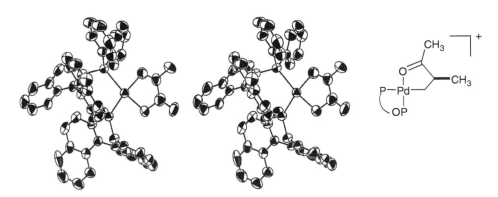

FIGURE 22.4 A stereoview for the major propylene-inserted intermediate (a) and a structural representation of the cation (b).

alkylpalladium species was isolated as the borate salt without β-hydride elimination, and its structure was determined by single-crystal X-ray analysis (Figure 22.4).[26] It should be noted that the five-membered ring chelation in the alkyl complex is essential to prevent it from undergoing β-hydride elimination.

22.4 SYNDIOTACTIC POLY(STYRENE-*ALT*-CO)

The first styrene/CO alternating copolymers were obtained using [Pd(2,2′-bipyridine)(S)$_2$][X]$_2$ or [Pd(1,10-phenanthroline)(S)$_2$][X]$_2$ complexes as catalysts where S is a solvent, mainly methanol, and X is a noncoordinating anion.[34–36] Remarkably, using these planar symmetrical ligands, the product poly(styrene-*ALT*-CO) was regioregular and prevailingly syndiotactic with a triad selectivity of *uu*:*ul*:*lu*:*ll* = 0.80:0.10:0.10:0.[13] The triad stereochemical composition can be estimated by ^{13}C NMR using the enchained styrene *ipso*-carbon peak at 136.1.[20] For copolymers of para-substituted styrenes with CO, the peaks due to the main chain carbons at δ 43.2 (CH$_2$), 54.0 (CH), 209.7 (C=O) are used to discuss the stereoregularity because there are two *ipso*-carbons for the substituted styrenes.

Successful chain-end control for the high syndiotacticity may be attributed to the exclusive 2,1-insertion of styrene, which is in sharp contrast to the predominant 1,2-insertion of aliphatic 1-alkenes such as propylene. The stereocontrol by the asymmetric center of the enchained styrene unit (located at the β-position to the metal after insertion of CO) seems rather surprising, because it seems too distant to control the direction of the next incoming styrene monomer. It is proposed that, the asymmetric center of the acylpalladium propagating species controls the orientation of the polymer chain carbonyl that is bound to palladium, and that this carbonyl interacts with the inserting styrene to control the stereochemistry of insertion (Scheme 22.5, **TS 1**).[35,7] Thus, mediated by the carbonyl, effective chain-end control appears to ensue. The chain growth process of styrene/CO copolymerization with a bipyridine acetyl palladium complex has been monitored by ^{1}H and ^{13}C NMR over the first three alternating insertion sequences, by using 4-*tert*-butylstyrene that exhibits more simple ^{13}C NMR peaks as a substrate (Scheme 22.5).[37] Insertion of 4-*tert*-butylstyrene gives a 3:1 mixture of σ-benzyl complex and π-benzyl complex. After the subsequent insertion of CO, 4-*tert*-butylstyrene, and then another CO, formation of a single diastereomer was confirmed at low

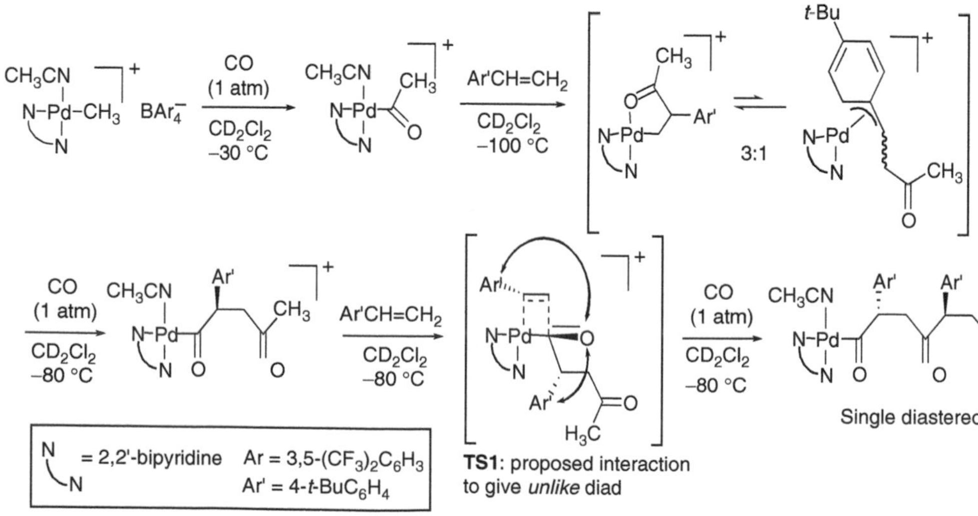

SCHEME 22.5 Stepwise chain growth using 4-*tert*-butylstyrene. TS 1 was not detected but it illustrates the proposed interactions that would result in the selective formation of an unlike diad.

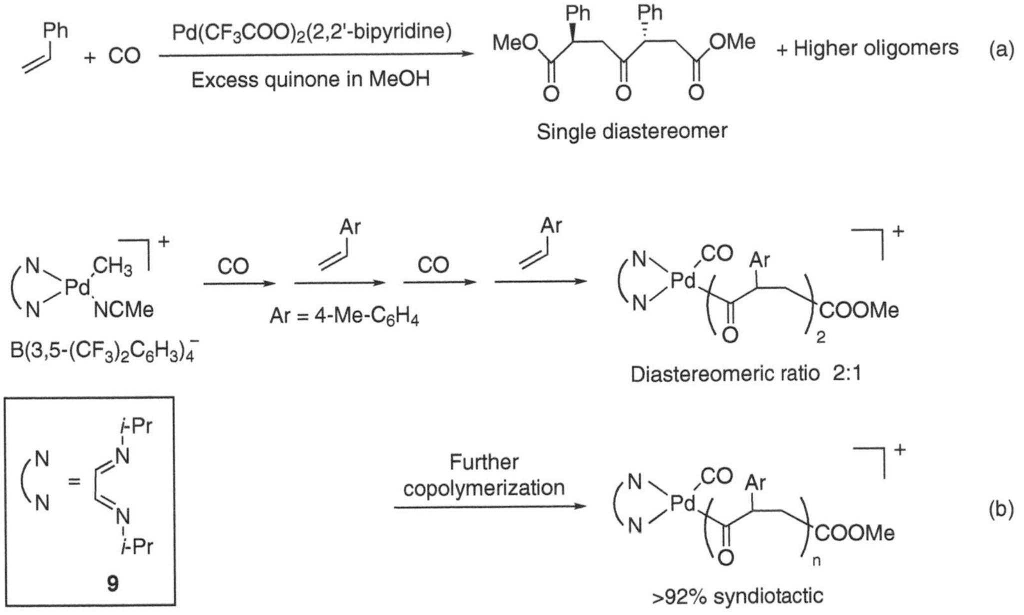

SCHEME 22.6 Stereochemical analysis for styrene/CO copolymers. (a) Highly diastereoselective diad formation with 2,2'-bipyridine. (b) Amplification of stereoselectivity occurs as the polymer chain grows using a diimine ligand **9**.

temperature for the diad. Although the relative configuration was not determined, one would expect it to be *unlike*.

Alternatively, lower molecular weight model compounds of poly(styrene-*ALT*-CO) were prepared in the presence of an excess oxidant using Pd(CF$_3$COO)$_2$/2,2'-bipyridine in methanol. The high concentration of oxidant accelerates the chain transfer so that shorter oligomers are obtained. Among various products, a dimer, dimethyl 2,5-diphenyl-4-oxoheptanedioate (Scheme 22.6a), was obtained in a diastereomerically pure form.[38] Interestingly, in contrast, the diastereoselectivity (~2:1) seen for the first two insertions of 4-methylstyrene in copolymerization with CO was much lower than the overall diastereoselectivity seen in the corresponding copolymer (~92% of *uu* triad, determined by ^{13}C NMR) when the catalyst system [Pd(H$_2$O)$_2$(**9**)](CF$_3$SO$_3$)$_2$ in methanol was employed (Scheme 22.6b).[39] Thus, the contribution of the growing chain to the stereocontrol is proposed in the latter case.

22.5 ISOTACTIC POLY(STYRENE-*ALT*-CO)

Brookhart and Wagner first reported the asymmetric alternating copolymerization of 4-*tert*-butylstyrene with carbon monoxide using [Pd(Me)(MeCN)(**10**)]$^+$[BAr$_4$]$^-$ (Ar = 3,5-(CF$_3$)$_2$C$_6$H$_3$), a methylpalladium catalyst containing a chiral bis(oxazoline) ligand (ligand **10**, Figure 22.5).[40] In this copolymerization, the enantioface of the olefin was selected by the chiral ligand instead of the chain end; as a result, the polymer was completely isotactic. Since one enantioface was discriminated against the other by the chiral catalyst, it is probable that the copolymer is also of high enantiopurity. The ligands employed for the isotactic copolymerization of styrenes with CO are summarized in Figure 22.5. Bidentate sp^2-nitrogen ligands, such as **11**, **12**, and **13** are most commonly used,[23,41–43] along with the phosphine-nitrogen bidentate ligand **14**[44,45] and phosphine-phosphite **7** (Figure 22.2).[25,26]

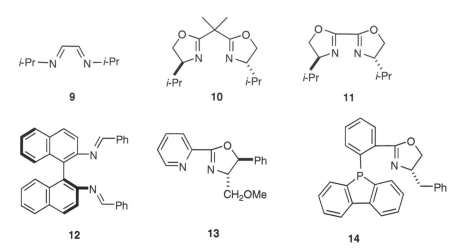

FIGURE 22.5 Ligands employed for the copolymerization of styrene (or its substituted analogues) with CO.

The enantioselectivities for styrene/CO copolymerizations were mostly estimated by molar optical rotation, $[\Phi]_D$, or CD, $\Delta\varepsilon$. Currently, $[\Phi]_D = -536$ for 4-*tert*-butylstyrene/CO using the catalyst system, $[Pd(Me)(MeCN)(\textbf{10})]^+[BAr_4]^-$,[40] and $\Delta\varepsilon = -11.75$ for styrene/CO using a catalyst system, $[Pd(Me)(MeCN)(\textbf{14})]^+[BAr_4]^-$,[46] are the highest reported values. A few studies involving the synthesis and characterization of oligomeric species have also been reported. Based on nuclear magnetic resonance (NMR) analysis, Consiglio and coworkers revealed that styrene insertion into the acylpalladium complex $[(\textbf{14})Pd(C(=O)Me)(MeCN)]^+[OTf]^-$ (OTf = OSO_2CF_3) is both completely regioselective for 2,1-insertion and enantioselective, producing $[(\textbf{14})Pd(CHPhCH_2C(=O)Me)]^+[OTf]^-$ as a single species.[47,48] This result is consistent with the fact that dimethyl (*R*)-2-phenylbutanedioate was obtained almost exclusively as the *S* enantiomer (95% ee) when styrene/CO copolymerization with $[(\textbf{14})Pd(Me)(MeCN)]^+[OTf]^-$ was carried out in the presence of a high concentration of the oxidant, benzoquinone.

It is noteworthy that only low molecular weight oligoketones can be obtained with bis(phosphine) ligands in alternating styrene/CO copolymerizations. Drent and Budzelaar have attributed this fact to the higher electron density on the Pd center when phosphine ligands rather than nitrogen ligands are used, because the growing styrene/CO copolymer has a higher tendency to terminate by β-hydride elimination than growing propylene/CO chains.[7] The rather unusual fact that high copolymer can be prepared with catalyst systems $[(\textbf{7})Pd(Me)(MeCN)]^+[B(3,5\text{-}(CF_3)_2\text{-}C_6H_3)_4]^-$ is proposed to be attributable to the steric demand of this bulky ligand, which causes styrene to undergo 1,2-insertion rather than the more typical 2,1-insertion.[49] The continuous 1,2-insertion provided the h-t polyketone with high enantiofacial selection ($[\Phi]_D = -451$).

22.6 TERPOLYMERS CONSISTING OF TWO KINDS OF OLEFINS AND CARBON MONOXIDE

Using a mixture of two kinds of olefins, a terpolymer can be generated by alternating olefin/CO copolymerization (Scheme 22.7). An olefin and carbon monoxide are incorporated in a completely alternating manner, and the order of the two olefins is mostly random. Asymmetric terpolymerization of styrene/ethylene/CO has been intensely studied by Consiglio.[41,46,50] Using $[(\textbf{14})Pd(CH_3)(MeCN)]^+[OTf]^-$ as a catalyst, ethylene was preferentially and randomly enchained in the terpolymer, in spite of the comparably higher reactivity of styrene for the copolymerization. The

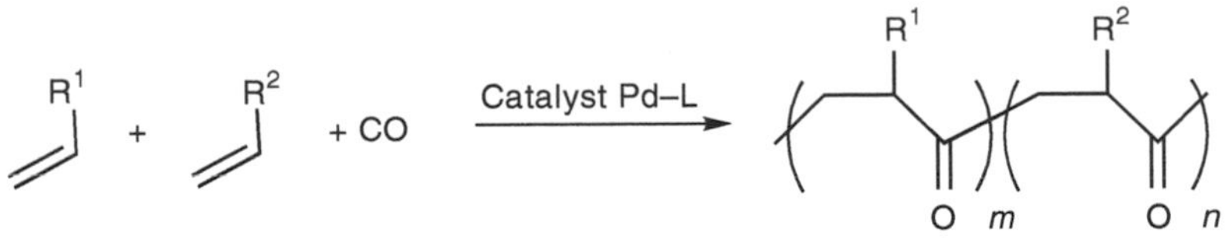

SCHEME 22.7 Reported terpolymerizations of two kinds of olefins with CO effected by Pd catalyst systems with N- and P-based ligands (L).

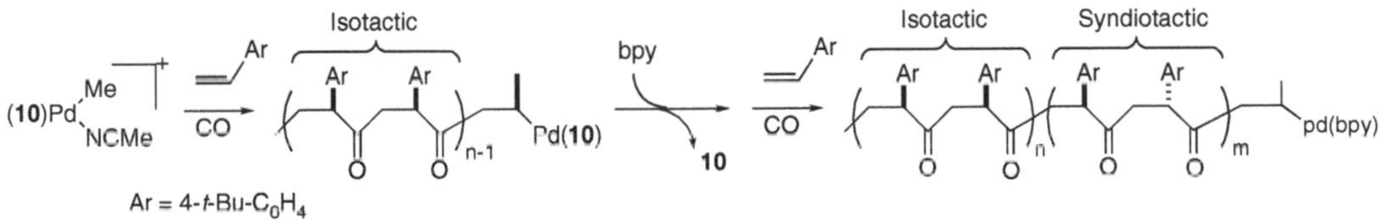

R^1	R^2	Employed L
Ph	H	14
Me	H	4, 7, or 8
4-*tert*-Bu-C$_6$H$_4$	CH$_3$	7
4-Me-C$_6$H$_4$	n-C$_8$H$_{17}$	10

Ar = 4-*t*-Bu-C$_6$H$_4$

SCHEME 22.8 Synthesis of a stereoblock 4-*tert*-butylstyrene/CO alternating copolymer by sequentially using two different ligands at the Pd center. (bpy = 2,2'-bipyridine).

enantioselectivity for styrene in the terpolymerization is as high as that observed for the styrene/CO copolymerization.

Asymmetric terpolymerization of propylene/ethylene/CO with Pd–bis(phosphines) such as Pd(OAc)$_2$/4/Ni(ClO$_4$)$_2$/1,4-naphthoquinone, [(7)Pd(Me)(MeCN)]$^+$[B(3,5-(CF$_3$)$_2$-C$_6$H$_3$)$_4$]$^-$ or [(8)Pd(Me)(MeCN)]$^+$[B(3,5-(CF$_3$)$_2$-C$_6$H$_3$)$_4$]$^-$ (ligands **4**, **7**, and **8**; Figure 22.2) also provide terpolymers in which the two olefins are incorporated in a random manner.[23,51] By changing the partial monomer pressures, the propylene/ethylene unit-ratio (CH$_2$CHMeC(=O))/(CH$_2$CH$_2$C(=O)) in the terpolymer was varied. The molar ellipticity ($\Delta\varepsilon$) per chiral unit, -CH$_2$CHMeC(=O)-, as measured by CD was essentially the same as for poly(propylene-*ALT*-CO).

Two examples have been reported of asymmetric 1-alkene/styrene/CO terpolymerizations, which involve catalyst systems incorporating either phosphine-phosphite ligand **7**[52] or bis(oxazoline) ligand **10** (ligand **7**, Figure 22.2; ligand **10**, Figure 22.5).[53] Propylene and 1-hexene were employed as the 1-alkene. Successive enantiofacial selection for both styrene and for 1-alkene gave the like-rich terpolymer. The stereoregularity was estimated by optical rotation values. With phosphorus-based [(7)Pd(Me)(MeCN)]$^+$[B(3,5-(CF$_3$)$_2$-C$_6$H$_3$)$_4$]$^-$ propylene was preferably incorporated into the terpolymer over styrene, whereas styrene was preferentially incorporated when nitrogen-based [(**10**)Pd(Me)(MeCN)]$^+$[B(3,5-(CF$_3$)$_2$-C$_6$H$_3$)$_4$]$^-$ was employed.

A stereoblock copolymer consisting of isotactic and syndiotactic 4-*tert*-butylstyrene/CO alternating copolymer was prepared by Brookhart.[54] First, copolymerization was initiated by using a Pd catalyst containing chiral bis(oxazoline) ligand **10**, to produce the isotactic block. Subsequently, addition of 2,2'-bipyridine to the system resulted in ligand replacement, so that the second block (formed by further copolymerization) was syndiotactic (Scheme 22.8).

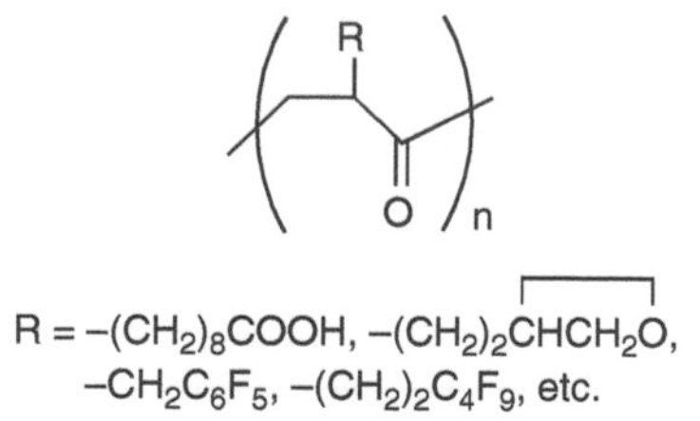

$$R = -(CH_2)_8COOH, -(CH_2)_2CHCH_2O,$$
$$-CH_2C_6F_5, -(CH_2)_2C_4F_9, \text{etc.}$$

FIGURE 22.6   Copolymers from functionalized olefins and CO.

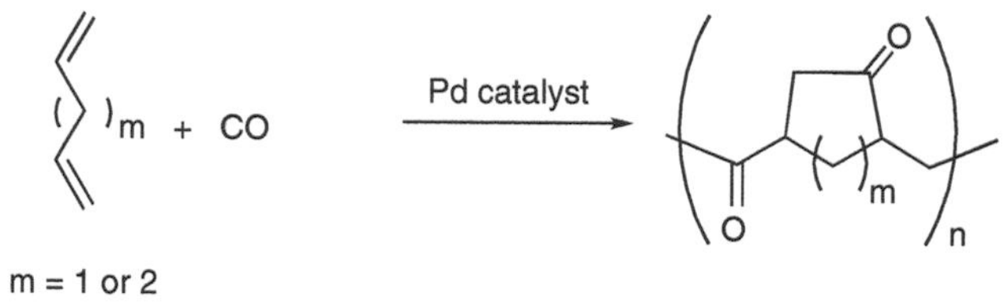

m = 1 or 2

SCHEME 22.9   Cyclocopolymerization of α, ω-olefin, and CO.

22.7 OTHER ALTERNATING COPOLYMERS PRODUCED BY OLEFIN/CO

Since palladium-catalysts are tolerant to various functional groups, alternating olefin/CO copolymers with polar-functionalized olefin units are accessible (Figure 22.6). Sen and coworkers have reported that alternating Pd-catalyzed copolymerization proceeds with alkenes having hydroxyl and carboxylic groups.[55] For the products, highly isotactic copolymer is suggested by ^{13}C NMR. Mono-epoxides of nonconjugated dienes are also employable as the olefin moiety.[56] The introduction of a fluorine atom into the olefin side chain, such as $CH_2=CH-CH_2-C_6F_5$ or $CH_2=CH-CH_2-C_4F_9$, affords fluorinated polyketones.[57–59] The stereoregularity of the fluorinated copolymers thus obtained is reported to be as high as that for with simple 1-alkene/CO copolymers. Fluorinated isotactic polyketones show a higher tendency to form polyspiroketals as compared to their hydrocarbon analogues.

α, ω-Dienes can also be used as the alkene comonomer for olefin/CO copolymerization; in this instance, cyclocopolymerization proceeds if the two C=C double bonds are separated by an adequate distance. Thus, cyclocopolymerization has been reported for 1,5-hexadiene[60] and 1,4-pentadiene (Scheme 22.9).[61] In both examples, complete cyclization has been achieved. When a chiral catalyst is employed for the copolymerization of CO with a cyclopolymerizable diene, the first olefin insertion takes place through catalyst-controlled enantiofacial selection. After a CO insertion, the second olefin insertion takes place in an intramolecular manner; the cis/trans relative configuration between the two substituents on the resultant cycloalkanone is not controlled completely.

More recently, the alternating copolymerization of vinyl acetate with carbon monoxide via a palladium catalyzed coordination insertion mechanism has been reported.[62] This is the first report of a nonradical pathway for vinyl acetate polymerization.

22.8 CONCLUSIONS

Intensive studies by Shell on catalyst development for the ethylene/propylene/CO terpolymerization provided the most effective catalyst represented as $[Pd(L^{\wedge}L')(S)_2][X]_2$, where $L^{\wedge}L'$ (L = or ≠ L') is a cis-chelating bis(phosphine) ligand, S is a solvent molecule, and X is an anion with low coordination capability. Ligand development leads to three patterns of stereoregular polymers; those are, isotactic poly(propylene-*ALT*-CO), syndiotactic, and isotactic poly(styrene-*ALT*-CO). Noteworthy is the fact that the isotactic polyketone does have asymmetric centers in the main chain. Thus, asymmetric synthesis, that is the production of one enantiomer of the isotactic polyketone by use of the corresponding enantiomer of a chiral catalyst, has been achieved.

REFERENCES AND NOTES

1. Reppe, W.; Magin, A. *Production of ketonic bodies*. U.S. Patent 2,577,208, December 4, 1951.
2. Drent, E., *Process for the preparation of polyketones*. European Patent 0121965 B1 (Shell Internationale Research Maatschappij B.V.), December 27, 1989.
3. Sen, A. Mechanistic aspects of metal-catalyzed alternating copolymerization of olefins with carbon monoxide. *Acc. Chem. Res.* **1993**, *26*, 303–310.
4. Drent, E.; van Broekhoven, J. A. M.; Doyle, M. J. Efficient palladium catalysts for the copolymerization of carbon monoxide with olefins to produce perfectly alternating polyketones. *J. Organomet. Chem.* **1991**, *417*, 235–251.
5. Milani, B.; Corso, G.; Mestroni, G.; Carfagna, C.; Formica, M.; Seraglia, R. Highly efficient catalytic system for the CO/styrene copolymerization: Toward the stabilization of the active species. *Organometallics* **2000**, *19*, 3435–3441.
6. Sen, A. Ed. *Catalytic Synthesis of Alkene-Carbon Monoxide Copolymers and Cooligomers*, Kluwer Academic Publishers: Dordrecht, 2003.
7. Drent, E.; Budzelaar, P. H. M. Palladium-catalyzed alternating copolymerization of alkenes and carbon monoxide. *Chem. Rev.* **1996**, *96*, 663–682.
8. Sommazzi, A; Garbassi, F. Olefin-carbon monoxide copolymers. *Prog. Polym. Sci.* **1997**, *22*, 1547.
9. Nozaki. K.; Hiyama, T. Stereoselective alternating copolymerization of carbon monoxide with alkenes. *J. Organomet. Chem.* **1998**, *576*, 248–253.
10. Abu-Surrah, A. S.; Rieger, B. High molecular weight 1-olefin/carbon monoxide copolymers: A new class of versatile polymers. *Top. Catal.* **1999**, *7*, 165–177.
11. Bianchini, C.; Meli, A. Alternating copolymerization of carbon monoxide and olefins by single-site metal catalysis. *Coord. Chem. Rev.* **2002**, *225*, 35–66.
12. Robertson, R. A. M.; Cole-Hamilton, D. J. The production of low molecular weight oxygenates from carbon monoxide and ethane. *Coord. Chem. Rev.* **2002**, *225*, 67–90.
13. About the nomenclature of *like* and *unlike*: The *meso* and *racemo* nomenclature commonly used for vinyl monomer diads, are not applicable to the head-to-tail polyketone because the junction unit between the two stereocenters, –CH$_2$-C(=O)–, is not symmetric. Poly(α-amino acid)s and poly(propylene oxide) are other examples of polymers with asymmetric (W. V. Metanomski, Compendium of Macromolecular Nomenclature, IUPAC Macromolecular division). Accordingly, the words *like* and *unlike*, which are used in organic chemistry, are applied; *like* (*l*) is used for the diad consisting of the same configuration (analogous to *meso*) and *unlike* (*u*) for the opposite (analogous to *racemo*).
14. Bronco, S.; Consiglio, G.; Hutter, R.; Batistini, A.; Suster, U. W. Regio-, stereo-, and enantioselective alternating copolymerization of propene with carbon monoxide. *Macromolecules* **1994**, *27*, 4436–4440.
15. Sesto, B.; Consiglio, G. Regioselectivity of the insertion of propene with achiral Pd(II) catalysts to highly isotactic poly[1-oxo-2-methylpropane-1,3-diyl]. Is the syndiotactic structure accessible? *J. Chem. Soc., Chem. Commun.* **2000**, 1011–1012.
16. Camalli, M.; Caruso, F.; Chaloupka, S.; Leber, E. M.; Rimml, H.; Venanzi, L. M. The coordination chemistry of 1,2-bis(diphenylphosphino)methyl]benzene with nickel(II), palladium(II), platinum(II), and platinum(0) and the X-ray crystals structure of ethylene[1,2-bis[(diphenylphosphino)methyl]benzene]platinum. *Helv. Chim. Acta* **1990**, *73*, 2263–2274.
17. Okamoto, Y.; Nakano, T. Asymmetric polymerization. *Chem. Rev.* **1994**, *94*, 349–372.
18. Nakano, K.; Kosaka, N.; Hiyama, T.; Nozaki, K. Metal-catalyzed synthesis of stereoregular polyketones, polyesters, and polycarbonates. *Dalton Trans.* **2003**, 4039–4056.
19. Batistini, A.; Consiglio, G.; Suter, U. W. Regioselectivity control in the palladium-catalyzed copolymerization of propylene with carbon monoxide. *Angew. Chem., Int. Ed. Engl.* **1992**, *31*, 303–305.

20. Barsacchi, M.; Batistini, A.; Consiglio, G.; Suter, U. W. Stereochemistry of alternating copolymerization of vinyl olefins with carbon monoxide. *Macromolecules* **1992**, *25*, 3604–3606.

21. Wong, P. K. *Copolymers of carbon monoxide.* European Patent 0384517 B1 (Shell Internationale Research Maatschappij B.V.), January 18, 1995.

22. Bronco, S.; Consiglio, G. Regio- and stereoregular copolymerisation of propene with carbon monoxide catalysed by palladium complexes containing atropisomeric diphosphine ligands. *Macromol. Chem. Phys.* **1996**, *197*, 355–365.

23. Jaing, Z.; Adams, S. E.; Sen, A. Stereo- and enantioselective alternating copolymerization of α-olefins with carbon monoxide. *Macromolecules* **1994**, *27*, 2694–2700.

24. Jiang, Z.; Sen, A. Palladium(II)-catalyzed isospecific alternating copolymerization of aliphatic α-olefins with carbon monoxide and isospecific alternating cooligomerization of 1,2-disubstituted olefin with carbon monoxide. Synthesis of novel, optically active, isotactic 1,4- and 1,5-polyketones. *J. Am. Chem. Soc.* **1995**, *117*, 4455–4467.

25. Nozaki, K.; Sato, N.; Takaya, H. Highly enantioselective alternating copolymerization of propene with carbon monoxide catalyzed by a chiral phosphine-phosphite complex of palladium(II). *J. Am. Chem. Soc.* **1995**, *117*, 9911–9912.

26. Nozaki, K.; Sato, N.; Tonomura, Y.; Yasutomi, M.; Takaya, H.; Hiyama, T.; Matsubara, T; Koga, N. Mechanistic aspects of the alternating copolymerization of propylene with carbon monoxide catalyzed by Pd(II) complexes of unsymmetrical phosphine-phosphite ligands. *J. Am. Chem. Soc.* **1997**, *119*, 12779–12795.

27. Gambs, C.; Chaloupka, S.; Consiglio, G.; Togni, A. Ligand electronic effect in enantioselective palladium-catalyzed copolymerization of carbon monoxide and propene. *Angew. Chem., Int. Ed.* **2000**, *39*, 2486–2488.

28. Gambs, C.; Consiglio, G.; Togni, A. Structural aspects of palladium and platinum complexes with chiral diphosphinoferrocenes relevant to the regio- and stereoselective copolymerization of CO with propene. *Helv. Chim. Acta* **2001**, *84*, 3105–3126.

29. Sperrle, M.; Consiglio, G. Diastereo- and enantioselectivity in the Co-oligomerization of propene and carbon monoxide to dimethyl-4-oxoheptanedioates. *J. Am. Chem. Soc.*, **1995**, *117*, 12130–12136.

30. Kosaka, N.; Nozaki, K.; Hiyama, T.; Fujiki, M.; Tamai, N.; Matsumoto, T. Conformational studies on an optically active 1,4-polyketone in solution. *Macromolecules* **2003**, *36*, 6884–6887.

31. Van Doorn, J. A.; Wong, P. K.; Sudmeijer, O. *Polymers of carbon monoxide with one or more α-olefins and their manufacture.* European Patent 376364. B1 (Shell Internationale Research Maatschappij B.V.), March 6, 1996.

32. Batistini, A.; Consiglio, G. Mechanistic aspects of the alternating copolymerization of carbon monoxide with olefins catalyzed by cationic palladium complexes. *Organometallics* **1992**, *11*, 1766–1769.

33. Kosaka, N.; Oda, T.; Hiyama, T.; Nozaki, K. Synthesis and photoisomerization of optically active 1,4-polyketones substituted by azobenzene side chains. *Macromolecules* **2004**, *37*, 3159–3164.

34. Drent, E., *Novel catalyst compositions and process for the copolymerization of ethane with carbon monoxide.* European Patent 0229408 B1 (Shell Internationale Research Maatschappij B.V.), January 9, 1991.

35. Barsacchi, M.; Consiglio, G.; Medici, L.; Petrucci, G.; Suter, U. W. Syndiotactic poly(1-oxo-2-phenyltrimethylene): Mechanism of palladium-catalyzed polymerization. *Angew. Chem., Int. Ed. Engl.* **1991**, *30*, 989–991.

36. Consiglio, G.; Milani, B. Stereochemical Aspects of Cooligomerization and Copolymerization In *Catalytic Synthesis of Alkene-Carbon Monoxide Copolymers and Cooligomers*, Kluwer Academic Publishers: Dordrecht, 2003.

37. Brookhart, M.; Rix, F. C.; DeSimone, J. M.; Barborak, J. C. Palladium(II) catalysts for living alternating copolymerization of olefins and carbon monoxide. *J. Am. Chem. Soc.* **1992**, *114*, 5894–5895.

38. Pisano, C.; Nefkens, S. C. A.; Consiglio, G. Stereochemistry of the dicarbonylation of olefins using styrene as the model compound. *Organometallics* **1992**, *11*, 1975–1978.

39. Carfagna, C.; Gatti, G.; Martini, D.; Pettinari, C. Organic syntheses via transition-metal complexes. A convenient regio- and stereoselective approach to the angular allylation of bicyclic cyclopentadienes generated by the (1-Alkynyl)carbene complex route. *Organometallics* **2001**, *20*, 2175–2182.

40. Brookhart, M.; Wagner, M. I. Polymers with main-chain chirality. Synthesis of highly isotactic, optically active poly(4-*tert*-butylstyrene-*ALT*-CO) using Pd(II) catalysts based on C_2-symmetric bisxazoline ligands. *J. Am. Chem. Soc.* **1994**, *116*, 3641–3642.

41. Aeby, A. Consiglio, G. P N versus N N ligands for the palladium-catalyzed alternating copolymerization of styrene and carbon monoxide. *Inorg. Chim. Acta* **1999**, *296*, 45–51.

42. Bartolini, S.; Carfagna, C.; Musco, A. Enantioselective isotactic alternating copolymerization of styrene and 4-methylstyrene with carbon monoxide catalyzed by a cationic bisoxazoline Pd(II) complex. *Makromol. Chem., Rapid. Commun.* **1995**, *16*, 9–14.

43. Reetz, M. T.; Haderlein, G.; Angermund, K. Chiral dikektimines as ligands in Pd-catalyzed reactions: Prediction of catalyst Activity by the AMS Model. *J. Am. Chem. Soc.* **2000**, *122*, 996–997.

44. Sperrle, M.; Aeby, A.; Consiglio, G.; Pfaltz, A. Isotactic and atactic copolymerization of styrene and carbon monoxide using cationic palladium-phosphino(dihydrooxazole) complexes. *Helv. Chim. Acta* **1996**, *79*, 1387–1392.

45. Aeby, A.; Gsponer, A.; Consiglio, G. From regiospecific to regioirregular alternating styrene/carbon monoxide copolymerization. *J. Am. Chem. Soc.* **1998**, *120*, 11000–11001.

46. Aeby, A.; Consiglio, G. Ethene and styrene insertion into the Pd–acyl bond of [Pd(COMe)(P N) (solv)]O₃SCF₃ and its role in the copolymerization of olefins with carbon monoxide. *J. Chem. Soc., Dalton Trans.* **1999**, 655–656.

47. Aeby, A.; Bangerter, F.; Consiglio, G. Enantioselective dicarbonylation of styrene to isotactic poly[1-oxo-2-phenylpropane-1,3-diyl] with phosphinodihydrooxazole-palladium(II) complexes: Model studies for enantioface selection. *Helv. Chim. Acta.* **1998**, *81*, 764–769.

48. Aeby, A.; Gsponer, A.; Sperrle, M.; Consiglio, G. Enantioface discriminating carbonylations of styrene with cationic palladium(II) catalyst precursors, *J. Organomet. Chem.* **2000**, *603*, 122–127.

49. Nozaki, K.; Komaki, H.; Kawashima, Y.; Hiyama, T.; Matsubara, T. Predominant 1,2-insertion of styrene in the Pd-catalyzed alternating copolymerization with carbon monoxide. *J. Am. Chem. Soc.* **2001**, *123*, 534–544.

50. Aeby, A.; Consiglio, G. Enantioselective alternating terpolymerization of styrene and ethene with carbon monoxide. *Helv. Chim Acta* **1998**, *81*, 35–39.

51. Sesto, B.; Bronco, S.; Gindro, E. L.; Consiglio, G. Enantioselective alternating terpolymerization of propene and ethene with carbon monoxide. *Macromol. Chem. Phys.* **2001**, *202*, 2059–2064.

52. Nozaki, K.; Kawashima, Y.; Nakamoto, K.; Hiyama, T. Asymmetric terpolymerization of propene, vinylarene, and carbon monoxide. *Macromolecules*, **1999**, *32*, 5168–5170.

53. Muellers, B. T.; Park, J.-W.; Brookhart, M.; Green, M. M. Glassy state and secondary structures of chiral macromolecules: Polyisocyanates and polyketones. *Macromolecules* **2001**, *34*, 572–581.

54. Brookhart, M.; Wagner, M. I. Synthesis of steroblock polyketone through ancillary ligand exchange. *J. Am. Chem. Soc.* **1996**, *118*, 7219–7220.

55. Kacker, S.; Jiang, Z.; Sen, A. Alternating copolymers of functional alkenes with carbon monoxide. *Macromolecules* **1996**, *29*, 5852–5858.

56. Lee, J. T.; Alper, H. Copolymers of vinyl epoxides with carbon monoxide. *J. Chem. Soc., Chem. Commun.* **2000**, 2189–2190.

57. Murtuza, S.; Harkins, S. B.; Sen, A. Palladium(II)-catalyzed synthesis of alternating fluoroalkene-carbon monoxide copolymers. *Macromolecules* **1999**, *32*, 8697–8720.

58. Nozaki, K.; Shibahara, F.; Elzner, S.; Hiyama, T. Alternating copolymerization of w-perfluoroalkyl-1-alkenes with carbon monoxide catalyzed by homogeneous and polymer-supported Pd-complexes. *Can. J. Chem.* **2001**, *79*, 593–597.

59. Fujita, T.; Nakano, K.; Yamashita, M.; Nozaki, K. Alternating copolymerization of fluoroalkenes with carbon monoxide. *J. Am. Chem. Soc.* **2006**, *128*, 1968–1975.

60. Borkowsky, S. L. W.; Waymouth, R. M. Pd²⁺-catalyzed cyclocopolymerization of 1,5-hexadiene and CO: Regioselectivity of olefin insertion. *Macromolecules* **1996**, *29*, 6377–6382.

61. Nozaki, K.; Sato, N.; Nakamoto, K.; Takaya, H. Cyclocopolymerization of α, ω-dienes with carbon monoxide catalyzed by (*R*,*S*)-BINAPHOS Pd(II). *Bull. Chem. Soc. Jpn.* **1997**, *70*, 659.

62. Kochi, T.; Nakamura, A.; Ida, H.; Nozaki, K. Alternating copolymerization of vinyl acetate with carbon monoxide. *J. Am. Chem. Soc.* **2007**, *129*, 7770–7771.

23 Stereoselective Acrylate Polymerization

Edward L. Marshall and Vernon C. Gibson

CONTENTS

23.1 INTRODUCTION

Poly(acrylates) and poly(methacrylates) are commercially important polymers with a myriad of uses, including paper and textile coatings, adhesives, caulks and sealants, plasticizers, paint and ink additives, and optical components for computer displays. Since they are derived from monosubstituted and unsymmetrical 1,1-disubstituted vinyl monomers, poly(acrylate) and poly(methacrylate) products with a spectrum of tacticities and thereby mechanical properties are potentially accessible. To date, however, industrially produced materials are generated using free-radical polymerization technology, which offers limited scope for tacticity control. Therefore, there has been much interest in the development of metal-catalyzed routes to these polymers where the coordination environment of the metal offers the potential to influence tacticity.

One of the main challenges in the use of metal catalysts is the suppression of undesirable termination reactions, which, for reasons described below, tend to be more prevalent for acrylate systems. Most of the recent developments have thus been seen in the polymerization of methacrylate monomers, especially commercially significant methyl methacrylate (MMA), which is polymerized to the commodity material poly(methyl methacrylate) (PMMA), known by its trademark names Perspex[TM], Plexiglass[TM], and Lucite[TM] (Scheme 23.1).[1] Much of this chapter, therefore, focuses on tacticity control in the production of PMMA using coordination polymerization catalysts. Though little has been reported on stereocontrolled acrylate polymerization, the basic principles established

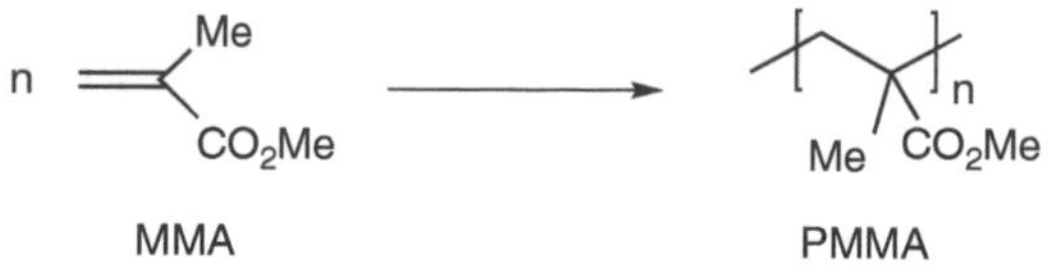

SCHEME 23.1 The polymerization of MMA.

Alkyl acrylates, RA   Alkyl methacrylates, RMA

FIGURE 23.1 Acrylate and methacrylate monomers.

Isotactic (*mm* triad)   Syndiotactic (*rr* triad)   Heterotactic (*mr* triad)

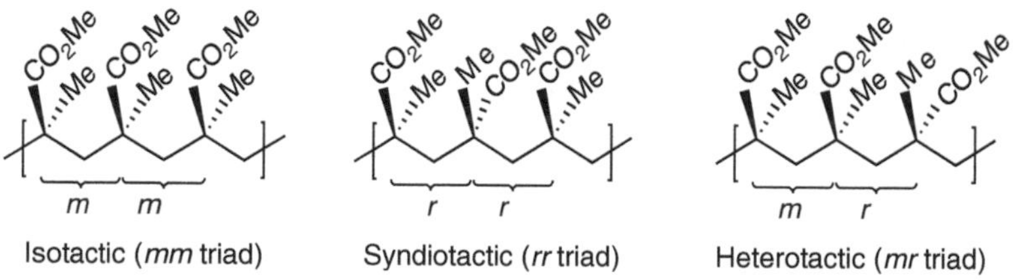

FIGURE 23.2 Stereoregular PMMA microstructures.

for PMMA are potentially applicable to poly(acrylate) materials, provided unwanted termination reactions can be eliminated. More recently, controlled free-radical polymerization methodologies, exploiting transition metal catalysts, have provided more encouraging indications of tacticity control for poly(acrylate) materials. The first part of this chapter reviews the effects of tacticity on the properties of PMMA and the commonly proposed coordinative polymerization mechanism. After a brief review of classical (poorly defined) initiators, the bulk of the chapter is devoted to recent advances in transition metal, rare earth, and main group single-site initiator systems. Finally, recent attempts to control stereoselectivity through metal-mediated free-radical polymerizations are described.

General formulae of acrylate and methacrylate monomers of relevance to the studies described in this chapter are shown in Figure 23.1 along with commonly used abbreviations. For the acrylates, the various alkyl esters are depicted by RA, (e.g., MA = methyl acrylate), while methacrylates are termed RMA, (e.g., MMA = methyl methacrylate).

23.1.1 The Properties of Stereoregular Poly(methyl Methacrylate)

There are three major stereoregular forms of PMMA: isotactic, syndiotactic, and heterotactic (Figure 23.2). However, commercially available samples prepared through free-radical initiators tend to be in the region of 60%–70% syndiotactic, the precise syndio content being dependent upon the reaction temperature (typically 65%–67% *rr* triad content at room temperature).[2] It is generally accepted (as a result of electron spin resonance (ESR) studies[3,4]) that the propagating species is a carbon-based radical chain end and thus the bias toward a syndiotactic microstructure, that is, with an alternating placement of the ester and methyl groups along the polymer backbone, arises from steric repulsion between the bulky ester groups of the radical chain end and the incoming monomer (Figure 23.3). Any deviation from such levels of syndiotacticity is often presented as good evidence for a nonradical mechanism, although workers in this field often use the term "stereospecific" to describe what, in reality, are relatively modest stereoselectivities.

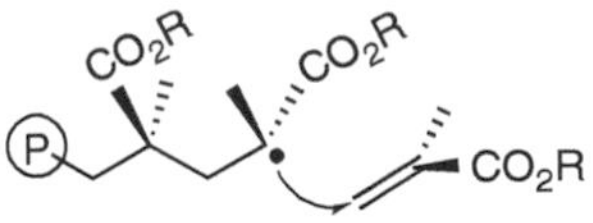

FIGURE 23.3 The proposed origin of syndiotactic bias in the radical polymerization of MMA.

Despite the inherent inclination toward a syndiotactic microstructure, the convenient production of very highly syndiotactic PMMA has proved highly challenging. Such a material has been recognized as offering potential technological advantages over less syndio-rich materials since high syndiotacticity imparts a higher glass transition temperature (T_g) to the material,[5] which may have advantages in higher temperature applications. Thus, the T_g of highly isotactic PMMA is ca. 48–50 °C,[6] significantly lower than its syndiotactic counterpart (ca. 130 °C)[7]. It has been reported that T_g values increase almost linearly with increasing r diad content.[8] Accordingly, highly heterotactic PMMA (96% rm content) displays an intermediate T_g value of 91 °C. All three stereoregular PMMA isomers exhibit some crystallinity, with melting points (T_ms) decreasing in the order heterotactic ($T_m = 166$ °C) > syndiotactic (159 °C) > isotactic (150 °C).[8]

The large majority of stereoselective acrylate polymerizations afford either isotactic or syndiotactic-biased material and since heterotactic polymers have only been prepared on rare occasions,[9] they are not covered in depth in this chapter. The principal initiating system for the production of heterotactic poly(alkyl methacrylates) is a stoichiometric 1:1 mixture of *tert*-butyllithium and bis(2,6-di-*tert*-butylphenoxy)methylaluminum.[9] The heterotactic content is particularly pronounced at low temperatures and increases in the order methyl < ethyl < propyl < allyl methacrylates. For example, at −95 °C, poly(allyl methacrylate) with 95.8% mr triad content is obtained; after reduction to poly(methacrylic acid) and subsequent methylation, highly heterotactic PMMA results.[10]

The microstructure of PMMA is readily examined using nuclear magnetic resonance (NMR) spectroscopy. The C(CH_3)CO$_2$Me region of the ^{1}H NMR spectrum in CDCl$_3$ consists of three resonances, at ca. δ 1.2, 1.0, and 0.8, assigned to the mm, mr, and rr triad repeat units respectively. Thus, much of the literature on PMMA describes microstructure in terms of triad contents, although occasionally dyad percentages may be used (even though these are calculated from the triad measurements, e.g., $\%m = \%mm + (0.5)(\%mr)$). More detailed information may be obtained from the carbonyl region of the ^{13}C NMR spectrum, which is sensitive to the pentad[11–14] and possibly even heptad level.[15] The assignment of NMR resonances to microstructures has also been performed for several other poly(methacrylate)s and poly(acrylate)s, including poly(ethyl methacrylate) (PEtMA),[16] poly(*n*-butyl methacrylate) (PnBMA),[17] poly(methyl acrylate) (PMA),[18,19] poly(isopropyl acrylate),[20] poly(*tert*-butyl acrylate),[21] poly(1-naphthyl acrylate),[22] poly(methyl α-(phenoxymethyl)acrylate) and poly(ethyl α-(phenoxymethyl)acrylate),[23] poly(benzyl α-(methoxymethyl)acrylate),[24] poly(methyl α-benzyl acrylate)s,[25] poly(ethyl α-benzoyloxymethylacrylate),[26] and poly(di-*n*-butyl itaconate).[27]

23.1.2 THE MECHANISM OF COORDINATIVE ACRYLATE POLYMERIZATION

Historically, the earliest reported stereoselective polymerizations of MMA employed a variety of main group organometallic catalysts, particularly those based upon lithium and magnesium, and to a lesser extent, aluminum. These initiators were considered to operate through anionic mechanisms, and the development of well-defined coordinative polymerization initiators may be considered a logical extension of this work. Both anionic and coordinative pathways are proposed to propagate through metal enolate species, and the difference between the two mechanisms may be simply considered in terms of the degree of dissociation of the enolate ligand/anion.

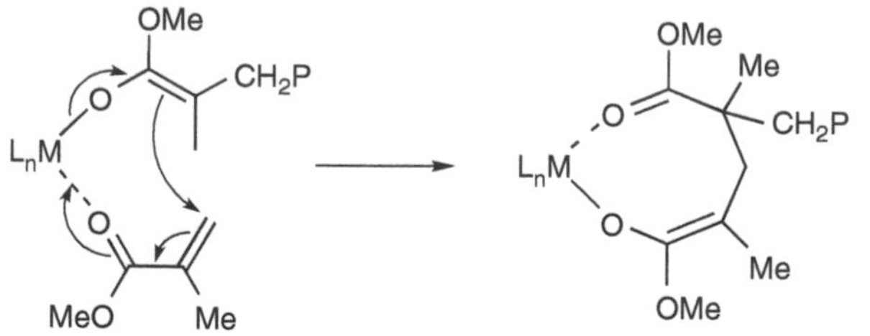

SCHEME 23.2 Addition of the propagating PMMA enolate to MMA monomer (M = transition or main group metal, L = spectator ligand and P = PMMA chain).

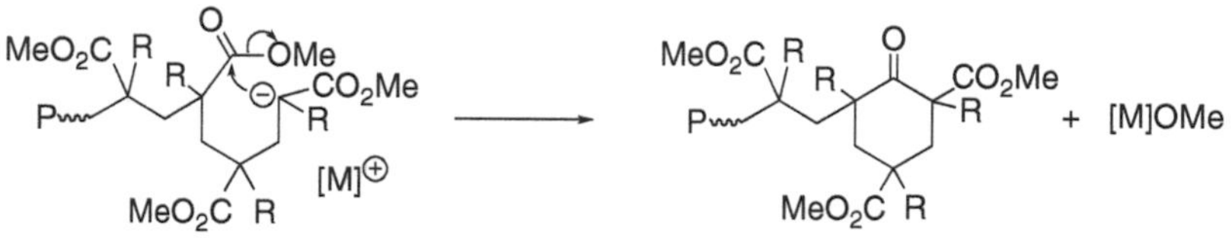

SCHEME 23.3 Intramolecular cyclization for PMA (R = H) and PMMA (R = Me) (P = polymer chain and M = metal cation).

As shown in Scheme 23.2, oxygen-metalated enolate species (as opposed to carbon-bound keto tautomers) undergo 1,4-addition of a coordinated monomer, in a manner often likened to the Michael reaction, leading to the formation of the backbone carbon–carbon bond and the generation of a new enolate ligand capable of repeating the insertion process. As described later in this chapter, it has also been proposed that a bimetallic intermolecular version of this process might operate with certain zirconocene initiators, whereby attack occurs by a metal enolate at a monomer bound to a second metal.

The most prevalent known termination process is an intramolecular cyclization (Scheme 23.3) with concomitant formation of an inactive metal methoxide species.[28] This is shown as an anionic process in Scheme 23.3 but may also occur with the metal covalently bonded to the polymer chain end. Cyclization would be expected to occur most readily for sterically unhindered chains, explaining the tendency of poly(acrylate)s such as PMA to be particularly susceptible to this form of termination.[29]

23.2 CLASSICAL (POORLY DEFINED) INITIATORS

A very large number of organometallic compounds of groups 1–3 have been shown to polymerize MMA, especially organolithium and organomagnesium species. These have been the subject of extensive reviews and are not discussed in detail here. Most initiate non-living polymerizations (especially at ambient temperatures) that afford little control over chain length, and molecular weight distributions are typically broad, consistent with multiple propagating species. Where living-like behavior has been documented, a rational approach to the development of stereospecific analogues has been hampered by the ill-defined nature of the catalysts. Stereoselective polymerizations using these initiators are known, but their discoveries have largely been the result of trial and error. For instance, highly iso- and syndiotactic PMMA may be prepared by Grignard reagents (*vide infra*), but identification of the active site(s) is complicated by processes such as aggregation, solvation effects, and ligand exchange (i.e., through the Schlenk equilibrium). Furthermore, high stereoselectivity is only observed at very low temperatures where propagation is consequently slow, and high molecular weight (>20,000 g/mol) products are rarely obtained.

23.2.1 CLASSICAL ORGANOLITHIUM INITIATORS

The most widely studied catalyst family for the anionic polymerization of MMA is that of lithium alkyls. For systems containing relatively non-bulky substituents, for example, n-BuLi, the polymerization is plagued by side reactions.[30] These may be avoided if bulkier initiators such as 1,1-diphenylhexyllithium[31] and low temperatures (typically $-78\ ^\circ C$) are employed. Under such conditions these systems often display living characteristics, including narrow molecular weight distributions (M_w/M_n; M_w = weight average molecular weight, M_n = number average molecular weight), and their use in the preparation of block copolymers has been reported.[32–34] Examples of stereoselective organolithium reagents include fluorenyllithium; with this system, at $-60\ ^\circ C$ highly isotactic PMMA is prepared using 95% toluene/5% Et_2O solvent mixtures, whereas syndiotactic material is obtained from 85% toluene/15% tetrahydrofuran (THF).[35]

Alkali metal alkoxides have also been examined in some detail.[36] The addition of bulky lithium alkoxides to alkyllithium initiators retards the rate of intramolecular cyclization, thus allowing the polymerization temperature to be raised. LiCl has been used to similar effect, allowing the preparation of monodisperse PMMA (M_w/M_n = 1.2) at $-20\ ^\circ C$.[37] Sterically bulky lithium aluminum alkyls have been used to similar effect, with good chain length control retained even at ambient temperature.[38] However, these approaches have not yet been successfully applied to stereoselective polymerizations.

23.2.2 CLASSICAL ORGANOMAGNESIUM INITIATORS

Some of the highest levels of stereocontrol in the polymerization of MMA have been reported using Grignard reagents as initiators at very low temperatures. For example, t-BuMgBr polymerizes MMA with almost 100% initiator efficiency at $-78\ ^\circ C$ in toluene, affording isotactic PMMA ($>96\%\ mm$ triads). The bulky alkyl groups are believed to prevent side reactions, and narrow polydispersities result. By contrast, m-vinylbenzyl magnesium chloride affords monodisperse ($M_w/M_n < 1.2$), highly syndiotactic PMMA (97% rr triad content) in THF at $-110\ ^\circ C$.[39] In such cases, the mechanistic origin of the stereocontrol remains unknown, a consequence of the ill-defined nature of the active site (i.e., its composition and geometry) and the role of the solvent.

23.3 WELL-DEFINED INITIATORS

23.3.1 GROUP 3 AND RARE EARTH METAL INITIATORS

Some of the most defining work carried out on the coordinative polymerization of MMA was reported in the early 1990s by Yasuda,[40,41] who described the living polymerization of MMA by lanthanocene catalysts. The complex $[Cp_2^*Sm(\mu\text{-}H)]_2$ ($Cp^* = C_5Me_5$) (**1**, Scheme 23.4) was shown to afford high molecular weight PMMA with very low polydispersities ($M_w/M_n \leq 1.05$).[42] At $-95\ ^\circ C$ the polymer was found to be highly syndiotactic (95% rr triads), and even at $0\ ^\circ C$ quite good stereoselectivity was exhibited (82% rr). Isolation and X-ray analysis of the 1:2 complex of **1** and MMA provided evidence for the participation of a metal-enolate as the active species. Complex **2** behaves in an identical manner to the hydride precursor, converting 100 equivalents of MMA to polymer with $M_n = 11,000\ g/mol$ and $M_w/M_n = 1.03$.[43] The intermediacy of the enolate structure confirmed the accuracy of proposed intermediates described decades earlier by Cram and Kopecky[44] and by Bawn and Ledwith.[45] The structure of **2** dramatically illustrates a striking difference between acrylate and olefin polymerization, with chelation of the propagating chain to the active site very probable for the former class of monomer. Much research is currently being performed to establish the degree to which the cyclic enolate structure affects the stereochemistry of insertion (a factor obviously not relevant to stereoselective α-olefin polymerization systems).

The samarocene and yttrocene alkyl complexes **3** and **4**[43] (Figure 23.4) are also highly active initiators for MMA polymerization and, since they bear the same supporting Cp* ligands, they are

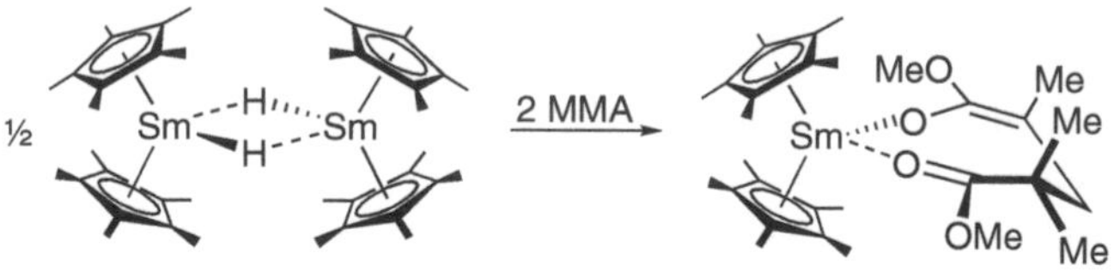

SCHEME 23.4   The isolation of the 1:2 complex of $[Cp_2^*Sm(\mu\text{-}H)]_2$ and MMA.

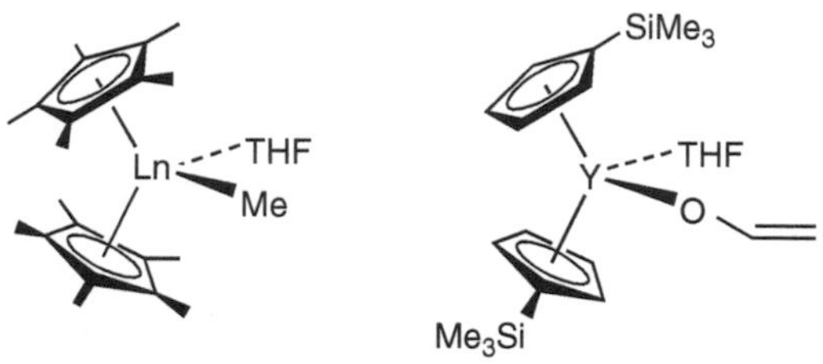

Ln = Sm, **3**
Ln = Yb, **4**

FIGURE 23.4   Sturctural formulae of complexes **3–5**.

also syndioselective, affording 82%–85% rr PMMA at 0 °C with high initiator efficiencies and narrow molecular weight distributions. Organic acids such as ketones and thiols act as chain transfer agents.[46] The range of monomers that can be polymerized in a well-controlled manner using the lanthanocene initiators includes ethyl methacrylate (EtMA), isopropyl methacrylate (iPrMA), and *tert*-butyl methacrylate (tBMA). Reactivity decreases in the order MMA ≈ EtMA > iPrMA > tBMA as may be anticipated on steric grounds. However, syndiotacticity is also greatest for the smaller alkyl substituents, an observation that remains to be understood.

Complexes **3** and **4** also rapidly polymerize acrylate monomers at 0 °C.[47,48] Both initiators give monodisperse poly(acrylate)s ($M_w/M_n = 1.02$–1.07) with molecular weights of less than 5×10^5 g/mol, although the polymers obtained are largely atactic. The polymerizations are exceedingly rapid with 500 equivalents of *n*-butyl acrylate (nBA) and ethyl acrylate (EtA) being fully consumed in <5 s at 0 °C; MA is somewhat slower, requiring 300 s for similar levels of conversion. An enolate is again believed to be the active propagating species since the model complex **5** (Figure 24.3) was shown to initiate the polymerization of MA in a near-identical manner to **4**. However, the synthesis of block copoly(acrylate)s is hampered by rapid decomposition of the active site following consumption of the first monomer, with lifetimes of less than 60 s in toluene at 0 °C for PMA, though slightly longer lifetimes are observed in THF (70% chain survival after 5 min).

The lanthanocene complexes **3–5** are achiral and therefore the source of stereocontrol is assumed to be chain-end control, though no detailed mechanism has been proposed. Marks and coworkers investigated whether the polymerization of MMA might also be subject to enantiomorphic site control by employing C_1-symmetric *ansa*-lanthanocenes bearing menthyl substituents.[49] When the (+)-neomenthyl catalyst **6** (Figure 23.5) is used, highly isotactic PMMA is produced (94% mm at −35 °C), whereas the (-)-menthyl derivative **7** affords syndio-rich PMMA (73% rr at 25 °C). However, NMR statistical analysis indicated that the PMMA microstructure could not arise exclusively either from chain-end or enantiomorphic site control. Instead, the observed stereoselectively was rationalized on the basis of competing insertion/isomerization pathways. As outlined in Scheme 23.5, fast addition of the propagating enolate to the monomer (relative to the rate of enolate isomerization) leads to high syndioselectivity. However, if enolate isomerization is rapid, and the diastereomeric equilibrium lies towards the side of the newly formed isomer (a consequence of the cyclopentandienyl R* substituent),

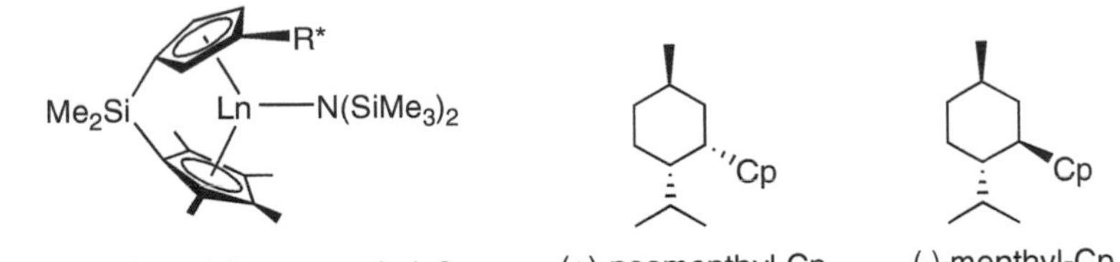

Ln = La; R* = (+)-neomenthyl, **6**   (+)-neomenthyl-Cp   (-)-menthyl-Cp
Ln = Lu; R* = (-)-menthyl, **7**

FIGURE 23.5   Structural formulae of complexes **6** and **7**.

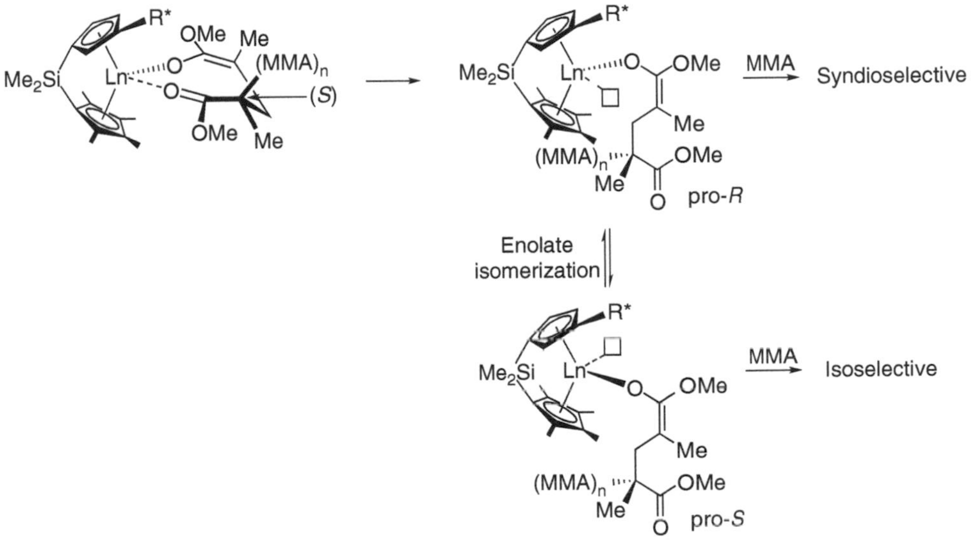

SCHEME 23.5   Proposed mechanism controlling the formation of isotactic versus syndiotactic PMMA from C_1-symmetric chiral lanthanocene initiators **6** and **7** (R* = (+)-neomenthyl or (-)-menthyl and □ = vacant coordination site).

then an isotactic polymer should be formed. It has been alternatively suggested that the menthyl-functionalized initiator **7** (Figure 23.5) generates syndiotactic PMMA from a cyclic eight-membered ring intermediate (akin to **2**) whereas complex **6** propagates through a nonchelated linear PMMA chain.[40]

Before the discovery of controlled MMA polymerization by lanthanocene initiators, several research groups had reported that lanthanocene alkyls were active for the polymerization of ethylene.[50–53] Yasuda reported that both $Cp_2^*SmMe(THF)$ and $[Cp_2^*Sm(\mu\text{-}H)]_2$ catalyzed the block copolymerization of ethylene with MMA (as well as the copolymerization of ethylene with other polar monomers, including MA, EtA, and lactones).[54] The scope of this system is limited somewhat by the need to keep the polyethylene block soluble (the olefin must be polymerized first), thus restricting the M_n of this prepolymer to ca. 12,000 g/mol. Reversal of the order of monomer addition, that is, polymerizing MMA first and then adding ethylene, affords PMMA homopolymer only. Other lanthanide complexes have since been reported to catalyze the preparation of this diblock material[55–57] as well as copolymers of MMA with higher olefins such as 1-pentene and 1-hexene.[58]

The divalent analogues **8–10** (Figure 23.6) also generate syndiotactic PMMA,[43] but exhibit much lower initiator efficiencies of typically 30%–40% (evaluated by comparing observed polymer molecular weights to calculated molecular weights). Yasuda had earlier proposed that initiation in such systems occurs through in situ hydride formation, but Boffa and Novak[59,60] presented evidence

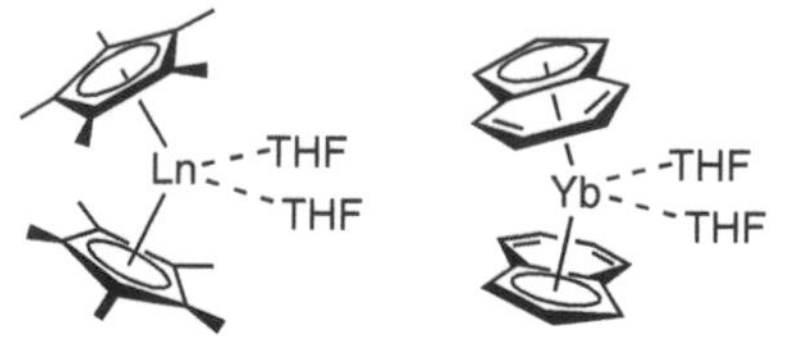

Ln = Sm, **8**
Ln = Yb, **9**   **10**

FIGURE 23.6   Divalent lanthanocene initiators **8–10**.

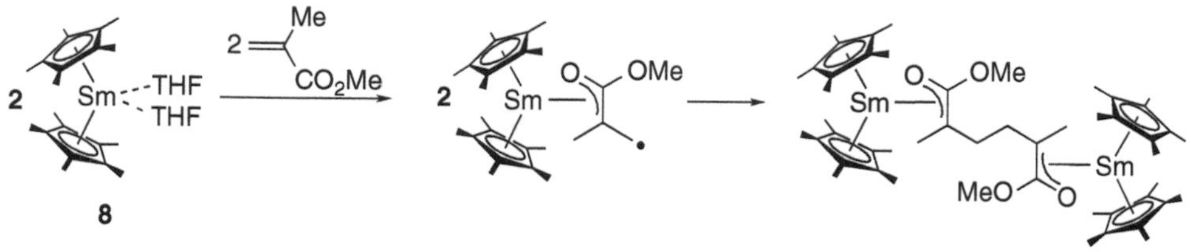

SCHEME 23.6   Formation of a bimetallic bis(enolate) intermediate during MMA polymerization using a divalent lanthanocene initiator.

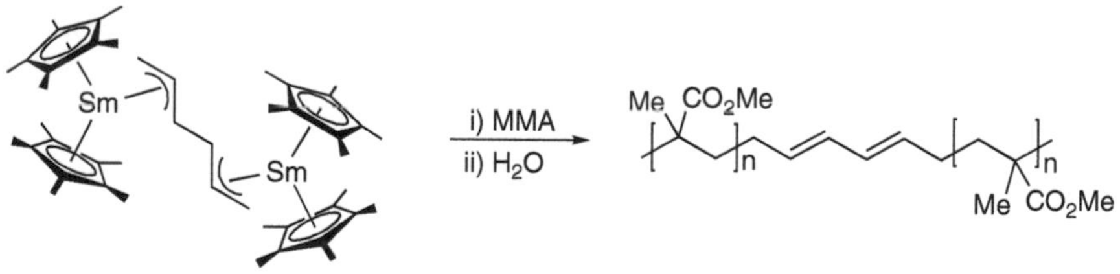

SCHEME 23.7   Synthesis of PMMA using a bimetallic lanthanide initiator.

for a bimetallic bis(enolate) intermediate, arising from the dimerization of a radical anion[61] (Scheme 23.6) that would lead to a maximum initiator efficiency of 50%. This system was then exploited to prepare unusual triblock materials: the "middle" segment of the triblock is formed initially by homopolymerization of the first monomer, and the two outer triblock segments are then grown simultaneously upon addition of the second monomer. Triblock copolymers having both methacrylate and acrylate segments have also been prepared.[62,63] Using a similar "double-ended" chain growth process, the bimetallic bis(allyl) complex **11** polymerizes MMA in a living fashion ($M_w/M_n \approx 1.1$) (Scheme 23.7).

Many other lanthanocene-based initiators for MMA polymerization have been described, especially amides;[64–68] a selection (**12–14**) is shown in Figure 23.7. Additional work in this field parallels contemporary research into new olefin polymerization catalysts, with rapid growth in interest in non-cyclopentadienyl-containing systems (**15–23**, Figure 23.8)[69–78] (mainly owing to the heavily patented nature of metallocene catalysts), and a variety of PMMA microstructures have proved accessible. The use of bulky ancillary ligands holds particular promise, as they may allow the approach of monomer and the orientation of the propagating enolate ligand (and hence the stereochemistry of insertion into the polymer chain) to be better understood and thence controlled. As an example, the bis(pyrrolylaldiminato) samarium alkyl complex **19** was found to polymerize MMA in a highly isoselective manner at room temperature (94.8% *mm* triads).

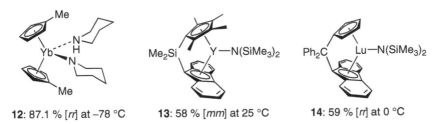

12: 87.1 % [*rr*] at −78 °C   **13**: 58 % [*mm*] at 25 °C   **14**: 59 % [*rr*] at 0 °C

FIGURE 23.7 Lanthanocene amides **12–14** used to polymerize MMA and the resultant PMMA microstructures.

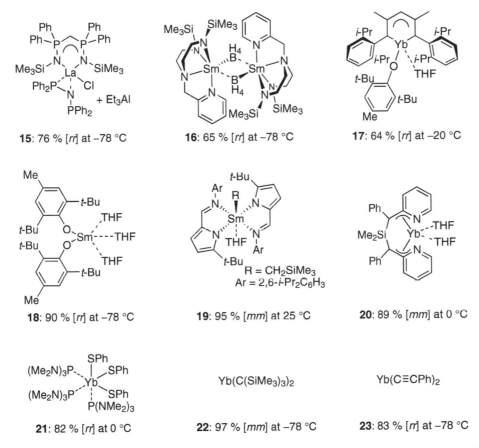

15: 76 % [*rr*] at −78 °C   **16**: 65 % [*rr*] at −78 °C   **17**: 64 % [*rr*] at −20 °C

18: 90 % [*rr*] at −78 °C   **19**: 95 % [*mm*] at 25 °C   **20**: 89 % [*mm*] at 0 °C

21: 82 % [*rr*] at 0 °C   **22**: 97 % [*mm*] at −78 °C   **23**: 83 % [*rr*] at −78 °C

FIGURE 23.8 A selection of nonmetallocene lanthanide-based initiators used for MMA polymerization and the resultant PMMA microstructures.

23.3.2 Group 4 Initiators

Although the polymerization of MMA using a group 4 metallocene initiator was first reported several decades ago,[79] it was not until more recent times that interest in this chemistry resurfaced. A patent filed in 1988 by workers at du Pont disclosed that MMA could be polymerized in a living manner using $Cp_2ZrCl(OC(OMe)=CMe_2)$ ($Cp = C_5H_5$).[80]

Subsequent work by Collins and Ward confirmed this observation; they showed that an equimolar mixture of the cationic alkyl complex $[Cp_2ZrMe(THF)]^+[BPh_4]^-$ (**24**, Scheme 23.8) and neutral Cp_2ZrMe_2 (**25**) generates low polydispersity PMMA (M_w/M_n = 1.2–1.4) with a syndiotactic bias comparable to free-radical MMA polymerizations (80% *r* dyads at 0 °C) and a microstructure

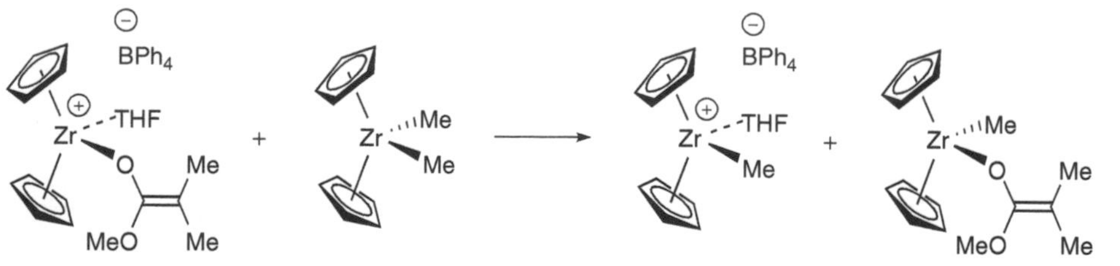

SCHEME 23.8 Generation of an active bimetallic initiator system for the polymerization of MMA from the (poorly active) cationic enolate complex **26**.

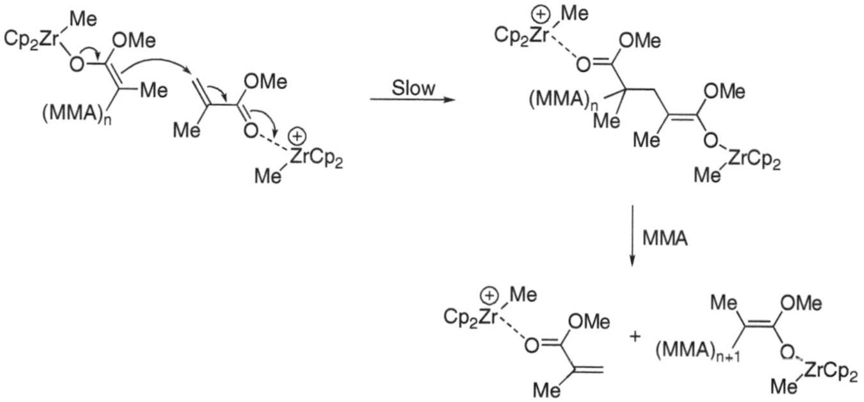

SCHEME 23.9 Bimetallic mechanism proposed by Collins and coworkers.

consistent with a chain-end controlled mechanism.[81] Initial support for the intermediacy of a cationic enolate[81] diminished when an isolable example of such a complex (**26**) was shown to be a poor initiator.[82] However, the 1:1 reaction of **26** with **25** to give **24** and the neutral alkyl enolate, [Cp$_2$ZrMe(OC(OMe)=CMe$_2$)] (**27**) (Scheme 23.8), was found to be an active initiator system. This mixture displays first order kinetics in both the neutral and cationic Zr species but is zero order in MMA. Consequently, a bimetallic group transfer polymerization (GTP) type mechanism[83,84] was proposed, the rate-limiting step of which involves intermolecular Michael addition of the propagating enolate to activated monomer (Scheme 23.9).[85] This system is highly moisture-sensitive, and trialkylaluminum compounds have been employed in situ to remove traces of water. However, chain transfer to the Al center occurs, unless the alkyl substituents are sufficiently bulky, for example, i-Bu$_3$Al.

The ability of the bimetallic system to polymerize the acrylate monomer nBA has also been examined.[85] Under similar conditions to those used for the polymerization of MMA (i.e., a 1:1 mixture of [Cp$_2$ZrMe(THF)]$^+$[BPh$_4$]$^-$ and [Cp$_2$ZrMe(OC(Ot-Bu)=CMe$_2$)] in CH$_2$Cl$_2$), the polymerization of nBA proceeded poorly at 0 °C. However, at −78 °C poly(n-butyl acrylate) (PnBA) was produced. Polydispersities were relatively narrow ($M_w/M_n < 1.3$) but broadened at higher temperatures owing to termination through backbiting cyclization (Scheme 23.10).

In the light of Yasuda's successful use of neutral lanthanocene alkyl complexes as initiators, several authors, including Collins et al.,[82] Soga et al.,[86] Gibson et al.,[87] and Höcker et al.,[88] rationalized that isolobal single-component, cationic zirconocene alkyl complexes should also polymerize MMA without the need for a second Zr center. This is indeed the case, and nearly all contemporary reports in this area now employ single-site cationic alkyl and cationic enolate initiators. Figure 23.9 collects many of the complexes examined to date and summarizes the resultant PMMA tacticities.

SCHEME 23.10 Intramolecular cyclization in the zirconocene-mediated polymerization of nBA (P = PnBA chain).

24/25: 80 % [r] at 0 °C

28: 67 % [rr] at 25 °C; 74 % [rr] at 0°C[a]

29: Inactive

30: 95 % [mm] at 25 °C; 96.5 % [mm] at 0 °C[a]

31: 96.7 % [mm] at 25 °C

32: 94.4 % [mm] at 0 °C[a]

33: 86.8 % [mm] at 0 °C[a]

34: 81 % [mm] at 25 °C

35: 54 % [mm] at 25 °C

36: Inactive

37: 72 % [rr] at 0 °C[a]; 74 % [rr] at −40 °C[a]

38: 69 % [rr] at 0 °C[a]; 78 % [rr] at −40 °C[a]

39: 67 % [rr] at 25 °C (CH₂Cl₂); 44 % [rr] at 25 °C (PhMe)

40: 89 % [rr] at −45 °C; 79 % [rr] at 0 °C

41: 60 % [rr] at 0 °C

42: 41 % [mm] at 0 °C

43: 64 % [mm] at 0 °C

44: 87 % [mm] at 0 °C

45: 84 % [mm] at 0 °C

46: 40 % [rr] at 0 °C

FIGURE 23.9 Cationic zirconocene initiators for MMA polymerization and the resultant PMMA microstructures ([a] = in the presence of ZnEt₂).

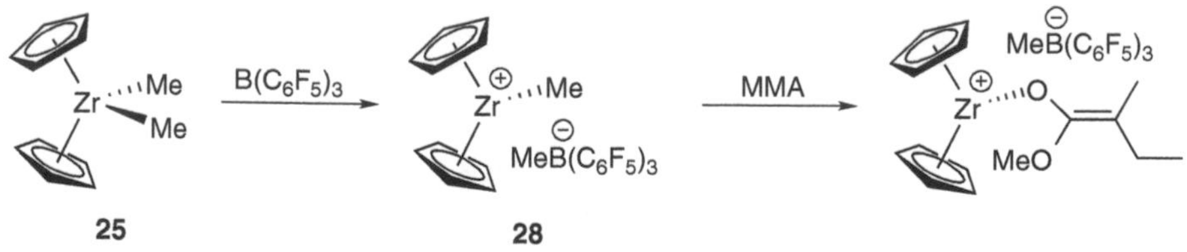

SCHEME 23.11   Proposed initiation step in the polymerization of MMA by **28**.

For example, $[Cp_2ZrMe]^+[MeB(C_6F_5)_3]^-$ (**28**) initiates the polymerization of MMA in a rapid and controlled manner,[87] with a similar syndioselectivity to that observed for the corresponding bimetallic system (**24/25**). Nucleophilic addition of the methyl ligand from the cationic zirconium center to a coordinated MMA molecule is believed to generate a cationic ester enolate, the postulated active propagating species (Scheme 23.11). This mechanism differs significantly from the bimetallic pathway outlined above in Scheme 23.9 since the PMMA chain only propagates from one zirconium center. Good supporting evidence for this mechanism has recently been presented by Chen and Bolig who have shown that the cationic zirconocene ester enolate complex **31** readily initiates the polymerization of MMA.[89] This observation may appear to conflict with previous observations by Collins, who found that the ester enolate complex **26** exhibits only low activity,[82] but this may be ascribed to the less coordinating nature of the anion in complex **31**, and to the greater reactivity of the bridged *rac*-bis(indenyl)zirconocenium fragment (*vide infra*).

Cationic initiators such as **28** are typically generated in situ by reaction of the dimethyl precursor with $B(C_6F_5)_3$ (Scheme 23.11).[87] In initial studies performed by Soga et al., the borane was added to the MMA monomer feedstock and this mixture was then treated with Cp_2ZrMe_2, but using such a procedure no activity is observed unless a large excess of $ZnEt_2$ is also added.[86] However, work in our laboratories indicates that the primary function of the organozinc reagent is to displace the monomer from the Lewis acidic borane, allowing the latter to extract a methyl ligand from the zirconium center and thus create the active cationic site.[87] By generating the cationic alkyl complex first, then treating with MMA, the need for the $ZnEt_2$ cocatalyst is obviated, and very fast propagation is observed (e.g., **28** consumes 200 equivalents MMA in under 2 min at 25 °C). $ZnEt_2$ is still used to activate bulky methacrylate monomers toward nucleophilic attack by the propagating enolate species.[90] However, the presence of $ZnEt_2$ has no appreciable effect upon the stereoselectivity of the propagation process (e.g., complex **30** is highly isoselective both in the presence and absence of the Lewis acid).

The stereoselectivity of MMA polymerization is obviously highly dependent upon the nature of the zirconocene fragment, with microstructures ranging from very highly isotactic to moderately syndiotactic (Figure 23.9). The nature of the Cp-based ligands also plays a major role in determining polymerization activity. Activities are additionally dependent upon the coordinating nature of the counteranion used. For example, the $[BPh_4]^-$ analogue of **28** (complex **29**) is inactive for MMA polymerization,[88] and counterparts of complex **45** featuring more strongly coordinating *N*-pyrrolyl-based borate anions have been found to be inactive under identical reaction conditions.[91] It has generally been assumed that the counteranion does not play a significant role in influencing the stereochemistry of monomer insertion, but the contrasting behavior of complexes **44** and **46** recently disclosed by Erker and coworkers suggests that this may not be the case.[91]

One conclusion to arise from these studies is the propensity of C_2-symmetric *ansa*-zirconocenium complexes (such as **30–33**) to exhibit high isoselectivities, whereas unbridged zirconocene initiators tend to favor syndioselective insertion. For example, in toluene or CH_2Cl_2, high levels of isotacticity (95% *mm*) are afforded by the *rac ansa*-bis(indenyl) complex **30** and its tetrahydroindenyl analogue **32**.[88,92] The analogous cationic enolate (**31**) behaves in a similar manner (consuming 400 equivalents MMA in under 10 min and giving PMMA with $M_w/M_n = 1.03$ and $mm = 96.7\%$[89]), and enolate end-groups have been detected in PMMA prepared with **31** using MALDI-TOF mass

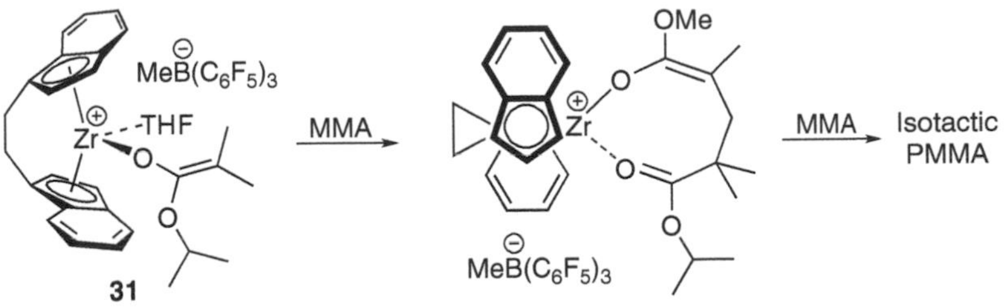

SCHEME 23.12 Proposed mechanism for the polymerization of MMA by a cationic zirconocene ester enolate initiator.

spectroscopy.[93] This polymerization exhibits a first-order dependence upon the cationic enolate and upon the monomer, and therefore the propagation step has been deduced to be a monometallic intramolecular Michael addition (Scheme 23.12).[93] By synthesizing a zirconocene-based model of the decamethylsamarocene cyclic enolate **2**, the same authors further concluded that the rate determining step of insertion is the associative displacement of the coordinated ester of an eight-membered cyclic intermediate by incoming MMA monomer. Microstructural analysis of the isotactic PMMA formed with this system indicated that the isotacticity most likely originates through enantiomorphic site control. Complex **31** has also been shown to generate very highly isotactic (>99% *mm*) PnBMA. The isotactic polymerization of *tert*-butyl acrylate has also been achieved with an *ansa*-bis(indenyl) zirconocene initiator.[94]

Highly isotactic PMMA has also been reported for a family of C_2-symmetric monoalkyl-substituted *ansa*-zirconocenes (**42–45**).[91] The isotactic triad content in the PMMA increases with the size of the ligand alkyl substituent, whereas the parent unsubstituted analogue **41** exhibits moderate syndioselectivity. However, C_2-symmetry is not a prerequisite feature of an isoselective group 4 initiator, as shown by the behavior of the C_1-symmetric *ansa*-metallocene complex **34**,[87] the [BPh$_4$]$^-$ variant of which has also been reported to afford PMMA with 94.7% *mm* triad content at −30 °C.[88]

Attempts to prepare highly syndiotactic PMMA using zirconocene catalysts have met with little success, despite the initial observation of *r* dyad selectivity using **28** and the report of modest syndioselectivities for a range of other methacrylate monomers using this initiator, including hexyl, decyl, and stearyl methacrylates.[95] The only notable exception is complex **40**, which produces 89% *rr* PMMA at −45 °C with $M_w/M_n = 1.31$.[96] However, Chen has shown that syndiotactic PMMA may be prepared using half-sandwich titanium-based initiators. For example, the cationic cyclopentadienyl-amido complex **47** (Figure 23.10) effects the controlled polymerization of methacrylates at room temperature, affording 82% *rr* PMMA and 89% *rr* PnBMA with $M_w/M_n < 1.1$ in both cases.[97] Syndiotactic block and random copolymers of these two monomers have also been prepared. The role of the metal in influencing the selectivity of monomer insertion is dramatically demonstrated by comparing the behavior of complexes **48** and **49** (Figure 23.10); the cationic titanium ester enolate complex **48** affords syndiotactic PMMA, whereas its isostructural zirconium counterpart **49** is isoselective (*mm* = 95.5% at −40 °C and 80.5% at −20 °C in CH$_2$Cl$_2$). However, the reason for this difference in behavior has yet to be elucidated.

In order to shed some light upon the mechanism of enolate addition and upon the origin of stereoselectivity with the zirconocene-based initiators, a series of computational studies have been undertaken. In one of the earliest such studies Sustmann et al. examined both the bimetallic GTP-type mechanism and the monometallic cationic alkyl/enolate pathway and showed that both were plausible (a third hypothetical possibility operating through neutral intermediates was also considered).[98] In an attempt to deduce why the cationic bridged *ansa*-zirconocene complex **40** is a highly active MMA polymerization initiator, while the closely related unbridged analogue **29** is inactive, Höcker and coworkers carried out a series of density functional theory (DFT) calculations on stationary points along the reaction coordinate (Scheme 23.13).[99] The difference in reactivity between **40** and **29**

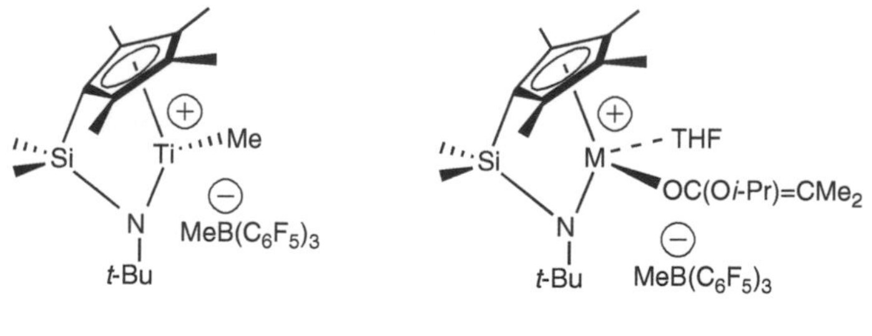

FIGURE 23.10   Structural formulae of complexes **47–49**.

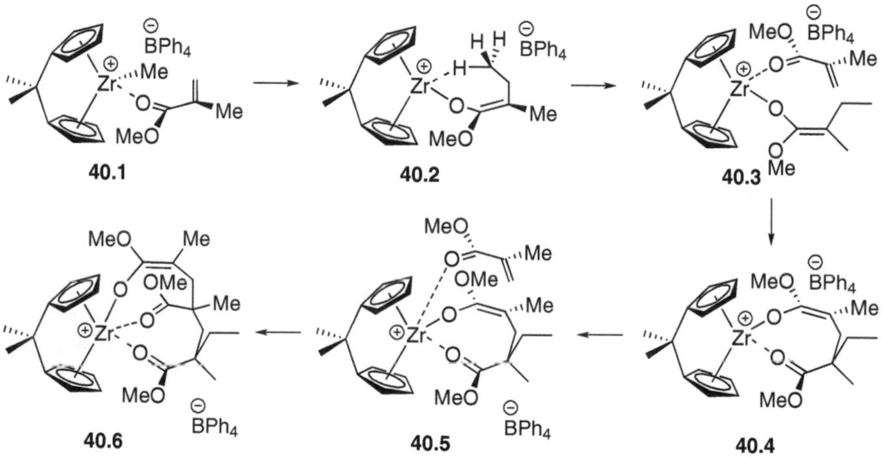

SCHEME 23.13   Computed key intermediates during the polymerization of MMA by complex **40**.

was then ascribed to the lower activation energies encountered along the reaction coordinate for **40** ($+20.52$ and $+11.48$ kcal/mol for [**40.1** → **40.2**]‡ and [**40.5** → **40.6**]‡ versus $+24.16$ and $+16.31$ kcal/mol for unbridged analogue **29**, respectively). This supposition was then rationalized in terms of how readily the enolate ligand and the coordinated monomer may closely approach each other before C–C bond formation. With bridged zirconocene fragments the metal center is more accessible (angle between the Cp rings in **40** = 127.5° versus 115.2° in **29**), thus allowing easier approach of the enolate ligand to the bound MMA molecule (hence *ansa*-zirconocene initiators should be more active than their unbridged counterparts). Furthermore, in the unbridged system, orbital overlap of the two Cp rings with the Zr center is stronger than in **40**; consequently the enolate ligand in **29** is bound less strongly to the metal, which leads to greater separation of the enolate from the coordinated monomer. Third, it was proposed that the eclipsed nature of the Cp rings in **40** leads to reduced steric interactions between the Cp hydrogen atoms and the enolate ligands compared to unbridged analogue **29**.

The same workers have also addressed the issue of stereoselectivity by computing competing pathways for MMA polymerization initiated by the isoselective cyclopentadienyl-*ansa*-indenyl complex **34**.[100] In the first phase of this study, the geometry of the metallacyclic enolate and the direction of monomer approach were examined. The authors concluded that the incoming MMA monomer prefers to approach the metal center from the same side as the indenyl unit (i.e., approach modes **34.1** and **34.3** in Figure 23.11) since this forces the propagating metallacyclic enolate to move away from the more hindered side of the *ansa*-metallocene ligand. In addition, in order for facile C–C bond formation to occur with the monomer, the enolate C=C bond must be proximal, that is, also on

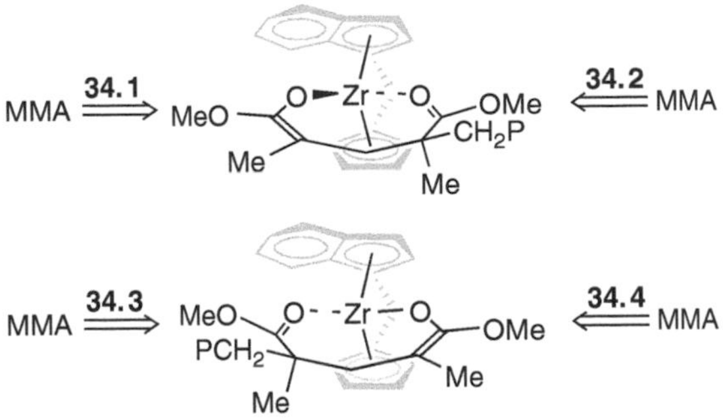

FIGURE 23.11 Possible MMA monomer approach pathways to the propagating metallacylic enolate derived from complex **34** (P = PMMA chain).

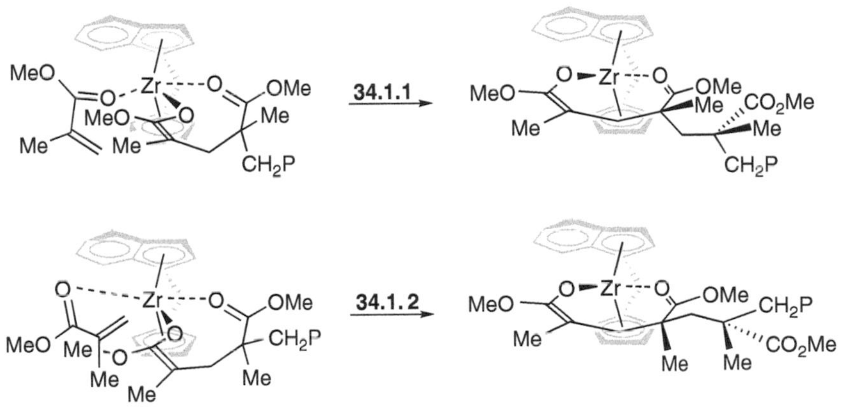

SCHEME 23.14 Computed MMA addition modes for complex **34** (P = PMMA chain).

the same side of the complex as the indenyl ring (as in **34.1** and **34.2** of Figure 23.11). Consequently, of the four possibilities shown in Figure 23.11, the approach mode depicted by **34.1** was computed to be the energetically most favorable.

Given the approach pathway outlined above (i.e., **34.1**), the stereoselectivity can then be rationalized by considering the way in which the MMA is oriented before C–C bond formation. As illustrated in Scheme 23.14, the monomer may approach the enolate with its olefinic CH_2 substituent either below (**34.1.1**) or above (**34.1.2**) the plane defined by the O-Zr-O atoms (a second pair of alternatives were also located but for the sake of brevity are not discussed here). Calculations reveal that the addition of MMA from below the O-Zr-O plane (**34.1.1**) is favored by ca. 10 kcal/mol. Furthermore, once C–C bond formation has taken place, the enolate C=C double bond is retained on the side of the complex, that is, immediately below the indenyl ring. As a consequence, subsequent monomer additions occur in exactly the same manner, leading to the continued formation of stereocenters with the same configuration (i.e., isotactic PMMA).

The block copolymerization of MMA with ethylene has been described using **34**/B(C$_6$F$_5$)$_3$.[101] Ethylene must be polymerized first (the reverse order of addition simply produces homo-PMMA) and the diblock nature of the products was inferred from solubility behavior. ^{1}H NMR spectroscopy confirmed that the PMMA block was highly isotactic, as expected for this initiator.

A block copolymer of isotactic-PP-co-isotactic-PMMA (PP = polypropylene) has also been synthesized using complex **40** (again, the olefin must be added first). Switching the initiator to the C$_s$-symmetric titanium complex **47** afforded an atactic-PP-co-syndiotactic-PMMA block copolymer.[102]

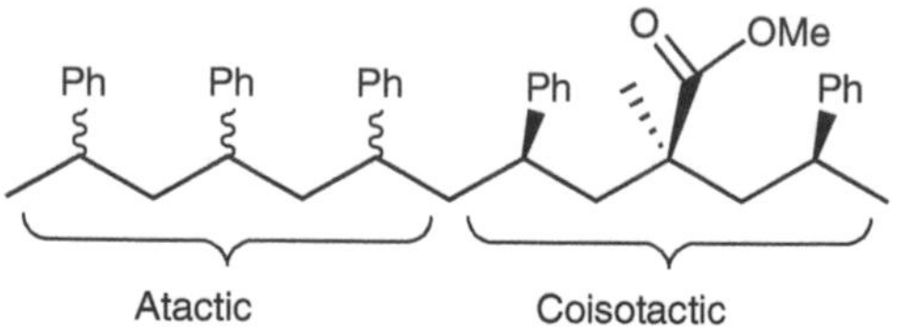

FIGURE 23.12   Coisoselective enchainment of MMA in a PS-PMMA copolymer.

Marks and coworkers have recently reported a low-valent titanium catalyst (generated in situ from Zn reduction of the product of the reaction of Cp*TiMe$_3$ with [Ph$_3$C$^+$][B(C$_6$F$_5$)$_4$]$^-$), which is capable of initiating the isoselective random copolymerization of styrene and MMA, although incorporation of the methacrylate monomer is low ($\leq$4%).[103] The polymer obtained is best described as containing atactic sequences of polystyrene (PS) with coisoselectively (ca. 80%) enchained MMA units (Figure 23.12).

23.3.3 GROUP 5 INITIATORS

In theory, the cations of V, Nb, and Ta should be more tolerant towards polar functionalities than their more oxophilic Group 4 counterparts. However, only a few reports of Group 5 MMA polymerization initiators have appeared. Mashima and coworkers have synthesized 1,4-diaza-1,3-diene tantalum complexes and found that in the presence of Me$_3$Al, MMA may be polymerized in a living manner at -20 °C to -30 °C ($rr = 71$–78%).[104] Chen has also shown that Cp$_2$TaMe$_3$ polymerizes MMA when activated with two equivalents of AlMe$_3$. However, initiator efficiencies are low and molecular weight distributions are broad.[105]

23.3.4 WELL-DEFINED MAGNESIUM AND LITHIUM INITIATORS

Given the high levels of stereocontrol for MMA polymerization exhibited by traditional Grignard reagents, albeit at very low temperatures, there is much promise in the development of single-site main group initiator systems capable of operating at higher reaction temperatures and influencing tacticity through their ancillary ligand(s). Furthermore, steric protection of the metal center by particularly large ancillary ligand sets is expected to help reduce (or even eliminate) the unwanted side reactions commonly found in anionic polymerization systems.

To this end, we have recently found that low-coordinate magnesium alkyl complexes can be stabilized by bulky N,N'-diisopropylphenyl β-diketiminate ligands, and that these compounds initiate a rapid polymerization of MMA. The alkyl complexes are prepared either by reaction of the diketimine with dialkyl precursors such as MgMe$_2$[106] or by treatment of RMgX (R = Me, i-Pr, t-Bu, Ph; X = Cl, Br) with the lithium salt of the ligand (Scheme 23.15).[107]

The methyl and *tert*-butyl complexes **50** and **51** give very high syndioselectivities ($rr > 95$%) for MMA polymerizations carried out at -78 °C; significantly, even at -30 °C the syndiotacticity remains high ($rr = 92$%).[108] However, chain length control using these two initiators is relatively poor, with much higher PMMA molecular weights obtained compared to those expected based on the monomer: initiator stoichiometries (Table 23.1). This behavior is attributed to a low efficiency of initiation (i.e., a low percentage of Mg–alkyl bonds actually undergoing insertion of MMA), with the steric bulk of the t-Bu initiating group implicated for **51**, and the bridging nature of the methyl groups (which is retained in solution) implicated for **50**.

Although **50** and **51** were shown to be poorly controlled initiators for MMA polymerization, it was reasoned that an alkyl group intermediate in size between a methyl and a *tert*-butyl might produce a mononuclear complex, yet still be sufficiently small enough to allow facile nucleophilic attack on a molecule of MMA (and therefore initiate polymerization more efficiently). In previous studies,[107]

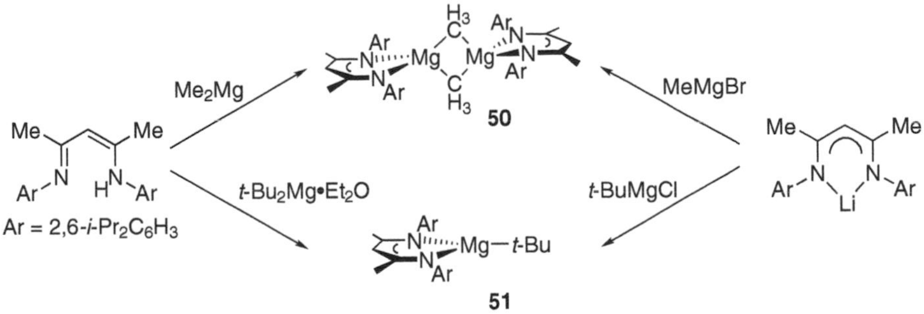

SCHEME 23.15 Synthetic routes to β-diketiminate magnesium alkyl complexes.

TABLE 23.1
Polymerizations of MMA Using β-Diketiminate Magnesium Initiators[a]

Complex	Time (min)	% Monomer[b] Conversion	M_n[c]	M_n (calculated)[d]	M_w/M_n[c]	% rr[e]
50	120	67	674,000	13,400	1.36	90
51	30	97	198,000	19,400	1.59	90
52	1.5	98	20,200	19,600	1.04	89
53	3.0	94	31,600	18,800	1.33	90
54	1.5	94	32,000	18,800	1.05	89

[a]Toluene, −30 °C, 200 equivalents MMA, [Mg] = 0.01 M.
[b]Determined by ^{1}H NMR spectroscopy.
[c]In g/mol; determined by gel permeation chromatography in chloroform at 25 °C (molecular weights quoted relative to polystyrene standards).
[d]Calculated from [MMA]$_0$/[Mg] × molecular weight of MMA (100.117 g/mol) × conversion.
[e]Determined by integrating the [CH$_2$C(CH$_3$)CO$_2$Me] resonances in CDCl$_3$ on a reaction aliquot.

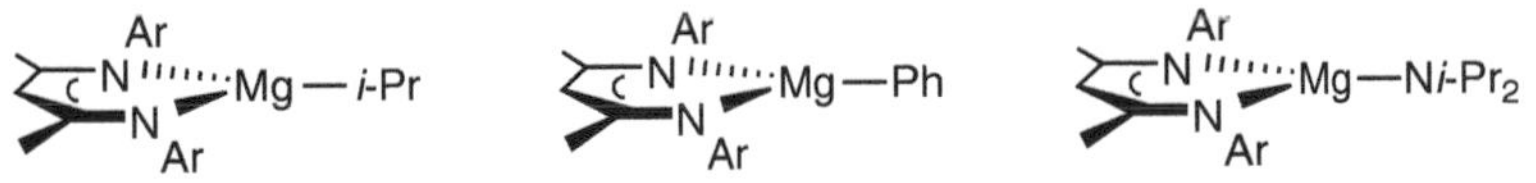

FIGURE 23.13 Structural formulae of complexes **52–54** (Ar = 2,6-i-Pr$_2$C$_6$H$_3$).

it had been shown that the isopropyl and phenyl analogues (**52** and **53**, respectively, Figure 23.13) are structurally similar to the t-Bu complex **51**, existing as three-coordinate mononuclear species. As summarized in Table 23.1, both **52** and **53** are also highly active and syndioselective initiators for the polymerization of MMA, but produce far more controlled molecular weights than when the methyl and $tert$-butyl complexes **50** and **51** are employed. The isopropyl complex **52** delivers chain lengths particularly close to those predicted from the initial monomer: initiator stoichiometry with very narrow molecular weight distributions, consistent with a living process. Further investigations with **52** confirmed these preliminary findings: the molecular weight of the PMMA formed by **52** increases linearly with monomer consumption, and polydispersities remain exceptionally narrow throughout the polymerization (Figure 23.14).

The effect of temperature on the stereoselectivity of MMA polymerization using **52** has also been investigated; the results are collected in Table 23.2. As might be anticipated, at higher temperatures

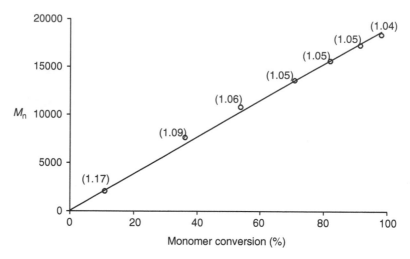

FIGURE 23.14   A plot of M_n versus monomer conversion for the polymerization of MMA initiated by complex **52** at -30 °C (in toluene, $[MMA]_0/[52] = 200$; M_w/M_n values in parentheses).

TABLE 23.2
Polymerizations of MMA Using Complex 52[a]

Temperature (°C)	Time (s)	% Monomer[b] Conversion	M_n[c]	M_n (calculated)[d]	M_w/M_n[c]	% rr[e]
-30	90	98	20,200	19,600	1.04	88.5
-20	60	97	18,500	19,400	1.05	84.9
-10	45	97	18,000	19,400	1.06	81.7
0	35	99	18,400	19,800	1.05	78.1
23	20	94	18,800	18,800	1.10	75.2

[a]Toluene, 200 equivalents MMA, $[Mg] = 0.01$ M.
[b]Determined by ^{1}H NMR spectroscopy.
[c]In g/mol; determined by gel permeation chromatography in chloroform at 25 °C (molecular weights quoted relative to polystyrene standards).
[d]Calculated from $[MMA]_0/[Mg]$ × molecular weight of MMA (100.117) × conversion.
[e]Determined by integrating the $[CH_2C(CH_3)CO_2Me]$ resonances in $CDCl_3$ on a reaction aliquot.

---

the stereoselectivity diminishes. Nonetheless, even at ambient temperature, this system still produces PMMA of notably greater syndiotacticity than would be expected by free-radical techniques, suggesting a coordinative mechanism subject to ligand-assisted chain-end controlled insertion.

By analogy to the work described above on lanthanide and zirconocene initiators, it is believed that initiation by the magnesium alkyl complexes occurs through addition of the alkyl ligand to one equivalent of MMA monomer to afford a magnesium ester enolate species (Scheme 23.16). Propagation should then occur in a manner similar to that outlined in Scheme 23.2. As magnesium enolates are often prepared through Hauser base precursors (i.e., magnesium amides), it was envisaged that the well-defined amide complex **54** (Figure 23.13) should also react with MMA to form its ester enolate and hence also function as a polymerization initiator. This indeed proves to be the case with **54** consuming 200 equivalents of MMA in just 90 s at -30 °C, again yielding highly syndiotactic, monodisperse polymer (Table 23.1).[108]

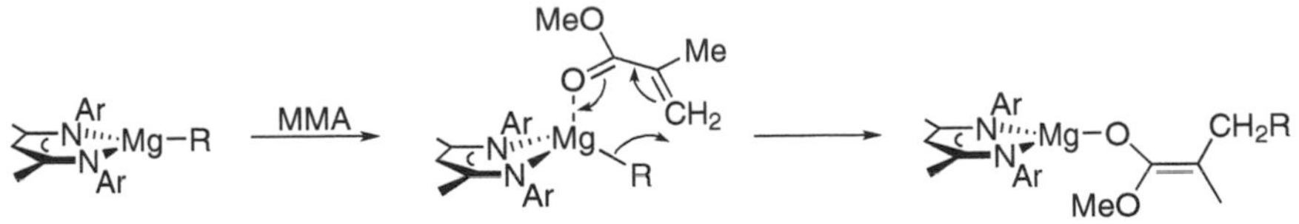

SCHEME 23.16 Postulated formation of the propagating enolate species in MMA polymerization with magnesium alkyl initiators.

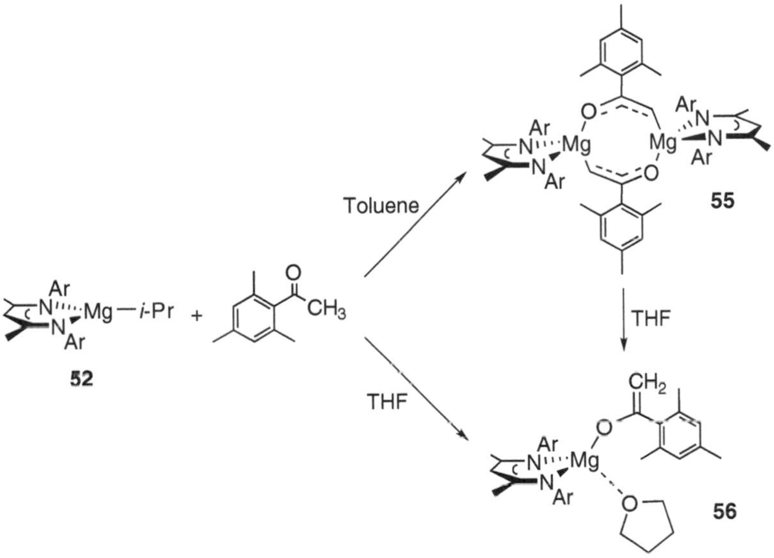

SCHEME 23.17 Synthesis of magnesium enolate complexes **55** and **56** (Ar = 2,6-i-Pr$_2$C$_6$H$_3$).

To investigate the initiation process further, we have also prepared a well-defined magnesium enolate species, **55**, by treatment of the isopropyl complex **52** with 2′,4′,6′-trimethylacetophenone (Scheme 23.17).[109] X-ray diffraction studies revealed that complex **55** exists as an enolate-bridged dimer in the solid state, with bond lengths indicative of oxygen metallation. Accordingly, it dissolves in THF to give the monomeric four-coordinate oxygen-bound enolate complex, **56**, featuring a terminal methylidene unit. Alternatively, complex **56** may be prepared directly by treating **52** with 2′,4′,6′-trimethylacetophenone in THF. It has also been found that both magnesium enolate complexes **55** and **56** can be prepared in high yields using a convenient, one-pot methodology whereby the free β-diketimine ligand is reacted with Bu$_2$Mg (a commercially available solution of equimolar amounts of n- and sec-butyl species) followed by treatment with 2′,4′,6′-trimethylacetophenone.

A three-coordinate, base-free analogue of **56** containing a terminal enolate ligand (**57**, Scheme 23.18) can be isolated by employing a bulkier β-diketiminate ligand.[108] The $tert$-butyl backbone substituents of the β-diketiminate ligand force the aryl groups to project further forward,[110] thereby increasing the steric protection of the magnesium center and stabilizing the lower coordination number. The molecular structure of **57** (Figure 23.15) confirms that the enolate ligand is indeed oxygen-metallated. MMA polymerization studies using the well-defined enolate initiators afforded excellent control over the molecular weight and narrow molecular weight distributions ($M_w/M_n < 1.1$).

Importantly, the high levels of syndiotacticity observed when the alkyl initiators **50–53** were employed for polymerization are retained with the enolate species. When MMA polymerization is

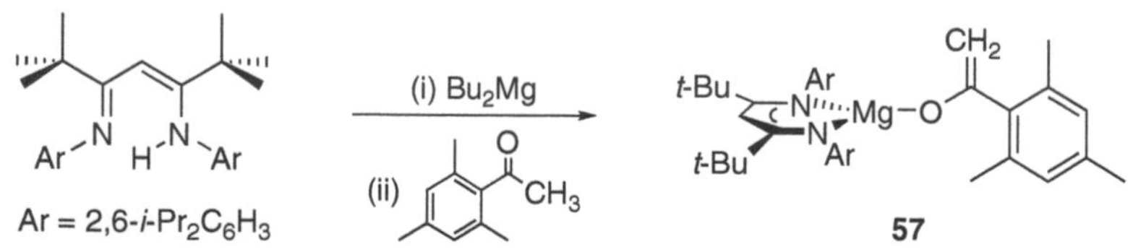

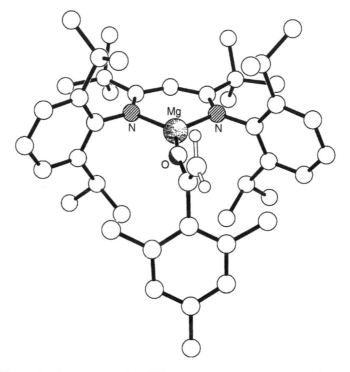

SCHEME 23.18   Synthesis of complex **57**.

FIGURE 23.15   The molecular structure of [HC{C(*t*-Bu)NAr}]Mg(OC(=CH$_2$)Ar′) (Ar = 2,6-*i*-Pr$_2$C$_6$H$_3$; Ar′ = 2,4,6-Me$_3$C$_6$H$_2$) (**57**); H atoms (except for those of the terminal methylidene group) are omitted for clarity.

performed at −78 °C using either **55** or **56** in toluene, ^{1}H NMR spectroscopy reveals a syndiotactic *rr* content of 97%, which drops only slightly (to 92%) when the polymerization is carried out at −30 °C.

The origin of stereoselectivity in this system has been investigated by synthesizing magnesium alkyl and enolate complexes bearing alternative β-diketiminate *N*-aryl substitution patterns.[108] It was found that reducing the size of the *ortho*-aryl substituents leads to a significant loss of stereocontrol. For example, in changing from 2,6-diisopropylphenyl to 2,6-diethylphenyl and 2,6-dimethylphenyl *N*-aryl groups, the syndiotactic *rr* triad contents of the resultant PMMAs prepared at −30 °C fall from 92% to 78% and 73%, respectively.[108] Clearly, the tacticity is highly sensitive to the steric environment presented by the β-diketiminate ligands, and has parallels to findings for related zinc β-diketiminate initiators used for lactide polymerization.[111,112]

Ester enolates of lithium are notoriously unstable. When they have proved isolable and have been structurally characterized, they have been found to adopt highly aggregated structures.[113] Nonetheless, the first example of a single-site mononuclear lithium initiator for the polymerization of MMA has recently been reported by Chen and Rodriguez-Delgado, who found that the sterically bulky Lewis acid bis(2,6-di-*tert*-butyl-4-methylphenoxy)aluminum methyl (**58**, Figure 23.16)

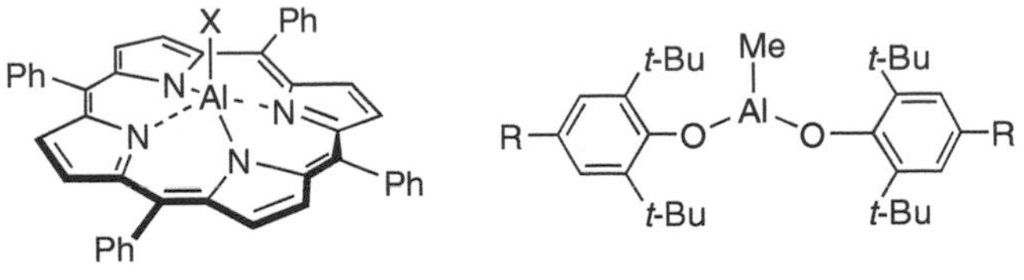

FIGURE 23.16 Porphyrinato-aluminum initiators for methacrylate polymerization (**61–63**) and Lewis acidic activators (**58** and **64**).

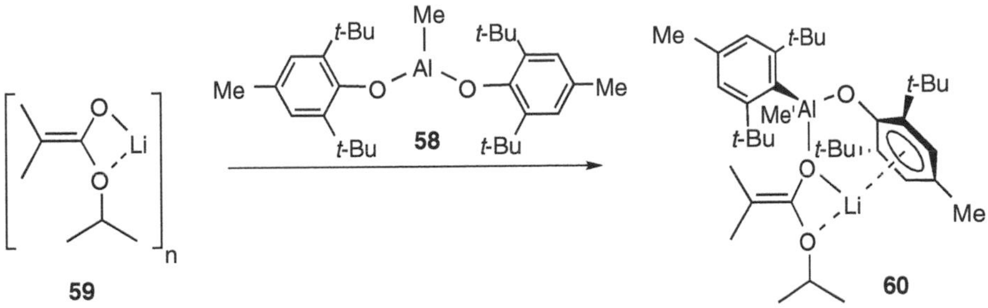

SCHEME 23.19 Synthesis of a mono-lithium ester enolate complex.

deaggregates hexameric isopropyl α-lithioisobutyrate **59** and helps stabilize a monolithium product, **60** (Scheme 23.19).[114] The difference in the ability of **59** and **60** to initiate the polymerization of MMA is marked: the aggregated enolate species affords isotactic PMMA in toluene at room temperature ($mm = 76\%$) with a broad polydispersity ($M_w/M_n = 6.8$), whereas the single-site enolate **60** generates monodisperse, syndiotactic PMMA ($M_w/M_n = 1.20$, $rr = 73\%$). Addition of a second equivalent of **58** serves to activate the monomer (coordination of MMA to the Al center through its carbonyl oxygen increases the δ^+ charge on its olefinic =CH$_2$ fragment, thus rendering it more susceptible to nucleophilic attack from the lithium enolate) and allows the quantitative consumption of 200 equivalents of MMA to be achieved within 1 h. This system has also been examined with nBMA, and is again well-controlled and stereoselective, producing monodisperse ($M_w/M_n = 1.18$), syndiotactic ($rr = 87\%$) PnBMA at 0 °C.

23.3.5 Well-Defined Aluminum Initiators

In 1987, Inoue and coworkers reported that irradiation of (TPP)AlCH$_3$ (**61**, TPP = tetraphenylporphyrinato, Figure 23.16) initiates the slow but well-controlled polymerization of MMA,[115] but the stereoselectivity of the process was not described. Photoactivation is required for polymerization to proceed; in the dark the system is inactive. Polymerizations of MMA with **61** and other methacrylate monomers all exhibit living characteristics, and ^{1}H NMR analysis of propagating oligomers of tBMA provided support for an oxygen-metallated enolate as the active site.[116] Successful diblock copolymerizations of MMA with nBMA,[77] and of MMA with epoxides,[117] are further testament to the living nature of this initiator. The analogous thiolate initiators, (TPP)AlSR (**62**, R = n-Pr; **63**, R = Ph, Figure 23.16), do not require photoexcitation to initiate polymerization, and consume 200 equivalents MMA in 18 h at 35 °C ($M_n = 22,000$ g/mol, M_n (calc) = 20,000 g/mol, $M_w/M_n = 1.12$).[118] The propagating species is again believed to be an enolate.[119]

The rate of polymerization may be increased by addition of a bulky Lewis acid such as the bis(aryloxy)aluminum methyl compounds **58** and **64** (Figure 23.16). Coordination of the Lewis acid to the carbonyl oxygen atom of the methacrylate monomer serves to increase the electrophilicity of the methylenic ($=CH_2$) carbon, thus activating it towards Michael addition by the propagating chain. For example, addition of **64** to a sample of living PMMA generated by irradiation of **61**/MMA caused a dramatic acceleration of the polymerization rate by a factor of over 45,000.[120] The dual-component systems consisting of **61/64** and **61/58** have proved particularly adept for preparing monodisperse, ultrahigh molecular weight samples of PMMA ($M_n > 10^6$ g/mol, $M_w/M_n = 1.2$).[121] Propagation from the thiolate initiators is also expedited by addition of **58**, with 100 equivalents of MMA requiring just 90 s for full conversion. A detailed study into the mechanism of rate enhancement revealed that use of a sterically bulky Lewis acid is imperative, in order to prevent scrambling of the propagating enolate between the two aluminum centers (of the porphyrin initiator and the Lewis acid).[122] Hence, for aluminum diphenolate Lewis acid additives, *ortho*-substitution is essential.

Complexes **61–63** may also be used to polymerize acrylates[123,124] in a living manner, although, unlike the thiolate complexes, the methyl initiator again requires photoexcitation. Acrylate monomers polymerize faster than methacrylates, with even sterically encumbered *tert*-butyl acrylate being consumed more rapidly than MMA. As for MMA, no information concerning the stereoselectivity of these polymerizations has been disclosed.

In theory, initiators based on tetradentate salicylaldiminato (salen) ligands (e.g., **65**, Figure 23.17) should offer greater potential for stereocontrol than their porphyrin relatives owing to the possibility of incorporating chiral backbones, yet complexes such as **65** do not initiate MMA polymerization under thermal or photochemical activation conditions, nor in the presence of a Lewis acid activator such as **58**. However, Gibson and coworkers discovered that addition of a nickel source, such as Ni(acac)$_2$ or Ni(COD)$_2$ (acac = acetylacetonate (2,4-pentanedionate), COD = 1,5-cyclooctadiene), catalyzes the formation of the active enolate species. Thus, using a three-component initiator system comprising **65**, **58**, and Ni(acac)$_2$, 200 equivalents MMA were polymerized within 2 min at room temperature ($M_n = 24,700$ g/mol, M_n (calc) = 20,000 g/mol, $M_w/M_n = 1.17$).[125] At this polymerization temperature the PMMA displays a slightly higher syndiotacticity than samples generated by free-radical methods, with $rr = 68\%$–72% rising to 84% rr at −20 °C. The living nature of the system was demonstrated by a linear dependence of M_n upon monomer conversion and the synthesis of a PMMA-*b*-PnBMA diblock by sequential monomer addition. The bulky Lewis acid **58** is proposed to serve two functions. The first is activation of the monomer as described by Inoue for the (TPP)Al-based initiator system. The second is generation of a nickel catalyst capable of catalyzing the formation of the aluminum enolate. This is proposed to involve methylation of the Ni species, which then inserts MMA to generate a carbon-bonded nickel enolate (Scheme 23.20). The enolate group of the latter species is then transferred to the aluminum center with concomitant exchange of the Al-methyl group, thereby regenerating the Ni catalyst and affording an oxygen-bonded aluminum enolate product. An analogous aluminum enolate has been prepared by treatment of LiOC(O*t*-Bu)=CMe$_2$ with the chloro analogue of complex **65** and shown to initiate the controlled polymerization of MMA.[126]

We have studied a wide range of alternative salen ligands, varying both the diimino backbone and the substituents on the phenoxy rings, and have found that in nearly all cases the stereoselectivity

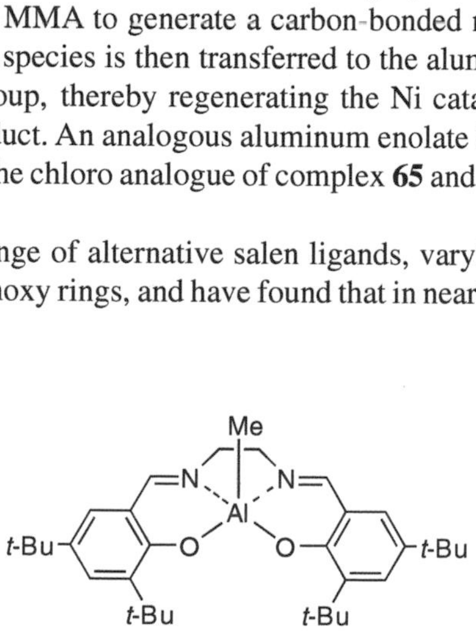

65

FIGURE 23.17 Structural formula of complex **65**.

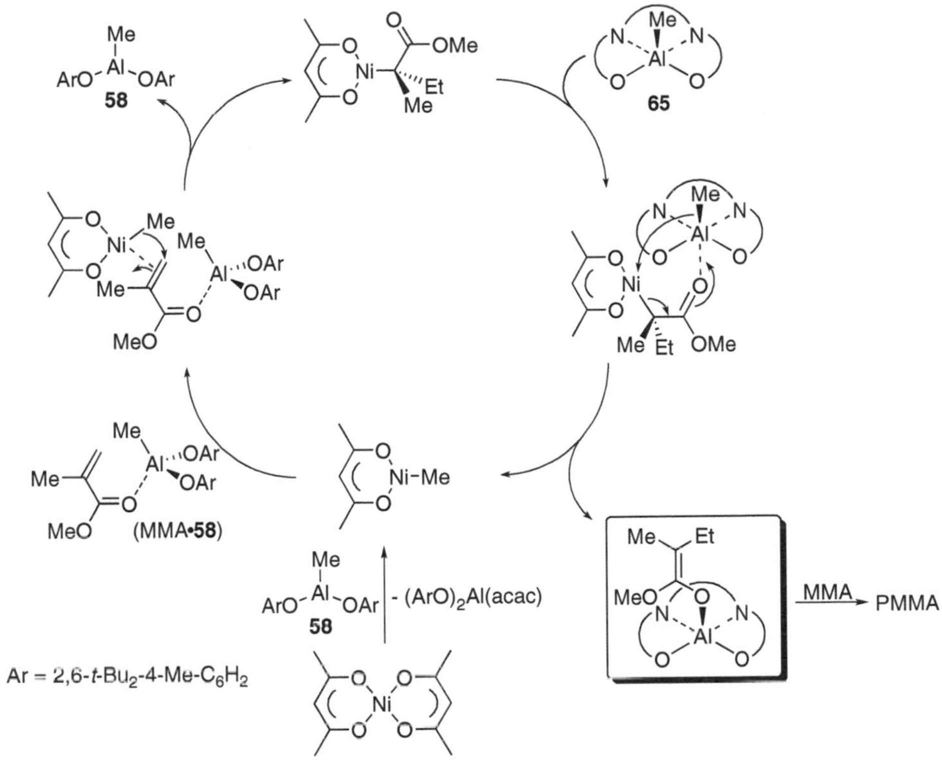

SCHEME 23.20 Polymerization of MMA using the **58/65**/Ni(acac)₂ system.

of the PMMA formed is invariant from that observed with **65**. For example, when [*N*,*N'*-bis(3,5-di-*tert*-butylsalicylidene)-*trans*-1,2-cyclohexanediamino]aluminum methyl is used in conjunction with **58** and Ni(acac)₂, a syndiotactic content of 69% *rr* is observed at room temperature in CH₂Cl₂.

Another family of aluminum initiators has arisen from work on the cationic zirconocene systems described in Section 23.3.2. Upon activation of the bis(indenyl)zirconocene complex **66** using the alane Al(C₆F₅)₃ (Scheme 23.21; as opposed to complex **30** derived from activation with B(C₆F₅)₃), Chen and Bolig[127] found moderate (60.4% *rr* at 23 °C) syndioselectivity for MMA polymerization (cf. 92.7% *mm* at 23 °C for **30**). Furthermore, whereas the stereoselectivity of MMA polymerization with borate-activated zirconocene initiators is highly dependent upon the metallocene ligand geometry (Figure 23.9), C_2-, C_{2v}-, and C_s-symmetric zirconocenes activated with the alane all afforded PMMA of similar syndiotacticity. ¹⁹F NMR studies revealed that initial generation of the MeAl(C₆F₅)₃ anion was followed by formation of a MMA→Al(C₆F₅)₃ adduct and several different propagating enolaluminate anions. The authors therefore proposed a polymerization mechanism whereby the MMA propagates from the aluminum center, with the role of the zirconocene component being initial generation of the active enolate ligand (Scheme 23.21) and subsequent monomer activation.[128] Good evidence for such a mechanism is the observation that the zirconocene complex may be replaced by other Michael nucleophiles (i.e., enolate precursors). For example, at −78 °C *t*-BuLi initiates slow, poorly controlled isoselective polymerization of MMA (75.0% *mm*; $M_w/M_n = 14.4$), but under similar conditions, a 2:1 mixture of Al(C₆F₅)₃ (one equivalent to generate the enolaluminate and one equivalent to activate the monomer) and *t*-BuLi produces 95.0% *rr* PMMA with a relatively narrow molecular weight distribution ($M_w/M_n = 1.33$). As above, it is believed that the organolithium reagent is only involved in enolate generation, with propagation occurring exclusively at the aluminum centers.

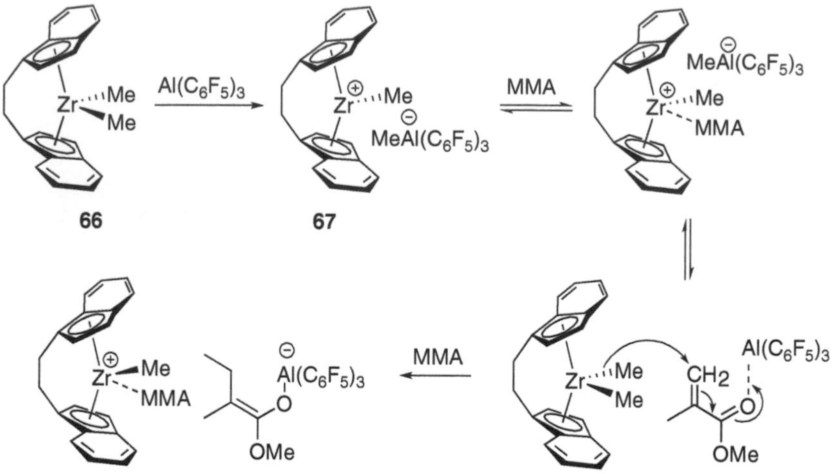

SCHEME 23.21   Generation of an enolaluminate initiator from reaction of **66** with Al(C$_6$F$_5$)$_3$.

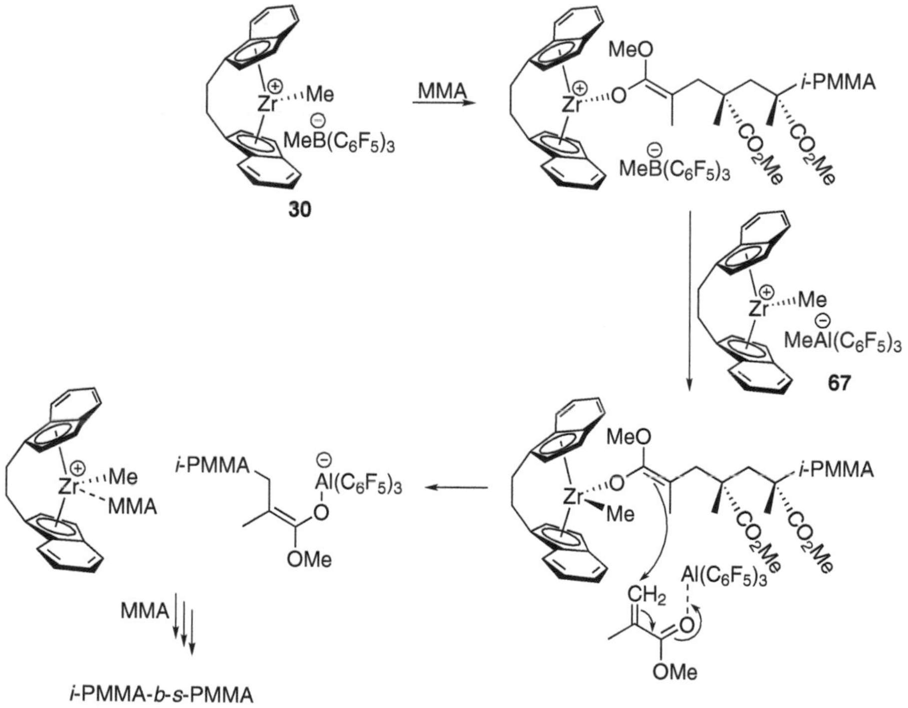

SCHEME 23.22   Synthesis of an isotactic-*b*-syndiotactic PMMA stereoblock.

Chen has further exploited this reaction to produce an isotactic-*b*-syndiotactic PMMA stereo-block copolymer through the route outlined in Scheme 23.22.[129] The isotactic block (*i*-PMMA) is generated using the borate-activated zirconocene complex **30** at room temperature, before the addition of **67** serves to convert the Zr-terminated propagating PMMA chain into an aluminate-terminated analogue. Cooling to −78 °C followed by addition of a second feed of MMA then affords the syndiotactic block (*s*-PMMA).

If the polymerization is performed at room temperature (or even at 0 °C) and the MMA is initially added to a 1:1 mixture of **30** and **67**, then a stereomultiblock PMMA material is produced.[130] At this temperature, highly isotactic PMMA is produced by **30** while moderately syndiotactic PMMA is formed at the enolaluminate derived from **67**. The two propagating PMMA chains may exchange between the cationic zirconium center and the anionic aluminum center, and since the rate of exchange is comparable to propagation, the PMMA ultimately produced consists of several isotactic and syndiotactic blocky regions.

23.4 FREE-RADICAL POLYMERIZATION

There has been much interest in the development of controlled free-radical processes for acrylate and methacrylate polymerization[4] and, recently, in the possibilities for stereocontrolled radical polymerizations. One of the most commonly exploited controlled radical polymerization methodologies is atom transfer radical polymerization (ATRP), pioneered by the groups of Sawamoto[131] and Matyjaszewski.[132]

ATRP has been used to successfully polymerize many acrylate and methacrylate monomers. For methacrylates, transition metal catalysts based upon Ru,[133,134] Cu,[135,136] Ni,[137–139] Fe,[140–143] Pd,[144] and Rh[145] have all been reported. Generally, polymerizations are performed at elevated temperatures (70–90°C) in toluene. PMMA samples generated by ATRP typically possess syndio-rich atactic structures consistent with a free-radical propagation mechanism (Figure 23.3), and may be prepared with relatively narrow polydispersities ($M_w/M_n = 1.1$–1.3).

Although the free-radical polymerization of MMA typically exhibits a syndiotactic bias (*rr* triad content = 60%–70%), it has long been known that the stereochemical interactions between the chain-end radical and vinyl monomers in free-radical polymerization can be modified by using chiral protecting groups. For example, the 80 °C AIBN-initiated polymerization (AIBN = 2,2'-azobisisobutyronitrile) of oxazolidine acrylamides based on valine and *tert*-leucine ultimately yields highly isotactic (92% *m* dyad content) poly(acrylic acid) and PMA after chemical modification (Scheme 23.23).[146]

Recently, more general approaches to stereoselective free-radical polymerizations have been reported, although the levels of stereocontrol are still modest compared to what is achievable using anionic or coordinative polymerization systems. For example, the free-radical polymerization of vinyl acetate (VA) is known to give a syndiotactic-biased polymer when performed in phenolic solvent, owing to hydrogen bonding between the acetyl group and the phenol.[147] Such effects are greatly increased in fluoroalcohol solvents, with $(CF_3)_3COH$ allowing the synthesis of 62% *r* dyad poly(vinyl acetate) (PVA) at 20 °C, rising to 72% *r* at −78 °C (PVA is atactic when prepared through traditional free-radical methods).[148] Similar studies with methacrylate monomers in fluoroalcohol solvents allowed the preparation of PMMA, PEtMA, and PtBMA at −40 °C with *rr* triad contents of 82.9%, 86.9%, and 74.4%, respectively (PEtMA = poly(ethyl methacrylate); PtBMA = poly(*tert*-butyl methacrylate).[149]

The role of the fluorinated alcohol in effecting stereocontrol has not been fully elucidated, but it is believed to coordinate to the ester groups of the methacrylate monomer and polymer through hydrogen bonding. Earlier studies from the 1960s have shown similar effects with MMA complexed to stoichiometric quantities of nonfluorinated Lewis acids.[150,151] As depicted in Scheme 23.24, this

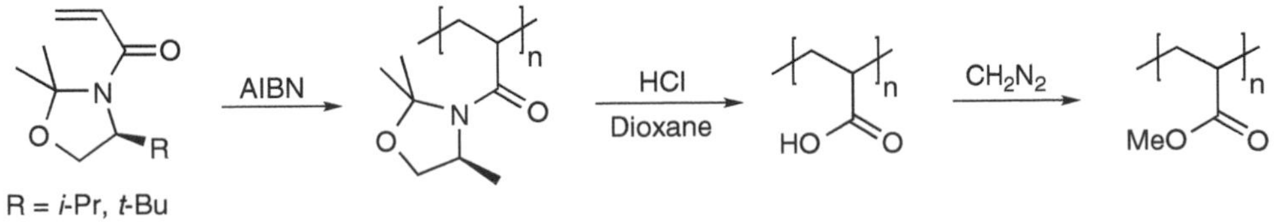

R = *i*-Pr, *t*-Bu

SCHEME 23.23 The free-radical polymerization of oxazolidine acrylamide monomers and conversion of the polymers into highly isotactic poly(acrylic acid) and PMA.

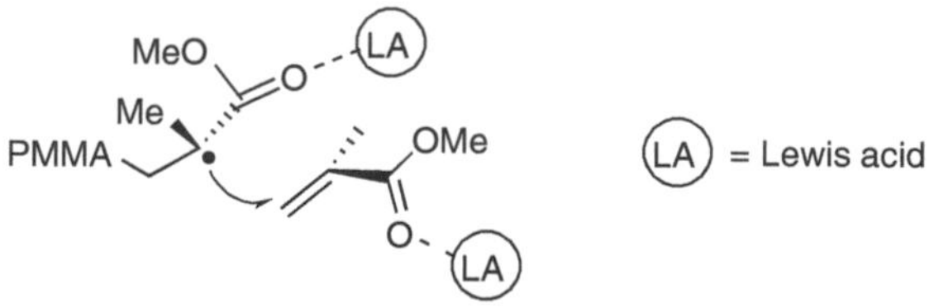

SCHEME 23.24   Proposed origin of increased syndioselectivity in the free-radical polymerization of MMA using bulky Lewis acid additives.

would lead to increased steric hindrance at one or both of the reacting centers, and hence presumably greater discrimination toward racemic (i.e., syndioselective) enchainment.

Such an effect should also be applicable to other bulky Lewis acids, and several workers have shown that this is indeed the case. For example, it has been shown that the stereoselectivity of free-radical MMA polymerization (carried out in nonfluorinated solvents) can be affected by addition of $MgBr_2 \cdot Et_2O$.[152] It has also been reported that the microstructures of poly(acrylamide)s may be significantly altered from those obtained through traditional free-radical methods by addition of $Sc(OTf)_3$ ($Tf = SO_2CF_3$) to the polymerization system.[153–156]

Increased levels of syndioselectivity have been reported for the polymerization of benzyl (α-methoxymethylacrylate); typical free-radical samples are atactic ($m:r = 54:46$), but syndio-bias is observed in the presence of $ZnBr_2$ ($r = 71\%$) as may be expected from the pathway outlined in Scheme 23.24.[157] However, when $Sc(OTf)_3$ was used instead of $ZnBr_2$, a polymer with increased isotactic content was obtained ($m = 69\%$).[158]

When the Lewis acids $HfCl_4$, $Zn(OTf)_2$, $ScCl_3$, and $Ln(OTf)_3$ ($Ln = Sc, Y, La, Sm, Eu, Yb, Lu$) were added to MMA polymerizations, the effects were negligible. For example, in the absence of Lewis acid, PMMA with $rr = 64\%$ was obtained, whereas typical rr values obtained in the presence of these Lewis acids fall in the range 60%–63%.[159] Interestingly, the largest deviations from typical free-radical tacticities, observed with $Sc(OTf)_3$ and $Yb(OTf)_3$, resulted in notably increased isotactic content (e.g., with $Sc(OTf)_3$ $mm:mr:rr = 14:46:40$), in contradiction to the proposed mode of action outlined in Scheme 23.24.[160] Similarly Matyjaszewski and coworkers found that addition of $Sc(OTf)_3$ to CuCl/bipyridine or CuCl/N,N,N',N'',N''-pentamethyldiethylene triamine ATRP MMA polymerization systems afforded increased isotactic contents in the resultant PMMAs ($mm:mr:rr = 21:45:33$). These results strongly suggest that the generalized mechanism shown in Scheme 23.24 has severe limitations.

23.5   CONCLUSIONS

The development of well-defined metal coordination catalysts for the polymerization of methacrylates and acrylates has led to significant advances in controlling the stereochemical outcome of these polymerizations. To date, by far the most extensive work has been carried out on the industrially relevant monomer MMA, which has provided much needed insight not only into the factors influencing stereocontrol, but also into the intricacies of the propagation mechanism and deactivation processes. Indeed, well-defined systems now exist that span the lanthanides, the early transition metals, and main group metals. Many of these have been shown to facilitate the living polymerization of methacrylate monomers, thereby providing access to tailored diblock and multiblock poly(methacrylate) materials, as well as copolymers with other technologically significant classes of polymers such as polyolefins. These studies on methacrylates have also provided a platform to expand understanding to the more challenging acrylate monomers, which tend to be more susceptible to termination processes, and for which stereocontrol has proved more difficult to harness.[161]

As more advanced catalyst systems are developed that do not readily deactivate, highly tailored materials that incorporate stereoregular poly(acrylates) can be confidently anticipated.

ACKNOWLEDGMENTS

The authors are grateful to the numerous talented postgraduate and postdoctoral researchers who have contributed to our zirconium, magnesium, and aluminum studies described herein, and to industrial collaborators (ICI Acrylics, now Lucite International, and BASF, in particular) who have encouraged our stereoselective acrylate polymerization catalyst program at various stages.

REFERENCES AND NOTES

1. Chisholm, M. Plastic fantastic. *Chem. Br.* April **1998**, *34*, 33–36.
2. Hatada, K.; Kitayama, T.; Ute, K. Stereoregular polymerization of α-substituted acrylates. *Prog. Polym. Sci.* **1988**, *13*, 189–276.
3. Bowden, M. J.; O'Donnell, J. H. Radiation-induced solid-state polymerization of derivatives of methacrylic acid. IV. Electron spin resonance spectra of barium methacrylate dihydrate. *J. Phys. Chem.* **1968**, *72*, 1577–1582.
4. Matsumoto, A.; Giese, B. Conformational structure of methacrylate radicals as studied by electron spin resonance spectroscopy: From small molecule radicals to polymer radicals. *Macromolecules* **1996**, *29*, 3758–3772.
5. Karasz, F. E.; MacKnight, W. J. The influence of stereoregularity on the glass transition temperatures of vinyl polymers. *Macromolecules* **1968**, *1*, 537–540.
6. Ute, K.; Miyatake, N.; Hatada, K. Glass transition temperature and melting temperature of uniform isotactic and syndiotactic poly(methyl methacrylate)s from 13 mer to 50 mer. *Polymer* **1995**, *36*, 1415–1419.
7. Kitayama, T.; Masuda, E.; Yamaguchi, M.; Nishiura, T.; Hatada, K. Syndiotactic-specific polymerization of methacrylates by tertiary phosphine triethylaluminum. *Polym. J.* **1992**, *24*, 817–827.
8. Hatada, K. Stereoregular uniform polymers. *J. Polym. Sci., Part A: Polym. Chem.* **1999**, *37*, 245–260.
9. Hatada, K.; Kitayama, T. Structurally controlled polymerizations of methacrylates and acrylates. *Polym. Int.* **2000**, *49*, 11–47.
10. Hirano, T.; Yamaguchi, H.; Kitayama, T.; Hatada, K. Heterotactic polymerization of methacrylates having C-3 ester group. *Polym. J.* **1998**, *30*, 767–769.
11. Kotyk, J. J.; Berger, P. A.; Remsen, E. E. Microstructural characterization of poly(methyl methacrylate) using proton-detected heteronuclear shift-correlated NMR spectroscopy. *Macromolecules* **1990**, *23*, 5167–5169.
12. Peat, I. R.; Reynolds, W. F. ^{13}C NMR spectra of stereoregular poly(methyl methacrylates). *Tetrahedon Lett.* **1972**, *13*, 1359–1362.
13. Spěváček, J.; Schneider, B.; Straka, J. High-resolution carbon-13 nuclear magnetic resonance spectra and structure of amorphous and crystalline forms of stereoregular poly(methyl methacrylate). *Macromolecules* **1990**, *23*, 3042–3051.
14. Spěváček, J. Application of ^{13}C NMR magnetic resonance spectroscopy to the determination of the tacticity of stereoregular poly(methyl methacrylate). *J. Mol. Struct.* **1985**, *129*, 175–177.
15. Nguyen, G.; Matlengiewicz, M.; Nicole, D. Incremental method for determination of sequence distribution of poly(methyl methacrylate) by ^{13}C NMR spectroscopy. *Analusis* **1999**, *27*, 847–853.
16. Kitayama, T.; Janco, M.; Ute, K.; Niimi, R.; Hatada, K. Analysis of poly(ethyl methacrylate)s by on-line hyphenation of liquid chromatography at the critical adsorption point and nuclear magnetic resonance spectroscopy. *Anal. Chem.* **2000**, *72*, 1518–1522.
17. Varshney, S. K.; Gao, Z.; Zhong, X. F.; Eisenberg, A. Effect of lithium chloride on the "living" polymerization of *tert*-butyl methacrylate and polymer microstructure using monofunctional initiators. *Macromolecules* **1994**, *27*, 1076–1082.
18. Brar, A. S.; Kumar, R.; Kaur, M. Poly(methyl acrylate): Spectral assignment by two-dimensional NMR spectroscopy. *Appl. Spectrosc.* **2002**, *56*, 1380–1382.

19. Matsuzaki, K.; Kanai, T.; Kawamura, T.; Matsumoto, S.; Urtu, Y. The C-13 nuclear magnetic resonance spectra of polyacrylates and their model compounds. *J. Polym. Sci., Part A1: Polym. Chem.* **1973**, *11*, 961–969.

20. Matsuzaki, K.; Okada, M.; Hosonuma, K. The stereoregularity of poly(isopropyl acrylate) and its racemization during hydrolysis. *J. Polym. Sci., Part A1: Polym. Chem.* **1972**, *10*, 1179–1186.

21. Suchopárek, M.; Spěváček, J. Characterization of the stereochemical structure of poly(*tert*-butyl acrylate) by one- and two-dimensional NMR spectroscopy. *Macromolecules* **1993**, *26*, 102–106.

22. Spyros, A.; Dais, P. Structure and dynamics of poly(1-naphthyl acrylate) in solution by ^{13}C NMR spectroscopy. *Macromolecules* **1992**, *25*, 1062–1067.

23. Lenz, R. W.; Saunders, K.; Balakrishnan, T.; Hatada, K. Synthesis and polymerization of alkyl α-(phenoxymethyl)acrylates. *Macromolecules* **1979**, *12*, 392–394.

24. Habaue, S.; Yamada, H.; Uno, T.; Okamoto, Y. Stereospecific polymerization of benzyl α-(alkoxymethyl)acrylates. *J. Polym. Sci., Part A: Polym. Chem.* **1997**, *35*, 721–726.

25. San Román, J.; Madruga, E. L.; Lavia, M. A. Stereochemical configuration of poly(methyl α-benzylacrylate) synthesized by radical polymerization. *Macromolecules* **1984**, *17*, 1762–1764.

26. Cuervo-Rodríguez, R.; Fernández-Monreal, M. C.; Fernández-García, M.; Madruga, E. L. Determination of the sequence distribution and stereoregularity of ethyl α-benzoyloxymethacrylate-methyl methacrylate copolymers by means of proton and carbon-13 NMR. *J. Polym. Sci., Part A: Polym. Chem.* **1997**, *35*, 3483–3493.

27. Hirano, T.; Tateiwa, S.; Seno, M.; Sato, T. Temperature dependence of stereospecificity in the radical polymerization of di-*n*-butyl itaconate in bulk. *J. Polym. Sci., Part A: Polym. Chem.* **2000**, *38*, 2487–2491.

28. Davis, T. P.; Haddleton, D. M.; Richards, S. N. Controlled polymerization of acrylates and methacrylates. *J. Macromol. Sci., Rev. Macromol. Chem. Phys.* **1994**, *C34*, 243–324.

29. Kitano, T.; Fujimoto, T.; Nagasawa, M. Anionic polymerization of *tert*-butyl acrylate. *Polym. J.* **1977**, *9*, 153–159.

30. Bywater, S. Polymerization initiated by lithium and its compounds. *Adv. Polym. Sci.* **1965**, *4*, 66–110.

31. Quirk, R. P.; Yoo, T.; Lee, Y.; Kim, J.; Lee, B. Applications of 1,1-diphenylethylene chemistry in anionic synthesis of polymers with controlled structures. *Adv. Polym. Sci.* **2000**, *153*, 67–162.

32. Anderson, B. C.; Andrews G. D.; Arthur, P.; Jacobson, H. W.; Melby, L. R.; Playtis, A. J.; Sharkey, A. H. Anionic polymerization of methacrylates. Novel functional polymers and copolymers. *Macromolecules* **1981**, *14*, 1599–1601.

33. Lutz, P.; Masson, P.; Beinert, G.; Rempp, P. Synthesis and characterisation of polyalkylmethacrylate macromonomers. *Polym. Bull.* **1984**, *12*, 79–85.

34. Mori, H.; Müller, A. H. E. New polymeric architectures with (meth)acrylic acid segments. *Prog. Polym. Sci.* **2003**, *28*, 1403–1439.

35. Glusker, D. L.; Galluccio, R. A.; Evans, R. A. The mechanism of the anionic polymerization of methyl methacrylate. III. Effects of solvents upon stereoregularity and rates in fluorenyllithium-initiated polymerizations. *J. Am. Chem. Soc.* **1964**, *86*, 187–196.

36. Viguier, M.; Collet, A.; Schue, F. Homogenous polymerization of methyl methacrylate initiated by potassium salts of methanol and *tert*-butyl alcohol in the presence of cryptand [222]. *Polym. J.* **1982**, *14*, 137–141.

37. Fayt, R.; Forte, R.; Jacobs, C.; Jérôme, R.; Ouhadi, T.; Teyssié, Ph.; Varshney, S. K. New initiator system for the living anionic polymerization of *tert*-alkyl acrylates. *Macromolecules* **1987**, *20*, 1442–1444.

38. Ballard, D. G. H.; Bowles, R. J.; Haddleton, D. M.; Richards, S. N.; Sellens, R.; Twose, D. L. Controlled polymerization of methyl methacrylate using lithium aluminum alkyls. *Macromolecules* **1992**, *25*, 5907–5913.

39. Hatada, K.; Nakanishi, H.; Ute, K.; Kitayama, T. Studies on *p*-vinylbenzylmagnesium and *m*-vinylbenzylmagnesium as initiators and monomers—preparations of macromers and poly(Grignard reagent)s. *Polym. J.* **1986**, *18*, 581–591.

40. Yasuda, H. Organo-rare-earth-metal initiated living polymerizations of polar and nonpolar monomers. *J. Organomet. Chem.* **2002**, *647*, 128–138.

41. Yasuda, H. Rare-earth-metal-initiated polymerizations of (meth)acrylates and block copolymerizations of olefins with polar monomers. *J. Polym. Sci., Part A: Polym. Chem.* **2001**, *39*, 1955–1959.

42. Yasuda, H.; Yamamoto, H.; Yokota, K.; Miyake, S.; Nakamura, A. Synthesis of monodispersed high molecular weight polymers and isolation of an organolanthanide(III) intermediate coordinated by a penultimate poly(MMA) unit. *J. Am. Chem. Soc.* **1992**, *114*, 4908–4910.

43. Yasuda, H.; Yamamoto, H.; Yamashita, M.; Yokota, K.; Nakamura, A.; Miyake, S.; Kai, Y.; Kanehisa, N. Synthesis of high molecular weight poly(methyl methacrylate) with extremely low polydispersity by the unique function of organolanthanide(III) complexes. *Macromolecules* **1993**, *26*, 7134–7143.

44. Cram, D. J.; Kopecky, K. R. Studies in stereochemistry. XXX. Models for steric control of asymmetric induction. *J. Am. Chem. Soc.* **1959**, *81*, 2748–2755.

45. Bawn, C. E. H.; Ledwith, A. Stereoregular addition polymerisation. *Quart. Rev. Chem. Soc.* **1962**, *16*, 361–434.

46. Nodono, M.; Tokimitsu, T.; Tone, S.; Makino, T.; Yanagase, A. Chain transfer polymerization of methyl methacrylate initiated by organolanthanide complexes. *Macromol. Chem. Phys.* **2000**, *201*, 2282–2288.

47. Ihara, E.; Morimoto, M.; Yasuda, H. Living polymerizations and copolymerizations of alkyl acrylates by the unique catalysis of rare earth metal complexes. *Macromolecules* **1995**, *28*, 7886–7892.

48. Kawaguchi, Y.; Yasuda, H. Synthesis of monodisperse poly(butyl acrylate) initiated by SmMe(C$_5$Me$_5$)$_2$(THF) and its adhesive properties after crosslinking by electron-beam irradiation. *J. Appl. Polym. Sci.* **2001**, *80*, 432–437.

49. Giardello, M. A.; Yamamoto, Y.; Brand, L.; Marks, T. J. Stereocontrol in the polymerization of methyl methacrylate mediated by chiral organolanthanide metallocenes. *J. Am. Chem. Soc.* **1995**, *117*, 3276–3277.

50. Ballard, D. G. H.; Courtis, A.; Holton, J.; McMeeking, J.; Pearce, R. Alkyl bridged complexes of the group 3A and lanthanoid metals as homogenous ethylene polymerisation catalysts. *J. Chem. Soc., Chem. Commun.* **1978**, 994–995.

51. Jeske, G.; Lauke, H.; Mauermann, H.; Swepston, P. N.; Schumann, H.; Marks, T. J. Highly reactive organolanthanides. Systematic routes to and olefin chemistry of early and late bis(pentamethylcyclopentadienyl) 4f hydrocarbyl and hydride complexes. *J. Am. Chem. Soc.* **1985**, *107*, 8091–8103.

52. Watson, P. L.; Parshall, G. W. Organolanthanides in catalysis. *Acc. Chem. Res.* **1985**, *18*, 51–56.

53. Watson, P. L. Ziegler–Natta polymerization: The lanthanide model. *J. Am. Chem. Soc.* **1982**, *104*, 337–339.

54. Yasuda, H.; Furo, M.; Yamamoto, H.; Nakamura, A.; Miyake, S.; Kibino, N. New approach to block copolymerizations of ethylene with alkyl methacrylates and lactones by unique catalysis with organolanthanide complexes. *Macromolecules* **1992**, *25*, 5115–5116.

55. Tanaka, K.; Furo, M.; Ihara, E.; Yasuda, H. Unique dual function of La(C$_5$Me$_5$)[CH(SiMe$_3$)$_2$]$_2$(THF) for polymerizations of both nonpolar and polar monomers. *J. Polym. Sci., Part A: Polym. Chem.* **2001**, *39*, 1382–1390.

56. Desurmont, G.; Tanaka, M.; Li, Y.; Yasuda, H.; Tokimitsu, T.; Tone, S.; Yanagase, A. New approach to block copolymerization of ethylene with polar monomers by the unique catalytic function of organolanthanide complexes. *J. Polym. Sci., Part A: Polym. Chem.* **2000**, *38*, 4095–4109.

57. Gromada, J.; Chenal, T.; Mortreaux, A.; Leising, F.; Carpentier, J. F. Homogeneous and heterogeneous alkyl-alkoxo-lanthanide type catalysts for polymerization and block-copolymerization of ethylene and methyl methacrylate. *J. Mol. Catal. A: Chem.* **2002**, *182/183*, 525–531.

58. Desurmont, G.; Tokimitsu, T.; Yasuda, H. First controlled block copolymerizations of higher 1-olefins with polar monomers using metallocene type single component lanthanide initiators. *Macromolecules* **2000**, *33*, 7679–7681.

59. Boffa, L. S.; Novak, B. M. Bimetallic samarium(III) catalysts via electron transfer initiation: The facile synthesis of well-defined (meth)acrylate triblock copolymers. *Tetrahedron* **1997**, *53*, 15367–15396.

60. Boffa, L. S.; Novak, B. M. "Link-functionalized" and triblock polymer architectures through bifunctional organolanthanide initiators: A review. *J. Mol. Catal. A: Chem.* **1998**, *133*, 123–130.

61. Evans, W. J.; Gonzales, S. L.; Ziller, J. W. Reactivity of decamethylsamarocene with polycyclic aromatic hydrocarbons. *J. Am. Chem. Soc.* **1994**, *116*, 2600–2608.

62. Boffa, L. S.; Novak, B. M. Bimetallic samarium(III) initiators for the living polymerization of methacrylates and lactones. A new route into telechelic, triblock, and "link-functionalized" polymers. *Macromolecules* **1994**, *27*, 6993–6995.

63. Boffa, L. S.; Novak, B. M. "Link-functionalized" polymers: An unusual macromolecular architecture through bifunctional initiation. *Macromolecules* **1997**, *30*, 3494–3506.

64. Mao, L.; Shen, Q.; Sun, J. Synthesis of bis(methylcyclopentadienyl)(piperidino)lanthanoids and their catalytic behavior for polymerization of methyl methacrylate. *J. Organomet. Chem.* **1998**, *566*, 9–14.

65. Mao, L.; Shen, Q. (Diisopropylamido)bis(methylcyclopentadienyl) lanthanides as single-component initiators for polymerization of methyl methacrylate. *J. Polym. Sci., Part A: Polym. Chem.* **1998**, *36*, 1593–1597.

66. Lee, M. H.; Hwang, J.-W.; Kim, Y.; Kim, J.; Han, Y.; Do, Y. The first fluorenyl *ansa*-yttrocene complexes: Synthesis, structures, and polymerization of methyl methacrylate. *Organometallics* **1999**, *18*, 5124–5129.

67. Qian, C.; Nie, W.; Chen, Y.; Sun, J. Synthesis and crystal structure of one carbon-atom bridged lutetium complex [Ph$_2$C(Flu)(Cp)]LuN(TMS)$_2$ and catalytic activity for polymerization of polar monomers. *J. Organomet. Chem.* **2002**, *645*, 82–86.

68. Shen, Q.; Wang, Y.; Zhang, K.; Yao, Y. Lanthanocene amide complexes as single-component initiators for the polymerization of (dimethylamino)ethyl methacrylate. *J. Polym. Sci., Part A: Polym. Chem.* **2002**, *40*, 612–616.

69. Britovsek, G. J. P.; Gibson, V. C.; Wass, D. F. The search for new-generation olefin polymerization catalysts: Life beyond metallocenes. *Angew. Chem., Int. Ed. Engl.* **1999**, *38*, 428–427.

70. Gibson, V. C.; Spitzmesser, S. K. Advances in non-metallocene olefin polymerization catalysis. *Chem. Rev.* **2003**, *103*, 283–315.

71. Gamer, M. T.; Rastätter, M.; Roesky, P. W.; Steffens, A.; Glanz, M. Yttrium and lanthanide complexes with various P,N ligands in the coordination sphere: Synthesis, structure, and polymerisation studies. *Chem. Eur. J.* **2005**, *11*, 3165–3172.

72. Bonnet, F.; Hillier, A. C.; Collins, A.; Dubberly, S. R.; Mountford, P. Lanthanide mono(borohydride) complexes of diamide-diamine donor ligands: Novel single site catalysts for the polymerisation of methyl methacrylate. *Dalton Trans.* **2005**, 421–423.

73. Yao, Y.; Zhang, Y.; Zhang, Z.; Shen, Q.; Yu, K. Synthesis and structural characterization of divalent ytterbium complexes supported by β-diketiminate ligands and their catalytic activity for the polymerization of methyl methacrylate. *Organometallics* **2003**, *22*, 2876–2882.

74. Nodono, M.; Tokimitsu, T.; Makino, T. Polymerization of methyl methacrylate initiated by a divalent samarium phenoxide complex with an alkyl aluminum compound. *Macromol. Chem. Phys.* **2003**, *204*, 877–884.

75. Cui, C. M.; Shafir, A.; Reeder, C. L.; Arnold, J. Highly isospecific polymerization of methyl methacrylate with a bis(pyrrolylaldiminato)samarium hydrocarbyl complex. *Organometallics* **2003**, *22*, 3357–3359.

76. Ihara, E.; Koyama, K.; Yasuda, H.; Kanehisa, N.; Kai, Y. Catalytic activity of allyl-, azaallyl- and diaza-pentadienyllanthanide complexes for polymerization of methyl methacrylate. *J. Organomet. Chem.* **1999**, *574*, 40–49.

77. Nakayama, Y.; Shibahara, T.; Fukumoto, H.; Nakamura, A. Syndiospecific polymerization of methyl methacrylate catalyzed by lanthanoid thiolate complexes bearing a hexamethylphosphoric triamide ligand. *Macromolecules* **1996**, *29*, 8014–8016.

78. Qi, G.; Nitto, Y.; Saiki, A.; Tomohiro, T.; Nakayama, Y.; Yasuda, H. Isospecific polymerizations of alkyl methacrylates with a bis(alkyl)Yb complex and formation of stereocomplexes with syndiotactic poly(alkyl acrylates). *Tetrahedron* **2003**, *59*, 10409–10418.

79. Simionescu, Cr.; Asandei, N.; Benedek, I.; Ungurenasu, C. La copolymerisation du système binaire acrylonitrile-methylmethacrylate a l'aide des promoteurs du type soluble Ziegler–Natta constitués par le complexe: Dichlorure du biscyclopentadienyl-titane-triethyle aluminium. *Eur. Polym. J.* **1969**, *5*, 449–462.

80. Farnham, W. B.; Hertler, W. R. Titanium, zirconium- and hafnium containing initiators in the polymerization of acrylic monomers to "living" polymers. U.S. Patent 4,728,706 (du Pont), March 1, 1988.

81. Collins, S.; Ward, D. G. Group-transfer polymerization using cationic zirconcene compounds. *J. Am. Chem. Soc.* **1992**, *114*, 5460–5462.

82. Collins, S.; Ward, D. G.; Suddaby, K. H. Group-transfer polymerization using metallocene catalysts: Propagation mechanisms and control of polymer stereochemistry. *Macromolecules* **1994**, *27*, 7222–7224.

83. Webster, O. W.; Hertler, W. R.; Sogah, D. Y.; Farnham, W. B.; RajanBabu, T. V. Group-transfer polymerization. 1. A new concept for addition polymerization with organosilicon initiators. *J. Am. Chem. Soc.* **1983**, *105*, 5706–5708.

84. Sogah, D. Y.; Hertler, W. R.; Webster, O. W.; Cohen, G. M. Group transfer polymerization. Polymerization of acrylic monomers. *Macromolecules* **1987**, *20*, 1473–1488.

85. Li, Y.; Ward, D. G.; Reddy, S. S.; Collins, S. Polymerization of methyl methacrylate using zirconocene initiators: Polymerization mechanisms and applications. *Macromolecules* **1997**, *30*, 1875–1883.

86. Soga, K.; Deng, H.; Yano, T.; Shiono, T. Stereospecific polymerization of methyl methacrylate initiated by dimethylzirconocene/$B(C_6F_5)_3$ (or $Ph_3CB(C_6F_5)_4$/$Zn(C_2H_5)_2$). *Macromolecules* **1994**, *27*, 7938–7940.

87. Cameron, P. A.; Gibson, V. C.; Graham, A. J. On the polymerization of methyl methacrylate by group 4 metallocenes. *Macromolecules* **2000**, *33*, 4329–4335.

88. Stuhldreier, T.; Keul, H.; Höcker, H. A cationic bridged zirconocene complex as the catalyst for the stereospecific polymerization of methyl methacrylate. *Macromol. Rapid Commun.* **2000**, *21*, 1093–1098.

89. Bolig, A. D.; Chen, E. Y.-X. *ansa*-Zirconocene ester enolates: Synthesis, structure, reaction with organo-Lewis acids, and application to polymerization of methacrylates. *J. Am. Chem. Soc.* **2004**, *126*, 4897–4906.

90. Kostakis, K.; Mourmouris, S.; Pitsikalis, M.; Hadjichristidis, N. Polymerization of acrylates and bulky methacrylates with the use of zirconocene precursors: Block copolymers with methyl methacrylate. *J. Polym. Sci., Part A: Polym. Chem.* **2005**, *43*, 3337–3348.

91. Strauch, J. W.; Fauré, J.-L.; Bredeau, S.; Wang, C.; Kehr, G.; Fröhlich, R.; Luftmann, H.; Erker, G. (Butadiene)metallocene/$B(C_6F_5)_3$ pathway to catalyst systems for stereoselective methyl methacrylate polymerization: Evidence for an anion dependent metallocene catalyzed polymerization process. *J. Am. Chem. Soc.* **2004**, *126*, 2089–2104.

92. Deng, H.; Shiono, T.; Soga, K. Isospecific polymerization of methyl methacylate initiated by chiral zirconocenedimethyl/$Ph_3CB(C_6F_5)_4$ in the presence of Lewis acid. *Macromolecules* **1995**, *28*, 3067–3073.

93. Rodriguez-Delgado, A.; Chen, E. Y.-X. Mechanistic studies of stereospecific polymerization of methacrylates using a cationic, chiral *ansa*-zirconocene ester enolate. *Macromolecules* **2005**, *38*, 2587–2594.

94. Deng, H.; Soga, K. Isotactic polymerization of *tert*-butyl acrylate with chiral zirconocene. *Macromolecules* **1996**, *29*, 1847–1848.

95. Karanikolopoulos, G.; Batis, C.; Pitsikalis, M.; Hadjichristidis, N. The influence of the nature of the catalytic system on zirconocene-catalyzed polymerization of alkyl methacrylates. *Macromol. Chem. Phys.* **2003**, *204*, 831–840.

96. Frauenrath, H.; Keul, H.; Höcker, H. Stereospecific polymerization of methyl methacrylate with single-component zirconocene complexes: Control of stereospecificity via catalyst symmetry. *Macromolecules* **2001**, *34*, 14–19.

97. Rodriguez-Delgado, A.; Mariott, W. R.; Chen, E. Y.-X. Living and syndioselective polymerization of methacrylates by constrained geometry titanium alkyl and enolate complexes. *Macromolecules* **2004**, *37*, 3092–3100.

98. Sustmann, R.; Sicking, W.; Bandermann, F.; Ferenz, M. On the mechanism of polymerization of acrylates by zirconocene complexes, an *ab initio* and density functional theory MO study. *Macromolecules* **1999**, *32*, 4204–4213.

99. Hölscher, M.; Keul, H.; Höcker, H. Explanation of the different reaction behaviors of bridged and unbridged cationic single component zirconocene catalysts in MMA polymerizations: A DFT study. *Macromolecules* **2002**, *35*, 8194–8202.

100. Hölscher, M.; Keul, H.; Höcker, H. Postulation of the mechanism of the selective synthesis of isotactic poly(methyl methacrylate) catalysed by [Zr{(Cp)(Ind)CMe$_2$}(Me)(thf)](BPh$_4$): A Hartree-Fock, MP2 and density functional study. *Chem. Eur. J.* **2001**, *7*, 5419–5426.

101. Frauenrath, H.; Balk, S.; Keul, H.; Höcker, H. First synthesis of an AB block copolymer with polyethylene and poly(methyl methacrylate) blocks using a zirconocene catalyst. *Macromol. Rapid Commun.* **2001**, *22*, 1147–1151.

102. Jin, J.; Chen, E. Y.-X. Stereoblock copolymerization of propylene and methyl methacrylate with single-site metallocene catalysts. *Macromol. Chem. Phys.* **2002**, *203*, 2329–2333.

103. Jensen, T. R.; Yoon, S. C.; Dash, A. K.; Luo, L.; Marks, T. J. Organotitanium-mediated stereoselective coordinative/insertive homopolymerizations and copolymerizations of styrene and methyl methacrylate. *J. Am. Chem. Soc.* **2003**, *125*, 14482–14494.

104. Matsuo, Y.; Mashima, K.; Tani, K. Half-metallocene tantalum complexes bearing methyl methacrylate (MMA) and 1,4-diaza-1,3-diene ligands as MMA polymerization catalysts. *Angew. Chem., Int. Ed. Engl.* **2001**, *40*, 960–962.

105. Feng, S.; Roof, G. R.; Chen, E. Y.-X. Tantalum(V)-based metallocene, half-metallocene, and non-metallocene complexes as ethylene-1-octene copolymerization and methyl methacrylate polymerization catalysts. *Organometallics* **2002**, *21*, 832–839.

106. Gibson V. C.; Segal J. A.; White A. J. P.; Williams D. J. Novel mono-alkyl magnesium complexes stabilized by a bulky β-diketiminate ligand: Structural characterization of a coordinatively unsaturated trigonal system. *J. Am. Chem. Soc.* **2000**, *122*, 7120–7121.

107. Dove, A. P.; Gibson, V. C.; Hormnirun, P.; Marshall, E. L.; Segal, J. A.; White, A. J. P.; Williams, D. J. Low coordinate magnesium chemistry supported by a bulky β-diketiminate ligand. *Dalton Trans.* **2003**, 3088–3097.

108. Dove, A. P.; Gibson, V. C.; Hormnirun, P.; Marshall, E. L.; Rzepa, H. S.; Segal, J. A.; White, A. J. P.; Williams, D. J. β-Diketiminate magnesium alkyl, amide and enolate complexes for the living, syndioselective polymerization of methyl methacrylate: A synthetic and theoretical study. Manuscript in preparation.

109. Dove, A. P.; Gibson, V. C.; Marshall, E. L.; White, A. J. P.; Williams, D. J. A well-defined magnesium enolate initiator for the living and highly syndioselective polymerisation of methyl methacrylate. *Chem. Commun.* **2002**, 1208–1209.

110. Budzelaar, P. H. M.; van Oort, A. B.; Orpen, A. G. β-Diiminato complexes of V^{III} and Ti^{III} – formation and structure of stable paramagnetic dialkylmetal compounds. *Eur. J. Inorg. Chem.* **1998**, 1485–1494.

111. Chamberlain, B. M.; Cheng, M.; Moore, D. R.; Ovitt, T. M.; Lobkovsky, E. B.; Coates, G. W. Polymerization of lactide with zinc and magnesium diiminate complexes: Stereocontrol and mechanism. *J. Am. Chem. Soc.* **2001**, *123*, 3229–3238.

112. Marshall, E. L.; Gibson, V. C.; Rzepa, H. S. A computational analysis of the ring-opening polymerization of *rac*-lactide initiated by single-site β-diketiminate metal complexes: Defining the mechanistic pathway and the origin of stereocontrol. *J. Am. Chem. Soc.* **2005**, *127*, 6048–6051.

113. Seebach, D. Structure and reactivity of lithium enolates. From pinacolone to selective *C*-alkylations of peptides. Difficulties and opportunities afforded by complex structures. *Angew. Chem., Int. Ed. Engl.* **1988**, *27*, 1624–1654.

114. Rodriguez-Delgado, A.; Chen, E. Y. X. Single-site anionic polymerization. Monomeric ester enolaluminate propagator synthesis, molecular structure and polymerization mechanism. *J. Am. Chem. Soc.* **2005**, *127*, 961–974.

115. Kuroki, M.; Aida, T.; Inoue, S. Novel photoinduced carbon–carbon bond formation via metal-alkyl and -enolate porphyrins — visible light-mediated polymerization of alkyl methacrylate catalyzed by aluminum porphyrin. *J. Am. Chem. Soc.* **1987**, *109*, 4737–4738.

116. Murayama, H.; Inoue, S. Photochemical activation of aluminum–alkyl bond in aluminum porphyrin—conjugate addition-reaction with vinyl ketone. *Chem. Lett.* **1985**, 1377–1380.

117. Kuroki, M.; Nashimoto, S.; Aida, T.; Inoue, S. Sequential addition-ring-opening living polymerizations by aluminum porphyrin. Synthesis of alkyl methacrylate-epoxide and -lactone block copolymers of controlled molecular weight. *Macromolecules* **1988**, *21*, 3114–3115.

118. Adachi, T.; Sugimoto, H.; Aida, T.; Inoue, S. Aluminum thiolate complexes of porphyrin as excellent initiators for Lewis acid-assisted high-speed living polymerization of methyl methacrylate. *Macromolecules* **1993**, *26*, 1238–1243.

119. Arai, T.; Sato, Y.; Inoue, S. Stereoselective formation of aluminum enolate on capped porphyrin. *Chem. Lett.* **1990**, 1167–1170.

120. Kuroki, M.; Watanabe, T.; Aida, T.; Inoue, S. Steric separation of nucleophile and Lewis acid providing dramatically accelerated reaction. High-speed polymerization of methyl methacrylate with enolate-aluminum porphyrin/sterically crowded organoaluminum systems. *J. Am. Chem. Soc.* **1991**, *113*, 5903–5904.

121. Adachi, T.; Sugimoto, H.; Aida, T.; Inoue, S. Controlled synthesis of high molecular weight poly(methyl methacrylate) based on Lewis acid-assisted high-speed living polymerization initiated with aluminum porphyrin. *Macromolecules* **1992**, *25*, 2280–2281.

122. Sugimoto, H.; Kuroki, M.; Watanabe, T.; Kawamura, C.; Aida, T.; Inoue, S. High-speed living anionic polymerization of methacrylic esters with aluminum porphyrin initiators. Organoaluminum compounds as Lewis acid accelerators. *Macromolecules* **1993**, *26*, 3403–3410.

123. Hosokawa, Y.; Kuroki, M.; Aida, T.; Inoue, S. Controlled synthesis of poly(acrylic esters) by aluminum porphyrin initiators. *Macromolecules* **1991**, *24*, 824–829.

124. Sugimoto, H.; Saika, M.; Hosokawa, Y.; Aida, T.; Inoue, S. Accelerated living polymerization of methacrylonitrile with aluminum porphyrin initiators by activation of monomer or growing species. Controlled synthesis and properties of poly(methyl methacrylate-*b*-methacrylonitrile)s. *Macromolecules* **1996**, *29*, 3359–3369.

125. Cameron, P. A.; Gibson, V. C.; Irvine, D. J. Nickel-catalyzed generation of Schiff base aluminum enolate initiators for controlled methacrylate polymerization. *Angew. Chem., Int. Ed. Engl.* **2000**, *39*, 2141–2144.

126. Cameron, P. A. New initiating systems for the polymerizations of acrylates. Ph.D. Thesis, Imperial College London, 1998.

127. Bolig, A. D.; Chen, E. Y.-X. Reversal of polymerization stereoregulation in anionic polymerization of MMA by chiral metallocene and non-metallocene initiators: A new reaction pathway for metallocene-initiated MMA polymerization. *J. Am. Chem. Soc.* **2001**, *123*, 7943–7944.

128. Chen, E. Y.-X. Stereospecific polymerization of methacrylates by metallocene and related catalysts. *J. Polym. Sci., Part A: Polym Chem.* **2004**, *42*, 3395–3403.

129. Bolig, A. D.; Chen, E. Y.-X. Isotactic-*b*-syndiotactic stereoblock poly(methyl methacrylate) by chiral metallocene/Lewis acid hybrid catalysts. *J. Am. Chem. Soc.* **2002**, *124*, 5612–5613.

130. Chen, E. Y.-X.; Cooney, M. J. Amphicatalytic polymerization: Synthesis of stereomultiblock poly(methyl methacrylate) with diastereospecific ion pairs. *J. Am. Chem. Soc.* **2003**, *125*, 7150–7151.

131. Kamigaito, M.; Ando, T.; Sawamoto, M. Metal-catalyzed living radical polymerization. *Chem. Rev.* **2001**, *101*, 3689–3745.

132. Matyjaszewski, K.; Xia, J. Atom transfer radical polymerization. *Chem. Rev.* **2001**, *101*, 2921–2990.

133. Kato, M.; Kamigato, M.; Sawimoto, M.; Higashimura, T. Polymerization of methyl methacrylate with the carbon tetrachloride/dichlorotris-(triphenylphosphine)ruthenium(II)/methylaluminum bis(2,6-di-*tert*-butylphenoxide) initiating system: Possibility of living radical polymerization. *Macromolecules* **1995**, *28*, 1721–1723.

134. Simal, F.; Demonceau, A.; Noels, A. F. Highly efficient ruthenium-based catalytic systems for the controlled free-radical polymerization of vinyl monomers. *Angew. Chem., Int. Ed. Engl.* **1999**, *38*, 538–540.

135. Grimaud, T.; Matyjaszewski, K. Controlled/"living" radical polymerization of methyl methacrylate by atom transfer radical polymerization. *Macromolecules* **1997**, *30*, 2216–2218.

136. Haddleton, D. M.; Jasieczek, C. B.; Hannon, M. J.; Shooter, A. J. Atom transfer radical polymerization of methyl methacrylate initiated by alkyl bromide and 2-pyridinecarbaldehyde imine copper(I) complexes. *Macromolecules* **1997**, *30*, 2190–2193.

137. Granel, C.; Dubois, Ph.; Jérôme, R.; Teyssié, Ph. Controlled radical polymerization of methacrylic monomers in the presence of a bis(*ortho*-chelated) arylnickel(II) complex and different activated alkyl halides. *Macromolecules* **1996**, *29*, 8576–8582.

138. Uegaki, H.; Kotani, Y.; Kamigaito, M.; Sawamoto, M. NiBr$_2$(Pn-Bu$_3$)$_2$-mediated living radical polymerization of methacrylates and acrylates and their block or random copolymerizations. *Macromolecules* **1998**, *31*, 6756–6761.

139. Moineau, G.; Minet, M.; Dubois, Ph.; Teyssié, Ph.; Senninger, T.; Jérôme, R. Controlled radical polymerization of (meth)acrylates by ATRP with NiBr$_2$(PPh$_3$)$_2$ as catalyst. *Macromolecules* **1999**, *32*, 27–35.

140. Matyjaszewski, K.; Wei, M.; Xia, J.; McDermott, N. E. Controlled/"living" radical polymerization of styrene and methyl methacrylate catalyzed by iron complexes. *Macromolecules* **1997**, *30*, 8161–8164.

141. Ando, T.; Kamigaito, M.; Sawamoto, M. Iron(II) chloride complex for living radical polymerization of methyl methacrylate. *Macromolecules* **1997**, *30*, 4507–4510.

142. Louie, J.; Grubbs, R. H. Highly active iron imidazolylidene catalysts for atom transfer radical polymerization. *Chem. Commun.* **2000**, 1479–1480.

143. Gibson, V. C.; O'Reilly, R. K.; Wass, D. F.; White, A. J. P.; Williams, D. J. Polymerization of methyl methacrylate using four-coordinate (α-diimine)iron catalysts: Atom transfer radical polymerization vs. catalytic chain transfer. *Macromolecules* **2003**, *36*, 2591–2593.

144. Lecomte, P.; Drapier, I.; Dubois, Ph.; Teyssié, Ph.; Jérôme, R. Controlled radical polymerization of methyl methacrylate in the presence of palladium acetate, triphenylphosphine, and carbon tetrachloride. *Macromolecules* **1997**, *30*, 7631–7633.

145. Moineau, G.; Granel, C.; Dubois, Ph.; Jérôme, R.; Teyssié, Ph. Controlled radical polymerization of methyl methacrylate initiated by an alkyl halide in the presence of the Wilkinson catalyst. *Macromolecules* **1998**, *31*, 542–544.

146. Porter, N. A.; Allen, T. R.; Breyer, R. A. Chiral auxiliary control of free-radical polymerization. *J. Am. Chem. Soc.* **1992**, *114*, 7676–7683.

147. Imai, K.; Shiomi, T.; Oda, N.; Otsuka, H. Effect of solvent for vinyl-acetate polymerization on microstructure of poly(vinyl-alcohol). *J. Polym. Sci., Part A: Polym. Chem.* **1986**, *24*, 3225–3231.

148. Yamada, K.; Nakano, T.; Okamoto, Y. Stereospecific free radical polymerization of vinyl esters using fluoroalcohols as solvents. *Macromolecules* **1998**, *31*, 7598–7605.

149. Isobe, Y.; Yamada, K.; Nakano, T.; Okamoto, Y. Stereospecific free radical polymerization of methacrylates using fluoroalcohols as solvents. *Macromolecules* **1999**, *32*, 5979–5981.

150. Otsu, T.; Yamada, B.; Imoto, M. Effect of zinc chloride on the stereoregularity of the radical polymerization of methyl methacrylate. *J. Macromol. Chem.* **1966**, *1*, 61–74.

151. Okazawa, S.; Hirai, H.; Makishima, S. Polymerization of coordinated monomers. I. Stereoregulation in the free-radical polymerization of methyl methacrylate-zinc chloride or methyl methacrylate-stannic chloride complexes. *J. Polym. Sci., Part A1: Polym. Chem.* **1969**, *7*, 1039–1053.

152. Matsumoto, A.; Nakamura, S. Radical polymerization of methyl methacrylate in the presence of magnesium bromide as the Lewis acid. *J. Appl. Polym. Sci.* **1999**, *74*, 290–296.

153. Mero, C. L.; Porter, N. A. Free-radical polymerization and copolymerization of acrylimides: Homopolymers of oxazolidinone acrylimide and control of 1,5-stereochemistry in copolymers derived from isobutylene and an oxazolidinone acrylimide. *J. Org. Chem.* **2000**, *65*, 775–781.

154. Suito, Y.; Isobe, Y.; Habaue, S.; Okamoto, Y. Isotactic-specific radical polymerization of methacrylamides in the presence of Lewis acids. *J. Polym. Sci., Part A: Polym. Chem.* **2002**, *40*, 2496–2500.

155. Isobe, Y.; Suito, Y.; Habaue, S.; Okamoto, Y. Stereocontrol during the free-radical polymerization of methacrylamides in the presence of Lewis acids. *J. Polym. Sci., Part A: Polym. Chem.* **2003**, *41*, 1027–1033.

156. Isobe, Y.; Fujioka, D.; Habaue, S.; Okamoto, Y. Efficient Lewis acid-catalyzed stereocontrolled radical polymerization of acrylamides. *J. Am. Chem. Soc.* **2001**, *123*, 7180–7181.

157. Habaue, S.; Uno, T.; Baraki, H.; Okamoto, Y. Stereospecific radical polymerization of α-(alkoxymethyl)acrylates controlled by a catalytic amount of zinc halides. *Macromolecules* **2000**, *33*, 820–824.

158. Baraki, H.; Habaue, S.; Okamoto, Y. Stereospecific radical polymerization of α-(alkoxymethyl)acrylates controlled by Lewis acids: Mechanistic study and effect of amino alcohols as ligand for zinc bromide. *Macromolecules* **2001**, *34*, 4724–4729.

159. Isobe, Y.; Nakano, T.; Okamoto, Y. Stereocontrol during the free-radical polymerization of methacrylates with Lewis acids. *J. Polym. Sci., Part A: Polym. Chem.* **2001**, *39*, 1463–1471.

160. Lutz, J. F.; Jakubowski, W.; Matyjaszewski, K. Controlled/living radical polymerization of methacrylic monomers in the presence of Lewis acids: Influence on tacticity. *Macromol. Rapid Commun.* **2004**, *25*, 486–492.

161. Liu, W. H.; Nakano, T.; Okamoto, Y. Stereochemistry of acrylate polymerization in toluene using *n*-BuLi. *Polym. J.* **1999**, *31*, 479–481.

24 Discrete Catalysts for Stereoselective Epoxide Polymerization

Hiroharu Ajiro, Scott D. Allen, and Geoffrey W. Coates

CONTENTS

24.1 INTRODUCTION

24.1.1 BACKGROUND

Near the time of Natta's discovery of stereoselective alkene polymerization, Baggett and Pruitt[1–3] reported that iron (III) chloride was capable of forming poly(propylene oxide) that could be divided into amorphous as well as crystalline materials using solvent fractionation.[4] Soon thereafter, Natta et al.[5] and Price et al.[6,7] provided evidence that the crystalline material was isotactic poly(propylene oxide), in which the main-chain methyl substituents were of the same relative configuration. This finding marked the first discovery of a stereoselective catalyst for epoxide polymerization. Since that time, significant advances have been made in stereoselective epoxide polymerization, which is the topic of this chapter.

24.1.2 SCOPE of CHAPTER

Many of the important contributors to the field of stereoselective epoxide polymerization have written accounts of their research,[8-20] but an up-to-date review does not exist. *This chapter covers discrete catalysts for the stereoselective polymerization of epoxides.* Due to space limitations, catalysts bearing at lease one ancillary ligand that is not likely to react with epoxides are discussed; those catalysts only bearing ligands that are well-known to react with epoxides are not covered.[21] A discussion of strategies for controlling the relative configuration of main-chain stereogenic centers of epoxide polymers is included. Because the emphasis is on stereochemical control of polymerization by the catalyst or initiator, the polymerization of optically active epoxides is not covered.

24.2 BASIC CONCEPTS IN STEREOSELECTIVE EPOXIDE POLYMERIZATION

24.2.1 REGIOCHEMISTRY

Both the ancillary ligands (L_n) surrounding the active metal center (M) and the growing polymer chain (OR) influence the regiochemistry and stereochemistry of epoxide polymerization.[22] When the epoxide is unsymmetrically substituted (i.e., propylene oxide), enchainment can occur in two ways: (1) attack by the polymer's alkoxide chain-end at the methylene with retention of the stereochemistry of the substituted carbon to give a secondary metal alkoxide, or (2) attack at the methine with inversion of stereochemistry to give a primary metal alkoxide (Scheme 24.1). The polymer is regioregular when only one process dominates; the polymer is regioirregular when both processes occur. The regiochemistry of a polyepoxide (such as poly(propylene oxide)) can be readily determined by ^{13}C NMR spectroscopy.[23-25]

24.2.2 CHAIN-END CONTROL AND ENANTIOMORPHIC-SITE CONTROL OF STEREOCHEMISTRY

In a chain-growth polymerization reaction, the end of the polymer chain remains at the active metal center during monomer enchainment. Thus, the stereogenic center in the polymer chain

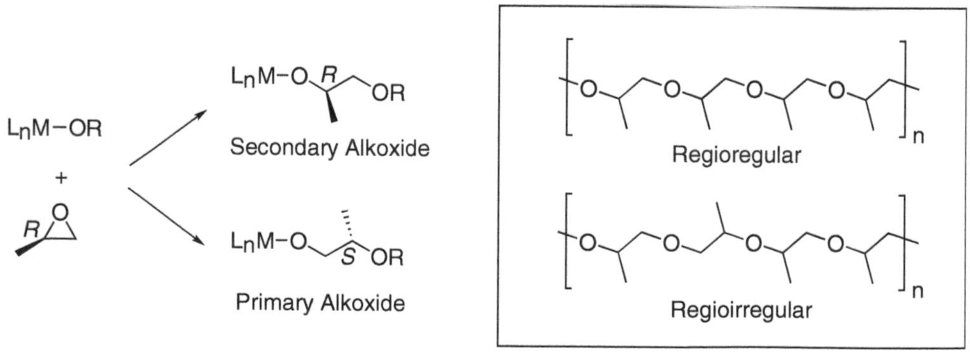

SCHEME 24.1 Regiochemistry and stereochemistry of epoxide polymerization.

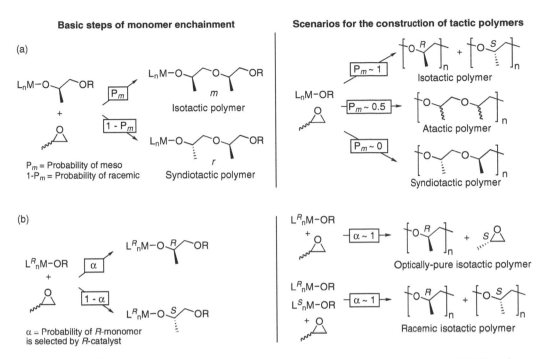

SCHEME 24.2 Chain-end (a) and enantiomorphic-site (b) mechanisms of stereocontrol in epoxide polymerization (L_n^RM-OR is an enantiomerically pure catalyst that prefers R-monomer).

from the last enchained monomer unit will have an influence on the stereochemistry of monomer enchainment. If this influence is significant, the mode of stereochemical regulation is referred to as "polymer chain-end control" (Scheme 24.2a). If the active site is chiral and overrides the influence of the polymer chain end, the mechanism of stereochemical direction is termed "enantiomorphic-site control" (Scheme 24.2b).[9–11] In the former mechanism, a stereochemical error is propagated (to give a polymer with ... *mmmrmm* ... sequences), but in the latter a correction occurs because the ligands direct the stereochemical events, leading to an isolated stereoerror (to give a polymer with ... *mmmrrmmm* ... sequences). (For additional information on chain-end control and enantiomorphic-site control, see Chapter 1.)

Optically active catalysts can kinetically resolve racemic monomers (Scheme 24.2b), producing optically active polymers as well as enantiomerically-enriched monomers since the nonpreferred enantiomer remains behind as monomer after the preferred monomer has been converted into polymer. The quantitative measure of stereocontrol in such a system is given by the parameter r, which is the ratio of the rate constants for the polymerization of the R- and S-enantiomers of the monomer $[r = k_R/k_S = \alpha/(1 - \alpha)]$.

24.2.3 ANALYSIS OF POLYMER STEREOCHEMISTRY

The most useful method for determining polymer tacticity and quantifying stereochemical purity is nuclear magnetic resonance (NMR).[26,27] In many cases the chemical shifts for the various polymer nuclei are sensitive to adjacent stereogenic centers, resulting in fine structure that can provide quantitative information about the polymer microstructure once the shift identities are assigned. For example, the methyl, methylene, and methine regions of a high-resolution ^{13}C NMR spectrum of atactic poly(propylene oxide) display several peaks, each of which represents a different set of consecutive stereocenters. Because the position of each peak in the spectrum has been assigned,[23,24,28] a routine ^{13}C NMR experiment can reveal both the tacticity and the degree of stereoregularity of a sample of poly(propylene oxide). The ratio of the peaks can also be used to determine the mechanism of stereocontrol because the spectra can be simulated using statistical models based on the various propagation mechanisms depicted in Scheme 24.2 and characteristic stereoerror sequences associated with each mechanism.

24.3 STEREOSELECTIVE EPOXIDE POLYMERIZATION

The vast majority of papers reporting stereoselective epoxide polymerization focus on isospecific propylene oxide polymerization. For clarity, this chapter is organized by the type of metal of the catalyst active center. The three most commonly used metals for discrete stereoselective epoxide polymerization catalysts are aluminum, zinc, and cobalt, and research using these metals forms the foundation of this chapter.

24.3.1 ALUMINUM-BASED CATALYSTS

Although aluminum alkoxide- and aluminoxane-based catalysts have shown promise for the isospecific polymerization of epoxides, the poorly defined nature of these species has significantly hampered their use in such polymerizations.[29] Some examples of discrete aluminum complexes have been reported, however.

24.3.1.1 Aluminum–Acetylacetonate Catalysts

The prevailing theory for the mechanism of epoxide polymerization has been that epoxide coordination to the aluminum precedes insertion. To support this mechanism, Vandenberg proposed that the addition of chelating agents such as acetylacetone (acacH) to an R_3Al/H_2O (R = alkyl) polymerization system would block potential coordination sites on the metal center, thus hindering the reaction. Instead, these additives enhanced the polymerization rate and ushered in a new class of versatile and highly active catalysts.[12,30–32] The presumed structure of the active aluminum-acac catalyst (acac = acetylacetonate) is shown in Scheme 24.3. Although the precise structure of the active catalyst is unknown, a few structural features have been determined: (1) an oxygen atom bridges two aluminum centers (although the presence of multiple linkages, such as those in oligomeric aluminoxanes, cannot be ruled out); (2) alkyl groups are present on the aluminum atoms; and (3) the acac ligand is chelated to the aluminum center.[20]

Tuning the $AlR_3/H_2O/acac$ catalyst (R = alkyl) composition by varying the R groups and the ratio of components creates systems that conduct epoxide polymerizations to give high conversions, in many instances achieving >90% conversion to give high-molecular-weight, ether-insoluble polyethers. The fraction of acetone-insoluble, isotactic polymer produced varies according to the exact qualities of the catalyst system used, and is generally around 30% of the total mass of the ether-insoluble material while the remaining 70% is acetone-soluble polymer that is generally atactic.[31,32]

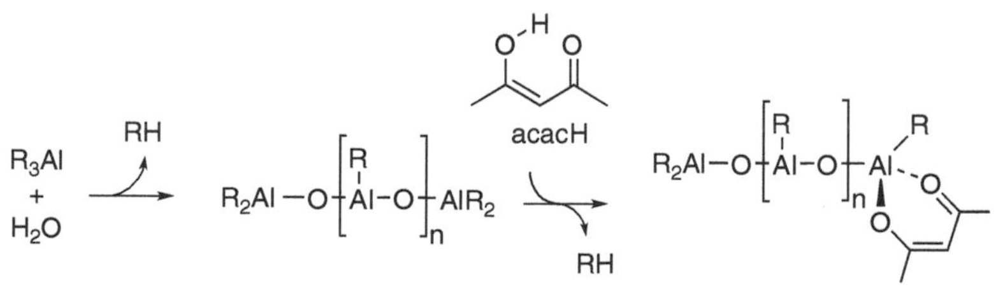

SCHEME 24.3 Synthesis of $AlR_3/H_2O/acacH$ epoxide polymerization catalysts (R = alkyl).

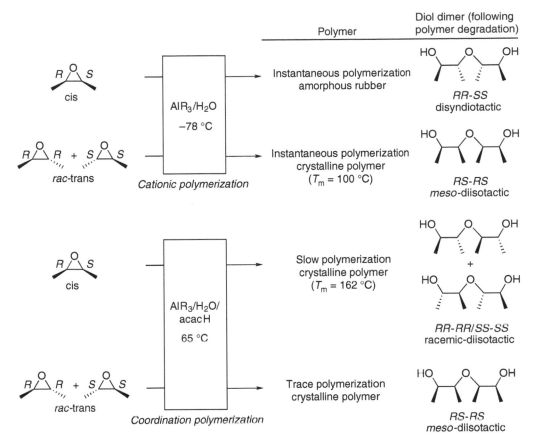

SCHEME 24.4 Stereoselectivity of polymerization of *cis*- and *trans*-2,3-epoxybutane using AlR$_3$/H$_2$O catalysts (R = alkyl).

In work investigating the mechanism of this system, Vandenberg used AlR$_3$/H$_2$O catalysts to polymerize *cis*- and *trans*-2,3-epoxybutane. From the properties of the resultant polymers and the examination of the diol decomposition products, mechanistic information was obtained. These results are summarized in Scheme 24.4.[12,20,33–35]

The AlR$_3$/H$_2$O catalysts polymerize both *cis*- and *trans*-2,3-epoxy butane instantaneously at −78 °C, consistent with a cationic process for monomer enchainment. The polymer isolated from the cis isomer is an amorphous rubber, whereas the polymer isolated from the trans isomer is crystalline with a T_m of 100 °C. This finding is in contrast to the AlR$_3$/H$_2$O/acacH system, which only slowly polymerizes the same monomers at 65 °C, presumably through a much slower coordination–insertion mechanism. The coordination polymerization of the cis isomer yields a crystalline polymer with a T_m of 162 °C, whereas the trans isomer polymerizes extremely slowly, producing only trace amounts of a crystalline polymer with a T_m similar to that obtained with the cationic polymerization. The extremely slow polymerization of the trans isomer is attributed to its increased steric bulk compared to the cis isomer. The steric bulk hinders the required precoordination to the metal center for monomer insertion.

Through the controlled degradation of the polyethers to diol dimers using *n*-butyl lithium, the stereochemistry of the monomer units in the polymer chain was determined.[33] The decomposition of all four polymers showed that inversion of stereochemistry at the site of attack on the epoxide ring occurred in both polymerization mechanisms. The cis epoxides (*RS* stereocenters) produce monomeric units in the polymer chain with *RR* and *SS* stereocenters, and the trans epoxides (either

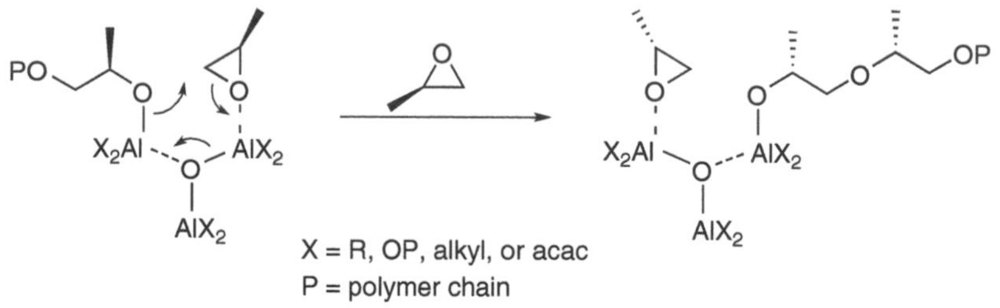

X = R, OP, alkyl, or acac
P = polymer chain

SCHEME 24.5 Proposed bimetallic enchainment of epoxides using (acac)Al complexes (acac = acetylacetonate, P = polymeryl).

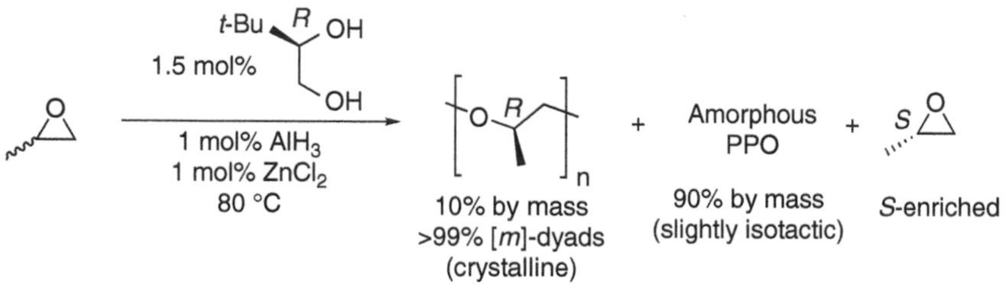

SCHEME 24.6 Polymerization of propylene oxide with $[(R\text{-dmbd})_{1.5}\text{Al}]_n/\text{ZnCl}_2$ catalysts (dmbd = 3,3-dimethyl-1,2-butanediolate).

RR or *SS*) produce monomeric units in the polymer chain with only *RS* units. This observation shows that inversion of configuration occurs at the site of attack, regardless of the polymerization mechanism.

To obtain the geometry required for a S_N2 attack in a coordination–insertion mechanism, Vandenberg proposed the transition-state structure shown in Scheme 24.5. In this scheme, an epoxide is activated by coordination to one aluminum center, while an adjacent aluminum center delivers the growing polymer chain. During this process, coordination bonds are exchanged to keep the charges balanced. Despite the evidence for bimetallic epoxide ring-opening events in other systems,[36–38] there is little evidence to show that it occurs in these heterogeneous aluminum-based systems.

24.3.1.2 Aluminum Catalysts Featuring Chiral Alkoxides

Haubenstock et al. synthesized chiral aluminum alkoxides for the stereoelective polymerization of propylene oxide.[39] Addition of 1.5 equivalents of (*R*)-(−)-3,3-dimethyl-1,2-butanediol ((*R*-dmbd)H$_2$) to AlH$_3$ generated the active complex $[(R\text{-dmbd})_{1.5}/\text{Al}]_n$. Alone, $[(R\text{-dmbd})_{1.5}/\text{Al}]_n$ had a very low activity for the polymerization of propylene oxide, achieving 85% conversion in 3 weeks with negligible optical activity in the unreacted monomer. The addition of ZnCl$_2$ (Al:Zn = 1:1) to the $[(R\text{-dmbd})_{1.5}/\text{Al}]_n$ initiator generated a much more active catalyst, as shown in Scheme 24.6. Furthermore, the optical activity of the unreacted propylene oxide was observed to increase with increasing conversion to polymer, and based on the rotation of the unreacted monomer, Haubenstock et al. determined that the catalyst system preferentially reacted with (*R*)-propylene oxide because the solution became enriched in the (*S*) isomer, although with modest selectivity (*r* = 1.05).[39] On fractionation, 10% of the total mass of the isolated polymer was acetone-insoluble and highly isotactic (>99% m-dyads), whereas the remainder of the polymer was acetone-soluble and atactic. Although a slight enantiomeric enrichment of monomer was achieved, this system did not significantly improve the yield of isotactic poly(propylene oxide) compared to similar systems,[40–42] and it is unclear whether the selective insertion of the *S*-enantiomer was contributing to the formation

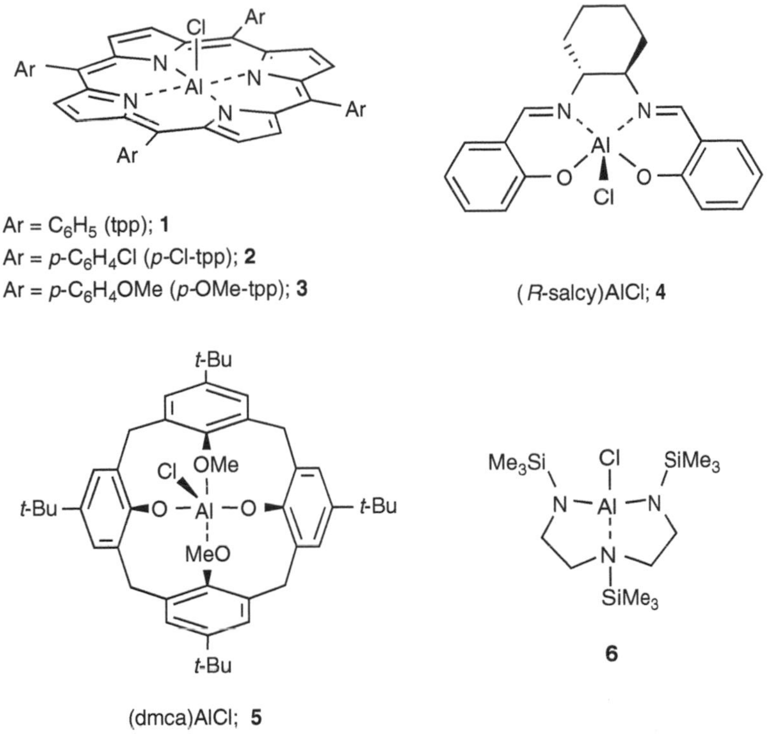

Ar = C_6H_5 (tpp); **1**
Ar = p-C_6H_4Cl (p-Cl-tpp); **2**
Ar = p-C_6H_4OMe (p-OMe-tpp); **3**

(R-salcy)AlCl; **4**

(dmca)AlCl; **5**

6

FIGURE 24.1 Well-defined complexes for the polymerization of epoxides (tpp = 5,10,15,20-tetraphenylporphyrin; salcy = N,N'-bis(2-hydroxybenzylidene)-(1R,2R)-1,2-cyclohexanediamine); dmca = dimethylcalixarene.

of the 10% of crystalline polymer as opposed to its arising from some other microstructural feature.

24.3.1.3 Aluminum–Porphyrin Catalysts

Inoue first reported that 5,10,15,20-tetraphenylporphyrin (tpp) aluminum chloride (Figure 24.1, **1**) was active for the living polymerization of propylene oxide.[43–45] Although the polymer microstructure was not studied in great detail, ^{13}C NMR spectra showed the polymers to be highly regioregular and slightly isotactic.[43,44] The activity of (tpp)AlCl was relatively low, requiring 6 days to achieve 100% conversion. The addition of Cl or OMe substituents on the porphyrin ligand, as in (p-Cl-tpp)AlCl (**2**) and (p-OMe-tpp)AlCl (**3**), increased the activity by a factor of two, but the stereochemical consequences for the resulting polymer were not discussed.[44] More detailed ^{13}C NMR analyses by Le Borgne et al. showed that the poly(propylene oxide) derived from **1** was slightly isotactic, with an m-dyad content of 69% and an mm-triad content of 45%, confirming the initial results reported by Inoue.[24]

24.3.1.4 Aluminum–Schiff-Base Catalysts

Well-defined [N,N'-bis(2-hydroxybenzylidene)-(1R,2R)-1,2-cyclohexane diamine] (R-salcy) aluminum complexes (e.g., Figure 24.1, **4**) have been used as stereoselective epoxide polymerization catalysts.[46–48] Polymerization of racemic propylene oxide in the presence of 5 mol% **4** yields ~70% conversion to poly(propylene oxide) after 62 h. The remaining unreacted monomer exhibits an optical rotation of +1.85°, which corresponds to an ee of 15% (Scheme 24.7). The modest r-factor

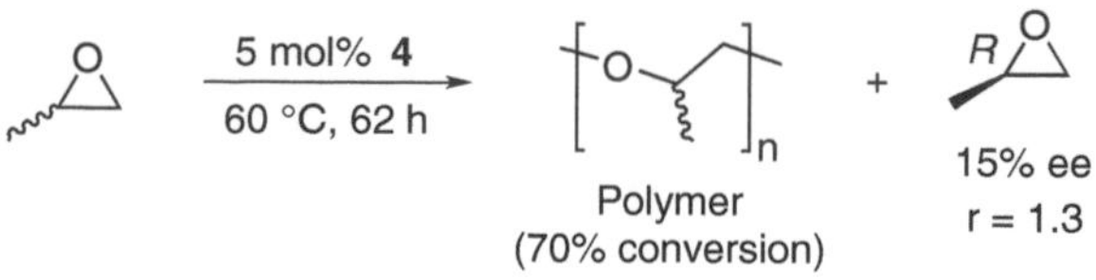

SCHEME 24.7   Attempted kinetic resolution of racemic propylene oxide using **4**.

of 1.3 obtained in this system is slightly higher than that observed for the previously discussed heterogeneous aluminum systems.[49] Examination of the isolated polymer reveals both chloro and hydroxyl end groups, suggesting that each metal center produces a single polymer chain since each chain bears a Cl atom from initiation and an OH group from termination.[47]

24.3.1.5 Aluminum–Calixarene Catalysts

Kuran et al. synthesized a dimethylcalixarene-based system (Figure 24.1, **5**) that was moderately active for propylene oxide polymerization.[50] The polymerization of propylene oxide produced predominantly isotactic poly(propylene oxide) with an m-dyad content of $\sim$74%.

24.3.1.6 Other Well-Defined Aluminum Catalysts

N,N',N''-Tris(trimethylsilyl)diethylenetriamine complexes of aluminum have been shown to be active oligomerization catalysts for propylene oxide.[51] Over the course of 2 days, **6** (Figure 24.1) produces low-molecular-weight poly(propylene oxide) ($M_n < 500$) with predominantly head-to-tail linkages and an m-dyad content of 60%.

24.3.2 Zinc-Based Catalysts

During research on aluminum catalysts for the stereospecific polymerization of epoxides, it was discovered that the addition of zinc cocatalysts to these systems greatly enhanced catalyst activity.[40] These enhancements prompted a number of studies focusing on the design of zinc-based catalyst systems.

24.3.2.1 Zinc Alkoxide Catalysts

Furukawa et al. explored the use of methanol and ethanol as additives for diethylzinc-based epoxide polymerization systems,[52] and found that both the yield and crystallinity of the resulting polymers were inferior to those for polymers synthesized with the $ZnEt_2/H_2O$ system. The use of achiral alcohols as cocatalysts was revisited in 1994 when Kuran and Listos reported the polymerization of propylene oxide and cyclohexene oxide (a *meso* molecule) with $ZnEt_2$/polyhydric phenol (such as 4-*tert*-butyl-catechol), phenol, or 1-phenoxy-2-propanol.[53] The poly(propylene oxide) formed from these systems contained mostly isotactic dyads (72% m), whereas the poly(cyclohexene oxide) contained mostly syndiotactic dyads (80% r) (Scheme 24.8).

24.3.2.2 Chiral Zinc Alkoxide Catalysts

Sigwalt and coworkers noted higher stereoselectivity in the polymerization of propylene sulfide using a (R)-3,3-dimethyl-1,2-butanediol/$ZnEt_2$ system when compared to a similar chiral alcohol/$ZnEt_2$ system. Based on these results, Sigwalt as well as others have applied this system to propylene oxide polymerization; however, the observed stereoselectivity was actually lower than that for the polymerization of propylene sulfide.[16,54,55] This lower stereoselectivity was presumably due to the

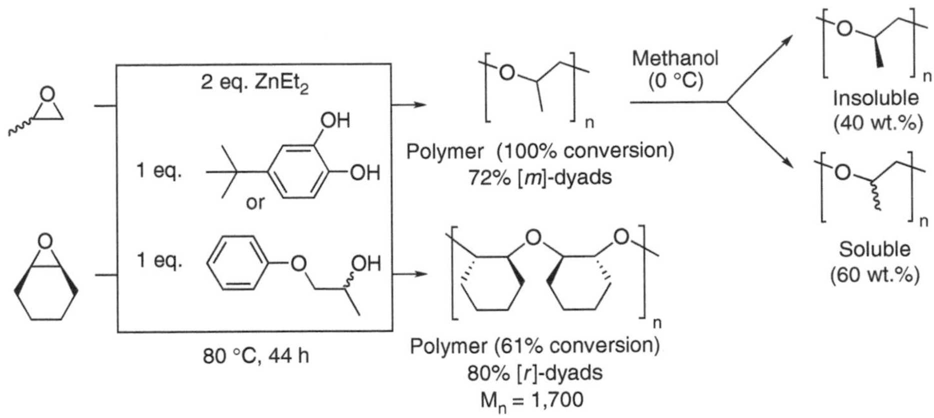

SCHEME 24.8 Polymerization of epoxides with ZnEt₂/1-phenoxy-2-propanol/4-*tert*-butylcatechol.

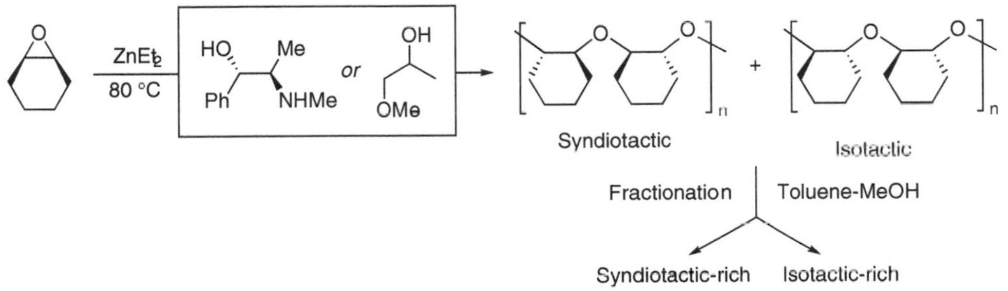

SCHEME 24.9 Preparation of a mixture of isotactic and syndiotactic poly(cyclohexene oxide) using chiral zinc alkoxide catalysts.

weaker coordination of the "harder" epoxide oxygen atom to zinc, as compared to the "softer" coordination of the episulfide sulfur atom.

Sepulchre et al. investigated the polymerization of cyclohexene oxide using ZnEt₂ activated with water, alcohols, and chiral alcohols. In their study, a mixture of ZnEt₂ and 1-methoxy-2-propanol or (1S,2R)-ephedrine simultaneously afforded a mixture of isotactic and syndiotactic poly(cyclohexene oxide) that was characterized using ^{1}H and ^{13}C NMR spectroscopy (Scheme 24.9).[56,57] They proposed a "flip-flop" mechanism (similar to that proposed by Vandenberg as shown in Scheme 24.5) involving neighboring zinc centers to explain this observation.[20]

24.3.2.3 Zinc Alkoxide Cluster Catalysts

Tsuruta and coworkers synthesized and investigated the epoxide polymerization activity of several well-defined zinc clusters (Figure 24.2).[58–67] Complexes [Zn(OMe)₂(EtZnOMe)₆] (**7**), [Zn(OCH₂CH₂OMe)₂(EtZnOCH₂CH₂OMe)₆] (**8**), and [{CH₃OCH₂CH(Me)OZnOCH(Me)CH₂-OCH₃}₂{EtZnOCH(Me)CH₂OCH₃}₂] ([Zn-MP]₂,₂) (**9**), were synthesized by the dropwise addition of 1.1 equivalents of the corresponding alkoxyalcohols to ZnEt₂ in heptanes at 5 °C, followed by heating the resultant solution for 1 h at 50–80 °C and then cooling it to 5 °C. Upon cooling, crystals suitable for X-ray analysis were isolated. Each complex was crystalline, and its molecular structure was determined using X-ray crystallography.[58,61,67]

The propylene oxide polymerization activity for each complex is shown in Scheme 24.10. Surprisingly, isostructural complexes **7** and **8** had significantly different polymerization activities; **7** achieved

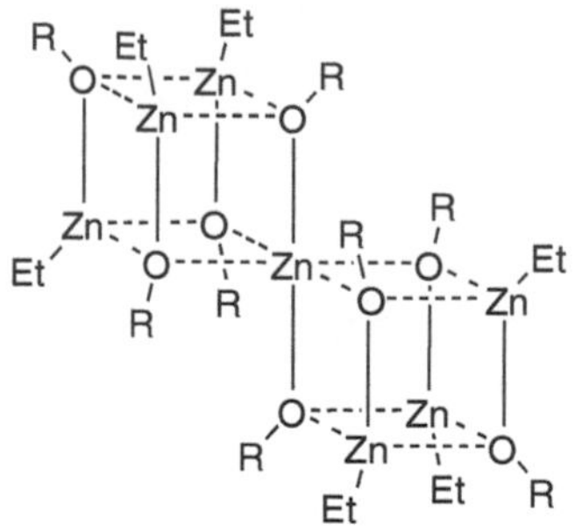

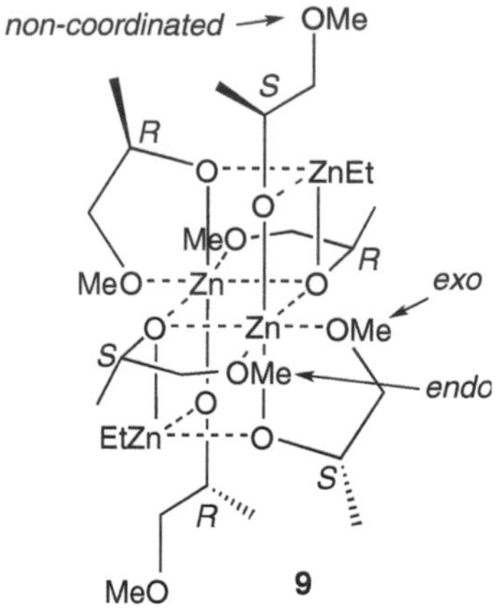

FIGURE 24.2   Structures of zinc cluster catalysts **7**, **8**, and **9**.

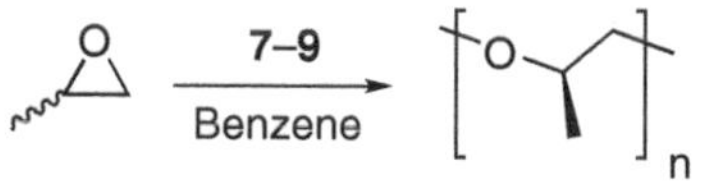

Catalyst	Temp. (°C)	Time (h)	Yield (%)	Dyads ([m]:[r])
7	80	216	91	63:37
8	80	240	22	59:41
9	80	100	93	79:21
9	35	240	9	81:19

SCHEME 24.10   Polymerization of racemic propylene oxide with zinc alkoxide cluster catalysts **7–9**.

91% conversion in 216 h, but **8** only attained 22% conversion in 240 h. Catalyst **9**, however, was twice as active as **7**. This catalyst has a different molecular structure, as seen in Figure 24.2. This complex bears six methoxy isopropyl groups in a chair-like structure in three different coordination environments; the methoxy groups are either endo- and exo-coordinated to the central zinc atoms through dative bonds, or non-coordinated.

Studies using a deuterated version of catalyst **9** revealed that the non-coordinating methoxyisopropyl groups initiated the polymerization by attack on the propylene oxide monomer whereas the coordinated methoxyisopropyl groups provided the chiral structure, which remained unchanged during the polymerization.[62] When these complexes were screened for cyclohexene oxide polymerization, only **9** was found to be active.[68,69] Through ^{1}H NMR analysis of polymer decomposition products (using Vandenberg's method[33]), Tsuruta and coworkers determined that the poly(cyclohexene oxide) obtained was syndiotactic (Scheme 24.11).

24.3.2.4 Zinc Porphyrin–Based Catalysts

Inoue and Takeda reported that the polymerization of propylene oxide at 20 °C with the zinc porphyrin catalyst (Et$_2$Zn/N-methyl-5,10,15,20-tetraphenylporphyrin, **10**) produced syndiotactic poly(propylene oxide) ($M_w = 31,000$, 60% r) (Scheme 24.12). This result was in contrast to those seen for all other zinc-based systems, which afford isotactic poly(propylene oxide). The authors attributed the unexpected syndiotactic microstructure of the polymer to the planar ligand and the isolated nature of the zinc center, which is different than that present in most other zinc aggregate systems.[43]

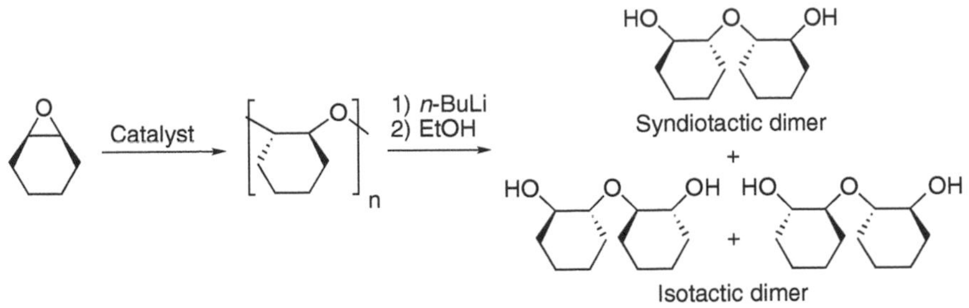

SCHEME 24.11 Syndioselective polymerization of cyclohexene oxide with **9** and subsequent degradation.

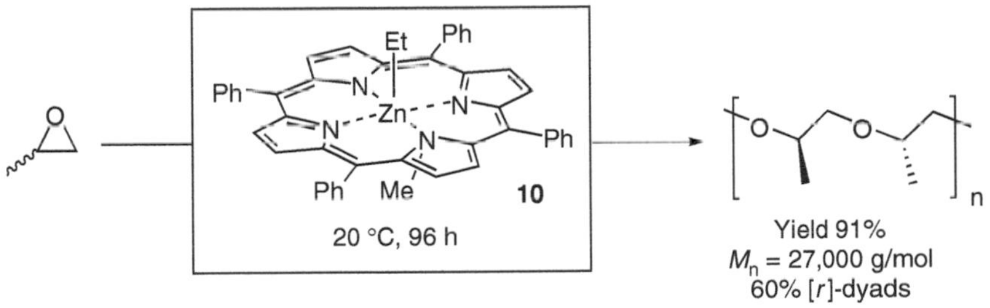

Catalyst	Syndiotactic dimer (%)	Isotactic dimer (%)
ZnEt$_2$	66	34
(EtZnOMe)$_4$	66	34
9	81	19

SCHEME 24.12 Syndiotactic poly(propylene oxide) prepared with **10**.

24.3.2.5 Zinc Catalysts for Asymmetric Cyclohexene Oxide/CO$_2$ Copolymerization

There is significant interest in controlling the absolute stereochemistry of ring-opening in epoxide/CO$_2$ copolymerization. Cyclohexene oxide, a *meso* molecule, is an ideal substrate for desymmetrization using chiral catalysts. In 1999, Nozaki et al. reported that a 1:1 mixture of ZnEt$_2$ and (S)-α,α-diphenylpyrrolidine-2-yl-methanol (**11**) (Scheme 24.13) was active for stereoselective cyclohexene oxide/CO$_2$ copolymerization at 40 °C and 30 atm CO$_2$ (Scheme 24.14).[70] The resultant polycarbonate contained 100% carbonate linkages, had an M_n of 8400 g/mol, and had a M_w/M_n of 2.2. Hydrolysis of this poly(cyclohexene carbonate) with base produced the corresponding *trans*-cyclohexane-1,2-diol with 73% ee. ^{13}C NMR spectroscopy studies of model polycarbonate oligomers afforded spectral assignments for the isotactic (153.7 ppm) and syndiotactic dyads (153.3–153.1 ppm) of poly(cyclohexene oxide),[71] which agreed with those proposed by Coates and coworkers.[72] Finally, the ring-opening polymerization proceeded via complete inversion of configuration (S$_N$2 mechanism); hence, no *cis*-cyclohexane-1,2-diol was observed upon base-catalyzed hydrolysis of the polycarbonate.

In a recent report, Nozaki and coworkers isolated presumed intermediates in the asymmetric alternating copolymerization of cyclohexene oxide with CO$_2$.[73] Reaction of a 1:1 mixture of ZnEt$_2$ and (S)-α, α-diphenylpyrrolidine-2-yl-methanol (**11**, Scheme 24.13) yielded dimeric **12**, which was structurally characterized by X-ray diffraction studies. At 40 °C and 30 atm CO$_2$, **12** catalyzed

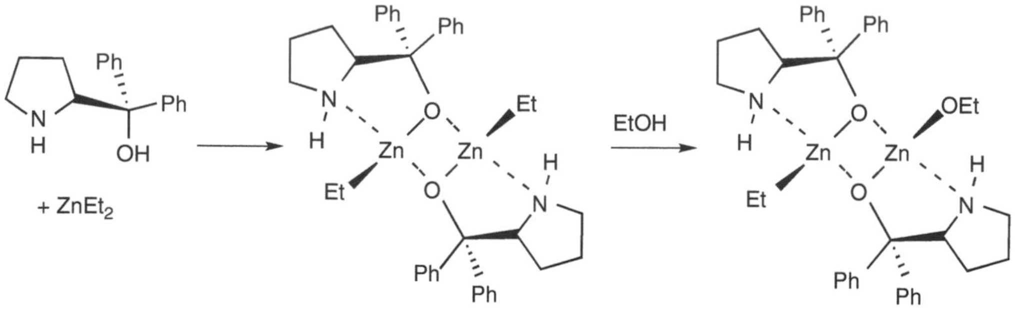

SCHEME 24.13 Chiral zinc catalysts for the asymmetric, alternating copolymerization of cyclohexene oxide and CO_2.

Isotactic
Poly(cyclohexene carbonate)

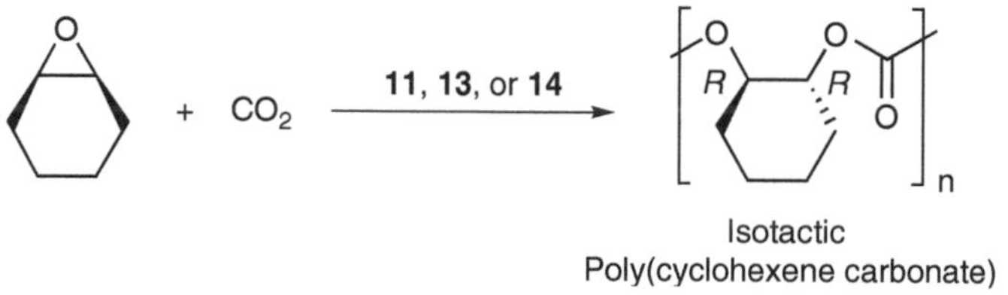

SCHEME 24.14 Enantioselective copolymerization of cyclohexene oxide and CO_2 with **11**, **13**, or **14**.

the formation of isotactic poly(cyclohexene carbonate) ($M_n = 11,800$ g/mol, $M_w/M_n = 15.7$) with a turnover frequency of 0.6 h^{-1}). Hydrolysis of the resulting poly(cyclohexene carbonate) yielded the *trans*-cyclohexane-1,2-diol of 49% ee, which was lower than that seen with catalyst **11**. When copolymerization was attempted using a catalyst system consisting of **12** and 0.2–1.0 equivalents EtOH (**12**/EtOH), enantioselectivities of the hydrolyzed cyclohexane diol increased up to 80% ee, and better control of polymer molecular weights and molecular weight distributions resulted as compared to polymerization using only **12**. Compound **13** (Scheme 24.13) was proposed to be the active initiating species in this polymerization. End-group analysis of the poly(cyclohexene carbonate)s prepared with **12** and **12**/0.2 EtOH by matrix-assisted laser desorption/ionization time-of-flight mass spectrometry revealed that in the absence of ethanol or in the presence of 0.2 equivalents ethanol, end-group signals assignable to an aminoalcohol-initiated polymerization were identified. However, as EtOH addition was increased from 0.2 to 1.0 equivalent, signals corresponding to the aminoalcohol-initiated polycarbonate disappeared as peaks corresponding to end-group structures for EtOH-initiated poly(cyclohexene carbonate) emerged. This result was further confirmed by end-group analysis using ^{1}H NMR spectroscopy. Finally, mechanistic studies suggested that the dimeric form of the catalyst, **13**, was in fact the active species.

In 2000, Coates and coworkers developed C_1-symmetric imine-oxazoline-ligated zinc bis(trimethylsilyl)amido compounds (Figure 24.3, **14**) for the stereoselective, alternating copolymerization of cyclohexene oxide and CO_2 (Scheme 24.14).[72] Through multiple electronic and steric manipulations of the imine-oxazoline ligand framework, compound **14** was found to exhibit the highest enantioselectivity for polymerization (*RR:SS* ratio in polymer = 86:14; 72% ee). Poly(cyclohexene carbonate) prepared with this catalyst possessed 100% carbonate linkages, an M_n of 14,700 g/mol, a M_w/M_n of 1.35, a glass transition temperature (T_g) of 120 °C, and a melting temperature (T_m) of 220 °C. Furthermore, stereocontrol was also achieved in the alternating copolymerization of cyclopentene oxide and CO_2, producing poly(cyclopentene carbonate) with an *RR:SS* ratio of 88:12 (76% ee). As revealed by ^{13}C NMR spectroscopy, the experimental carbonyl tetrad

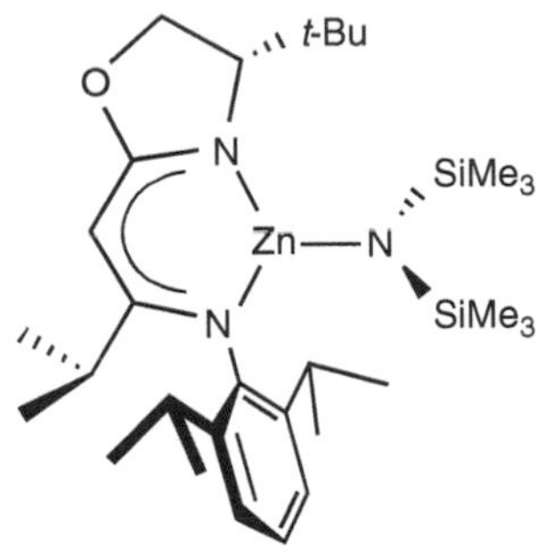

14

FIGURE 24.3 Chiral zinc catalyst **14** for the asymmetric, alternating copolymerization of cycloalkene oxides and CO_2.

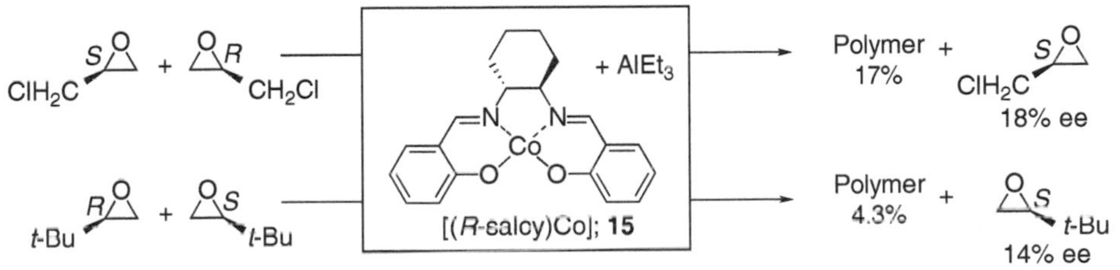

SCHEME 24.15 Kinetic resolution of racemic *tert*-butyl ethylene oxide and epichlorohydrin using **15**/AlEt$_3$.

concentrations of this material matched the predicted tetrad concentrations for an enantiomorphic-site control mechanism.[72]

24.3.3 COBALT–BASED CATALYSTS

Tsuruta found that the optically pure complex [(R-salcy)Co] (**15**) was active for epoxide polymerization (Scheme 24.15) when activated with AlEt$_3$. Although the system exhibited no enantioselectivity for the polymerization of propylene oxide, it was moderately selective (r = ~1.5) for the kinetic resolutions of *tert*-butyl ethylene oxide and epichlorohydrin (Scheme 24.15).

Coates and coworkers reported that [(R-salcy-t-Bu)CoOAc] (**16**) copolymerizes propylene oxide and CO_2 (Scheme 24.16).[74] Novel features of the catalyst are high regioregularity and alternation, coupled with high selectivity for polycarbonate formation (propylene carbonate is not formed). (S)-Propylene oxide is consumed faster than (R)-propylene oxide with a modest r-factor of 2.8. Given the same absolute monomer configuration and similar r-factor observed by Jacobsen and coworkers for the cobalt-catalyzed ring-opening of aliphatic epoxides with benzoic acid,[38] a related mechanism was proposed to occur for polymerization with [(R-salcy-t-Bu)CoOAc], giving an active alkoxide active site with the regioregular structure shown in Scheme 24.16. Lu and Wang found that the addition of quaternary ammonium salts to this system increased the r-factor to 3.5.[75] More recently, Coates and Cohen have reported that a combination of complex **17** and [Ph$_3$P=N=PPh$_3$]Cl exhibits an r-factor of 9.7 for the copolymerization of propylene oxide and CO_2 at –20 °C (Scheme 24.16).[76]

Coates and coworkers recently reported the first syndioselective copolymerization of cyclohexene oxide and CO_2 (Scheme 24.17).[77] Using complex [rac-(salpr-t-Bu)CoBr] (**18**), poly(cyclohexene carbonate) was formed with 80% [r]-centered tetrads, as determined by ^{13}C NMR spectroscopy.

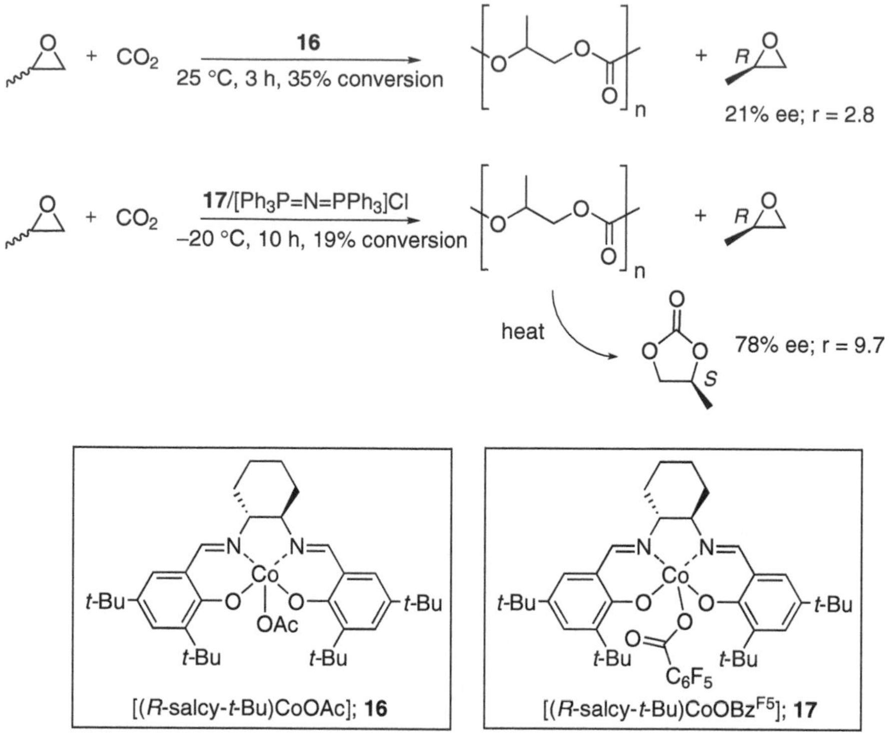

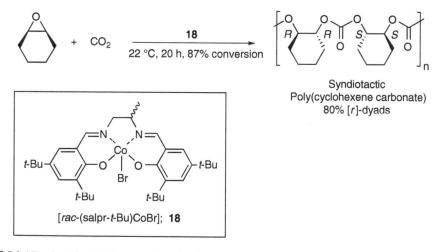

SCHEME 24.16   Kinetic resolution of racemic propylene oxide using **16** and **17**/[Ph$_3$P=N=PPh$_3$]Cl (B$_z$ = C(= O)C$_6$F$_5$.)

SCHEME 24.17   Syndioselective copolymerization of cyclohexene oxide and CO$_2$ with **18**.

The carbonyl and methylene regions were best simulated using Bernoullian statistical methods, supporting a chain-end stereochemical control mechanism.

In 2005, Coates and coworkers reported the first highly active and selective catalyst system for the isoselective polymerization of racemic propylene oxide (Scheme 24.18).[78] The complex [(salph-*t*-Bu)CoOAc] (**19**) exhibits excellent activity, regioselectivity, and stereoselectivity for the formation of isotactic poly(propylene oxide). This is the first example of isotactic poly(propylene oxide) generation from racemic propylene oxide without concomitant atactic byproduct. Notably,

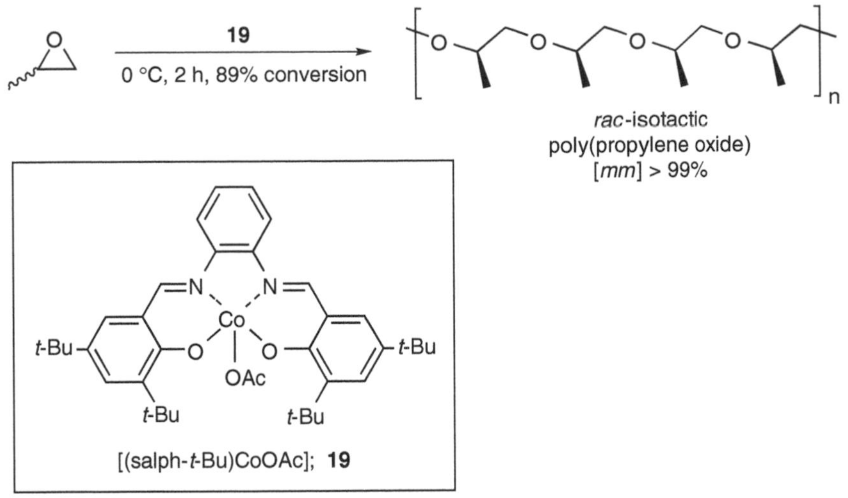

SCHEME 24.18   Isoselective polymerization of racemic propylene oxide with **19**.

the complex itself is not chiral, suggesting the active catalyst adopts a chiral arrangement different than that depicted in Scheme 24.18. Complex **19** also converts racemic butene and hexene oxides to their respective isotactic polyethers.

24.4 CONCLUSIONS

Although significant advances in stereoselective epoxide polymerization have been achieved over the last half-century, few known catalysts are capable of excellent levels of stereocontrol. Historically, most catalysts for epoxide polymerization have been of the heterogeneous variety and have exhibited poor selectivity. It is our opinion that the most fertile area for future catalyst exploration involves homogeneous, discrete catalysts that are capable of involving multiple metal centers in the polymerization mechanism. If the spatial environment of the active catalyst is precisely controlled, new generations of stereoselective epoxide polymerization catalysts will become available. Our current research focuses on the search for such catalysts.

REFERENCES AND NOTES

1. Pruitt, M. E.; Baggett, J. M. Catalysts for the polymerization of olefin oxides. U.S. Patent 2,706,181 (Dow Chemical Co.), 1955.
2. Pruitt, M. E.; Baggett, J. M.; Bloomfield, R. J.; Templeton, J. H. Catalysts for the polymerization of olefin oxides. U.S. Patent 2,706,182 (Dow Chemical Co.), 1955.
3. Pruitt, M. E.; Baggett, J. M. Solid polymers of propylene oxide. U.S. Patent 2,706,189 (Dow Chemical Co.), 1955.
4. Booth, C.; Jones, M. N.; Powell, E. Fractionation of partially isotactic poly(propylene oxide). *Nature* **1962**, *196*, 772–773.
5. Natta, G.; Corradini, P.; Dall'Asta, G. Structure of the crystalline poly(propylene oxide) chain. *Atti acad. nazl. Lincei Rend., Classe sci., fis., mat. e nat.* **1956**, *20*, 408–413.
6. Price, C. C.; Osgan, M. Polymerization of l-propylene oxide. *J. Am. Chem. Soc.* **1956**, *78*, 4787–4792.
7. Price, C. C.; Osgan, M.; Hughes, R. E.; Shambelan, C. The polymerization of l-propylene oxide. *J. Am. Chem. Soc.* **1956**, *78*, 690–691.
8. Vandenberg, E. J. Award address: Reflections on 50 years of polymer chemistry. *Polymer* **1994**, *35*, 4933–4939.

9. Spassky, N.; Dumas, P.; Le Borgne, A.; Momtaz, A.; Sepulchre, M. Asymmetric ring-opening polymerization. *Bull. Soc. Chim. Fr.* **1994**, *131*, 504–514.

10. Spassky, N.; Momtaz, A.; Kassamaly, A.; Sepulchre, M. Asymmetric synthesis polymerization of *meso* oxiranes and thiiranes. *Chirality* **1992**, *4*, 295–299.

11. Spassky, N. Stereospecific and anionic ring-opening polymerization. *Makromol. Chem., Macromol. Symp.* **1991**, *42/43*, 15–49.

12. Vandenberg, E. J. Coordination Polymerization. In *Coordination Polymerization.* Vandenberg, E. J., Price, C. C., Eds.; Plenum Publishing Corp.: New York, 1983; pp 11–44.

13. Tsuruta, T. Structure-reactivity relationship of catalysts for ring-opening polymerization of some oxiranes. *Pure Appl. Chem.* **1981**, *53*, 1745–1751.

14. Spassky, N.; Leborgne, A.; Sepulchre, M. Stereochemistry of the polymerization of three-membered and four-membered ring monomers using chiral initiators. *Pure Appl. Chem.* **1981**, *53*, 1735–1744.

15. Spassky, N. Stereoselective and stereoelective polymerization of oxiranes and thiiranes. *ACS Symposium Series* **1977**, *59*, 191–209.

16. Sigwalt, P. Stereoelection and stereoselection in the ring-opening polymerization of epoxides and episulfides. *Pure Appl. Chem.* **1976**, *48*, 257–266.

17. Price, C. C. Polyethers. *Acc. Chem. Res.* **1974**, *7*, 294–301.

18. Tsuruta, T. Stereoselective and asymmetric-selective (or stereoelective) polymerizations. *J. Poly. Sci., Part D: Macromol. Rev.* **1972**, *6*, 179–250.

19. Duda, A. Stereocontrolled polymerization of chiral heterocyclic monomers. *Polimery* **2004**, *49*, 469–478.

20. Vandenberg, E. J. Epoxide polymers: Synthesis, stereochemistry, structure, and mechanism. *J. Poly. Sci., Poly. Chem. Ed.* **1969**, *7*, 525–567.

21. A comprehensive review on the field of stereoselective epoxide polymerization is currently being prepared (Ajiro, H; Allen, S. D.; Coates, G. W., manuscript in preparation).

22. Farina, M. The stereochemistry of linear macromolecules. *Top. Stereochem.* **1987**, *17*, 1–111.

23. Schilling, F. C.; Tonelli, A. E. Carbon-13 NMR determination of poly(propylene oxide) microstructure. *Macromolecules* **1986**, *19*, 1337–1343.

24. Le Borgne, A.; Spassky, N.; Jun, C. L.; Momtaz, A. Carbon-13 NMR study of the tacticity of poly(propylene oxide)s prepared by polymerization with $\alpha, \beta, \gamma, \delta$-tetraphenylporphyrin/AlEt$_2$Cl as initiator system: An example of first-order Markovian statistics in ring-opening polymerization. *Makromol. Chem.* **1988**, *189*, 637–650.

25. Ugur, N.; Alyuruk, K. Stereoregularity of fractionally crystallized poly(propylene oxide) samples by carbon-13 NMR spectroscopy. *J. Poly. Sci., Part A: Polym. Chem.* **1989**, *27*, 1749–1761.

26. Cheng, H. N., NMR Characterization of Polymers. In *Modern Methods of Polymer Characterization*, Barth, H. G., Mays, J. W., Eds.; John Wiley & Sons: New York, 1991; pp 409–493.

27. Bovey, F. A.; Mirau, P. A. *NMR of Polymers.* Academic Press: San Diego, 1996.

28. Chisholm, M. H.; Navarro-Llobet, D. NMR assignments of regioregular poly(propylene oxide) at the triad and tetrad level. *Macromolecules* **2002**, *35*, 2389–2392.

29. Wu, B.; Harlan, C. J.; Lenz, R. W.; Barron, A. R. Stereoregular polymerization of (R,S)-propylene oxide by an aluminoxane-propylene oxide complex. *Macromolecules* **1997**, *30*, 316–318.

30. Vandenberg, E. J. Organometallic catalysts for polymerizing monosubstituted epoxides. *J. Poly. Sci.* **1960**, *47*, 486–489.

31. Vandenberg, E. J. Catalysts for polymerization of epoxides. U.S. Patent 3,135,705 (Hercules Powder Co.), 1964.

32. Vandenberg, E. J. Catalysts for polymerization of epoxides. U.S. Patent 3,219,591 (Hercules Powder Co.), 1965.

33. Vandenberg, E. J. Mechanism aspects of epoxide polymerization. Stereochemical structure of the crystalline polymer from 2,3-epoxybutanes. *J. Poly. Sci., Part B: Poly. Lett.* **1964**, *2*, 1085–1088.

34. Vandenberg, E. J. Crystalline polymers of the 2,3-epoxybutanes-structure and mechanism aspects. *J. Am. Chem. Soc.* **1961**, *83*, 3538–3539.

35. Vandenberg, E. J. High polymers from symmetrical disubstituted epoxides. *J. Poly. Sci.* **1960**, *47*, 489–491.

36. Braune, W.; Okuda, J. An efficient method for controlled propylene oxide polymerization: The significance of bimetallic activation in aluminum Lewis acids. *Angew. Chem., Int. Ed.* **2003**, *42*, 64–68.

37. Moore, D. R.; Cheng, M.; Lobkovsky, E. B.; Coates, G. W. Mechanism of the alternating copolymerization of epoxides and CO_2 using beta-diiminate zinc catalysts: Evidence for a bimetallic epoxide enchainment. *J. Am. Chem. Soc.* **2003**, *125*, 11911–11924.

38. Tokunaga, M.; Larrow, J. F.; Kakiuchi, F.; Jacobsen, E. N. Asymmetric catalysis with water: Efficient kinetic resolution of terminal epoxides by means of catalytic hydrolysis. *Science* **1997**, *277*, 936–938.

39. Haubenstock, H.; Panchalingam, V.; Odian, G. Stereoelective polymerization of propylene oxide with a chiral aluminum alkoxide initiator. *Makromol. Chem.* **1987**, *188*, 2789–2799.

40. Osgan, M.; Price, C. C. New catalysts for preparation of isotactic polypropylene oxide. *J. Poly. Sci.* **1959**, *34*, 153–156.

41. Kasperczyk, J.; Dworak, A.; Jedlinski, Z. Polymerization of propylene oxide using chiral, aluminum-containing initiators. *Makromol. Chem., Rapid Commun.* **1981**, *2*, 663–666.

42. Dworak, A.; Jedlinski, Z. Investigation of the structure of poly(p-chlorophenyl glycidyl ether) by the carbon-13 NMR technique: Tacticity and addition isomerism. *Polymer* **1980**, *21*, 93–96.

43. Takeda, N.; Inoue, S. Polymerization of 1,2-epoxypropane and copolymerization with carbon dioxide catalyzed by metalloporphyrins. *Makromol. Chem.* **1978**, *179*, 1377–1381.

44. Aida, T.; Inoue, S. Living polymerization of epoxide catalyzed by the porphyrin-chlorodiethylaluminum system. Structure of the living end. *Macromolecules* **1981**, *14*, 1166–1169.

45. Aida, T.; Inoue, S. Living polymerization of epoxides with metalloporphyrin and synthesis of block copolymers with controlled chain lengths. *Macromolecules* **1981**, *14*, 1162–1166.

46. Vincens, V.; Le Borgne, A.; Spassky, N. Stereoelective oligomerization of methyloxirane with a chiral aluminum complex of a Schiff's base as initiator. *Makromol. Chem., Rapid Commun.* **1989**, *10*, 623–628.

47. Vincens, V.; Le Borgne, A.; Spassky, N. Aluminum complex of a Schiff base as new initiator for oligomerization of heterocycles. *Makromol. Chem., Macromol. Symp.* **1991**, *47*, 285–291.

48. Le Borgne, A.; Vincens, V.; Jouglard, M.; Spassky, N. Ring-opening oligomerization reactions using aluminum complexes of Schiff's bases as initiators. *Makromol. Chem., Macromol. Symp.* **1993**, *73*, 37–46.

49. Matsuura, K.; Inoue, S.; Tsuruta, T. Asymmetric-selective polymerization of DL-propylene oxide with triethylaluminum-*N*-carboxy-L(+)-alanine anhydride system. *Makromol. Chem.* **1965**, *86*, 316–319.

50. Kuran, W.; Listos, T.; Abramczyk, M.; Dawidek, A. Epoxide polymerization and copolymerization with carbon dioxide using diethylaluminum chloride-25,27-dimethoxy-26,28-dihydroxy-*p-tert*-butylcalix[4]arene system as a new homogeneous catalyst. *J. Macromol. Sci., Pure Appl. Chem.* **1998**, A35, 427–437.

51. Emig, N.; Nguyen, H.; Krautscheid, H.; Reau, R.; Cazaux, J.-B.; Bertrand, G. Neutral and cationic tetracoordinated aluminum complexes featuring tridentate nitrogen donors: Synthesis, structure, and catalytic activity for the ring-opening polymerization of propylene oxide and (D,L)-lactide. *Organometallics* **1998**, *17*, 3599–3608.

52. Furukawa, J.; Tsuruta, T.; Sakata, R.; Saigusa, T.; Kawasaki, A. Polymerization of propylene oxide by diethylzinc in the presence of cocatalysts. *Makromol. Chem.* **1959**, *32*, 90–94.

53. Kuran, W.; Listos, T. Polymerization of 1,2-epoxypropane and 1,2-epoxycyclohexane by diethylzinc-polyhydric phenol and/or phenol or 1-phenoxy-2-propanol as catalysts. *Macromol. Chem. Phys.* **1994**, *195*, 401–411.

54. Coulon, C.; Spassky, N.; Sigwalt, P. Stereoelective polymerization of racemic propylene oxide using a diethylzinc-chiral-1,2-diol system as initiator. *Polymer* **1976**, *17*, 821–827.

55. Kassamaly, A.; Sepulchre, M.; Spassky, N. Enantioasymmetric polymerization of racemic styrene oxide. *Polym. Bull.* **1988**, *19*, 119–122.

56. Sepulchre, M.; Kassamaly, A.; Spassky, N. Stereospecific polymerization of cyclohexene oxide. *Poly. Prepr. (Am. Chem. Soc., Div. Polym. Chem.)* **1990**, *31*, 91–92.

57. Sepulchre, M.; Kassamaly, A.; Spassky, N. Stereospecific polymerization of cyclohexene oxide. *Makromol. Chem., Macromol. Symp.* **1991**, *42/43*, 489–500.

58. Ishimori, M.; Hagiwara, T.; Tsuruta, T.; Kai, Y.; Yasuoka, N.; Kasai, N. The structure and reactivity of $[Zn(OMe)_2(EtZnOMe)_6]$. *Bull. Chem. Soc. Jpn.* **1976**, *49*, 1165–1166.

59. Tsuruta, T. J. Mechanism of stereoselective polymerization of oxirane. *Poly. Sci., Poly. Symp.* **1980**, *67*, 73–82.
60. Tsuruta, T. Stereoselective and stereoelective polymerizations of some oxiranes. *Macromol. Chem. Phys., Suppl.* **1981**, *5*, 230–233.
61. Kageyama, H.; Miki, K.; Tanaka, N.; Kasai, N.; Ishimori, M.; Heki, T.; Tsuruta, T. Molecular structure of [Zn(OCH₂CH₂OMe)₂(EtZnOCH₂CH₂OMe)₆]. An enantiomorphic catalyst for the stereoselective polymerization of methyloxirane. *Makromol. Chem., Rapid Commun.* **1982**, *3*, 947–951.
62. Hasebe, Y.; Tsuruta, T. Mechanism of stereoselective polymerization of propylene oxide with [{MeOCH₂CH(Me)OZnOCH(Me)CH₂OMe}₂{EtZnOCH(Me)CH₂OMe}₂] as initiator. *Makromol. Chem.* **1988**, *189*, 1915–1926.
63. Yoshino, N.; Suzuki, C.; Kobayashi, H.; Tsuruta, T. Some features of a novel organozinc complex, [{MeOCH₂CH(Me)OZnOCH(Me)CH₂OMe}₂{EtZnOCH(Me)CH₂OMe}₂], as an enantiomorphic catalyst for stereoselective polymerization of propylene oxide. *Makromol. Chem.* **1988**, *189*, 1903–1913.
64. Tsuruta, T. Stereochemical behavior of oxirane and bioxirane in their polymerizations initiated with an organozinc complex having chair-type structure. *Makromol. Chem., Macromol. Symp.* **1991**, *47*, 277–283.
65. Tsuruta, T.; Hasebe, Y. Stereocontrol mechanism of an organozinc complex having chain-type structure in the polymerization of *tert*-butylethylene oxide; comparison with propylene oxide. *Macromol. Chem. Phys.* **1994**, *195*, 427–438.
66. Tsuruta, T. Molecular level approach to the origin of the steric control in organozinc catalyst for oxirane polymerization. *Makromol. Chem., Macromol. Symp.* **1986**, *6*, 23–31.
67. Kageyama, H.; Kai, Y.; Kasai, N.; Suzuki, C.; Yoshino, N.; Tsuruta, T. Molecular structure of [{MeOCH₂CH(Me)OZnOCH(Me)CH₂OMe}₂{EtZnOCH(Me)CH₂OMe}₂]. An enantiomorphic catalyst for the stereoselective polymerization of methyloxirane. *Makromol. Chem., Rapid Commun.* **1984**, *5*, 89–93.
68. Hasebe, Y.; Tsuruta, T. Structural studies of poly(1,2-cyclohexene oxide) prepared with well-defined organozinc compounds. *Makromol. Chem.* **1987**, *188*, 1403–1414.
69. Hasebe, Y.; Izumitani, K.; Torii, M.; Tsuruta, T. Studies on elementary reactions in the polymerization of 1,2-epoxycyclohexane with organozinc compounds as initiators. *Makromol. Chem.* **1990**, *191*, 107–119.
70. Nozaki, K.; Nakano, K.; Hiyama, T. Optically active polycarbonates: Asymmetric alternating copolymerization of cyclohexene oxide and carbon dioxide. *J. Am. Chem. Soc.* **1999**, *121*, 11008–11009.
71. Nakano, K.; Nozaki, K.; Hiyama, T. Spectral assignment of poly[cyclohexene oxide-*alt*-carbon dioxide]. *Macromolecules* **2001**, *34*, 6325–6332.
72. Cheng, M.; Darling, N. A.; Lobkovsky, E. B.; Coates, G. W. Enantiomerically-enriched organic reagents via polymer synthesis: Enantioselective copolymerization of cycloalkene oxides and CO₂ using homogeneous, zinc-based catalysts. *Chem. Commun.* **2000**, 2007–2008.
73. Nakano, K.; Nozaki, K.; Hiyama, T. Asymmetric alternating copolymerization of cyclohexene oxide and CO₂ with dimeric zinc complexes. *J. Am. Chem. Soc.* **2003**, *125*, 5501–5510.
74. Qin, Z. Q.; Thomas, C. M.; Lee, S.; Coates, G. W. Cobalt-based complexes for the copolymerization of propylene oxide and CO₂: Active and selective catalysts for polycarbonate synthesis. *Angew. Chem., Int. Ed.* **2003**, *42*, 5484–5487.
75. Lu, X. B.; Wang, Y. Highly active, binary catalyst systems for the alternating copolymerization of CO₂ and epoxides under mild conditions. *Angew. Chem., Int. Ed.* **2004**, *43*, 3574–3577.
76. Cohen, C. T.; Coates, G. W. Copolymerization of propyelene oxide and carbon dioxide with highly efficient and selective (salen)Co(III) catalysts: Effect of ligand and cocatalyst variation. *J. Poly. Sci., Part A* **2006**, *44*, 5182–5191.
77. Cohen, C.; Thomas, C.; Peretti, K.; Lobkovsky, E.; Coates, G. Copolymerization of cyclohexene oxide and carbon dioxide using (salen)Co(III) complexes: Synthesis and characterization of syndiotactic poly(cyclohexene carbonate). *Dalton Trans.* **2006**, 237–249.
78. Peretti, K.; Ajiro, H.; Cohen, C.; Lobkovsky, E.; Coates, G. A highly active, isospecific cobalt catalyst for propylene oxide polymerization. *J. Am. Chem. Soc.* **2005**, *127*, 11566–11567.

25 Stereoselective Polymerization of Lactide

Malcolm H. Chisholm and Zhiping Zhou

CONTENTS

25.1 INTRODUCTION

Poly(lactic acids) (PLAs) were first synthesized by heating lactic acid and driving off water (Scheme 25.1). The product formed by this process is a gummy, tar-like material of ill-defined molecular weight and microstructure. If this polymer is heated in the presence of zinc oxide, the volatile cyclic condensation product of two lactic acid units is distilled from the polymeric mass. This product is lactide (LA), 3,6-dimethyl-1,4-dioxane-2,5-dione.[1,2]

There are three isomers of lactide: L-, D-, and *meso* (Figure 25.1; L-LA, D-LA, and *meso*-LA). The 50:50 racemic mixtures of L and D is known as *rac*-lactide (*rac*-LA). Lactide, in particular L-LA, can also be prepared by an enzymatic process from biomass; this is currently a major commodity material produced by Cargill.[3–6] Ring-opening polymerization of lactides can be achieved by a wide variety of catalysts, many of which are metal-containing coordination complexes that are better considered as catalyst precursors.[7–10] Among the most common of these are tin(II) bis(2-ethylhexanoate) (Sn(Oct)$_2$), zinc lactate, and aluminum tris(acetylacetonate).[11–16] These catalysts are typically employed in high-temperature ($\sim$140 °C) melt polymerizations. Many more well-defined single-site metal alkoxide complexes are now known that can effect LA polymerization at room temperature. A number of these are discussed in detail in this chapter since they can bring about living polymerizations with control of both polymer molecular weight and stereochemistry. The reaction sequence can be described by Scheme 25.2.[17]

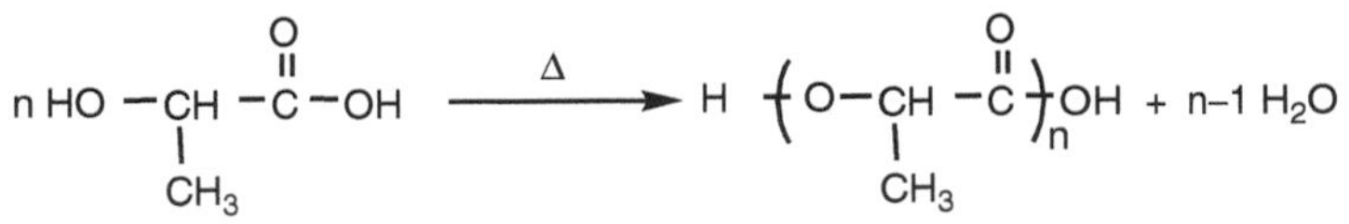

SCHEME 25.1 Poly(lactic acid) formed by condensation polymerization of lactic acids.

645

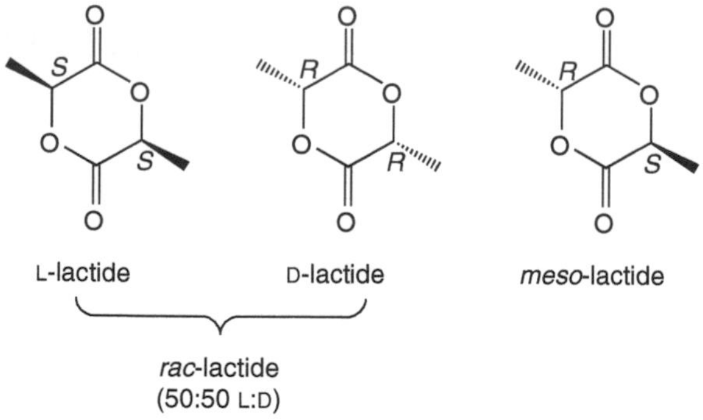

FIGURE 25.1   Stereoisomers of lactide.

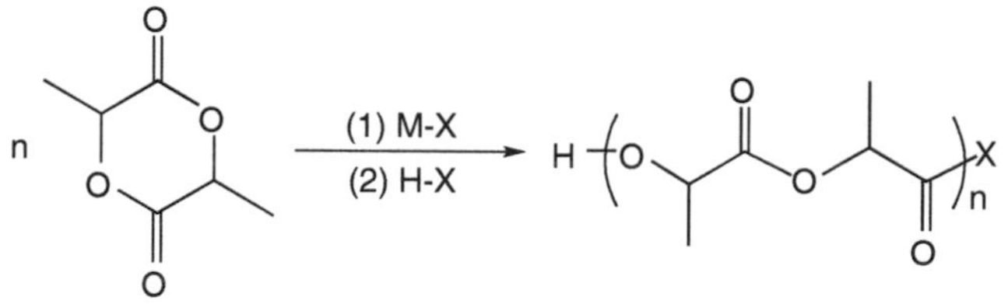

SCHEME 25.2   PLA formed by ring-opening polymerization of lactides (M = metal; X = OH or OR where R = alkyl or aryl).

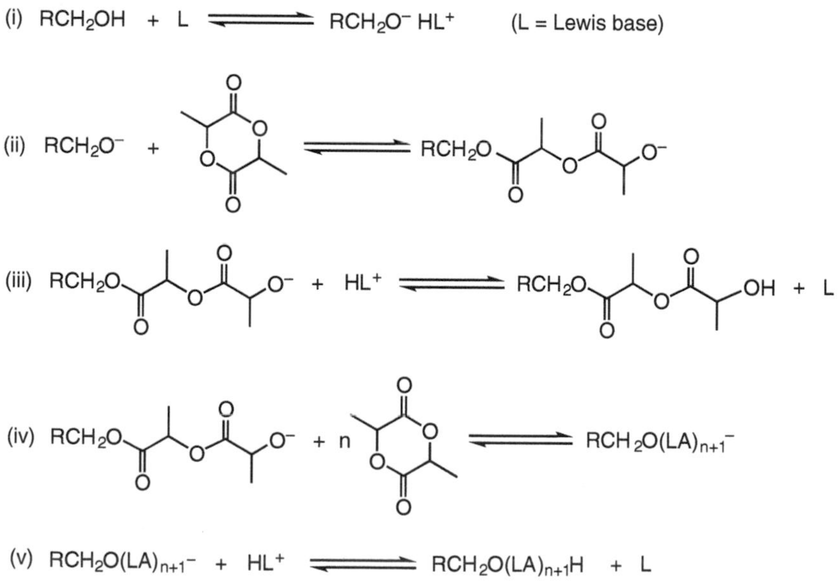

SCHEME 25.3   Ring-opening polymerization of LA by Lewis bases (LA = enchained lactide unit).

Ring-opening polymerization of lactides can also be brought about by strong Lewis bases such as 4-(dimethylamino)pyridine (DMAP), or a N-heterocyclic carbene in the presence of a primary alcohol initiator (Scheme 25.3).[18–23] These polymerizations are living, and polymer molecular weight can be controlled by the amount of the primary alcohol introduced.[8]

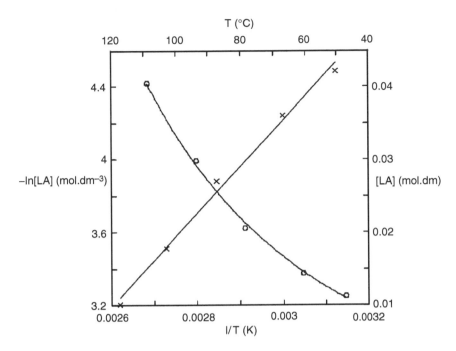

FIGURE 25.2 Dependence of the equilibrium concentration of LA monomer in ring-opening polymerization on temperature (crosses = ln[LA] versus 1/T; circles = [LA] versus 1/T). (From Chisholm, M. H.; Delbridge, E. E. *New J. Chem.* **2003**, *27*, 1177–1183. Reproduced by permission of The Royal Society of Chemistry on behalf of the Centre National de la Recherche Scientifique.)

A final general point of note is that ring-opening LA polymerization is an equilibrium process. Formation of the polymer is favored on enthalpic grounds ($\Delta H°$ for ring opening is approximately -23 kJ/mol), but the cyclic dione is entropically favored.[24] At low temperatures the equilibrium lies greatly in favor of the polymer, but at higher temperatures the equilibrium concentration of monomer becomes significant; that is, at 100 °C, ca. 15% LA monomer is present (Figure 25.2).[24,25] Depolymerization of the polymer occurs by intrachain transesterification (Scheme 25.4a), and this process must be avoided if stereocontrol is to be achieved. Intermolecular transesterification (Scheme 25.4b) also becomes significant at higher temperatures, leading to a loss of both stereoselectivity and the narrow molecular distributions characteristic of living polymerization.[25]

25.2 STEREOSEQUENCES

Polymerization of L-LA, which is the most readily available enantiomerically pure form of lactide, leads to the formation of poly(L-LA) providing that no epimerization of the monomer stereocenter occurs. Epimerization can be avoided by use of the majority of metal coordination catalysts (Scheme 25.2) and by using the Lewis base/alcohol process (Scheme 25.3). However, in the polymerization of *rac*-LA or *meso*-LA, stereorandom polymers are generally produced. Figure 25.3 shows polylactides with various tacticities. Epimerization occurs only with the most highly active of catalysts, such as organolithium and organomagnesium initiators.

Stereosequences in PLAs are routinely determined by the examination of nuclear magnetic resonance (NMR) spectra, particularly with regard to the ^{13}C NMR methine resonance and the proton decoupled ^{1}H NMR methine resonance. Figure 25.4 shows the possible PLA tetrad stereosequences (*i* = isotactic dyad; *s* = syndiotactic dyad). The original (^{13}C and ^{1}H) NMR peak assignments for tactic PLAs were put forward at the tetrad level of sensitivity by Kricheldorf et al. based

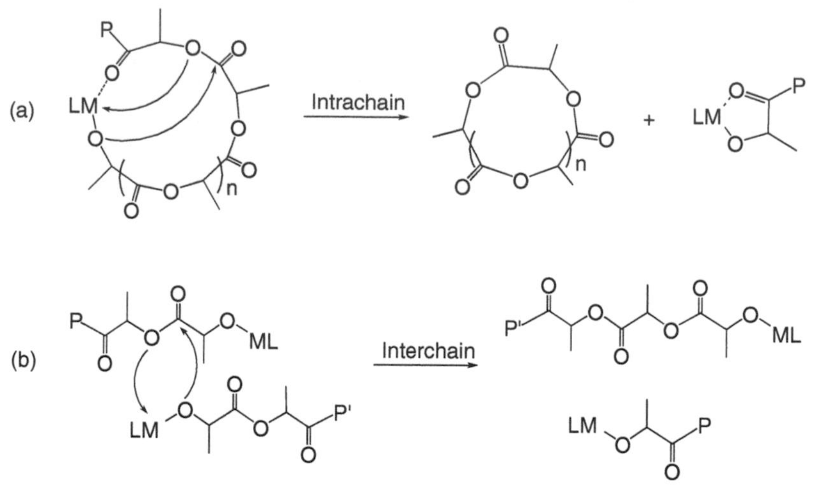

SCHEME 25.4 (a) Intrachain and (b) interchain transesterification reactions occurring in LA polymerization (P, P′ = polymer chain; M = metal; L = ligand).

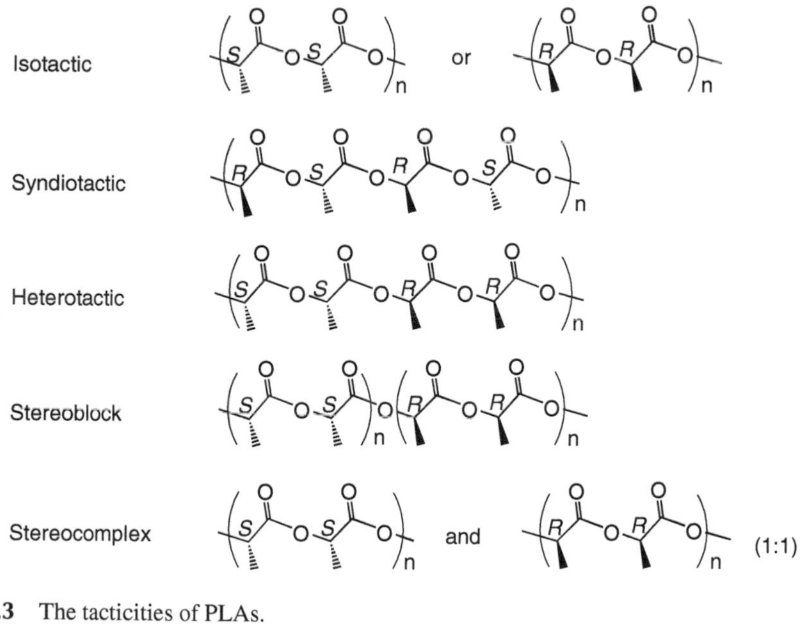

FIGURE 25.3 The tacticities of PLAs.

on an examination of PLAs formed from varying concentrations of L-LA, D-LA, and *meso*-LA monomers.[26,27] Chisholm et al. challenged these assignments based on heteronuclear correlation (HETCOR) spectra,[28,29] but this challenge was quickly refuted. Subsequently, more detailed studies revealed that the stereosequence of a PLA tetrad is sensitive in the following manner: The ^{1}H signal of each enchained LA methine moiety senses two stereocenters to the right and one to the left, whereas the methine ^{13}C signal is sensitive in the opposite manner (Figure 25.5).[30] At high fields (>500 MHz for ^{1}H), assignments can be extended to the hexad level, but it is generally sufficient and customary to employ the tetrad sensitivity when reporting stereoselectivities in LA polymerization reactions.[31–33]

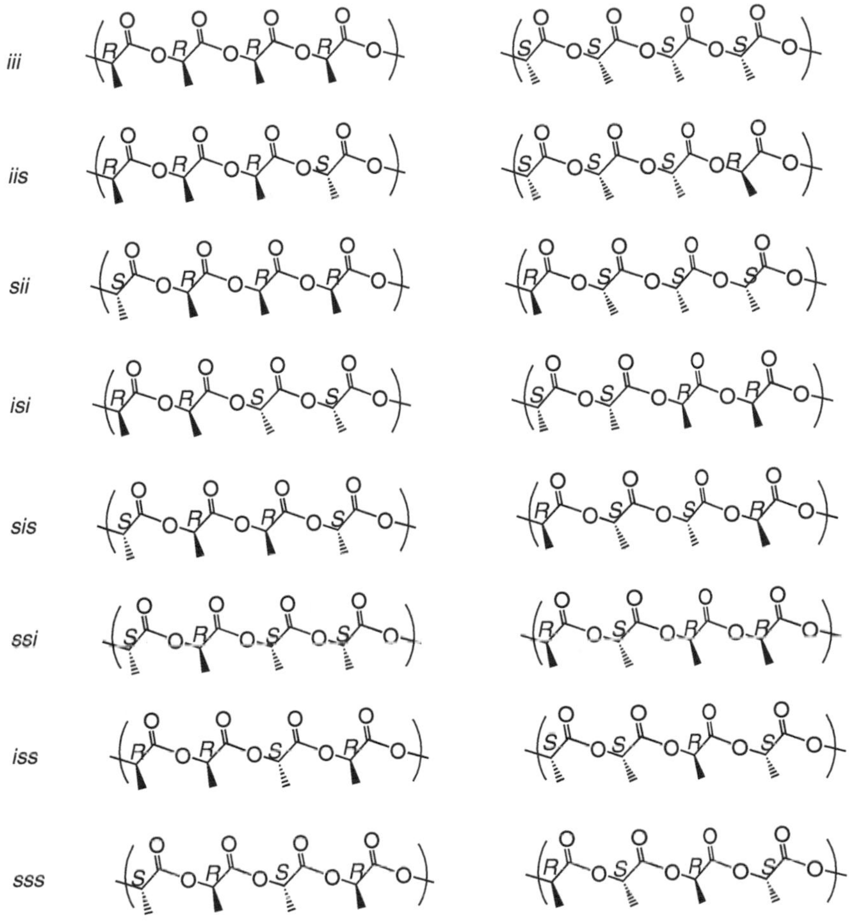

FIGURE 25.4 The possible stereosequences in PLA at the tetrad level (i = isotactic dyad; s = syndiotactic dyad).

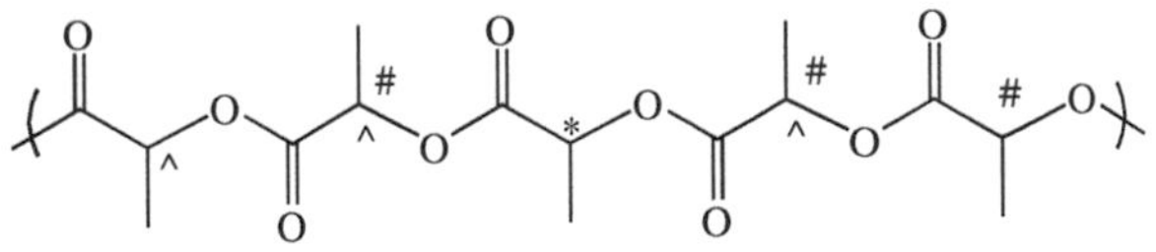

FIGURE 25.5 Tetrad sensitivity of PLA methine group in ^{1}H and ^{13}C NMR: each methine (represented by unit with *) is split in the ^{1}H by two neighboring units to the right and one to the left (units with #), and in the ^{13}C by one neighboring unit to the right and two to the left (units with ∧).

Polymerization of L-LA (or D-LA) leads to an isotactic polymer that is assigned as *iii* at the tetrad level. The stereorandom polymerization of *rac*-LA leads to five possible tetrads: *iii, iis, sii, isi*, and *sis*. Because each monomer unit in *rac*-LA contains an isotactic junction, there can be no *ss* junctions. Similarly, in the polymerization of *meso*-LA (which contains an *s* junction) there can be only five tetrads, namely, *sss, ssi, iss, sis*, and *isi*. No *ii* junctions are possible for poly(*meso*-LA). However, if a mixture of L-, D-, and *meso*-LA is polymerized, or if pure L-, D-, or *meso*-LA is polymerized with concomitant epimerization (for L-LA or D-LA) or transesterification (for *rac*-LA or

TABLE 25.1
Tetrad Probabilities in Stereorandom Polymerization of rac-LA and meso-LA Based on Bernoullian Statistics

Tetrads	Probability	
	rac-LA	meso-LA
iii	0.375	0
iis	0.125	0
sii	0.125	0
isi	0.250	0.125
sis	0.125	0.250
ssi	0	0.125
iss	0	0.125
sss	0	0.375

meso-LA), then all eight tetrad sequences are possible. A polymer containing a random distribution of all eight tetrads is said to be atactic. A polymerization of either rac-LA or meso-LA that gives a statistical distribution of the respective possible five tetrads is a stereorandom polymerization. These tetrad sequences are shown in Table 25.1 along with their statistical abundance for stereorandom polymerization.[34,35] Note that in the polymerization of rac-LA, where each monomer contains an i junction, the formation of a heterotactic polymer is equivalent to the formation of a syndiotactic polyolefin.

The stereoselective polymerization of meso-LA could give either syndiotactic PLA (with sss tetrads) or heterotactic PLA (with isi/sis tetrads). Stereoselective polymerization of rac-LA can give either heterotactic (isi/sis) or isotactic (iii) PLA, the latter being composed of a 1:1 mixture of chains of poly(L-LA) and poly(D-LA) that form a stereocomplex polymer. This polymer has a higher melting point (230 °C) than poly(L-LA) (175 °C), and as such is more attractive for certain processing procedures and applications.[36–38] In general, the observed tetrad populations from polymerization may be analyzed by NMR spectroscopy and compared to statistically calculated distributions (Table 25.1). The polymerization may then be mathematically assigned a stereoselectivity of P_i (= the fractional probability of forming a new isotactic tetrad) or P_h (= the fractional probability of forming a new heterotactic tetrad) where $P_h + P_i = 1.0$[34,39] (Equations 25.1 and 25.2). Selected PLA spectra are shown in Figure 25.6.

$$P_i = \frac{\text{observed} - \text{statistical}}{1 - \text{statistical}} \tag{25.1}$$

$$P_h = \frac{\text{observed} - \text{statistical}}{1 - \text{statistical}} \tag{25.2}$$

At this point, it is worth noting that in some places in the literature the nomenclature for the stereochemistry of polylactides is based on the use of m (meso) and r (racemo) dyads rather than i and s dyads. This is common for poly(α-olefins), for which regioregular polymers can have mirror symmetry with respect to the methylene carbon (Figure 25.7).[35] Hence, an isotactic junction is designated m and a syndiotactic junction r, and an isotactic tetrad is designated mmm rather than iii and a syndiotactic tetrad rrr rather than sss. However, PLA is not a symmetric polymer, and as noted previously, at the tetrad level the sensing of the ^1H and ^{13}C signals are in the opposite direction as shown in Figure 25.5. For this reason, we prefer the nomenclature based on i and s, but note that others may use m or r instead. Similarly, in referring to Equations 25.1 and 25.2, in this case $P_m = P_i$ and $P_r = P_h$.

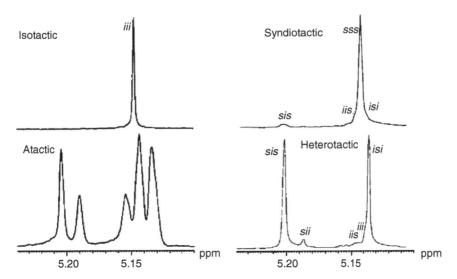

FIGURE 25.6 Homodecoupled ^{1}H NMR spectra (methine proton region) of PLAs with various tacticities. (Reprinted with permission from Chamberlain, B. M.; Cheng, M.; Moore, D. R.; Ovitt, T. M.; Lobkovsky, E. B.; Coates, G. W. *J. Am. Chem. Soc.* **2001**, *123*, 3229–3238; Radano, C. P.; Baker, G. L.; Smith, M. R. III. *J. Am. Chem. Soc.* **2000**, *122*, 1552–1553; Ovitt, T. M.; Coates, G. W. *J. Am. Chem. Soc.* **2002**, *124*, 1316–1326. Copyright 2000, 2001, 2002 American Chemical Society.)

FIGURE 25.7 Tetrad stereosequences in polypropylene.

25.3 STEREOSELECTIVE POLYMERIZATIONS

25.3.1 METAL SITE CONTROL

The first attempts to control the stereoselective polymerization of *rac*-LA were carried out by Spassky et al. in 1996.[40] Working with a chiral Schiff-base salen (bis(salicylaldehyde)-ethylenediamine) aluminum complex (*R*-1, Figure 25.8), he found that D-LA was preferentially polymerized during the early course of *rac*-LA polymerization. This was determined by examining the optical activity of the residual LA monomer. Furthermore, he estimated that the ratio of the polymerization rate constants for L-LA and D-LA, k_D:k_L, was 20:1. Following this work, many other researchers employed chiral salen aluminum initiators (**2**, **4–6**, Figure 25.8) with varying results. Reactions employing (*R*,*R*-salen)Al(O-*i*-Pr) ((***R,R***)-**4**) were found by Feijen and coworkers to preferentially ring-open L-LA with a k_L:k_D of ~14:1.[39,41]

In studies of the initial ring-opening event, it was found that (*R*,*R*-salen)AlOEt ((***R,R***)-**5**) preferentially reacts with L-LA in benzene, toluene, and O-donor solvents such as tetrahydrofuran (THF). However, in CHCl$_3$, this preference is reversed in favor of D-LA (Scheme 25.5 and Table 25.2).[42] Equally surprising is the fact that (*R*,*R*-salen)AlOCH$_2$CH(*S*)MeCl ((***R,R***)-**6**) preferentially ring opens D-LA in all solvents (in contrast to the analogous ethoxide initiator, (***R,R***)-**5**) (Table 25.2). These results point to complexity of stereocontrol in a ring-opening polymerization where both chain-end control and enantiomorphic site control are involved.

Based on Spassky's results, Baker and coworkers employed a racemic salen aluminum catalyst (***rac*-2**, Figure 25.8) and obtained a polymer proposed to be ~90% poly(L-LA) and poly(D-LA), that is, a stereocomplex polymer.[43] However, Coates and Ovitt reexamined this work, and based on the

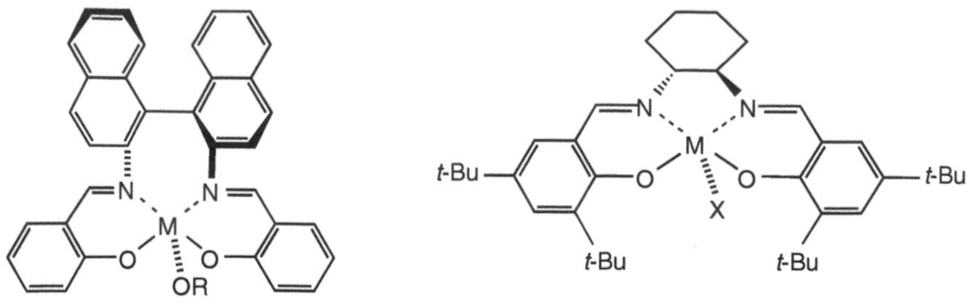

1 : M = Al, R = Me
2 : M = Al, R = i-Pr
3 : M = Y, R = (CH$_2$)$_2$NMe$_2$

4 : M = Al, X = O-i-Pr
5 : M = Al, X = OEt
6 : M = Al, X = OCH$_2$CH(S)MeCl
7 : M = Y, X = N(SiHMe$_2$)$_2$

FIGURE 25.8   Chiral salen metal catalysts used for LA polymerization (for **1–3**, R-enantiomers are shown; for **4–7**, R,R-enantiomers are shown).

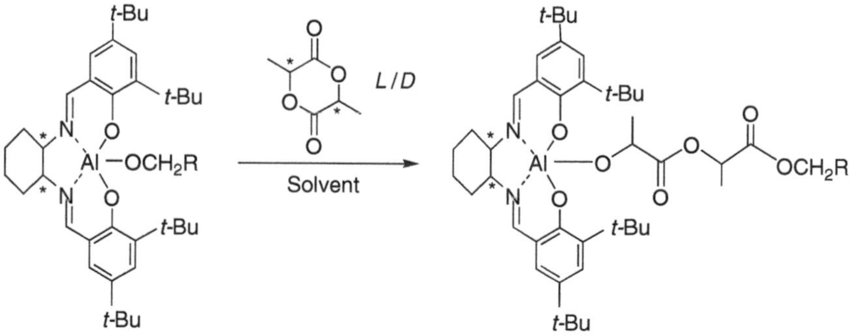

R = CH$_3$, CHMe(S)Cl

SCHEME 25.5   1:1 Reactions of chiral salen Al complexes (**5**, **6**) and rac-LA in various solvents.

TABLE 25.2

Stereoselectivity of Products of 1:1 Reactions of Chiral Salen Al Complexes and rac-LA in Various Solvents (per Scheme 25.5)

Solvent	L - D (de, %)[a]			
	(R,R)-5	(S,S)-5	(R,R)-6	(S,S)-6
Benzene	20	−18	−31	−40
Toluene	12	−17	−33	−37
THF	22	−20	−33	−13
S-Propylene oxide	17	−13	−32	−12
Chloroform	−16	14	−30	−3

[a] Diastereomeric excess of L-LA vs. D-LA ring-opened products.

Source: Chisholm, M. H.; Patmore, N. J.; Zhou, Z. *Chem. Commun.* **2005**, 127–129. Reproduced by permission of The Royal Society of Chemistry

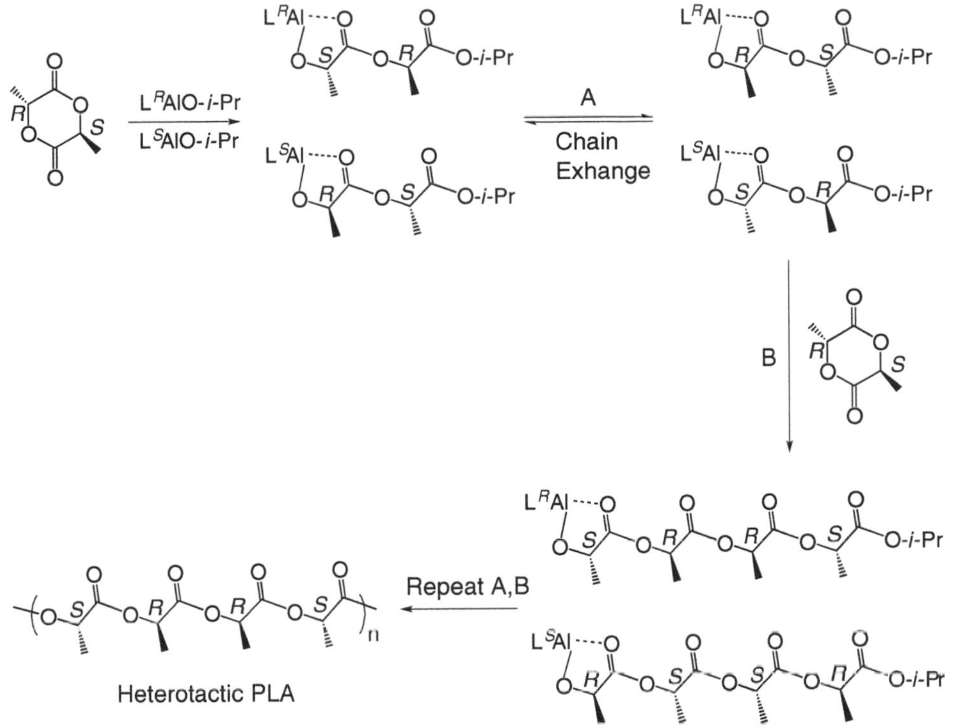

SCHEME 25.6 Coates' proposed chain exchange mechanism of heterotactic PLA formation by *rac*-2 salen Al catalyst (L = salen ligand). (Reprinted with permission from Ovitt, T. M.; Coates, G. W. *J. Am. Chem. Soc.* **2002**, *124*, 1316–1326. Copyright 2002 American Chemical Society.)

"mistakes" in the polymer microstructure, proposed that this material was in fact a blocky polymer consisting of (L-LA)$_n$-(D-LA)$_n$ segments where n ~11.[44] This distinction was made based on the presence of the *iis*, *sii*, and *isi* tetrads.

Employing the enantiomerically pure catalyst of **R-2**, Coates and Ovitt prepared the first samples of syndiotactic PLA (96% *sss*) from the ring-opening polymerization of *meso*-LA.[45,46] In some ways this is puzzling, because **R-2** was shown to preferentially open L-LA having *S,S*-configuration. Thus, it might have been anticipated that by chain-end control, an Al-(*S,R*) growing chain would preferentially ring-open *meso*-LA at the *S* ketonic carbon generating a Al-(*R,S*)(*S,R*) unit having an *i* junction. Obviously, *meso*-LA has two stereocenters of different configuration (*R* and *S*), and both individually influence the rate of ring-opening polymerization as compared to the two identical *S* stereocenters in L-LA. Even more puzzling was Coates' finding that the racemic catalyst **rac-2** polymerized *meso*-LA to give heterotactic PLA having 80% *isi* and *sis* junctions. To account for this observation, Coates and Ovitt suggested a chain exchange mechanism (Scheme 25.6).[46] For this mechanism to work effectively, the rate of chain exchange must be much faster than the rate of ring opening (propagation). It should be noted here that these aluminum (III) complexes are very slow polymerization catalysts, typically requiring several days at 70 °C to polymerize 100 equivalents of LA to ~90% conversion.

Chiral salen yttrium complexes were also examined in the ring-opening polymerization of LA (**R-3**, (*S,S*)-**7**, Figure 25.8). Although the yttrium catalyst **R-3** showed higher kinetic activity (faster polymerization rate) than Al catalyst **R-2**, unlike **R-2**, it produced atactic PLA from *meso*-LA.[46] Similarly, atactic PLA was formed from *rac*-LA by the (*S,S*-salen) yttrium catalyst (*S,S*)-**7**.[47,48]

Chiral Mg and Zn trisindazolylborate metal complexes (**8** and **9**, Figure 25.9) have been synthesized for use as single-site catalysts for the stereoselective ring-opening polymerization of LA by

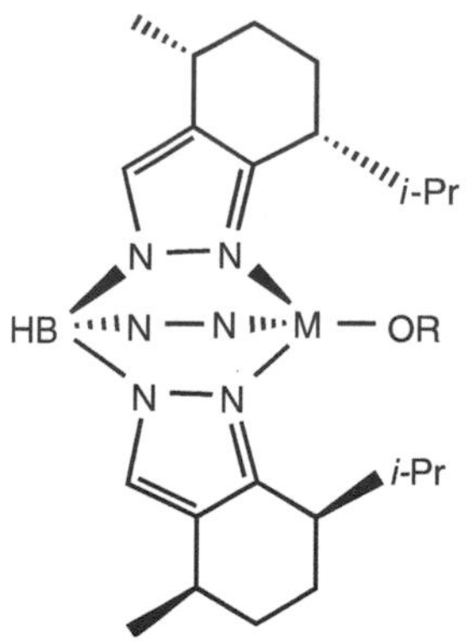

FIGURE 25.9 Chiral trisindazolylborate metal catalysts used for LA polymerization (third indazolyl ligand represented as N–N for clarity).

Chisholm et al.[49] In the copolymerization of a 1:1 mixture of *meso:rac* lactide, the chiral magnesium complex **8** showed a marked preference for the polymerization of *meso*-LA over *rac*-LA, and a modest preference for the formation of syndiotactic junctions ($\sim$60% *sss*). However, the chiral zinc complex **9** showed a much lesser degree of diastereoselectivity.

25.3.2 Chain-End Control

When LA is ring opened by a M-X polymerization catalyst (M = metal, X = alkoxide, hydroxide, amide, etc.), a chiral alkoxide ligand is formed at the metal center. This ligand may then impose its influence in selecting the next LA molecule for ring opening, and so forth. For example, the initiator LiO-t-Bu was reported to produce predominantly heterotactic polymer from *rac*-LA.[31,50] Here, the initial ring opening of L-LA would be followed by D-LA, and the initial ring-opening of D-LA followed by L-LA, such that *isi* and *sis* tetrads would be formed. Such a chain-end control mechanism has been exclusively studied by employing metal complexes with achiral auxiliary ligands.

Again, Schiff-base type salen aluminum catalysts (**10–13**, Figure 25.10) were used as mechanistic probes. In 1997, Spassky and coworkers reported that when achiral salen Al catalyst **10** was employed in the ring-opening polymerization of *rac*-LA, it offered a moderate stereoselectivity and produced PLA with some enhancement in isotactic junctions.[51] Nomura et al. prepared a series of achiral salen Al catalysts (**11a–12e**); highly isotactic PLA ($P_i = 0.9$) was obtained from *rac*-LA using catalyst **12e** bearing bulky substituent groups.[52] Chen and coworkers also reported that achiral salen Al catalyst **13** polymerized *rac*-LA to give PLA with predominantly isotactic junctions ($P_i = 0.9$).[53] Gibson and coworkers recently prepared discrete salan-type (bis-*o*-hydroxybenzyl)-1,2-diaminoethane) Al catalysts (**14a–15d**, Figure 25.11) that provide a wide range of PLA tacticities from *rac*-LA, from highly isotactic to highly heterotactic ($P_h:P_i = 0.21:0.79$ to $0.96:0.04$). A subtle change in substituents resulted in a dramatic switch in selectivity.[54]

Achiral tris(pyrazolyl)borate (Tp) metal complexes of Mg, Zn, and Ca (**16–18**, Figure 25.12) were first prepared by Chisholm et al. as initiators for the ring-opening polymerization of lactides.[49,55,56] In the copolymerization of a 1:1 mixture of *meso*-LA and *rac*-LA, the magnesium and zinc complexes (**16** and **17**) showed significant preferences for the polymerization of *meso*-LA over L- and D-lactides in CH$_2$Cl$_2$.[49] Remarkably, polymerization with the calcium complex (**18**) is extremely rapid, with >90% conversion achieved within one minute at room temperature and a 200:1 monomer:catalyst ratio; this catalyst selectively produces 90% heterotactic PLA (*isi/sis*) from *rac*-LA in THF solvent.[56] In CH$_2$Cl$_2$, less selectivity is observed, which implicates the role of solvent THF in binding to the metal center.

10

13

11: n = 2; **12**: n = 3
a: $R_1 = R_2 = H$
b: $R_1 = R_2 = Me$
c: $R_1 = i\text{-Pr}, R_2 = H$
d: $R_1 = Ph, R_2 = H$
e: $R_1 = R_2 = t\text{-Bu}$

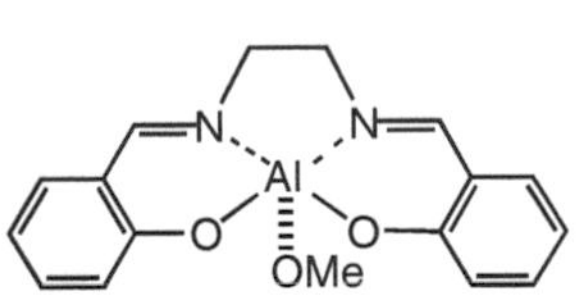

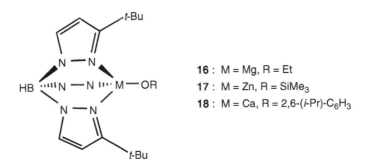

FIGURE 25.10 Achiral salen aluminum catalysts used for LA polymerization.

14: $R_1 = Me$
15: $R_1 = CH_2Ph$
a: $R_2 = H$
b: $R_2 = Me$
c: $R_2 = t\text{-Bu}$
d: $R_2 = Cl$

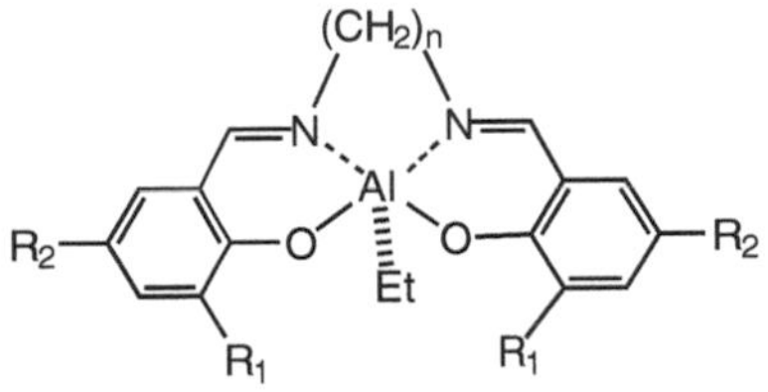

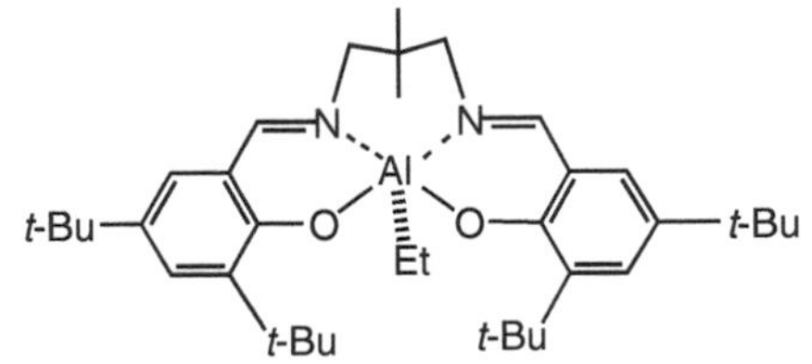

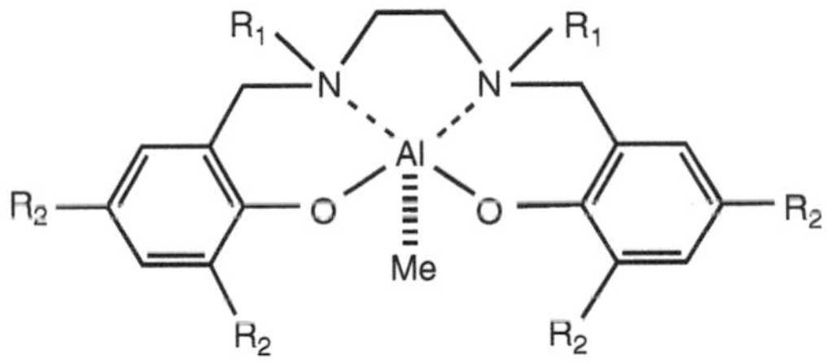

FIGURE 25.11 Salan aluminum catalysts used for LA polymerization.

16 : M = Mg, R = Et
17 : M = Zn, R = $SiMe_3$
18 : M = Ca, R = 2,6-(i-Pr)-C_6H_3

FIGURE 25.12 Achiral tris(pyrazolyl)borate (Tp) metal catalysts used for LA polymerization (third pyrazolyl ligand represented as N–N for clarity).

Significant stereocontrol in the ring-opening polymerization of lactides was achieved by Coates' group by using achiral β-diiminate (BDI) metal complexes (**19–22**, Figure 25.13).[34,57] With the bulky BDI ligand bound to zinc, the complex [(BDI)Zn(O-*i*-Pr)]$_2$ (**21**) is effective in polymerizing *rac*-LA to give heterotactic PLA (94% *isi*/*sis*) and *meso*-LA to give syndiotactic PLA (76% *sss*) in CH_2Cl_2. Here, chain-end control is operative as shown in Scheme 25.7. Also, Chisholm et al. have prepared monomeric bulky BDI zinc complexes (**23**, **24**, **26**, **27**, Figure 25.13) that similarly polymerize *rac*-LA to heterotactic PLA (~90% *isi*/*sis*).[58–60] Interestingly, with analogous magnesium BDI complexes (**22**, **25**, **28**, Figure 25.13), neither the Coates nor Chisholm groups observed stereoselective polymerization of lactides in CH_2Cl_2, although the Mg-catalyzed polymerization reactions were much more rapid than the Zn-catalyzed polymerizations. Even more interestingly, Chisholm

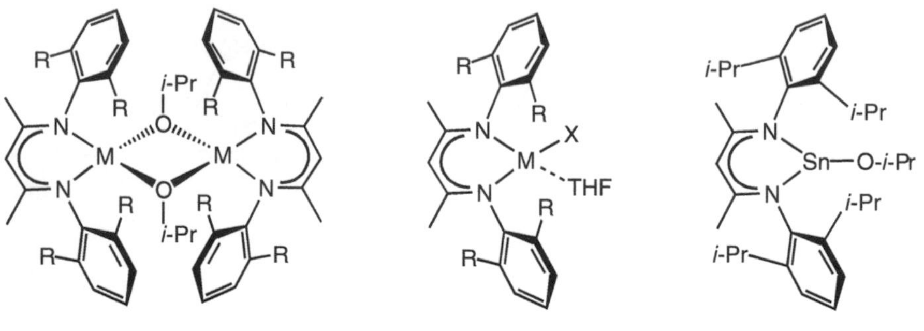

19 : M = Zn, R = Et
20 : M = Zn, R = n-Pr
21 : M = Zn, R = i-Pr
22 : M = Mg, R = i-Pr

23 : M = Zn, R = i-Pr, X = OSiPh$_3$
24 : M = Zn, R = i-Pr, X = O-t-Bu
25 : M = Mg, R = i-Pr, X = O-t-Bu
26 : M = Zn, R = t-Bu, X = O-t-Bu
27 : M = Zn, R = t-Bu, X = N(i-Pr)$_2$
28 : M = Mg, R = t-Bu, X = N(i-Pr)$_2$

29

FIGURE 25.13 β-Diiminate (BDI) metal catalysts used for LA polymerization.

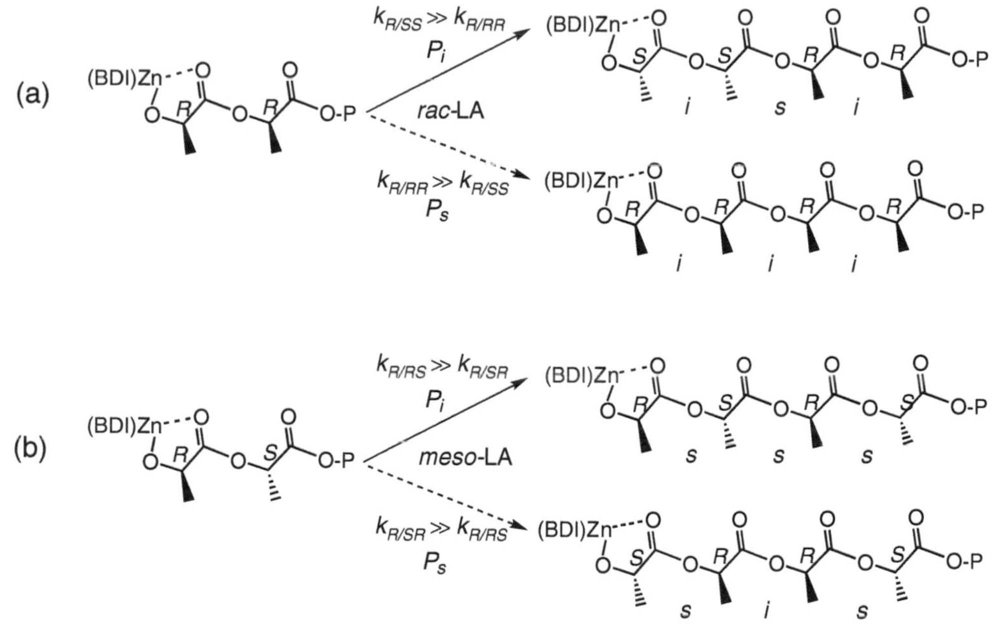

SCHEME 25.7   Chain-end control stereoselectivity in the ring-opening polymerizations of (a) rac-LA and (b) meso-LA by the initiator [(BDI)Zn(O-i-Pr)]$_2$ (**21**) (P = polymer chain; $k_{R/SS}$ = rate constant of ring opening of L-LA (S,S) by a R chain end). (Reprinted with permission from Chamberlain, B. M.; Cheng, M.; Moore, D. R.; Ovitt, T. M.; Lobkovsky, E. B.; Coates, G. W. J. Am. Chem. Soc. **2001**, 123, 3229–3238. Copyright 2001 American Chemical Society.)

et al. found that working in THF as the polymerization solvent restored the stereoselectivity ($\sim$90% isi/sis) of the magnesium catalyst system back to that of the zinc system without significant loss of rate.[59,60] This again implicates the role of the solvent THF as a ligand. Gibson and coworkers prepared (BDI)Sn(II)O-i-Pr (**29**, Figure 25.13) and obtained a modest preference for heterotacticity when polymerizing rac-LA in toluene.[61]

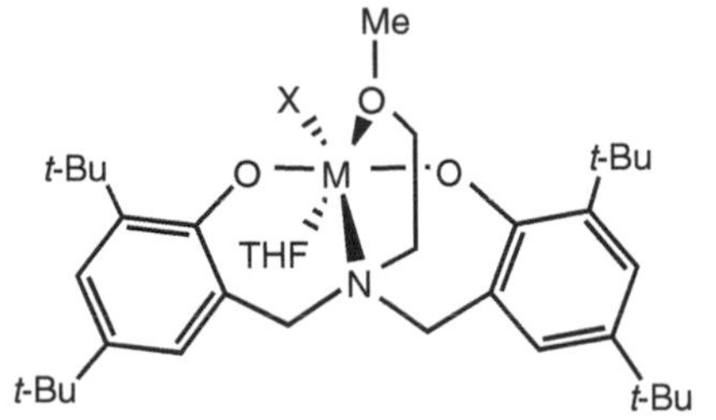

30: M = Y, X = CH$_2$SiMe$_3$
31: M = Y, X = N(SiHMe$_2$)$_2$
32: M = La, X = N(SiHMe$_2$)$_2$

FIGURE 25.14 Group 3 alkoxy-amino-bis(phenolate) catalysts used for LA polymerization.

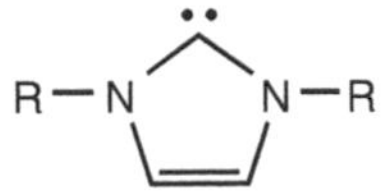

33: R = 2,4,6-trimethylphenyl
34: R = 2,6-diisopropylphenyl
35: R = tert-butyl

FIGURE 25.15 Carbenes used as catalysts for LA polymerization.

Carpentier and coworkers recently synthesized yttrium and lanthanide LA polymerization catalysts with tetradentate alkoxy-amino-bis(phenolate) ligands (**30–32**, Figure 25.14).[62] Yttrium compounds **30** and **31** polymerized *rac*-LA to give predominantly heterotactic polymer ($P_h = 0.8$) in THF, showing less selectivity in toluene ($P_h = 0.6$). The heterotacticity of PLA formed by lanthanide complex **32** from *rac*-LA in THF is significantly decreased ($P_h = 0.64$) compared to its yttrium analogue **31**.

In the organic catalyst systems used for LA polymerization, such as carbene/primary alcohol, usually little stereocontrol can be obtained. However, Hillmyer and Tolman have reported that the carbenes **33–35** (Figure 25.15), in the presence of *n*-butanol, produced isotactic-enriched PLA ($P_i = 0.6$) from *rac*-LA in CH$_2$Cl$_2$. Even higher selectivity ($P_i = 0.75$) was obtained when the polymerization was carried out at $-20\,^\circ$C.[63]

25.4 CONCLUSIONS

It is clear that within the past decade, considerable progress has been made in the stereoselective polymerizations of lactides. However, it is also fair to state that the success to date has largely been from empirical studies. Unlike α-olefins, lactide monomers contain two stereocenters, and the way in which these interact with the chiral end of the growing polymer chain and the chiral metal center is far from understood. Moreover, the solvent may play a critical and somewhat surprising role. This was seen, for example, with the highly active (BDI)MgOR and TpCaOR catalyst systems (**18**, **25**, **28**) that convert *rac*-LA to heterotactic PLA in the O-donor solvent THF, which can act as a ligand to these oxophilic metals. The complex nature of the role of the solvent and the alkoxide ligand is also emphasized by the studies of the initial ring-opening event. Clearly a good deal needs to be learned before catalysts can be designed for the stereoselective ring-opening polymerization of lactides, and computational efforts underway in various groups may assist in these matters.[64,65]

ACKNOWLEDGMENTS

We thank the US Department of Energy, Office of Basic Sciences, Chemistry Division, for support of research at The Ohio State University.

REFERENCES AND NOTES

1. Kricheldorf, H. R. Syntheses and application of polylactides. *Chemosphere* **2001**, *43*, 49–54.
2. Kharas, G. B.; Sanchez-Riera, F.; Severson, D. K. Polymers of lactic acid. *Plast. Microbes* **1994**, 93–137.
3. Gruber, P.; O'Brien, M. Polylactides. Natureworks PLA. *Biopolymers* **2002**, *4*, 235–250.
4. Lunt, J.; Bone, J. Properties and dyeability of fibers and fabrics produced from polylactide (PLA) polymers. *AATCC Review* **2001**, *1*, 20–23.
5. Dartee, M.; Lunt, J.; Shafer, A. Natureworks PLA: Sustainable performance fiber. *Chem. Fibers Int.* **2000**, *50*, 546–551.
6. Drumright, R. E.; Gruber, P. R.; Henton, D. E. Polylactic acid technology. *Adv. Mater.* **2000**, *12*, 1841–1846.
7. O'Keefe, B. J.; Hillmyer, M. A.; Tolman, W. B. Polymerization of lactide and related cyclic esters by discrete metal complexes. *J. Chem. Soc., Dalton Trans.* **2001**, 2215–2224.
8. Chisholm, M. H.; Zhou, Z. New generation polymers: The role of metal alkoxides as catalysts in the production of polyoxygenates. *J. Mater. Chem.* **2004**, *14*, 3081–3092.
9. Nakano, K.; Kosaka, N.; Hiyama, T.; Nozaki, K. Metal-catalyzed synthesis of stereoregular polyketones, polyesters, and polycarbonates. *J. Chem. Soc., Dalton Trans.* **2003**, 4039–4050.
10. Dechy-Cabaret, O.; Martin-Vaca, B.; Bourissou, D. Controlled ring-opening polymerization of lactide and glycolide. *Chem. Rev.* **2004**, *104*, 6147–6176.
11. Penczek, S.; Duda, A.; Kowalski, A.; Libiszowski, J.; Majerska, K.; Biela, T. On the mechanism of polymerization of cyclic esters induced by tin(II) octoate. *Macromol. Symp.* **2000**, *157*, 61–70.
12. Schwach, G.; Coudane, J.; Engel, R.; Vert, M. Zn lactate as initiator of DL-lactide ring opening polymerization and comparison with Sn octoate. *Polym. Bull.* **1996**, *37*, 771–776.
13. Bero, M.; Kasperczyk, J.; Adamus, G. Coordination polymerization of lactides. 3. Copolymerization of L,L-lactide and ε-caprolactone in the presence of initiators containing zinc and aluminum. *Makromol. Chem.* **1993**, *194*, 907–912.
14. Kasperczyk, J. Microstructural analysis of poly[(L-lactide)-*co*-(glycolide)] by ^{1}H and ^{13}C NMR spectroscopy. *Polymer* **1996**, *37*, 201–203.
15. Schwach, G.; Coudane, J.; Engel, R.; Vert, M. Ring opening polymerization of DL-lactide in the presence of zinc metal and zinc lactate. *Polym. Int.* **1998**, *46*, 177–182.
16. Kricheldorf, H. R.; Damrau, D. O. Polylactones. Part 37. Polymerizations of L-lactide initiated with Zn(II) L-lactate and other resorbable Zn salts. *Macromol. Chem. Phys.* **1997**, *198*, 1753–1766.
17. Chisholm, M. H.; Delbridge, E. E. A study of the ring-opening of lactides and related cyclic esters by Ph_2SnX_2 and Ph_3SnX compounds (X = NMe_2, OR). *New J. Chem.* **2003**, *27*, 1167–1176.
18. Nederberg, F.; Connor, E. F.; Glausser, T.; Hedrick, J. L. Organocatalytic chain scission of poly(lactides): A general route to controlled molecular weight, functionality and macromolecular architecture. *Chem. Commun.* **2001**, 2066–2067.
19. Nederberg, F.; Connor, E. F.; Moeller, M.; Glauser, T.; Hedrick, J. L. New paradigms for organic catalysts: The first organocatalytic living polymerization. *Angew. Chem., Int. Ed. Engl.* **2001**, *40*, 2712–2715.
20. Connor, E. F.; Nyce, G. W.; Myers, M.; Moeck, A.; Hedrick, J. L. First example of N-heterocyclic carbenes as catalysts for living polymerization: Organocatalytic ring-opening polymerization of cyclic esters. *J. Am. Chem. Soc.* **2002**, *124*, 914–915.
21. Myers, M.; Connor, E. F.; Glauser, T.; Mock, A.; Nyce, G.; Hedrick, J. L. Phosphines: Nucleophilic organic catalysts for the controlled ring-opening polymerization of lactides. *J. Polym. Sci., Part A: Polym. Chem.* **2002**, *40*, 844–851.
22. Nyce, G. W.; Connor, E. F.; Hedrick, J. L. In-situ formation of N-heterocyclic carbenes as organic catalysts for living polymerization. *Polym. Mater. Sci. Eng.* **2002**, *87*, 213–214.
23. Nyce, G. W.; Glauser, T.; Connor, E. F.; Moeck, A.; Waymouth, R. M.; Hedrick, J. L. In situ generation of carbenes: A general and versatile platform for organocatalytic living polymerization. *J. Am. Chem. Soc.* **2003**, *125*, 3046–3056.
24. Duda, A.; Penczek, S. Thermodynamics of L-lactide polymerization. Equilibrium monomer concentration. *Macromolecules* **1990**, *23*, 1636–1639.

25. Chisholm, M. H.; Delbridge, E. E. A study of the ring-opening polymerization (ROP) of L-lactide by Ph$_2$SnX$_2$ precursors (X = NMe$_2$, OPr^i): The notable influence of initiator group. *New J. Chem.* **2003**, *27*, 1177–1183.

26. Kricheldorf, H. R.; Boettcher, C.; Toennes, K. U. Polylactones. 23. Polymerization of racemic and *meso* D,L-lactide with various organotin catalysts – stereochemical aspects. *Polymer* **1992**, *33*, 2817–2824.

27. Kricheldorf, H. R.; Kreiser-Saunders, I. Polylactides—synthesis, characterization and medical application. *Macromol. Symp.* **1996**, *103*, 85–102.

28. Chisholm, M. H.; Iyer, S. S.; Matison, M. E.; McCollum, D. G.; Pagel, M. Concerning the stereochemistry of poly(lactide), PLA. Previous assignments are shown to be incorrect and a new assignment is proposed. *Chem. Commun.* **1997**, 1999–2000.

29. Chisholm, M. H.; Iyer, S. S.; McCollum, D. G.; Pagel, M.; Werner-Zwanziger, U. Microstructure of polylactide. Phase-sensitive HETCOR spectra of poly-*meso*-lactide, poly-*rac*-lactide, and atactic polylactide. *Macromolecules* **1999**, *32*, 963–973.

30. Zell, M. T.; Padden, B. E.; Paterick, A. J.; Thakur, K. A. M.; Kean, R. T.; Hillmyer, M. A.; Munson, E. J. Unambiguous determination of the ^{13}C and ^{1}H NMR stereosequence assignments of polylactide using high-resolution solution NMR spectroscopy. *Macromolecules* **2002**, *35*, 7700–7707.

31. Kasperczyk, J. E. Microstructure analysis of poly(lactic acid) obtained by lithium *tert*-butoxide as initiator. *Macromolecules* **1995**, *28*, 3937–3939.

32. Kasperczyk, J. E. HETCOR NMR study of poly(*rac*-lactide) and poly(*meso*-lactide). *Polymer* **1999**, *40*, 5455–5458.

33. Coudane, J.; Ustariz-Peyret, C.; Schwach, G.; Vert, M. More about the stereodependence of DD and LL pair linkages during the ring-opening polymerization of racemic lactide. *J. Polym. Sci., Part A: Polym. Chem.* **1997**, *35*, 1651–1658.

34. Chamberlain, B. M.; Cheng, M.; Moore, D. R.; Ovitt, T. M.; Lobkovsky, E. B.; Coates, G. W. Polymerization of lactide with zinc and magnesium β-diiminate complexes: Stereocontrol and mechanism. *J. Am. Chem. Soc.* **2001**, *123*, 3229–3238.

35. Bovey, F. A.; Mirau, P. A. Physical properties of synthetic high polymers. In *NMR of Polymers*; Academic Press: San Diego, CA, 1996; p 459.

36. Spinu, M.; Jackson, C.; Keating, M. Y.; Gardner, K. H. Material design in poly(lactic acid) systems: Block copolymers, star homo- and copolymers, and stereocomplexes. *J. Macromol. Sci., Pure Appl. Chem.* **1996**, *A33*, 1497–1530.

37. Yui, N.; Dijkstra, P. J.; Feijen, J. Stereo block copolymers of L- and D-lactides. *Makromol. Chem.* **1990**, *191*, 481–488.

38. Tsuji, H.; Ikada, Y. Stereocomplex formation between enantiomeric poly(lactic acid)s. XI. Mechanical properties and morphology of solution-cast films. *Polymer* **1999**, *40*, 6699–6708.

39. Zhong, Z.; Dijkstra, P. J.; Feijen, J. Controlled and stereoselective polymerization of lactide: Kinetics, selectivity, and microstructures. *J. Am. Chem. Soc.* **2003**, *125*, 11291–11298.

40. Spassky, N.; Wisniewski, M.; Pluta, C.; Le Borgne, A. Highly stereoelective polymerization of *rac*-(D,L)-lactide with a chiral Schiff's base/aluminum alkoxide initiator. *Macromol. Chem. Phys.* **1996**, *197*, 2627–2637.

41. Zhong, Z.; Dijkstra, P. J.; Feijen, J. [(salen)Al]-Mediated, controlled and stereoselective ring-opening polymerization of lactide in solution and without solvent: Synthesis of highly isotactic polylactide stereocopolymers from racemic D,L-lactide. *Angew. Chem., Int. Ed. Engl.* **2002**, *41*, 4510–4513.

42. Chisholm, M. H.; Patmore, N. J.; Zhou, Z. Concerning the relative importance of enantiomorphic site vs. chain end control in the stereoselective polymerization of lactides: Reactions of (*R,R*-salen)- and (*S,S*-salen)-aluminium alkoxides LAlOCH$_2$R complexes (R = CH$_3$ and *S*-CHMeCl). *Chem. Commun.* **2005**, 127–129.

43. Radano, C. P.; Baker, G. L.; Smith, M. R. III. Stereoselective polymerization of a racemic monomer with a racemic catalyst. Direct preparation of the poly(lactic acid) stereocomplex from racemic lactide. *J. Am. Chem. Soc.* **2000**, *122*, 1552–1553.

44. Ovitt, T. M.; Coates, G. W. Stereoselective ring-opening polymerization of *rac*-lactide with a single-site, racemic aluminum alkoxide catalyst: Synthesis of stereoblock poly(lactic acid). *J. Polym. Sci., Part A: Polym. Chem.* **2000**, *38*, 4686–4692.

45. Ovitt, T. M.; Coates, G. W. Stereoselective ring-opening polymerization of *meso*-lactide: Synthesis of syndiotactic poly(lactic acid). *J. Am. Chem. Soc.* **1999**, *121*, 4072–4073.

46. Ovitt, T. M.; Coates, G. W. Stereochemistry of lactide polymerization with chiral catalysts: New opportunities for stereocontrol using polymer exchange mechanisms. *J. Am. Chem. Soc.* **2002**, *124*, 1316–1326.

47. Lin, M.-H.; RajanBabu, T. V. Ligand-assisted rate acceleration in transacylation by a yttrium-salen complex. Demonstration of a conceptually new strategy for metal-catalyzed kinetic resolution of alcohols. *Org. Lett.* **2002**, *4*, 1607–1610.

48. Chisholm, M. H.; Zhou, Z. The Ohio State University, Columbus, OH. Unpublished results, 2004.

49. Chisholm, M. H.; Eilerts, N. W.; Huffman, J. C.; Iyer, S. S.; Pacold, M.; Phomphrai, K. Molecular design of single-site metal alkoxide catalyst precursors for ring-opening polymerization reactions leading to polyoxygenates. 1. Polylactide formation by achiral and chiral magnesium and zinc alkoxides, $(\eta^3$-L)MOR, where L = trispyrazolyl- and trisindazolylborate ligands. *J. Am. Chem. Soc.* **2000**, *122*, 11845–11854.

50. Bero, M.; Dobrzynski, P.; Kasperczyk, J. Synthesis of disyndiotactic polylactide. *J. Polym. Sci., Part A: Polym. Chem.* **1999**, *37*, 4038–4042.

51. Wisniewski, M.; Le Borgne, A.; Spassky, N. Synthesis and properties of (D)- and (L)-lactide stereocopolymers using the system achiral Schiff's base/aluminum methoxide as initiator. *Macromol. Chem. Phys.* **1997**, *198*, 1227–1238.

52. Nomura, N.; Ishii, R.; Akakura, M.; Aoi, K. Stereoselective ring-opening polymerization of racemic lactide using aluminum-achiral ligand complexes: Exploration of a chain-end control mechanism. *J. Am. Chem. Soc.* **2002**, *124*, 5938–5939.

53. Tang, Z.; Chen, X.; Pang, X.; Yang, Y.; Zhang, X.; Jing, X. Stereoselective polymerization of *rac*-lactide using a monoethylaluminum Schiff base complex. *Biomacromolecules* **2004**, *5*, 965–970.

54. Hormnirun, P.; Marshall, E. L.; Gibson, V. C.; White, A. J. P.; Williams, D. J. Remarkable stereocontrol in the polymerization of racemic lactide using aluminum initiators supported by tetradentate aminophenoxide ligands. *J. Am. Chem. Soc.* **2004**, *126*, 2688–2689.

55. Chisholm, M. H.; Lin, C.-C.; Gallucci, J. C.; Ko, B.-T. Binolate complexes of lithium, zinc, aluminium, and titanium; preparations, structures, and studies of lactide polymerization. *J. Chem. Soc., Dalton Trans.* **2003**, 406–412.

56. Chisholm, M. H.; Gallucci, J. C.; Phomphrai, K. Well-defined calcium initiators for lactide polymerization. *Inorg. Chem.* **2004**, *43*, 6717–6725.

57. Cheng, M.; Attygalle, A. B.; Lobkovsky, E. B.; Coates, G. W. Single-site catalysts for ring-opening polymerization: Synthesis of heterotactic poly(lactic acid) from *rac*-lactide. *J. Am. Chem. Soc.* **1999**, *121*, 11583–11584.

58. Chisholm, M. H.; Huffman, J. C.; Phomphrai, K. Monomeric metal alkoxides and trialkyl siloxides: (BDI)Mg(O^{t}Bu)(THF) and (BDI)Zn(OSiPh$_3$)(THF). Comments on single site catalysts for ring-opening polymerization of lactides. *J. Chem. Soc., Dalton Trans.* **2001**, 222–224.

59. Chisholm, M. H.; Gallucci, J.; Phomphrai, K. Coordination chemistry and reactivity of monomeric alkoxides and amides of magnesium and zinc supported by the diiminato ligand CH(CMeNC$_6$H$_3$-2,6-^{i}Pr$_2$)$_2$. A comparative study. *Inorg. Chem.* **2002**, *41*, 2785–2794.

60. Chisholm, M. H.; Phomphrai, K. Conformational effects in β-diiminate ligated magnesium and zinc amides. Solution dynamics and lactide polymerization. *Inorg. Chim. Acta.* **2003**, *350*, 121–125.

61. Dove, A. P.; Gibson, V. C.; Marshall, E. L.; White, A. J. P.; Williams, D. J. A well defined tin(II) initiator for the living polymerisation of lactide. *Chem. Commun.* **2001**, 283–284.

62. Cai, C.-X.; Amgoune, A.; Lehmann, C. W.; Carpentier, J.-F. Stereoselective ring-opening polymerization of racemic lactide using alkoxy-amino-bis(phenolate) group 3 metal complexes. *Chem. Commun.* **2004**, 330–331.

63. Jensen, T. R.; Breyfogle, L. E.; Hillmyer, M. A.; Tolman, W. B. Stereoselective polymerization of D,L-lactide using N-heterocyclic carbene based compounds. *Chem. Commun.* **2004**, 2504–2505.

64. von Schenck, H.; Ryner, M.; Albertsson, A.-C.; Svensson, M. Ring-opening polymerization of lactones and lactides with Sn(IV) and Al(III) initiators. *Macromolecules* **2002**, *35*, 1556–1562.

65. Marshall, E. L.; Gibson, V. C.; Rzepa, H. S. A computational analysis of the ring-opening polymerization of *rac*-lactide initiated by single-site β-diketiminate metal complexes: Defining the mechanistic pathway and the origin of stereocontrol. *J. Am. Chem. Soc.* **2005**, *127*, 6048–6051.

Glossary

Ψ	stereoregularity
Ψ_{abs}	absolute stereoregularity
Ψ_{rel}	relative stereoregularity
μ or $[\eta]$	polymer viscosity
a or a-	atactic
AA	acrylic acid or aluminum-activated
acac	acetylacetonate
acacH	acetylacetone
Adam	adamantyl
ADMET	acyclic diene metathesis polymerization
AFM	atomic force microscopy
AIBN	2,2′-azobisisobutyronitrile
An	acenaphthyl
ANiBr	bis(allylnickel bromide)
ANiTFA	bis[(η^3-allyl)(trifluoroacetato)nickel(II)]
APT	Approximate Pair Theory
ATRP	atom transfer radical polymerization
B3LYP	3-Parameter Becke fit + Lee Yang Parr correlation function
9-BBN	9-borabicyclo[3.3.1]nonane
BD	1,3-butadiene
BDI	β-diiminate
BenzInd	benz[e]indenyl
BINOL	1,1′-bi-2-naphtholate
bipy or bpy	2,2′-bipyridyl or 2,2′-bipyridine
Bn	benzyl
Bz^{F5}	perfluorobenzoyl
$C_2^=$	ethylene
$C_3^=$	propylene
$C_8^=$	1-octene
CAS	Chemical Abstracts Service
CD	circular dichroism
CE	chain end
C_6F_5	pentafluorophenyl
CGC	constrained geometry catalyst
CHD	1,3-cyclohexadiene
CM	cross metathesis
COC	cyclic olefin copolymer or cycloolefin copolymer

COD or cod	cyclooctadiene
coe	cyclooctene
COSY	correlation spectroscopy
Cp	cyclopentadienyl (C_5H_5)
Cp*	pentamethylcyclopentadienyl (C_5Me_5)
Cp′	cyclopentadienyl ligand (substituted or unsubstituted)
Cp^R	substituted cyclopentadienyl ligand
Cp_{cent}	cyclopentadienyl centroid
CPE	cyclopentene
CPO	chlorinated maleated isotactic polypropylene
CP MAS or CPMAS	cross-polarization magic-angle spinning
CTA	chain transfer agent
Cy	cyclohexyl
Da	Dalton
D/A	donor–acceptor
DABCO	1,4-diazabicyclo[2.2.2]octane
DEPT	distortionless enhancement by polarization transfer
DFT	density functional theory
DMAP	4-(N,N-dimethylamino)pyridine
DMB	2,3-dimethyl-1,3-butadiene
dmbd	3,3-dimethyl-1,2-butanediolate
$dmbdH_2$	3,3-dimethyl-1,2-butanediol
dmca	dimethylcalixarene
dme	1,2-dimethoxyethane
3,4-DMP1	3,4-dimethyl-1-pentene
dmpe	1,2-bis(dimethylphosphino)ethane
dpm	2,2,6,6-tetramethyl-3,5-heptanedionate
dppp	1,3-propanediyl-bis(diphenylphosphine)
DSC	differential scanning calorimetry
E	ethylene
ECP	ethylene-1,2-cyclopentane
ee	enantiomeric excess
EHPP	elastomeric homopolypropylene
ENB	5-ethylidene-2-norbornene
EP, EPM	ethylene/propylene copolymer
EPNB	ethylene/propylene/norbornene terpolymer
EPDM	ethylene/propylene/diene monomer terpolymer
ES	enantiomorphic sites
ESR	electron spin resonance
EtMA	ethyl methacrylate
EtA	ethyl acrylate
FI Catalyst	bis(phenoxy-imine) group 4 complex
Flu	fluorenyl
GPC	gel permeation chromatography
GTP	group transfer polymerization
1,5-HD	1,5-hexadiene
HETCOR	heteronuclear correlation spectroscopy

H_f, ΔH_f, ΔH	heat of fusion or heat of transition
hfacac	hexafluoroacetylacetonate
HFIP	1,1,1,3,3,3-hexafluoropropan-2-ol
h-h	head-to-head
h-t	head-to-tail
H_4Ind	4,5,6,7-tetrahydroindenyl
i or i-	isotactic
Ind	indenyl
iPP	isotactic polypropylene
iPrMA	isopropyl methacrylate
iPS	isotactic polystyrene
IR	infrared
l	*like* diad
L^L′	*cis*-chelating bidentate ligand
LA	lactide
m	*meso*
MA	methyl acrylate
MAO	methylaluminoxane or methylalumoxane
MAO9	anionic (AlMeO)$_9$ cluster counterion model
MCH	methylene-1,3-cyclohexane
MCP	methylene-1,3-cyclopentane
mCPBA	*meta*-chloroperoxybenzoic acid
2-Me-4-PhInd	2-methyl-4-phenylindenyl
3-MH1	3-methyl-1-hexene
4-MH1	4-methyl-1-hexene
mid	*N*-methylimidazole
MM	molecular mechanics
MMA	methyl methacrylate
MMAO	modified methylaluminoxane
M_n	number average molecular weight
MOD	7-methyl-1,6-octadiene
3-MP1	3-methyl-1-pentene
3-MPD	3-methyl-1,3-pentadiene
4-MPD	4-methyl-1,3-pentadiene
M_v, M_η	viscosity average molecular weight
M_w	weight average molecular weight
NA or na	not applicable or not analyzed
Naph	naphthyl
α-naph	α-nitroacetophenonate
NB	norbornene
NBD or nbd	norbornadiene
nBA	*n*-butyl acrylate
nBMA	*n*-butyl methacrylate
ND or nd	not determined or not detected
NMR	nuclear magnetic resonance
nr	not reported

OAc	acetate
1,7-OD	1,7-octadiene
Oct	octamethyloctahydrodibenzofluorenyl
P	propylene or polymeryl or polymer chain
PBD	polybutadiene
PCL	poly(ε-caprolactone)
PCPE	poly(cyclopentene)
PD	1,3-pentadiene
PDI	polydispersity index (also abbreviated PI)
PE	polyethylene
PECP	poly(ethylene-1,2-cyclopentane)
PEO	poly(ethylene oxide)
PEtMA	poly(ethyl methacrylate)
phen	1,10-phenanthroline
PI	polyisoprene or polydispersity index (also abbreviated PDI)
PLA	polylactide or poly(lactic acid)
PMA	poly(methyl acrylate)
PMCH	poly(methylene-1,3-cyclohexane)
PMCP	poly(methylene-1,3-cyclopentane)
PMMA	poly(methyl methacrylate)
PMP2	Projected Second Order Møller Plesset perturbation theory
p-MS	*para*-methylstyrene
P_n	number average degree of polymerization
PNB	polynorbornene (also abbreviated poly(NB))
PnBA	poly(*n*-butyl acrylate)
PnBMA	poly(*n*-butyl methacrylate)
poly(NB)	polynorbornene (also abbreviated PNB)
poly(NBD)	poly(norbornadiene)
POM	polarized optical microscopy
PP	polypropylene
PPA	poly(phenylacetylene)
PPP	poly(*p*-phenylene)
PPD	poly(1,3-pentadiene)
PS	polystyrene
PtBMA	poly(*t*-butyl methacrylate)
PVA	poly(vinyl acetate)
QM	quantum mechanics
QM-MM	hybrid quantum mechanical-molecular mechanics method
r	*racemo* (racemic)
RCM	ring closing metathesis
RFF	reactive force field
RMS	root mean square
ROMP	ring opening metathesis polymerization
s or *s*-	syndiotactic
S	styrene or solvent
salcy	N,N'-bis-(2-hydroxybenzylidene)-1,2-cyclohexanediamine

salan	bis(*o*-hydroxybenzyl)-1,2-diaminoethane ligand
salen	N,N'-bis-(2-hydroxybenzylidene)-1,2-ethanediamine
salph	N,N'-bis-(2-hydroxybenzylidene)-1,2-benzenediamine
salpr	N,N'-bis-(2-hydroxybenzylidene)-1,2-propanediamine
SEM	scanning electron microscopy
SFM	scanning force microscopy
SHOP	Shell Higher Olefins Process
sPP	syndiotactic polypropylene
sPS	syndiotactic polystyrene
SSC	single-site catalyst
*t*BMA	*t*-butyl methacrylate
TEA	triethylaluminum
TEM	transmission electron spectroscopy
Tf	triflate (trifluoromethanesulfonate)
T_g	glass transition temperature
t-h	tail-to-head
THF	tetrahydrofuran
TIBAL or TIBA	triisobutylaluminum
T_m, T_M	melting temperature
TMA	trimethylaluminum
TMEDA	N,N,N',N'-tetramethylethylenediamine
3,5,5-TMH1	3,5,5-trimethyl-1-hexene
TM-SFM	tapping mode scanning force microscopy
TOA	trioctylaluminum
T_p	polymerization temperature
Tp	tris(pyrazolyl)borate
TPE	thermoplastic elastomer
TPE-PP	thermoplastic elastic (elastomeric) polypropylene
TPP	5,10,15,20-tetraphenylporphyrinato
TREF	temperature rising elution fractionation
TS	transition state
TsH	*p*-toluenesulfonhydrazide
t-t	tail-to-tail
u	*unlike* diad
UHMW	ultra high molecular weight
UV	ultraviolet
V	variation coefficient
VA	vinyl acetate
VNB	5-vinyl-2-norbornene
VTM	1- or 3-vinyl tetramethylene
WAXD	wide-angle X-ray diffraction
WAXS	wide angle X-ray scattering
x	crystallinity
X_n	degree of polymerization
XRD	X-ray diffraction

Index